PRINCIPLES OF ELECTRONIC DEVICES

William D. Stanley
Old Dominion University

Prentice Hall
Englewood Cliffs, New Jersey | Columbus, Ohio

Library of Congress Cataloging-in-Publication Data

Stanley, William D.

        Principles of electronic devices / William D. Stanley.

            p.   cm.

      Includes index.

      ISBN 0-02-415560-8

      1. Electronics.   I. Title.

TK7816.S688 1995

621.381—dc20   94-377

                                               CIP

Cover photo: FPG International

Editor: Dave Garza

Developmental Editor: Carol Hinklin Robison

Production Editor: Stephen C. Robb

Cover Designer: Gryphon III

Production Buyer: Patricia A. Tonneman

Production Coordination: Spectrum Publisher Services

Illustrations: Academy ArtWorks, Inc.

This book was set in Times Roman and Univers by Bi-Comp, Incorporated and was printed and bound by R. R. Donnelley & Sons Company. The cover was printed by Phoenix Color Corp.

 © 1995 by Prentice-Hall, Inc.
A Simon & Schuster Company
Englewood Cliffs, New Jersey 07632

Earlier edition, entitled *Electronic Devices: Circuits and Applications,* © 1989 by Prentice-Hall, Inc.

Printed in the United States of America

10 9 8 7 6 5 4 3 2 1

ISBN: 0-02-415560-8

Prentice-Hall International (UK) Limited, *London*
Prentice-Hall of Australia Pty. Limited, *Sydney*
Prentice-Hall of Canada, Inc., *Toronto*
Prentice-Hall Hispanoamericana, S.A., *Mexico*
Prentice-Hall of India Private Limited, *New Delhi*
Prentice-Hall of Japan, Inc., *Tokyo*
Simon & Schuster Asia Pte. Ltd., *Singapore*
Editora Prentice-Hall do Brasil, Ltda., *Rio de Janeiro*

PREFACE

The primary objective of this book is to provide a comprehensive treatment of electronic devices and circuits at a level suitable for first- and second-year students in electronic and electrical technology curricula. The large number of example problems and exercises should also make it very useful for self-study, review, and in support of more advanced courses. The approach in this book differs somewhat from that of other books and is believed to offer some significant advantages.

Since the vast majority of electronic technicians and technologists are *users* of electronic devices, and only a few are actually involved in device fabrication, the treatment of semiconductor physics here is limited to a few sections. Rather, the major focus throughout the book is on device *modeling*, circuit *operation* and *analysis*, and applied *design*.

Basic electronic devices are introduced in the first third or so of the book, and these include diodes, bipolar junction transistors (BJTs), and field-effect transistors (FETs). In each case, a chapter dealing with the particular device category and its models is first presented, and then a chapter dealing with simplified applications follows. The applications at this point include *only resistive components,* and circuits requiring reactive components are delayed until the reader has developed more sophistication. Included in the applications, however, are such important circuits as electronic switches, the emitter follower, the source follower, voltage-controlled resistors, and stable biasing circuits. Note that field-effect transistors are introduced in parallel with bipolar junction transistors, an approach that differs from most books at this level.

Prior to developing transistor amplifier circuits, a general development of amplifiers from a block diagram point of view is given. Concepts such as gain, input resistance, output resistance, stage loading, and interaction are developed with simpler models before the student becomes involved with specific circuit details.

Small-signal amplifiers, including all basic BJT and FET types, are developed in detail. The emphasis in circuit modeling is that of the simplified low-frequency *hybrid-pi model*. This represents a significant departure from most books at this level. However, it is believed by this author to be the best and most timely approach. It has several distinct advantages: (1) The forms of the BJT and FET models are essentially the same; (2) the basic model can readily be extended to cover high-frequency circuits; and (3) it can readily be adapted to computer modeling. With regard to advantage (1), the various equations developed first for BJT amplifiers are quickly adapted to the corresponding FET amplifiers, and this results in a significant correlation between the two types of amplifier circuits.

A few additional comments about small-signal models of transistors are in order. Most books at this level employ either the standard hybrid parameter model or the r'_e model. It is this author's strong opinion that early introduction of hybrid parameters acts more to confuse beginning students than to serve any real educational purpose. Their use is meaningful only after considerable intuition for the practical operation of electronic amplifiers is established. For instructors wishing to cover hybrid parameters, Appendix C is devoted to their coverage.

The r'_e model is quite simple and easy to apply, but it has the disadvantage that the approach is more formula oriented than circuit-model oriented. In fact, the value of r'_e is the reciprocal of g_m as used in the hybrid-pi model. Consequently, from one point of view, the approach of this text is nearly equivalent to that of the r'_e model if interpreted properly. However, the hybrid-pi model can be readily adapted to computer modeling and to high-frequency effects.

A complete chapter is devoted to the differential amplifier and its integrated circuit evolution to the operational amplifier. This sets the stage for the next several chapters, which are predominantly oriented toward integrated-circuit applications.

Negative feedback is developed in detail through the application of op-amp circuits. Various op-amp circuit applications (both linear and nonlinear) are considered. Included are numerous amplifier circuits, holding circuits, clamping circuits, and comparators.

Other topics include positive feedback and oscillators, timers, high-frequency amplifiers, power amplifiers, and power supplies. Placing the chapter on power supplies near the end of the book represents a significant departure from most other texts. However, this author believes it to be appropriate. The usual treatment of power supplies in a very early chapter causes difficulty with many students because of the variety of waveform concepts, definitions, and nonlinear functions encountered.

An entire chapter (Chapter 21) is devoted to the important topic of active filters. The approach here is aligned with that of many industry-wide design references in providing *normalized* filter design data from which the student can quickly learn the process of adjusting the frequency and impedance levels to achieve workable filters. Data are provided for low-pass, high-pass, band-pass, and band-rejection filters. The circuit configurations include some of the most common active designs using op-amps as well as state-variable forms.

The last chapter (Chapter 22) is devoted to the very timely subjects of analog-to-digital and digital-to-analog conversion. The basic concepts for implementing these circuits are covered in some detail. However, since most users will employ off-the-shelf converters, a major emphasis is devoted to understanding the numerous specifications and translating these results to predict circuit performance.

The book is suitable for a two-semester sequence in electronic devices and circuits. Such a sequence could start after the first course in circuits or basic electricity, or it could be taught in parallel with minimal modification in teaching style. The first chapter of the book provides a ''minicourse'' in basic dc circuit analysis, with specific emphasis on techniques useful in electronic circuit analysis. This chapter could be covered very quickly for students having at least one term in dc circuits. An important feature is that *ac circuit analysis is not required until about the middle of the book.*

PSPICE EXAMPLES

A major feature of this textbook is the presence of approximately 60 completely worked-out PSPICE examples distributed throughout the text. Each chapter has from one to six examples with an average of about three per chapter. A brief discussion of the philosophy surrounding this phase of the text follows.

It is the author's opinion that students, at the level intended for this text, should first be taught to deal with the circuits through the simple intuitive and algebraic processes involved before becoming engrossed with computer analysis. It is through such drills that students develop the proper insight for circuit behavior and modeling. Consequently, all PSPICE examples in a given chapter have been grouped into a single section at the end of the chapter. There is no reference in any other section of the chapter to the PSPICE section. Therefore, instructors wishing to bypass this material may do so without any loss of continuity.

However, the trend throughout industry is on computer modeling for advanced analysis and design, and the PSPICE sections should provide a significant supplement to prepare students for that objective. A supplementary appendix (Appendix E) provides the primary concepts for developing circuit codes to support the PSPICE examples. In addition, each PSPICE example provides a thorough discussion of new concepts as they are introduced.

DISCUSSION OF END-OF-CHAPTER PROBLEMS

At the end of most chapters, there are four sets of problems. These problems are designated as Drill Problems, Derivation Problems, Design Problems, and Troubleshooting Problems. The nature of these categories is discussed here.

Drill Problems

This set usually has more problems than the other categories and will represent the major focus for much of the material. Drill problems have been designed to solidify and reinforce the major principles presented in the text. Drill problems range from very simple types to more challenging problems representing situations somewhat different from example problems in the text.

Derivation Problems

The derivation problems are those in which the reader is asked to prove or develop some result in general terms. In some cases, these are simply verifications of rules or theorems given in the text, while in other cases, they are extensions or generalizations of the text results. The outcomes of most derivation problems are formulas or equations rather than numerical values. A few of the derivation problems require calculus, but such problems are identified as such. Such exercises represent the only points in the book where calculus is required and may be bypassed without loss of continuity.

Design Problems

Design problems are selected exercises in which the reader is expected to determine a specific circuit to meet a prescribed objective. In most cases, the actual circuit configuration will be one of those studied in the chapter, but the design aspect of the problem will involve the determination of element values for the various components. In some cases, it will be expected that component values can be rounded to standard element values as tabulated in Appendixes A and B. Design problems do not necessarily have a single correct solution.

Troubleshooting Problems

Troubleshooting problems are exercises in which the reader is expected to display judgment in selecting the probable cause of a malfunction or in predicting the expected response of the instrumentation to typical faults. Because a variety of faults can be responsible for an observed malfunction, most of the troubleshooting exercises have multiple-choice answers based on the *most likely* problem.

SEMICONDUCTOR DATA SHEETS

A concept heavily promoted throughout this book is the use of manufacturers' data or specifications sheets for obtaining information about semiconductor devices. To assist in that endeavor, samples of typical data sheets are provided in Appendix D. These sheets will be referred to throughout the book as the subject matter is developed and as individual specifications become relevant.

Device specifications are necessary in performing any type of design or special system implementation where components must be selected for a particular application. Without such specifications, one is "working in the dark," and what may appear to be a satisfactory design could be totally unworkable for the given set of conditions. Worse yet, if the maximum ratings of components are exceeded, the components may be destroyed.

The preceding discussion undoubtedly seems logical to the perceptive reader, but a few further comments are in order. First, there are literally thousands of different types of semiconductor devices available from a large number of different manufacturers. Some common devices are made by a number of different companies, and the identifying number may be the same or it may be different. The situation is often confusing even to experienced engineers and technicians, so it can appear to be overwhelming to the uninitiated.

The best advice to give to the beginner in this regard is not to worry about the fact that there are thousands of such devices in the field. In your entire career, you will probably have a direct encounter with only a small percentage of the available semiconductor devices, and you will tend to become quite familiar with these. When new devices are encountered, the first thing to do is to obtain data and/or applications sheets for these devices, and you can study these for further information. This author is personally acquainted with many outstanding engineers and technicians, and they all feel overwhelmed by the same situation. The fact is that we really do not need so many available types of devices, but the semiconductor industry has proliferated and produced the existing situation.

The next comment is in regard to the availability of data and applications sheets on individual components. Many semiconductor manufacturers publish one or more data manuals providing compilations of data sheets on devices. Such data manuals are constantly being revised with new devices added and older devices discontinued. As time passes and integrated circuits have become the major focus of most design, more diodes and transistors are being discontinued, and it may be difficult to find specifications on certain components. Some services provide coverage "across the board" and include data from a number of manufacturers. Of course, you need not be too concerned about this now, while you are learning the basics of electronics. Once you are on the job and have the need to know, help should be available in breaking through the information barrier.

The final comment is in regard to the specifications themselves. It is an interesting irony that a simple-looking three-terminal transistor may have two or more pages of specifications with dozens of graphs and several tables of data. This is enough to scare the impressionable person right out of the field! Fear not, however, for all is not yet lost. Such detailed specifications tend to provide about everything one would ever want to know about the device (and perhaps more!). Most general applications require the detailed consideration of only a fraction of the available information. The additional information is there for special applications and situations in which critical specifications are being "stretched to the limit."

In virtually any application, some of the most important data are the *maximum ratings*. Depending on the device, the maximum permitted values of such basic quantities as voltage, current, and power must be carefully monitored. Additional specifications relate to the types of applications to be performed. Throughout the book, we frequently extract pertinent data and apply it as the need arises.

It is impossible to acknowledge all of the people who have contributed to the development of this text, but I will mention a few. First, I wish to thank Dave Garza, Carol H. Robison, and Stephen C. Robb of Merrill for their strong support and assistance throughout this project. Second, I wish to thank Kelly Ricci and

Kristin Miller of Spectrum Publisher Services for outstanding cooperation during the production phase. Third, I wish to thank a fellow colleague, John R. Hackworth, who has taught from the manuscript and provided valuable suggestions. Finally, I would like to express my sincere gratitude to Ms. Michele Boulden for typing a significant portion of the text.

William D. Stanley
Virginia Beach, Virginia

CONTENTS

1

REVIEW OF CIRCUIT ANALYSIS

OBJECTIVES

After completing this chapter, the reader should be able to:

- Discuss the various circuit variables, their symbols, units, and abbreviations.
- Define the two ideal active circuit element models.
- Define the three ideal passive circuit element models.
- State Ohm's law, Kirchhoff's voltage law, and Kirchhoff's current law.
- Solve for all the variables in a series circuit.
- Solve for all the variables in a parallel circuit.
- State the voltage-divider and current-divider rules.
- Solve for the variables contained in a dc resistive circuit containing one source and various series-parallel combinations.
- For a given circuit at its external terminals, determine the Thévenin and Norton equivalent circuits.
- Apply source transformation to change the forms of source models.
- Approximate realistic sources as ideal source models, and identify conditions in which these approximations may be used.
- Define the four types of controlled or dependent sources.
- Apply the principle of superposition in the analysis of linear circuits.
- Define the notation used for voltages between two points, for voltages with respect to ground, and for current.
- Apply the concept of the voltmeter loop to determine the voltage between two points.
- Analyze some of the circuits in the chapter with PSPICE.

1-1 OVERVIEW

Much of the analysis, design, and troubleshooting of electronic circuits involves the systematic application of basic circuit laws and theorems. This introductory chapter will cover these. It is expected that most readers will already be familiar with much of this material, but it is recommended that it be reviewed completely before pursuing the remainder of the book. Chances are that some new perspectives and interpretations will result.

1-2 DEFINITIONS AND UNITS

The two circuit variables used most often in analyzing electronic circuits are voltage and current. **Voltage** represents an electrical pressure or potential difference and is always measured *across an element* or *between two points* in a circuit. **Current** represents a rate of flow of electricity and is always measured *through an element* or *a conductor*.

Refer to Figure 1–1, in which some arbitrary circuit element is shown. The symbol v is most often used to represent voltage, and it is always identified by + and − signs. The + sign is the positive reference, and the − sign is the negative reference. This means that the electrical pressure or potential at the top is *assumed to be* higher than the value at the bottom. (Some references use the symbol e for voltage.)

FIGURE 1-1
Arbitrary circuit element with voltage across it and current through it.

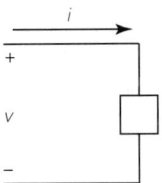

The symbol i is used to represent current, and it is always identified by an arrow. The arrow is assumed to be pointed in the direction of conventional positive current flow.

The unit of voltage is the **volt,** and it is abbreviated as V. The unit of current is the **ampere,** and it is abbreviated as A.

In general, voltage and current vary with time in an electronic circuit. To emphasize these time-varying quantities, both variables are often written in **functional notation** as $v(t)$ and $i(t)$, respectively, where t denotes the time. Such forms are also referred to as **instantaneous** voltage and current forms. For the most part, our practice throughout the book will be to omit the functional notation unless necessary to emphasize a point.

Although there are exceptions, lowercase symbols are most often used when the variables are or could be time-varying, and uppercase symbols are used for dc or fixed quantities such as peak voltage values and so on. For example V_1

would indicate some constant voltage value, and I_p would indicate some constant current value.

Another important variable is **power,** which is the rate of performing work or energy. The symbol for power is p (or P when it is constant), and the unit is the **watt** (abbreviated as W). For any element with voltage v across it and current i through it, the instantaneous power p being generated is

$$p = vi \qquad\qquad (1\text{--}1)$$

If current is *flowing into the positive terminal* of the device, the device is *absorbing power,* but if current is *flowing out of the positive terminal,* it is *delivering power.*

The most common quantities used in the analysis of electronic circuits are summarized in Table 1–1. Only a few of these quantities will appear in this chapter, but the others are included for later reference.

TABLE 1–1
Most common quantities used in electronic circuit analysis

Quantity	Symbols	Unit	Abbreviation of Unit
Time	t, T	second	s
Voltage	v, V, e, E	volt	V
Current	i, I	ampere	A
Power	p, P	watt	W
Resistance	R	ohm	Ω
Conductance	G	siemens	S
Capacitance	C	farad	F
Inductance	L	henry	H
Impedance	$\overline{Z}$	ohm	Ω
Reactance	X	ohm	Ω
Admittance	$\overline{Y}$	siemens	S
Susceptance	B	siemens	S
Frequency (cyclic)	f	hertz	Hz
Frequency (radian)	ω	radian/second	rad/s

The most common prefixes used for electrical units are summarized in Table 1–2. By preceding the units with properly chosen prefixes, a wide range of values may be easily listed without the necessity of using a large number of zeros.

EXAMPLE 1–1

The value of a certain high voltage is 18,100 V. Express in **(a)** kV, and **(b)** MV.

Solution

(a) Since $1 \text{ kV} = 10^3 \text{ V} = 1000 \text{ V}$, there will be fewer kV required to express the voltage by a factor of 1/1000. We thus move the decimal point three places to the left, and an equivalent value is 18.1 kV.

TABLE 1–2
Most common prefixes used in electronic
circuit analysis

Value	Prefix	Abbreviation
10^{-15}	femto	f
10^{-12}	pico	p
10^{-9}	nano	n
10^{-6}	micro	μ
10^{-3}	milli	m
10^{3}	kilo	k
10^{6}	mega	M
10^{9}	giga	G

(b) Since 1 MV $= 10^6$ V, we continue to move the decimal three more places
to the left, and the value is 0.0181 MV.

**EXAMPLE
1–2**
The value of a certain current is 2×10^{-4} A. Express in **(a)** mA, and **(b)** μA.

Solution
(a) Since 1 mA $= 10^{-3}$A, there will more mA required to express the current
by a factor of 1000. We thus multiply the value by $1000 = 1 \times 10^3$, which is
equivalent to moving the decimal point three places to the right. The proper value
is thus 2×10^{-1} mA $= 0.2$ mA.

(b) Since 1 μA $= 10^{-6}$ A, we multiply the value again by $1000 = 1 \times 10^3$ and
obtain 200 μA as the proper form.

**EXAMPLE
1–3**
The two elements in Figure 1–2 are portions of larger circuits not shown. Determine
the power for each element and whether it is being absorbed or delivered.

FIGURE 1–2
Circuits for Example 1–3.

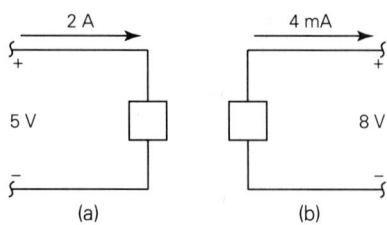

(a)   (b)

Solution
(a) The power associated with Figure 1–2(a) is

$$P = 5 \text{ V} \times 2 \text{ A} = 10 \text{ W} \qquad\qquad \textbf{(1–2)}$$

Since the current is flowing *into* the positive terminal, this power is being *absorbed* by the device.

(b) The power associated with Figure 1–2(b) is

$$P = 8\text{ V} \times 4 \times 10^{-3}\text{ A} = 32 \times 10^{-3}\text{ W} = 32\text{ mW} \qquad \text{(1–3)}$$

Since the current is flowing *out* of the positive terminal, this power is being *delivered* by the device.

1–3 CIRCUIT ELEMENTS

Electrical and electronic circuits are studied extensively through the use of so-called circuit **models.** These are elements that represent various circuit effects in a manner that can be readily predicted. We will consider later in the text many models of semiconductor devices, but at this point, we will introduce only five basic circuit element models. The reader is likely already familiar with these elements, but a brief review is warranted.

Active elements are those that are capable of delivering power to some external load. The two ideal active elements are (a) the **ideal voltage source,** and (b) the **ideal current source.** The source models are identified by the symbols shown in Figure 1–3.

FIGURE 1–3

Circuit models for (a) ideal voltage source and (b) ideal current source.

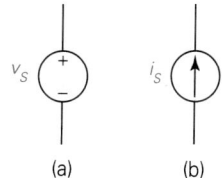

(a)   (b)

The ideal voltage source of (a) is assumed to maintain a voltage v_s across its terminals that is independent of the load connected to it or the current flowing through it. The voltage may be dc or it may be time-varying. Examples of sources that may usually be represented by ideal voltage sources are well-charged batteries, regulated dc voltage power supplies, and the commercial ac voltage supply.

The ideal current source of (b) is assumed to maintain a current flow i_s through its terminals that is independent of the load connected to it or the voltage across it. Ideal current sources are not as obvious in everyday life as ideal voltage sources, but many semiconductor devices behave like ideal current sources, as we will see.

Passive elements are those that are not capable of generating power, but some may actually store energy. The three passive elements are (a) resistance, R, (b) capacitance, C, and (c) inductance, L. The element models are identified

FIGURE 1-4

Circuit models for (a) ideal resistance, (b) ideal capacitance, and (c) ideal inductance.

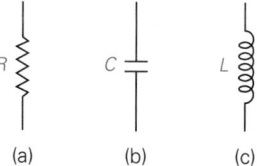

(a)　　(b)　　(c)

by the symbols shown in Figure 1–4. Resistance can only dissipate energy, while capacitance and inductance are both capable of storing energy and releasing it later to the circuit.

Of the three passive elements, only resistance appears in the discussion of electronic circuits in the first seven chapters. The effect of capacitance and inductance will be considered later.

1–4 OHM'S LAW

Probably the most fundamental of all circuit laws is Ohm's law. Consider the circuit of Figure 1–5 that contains a **resistor** (a body of resistance) in which the resistance has a value of R. The unit of resistance is the **ohm,** which is abbreviated as Ω. The voltage v across the resistance and the current i through it are related by Ohm's law:

$$v = Ri \tag{1-4}$$

The positive terminal for the voltage is the terminal at which the current enters in such a way that power is always *absorbed* by a resistor.

FIGURE 1-5

Circuit form of resistor used to define Ohm's law.

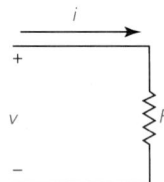

An alternate way to describe resistance is through the conductance parameter, G, which is related to R by

$$G = \frac{1}{R} \tag{1-5}$$

The unit of conductance is the **siemens,** which is abbreviated as S.

The power dissipated in a resistance may be expressed three ways by the use of Ohm's law:

$$p = vi = i^2 R = \frac{v^2}{R} \tag{1-6}$$

Current Sampling Resistor

We will discuss a useful and practical property of Ohm's law for instantaneous voltage and current. From (1–4), the voltage is always a constant times the current, so the waveforms of voltage and current for a resistor have the same forms. Frequently, it is desired to monitor the waveform of a certain time-varying current on an oscilloscope. However, an oscilloscope is a voltage-measuring device. If the circuit contains a resistance (normally with one terminal grounded) through which the current of interest flows, the voltage across it will have the same waveform as the current through it, and the oscilloscope can be made to respond to current indirectly. In some cases, a small sampling resistor may be inserted in the circuit for this purpose. Obviously, the resistance must be sufficiently small that it does not disturb the circuit. The oscilloscope voltage scale can be calibrated so that the current level can be determined by a simple application of Ohm's law.

EXAMPLE 1–4

It is desired to monitor the waveform of current in a certain electronic circuit by means of a standard oscilloscope. It is determined that the insertion of a 10-Ω precision sampling resistor presents negligible disturbance to the circuit. The resistor is inserted at a point where one side is connected to the common ground, so that the grounded lead of the oscilloscope may be maintained at the circuit common ground. A satisfactory display is obtained when the oscilloscope's vertical calibration factor is set at 20 mV/cm. Determine the equivalent calibration factor for the measured current.

Solution
By Ohm's law with $R = 10\ \Omega$ and $v = 20$ mV, the current is $i = 0.02$ V/10 Ω = 2×10^{-3} A = 2 mA. The calibration factor on this scale for current is thus 2 mA/cm.

1–5 KIRCHHOFF'S VOLTAGE LAW

Kirchhoff's voltage law deals with a closed loop in a circuit, of which an example is shown in Figure 1–6. This law can be stated as

$$\text{Algebraic sum of loop voltages} = 0 \tag{1–7}$$

FIGURE 1–6
Circuit used to illustrate Kirchhoff's voltage law.

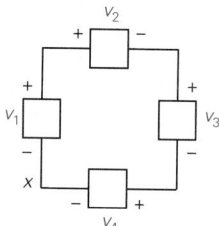

To apply this law, an algebraic pattern is established in which a voltage drop is given one sign and a rise is given the opposite sign. In most circuits, there are more drops than rises, so we will follow the convention that a drop is considered as positive and a rise is considered as negative. For the circuit of Figure 1–6, a clockwise loop is formed starting at point x, and we have

$$-v_1 + v_2 + v_3 + v_4 = 0 \qquad (1\text{–}8)$$

EXAMPLE 1–5 A portion of some dc circuit of interest is shown in Figure 1–7. Determine the voltage V and the current I. (Uppercase notation is used since the circuit is a dc circuit.)

FIGURE 1–7
Circuit of Example 1–5.

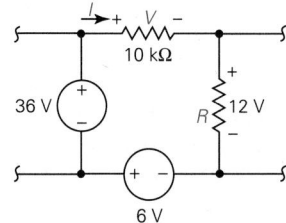

Solution
We do not see the complete circuit, nor do we know the values of all components shown. However, after some initial inspection, we do recognize that there is a closed loop shown for which three of the four voltages in the loop are known. Thus, by KVL, we should be able to determine the remaining voltage V.

Starting at the negative terminal of the 36-V source, a clockwise KVL loop reads

$$-36 + V + 12 - 6 = 0 \qquad (1\text{–}9)$$

Solving for V, we obtain

$$V = 30 \text{ V} \qquad (1\text{–}10)$$

Since I is the current through the 10-kΩ resistor, for which the voltage is now known, it may be determined by Ohm's law. Thus,

$$I = \frac{30 \text{ V}}{10 \times 10^3 \text{ }\Omega} = 3 \times 10^{-3} \text{ A} = 3 \text{ mA} \qquad (1\text{–}11)$$

1-6 KIRCHHOFF'S CURRENT LAW

Kirchhoff's current law deals with a **node,** which is a junction point between two or more branches as illustrated in Figure 1–8. This law can be stated as

$$\text{Algebraic sum of all currents at a node} = 0 \qquad \textbf{(1–12)}$$

FIGURE 1–8
Circuit used to illustrate Kirchhoff's current law.

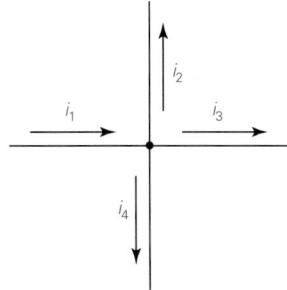

To apply this law, an algebraic pattern is established in which a current leaving is given one sign and a current entering is given the opposite sign. We will follow the convention that a current leaving is considered as positive and a current entering is considered as negative. For the circuit of Figure 1–8, we have

$$-i_1 + i_2 + i_3 + i_4 = 0 \qquad \textbf{(1–13)}$$

EXAMPLE 1–6 A portion of some dc circuit of interest is shown in Figure 1–9. Determine the current I and the resistance R.

FIGURE 1–9
Circuit of Example 1–6.

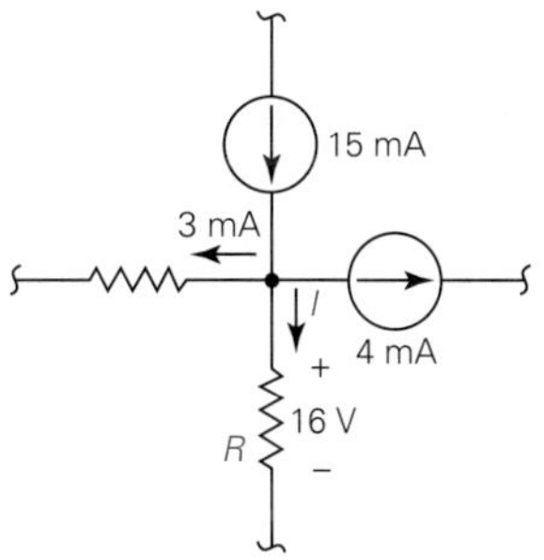

Solution
As in Example 1–5, not all the circuit is evident. However, in this case, there is a node shown for which three of the four currents are known. Thus, by KCL, we should be able to determine the remaining current.

Assuming positive currents leaving the node, the KCL equation is

$$3 \times 10^{-3} - 15 \times 10^{-3} + 4 \times 10^{-3} + I = 0 \qquad \textbf{(1–14)}$$

Solving for I, we obtain

$$I = 8 \times 10^{-3}\,\mathrm{A} = 8\,\mathrm{mA} \qquad \textbf{(1–15)}$$

Since I is the current through R, and the voltage across R is known, the value of R can be determined by Ohm's law as

$$R = \frac{16\,\mathrm{V}}{8 \times 10^{-3}\,\mathrm{A}} = 2 \times 10^{3}\,\Omega = 2\,\mathrm{k}\Omega \qquad \textbf{(1–16)}$$

1–7 SERIES CIRCUITS

One of the simplest configurations is the **series circuit** whose basic form is illustrated in Figure 1–10. It is also called a **single-loop circuit.** Since the current entering any node within the circuit must equal the current leaving the node, we can conclude that *the current within the loop must be the same at all points.*

FIGURE 1–10
Form of series (or single-loop) circuit.

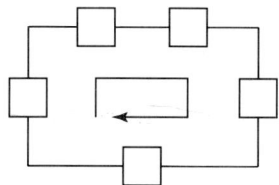

Assume that one or more voltage sources and one or more resistors are present in a series circuit and that their values are known. The current i can then be determined by applying KVL to the loop and expressing all voltage drops in terms of the unknown current. The resulting equation can then be solved for the current, as will be illustrated in Example 1–7.

If there is a single ideal current source within the loop, the current will automatically assume the value of this current. A situation of this type will be illustrated in Example 1–8.

A special situation of interest arises when there is one voltage source and a number of resistances as illustrated in Figure 1–11. Although KVL could be applied, the simplest approach in this case is to first determine the **equivalent resistance** R_{eq} of the series combination as "seen" by the source. The value of this resistance is

$$R_{eq} = R_1 + R_2 + R_3 + \cdots + R_n \qquad \textbf{(1–17)}$$

FIGURE 1–11

Series circuit with one voltage source
and a number of resistors.

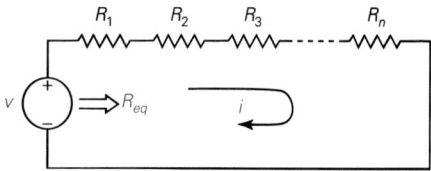

The current i can be determined by Ohm's law as

$$i = \frac{v}{R_{eq}}$$ (1–18)

EXAMPLE 1-7

For the dc resistive circuit of Figure 1–12(a), determine **(a)** the loop current, **(b)** the voltage drops across all resistors, and **(c)** the power dissipated in the three resistors.

FIGURE 1–12

Circuit of Example 1–7.

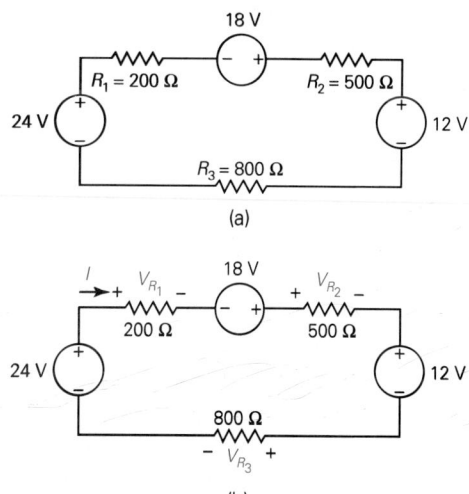

Solution

As given initially, the circuit has no labels for the unknown current and voltages desired in order that such labels may be determined as part of the analysis. First, intuition tells us that the loop current will be clockwise, since both the 24-V source and 18-V source would tend to cause current to flow clockwise, and their combined effect is more powerful than that of the 12-V source. Thus, a clockwise current I is assumed. Once the current direction is assumed, the various voltage drops across resistors are chosen in the proper directions as indicated in Figure 1–12(b). For convenience, each voltage is also labeled with an appropriate defining subscript. (*Note:* Uppercase symbols are used for the current and voltages, since this is a dc resistive circuit.)

(a) Starting at a point just below the 24-V source and moving clockwise, the KVL equation is

$$-24 + 200\,I - 18 + 500I + 12 + 800I = 0 \qquad\text{(1–19)}$$

This simplifies to

$$1500I = 30 \qquad\text{(1–20)}$$

or

$$I = 0.02\ \text{A} = 20\ \text{mA} \qquad\text{(1–21)}$$

(b) Once the current is known, the voltage drops across the resistors are determined as follows:

$$V_{R_1} = R_1 I = 200 \times 20 \times 10^{-3} = 4\ \text{V} \qquad\text{(1–22)}$$
$$V_{R_2} = R_2 I = 500 \times 20 \times 10^{-3} = 10\ \text{V} \qquad\text{(1–23)}$$
$$V_{R_3} = R_3 I = 800 \times 20 \times 10^{-3} = 16\ \text{V} \qquad\text{(1–24)}$$

(c) Let P_{R_1}, P_{R_2}, and P_{R_3} represent the three power levels dissipated in R_1, R_2, and R_3, respectively.
We have

$$P_{R_1} = R_1 I^2 = 200(20 \times 10^{-3})^2 = 0.08\ \text{W} = 80\ \text{mW} \qquad\text{(1–25)}$$
$$P_{R_2} = R_2 I^2 = 500(20 \times 10^{-3})^2 = 0.2\ \text{W} = 200\ \text{mW} \qquad\text{(1–26)}$$
$$P_{R_3} = R_3 I^2 = 800(20 \times 10^{-3}) = 0.32\ \text{W} = 320\ \text{mW} \qquad\text{(1–27)}$$

Incidentally, the 24-V and 18-V sources are delivering power, and the 12-V source is absorbing power. The net power delivered must equal the net power absorbed, and the reader is invited to verify this fact.

EXAMPLE 1–8

For the circuit of Figure 1–13, determine **(a)** the current I, **(b)** the voltage V_R, and **(c)** the voltage V_O across the current source. **(d)** Determine also the power associated with each element and whether it is delivered or absorbed.

FIGURE 1–13
Circuit of Example 1–8.

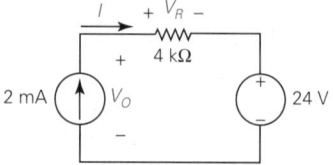

Solution
(a) Since the single loop contains an ideal current source, the value of the current is immediately deduced as

$$I = 2\ \text{mA} \qquad\text{(1–28)}$$

(b) The voltage V_R across the resistor is determined from Ohm's law as

$$V_R = RI = 4 \times 10^3 \times 2 \times 10^{-3} = 8 \text{ V} \qquad (1\text{--}29)$$

(c) The voltage V_O across the current source is determined from KVL by forming a loop around the circuit. Starting below the current source and moving clockwise, we have

$$-V_O + 8 + 24 = 0 \qquad (1\text{--}30)$$

Solving for V_O, we obtain

$$V_O = 32 \text{ V} \qquad (1\text{--}31)$$

It can be deduced that changing either the resistance or the value of the voltage source will change the value of V_O. In other words, the voltage across the current source depends as much on the remainder of the circuit as it does on the current source itself.

(d) The current source is *delivering* power P_I as given by

$$P_I = 32 \text{ V} \times 2 \times 10^{-3} \text{ A} = 64 \text{ mW} \qquad (1\text{--}32)$$

The resistor is *absorbing* power P_R of value

$$P_R = (2 \times 10^{-3})^2 \times 4 \times 10^3 = 16 \text{ mW} \qquad (1\text{--}33)$$

The voltage source is also *absorbing* power P_V with value

$$P_V = 24 \text{ V} \times 2 \times 10^{-3} \text{ A} = 48 \text{ mW} \qquad (1\text{--}34)$$

In general, when there are several sources in a circuit, at least one must be delivering power, but others may be absorbing or delivering. Note that 16 + 48 = 64; that is, the net power delivered must equal the net power absorbed.

1–8 PARALLEL CIRCUITS

A second simple circuit configuration is the **parallel circuit,** whose basic form is illustrated in Figure 1–14. It is also called a **single node-pair** as evidenced by the presence of two nodes. A fundamental constraint for a parallel circuit is that *the voltage across each element be the same.*

FIGURE 1–14
Form of parallel (or single node-pair) circuit.

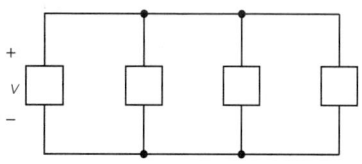

FIGURE 1–15
Parallel circuit with one voltage source
and a number of resistors.

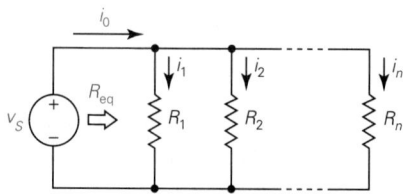

Consider the case where a single voltage source v_s is connected across a number of resistors as illustrated in Figure 1–15. The presence of the parallel voltage source forces the voltage across each branch to be v_s. By Ohm's law, the current through the individual resistances can be determined as follows:

$$i_1 = \frac{v_s}{R_1} \tag{1–35}$$

$$i_2 = \frac{v_s}{R_2} \tag{1–36}$$

$$i_n = \frac{v_s}{R_n} \tag{1–37}$$

By KCL, the current i_o leaving the voltage source is

$$i_o = i_1 + i_2 + \cdots + i_n \tag{1–38a}$$

$$= \frac{v_s}{R_1} + \frac{v_s}{R_2} + \cdots + \frac{v_s}{R_n} \tag{1–38b}$$

A different way to view the preceding situation is to determine the equivalent resistance R_{eq} "seen" by the source. The equivalent resistance of n resistances in parallel is given by the formula

$$\frac{1}{R_{eq}} = \frac{1}{R_1} + \frac{1}{R_2} + \cdots + \frac{1}{R_n} \tag{1–39}$$

The net current can then be expressed as

$$i_o = \frac{v_s}{R_{eq}} \tag{1–40}$$

If one is working with conductances, the equivalent conductance is given by

$$G_{eq} = G_1 + G_2 + \cdots + G_n \tag{1–41}$$

where $G_{eq} = 1/R_{eq}$, $G_1 = 1/R_1$, $G_2 = 1/R_2$, and so on.

For two resistances in parallel the general formula reduces to

$$R_{eq} = \frac{R_1 R_2}{R_1 + R_2} \qquad\qquad (1\text{–}42)$$

This is easily remembered as the product divided by the sum.

Frequently, in working with electronic circuits there will be two or more resistances in parallel, in which the resulting expression may be somewhat bulky. A common notation is to express the equivalent parallel resistance as $R_1 \| R_2$. Likewise, for three resistances, the notation $R_1 \| R_2 \| R_3$ may be used to represent the equivalent parallel resistance, and so on.

Consider now the case where there is no voltage source, but in which there is one or more current sources in parallel with several resistances. In this case KCL may be applied at the upper node by expressing all currents through resistances in terms of the unknown voltage. This process will be illustrated in Example 1–10.

EXAMPLE 1–9

For the dc resistive circuit of Figure 1–16(a), determine **(a)** the node-pair voltage, **(b)** the currents through the resistors, and **(c)** the current through the voltage source.

FIGURE 1–16
Circuit of Example 1–9.

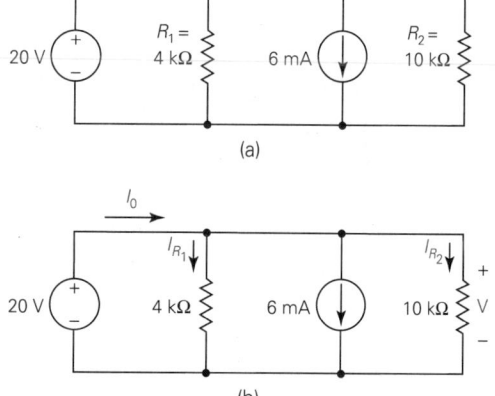

(a)

(b)

Solution

The circuit with appropriate labels is shown in Figure 1–16(b). Note that a current I_o is shown through the voltage source.

(a) The voltage V is simply the value of the voltage source and is

$$V = 20\ \text{V} \qquad\qquad (1\text{–}43)$$

(b) The currents through the resistors are

$$I_{R_1} = \frac{V}{R_1} = \frac{20}{4 \times 10^3} = 5 \times 10^{-3} = 5 \text{ mA} \qquad \textbf{(1–44)}$$

$$I_{R_2} = \frac{V}{R_2} = \frac{20}{10 \times 10^3} = 2 \times 10^{-3} = 2 \text{ mA} \qquad \textbf{(1–45)}$$

(c) To determine the current I_o, KCL is applied to the upper node. We have

$$-I_o + I_{R_1} + 6 \times 10^{-3} + I_{R_2} = 0 \qquad \textbf{(1–46)}$$

This leads to

$$\begin{aligned} I_o &= I_{R_1} + I_{R_2} + 6 \times 10^{-3} = 5 \times 10^{-3} + 2 \times 10^{-3} + 6 \times 10^{-3} \\ &= 13 \times 10^{-3} \text{ A} = 13 \text{ mA} \end{aligned} \qquad \textbf{(1–47)}$$

EXAMPLE 1–10 For the dc resistive circuit of Figure 1–17, determine **(a)** the node-pair voltage, and **(b)** the currents through all resistors.

FIGURE 1–17
Circuit of Example 1–10.

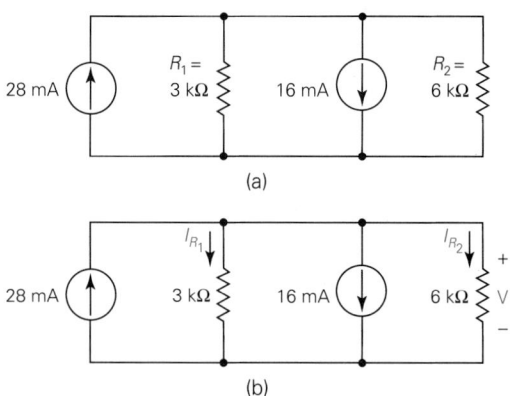

(a)

(b)

Solution
As given initially, the circuit has no labels for the unknown voltage and currents desired in order that such labels may be determined as part of the analysis. First, intuition tells us that the more positive terminal of the voltage is at the top, since the 28-mA source is more powerful than the 16-mA source. With this direction of V chosen, the directions of the two resistive branch currents are then determined in accordance with the standard sign convention as shown in Figure 1–17(b). For convenience, each current is labeled with appropriate defining subscripts.

(a)   A KCL equation may be written at the upper node, and it is

$$-28 \times 10^{-3} + \frac{V}{3 \times 10^3} + 16 \times 10^{-3} + \frac{V}{6 \times 10^3} = 0 \qquad \textbf{(1–48)}$$

This simplifies to

$$5 \times 10^{-4}V = 12 \times 10^{-3} \qquad \textbf{(1–49)}$$

or

$$V = 24 \text{ V} \qquad \textbf{(1–50)}$$

(b)   Once the voltage is known, the currents through the two resistors are determined as follows:

$$I_{R_1} = \frac{V}{R_1} = \frac{24}{3 \times 10^3} = 8 \times 10^{-3} \text{ A} = 8 \text{ mA} \qquad \textbf{(1–51)}$$

$$I_{R_2} = \frac{V}{R_2} = \frac{24}{6 \times 10^{-3}} = 4 \times 10^{-3} \text{ A} = 4 \text{ mA} \qquad \textbf{(1–52)}$$

1–9   VOLTAGE- AND CURRENT-DIVIDER RULES

Two simple rules that may be employed in the analysis of many circuits are (a) the **voltage-divider rule,** and (b) the **current-divider rule.**

Voltage-Divider Rule

Consider the circuit of Figure 1–18. Assume that v_i is the input voltage and that the drop v_o across R_o is to be determined in terms of v_i. It is assumed that v_i either is a source or is a known voltage produced by a source elsewhere in a circuit to the left that is not shown. This configuration represents a voltage divider, and the voltage v_o may be expressed as

$$v_o = \frac{R_o}{R_o + R_1} v_i \qquad \textbf{(1–53)}$$

It can be deduced that v_o is always less than v_i, since some of the available voltage is dropped across R_1. However, when $R_1 \ll R_o$, the voltage v_o is nearly

FIGURE 1–18
Circuit used to illustrate voltage-divider rule.

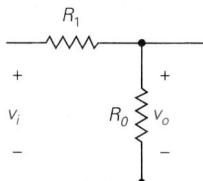

equal to v_i. As an easily remembered rule of thumb, it can be readily shown that when $R_1 \leq 0.01R_o$, the voltage across R_o is greater than 99% of the available voltage.

The voltage-divider rule is readily extended to three or more resistors in series by adding the additional resistances in the denominator of (1–53). However, the voltage across any one resistor is determined by placing that resistance in the numerator of (1–53).

Current-Divider Rule

Consider next the circuit of Figure 1–19. Assume that i_i is a source or is a known current produced by a source elsewhere in a circuit to the left that is not shown. this circuit represents a current divider, and the current i_o may be expressed as

$$i_o = \frac{R_1}{R_1 + R_o} i_i \qquad (1-54)$$

FIGURE 1–19
Circuit used to illustrate current-divider rule.

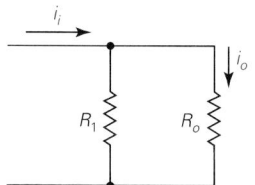

It can be deduced that i_o is always less than i_i, since some of the available current flows through R_1. However, when $R_1 \gg R_o$, the current i_o is nearly equal to i_i. As an easily remembered rule of thumb, it can be shown that when $R_1 \geq 100R_o$, the current through R_o is greater than 99% of the available current.

When there are more than two resistances in parallel, all resistances except R_o are reduced to an equivalent single resistance, and this is the value used for R_1 in (1–54).

EXAMPLE 1–11   Use the voltage-divider rule to determine the three voltages V_1, V_2, and V_3 in Figure 1–20.

FIGURE 1–20
Circuit of Example 1–11.

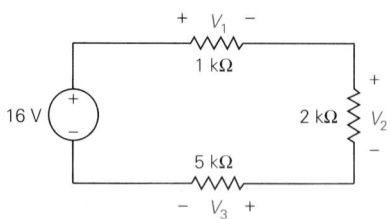

Solution

The relationship of (1–53) is extended to three resistances as discussed in the text. The various voltages are

$$V_1 = \frac{1000}{1000 + 2000 + 5000} \times 16 = \frac{1000}{8000} \times 16 = 2 \text{ V} \qquad \textbf{(1–55)}$$

$$V_2 = \frac{2000}{8000} \times 16 = 4 \text{ V} \qquad \textbf{(1–56)}$$

$$V_3 = \frac{5000}{8000} \times 16 = 10 \text{ V} \qquad \textbf{(1–57)}$$

It can be readily verified that the sum of these three voltages is 16 V, the value of the source voltage.

EXAMPLE 1–12 Use the current-divider rule to determine the current I_o in Figure 1–21(a).

FIGURE 1–21
Circuit of Example 1–12.

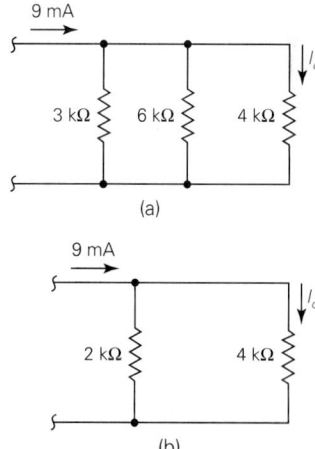

(a)

(b)

Solution

Since there are three resistors, to find the current through the 4-kΩ resistance, it is first necessary to determine the equivalent resistance of the 3-kΩ resistance and the 6-kΩ resistance in parallel. This equivalent resistance is readily determined to be 2 kΩ, and a modified equivalent circuit is shown in Figure 1–21(b). Application of the current-divider rule yields

$$I_o = \frac{2000}{2000 + 4000} \times 9 \text{ mA} = \frac{1}{3} \times 9 \text{ mA} = 3 \text{ mA} \qquad \textbf{(1–58)}$$

1–10 MORE COMPLEX CIRCUITS

Complex circuits involving several loops or nodes may be solved by the methods of mesh current analysis or node voltage analysis, which are covered in books devoted to circuit or network analysis. Such methods should be learned as the reader progresses to the level of complex electronic circuits. At this point, however, our focus will be on certain structures that can be easily handled with the simple circuit laws of the preceding few sections.

The circuits that will be considered are some orderly combinations of series and parallel arrangements that have a single voltage or current source at the input. The equivalent resistance seen by the source is first determined and all unknown quantities at the input terminals are then calculated. By progressively working back toward the opposite end of the circuit in a step-by-step fashion, simple circuit laws may be used to determine all unknown variables. This process is best illustrated with some examples.

EXAMPLE 1–13

Using successive series and parallel reduction, determine the equivalent resistance at the input terminals for the circuit of Figure 1–22.

FIGURE 1–22
Circuit of Example 1–13.

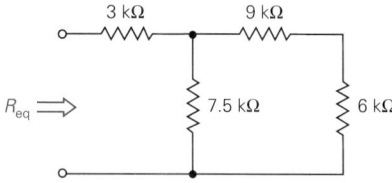

Solution

The process consists of starting on the right-hand side of the circuit and working back toward the input, making appropriate series and parallel combinations at each step. The various steps are illustrated by the different parts of Figure 1–23. The calculations will not be given here, since they consist of simple applications of the formulas of this chapter, but the reader may want to verify the results.

First, the series combination of the 9-kΩ and 6-kΩ resistances is determined as shown in Figure 1–23(a), and this result is 15 kΩ. Next, the parallel combination of the 15-kΩ equivalent resistance and the 7.5-kΩ resistor is determined, and the result is 5 kΩ, as shown in Figure 1–23(b). Finally, the series combination of the 5-kΩ equivalent resistance and the 3-kΩ resistor is determined, as shown in Figure 1–23(c). The result is the equivalent resistance, and thus R_{eq} = 8 kΩ.

EXAMPLE 1–14

Consider the dc resistive circuit of Figure 1–24. The passive part of the circuit is the same as that of Example 1–13, but a 48-V dc source has been connected to the input. Using a step-by-step approach, determine all branch voltages and currents in the circuit.

FIGURE 1–23
Successive steps in determining the equivalent
resistance of the circuit of Example 1–13.

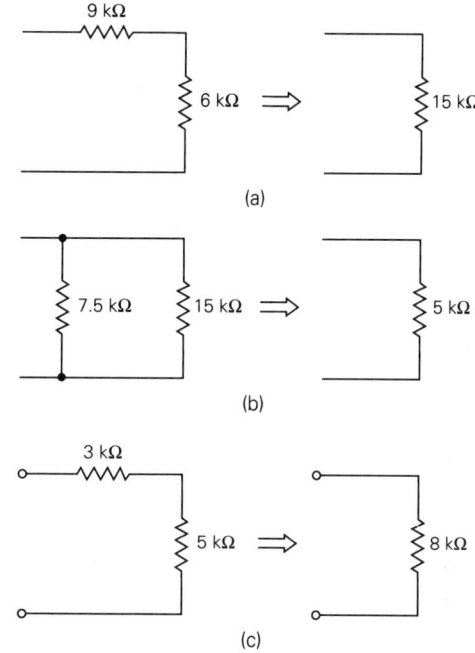

(a)

(b)

(c)

FIGURE 1–24
Circuit of Example 1–14.

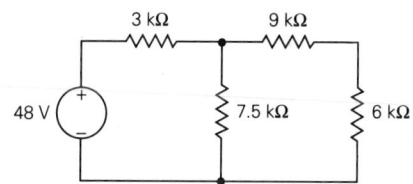

Solution
The equivalent resistances at various points in the circuit were calculated in
Example 1–13, so the reader may wish to refer to that example (including Figure
1–23). In addition, some equivalent circuits with certain voltages and currents
identified are shown in Figure 1–25 to aid in the steps that follow.

First, the equivalent resistance "seen" by the voltage source is 8 kΩ, so the
current leaving the 48-V voltage source is readily determined by Ohm's law as
48 V/8 × 10³ Ω = 6 × 10⁻³ A = 6 mA. This is illustrated in Figure 1–25(a), with
the resistance represented as the series combination of the 3-kΩ resistor and the
additional 5-kΩ equivalent resistance. The voltage across the 3-kΩ resistor is
determined as 3 × 10³ × 6 × 10⁻³ A = 18 V, and the voltage across the circuit
to the right of the 3-kΩ resistor is determined as 5 × 10³ Ω × 6 × 10⁻³ A = 30
V. Alternatively, the voltage-divider rule could have been used twice to determine
the two voltages.

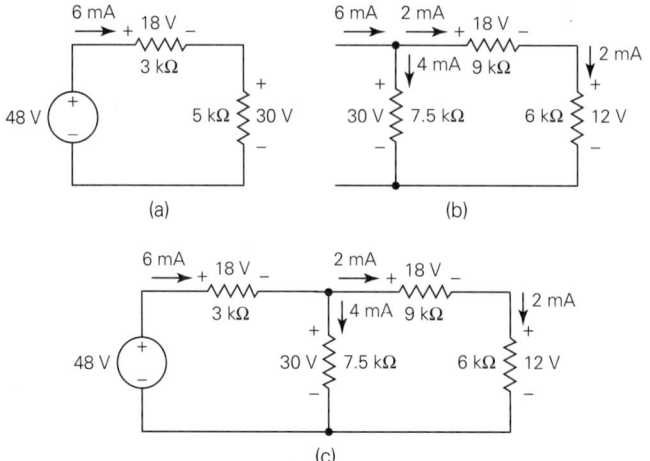

(a)   (b)

(c)

FIGURE 1–25
Steps involved in analyzing circuit of Example 1–14.

With the 30-V voltage across the 5-kΩ equivalent resistance known, this resistance is next represented as the parallel combination of the 7.5-kΩ resistance and the additional two resistors in series, as shown in Figure 1–25(b). The current through the 7.5-kΩ resistor is determined as $30 \text{ V}/7.5 \times 10^3 \ \Omega = 4 \times 10^{-3} \text{ A} = 4$ mA. The current through the series combination of the two remaining resistors is then determined as $30 \text{ V}/15 \times 10^3 \ \Omega = 2 \times 10^{-3} \text{ A} = 2$ mA. Alternatively, the current-divider rule could have been used twice to determine the division of the 6-mA current through the two paths.

With the 2-mA current to the right now known, the two remaining voltages may be calculated by Ohm's law as $9 \times 10^3 \ \Omega \times 2 \times 10^{-3} \text{ A} = 18$ V and 6×10^3 $\Omega \times 2 \times 10^{-3} \text{ A} = 12$ V.

All the currents and voltages calculated in the preceding steps are identified on the original circuit diagram in Figure 1–25(c). The reader may readily verify that KVL is satisfied around all loops and that KCL is satisfied at all nodes.

The perceptive reader may have observed some possible variations in some of the analysis steps that could have been taken in "moving" through the circuit. The key with circuits of this type is to be able to simplify the circuit successively at different points so that the intuitive rules may be applied. To apply either of the divider rules, the loading effects of all resistors must be considered in the form of equivalent resistances at different points.

1–11 THÉVENIN'S AND NORTON'S THEOREMS

Two of the most useful techniques in analyzing and simplifying electronic circuits are **Thévenin's theorem** and **Norton's theorem**.

Thévenin's Theorem

Refer to Figure 1–26. Thévenin's theorem states that any effects of a complex circuit produced external to the circuit may be predicted from a single voltage source v_t in series with a resistance R_{eq}. The combination of the voltage source and the series resistance is referred to as the **Thévenin equivalent circuit.** Once the Thévenin equivalent circuit is known, it may be substituted for the original circuit for modeling and analysis purposes.

FIGURE 1–26

Thévenin model of an electrical circuit at a set of reference terminals.

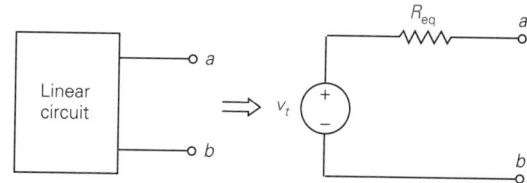

Norton's Theorem

Refer to Figure 1–27. Norton's theorem states that any effects of a complex circuit produced external to the circuit may be predicted from a single current source i_n in parallel with a resistance R_{eq}. The combination of the current source and the parallel resistance is referred to as the **Norton equivalent circuit.** Once the Norton equivalent circuit is known, it may be substituted for the original circuit for modeling and analysis purposes.

FIGURE 1–27

Norton model of an electrical circuit at a set of reference terminals.

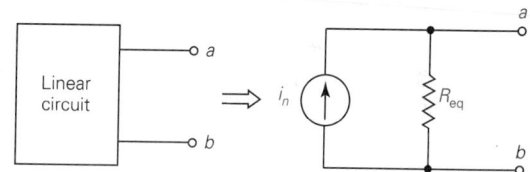

Determining the Thévenin and Norton Models

For each of the equivalent circuits, there are two quantities to be determined. However, the resistance R_{eq} is the same for both circuits. For circuits containing only independent sources, R_{eq} is the equivalent resistance seen looking back from the output terminals when all internal sources are deenergized. To deenergize an ideal voltage source, replace it by a short circuit, and to deenergize an ideal current source, replace it by an open circuit. To deenergize nonideal sources, replace them by their internal resistances.

A word of caution is in order here. When reference is made to replacing an ideal voltage source by a short circuit or, as we will see, measuring short circuit current, we are primarily discussing "paper exercises" for analysis purposes. It

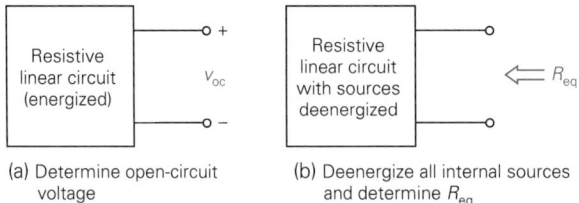

(a) Determine open-circuit   (b) Deenergize all internal sources
    voltage   and determine R_{eq}

FIGURE 1–28
General approach for determining Thévenin equivalent of resistive current when circuit contains no dependent sources.

may be quite impractical and even dangerous to attempt such practices in the laboratory. There are procedures that may be applied in the laboratory, but our focus here is on the analytical modeling.

The Thévenin voltage v_t is the **open-circuit** voltage measured at the external terminals. The Norton current i_n is the **short-circuit** current at the terminals; that is, the current measured through a short (on paper) connected to the external terminals. These procedures are summarized in Figures 1–28 and 1–29.

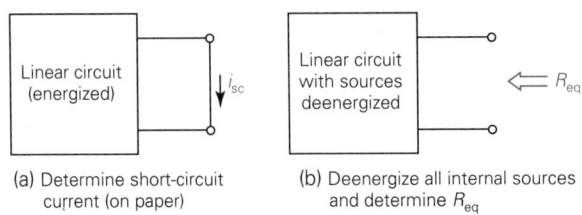

(a) Determine short-circuit   (b) Deenergize all internal sources
    current (on paper)   and determine R_{eq}

FIGURE 1–29
General approach for determining Norton equivalent of resistive circuit when circuit contains no dependent sources.

Source Transformations

If either the Thévenin equivalent circuit or the Norton equivalent circuit is known, the other may be readily determined. In Figure 1–30, the Thévenin equivalent circuit on the left is known. The current i_{sc} that would flow through a short connected to the terminals is simply

$$i_{sc} = i_n = \frac{v_t}{R_s} \qquad \textbf{(1–59)}$$

The corresponding Norton equivalent circuit is shown on the right.

FIGURE 1–30
Transformation of voltage source to current source.

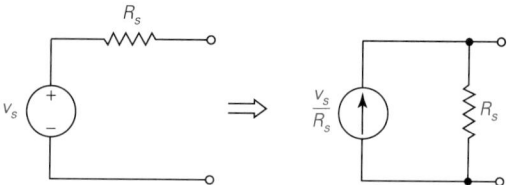

In Figure 1–31, the Norton equivalent circuit on the left is known. The open-circuit voltage v_{oc} at the terminals is simply

$$v_{oc} = v_t = R_{eq}i_n \tag{1–60}$$

FIGURE 1–31
Transformation of current source to voltage source.

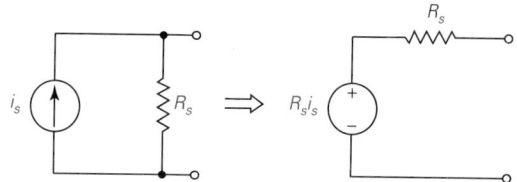

The corresponding Thévenin equivalent circuit is shown on the right. These two special conversions are often referred to as **source transformations.** They permit a voltage source to be replaced with a current source that could produce the same external effect. The only requirement to permit this conversion is that there be an internal resistance present.

EXAMPLE 1–15

Determine the Thévenin equivalent circuit for the circuit of Figure 1–32(a) at the terminals $a-a'$.

Solution

This "L" circuit occurs so often in practical problems that it warrants special attention. First, we calculate the open-circuit voltage V_{oc} as indicated in Figure 1–32(b) using the voltage-divider rule. We have

$$V_{oc} = \frac{6\ k\Omega}{6\ k\Omega + 12\ k\Omega} \times 60 = \frac{1}{3} \times 60 = 20\ V \tag{1–61}$$

Next, we deenergize the voltage source by replacing it with a short circuit as shown in Figure 1–32(c). We then look back from the external terminals and note that the two resistors are in parallel. Thus,

$$R_{eq} = 6\ k\Omega \| 12\ k\Omega = 4\ k\Omega \tag{1–62}$$

The Thévenin equivalent circuit is shown in Figure 1–32(d).

A common error made by beginners with this circuit configuration is to consider R_{eq} as the sum of the two resistances. Indeed, the resistors are in series as far as the *internal* form of the circuit is concerned. However, the Thévenin equivalent resistance is that value obtained by looking in from the *external* terminals, and its value is the parallel combination.

FIGURE 1–32

Circuit of Example 1–15.

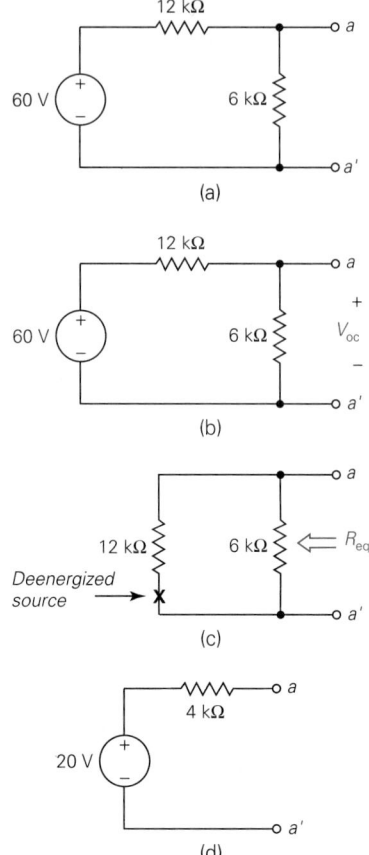

(a)

(b)

(c)

(d)

EXAMPLE 1–16

Determine the Thévenin equivalent circuit for the circuit of Figure 1–33(a) at the terminals a–a'.

Solution

The "T" circuit configuration also occurs quite frequently. In this example, the circuit up to the last resistor is the same as that of Example 1–15 in order to make a point. The open-circuit voltage V_{oc} determined in Figure 1–33(b) will be the same as that in Example 1–15, since no current flows through the 1-kΩ resistor. Remember that an ideal voltmeter does not draw any current, so there is no voltage drop across the 1-kΩ resistor. Thus,

$$V_{oc} = \frac{6\,k\Omega}{6\,k\Omega + 12\,k\Omega} \times 60 = 20\text{ V} \tag{1–63}$$

FIGURE 1–33
Circuit of Example 1–16.

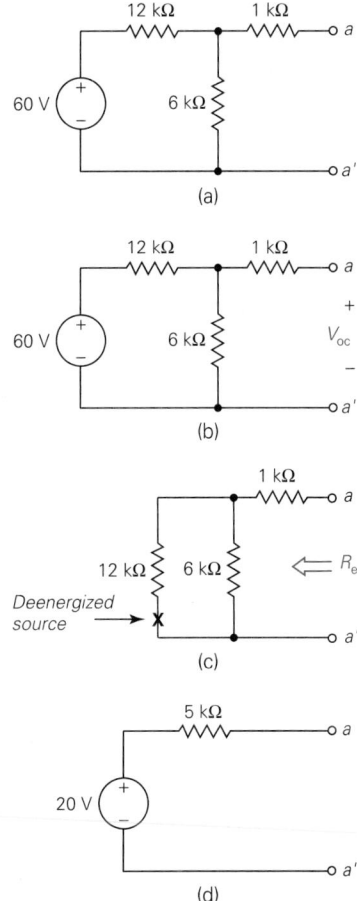

(a)

(b)

(c)

(d)

Again, in determining R_{eq}, the source is replaced by a short circuit as shown in Figure 1–33(c), and the equivalent resistance viewed from a–a' is determined. In this case, we have

$$R_{eq} = 1 \text{ k}\Omega + 6 \text{ k}\Omega \parallel 12 \text{ k}\Omega = 1 \text{ k}\Omega + 4 \text{ k}\Omega = 5 \text{ k}\Omega \qquad \textbf{(1–64)}$$

The Thévenin equivalent circuit is shown in Figure 1–33(d). The Thévenin voltage is the same as in Example 1–15, but the equivalent resistance is different.

Both the "L" configuration and the "T" configuration occur frequently in electronic circuits, so the reader should carefully note the forms involved.

EXAMPLE 1–17   Determine the Norton equivalent circuit of the circuit of Example 1–16.

Solution

Although we could start all over if desired, the simplest approach is to determine the Norton equivalent circuit directly from the Thévenin model determined in Example 1–16. Referring to Figure 1–34(a), we simply measure the current I_{sc} that would flow through a short connected across the terminals, and we have

$$I_{sc} = \frac{20\ \text{V}}{5\ \text{k}\Omega} = 4\ \text{mA} \tag{1–65}$$

The resulting Norton equivalent circuit is shown in Figure 1–34(b). This amounts to using a source transformation.

FIGURE 1–34
Circuit of Example 1–17.

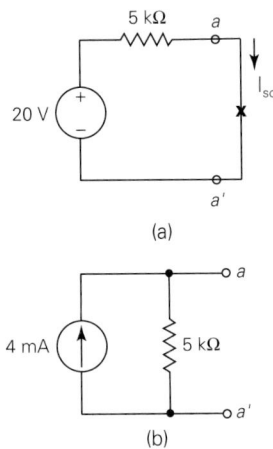

(a)

(b)

A basic point to deduce is that once either the Thévenin or the Norton equivalent circuit is known, the other may be easily determined.

EXAMPLE 1–18   Determine the Thévenin equivalent circuit at the terminals *x–y* in Figure 1–35.

FIGURE 1–35
Circuit of Example 1–18.

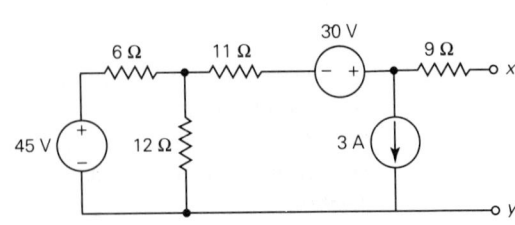

Solution

The process of determining V_{oc} directly with this circuit would require a somewhat detailed evaluation, so we will use a step-by-step approach using source transformations. In the steps that follow, reference should be made to the circuit diagrams of Figure 1–36.

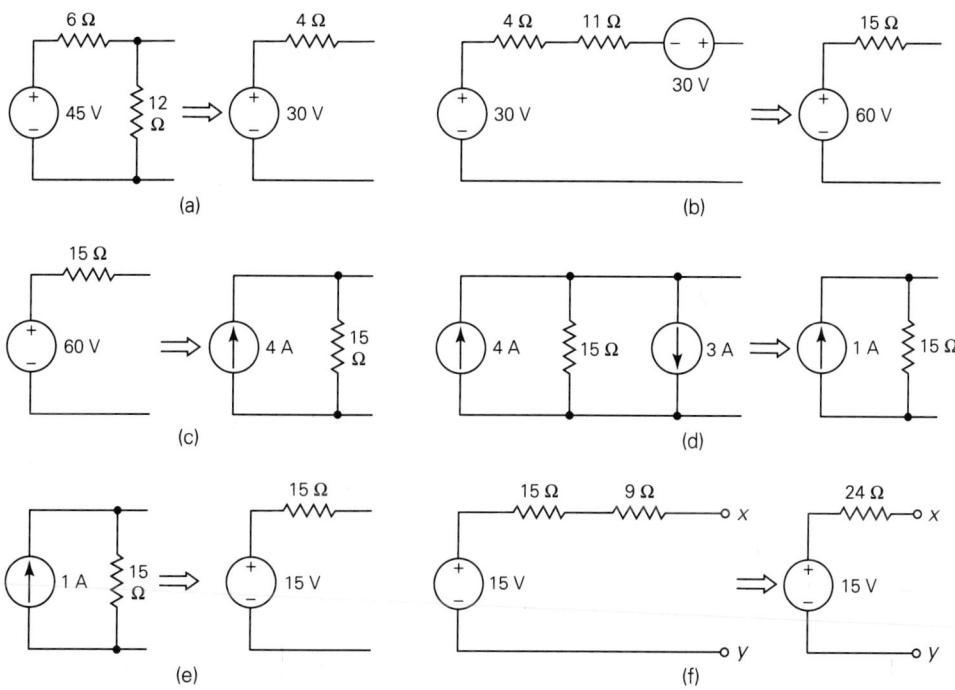

FIGURE 1–36
Successive reduction of the circuit of Example 1–18.

First, the circuit to the left of the 11-Ω resistor in Figure 1–35 is reduced to a 30-V source in series with 4 Ω, as shown in Figure 1–36(a). The details of this reduction are not shown, but it can be achieved by calculating the open-circuit voltage across the 12-Ω resistor and then determining the parallel combination of the two resistors after the source is deenergized. Next, the 11-Ω resistor and 30-V source are combined with the results of the preceding step, and the circuit is simplified as shown in Figure 1–36(b).

To combine this circuit with the 3-A shunt current source in Figure 1–36(d), the voltage sources and series resistor are first converted to a current source in parallel with a resistor as shown in Figure 1–36(c). Then, when the shunt current source is added, the two current sources may be combined as shown in Figure 1–36(d).

The next step consists of converting back to a voltage source form, as shown in Figure 1–36(e). The additional 9-Ω resistance is included and combined with the 15-Ω resistance as shown in Figure 1–36(f), which is the desired Thévenin form.

1–12 PRACTICAL SOURCE MODELS

All sources contain some internal resistance, and none are completely ideal. How does one know when a real source can be approximated by an ideal source model? We will investigate that question in this section.

First, consider the source of Figure 1–37(a) and assume that its internal resistance R_s is much smaller than any load R_L to which it is to be connected. Although either the Thévenin or the Norton model of the source could be used, for reasons that will be clear shortly, we will use the Thévenin model. By the voltage-divider rule, the load voltage v_L is given by

$$v_L = \frac{R_L}{R_L + R_S} v_S \qquad (1–66)$$

From previous work, we recall that if $R_s \leq 0.01\, R_L$, the error in approximating v_L by v_S is less than 1%. Thus, *a real source can be closely approximated by an ideal voltage source when its internal resistance is very small in comparison to the load resistance.*

Next, consider the source of Figure 1–37(b), and assume that its internal resistance R_s is much greater than any load R_L to which it is to be connected. In this case the Norton model is used. By the current-divider rule, the load current i_L is given by

$$i_L = \frac{R_s}{R_s + R_L} i_s \qquad (1–67)$$

FIGURE 1–37
Realistic source approximations for very low and very high internal resistances.

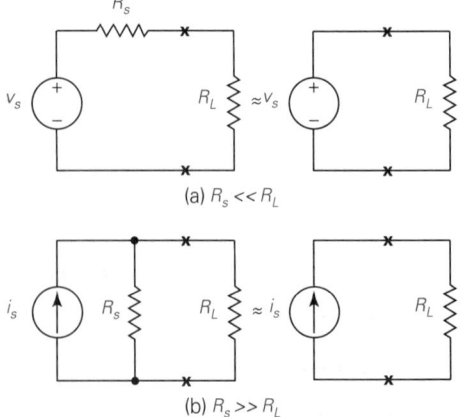

(a) $R_s \ll R_L$

(b) $R_s \gg R_L$

From previous work, we recall that if $R_s \geq 100\, R_L$, the error in approximating i_L by i_s is less than 1%. Thus *a real source can be closely approximated by an ideal current source when its internal resistance is very large in comparison to the load resistance.*

From the preceding two cases, we can say that *an ideal voltage source has zero internal resistance and that an ideal current source has infinite internal resistance.*

1–13 DEPENDENT SOURCES

All voltage and current sources considered thus far have been assumed to be **independent sources.** This means that their values are in no way dependent on other voltages and currents in the circuit.

The complex phenomena of semiconductor and vacuum electronic devices produce voltages and currents that are dependent on other voltages and currents within the circuit. The models for predicting the behavior of such devices use so-called **dependent** or **controlled sources.**

There are four possible types of dependent or controlled sources: (a) voltage-controlled voltage source, (b) voltage-controlled current source, (c) current-controlled voltage source, and (d) current-controlled current source. All of these arise in various electronic devices and circuits, as we will see throughout the text.

Voltage-Controlled Voltage Source

The model for the voltage-controlled voltage source (VCVS) is shown in Figure 1–38(a). An independent control voltage v_i is assumed to exist across certain control terminals. The voltage controls a dependent voltage whose value is Av_i. The quantity A is dimensionless and corresponds to a **voltage gain** when the model represents a voltage amplifier.

Voltage-Controlled Current Source

The model for the voltage-controlled current source (VCIS) is shown in Figure 1–38(b). An independent control voltage v_i is assumed to exist across certain control terminals. This voltage controls a dependent current whose value is $g_m v_i$. The quantity g_m, relating dependent current to the controlling voltage, has the dimensions of siemens and is called the **transconductance.**

Current-Controlled Voltage Source

The model for the current-controlled voltage source (ICVS) is shown in Figure 1–38(c). An independent control current i_i is assumed to be flowing in certain control terminals. This current controls a dependent voltage whose value is $R_m i_i$. The quantity R_m, relating the dependent voltage to the controlling current, has the dimensions of ohms and is called the **transresistance.**

FIGURE 1–38
Four possible models of ideal controlled
(or dependent) sources in electronic circuits.

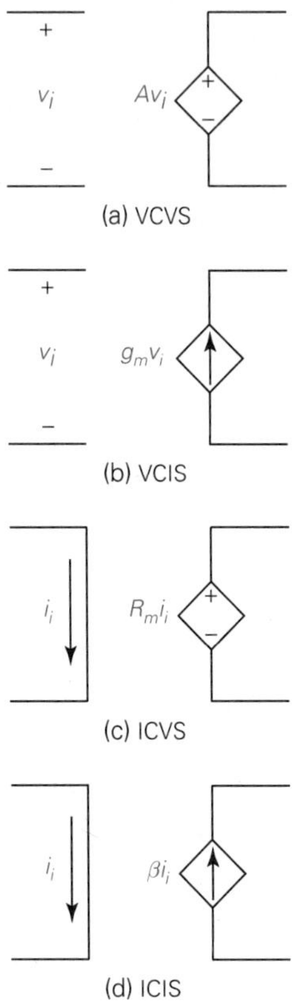

(a) VCVS

(b) VCIS

(c) ICVS

(d) ICIS

Current-Controlled Current Source

The model for the current-controlled current source (ICIS) is shown in Figure
1–38(d). An independent control current i_i is assumed to be flowing in certain
control terminals. This current controls a dependent current whose value is βi_i.
The quantity β is dimensionless and corresponds to a **current gain** when the model
represents a current amplifier.

1–14 THÉVENIN'S AND NORTON'S THEOREMS
WITH DEPENDENT SOURCES

The models used for predicting the performance of active electronic devices often
involve dependent sources. It is frequently necessary to apply Thévenin's and/
or Norton's theorems to such circuits to determine the gain and equivalent output
resistance. This analysis can be a bit tricky, and even experienced engineers and
technologists can become confused when working with such circuits.

In this section, we will discuss a procedure for working with such circuits.
We will call the procedure the **external generator method.** It turns out that this

procedure will actually work with virtually all circuits, but for a circuit without a dependent source, it is usually a bit of an overkill.

The external generator method is illustrated in Figure 1–39. First, the open-circuit voltage v_{oc} is determined as in Figure 1–39(a), as usual with a Thévenin analysis. The primary difference in this case is that some algebraic manipulation may be required, since the expression determined may be a function of itself. (This will be illustrated in Example 1–19.)

FIGURE 1–39

Steps in applying external generator method for determining Thévenin equivalent circuit.

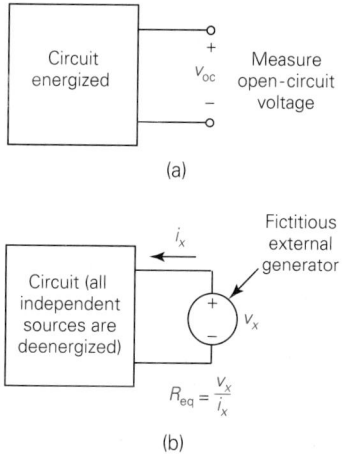

(a)

(b)

The part of the procedure requiring the most care is shown in Figure 1–39(b). First, all *independent* sources inside are deenergized and a fictitious generator v_x is applied to the external terminals. While independent sources inside will have zero values, there may indeed be *dependent* sources inside, since the presence of the external generator may keep some of them alive.

The circuit is analyzed by whatever method is appropriate to determine i_x as a function of v_x. The expression obtained will be proper when and only when i_x is a constant times v_x, that is, a linear function of v_x. If any other voltages or currents appear in the expression, the analysis is not complete.

Once the preceding operation is complete, the equivalent resistance is determined simply as

$$R_{eq} = \frac{v_x}{i_x} \qquad (1\text{–}68)$$

Not much more can be said about this procedure, since every case is a little different. Fortunately, this procedure will be necessary for only a few circuits in the text.

EXAMPLE 1–19

The circuit of Figure 1–40(a) represents the model of a certain electronic device that has negative feedback. The voltage v_i is an *independent* input signal voltage,

FIGURE 1–40
Circuit of Example 1–19.

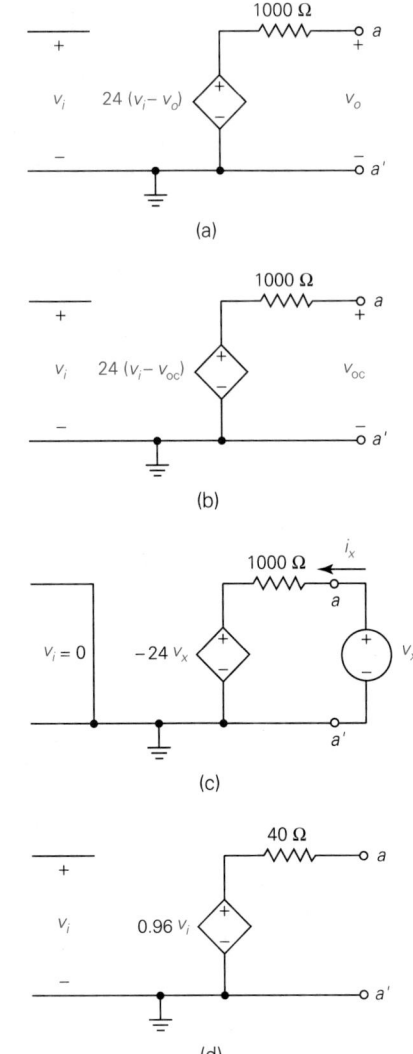

(a)

(b)

(c)

(d)

and v_o is the output voltage. Determine the Thévenin equivalent circuit at the output terminals $a-a'$.

Solution

In one sense, the output is already in the form of a Thévenin equivalent circuit. However, this is misleading, because the Thévenin voltage is a function of the input independent voltage and the output dependent voltage. To show the true Thévenin form, the effect of the dependent voltage v_o must be eliminated.

First, the open circuit output voltage v_{oc} should be calculated, and this process is illustrated in Figure 1–40(b). Note that since the output voltage for this measurement is v_{oc}, this value is substituted for v_o in the expression for the dependent generator. The value of v_{oc} is

$$v_{oc} = 24\,(v_i - v_{oc}) = 24v_i - 24v_{oc} \tag{1-69}$$

This expression seems strange in that v_{oc} is expressed in terms of itself, which does not truly demonstrate the relationship between v_i and v_{oc}. This problem is circumvented by moving the term $-24v_{oc}$ to the left side of the equation and combining with v_{oc}. This leads to

$$v_{oc} = \frac{24}{25}v_i = 0.96v_i \tag{1-70}$$

The Thévenin equivalent resistance is determined from the circuit shown in Figure 1–40(c). First, we deenergize the independent source by setting $v_i = 0$. Next, we apply a fictitious external generator v_x. Observe that these two steps change the value of the dependent source to $24\,(v_i - v_o) = 24\,(0 - v_x) = -24v_x$. A counterclockwise loop for the output circuit reads

$$-v_x + 1000\,i_x - 24v_x = 0 \tag{1-71}$$

Regrouping v_x terms on one side of the equation and i_x on the other side, we have

$$25v_x = 1000i_x \tag{1-72}$$

or

$$R_{eq} = \frac{v_x}{i_x} = \frac{1000}{25} = 40\,\Omega \tag{1-73}$$

The final Thévenin equivalent circuit is shown in Figure 1–40(d).

1–15 SUPERPOSITION

The principle of **superposition** states that any voltage or current response in a linear circuit resulting from several voltage and/or current sources may be determined by first considering the response produced by each source individually and then algebraically combining the individual responses. As the effect of each source is considered, all other sources are deenergized. To deenergize an ideal *voltage source*, it is replaced by a *short circuit*. To deenergize an ideal *current source*, it is replaced by an *open circuit*. Any internal resistances assumed in the source models are retained in the circuit when the ideal sources are deenergized. We illustrate the use of superposition with an example.

EXAMPLE 1–20

Consider the circuit of Example 1–18 (Figure 1–35), for which Thévenin and Norton equivalent circuits were established at terminals x–y. The circuit was simplified in Example 1–18 by using successive source transformations. Use superposition to compute V_{oc} from the original circuit.

Solution

For convenience, the circuit is repeated in Figure 1–41(a), and the desired variable V_{oc} is indicated. Since there are three sources in the circuit, there are three contributions to the voltage. Let

$$V_{oc} = V'_{oc} + V''_{oc} + V'''_{oc} \tag{1-74}$$

where V'_{oc} is the response due to the 45-V source, V''_{oc} is the response due to the 30-V source, and V'''_{oc} is the response due to the 3-A source.

To determine V'_{oc}, we deenergize the 30-V source and the 3-A source as shown in Figure 1–41(b). Note that the 30-V source is replaced by a short circuit and the 3-A source is replaced by an open circuit. Since no current flows through

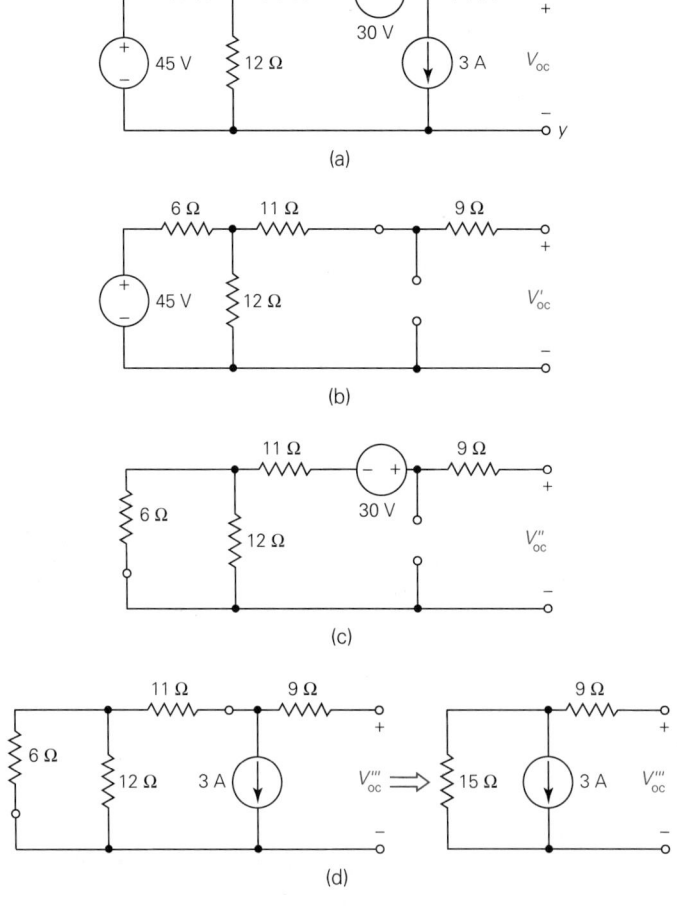

(a)

(b)

(c)

(d)

FIGURE 1–41
Circuit of Example 1–20 used to illustrate superposition.

the 11-Ω and 9-Ω resistors, V'_{oc} is the voltage across the 12-Ω resistor, and this is determined by the voltage-divider rule as follows:

$$V'_{oc} = \frac{12}{12 + 6} \times 45 = 30 \text{ V} \tag{1-75}$$

To determine V''_{oc}, we deenergize the 45-V source and the 3-A source as shown in Figure 1–41(c). No current flows in this case, so the open-circuit voltage is the same as the source, that is,

$$V''_{oc} = 30 \text{ V} \tag{1-76}$$

To determine V'''_{oc}, we deenergize the 45-V source and the 30-V voltage as shown on the left in Figure 1–41(d). To simplify this circuit, the resistance in parallel with the 3-A source is determined as 15 Ω, as shown on the right. The open-circuit voltage will have its more positive terminal at the bottom, and V'''_{oc} is

$$V'''_{oc} = -3 \times 15 = -45\text{V} \tag{1-77}$$

The net voltage from (1–74) is

$$V_{oc} = 30 + 30 - 45 = 15 \text{ V} \tag{1-78}$$

This result is in agreement with the outcome of Example 1–18, as shown in Figure 1–36.

1–16 VOLTAGE AND CURRENT VARIABLES AND NOTATION

We establish some basic conventions concerning voltage and current in this section. In a relatively simple circuit with only a few voltages and currents, no special notation may be necessary. However, as the circuit complexity increases, special notation and conventions are necessary, so that one can keep track of the many variables within the circuit. The convention will be established using uppercase symbols, which generally imply dc quantities. However, the concept may be applied to time-varying quantities as well. In the discussion that follows, the term *point* in a circuit will usually denote the same meaning as the term *node*.

Voltages between Nodes

Consider two nodes designated as node A and node B, respectively. The two-point notation V_{AB} is defined as the voltage at node A with respect to node B. Thus, node A is considered as the **positive reference** for the voltage. However, V_{AB} may be either a positive or a negative value. If V_{AB} is positive, the voltage at node A is indeed more positive than the voltage at node B. If, however, V_{AB} is negative, the voltage at node A is actually lower than the voltage at node B. In some cases it may be desirable to reverse the positive reference in the latter case.

If we move the positive reference node to B, the voltage should be expressed as V_{BA}. The two voltages are related by

$$\boxed{V_{BA} = -V_{AB}}$$

(1–79)

This concept is illustrated by the examples of Figure 1–42. The blocks could represent any circuit components. In (a), the voltage at node A is 5 V higher than the voltage at node B. In this case, $V_{AB} = 5V$. In (b), the voltage at node A is 6 V lower than the voltage at node B, so $V_{AB} = -6$ V. Alternatively, we can reverse the reference and say that $V_{BA} = 6$ V.

FIGURE 1–42
Examples of two-point voltage notation.

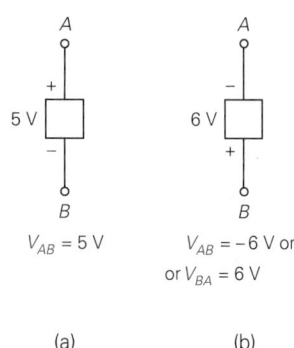

$V_{AB} = 5$ V

$V_{AB} = -6$ V or
or $V_{BA} = 6$ V

(a)   (b)

Voltages to Ground

Most electronic circuits contain a common ground from which most of the voltages are referred. Schematic diagrams often show many of the typical voltages with respect to ground that can be measured if the equipment is functioning properly. This allows one to determine probable faulty components when troubleshooting. The voltmeter common terminal is connected to the system common ground point, and the other lead is simply moved around to various nodes. In many cases, the common circuit ground is also connected to the chassis of the unit.

One could label the ground node as G and express the voltage at a point A with respect to ground as V_{AG}, if desired. However, when two-point quantities are being used in an analysis, we will usually simplify the notation by omitting the G subscript and having it understood. Thus a voltage V_A will be interpreted as the voltage at point A with respect to ground unless otherwise indicated.

This concept is illustrated by the circuit of Figure 1–43. This circuit is assumed to be part of a larger circuit, and the blocks could represent any components. Note that two ground symbols are shown, but they are at the same potential. To simplify schematic diagrams, ground symbols are typically shown at many points. In the actual wiring of the circuits, the grounds may be connected at different points, or in some critical cases, they may be brought to a single point to minimize difficulties with a phenomenon called **ground loops.**

FIGURE 1–43
Examples of voltages with respect to ground.

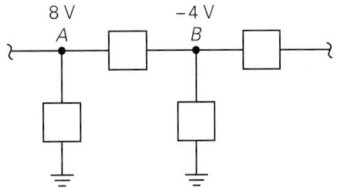

$V_A = 8$ V (with respect to ground)
$V_B = -4$ V (with respect to ground)

From the diagram, point A is 8 V above ground, which can be expressed as $V_A = 8$ V. Point B is 4 V below ground, so it can be expressed as $V_B = -4$ V. If it were necessary to express the latter voltage as a positive value, the ground symbol G would have to be introduced, and we could write $V_{GB} = 4$ V.

Currents

Consider a single branch connected between nodes A and B, and assume that there are no separate paths for current between A and B. This restriction assures that there is only a single current to consider. The current I_{AB} is defined as the current leaving node A and entering B through the branch. The actual current may be flowing from A to B or from B to A. If I_{AB} is positive, the current is actually flowing from A to B. If, however, I_{AB} is negative, the current is actually flowing from B to A.

In some cases it may be desirable to reverse the assumed direction. If we reverse the assumed direction, the current should be expressed as I_{BA}. The two currents are related by

$$I_{BA} = -I_{AB} \qquad (1\text{--}80)$$

This concept is illustrated by the examples of Figure 1–44. The blocks could represent any circuit components. In (a), the current is actually flowing from A to B, so $I_{AB} = 3$ A. In (b), the current is flowing from B to A, so $I_{AB} = -2$ A. Alternatively, we can say that $I_{BA} = 2$ A.

The general guidelines given here will be followed as necessary and appropriate throughout the book. The node symbols will often be those that represent certain terminals of a semiconductor device. For example, a bipolar junction

FIGURE 1–44
Examples of two-point current notation.

$A \xrightarrow{\quad 3\text{ A}\quad} B$

$I_{AB} = 3$ A

(a)

$A \xleftarrow{\quad 2\text{ A}\quad} B$

$I_{AB} = -2$ A or
or $I_{BA} = 2$ A

(b)

transistor (BJT) has emitter, base, and collector terminals, so the corresponding node symbols will be *E, B,* and *C*. With many transistor currents, however, it may be desirable to employ only one subscript.

1–17 USEFUL VOLTAGE RELATIONSHIPS

We discuss two useful concepts for determining voltages between points in a circuit in this section. Both of these concepts are special applications of Kirchhoff's voltage laws. However, they are sufficiently useful that they desire special consideration.

Voltmeter Loop

Frequently, it is necessary to determine the voltage between two points in a circuit separated by a number of branches, for which the voltage across each branch is known. As a simple example, consider the series connection of Figure 1–45(a), in which the voltage V_{AB} across the combination is desired. The two branch voltages are 10 V and 5 V, and the polarities are aiding. Most readers will probably deduce immediately by an intuitive process that the net voltage between the two points is 15 V and that point *A* is the most positive node.

FIGURE 1–45
Technique for forming a voltmeter loop
with a simple circuit.

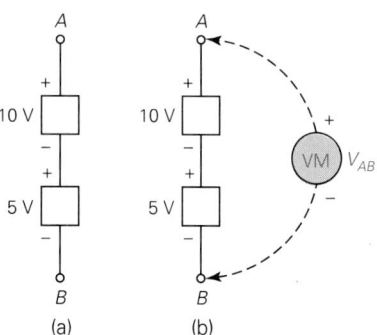

If that example seems too easy, look at Figure 1–46, in which the voltage between points *A* and *B* is to be determined. Some readers might be able to deal with this circuit on an intuitive basis, but the chances of error in determining the magnitude and the sign of the net voltage are much higher in this case.

A technique that we refer to as "forming a voltmeter loop" is the best way to deal with this situation. To apply this concept, there must be at least one path between the two points along which all branch voltages are known. We then assume that an ideal voltmeter is connected between the two points. The voltage measured by the voltmeter is the unknown voltage to be determined. The combination of the path mentioned previously and the voltmeter constitute a closed loop, and KVL may be applied.

FIGURE 1–46
Voltmeter loop formed in a more complex circuit.

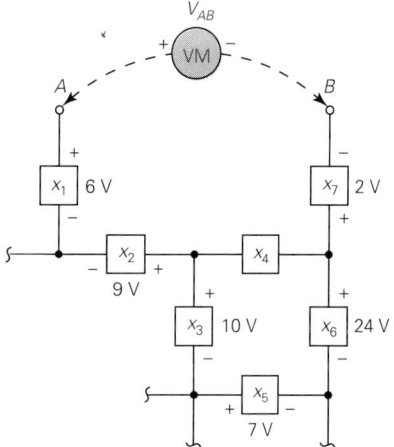

We illustrate the concept first by the simple circuit of Figure 1–45(a). An
ideal voltmeter is assumed to be connected between points A and B, as shown in
Figure 1–45(b). The positive reference terminal of the voltmeter is connected to
the assumed positive reference terminal of the voltage. Thus, to determine V_{AB},
the positive voltmeter terminal is connected to node A, and the negative voltmeter
terminal is connected to node B. The voltage across the voltmeter is labeled as
V_{AB}. An ideal voltmeter draws no current, so none of the branches are assumed
to be disturbed by the voltmeter's presence.

The closed loop that results is now evident, and KVL may be applied.
Starting at B and moving clockwise, we have

$$-5 - 10 + V_{AB} = 0 \qquad \qquad \textbf{(1–81)}$$

or

$$V_{AB} = 15 \text{ V} \qquad \qquad \textbf{(1–82)}$$

which agrees with our earlier intuitive observation.

Consider next the more complex circuit of Figure 1–46. Observe that the
path from A to B through x_1, x_2, x_3, x_5, x_6, and x_7 has all branch voltages known.
Note that the voltage across branch x_4 is not known without a separate calculation,
so that branch will be avoided in the path. (The voltage across x_4 could be calculated
if desired. Can the reader supply this value?)

Consider that point A is the assumed positive reference. In this case we start
at point A and move clockwise. Application of KVL yields

$$V_{AB} - 2 + 24 - 7 - 10 + 9 - 6 = 0 \qquad \qquad \textbf{(1–83)}$$

or

$$V_{AB} = -8 \text{ V} \qquad \qquad \textbf{(1–84)}$$

This result means that the voltage at point A is actually lower than the voltage at point B, so we write

$$V_{BA} = 8 \text{ V} \tag{1–85}$$

A good practical measurement reminder will be inserted here. Assume that this was an actual circuit and that the voltmeter had been connected as shown. If the voltmeter had been a digital meter, the negative or minus indication would have been generated, and the meter could remain as originally connected to take the reading. However, an analog voltmeter would have experienced a negative deflection, and possible damage could result. When the polarity and approximate level of a voltage are unknown, start with the highest available meter scale. A quick initial check can be made to determine the polarity. If the deflection is negative, the meter terminals can be reversed so that a positive deflection results. The range switch can then be adjusted downward until an accurate measurement can be made.

Voltage between Two Ground-Referred Points

Frequently, it is desired to determine the voltage between two points in a circuit, for which the voltage is known at each point with respect to ground. Consider the two nodes A and B in Figure 1–47(a), and assume that the respective voltages with respect to ground are V_A and V_B.

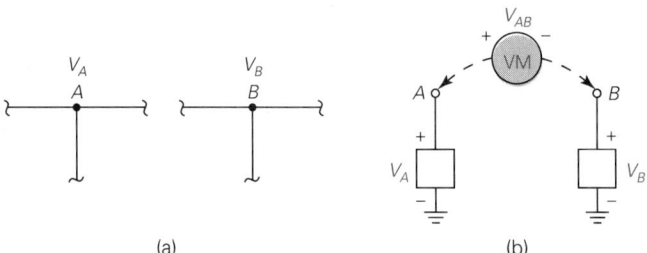

(a)   (b)

FIGURE 1–47
Process of determining the voltage between two ground-referred points using a voltmeter loop.

To determine the voltage between A and B, we can model the circuit as shown in Figure 1–47(b). Although there can be any number of branches between points A and B and ground, the voltages V_A and V_B can be visualized as shown. An ideal voltmeter is connected between A and B to form a loop according to the procedure discussed. Note that the positive reference of V_A and V_B are at nodes A and B, respectively. However, either V_A, V_B, or both could be negative.

Starting at ground and moving in a clockwise direction through the left-hand part of the circuit, the KVL equation is

$$-V_A + V_{AB} + V_B = 0 \qquad \textbf{(1–86)}$$

or

$$\boxed{V_{AB} = V_A - V_B} \qquad \textbf{(1–87)}$$

Once understood, this relationship is very simple to apply. The voltage between any two points is the voltage of the assumed positive reference node with respect to ground minus the voltage of the assumed negative reference node with respect to ground.

As an example, refer to Figure 1–43, in which $V_A = 8$ V and $V_B = -4$ V. The voltage V_{AB} between points A and B, with A the assumed positive reference, is

$$V_{AB} = V_A - V_B = 8 - (-4) = 12 \text{ V} \qquad \textbf{(1–88)}$$

1–18 PSPICE EXAMPLES

Beginning with this chapter and continuing throughout the text, we will present one or more PSPICE examples as the last section of each chapter. The sections are optional and may be omitted without loss of continuity in the remainder of the text. Nowhere are references made elsewhere in the chapters to these examples. However, they are offered as supplements to illustrate the power and utility of computer modeling.

Many of the examples involve adapting various text examples to PSPICE analysis. Depending on the circuits and the complexity of the models involved, the results may turn out exactly the same as the text examples or they may be slightly different. This is to be expected, since there are many approximations involved in electronic circuit analysis. In this chapter, the results should turn out to be very accurate, since they are based on exact circuit models, but later chapters will reveal various differences.

We will present PSPICE examples as if the reader is familiar with the general process of PSPICE analysis. However, readers who do not have this background should carefully study Appendix E first. A sufficient description of the process is given there to acquaint readers with enough background to follow the examples. Further explanations will be given as they arise in examples.

The notation involved with PSPICE varies somewhat from conventional circuit analysis, so the practice of presenting circuit diagrams adapted to the notation will be followed extensively. For example, subscripted variables are not easily expressed in PSPICE, so a circuit quantity such as I_E will be expressed as IE, and references in the example discussion will sometimes follow the second form. Although lowercase and uppercase symbols are usually interchangeable with PSPICE, we prefer uppercase symbols. Thus, a circuit variable such as i_b could be expressed in PSPICE as ib, but in this text we will use IB.

PSPICE employs a number of prefixes on values, and some readers may find them useful. However, some differ from standard SI notation. For example, either m or M could represent 10^{-3} in PSPICE, while the latter represents 10^6 in SI notation. To avoid possible misinterpretation, *all values will be expressed in their basic units in PSPICE circuits and files, and scientific notation in floating point form will be used extensively*. Thus, a current of 5 μA will be expressed as 5E−6, and a resistance of 2 kΩ will be expressed as 2E3. Basic units will be so understood in all cases unless indicated otherwise.

PSPICE EXAMPLE 1–1

Develop a PSPICE program to determine the voltages across the three resistors and the current in the circuit of Example 1–7 (Figure 1–12).

Solution

For all circuits to be solved by PSPICE, the first step should always be to redraw the circuit and label it in PSPICE notation. This should be considered as a mandatory step, for without it the coding can become very unwieldy.

The circuit is redrawn for PSPICE analysis in Figure 1–48(a). The discussion that follows will explain the various conventions and symbols.

The next step is to label all of the nodes, that is, all of the junction points between branches. All PSPICE circuits *must* contain a ground node, and this *node*

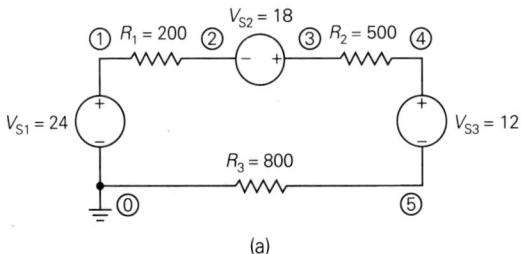

(a)

```
PSPICE EXAMPLE 1-1
VS1 1 0 DC 24
R1 1 2 200
VS2 3 2 18
R2 4 3 500
VS3 4 5 12
R3 5 0 800
.DC VS1 24 24 1
.PRINT DC V(1,2) V(3,4) V(5) V(R1) V(R2)
.PRINT DC I(R1) I(R2) I(R3) I(VS1) I(VS3)
.OPTIONS NOPAGE
.END
```

(b)

FIGURE 1–48
Circuit and code of PSPICE Example 1–1.

must be denoted as node 0. Since the original circuit does not contain a ground node, the negative terminal of the 24-V source will be arbitrarily selected as the ground reference. The numbers for the other nodes are arbitrary, but for ease in checking, a systematic scheme is recommended. In this circuit, the other five nodes are numbered as 1 through 5 in a clockwise manner. Throughout the text, node numbers will be circled as shown.

All elements must have names associated with the types of elements, and since V is the required first symbol for a voltage source, the three sources have been designated as V_{S1}, V_{S2}, and V_{S3}, respectively. All values are expressed in their basic units, so $R_1 = 200$ refers to 200 Ω, and $V_{S1} = 24$ refers to 24 V, and so on.

The code is shown in Figure 1–48(b). Each PSPICE program must begin with a title line, which in this case is listed as PSPICE Example 1–1.

The next six lines constitute the circuit block and describe to the computer the exact connections within the circuit. Each line corresponds to one element or branch in the circuit. The order is arbitrary, but in the same sense as the labeling, it should follow some logical order.

Let us begin with the 24-V source and continue in a clockwise fashion. The first line in the circuit block reads

```
VS1 1 0 DC 24
```

The name of the source is the first entry, and it is VS1. The next two entries are the nodes between which the element is connected. The order for a voltage source is very important in that the first number (1) is the positive terminal of the source, and the second number (0) is the negative terminal. The next entry DC indicates that the source is a dc source. The last entry 24 is the value of the voltage source in volts.

The next line reads

```
R1 1 2 200
```

The name of the resistor is the first entry, and it is R1. The next two entries 1 2 are the nodes between which the resistor is connected. The order may or may not be important depending on the desired output. If current is not to be measured, the order is usually unimportant. However, if current is to be measured, the order of the nodes defines the positive reference direction for the current from the first node number to the second node number. Thus, the positive reference direction for current in R1 is from node 1 to node 2.

The next line defines VS2 and it has the same form as that of VS1. In this case, the first node number 3 is the positive terminal of that source.

To make a point for later discussion in the data printout, the positive reference direction of current flow in R2 is from node 4 to node 3, which is opposite to that of R1.

The reader should be able to follow the remaining code lines for the circuit block. Note that the positive current direction for R3 is consistent with that of R1 in that clockwise current flow would be positive.

The circuit description is now concluded. We must now tell PSPICE what types of analysis and output data are desired.

The first statement in the control block reads

```
.DC VS1 24 24 1
```

This command looks a little clumsy and needs to be explained. The command .DC tells PSPICE to perform a dc analysis. In general, the format for a dc analysis is based on a sweep of one or more of the variables. For a single-point analysis, one of the variables must be selected for the sweep, and this variable has been arbitrarily selected as VS1 in this case. The first value identifies the starting value, and the second value indicates the ending value. The third value indicates the step size. For a single-point analysis, the starting and ending values are listed as the same (24 V in this case). The third value may be any nonzero value, and the simplest choice is the number 1. This type of statement will appear often in the text, so the form should be carefully noted.

Following the dc analysis statement, one or more print statements are listed. Up to five dc variables may be listed in a single block of output data. A brief discussion of the format for output print variables will now be made.

The most general form for indicating the voltage desired between two points in a circuit is V(N, M), where N and M are the two node numbers across which the voltage is desired. The first number N is the positive reference node, and M is the negative reference. For example, V(3, 5) is the voltage between nodes 3 and 5 with 3 being the positive reference.

Any voltage referred to ground could, by the preceding scheme, be indicated as V(N, 0). However, PSPICE allows the 0 to be omitted whenever the negative reference is node 0. Thus, V(3) is the same thing as V(3, 0), and the shorter form will be employed throughout the text.

The preceding voltage rules are actually the most general and can be employed in all circuits when desired. However, there are other rules that can be employed when it is more convenient. For example, the voltage across a resistance R1 can be indicated as V(R1) if desired. The positive reference in this case is the first node number used in defining the branch for R1. As we will see, many active devices have special codes referring to specific terminal names.

The rules for measuring current are somewhat more restricted than for voltages. There are no formats for measuring currents based on node numbers, for example. For the most part, currents are designated by I ( ) with the particular *element name* enclosed within the parentheses. For example, if the current through R1 is desired, it would be listed as I(R1). The positive reference for the current flow is from the first node number to the second node number in the defining branch. This results in a slightly awkward interpretation for the current through a voltage source. For example, I(VS) refers to the current through a voltage source VS. Since in the branch description for VS, the first node number is the positive reference, this implies that the positive reference flow for current is *into* the positive terminal for the voltage source. While awkward, it need not cause any problem if properly understood, and it is consistent with the positive flow for all other devices. In the same manner as for voltages, there will be special designations for certain semiconductor terminals that we will encounter later.

To show some of these various forms in the output data, two print statements have been used. Obviously, there is only one current in the circuit, but the second print statement will show how several variations of this variable can be listed.

The output data are shown in Figure 1–49. The first block of data indicated as "DC Transfer Curves" is associated with the first print statement, and the sweep variable VS1 is listed as 24 V. The next five quantities represent the five voltages requested in the printout. Note from Figure 1–44 that V(1, 2) is the same as V(R1); that is, V(1, 2) = V(R1) = 4 V. The voltage V(3, 4) would be the same as V(R2) if R2 had been defined from 3 to 4. However, since R2 was defined from 4 to 3, V(R2) = −10 V, while V(3, 4) = 10 V. Finally, V(5) is the same as V(5, 0) and is V(5) = 16 V. This voltage could also be indicated as V(R3) if desired.

The second block of data is associated with the second print statement, and the sweep variable VS1 is listed again. The next five quantities represent different ways of expressing the single loop current. The currents I(R1) and I(R3) are based

```
PSPICE EXAMPLE 1-1

****      CIRCUIT DESCRIPTION

*********************************************************************

VS1 1 0 DC 24
R1 1 2 200
VS2 3 2 18
R2 4 3 500
VS3 4 5 12
R3 5 0 800
.DC VS1 24 24 1
.PRINT DC V(1,2) V(3,4) V(5) V(R1) V(R2)
.PRINT DC I(R1) I(R2) I(R3) I(VS1) I(VS3)
.OPTIONS NOPAGE
.END
```

****	DC TRANSFER CURVES			TEMPERATURE =	27.000 DEG C
VS1	V(1,2)	V(3,4)	V(5)	V(R1)	V(R2)
2.400E+01	4.000E+00	1.000E+01	1.600E+01	4.000E+00	−1.000E+01

****	DC TRANSFER CURVES			TEMPERATURE =	27.000 DEG C
VS1	I(R1)	I(R2)	I(R3)	I(VS1)	I(VS3)
2.400E+01	2.000E−02	−2.000E−02	2.000E−02	−2.000E−02	2.000E−02

```
        JOB CONCLUDED

        TOTAL JOB TIME        1.09
```

FIGURE 1–49
Computer printout of PSPICE Example 1–1.

on the defining branches for R1 and R3, which were in the positive direction of flow and I(R1) = I(R3) = 20 mA. The current I(R2) = −20 mA is a result of that branch being defined from node 4 to node 3. The current is flowing out of the positive terminal of VS1 so I(VS1) = −20 mA. However, the current is flowing into the positive terminal of VS3, and I(VS3) = 20 mA. All of the results are in perfect agreement with those of Example 1–7.

The last statement before the .END statement reads

```
.OPTIONS NOPAGE
```

The OPTIONS line allows several format conventions to be listed. The NOPAGE option is one that suppresses the repetition of certain header information on each page of the printout. While not required, its use generally reduces the volume of paper in the printout.

PSPICE EXAMPLE 1–2

Use PSPICE to analyze the single node-pair circuit of Example 1–10 (Figure 1–17).

Solution

The circuit adapted to the PSPICE format is shown in Figure 1–50(a), and the code is shown in Figure 1–50(b). Following the title line, the description of the current source IS1 reads as follows:

```
IS1 0 1 DC 28E-3
```

The first letter I identifies the source as a current source, and the additional characters are identifiers.

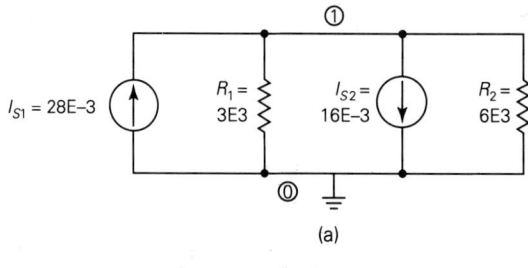

(a)

```
PSPICE EXAMPLE 1-2
IS1 0 1 DC 28E-3
R1 1 0 3E3
IS2 1 0 16E-3
R2 1 0 6E3
.DC IS1 28E-3 28E-3 1
.PRINT DC V(1) I(R1) I(R2)
.OPTIONS NOPAGE
.END
```

(b)

FIGURE 1–50
Circuit and code of PSPICE Example 1–2.

The node numbers 0 1 that follow indicate the terminals of the current source. The order is very important in that the first node number (0) is the node at which the current enters the source, and the second node number (1) is the node at which the current leaves the source. Note that the order of nodes for IS2 is opposite to that of IS1, since that current is directed downward.

The remainder of the code should be self-explanatory from the work of the preceding example. The data desired in the printout are the voltage V(1) at node 1 and the currents through the two resistors.

The output data are provided in Figure 1–51. The results are in perfect agreement with those of Example 1–10.

```
PSPICE EXAMPLE 1-2

****      CIRCUIT DESCRIPTION


*********************************************************************

IS1 0 1 DC 28E-3
R1 1 0 3E3
IS2 1 0 16E-3
R2 1 0 6E3
.DC IS1 28E-3 28E-3 1
.PRINT DC V(1) I(R1) I(R2)
.OPTIONS NOPAGE
.END


****      DC TRANSFER CURVES              TEMPERATURE = 27.000 DEG C


 IS1          V(1)        I(R1)        I(R2)
  2.800E-02    2.400E+01    8.000E-03   4.000E-03


      JOB CONCLUDED

      TOTAL JOB TIME            .77
```

FIGURE 1–51
Computer printout of PSPICE Example 1–2.

PSPICE EXAMPLE 1–3

Use PSPICE to analyze the dc resistive circuit of Example 1–14 (Figure 1–24).

Solution
The circuit adapted to the PSPICE format is shown in Figure 1–52(a), and the code is shown in Figure 1–52(b). The only new statement in this example is the use of the NOECHO option. This option suppresses the circuit file from appearing on the page containing the output data file. As in the case of the NOPAGE option, this will usually reduce the volume of the printout. The NOECHO option should appear in the second line, so that the remaining part of the code will be suppressed. The NOPAGE option can be placed on the same line.

FIGURE 1–52
Circuit and code of PSPICE Example
1–3.

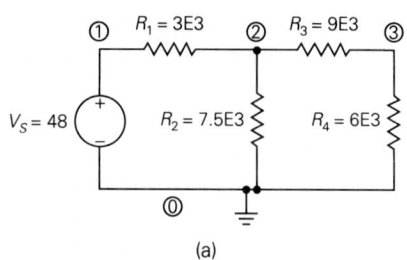

(a)

```
PSPICE EXAMPLE 1-3
.OPTIONS NOPAGE NOECHO
VS 1 0 DC 48
R1 1 2 3E3
R2 2 0 7.5E3
R3 2 3 9E3
R4 3 0 6E3
.DC VS 48 48 1
.PRINT DC V(R1) V(R2) V(R3) V(R4)
.PRINT DC I(R1) I(R2) I(R3) I(R4)
.END
```

(b)

```
PSPICE EXAMPLE 1-3

****      CIRCUIT DESCRIPTION

*************************************************************************

.OPTIONS NOPAGE NOECHO

****      DC TRANSFER CURVES              TEMPERATURE =   27.000 DEG C

VS           V(R1)         V(R2)        V(R3)         V(R4)

 4.800E+01    1.800E+01     3.000E+01    1.800E+01     1.200E+01

****      DC TRANSFER CURVES              TEMPERATURE =   27.000 DEG C

VS            I(R1)         I(R2)        I(R3)         I(R4)

 4.800E+01    6.000E-03     4.000E-03    2.000E-03     2.000E-03

          JOB CONCLUDED

          TOTAL JOB TIME            1.09
```

FIGURE 1–53
Computer printout of PSPICE Example 1–3.

The computer output data are shown in Figure 1–53. The results are in perfect agreement with those of Example 1–14.

PSPICE
EXAMPLE
1–4

Use the transfer function command to determine the Thévenin equivalent circuit of Example 1–15 (Figure 1–32).

Solution

The transfer function command (.TF) is a dc function that determines for a given circuit the following three parameters: (a) input resistance, (b) output resistance, and (c) ratio of an output variable to an input variable. Careful use of parts (b) and (c) can lead to the determination of a Thévenin equivalent circuit, as we will see shortly.

The circuit adapted to the PSPICE format is shown Figure 1–54(a), and the code is shown in Figure 1–54(b). To determine a Thévenin equivalent circuit, the output variable should be the voltage across the reference output terminals, which in this case is V(2). The transfer function command is written as

```
.TF V(2) VS
```

The variable following .TF is the output voltage V(2), and the second variable is the assumed input VS.

FIGURE 1–54
Circuit and code of PSPICE Example 1–4.

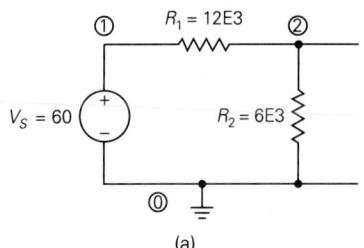

(a)

```
PSPICE EXAMPLE 1-4
.OPTIONS NOPAGE NOECHO
VS 1 0 DC 60
R1 1 2 12E3
R2 2 0 6E3
.TF V(2) VS
.END
```

(b)

The computer printout is shown in Figure 1–55. Under "Small-Signal Characteristics," the output resistance is listed as 4000 Ω. Since V(2)/VS = 0.3333, the open-circuit voltage can be calculated as V(2) = 0.3333 × 60 = 20 V. However, this voltage may also be read from the "Small-Signal Bias Solution."

```
PSPICE   EXAMPLE 1-4


 ****      CIRCUIT DESCRIPTION


********************↷*********************************************************

.OPTIONS NOPAGE NOECHO

 ****      SMALL SIGNAL BIAS SOLUTION      TEMPERATURE =  27.000 DEG C

 NODE   VOLTAGE       NODE   VOLTAGE      NODE  VOLTAGE      NODE   VOLTAGE

(    1)   60.0000  (    2)   20.0000

        VOLTAGE SOURCE CURRENTS
        NAME           CURRENT

        VS             -3.333E-03

        TOTAL POWER DISSIPATION   2.00E-01 WATTS

 ****      SMALL-SIGNAL CHARACTERISTICS

        V(2)/VS =  3.333E-01

        INPUT RESISTANCE AT VS =  1.800E+04

        OUTPUT RESISTANCE AT V(2) =  4.000E+03


            JOB CONCLUDED

        TOTAL JOB TIME             .99
```

FIGURE 1–55
Computer printout of PSPICE Example 1–4.

PSPICE EXAMPLE 1–5

Use the transfer function command to determine the Thévenin equivalent circuit of Example 1–19 (Figure 1–40).

Solution

The circuit adapted to the PSPICE format is shown in Figure 1–56(a), and the code is shown in Figure 1–56(b). PSPICE will not allow a dangling node (except where it is a control variable), and thus it is necessary to place a resistor from node 3 to ground. The value 1E9 Ω should be sufficiently large that its effect will be negligible.

The circuit contains a voltage-controlled voltage source, for which the PSPICE code must begin with the letter E. This particular branch reads

```
E 2 0 1 3 24
```

FIGURE 1–56
Circuit and code of PSPICE Example 1–5.

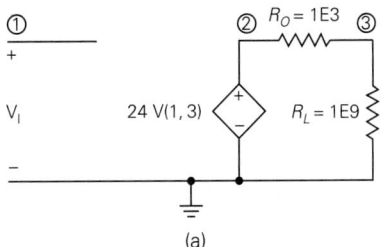

(a)

```
PSPICE EXAMPLE 1-5
.OPTIONS NOPAGE NOECHO
VI 1 0 DC 1
E 2 0 1 3 24
RO 2 3 1E3
RL 3 0 1E9
.TF V(3) VI
.END
```

(b)

```
PSPICE EXAMPLE 1-5

****      CIRCUIT DESCRIPTION


*************************************************************

.OPTIONS NOPAGE NOECHO

****      SMALL SIGNAL BIAS SOLUTION       TEMPERATURE =  27.000 DEG C

NODE   VOLTAGE     NODE   VOLTAGE     NODE   VOLTAGE     NODE   VOLTAGE

(   1)    1.0000  (   2)     .9600  (   3)     .9600


       VOLTAGE SOURCE CURRENTS
       NAME            CURRENT

       VI            0.000E+00

       TOTAL POWER DISSIPATION   0.00E+00   WATTS

****      SMALL SIGNAL CHARACTERISTICS

          V(3)/VI = 9.600E-01

          INPUT RESISTANCE AT VI = 1.000E+20

          OUTPUT RESISTANCE AT V(3) = 4.000E+01


          JOB CONCLUDED

          TOTAL JOB TIME              .71
```

FIGURE 1–57
Computer printout of PSPICE Example 1–5.

The first two integers 2 0 represent the output nodes of the source with the first (2) representing the positive terminal. The next two integers 1 3 represent the nodes across which the controlling voltage is defined, with the first (1) representing the positive control reference. Thus, the controlling voltage is $V(1) - V(3) = V(1, 3)$. The last number 24 is the gain factor; that is, the value of the dependent source is $24V(1, 3)$.

The output data are shown in Figure 1–57. These results for output resistance and $V(3)/VI$ are in perfect agreement with those of Example 1–19.

PROBLEMS

Drill Problems

1–1. For each of the branches shown in Figure P1–1, determine the power and whether it is being delivered or absorbed. (The components are not identified.)

FIGURE P1–1

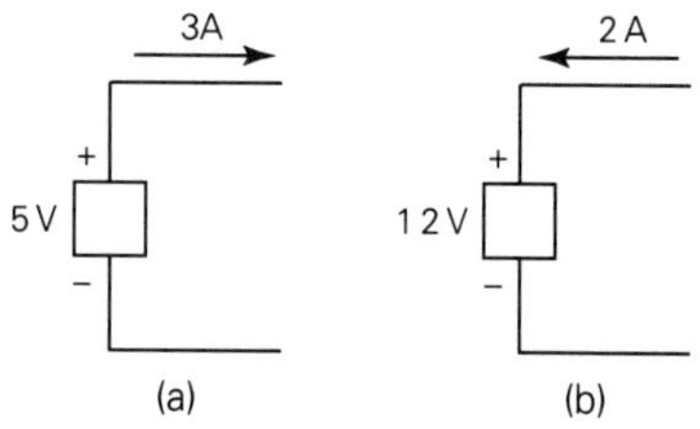

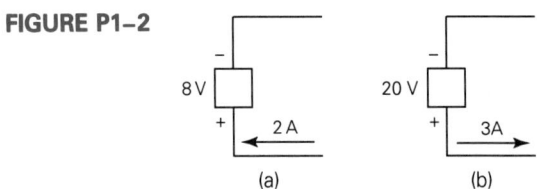

(a)   (b)

1–2. For each of the branches shown in Figure P1–2, determine the power and whether it is being delivered or absorbed. (The components are not identified.)

FIGURE P1–2

(a)   (b)

1–3. The dc voltage across a 2.7-kΩ resistor is 12.6 V. Determine the dc current and the power.

1–4. The dc voltage across a 68-kΩ resistor is 30 V. Determine the dc current and the power.

1–5. The dc current through a 3-kΩ resistor is 12 mA. Determine the dc voltage and the power.

1–6. The dc current through a 150-kΩ resistor is 8 μA. Determine the dc voltage and the power.

1–7. The dc voltage and current associated with a certain resistor are 30 V and 2 mA. Determine the resistance and the power.

1–8. The dc voltage and current associated with a certain resistance are 60 V and 3 μA. Determine the resistance and the power.

1–9. A certain resistor has a resistance of 2 kΩ. Determine the conductance.

1–10. A certain resistor has a conductance of 8 μS. Determine the resistance.

1–11. A certain resistor dissipates 0.4 W when the voltage is 20 V. Determine the resistance.

1–12. A certain resistor dissipates 0.2 W when the current is 50 mA. Determine the resistance.

1–13. In the circuit of Figure P1–13, the source voltage and the voltage at the right-hand node with respect to ground are known. Two resistance values are known. Determine V_1, I_1, I_2, and R.

FIGURE P1–13

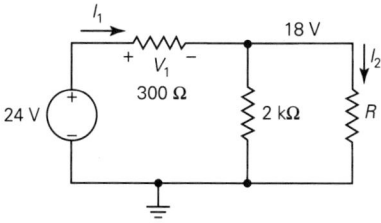

1–14. In the circuit of Figure P1–14, the two source voltages and the voltage across one resistor are known. Determine I_1, I_2, and R.

FIGURE P1–14

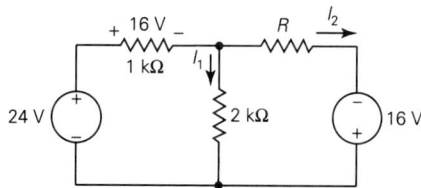

1–15. For the circuit of Figure P1–15, determine V_o and R.

FIGURE P1–15

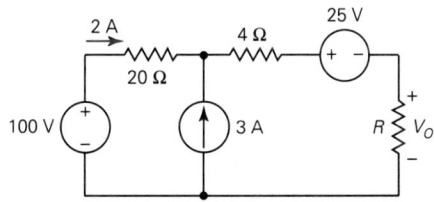

1–16. For the circuit of Figure P1–16, determine I_1, I_2, and V_o.

FIGURE P1–16

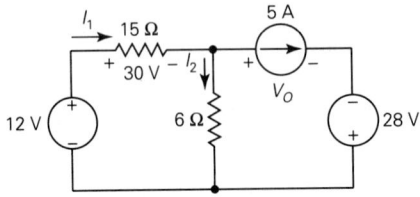

1–17. Using successive series and parallel reduction, determine the equivalent resistance at the input terminals for the circuit of Figure P1–17.

FIGURE P1–17

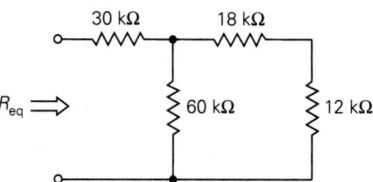

1–18. Using successive series and parallel reduction, determine the equivalent resistance at the input terminals for the circuit of Figure P1–18.

FIGURE P1–18

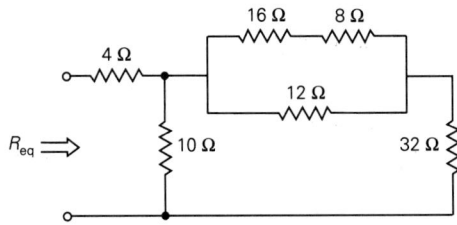

1–19. For the circuit of Figure P1–19, determine **(a)** the loop current I, **(b)** the voltage drops across all resistors, and **(c)** the power associated with each element and whether it is absorbed or delivered.

FIGURE P1–19

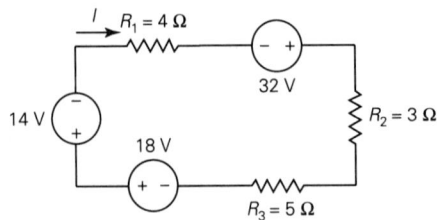

1-20. For the circuit of Figure P1–20, determine (a) the loop current I, (b) the voltage drops all resistors, and (c) the power associated with each element and whether it is absorbed or delivered.

FIGURE P1–20

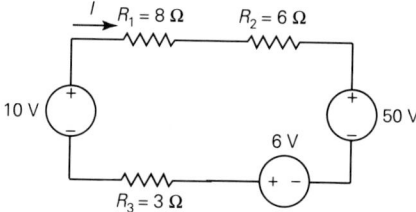

1-21. For the circuit of Figure P1–21, determine (a) the loop current I, (b) the voltage drops across all resistors, and (c) the voltage V_o across the current source.

FIGURE P1–21

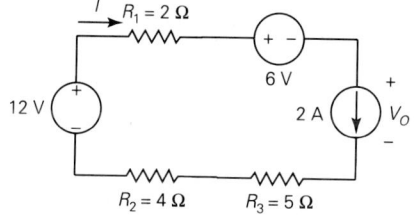

1-22. For the circuit of Figure P1–22, determine (a) the loop current I, (b) the voltage drops across all resistors, and (c) the voltage V_o across the current source.

FIGURE P1–22

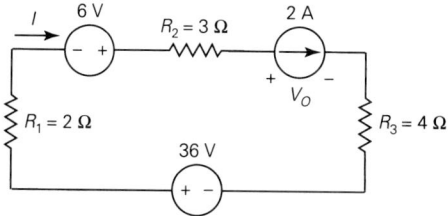

1-23. For the circuit of Figure P1–23, determine (a) the node-pair voltage V, and (b) the currents through all resistors.

FIGURE P1–23

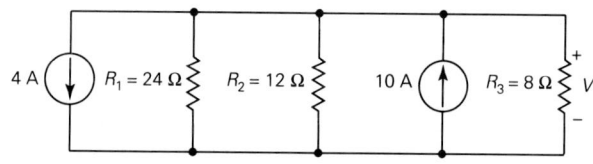

1–24. For the circuit of Figure P1–24, determine **(a)** the node-pair voltage V, and **(b)** the currents through all resistors.

FIGURE P1–24

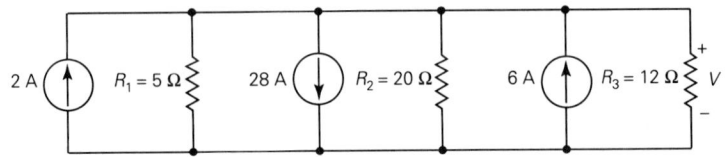

1–25. For the circuit of Figure P1–25, determine **(a)** the node-pair voltage V, **(b)** the currents through all resistors, and **(c)** the current I_o through the voltage source.

FIGURE P1–25

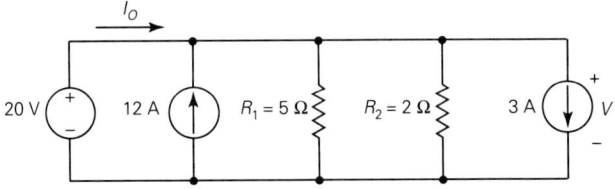

1–26. For the circuit of Figure P1–26, determine **(a)** the node-pair voltage V, **(b)** the currents through all resistors, and **(c)** the current I_o through the voltage source.

FIGURE P1–26

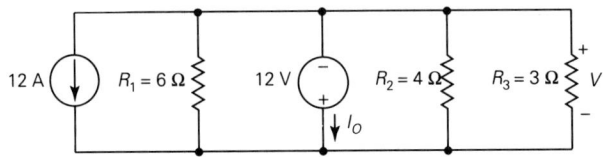

1–27. Use the voltage-divider rule to determine the voltages V_1 and V_2 in Figure P1–27.

FIGURE P1–27

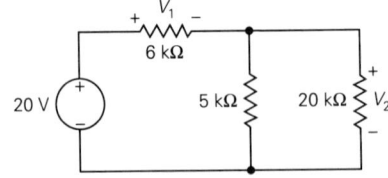

1–28. Use the voltage-divider rule to determine the voltages V_1 and V_2 in Figure P1–28.

FIGURE P1–28

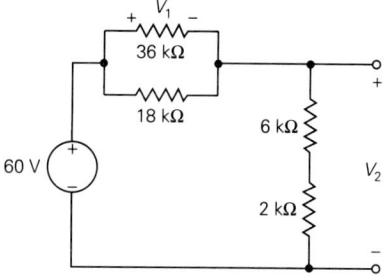

1–29. Use the current-divider rule to determine the currents I_1 and I_2 in Figure P1–29.

FIGURE P1–29

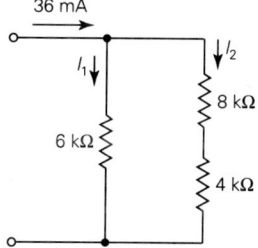

1–30. Use the current-divider rule to determine the current I_1 in Figure P1–30.

FIGURE P1–30

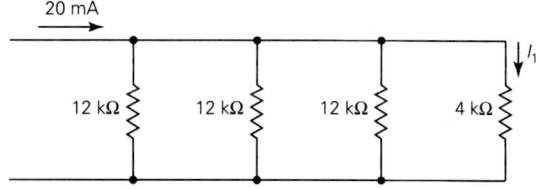

1–31. Assume that a 100-V dc source is connected across the input terminals in the circuit of Figure P1–17. Determine all branch voltages and currents.

1–32. Assume that a 120-V dc source is connected across the input terminals in the circuit of Figure P1–18. Determine all branch voltages and currents.

1–33. Determine the voltage V_{AB} in Figure P1–33.

FIGURE P1–33

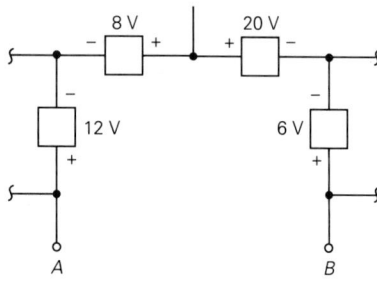

1–34. Determine the voltage V_{AB} in Figure P1–34.

FIGURE P1–34

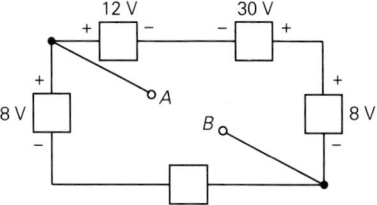

1–35. The voltages referred to ground at two points A and B are $V_A = -6$ V and $V_B = -18$ V. Determine V_{AB}.

1–36. The voltages referred to ground at two points A and B are $V_A = 6$ V and $V_B = -14$ V. Determine V_{AB}.

1–37. Perform a source transformation on the circuit of Figure P1–37 to obtain an equivalent current source.

FIGURE P1–37

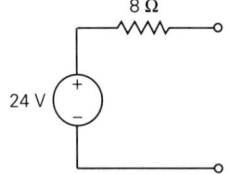

1–38. Perform a source transformation on the circuit of Figure P1–38 to obtain an equivalent current source.

FIGURE P1–38

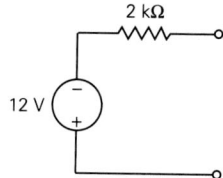

1–39. Perform a source transformation on the circuit of Figure P1–39 to obtain an equivalent voltage source.

FIGURE P1–39

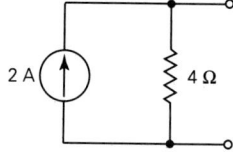

1–40. Perform a source transformation on the circuit of Figure P1–40 to obtain an equivalent voltage source.

FIGURE P1–40

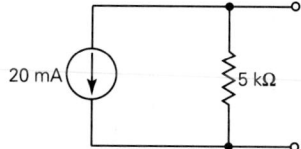

1–41. The load to be connected to the circuit of Figure P1–41 will always be greater than 5 kΩ. Based on a maximum 1% error, determine the form and value of an ideal source approximation.

FIGURE P1–41

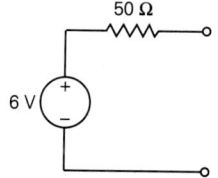

1–42. The load to be connected to the circuit of Figure P1–42 will always be less than 2 kΩ. Based on a maximum 1% error, determine the form and value of an ideal source approximation.

FIGURE P1–42

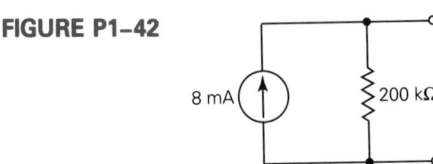

8 mA   200 kΩ

1–43. The load to be connected to the circuit of Figure P1–43 will always be less than 5 kΩ. Based on a maximum 1% error, determine the form and value of an ideal source approximation.

FIGURE P1–43

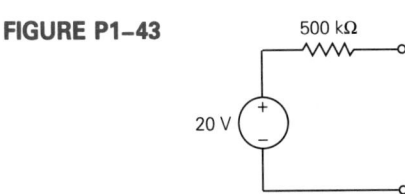

500 kΩ

20 V

1–44. The load to be connected to the circuit of Figure P1–44 will always be greater than 200 kΩ. Based on a maximum 1% error, determine the form and value of an ideal source approximation.

FIGURE P1–44

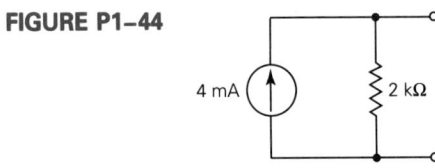

4 mA   2 kΩ

1–45. Determine the Thévenin equivalent circuit for the dc circuit of Figure P1–45 at the terminals x–y.

FIGURE P1–45

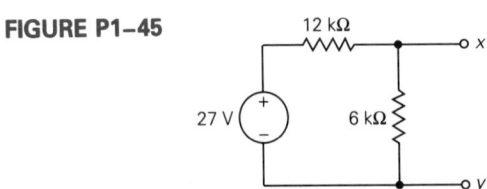

12 kΩ x

27 V   6 kΩ

y

1–46. Determine the Thévenin equivalent circuit for the dc circuit of Figure P1–46 at the terminals x–y.

FIGURE P1–46

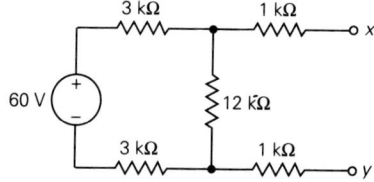

1–47. Determine the Thévenin equivalent circuit for the circuit of Figure P1–47 at the terminals x–y.

FIGURE P1–47

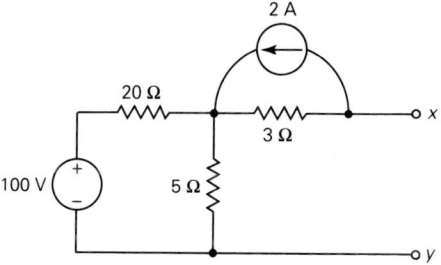

1–48. Determine the Thévenin equivalent circuit for the circuit of Figure P1–48 at the terminals x–y.

FIGURE P1–48

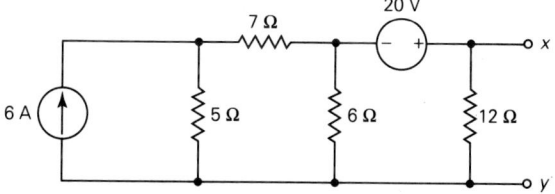

1–49. The open-circuit ouput voltage of a certain electronic circuit is 8 V. When a resistance of 5 kΩ is connected across the output, the voltage drops to 4 V. Determine the Thévenin equivalent circuit. (*Hint:* Look ahead to Problem 1–55.)

1–50. The open-circuit output voltage of a certain electronic circuit is 6 V. When a 120-Ω load is connected across the output, the voltage drops to 5.5 V. Determine the Thévenin equivalent circuit. (*Hint:* Look ahead to Problem 1–56.)

1–51. Use the principle of superposition to determine the open-circuit voltage in the circuit of Figure P1–47 by considering the effects of the individual sources separately.

1–52. Use the principle of superposition to determine the open-circuit voltage in the circuit of Figure P1–48 by considering the effects of the individual sources separately.

Derivation Problems

1–53. Consider the voltage-divider circuit of Figure 1–21(a). Verify that if $R_o = 100R_1$, $v_o/v_i > 0.99$, which means that the loss in input voltage through the divider is less than 1% of the available input voltage.

1–54. Consider the current-divider circuit of Figure 1–23. Verify that if $R_1 = 100R_o$, $i_o/i_i > 0.99$, which means that the loss in input current through the divider is less than 1% of the available input current.

1–55. The determination of the Thévenin or Norton equivalent circuit of a signal source in the laboratory usually requires a procedure completely different from the "paper" procedures discussed in the chapter. A procedure that works well for small-signal sources with moderate to high values of internal resistance is illustrated in Figure P1–55. First, the open-circuit voltage v_{oc} is measured, and we know that $v_t = v_{oc}$. Next a variable resistance R_L is connected across the output, and the resistance is adjusted until the load voltage v_L is

$$v_L = \frac{v_{oc}}{2}$$

At this point R_L is disconnected and measured. Show that the equivalent resistance R_{eq} is

$$R_{eq} = R_L$$

FIGURE P1–55

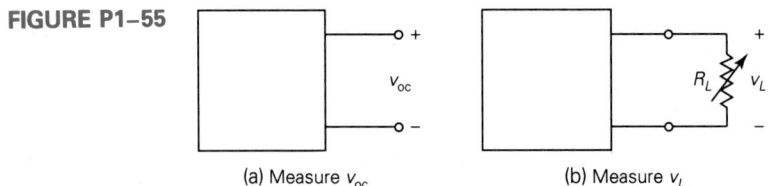

(a) Measure v_{oc}　　　　　　(b) Measure v_L

1–56. The procedure of Problem 1–55 for determining the equivalent resistance of a source is generally not suitable for low internal resistance sources due to the excessive loading when $R_L = R_{eq}$. Referring again to Figure P1–55, an alternative procedure is to adjust R_L for only a moderate decrease in the terminal voltage. For a terminal voltage v_L with load, show that the value of R_{eq} is

$$R_{eq} = \frac{(v_{oc} - v_L)R_L}{v_L}$$

1–57. A requirement that frequently arises in laboratory testing is to provide a simple voltage divider to attenuate (reduce) the level of a signal without changing the output resistance. Referring to Figure P1–57, assume that a given laboratory generator has

FIGURE P1–57

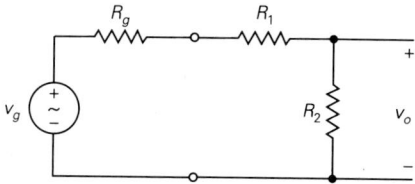

an open-circuit voltage v_g and an output resistance R_g. The resistances R_1 and R_2 are added externally, and the following requirements are imposed:

$$\frac{v_o}{v_g} = \alpha \qquad (\alpha < 1)$$

R_{eq} = Thévenin equivalent resistance looking back from output = R_g. Show that the resistances R_1 and R_2 are

$$R_1 = \frac{1 - \alpha}{\alpha} R_g$$

$$R_2 = \frac{R_g}{1 - \alpha}$$

1–58. An ideal voltmeter would have infinite resistance and would present no loading to a circuit. Practical voltmeters have a finite resistance, of course. In general, the resistance should, if possible, be very large compared with the Thévenin resistance seen at the point of measurement. With passive voltmeters not containing electronic isolation, this is sometimes difficult to achieve with high-resistance circuits.

The actual resistance of many passive voltmeters is a function of the scale range. Assume that two voltage measurements, based on two different ranges, are made. On a range in which the voltmeter resistance is R_1, the voltage is measured as V_1, and on a scale in which the voltmeter resistance is R_2, the voltage is measured as V_2. Show that the true open-circuit voltage V_{oc} and Thévenin equivalent resistance R_{eq} are given by

$$R_{eq} = \frac{R_1 R_2 (V_2 - V_1)}{V_1 R_2 - V_2 R_1}$$

$$V_{oc} = \frac{V_1 V_2 (R_2 - R_1)}{V_1 R_2 - V_2 R_1}$$

(*Hint:* A voltage-divider model may be constructed in which the measured voltage is the drop across the voltmeter resistance.)

Design Problems

1–59. The range of an ammeter may be extended by placing a shunt resistance R_s in parallel with the ammeter. The resistance is selected to absorb the current in excess of full-scale meter current. Assume that a given 0–1 mA ammeter is to be extended to cover the range 0 to 10 mA. If the internal resistance is 50 Ω, determine the required value of the shunt resistance.

1–60. A classical approach to designing dc voltmeters is illustrated in Figure P1–60. A basic D'Arsonval galvanometer is used as the indicating device. The series resistances are

FIGURE P1-60

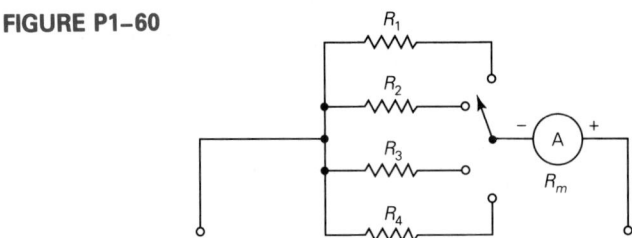

chosen such that the full-scale current I_{fs} flows when the maximum voltage for a given range is applied. For the circuit illustrated, assume that the four full-scale desired voltages are 3, 10, 30, and 100 V. Assume that I_{fs} = 100 μA and that the meter contains an internal series resistance R_m = 1 kΩ. Determine the required design values of the four resistances R_1, R_2, R_3, and R_4.

1–61. Using the results of Problem 1–57, assume that a given laboratory generator has an output resistance of 50 Ω and that it is desired to reduce the output voltage by a 10:1 ratio (i.e., α = 0.1). Design the attenuation circuit.

1–62. Repeat problem 1–61 with α = 0.01.

Troubleshooting Problems

1–63. Assume that the circuit of Figure P1–27 is not working as expected and that you measure the following voltages: V_1 = 4.6 V and V_2 = 15.4 V. Which one of the following is the likely trouble? **(a)** 6-kΩ resistor open; **(b)** 6-kΩ resistor shorted; **(c)** 5-kΩ resistor open; **(d)** 20-kΩ resistor open; **(e)** incorrect value of dc input voltage.

1–64. Assume that the circuit of Figure P1–28 is not working as expected and that you measure the following voltages: V_1 = 51.4 V and V_2 = 8.6 V. Which one of the following is the likely trouble? **(a)** 18-kΩ resistor open; **(b)** 36-kΩ resistor open; **(c)** 6-kΩ resistor open; **(d)** 6-kΩ resistor shorted; **(e)** 2-kΩ resistor open; **(f)** 2-kΩ resistor shorted.

2

SEMICONDUCTOR DIODES

OBJECTIVES

After completing this chapter, the reader should be able to:

- State the three classifications of materials in terms of their relative conductivity.
- State the two most common types of semiconductors.
- State the two types of charge carriers in semiconductors.
- Explain briefly the concepts of conduction band and valence band.
- Classify the types of impurities and the types of charge carriers they produce.
- Explain the difference between p-type material and n-type material.
- Discuss the construction of a pn junction diode.
- Discuss the depletion layer and its effect.
- Draw the symbol of a diode, and identify the anode and cathode terminals.
- Show the directions of "easy" and "hard" current flow.
- Explain the significance of a terminal characteristic, and indicate the independent and dependent variables.
- Sketch the form of the terminal characteristic of a pn junction diode.
- Discuss the properties of the diode terminal characteristic for the forward-biased region and for the reverse-biased region.
- Construct the circuit form and the terminal characteristic of the ideal diode model.
- Construct the circuit form and the terminal characteristic of the constant voltage model for the diode.
- Construct the circuit form and the terminal characteristic of the constant voltage source plus resistance model of the diode.
- Analyze simple dc circuits containing a diode using one of the models of the preceding three objectives.
- Use specification sheets to determine diode parameters.

67

- Define *peak reverse voltage* (or *peak inverse voltage*).
- Explain the behavior of a zener diode and draw its schematic diagram.
- Define for a zener diode the *knee current,* the *maximum reverse current*, and the *test current.*
- Define the zener impedance, and apply it to predict voltage variations.
- Define the temperature coefficient of the zener voltage, and apply it to predict voltage variation.
- Discuss the general operation of optoelectronic diodes, including light-emitting diodes, photodiodes, and optocouplers.
- Discuss briefly the general operation of other miscellaneous diodes, such as Schottky diodes, varactor diodes, constant current diodes, tunnel diodes, varistors, step-recovery diodes, and back diodes.
- Explain the procedure for testing a diode with an ohmmeter.

2–1   APPROACH

An initial discussion of the approach for dealing with electronic devices in this chapter and in succeeding chapters is appropriate. For the purpose of this discussion, let us classify the behavior of electronic devices at two levels: (1) internal device physics, and (2) external terminal characteristics.

At the most fundamental level, the internal device physics are very important and must be understood by those who design and manufacture semiconductor devices and integrated circuits. However, this segment of the electronics industry is much smaller than those segments that use electronic devices in the numerous application areas. There are books, courses, and complete college curricula devoted to semiconductor theory at a level appropriate for component and integrated-circuit design. Because the approach at that level is quite different from that of circuit design and applications, books attempting to cover all aspects of the subject equally well are quite rare.

The vast majority of engineers, technologists, technicians, and instrumentation personnel tend to be "component users"; that is, they purchase components or integrated circuits from a manufacturer and adapt them to fit their needs. In most cases, they are not particularly concerned with what is happening inside the component, but they want to know about the terminal characteristics and how it relates to their requirements.

What we are going to do in this book is concentrate on the terminal characteristics of devices and how they can be modeled for analysis and design. The internal mechanisms will be considered only qualitatively and to the extent that they aid in the visualization of circuit operation and in reading specifications. This is the approach taken by the majority of electronic component and integrated-circuit users in their day-to-day encounters with the complex world of electronics.

Although no substantive treatment of device physics is given, it will be necessary to provide an introduction to the concepts of semiconductor theory, and that is the basis for the next section.

2–2 SEMICONDUCTOR CONCEPTS

A brief overview of some basic concepts of semiconductor theory are given. As explained earlier, the purpose is to provide enough insight to visualize the general phenomena and to understand the terminology used in the classification of materials and components.

From the standpoint of electrical conductivity, materials may be classified as: (1) **conductors,** (2) **insulators** (or **nonconductors**), and (3) **semiconductors.** *Conductors* are materials having very low resistance or high conductivity. Examples are copper and aluminum. *Insulators* are materials having very high resistance or very low conductivity. Examples are wood and plastics.

Semiconductors are materials that lie between conductors and insulators; that is, current can flow but not as easily as in conductors. The two most common examples of semiconductors are *silicon* and *germanium*. Silicon is the most widely used at the present, but there are still some applications for germanium devices. Semiconductors are the basis for most electronic devices because their atomic structures may be modified to create various types of voltage and current control functions. To understand the behavior of semiconductors, it is necessary to focus on some of the basic concepts of atomic theory.

Atomic Theory

All matter is composed of *atoms*. Each of the basic chemical elements has a unique atomic structure. The classical Bohr model of the atom utilizes a planetary type of structure that can be visualized by the form shown in Figure 2–1. The particular atom considered here is copper.

An atom consists of three types of basic particles: (1) **electrons,** (2) **protons,** and (3) **neutrons.** An electron is negatively charged, and a proton is positively charged. However, the neutron is neutral and does not carry a charge. The nucleus contains the protons and the electrons, and the electrons revolve around the nucleus in fixed orbits. The charge of an electron and the charge of a proton are equal in magnitude.

In the stable state of an atom, the number of electrons is equal to the number of protons, and the net charge is zero. However, if the atomic structure is altered so that the number of electrons is not equal to the number of protons, the resulting charged atom is referred to as an **ion.**

Elements are arranged in a periodic table according to their **atomic number.** The *atomic number* is the number of electrons in a neutral atom. Protons and neutrons are much heavier than electrons, and the **atomic weight** is approximately equal to the number of protons plus the number of neutrons.

Electrons revolve around the nucleus in fixed orbits. Electrons in orbits close to the nucleus have lower energy levels than those in outer orbits. The orbits are considered to be grouped into energy bands known as **shells.** The shells correspond to discrete energy levels, and only certain levels are permissible in a given atom. A given shell has a maximum number of electrons that can exist in

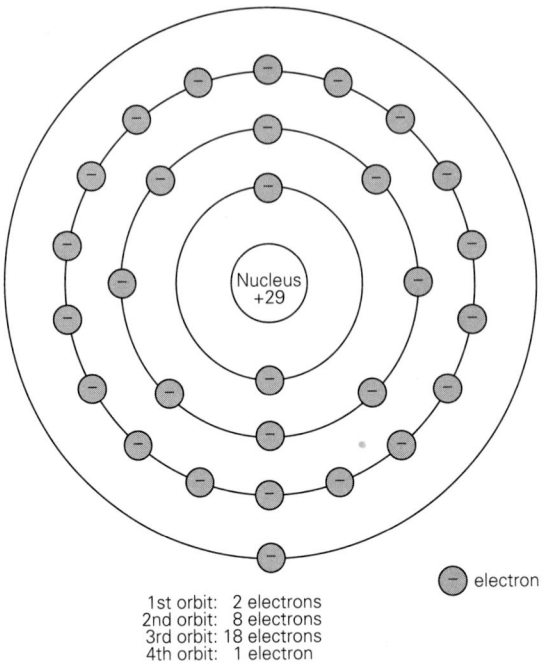

1st orbit: 2 electrons
2nd orbit: 8 electrons
3rd orbit: 18 electrons
4th orbit: 1 electron

FIGURE 2–1
Bohr model of the atom for copper (atomic number = 29, valence = 1).

that particular shell. Since the force of attraction decreases as the radius from the center of the atom increases, electrons in the distant orbits are not bound as tightly to the atom as those in the inner shells.

The outer shell of an atom is known as the **valence band,** and the number of electrons in that band is called the **valence** of the element. The valence electrons play a major role in the chemical processes leading to compounds and chemical reactions.

Semiconductor Structures

The atomic models of silicon and germanium, the two most commonly used semiconductors, are shown in Figure 2–2. The silicon atom of (a) has 14 electrons, while the germanium atom of (b) has 32 electrons. Each, however, has a valence of four, which means that there are four electrons in the outer orbit.

Semiconductor materials form **crystals** in which the electrons of the outer orbit are shared with other atoms. This process is referred to as **covalent bonding** and is illustrated in Figure 2–3.

Intrinsic semiconductors are those that have been carefully constructed to reduce impurities to a very low level. However, intrinsic semiconductors are still relatively poor conductors.

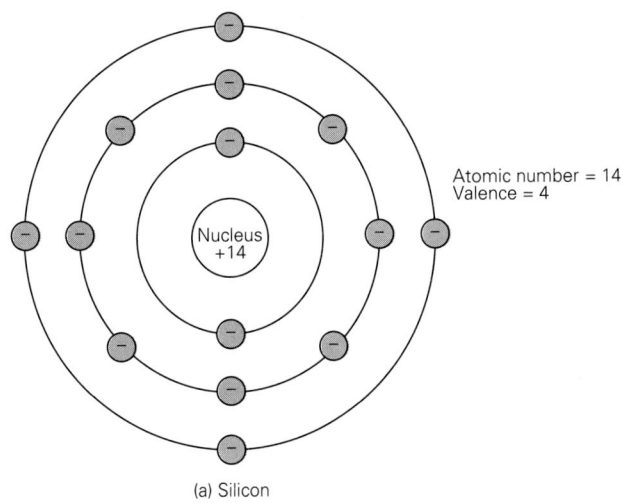

Atomic number = 14
Valence = 4

(a) Silicon

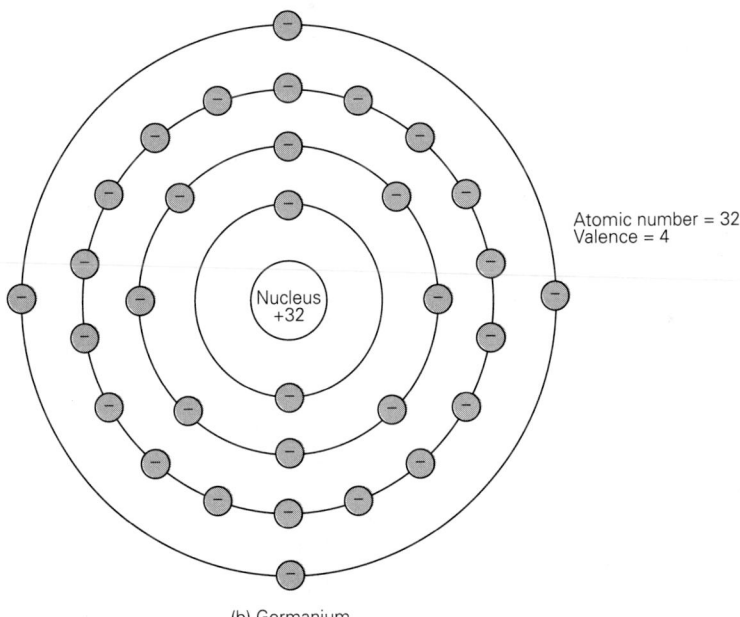

Atomic number = 32
Valence = 4

(b) Germanium

FIGURE 2–2
Atomic models for (a) silicon and (b) germanium.

FIGURE 2–3
Covalent bond with shared valence electrons.

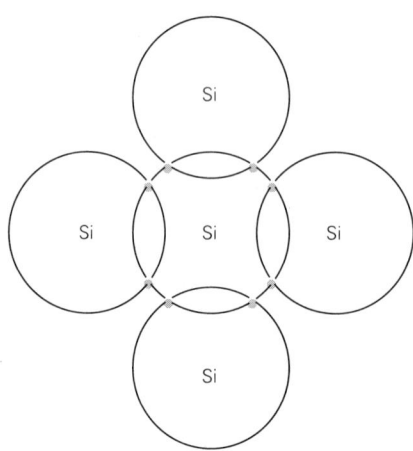

Two electrons of one atom fill two holes of adjacent atom.

Extrinsic semiconductors are those that have been altered by the addition of certain impurities to the material. The process of adding the impurities is called **doping.** Even the addition of impurities of the order of one part in a million may result in major changes in the property of the material.

n and p Materials

The two types of semiconductors formed by doping are called **n-type** materials and **p-type** materials. The *n-type* materials are formed by the addition of impurities that have a valence of five, that is, *five* electrons in the outer orbit. This type of impurity is called a **pentavalent** material; some examples are **antimony, arsenic,** and **phosphorus.** The bonding structure with such an impurity is shown in Figure 2–4. This process generates free electrons in the material.

The four covalent bonds are present in the *n*-type material, but there is a fifth electron provided by each impurity atom, which is not associated with any particular bond. This electron is relatively free to move within the *n*-type material. Since additional electrons have been given to the structure, these materials with five valence electrons are called **donors.**

The *p-type* materials are formed by the addition of impurities that have a valence of three, that is, *three* electrons in the outer orbit. This type of impurity is called a **trivalent** material, and some examples are **boron, gallium,** and **indium.** The lattice structure with such an impurity is shown in Figure 2–5. In the case of *p*-type materials, there are not enough electrons to complete the bonds of the lattice. The result is the creation of a **hole** or vacancy in the bond. The absence of a negative charge is electrically equivalent to the addition of a positive charge so holes may be considered, in a sense, as positive charges. Since these holes are

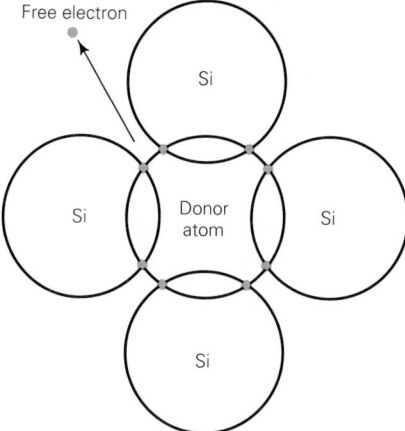

The extra electron from the donor atom becomes free from the bond.

FIGURE 2–4
Donor impurity atom in a silicon crystal.

capable of accepting free electrons, these materials with five valence electrons are called **acceptors.**

Conduction is a result of the movement of either electrons or holes. Electron conduction is, of course, the process that occurs in ordinary conductors. The process of hole conduction, however, results when a valence electron acquires sufficient energy to break its covalent bond, in which case it fills the void created

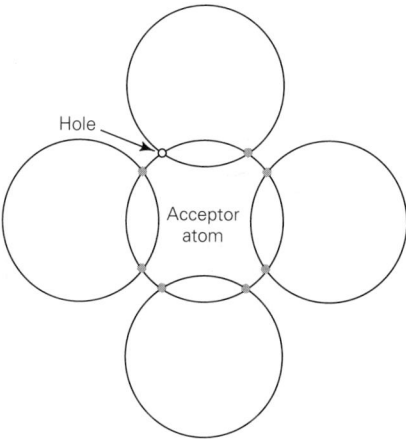

The lack of sufficient electrons in the acceptor atom results in the creation of a hole.

FIGURE 2–5
Acceptor impurity atom in a silicon crystal.

by an electron. This transfer of an electron in one direction is electrically equivalent to the transfer of a hole in the opposite direction. The resulting direction is that of the positive or conventional direction of current flow. In a sense then, the movement of holes is equivalent to a flow of positive charges in the conventional direction of current flow.

In an *n*-type material most of the current flow results from the flow of electrons. Thus, *electrons* are referred to as the **majority carriers,** and holes are referred to as the **minority carriers.** However, in a *p*-type material, most of the conduction results from the movement of holes. Thus, for *p*-type materials, *holes* are the *majority* carriers, and electrons are the *minority* carriers.

pn Junction

A *pn* junction, which is the basis of a diode, is formed by combining a region of *p*-type material with a region of *n*-type material as illustrated in Figure 2–6(a). Initially, most of the charge carriers in the *p*-material are holes, and most of the carriers in the *n*-material are electrons as shown in Figure 2–6(b). However, each contains some minority carriers resulting from thermal energy causing electrons to break away from the bonds.

FIGURE 2–6
Formation of a *pn* junction by combining *p*- and *n*-type materials.

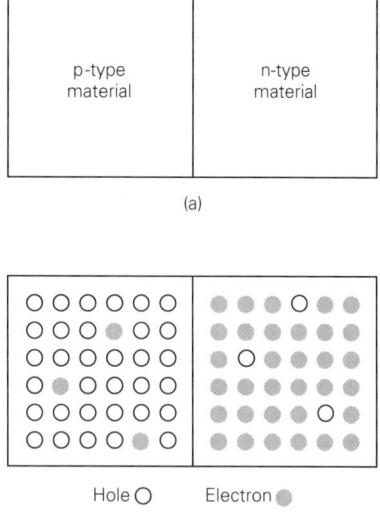

Depletion Region

When a junction is formed between *p*-material and *n*-material, free electrons on the *n*-side diffuse across the junction into the *p*-region. Since an electron represents a minority carrier on the *p*-side, it will quickly fall into a hole and become a valence electron. The result is the formation of two ions: (1) a positive ion on the *n*-side resulting from the loss of the electron, and (2) a negative ion on the *p*-side for the atom that captures the electron. As this process builds, it depletes the carriers in the neighborhood of the junction. The resulting region is called the **depletion region** and is illustrated in Figure 2–7.

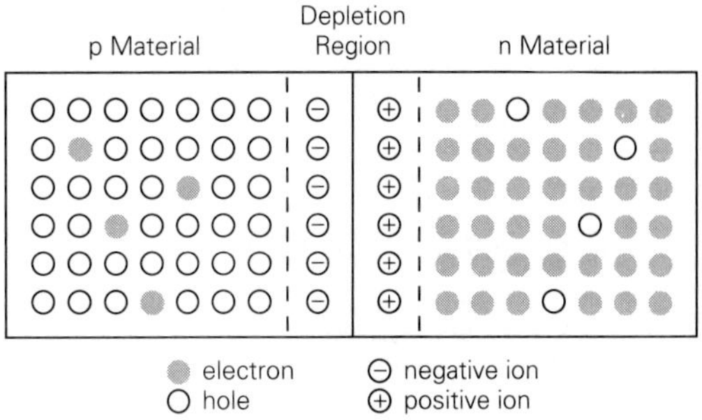

FIGURE 2–7
Equilibrium condition for *pn* junction with depletion region.

The effect of the depletion region is to create a **barrier potential.** For silicon diodes, the barrier potential is about 0.7 V, and for germanium diodes, it is about 0.3 V. This potential cannot be measured with a voltmeter unless an appropriate external source is applied.

Forward Bias

Assume that a **forward** voltage is applied to the *pn* junction by connecting the positive external voltage to the *p*-type material and the negative terminal to the *n*-type material as shown in Figure 2–8. This voltage will force electrons in the *n*-type material and holes in the *p*-type material to combine with ions near the boundary. The effect is to reduce the width of the depletion region. This reduction results in a heavy flow of majority carriers across the junction. An electron in the *n*-type material now experiences a strong attraction to the positive external voltage. The depletion region continues to reduce, and the current flow continues to increase as the voltage continues to increase.

Reverse Bias

Assume now that the external source is reversed with the positive terminal connected to the *n*-side and the negative terminal connected to the *p*-side. The negative terminal of the source attracts the holes, and the positive terminal attracts the electrons. Free electrons and holes thus move away from the junction, and the depletion layer widens.

A very small curent actually flows with reverse bias. This current results from minority carriers created by free electrons and holes. The resulting current is called **reverse saturation current.**

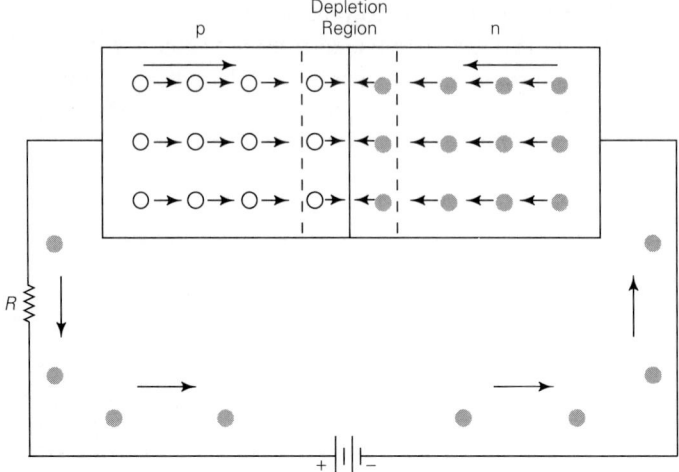

FIGURE 2–8
Current flow in a forward-biased *pn* junction diode.

Energy Bands

The concept of energy bands is best illustrated by an **energy band diagram,** whose basic form is illustrated in Figure 2–9 for silicon. As mentioned earlier, only discrete energy levels can exist in an atom. Each energy band corresponds to a given shell within the atom. When an electron in the valence band acquires suffi-

FIGURE 2–9
Energy band diagram for silicon with no electrons in conduction band.

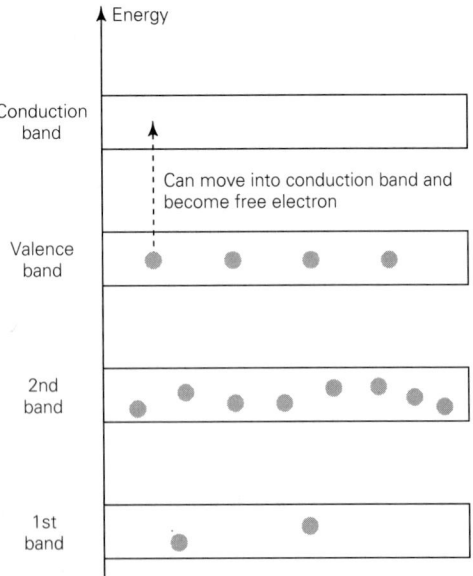

cient energy, it can dislodge from the valence band and become a free electron in the conduction band as illustrated. This leaves a hole in the valence band.

An energy band diagram is useful for explaining the difference between the three types of materials as illustrated in Figure 2–10. The energy diagram of (a) shows an insulator, in which there is a very wide energy gap between the valence band and the conduction band.

For the semiconductor of (b), the energy gap is much narrower, and as seen earlier, allows appropriate doping to achieve controlled conduction.

FIGURE 2–10
Energy band diagrams for three types of materials.

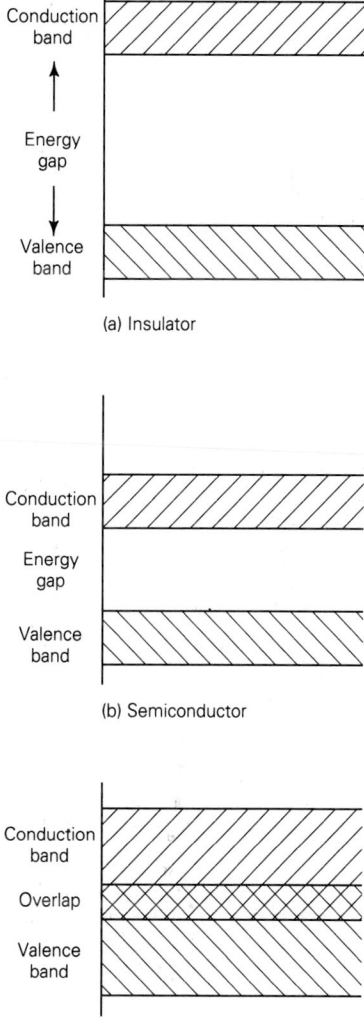

(a) Insulator

(b) Semiconductor

(c) Conductor

For the conductor of (c), however, there is an overlap between the conduction band and the valence band, for which electrons are easily freed from the valence band.

2–3 JUNCTION DIODE

As explained earlier, a basic semiconductor *pn* **junction diode** is formed from the combination of a section of *p*-type semiconductor material with a section of *n*-type semiconductor material, as illustrated in Figure 2–11(a) and throughout the chapter. It is difficult for current to flow in the direction shown by the dashed line in Figure 2–11(a). However, a positive voltage applied to the *p*-terminal with respect to the *n*-terminal is capable of overcoming the effect of the depletion layer, and the result is possible current flow in the direction of the solid line in Figure 2–11(a). Thus, the junction diode acts as a type of one-way valve for current flow.

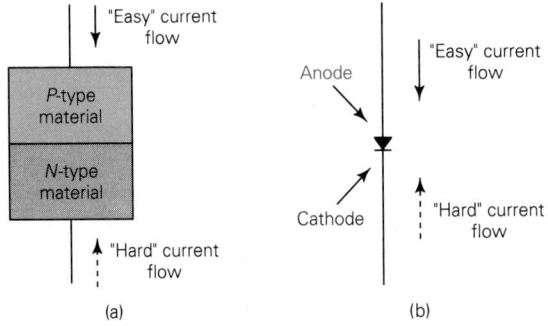

FIGURE 2–11
Composition of *pn* junction diode and the circuit symbol.

The schematic symbol for the diode is shown in Figure 2–11(b). Note that the arrow on the symbol points in the direction of "easy" positive conventional current flow. The terminal on the *p*-side is called the **anode,** and the terminal on the *n*-side is called the **cathode.** These terms date back to the era of the vacuum-tube diode, but they are still widely used. In summary, current flows easily from anode to cathode but not from cathode to anode. Diodes come in a variety of sizes, with current levels ranging from a fraction of an ampere to thousands of amperes.

2–4 TERMINAL CHARACTERISTICS

Before developing circuit models for the diode, it is important to understand and be comfortable with certain basic concepts of linear and nonlinear relationships. Consider the block shown in Figure 2–12, having a set of terminals. This set of

FIGURE 2–12
Block used to discuss terminal character-istic.

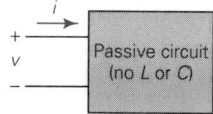

terminals is also called a **port.** An arbitrary voltage v is assumed across the terminals, and a current i is assumed to flow into the upper terminal (which means that i will be flowing out of the lower terminal, but it is usually not shown on the lower terminal). For the purpose at hand, we assume that the port has no reactive elements (i.e., no inductance or capacitance) and is passive (i.e., there are no energy sources inside).

Assume that the voltage v is varied through some range of values and that i is measured at each value of v specified. In a sense, we are considering v as the **independent variable** and i as the **dependent variable.** However, we could also vary i through a range of values and measure v at corresponding points. In this sense we would be considering i as the independent variable and v as the dependent variable. A curve showing i as a function of v (or v as a function of i) is called a **terminal characteristic.** In many cases, it is also possible to express the terminal characteristic as an equation relating i to v, or vice versa.

Let us illustrate the concept by observing the terminal characteristic of a simple resistor. Consider the resistance of Figure 2–13(a) with voltage v and current i. If we consider v as the independent variable and i as the dependent variable, we have

$$i = \frac{v}{R} = \frac{1}{R}v \qquad (2\text{–}1)$$

The terminal characteristic is shown in Figure 2–13(b). Observe that the independent variable is shown as the horizontal axis (abscissa) and the dependent variable is shown as the vertical axis (ordinate). The slope of this curve is $1/R$.

If current is considered as the independent variable, we have

$$v = Ri \qquad (2\text{–}2)$$

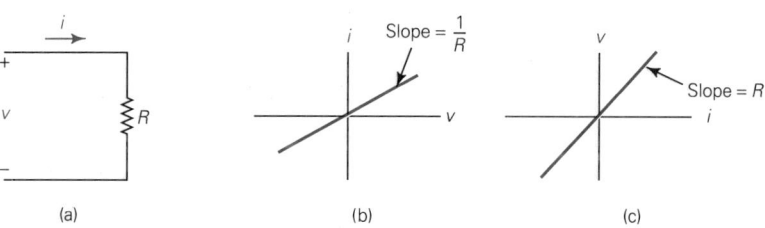

(a) (b) (c)

FIGURE 2–13
Ideal linear resistor and two forms of linear terminal characteristics.

This curve is shown in Figure 2–13(c), and its slope is R. In both curves we are assuming that the resistor is operated below its maximum power rating for the curve to be valid.

Either form of the terminal characteristic of an ideal resistor is a **linear characteristic;** that is, the dependent variable changes linearly with changes in the independent variable. Although this rather trivial example has not taught us anything new, it should help us to visualize the concept of terminal characteristics before proceeding to more complex situations. Furthermore, by visualizing the characteristic for an ideal linear resistor, we can use it as a basis of comparison with other types of characteristics.

As an example of a nonlinear characteristic, consider the circuit of Figure 2–14(a), in which a nonlinear resistor is shown. Although there are a number of different types of nonlinear resistors, this particular one is assumed to have a resistance that increases with increasing voltage. The resulting terminal characteristic might be of the form shown in Figure 2–14(b). Many nonlinear characteristics are not easily described by equations, so the graphical form is often the only way to characterize the device.

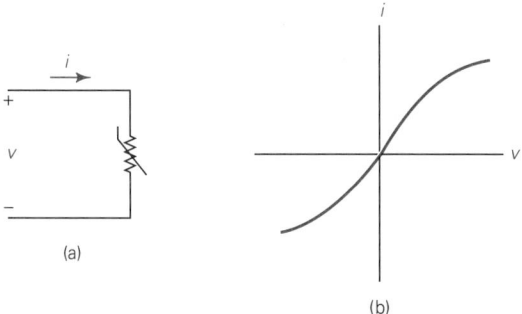

(a)

(b)

FIGURE 2–14
Nonlinear resistor and possible terminal characteristic.

Diode Terminal Characteristic

The terminal characteristic of the junction *pn* diode will now be investigated. Consider the diode shown in Figure 2–15(a) with voltage V_D and current I_D as labeled. The assumed voltage polarity and current direction are referred to throughout as the **positive reference direction** for the diode.

Assume that the diode is connected in the circuit of Figure 2–15(b). An experiment is to be performed for the purpose of obtaining a terminal characteristic. By varying the dc voltage source, the voltage V_D across the diode and the corresponding current I_D through the diode can be measured so that enough points can be determined to plot a curve. The voltmeter measures the voltage across the

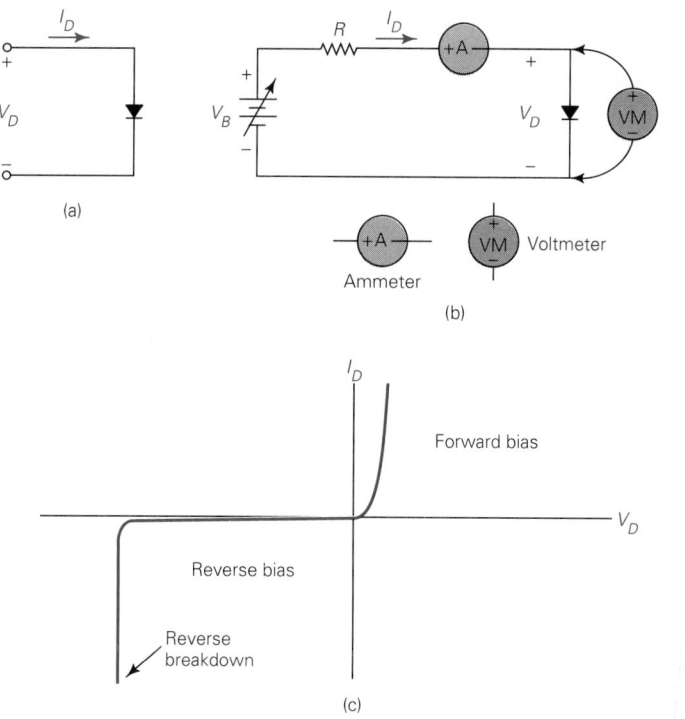

FIGURE 2–15
Semiconductor diode and the form of the terminal characteristic.

diode, and the ammeter measures the current through it. The only purpose of the resistor is to limit the maximum current through the diode. As we will see shortly, slight changes in the diode voltage in one direction can result in large changes in current.

Refer to Figure 2–15(c) for the discussion that follows. This curve represents a typical junction diode terminal characteristic, which is discussed in relationship to the experiment. First, the portion of the characteristic corresponding to positive voltage and positive current is considered. This condition is called **forward bias.** As the power supply voltage is varied in positive steps from zero, there is little or no measurable current flow for V_D less than a few tenths of a voltage. This is the range in which the depletion layer within the diode is blocking current flow. However, once V_D reaches a certain voltage level (to be quantified later), the current starts to flow and increases rapidly for small changes in voltage, as shown. Without the resistor it would be difficult to adjust the voltage without creating an excessive current.

The dc voltage is next reduced to zero, and arrangements are made to obtain the portion of the characteristic corresponding to negative voltage and negative current. This can be achieved by reversing the terminals of the power supply and the polarities of the ammeter and voltmeter. (Alternatively, one can simply reverse the diode.) All subsequent measurements will then be based on negative voltage and current with respect to the assumed positive direction.

The condition of negative voltage and current with respect to the original direction is called **reverse bias.** The voltage and current in this condition are also called **reverse voltage** and **reverse current.**

As the reverse voltage is increased from zero in the reverse-biased region, the current would ideally be zero. In practice, a very small *reverse current* does flow, but it is normally much smaller than the forward current measured previously. In many applications, the reverse current will be ignored, but some discussion of this phenomenon is included in Section 2–6.

If the magnitude of reverse voltage were allowed to reach a certain level, a **reverse avalanche breakdown** would occur; then the current magnitude would experience a pronounced increase, as shown. **Zener** diodes are designed to use the nearly constant voltage properties of this region, and such diodes are considered in Section 2–7. However, ordinary diodes should *not* be operated in the reverse breakdown region, since they may easily be damaged.

2–5 DIODE CIRCUIT MODELS

We consider next the process of modeling diodes for the purpose of analyzing their behavior in circuits. The diode is basically a nonlinear device, and an exact analysis is rather difficult. However, some very good approximations have been developed that permit simplified analysis to be performed.

Ideal Diode Model

The first and simplest approximation is the **ideal diode model.** Consider the diode shown in Figure 2–16(a) with the positive references for voltage and current indicated. The form of the ideal diode model is shown by the two circuits of Figure 2–16(b). This model assumes that the diode is either a perfect short circuit or a perfect open circuit, as shown. It is considered as a perfect short circuit if the

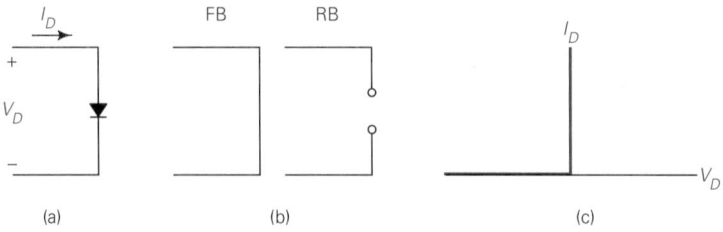

FIGURE 2–16
Ideal diode model and terminal characteristic.

direction of current flow is positive (in the direction of the arrow), and it is considered as a perfect open circuit if the voltage is negative (i.e., the cathode is more positive than the anode). The perfect short circuit corresponds to ideal **forward bias** (indicated as FB), and the perfect open circuit corresponds to ideal **reverse bias** (indicated as RB).

The terminal characteristic of the ideal diode model is shown in Figure 2–16(c). In effect, the characteristic becomes the positive vertical axis and the negative horizontal axis. This means that the voltage cannot become positive and the current cannot become negative under these ideal conditions. This curve is a graphical representation of the following conditions: *The voltage is zero when the current is positive, and the current is zero when the voltage is negative.*

Constant Voltage Model

The second model to be considered will be referred to as the **constant voltage model.** It is based on the fact that when the diode is forward biased, very little current flows until the forward voltage reaches a few tenths of a volt. (Refer to Figure 2–15(c), if necessary.) However, once current starts flowing, very little change in voltage occurs over the much larger range of operating current. As a very good approximation for many purposes, therefore, the forward-biased diode can be approximated as an ideal voltage source.

Refer to Figure 2–17, in which the diode references are shown again in (a). The constant voltage source model is shown in (b). The voltage V_o is denoted as the "turn-on" voltage, the "cut-in" voltage, or "knee" voltage, depending on the reference.

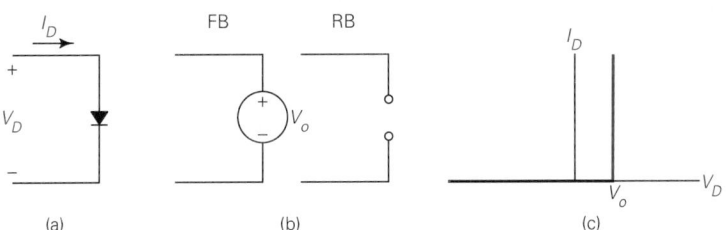

(a) (b) (c)

FIGURE 2–17
Constant voltage diode model and terminal characteristic.

The value of the knee voltage V_o depends on whether the diode material is germanium or silicon. The value for germanium diodes typically varies from about 0.2 to 0.3 V, and the value for silicon diodes varies from about 0.6 to 0.7 V for small-signal diodes. For diodes designed to handle large current levels, such as in-power supplies, the voltage may be of the order of 1 V or so.

In most circuit analysis problems throughout the text, the value $V_o = 0.7$ V will be assumed. This will represent a sort of "standard" valued based on silicon diodes. However, the reader should understand that it is not an exact value and will vary both with operating current and from one diode to another.

Understand that although the circuit model shows a voltage source, the diode is a passive device and cannot cause current to flow. The voltage source simply accounts for the forward-biased voltage drop, and this source does not "come alive" until the diode is forward biased. The terminal characteristic of the constant voltage model is shown in Figure 2–17(c).

Constant Voltage Source plus Resistance Model

The third model that we will consider is the **constant voltage source plus resistance model.** For the diode shown in Figure 2–18(a), the pertinent model is shown in (b). In this model, a resistance r_D is added to the constant voltage source model. This resistance allows the net voltage across the diode to increase as the current increases, which makes the result closer to a real situation. The corresponding terminal characteristic is shown in Figure 2–18(c).

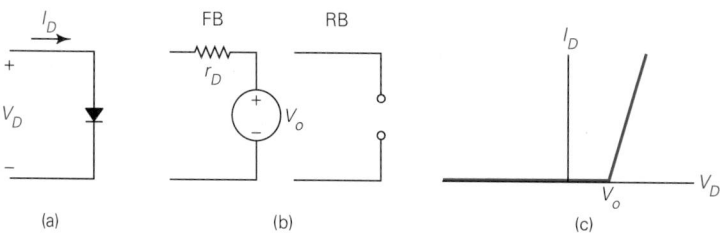

FIGURE 2–18
Constant voltage source plus resistance diode model and terminal characteristic.

The major difficulty with this model is in determining the value of the resistance to be used in the model. There is no simple value that will work over the entire range; that is, the resistance is heavily dependent on the operating range. For that reason, the constant voltage source plus resistance model has limited value in practical circuit analysis.

Shockley Model

The last method to be considered will be mentioned briefly, but is not considered within the scope of this book. This model utilizes the so-called **Shockley equation,** in which an exact mathematical voltage–current relationship using an exponential function for the diode is provided. This method is capable of providing virtually exact results, but it cannot be adapted easily to a simple circuit model.

The reader is no doubt confused at this point and is asking: "Which model do I use?" The answer is that any of the models can be used provided that the answers are interpreted in terms of their real-life meaning. We must understand that with the exception of the Shockley equation, none provide *exact* answers. However, acceptable answers can be obtained from these models, and they can produce meaningful results if interpreted properly.

The ideal diode model is used when a very quick analysis of a circuit is desired and where high accuracy is not required. For example, when one is making a quick overview of a circuit and ballpark estimates are being used, ideal diode models are frequently assumed.

The constant voltage drop model is probably the most widely employed model, and it is capable of providing results sufficiently close for a variety of applications. Although one does not always know the exact value of voltage drop, an assumed value of 0.7 V for silicon or 0.3 V for germanium usually results in acceptable answers for most routine applications. This model is used extensively throughout the book.

As mentioned, the ideal voltage source plus resistance model is more difficult to use because of the possible wide variation in resistance, so we will not use it very often. We illustrate its application in Example 2–3.

Forward or Reverse Bias?

One basic question that must always be answered in analyzing a circuit with a diode is to determine whether the diode is forward biased (FB) or reverse biased (RB). When there is only one diode in a circuit, it is usually straightforward to predict the condition. Based on your intuition (which will develop with practice), assume momentarily one of the two possible conditions (FB or RB). If you assumed FB, replace the diode with the appropriate FB model (usually the constant voltage source), and calculate the current through the diode. If the current turns out to be in the direction of the diode arrow, your guess was correct, and the assumed model may be retained. If, however, the current turns out in the opposite direction (which the diode won't allow), your guess was wrong and you must replace the model with the RB model.

Suppose your intuition told you initially that the diode was probably reverse biased. In that case, replace the diode with the RB model (open circuit), and calculate the voltage across the diode. If the voltage at the cathode turns out to be more positive than the voltage at the anode, your guess was correct, and the assumed model may be retained. If, however, the voltage turns out in the opposite direction, your guess was wrong, and you must replace the model with the FB model.

With a little practice, the reader should develop a quick intuitive approach for dealing with diode circuits having one diode. However, when the circuit contains more than one diode, the situation starts to get complicated, particularly when there are several diodes. The reason is that the assumption of one diode condition can affect the condition of other diodes. We defer this situation for a while.

The examples that follow are designed to illustrate the various levels of modeling based on the preceding developments.

EXAMPLE 2-1

Consider the circuit shown in Figure 2–19(a). Verify that the diode is forward biased. Calculate the current I and the two voltages V_D and V_R based on **(a)** the ideal diode model, and **(b)** the constant voltage drop model. The diode is assumed to be a silicon type with $V_o = 0.7$ V.

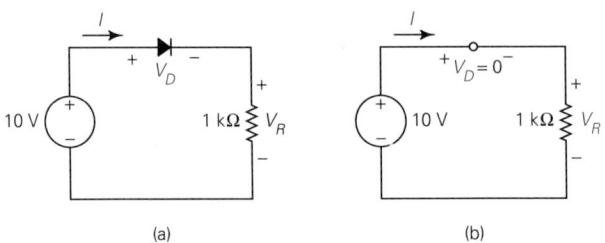

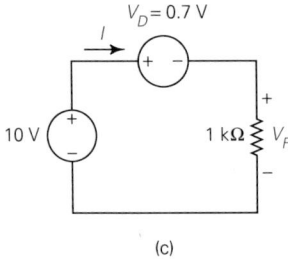

FIGURE 2–19
Circuit of Example 2–1.

Solution

Forward bias is readily verified by observing that the direction of current flow would have to be clockwise in the loop, that is, out of the positive terminal of the source and back into its negative terminal. This current flow is in the direction of the diode arrow, so the diode must necessarily be forward biased. With a little practice, the direction of forward bias can be quickly determined in a simple situation such as this.

(a) The equivalent circuit with the ideal diode is shown in Figure 2–19(b). In this case, the diode is replaced by a short circuit, and we readily note that

$$V_D = 0 \qquad (2\text{--}3)$$

Thus the entire 10-V source would appear across the resistance in this case, and

$$V_R = 10 \text{ V} \qquad (2\text{--}4)$$

The current is determined by Ohm's law as

$$I = \frac{10 \text{ V}}{1 \text{ k}\Omega} = 10 \times 10^{-3} \text{ A} = 10 \text{ mA} \qquad (2\text{--}5)$$

(b) The equivalent circuit with the constant voltage diode model is shown in Figure 2–19(c). We immediately note, of course, that

$$V_D = 0.7 \text{ V} \tag{2-6}$$

By KVL we have

$$-10 + V_D + V_R = 0 \tag{2-7}$$

or

$$V_R = 10 - V_D = 10 - 0.7 = 9.3 \text{ V} \tag{2-8}$$

The current is

$$I = \frac{9.3 \text{ V}}{1 \text{ k}\Omega} = 9.3 \times 10^{-3} \text{ A} = 9.3 \text{ mA} \tag{2-9}$$

As explained earlier, neither solution is necessarily correct, but the second solution is probably quite close to the actual result and *could* be the exact solution. (It *could be* because there is a point in which the voltage drop across the diode is *exactly* 0.7 V, but we do not usually know when this condition is met.)

EXAMPLE 2–2   Consider the circuit shown in Figure 2–20(a). Verify that the diode is reverse biased, and determine the voltage across the diode.

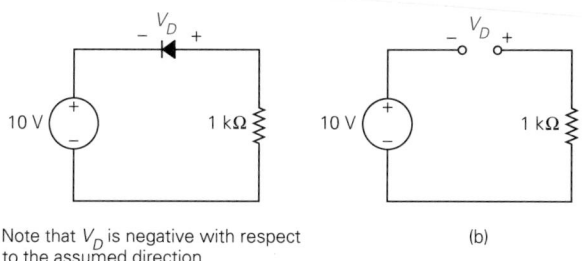

(a) Note that V_D is negative with respect to the assumed direction

(b)

FIGURE 2–20
Circuit of Example 2–2.

Solution
Current flow around the circuit *would* be clockwise (if it could flow), and this would be opposite to the diode arrow, which is impossible. Thus, the diode is reverse biased, and it is replaced by an open circuit as shown in Figure 2–20(b). Observe that we have chosen the positive reference terminal of the voltage across the diode in accordance with the basic convention established earlier. Since no

current flows in the circuit, there is no voltage drop across the resistor, and an application of KVL gives

$$-10 - V_D + 0 = 0 \qquad (2\text{--}10)$$

or

$$V_D = -10 \text{ V} \qquad (2\text{--}11)$$

Thus, the voltage across the diode with respect to the reference direction is negative, meaning that the cathode (left-hand terminal) is actually more positive than the anode (right-hand terminal). This validates our assumption of reverse bias.

We could, of course, have assumed the diode voltage drop as plus to minus from left to right, and the answer would have been positive. Later, we will take more liberties in choosing directions, but for the moment, we will adhere to the basic reference directions established earlier. With that convention, a reverse-biased junction will always yield a negative voltage.

It should be emphasized that although there is no current flow, the reverse voltage clearly exists across the diode. In circuit design it is sometimes easy to overlook the fact that very large negative voltages may appear across reverse-biased diodes. From the actual diode characteristic of Figure 2–15, recall that junction breakdown may occur if this reverse voltage is too large. More will be said about this later.

EXAMPLE 2–3

This example illustrates the effect resulting from the constant voltage source plus resistance model. Consider the circuit of Figure 2–21(a), and assume that the diode is represented by a constant voltage source of value 0.6 V in series with 50 Ω as shown in (b). Calculate the current I, the diode voltage drop V_D, and the resistive voltage drop V_R for each of the following values of V_i: **(a)** 10 V; **(b)** 20 V.

Solution

(a) The equivalent circuit with $V_i = 10$ V is shown in Figure 2–21(c). The KVL equation for the loop is

$$-10 + 50I + 0.6 + 10^4 I = 0 \qquad (2\text{--}12)$$

Solving for I, we obtain

$$I = \frac{9.4}{10,050} = 0.9353 \times 10^{-3} \text{A} = 0.9353 \text{ mA} \qquad (2\text{--}13)$$

The resistive voltage V_R is simply

$$V_R = 10^4 I = 10^4 \times 0.9353 \times 10^{-3} = 9.353 \text{ V} \qquad (2\text{--}14)$$

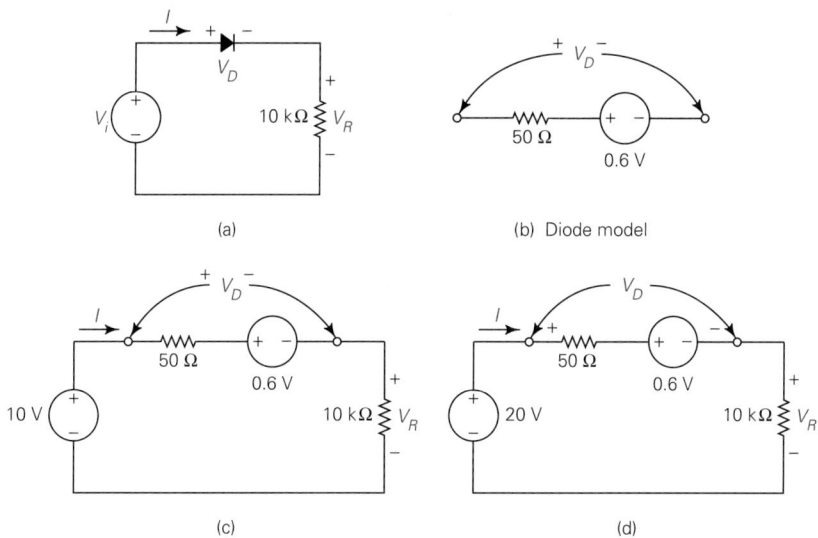

FIGURE 2–21
Circuit of Example 2–3 for different input conditions.

The diode voltage V_D consists of a contribution from the assumed internal resistance plus the constant value of 0.6 V. By forming a "voltmeter loop," we obtain

$$V_D = 0.6 + 50I = 0.6 + 50 \times 0.9353 \times 10^{-3} = 0.647 \text{ V} \qquad \textbf{(2–15)}$$

Alternatively, the diode voltage is simply the difference between the source voltage and the resistive voltage.

(b) The equivalent circuit with $V_i = 20$ V is shown in Figure 2–21(d). The KVL equation is now

$$-20 + 50I + 0.6 + 10^4 I = 0 \qquad \textbf{(2–16)}$$

Solving for I, we obtain

$$I = \frac{19.4}{10,050} = 1.930 \text{ mA} \qquad \textbf{(2–17)}$$

The voltages V_R and V_D are now determined as

$$V_R = 10^4 \times 1.930 \times 10^{-3} = 19.30 \text{ V} \qquad \textbf{(2–18)}$$

and

$$V_D = 0.6 + 50 \times 1.930 \times 10^{-3} = 0.697 \text{ V} \qquad \textbf{(2–19)}$$

Let us see if this example can serve to illustrate a few points. The source voltage in (b) is twice the value in (a). In a perfectly passive linear resistive circuit, all voltages and currents would also double. However, the current actually increased by a factor of $1.930/0.9353 = 2.064$. What is happening here is that the diode drop has less effect at the higher source voltage, so the current more than doubles. The resistive voltage V_R increases in the same proportion as the current, since that voltage drop varies linearly with current. The diode voltage, however, only increases from 0.647 V to 0.697 V. Of course, if the simpler constant voltage source model had been used, the result would have shown no increase in diode voltage. A diode voltage actually does increase with increased current, so this model predicts the actual behavior more closely. Yet, it is still approximate, because the assumption of a constant voltage plus a fixed resistance is usually valid only over a very narrow range of operation.

2-6 READING DIODE DATA SHEETS

A typical set of specifications for some representative diodes is provided on the first page of Appendix D. Although we are not yet able to relate to all these specifications, some of the information may be understood at this point. This particular sheet provides specifications on the family consisting of seven diodes, denoted as 1N4001 through 1N4007. Note that they are called **rectifiers** on this sheet. As we will see in Chapter 3, a common application of a diode is as a rectifier, so the term is often used interchangeably with diode. Observe the band on one end of the diode, which represents the *cathode* side. This is a widely employed convention.

The various maximum ratings are provided in a table. The significance of some of these ratings may not be clear at this point, but a few will be meaningful. The first two maximum ratings relate to reverse bias and provide data on the **peak reverse voltage,** which is also referred to in some references as the **peak inverse voltage.** The first rating is based on a repetitive situation, which means that the diode would be reverse biased on a regularly occurring basis. Note that in all cases, this voltage rating is less than that of a nonrepetitive situation. Note also that the peak reverse ratings vary considerably for different diodes in the family. The maximum value of the average rectified foward current is given as 1 A. However, the diodes can withstand a surge of 30 A for 1 cycle.

A table providing certain electrical characteristics is also given. A typical value of forward voltage drop is given as 0.93 V at 1 A. Note that this is somewhat greater than our assumption of 0.7 V. However, the value of 0.7 V is more accurate at the much lower current levels typical of electronic signal levels. The value of 1 A used here is more typical of certain power supply levels.

A typical reverse current is indicated as 0.05 μA at a junction temperature (T_J) of 25°C. However, the typical value increases to 1 μA for a junction temperature of 100°C. The maximum values for the two cases are 10 μA and 50 μA, respectively.

2–7 ZENER DIODES

In the discussion of the semiconductor diode characteristics in Section 2–4, reference was made to the reverse avalanche breakdown process, which can occur in any diode if the reverse voltage is excessive. By increasing the doping level, it is possible to decrease the magnitude of the reverse voltage and to control the process in an orderly fashion.

In the higher range of voltages, the breakdown is due to the reverse avalanche effect, as already noted. In the lower range of voltages, a separate effect called **zener breakdown** occurs. However, all diodes in which specified reverse breakdown voltage levels are established are collectively referred to as **zener diodes.** Thus, whereas reverse breakdown is undesirable for an ordinary diode, it is the primary objective sought in a zener diode. The variety of available zener diodes is illustrated in Figure 2–22.

The schematic symbol of a zener diode is shown in Figure 2–23(a). The normal positive reference voltage and current for a diode are labeled in Figure 2–23(b). Although intended operation of the zener diode is in the reverse region, the normal voltage and current convention is compatible with standard semiconductor diodes.

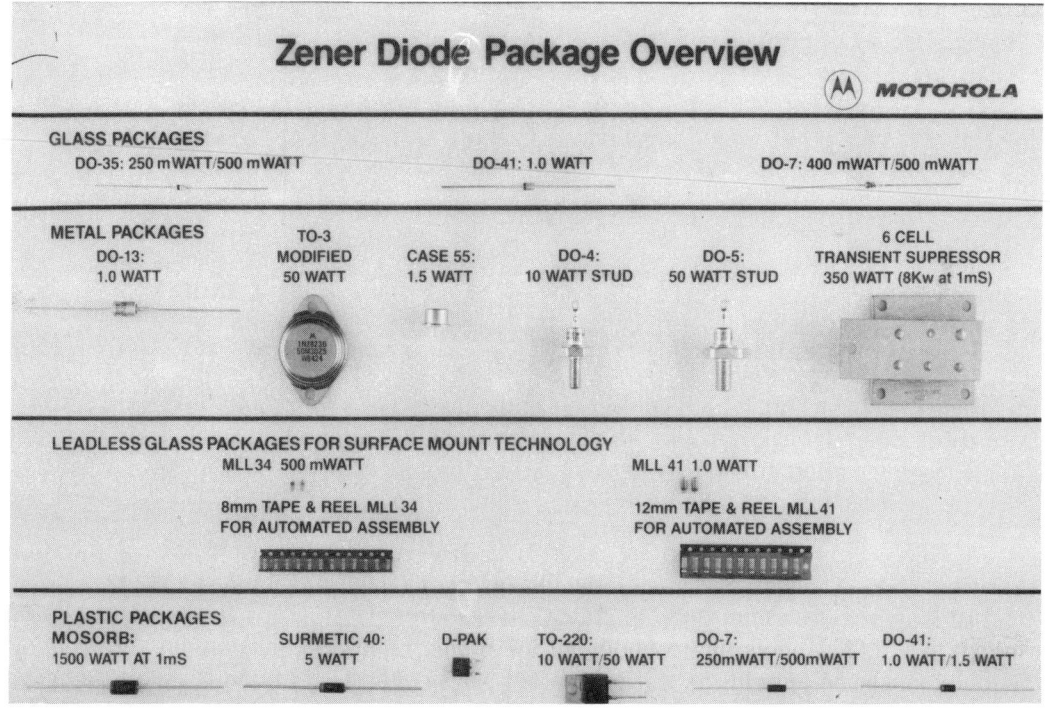

FIGURE 2–22
Typical packages for zener diodes. (Courtesy of Motorola, Inc.)

FIGURE 2–23
Schematic symbol and variables for zener diode.

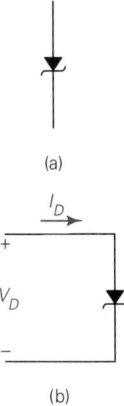

(a)

(b)

The terminal characteristic of a typical zener diode is shown in Figure 2–24. Generally, it has the same form as an ordinary diode, as given in Figure 2–15. In the forward-bias region, a zener diode behaves like an ordinary diode, and the same circuit models used earlier may be applied to the zener diode in that region.

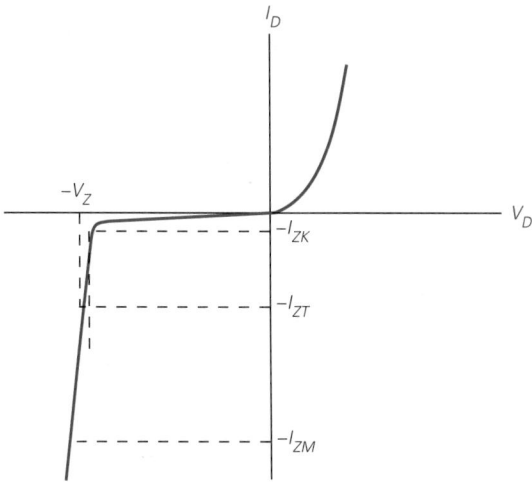

FIGURE 2–24
Zener characteristic with parameters identified.

As the zener characteristics are studied in the remainder of this section, it may be helpful for the reader to refer to Appendix D. The second set of specifications provided there relate to several families of zener diodes. Included are the 1N746 through the 1N759, the 1N957A through the 1N986A, and the 1N4370 through the 1N4372. These are all 500-mW diodes. However, these are only a small sample of the available zener types, and the available power levels vary widely.

The notation on Figure 2–24, along with other parameters to be discussed, has been formulated to be compatible with that of the zener diode specifications in Appendix D.

The distinguishing feature of the zener diode is that reverse breakdown occurs at a specified voltage $-V_Z$, and the value of V_Z is usually much smaller than the reverse breakdown voltage of ordinary diodes. Once breakdown is reached, the voltage varies little over a wide range of current. The zener is thus capable of providing voltage regulation and in establishing voltage references.

Some important parameters of a zener diode are identified in Figure 2–24. Strictly speaking, some of these quantities have negative values when referred to the basic voltage and current convention of Figure 2–23(b). However, as is customary with data manuals and with references, definitions are given in terms of positive values.

The nominal **zener voltage** is designated as V_Z. Since V_Z does vary slightly, it is customary in data manuals to state the value of V_Z at some reference **test current,** which is denoted as I_{ZT}.

An important parameter for a zener diode is the maximum power rating. In general, the power dissipation P_D is given by

$$P_D = V_Z I_Z$$

(2–20)

where I_Z is the zener current. Since the value of V_Z is known, the value of I_Z corresponding to the maximum power dissipation could be computed. However, since V_Z will vary somewhat from one unit to another and with different operating conditions, a separate maximum current rating is provided for a given zener type. The maximum current rating provides some leeway to ensure operation below the maximum power level and will be discussed shortly.

The current I_{ZK} is referred to as the **knee current.** It is the smallest value of current at which regulation of the voltage can be assumed to occur.

The current I_{ZM} is the *maximum* reverse current at which the diode can operate safely, and it provides sufficient leeway to ensure operation below the maximum power, Thus, all circuit designs must be checked to ensure that operation is between I_{ZK} and I_{ZM} under all conditions.

Since there is some variation of voltage in the zener region, the **zener impedance** is specified. It is actually a resistance, but a common symbol for the quantity is Z_{ZT}. Let ΔI_Z represent a change in current in the zener region, and let ΔV_Z represent the corresponding change in voltage. Then Z_{ZT} is defined as

$$Z_{ZT} = \frac{\Delta V_Z}{\Delta I_Z}$$

(2–21)

Assume that the voltage V_Z and impedance Z_{ZT} at a given current I_Z are known. If the current then changes to $I_Z + \Delta I_Z$, the corresponding zener voltage changes to $V_Z + \Delta V_Z$, and

$$\Delta V_Z = Z_{ZT}\Delta I_Z \qquad (2\text{--}22)$$

The new value of zener voltage V_Z' can then be estimated as

$$V_Z' = V_Z + \Delta V_Z = V_Z + Z_{ZT}\Delta I_Z \qquad (2\text{--}23)$$

The value of Z_{ZT} varies with the operating point. From the zener data of Appendix D, note that the maximum value of Z_{ZT} is specified at the test current I_{ZT}. For some of the diodes, the maximum value of Z_{ZT} is also specified at the knee current I_{ZK}. The value of the impedance at the knee current is quite large, indicating that regulation is very poor at that level of current. The value of the impedance at the test current is more typical of the impedance in the desired operating range.

The variation of the zener voltage with temperature can be determined from the **temperature coefficient of the zener voltage,** which is denoted in Appendix D as ΘV_Z. This quantity is usually expressed in mV/°C and may be positive or negative, depending on the breakdown process. In general, at zener voltages below about 6 V, breakdown is due to the zener effect, and the temperature coefficient is *negative*. However, at zener voltages above about 6 V, breakdown is caused by the avalanche effect, and the temperature coefficient is *positive*.

Assume that V_Z and Θ_{VZ} are known at a certain reference temperature. If the temperature then changes by ΔT, the change in voltage ΔV_Z is

$$\Delta V_Z = \Theta_{VZ}\Delta T \qquad (2\text{--}24)$$

The new value of voltage V_Z' is then estimated as

$$V_Z' = V_Z + \Delta V_Z = V_Z + \Theta_{VZ}\Delta T \qquad (2\text{--}25)$$

Note that Θ_{VZ} may be either positive or negative, so V_Z' may be larger or smaller than V_Z.

On the fourth page of the zener specifications in Appendix D, curves providing temperature coefficients are provided. Note that the two curves represent a *range* of values, so the actual coefficient for a given diode could lie anywhere within the range. Note also how the temperature coefficient changes from a negative value at low voltages to a positive increasing value as the voltage increases.

EXAMPLE 2–4   A 1N753 zener diode has a specified value of $V_Z = 6.2$ V at $I_{ZT} = 20$ mA. The maximum zener impedance is specified as 7 Ω. Estimate the maximum value of zener voltage at $I_Z = 60$ mA.

Solution
The change in current between the reference point and the point of interest is

$$\Delta I_Z = 60\ \text{mA} - 20\ \text{mA} = 40\ \text{mA} = 0.04\ \text{A} \qquad (2\text{--}26)$$

The change in voltage is

$$\Delta V_Z = Z_{ZT}\Delta I_Z = 7 \times 0.04 = 0.28 \text{ V} \qquad \text{(2-27)}$$

The new voltage is

$$V'_Z = V_Z + \Delta V_Z = 6.2 + 0.28 = 6.48 \text{ V} \qquad \text{(2-28)}$$

Note that the basic units of ohms and amperes were used in the multiplication of (2–27).

EXAMPLE 2–5   A possible value of the temperature coefficient of a 1N753 zener diode is +3 mV/°C. If $V_Z = 6.2$ V at 25°C, estimate the zener voltage at 75°C.

Solution
The temperature difference is

$$\Delta T = 75 - 25 = 50°\text{C} \qquad \text{(2-29)}$$

The change in voltage is

$$\Delta V_Z = 3 \times 10^{-3} \times 50 = 0.15 \text{ V} \qquad \text{(2-30)}$$

The new voltage is

$$V'_Z = 6.2 + 0.15 = 6.35 \text{ V} \qquad \text{(2-31)}$$

Note that the temperature coefficient was expressed in the basic units of volts/°C in (2–30).

2–8 OPTOELECTRONIC DIODES

The field of **optoelectronics** is a rapidly developing segment of the electronics industry. This term refers to those devices and applications that combine *optical* and *electronic* technology. Some of the most common optoelectronic devices combine the action of a *pn* junction diode with light-sensitive materials. In this section we consider three of the most common optoelectronic devices: light-emitting diodes, photodiodes, and optocouplers.

Light-Emitting Diodes

The **light-emitting diode** (widely referred to as an LED) is a special *pn* junction diode that emits light when forward biased. In an ordinary *pn* junction diode, charge carriers crossing the junction radiate energy in the form of heat. In the LED, special materials are used that convert a portion of this energy to light. Some of the elements used are gallium, arsenic, and phosphorus. A variety of colors are available in LEDs. In addition to visible radiation with common colors, some LEDs produce invisible infrared radiation. Invisible radiation could be useful in applications such as burglar alarm systems, for example.

The schematic symbol for an LED is shown in Figure 2–25. The arrows obviously indicate light *leaving the diode*.

FIGURE 2–25
LED schematic symbol.

The voltage drop for a forward-biased LED is less predictable than that of an ordinary diode. Typical values range from 1.5 to 2.5 V, and the current range is typically from 10 to 50 mA. A reasonable estimate for the voltage when the exact value is not known is 2 V.

A number of LEDs can be combined to form an electronic display. One of the most common electronic displays is the **seven-segment indicator.** The form of the seven-segment display is shown in Figure 2–26(a), and the schematic diagram is shown in Figure 2–26(b). The seven segments of the display are denoted as A through G, respectively, and each represents a separate diode and associated dropping resistor. The form shown in Figure 2–26 is referred to as a **common-anode** form, since all the anodes are connected to a common point. If a positive voltage with respect to ground is connected to the common-anode form, an individual segment is activated by grounding the appropriate branch. These indications are also available in **common-cathode** form.

By activating various combinations of the seven segments, all integer numbers can be formed. Certain letter combinations can be formed as well.

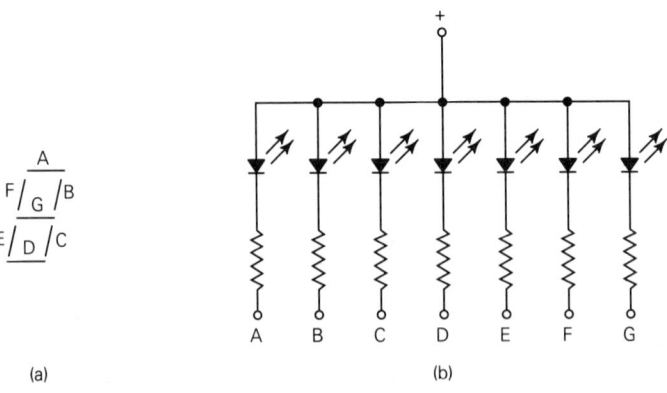

(a)

(b)

FIGURE 2–26
Seven-segment LED indicator and schematic.

Photodiode

The **photodiode** is opposite to the LED in that it responds to impinging light. The schematic symbol for a photodiode is shown in Figure 2–27. In contrast to the symbol for the LED, the arrows point *toward* the diode in the photodiode symbol.

FIGURE 2–27
Photodiode schematic symbol.

The photodiode employs the principle that the reverse current in a reverse-biased *pn* junction is a function of any light radiation received by the junction. The energy received from the light increases the reverse current. A photodiode is built to enhance this effect, and the light is channeled to the junction by a special window. Thus, the normal mode of operation of a photodiode is with reverse bias.

Optocoupler

An **optocoupler** is a combination of an LED and an optically aligned photodiode in a single package. The schematic diagram is shown in Figure 2–28. The LED is on the left, and the photodiode is on the right.

FIGURE 2–28
Optocoupler schematic symbol.

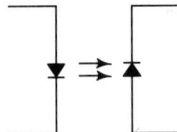

Assume that a signal is connected on the left and that a load is connected on the right. Both portions of the circuit will require appropriate biasing circuitry, of course. The signal will cause variations in the light intensity produced by the LED, and this will result in corresponding variations in the current through the photodiode. The result is that the signal and source are completely isolated from each other electrically. The signals are, instead, said to be **optically coupled.**

This type of application is particularly useful where a small signal is to be transferred between two points without the presence of undesired electrical coupling such as high voltages or ground loops.

The term *optocoupler* includes similar coupling components employing transistors instead of diodes. Larger output levels are achieved with transistors as a result of the amplification process (to be discussed later).

EXAMPLE 2–6

An LED is to be connected through a resistor to a 15-V dc supply as shown in Figure 2–29(a). If the LED voltage is known to be 2 V and the desired current is 20 mA, determine the required resistance R.

FIGURE 2–29
Circuit of Example 2–6 and equivalent loop.

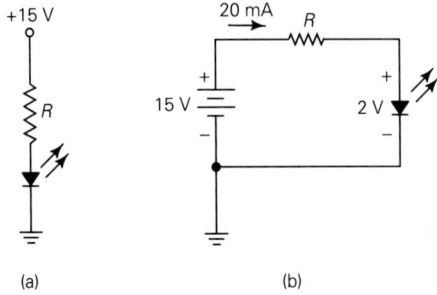

(a)   (b)

Solution
The circuit form shown in Figure 2–29(a) is the form that probably appears on a schematic diagram. The circuit is redrawn in an equivalent single-loop form in Figure 2–29(b), in which the 15-V supply is represented as a battery. An application of KVL for the loop reads

$$-15 + 20 \times 10^{-3}R + 2 = 0 \qquad (2\text{–}32)$$

or

$$R = \frac{15 - 2}{20 \times 10^{-3}} = \frac{13}{20 \times 10^{-3}} = 650 \ \Omega \qquad (2\text{–}33)$$

In the expression of (2–33), note that the value $15 - 2 = 13$ V represents the voltage absorbed by the resistance, and this value is divided by the known current in accordance with Ohm's law. With reasonable practice, a simple circuit such as this can be analyzed directly. However, one can always return to the more "formal" approach when necessary.

2–9 OTHER TYPES OF DIODES

In this section, several other types of diodes are discussed. These include Schottky diodes, varactor diodes, and others.

Schottky Diodes

When an ordinary *pn* junction diode is forward biased, holes are injected from the *p*-type material across the junction into the *n*-type material, and electrons from the *n*-type material are injected into the *p*-type material. These charges on both sides are momentarily stored before recombining with charges of the opposite side. To change the forward current or voltage, it is necessary that the stored charge change. This phenomenon is a capacitive effect and is referred to as **diffusion capacitance.**

Because of the stored charge, it takes some time to terminate current flow when the diode is changed from forward to reverse bias. This time is called the **reverse recovery time.** It is typically of the order of a few nanoseconds for a small-signal diode. Its effect is negligible at low frequencies, but it can become significant above 10 MHz or so.

The **Schottky diode** is of a special construction that reduces the forward diffusion capacitance to a negligible value, which in turn reduces the reverse recovery time considerably. This is achieved by using a thin layer of aluminum on lightly doped *n*-type silicon, which forms a rectifying junction between the metal and the semiconductor.

The schematic symbol of a Schottky diode is shown in Figure 2–30. These diodes are usable well into the microwave frequency range. They are also used in high-speed digital switching circuits.

FIGURE 2–30
Schematic symbol for Schottky diode.

Varactor Diodes

All junction diodes exhibit a capacitance effect when the diode is reverse biased. This effect is referred to as **reverse depletion capacitance.** This capacitance varies with the magnitude of the reverse voltage across the junction.

Diodes manufactured specifically to employ the capacitive effect are referred to as **varactor diodes.** The term *varactor* refers to **voltage-variable capacitance.** Such diodes are also called **varicaps.** Among the applications of varactors are the tuning circuits of many high-frequency receivers. The schematic symbol is shown in Figure 2–31.

FIGURE 2–31
Schematic symbol for varactor diode.

Other Diodes

Other diodes include constant current diodes, tunnel diodes, varistors, step-recovery diodes, and back diodes.

The **constant current diode** is capable of providing a constant current to a variable load. It is thus opposite in its effect to a zener diode. Constant current diodes are related very closely to field-effect transistors (discussion is deferred to Chapter 7).

A **tunnel** or **Esaki diode** exhibits a **negative resistance** region in its terminal characteristic. This means that over a portion of the terminal characteristic, the current decreases as the voltage increases. Tunnel diodes are used in certain high-frequency and microwave circuits.

A **varistor** is similar to two zener diodes connected in series with opposite directions and with a high breakdown voltage in either direction. These devices are used as transient suppressors in power systems. Sudden line voltage changes, switching loads, lighting, and other effects can produce devastating overvoltages in some power systems. Varistors are used to suppress these effects.

A **step-recovery diode** has a doping profile that results in a continued current flow for a portion of a reverse-biased interval, followed by a sudden termination of the current. This effect is called **reverse snapoff,** and step-recovery diodes are sometimes referred to as **snap diodes.** Step-recovery diodes are used in frequency-multiplier circuits to generate frequencies that are integer multiples of the input frequency.

A **back diode** is one in which the doping level is increased to the point where reverse breakdown occurs at approximately -0.1 V. By employing this diode in a reverse direction and for signal levels below 0.7 V, the output is reduced only by the 0.1-V drop.

2–10 TESTING A DIODE WITH AN OHMMETER

An **ohmmeter** may be used to provide a quick check on the probable condition of a diode. Current actually flows out of one of the terminals of an ohmmeter through the resistance being measured and back into the other terminal. Check the manual for the ohmmeter so that you will know the actual positive direction of current flow through an external load. *It is absolutely essential that all power sources be turned off for components being tested.* An ohmmeter can easily be damaged when connected in a "hot" circuit.

The diode being tested should not be shunted by any additional resistive load. Consequently, it may be necessary to disconnect at least one of the diode terminals from any circuit in which it is connected.

Connect the ohmmeter leads to the diode in the direction that will result in forward bias, and note the approximate value of resistance. Next, reverse the ohmmeter leads so that reverse bias is established, and note the resistance. A good diode should exhibit a large reverse resistance compared to the forward resistance. A ratio of 1000:1 is typical. In fact, on some ohmmeter scales, the reverse resistance of a good diode appears to be an open circuit ($R = \infty$).

The actual value of forward resistance that would be read on a typical ohmmeter depends on the scale and can vary from a few hundred ohms to thousands of ohms. The resistance read on the ohmmeter will change as the scale is changed. Consequently, this process is more of a quick check than a quantitative measure.

Some ohmmeters can produce enough current on the very lowest ranges to destroy a small-signal diode. Consequently, when testing a very small diode, avoid using the lowest ranges.

A final point is that not all ohmmeter scales of many modern multimeters necessarily provide sufficient voltage to forward-bias diodes. With such instruments, it is intended that only certain scales be used to check diodes. Sometimes these ohmmeter scales are identified on the instrument. In any event, the operating manual will be your best guide.

2–11 PSPICE EXAMPLES

PSPICE EXAMPLE 2–1

Obtain the forward voltage–current characteristic curve of the PSPICE basic diode model.

Solution

PSPICE contains many models of diodes and transistors that correspond very closely with commercially available units, and we will investigate some of these throughout the text. In this case, we will use the D model, which is not intended to represent any particular device, but which is representative of a variety of different types. Henceforth, we refer to this model as the **generic diode model.**

The circuit model used to obtain the characteristic curve is shown in Figure 2–32(a), and the PSPICE code is shown in Figure 2–32(b). Note that the test voltage VD is connected directly across the diode. A real diode should never be connected directly across a power supply like this without a limiting resistor, since the diode could be easily damaged. However, "software diodes" are rugged and cannot be burned out!

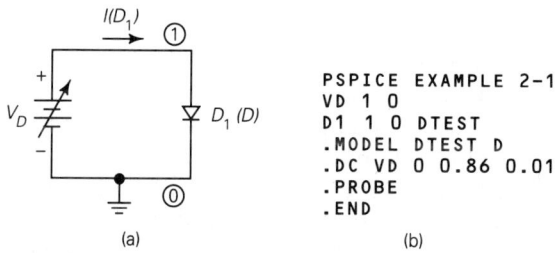

```
PSPICE EXAMPLE 2-1
VD 1 0
D1 1 0 DTEST
.MODEL DTEST D
.DC VD 0 0.86 0.01
.PROBE
.END
```

(a)   (b)

FIGURE 2–32

Circuit and code of PSPICE Example 2–1.

Following the title line, the voltage is expressed as

```
VD 1 0
```

This has fewer elements than we have previously encountered, and it needs some explanation. The numbers 1 and 0 refer to the node connections, with 1 identified as the positive node; however, neither a voltage type nor a value is given. It turns out that when no designation for the type of source is given, PSPICE interprets the source as a dc source. For clarity, we have shown it earlier in the text and we will use it often as we progress for emphasis. However, when there is little chance of confusion (e.g., when there is only a dc source present) it may be omitted.

The absence of a voltage value is interpreted by PSPICE as a value of 0 V. In some situations, this is useful for measuring or monitoring current, but for now we need a range of values of voltage. The range of voltages is identified in the control block by the sweep statement:

```
.DC VD 0 0.86 0.01
```

This statement tells PSPICE to perform a dc analysis and to step the voltage source VD from 0 V to 0.86 V in steps of 0.01 V. This corresponds to 87 steps for the source voltage. When a dc sweep is performed, the values in the .DC line override any value on the VD line. Thus, any value could have been placed on the VD line, and the simplest choice is to omit it, which causes a default value of 0.

Back to the circuit description, the generic diode model D is being employed, and this requires two lines. The forms for the two lines are as follows:

```
D1 1 0 DTEST
.MODEL DTEST D
```

The first line appears in the circuit description and must begin with the letter D (for diode). Additional letters or numbers without a space are identifiers. In this case, the diode is denoted simply as D1. The integers 1 and 0 define the two terminals of the diode. The order is very important in that *the first number represents the anode node, and the second number represents the cathode node*. Said differently, the *easy* direction of positive current flow is from the first node number to the second node number. The name DTEST is any arbitrary name that will be matched to the model definition, as we will see shortly.

The model identification line must begin with .MODEL. The next entry on the line is DTEST, which must correspond to the name provided on the circuit element line. Finally, the generic name of the model must be given, and this is simply D. Note that there is some choice with the other names, but this latter description is unique and identifies the particular model.

The preceding codes may seem cumbersome and redundant, but this is a simple circuit containing only one diode. That is, when a circuit contains several diodes with different characteristics, the labeling scheme will appear more logical.

For virtually all of the work in this text, the diode and transistor models are used exactly as provided by PSPICE. For advanced work beyond the scope of this text, however, it is possible to modify some of the diode characteristics by providing more optional information on the .MODEL line.

If we so desired, we could execute a dc print statement and send a lot of numerical data to a printer. But our purposes are better served by letting PSPICE provide a plot for us. This is achieved by the use of the .PROBE statement; we use this powerful tool extensively.

When the program is executed, a coordinate system is established on the screen by .PROBE with VD as the independent variable. Several different quantities could be plotted, but the only one of interest to us at this time is the diode current, which is indicated by .PROBE as I(D1). Although a full discussion of .PROBE is beyond the scope of this text, we should mention that .PROBE is menu driven and very easy to use. Much of its utility can be discovered by simple trial-and-error procedures, but the inquisitive reader is referred to one of the several books on PSPICE for further details.

The plot of diode forward current versus diode forward voltage is shown in Figure 2–33. Negligible current is observed until the diode voltage approaches about 0.7 V. The current then increases sharply, as expected.

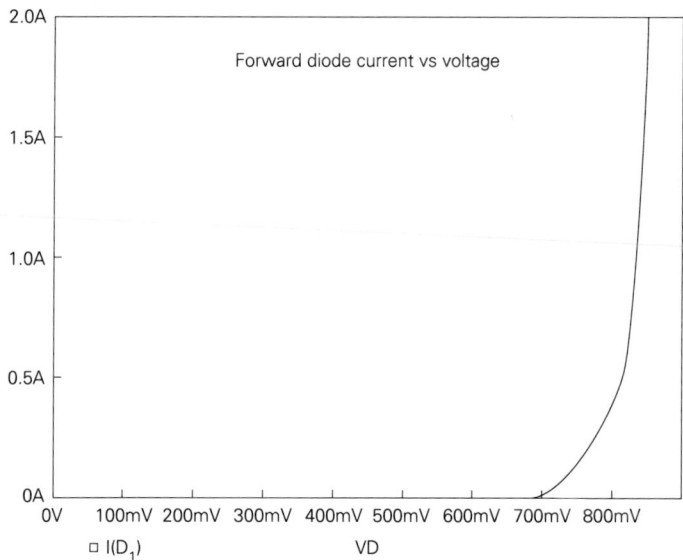

FIGURE 2–33
Forward characteristic of generic diode model in Example 2–1.

PSPICE EXAMPLE 2–2

Obtain the forward voltage–current characteristic curve of the PSPICE 1N4148 library model.

Solution

The circuit model is shown in Figure 2–34(a), and the PSPICE code is shown in Figure 2–34(b). In most ways, this example is the same as PSPICE Example 2–1,

FIGURE 2–34
Circuit and code of PSPICE Example 2–2.

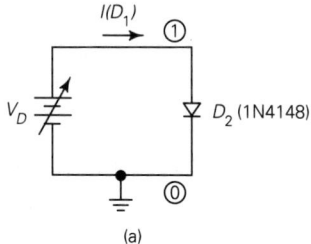

(a)

```
PSPICE EXAMPLE 2-2
VD 1 0
D1 1 0 D1N4148
.LIB EVAL.LIB
.DC VD 0 0.9 0.01
.PROBE
.END
```

(b)

but with two differences. The characteristics correspond closely to those of the real diode designated as a 1N4148, and the characteristics must be obtained from the PSPICE model library.

The two lines in the code that differ in form from those of the preceding example are as follows:

```
D1 1 0 D1N4148
.LIB EVAL.LIB
```

The first line begins in the same fashion as the preceding example with D1 representing the identifier and 1 and 0 representing the anode and cathode nodes, respectively. However, the next entry is D1N4148. The first letter is D, and the remainder of the name is the diode type to be called from the library.

When a library model is to be used, a control line beginning with .LIB must be included. The name EVAL.LIB represents the evaluation version library, which was used for this simulation. The reader should compare the *library* form from this example and the *model* form from the preceding example, so that their coding differences are understood.

The forward current versus voltage characteristic as obtained from .PROBE is shown in Figure 2–35. Note that the form is very similar to that of the generic diode, but the current does not increase as sharply above the knee voltage as for the generic model.

PSPICE EXAMPLE 2–3   Obtain the voltage–current characteristic curve of the 1N750 PSPICE diode over the range from zener breakdown to forward conduction.

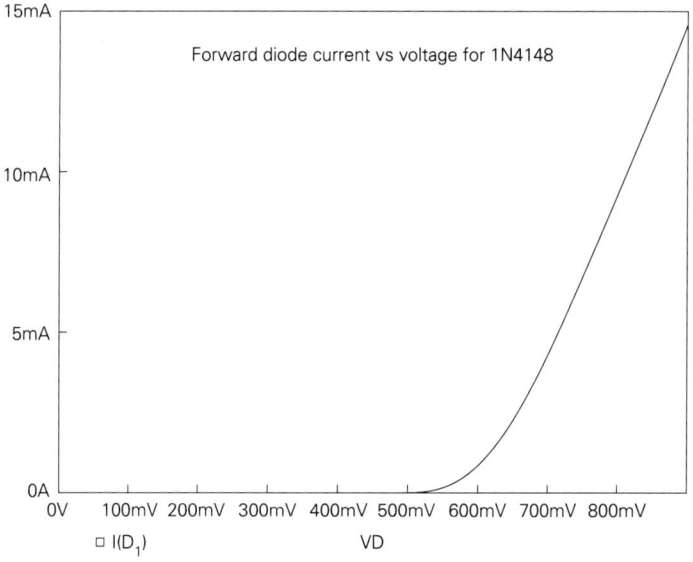

FIGURE 2–35
Forward characteristic of 1N4148 diode model in Example 2–2.

Solution

The circuit model is shown in Figure 2–36(a), and the PSPICE code is shown in Figure 2–36(b). This example is very similar to that of PSPICE Example 2–2 in that we use a library model, but the present diode will be swept over both positive and negative voltages.

FIGURE 2–36
Circuit and code of PSPICE Example 2–3.

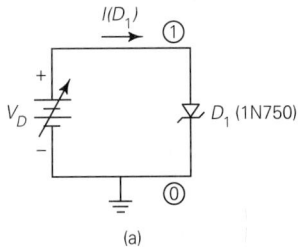

(a)

```
PSPICE EXAMPLE 2-3
VD 1 0
D1 1 0 D1N750
.LIB EVAL.LIB
.DC VD -4.8 0.8 0.001
.PROBE
.END
```

(b)

Based on the preceding example, the code should be self-evident for the most part. The diode designation on the D1 line in this case is D1N750. The normal zener voltage for this diode is about 4.7 V, so the sweep range has been selected as −4.8 V to 0.8 V (i.e., from just below the zener breakdown level to the level for forward conduction).

The voltage–current characteristic obtained from .PROBE is shown in Figure 2–37. Note how reverse breakdown starts to occur near −4.7 V and how the voltage changes very little over a wide range of current. In fact, the current level on the curve actually exceeds the maximum current rating for this particular diode, which indicates that a voltage of −4.8 V across the real diode would be excessive.

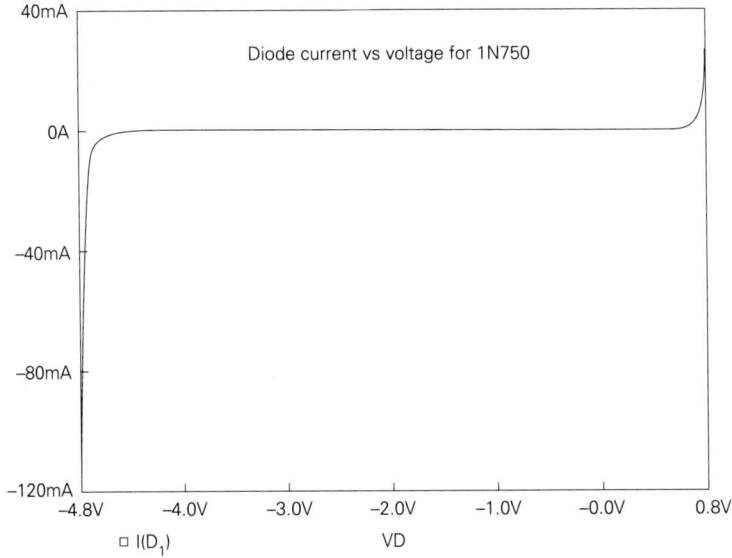

FIGURE 2–37
Characteristic of zener diode 1N750 in Example 2–3.

PROBLEMS

Drill Problems

2–1. Consider the circuit shown in Figure P2–1. Verify that the diode is forward biased. Calculate the current I and the two voltages V_D and V_R based on **(a)** the ideal diode model and **(b)** the constant voltage drop model. The diode is assumed to be a silicon type with $V_o = 0.7$ V.

FIGURE P2-1

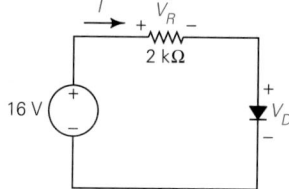

2-2. Consider the circuit shown in Figure P2-2. Verify that the diode is forward biased. Calculate the current I and the two voltages V_D and V_R based on **(a)** the ideal diode model and **(b)** the constant voltage drop model. The diode is assumed to be a silicon type with $V_o = 0.7$ V.

FIGURE P2-2

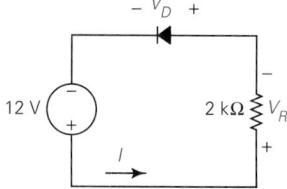

2-3. Consider the circuit shown in Figure P2-3. Verify that the diode is reverse biased, and determine the voltage across the diode.

FIGURE P2-3

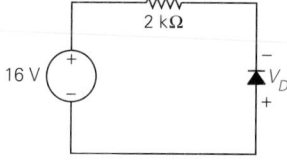

2-4. Consider the circuit shown in Figure P2-4. Verify that the diode is reverse biased, and determine the voltage across the diode.

FIGURE P2-4

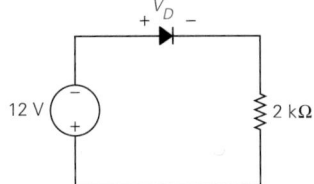

2-5. For the circuit of Problem 2-1 (Figure P2-1), assume that the diode is represented by a constant voltage source of 0.6 V in series with a resistance of 10 Ω. Determine I, V_D, and V_R.

2–6. For the circuit of Problem 2–2 (Figure P2–2), assume that the diode is represented by a constant voltage source of 0.56 V in series with a resistance of 20 Ω. Determine I, V_D, and V_R.

2–7. A 1N755 zener diode has a specified value of $V_Z = 7.5$ V at $I_{ZT} = 20$ mA. The maximum zener impedance is specified as 6 Ω. Estimate the value of zener voltage at $I_Z = 50$ mA.

2–8. A 1N759 zener diode has a specified value of $V_Z = 12$ V at $I_{ZT} = 20$ mA. The maximum zener impedance is specified as 30 Ω. Estimate the value of zener voltage at $I_Z = 30$ mA.

2–9. A possible value of the temperature coefficient of a 1N755 zener diode is +4 mV/°C. If $V_Z = 7.5$ V at 25°C, estimate the zener voltage at 75°C.

2–10. A possible value of the temperature coefficient of a 1N759 zener diode is +8 mV/°C. If $V_Z = 12$ V at 25°C, estimate the zener voltage at 75°C.

2–11. A possible value of the temperature coefficient of a 1N4372 zener diode is −2 mV/°C. For $V_Z = 3$ V at 25°C, estimate the zener voltage at 75°C.

2–12. A certain zener diode has a specified value of $V_Z = 10$ V at 25°C and $I_{ZT} = 20$ mA. The zener impedance is 20 Ω, and the temperature coefficient is +6 mV/°C. If the impedance and temperature effects are assumed to be additive, estimate the zener voltage for $I_Z = 30$ mA and at a temperature of 75°C.

2–13. In Example 2–6, the resistance was determined on the assumption that the diode drop was 2 V. If the diode drop actually turns out to be 1.5 V, determine the actual LED current that would result from the use of the 650-Ω resistance.

2–14. In Example 2–6, the resistance was determined on the assumption that the diode drop was 2 V. If the diode drop actually turns out to be 2.5 V, determine the actual LED current that would result from the use of the 650-Ω resistance.

2–15. **(a)** For the LED indicator of Figure 2–26, indicate which segments should be grounded to form the number 2. **(b)** If the current in each segment is 20 mA, what is the total current for this character?

2–16. **(a)** For the LED indicator of Figure 2–26, indicate which segments should be grounded to form the letter A. **(b)** If the current in each segment is 20 mA, what is the total current for this character?

Design Problems

2–17. An LED is to be connected through a resistor to a 5-V dc power supply as shown in Figure P2–17. **(a)** If the LED voltage is estimated as 2 V and the desired current is 20 mA, determine the required resistance R. **(b)** Based on the value of R determined in **(a)**, estimate the possible range of current if the actual LED voltage can range from 1.5 to 2.5 V.

FIGURE P2–17   +5 V

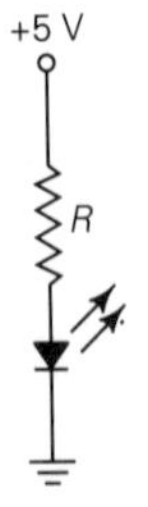

109 | PROBLEMS

2–18. An LED is to be powered from a power supply whose voltage is -15 V with respect to ground. The LED voltage is estimated as 2 V, and the desired current is 15 mA. **(a)** Draw the schematic diagram and determine the value of a dropping resistance R. **(b)** Based on the value of R determined in **(a)**, estimate the possible range of current if the actual LED voltage can range from 1.5 to 2.5 V.

Troubleshooting Problems

2–19. The voltages across the resistor and diode in Figure P2–1 are measured as $V_R = 0$ and $V_D = 16$ V. What is the likely problem?

2–20. The voltages across the resistor and diode in Figure P2–1 are measured as $V_R = 16$ V and $V_D = 0$. What is the likely problem?

3

APPLICATIONS OF DIODE-RESISTOR CIRCUITS

OBJECTIVES

After completing this chapter, the reader should be able to:

- Describe and illustrate the form of a transfer characteristic.
- Show the difference between a linear and a nonlinear transfer characteristic.
- Define the process of rectification.
- Draw the circuit forms and transfer characteristics, explain the operation of, and perform the analysis of several half-wave rectifier circuits.
- Explain the process of amplitude limiting.
- Explain the function of a clipping circuit.
- Explain the operation of an OR gate employing diodes and resistors.
- Explain the operation of an AND gate employing diodes and resistors.
- Draw the circuit form, explain the operation of, and perform the analysis of a zener regulator circuit.
- Perform worst-case computations in a zener regulator circuit.
- Design a zener regulator circuit to derive a fixed reference voltage from a larger available voltage.
- Apply PSPICE to analyze circuits in the chapter.

3-1 OVERVIEW

We introduced the basic semiconductor diode and several special-purpose diodes in Chapter 2. Methods of analyzing circuits containing diodes were developed. However, up to this point we have considered no specific applications of these devices.

In this chapter we cover a number of basic applications of diodes in practical circuits. The intent is to provide immediate application of the analysis skills of Chapter 2 to some useful circuits. The reader will also begin to start developing a "mental collection" of the types of circuits used in the field.

Of the three basic linear circuit parameters (R, L, and C), only resistance is considered in this chapter. The resulting circuits containing resistors and diodes will be classified as **diode-resistor** circuits. Thus, circuits involving capacitance and/or inductance will be delayed until the reader's circuit analysis and mathematical skills are sufficiently developed. These diode-resistor circuits may contain fixed dc voltages as required, and their application with time-varying signal sources is also considered.

The specific types of applications considered include rectifiers, limiters, clippers, logic gates, and zener regulators. We present some of the fundamental circuits in each category.

Some basic advice to the reader is in order. This advice applies to this chapter and many others throughout the book. As you start looking at the various circuits, you will soon realize that there are many different types of functional operations that can be performed with electronic components, and there are often many ways to achieve the same operation. No one (including this author) can store the complete details of all these circuits in his or her memory.

Obviously, a reasonable amount of memory work is required when learning a new subject, and periodic review is in order. However, it is more important to develop the necessary approach to analyze or develop the form of a given circuit than to rigidly memorize the circuit details without understanding. As you progress through various sections, it is hoped that you will see the similarity in analysis procedures as they are applied to individual circuits.

3–2 TRANSFER CHARACTERISTICS

Early in Chapter 2 we introduced the concepts of both linear and nonlinear **terminal characteristics** relating voltage and current at a set of terminals. Consider now a circuit with a set of **input terminals** (also called the **input port**) and a set of **output terminals** (also called the **output port**) as illustrated in block form in Figure 3–1. In many circuits it is possible to obtain an equation and/or a graph relating an output variable to an input variable. Such a relationship is called an **input–output characteristic**, or **transfer characteristic**. Both terms are used in this book.

FIGURE 3–1
Circuit with input and output ports.

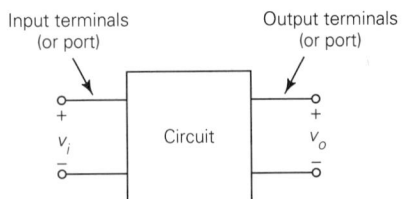

Depending on the circuit and the application, either the voltage or the current at the input port could be considered as the input variable, and either the voltage or the current at the output port could be considered as the output variable. In most situations the *input variable* is considered as the **independent variable** and is shown on the *horizontal* axis, while the *output variable* is considered as the **dependent variable** and is shown on the *vertical* axis.

The most common type of application is where the input voltage is considered as the independent variable and the output voltage is assumed as the dependent variable. All applications in this chapter assume that situation. The most common symbol for the instantaneous input voltage is v_i, and the corresponding output voltage is denoted as v_o, as illustrated in Figure 3–1. In a few places, however, it will be necessary to modify that notation.

First, we consider a **linear transfer** characteristic. A simple example of a circuit that will produce a linear transfer characteristic is the **voltage-divider,** or **attenuator,** circuit of Figure 3–2, which can be used to *attenuate,* or reduce, the level of a signal. The output voltage v_o is readily expressed in terms of the input voltage as

$$v_o = \frac{R_2}{R_1 + R_2} v_i \tag{3–1}$$

This equation describes a linear equation with slope $R_2/(R_1 + R_2)$ and zero intercept. The term *zero intercept* means that the line passes through the origin (i.e., $v_o = 0$ when $v_i = 0$). The transfer characteristic is shown in Figure 3–2(b). It is assumed, of course, that operation is below the maximum power ratings of the resistors.

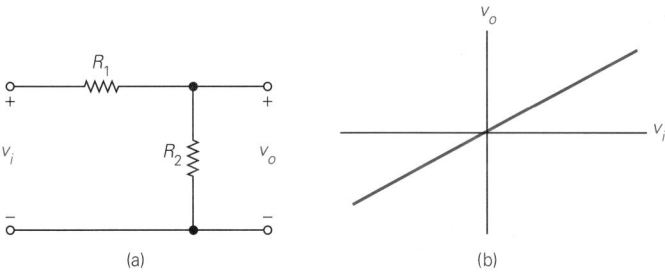

FIGURE 3–2
Simple attenuator circuit and transfer characteristic.

A basic property of a linear transfer characteristic with zero intercept is that the shape of any signal is preserved (i.e., no **distortion** occurs). For this attenuator circuit, the level would be reduced. For example, assume that $R_1 = R_2 = 10 \text{ k}\Omega$, which results in $R_2/(R_1 + R_2) = \frac{1}{2}$. For the waveform of Figure 3–3(a), the output

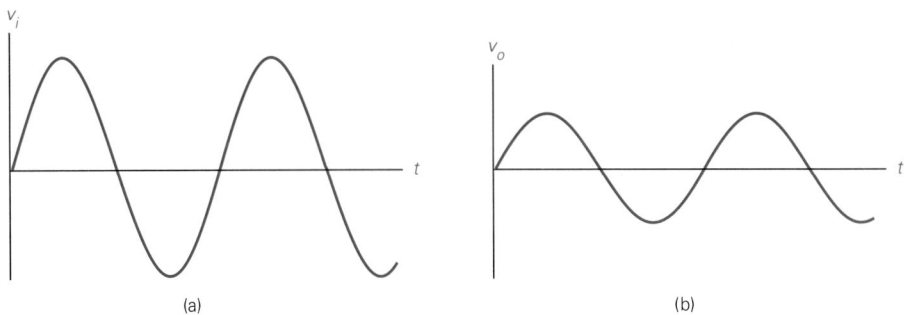

FIGURE 3–3
Input and output waveforms of attenuator circuit with $R_1 = R_2 = 10 \ \text{k}\Omega$.

would be of the form shown in (b), but the exact shape is preserved, so that no distortion occurs.

An example of some arbitrary nonlinear transfer characteristic is shown in Figure 3–4. A signal passing through a circuit such as this would undergo a modification; that is, the output would be **distorted** with respect to the input. Distortion is undesirable in amplifiers, as we will see later, but in some circuits, such as those in this chapter, possible alteration of the signal shape may be a design goal.

FIGURE 3–4
Example of a nonlinear transfer characteristic.

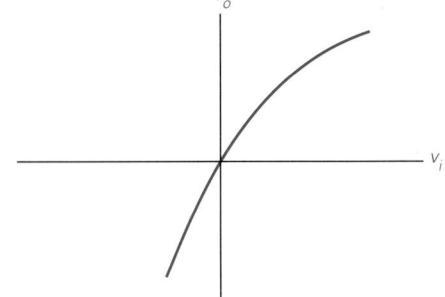

Piecewise Linear Characteristics

Many of the circuit applications involving diodes can be classified as having approximately **piecewise linear** characteristics. An example of a piecewise linear transfer characteristic is shown in Figure 3–5. Strictly speaking, piecewise linear characteristics are nonlinear. However, operation in certain regions of some piecewise linear characteristics may result in linear input–output conditions. With the characteristic of Figure 3–5, for example, a signal whose peak level is sufficiently low to keep operation along segment A would result in a linear input–output

FIGURE 3–5

Example of a piecewise linear transfer
characteristic.

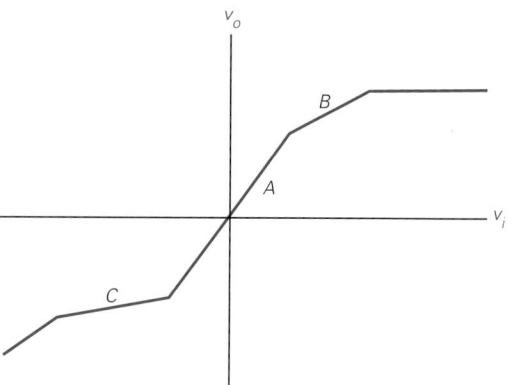

relationship, and no distortion would occur. However, if the peak signal moved
to segments B and C, nonlinear operation and distortion would occur.

Linear Equations

Many transfer characteristics of electronic circuits may be closely approximated
by linear or piecewise linear functions, as noted. To model or analyze such circuits,
it is important to understand the mathematical and graphical forms of straight-
line functions.

To relate this material to the reader's basic background, let us first review
the form of a **straight-line** or **linear equation** as studied in most mathematics texts.
Assuming x as the independent variable and y as the dependent variable, the most
common form of a linear equation is

$$y = mx + b \qquad\qquad (3\text{--}2)$$

where m is in the **slope.**

A representative graph of a linear function is shown in Figure 3–6. When
$x = 0$, $y = b$, and this value of y is called the **vertical intercept.** In this particular
example, the vertical intercept is positive. When $y = 0$, $x = -b/m$, and this value
of x is called the **horizontal intercept.** In this particular example, the horizontal
intercept is negative. Note that in the common form of the straight-line equation
as given by (3–2), the vertical intercept appears explicitly, but the horizontal
intercept must be determined by setting $y = 0$ and solving for x.

The **slope** m provides a measure of how fast y changes as x changes. If m
is very small, the linear function is nearly horizontal and displays very little change
in y as x changes. Conversely, if m is very large, the linear function approaches
a vertical characteristic, and y changes rapidly as x changes. If m is positive, y
increases as x increases, while if m is negative, y decreases as x increases. In the
example of Figure 3–6, the slope is positive.

FIGURE 3–6
Basic linear (straight-line) function.

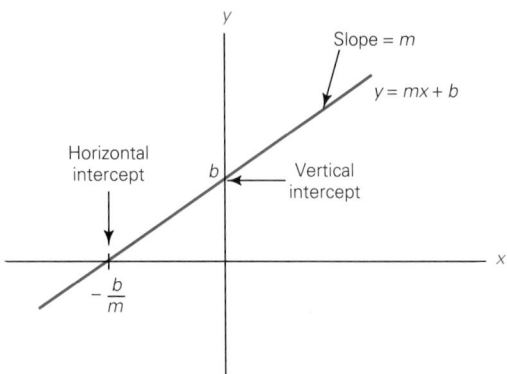

Common Linear Forms

We now observe some of the possible forms that linear functions may assume as transfer characteristics for electronic circuits. The notation will change to correspond to typical electronic circuit terminology. Specifically, the input (independent) variable will be assumed to be an instantaneous voltage v_i, and the output (dependent) variable will be assumed to be an instantaneous voltage v_o.

A perfectly linear transfer characteristic with zero intercept is shown in Figure 3–7(a). This function can be expressed as

$$v_o = Av_i \qquad\qquad \text{(3–3)}$$

where A is the slope. Any purely resistive circuit with no sources could be described in this manner. For example, the voltage-divider circuit of Figure 3–2 has a characteristic of this form, and from (3–1), the slope is $A = R_2/(R_1 + R_2)$. In this case, $A < 1$. As we will see, ideal amplifiers also have a linear characteristic with zero intercept. However, most amplifier characteristics have slopes whose magnitudes are greater than unity.

Some electronic circuits display an offset in the output voltage. An example is shown in Figure 3–7(b), and this characteristic is expressed as

$$v_o = Av_i + V_o \qquad\qquad \text{(3–4)}$$

The output offset V_o is also the vertical intercept. Thus $v_o = V_o$ when $v_i = 0$.

Certain circuits display a constant value of output voltage when the input is in one range and display a different behavior when the input is in a different range. A representative example is shown in Figure 3–7(c). This characteristic is a piecewise linear function. For this example, the output v_o can be expressed as

$$v_o = \begin{cases} V_o & \text{for } v_i < 0 \\ V_o + Av_i & \text{for } v_i > 0 \end{cases} \qquad\qquad \text{(3–5)}$$

Note that the slope is zero for $v_i < 0$, and the slope is A for $v_i > 0$.

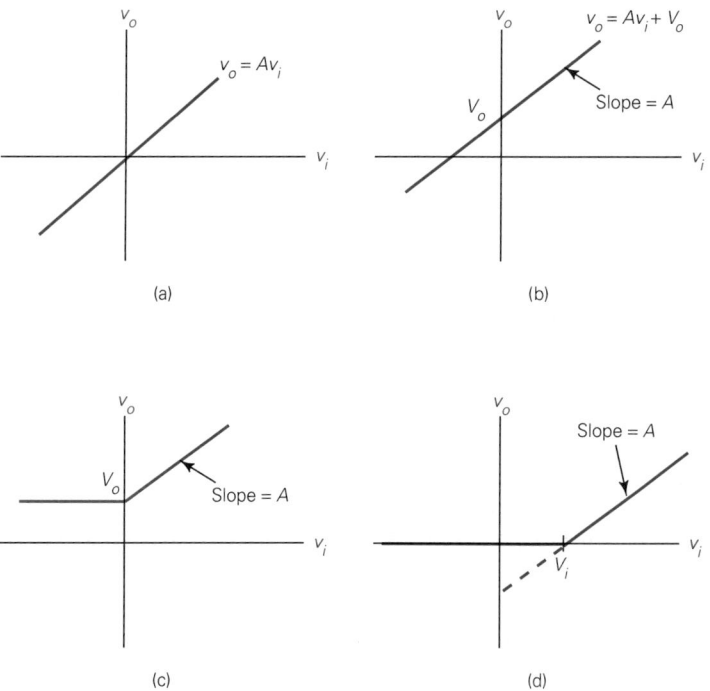

FIGURE 3–7
Some examples of linear and piecewise linear functions.

In the preceding example, the vertical intercept for the right-hand segment was immediately evident, so the linear equation could be written by inspection. A different type of piecewise linear function in which the horizontal intercept is immediately evident is shown in Figure 3–7(d). In this case $v_o = 0$ for $v_i < V_i$ by inspection. To determine the straight-line equation applicable for $v_i > V_i$, we write

$$v_o = Av_i + b \qquad\qquad (3\text{–}6)$$

where b is to be determined. We then substitute $v_o = 0$ and $v_i = V_i$ in (3–6) and solve for b. It is readily determined that $b = -AV_i$. Putting the preceding information together, we can write

$$v_o = \begin{cases} 0 & \text{for } v_i < V_i \\ Av_i - AV_i = A(v_i - V_i) & \text{for } v_i > V_i \end{cases} \qquad (3\text{–}7)$$

Note in the second form of the second equation of (3–7) how the quantity $v_i - V_i$ appears as a multiplier for the slope. Some persons remember this form when the horizontal intercept is given.

The preceding linear and piecewise linear characteristics represent only a few of the possible types that we will encounter. However, the basic straight-line equation form is always applicable to each segment of a piecewise linear function. Different equations are determined for the various segments, and the ranges of independent variables applicable to each are noted.

3-3 RECTIFIERS

The first application that we consider for diode-resistor circuits is that of **rectification.** Rectification is the process of converting a voltage with both positive and negative polarities at different times to one with only one polarity. A circuit that performs rectification is called a **rectifier.** As noted in Chapter 2, diodes are sometimes called rectifiers.

Rectification is a fundamental process within a power supply, in which ac input is converted to dc output. However, a power supply requires a transformer, rectifiers, filter components, and fusing. A full treatment of power supplies is given in Chapter 20. The simple rectifier circuits we consider here are those intended to function with electronic signals, and various voltage, current, and power levels are chosen on that basis.

We consider only simple half-wave rectifier circuits in this section. A **half-wave** rectifier circuit is one that passes the signal when it has one of the two possible polarities but rejects the signal when it has the opposite polarity.

The first rectifier to be considered is a **series diode half-wave rectifier** circuit, of which one form is shown in Figure 3–8(a). When the input voltage v_i is less than about 0.7 V, the diode is reverse biased, and the equivalent circuit is as shown in Figure 3–8(b). There is an open circuit between the input and output under this condition, and the negative portion of the signal does not appear at the output. (It is assumed here that the peak value of the negative voltage is less than the reverse breakdown voltage for the particular diode used.)

When the input voltage exceeds about 0.7 V, the diode is forward biased, and the equivalent circuit is as shown in Figure 3–8(c). A loop equation reads

$$-v_i + 0.7 + v_o = 0 \tag{3-8}$$

or

$$v_o = v_i - 0.7 \tag{3-9}$$

The complete operation of the circuit can now be expressed in idealized form as

$$v_o = \begin{cases} 0 & \text{for } v_i < 0.7 \text{ V} \tag{3-10} \\ v_i - 0.7 & \text{for } v_i > 0.7 \text{ V} \tag{3-11} \end{cases}$$

The transfer characteristic curve is of the form shown in Figure 3–8(d). The ideal transfer characteristic of (3–11) is represented by the solid line. In the vicinity of the knee voltage, the actual behavior is more like what is shown by the dashed curve. Based on the assumed reference directions, the active region corresponds to positive voltage and positive current, so this circuit can be described as one with **first-quadrant** active operation.

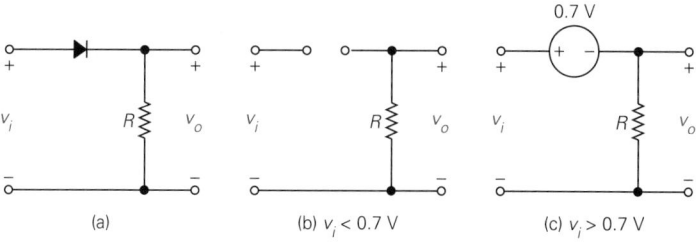

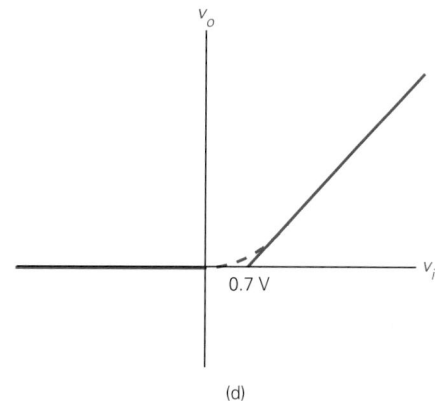

FIGURE 3–8
Series diode half-wave rectifier for first-quadrant operation, two equivalent circuits, and transfer characteristic.

The circuit can readily be converted to one in which the positive portion of the signal is rejected simply by reversing the diode, and this circuit is shown in Figure 3–9(a). For $v_i > -0.7$ V (more positive than -0.7 V), the diode is reverse biased, and the equivalent circuit is shown in Figure 3–9(b). When $v_i < -0.7$ V, the diode is forward biased, and the equivalent circuit is shown in Figure 3–9(c). The loop equation now reads

$$-v_i - 0.7 + v_o = 0 \qquad \text{(3–12)}$$

or

$$v_o = v_i + 0.7 \qquad \text{(3–13)}$$

The complete operation can be expressed in idealized form as

$$v_o = \begin{cases} 0 & \text{for } v_i > -0.7 \text{ V} \qquad \text{(3–14)} \\ v_i + 0.7 & \text{for } v_i < -0.7 \text{ V} \qquad \text{(3–15)} \end{cases}$$

The transfer characteristic curve is of the form shown in Figure 3–9(d). Active operation is now in the third quadrant (negative input and negative output).

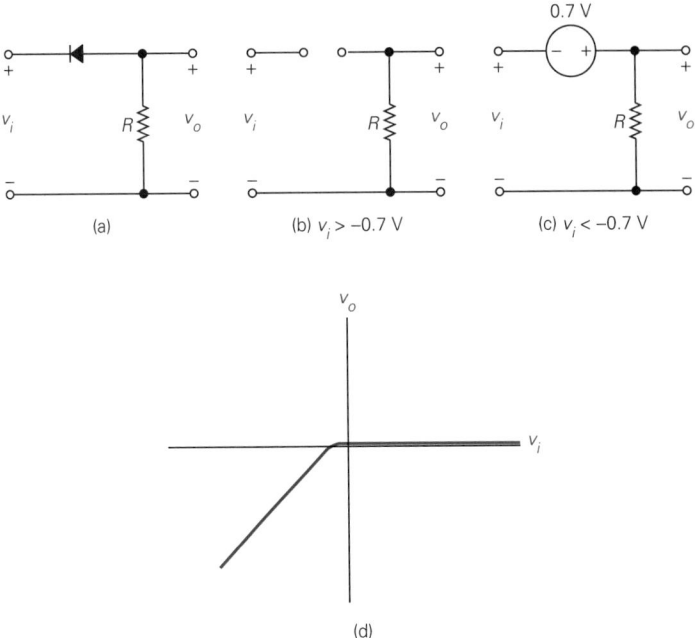

FIGURE 3–9
Series diode half-wave rectifier for third-quadrant operation, two equivalent circuits, and transfer characteristic.

The value $+0.7$ in (3–15) may bother some readers, since we know that the output voltage magnitude is less than the input voltage magnitude. This is caused by the fact that v_i is negative in the third quadrant, and this forces the magnitude of v_o to be less than the magnitude of v_i.

The operation and certain limitations of the preceding two rectifier circuits will be illustrated with an assumed signal chosen to illuminate certain points. Consider the signal shown in Figure 3–10(a) with the assumption that both positive and negative peaks are much larger in magnitude than 0.7 V. If the signal is applied as the input of the half-wave rectifier of Figure 3–8, the output would have a form similar to that of Figure 3–10(b). There is a cut-in effect near zero resulting from the fact that the input voltage must be close to 0.7 V to produce an output. In addition, there is a loss of about 0.7 V on the peaks. If the signal is applied to the rectifier circuit of Figure 3–9, the output is of the form shown in Figure 3–10(c), and similar effects are noted there.

The relative effects of these degradations are much more significant at low signal levels. In some applications, they are unimportant, and in other applications, they may be unacceptable.

A different way to produce approximate rectification for some applications is the **shunt diode half-wave rectifier** shown in Figure 3–11(a). When $v_i < -0.7$

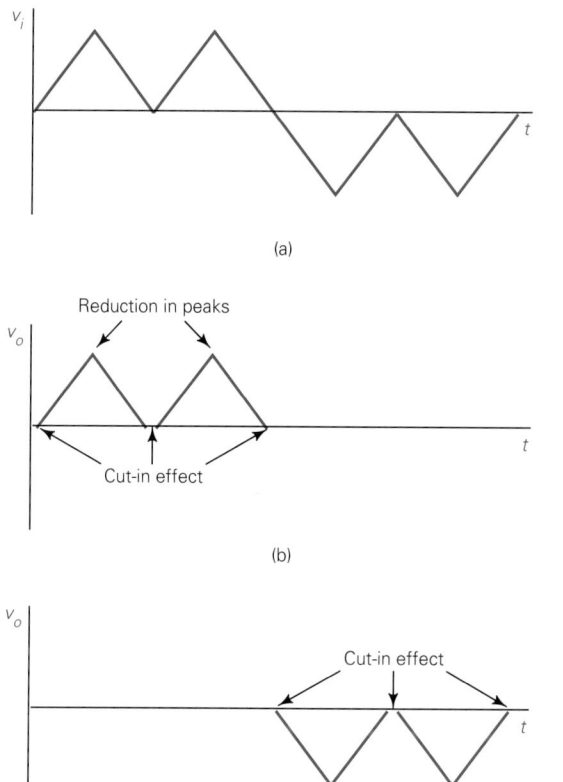

(a)

(b)

(c)

FIGURE 3–10
Input signal and outputs of the two rectifiers of Figures 3–8 and 3–9.

V, the diode is forward biased, and the equivalent circuit is shown in Figure 3–11(b). The output is $v_o \simeq -0.7$ V in this case. When $v_i > -0.7$ V, the diode is reverse biased, and the equivalent circuit is shown in Figure 3–11(c). The output without any load is $v_o = v_i$. As in previous circuits, it is assumed here that the reverse breakdown voltage of the diode is not reached.

The complete operation of the circuit can be expressed in idealized form as

$$v_o = \begin{cases} -0.7 \text{ V} & \text{for } v_i < -0.7 \text{ V} \qquad \textbf{(3–16)} \\ v_i & \text{for } v_i > -0.7 \text{ V} \qquad \textbf{(3–17)} \end{cases}$$

The transfer characteristic curve is shown in Figure 3–11(d), and first-quadrant active operation is observed. This circuit appears to have a very ideal relationship for positive voltage. However, the presence of the −0.7-V output for negative input voltages is a source of error and may not be acceptable for many applications.

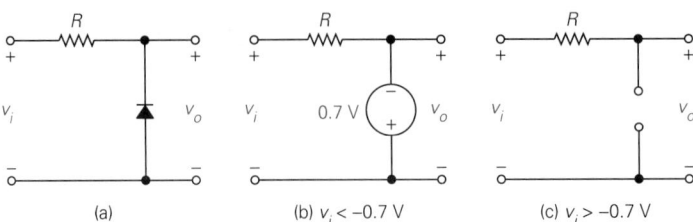

(a)   (b) $v_i < -0.7$ V   (c) $v_i > -0.7$ V

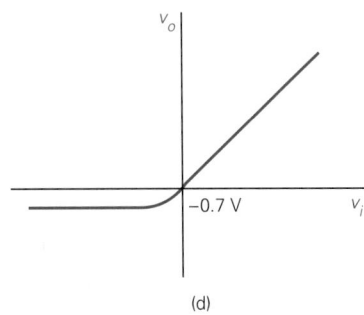

(d)

FIGURE 3–11
Shunt diode half-wave rectifier for first-quadrant operation, two equivalent circuits, and transfer characteristic.

Furthermore, any load on the ouput will affect the slope of the transfer characteristic (see Problems 3–11 and 3–12).

The shunt diode converted to third-quadrant operation is shown in Figure 3–12(a), and the transfer characteristic is shown in Figure 3–12(b). Analysis of this circuit is left as an exercise for the reader, and the idealized results are

$$v_o = \begin{cases} 0.7 \text{ V} & \text{for } v_i > 0.7 \text{ V} \quad\quad\quad\quad\text{(3–18)} \\ v_i & \text{for } v_i < 0.7 \text{ V} \quad\quad\quad\quad\text{(3–19)} \end{cases}$$

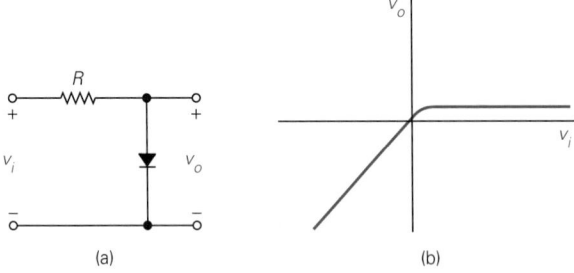

(a)   (b)

FIGURE 3–12
Shunt diode half-wave rectifier for third-quadrant operation and transfer characteristic.

All the basic half-wave rectifier circuits studied thus far are subject to errors, as we have noted. However, they are useful for noncritical applications. Highly linearized small-signal rectifier characteristics may be obtained with the use of an operational amplifier, one of the topics of Chapter 16.

The examples that follow illustrate some of the types of calculations required in troubleshooting, analyzing, or designing these simple circuits.

EXAMPLE 3–1

Consider the rectifier circuit of Figure 3–13(a). A source with time-varying voltage v_s and an internal resistance of 250 Ω is connected to the input.

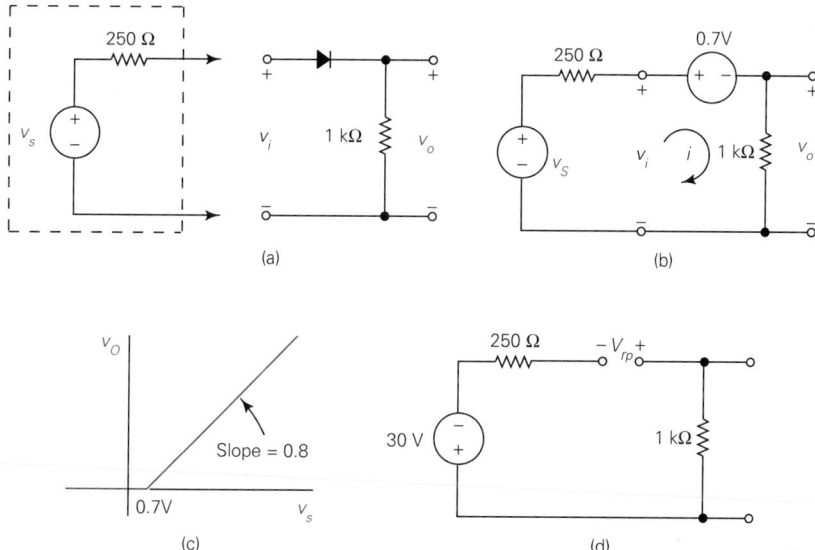

(a) (b)

(c) (d)

FIGURE 3–13
Circuit of Example 3–1, equivalent circuits, and transfer characteristic.

(a) Determine a set of equations describing the relationship between the source voltage and output voltage, and sketch the transfer characteristic. **(b)** If v_s varies from −30 to 30 V, determine the peak values of the output voltage, the diode current, and the reverse voltage across the diode.

Solution
The rectifier circuit is exactly like that of Figure 3–8, and analysis would be exactly like that performed earlier if the source had no internal resistance. In fact, the relationship between v_i and v_o for the diode circuit alone would be of the form of (3–10) and (3–11). However, the source has an internal resistance, and we are interested in a relationship between the open-circuit source voltage v_s and the output voltage in this case. As we will see shortly, the internal source resistance has a significant effect on the output voltage.

(a) For $v_s > 0.7$ V, the equivalent circuit is shown in Figure 3–13(b). The circuit has the form of a single loop, and a current i has been chosen as the variable in the analysis. A clockwise loop equation reads

$$-v_s + 250i + 0.7 + 1000i = 0 \qquad \textbf{(3–20)}$$

Solution for i yields

$$i = -0.56 \times 10^{-3} + 0.8 \times 10^{-3} v_s \qquad \textbf{(3–21)}$$

The output voltage is

$$v_o = 1000i \qquad \textbf{(3–22)}$$

Substitution of (3–21) in (3–22) yields

$$v_o = -0.56 + 0.8 v_s \qquad \text{for } v_s > 0.7 \text{ V} \qquad \textbf{(3–23)}$$

Of course, we can also state that

$$v_o = 0 \qquad \text{for } v_s < 0.7 \text{ V} \qquad \textbf{(3–24)}$$

The ideal transfer characteristic relating v_s to v_o is sketched in Figure 3–13(c). Observe that the independent variable is now v_s rather than v_i. From either (3–23) or Figure 3–13(c), observe that the slope of the curve is now 0.8, a result of the fact that a fraction of the source voltage is dropped in the internal resistance. The value -0.56 in (3–23) *would be* the vertical intercept if the straight line in the transfer characteristic were extended downward.

(b) The peak values of the output voltage and diode current occur when the source voltage reaches its positive peak value. Substituting $v_s = 30$ V in (3–23) and (3–21), we obtain the following *peak* values:

$$v_o = -0.56 + 0.8 \times 30 = 23.44 \text{ V} \qquad \textbf{(3–25)}$$
$$\begin{aligned} i &= -0.56 \times 10^{-3} + 0.8 \times 10^{-3} \times 30 \\ &= 23.44 \times 10^{-3} \text{ A} = 23.44 \text{ mA} \end{aligned} \qquad \textbf{(3–26)}$$

The peak reverse diode voltage occurs at the negative peak of the input voltage. An equivalent circuit based on this condition is shown in Figure 3–13(d). For convenience, the source voltage is shown as an inverted positive voltage, and the corresponding reverse voltage V_{rp} across the diode has been labeled so that the value is positive. Since no current flows, we note that

$$V_{rp} = 30 \text{ V} \qquad \textbf{(3–27)}$$

An interesting point is that while the peak positive output voltage is reduced to 23.44 V, the peak diode reverse voltage is equal to the full negative peak value of the source voltage.

The peak diode current and the peak reverse diode voltage are important parameters in selecting a diode for a given application. These values must not exceed the corresponding maximum ratings for the diode.

3-4 LIMITERS

Diodes may be used to limit the allowable range of a variable signal. Many time-varying signals have a range of magnitudes that are generally predictable and that the associated circuitry has been designed to accommodate. However, most signal sources may occasionally produce peaks or bursts outside the normal operating range. In addition, sudden noise spikes or stray pickup may result in overload conditions. Finally, the human element is always a factor, and an excessive level may accidentally be set during an adjustment or system connection.

Most electronic circuits are vulnerable to excessive signal levels and may accidentally be damaged or destroyed if safe levels are exceeded. In amplifiers designed to drive commercial broadcasting transmitters, a condition known as **overmodulation** can occur if the signal levels are excessive.

Circuits designed to limit the allowable range of a signal are known as **limiting circuis,** or simply **limiters.** The limiting action may be gradual or abrupt. At this point we will study some of the simpler limiting circuits implemented with diodes, voltage sources, and possibly resistors. The particular circuits to be considered at this point are **nearly abrupt.** (They would be completely abrupt if the diodes were ideal.)

Abrupt limiting circuits are also referred to as **clamping circuits,** for reasons that will be clear shortly. However, there is another class of circuits called clamping circuits that perform a shifting level at the same time, and these circuits are studied in Chapter 16. To avoid confusion, we reserve the term *clamping circuits* for the group of circuits to be studied in Chapter 16.

In the analysis of the circuits here and in many other sections to follow, the schematic diagrams will often use the battery symbol in showing typical dc voltage for ease of illustration. In most applications, however, such voltages are obtained directly from available dc system voltages or indirectly through voltage reference circuits.

The basic form of a limiter circuit designed for positive voltage limiting is shown in Figure 3–14(a). The voltage v_i is the input signal voltage, and the voltage v_o is the output signal voltage.

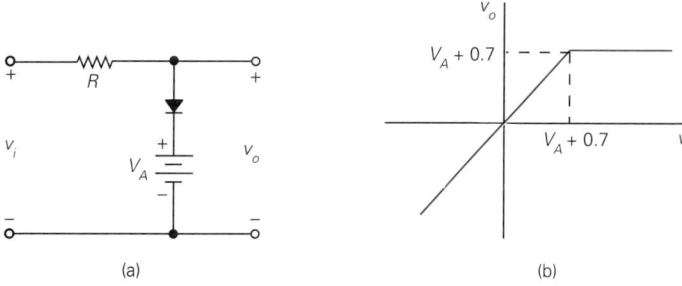

(a) (b)

FIGURE 3–14
Limiter circuit with positive voltage limiting.

The reader who has carefully studied the text up to this point should, it is hoped, be able to infer forward- and reverse-bias conditions in this circuit without being "too formal" in the mathematical development. The algebraic sum of the sources around the loop tending to cause current flow in the forward-bias (FB) direction is $v_i - V_A$, so when $v_i - V_A < 0.7$, or $v_i < V_A + 0.7$, the diode will be reverse biased. For this condition, there is no current flow around the loop, and $v_o = v_i$. However, when $v_i - V_A > 0.7$, or $v_i > V_A + 0.7$, the diode becomes forward biased. In this case the output voltage is the sum of the 0.7-V drop plus the dc voltage V_A, or $v_o = V_A + 0.7$.

A mathematical summary of the preceding conditions is

$$v_o = \begin{cases} v_i & \text{for } v_i < V_A + 0.7 & \textbf{(3–28)} \\ V_A + 0.7 & \text{for } v_i > V_A + 0.7 & \textbf{(3–29)} \end{cases}$$

The ideal transfer characteristic is shown in Figure 3–14(b).

As long as the positive peak of the input signal is less than about $V_A + 0.7$, the diode and dc voltage are isolated from the line and have no effect. However, when the signal level reaches about $V_A + 0.7$, the dc voltage and diode "take over" and prevent the output signal from increasing further.

Understand that the limiting action of this circuit, along with others to be considered in this section, is not precise because of the uncertainty of the forward diode drop. The limiting actually starts at a voltage slightly less than $V_A + 0.7$, and there is a slight increase in the output as the input increases. However, many applications of limiters do not require precise levels of the limiting value, and this type of circuit may be perfectly suitable. More accurate limiting levels may be established with operational amplifiers.

The corresponding form of a limiter circuit designed for negative voltage limiting is shown in Figure 3–15(a). The algebraic sum of the sources around the loop tending to cause forward bias (counterclockwise current flow) is $-v_i - V_B$, so when $-v_i - V_B < 0.7$ or $v_i > -V_B - 0.7$, the diode will be reverse biased. In this case, $v_o = v_i$. However, when $-v_i - V_B > 0.7$ or $v_i < -V_B - 0.7$, the diode becomes forward biased. In this case the output voltage is $v_o = -V_B - 0.7$.

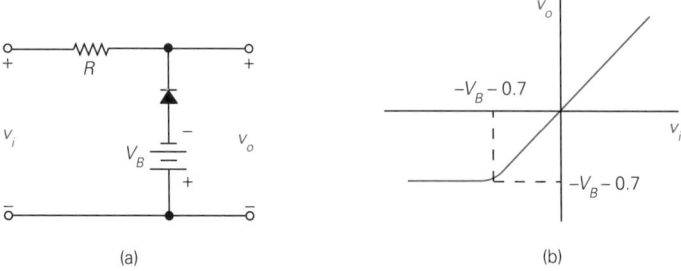

(a) (b)

FIGURE 3–15
Limiter circuit with negative voltage limiting.

A summary of the preceding conditions is

$$v_o = \begin{cases} v_i & \text{for } v_i > -V_B - 0.7 & \textbf{(3–30)} \\ -V_B - 0.7 & \text{for } v_i < -V_B - 0.7 & \textbf{(3–31)} \end{cases}$$

The ideal transfer characteristic is shown in Figure 3–15(b).

The preceding two circuits may be effectively combined to produce a limiter circuit capable of limiting on both positive and negative levels, and the resulting circuit is shown in Figure 3–16(a). Although the circuit now contains two diodes, it is easy to deduce from the earlier simpler circuits that the left-hand shunt path will conduct only for positive voltages and that the right-hand shunt path will conduct only for negative voltages. Thus, when either shunt path is conducting, the other is necessarily reverse biased and has no effect (unless, of course, the reverse breakdown voltage of the diode is inadvertently exceeded). A summary for this circuit is

$$v_o = \begin{cases} v_i & \text{for } -V_B - 0.7 < v_i < V_A + 0.7 & \textbf{(3–32)} \\ -V_B - 0.7 & \text{for } v_i < -V_B - 0.7 & \textbf{(3–33)} \\ V_A + 0.7 & \text{for } v_i > V_A + 0.7 & \textbf{(3–34)} \end{cases}$$

The ideal transfer characteristic is shown in Figure 3–16(b).

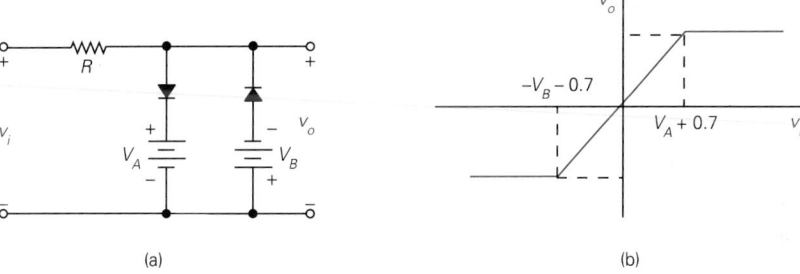

(a)   (b)

FIGURE 3–16
Limiter circuit with both positive and negative voltage limiting.

If the signal level is very low, and if limiting in the range of ±0.7 V or so is acceptable, it is possible to limit directly with a pair of parallel diodes, as shown in Figure 3–17(a). In this case it is necessary to scrutinize the constant voltage diode model carefully, since operation will be in the vicinity of the knee voltage. The transfer characteristic is shown in Figure 3–17(b). The dashed line represents the characteristic based on the constant voltage drop assumption, and the solid line represents a typical realistic characteristic. In practice, the peak value of the normal signal should be well below 0.7 V in order not to be affected by the diodes. This type of circuit is often used on integrated-circuit chips for protection.

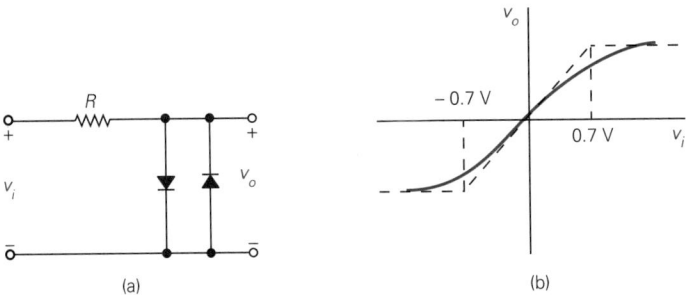

FIGURE 3–17
Limiter circuit with positive and negative voltage limiting at very low signal levels.

A few additional comments about all the circuits considered in this section are in order. First, the resistance R in either of the circuits may represent an actual circuit resistance, a source internal resistance, or a combination of both. However, there must be a minimum resistance for the circuit to function properly. Imagine, for example, that one of these limiting circuits were placed directly across a nearly ideal voltage source. As the input voltage started to increase above the level required to forward-bias the diode, there would be no resistance to separate v_i from the output, and the diode would attempt to absorb the voltage difference. The result would be a sharp increase in current, and the diode could be destroyed quickly. The resistance required is a function of the maximum loop current permitted, and this is limited by either the diode or the reference voltage.

If the Thévenin equivalent resistance of the source is sufficiently high, it may serve as the resistance R. Thus, it may be necessary to know something about the nature of the source before implementing these limiting circuits.

Frequently, a schematic diagram will be encountered in which one or more diodes (together with possible series voltage references) appear to be connected directly across a source. In such a case, the source is assumed to contain sufficient internal resistance to limit the current, as discussed.

The analysis of the circuits has assumed no loading at the output. In effect, this means that any additional circuitry following the limiting circuit will have sufficiently high input resistance that the open-circuit voltage v_o will remain the same with the load added. If this condition is not met, some attenuation of the signal level will occur, and the limiting levels will be reduced.

Clippers

Most of the limiter circuits considered here may also be used as **clipping circuits,** or **clippers.** A clipping circuit is one in which one or both peaks of the input signal are intentionally operated above the limiting level, so that the peaks are deliberately suppressed or "clipped." Thus, the major difference between a *limiter* and a *clipper* is intent rather than circuit function.

Clipping circuits can be used to remove undesirable amplitude noise from certain types of signals. One must be cautious, however, to ensure that the signal itself is not severely altered in the process. Examples of signals that can be clipped for reducing noise levels are frequency-modulated signals and many types of digital data signals.

Clippers can also be used to reshape certain types of waveforms. For example, consider the sine wave of Figure 3–18(a), and assume that it is applied as the input to the circuit of Figure 3–16(a). Assume that $V_A = V_B$ and that the peak sine wave voltage is much larger than the limiting voltage. In this case the output would appear in the form shown in Figure 3–18(b). The result is a flat-top signal that has some of the qualities of a square wave. (There are much better ways of generating a good square wave from a sine wave, as we will see. This example was chosen simply to illustrate wave shaping.)

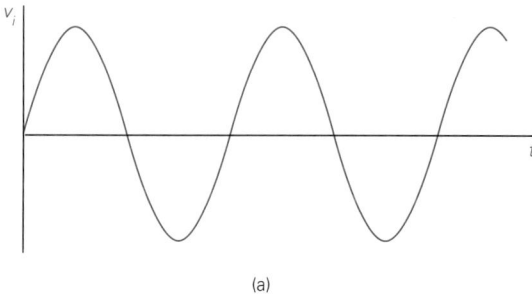

(a)

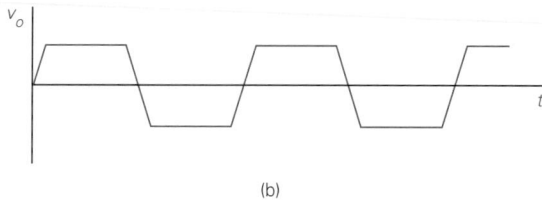

(b)

FIGURE 3–18
Application of circuit of Figure 3–16 as a clipper.

Zener Limiters

Zener diodes may be used as limiters in certain configurations. In fact, the use of a properly selected zener diode may eliminate the need for a separate reference voltage. However, since a zener diode acts like a regular junction diode when forward biased, it is necessary to consider this fact when analyzing a circuit containing a zener.

A zener circuit useful for limiting positive voltages (and very small negative voltages) is shown in Figure 3–19(a). For $0 < v_i < V_Z$, the diode is reverse biased

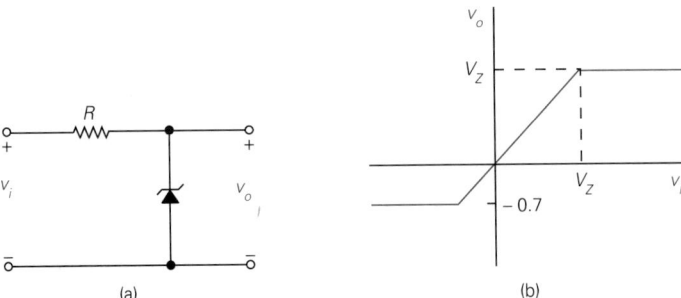

FIGURE 3–19
Zener diode used to limit positive voltage.

and is less than the zener level, so the diode appears as an open circuit. There is no current flow, and the output voltage without load is simply $v_o = v_i$. For $v_i > V_Z$, the zener conducts and holds the output voltage at V_Z for any subsequent increase of the input.

When $v_i < 0$, the diode behavior is that of a conventional diode. Thus, very little conduction occurs until v_i reaches a level of about -0.7 V, and the diode then becomes forward biased and holds the output to that level.

A complete summary of the operation of this circuit is

$$v_o = \begin{cases} v_i & \text{for } -0.7 \text{ V} < v_i < V_Z & \textbf{(3–35a)} \\ -0.7 \text{ V} & \text{for } v_i < -0.7 \text{ V} & \textbf{(3–35b)} \\ V_Z & \text{for } v_i > V_Z & \textbf{(3–35c)} \end{cases}$$

The transfer characteristic is shown in Figure 3–19(b).

This circuit is primarily useful for limiting a voltage that is generally expected to be always positive or zero but for which a negative output up to about -0.7 V should be acceptable.

The circuit is readily modified to accommodate a signal that is generally expected to be negative simply by reversing the diode, and this circuit is shown in Figure 3–20(a). Without going through all the details, the output can be expressed as

$$v_o = \begin{cases} v_i & \text{for } -V_Z < v_i < 0.7 \text{ V} & \textbf{(3–36a)} \\ -V_Z & \text{for } v_i < -V_Z & \textbf{(3–36b)} \\ 0.7 \text{ V} & \text{for } v_i > 0.7 \text{ V} & \textbf{(3–36c)} \end{cases}$$

Back-to-Back Zener Limiter

A zener configuration that can be used to provide limiting for both positive and negative voltages is shown in Figure 3–21. The two diodes are connected in what is called a **back-to-back** connection; that is, they are in series and oriented in opposite directions. To provide some generality here, diode D1 is assumed to

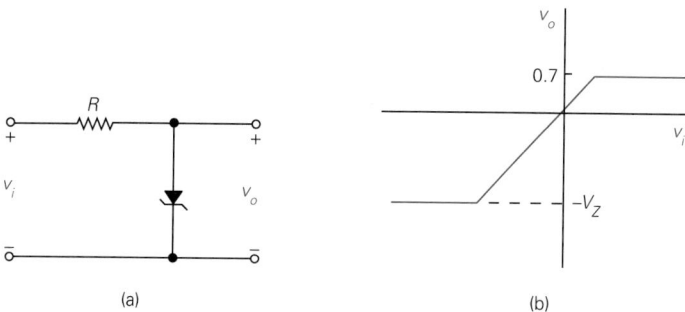

FIGURE 3-20
Zener diode used to limit negative voltage.

have a zener voltage V_{Z1}, and diode D2 is assumed to have a zener voltage V_{Z2}. For symmetrical signals, the zener levels are usually chosen to be the same.

A key point in the analysis of the circuit is that for either polarity of input signal, one diode will be forward biased and one will be reverse biased. Thus, the critical transition voltage magnitudes are $V_{Z1} + 0.7$ and $V_{Z2} + 0.7$. The reader may wish to draw more detailed equivalent circuits to illuminate the development that follows.

For $0 < v_i < V_{Z1} + 0.7$, D1 is reverse biased and acts as an open circuit. Similarly, for $-V_{Z2} - 0.7 < v_i < 0$, D2 is reverse biased. For $v_i > V_{Z1} + 0.7$, zener reverse breakdown occurs in D1, and the output voltage is the sum of V_{Z1} and the forward-biased drop of D2 (i.e., $v_o = V_{Z1} + 0.7$). Similarly, for $v_i < -V_{Z2} - 0.7$, zener reverse breakdown occurs in D2, and $v_o = -V_{Z2} - 0.7$.

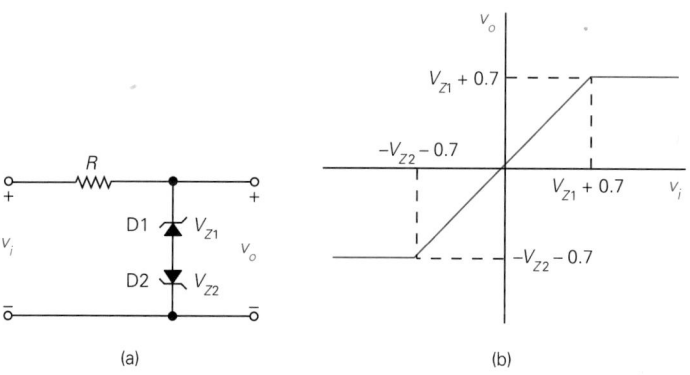

FIGURE 3-21
Back-to-back zener diode limiter circuit.

A summary of the preceding results follows.

$$v_o = \begin{cases} -V_{Z2} - 0.7 & \text{for } v_i < -V_{Z2} - 0.7 & \text{(3–37a)} \\ v_i & \text{for } -V_{Z2} - 0.7 < v_i < V_{Z1} + 0.7 & \text{(3–37b)} \\ V_{Z1} + 0.7 & \text{for } v_i > V_{Z1} + 0.7 & \text{(3–37c)} \end{cases}$$

The transfer characteristic is shown in Figure 3–21(b).

EXAMPLE 3–2

A certain signal source is connected to the input of an amplifier, and to protect the amplifier from possible excess signal levels, the dual limiting circuit of Figure 3–22(a) is connected across the line. Two reference voltages of ±6 V are used to establish the limiter levels. The Thévenin output resistance of the source is 500 Ω, and no additional series resistance is added. The input to the amplifier is essentially an open circuit. **(a)** Determine the approximate levels of the input voltage at which limiting action occurs. **(b)** Write equations describing the transfer characteristic. **(c)** Assume that a positive "spike" of 15 V appears at the source. Determine the conducting diode current based on this condition. **(d)** For the same condition as **(c)**, determine the reverse voltage across the nonconducting diode.

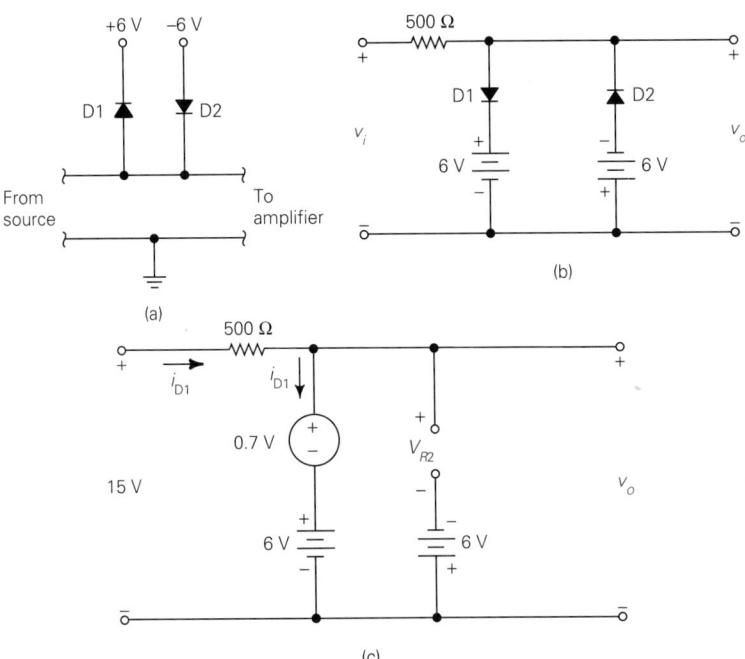

FIGURE 3–22
Limiter circuit of Example 3–2 shown two ways and equivalent circuit when $v_i = 15$ V.

Solution

(a) The circuit as shown in Figure 3–22(a) is representative of the way it might appear on a circuit diagram. However, to be analyzed, a circuit model showing the effective series resistance can be drawn as in Figure 3–22(b). The two voltage references have also been drawn as "batteries," so that the equivalence of this circuit to that of Figure 3–16(a) is clearly demonstrated.

The level of limiting for a positive input corresponds to the +6-V reference plus the forward-biased drop of 0.7 V for D1, or 6.7 V. By symmetry, the corresponding negative limiting value is −6.7 V. Thus, for −6.7 V < v_i < 6.7 V, the signal is coupled from the source to the amplifier, and the diodes ideally have no effect.

(b) The equations may be expressed as

$$v_o = \begin{cases} v_i & \text{for } -6.7 \text{ V} < v_i < 6.7 \quad \text{(or } |v_i| < 6.7 \text{ V)} & \textbf{(3–38)} \\ -6.7 \text{ V} & \text{for } v_i < -6.7 \text{ V} & \textbf{(3–39)} \\ 6.7 \text{ V} & \text{for } v_i > 6.7 \text{ V} & \textbf{(3–40)} \end{cases}$$

The transfer characteristic is of the form shown in Figure 3–16(b), with the two voltages labeled as +6.7 V and −6.7 V.

(c) With an assumed input voltage of 15 V, the equivalent circuit is shown in Figure 3–22(c). Note that the constant voltage model for D1 is assumed, but D2 is assumed to be an open circuit.

The current in diode D1 is i_{D1}, and since D2 is open, this current also flows through the 500-Ω resistance. A KVL equation for the left-hand loop is

$$-15 + 500 i_{D1} + 0.7 + 6 = 0 \qquad \textbf{(3–41)}$$

or

$$i_{D1} = \frac{8.3}{500} = 16.6 \text{ mA} \qquad \textbf{(3–42)}$$

(d) Referring again to Figure 3–22(c), note that the reverse voltage v_{R2} across D2 has been shown in the direction to yield a positive value. This voltage can readily be determined by a loop encompassing the two 6-V sources, the 0.7-V drop, and the desired voltage v_{R2}. Starting below the left-hand 6-V source and moving clockwise, we have

$$-6 - 0.7 + V_{R2} - 6 = 0 \qquad \textbf{(3–43)}$$

or

$$V_{R2} = 12.7 \text{ V} \qquad \textbf{(3–44)}$$

By symmetry, if the circuit were analyzed for $v_i = -15$ V, the magnitudes of diode current and reverse voltage would be the same, but the roles of the two diodes would be interchanged.

3–5 LOGIC GATES

The use of diodes and resistors to implement simple combinational logic gates is illustrated in this section. A logic gate is a circuit that has two or more possible output states (levels and/or conditions) and in which the existing state is a function of the possible input states. Combinational logic circuits are among the most basic of digital circuits.

Integrated-circuit logic functions are widely available at low cost, so the discrete circuits discussed here have limited utility in modern technology. However, they are included for two purposes: (1) they provide good drill in analyzing diode circuits that have several states, and (2) they orient the reader to think in terms of logic levels and conditions.

For digital logic circuits, binary variables are employed. A binary variable is one with only two levels. The two levels may be denoted as 0 and 1, *low* and *high,* or *off* and *on.* Since all voltage levels are subject to variation because of losses, power-line variations, component tolerances, and so on, a range is usually specified for each binary variable, and a "forbidden" region is provided as a buffer zone. For example, a voltage level between 0 and 0.8 V may be defined as a binary *0* or *low* level, and a voltage level between 2 V and 5 V may be defined as a binary *1* or *high* level. The range between 0.8 and 2 V would then be undefined.

The logic gates discussed here assume a voltage level equal to or near zero for a logical 0 and a positive voltage (well above 0.7 V) for a logical 1.

OR Gate

A two-input **OR gate** implemented with diodes and a resistor is shown in Figure 3–23. The two inputs are v_1 and v_2, and the output is v_o. The output v_o is high if *either* v_1 *or* v_2 is high. We will analyze this circuit in detail in Example 3–3.

FIGURE 3–23
Simple two-input diode logic OR gate.

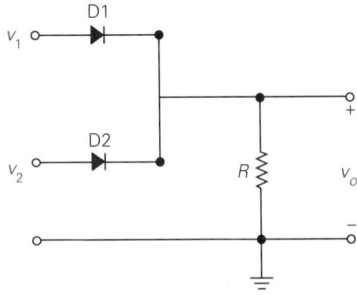

AND Gate

A two-input **AND gate** implemented with diodes and a resistor is shown in Figure 3–24. The two inputs are v_1 and v_2, and the output is v_o. The output v_o is high only if v_1 *and* v_2 are high. We will analyze this circuit in detail in Example 3–4.

FIGURE 3–24
Simple two-input diode AND gate.

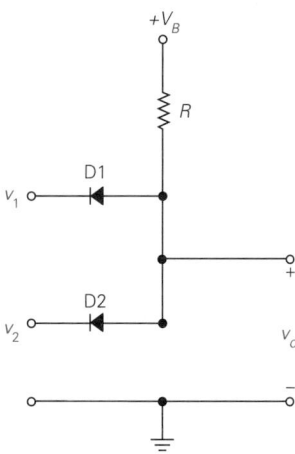

EXAMPLE 3–3

Consider the OR gate of Figure 3–25(a), in which logic levels of 0 and 5 V are employed. Determine the output voltage for each of the four combinations of the two input voltages, and verity that the OR function is performed.

Solution

As stated in the problem, there are four combinations of input variables. These correspond to (1) $v_1 = v_2 = 0$; (2) $v_1 = 5$ V, $v_2 = 0$; (3) $v_1 = 0$ V, $v_2 = 5$ V; and (4) $v_1 = v_2 = 5$ V. In general, for n inputs, there are 2^n different combinations of input variables.

Since the circuit contains two diodes, for any combination of input variables, it might be necessary to guess at the four possible sets of diode conditions to determine the correct one. However, we use our intuition to infer the correct one in each case and verify that it is correct. Let us consider each of the four input combinations.

(1) $v_1 = v_2 = 0$. We can analyze this one by inspection. This circuit is completely passive with both input voltages zero, so obviously $v_o = 0$.

(2) $v_1 = 5$ V, $v_2 = 0$. An equivalent circuit to be verified is shown in Figure 3–25(b). It is assumed that D1 is forward biased and D2 is reverse biased. The first condition is verified by noting that the net potential around the loop would result in clockwise current. The second condition is verified by noting that the net potential across D2 is more positive at the cathode than at the anode.

A loop equation reads

$$-5 + 0.7 + 1000i_R = 0 \qquad (3\text{–}45)$$

This yields

$$i_R = \frac{4.3}{1000} = 4.3 \text{ mA} = i_{D1} \qquad (3\text{–}46)$$

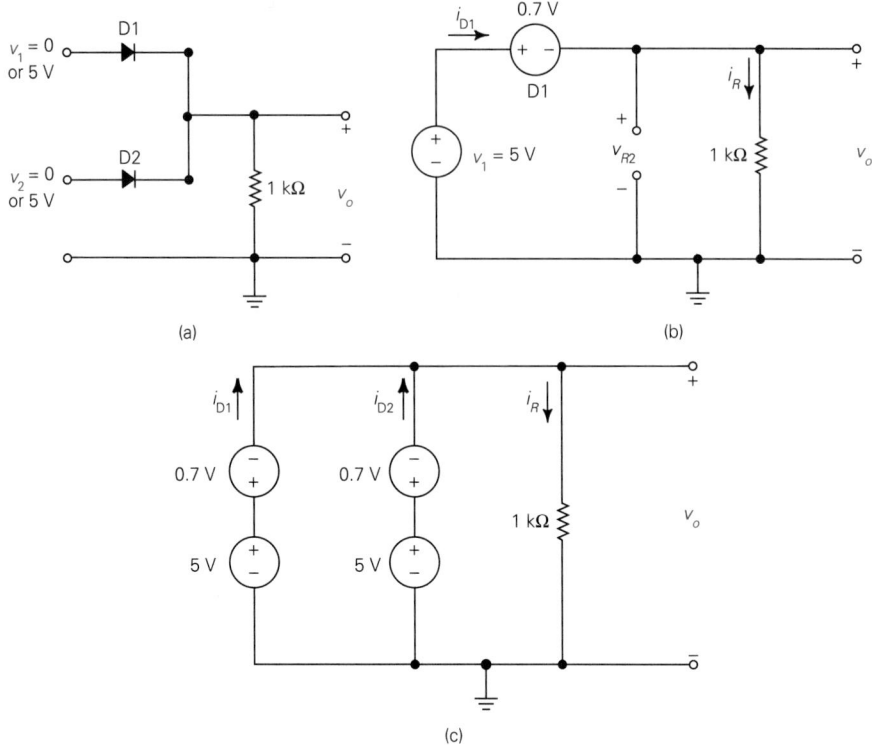

FIGURE 3–25
OR gate of Example 3–3 and equivalent circuits of two states.

The output voltage and reverse diode voltage v_{R2} (for D2) are the same.

$$v_o = v_{R2} = 1000i_R = 4.3 \text{ V} \tag{3–47}$$

Note that the output is high when one of the variables is high. Do not be disturbed by the fact that this voltage is 4.3 V rather than 5 V. The levels of digital variables frequently change through a circuit, but as long as they remain within a defined range, they can be interpreted properly.

(3) $v_1 = 0$, $v_2 = 5$ V. By the symmetry of the circuit, the condition is the same as Figure 3–25(b), but with the roles of D1 and D2 interchanged. Thus, the output is high again.

(4) $v_1 = v_2 = 5$ V. An equivalent circuit to be verified is shown in Figure 3–25(c). Both diodes are forward biased in this case. This is readily verified by noting that current will flow upward through both of these branches, combine, and flow down through the load.

A loop equation around either loop that includes the 1-kΩ resistor is

$$-5 + 0.7 + 1000i_R = 0 \tag{3–48}$$

This yields

$$i_R = \frac{4.3}{1000} = 4.3 \text{ mA} \tag{3-49}$$

and

$$v_o = 4.3 \text{ V} \tag{3-50}$$

By symmetry, each of the two source branches should supply half of this current.

$$i_{D1} = i_{D2} = \frac{4.3}{1000} = 2.15 \text{ mA} \tag{3-51}$$

The results of this problem can be summarized in the table that follows:

v_2	v_1	v_0
Low	Low	Low
Low	High	High
High	Low	High
High	High	High

The output is thus high if *either* v_1 *or* v_2 is high. The preceding table is called a **truth table.**

EXAMPLE 3-4 Consider the AND gate of Figure 3–26, in which logic levels of 0 and 5 V are employed. Determine the output voltage for each of the four combinations of the two input voltages, and verify that the AND function is performed.

Solution
The solution proceeds in the same manner as that of the OR gate in Example 3–3, and the same comments concerning possible combinations of input variables apply here. We will consider each of the four combinations.

(1) $v_1 = v_2 = 0$. The equivalent circuit for this condition is shown in Figure 3–26(b). The circuit has been redrawn for ease of analysis, and the 5-V supply has been shown as a battery. The net voltage around either loop containing this voltage and a diode is such that current flows in the direction of i_{D1} and i_{D2}, so both diodes are forward biased. A loop equation including the voltage source and either diode is

$$-5 + 1000 i_R + 0.7 = 0 \tag{3-52}$$

and

$$i_R = \frac{4.3}{1000} = 4.3 \text{ mA} \tag{3-53}$$

By symmetry, this current will divide evenly between the diodes and

$$i_{D1} = i_{D2} = \frac{4.3}{2} = 2.15 \text{ mA} \tag{3-54}$$

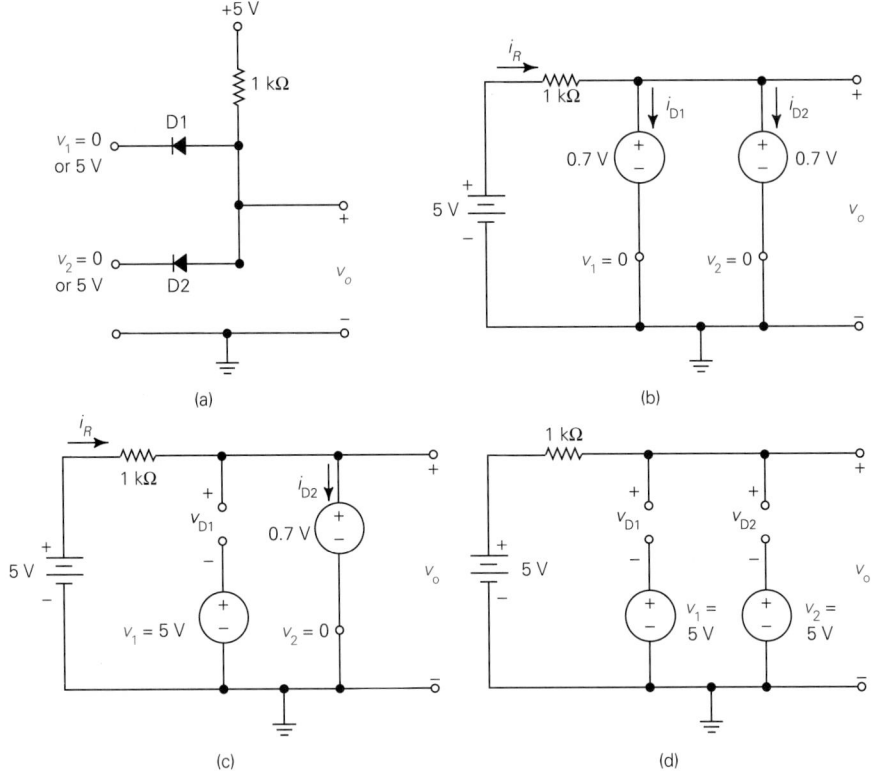

FIGURE 3–26
AND gate of Example 3–4 and equivalent circuits of two states.

By direct inspection,

$$v_o = 0.7 \text{ V} \qquad (3\text{–}55)$$

This voltage must be interpreted as a *low* condition, so the output is low when both inputs are low.

(2) $v_1 = 5$ V, $v_2 = 0$. The equivalent circuit is shown in Figure 3–26(c). In this case D2 is assumed to be forward biased, and D1 is assumed to be reverse biased. The first condition is verified by noting that the net voltage in the loop encompassing the 5-V voltage source and that D2 is such that current must flow in the direction of i_{D2}. A loop equation in this case reads

$$-5 + 1000i_R + 0.7 = 0 \qquad (3\text{–}56)$$

Solution for i_R yields

$$i_R = \frac{4.3}{1000} = 4.3 \text{ mA} = i_{D2} \qquad (3\text{–}57)$$

To verify the reverse-biased condition for v_{D1}, a loop equation is written around the loop containing the 5-V source and the 0.7-V drop across D2. We have

$$-5 - v_{D1} + 0.7 = 0 \qquad \textbf{(3–58)}$$

This yields

$$v_{D1} = -4.3 \text{ V} \qquad \textbf{(3–59)}$$

Since the assumed reference was positive from anode to cathode, but the value turns out to be negative, this diode is reverse biased.

The output voltage is simply

$$v_o = 0.7 \text{ V} \qquad \textbf{(3–60)}$$

which is a *low* condition, as in the previous case.

(3) $v_1 = 0$, $v_2 = 5$ V. By the symmetry of the circuit, the condition is the same as Figure 3–26(c), but with D1 and D2 reversed. Thus, the output is low again.

(4) $v_1 = v_2 = 5$ V. The equivalent circuit is shown in Figure 3–26(d). Both diodes are reverse biased in this case. This is verified by writing a loop equation around either loop containing the 5-V source and one of the inputs. No current is assumed to flow, so there is no voltage drop across the 1-kΩ resistor. For the loop containing v_{D1} and the 5-V source, we have

$$-5 + v_{D1} + 5 = 0 \qquad \textbf{(3–61)}$$

which results in

$$v_{D1} = 0 \qquad \textbf{(3–62)}$$

By symmetry, we deduce that

$$v_{D2} = 0 \qquad \textbf{(3–63)}$$

Since these voltages are below the knee voltages, both diodes are reverse biased. The output voltage in this case is

$$v_o = 5 \text{ V} \qquad \textbf{(3–64)}$$

The results of this problem can be summarized in the following truth table:

v_2	v_1	v_0
Low	Low	Low
Low	High	Low
High	Low	Low
High	High	High

The output is thus high only if v_1 *and* v_2 are high.

3–6 ZENER DIODE REGULATOR CIRCUITS

A general treatment of power supplies, including the process of ac-to-dc conversion, filtering, and electronic regulation, will be given in Chapter 20 after the reader has developed sufficient background for full comprehension. However, enough material has been covered already to permit the analysis and design of dc zener regulator circuits. These regulator circuits are discussed in this section to provide some common applications of zener diodes.

Zener regulator circuits are used to obtain a reasonably regulated dc voltage from a larger available dc voltage. Since electronic devices require different voltage levels for operation, it is possible to employ a few fixed power supply voltages and to obtain the required fixed voltages by employing various zener circuits. A fundamental requirement at the outset is that the available dc voltage must be *larger* in magnitude than the desired voltage to be derived.

Although the analysis and design computations for zener circuits involve simple dc calculations and equally simple circuits, many beginners find the process a bit tricky, particularly when input and/or load conditions are subject to variations. The difficulty is in understanding what happens to certain circuit variables when a change occurs in a different variable.

The basic form of a zener regulator circuit arranged for positive voltages is shown in Figure 3–27. A dc voltage V_1 is available, and it is desired to derive a stable smaller dc voltage V_L by means of a zener diode. The diode is placed across the output in reverse-biased form, so the zener voltage V_Z is the desired load voltage (i.e., $V_Z = V_L$). The resistance R_L represents some load of interest, and the load current is I_L. Since V_Z appears across R_L, Ohm's law indicates that

$$I_L = \frac{V_L}{R_L} = \frac{V_Z}{R_L} \qquad (3\text{–}65)$$

FIGURE 3–27
Basic form of zener regulator circuits.

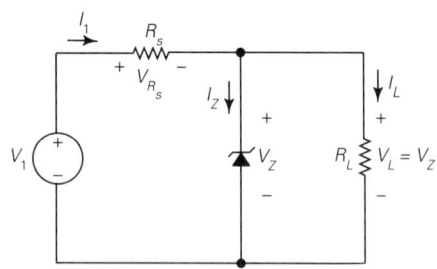

The current through the diode is I_Z, and the current flowing from the source is I_1. From KCL applied to the node above the zener, we have

$$-I_1 + I_Z + I_L = 0 \qquad (3\text{–}66)$$

or

$$I_1 = I_Z + I_L$$
(3–67)

Stated in words, the source must supply a current equal to the sum of the zener current and the load current.

A KVL equation for the left-hand loop reads

$$-V_1 + R_s I_1 + V_Z = 0$$
(3–68)

or

$$V_1 = R_s I_1 + V_Z$$
(3–69)

The input voltage must thus equal the sum of the voltage drop across R_s plus the zener (or output) voltage.

The preceding equations are all straightforward based on the conditions given. Reflecting on the assumptions, suppose that the variables in the preceding equations were all fixed at constant values. If this were true, the situation would be quite simple. The utility of the zener diode is that it can maintain a reasonably constant voltage over a range of circuit parameter variations.

It was stated in Section 2–7 that the voltage V_Z does vary somewhat with bias current and that the effect can be modeled by a dynamic zener impedance Z_{ZT}. Let ΔI_Z represent a change in zener current, and let ΔV_Z represent the corresponding change in voltage. Recall that those quantities are related by

$$\Delta V_Z = Z_{ZT} \Delta I_Z$$
(3–70)

The impedance is positive, so an increase in current results in an increase in voltage.

In the development that follows, a reasonable approximation will be made that might bother some readers, but it is typical of the types of assumptions made in "real-life" electronics. In studying the effects of various changes for input and load, we assume initially that V_Z *remains fixed*. Once these computations are made, we can then observe the net change in current and employ (3–70) to estimate the net change in zener voltage. This process is justified, because ΔV_Z is typically much smaller than V_Z and usually can be ignored for most computations outside of the zener itself.

Based on the preceding assumption, a basic concept will be emphasized at the outset. *As long as the zener diode is biased in the breakdown region, it will automatically adjust the current I_Z that it absorbs in order to attempt to maintain the voltage V_Z at a nearly constant level.* This process is fundamental to the analysis of a circuit and will appear in various forms in the discussion that follows.

Changes in Input Voltage

First, assume that the input voltage increases from V_1 to a new value V'_1 as shown in Figure 3–28(a). If V_Z is to remain constant, the voltage across R_s must increase, since the resistance must absorb the difference between V_1 and V_Z. Since R_s is fixed, the only way the resistive drop $R_s I_1$ can increase is for I_1 to increase. I_1 is composed of two components: I_Z and I_L. I_L cannot increase, since it is fixed by V_Z and R_L. Thus, I_Z must increase; that is, the zener absorbs more current when the input voltage increases. The opposite pattern occurs when V_1 decreases (i.e., I_Z decreases).

To summarize: *For fixed load current, the zener current increases as the input voltage increases, and vice versa.*

Changes in Load Current

Next, assume that the load current increases from a value I_L to a new value I'_L as shown in Figure 3–28(b). (Note that this corresponds to a decrease in R_L.) If

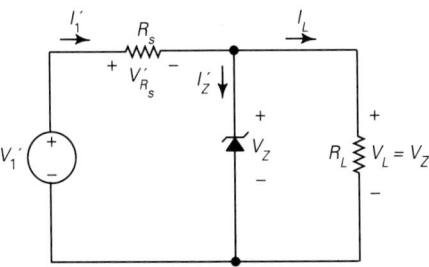

(a) Input voltage increases: if V_1 increases to V_1', I_Z increases to I_Z', I_1 increases to I_1', and V_{R_s} increases to V'_{R_s} so as to keep V_Z and I_L constant.

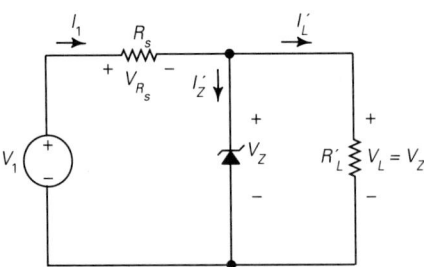

(b) Load current increases: if I_L increases to I_L', I_Z decreases to I_Z' so as to keep I_1, V_{R_s}, and V_Z constant.

FIGURE 3–28
Separate effects of increased input voltage and increased load current.

V_Z is to remain constant, the voltage across R_s must remain constant, since V_1 has not changed. Since the resistive voltage drop is $R_s I_1$ and R_s is fixed, I_1 cannot change. However, $I_1 = I_Z + I_L$, and since I_L has increased, I_Z must decrease to maintain a constant sum. The opposite pattern occurs when I_L decreases (i.e., I_Z increases).

To summarize: *For fixed input voltage, the zener current decreases as the load current increases, and vice versa.*

Changes in Input Voltage and Load Current

When both input voltage and load current are subject to variations, the variation in zener current is a combination of the two separate effects discussed previously. Look at each combination of changes, and determine what adjustment in zener current must occur to keep the voltage across the diode constant.

Worst-Case Effects

Refer to the zener diode characteristic shown in Figure 3–29. For proper circuit operation, the zener current must always be in the range from I_{ZK} to I_{ZM}. If the current magnitude ever drops below I_{ZK}, the circuit loses its ability to regulate, and the voltage magnitude drops drastically. Conversely, if the current magnitude exceeds I_{ZM}, the diode will be destroyed.

FIGURE 3–29
Allowable operating range for zener regulator circuits.

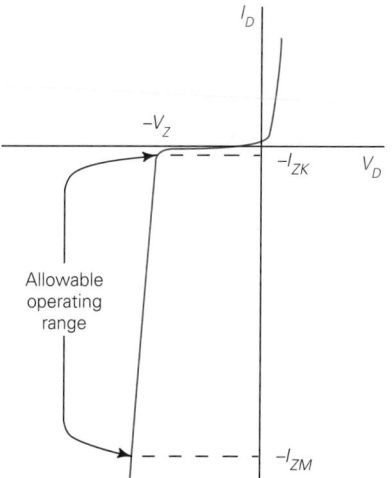

In any practical design, it is important that some investigation be made to ensure that the expected range of diode current always remain in the allowable range. For this purpose, estimated values of the minimum zener current $I_{Z(\min)}$ and the maximum zener current $I_{Z(\max)}$ should be made.

Assume first that the minimum and maximum expected values of V_1 and I_L are somehow estimated. Denote these values as $V_{1(min)}$, $V_{1(max)}$, $I_{L(min)}$, and $I_{L(max)}$, respectively. Even when the load current is expected to remain nearly constant, it could be disconnected accidentally, or an open circuit in the load could occur. Since this represents an increase in zener current, a good worst-case choice for the minimum-load current is $I_{L(min)} = 0$ for many situations.

In general, from (3–67),

$$I_Z = I_1 - I_L \tag{3–71}$$

From (3–68),

$$I_1 = \frac{V_1 - V_Z}{R_s} \tag{3–72}$$

Substituting (3–72) in (3–71) results in

$$I_Z = \frac{V_1 - V_Z}{R_s} - I_L \tag{3–73}$$

The minimum value of I_Z corresponds to the minimum value of V_1 and the maximum value of I_L.

$$I_{Z(min)} = \frac{V_{1(min)} - V_Z}{R_s} - I_{L(max)} \tag{3–74}$$

The maximum value of I_Z corresponds to the maximum value of V_1 and the minimum value of I_L.

$$I_{Z(max)} = \frac{V_{1(max)} - V_Z}{R_s} - I_{L(min)} \tag{3–75}$$

Although these formulas will allow us to "plug in" the different values and obtain the results directly, the reader is encouraged to think through the process in each case rather than memorize formulas.

Design of a Regulator Circuit

In designing a regulator circuit, a diode with a value of V_Z equal to the desired load voltage is used. Assume that the nominal values of I_L and V_1 are given. A nominal zener operating current I_Z must be selected, and the design then reduces to determining R_s. From (3–72), R_s is selected as

$$R_s = \frac{V_1 - V_Z}{I_1} = \frac{V_1 - V_Z}{I_Z + I_L} \tag{3–76}$$

The value of I_Z is selected from available zener diode data. If both I_L and V_1 are expected to remain nearly constant, the value of data test current I_{ZT} may be a good choice.

Worst-case variations of V_1 and I_L should then be estimated. As indicated earlier, the value $I_{L(min)} = 0$ may be a good choice for the lower worst-case value of the load current. The worst-case range of I_Z should then be calculated as discussed earlier. If $I_{Z(max)}$ exceeds I_{ZM} or if $I_{Z(min)}$ is lower than I_{ZK}, the design cannot be achieved with the conditions assumed.

EXAMPLE 3–5

Consider the zener regulator circuit in Figure 3–30. Assume initially that $V_1 = 15$ V, $R_L = 100\ \Omega$, and $V_Z = 10$ V for these conditions. **(a)** Determine I_L, I_1, and I_Z. **(b)** Assume that V_1 changes to 16 V but that all other parameters remain the same. Repeat the computations of **(a)**. **(c)** With $V_1 = 16$ V, assume that R_L is changed to 125 Ω. Repeat the computations of **(a)**. **(d)** If the zener impedance is $Z_{ZT} = 5\ \Omega$, estimate the final zener voltage in **(c)**.

FIGURE 3–30
Circuits of Example 3–5.

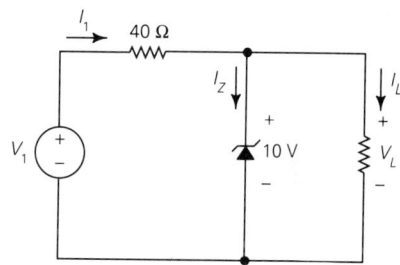

Solution
(a) For $R_L = 100\ \Omega$ and $V_L = V_Z = 10$ V, the load current is

$$I_L = \frac{10}{100} = 0.1\ \text{A} = 100\ \text{mA} \tag{3–77}$$

The current I_1 is

$$I_1 = \frac{V_1 - V_Z}{40} = \frac{15 - 10}{40} = 0.125\ \text{A} = 125\ \text{mA} \tag{3–78}$$

The zener current I_Z is then

$$I_Z = I_1 - I_L = 125 - 100 = 25\ \text{mA} \tag{3–79}$$

(b) When V_1 changes to 16 V, the voltage across R_s will increase to maintain V_Z close to 10 V. Actually, V_Z changes slightly, but that change will be ignored for the present analysis. The new value of I_1 is

$$I_1 = \frac{16 - 10}{40} = 0.15\ \text{A} = 150\ \text{mA} \tag{3–80}$$

Since V_L and R_L are both assumed as constants, we have

$$I_L = 100\ \text{mA} \tag{3–81}$$

The zener current changes to

$$I_Z = 150 - 100 = 50 \text{ mA} \tag{3-82}$$

(c) When R_L is changed to 125 Ω, the load current is

$$I_L = \frac{10}{125} = 0.08 \text{ A} = 80 \text{ mA} \tag{3-83}$$

As in **(b)**, we have assumed that V_Z remains at 10 V. Since $V_1 = 16$ V, the current I_1 is the same as in **(b)**; that is,

$$I_1 = 150 \text{ mA} \tag{3-84}$$

The zener current is now

$$I_Z = 150 - 80 = 70 \text{ mA} \tag{3-85}$$

Reviewing the problem thus far, we see that the zener current increased when the input voltage increased and again when the load current decreased.

(d) Over the range of this problem, the zener current changed from 25 mA initially to 70 mA in **(c)**. Thus $\Delta I_Z = 70 - 25 = 45$ mA. The change ΔV_Z in zener voltage can be estimated as

$$\Delta V_Z = Z_{ZT} \Delta I_Z = 5 \times 45 \times 10^{-3} = 0.225 \text{ V} \tag{3-86}$$

The final voltage V'_Z is then

$$V'_Z = 10 + 0.225 = 10.225 \text{ V} \tag{3-87}$$

One could correctly argue that this value when applied in **(c)** would result in a different value of current than was obtained in that part. Similarly, an adjustment in **(b)** would result in a different value. Such adjustments can be made when it is necessary to be as precise as possible. However, the approximate nature of the data involved usually makes the approach here sufficient for most purposes.

EXAMPLE 3-6   A zener regulator circuit of the form shown in Figure 3–27 is to be designed to establish a 6.8-V reference from a fixed 10-V supply. The load current requirement is fixed at 100 mA. The zener is to be biased at 20 mA. **(a)** Determine the value of R_s required. **(b)** If the load is accidentally removed from the zener, determine the maximum diode current.

Solution
(a)   The net current I_1 through R_s is the load current plus the zener current, that is,

$$I_1 = I_L + I_Z = 100 + 20 = 120 \text{ mA} \tag{3-88}$$

The voltage V_{R1} across R_s must be

$$V_{R1} = V_1 - V_Z = 10 - 6.8 = 3.2 \text{ V} \qquad (3\text{--}89)$$

The value of R_s is then

$$R_s = \frac{V_1}{I_1} = \frac{3.2}{120 \times 10^{-3}} = 26.7 \ \Omega \qquad (3\text{--}90)$$

(b) If the load is accidentally removed, the zener will absorb the entire current. Thus,

$$I_{Z(\text{max})} = 120 \text{ mA} \qquad (3\text{--}91)$$

The maximum diode current rating should be based on this possible worst-case effect.

3–7 PSPICE EXAMPLES

PSPICE EXAMPLE 3-1   Use PSPICE and the generic diode model to determine the transfer characteristic for the series diode rectifier circuit of Figure 3–8(a) over the range $-10 \text{ V} < v_i < 10 \text{ V}$ with $R = 10 \text{ k}\Omega$.

Solution
The circuit adapted to the PSPICE format is shown in Figure 3–31(a), and the code is shown in Figure 3–31(b). Based on the discussion of diode circuits in

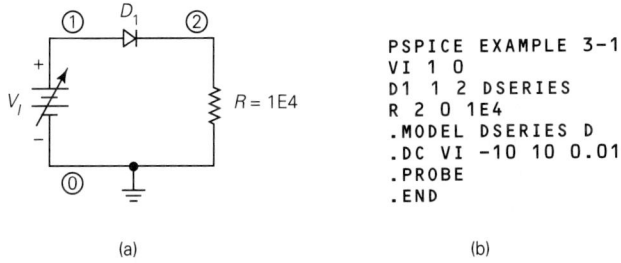

```
PSPICE EXAMPLE 3-1
VI 1 0
D1 1 2 DSERIES
R 2 0 1E4
.MODEL DSERIES D
.DC VI -10 10 0.01
.PROBE
.END
```

(a)   (b)

FIGURE 3–31
Circuit and code of PSPICE Example 3–1.

Chapter 2, the code should be self-evident. The name employed for the diode is DSERIES. Note that we have chosen to sweep the circuit in steps of 0.01 V. A graphic presentation with .PROBE is employed.

A plot of V(2) versus VI as obtained from .PROBE is shown in Figure 3–32. Observe that there is negligible output for very small positive input voltages

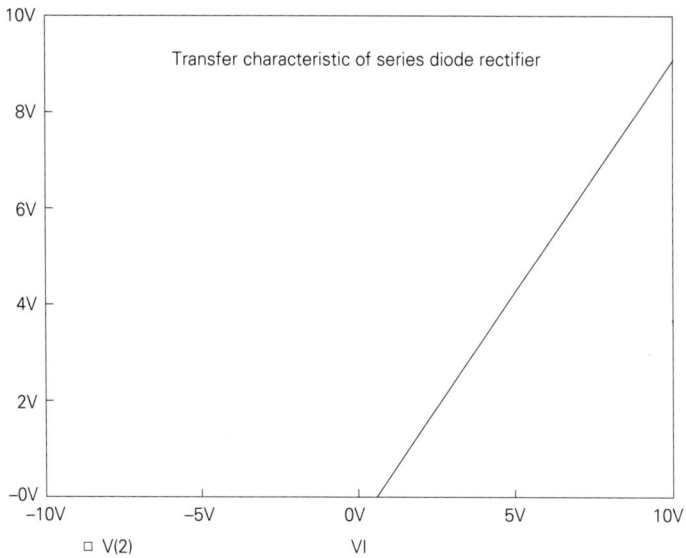

FIGURE 3–32
Transfer characteristic of PSPICE Example 3–1.

because of the cut-in effect. Once conduction starts, the transfer characteristic appears linear. However, when the input reaches a value of 10 V, the output is around 9.3 V, as expected.

PSPICE EXAMPLE 3–2   Demonstrate the operation of the rectifier circuit of PSPICE Example 3–1 by applying a sine wave as the input and employing *transient* analysis. Use a peak value of 5 V and a frequency of 1000 Hz.

Solution
We are momentarily delving into an area that could easily go beyond the present scope of material, but explanations will be carefully provided to keep the reader from getting lost. In general, transient analysis is the form of analysis that deals with the instantaneous behavior of voltages and currents within a circuit following the application of an excitation to the input. Whenever inductance or capacitance is present, this involves the solution of differential equations, that is, terms that contain calculus forms. This circuit does not contain inductance or capacitance, but it does contain a nonlinear element, for example, the diode. PSPICE can readily solve linear or nonlinear problems in the transient mode. The user need not understand how to solve either differential equations or nonlinear equations to make use of this powerful tool. However, some judgment in guiding the program in the right direction is required, and this facility develops with experience. Transient analysis has more chances of going astray than simple dc or ac analysis, so care must be exercised in its usage.

Transient analysis provides a variety of waveforms of which the most common one is the sinusoidal function. Sinusoidal functions may have up to six parameters in the PSPICE code, but most of our work can be achieved with the simplest form of the code, which is satisfied with *three* parameters. This simplified code for a voltage is represented as follows

```
SIN (VDC VP FREQUENCY)
```

The use of parentheses is optional, but it is often preferred for clarity. Commas may also be inserted between the parameters if desired, but spaces will suffice. The parameters are defined as follows:

VDC = dc value or offset added to sine wave

VP = peak value of sinusoid

FREQUENCY = number of complete cycles per second

The time duration of one cycle is called the period, and it is the reciprocal of the frequency; that is,

$$\text{PERIOD} = \frac{1}{\text{FREQUENCY}} \tag{3-92}$$

Thus, a frequency of 1000 Hz would correspond to a period of 1/1000 = 0.001 seconds = 1 ms. A sine wave of this frequency with a peak value of 5 V and no offset could be described as

```
SIN (0 5 1000)
```

This description should be placed on the line with the source connection code, as we will see shortly.

The code for the circuit is shown in Figure 3–33. Following the title line, the input voltage is described as

```
VI 1 0 SIN (0 5 1000)
```

This voltage is connected between nodes 1 and 0 with the positive reference at 1. Based on the preceding discussion, the source is a sinusoidal form that has no dc offset, a peak value of 5 V, and a frequency of 1000 Hz.

FIGURE 3–33
Code of PSPICE Example 3–2.

```
PSPICE EXAMPLE 3-2
VI 1 0 SIN (0 5 1000)
D1 1 2 DSERIES
R 2 0 1E4
.MODEL DSERIES D
.TRAN 2E-3 2E-3
.PROBE
.END
```

The next three lines describe the circuit configuration and the diode model, which are the same as in the preceding example and require no further discussion. The next line requiring a major discussion is the one beginning with .TRAN. This command informs PSPICE to perform a **transient analysis.** There could be as many as five entries following this name. However, the minimum number is *two,* and for most of the work that we will do, this number will suffice. The transient analysis will always begin at a reference time of $t = 0$. The second of the two values is the easiest to understand, and it is the **final time,** that is, the point at which the analysis is terminated.

For the case at hand, the command reads

```
.TRAN 2E-3 2E-3
```

The second value is 2E-3, or 2×10^{-3} s = 2 ms. For a sine wave frequency of 1000 Hz, the period is 1/1000 = 1×10^{-3} s = 1 ms. Thus, the analysis is to be performed over *two* complete cycles of the input.

Now let us explain the first number. If we wanted to generate a table of exact values for the output, this would represent the spacing between the output tabulated data values. In general, this is *not* the same as the internal time steps that PSPICE employs in the actual computation. The complex mathematical process in the program selects time steps sufficiently small to ensure accurate output. However, the points at which the output values are to be printed depend on the user's needs, and this can be controlled by the choice of this value.

In most of our work we employ graphical output using .PROBE. It turns out that .PROBE will actually select time steps for this purpose, but a value must be inserted here to satisfy the PSPICE code. The simplest choice is the same value as the final time. Thus, *for many graphical transient solutions, we will select both values following .PROBE to be the final time value.*

Plots of the input voltage V(1) and the output voltage V(2) are shown in Figure 3–34. Observe that V(1) is a sinusoid oscillating between 5 V and −5 V, as expected. The period is 1 ms, and two cycles are shown. The output voltage

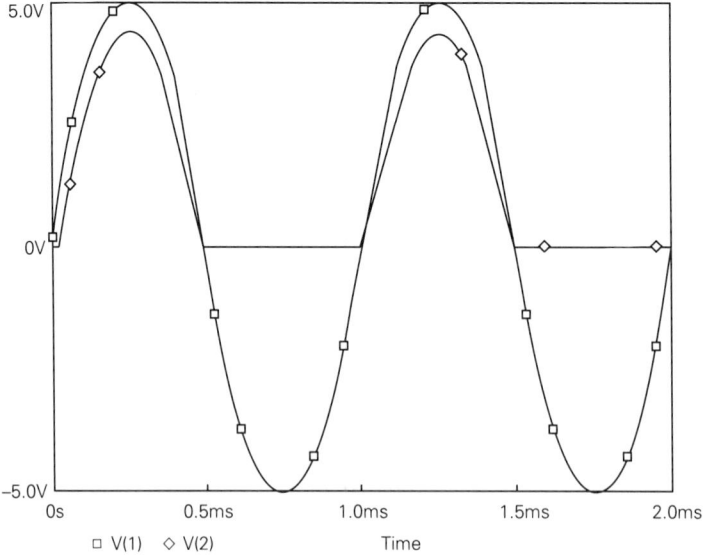

FIGURE 3–34
Waveforms of PSPICE Example 3–2.

V(2) follows the input when the input is positive but is zero when the input is negative. However, there is an error for positive input because of the diode drop. Thus, the maximum value of the output is close to 4.3 V, as expected.

Although the preceding plots are sufficient for many applications, there is a way of obtaining better resolution by adjusting the transient time step.* It turns out that the time steps taken by PSPICE are based on a numerical accuracy criterion, but in some cases, the resulting plot may be a little jagged. To see how a better plot can be obtained, we need to extend the .TRAN statement to include two more parameters. A new code containing two more parameters in the .TRAN statement is shown in Figure 3–35. This statement reads

```
.TRAN 2E-3 2E-3 0 1E-6
```

The first two values have the same meaning as before and have not been changed. When more than two parameters are employed, *the third value represents the time at which observation of the transient is to begin.* When this is omitted,

FIGURE 3–35
Modified code of PSPICE Example 3–2 with .TRAN statement adjusted for higher resolution.

```
PSPICE EXAMPLE 3-2
VI 1 0 SIN (0 5 1000)
D1 1 2 DSERIES
R 2 0 1E4
.MODEL DSERIES D
.TRAN 2E-3 2E-3 0 1E-6
.PROBE
.END
```

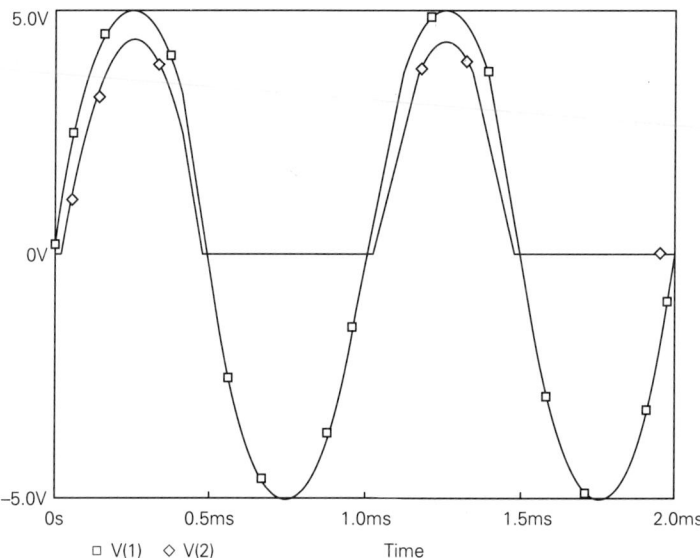

□ V(1) ◇ V(2)   Time

FIGURE 3–36
Waveforms of PSPICE Example 3–2 with higher resolution.

---

* Since these figures were transferred to the text with an electronic scanner, the differences here are not too noticeable.

observation begins at $t = 0$, as was evident in the preceding plot. This option is useful when there is a long settling time in a circuit, which is not of interest to us, and we may skip this interval by an appropriately chosen value here. Note that this value does not affect the computation time, which always begins at $t = 0$; it only affects when we see the response. We still want to see the response beginning at $t = 0$, so we must insert the value 0. Our main interest here is the fourth parameter, but the third parameter must be present in this case.

The fourth parameter allows us to establish a maximum time step that PSPICE will never exceed. One must be careful, of course, because an extremely small time step would increase the computation time markedly. In this case, we have chosen 1E-6 = 1 μs as the maximum time step. This choice will result in 1000 steps per cycle, which should be more than adequate.

The resulting plot is shown in Figure 3-36. Observe that both functions are smoother than in the preceding case.

PSPICE EXAMPLE 3-3
Use PSPICE and the generic diode model to determine the transfer characteristic for the shunt diode rectifier circuit of Figure 3-11(a) over the range $-10 < v_i < 10$ V with $R = 10$ kΩ.

Solution
The circuit adapted to the PSPICE format is shown in Figure 3-37(a), and the code is shown in Figure 3-37(b). The name employed for the diode is DSHUNT.

A plot of V(2) versus VI is shown in Figure 3-38. Note that the characteristic is very linear in the first quadrant, but there is a very small negative output voltage when the input is negative because of the diode drop.

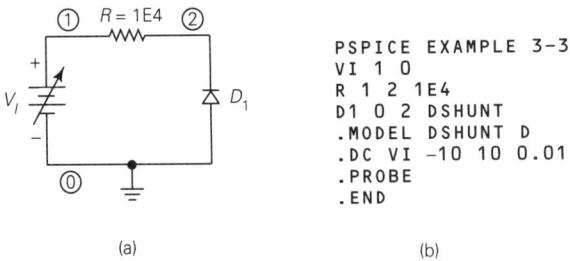

```
PSPICE EXAMPLE 3-3
VI 1 0
R 1 2 1E4
D1 0 2 DSHUNT
.MODEL DSHUNT D
.DC VI -10 10 0.01
.PROBE
.END
```

(a)   (b)

FIGURE 3-37
Circuit and code of PSPICE Example 3-3.

PSPICE EXAMPLE 3-4
Prepare a PSPICE program to simulate the limiting circuit of Example 3-2. Demonstrate the operation of the circuit with sinusoidal inputs that have peak values of 5 V and 15 V by plotting both input and output voltages. For the latter case, obtain separate plots for the diode current of D_1 and the voltage across D_2.

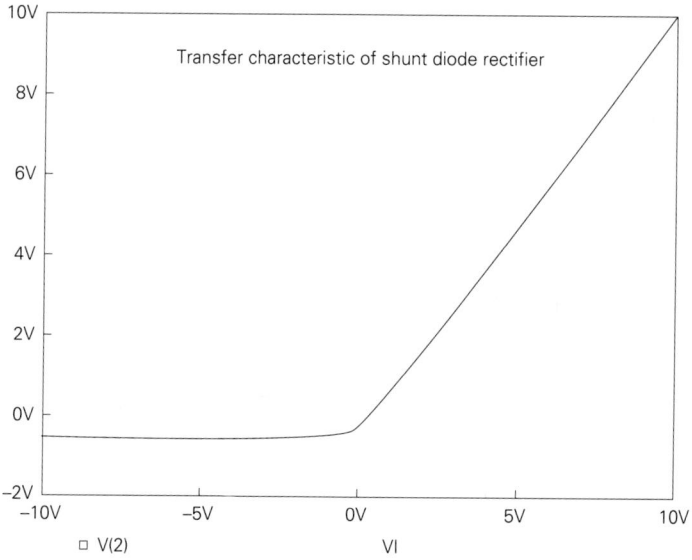

FIGURE 3–38
Transfer characteristic of PSPICE Example 3–3.

Solution

The circuit adapted to the PSPICE format is shown in Figure 3–39(a), and the codes for a 5-V input and a 15-V input are shown in Figure 3–39(b) and 3–39(c), respectively.

The input and output waveforms for a peak input of 5 V are shown in Figure 3–40. The two curves blend completely together, since they are identical; that is, no limiting has occurred.

When the input is raised to a peak value of 15 V, the output starts to clip at levels of about ±6 V as shown in Figure 3–41. For this condition, the current waveform of D_1 is shown in Figure 3–42. In this analysis we used a special feature

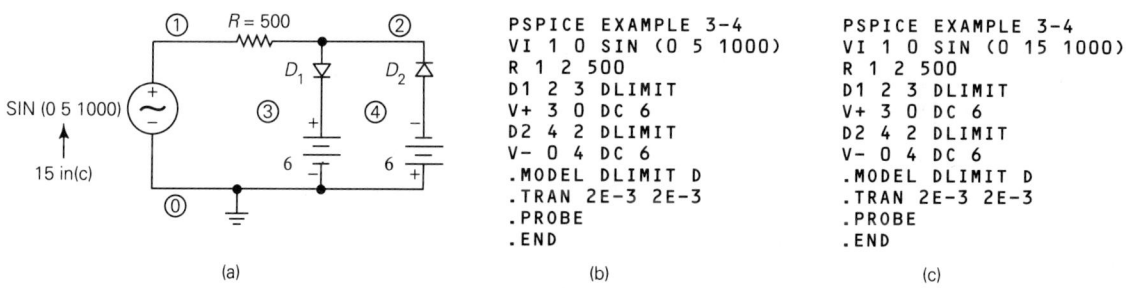

(a)

```
PSPICE EXAMPLE 3-4
VI 1 0 SIN (0 5 1000)
R 1 2 500
D1 2 3 DLIMIT
V+ 3 0 DC 6
D2 4 2 DLIMIT
V- 0 4 DC 6
.MODEL DLIMIT D
.TRAN 2E-3 2E-3
.PROBE
.END
```

(b)

```
PSPICE EXAMPLE 3-4
VI 1 0 SIN (0 15 1000)
R 1 2 500
D1 2 3 DLIMIT
V+ 3 0 DC 6
D2 4 2 DLIMIT
V- 0 4 DC 6
.MODEL DLIMIT D
.TRAN 2E-3 2E-3
.PROBE
.END
```

(c)

FIGURE 3–39
Circuit and codes of PSPICE Example 3–4.

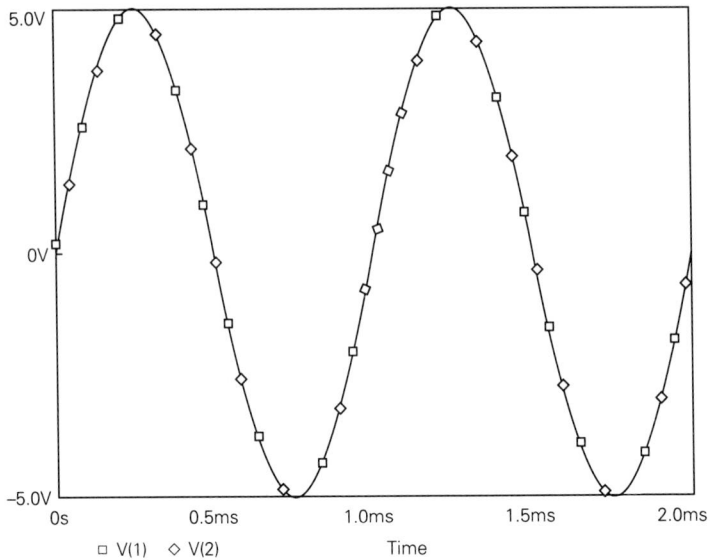

FIGURE 3–40
Input and output waveforms of PSPICE Example 3–4 for a peak input of 5 V. (The two waveforms are identical.)

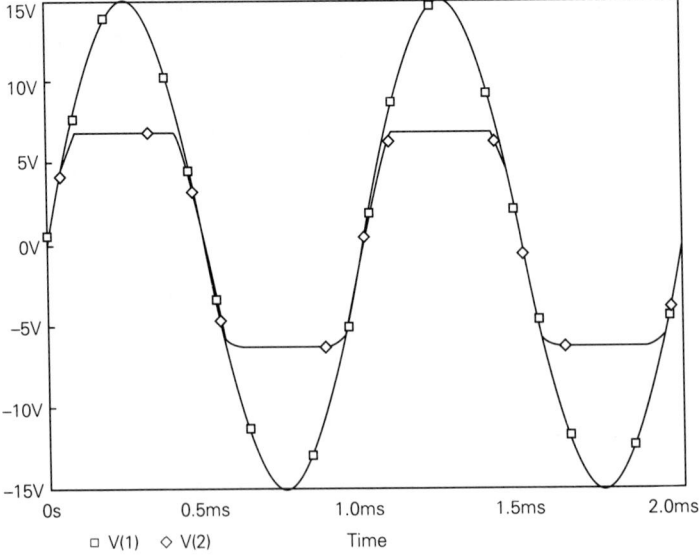

FIGURE 3–41
Input and output waveforms of PSPICE Example 3–4 for a peak input of 15 V.

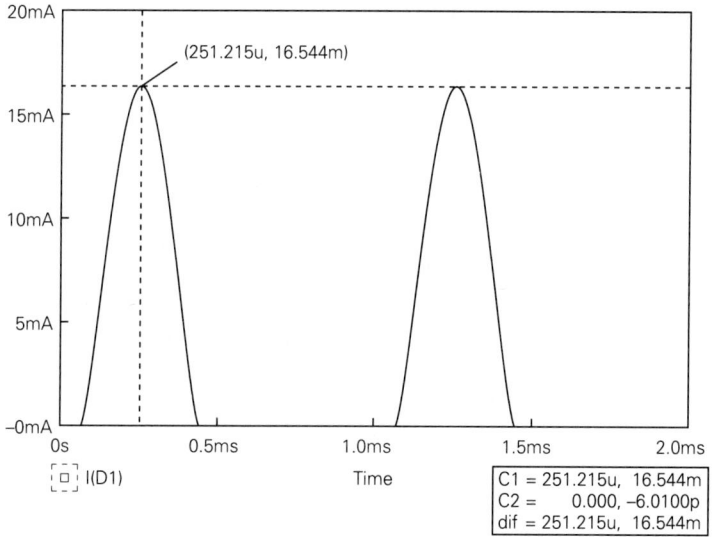

FIGURE 3–42
Diode D_1 current for PSPICE Example 3–4 for a peak input of 15 V.

of .PROBE allowing the maximum value of a waveform to be determined. This peak value was determined to be 16.544 mA, which agrees very favorably with the calculated value of 16.6 mA.

The voltage across diode D_2 is shown in Figure 3–43. Note that this voltage

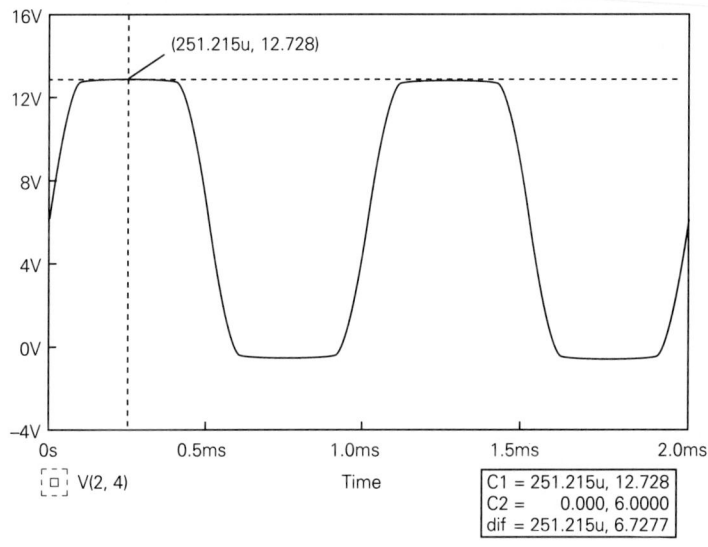

FIGURE 3–43
Diode D_2 voltage for PSPICE Example 3–4 with a peak input of 15 V.

is denoted as V(2, 4) with the positive reference at the cathode. The positive peak value here represents the peak inverse voltage, which is indicated to be 12.728 V. This compares favorably with the calculated value of 12.7 V.

PSPICE EXAMPLE 3–5

The zener regulator circuit of Figure 3–44(a) was designed to obtain a reference voltage very close to 4.7 V from an unregulated voltage whose nominal value is about 6.3 V using a 1N750 diode. The expected value of the load current under nominal conditions is 40 mA. Develop a PSPICE program with suitable modifications to perform the following studies: **(a)** Determine the values of the load voltage, diode current, and load current for $v_1 = 6.3$ V. **(b)** Determine the values of **(a)** when the input voltage is swept from 6 to 6.7 V. **(c)** Determine the values of **(a)** when the load is removed. **(d)** Determine the values of **(a)** when the load changes to 60 Ω.

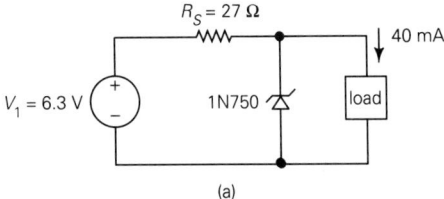

(a)

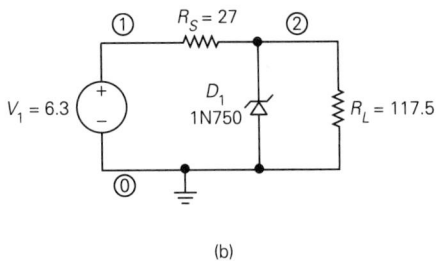

(b)

```
PSPICE EXAMPLE 3-5
.OPTIONS NOECHO
V1 1 0 6.3
RS 1 2 27
D1 0 2 D1N750
RL 2 0 117.5
.LIB EVAL.LIB
.DC V1 LIST 6.3
.PRINT DC V(2) I(D1) I(RL)
.OPTIONS NOPAGE
.END
```

(c)

FIGURE 3–44
Circuit and code of PSPICE Example 3–5.

Solution

If the zener voltage were exactly 4.7 V, the current I_1 through the dropping resistor would be

$$I_1 = \frac{6.3 - 4.7}{27} = 59.26 \text{ mA} \tag{3–93}$$

The zener current I_Z would then be

$$I_Z = 59.26 - 40 = 19.26 \, \text{mA} \qquad \text{(3–94)}$$

Now let us see what the PSPICE model predicts.

(a) On the assumption that the 40-mA load corresponds to a fixed resistance, the value of this load R_L is

$$R_L = \frac{4.7}{40 \times 10^{-3}} = 117.5 \, \Omega \qquad \text{(3–95)}$$

This circuit adapted to the PSPICE format is shown in Figure 3–44(b), and the code is shown in Figure 3–44(c).

The printout for the analysis is shown in Figure 3–45. Don't be intimidated by the bewildering array of diode model parameters listed. They relate to the complex mathematical model used in the software. They are intended for the

```
PSPICE EXAMPLE 3-5

****    CIRCUIT DESCRIPTION

*************************************************************************

.OPTIONS NOECHO

*****    DIODE MODEL PARAMETERS

              D1N750
        IS  880.500000E-18
       ISR    1.859000E-09
        BV    4.7
       IBV     .020245
       NBV    1.6989
      IBVL    1.955600E-03
      NBVL   14.976
        RS     .25
       CJO  175.000000E-12
        VJ     .75
         M     .5516
      TBV1   21.277000E-06


****  DC TRANSFER CURVES     TEMPERATURE = 27.000 DEG C

V1          V(2)          I(D1)          I(RL)

 6.300E+00    4.698E+00   -1.934E-02    3.998E-02

             JOB CONCLUDED

             TOTAL JOB TIME          1.26
```

FIGURE 3–45
Printout of PSPICE Example 3–5 at nominal operating point.

specialist who may need to make precise adjustments on the actual characteristic. The average user need not understand the meanings of all these terms.

The output data consist of the four values listed under "DC Transfer Curves." The input voltage is listed as V1 = 6.3 V, as expected. The zener voltage is V(2), and its value is V(2) = 4.698 V, which differs by less than 0.5% from the design goal of 4.7 V. (A slight adjustment in R_s could be made to compensate if desired.) The load current is I(RL) = 39.98 mA, which also differs by the same percentage from the expected load current of 40 mA. Finally, the diode current is I(D1) = −19.34 mA. The negative sign relates to the fact that the positive reference of current flow is from node 0 to node 2, while the actual current flow is in the opposite direction.

(b) The program modified for a dc sweep is shown in Figure 3–46. In order to suppress the listing of the diode model parameters, a new option NOMOD ("no model parameters") has been added in the options line. Note that the input voltage is being stepped from 6 to 6.7 V in steps of 0.1 V.

FIGURE 3–46
Code of PSPICE Example 3–5 modified for a dc sweep.

```
PSPICE EXAMPLE 3-5
.OPTIONS NOECHO NOPAGE NOMOD
V1 1 0 6.3
RS 1 2 27
D1 0 2 D1N750
RL 2 0 117.5
.LIB EVAL.LIB
.DC V1 6 6.7 0.1
.PRINT DC V(2) I(D1) I(RL)
.END
```

```
PSPICE EXAMPLE 3-5

****   CIRCUIT DESCRIPTION

*************************************************************************

.OPTIONS NOECHO NOPAGE NOMOD

****  DC TRANSFER CURVES     TEMPERATURE = 27.000 DEG C

V1          V(2)          I(D1)          I(RL)

6.000E+00   4.662E+00   -9.868E-03   3.968E-02
6.100E+00   4.677E+00   -1.292E-02   3.980E-02
6.200E+00   4.688E+00   -1.608E-02   3.990E-02
6.300E+00   4.698E+00   -1.935E-02   3.998E-02
6.400E+00   4.707E+00   -2.267E-02   4.006E-02
6.500E+00   4.714E+00   -2.603E-02   4.012E-02
6.600E+00   4.721E+00   -2.943E-02   4.018E-02
6.700E+00   4.727E+00   -3.286E-02   4.023E-02

        JOB CONCLUDED

        TOTAL JOB TIME          1.27
```

FIGURE 3–47
Data obtained from dc sweep in PSPICE Example 3–5.

The computer printout is shown in Figure 3–47. Obviously, the diode is not a perfect regulator, since the load voltage varies from 4.662 to 4.727 V over the range given. However, this variation is less than 1.4%, which compares with an input variation of more than 11%. Note, however, that the zener current magnitude varies all the way from 9.868 mA to 32.86 mA, which is more than a 3 : 1 ratio.

(c) To remove the load, the resistance R_L is eliminated, and this code is shown in Figure 3–48. The corresponding data printout is shown in Figure 3–49. While the output voltage is still close to 4.7 V, the diode current magnitude has now increased to 71.44 mA. This is near the maximum current rating of the data, and this illustrates the potential difficulty in a zener regulator circuit when the load is removed.

FIGURE 3–48
Code for PSPICE Example 3–5 when load is removed.

```
PSPICE EXAMPLE 3-5
.OPTIONS NOECHO NOPAGE NOMOD
V1 1 0 6.3
RS 1 2 27
D1 0 2 D1N750
.LIB EVAL.LIB
.DC V1 LIST 6.7
.PRINT DC V(2) I(D1)
.END
```

```
PSPICE EXAMPLE 3-5

****    CIRCUIT DESCRIPTION

**********************************************************************

.OPTIONS NOECHO NOPAGE NOMOD

**** DC TRANSFER CURVES      TEMPERATURE = 27.000 DEG C

V1          V(2)          I(D1)

6.700E+00 4.772E+00 -7.145E-02



        JOB CONCLUDED

    TOTAL JOB TIME          .83
```

FIGURE 3–49
Data obtained from PSPICE Example 3–5 when load is removed.

(d) The code for the load current requirement increased with $R_L = 60\ \Omega$ is shown in Figure 3–50. The corresponding data printout is shown in Figure 3–51.

The load voltage has now dropped to 4.331 V, which is nearly 8% below the intended voltage. The problem is noted by observing that the diode current

FIGURE 3–50

Code for PSPICE Example 3–5 when load current requirement is increased.

```
PSPICE EXAMPLE 3-5
.OPTIONS NOECHO NOPAGE NOMOD
V1 1 0 6.3
RS 1 2 27
D1 0 2 D1N750
RL 2 0 60
.LIB EVAL.LIB
.DC V1 LIST 6.3
.PRINT DC V(2) I(D1) I(RL)
.END
```

```
PSPICE EXAMPLE 3-5

****   CIRCUIT DESCRIPTION

*********************************************************************

.OPTIONS NOECHO NOPAGE NOMOD

****   DC TRANSFER CURVES    TEMPERATURE = 27.000 DEG C

V1        V(2)        I(D1)        I(RL)

6.300E+00   4.331E+00   -7.579E-04   7.218E-02



    JOB CONCLUDED

    TOTAL JOB TIME        1.21
```

FIGURE 3–51

Data obtained from PSPICE Example 3–5 when load current requirement is increased.

magnitude is now only 0.7579 mA, which is well below the required level of bias current for the 1N750. Thus, the diode has moved out of the optimum regulator range of operation.

PROBLEMS

Drill Problems

For all problems except those involving zener diodes, assume that reverse breakdown does not occur within the range of signal voltages given. Assume a 0.7-V drop for forward-biased diodes unless otherwise indicated.

3–1. Consider the rectifier circuit of Figure 3–8(a) with $R = 1$ kΩ. An ideal time-varying voltage source is applied across the input terminals. If the source voltage varies from -20 to 20 V, determine the peak values of **(a)** the output voltage, **(b)** the diode current, and **(c)** the reverse voltage across the diode.

3–2. Consider the rectifier circuit of Figure 3–9(a) with $R = 2$ kΩ. An ideal time-varying voltage source is applied across the input terminals. If the source voltage varies from -15 to 15 V, determine the peak values of **(a)** the output voltage, **(b)** the diode current, and **(c)** the reverse voltage across the diode.

3–3. Repeat Problem 3–1 if the source voltage is nonsymmetrical and varies from -30 to 25 V.

3–4. Repeat Problem 3–2 if the source voltage is nonsymmetrical and varies from -15 to 20 V.

3–5. Consider the rectifier circuit of Figure 3–11(a) with $R = 1$ kΩ. An ideal time-varying voltage source is applied across the input terminals. If the source voltage varies from -20 to 20 V, determine the peak values of **(a)** the output voltage, **(b)** the diode current, and **(c)** the reverse voltage across the diode.

3–6. Consider the rectifier circuit of Figure 3–12(a) with $R = 2$ kΩ. An ideal time-varying voltage source is applied across the input terminals. If the source voltage varies from -15 to 15 V, determine the peak values of **(a)** the output voltage, **(b)** the diode current, and **(c)** the reverse voltage across the diode.

3–7. Repeat Problem 3–5 if the source voltage is nonsymmetrical and varies from -30 to 25 V.

3–8. Repeat Problem 3–6 if the source voltage is nonsymmetrical and varies from -15 to 20 V.

3–9. Consider the rectifier circuit of Figure P3–9. A source with time-varying voltage v_s and an internal resistance of 500 Ω is connected to the input. **(a)** Determine a set of equations describing the relationship between the source voltage and output voltage, and sketch the transfer characteristic. **(b)** If v_s varies from -30 to 30 V, determine the peak values of the output voltage, the diode current, and the reverse voltage across the diode.

FIGURE P3–9

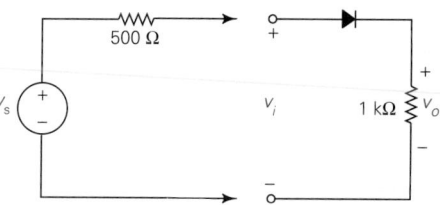

3–10. Consider the rectifier circuit of Figure P3–10. A source with time-varying voltage v_s and an internal resistance of 300 Ω is connected to the input. **(a)** Determine a set of equations describing the relationship between the source voltage and output voltage, and sketch the transfer characteristic. **(b)** If v_s varies from -18 to 18 V, determine the peak values of the output voltage, the diode current, and the reverse voltage across the diode.

FIGURE P3–10

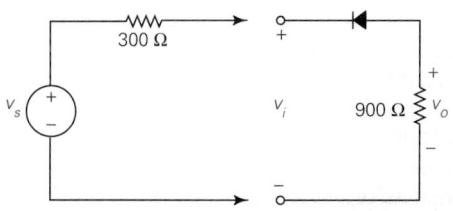

3–11. Consider the rectifier circuit of Figure P3–11. Determine a set of equations describing the relationship between v_i and v_o. (*Hint:* Redraw the parallel combination of the diode and resistor and place the resistor on the left. Then, apply Thévenin's theorem looking back from the diode.)

FIGURE P3–11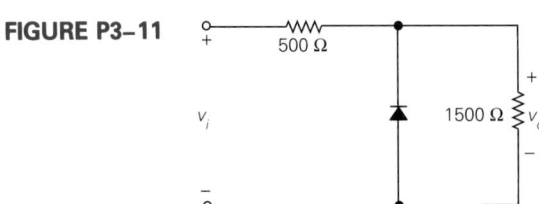

3–12. Consider the rectifier circuit of Figure P3–12. Determine a set of equations describing the relationship between v_i and v_o. (See hint for Problem 3–11.)

FIGURE P3–12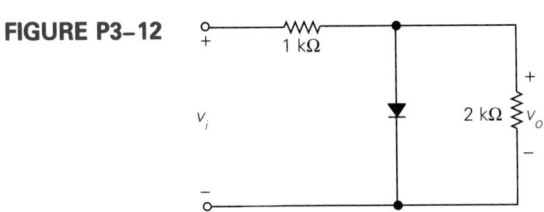

3–13. A certain circuit is designed to limit positive voltages and is of the form shown in Figure 3–14(a) with $R = 1$ kΩ and $V_A = 6$ V. **(a)** Write equations for the transfer characteristics, and sketch the form. **(b)** Determine the output voltage for each of the following values of input voltage: $v_i = -12, -5, 5,$ and 12 V. **(c)** If a "spike" of 30 V appears on the input, determine the diode current i_D.

3–14. A certain circuit is designed to limit negative voltages and is of the form shown in Figure 3–15 with $R = 2$ kΩ and $V_B = 12$ V. **(a)** Write equations for the transfer characteristic, and sketch the form. **(b)** Determine the output voltage for each of the following values of input voltage: $v_i = -20, -6, 6,$ and 20 V. **(c)** If a "spike" of -50 V appears on the input, determine the diode current i_D.

3–15. A certain circuit is designed to limit both positive and negative voltages and is of the form shown in Figure P3–15. The Thévenin output resistance of the source is 1000 Ω, and no additional series resistance is added. The load is an amplifier input open circuit. **(a)** Determine the approximate levels of the input voltage at which limiting action occurs. **(b)** Write equations describing the transfer characteristic, and sketch it. **(c)** Assume that a positive spike of 30 V appears at the source. Determine the conducting diode current based on this condition. **(d)** For the same condition as **(c)**, determine the reverse voltage across the nonconducting diode.

FIGURE P3–15

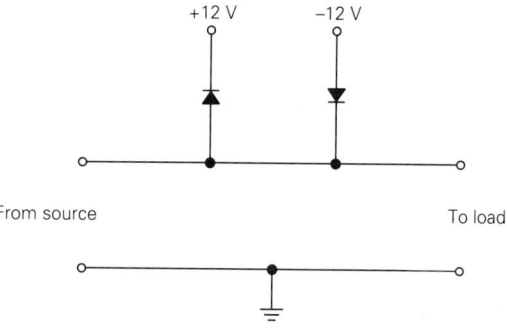

From source   To load

3–16. Repeat the analysis of Problem 3–15 for the circuit of Figure P3–16. All conditions not shown in the figure are the same as in Problem 3–15.

FIGURE P3–16

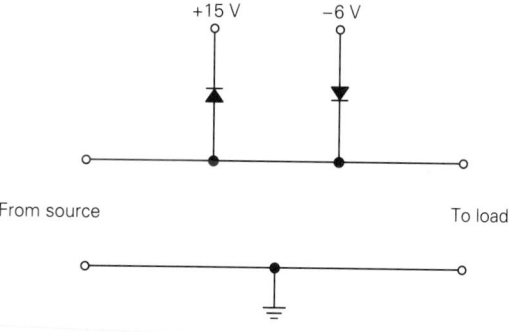

From source   To load

3–17. A zener limiting circuit of the form shown in Figure 3–19 has $V_Z = 6.2$ V and $R = 200 \ \Omega$. **(a)** Write equations for the transfer characteristic, and sketch it. **(b)** If $v_i = 20$ V, determine v_o and the diode current. Assume that any load connected across the output draws negligible current.

3–18. A zener limiting circuit of the form shown in Figure 3–20 has $V_Z = 5.1$ V and $R = 300 \ \Omega$. **(a)** Write equations for the transfer characteristic, and sketch it. **(b)** If $v_i = -25$ V, determine v_o and the diode current. Assume that any load connected across the output draws negligible current.

3–19. A zener limiting circuit of the form shown in Figure 3–21 has $V_{Z1} = V_{Z2} = 6.2$ V and $R = 200 \ \Omega$. **(a)** Write equations for the transfer characteristic, and sketch it. **(b)** If $v_i = 20$ V, determine v_o and the diode current. Assume that any load connected across the output draws negligible current.

3–20. A zener limiting circuit of the form shown in Figure 3–21 has $V_{Z1} = 5.1$ V, $V_{Z2} = 3.3$ V, and $R = 300 \ \Omega$. **(a)** Write equations for the transfer characteristic, and sketch it. **(b)** If $v_i = 25$ V, determine v_o and the diode current. **(c)** If $v_i = -25$ V, determine v_o and the diode current.

3–21. In the circuit of Example 3–5 (Figure 3–30), assume that V_1 changes to 17 V but that all other parameters remain the same. **(a)** Determine I_L, I_1, and I_Z. **(b)** With $V_1 = 17$ V, assume that R_L changes to 80 Ω. Repeat the computations of **(a)**.

3–22. In the circuit of Example 3–5 (Figure 3–30), assume that V_1 changes to 14.6 V but that all other parameters remain the same. **(a)** Determine I_L, I_1, and I_Z. **(b)** With $V_1 = 14.6$ V, assume that R_L changes to 125 Ω. Repeat the computations of **(a)**.

3–23. Consider the zener regulator circuit of Figure P3–23. Assume initially that $V_1 = 18$ V, $R_L = 150\ \Omega$, and $V_Z = 15$ V. **(a)** Determine I_L, I_1, and I_Z. **(b)** Assume that V_1 changes to 18.5 V but that all other parameters remain the same. Repeat the computations of **(a)**. **(c)** With $V_1 = 18.5$ V, assume that R_L changes to 170 Ω. Repeat the computations of **(a)**. **(d)** If the zener impedance is 12 Ω, estimate the final zener voltage in **(c)**.

FIGURE P3–23

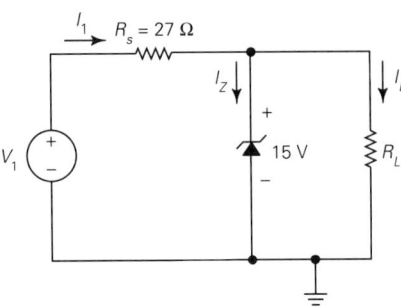

3–24. Consider the negative voltage zener regulator circuit of Figure P3–24. Assume initially that $V_1 = 5$ V, $R_L = 40\ \Omega$, and $V_Z = 3.0$ V. **(a)** Determine I_L, I_1, and I_Z. **(b)** Assume that V_1 changes to 5.5 V but that all other parameters remain the same. Repeat the computations of **(a)**. **(c)** With $V_1 = 5.5$ V, assume that R_L changes to 60 Ω. Repeat the computations of **(a)**. **(d)** If the zener impedance is 6 Ω, estimate the final zener voltage in **(c)**.

FIGURE P3–24

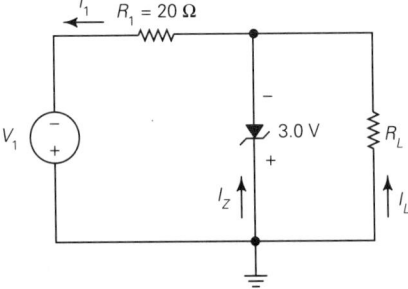

Design Problems

3–25. Design a simple half-wave rectifier circuit to meet the following specifications:

Positive output for positive input

Open-circuit signal voltage range from −30 to 30 V

Peak diode current ≤ 20 mA

Ideal voltage source input

Negligible loading on output

Draw a schematic diagram, and specify the *minimum* value of the resistance R and the diode peak reverse voltage rating. (In practice, one should select ratings well above the minimum values.)

3–26. In the design of Problem 3–25, assume that the voltage source has an internal resistance of 600 Ω. Specify again the *minimum* values of the *external* resistance R and the diode peak reverse voltage rating.

3–27. Design a limiter circuit employing standard junction diodes to meet the following specifications:

Symmetrical limiting at levels of about ±15.7 V

Worst-case calculations to be based on assumed peak values of ±50 V

Peak diode current ≤ 40 mA

Ideal voltage source input

Negligible loading on output

Assume that the system contains $+15$-V and -15-V power supplies. Draw a schematic diagram, and specify the *minimum* values of the resistance R and the diode peak reverse voltage rating.

3–28. In the design of Problem 3–27, assume that the voltage source has an internal resistance of 600 Ω. Specify again the *minimum* values of the *external* resistance R and the diode peak reverse voltage rating.

3–29. Design a limiter circuit using two zener diodes to meet the following specifications:

Symmetrical limiting at levels of about ±5.8 V

Worst-case calculations to be based on assumed peak values of ±50 V

Peak diode current ≤ 40 mA

Ideal voltage source input

Negligible loading on output

Draw a schematic diagram, and specify the values of the two zener voltages. Specify the *minimum* value of the resistance R.

3–30. Assume in Problem 3–29 that nonsymmetrical limiting at levels of $+12.7$ V and -4 V is desired. All other conditions and assumptions are unchanged. Repeat the design and the design calculations of Problem 3–29.

3–31. Design a three-input circuit to satisfy the following requirements:

Output is low if all three inputs are low.

Output is high if either input or any input combination is high.

Assume at the input that a high level is 5 V and a low level is 0. However, an output high level can be from 4 to 5 V. Diode current is not to exceed 2.5 mA. Assume ideal sources at input and no loading at output.

3–32. Design a three-input circuit to satisfy the following requirements:

Output is high if all three inputs are high.

Output is low if either input is low.

Assume at the input that a high level is 5 V and a low level is 0. However, an output low level can be from 0 to 1 V. Diode current is not to exceed 2.5 mA. Assume ideal sources at input and no loading at output.

3–33. Design a zener regulator circuit to establish a 5.1-V reference from a 15-V supply. The load current requirement is 50 mA, and the zener is to be biased at 20 mA. If the load is accidentally removed from the zener, determine the maximum diode current.

3–34. Design a zener regulator circuit to establish a −12-V reference from a −15-V supply. The load resistance is 500 Ω, and the zener is to be biased at 20 mA. If the load is accidentally removed from the zener, determine the maximum diode current.

Troubleshooting Problems

3–35. A circuit of the form shown in Figure P3–9 is not working properly. The following dc voltages are measured:

$v_s = 10$ V (open-circuit voltage with load temporarily disconnected)

$v_i = 10$ V

$v_o = 0$

Indicate the likely trouble.

3–36. A circuit of the form shown in Figure P3–10 is not working properly. The following dc voltages are measured:

$v_s = 12$ V (open-circuit voltage with load temporarily disconnected)

$v_i = 9$ V

$v_o = 9$ V

Indicate the likely trouble.

3–37. A circuit of the form shown in Figure P3–11 is not working properly. The following dc voltages are measured:

$v_i = 12$ V

$v_o = 12$ V

Identify two *possible* circuit faults.

3–38. A circuit of the form shown in Figure P3–11 is implemented without checking diode specifications and worst-case conditions. When the circuit fails to function properly, the following dc conditions are noted:

When $v_i = 16$ V, $v_o = 12$ V.

When $v_i = 32$ V, $v_o = 24$ V.

When $v_i = 40$ V, $v_o = 24$ V.

Indicate the likely cause of the problem.

4

BIPOLAR JUNCTION TRANSISTORS

OBJECTIVES

After completing this chapter, the reader should be able to:

- Identify the two major classes of transistors.
- Draw a simplified layout representing the construction of a bipolar junction transistor (BJT).
- Identify the two types of BJTs according to the polarities of the semiconductor materials.
- Draw the schematic of and identify the three terminals of a BJT.
- State the three operating regions of a BJT, and identify the bias conditions for each.
- Draw equivalent circuit models for each of the three BJT operating regions.
- Sketch the form of the collector characteristics of a typical BJT, and label all variables involved.
- Define the dc current gain of a BJT.
- State the relationship between base current, emitter current, and collector current.
- Discuss the difference between a small-signal and a power transistor.
- Inspect the specification sheets of typical BJTs to determine pertinent data.
- Apply the derating factor to determine the power rating of a BJT at a specified temperature.
- Apply PSPICE to analyze circuits in the chapter.

4–1 OVERVIEW

The development of the **bipolar junction transistor** (BJT) in the late 1940s had a most significant impact on the future of electronics. Prior to that time, virtually all electronic control and amplification functions were performed with vacuum tubes. Over the next two decades, a major transition took place, with transistors and related semiconductor devices replacing vacuum tubes. Transistors had the advantages of reliability, smaller size, lower power dissipation, and lower costs.

The transition from vacuum tubes to transistors was only the beginning of electronics as we know it today. In the 1960s, semiconductor manufacturers developed the **integrated circuit (IC),** in which the equivalent of a number of transistors was contained on a single chip. This process continued through the levels of **large-scale integration (LSI)** and **very-large-scale integration (VLSI).** At the time of this writing, integrated circuits containing the equivalent of hundreds of thousands of transistors are available, and this trend will doubtless continue.

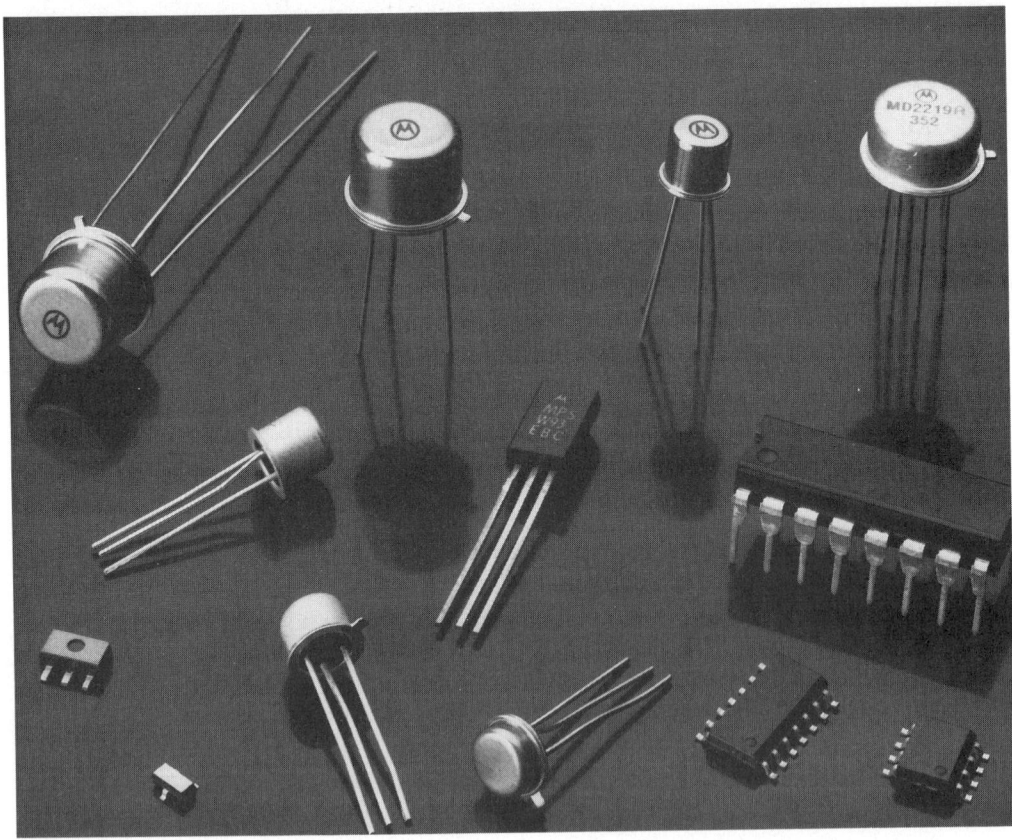

FIGURE 4–1
Motorola Semiconductor Products Sector offers a vast selection of discrete small-signal transistors in a variety of package types. (Courtesy of Motorola, Inc.)

Despite these major developments, the serious electronics specialist must still understand how to deal with an individual (or **discrete**) transistor. The state of the art has never developed and probably never will develop to the point where the user can blindly connect integrated-circuit chips without regard to interactions and interface problems. In dealing with such problems, the need for considering transistor action at the input and output terminals of the chip is essential. In many situations, the voltage, current, and power levels are different, and discrete transistors may be required to match properly the levels between chips. The control of external devices from IC chips often requires transistors to serve as buffers or drivers. Finally, integrated circuits are not available to perform many functions at higher frequencies, and there is still heavy use of discrete transistors for such applications.

Transistors and integrated circuits are available in a variety of packages, depending on the power level and other factors. Some of the different types of transistor and integrated-circuit packages are illustrated in Figure 4–1. The transistors in this photograph are of the **small-signal** type (to be defined later). Integrated circuits contain the equivalent of many different transistors and may have many external connections (referred to as **pins**).

There are two major classes of transistors: (1) the **bipolar junction transistor (BJT),** and (2) **the field-effect transistor (FET).** There are various categories within each individual class. Integrated circuits use both classes of transistors.

When the term *transistor* is used without a modifier in this book, it will be interpreted as either the major type being discussed at that point in the text, or the point of discussion could apply to either type. Where specific properties of individual types are discussed, the acronyms BJT and FET will often be used.

This chapter deals with the BJT and its operating characteristics. The major focus is on the external characteristics relating voltage and current at specific sets of terminals.

4–2 BIPOLAR JUNCTION TRANSISTOR

The BJT in its basic form consists of three sections of semiconductor material. The three sections are denoted as the **emitter, base,** and **collector.** There are two possible configurations of this basic form, and they are denoted as (1) the *npn* type, and (2) the *pnp* type. The choice of these designations will be clear shortly.

A simplified layout form of an *npn* BJT is shown in Figure 4–2(a). The emitter and collector constitute the outer sections of the BJT, and they are both composed of *n*-type semiconductor material. The base is a thin layer of *p*-type material sandwiched between the emitter and collector, and it is very lightly doped. The emitter is heavily doped, and the collector has an intermediate level of doping. Another distinction between the emitter and collector is that the collector region is larger and can dissipate more heat.

The schematic symbol for the *npn* transistor is shown in Figure 4–2(b). As far as a user is concerned, the contact with the "outside world" is through the three terminals, with *E* denoting *emitter, B* denoting *base,* and *C* denoting *collector.*

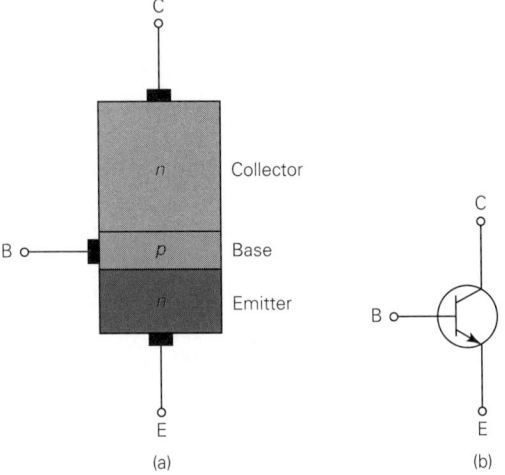

FIGURE 4–2

Physical layout of *npn* bipolar junction transistor and schematic symbol.

Later we will see that the arrow in the emitter of the schematic symbol is pointing in the direction of positive conventional current flow for that terminal.

A simplified layout form of a *pnp* BJT is shown in Figure 4–3(a). The various doping properties discussed apply to the *pnp* transistor, but its emitter and collector consist of *p*-type semiconductor material, and its base is composed of a thin layer of *n*-type material.

For either transistor type, the semiconductor junction between the base and emitter is called the **base–emitter junction,** or **base–emitter diode.** The junction between the base and collector is called the **base–collector junction,** or **base–collector diode.**

In the sense of the definitions given and bias conditions to be discussed, it

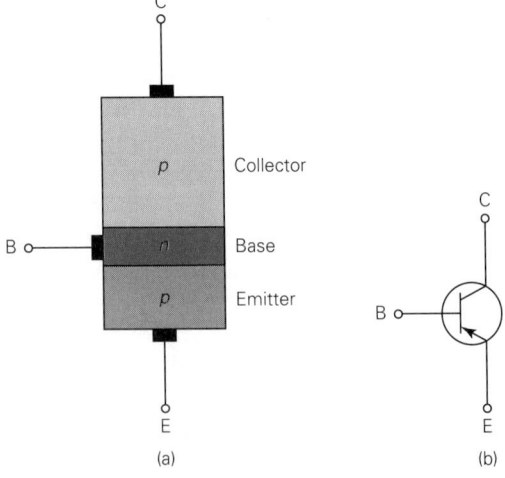

FIGURE 4–3

Physical layout of *pnp* bipolar junction transistor and schematic symbol.

is helpful to visualize a BJT as two diodes connected in series, with the base connection at the midpoint. Such forms for both the *npn* and *pnp* transistors are shown in Figure 4–4. These diagrams are particularly useful in deducing forward- and/or reverse-bias conditions for the two junctions. However, it must be stressed that *the BJT is much more than the simple connection of two diodes back to back.* The connection to the base terminal is shown as a dashed line to remind us that this is not a simple connection of two diodes.

FIGURE 4–4
Visualization of bipolar junction transistor as two diodes.

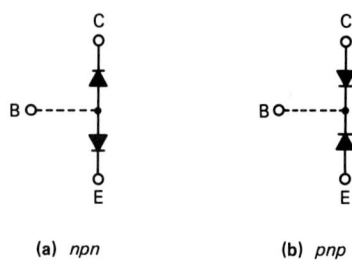

(a) *npn*   (b) *pnp*

As a result of some complex semiconductor phenomena, a small current flow in the lightly doped base can have a very large effect on current flow from collector to emitter (*npn*) or from emitter to collector (*pnp*). Since our objective in this book is circuit analysis, troubleshooting, and design, we deal with this phenomenon through the external charcteristics rather than the internal physics. The remainder of this chapter is devoted to a development of the properties of such external characteristics.

4–3 BJT OPERATING REGIONS

Since the BJT contains a base–emitter junction and a base–collector junction, each conceivably could be forward biased or reverse biased for a given set of circuit conditions. With two junctions, there are four possible combinations of bias conditions: (1) both junctions reverse biased, (2) both junctions forward biased, (3) base–emitter junction forward biased and base–collector junction reverse biased, and (4) base–emitter junction reverse biased and base–collector junction forward biased. Of these four conditions, the last one is not used often. When condition 4 does arise, it can be viewed in a manner similar to condition 3, but with the collector effectively becoming the emitter and the emitter becoming the collector. The characteristics are usually not optimum in that form, since the emitter and collector are constructed differently. We will eliminate condition 4 from consideration and focus on the other three important cases. A summary of these three cases and the designations of the regions of operation are given in Table 4–1. Each condition will now be discussed in detail.

TABLE 4–1
Transistor operating regions[a]

Base Emitter	Base Collector	Region
RB	RB	Cutoff
FB	FB	Saturation
FB	RB	Active

[a] RB, reverse bias; FB, forward bias.

Cutoff

This condition corresponds to *reverse bias* for *both base–emitter* and *base–collector junctions*. In effect, both diodes act nearly like open circuits under these conditions. An ideal equivalent circuit is shown in Figure 4–5.

FIGURE 4–5
Ideal model of BJT in *cutoff* region.

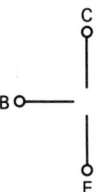

    Actually, there are some small temperature-dependent reverse leakage currents that will exist under reverse-bias conditions. For normal temperature conditions and stable circuit designs, these currents are usually negligible and will be ignored at the present time. However, these currents may be significant in some applications.

Saturation

This condition corresponds to *forward bias* for *both base–emitter* and *base–collector junctions*. As was the case for an individual junction diode in Chapter 3, various models may be applied for the two effective diodes in the saturated transistor. The most ideal model is to assume a perfect short circuit for both the base–emitter and base–collector diodes. This ideal model, shown in Figure 4–6, corresponds

FIGURE 4–6
Ideal model of BJT in *saturation* region.

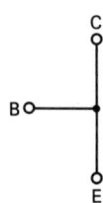

to the ideal diode model of Chapter 2. In effect, this ideal model indicates that all three transistor terminals would "come together" and assume the same potential under conditions of ideal saturation. This is in contrast to ideal cutoff, where all terminals are uncoupled from each other. The assumption of ideal saturation is used where rough checks of saturation conditions are made and where high accuracy is not required.

Recall that the constant voltage source model is the most widely used form for representing a single diode. In the same sense, a constant voltage source model could be used to represent both the base–emitter and base–collector junctions. Such a model would have the form shown in Figure 4–7, where V_{BE} and V_{BC} are the respective forward-biased constant voltage drops for the diodes. The polarities shown here represent an *npn* transistor, and both would be reversed for a *pnp* unit. If V_{BE} and V_{BC} each had a value of 0.7 V, the net voltage drop between collector and emitter would be zero, since the two voltages are opposite in direction.

FIGURE 4–7
Possible form for a constant voltage source saturation model of an *npn* transistor.

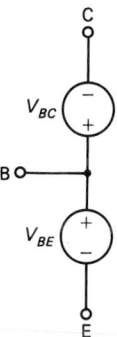

In practice, the model of Figure 4–7 as given is somewhat clumsy to apply and is not used often. Under most conditions of saturation, the collector current is much greater than the base current, and the two sources in Figure 4–7 would have to be made slightly different to account for the small collector–emitter saturation voltage drop. We will drop this model from further consideration.

A better model for saturation is shown in Figure 4–8, with (a) representing an *npn* transistor and (b) representing a *pnp* transistor. In this model the standard silicon forward-biased voltage drop of 0.7 V for the base–emitter junction is assumed, with the polarity a function of the forward-biased direction of that junction. The base current is generally small compared with the collector current, and the constant voltage drop assumption is reasonably valid. However, we need to remind ourselves that the value 0.7 V is an approximation and may vary with different transistors and operating conditions.

To account for the fact that the voltage between collector and emitter saturation is not exactly zero, a small voltage source, denoted as $V_{CE(sat)}$, is placed between these terminals. The value of $V_{CE(sat)}$ typically varies from less than 0.1

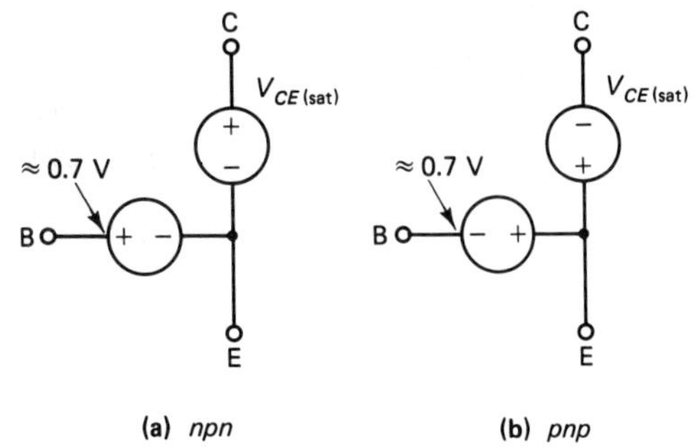

FIGURE 4–8
Best constant voltage source models of BJT in saturation region.

(a) *npn*   (b) *pnp*

V at small saturation currents to as much as 1 V or more at very large saturation currents.

In many applications involving saturation, the value of $V_{CE(sat)}$ is not known and cannot easily be determined. In such applications it may be necessary to assume that $V_{CE(sat)} = 0$. For such an assumption, the $V_{CE(sat)}$ sources in Figure 4–8 are replaced by short circuits. If, however, the application requires that the exact collector-to-emitter saturation voltage be determined, these simplified models predict invalid results.

Active Region

This condition corresponds to *forward bias* for the *base–emitter junction* and *reverse bias* for the *base–collector junction*. This means that for an *npn* transistor, the base is more positive than the emitter, and the collector is more positive than the base. The sense of polarities is reversed for a *pnp* transistor.

As transistors are connected in various circuits for specific applications, any one of the three terminals may be considered as a **common terminal.** The other two terminals may then be considered as input and output terminals, respectively.

The most widely employed common terminal is the emitter terminal, and any circuit connected in such a fashion is referred to as a **common-emitter** configuration. The most widely available transistor data are obtained through common-emitter measurement circuits and references.

An *npn* transistor, with three possible voltages labeled, is shown in Figure 4–9. Using the reference directions shown, the following quantities are defined:

V_{BE} = base–emitter voltage

V_{BC} = base–collector voltage

V_{CE} = collector–emitter voltage

In all cases, the first subscript refers to the assumed positive reference. Thus, if a given voltage is positive, the terminal represented by the first subscript is more positive than the other terminal. On the other hand, if a given voltage is negative, the terminal represented by the first subscript is more negative than the terminal

FIGURE 4-9
Three possible voltages for an *npn* transistor.

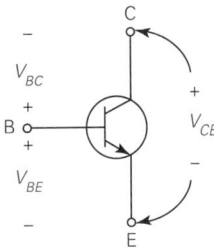

represented by the second subscript. In that case it is often more convenient to rewrite the voltage as a positive value and reverse the order of the subscript. The reference + and − signs are also reversed in the process. For example, if $V_{CE} = -10$ V, the assumed positive reference terminal (collector) is actually more negative than the emitter terminal. This could be rewritten as $V_{EC} = 10$ V, where the reference positive terminal is now identified as the emitter terminal. In general, for any terminals A and B,

$$V_{AB} = - V_{BA} \qquad (4-1)$$

For active-region operation, the base–emitter junction must be forward biased, and the base–collector junction must be reverse biased. For an *npn* transistor, the first condition implies that $V_{BE} \approx 0.7$ V. The assumed reference direction for V_{BE} with an *npn* transistor is thus compatible with the actual polarity for forward bias.

The second requirement for active-region operation (reverse-biased base–collector junction) needs some further clarification. Although it is the base–collector junction that establishes the condition, the collector–emitter voltage is more convenient as a reference in the common-emitter configuration. A relationship between base–collector voltage and collector–emitter voltage can be determined from a loop equation.

Referring to Figure 4–9, a KVL equation can be written by forming a loop from the emitter to the base, on to the collector, and back to the emitter. This equation reads

$$-V_{BE} + V_{BC} + V_{CE} = 0 \qquad (4-2)$$

Solving for V_{BC}, we obtain

$$\boxed{V_{BC} = V_{BE} - V_{CE}} \qquad (4-3)$$

If the base–collector junction in an *npn* transistor is to be reverse biased, we require that $V_{BC} < 0.7$ V. This constraint gives

$$V_{BE} - V_{CE} < 0.7 \qquad (4-4)$$

or

$$V_{CE} > V_{BE} - 0.7 \qquad (4-5)$$

If we assume that $V_{BE} = 0.7$ and substitute this value in (4–5), the condition for V_{CE} is determined as

$$\boxed{V_{CE} > 0} \tag{4–6}$$

The assumed positive reference direction for V_{CE} in Figure 4–9 for an *npn* transistor is thus compatible with the actual polarity for reverse bias.

We have seen earlier that the saturation condition results in a small positive voltage (typically about 0.1 V), so there is a transition region of a few tenths of a volt between saturation and full active-region operation. This transition region will be evident on the characteristic curves to be presented. For simplicity, the expression of (4–6) will be retained as given, but it should be understood that some realistic positive voltage for an *npn* transistor is required for full active-region operation.

The conditions for active-region operation, with V_{CE} used as a criterion, are summarized for both *npn* and *pnp* transistors in Table 4–2. Note that for *pnp* transistors, the voltages are stated two ways. In the first way, the assumed positive reference terminals are the same as for the *npn* transistor, which results in negative voltages for both V_{BE} and V_{CE}. However, by reversing the subscripts and assuming the emitter terminal as the positive reference, the values become positive.

TABLE 4–2
Conditions for active-region operation

Condition	*NPN*	*PNP*
Base–emitter forward bias	$V_{BE} \approx 0.7$ V	$V_{BE} \approx -0.7$ V or $V_{EB} \approx 0.7$ V
Base–collector reverse bias	$V_{CE} > 0$	$V_{CE} < 0$ or $V_{EC} > 0$

A schematic summary of both *npn* and *pnp* transistors in the forms most appropriate for common-emitter reference is given in Figure 4–10. As shown, all voltages will be positive for active-region operation. In this book, the "upside-down" positive references will be assumed in most cases when working with *pnp* transistors.

FIGURE 4–10
Reference polarities of voltages for *npn* and *pnp* transistors.

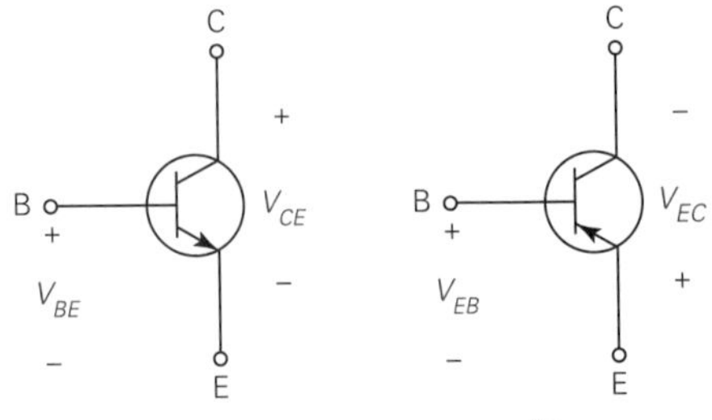

(a) *npn*   (b) *pnp*

EXAMPLE
4–1
Classify the region of operation—*cutoff, saturation,* or *active*—for each of the following transistor types and conditions: **(a)** *npn:* $V_{BE} = 0.7$ V, $V_{CE} = 10$ V; **(b)** *npn:* $V_{BE} = 0.7$ V, $V_{CE} = 0$ V; **(c)** *npn:* $V_{BE} = -1$ V, $V_{CE} = 5$ V; **(d)** *pnp:* $V_{BE} = -0.7$ V, $V_{CE} = -10$ V; **(e)** *pnp:* $V_{EB} = 0.7$ V, $V_{EC} = 10$ V; **(f)** *pnp:* $V_{EB} = 0.7$ V, $V_{EC} = 0$ V; **(g)** *pnp:* $V_{BE} = 0.7$ V, $V_{EC} = 6$ V.

Solution
As the reader develops a "feel" for this process, it may be helpful to refer to the cutoff and saturation models, Table 4–1 (p. 172), and, particularly, Table 4–2. However, one should strive to develop the ability to predict the region of operation from the definition and the stated condition. Each part will now be considered, and comments will be made as appropriate.

(a) For an *npn* transistor, $V_{BE} = 0.7$ V and $V_{CE} = 10$ V imply *active-region* operation.

(b) For an *npn* transistor, $V_{BE} = 0.7$ V and $V_{CE} = 0$ imply *saturation.*

(c) For an *npn* transistor, $V_{BE} = -1$ V (or $V_{EB} = 1$ V) implies reverse bias for the base–emitter junction. Coupled with this is $V_{CE} = 5$ V, so the transistor is in *cutoff.*

(d) For a *pnp* transistor, the conditions $V_{BE} = -0.7$ V and $V_{CE} = -10$ V imply the *active region.*

(e) For a *pnp* transistor, the conditions $V_{EB} = 0.7$ V and $V_{EC} = 10$ V imply the *active region.* Note that this case is actually the same as **(d),** but the intent was to illustrate different ways of specifying the voltages.

(f) For a *pnp* transistor, $V_{EB} = 0.7$ V and $V_{EC} = 0$ V imply *saturation.*

(g) For a *pnp* transistor, $V_{BE} = 0.7$ V implies *reverse bias* for the base–emitter junction. (Do not be misled by the value 0.7 V; it could have been any positive value below the reverse breakdown level. This value was selected to test the reader's alertness!) Coupled with this is $V_{EC} = 6$ V, so the transistor is in *cutoff.*

4–4 TRANSISTOR CHARACTERISTIC CURVES

We will now explain the operating characteristics of a transistor by discussing a laboratory experiment that can be used to obtain certain characteristic curves for the transistor. Although the transistor to be discussed is hypothetical and its characteristics are somewhat idealized for clarity, the procedure can be readily verified in an electronics laboratory.

The circuit to be used in the discussion is shown in Figure 4–11. This circuit

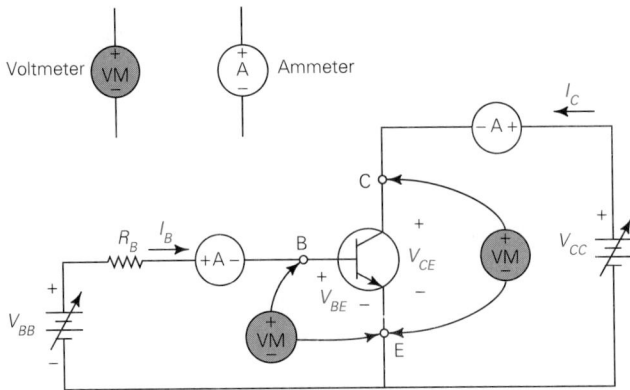

FIGURE 4–11
Measurement circuit for obtaining transistor characteristics.

contains an *npn* transistor, two adjustable dc voltage sources, a resistance to ease adjustment of and limit the maximum current in the base circuit, two voltmeters, and two ammeters. The voltage source on the left will be referred to as the **base bias supply,** and the source on the right will be referred to as the **collector bias supply.**

Our major focus will be on the active region, since many more data are required to characterize that condition, but cutoff and saturation conditions will also be evident. Note that the base and collector bias voltage sources have the proper polarities for active-region operation with an *npn* transistor. If the transistor were a *pnp* type, the polarities of both sources (along with voltmeter and ammeter terminals) would have to be reversed. Observe that the transistor is connected in a common-emitter configuration, since the emitter is common between the input (base–emitter circuit) and output (collector–emitter circuit).

Base Characteristics

First, we concentrate on a set of measurements to be made on the base side of the circuit of Figure 4–11. The curves obtained are referred to as the **base,** or **input, characteristics** of the BJT.

Assume that the collector–emitter voltage V_{CE} is set at a fixed nominal positive voltage (e.g., 10 V). Assume then that the base bias voltage source is varied from zero to some positive value. At a number of points, the base–emitter voltage V_{BE} and the base current I_B are measured.

At very low values of V_{BE} (well below 0.1 V), there is no measurable base current. However, as V_{BE} approaches 0.7 V, the base current increases rapidly, and the resulting curve would have the form shown in Figure 4–12.

We recognize immediately that this curve has the same form as the current-versus-voltage curve for a forward-biased junction diode. Indeed, the knee voltage is at about the same level (i.e., about 0.7 V for silicon). There is, however, a

FIGURE 4–12
Form of base characteristic of BJT with V_{CE} set to a constant value.

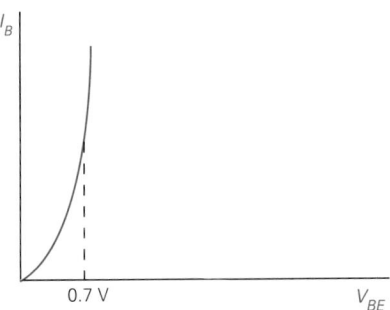

major difference. *For a given forward voltage, the base current in the transistor is much smaller than the current for a junction diode.* On the hypothetical transistor curve shown, the current values are likely in the range from about 20 to 100 μA. A typical junction diode might have current values 100 times as large for the same forward voltage.

Recall that the collector–emitter voltage V_{CE} is being held constant. Suppose that we change V_{CE} to a new value and obtain a new curve. The new curve will have the same shape but will be shifted slightly. The shift is so slight that there is very little need to obtain more than about two curves. Instead, Figure 4–13 shows the form of extreme cases. The one on the left corresponds to setting the collector bias voltage supply to zero (i.e., $V_{CE} = 0$). The right curve is for some value $V_{CE} \gg 0$ (but below the collector–base junction breakdown voltage, of course).

FIGURE 4–13
Forms of base characteristics of BJT for two extreme cases of V_{CE}.

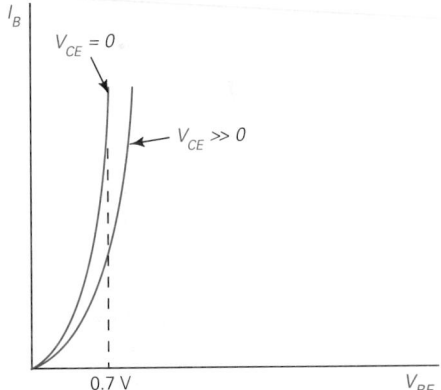

The shift of the input characteristic with the change in collector–emitter voltage is called the **Early effect.** The direction of the shift in the curve is to the right as V_{CE} increases. If V_{CE} increases and V_{BE} is fixed, I_B decreases. Conversely, if V_{CE} increases and I_B is fixed, V_{BE} increases.

For many applications, the shift in the input characteristics as V_{CE} is changed is insignificant. Unless otherwise indicated, it will be ignored in most of our subsequent work.

As in the case of a diode, much analysis with transistors uses the constant voltage drop assumption for a forward-biased base–emitter junction; that is, $V_{BE} = 0.7$ V for an *npn* transistor and $V_{EB} = 0.7$ V for a *pnp* transistor.

Collector Characteristics

We are now ready to discuss some very important measurements performed on the collector side of the circuit of Figure 4–11. The set of curves to be obtained are referred to as the **collector, or output, characteristics.**

To enhance the discussion, we assume some numerical values that have been chosen as representative of many small-signal transistors. However, these values simply illustrate the process and are not intended to suggest any general results. Actual values vary widely from one transistor type to another.

Basically, the goal is to obtain a plot of collector current I_C as a function of collector–emitter voltage V_{CE}. However, since base current I_B has a large controlling effect on collector current, all three variables must be considered. This could suggest a three-dimensional plot, which is hardly desirable.

In a situation involving three variables, the standard approach is to set one of the variables at a fixed value and obtain a two-dimensional curve involving the other two variables. The first variable is then set to a new fixed value, and a second curve is obtained. This process is repeated as many times as desired until a so-called family of curves is obtained. The variable that is fixed for a given curve is referred to as a **parameter** for this purpose.

The *collector characteristics* of a BJT are a family of curves displaying collector current versus collector–emitter voltage with base current as the parameter. The collector characteristics for our hypothetical transistor are shown in Figure 4–14. These curves will be discussed in the context of the measurement system of Figure 4–11.

First, $I_B = 0$ is established. This is achieved by either setting the dc voltage on the base side to zero or simply opening the base circuit. In this case $I_C = 0$ for most practical purposes as V_{CE} is varied. (There is a very small leakage current, as mentioned previously, but its effect will be ignored at this time.) This limiting case corresponds to the cutoff region, and it is a starting point for transition into the active region.

To move into the active region, the dc voltage on the base side is increased until forward bias is achieved in the base–emitter diode. The particular base current that is established is referred to as the **base bias current.** For our hypothetical transistor, a base bias current $I_B = 20$ μA is set. The collector voltage is then varied, and the collector current is measured.

At very small values of collector–emitter voltage, there is a sharp increase in collector current. Recall that there is a transition of a few tenths of a volt between saturation and active-region operation. Once full active-region operation

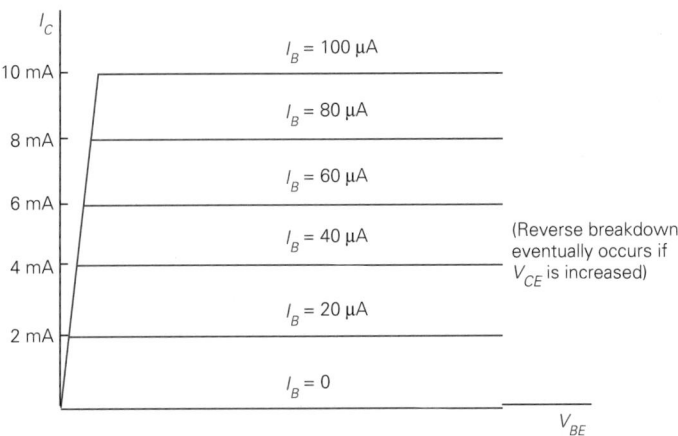

FIGURE 4–14
Collector characteristics of hypothetical BJT.

is attained, the collector current of this transistor levels off at about 2 mA for $I_B = 20$ μA. As V_{CE} increases over a fairly wide range, I_C increases slightly but not significantly. A good approximation for many transistors is to assume that it remains constant as V_{CE} varies.

We pause here to marvel at the process that is taking place. Earlier, the analogy of the transistor as two diodes connected in series was noted. However, if that model exactly described the transistor, there would be no collector current, since the base–collector junction is reverse biased. The special construction of the transistor permits a small base current flowing through the forward-biased base–emitter junction to control a much larger collector current through the reverse-biased collector–base junction.

Two assumptions need to be pointed out. First, V_{CE} cannot be increased indefinitely. Like all *pn* junctions, there is a reverse breakdown voltage for the collector–base junction at which the collector current would increase sharply, and this could destroy the transistor. We assume that V_{CE} is always kept below that level. Second, we have seen that the input characteristic is affected slightly by variations in V_{CE}. We will assume in our experiment that the base bias voltage is adjusted as necessary to keep I_B constant as V_{CE} is varied.

Next, assume that $I_B = 40$ μA is established by increasing the base bias voltage source. As V_{CE} is varied, the initial behavior is similar to that obtained in the previous curve. In this case, however, the collector current reaches a value of 4 mA at the leveling point.

This process is repeated here for several other values of base current, and the family of collector characteristics shown in Figure 4–14 is obtained. Observe that there is a slight shift to the right of the voltage at which the leveling occurs as higher values of bias current are established.

Look at Figure 4–15, and note the three regions of operation for a transistor.

FIGURE 4–15

Three regions of operation for BJT shown on collector characteristics.

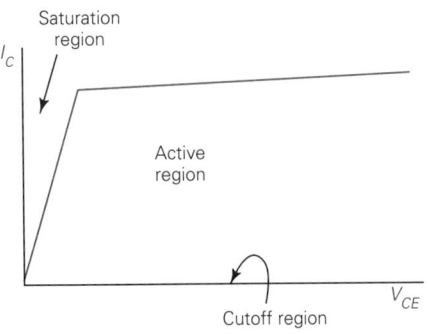

It is helpful to visualize these as transistor operation is discussed. Returning to Figure 4–14, some important properties of the active region can be deduced. In the active region, the collector voltage has very little effect on the collector current. In contrast, however, the *base current has a very large effect on the collector current*. In effect, the small base current controls the much larger collector current. It is this property that makes the transistor a very useful device. As we will see throughout the text, this type of control mechanism is fundamental to all useful electronic devices; that is, a small variable at the input controls a much larger variable at the output.

DC Current Gain

We will now define an important transistor parameter, referred to as the **dc current gain.** This quantity has traditionally been denoted as β_{dc}, particularly in textbooks. However, for certain reasons irrelevant to our present train of thought, it is denoted on specifications sheets as h_{FE}. Except where we are specifically using specifications sheets, we will use the symbol β_{dc}.

The definition of dc current gain is

$$\beta_{\text{dc}} = \frac{I_C}{I_B} \qquad (4\text{–}7)$$

Stated in words, the *dc current gain is the ratio of total dc collector current to total dc base current*. Both numerator and denominator of (4–7) must be expressed in the same units, but the resulting quantity has no units (i.e., it is *dimensionless*).

Let us see what β_{dc} is for our hypothetical transistor. Suppose that we first pick the curve for $I_B = 20 \ \mu\text{A}$ and note that $I_C = 2$ mA. Expressing both quantities in milliamperes, we obtain

$$\beta_{\text{dc}} = \frac{2 \ \text{mA}}{0.02 \ \text{mA}} = 100 \qquad (4\text{–}8)$$

If we used instead the curve for $I_B = 40~\mu A$, in which case $I_C = 4$ mA, we would obtain

$$\beta_{dc} = \frac{4~\text{mA}}{0.04~\text{mA}} = 100 \qquad\qquad (4\text{--}9)$$

which is obviously the same result as before. In effect, the collector current has doubled because the base current has doubled, but the ratio remains the same.

As indicated, these curves have been idealized for clarity. Most transistor curves are not quite as "nice." In general, the value of β_{dc} will vary somewhat with different values of base current at which it is calculated. Furthermore, as we move to the right, I_C usually increases slightly for a given I_B, so β_{dc} will also vary with V_{CE}.

The variation of β_{dc} with different conditions is discussed later, so let us not confuse the situation too much now. We will learn how to analyze and even design circuits under conditions where β_{dc} can vary widely. The main point now is to understand the definition and the all-important implications of this quantity.

β_{dc} provides a measure of how large the control property is between base and collector. The small base current has a very pronounced effect on the much larger collector current when β_{dc} is large. The value of β_{dc} for transistors can vary from less than 50 to more than 500. The value of 100 for our hypothetical transistor is typical for general-purpose transistors.

The characteristics that we have investigated have been for an *npn* transistor. A similar experiment could have been performed for a *pnp* transistor if all polarities of sources and instruments are reversed. In the strict sense of polarities and references as defined for the *npn* transistor, the various voltages and currents for a *pnp* transistor would appear in the third quadrant (negative current and negative voltage). Indeed, transistor curve tracers, which are instruments providing an oscilloscopic display of the characteristic curves, show "upside-down," or third-quadrant, displays for *pnp* transistors. However, for analysis purposes, positive values may be obtained for a *pnp* transistor by reversing the defined directions of currents and reversing subscripts and reference directions for voltages, as noted.

A summary of the assumed directions and polarity conventions to be employed in this text for both *npn* and *pnp* transistors is shown in Figure 4–16. The

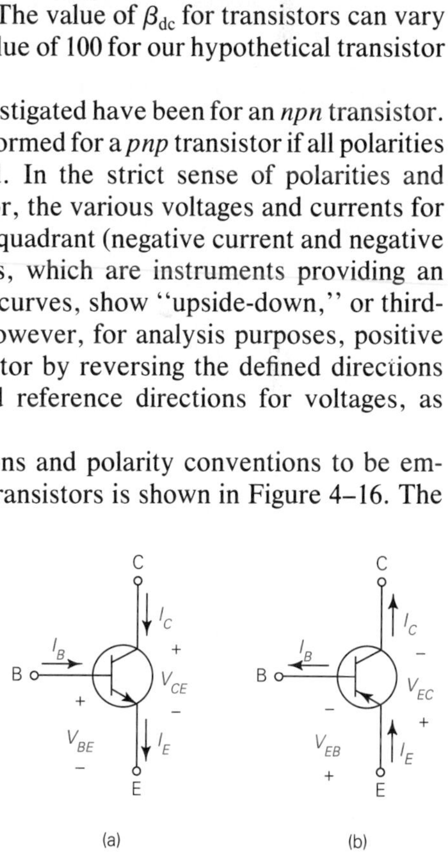

FIGURE 4–16
Recommended reference directions for *npn* and *pnp* transistors.

(a)

(b)

directions for the *npn* transistor of (a) are the same as for the hypothetical experiment discussed previously (although emitter current was not discussed). For the *pnp* transistor of (b), note that I_B is *out* of the base, I_E is *into* the emitter, and I_C is *out* of the collector. The pertinent voltages are now V_{EB} and V_{EC}. *With these directions and polarities, all quantities are positive for active-region operation.*

One more helpful point to remember is that *the arrow in the schematic symbol of a transistor has the direction of positive current flow in the emitter.*

EXAMPLE 4-2

Some measurements with a nonideal *npn* BJT have been made, and the results for three different conditions are summarized below:

	$I_B(\mu A)$	$I_C(mA)$	$V_{CE}(V)$
(a)	20	1.8	10
(b)	100	11	10
(c)	100	11.2	20

Determine the value of β_{dc} at each point.

Solution

For each point, the value of β_{dc} is

$$\beta_{dc} = \frac{I_C}{I_B} \tag{4-10}$$

The value of V_{CE} obviously does not appear in the calculation. However, the different values are given to substantiate the fact that operation is in the active region and to show that β_{dc} does typically vary with V_{CE}.

(a) For the first point, we have

$$\beta_{dc} = \frac{1.8 \times 10^{-3}}{20 \times 10^{-6}} = 90 \tag{4-11}$$

(b) For the second point, I_B is five times as great as for the first point, but V_{CE} is the same. We have

$$\beta_{dc} = \frac{11 \times 10^{-3}}{100 \times 10^{-6}} = 110 \tag{4-12}$$

(c) For the third point, I_B is the same as for the second point, but V_{CE} has been doubled. The value of β_{dc} is

$$\beta_{dc} = \frac{11.2 \times 10^{-3}}{100 \times 10^{-6}} = 112 \tag{4-13}$$

These results should be interpreted simply as illustrations of typical variations. For a given transistor *type*, the variation β_{dc} from one unit to another is even greater, as we will see.

EXAMPLE 4–3

Suppose that the same transistor for which measurements were performed in Example 4–2 is used in a circuit application in which $I_B = 90\ \mu$A and $V_{CE} = 12$ V. Estimate the approximate value of I_C expected.

Solution
It is immediately noted that the operating conditions do not correspond exactly to the conditions under which the test was performed. However, since only an estimate of the approximate value is desired, a reasonable approach is to use the value of β_{dc} closest to the operating point.

The given conditions correspond closely to those of **(b)**, for which it was determined that $\beta_{dc} = 110$. Employing that assumption, I_C is estimated as

$$I_C = \beta_{dc}I_B = 110 \times 90 \times 10^{-6} = 9.9\ \text{mA} \qquad \textbf{(4–14)}$$

This problem illustrates a typical estimation process of the type so commonly encountered in working with practical electronic circuits. It is possible with detailed equivalent circuits to extrapolate between points and, perhaps, obtain more accurate results. However, in many applications, the approach taken here is sufficient considering the uncertainty of the transistor characteristics.

4–5 ACTIVE-REGION DC MODEL

We consider next a dc model that can be used for predicting transistor operation in the active region. The adjective *dc* may be misleading and needs to be qualified. Much later in the text, some so-called **ac** or **small-signal** models are developed. Those models deal with *small changes* about some nominal operating point and, in general, have different forms from those discussed here. The terms **dc model** and **ac model** are widely used.

The dc model to be discussed here can be applied to certain signal-type situations as long as it is the *total* voltage and/or current that is of interest and as long as the transistor high-frequency and switching effects are negligible. Thus, in a number of early chapters, the dc model is used for a number of applications where time-varying signals are involved, but in which total voltage and current levels are used.

The basic dc models for the BJT are shown in Figure 4–17, with (a) representing the form for an *npn* transistor and (b) representing the form for a *pnp* transistor. The directions of all currents are labeled, so that they will be positive quantities as given. If the direction of a given variable is reversed in working a problem, a negative sign should be used. The assumed knee voltage for silicon (i.e., 0.7 V) has been assumed.

The base–emitter junction is represented by a constant voltage source model, since it is forward biased. For an *npn* transistor, the base is, therefore, about 0.7 V higher than the emitter terminal. For a *pnp* transistor, however, the emitter is about 0.7 V higher than the base voltage. The positive direction of base current is *into* the base for an *npn* transistor and *out* of the base for a *pnp* transistor.

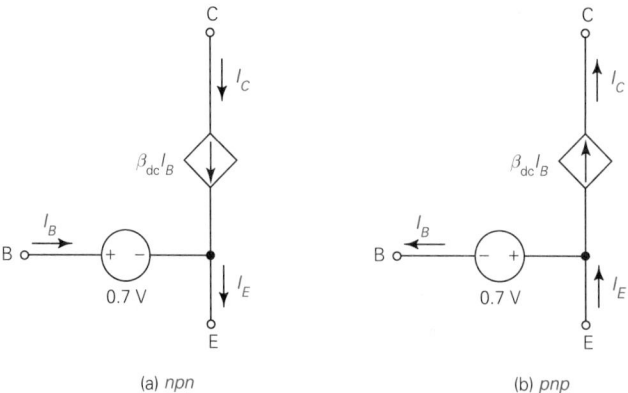

FIGURE 4–17

Active-region dc models for *npn* and *pnp* transistors. (Reverse leakage current is neglected.)

To account for the effect of the base control, a current source with value $\beta_{dc}I_B$ is placed between collector and base terminals. This source is a *dependent* or *controlled* source, since it is a function of a variable in a different part of the circuit. In effect, the collector current I_C is equal to this current source value, since there is no shunting resistance (i.e., $I_C = \beta_{dc}I_B$).

At the junction between collector branch, base branch, and emitter branch, an application of KCL yields

$$I_E = I_B + I_C \qquad\qquad (4\text{–}15)$$

Stated in words, *the emitter current is the sum of base current plus collector current.* Note that this property applies to both *npn* and *pnp* transistors as long as all quantities are expressed as positive numbers.

In many situations, the base current is much smaller than the collector current. As long as I_B is known, one might as well use the exact form of (4–15) when required. However, as we will see, in many circuits the base current may not be known accurately. In that event, the following rule is convenient: *In many active-region transistor circuits, the emitter current is only slightly larger than the collector current, and for many applications, the approximation $I_E \approx I_C$ is valid.*

It must be understood that the models of Figure 4–17 apply only if the transistor is based in the active region, which means that the base–emitter junction is forward biased and $V_{CE} > 0$ for an *npn* unit (or $V_{EC} > 0$ for a *pnp* unit). Suppose that one is analyzing a circuit problem and either of the two following conditions result: (1) the calculated base current *appears* to be negative, or (2) the calculated collector–emitter voltage for an *npn* unit *appears* to be negative (or the opposite situation for a *pnp* unit). The models given here do not provide reverse leakage effects, so both of the preceding possibilities are *impossible*. The model would, therefore, be invalid under those circumstances, and further analysis would be required to determine the exact region of operation. The conclusion is that this model is valid only for the active region.

EXAMPLE 4–4

Construct the active-region dc model for the hypothetical *npn* transistor whose characteristics were given in Figure 4–14.

Solution
It was determined earlier that $\beta_{dc} = 100$ for the transistor. The equivalent circuit will be of the form of Figure 4–17(a), and it is shown in Figure 4–18 for this particular transistor.

FIGURE 4–18
Active-region dc model for *npn* transistor of Example 4–4.

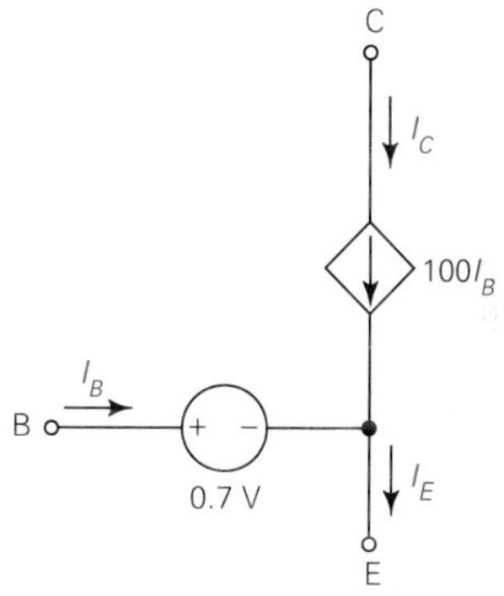

EXAMPLE 4–5

A certain *pnp* transistor has $\beta_{dc} = 120$. Construct the active-region dc equivalent circuit for the transistor.

Solution
The equivalent circuit will be of the form of Figure 4–17(b), and it is shown in Figure 4–19 for this particular transistor. Compare this equivalent circuit of the

FIGURE 4–19
Active-region dc model for *pnp* transistor of Example 4–5.

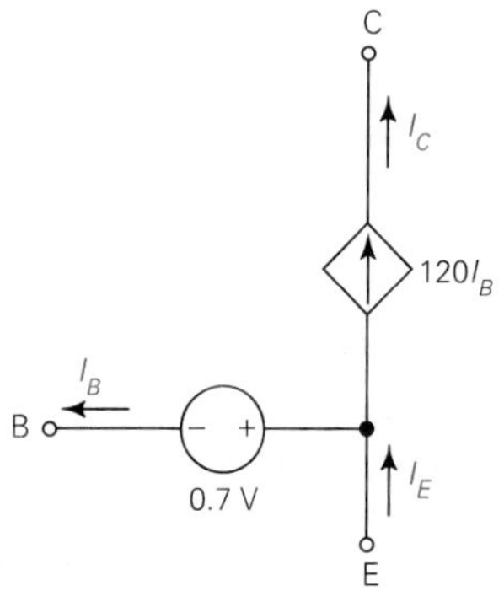

pnp transistor with that of the *npn* transistor of Example 4–4 (Figure 4–18) so that you are clear about the directions of the current sources and the base–emitter junction voltage drops.

EXAMPLE 4–6

Some conditions concerning one or more of the three transistor currents I_B, I_C, and I_E for different transistors are given in the different parts of this problem. Active-region operation is assumed in all cases. In each case, *determine exactly, if possible,* or *estimate, if necessary,* the values of the other currents. **(a)** $I_B = 50$ μA, $I_C = 3$ mA; **(b)** $I_B = 80$ μA, $I_E = 5$ mA; **(c)** $I_C = 12$ mA, $I_E = 12.15$ mA; **(d)** $I_E = 7$ mA.

Solution
The basic relationship to be employed is that of (4–15). However, in certain parts, we may be required to assume that $I_E \approx I_C$.

(a) $I_B = 50$ μA, $I_C = 3$ mA. In this case the value of I_E may be determined exactly as

$$I_E = 50 \times 10^{-6} + 3 \times 10^{-3} = 3.05 \times 10^{-3} \text{ A} = 3.05 \text{ mA} \qquad \textbf{(4–16)}$$

(b) $I_B = 80$ μA, $I_E = 5$ mA. In this case the value of I_C may be determined exactly as

$$I_C = I_E - I_B = 5 \times 10^{-3} - 80 \times 10^{-6} = 4.92 \times 10^{-3} \text{ A} = 4.92 \text{ mA} \quad \textbf{(4–17)}$$

(c) $I_C = 12$ mA, $I_E = 12.15$ mA. In this case the value of I_B may be determined exactly as

$$I_B = I_E - I_C = 12.15 \times 10^{-3} - 12 \times 10^{-3} = 0.15 \times 10^{-3} \text{ mA}$$
$$= 150 \ \mu\text{A} \qquad \textbf{(4–18)}$$

(d) $I_E = 7$ mA. All that can be done in this case is simply to estimate that

$$I_C \approx 7 \text{ mA} \qquad \textbf{(4–19)}$$

The value of I_B cannot be determined or even estimated closely. If one wanted to make a ballpark guess, an assumption of $\beta_{dc} = 100$ might be reasonable. This would lead to a guess of $I_B = 7 \times 10^{-3}/100 = 70$ μA.

In some design procedures studied later, an attempt is made to estimate the *worst-case* or *maximum* possible value of I_B. For active-region operation, the minimum expected value of β_{dc} for the transistor would then be used, since that value would predict the maximum possible base current. If, for example, the minimum value of β_{dc} for the particular transistor were 35, the maximum expected value of base current would be $I_B = 7 \times 10^{-3}/35 = 200$ μA. Although this type of process may bother some readers accustomed to exact numerical results, it is often the best that can be done with real-life electronic devices.

4–6 READING BJT DATA SHEETS

The next step in our study of BJTs is to observe some actual transistor specifications to see how they relate to the developments in the text. Any actual real-life circuit design and development must be performed in conjunction with a detailed study of the specifications of the devices used.

At the peak of discrete transistor use, there were several thousand different transistor types available. Many of these have been discontinued and will be found only in older equipment. Because of the close similarity between many different types, replacement charts have been developed to assist in replacing unavailable types with the available types. The major transistor manufacturers have transistor manuals that provide data on individual transistor characteristics.

Transistors are often loosely classified as either (1) **small-signal transistors,** or (2) **power transistors.** The lines between these categories are not rigidly defined, and some transistors could be classified as either type, depending on their application. As a reasonable guide, small-signal transistors are those in which the transistor power level is generally less than 0.5 W, and power transistors usually have power levels exceeding 0.5 W.

A major factor in the classification is the physical size and construction. Small-signal transistors usually have small physical sizes and no special provisions for heat transfer. In contrast, power transistors often have relatively large structures and provisions for **heat sinks,** which are devices used to facilitate heat transfer out of the device.

Power transistors are considered as a separate topic in Chapter 19. At this point we focus on some representative specifications of the so-called small-signal general-purpose transistors.

There are a number of small-signal transistors that are quite popular because of their wide availability, low cost, and versatility. We focus on one of these for the purpose of studying typical characteristics. We selected the 2N3904, for which a complete set of specifications is reproduced in Appendix D. We are prepared at this time to investigate only the dc characteristics. Later portions of the specifications will be studied in subsequent chapters.

Refer now to the specifications sheets of the 2N3903/2N3904 transistors in Appendix D. Manufacturers frequently group several similar semiconductor types together, as we noted for diodes in Chapter 2. We focus on the data for the 2N3904.

The first group of specifications we study are the **absolute maximum ratings.** First, the maximum values of three voltages, V_{CEO}, V_{CBO}, and V_{EBO}, are listed. The first two subscripts represent transistor terminals in accordance with our standard notation, but the third subscript O is new to us. It represents an **open condition** for the terminal not listed in the first two subscripts. Thus, V_{CEO} represents collector–emitter voltage with the base open, V_{CBO} represents collector–base voltage with the emitter open, and V_{EBO} represents emitter–base voltage with the collector open. The maximum values of these three voltages are 40, 60, and 6 V, respectively.

It should be noted that all three of these voltages represent reverse-bias conditions. If the emitter is open, the maximum collector–base voltage is 60 V.

However, in the more common situation where the emitter is not open, the maximum collector–emitter voltage (with base open) is 40 V. The maximum emitter–base voltage is only 6 V. In the active and saturation regions, the base–emitter junction is forward biased, and this need not be a concern. However, in applications where cutoff is encountered, care must be taken not to exceed this relatively low reverse-bias rating for the base–emitter junction. The maximum rating of the collector current is specified as I_C = 200 mA.

The total power absorbed by a transistor is denoted on some specifications sheets as P_D and on others by P_T. The latter symbol is used on the particular specifications sheets considered here.

The total power absorbed by a transistor consists of the power absorbed by the collector–emitter side of the circuit plus the power absorbed by the base–emitter side of the circuit. In many (but not all) applications, the second power level is considerably smaller than the first power level and may be neglected. For dc power analysis, let V_{CE} represent the dc collector–emitter voltage, and let I_C represent the dc collector current. Under conditions in which the base power is negligible, the total dc power may be approximated as

$$P_T = V_{CE}I_C \qquad (4\text{--}20)$$

For ambient temperatures of 25°C or less, the maximum value of P_T for the 2N3904 is specified as 350 mW.

Observe that a **derating factor** follows the maximum power specification. If the **ambient temperature** T_A exceeds 25°C, the amount of heat that can be transferred out of the transistor decreases. The change in power is a linear function of the increase in temperature.

Let D represent the derating factor. The basic unit is watts/°C (W/°C), but for small-signal transistors, the value is usually specified in mW/°C. The derating factor specified for the 2N3904 is 2.8 mW/°C.

Let $P_T(T_A)$ represent the maximum power dissipation at ambient temperature T_A, where T_A > 25°C, and let $P_T(25)$ represent the specified maximum power dissipation at T_A = 25°C. The following relationship is used to derate the power dissipation:

$$P_T(T_A) = P_T(25) - (T_A - 25)D \qquad (4\text{--}21)$$

It should be noted from the preceding data that there are separate maximum ratings for *voltage, current,* and *power,* and *all three* variables must be maintained below these ratings. For example, there are combinations of voltage and current that fall within the allowable range of each, but whose product exceeds the maximum power level. In the same fashion, some voltage and current combinations result in power levels below the maximum power level, but in which either the voltage or the current maximum is exceeded. For any design application, it is essential that *all three conditions be checked* before further consideration of the transistor is made.

The last set of maximum ratings are the **temperature ratings.** The symbol T_J represents the transistor junction temperature, and it is required to be in the range -55 to $135°C$. Storage and lead temperature ratings are also provided.

The next data to be considered are the **electrical characteristics.** The first three quantities listed are the reverse-biased **breakdown voltages.** The breakdown condition is denoted by BR. The remaining subscripts follow the pattern defined previously. As a matter of fact, the **minimum values** of the three voltages given represent the same voltages listed under **absolute maximum ratings** considered previously. This may seem confusing, but it is indicating that if the respective voltages are kept below the maximum ratings we considered, breakdown will not occur.

The next two data values represent **leakage currents** under **cutoff conditions.** The collector cutoff current I_{CEV} represents collector leakage current when the base–emitter junction is reverse biased. The base cutoff current I_{BEV} represents base current under the same condition. Both of the values are given as 50 nA.

The next complete block of data provides important information on the dc current gain. The symbol h_{FE} is used, and, as noted, it is the same as β_{dc}. However, do not confuse this with another symbol, h_{fe}, which appears at a different point in the specifications sheets and relates to the small-signal, or ac, data.

Values of dc current gain are specified at a number of different operating points. At one operating point ($V_{CE} = 1$ V, $I_C = 10$ mA), the minimum value is listed as 100, and the maximum value is listed as 300. This does not mean that h_{fe} for a *particular* 2N3904 will vary from 100 to 300. Rather, it means that out of a large stock of 2N3904 transistors, the values of β_{dc} could be expected to range from 100 to 300.

At other points, only minimum values are given. The wide variations shown here are typical and represent the kind of uncertainty associated with transistor specifications.

The next block of data is entitled **collector–emitter saturation voltage.** The maximum values of $V_{CE(sat)}$ are provided at two operating points. These values are 0.2 and 0.3 V, respectively.

The final block of static data is entitled **base–emitter saturation voltage,** and this quantity is denoted as $V_{BE(sat)}$. This value represents the usual forward-biased base–emitter junction drop. At one point, the minimum value is listed as 0.65 V, and the maximum value is listed as 0.85 V. At the second point, only the maximum value of 0.95 V is listed. We have been assuming and will continue to assume the common value of 0.7 V as representative for moderate bias conditions.

The large section of data on the third page, entitled **dynamic characteristics,** relates to small-signal, switching, and delay characteristics of the transistor and is discussed later in the book.

Some graphical data are provided and are useful in studying the dc properties of the transistor. At the top of the fifth page of the specifications, a family of curves of h_{FE} (or β_{dc}) for the 2N3904 as a function of I_C with T_A as a parameter is given. For a given temperature, h_{FE} shows a slight increase as I_C varies from 0.1 mA to slightly more than 10 mA, but then it begins to decrease with increased collector current.

Note the temperature dependency of h_{FE} as indicated by the different curves. The value of h_{FE} increases with increasing temperature. Finally, note that $V_{CE} = 1$ V for three of the curves, but $V_{CE} = 5$ V for the fourth. There is little difference in the two curves for the different values of V_{CE} at $T_A = 25°C$, indicating only minimal variation of h_{FE} with V_{CE}.

The variations of h_{FE} with I_C, T_A, and V_{CE} shown here are reasonably typical, but exact data for a given transistor should be consulted when such information is critical. For a reasonable range of operating conditions for many transistors, h_{FE} usually displays only a moderate change with I_C and V_{CE}, but the change with temperature is more pronounced. For a stable temperature, therefore, it is usually reasonable to use as an estimate the value of h_{FE} at the closest operating voltage and current for which data are available.

The bottom figure on the fifth page shows the base–emitter voltage as a function of collector current with temperature as a parameter. Note that V_{BE} *decreases* with *increasing* temperature. One rule of thumb is that V_{BE} *decreases* by about 2.2 mV for each *increase* of 1°C in the temperature.

The top figure on the sixth page of the specifications shows the collector–emitter saturation voltage as a function of the base current with collector current as a parameter. The steep vertical lines represent the values of base current at which saturation starts to occur. As the curves flatten out, the transistor is driven further into saturation, and the value of $V_{CE(sat)}$ is reduced.

The bottom figure on the sixth page shows the collector–emitter saturation voltage as a function of collector current with temperature as a parameter. Note that $V_{CE(sat)}$ increases with increasing temperature.

EXAMPLE 4–7 Determine the maximum power rating of a 2N3904 at an ambient temperature $T_A = 40°C$.

Solution

It was noted in the text that the value of power dissipation of 25°C is $P_T(25) = 350$ mW and the derating factor is $D = 2.8$ mW/°C. Since both quantities are given in milliwatts, the analysis can be performed directly with such units. The desired quantity is $P_T(40)$, and we have

$$P_T(40) = P_T(25) - (40 - 25) \times 2.8$$
$$= 350 - 15 \times 2.8 = 308 \text{ mW}$$

(4–22)

Note the significant reduction in power dissipation at the higher temperature.

4–7 TESTING TRANSISTORS

A number of different procedures can be used for testing transistors. This includes the use of general-purpose instruments such as an ohmmeter as well as specialized instruments designed specifically for transistors.

Ohmmeter Tests

An ohmmeter can provide some quick checks on a BJT in much the same way as for a diode. The power should be off, and the transistor should be removed from the circuit in which it is connected.

Since a BJT can be considered in a sense as two diodes in series, check the forward and reverse resistances of the base–emitter diode and the base–collector diode. The discussion of Section 2–10 concerning diodes is applicable here. Depending on whether the unit is an *npn* or a *pnp* type, the appropriate forward- and reverse-bias directions can be inferred.

In addition to the diode tests, check the resistance between collector and emitter with the ohmmeter leads in both combinations. The resistance should be high in both directions for this test, since there are two diodes in series with opposite directions. A common transistor trouble is a collector–emitter short, so if you read zero to a few thousand ohms between emitter and collector in either direction, the transistor is shorted.

Transistor Testers

If the transistor fails any of these quick checks, it is bad. However, it may pass the preceding tests and still have some faults. The ohmmeter only provides dc tests and does not indicate problems with gain or signal conditions. For example, the value of β_{dc} may be too low. Active transistor checks may be made with special transistor testers. These testers usually provide a measure of the value of β_{dc} for the transistor as well as other checks.

The most elaborate means for checking a transistor is with a *curve tracer*. These oscilloscope-based units actually display the collector characteristic curves on the screen. Curve tracers are used primarily in electronic circuit design and testing.

Tests Performed in a Circuit

When a transistor is energized within a circuit, various dc voltages may be measured as a check on operation. It is usually more convenient to measure all voltages with respect to ground. Voltages between terminals may then be determined by taking the differences between ground-referred voltages.

Based on your understanding of voltage levels for different transistor terminals, check the various voltages to ensure proper conditions. For example, if the base voltage with respect to ground is $V_B = 3.7$ V and the emitter voltage with respect to ground is $V_E = 3$ V, the value of $V_{BE} = 3.7 - 3 = 0.7$ V, and this indicates the forward-bias condition for the base–emitter junction of an *npn* transistor.

4-8 PSPICE EXAMPLES

PSPICE Example 4-1

Develop a PSPICE program to determine the collector characteristics of the generic *npn* BJT.

Solution

The procedure for obtaining collector characteristics in the laboratory was discussed in Section 4–4, and a test circuit was shown in Figure 4–11. One could, of course, develop a PSPICE code that would correspond exactly with that circuit configuration. However, for PSPICE analysis it is easier to bias the base directly with an ideal current source.

A circuit for obtaining the collector characteristics as adapted to PSPICE is shown in Figure 4–20(a), and the corresponding code is shown in Figure 4–20(b). We will illustrate in this case how to sweep both the base current source I_B and the collector voltage V_C. Consequently, no values are specified on those lines, since the range of values will be specified in a sweep control statement.

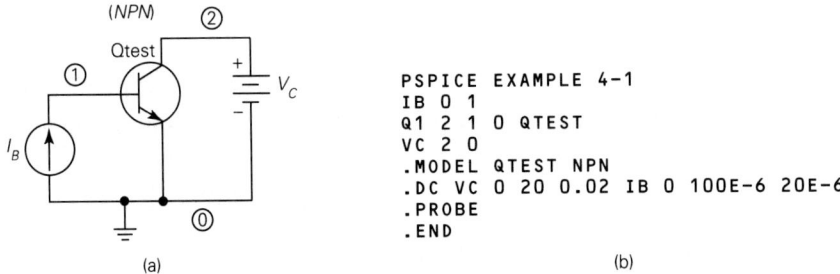

(a)

```
PSPICE EXAMPLE 4-1
IB 0 1
Q1 2 1 0 QTEST
VC 2 0
.MODEL QTEST NPN
.DC VC 0 20 0.02 IB 0 100E-6 20E-6
.PROBE
.END
```

(b)

FIGURE 4–20
Circuit and code of PSPICE Example 4–1.

Bipolar junction transistors must begin with Q, and additional letters or numbers may be added as required. Two lines are required, and they are indicated as follows:

```
Q1 2 1 0 QTEST
.MODEL QTEST NPN
```

The first line defines a BJT $Q1$, and the three integers that follow are the terminal connections. The order of the three numbers is very important, and it has the following pattern: *collector, base,* and *emitter.* Thus, the collector is connected to node 2, the base is connected to node 1, and the emitter is connected to node 0. The name of the transistor is arbitrary and is selected here as QTEST.

The second line begins with the .MODEL designation, after which the name chosen on the first line is given. Finally, the model type is identified as NPN, which is the required identifier for the NPN generic model. The reader will note that the general pattern is similar to that of the diode model D of Chapter 2.

The sweep of two variables is required, and the control statement for this operation is as follows:

```
.DC VC 0 20 0.02 IB 0 100E-6 20E-6
```

The coding following each of the variables follows the normal pattern. Thus, V_C is to be swept from 0 to 20 V in steps of 0.02 V, while I_B is to be swept from 0 to 100 μA in steps of 20 μA. The order is important in that the first variable V_C identifies the "inner loop," while the second variable I_B identifies the "outer loop." What this means is that the second variable I_B is set to its first value ($I_B = 0$), and the entire sweep is performed for the first variable V_C. Next, the second variable is set to its second value ($I_B = 20E$-6), and the entire sweep is repeated for the first variable. This process continues until all sweeps are generated. This means that six sweeps for V_C are obtained, each corresponding to a fixed value of I_B. The result is the family of curves representing the collector characteristics with I_B as a parameter.

The resulting collector characteristics of using .PROBE and the **label option** are shown in Figure 4–21. It turns out by coincidence that the generic characteristics are nearly the same as those of our hypothetical transistor considered in Figure 4–14.

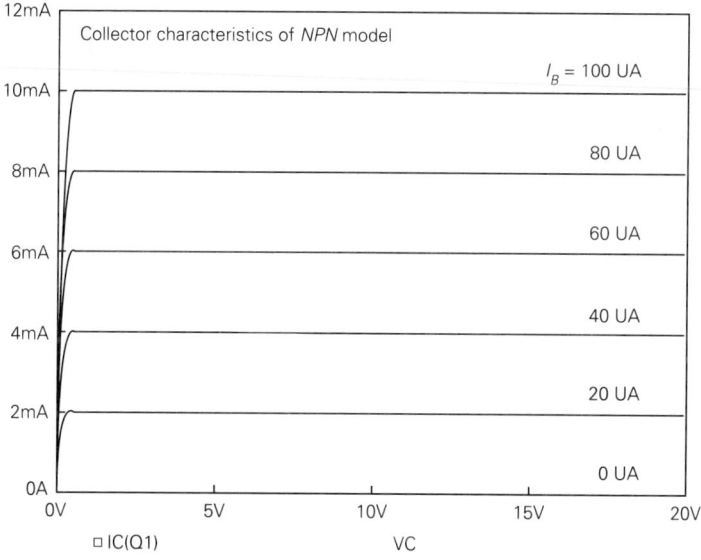

FIGURE 4–21

Collector characteristics of PSPICE generic *npn* transistor as determined in PSPICE Example 4–1.

PSPICE EXAMPLE 4–2

Develop a PSPICE program to determine the collector characteristics of the 2N3904 BJT library model.

Solution

This exercise is almost a repeat of PSPICE Example 4–1, but it is included for two reasons: (1) the characteristics have been tailored to correspond closely to those of a real transistor, and (2) the coding is slightly different because of the fact that the 2N3904 is a library model.

The circuit, which is the same form as that used for the preceding example, is shown in Figure 4–22(a), and the code is shown in Figure 4–22(b). The only lines that have a different form from the preceding example are the following:

```
Q1 2 1 0 Q2N3904
.LIB EVAL.LIB
```

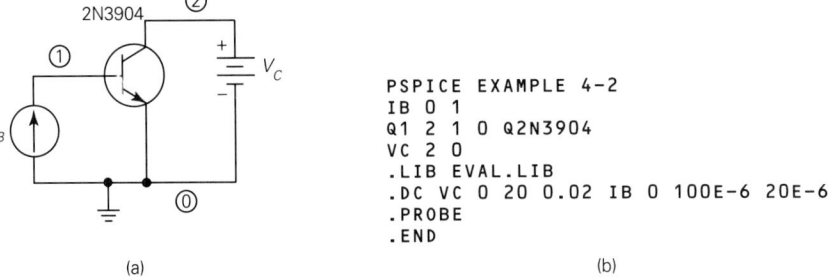

```
PSPICE EXAMPLE 4-2
IB 0 1
Q1 2 1 0 Q2N3904
VC 2 0
.LIB EVAL.LIB
.DC VC 0 20 0.02 IB 0 100E-6 20E-6
.PROBE
.END
```

(a)   (b)

FIGURE 4–22
Circuit and code of PSPICE Example 4–2.

The first line defines a BJT Q_1 with collector, base, and emitter terminals defined in consecutive order. However, the model type specified in this example is the Q2N3904. Note that the letter Q is added to the 2N3904 designation as the model name. The second line is the designation for the evaluation version of the library, as required.

The collector characteristics with additional labeling established by .PROBE are shown in Figure 4–23. These characteristics are more realistic than those of the generic *npn* model.

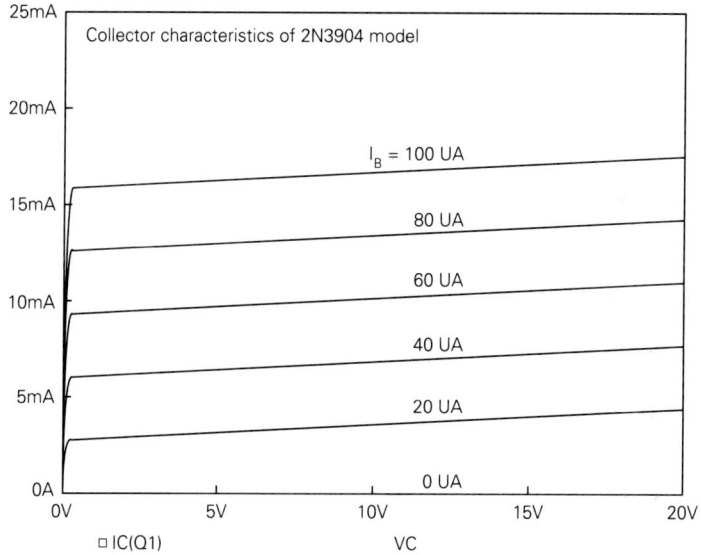

FIGURE 4–23
Collector characteristics of 2N3904 as determined in PSPICE Example 4–2.

PROBLEMS

Drill Problems

4–1. Classify the region of operation—*cutoff, saturation,* or *active*—for each of the following transistor types and conditions: **(a)** *npn:* $V_{BE} = -2$ V, $V_{CE} = 12$ V; **(b)** *npn:* $V_{BE} = 0.7$ V, $V_{CE} = 15$ V; **(c)** *npn:* $V_{BE} = 0.7$ V, $V_{CE} = 0$; **(d)** *pnp:* $V_{EB} = 0.7$ V, $V_{CE} = -15$ V; **(e)** *pnp:* $V_{BE} = 3$ V, $V_{EC} = 10$ V; **(f)** *pnp:* $V_{BE} = -0.7$ V, $V_{CE} = 0$.

4–2. Based on the dc voltage shown, classify the region of operation—*cutoff, saturation,* or *active*—for each of the transistors shown in Figure P4–2.

4–3. Determine V_{BC}, the voltage from base to collector, for each of the conditions in Problem 4–1.

4–4. Determine V_{BC}, the voltage from base to collector, for each of the circuits of Problem 4–2 (Figure P4–2).

4–5. For a certain transistor and operating conditions, $I_B = 40$ μA and $I_C = 10$ mA. Determine β_{dc} and I_E.

4–6. For a certain transistor and operating conditions, $I_B = 100$ μA and $I_E = 2$ mA. Determine β_{dc} and I_C.

4–7. For a certain transistor and operating conditions, $I_B = 30$ μA and $\beta_{dc} = 120$. Determine I_C and I_E.

4–8. For a certain transistor and operating conditions, $I_C = 50$ mA and $\beta_{dc} = 200$. Determine I_B and I_E.

4–9. The collector characteristics of a certain hypothetical *npn* BJT are shown in Figure P4–9. Determine the value of β_{dc}.

FIGURE P4-2

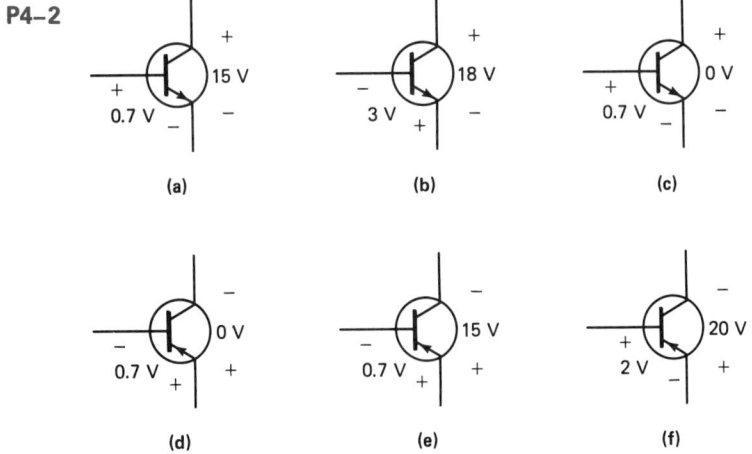

(a)　　　　　　　(b)　　　　　　　(c)

(d)　　　　　　　(e)　　　　　　　(f)

FIGURE P4-9

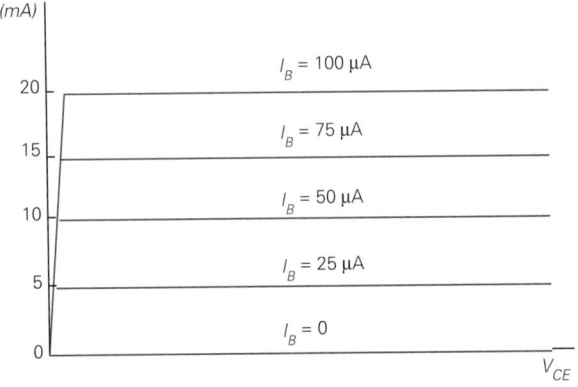

4-10. A certain idealized *npn* BJT has $\beta_{dc} = 250$ over a wide range of conditions. Sketch the collector characteristic curves corresponding to $I_B = 10\ \mu A$, $20\ \mu A$, and $30\ \mu A$. Label the collector current axis properly.

4-11. Construct the active-region dc model for the transistor of Problem 4-9.

4-12. A certain *pnp* transistor has $\beta_{dc} = 150$. Construct the active-region dc model.

4-13. A certain transistor has $I_B = 100\ \mu A$ and $I_E = 2\ mA$. Determine I_C.

4-14. A certain transistor has $I_B = 100\ \mu A$ and $I_C = 2\ mA$. Determine I_E.

4-15. A certain transistor has $I_C = 1.51\ mA$ and $I_E = 1.53\ mA$. Determine I_B.

4-16. A certain transistor has $I_C = 10\ mA$. The value of β_{dc} is specified to be in the range 50 to 200. Estimate I_E, and determine the maximum and minimum possible values of I_B.

4-17. Determine the maximum power rating of a 2N3904 at an ambient temperature $T_A = 30°C$.

4-18. Determine the maximum power rating of a 2N3904 at an ambient temperature $T_A = 100°C$.

4–19. A rule of thumb stated in the text is that V_{BE} decreases by about 2.2 mV for each increase of 1°C in the temperature. Refer to the bottom figure on the fifth page of the 2N3904 specifications. At $I_C = 1$ mA, note the value of $V_{BE(\text{sat})}$ at 25°C. Using the rule of thumb, calculate $V_{BE(\text{sat})}$ at $+125$°C, and compare with the value given on the figure.

4–20. Repeat the analysis of Problem 4–19 at $I_C = 10$ mA.

4–21. Look ahead to Problem 4–23, and note the definition and expression for α_{dc}. Calculate α_{dc} for the following values of β_{dc}: **(a)** 50; **(b)** 100; **(c)** 200; **(d)** 5000. What can you conclude about α_{dc}?

4–22. Look ahead to Problem 4–24, and note the expression for β_{dc} in terms of α_{dc}. Calculate β_{dc} for each of the following values of α_{dc}: **(a)** 0.9; **(b)** 0.99; **(c)** 0.999.

Derivation Problems

4–23. Early transistor references used an α_{dc} gain parameter, which is defined as

$$\alpha_{\text{dc}} = \frac{I_C}{I_E}$$

Show that α_{dc} can be expressed in terms of β_{dc} as

$$\alpha_{\text{dc}} = \frac{\beta_{\text{dc}}}{1 + \beta_{\text{dc}}}$$

4–24. From the results of Problem 4–23, show that β_{dc} can be expressed in terms of α_{dc} as

$$\beta_{\text{dc}} = \frac{\alpha_{\text{dc}}}{1 - \alpha_{\text{dc}}}$$

4–25. Show that emitter current can be expressed in terms of base current as

$$I_E = (1 + \beta_{\text{dc}})I_B$$

4–26. The transistor power dissipation P_T is $P_T = V_{CE}I_C$. For a given value of P_T, sketch the form of the curve of I_C versus V_{CE}. Since I_C and V_{CE} are limited to $I_{C(\text{max})}$ and $V_{CE(\text{max})}$, show how the curve and these limits define a safe operating region for the transistor.

Troubleshooting Problems

4–27. You have been assigned the task of checking circuit conditions for a number of BJTs in a complex circuit. Most of the BJTs have none of their terminals connected to the circuit common ground. However, the common terminal of your voltmeter is connected to the circuit ground, and all voltages indicated are referred to this ground. The instrument is a digital voltmeter, and both positive and negative voltages are indicated. For each of the following measurements and transistor types, identify the region of operation: *cutoff, saturation,* or *active*. **(a)** *npn:* $V_B = 6$ V, $V_E = 8$V, $V_C = 24$ V; **(b)** *npn:* $V_E = -3$ V, $V_C = 10$ V, $V_B = -2.3$ V; **(c)** *npn:* $V_C = -9$ V, $V_E = -9$ V, $V_B = -8.3$ V; **(d)** *pnp:* $V_C = 2$ V, $V_E = 20$ V, $V_B = 20.7$ V; **(e)** *pnp:* $V_E = 0$ V, $V_C = 0$ V, $V_B = -0.7$ V; **(f)** *pnp:* $V_E = 12$ V, $V_B = -12.7$ V, $V_C = -24$ V.

4–28. Repeat Problem 4–27 for the following transistor types and measurements: **(a)** *npn:* $V_E = 6$ V, $V_C = 12$ V, $V_B = 6.7$ V; **(b)** *npn:* $V_C = 0$ V, $V_E = 0$ V, $V_B = 0.7$ V;

(c) *npn:* $V_E = 10$ V, $V_C = 22$ V, $V_B = 9.3$ V; **(d)** *pnp:* $V_C = 12$ V, $V_E = 12$ V, $V_B = 11.3$ V; **(e)** *pnp:* $V_C = -8$ V, $V_E = -3$ V, $V_B = -2$ V; **(f)** *pnp:* $V_E = -10$ V, $V_C = -18$ V, $V_B = -10.7$ V.

4-29. While troubleshooting a circuit containing an *npn* transistor that *should* be operating in the active region, the following ground-referred voltages are measured: $V_B = 12$ V, $V_E = 11.3$ V, $V_C = 11.3$ V. What can you conclude about the transistor?

4-30. While troubleshooting a circuit containing a *pnp* transistor that *should* be operating in the active region, the following ground-referred voltages are measured: $V_B = 2$ V, $V_E = 0$ V, $V_C = -12$ V. What can you conclude about the transistor?

5

BJT-RESISTOR CIRCUIT APPLICATIONS

OBJECTIVES

After completing this chapter, the reader should be able to:

- Demonstrate the notational conventions for dc variables, total instantaneous variables, and time-varying or ac variables.
- Analyze various dc circuit forms containing a transistor, resistors, and a power supply.
- Analyze or design a circuit in which a BJT is used as a switch.
- Analyze or design a circuit in which a BJT is used as a current source.
- Analyze or design a basic emitter-follower circuit.
- Analyze a stable bias circuit.
- Design a stable bias current from operating point specifications.
- Construct a dc load line for either the base circuit or the collector circuit of a BJT.
- Demonstrate how changes in different transistor circuit parameters and variables affect the load line.
- Analyze some of the circuits in the chapter with PSPICE.

5–1 OVERVIEW

The bipolar junction transistor (BJT) was introduced in Chapter 4, and the terminal characteristics were discussed. We are now ready to start exploring some of the various circuit applications that use the BJT.

At this point in the book, all circuits considered will use only transistors and resistors (plus power supplies and, possibly, diodes). Primary focus will be on dc analysis and operating characteristics. However, many of these circuits can be used with time-varying inputs and outputs provided that instantaneous voltages and currents are considered.

Among the applications to be considered are BJT electronic switches, current sources, and emitter-follower circuits. These circuits find numerous applications in interfacing between integrated circuits, for buffers and drivers, and in switching loads.

The establishment of a stable dc bias circuit for a BJT is considered in detail. Both the analysis of existing circuits and the design of such circuits are discussed.

5–2 NOTATION AND SYMBOLS

The beginning electronics student is often bewildered by the many different symbols and subscripts encountered in the study of electronic devices and circuits. This author is convinced that the learning process is facilitated by an orderly process of notation, and this section is devoted to developing a basis for much of the remainder of the book. Although all possible situations cannot be covered this early, the general pattern is discussed here, and this should enhance all developments that follow.

The circuit variables that appear most often in electronic applications are voltages and currents (although power may also be considered as a circuit variable in some cases). Depending on the circuit and the desired purpose of analysis, these variables may appear in at least three forms: (1) **dc form,** (2) **total instantaneous form,** and (3) **time-varying signal (or ac) form.**

The **dc form** for variables is used when the primary objective in the analysis or design is to determine various dc voltages, currents, and power in a circuit or to characterize a fixed quantity. No variation of these quantities for the given set of conditions is normally assumed in a dc analysis. All voltages and currents could be measured with dc instruments, and oscilloscopic presentations of dc voltages would simply be horizontal lines. As a general rule, *dc or fixed variables are represented by uppercase symbols.* The subscripts may be either uppercase or lowercase. Examples of dc variables used in this chapter are I_C, V_{CE}, and V_{R_C}. The convention for constant values associated with a waveform (e.g., peak value) follows the same pattern as for dc values.

The **total instantaneous form** for variables is used when the primary objective in the analysis or design is to observe the combined effect of both a dc level and a variation about that level as a result of a signal. Oscilloscopic presentations of total instantaneous voltages must be done with the vertical channel set for dc (direct coupled) input. Measurements of total instantaneous variables made with instruments must be done with some care, since different types of instruments respond in different ways to complex waveforms. As a general rule, *total instantaneous variables are represented by lowercase symbols with uppercase subscripts.* Examples of total instantaneous variables are i_C, v_{CE}, and v_{BE}.

The **time-varying signal,** or **ac form,** for variables is used when the primary objective in the analysis or design is to study the effect of the circuit on signals.

In this case, the dc levels are ignored, and only time-varying variables are shown. Oscilloscopic presentations of time-varying voltages can be facilitated by setting the vertical channel for ac input. In this manner, the dc offsets are eliminated, and the viewer can optimize the display of the time-varying component. As a general rule, *time-varying signal variables are represented by lowercase symbols with lowercase subscripts*. Examples of time-varying instantaneous variables are i_c, v_{ce}, and v_{be}. The term *ac variable* is often used as a more casual and abbreviated description of a *time-varying signal variable,* and we use that term at many points in the text.

The difference between a total instantaneous variable and a time-varying signal variable is illustrated in Figure 5–1. The variable of interest here is the voltage between collector and emitter. The total instantaneous voltage is denoted as v_{CE}, and it is shown in Figure 5–1(a). Observe that this voltage varies up and down around a dc level. If the time axis were the baseline of an oscilloscope, the presentation would have the form shown with **dc coupling** on the input.

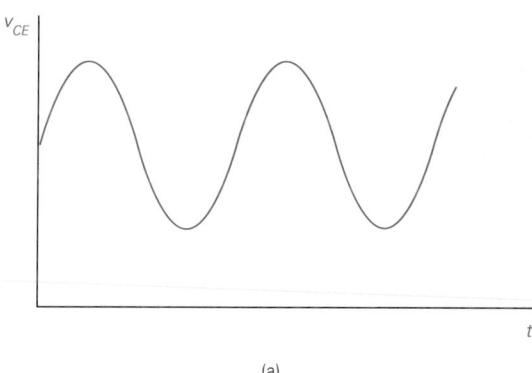

(a)

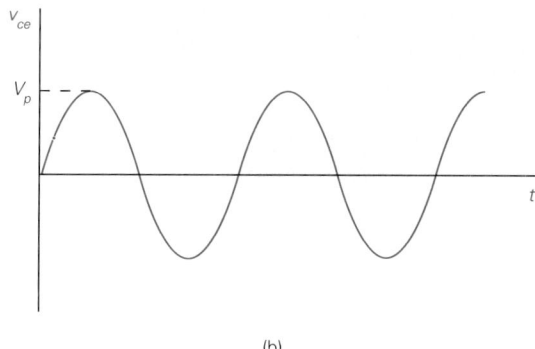

(b)

FIGURE 5–1
Typical instantaneous voltage and the time-varying (ac) portion of the voltage.

The time-varying signal form of the voltage is shown in Figure 5–1(b), and it is denoted as v_{ce} in accordance with the notation discussed previously. This form would result on an oscilloscopic display if the input were changed to **ac coupling.** Note that the peak value of this waveform is denoted as V_p. Although the signal itself is time varying and is denoted by a lowercase symbol, its peak value is a constant property of the waveform and is denoted by an uppercase symbol. This particular signal is a sine wave, but time-varying signals assume many different forms.

We discussed the pattern on subscripts in some detail in Section 1–16. We now give a review of this notation, along with some additional points relative to electronic circuit conventions.

The vast majority of electronic circuits contain a common circuit ground. *Voltage variables with a signal subscript represent voltages referred to the common circuit ground.* For example, V_C represents the dc collector voltage with respect to ground, v_C represents the total instantaneous collector voltage with respect to ground, and v_c represents the signal portion of the collector voltage with respect to ground.

Voltages with respect to circuit ground are the most useful in testing and troubleshooting circuits. In many test procedures, the ground terminal of the instrument is connected to the circuit ground, and the other instrument terminal is moved around from point to point in the circuit. In fact, through power supply safety grounding, it may be difficult and even hazardous with some instruments to measure directly voltages other than those with respect to ground. Thus there is a very practical association in the electrical circuit analysis of voltages referred to common ground.

Voltage variables with two parallel subscripts represent voltages between two points in a circuit, with the first subscript being the assumed positive reference point. For example, V_{CE} represents the dc voltage from collector to emitter, v_{CE} represents the total instantaneous voltage from collector to emitter, and v_{ce} represents the time-varying (or ac) part of the voltage from collector to emitter.

Voltages between two points in a circuit can always be expressed in terms of the differences between the ground-referred voltages. For example, if the ground-referred total instantaneous collector and emitter voltages are denoted as v_C and v_E, respectively, the collector-to-emitter total instantaneous voltage v_{CE} is

$$v_{CE} = v_C - v_E \qquad (5-1)$$

In some circuits it will be necessary to identify voltages across specific circuit components. In such cases the subscript often assumes the form of the circuit component for convenience in notation. In some cases, this may involve a "three-level" symbol, but the resulting variable will be more obvious in the process. For example, the dc voltage across a resistance R_C will be denoted as V_{R_C}.

It is customary to identify power supply voltages used to establish bias for transistors by a double subscript containing the letter twice representing the transistor terminal closest to the voltage. Examples are V_{CC} for a collector dc supply, V_{EE} for an emitter dc supply, and V_{BB} for a base dc supply.

The double-subscript notation can be used for currents when necessary, and this was explained in Section 1–16. As it turns out, however, the currents involved with most basic semiconductor devices can be adequately defined by a single subscript or by a subscript indicating the component through which the current is flowing. For example, I_C represents the dc collector current, and I_{R_1} represents the dc current flowing through a resistance R_1.

Let us now illustrate some of these conventions for a portion of some hypothetical transistor circuit. The reader need not be concerned about how the particular circuit works at this point, since the circuit is not even complete. Rather, the intent is simply to illustrate the terminology.

Consider the portion of some *npn* transistor circuit shown in Figure 5–2, which has been labeled with a number of dc voltages and currents. First, we note that the collector terminal is connected through a collector resistance R_C to a positive bias supply voltage V_{CC}. (The negative terminal of this bias voltage is connected to the reference ground.) The emitter terminal is connected through an emitter resistance R_E to a negative supply voltage $-V_{EE}$. (The positive terminal of this supply voltage is also connected to the reference ground.)

FIGURE 5–2
Illustration of a number of possible transistor dc voltages and currents.

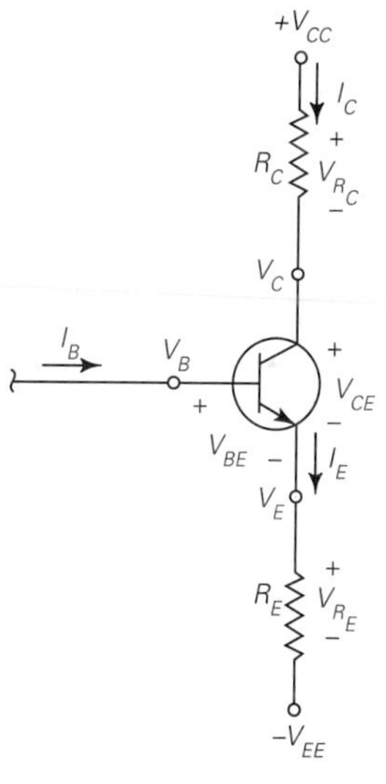

The base dc current, collector dc current, and emitter dc current are denoted as I_B, I_C, and I_E, respectively, and are shown in their normal positive directions. The dc voltages at the three transistor terminals, as referred to the circuit common ground, are denoted as V_B, V_C, and V_E. In addition, the dc voltage V_{BE} between base and emitter and the dc voltage V_{CE} between collector and emitter are also identified. Finally, the dc voltage V_{R_C} across R_C, and the dc voltage V_{R_E} across

R_E are identified. Both of the latter voltages are defined in the directions that will result in positive values.

Assume next that some external signal is applied to this circuit and it is desired to study the total instantaneous voltages and currents. The notation for this condition is shown in Figure 5–3. Not all the variables used in the dc analysis are considered in this case.

FIGURE 5–3
Circuit of Figure 5–2 with assumed signal and a number of possible transistor instantaneous voltages and currents.

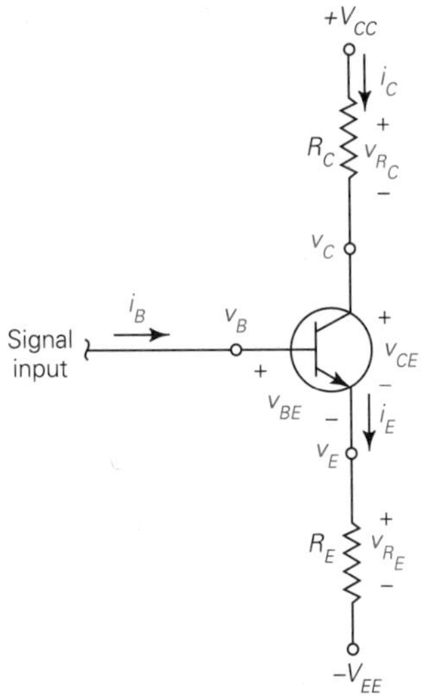

First, the power supplies $+V_{CC}$ and $-V_{EE}$ remain the same, since these are fixed quantities. The total instantaneous currents for the three terminals are now denoted as i_B, i_C, and i_E, respectively. The total instantaneous voltages at the three terminals are denoted as v_B, v_C, and v_E, respectively. Finally, the total instantaneous base–emitter and collector–emitter voltages are denoted as v_{BE} and v_{CE}, respectively.

Later in the text we will learn how to construct special circuit models for dealing only with the time-varying signal portions of the various voltages and currents. In such cases, we will use the notation introduced here for such variables (lowercase symbols and lowercase subscripts) extensively.

The reader may find it instructive to refer to the present section, along with Sections 1–16 and 1–17, and review these conventions as new devices and circuits are introduced throughout the book.

5–3 ACTIVE-REGION BJT CIRCUIT DRILL

We developed a dc equivalent circuit for the BJT in the active region in Chapter 4. In this section we study and analyze some relatively simple circuit configurations using this equivalent circuit in detail. The circuits considered at this point are not necessarily intended to represent complete circuits or particular applications. Rather, the intent is to provide drill in constructing circuit models and in computing various voltages and currents, so that the reader may develop a "feel" for the process. However, the concepts established from these drill circuits will arise extensively in later sections as we consider more complete circuit applications.

First, consider the circuit shown in Figure 5–4(a) containing an *npn* BJT. The base current I_B is assumed to be supplied from a circuit whose details are not yet specified. The emitter is connected to the circuit ground, and the collector is connected through a resistance R_C to a positive dc supply voltage V_{CC}. These "vertical" connections to $+V_{CC}$ above and ground below are used extensively in circuit diagrams. Remember that the negative terminal of the V_{CC} supply is also grounded.

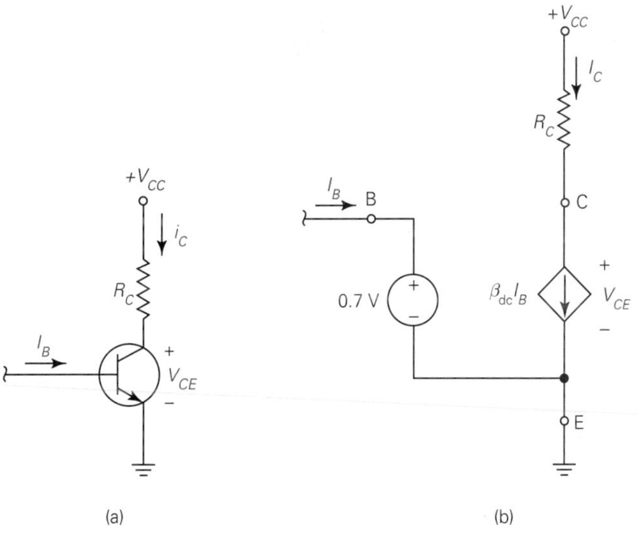

FIGURE 5–4
Basic *npn* BJT common-emitter circuit and active-region dc model used for drill circuit.

The equivalent circuit model in the linear active region is shown in Figure 5–4(b). The model for an *npn* transistor as given in Chapter 4 replaces the transistor between the three terminals, but the remainder of the circuit is not changed.

Before analyzing this circuit, we wish to emphasize that the portion of the circuit consisting of the collector and emitter terminals, the resistance, and the power supply constitutes a closed loop. The circuit is redrawn for this purpose in Figure 5–5(a), and the model oriented toward this new layout is shown in Figure 5–5(b). Be sure that you understand the equivalence of the forms of Figures 5–4 and 5–5. In most subsequent work, we will use the more standard forms of Figure 5–4.

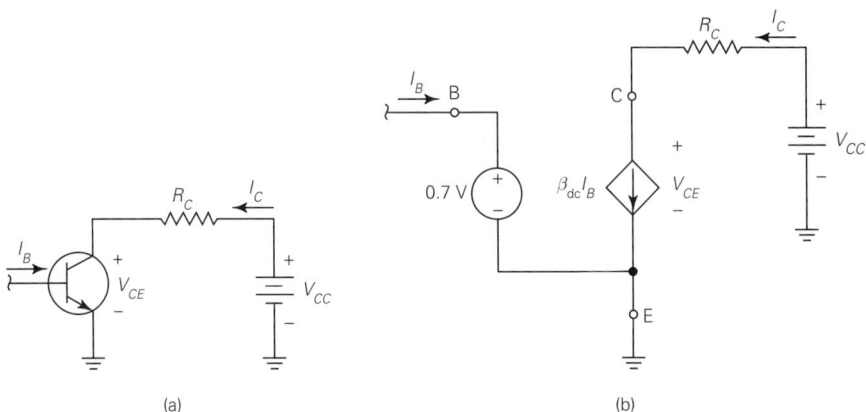

FIGURE 5–5
Alternative way to draw circuit of Figure 5–4 to illustrate collector loop.

Let us now write a KVL equation for the collector loop. Returning to the forms of Figure 5–4, follow this process: Start at ground, and move up through the power supply to V_{CC}. Next, move down through R_C to the collector, and then on down through the transistor back to ground. The KVL equation reads

$$-V_{CC} + R_C I_C + V_{CE} = 0 \qquad (5\text{--}2)$$

If Figure 5–5 had been used in formulating this equation, the KVL loop would be counterclockwise in direction. It turns out that counterclockwise loops are often more convenient for the collector circuits of *npn* transistors. Solving for V_{CE} in (5–2), we obtain

$$\boxed{V_{CE} = V_{CC} - R_C I_C} \qquad (5\text{--}3)$$

Before commenting on this result, we will consider the comparable situation with a *pnp* transistor. This circuit is shown in Figure 5–6(a), and the dc active-region model is shown in Figure 5–6(b). In this case, the power supply is a negative voltage $-V_{CC}$ with respect to ground. Note that conventions of voltage and currents are reversed in accordance with the suggestions of Chapter 4.

Before analyzing this circuit, an alternative circuit layout is shown in Figure 5–7(a), and the equivalent-circuit model that has a similar form is shown in Figure 5–7(b). This circuit and model are exactly the same as those of Figure 5–6. This "upside-down" form with emitter on the top and collector on the bottom is often used for *pnp* transistors. Many beginners are confused by changing circuit layouts to different forms, so learn to recognize this form when it occurs.

For the *pnp* transistor, the following KVL process is used: Start at ground, and move through the transistor to the collector and on through R_C to $-V_{CC}$.

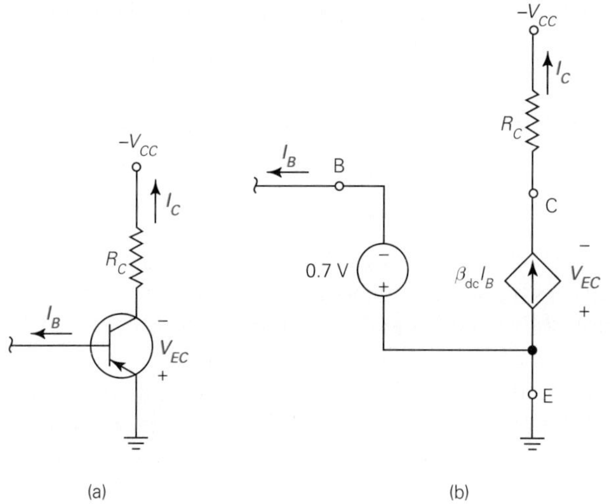

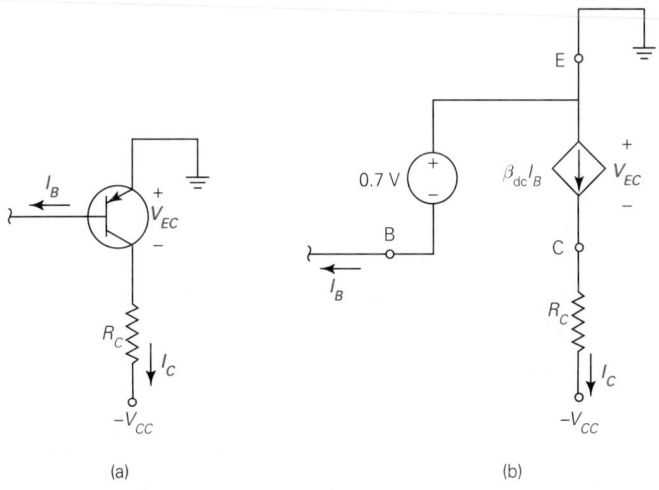

FIGURE 5–6

Basic *pnp* BJT common-emitter circuit and active-region dc model used for drill calculations.

FIGURE 5–7

"Upside-down" form of *pnp* circuit of Figure 5–6.

Finally, move through $-V_{CC}$ to ground (a voltage rise). The loop equation reads

$$V_{EC} + R_C I_C - V_{CC} = 0 \tag{5-4}$$

Solving for V_{EC}, we obtain

$$\boxed{V_{EC} = V_{CC} - R_C I_C} \tag{5-5}$$

Comparing (5–5) with (5–3), we note that the collector loop equations for *npn* and *pnp* transistors have the same form, but with V_{EC} for the *pnp* unit replacing V_{CE} for the *npn* unit. All variables and parameters in both equations have positive values, so the equations behave in the same manner.

Although one could and should be able to "walk through" this process when required, an intuitive point to remember is this: *The voltage across the transistor on the collector side is the power supply voltage minus the collector current times the collector resistance.*

The collector current I_C in the *npn* circuit of Figure 5–4 or the *pnp* current of Figure 5–6 is given by

$$\boxed{I_C = \beta_{dc} I_B} \tag{5-6}$$

Curves displaying idealized variations of I_C and V_{CE} (or V_{EC}) with I_B are shown in Figure 5–8. In the ideal case, I_C varies linearly with base current until saturation is reached. However, V_{CE} (or V_{EC}) decreases as I_B increases according to (5–3) or (5–5), which is deduced by substituting I_C in terms of I_B from (5–6). This decrease in voltage across the transistor results from an increasing portion of the supply voltage being dropped across R_C as I_C increases.

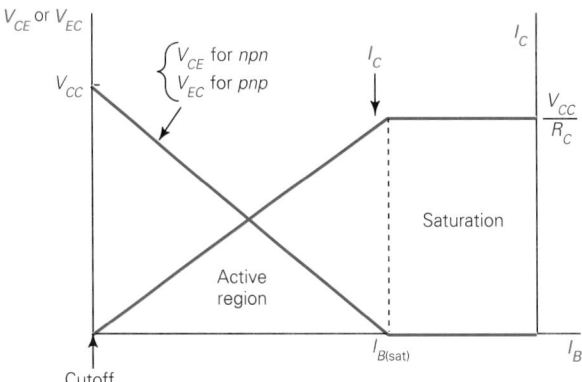

FIGURE 5–8
Ideal variation of collector–emitter (or emitter–collector) voltage and collector current with base current.

We will now relate the equations and Figure 5–8 to some properties of significance at cutoff and saturation.

Cutoff

In the cutoff region, $I_B = 0$ and $I_C = 0$. Since $I_C = 0$, there is no drop across R_C, and the full power supply voltage appears across the transistor. These results for cutoff are summarized as follows:

$$I_B = 0 \qquad (5\text{--}7)$$
$$I_C = 0 \qquad (5\text{--}8)$$
$$V_{CE} = V_{CC} \quad npn \qquad (5\text{--}9a)$$
$$V_{EC} = V_{CC} \quad pnp \qquad (5\text{--}9b)$$

This situation corresponds to the left-hand side of Figure 5–8.

Saturation

For this purpose we will assume at this time that $V_{CE(\text{sat})} = 0$ in the saturation model. As I_B increases, I_C increases, and the drop across R_C increases. The value of V_{CE} (or V_{EC}) decreases as shown in Figure 5–8 until it reaches zero. At that point, all the supply voltage V_{CC} appears across R_C, and there is none left for the transistor. The value of I_C at that point will be denoted as $I_{C(\text{sat})}$, and it is determined by setting $V_{CE} = 0$ in (5–3) or $V_{EC} = 0$ in (5–5). The result is

$$I_{C(\text{sat})} = \frac{V_{CC}}{R_C} \qquad (5\text{--}10)$$

This value of I_C is the maximum collector current that can flow in the circuit. For $I_C < I_{C(\text{sat})}$, the active-region model is applicable and $I_C = \beta_{\text{dc}} I_B$. However, once $I_{C(\text{sat})}$ is reached, further increases in I_B do not result in any additional increase in collector current. In effect, the active-region model is no longer valid, and the saturation-region model is applicable.

Let $I_{B(\text{sat})}$ represent the value of base current that will result in collector saturation. This value is determined as

$$I_{B(\text{sat})} = \frac{I_{C(\text{sat})}}{\beta_{\text{dc}}} \qquad (5\text{--}11)$$

The value of I_B can be made larger than $I_{B(\text{sat})}$, but I_C cannot increase beyond $I_{C(\text{sat})}$ for the given conditions. When $I_B > I_{B(\text{sat})}$, the control process between base and collector is lost, and I_C remains at $I_{C(\text{sat})}$. When $I_B > I_{B(\text{sat})}$, the base is said to be **overdriven.**

In actual practice, the value of V_{CE} at saturation is $V_{CE(\text{sat})}$. When it is necessary to consider this voltage, a more accurate value of the collector saturation current is

$$I_{C(\text{sat})} = \frac{V_{CC} - V_{CE(\text{sat})}}{R_C} \qquad (5\text{--}12)$$

The problem with using this more exact form is that the value of $V_{CE(\text{sat})}$ is a function of the collector current. In turn, the collector current depends on $V_{CE(\text{sat})}$. Thus, some estimating is involved in attempting an exact solution. For many noncritical applications, the assumption $V_{CE(\text{sat})} = 0$ is acceptable.

Along with the analysis procedures, the reader should strive to develop the following visualization process from this section: As the base current increases, the collector current increases, but the voltage magnitude between collector and emitter decreases. This process will arise frequently in later applications.

EXAMPLE 5–1

Consider the circuit of Figure 5–9 containing an *npn* transistor. For $\beta_{\text{dc}} = 50$, determine the values of I_C, I_B, and v_i that result in saturation. Assume that $V_{BE} = 0.7$ V and $V_{CE(\text{sat})} = 0$.

FIGURE 5–9
Circuit of Example 5–1.

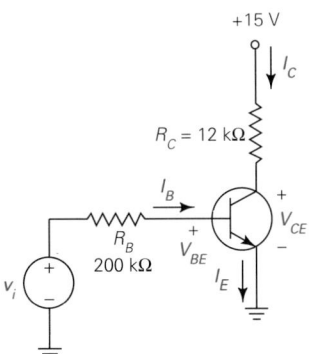

Solution
At saturation with $V_{CE(\text{sat})} = 0$, the full power supply voltage appears across R_C. Thus,

$$I_{C(\text{sat})} = \frac{V_{CC}}{R_C} = \frac{15 \text{ V}}{12 \times 10^3 \, \Omega} = 1.25 \text{ mA} \qquad (5\text{--}13)$$

The value of I_B that causes saturation is

$$I_{B(\text{sat})} = \frac{I_{C(\text{sat})}}{\beta_{\text{dc}}} = \frac{1.25 \times 10^{-3} \text{ A}}{50} = 25 \, \mu\text{A} \qquad (5\text{--}14)$$

Next, a clockwise loop equation is formed for the base circuit.

$$-v_i + R_B I_B + V_{BE} = 0 \qquad\qquad (5\text{--}15)$$

Solving for v_i and substituting $I_B = 25\ \mu\text{A}$, $V_{BE} = 0.7\ \text{V}$, and $R_B = 200\ \text{k}\Omega$, we obtain

$$v_i = R_B I_B + V_{BE} = 2 \times 10^5 \times 25 \times 10^{-6} + 0.7 = 5.7\ \text{V} \qquad (5\text{--}16)$$

EXAMPLE 5–2

Consider the circuit of Example 5–1 (Figure 5–9) again. With the same assumptions as in Example 5–1, determine I_B, I_C, I_E, and V_{CE} for each of the following values of v_i: **(a)** 0; **(b)** 2 V; **(c)** 4 V; **(d)** 10 V.

Solution
At the outset, it must be stressed that there is a questionable assumption in this problem. The value of β_{dc} in a typical transistor does vary with different operating points, and the assumption of a constant value $\beta_{dc} = 50$ at different operating points is not completely valid. Despite that, the results with such an assumption are often as best as can be done, so we will proceed on that basis.

    The active-region model is shown in Figure 5–10. The analysis that follows may be applied to this active-region model or to the original circuit.

(a) $v_i = 0$. In this case the base–emitter junction is reverse biased and cutoff exists. We thus have

$$I_B = I_C = I_E = 0 \qquad\qquad (5\text{--}17)$$

and

$$V_{CE} = 15\ \text{V} \qquad\qquad (5\text{--}18)$$

FIGURE 5–10
Active-region model of circuit of Figure 5–9 as used in Example 5–2.

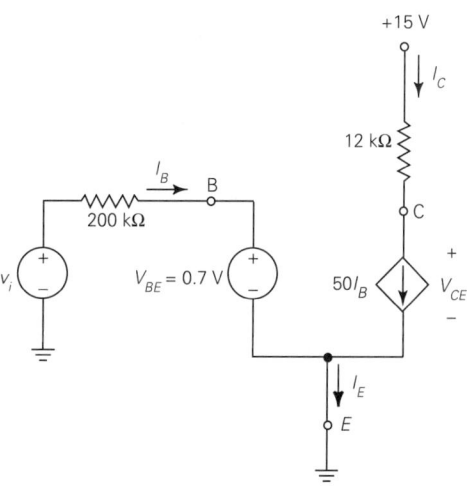

For the next several steps, a clockwise loop equation for the base circuit will be written as was done in Example 5–1. However, the intent in this case is to determine I_B as a function of v_i. We have

$$-v_i + R_B I_B + V_{BE} = 0 \qquad (5\text{--}19)$$

Solving for I_B and substituting pertinent values, we obtain

$$I_B = \frac{v_i - V_{BE}}{R_B} = \frac{v_i - 0.7}{2 \times 10^5} \qquad (5\text{--}20)$$

Note the form of (5–20) for the base loop. The voltage $v_i - 0.7$ is the net voltage across R_B. This voltage divided by R_B yields the base current.

Once I_B is known, the collector current in the active region is

$$I_C = 50\, I_B \qquad (5\text{--}21)$$

The emitter current is

$$I_E = I_B + I_C \qquad (5\text{--}22)$$

Finally, the collector–emitter voltage in the active region is

$$V_{CE} = V_{CC} - R_C I_C = 15 - 12 \times 10^3 I_C \qquad (5\text{--}23)$$

(b) $v_i = 2$ V. If this value is substituted in (5–20), we obtain

$$I_B = \frac{2 - 0.7}{2 \times 10^5} = 6.5\ \mu A \qquad (5\text{--}24)$$

The collector current from (5–21) is

$$I_C = 50 \times 6.5 \times 10^{-6} = 0.325\ \text{mA} \qquad (5\text{--}25)$$

The emitter current from (5–22) is

$$I_E = 0.325\ \text{mA} + 0.0065\ \text{mA} = 0.3315\ \text{mA} \qquad (5\text{--}26)$$

Note that I_E differs from I_C by only about 2%. The collector–emitter voltage from (5–23) is

$$V_{CE} = 15 - 12 \times 10^3 \times 0.325 \times 10^{-3} = 11.1\ \text{V} \qquad (5\text{--}27)$$

(c) $v_i = 4$ V. This step proceeds in the same manner as **(b)**, and only the calculations will be listed.

$$I_B = \frac{4 - 0.7}{2 \times 10^5} = 16.5\ \mu A \qquad (5\text{--}28)$$

$$I_C = 50 \times 16.5 \times 10^{-6} = 0.825\ \text{mA} \qquad (5\text{--}29)$$

$$I_E = 0.825\ \text{mA} + 0.0165\ \text{mA} = 0.8415\ \text{mA} \qquad (5\text{--}30)$$

$$V_{CE} = 15 - 12 \times 10^3 \times 0.825 \times 10^{-3} = 5.1\ \text{V} \qquad (5\text{--}31)$$

(d) $v_i = 10$ V. Assume that we proceed as in parts **(b)** and **(c)**, in which case I_B is calculated as

$$I_B = \frac{10 - 0.7}{2 \times 10^5} = 46.5 \ \mu A \tag{5-32}$$

If we were not alert, this value of I_B might lead us to calculate $I_C = 50 \times 46.5 \times 10^{-6} = 2.325$ mA, which might further lead us to calculate $V_{CE} = 15 - 12 \times 10^3 \times 2.325 \times 10^{-3} = -12.9$ V! Something is wrong, because V_{CE} cannot be negative in this circuit!

The answer, of course, is that we have reached the saturation point and the active-region model is no longer valid. In this circuit we can quickly refer to the results of Example 5–1 and note that $I_{B(\text{sat})} = 25 \ \mu A$ and $I_{C(\text{sat})} = 1.25$ mA. Thus, I_B from (5–32) exceeds the value of $I_{B(\text{sat})}$, and further calculations of I_C and V_{CE} based on the active-region model are meaningless. Thus, we conclude that

$$I_C = I_{C(\text{sat})} = 1.25 \text{ mA} \tag{5-33}$$
$$V_{CE} = 0 \tag{5-34}$$
$$I_E = 1.25 \text{ mA} + 0.0465 \text{ mA} = 1.2965 \text{ mA} \tag{5-35}$$

Some comments are in order. First, while I_C cannot exceed $I_{C(\text{sat})}$, I_B can continue to increase beyond the value of base current that causes saturation. As we will see in Section 5–4, a good design approach for circuits in which saturation is desired is to ensure that I_B is large enough always to cause saturation. Second, since I_B continues to increase after saturation occurs but I_C remains constant, the assumption that $I_E \simeq I_C$ becomes less valid for highly saturated circuits. The difference between I_C and I_E increases from about 2% in previous steps to nearly 4% in this example. In highly saturated circuits, the difference may be much greater.

EXAMPLE 5–3

The circuit of Figure 5–11(a) containing an *npn* transistor illustrates the simplest (but *not* the best) manner in which a single power supply may be used to establish active-region bias (or possibly saturation). Determine the values of I_C and V_{CE} for each of the following values of β_{dc}: **(a)** 50; **(b)** 100; **(c)** 200. Assume that $V_{BE} = 0.7$ V and $V_{CE(\text{sat})} = 0$.

Solution

The active-region model is shown in Figure 5–11(b). As in Example 5–2, both the circuit and its active-region model have been shown for clarity. In relatively simple circuit configurations, the complete analysis can often be performed directly on the original circuit configuration, and it may not be necessary to bother with the active-region model. However, the active-region model is another way of visualizing circuit operation, and it will be used as appropriate throughout the book.

Although there is only a single power supply, it is common for both the base

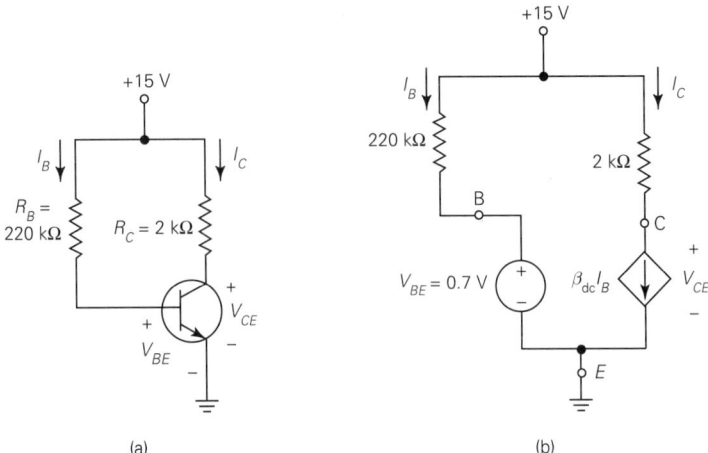

FIGURE 5-11
Circuit and active-region model for Example 5–3.

and collector circuits. Thus the loop equations for both base and collector circuits contain the 15-V dc supply.

To form the KVL loop for the base circuit, we will start at ground, move up through the power supply (voltage rise), and then move down through R_B and the base–emitter junction drop to ground. We have

$$-15 + 220 \times 10^3 I_B + 0.7 = 0 \qquad (5\text{--}36)$$

Solving for I_B, we obtain

$$I_B = \frac{15 - 0.7}{220 \times 10^3} = 65 \ \mu\text{A} \qquad (5\text{--}37)$$

The collector current in the active region is

$$I_C = \beta_{dc} I_B = 65 \times 10^{-6} \beta_{dc} \qquad (5\text{--}38)$$

The collector loop equation is formed in a manner similar to that of the base circuit, and we have

$$-15 + 2 \times 10^3 I_C + V_{CE} = 0 \qquad (5\text{--}39)$$

Solving for V_{CE}, we obtain

$$V_{CE} = 15 - 2 \times 10^3 I_C \qquad (5\text{--}40)$$

The results for each value of β_{dc} will be considered next.

(a) $\beta_{dc} = 50$. When this value is substituted in (5–38) and (5–40), we obtain

$$I_C = 65 \times 10^{-6} \times 50 = 3.25 \ \text{mA} \qquad (5\text{--}41)$$
$$V_{CE} = 15 - 2 \times 10^3 \times 3.25 \times 10^{-3} = 8.5 \ \text{V} \qquad (5\text{--}42)$$

(b) $\beta_{dc} = 100$. Proceeding as in part **(a)**, we obtain

$$I_C = 65 \times 10^{-6} \times 100 = 6.5 \text{ mA} \tag{5-43}$$
$$V_{CE} = 15 - 2 \times 10^3 \times 6.5 \times 10^{-3} = 2 \text{ V} \tag{5-44}$$

(c) $\beta_{dc} = 200$. If the procedure of the preceding two steps were followed without further consideration, the values of I_C and V_{CE} would be calculated as 13 mA and -11 V, respectively! However, these impossible results quickly tell us that saturation has been reached and that the active-region model is no longer valid.
    At saturation, and with $V_{CE(sat)} = 0$, the collector current is

$$I_C = I_{C(sat)} = \frac{V_{CC}}{R_C} = \frac{15}{2000} = 7.5 \text{ mA} \tag{5-45}$$

which is the maximum collector current for the given circuit. The value of V_{CE} is, of course,

$$V_{CE} = V_{CE(sat)} = 0 \tag{5-46}$$

This particular circuit configuration had a fixed value of base current in all three parts. From the results it can be concluded that as β_{dc} increases, I_C increases and V_{CE} decreases for a fixed base current. There is some value of β_{dc} that causes saturation—can the reader determine this value?—and the transistor would remain saturated for subsequent increases in β_{dc}.
    This problem illustrates the difficulty of establishing fixed values of I_C and V_{CE} in the active region when β_{dc} of a given transistor type varies widely. We deal with this process in Sections 5–7 and 5–8 and learn how a fixed operating point is established through proper stable circuit design.

EXAMPLE 5-4

Consider the circuit of Figure 5–12(a) containing a *pnp* transistor with $\beta_{dc} = 50$. Determine I_B, I_C, I_E, and V_{EC}. Assume that $V_{EB} = 0.7$ V.

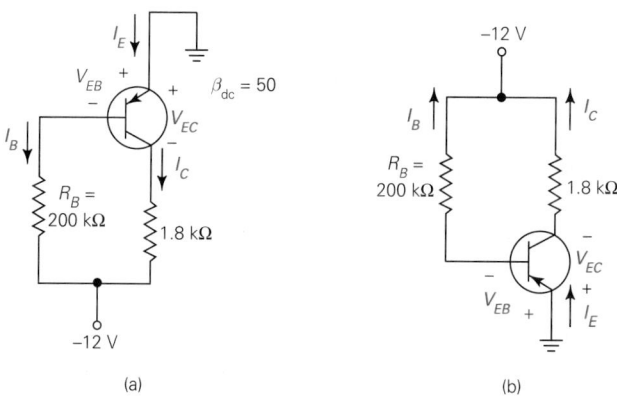

(a)   (b)

FIGURE 5–12
Circuit of Example 5–4 drawn two ways.

Solution

The circuit as given in Figure 5–12(a) is an "upside-down" form which, in certain ways, is more convenient for *pnp* transistors than the standard form, since most voltages have the more positive terminal above the negative terminal. However, to provide maximum insight, the circuit is redrawn with emitter at the bottom in Figure 5–12(b). Comparing this circuit with that of Example 5–3 as shown in Figure 5–11(a), it is noted that both circuits have identical structures (but with opposite polarities and directions, of course).

In this case we work directly with the actual circuit diagram rather than the model. To form the base loop, we start at ground and move on through the base–emitter junction and R_B to the power supply negative terminal. Moving from that point back to ground (a voltage rise) completes the loop. We have

$$V_{EB} + 200 \times 10^3 I_B - 12 = 0 \tag{5–47}$$

Solving for I_B and substituting $V_{EB} = 0.7$ V results in

$$I_B = \frac{12 - V_{EB}}{200 \times 10^3} = \frac{12 - 0.7}{200 \times 10^3} = 56.5 \ \mu\text{A} \tag{5–48}$$

The collector current is

$$I_C = \beta_{dc} I_B = 50 \times 56.5 \times 10^{-6} = 2.825 \ \text{mA} \tag{5–49}$$

To form the collector loop, we start again at ground, move to the collector, on to the power supply, and back to ground. The KVL equation is

$$V_{EC} + 1.8 \times 10^3 I_C - 12 = 0 \tag{5–50}$$

Solving for V_{EC} and substituting I_C from (5–49), we obtain

$$V_{EC} = 12 - 1.8 \times 10^3 I_C = 12 - 1.8 \times 10^3 \times 2.825 \times 10^{-3} = 6.915 \ \text{V} \tag{5–51}$$

The emitter current is

$$I_E = I_B + I_C = 56.5 \times 10^{-6} \ \text{A} + 2.825 \times 10^{-3} \ \text{A} = 2.8815 \ \text{mA} \tag{5–52}$$

5–4   BJT SWITCHES

One of the most common applications of the BJT is that of an electronic switch. This particular application of discrete transistors remains quite important even with the widespread existence of integrated circuits.

A typical application of a switch involves some load that is to be turned on or off by a control signal. The control signal might be a signal appearing at the output of a digital logic circuit or a microprocessor. The power level of the control signal will be very small and incapable of switching a load. However, the control signal may be capable of providing enough base drive to switch a transistor between two extreme levels, and the transistor may then serve to switch the load.

Most applications of the BJT as switches involve two levels of a control signal. When one level of the control signal is applied to the transistor, it operates in the cutoff region and acts as an open circuit. When the second level of the control signal is applied to the transistor, the state changes all the way to the saturation region and acts essentially as a short circuit.

One of the most common switching configurations is a **series switch** using an *npn* transistor, a positive dc supply voltage, and a control voltage with the levels zero and a positive voltage. The form of this circuit is shown in Figure 5–13(a).

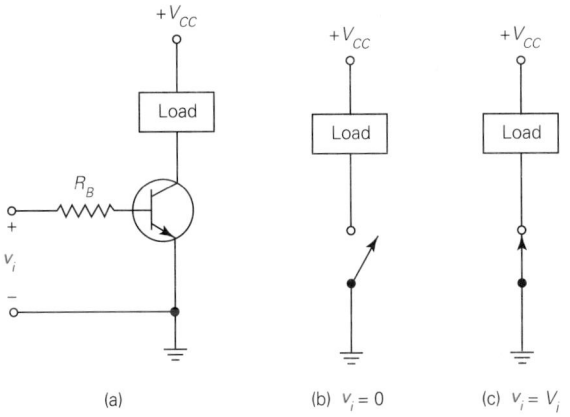

(a)   (b) $v_i = 0$ (c) $v_i = V_i$

FIGURE 5–13
Transistor switch using an *npn* transistor to close switch with positive voltage.

When $v_i = 0$, the base–emitter junction of the transistor is reverse biased, and the transistor is open. The equivalent circuit as viewed by the load is shown in Figure 5–13(b). Neglecting reverse leakage current, the switch is open.

Next assume that $v_i = V_i$, where V_i is a positive voltage of sufficient magnitude to produce saturation. In this case, the ideal equivalent circuit is shown in Figure 5–13(c). In the ideal case, the entire supply voltage appears across the load and the switch is closed. In practice, there is a small saturation voltage drop across the transistor switch, so the voltage across the load is slightly less than V_{CC}. This factor must be considered in the design if the load is very voltage sensitive.

In the design of the *npn* BJT switch, it is necessary that the positive input level of the control signal drive the transistor into saturation. If one knew exactly the value of β_{dc}, the base circuit could theoretically be designed to reach the exact value of saturation when the positive control voltage is applied. However, this minimum design approach is unsatisfactory for several reasons. First, the value of β_{dc} for a given transistor is subject to change with collector voltage, temperature, and other factors. Second, if one is designing for production, it is impractical to produce a "custom design" for each individual transistor. As pointed out earlier, the value of β_{dc} for a given transistor type in a large stock may vary by a $5:1$ range or more.

The key to this dilemma is to design around a *worst-case assumption*. It has been noted in several examples that once saturation is reached, this condition remains as base current is increased further. The approach is to select the most pessimistic conditions concerning β_{dc} (and any other circuit variables) and to ensure that base current is sufficient to produce saturation for those conditions. It follows that saturation will also occur for more favorable conditions.

As far as the base circuit is concerned, it is the *minimum* value of β_{dc} that must be used in the worst-case design. If there is sufficient base drive to ensure saturation with the minimum value of β_{dc}, all other values of β_{dc} must necessarily produce saturation.

There are several rules of thumb that have been used in this endeavor. One is to use the specified minimum value of β_{dc} and design the base circuit to produce a base drive current perhaps twice the value required (to provide additional leeway) for this value of β_{dc}.

A second rule of thumb is to select an arbitrarily low β_{dc}, well below a realistic minimum value of β_{dc}, and design around that value. The value $\beta_{dc} = 10$, for example, has been used in many designs. If the positive value of the control voltage v_i is subject to uncertainty, the worst-case design approach would be to select the *minimum* expected value of that voltage as the basis for the design.

Another factor that must be considered in the design is the additional effect of the output resistance of the control voltage on the base drive current. In the basic circuit of Figure 5–13, a signal voltage v_i and resistance R_B are shown. In practice, the Thévenin equivalent circuit of the control voltage will be in series with R_B. In some cases, this resistance will be very small in comparison with R_B and may be neglected. However, its value should always be estimated, and its possible effect in series with R_B must be considered.

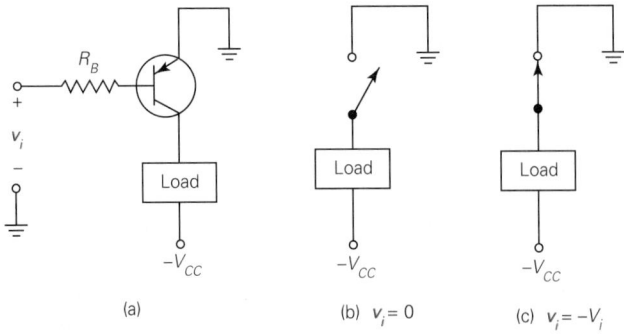

FIGURE 5–14
Transistor switch using a *pnp* transistor to close switch with negative voltage.

A different form of a *series switch* utilizing a *pnp* transistor, a negative dc supply voltage, and a control voltage with the levels zero and a negative voltage is shown in Figure 5–14(a). The resulting conditions for this circuit are shown in Figure 5–14(b) and (c). All conditions in this circuit are simply the negative mirror images of those in Figure 5–13.

EXAMPLE 5–5

A common relay switching circuit is shown in Figure 5–15. The switching operation is actually a two-step process in the following sense. First, the low power control signal turns on the transistor switch, which can sink (absorb to ground) sufficient current to turn on the relay. The relay switch contacts then close, and these mechanical contacts may be designed to handle a very large current (such as an industrial load or motor control circuit, etc.) The relay coil requires more power than the small control signal is capable of providing. The transistor in turn is incapable of handling the large external load current. However, the transistor can handle the relay control current, so it serves as a **buffer** or **interface** between the control signal and the relay coil.

FIGURE 5–15
Circuit of Example 5–5.

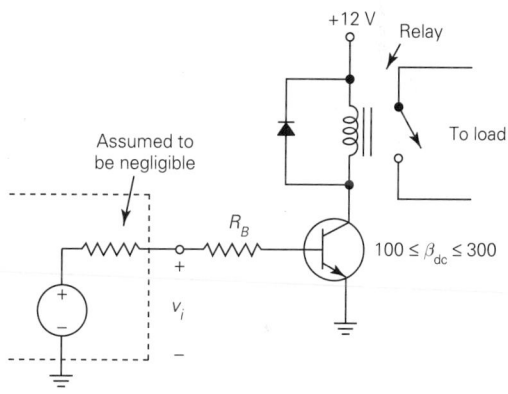

The diode across the coil is reverse biased during the period when the relay coil is excited, so its effect will be neglected in the analysis that follows. However, when current has been flowing through the coil and the transistor is turned off, the effect is to terminate current through an inductor. When an attempt is made abruptly to terminate current through an inductance, a large negative voltage is generated. Such a voltage may have destructive effects unless some means are provided for suppressing it. The diode serves this purpose, since it becomes forward biased under the conditions discussed previously and provides an inductive discharge path.

Assume that design for production is being performed and that the β_{dc} range for the transistor is from 100 to 300, as noted on the diagram. Assume that the control voltage has two levels: (1) $v_i = 0$ and (2) $v_i = 6$ V. Assume that the output resistance of the control circuit is small, so that it can likely be neglected in comparison with R_B. Assume that the relay coil operates at about 12 V and has a coil resistance of 100 Ω (not shown). Finally, assume that $V_{BE} = 0.7$ V and $V_{CE(sat)} = 0$. Using the criterion of twice the base current for worst-case conditions, complete the design by specifying the value of R_B.

Solution

Since the low value of the control voltage is $v_i = 0$, the transistor will be at cutoff and the relay will be open for that condition. Thus, the design calculations are based on $v_i = 6$ V.

First, the value of $I_{C(sat)}$ is determined. When the transistor is saturated, the 12-V supply appears across the relay coil, and since the relay coil has a dc resistance of 100 Ω, we have

$$I_{C(sat)} = \frac{12 \text{ V}}{100 \text{ } \Omega} = 0.12 \text{ A} \qquad (5\text{--}53)$$

The base drive required to create this collector current is first determined under the worst-case condition of $\beta_{dc} = 100$ (minimum value). This value, denoted as $I_{B(sat)}$, is

$$I_{B(sat)} = \frac{0.12}{100} = 0.0012 \text{ A} = 1.2 \text{ mA} \qquad (5\text{--}54)$$

Based on the criterion specified, we will choose

$$I_B = 2I_{B(sat)} = 2 \times 1.2 \text{ mA} = 2.4 \text{ mA} \qquad (5\text{--}55)$$

From the base circuit with $v_i = 6$ V, we can calculate R_B as follows:

$$R_B = \frac{6 - 0.7}{2.4 \times 10^{-3}} = 2208 \text{ } \Omega \qquad (5\text{--}56)$$

If the output resistance of the control source were an appreciable fraction of the required value of R_B, it should be subtracted from R_B. For example, assume that the control source has an output resistance of 600 Ω. The external resistance required would then be $R_B = 2206 - 600 = 1606 \text{ } \Omega$.

5–5 BJT CURRENT SOURCES

Bipolar junction transistors may be connected in a manner that will result in a nearly constant value of current for a variable load resistance. A circuit that supplies a current that is independent of the load resistance is, of course, a current source. The development here will be based on a single transistor current source.

A current source using an *npn* BJT is shown in Figure 5–16. At first glance, this circuit seems somewhat similar to a transistor switch. However, there are two major differences in the circuit layout and operation as compared with the switch. First, the current source contains a resistance R_E between emitter and

FIGURE 5-16

Single-transistor constant current source.

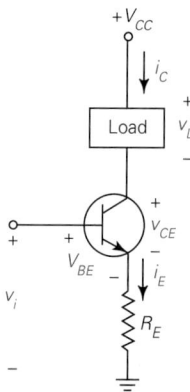

ground. This resistance will play a major role in establishing a different mode of operation, as we will see shortly. Second, the current source is operated in the active region, while the switch utilizes the cutoff and saturation regions.

We will now analyze the current source. Assume at this time that an ideal voltage source v_i appears between base and ground. If the emitter were grounded, as has been the case with most circuits considered thus far, it would be obvious that the transistor would start conducting for $v_i \approx 0.7$ V. However, the presence of R_E may puzzle the reader in determining the point where active-region operation begins.

To assist in developing a "feel" for this circuit, momentarily assume that $v_i < 0.7$ V, so that cutoff exists. In this case the emitter current is zero, and there is no voltage across R_E. This means that the emitter is at ground potential, and the initial value of v_i required to produce current is the same as if R_E were not there (i.e., $v_i = 0.7$ V).

When v_i reaches about 0.7 V, current starts to flow through the transistor, and a small drop appears across R_E. This drop is in opposition to v_i, so the net base–emitter voltage is decreased very slightly, but active-region operation remains. This is a case where we must alter our thinking to realize that the base–emitter voltage is not exactly 0.7 V but varies slightly on either side. If v_i is increased further, collector and emitter current increase further, but the drop across R_E also increases, again tending to stabilize the base–emitter voltage to a nearly constant value. Thus, while the knee voltage in this circuit is the same as if the emitter were grounded, the process of conduction is much more gradual as the input voltage increases.

A brief pause will be made here for an important comment. This circuit represents a form of **negative feedback,** a subject that will constitute a major topic later in the text. The emitter resistance is the significant factor and it can result in a highly stabilized circuit.

Let us now write a KVL clockwise loop equation for the base circuit. We have

$$-v_i + V_{BE} + R_E i_E = 0 \qquad (5\text{--}57)$$

Solving for i_E, we obtain

$$i_E = \frac{v_i - V_{BE}}{R_E} \qquad\qquad (5\text{--}58)$$

Although we must accept "slight" variations in V_{BE} for the circuit to function properly, from an overall circuit point of view we will assume that $V_{BE} = 0.7$ V. Furthermore, we will assume that β_{dc} is relatively large for the transistor, which means that $i_E \simeq i_C$. These assumptions result in

$$\boxed{i_C \simeq i_E = \frac{v_i - 0.7}{R_E}} \qquad\qquad (5\text{--}59)$$

The preceding equation gives us an exciting result! It says that the collector current is a function only of the input voltage and the emitter resistance, two quantities that are easily controlled (assuming, of course, that we disregard slight variations in the 0.7-V value). The very elusive parameter β_{dc} does not even appear in the equation! In effect, we have circumvented the problem with the wide β_{dc} variation through the use of a stabilizing emitter resistance.

This circuit forms the basis of bias stabilizing networks and is considered from that point of view in Sections 5–7 and 5–8. The focus at this point is on the implementation of a current source, and the discussion here addresses that application.

So far, most of the analysis has centered on the base–emitter part of the circuit. In effect, the collector current is determined from conditions established in the base–emitter circuit and the emitter resistance. However, the collector–emitter part of the circuit must be studied to determine the range of conditions that ensure active-region operation.

Let v_L represent the voltage across the load. If this were a simple resistive load, v_L would be related to i_C by Ohm's law. In general, however, the load could represent a nonlinear device (e.g., a diode). A KVL equation for the collector–emitter loop reads

$$-V_{CC} + v_L + v_{CE} + R_E i_E = 0 \qquad\qquad (5\text{--}60)$$

Solution for v_{CE} yields

$$v_{CE} = V_{CC} - v_L - R_E i_E \qquad\qquad (5\text{--}61\text{a})$$
$$= V_{CC} - v_L - R_E i_C \qquad\qquad (5\text{--}61\text{b})$$

where $i_C \simeq i_E$ has been substituted in the last form. Stated in words, the preceding equation indicates that the voltage across the transistor is the supply voltage minus the voltage drops of the load and the emitter resistance.

For active-region operation, $v_{CE} > 0$, and the circuit will function as a current source for that condition. In practice, since $V_{CE(sat)}$ is a few tenths of a volt, some additional allowance should be made for that condition.

Returning to the expression for collector current of (5–59), the design of a current source will depend on the objective and given conditions. If one particular fixed value of load current is desired, v_i and/or R_E are selected to produce the required i_C. In some applications it is desired to provide an adjustable current source. This could easily be accommodated by providing an adjustment for v_i. This situation results in a **voltage-controlled current source,** which will be studied in more general terms later. Examples 5–6 and 5–7 will illustrate some of the typical situations encountered in this type of circuit.

Thus far we have assumed that v_i is an ideal voltage source, which simplified the base–emitter circuit analysis. In practice, however, the source will usually have some output resistance in its Thévenin model. A circuit diagram containing an effective source resistance R_S is shown in Figure 5–17. A loop equation is

$$-v_i + R_S i_B + V_{BE} + R_E i_E = 0 \qquad (5\text{–}62)$$

Solving for i_E, letting $V_{BE} = 0.7$ V, and assuming that $i_C \simeq i_E$, there results

$$i_C = \frac{v_i - 0.7 - R_S i_B}{R_E} \qquad (5\text{–}63)$$

When this result is compared with (5–59), it is noted that the numerator of (5–63) contains the additional term $R_S i_B$. Since $i_C = \beta_{dc} i_B$, $i_B = i_C/\beta_{dc}$. If this relationship were substituted in (5–63), a more exact equation for i_C could be determined, with β_{dc} appearing as a parameter.

FIGURE 5–17

Single-transistor constant current source with effect of source resistance considered.

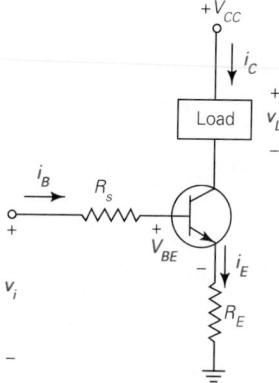

In practice, the more exact procedure suggested by the preceding paragraph may not be warranted. If β_{dc} is reasonably large and if R_S is not too large, the product $R_S i_B$ is usually quite small and may be neglected. The term "reasonably large" and "not too large" probably puzzle the reader, so the best way to resolve the dilemma is to make a *worst-case* calculation of $R_S i_B$ based on the worst-case value of base current $i_{B(\max)}$. The quantity $i_{B(\max)}$ is that base current that would be required to produce i_C if the value of β_{dc} were at the minimum value (i.e., $\beta_{dc(min)}$). This base current is

$$i_{B(max)} = \frac{i_C}{\beta_{dc(min)}} \tag{5-64}$$

If the product $R_S i_{B(max)}$ is small compared with 0.7 V, the term $R_S i_B$ in (5-63) may be ignored, and the simplified equation of (5-59) may be assumed.

EXAMPLE 5-6

The circuit of Figure 5-18 is designed to produce a nearly constant current through a variable resistive load R_L. An ideal 5-V source is used to establish the current. **(a)** Determine the value of the current I_C. **(b)** Determine the range of R_L over which the circuit will function properly. It is assumed that β_{dc} is sufficiently large to justify the approximations used in the text.

FIGURE 5-18
Circuit of Example 5-6.

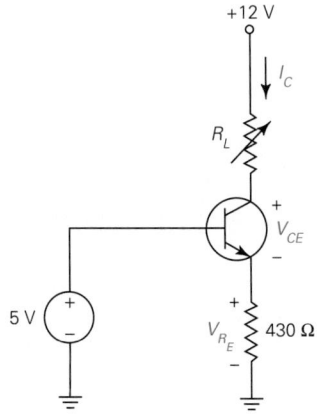

Solution

(a) Uppercase symbols are used for I_C and the various voltages because the input is no longer considered as an instantaneous voltage, but is a fixed 5-V source. The load current I_C is determined from (5-59) as

$$I_C = \frac{v_i - 0.7}{R_E} = \frac{5 - 0.7}{430} = 10 \text{ mA} \tag{5-65}$$

The voltage across the emitter resistance V_{R_E} is

$$V_{R_E} = 4.3 \text{ V} \tag{5-66}$$

This voltage remains constant as long as active-region operation occurs.

(b) When $R_L = 0$, the value of V_{CE} is simply

$$V_{CE} = 12 - 4.3 = 7.7 \text{ V} \tag{5-67}$$

which is certainly well within the active region. As R_L increases, however, its voltage increases, and V_{CE} decreases. Thus, there will be some maximum value of R_L at which active-region operation ceases.

In general,

$$V_{CE} = 12 - 4.3 - R_L I_C = 7.7 - R_L I_C \qquad (5\text{--}68)$$

The value of $R_{L(max)}$ is determined by setting $V_{CE} = 0$. Thus

$$0 = 7.7 - R_{L(max)} I_C$$

or

$$R_{L(max)} = \frac{7.7}{I_C} = \frac{7.7}{0.01} = 770 \ \Omega \qquad (5\text{--}69)$$

Thus the circuit functions as a current source provided that R_L is within the range $0 < R_L < 770 \ \Omega$. When $R_L \geq 770 \ \Omega$, the BJT is saturated.

EXAMPLE 5–7

The circuit of Figure 5–19 is to be used in a production design as a current source driver for the LED shown. The advantage of using a current source is that the light of an LED is a function of the current, and if the current is constant, the light intensity will be more uniform, even though the LED voltage will vary with different diodes. **(a)** For a desired LED current of 20 mA, determine the value of R_E required. **(b)** If the LED voltage drop can vary from 1.5 to 2.5 V, determine the expected range of V_{CE} in order to check active-region operation for all conditions.

FIGURE 5–19
LED driver circuit of Example 5–7.

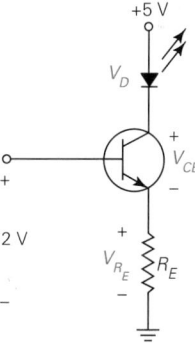

Solution

(a) The value of R_E is determined from (5–59) as

$$R_E = \frac{2 - 0.7}{0.02} = 65 \ \Omega \qquad (5\text{--}70)$$

(b) The collector–emitter voltage V_{CE} is

$$V_{CE} = 5 - V_{R_E} - V_D = 5 - 1.3 - V_D = 3.7 - V_D \qquad (5\text{--}71)$$

The maximum value of V_{CE}, denoted as $V_{CE(max)}$, occurs for the minimum value of V_D.

$$V_{CE(max)} = 3.7 - 1.5 = 2.2 \text{ V} \qquad (5\text{--}72)$$

The minimum value of V_{CE}, denoted as $V_{CE(min)}$, occurs for the maximum value of V_D.

$$V_{CE(min)} = 3.7 - 2.5 = 1.2 \text{ V} \qquad (5\text{--}73)$$

Thus, the range of V_{CE} is from 1.2 to 2.2 V, and active-region operation should be ensured under all conditions.

5–6 EMITTER FOLLOWER

A BJT circuit that has application both in discrete form and in integrated circuits is the **emitter follower.** The simplest dc coupled form of such a circuit using an *npn* transistor is shown in Figure 5–20(a). The collector terminal is connected directly to the positive power supply terminal, and the load is connected between the emitter and ground. In this simple form of the emitter follower, *it is assumed that both the input v_i and the load have dc paths to ground.*

The circuit is almost identical in form to the single transistor current source of Section 5–5. What distinguishes the emitter follower, however, is the fact that the desired output variable is the voltage v_o across the load as a function of the input voltage variable v_i.

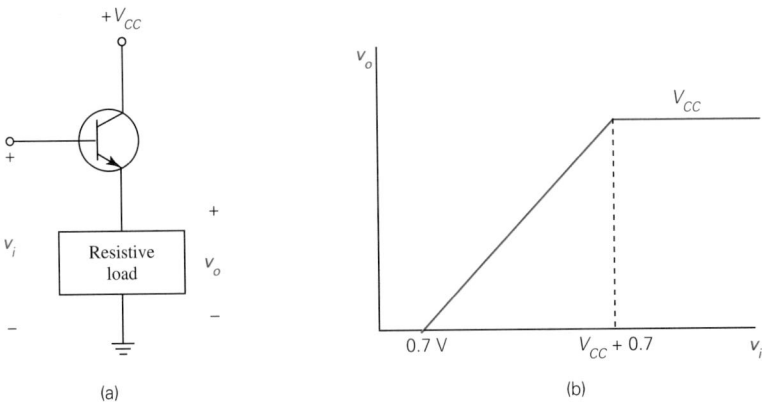

FIGURE 5–20
Basic *npn* dc-coupled emitter-follower circuit and ideal transfer characteristic.

For $v_i \geq 0.7$ V, the base–emitter junction is forward biased and current flows through the transistor to the load. Assuming a base–emitter voltage drop of 0.7 V, a loop equation for the base circuit yields

$$-v_i + 0.7 + v_o = 0 \qquad (5\text{--}74)$$

or

$$\boxed{v_o = v_i - 0.7} \qquad (5\text{--}75)$$

From (5–75) we see that as the input voltage increases, the output voltage also increases but differs from the input by 0.7 V. In effect, the output voltage at the emitter *follows* the input as suggested by the name of the circuit.

The transfer characteristic of the emitter follower is shown in Figure 5–20(b). The output is zero for $v_i < 0.7$ V, since the base–emitter junction will be reverse biased. For $v_i > 0.7$ V, the straight-line characteristic of (5–75) applies as long as active-region operation occurs. However, when v_i becomes sufficiently positive that the voltage between collector and emitter approaches zero, saturation occurs, and the output can increase no further. Ideally, this would occur when $v_i = V_{CC} + 0.7$ and $v_o = V_{CC}$, as shown. In practice, saturation occurs a few tenths of a volt below this level.

Since the output voltage is close to, but always smaller than, the input voltage, a beginner might logically question the value of this circuit. The value lies in the **isolation** that the circuit provides. A typical application might be as follows: A voltage close to a level v_i is required to furnish a moderate or large amount of current to a load. However, the source itself is incapable of supplying the amount of current required. A transistor with the necessary current capability is selected, and the emitter follower is used to separate the source from the load. All the source need do is to provide the much smaller base current drive, while the transistor supplies the required larger load current. The circuit thus serves as a **buffer.**

The actual base current drive is a function of the required load current and the value of β_{dc} for the transistor. Let i_o represent the required load current, which is, of course, the emitter current (i.e., $i_o = i_E$). Since $i_E = i_B + i_C$ and $i_C = \beta_{dc} i_B$, the following relationship is established between load current and base current:

$$i_o = i_B + i_C = i_B + \beta_{dc} i_B = (1 + \beta_{dc}) i_B \qquad (5\text{--}76)$$

Solving for i_B, we obtain

$$i_B = \frac{i_o}{1 + \beta_{dc}} \qquad (5\text{--}77)$$

The last result indicates how much smaller i_B is than the required load current, which provides the isolation inherent in the circuit.

Understand that the output voltage differs from the input voltage by the 0.7-V drop, so any application of this circuit in this simple form must be one in which this difference poses no problem. We will see how additional feedback can be used with integrated circuits to overcome this difference.

A *pnp* transistor can be used to implement an emitter follower for a negative input voltage, and this circuit is shown in Figure 5–21(a). Since the same standard positive input and output references are used here as for the previous circuit, the active region of operation is now for $v_i \leq -0.7$ V. In this region the output is given by

$$v_o = v_i + 0.7 \qquad (5\text{-}78)$$

Note that since $v_i < 0$, the magnitude of v_o will be less than the magnitude of v_i. The transfer characteristic for this circuit is shown in Figure 5–21(b). Saturation in this circuit occurs at an output level of $-V_{CC}$ in the ideal case.

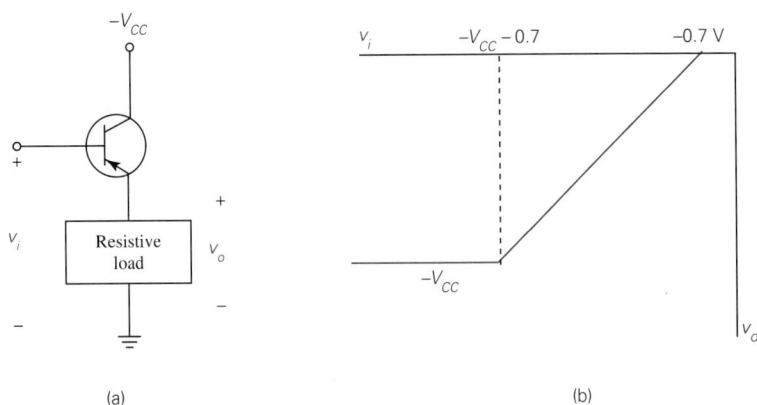

(a) (b)

FIGURE 5–21
Basic *pnp* emitter-follower circuit and ideal transfer characteristic.

EXAMPLE 5–8 The emitter-follower circuit of Figure 5–22 is used to isolate a source v_i from a 100-Ω load. The small difference between v_i and v_o is not critical for the particular application. **(a)** Assuming that $V_{BE} = 0.7$ V, determine v_o in terms of v_i. **(b)** Assuming that $V_{CE(\text{sat})} = 0$, determine the range of input voltage for which the circuit functions. **(c)** For $v_i = 12$ V and $\beta_{dc} = 50$, determine the value of base current i_B.

Solution

(a) From a base KVL loop equation or from (5–75), the output voltage is simply

$$v_o = v_i - 0.7 \qquad (5\text{-}79)$$

FIGURE 5-22
Emitter-follower circuit of Example 5-8.

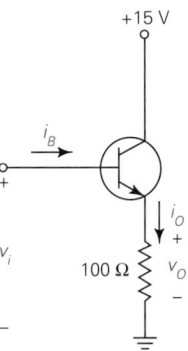

(b) The smallest value of the input at which the circuit starts to function is $v_i = 0.7$ V. The largest value corresponds to saturation. At saturation, $v_o = 15$ V and $v_i = 15.7$ V. Thus, the ideal range is $0.7\ \text{V} < v_i < 15.7\ \text{V}$. In practice, the upper value would be reduced slightly because of the realistic collector–emitter saturation voltage.

(c) For $v_i = 12$ V, the output voltage is

$$v_o = 12 - 0.7 = 11.3\ \text{V} \tag{5-80}$$

The output load current is

$$i_o = \frac{11.3\ \text{V}}{100\ \Omega} = 113\ \text{mA} \tag{5-81}$$

The value of i_B is

$$i_B = \frac{i_o}{1 + \beta_{dc}} = \frac{113\ \text{mA}}{51} = 2.216\ \text{mA} \tag{5-82}$$

The advantage of this circuit is clear from these results. While the required load current is 113 mA, the current supplied by the input source need only be 2.216 mA.

5-7 STABLE BIAS CIRCUITS WITH SINGLE POWER SUPPLY

Many applications of bipolar junction transistors require that a stable set of conditions be established at a specific point in the active region. A stable operating point is one in which the collector current and collector–emitter voltage should remain nearly constant and predictable, even though transistor characteristics of the given type vary widely from one unit to another.

If a given individual transistor were being used for a specific circuit, the operating point could be adjusted experimentally around the actual β_{dc}, and this might suffice for a limited special application. This approach is unsatisfactory for production, however, since transistor characteristics vary widely within a given type. As we have seen, the variation of β_{dc} for a given transistor type may swing the operating point all the way from well in the active region to saturation with a simple bias circuit.

In this section we will analyze and discuss a very popular stable bias circuit using a single power supply. The circuit is the most widely used for discrete BJT active-region applications. This circuit is a prelude for dealing with basic BJT amplifier configurations to be discussed in later chapters.

Our approach in this section is to analyze a given circuit to determine properties of the operating point established. The design of such circuits is considered in the next section, so all examples considered in this section are analyzed as they are given.

The key to the single power supply stable bias circuit is the concept of the constant current source as developed in Section 5–5. In this case, however, both the collector–emitter circuit and base–emitter circuits will be biased from a single power supply.

The basic stable bias circuit as applied to an *npn* transistor using a positive bias power supply (with respect to circuit ground) is shown in Figure 5–23(a). In general, the circuit includes the transistor, an emitter resistance R_E, a collector resistance R_C, and a voltage-divider network consisting of R_1 and R_2. In all circuit configurations that follow, R_1 will represent the resistance adjacent to the collector resistance, and R_2 will represent the resistance adjacent to the emitter resistance. This arrangement should be noted, because in some of the upside-down or *pnp* configurations, R_2 may be above R_1 on the schematic.

An intuitive approach may be formulated as follows: The voltage divider on the base side results in a positive voltage well below V_{CC} appearing at the base. This voltage in consort with R_E establishes an emitter current in the same manner as discussed for the current source concept of Section 5–5. Since emitter and collector currents are nearly the same, the collector current turns out to be dependent primarily on the dc voltage across R_2 and the emitter resistor R_E.

To illustrate the analysis, the circuit is first redrawn as shown in Figure 5–23(b). Although there is only one power supply, the circuit may be studied *as if* there were two supplies of equal value as shown, and this allows the base–emitter and collector–emitter circuits to be separated.

Next, a Thévenin equivalent circuit looking back from the base (point B) will be determined. The value of the Thévenin voltage V_{TH} is determined from the voltage-divider rule as

$$V_{TH} = \frac{R_2}{R_1 + R_2} V_{CC} \qquad\qquad \textbf{(5–83)}$$

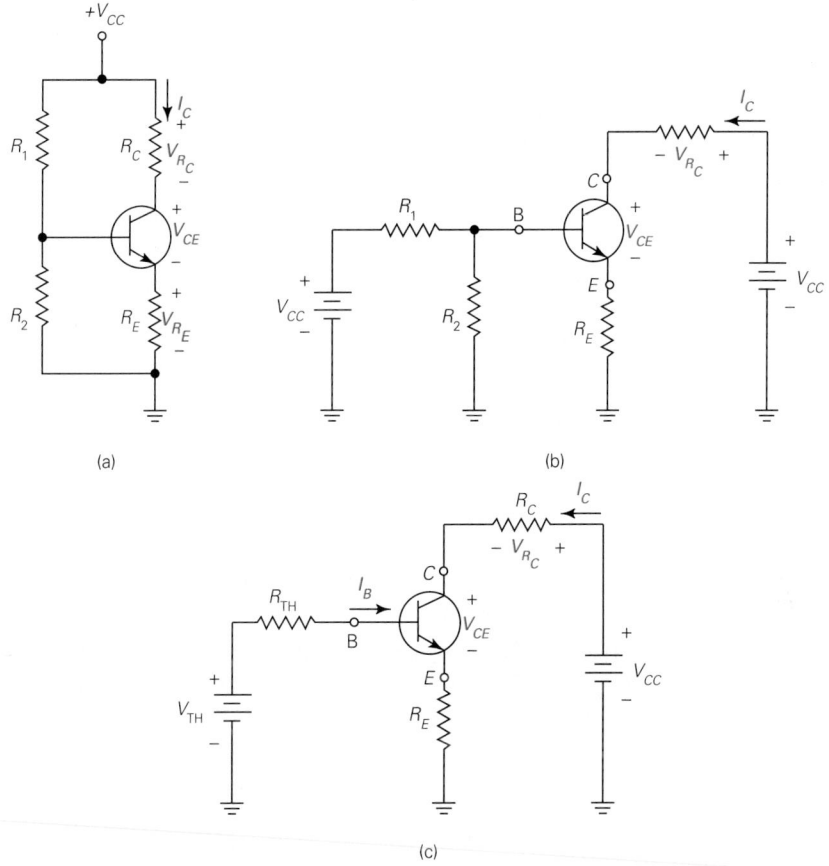

FIGURE 5–23
Single power supply stable bias for an *npn* transistor and equivalent circuits.

The Thévenin equivalent resistance is

$$R_{TH} = R_1 \| R_2 = \frac{R_1 R_2}{R_1 + R_2} \qquad (5\text{–}84)$$

Using this Thévenin equivalent circuit, a simplified circuit model is shown in Figure 5–23(c). The base current I_B may be considered as flowing from a source V_{TH} in series with R_{TH}.

From the base equivalent circuit, a clockwise KVL loop equation is written as follows:

$$-V_{TH} + R_{TH} I_B + V_{BE} + R_E I_E = 0 \qquad (5\text{–}85)$$

The current I_B is given by

$$I_B = \frac{I_C}{\beta_{dc}} \tag{5-86}$$

The largest value of this term corresponds to the minimum value of β_{dc}. In the practical design of the bias circuit (to be discussed in Section 5–8), the resistances R_1 and R_2 are chosen such that the largest value of this term is very small compared with the other terms in (5–85). It is convenient to relate this term specifically to the term $R_E I_E$, so the following inequality is required:

$$\frac{R_{TH} I_C}{\beta_{dc}} \ll R_E I_E \qquad \text{for a stable design} \tag{5-87}$$

Since $I_C \simeq I_E$, this inequality may be restated as

$$\frac{R_{TH}}{\beta_{dc}} \ll R_E \qquad \text{for a stable design} \tag{5-88}$$

or

$$R_{TH} \ll \beta_{dc} R_E \qquad \text{for a stable design} \tag{5-89}$$

Based on the assumed inequality, the second term in (5–85) may be neglected. The value of I_E (and I_C) may then be closely approximated as

$$I_C \approx I_E \approx \frac{V_{TH} - V_{BE}}{R_E} = \frac{V_{TH} - 0.7}{R_E} \tag{5-90}$$

The result is in agreement with the development of the current source in Section 5–5. In this case the voltage establishing the current is the open-circuit voltage V_{TH}. Once again, we see that the current is virtually independent of the transistor β_{dc}, provided, of course, that β_{dc} is large and that the inequality we discussed earlier is met.

On the collector side, we may now determine the voltages across the individual portions of the circuit. The voltage across R_C, denoted as V_{R_C}, is determined as

$$V_{R_C} = R_C I_C \tag{5-91}$$

The voltage across R_E, denoted as V_{R_E}, is

$$V_{R_E} = R_E I_E \approx R_E I_C \tag{5-92}$$

The voltage V_{CE} from collector to emitter is

$$V_{CE} = V_{CC} - V_{R_E} - V_{R_C} \qquad \text{(5–93)}$$

In effect, the dc supply voltage V_{CC} is divided into three parts: V_{R_E}, V_{R_C}, and V_{CE}.

The single supply bias circuit for a *pnp* circuit using a negative power supply with respect to ground is shown in Figure 5–24(a). This circuit layout is, as shown, in the same form as for the *npn* transistor. The corresponding "upside-down" layout with ground at the top for the same circuit is shown in Figure 5–24(b). Note that when the circuit is shown in this form, the resistors R_2 and R_E are at the top, and R_1 and R_C are at the bottom.

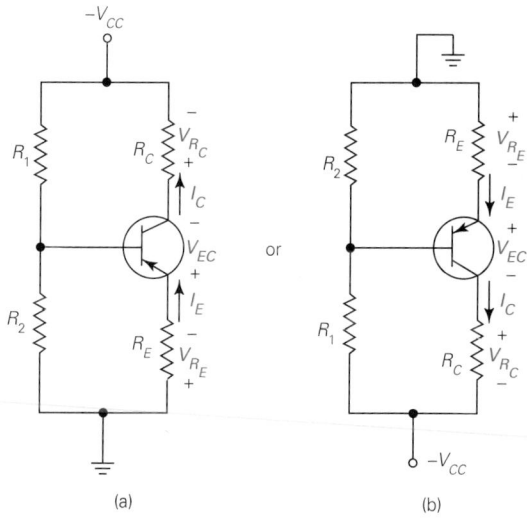

(a)   or   (b)

FIGURE 5–24
Single negative power supply stable bias for a *pnp* transistor shown two ways.

The expressions for V_{TH}, R_{TH}, and I_E or I_C are exactly the same as for the *npn* transistor. However, I_E and I_C are defined in the direction from emitter to collector, so that they are positive quantities. Similarly, the voltage across the transistor is defined as V_{EC}, which will be a positive quantity. The two resistive voltage drops retain the same notation as before, but the reference directions are reversed. The expressions for V_{R_E} and V_{R_C} will be the same as for the *npn* transistors (i.e., [5–91] and [5–92]). The expression for V_{EC} will be

$$V_{EC} = V_{CC} - V_{R_C} - V_{R_E} \qquad \text{(5–94)}$$

In some applications, there may be available only a positive dc power supply voltage with respect to ground. This is ideal if only *npn* transistors are available. However, suppose that one desires to use a *pnp* transistor in such a case. Although the basic approach would be to bias the collector to a negative level with respect to the emitter, *the same effect can be achieved by biasing the emitter to a positive level with respect to the collector.* A stable bias arrangement to achieve this result is shown in Figure 5–25. Basically, the voltage across the complete circuit has the same layout as for the *pnp* circuit of Figure 5–24(b). The two circuit forms differ only with respect to the ground point. Thus, the preceding discussion for the *pnp* circuits of Figure 5–24 applies to this circuit with respect to the currents and voltages across components. The voltages with respect to ground in the two circuits are, of course, different. As in the "upside-down" form of Figure 5–24(b), note that R_2 and R_E are at the top, and R_1 and R_C are at the bottom.

FIGURE 5–25
Method for establishing stable bias for a *pnp* transistor with a single positive power supply.

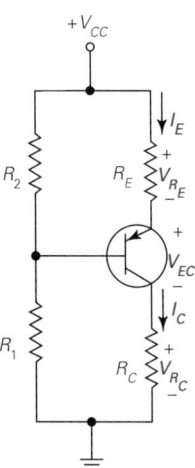

EXAMPLE 5–9

(a) For the circuit of Figure 5–26, determine I_E, I_C, V_{R_C}, V_{R_E}, and V_{CE}. (b) If the minimum value of β_{dc} for the transistor is 50, determine the maximum possible value of I_B. (c) Check the inequality of (5–89) as a means of assessing the relative stability of the bias point.

Solution
(a) The value of V_{TH} is

$$V_{TH} = \frac{4700}{4700 + 1 \times 10^4} \times 15 = 4.80 \text{ V} \qquad (5\text{–}95)$$

Assuming that $V_{BE} = 0.7$ V, we have

$$I_C \simeq I_E = \frac{4.8 - 0.7}{3900} = 1.05 \text{ mA} \qquad (5\text{–}96)$$

FIGURE 5–26
Circuit of Example 5–9.

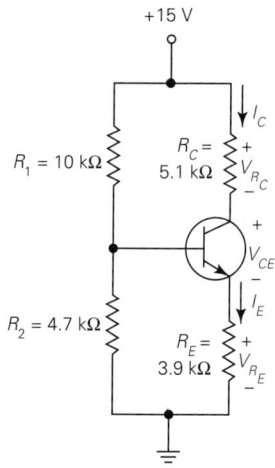

The three voltage drops around the transistor are

$$V_{R_E} = 3.9 \times 10^3 \times 1.05 \times 10^{-3} = 4.1 \text{ V} \tag{5–97}$$

$$V_{R_C} = 5.1 \times 10^3 \times 1.05 \times 10^{-3} = 5.36 \text{ V} \tag{5–98}$$

$$V_{CE} = 15 - 5.36 - 4.1 = 5.54 \text{ V} \tag{5–99}$$

(b) The maximum base current $I_{B(\text{max})}$ corresponds to $\beta_{\text{dc(min)}}$ and is

$$I_{B(\text{max})} = \frac{1.05 \times 10^{-3}}{50} = 21 \text{ }\mu\text{A} \tag{5–100}$$

(c) The value of R_{TH} is

$$R_{\text{TH}} = 4.7 \times 10^3 \| 10^4 = 3197 \text{ } \Omega \tag{5–101}$$

The value of $\beta_{\text{dc}} R_E$ is $50 \times 3.9 \times 10^3 = 195,000$. From the inequality of (5–89), we see that $3197 \ll 195,000$, and the circuit should be stable.

5–8 DESIGN FOR A STABLE BIAS POINT

In the preceding section, we performed the analysis of a stable biasing circuit, and we developed the relationships for predicting the approximate circuit bias conditions. In this section, we give an approach for actually designing such a circuit. The design of a stable bias operating point is a necessary part of an overall amplifier design using BJTs.

First, it must be understood that there is no one single procedure for a design such as this. Various restrictions may result in modifications of any design procedure. Furthermore, there is not necessarily any one final design that is "best"; a number of different designs could all be acceptable. Some individuals develop an intuitive approach with practice that may be superior to a formal approach. Nevertheless, there is a strong motivation for an approach that will guide the inexperienced person to an acceptable result.

The approach here is based on one provided by Hayt and Neudick* and is judged by this author to be one of the easiest methods to apply. It will be given initially as a step-by-step procedure. However, with some practice, you will understand the logic behind the steps, and it then becomes less "cookbook" in form.

Assumed Conditions

Refer to the circuit diagram of Figure 5–27 in the development that follows. Assume that the value of V_{CC} is given and that a desired value of I_C is to be established for a particular transistor type. The exact value of β_{dc} need not be known, since the intent of the circuit is to establish a specified value of I_C for any transistor of the given type. However, the minimum value of β_{dc}, denoted as $\beta_{dc(min)}$, at the given operated point should be estimated from specifications if possible. If this is not practical, even a pessimistic reasonable guess, such as $\beta_{dc(min)} = 25$, may be acceptable in some cases.

FIGURE 5–27
Circuit form of stable bias circuit with variables used in design.

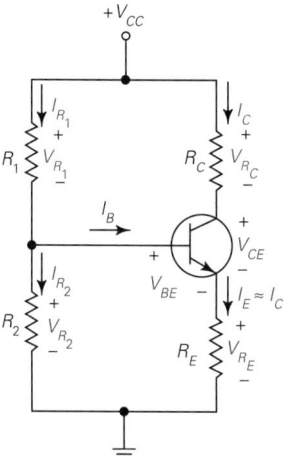

The power supply voltage V_{CC} is divided into three parts: V_{R_C}, V_{CE}, and V_{R_E}, in which

$$V_{CC} = V_{R_C} + V_{CE} + V_{R_E}$$

(5–102)

---

* W. H. Hayt, Jr., and G. W. Neudick, *Electronic Circuit Analysis and Design,* 2nd ed. (Boston: Houghton Mifflin, 1983).

The choice of the three voltages depends on the ultimate application of the circuit. For some amplifier designs, the two voltages V_{R_C} and V_{CE} are chosen to be the same (i.e., $V_{R_C} = V_{CE}$). In some cases, all three voltages are chosen to be equal; that is,

$$V_{R_C} = V_{CE} = V_{R_E} = \frac{V_{CC}}{3} \qquad \text{(5–103)}$$

The latter choice is quite good when the peak expected output signal voltage is well below $V_{CC}/3$ and when there is no additional loading on the output. However, it tends to be somewhat "wasteful" of the available dc voltage, as it is not necessary to use as much of the available voltage as $V_{CC}/3$ for V_{R_E} to achieve a stable design.

One approach is to use a 40–40–20 split, which amounts to

$$V_{R_C} = V_{CE} = 0.4 V_{CC} \qquad \text{(5–104)}$$

and

$$V_{R_E} = 0.2 V_{CC} \qquad \text{(5–105)}$$

Other possibilities may be used, and the reader will be guided on the approach at this point in the text.

As a general rule, the larger the value of V_{R_E}, the more stable the circuit, but the less the dynamic range of the output signal.

Once the assumptions of I_C and the three voltages are made, the design procedure is given as follows:

Design steps

1. Determine R_C as

$$R_C = \frac{V_{R_C}}{I_C} \qquad \text{(5–106)}$$

2. Assuming that $I_E \approx I_C$, determine R_E as

$$R_E = \frac{V_{R_E}}{I_E} \approx \frac{V_{R_E}}{I_C} \qquad \text{(5–107)}$$

3. Calculate

$$V_{R_2} = V_{R_E} + V_{BE} = V_{R_E} + 0.7 \qquad \text{(5–108)}$$

where $V_{BE} = 0.7$ V has been assumed.

4. Calculate V_{R_1} as

$$\boxed{V_{R_1} = V_{CC} - V_{R_2}}$$ (5–109)

5. Compute the maximum value of I_B, indicated as $I_{B(\max)}$, using the minimum value of β_{dc}, indicated as $\beta_{dc(\min)}$.

$$\boxed{I_{B(\max)} = \frac{I_C}{\beta_{dc(\min)}}}$$ (5–110)

6. Select a current I_{R_1} through the resistance R_1 to be much larger than the maximum value of I_B. This can be stated as

$$\boxed{I_{R_1} = K I_{B(\max)}}$$ (5–111)

where K is a constant to be selected. The choice of this constant is arbitrary, but a value between 10 and 100 is usually satisfactory. As a general rule, larger values of K result in a more stable circuit but increase loading on any signal source at the input (to be considered in amplifier configurations) and increase circuit power dissipation.

7. Calculate R_1 as

$$\boxed{R_1 = \frac{V_{R_1}}{I_{R_1}}}$$ (5–112)

8. Calculate I_{R_2} as

$$\boxed{I_{R_2} = I_{R_1} - I_{B(\max)}}$$ (5–113)

9. Calculate R_2 as

$$\boxed{R_2 = \frac{V_{R_2}}{I_{R_2}}}$$ (5–114)

This completes the preliminary design calculations.

In general, the four resistance values obtained from this process will usually not turn out to be standard values. One can then substitute the nearest standard values as a trial approach. In most cases it is best to round R_1 and R_2 in the same direction; that is, use the next *larger standard values* for *both,* or use the next *smaller standard values* for *both.* However, rounding in the same direction may not be the best approach for R_E and R_C, since the changes in both values accentuate the variation of V_{CE} from the design goal.

Once a tentative design is determined, return to the basic equation for I_C as given by (5–90), and determine I_C for the actual component values selected. Next, compute V_{R_C}, V_{R_E}, and V_{CE} based on this I_C. If all values are acceptable, the design may be considered complete as far as the work of this chapter is concerned. (Other considerations for amplifier applications may have to be studied.)

If this design seems unsatisfactory because of either an unacceptable I_C or unacceptable voltages in the collector loop, some additional work is required. Sometimes, the change of one or two of the four resistances may resolve the problem. In other cases it may be necessary to choose a different value of K and start all over.

EXAMPLE 5–10 Design a stable bias circuit for an *npn* transistor based on the following specifications and conditions:

Power supply voltage = 15 V

Desired collector current = 2 mA

Desired $V_{R_C} = V_{CE} = 6$ V

Desired $V_{R_E} = 3$ V

$\beta_{dc(min)} = 50$

Select $K = 30$. In addition to the idealized design, investigate the results of using standard values on I_C and the three voltages.

Solution

The procedure delineated in the text will be followed exactly in these steps:

$$R_C = \frac{V_{R_C}}{I_C} = \frac{6}{2 \times 10^{-3}} = 3\,\text{k}\Omega \tag{5–115}$$

$$R_E = \frac{V_{R_E}}{I_C} = \frac{3}{2 \times 10^{-3}} = 1.5\,\text{k}\Omega \tag{5–116}$$

$$V_{R_2} = V_{R_E} + 0.7 = 3 + 0.7 = 3.7\,\text{V} \tag{5–117}$$

$$V_{R_1} = V_{CC} - V_{R_2} = 15 - 3.7 = 11.3\,\text{V} \tag{5–118}$$

$$I_{B(max)} = \frac{I_C}{\beta_{dc(min)}} = \frac{2 \times 10^{-3}}{50} = 40\,\mu\text{A} \tag{5–119}$$

$$I_{R_1} = K I_{B(max)} = 30 \times 40\,\mu\text{A} = 1.2\,\text{mA} \tag{5–120}$$

$$R_1 = \frac{V_{R_1}}{I_{R_1}} = \frac{11.3}{1.2 \times 10^{-3}} = 9417\,\Omega \tag{5–121}$$

$$I_{R_2} = I_{R_1} - I_{B(max)} = 1.2 \times 10^{-3} - 40 \times 10^{-6} = 1.16\,\text{mA} \tag{5–122}$$

$$R_2 = \frac{V_{R_2}}{I_2} = \frac{3.7}{1.16 \times 10^{-3}} = 3190 \ \Omega \qquad (5\text{–}123)$$

The circuit based on these idealized values is shown in Figure 5–28. (Momentarily ignore the values in parentheses.) Let us now calculate the value of I_C based on these values and the idealized relationship of (5–90). First V_{TH} is calculated as

$$V_{TH} = \frac{3190}{3190 + 9417} \times 15 = 3.796 \ \text{V} \qquad (5\text{–}124)$$

The value of I_C is then

$$I_C = \frac{3.796 - 0.7}{1.5 \times 10^{-3}} = 2.064 \ \text{mA} \qquad (5\text{–}125)$$

FIGURE 5–28
Design of Example 5–10 (rounded standard values are in parentheses).

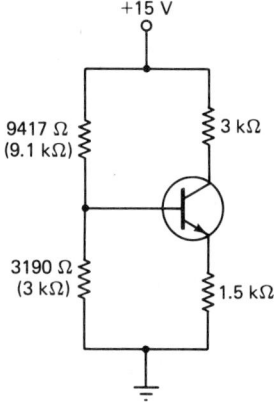

+15 V

9417 Ω
(9.1 kΩ)

3 kΩ

3190 Ω
(3 kΩ)

1.5 kΩ

The reader may be puzzled by the fact that even with the ideal resistance values, I_C differs somewhat (about 3%) from the design goal. This difference is a result of the approximate nature of the design procedure in assuming the worst-case value of I_B. In fact, this procedure compensates somewhat for the idealized nature of the result of (5–90) in the following manner: When (5–90) is employed, the drop resulting from $R_{TH}I_B$ is ignored. Said differently, (5–90) assumes that $\beta_{dc} = \infty$. Any finite value of β_{dc} will result in a smaller value of I_C than (5–90) predicts. Thus the actual value of I_C will be somewhat smaller than the value predicted by (5–125).

Now, let us explore the possibility of using standard values (see Appendix A). The closest standard 5% value to R_1 is 9.1 kΩ, and the closest standard 5% value to R_2 is 3 kΩ. Both of these are less than the ideal values, so we are rounding in the same direction, as suggested. The rounded values for R_1 and R_2 are shown in parentheses on Figure 5–28.

It turns out that the ideal values of R_E (1.5 kΩ) and R_C (3 kΩ) are standard values, so we are in good shape there. Based on using the preceding standard values of R_1 and R_2, the value of I_C will now be calculated. First V_{TH} is calculated as

$$V_{TH} = \frac{3 \times 10^3}{3 \times 10^3 + 9.1 \times 10^3} \times 15 = 3.719 \text{ V} \qquad \textbf{(5–126)}$$

The value of I_C is

$$I_C = \frac{3.719 - 0.7}{1.5 \times 10^3} = 2.013 \text{ mA} \qquad \textbf{(5–127)}$$

This value of I_C turns out to be closer to the design goal with standard rounded values than with the values obtained from the design process. However, this result is somewhat misleading, since, as pointed out earlier, the actual value of I_C will be somewhat smaller than the result of (5–127) suggests.

Next, the voltages are calculated based on the value of I_C of (5–127).

$$V_{R_E} = 2.013 \times 10^{-3} \times 1.5 \times 10^3 = 3.02 \text{ V} \qquad \textbf{(5–128)}$$
$$V_{R_C} = 2.013 \times 10^{-3} \times 3 \times 10^3 = 6.04 \text{ V} \qquad \textbf{(5–129)}$$

$$V_{CE} = 15 - 3.02 - 6.04 = 5.94 \text{ V} \qquad \textbf{(5–130)}$$

Although V_{R_C} and V_{CE} do not turn out to be exactly the same (they rarely do with real-life values), they differ by only about 1.7%. Considering the variations and tolerances involved, this is about as good as can usually be achieved with this type of process.

5–9 LOAD-LINE ANALYSIS

For most transistor dc analysis and design procedures, the various equivalent circuits considered throughout the chapter are the most direct means for solving problems. There is, however, a graphical procedure for determining transistor variables in certain circuits from the voltage–current characteristics. This procedure involves superimposing a linear circuit equation on the voltage–current characteristics and determining the point of intersection of the two functions. The linear circuit equation is called a **load line.**

Circuits that are directly amenable to load-line analysis are usually limited to those that can be reduced to simple series configurations for both collector and base circuits. It is also quite difficult to employ load-line analysis when there is a resistance between emitter and ground.

Before getting involved with the details, it should be stressed that the most powerful outcome of load-line analysis is the **visualization process** of operating conditions that it promotes. It is somewhat limited as an exact analysis and design tool, but one can gain tremendous insight into the effects of various circuit parameters and their variations by visualizing the effects on the load line.

We will develop the process of load-line analysis with the dc circuit of Figure 5–29(a). For this purpose, both base and collector circuits have been constructed as loops with power supplies assumed for each. Two load lines can be constructed corresponding to the respective sides of the circuit, and these will be designated as the **base circuit load line** and the **collector circuit load line.**

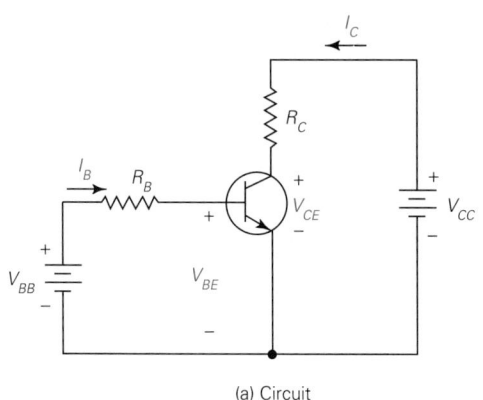

(a) Circuit

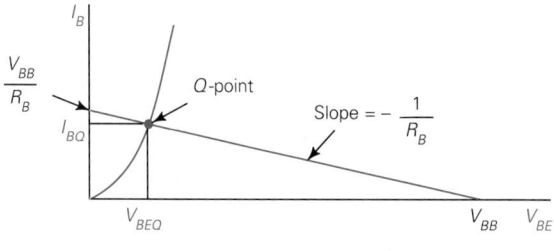

(b) Base circuit load line

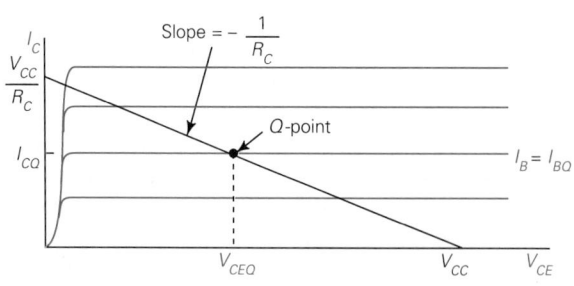

(c) Collector circuit load line

FIGURE 5–29
BJT circuit and input and output load lines.

Base Circuit Load Line

Assume that the BJT base characteristic providing a graph of I_B versus V_{BE} is given as shown in Figure 5–29(b). (Recall that there is some variation of this characteristic with V_{CE}, but we are ignoring that for this purpose.) A KVL equation for the base–emitter loop of Figure 5–29(a) reads

$$-V_{BB} + R_B I_B + V_{BE} = 0 \qquad (5\text{--}131)$$

Solving for I_B, we obtain

$$I_B = \left(-\frac{1}{R_B}\right) V_{BE} + \frac{V_{BB}}{R_B} \qquad (5\text{--}132)$$

This result has been expressed in a different form than earlier in the chapter for a special reason. The variables are assumed to be V_{BE} and I_B, with V_{BE} considered as the independent variable and I_B considered as the dependent variable.

When (5–132) is compared with the standard form of a linear equation $y = mx + b$, the following associations are readily noted:

$$m = \text{slope} = -\frac{1}{R_B} \qquad (5\text{--}133)$$

$$b = \text{vertical intercept} = \frac{V_{BB}}{R_B} \qquad (5\text{--}134)$$

Alternatively, the vertical intercept could be obtained by simply setting $V_{BE} = 0$ in (5–132) and noting that $I_B = V_{BB}/R_B$ for this case.

The horizontal intercept is readily obtained by setting $I_B = 0$ in (5–131) and solving for V_{BE}. The result is

$$\text{horizontal intercept} = V_{BB} \qquad (5\text{--}135)$$

The straight-line equation corresponding to (5–132) is the **base circuit load line,** and it can readily be superimposed on the base–emitter characteristics as shown in Figure 5–29(b). *The intersection of the base–emitter load line with the input characteristic determines the input operating conditions.*

One slight notational problem needs to be addressed before proceeding. Throughout the chapter, we have used notation such as I_B, V_{BE}, I_C, V_{CE}, and so on, to represent dc transistor currents and voltages for a given set of conditions, and this has not created any problems. For this graphical analysis, however, the characteristic curves are labeled with these same quantities for *any possible operating conditions* for the transistor. Yet, for a given set of circuit conditions, there will be only *one possible set of values* of the transistor variables that can exist. We need a separate notational form that distinguishes the *particular* set of values that actually exist from the *possible* range of all values that could exist from the characteristics.

We resolve this potential dilemma by first defining the widely used term **Q-point** as a reference to the particular set of variables that actually satisfy all circuit relationships. (The term *Q-point* is a contraction for **quiescent point,** which is significant in amplifier analysis, as we will see.) For the moment, we will simply refer to the *Q*-point as the actual operating point in which all circuit constraints are satisfied. From a notational point of view, the symbol *Q* will be added to the end of the subscript. Thus the values of I_B, V_{BE}, I_C, and V_{CE} at the *Q*-point will be denoted as I_{BQ}, V_{BEQ}, I_{CQ}, and V_{CEQ}, respectively.

At various points in the text, the *Q* subscript will be omitted when it does not contribute to the clarity of any analysis or when it could be misinterpreted. It will be used primarily where the concept of a designated operating point is to be delineated and where ambiguity in notation would result without it.

Returning now to the base–emitter load line of Figure 5–29(b), we see that the intersection of the load line with the base characteristic results in a base-to-emitter voltage V_{BEQ} and a base current I_{BQ}. While no actual values are shown, we would expect V_{BEQ} to be close to 0.7 V. The load line, however, indicates the actual value of V_{BEQ} that would result.

Collector Circuit Load Line

Assume that the BJT collector characteristics are of the form shown in Figure 5–29(c). A KVL equation for the collector–emitter loop of Figure 5–29(a) reads

$$-V_{CC} + R_C I_C + V_{CE} = 0 \qquad (5\text{--}136)$$

Solving for I_C, we have

$$I_C = \left(-\frac{1}{R_C}\right) V_{CE} + \frac{V_{CC}}{R_C} \qquad (5\text{--}137)$$

in which, again, the straight-line equation form $y = mx + b$ has been used. The independent variable in this case is V_{CE}, and the dependent variable is I_C.

From the basic straight-line equation properties, we have

$$\text{slope} = -\frac{1}{R_C} \qquad (5\text{--}138)$$

$$\text{vertical intercept} = \frac{V_{CC}}{R_C} \qquad (5\text{--}139)$$

By setting $I_C = 0$ in (5–136), the horizontal intercept is determined as

$$\text{horizontal intercept} = V_{CC} \qquad (5\text{--}140)$$

The straight-line equation corresponding to (5–137) is the **collector circuit load line,** and it is superimposed on the collector characteristics of Figure 5–29(c).

In this case there is a whole family of curves with which it could intersect. *The correct solution is the point on the particular curve $I_B = I_{BQ}$ with which the load line intersects.* Thus, the base current has pronounced influence on the collector circuit by determining which characteristic curve represents the applicable circuit condition.

For the characteristics of Figure 5–29(c), the Q-point is observed at the intersection of the curve corresponding to $I_B = I_{BQ}$ and the load line. The values of V_{CE} and I_C at this point are designated as V_{CEQ} and I_{CQ}, as shown.

Although load lines can be drawn for both base–emitter and collector–emitter circuits as we have shown, most references to the term *load line* without a modifier refer to the collector–emitter load line. The reason is that the collector–emitter load line usually provides much more information about a variety of circuit conditions. In the remainder of this section, we explore the use of the collector–emitter load line for this purpose.

In the remainder of this book, the term *load line* without a modifier will be interpreted to mean the collector–emitter load line for a BJT or an analogous output circuit load line for a field-effect transistor (FET) (drain–source load line, as we will see).

To observe the visualization process that the load line provides, we will consider a series of circuit element and parameter variations. No quantitative results will be considered, but representative loads will be shown and discussed qualitatively.

Operating Regions

First, the three BJT operating regions are illustrated by the three load lines of Figure 5–30. For the cutoff region in Figure 5–30(a), the Q-point is all the way to the right at the horizontal intercept. This condition corresponds to $I_B = 0$.

Second, the saturation region is illustrated in Figure 5–30(b). The Q-point in this case corresponds to the intersection of the load line with the saturation region of the characteristic curves.

Third, the active region is illustrated in Figure 5–30(c). In this case, the Q-point is established at some intersection between cutoff and saturation. In many applications, an intersection at or near the midpoint is desired.

Variation of Base Current

Consider the load line of Figure 5–31. The direction of the arrows indicate the manner in which base current variations affect the Q-point if all other circuit conditions are fixed. From any point on the load line, a decrease in base current causes the Q-point to move toward the cutoff point. This results in an increase in collector–emitter voltage and a decrease in collector current, as is readily noted from the projections of the Q-point to the two axes. Conversely, an increase in

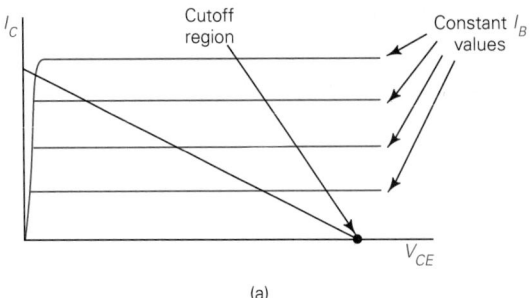

(a)

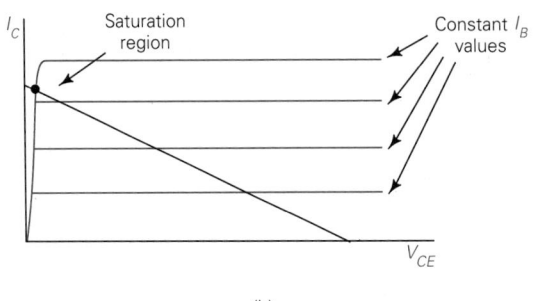

(b)

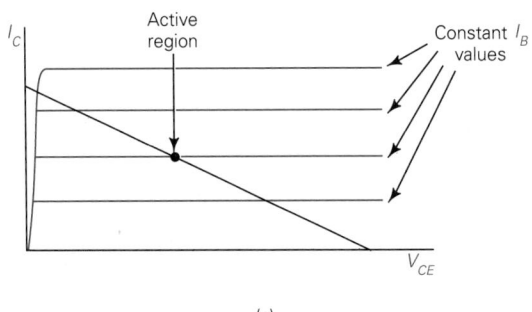

(c)

FIGURE 5–30
Load lines showing the three BJT operating regions.

FIGURE 5–31
Load line showing variation of base current.

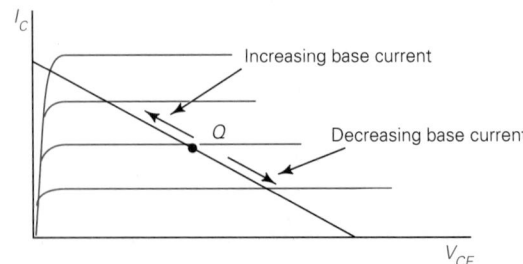

base current causes the Q-point to move toward saturation. This results in a decrease in collector–emitter voltage and an increase in collector current.

We will see in Chapter 10 that the process of movement along the load line is the basis of amplifier actions in a transistor. Such movement would be caused by the signal to be amplified.

Variation of Collector Bias Voltage

When the collector power supply bias voltage is changed and all other circuit parameters are fixed, the result is a completely new load line. Assume first a collector power supply voltage V_{CC1}, and then assume that this voltage is changed to a larger value V_{CC2}. Two possible load lines are shown in Figure 5–32. As long as R_C remains constant, the two load lines have the same slope. However, the respective horizontal intercepts (V_{CC1} and V_{CC2}) and the vertical intercepts (V_{CC1}/R_C and V_{CC2}/R_C) are different.

FIGURE 5–32
Load lines showing variation of collector power supply voltage.

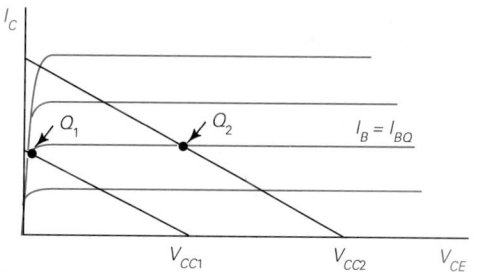

If the base current were fixed, the two Q-points would be determined from the intersections of the curve corresponding to $I_B = I_{BQ}$ with the two load lines. For the hypothetical curves of Figure 5–32, note that the first Q-point (denoted as Q_1) is almost in the saturation region, while the second Q-point (denoted as Q_2) is much closer to the middle of the active region. This illustrates the type of visualization that is very worthwhile with the load line.

Variation of Collector Resistance

When the collector resistance R_C is varied and all other circuit parameters are fixed, the result is a completely new load line. In this case, however, the slope varies.

Assume first a resistance R_{C1}, and then assume that this resistance is changed to a larger value R_{C2}. Two possible load lines are shown in Figure 5–33. As long as V_{CC} remains constant, the horizontal intercept remains constant. However,

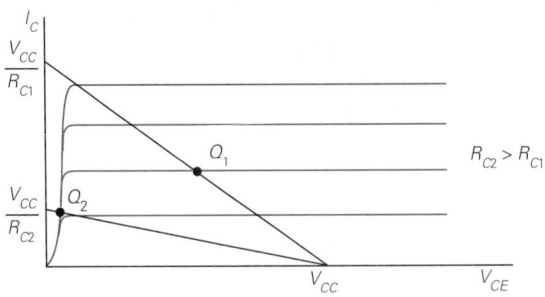

FIGURE 5–33

Load lines showing variation of collector resistance.

both the respective slopes $(-1/R_{C1}$ and $-1/R_{C2})$ and the vertical intercepts $(V_{CC}/R_1$ and $V_{CC}/R_2)$ are different.

　　If the base current were fixed, the two Q-points would be determined from the intersections of the curve corresponding to $I_B = I_{BQ}$ with the load lines. For the hypothetical curves of Figure 5–33, note that the first Q-point (Q_1) is close to the middle of the active region, while the second Q-point is at saturation.

EXAMPLE 5–11

A certain idealized *npn* transistor has the collector characteristics shown in Figure 5–34. The transistor is connected in the circuit of Figure 5–29(a) with the following parameter values: $V_{BB} = 6$ V, $V_{CC} = 20$ V, $R_B = 132.5$ kΩ, $R_C = 2$ kΩ. **(a)** Draw the dc load line and determine V_{CEQ} and I_{CQ}. **(b)** From the load line, determine the values of $I_{C(\text{sat})}$ and $I_{B(\text{sat})}$.

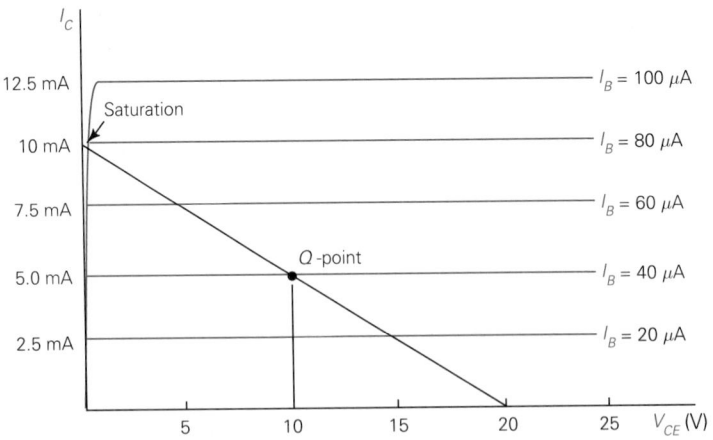

FIGURE 5–34

Collector characteristics for Example 5–11.

Solution

(a) The load line is most easily constructed by first determining the two intercepts. The horizontal intercept is simply $V_{CC} = 20$ V. The vertical intercept is $V_{CC}/R_C = 20/2000 = 0.01$ A $= 10$ mA. A straight line is drawn between these points, as shown in Figure 5–34.

Before determining the Q-point, the value of I_{BQ} must be calculated. This current is

$$I_{BQ} = \frac{V_{BB} - V_{BE}}{R_B} = \frac{6 - 0.7}{132.5 \times 10^3} = 40 \ \mu A \tag{5-141}$$

The Q-point is then located at the intersection of the load line with the curve corresponding to $I_B = 40 \ \mu A$ as shown in Figure 5–34. The values of collector–emitter voltage and collector current at this point are $V_{CEQ} = 10$ V and $I_{CQ} = 5$ mA.

(b) The saturation point is located on the left side of the load line where the characteristic curves intersect the load line. The value of collector current at this point is $I_{C(\text{sat})} = 10$ mA. The value of $I_{B(\text{sat})}$ is the *smallest* value of I_B that results in saturation and this value is $I_{B(\text{sat})} = 80 \ \mu A$. Note that there is a small value of collector–emitter voltage at this point (i.e., $V_{CE(\text{sat})}$), but it is very small for the idealized characteristics.

5–10 TTL INTEGRATED LOGIC CIRCUITS

In this section we will provide an introduction to logic operations performed with BJTs. The purpose is to demonstrate a very important application of BJTs as well as to relate the study of electronic devices in this book to the study of digital circuits in texts emphasizing digital electronics.

One of the most widely employed concepts in digital electronics is the so-called **transistor-transistor logic (TTL)** circuit family. A wide variety of integrated-circuit chips are available in TTL form, and a large percentage of digital application circuits employ TTL devices.

You will learn about designing, analyzing, and troubleshooting TTL circuits from a separate course or book devoted to the field of digital electronics. Our purpose here is to show how the basic circuit functions of such chips are directly related to the material covered in the preceding two chapters. Very few users of integrated-circuit modules understand all the details of the chip operation. However, it is helpful to visualize certain important input–output functions, so that appropriate drive, load, and interfacing conditions can be met.

NAND Circuit

We illustrate the TTL operation with a discussion of a two-input **NAND gate**, whose typical integrated circuit form is shown in Figure 5–35. The NAND designa-

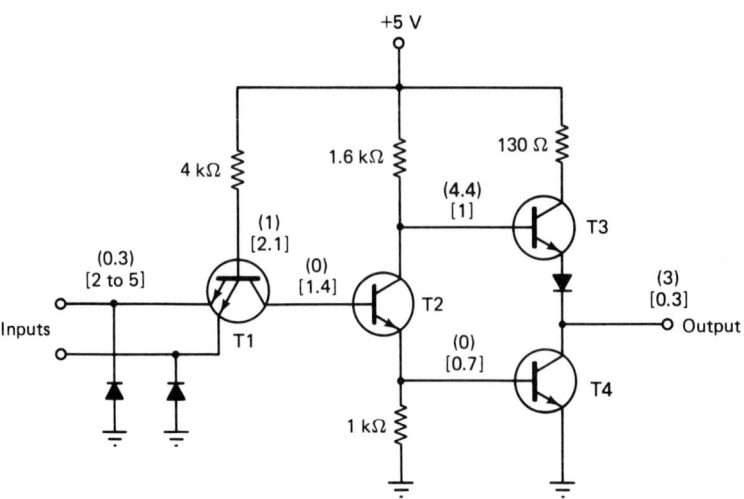

Values in parentheses and brackets are typical voltage levels
(in volts) for different input conditions (see text)

FIGURE 5–35
Form of a transistor-transistor logic (TTL) NAND gate.

tion is a contraction of the terms *NOT* and *AND*. A *NOT* operation is an inversion of the logical state (i.e., a logical 1 is changed to a logical 0, and vice versa).

The two-input circuit of Figure 5–35 contains four transistors and three diodes. The two input diodes on the left are limiting diodes used to prevent excessive voltage on the input transistor, and they do not contribute directly to circuit operation. The input transistor may seem strange in that it has two emitter terminals, but they may be considered essentially as parallel paths.

Above various points on the circuit diagrams are two numbers, one of which is in parentheses ( ), and the other of which is in brackets [ ]. The first number represents a *typical* value of voltage in volts when *either* or *both* input voltages are at a logical 0 level. The second number represents a *typical* value of voltage in volts when *both* input voltages are at a logical 1 level.

Logic 0 at Either Input

First, assume that either or both inputs have a logical 0 value. TTL specifications at the input stipulate that this is any voltage level between zero and 0.8 V. A typical value of 0.3 V is assumed in the analysis. For this condition, the base–emitter junction of T1 will be forward biased, and current will flow from the 5-V source, through the 4-kΩ resistor out through emitter (or emitters) of T1 and through the input (not shown) to ground. The input is said to be *sinking* the current, and for that reason, the TTL is referred to as **current sinking logic.** The base of T1 is about 0.7 V higher than the emitter or about 1 V to ground, as shown.

To forward bias the base–collector junction of T1, a base voltage of about 2.1 V would be required. This particular voltage represents the sum of the T1 0.7-V base–collector junction, the 0.7-V base–emitter drop of T2, and the 0.7-V base–emitter drop of T4. Since the base voltage is only 1 V, the base–collector junction of T1 is reverse biased, and no base drive is applied to either T2 or T4. Consequently, T2 and T4 are in the cutoff state.

Since no collector current flows in T2, the voltage at the bottom of the 1.6-kΩ resistor is high enough to forward bias T3, and this stage acts essentially as an emitter follower. Because of the presence of both base–emitter drop and a diode drop, the typical output shown is about 1.4 V below the base of T3 or about 3 V. This voltage level is within the range of a logical 1.

Logic 1 at Both Inputs

Next, assume that *both* inputs are at a logical 1 level. TTL specifications at the input stipulate that this is any voltage level between 2 and 5 V, as indicated. For this condition the base–emitter junction of T1 is reverse biased. The voltage at the base of T1 is then sufficient to forward-bias the combination of the base–collector junction of T1 and the base–emitter junctions of T2 and T4. The typical value of 2.1 V at the base of T1 is the sum of the base–collector forward drop of T1 plus the two base–emitter drops of T2 and T4 (i.e., 3×0.7 V $= 2.1$ V).

Since T2 is now forward biased, its collector voltage (1 V) is too low to forward-bias the series combination of the T3 base–emitter voltage plus the diode drop. Thus T3 is uncoupled from the output. However, T4 is driven into saturation by the emitter of T2, and the output voltage of T4 is the collector–emitter saturation voltage. A typical value of 0.3 V is shown. TTL specifications at the output stipulate a maximum level of 0.4 V for a logical 0.

A summary of TTL specifications is given in Table 5–1. A truth table depicting logic conditions for the NAND gate is given in Table 5–2.

TABLE 5–1
TTL logic-level specifications

Logical 0 level at input:	0 to 0.8 V
Logical 1 level at input:	2 to 5 V
Logical 0 level at output:	0 to 0.4 V
Logical 1 level at output:	2.4 to 5 V

TABLE 5–2
Truth table for NAND gate

Input 1	Input 2	Output
0	0	1
0	1	1
1	0	1
1	1	0

5–11 DARLINGTON COMPOUND CONNECTION

A circuit that may be used to achieve a very high value of β is the **Darlington compound connection.** This circuit is used both in discrete form and in integrated-circuit packages.

The basic form of the Darlington compound pair using *npn* transistors is shown in Figure 5–36. As far as the three *external* connections shown outside the circle are concerned, the pair acts like a single transistor, so those variables are labeled without subscripts. However, the individual transistor variables are labeled with subscripts 1 and 2 as noted.

FIGURE 5–36
Darlington compound connection with different variables labeled.

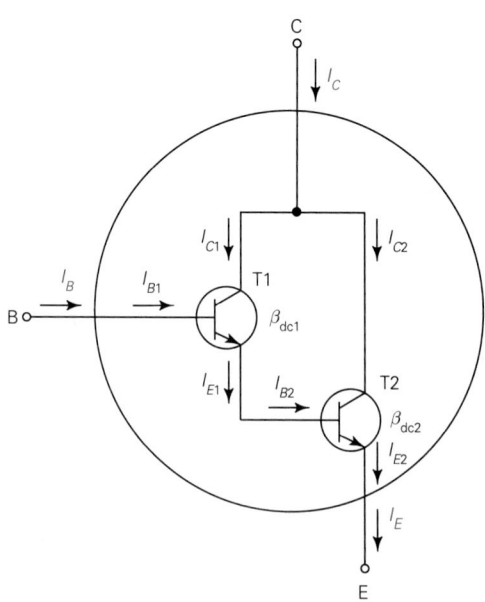

The collector current I_{C1} is related to the base current I_{B1} by

$$I_{C1} = \beta_{dc1}I_{B1} \tag{5-142}$$

The emitter current I_{E1} is

$$I_{E1} = I_{B1} + I_{C1} = I_{B1} + \beta_{dc1}I_{B1} = (1 + \beta_{dc1})I_{B1} \tag{5-143}$$

This current acts as the base current for the second transistor (i.e., $I_{B2} = I_{E1}$).
The collector current I_{C2} is

$$I_{C2} = \beta_{dc2}I_{B2} = \beta_{dc2}I_{E1} \tag{5-144}$$

Substitution of (5-143) in (5-144) yields

$$I_{C2} = \beta_{dc2}(1 + \beta_{dc1})I_{B1} \tag{5-145}$$

The total collector current I_C is

$$I_C = I_{C1} + I_{C2} \tag{5-146}$$

Substitution of (5-142) and (5-145) in (5-146) yields

$$I_C = \beta_{dc1}I_{B1} + \beta_{dc2}(1 + \beta_{dc1})I_{B1} \tag{5-147}$$
$$= (\beta_{dc1} + \beta_{dc2} + \beta_{dc1}\beta_{dc2})I_B \tag{5-148}$$

where the substitution $I_{B1} = I_B$ was made.
The net value of β, denoted simply as β_{dc}, is

$$\beta_{dc} = \frac{I_C}{I_B} = \beta_{dc1} + \beta_{dc2} + \beta_{dc1}\beta_{dc2} \tag{5-149}$$

Under relatively large individual values of β, this result is often estimated simply as

$$\beta_{dc} \approx \beta_{dc1}\beta_{dc2} \tag{5-150}$$

EXAMPLE 5-12

A Darlington compound connection of the form shown in Figure 5-36 has $\beta_{dc1} = 100$ and $\beta_{dc2} = 80$. **(a)** Determine the exact value of β_{dc} for the compound pair. **(b)** Determine an estimate of β_{dc} based on (5-150).

Solution

(a) The exact value of β_{dc} is

$$\beta_{dc} = \beta_{dc1} + \beta_{dc2} + \beta_{dc1}\beta_{dc2} = 100 + 80 + 100 \times 80 = 8180 \tag{5-151}$$

(b) The estimated value of β_{dc} based on (5-150) is

$$\beta_{dc} \approx \beta_{dc1}\beta_{dc2} = 100 \times 80 = 8000 \tag{5-152}$$

This value differs from the exact value by about 2.2%. This estimate is often adequate for quick checks considering the wide variation and uncertainty in parameter values.

Note the very large value of β_{dc} obtained through this process. Darlington compound pairs, with appropriately matched transistors, are available as stock items.

5–12 PSPICE EXAMPLES

PSPICE EXAMPLE 5–1

Develop a PSPICE program to simulate the transistor bias circuit of Example 5–9 (Figure 5–26), using a 2N3904 BJT.

Solution

The circuit adapted to PSPICE is shown in Figure 5–37(a), and the code is shown in Figure 5–37(b). Three options have been employed in the code: NOECHO, NOPAGE, and NOMOD. The first two options have been previously considered, but the last one (NOMOD) suppresses the very detailed printout of the device model parameters, which are beyond the scope of the present discussion.

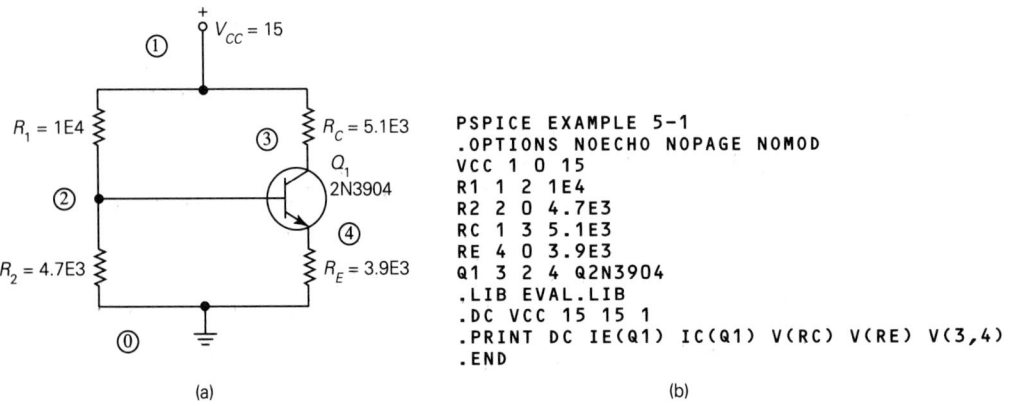

(a)

```
PSPICE EXAMPLE 5-1
.OPTIONS NOECHO NOPAGE NOMOD
VCC 1 0 15
R1 1 2 1E4
R2 2 0 4.7E3
RC 1 3 5.1E3
RE 4 0 3.9E3
Q1 3 2 4 Q2N3904
.LIB EVAL.LIB
.DC VCC 15 15 1
.PRINT DC IE(Q1) IC(Q1) V(RC) V(RE) V(3,4)
.END
```

(b)

FIGURE 5–37
Circuit and code of PSPICE Example 5–1.

The printout obtained from the output file is shown in Figure 5–38. Don't be disturbed by the negative sign associated with I_E. The positive reference directions for all currents in PSPICE transistor models are *into* the terminals. For I_E, this is opposite to our convention, and the value is thus negative.

The values generated by the computer model are compared with those obtained in Example 5–9 in the table on page 257. All are seen to be in very close agreement.

```
PSPICE EXAMPLE 5-1

****        CIRCUIT DESCRIPTION

****************************************************************************************

.OPTIONS NOECHO NOPAGE NOMOD

******      DC TRANSFER CURVES             TEMPERATURE = 27.000 DEG C


VCC         IE(Q1)      IC(Q1)      V(RC)       V(RE)       V(3,4)
1.500E+01   -1.054E-03  1.046E-03   5.333E+00   4.107E+00   5.560E+00



        JOB CONCLUDED

        TOTAL JOB TIME          1.27
```

FIGURE 5–38
Computer printout of PSPICE Example 5–1.

Variable	Calculated value	PSPICE value
I_E	1.05 mA	1.054 mA
I_C	1.05 mA	1.046 mA
V_{R_C}	5.36 V	5.333 V
V_{R_E}	4.1 V	4.107 V
V_{CE}	5.54 V	5.560 V

PSPICE EXAMPLE 5-2

Develop a PSPICE program to simulate the bias circuit designed in Example 5–10, employing a 2N3904 BJT and the rounded standard values.

Solution
The circuit adapted to PSPICE is shown in Figure 5–39(a), and the code is shown in Figure 5–39(b). The format is the same as in PSPICE Example 5–1 and needs no explanation.

The printout obtained from the output file is shown in Figure 5–40. The values generated by the computer are compared with the design target values and those obtained in Example 5–10 in the table that follows.

Variable	Design goal	Calculated value	PSPICE value
I_E	2 mA		2.005 mA
I_C	2 mA	2.013 mA	1.992 mA
V_{R_C}	6 V	6.04 V	5.976 V
V_{R_E}	3 V	3.02 V	3.007 V
V_{CE}	6 V	5.94 V	6.017 V

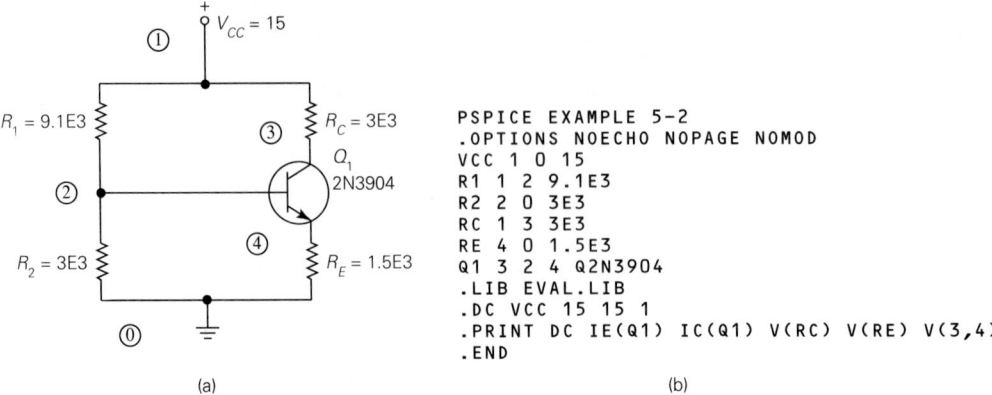

```
PSPICE EXAMPLE 5-2
.OPTIONS NOECHO NOPAGE NOMOD
VCC 1 0 15
R1 1 2 9.1E3
R2 2 0 3E3
RC 1 3 3E3
RE 4 0 1.5E3
Q1 3 2 4 Q2N3904
.LIB EVAL.LIB
.DC VCC 15 15 1
.PRINT DC IE(Q1) IC(Q1) V(RC) V(RE) V(3,4)
.END
```

(a)   (b)

FIGURE 5–39
Circuit and code of PSPICE Example 5–2.

```
PSPICE EXAMPLE 3-5

*****          CIRCUIT DESCRIPTION

******************************************************************************

.OPTIONS NOECHO NOPAGE NOMOD

*****          DC TRANSFER CURVES              TEMPERATURE = 27.000 DEG C

VCC          IE(Q1)      IC(Q1)      V(RC)       V(RE)       V(3,4)
1.500E+01  -2.005E-03  1.992E-03  5.976E+00  3.007E+00  6.017E+00


            JOB CONCLUDED

            TOTAL JOB TIME                      1.27
```

FIGURE 5–40
Computer printout of PSPICE Example 5–2.

PROBLEMS

Drill Problems

Note: Unless indicated otherwise, the following assumptions can be made in all problems:

1. $V_{BE} = 0.7$ V for *npn* transistors and $V_{EB} = 0.7$ V for *pnp* transistors with forward-biased base–emitter junctions.

2. $V_{CE(sat)} = 0$ for *npn* transistors and $V_{EC(sat)} = 0$ for *pnp* transistors.

5–1. Consider the circuit of Figure P5–1. If $\beta_{dc} = 80$, determine $I_{C(sat)}$, $I_{B(sat)}$, and the value of v_i that results in saturation.

FIGURE P5–1

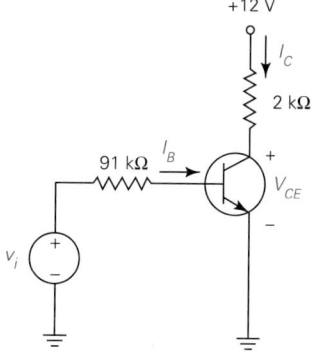

5–2. Consider the circuit of Figure P5–2. If $\beta_{dc} = 125$, determine $I_{C(sat)}$, $I_{B(sat)}$, and the value of v_i that results in saturation.

FIGURE P5–2

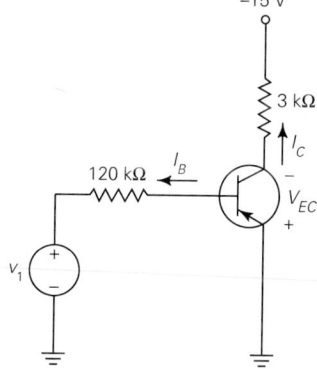

5–3. Consider the circuit of Problem 5–1 (Figure P5–1) again. With the same assumptions as in Problem 5–1, determine I_B, I_C, I_E, and V_{CE} for each of the following values of v_i: **(a)** 0; **(b)** 3 V; **(c)** 6 V; **(d)** 9 V.

5–4. Consider the circuit of Problem 5–2 (Figure P5–2) again. With the same assumptions as in Problem 5–2, determine I_B, I_C, I_E, and V_{EC} for each of the following values of v_i: **(a)** 0; **(b)** −2 V; **(c)** −4 V; **(d)** −6 V.

5–5. For the circuit of Figure P5–5, determine the values of I_C and V_{CE} for each of the following values of β_{dc}: **(a)** 50; **(b)** 100; **(c)** 150.

5–6. For the circuit of Figure P5–6, determine the values of I_C and V_{EC} for each of the following values of β_{dc}: **(a)** 50; **(b)** 100; **(c)** 150.

5–7. For the circuit of Figure P5–7, determine the values of I_C and V_{CE} for each of the following values of R_B: **(a)** 200 kΩ; **(b)** 390 kΩ; **(c)** 560 kΩ.

5–8. For the circuit of Figure P5–8, determine the values of I_C and V_{EC} for each of the following values of R_B: **(a)** 100 kΩ; **(b)** 200 kΩ; **(c)** 270 kΩ.

FIGURE P5–5

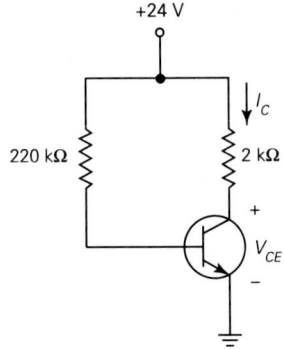

FIGURE P5–6

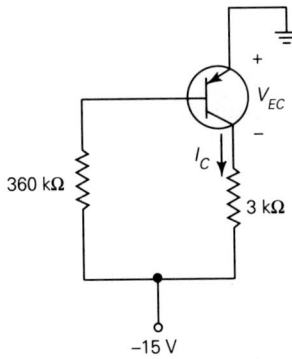

FIGURE P5–7

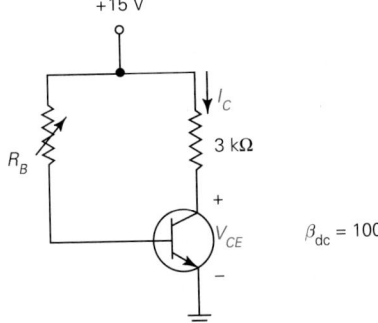

FIGURE P5–8

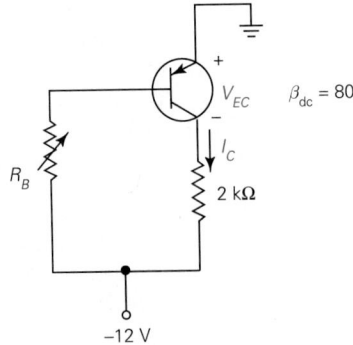

5–9. For the circuit of Figure P5–9, determine the values of I_C and V_{CE} for each of the following values of R_C: **(a)** 1 kΩ; **(b)** 2 kΩ; **(c)** 3 kΩ.

FIGURE P5–9

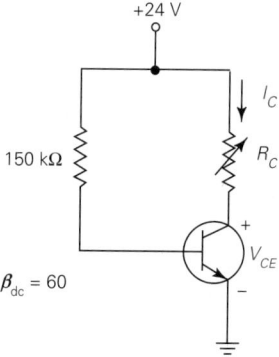

$\beta_{dc} = 60$

5–10. For the circuit of Figure P5–10, determine the values of I_C and V_{EC} for each of the following values of R_C: **(a)** 1 kΩ; **(b)** 2 kΩ; **(c)** 3 kΩ.

FIGURE P5–10

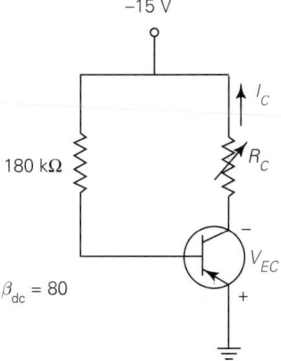

$\beta_{dc} = 80$

5–11. The circuit of Figure P5–11 is used to establish a constant current through a resistive load R_L. Determine **(a)** the current I_C, and **(b)** the range of R_L over which the circuit will function properly.

5–12. The circuit of Figure P5–12 is used to establish a constant current through a resistive load R_L. Determine **(a)** the current I_C, and **(b)** the range of R_L over which the current will function properly.

5–13. The emitter-follower circuit of Figure P5–13 is used to isolate a source v_i from a 200-Ω load. **(a)** Determine an expression for v_o in terms of v_i. **(b)** Determine the range of input voltage for which the circuit functions. **(c)** For $v_i = 20$ V and $\beta_{dc} = 60$, determine the value of base current i_B.

FIGURE P5–11

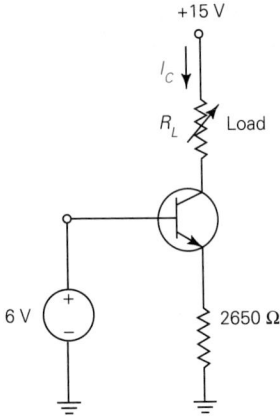

FIGURE P5–12

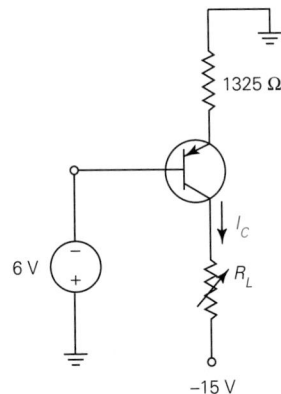

FIGURE P5–13

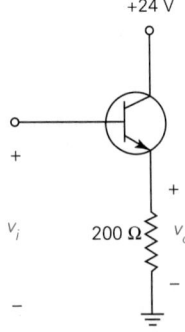

5–14. The emitter-follower circuit of Figure P5–14 is used to isolate a source v_i from a 120-Ω load. **(a)** Determine an expression for v_o in terms of v_i. **(b)** Determine the range of input voltage for which the circuit functions. **(c)** For $v_i = -12$ V and $\beta_{dc} = 40$, determine the value of the base current i_B.

FIGURE P5–14

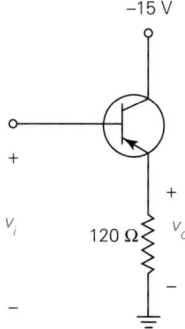

5–15. In the circuit of Figure P5–15, $V_{CC} = 24$ V, $R_1 = 20$ kΩ, $R_2 = 8.2$ kΩ, $R_C = 4.3$ kΩ, and $R_E = 3.3$ kΩ. **(a)** Determine I_C, I_E, V_{R_C}, V_{R_E}, and V_{CE}. **(b)** If the minimum value of β_{dc} for the transistor is 50, determine the maximum possible value of I_B. **(c)** Check the inequality (5–89) as a means of assessing the relative stability of the circuit.

FIGURE P5–15

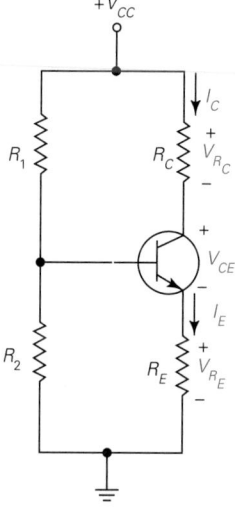

5–16. Repeat the analysis of Problem 5–15 if $V_{CC} = 30$ V, $R_1 = 27$ kΩ, $R_2 = 10$ kΩ, $R_C = 2$ kΩ, and $R_E = 1.5$ kΩ.

5–17. For the circuit of Figure P5–17, determine I_C, I_E, V_{R_C}, V_{R_E}, and V_{EC}.

FIGURE P5-17

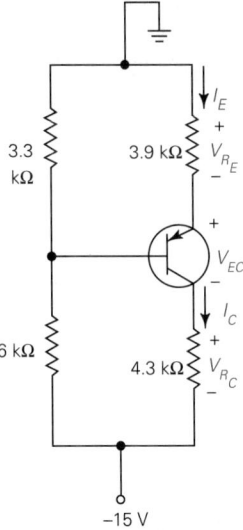

5-18. For the circuit of Figure P5-18, determine I_C, I_E, V_{R_C}, V_{R_E}, and V_{EC}.

FIGURE P5-18

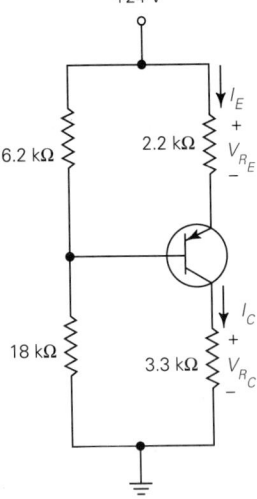

5-19. A certain idealized *npn* transistor has the collector characteristics shown in Figure P5-19. The transistor is connected in the circuit of Figure 5-29(a) with the following parameter values: $V_{BB} = 5$ V, $V_{CC} = 20$ V, $R_B = 430$ kΩ, $R_C = 4$ kΩ. **(a)** Draw the dc load line and determine V_{CEQ} and I_{CQ}. **(b)** From the load line, determine the values of $I_{C(\text{sat})}$ and $I_{B(\text{sat})}$.

5-20. Repeat the analysis of Problem 5-19 if the parameters are changed to the following values: $V_{BB} = 6$ V, $V_{CC} = 16$ V, $R_B = 1.06$ MΩ, $R_C = 8$ kΩ.

FIGURE P5–19

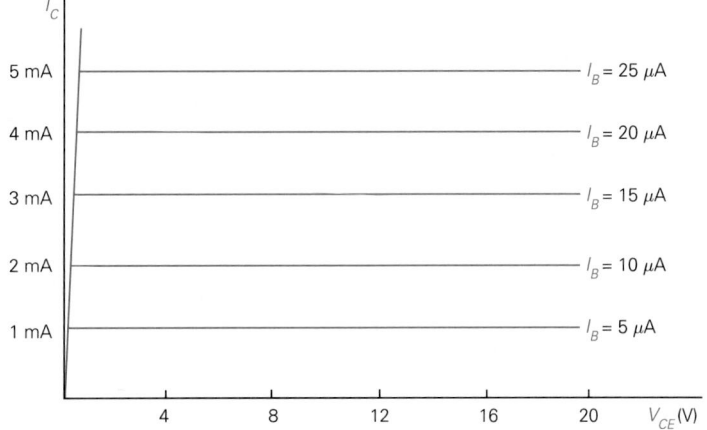

5–21. The circuit of Figure P5–21 is referred to as a **complementary common-emitter switch.** Assume that $V_{CE(sat)} = 0.1$ V for the *npn* transistor and $V_{EC(sat)} = 0.1$ V for the *pnp* transistor. Assume that $\beta_{dc(min)} = 25$. Determine v_o for **(a)** $v_i = 0$ V, and **(b)** $v_i = 5$ V.

FIGURE P5–21

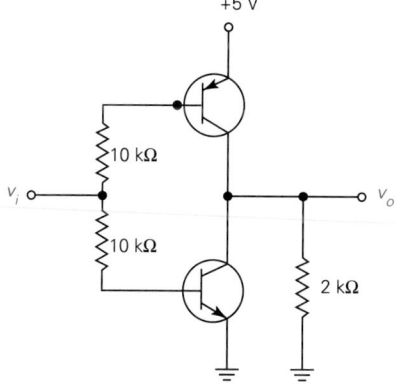

5–22. The circuit of Figure P5–22 can be used as an emitter-follower switch. Determine v_o for **(a)** $v_i = -0.7$ V, and **(b)** $v_i = 5.7$ V.

5–23. The form of a certain RTL gate is shown in Figure P5–23. For your analysis, assume idealized input values of 0 V for logical 0, 5 V for logical 1, and $V_{CE(sat)} = 0$. **(a)** Assume that both inputs are at a logical 0 level. Determine v_o. **(b)** Assume that either input is at a logical 1 level. Determine v_o. **(c)** Construct a truth table for the four possible inputs. Identify the type of circuit.

5–24. The form of a certain **diode-transistor logic (DTL)** gate is shown in Figure P5–24. For your analysis, assume *idealized* input values of 0 V for logical 0, 5 V for logical 1, and $V_{CE(sat)} = 0$. **(a)** Assume that either input is at a logical 0 level. Determine v_o. **(b)** Assume that both inputs are at a logical 1 level. Determine v_o. **(c)** What type of function is performed by this gate? (You may assume that the output transistor saturates when its base–emitter junction is forward biased.)

FIGURE P5–22

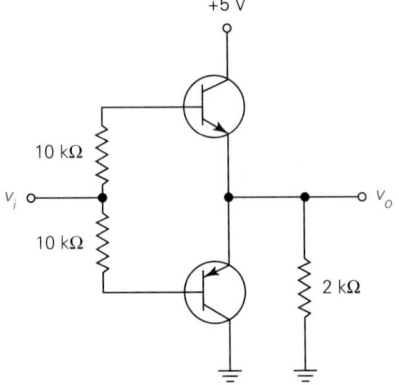

FIGURE P5–23
Form of resistor-transistor logic (RTL)
NOR gate.

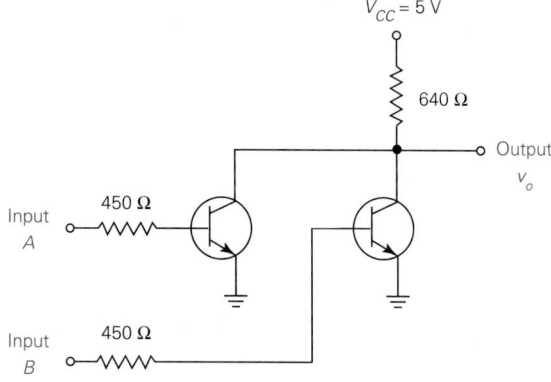

FIGURE P5–24
Form of a diode-transistor logic (DTL)
NAND gate.

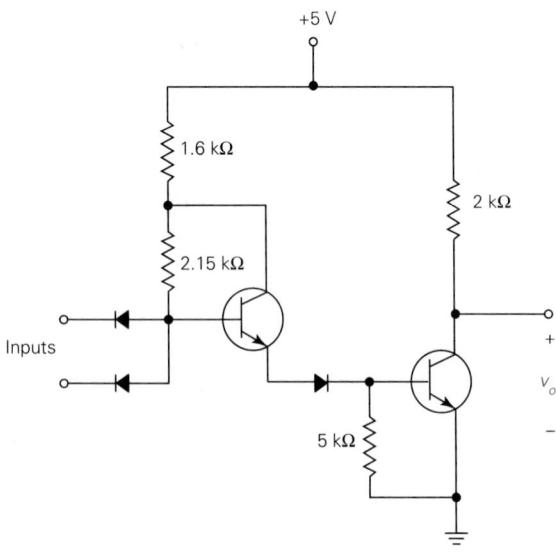

Derivation Problems

5-25. One relatively stable circuit sometimes used to bias a BJT is referred to as **collector-feedback bias** and is shown in Figure P5-25. Show that the collector current can be expressed as

$$I_C = \frac{V_{CC} - 0.7}{R_C + (R_B + R_C)/\beta_{dc}}$$

FIGURE P5-25

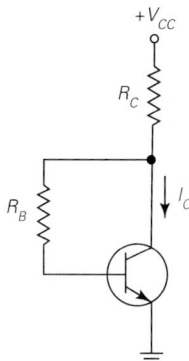

5-26. In this exercise, an *exact* expression for I_C for the stable bias circuit of Figure 5-23(a) will be derived with β_{dc} retained in the equation. **(a)** In the equivalent circuit of Figure 5-23(c), write a loop equation for the base–emitter loop, recognizing that the current through R_E is $I_E = I_B + I_C$. **(b)** By substituting $I_B = I_C/\beta_{dc}$, show that I_C can be expressed as

$$I_C = \frac{\beta_{dc}(V_{TH} - V_{BE})}{R_{TH} + (\beta_{dc} + 1)R_E}$$

Design Problems

5-27. The circuit of Figure P5-27 is to be used to switch the 24-V source across the load. Assume that the control voltage has two levels: (1) $v_i = 0$, and (2) $v_i = 12$ V. The load is resistive and has a value of 400 Ω. Assume that design for production is being performed and that the β_{dc} variation for the transistor type is indicated. Using the criterion of twice the base current for worst-case conditions, complete the design by determining the value of R_B.

5-28. The circuit of Figure P5-28 is to be used to switch the -24-V source across the load. Assume that the control voltage has two levels: (1) $v_i = 0$, and (2) $v_i = -6$ V. The load is resistive and has a value of 600 Ω. Assume that design for production is being performed and that the β_{dc} variation of the transistor type is indicated. Using the criterion of twice the base current for worst-case conditions, complete the design by determining the value of R_B.

FIGURE P5–27

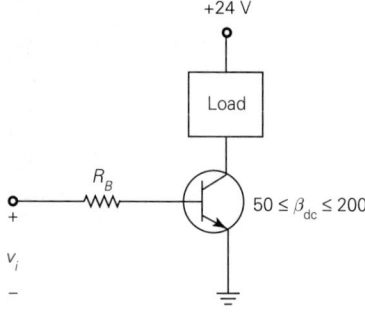

FIGURE P5–28

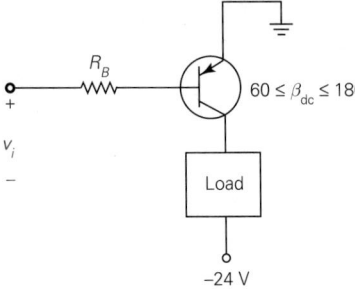

5–29. Design a current source using an *npn* transistor to provide a current of 1 mA through a variable load. The available power supply voltage is +15 V, and this voltage will be used for biasing both the collector and the base circuits.

5–30. Design a current source using a *pnp* transistor to provide a current of 3 mA through a variable load. The available power supply is −15 V, and this voltage will be used for biasing both the collector and the base circuits.

5–31. Design a stable bias circuit for an *npn* transistor based on the following specifications and conditions:

       Power supply voltage = 24 V

       Desired collector current = 1 mA

       Desired $V_{R_C} = V_{CE} = 10$ V

       Desired $V_{R_E} = 4$ V

       $\beta_{dc(min)} = 50$

Select $K = 20$.

5–32. Design a stable bias circuit for an *npn* transistor based on the following specifications and conditions:

       Power supply voltage = 15 V

       Desired collector current = 2 mA

       Desired $V_{R_C} = V_{CE} = V_{R_E} = 5$ V

       $\beta_{dc(min)} = 40$

Select $K = 20$.

5–33. Repeat the design of Problem 5–31 with $K = 80$.

5–34. Repeat the design of Problem 5–32 with $K = 80$.

Troubleshooting Problems

5–35. You are troubleshooting a circuit of the form shown in Figure P5–15, which should be operating in the active region, and you measure the following voltages with respect to ground: $V_{CC} = 15$ V, $V_C = 7.7$ V, $V_E (= V_{R_E}) = 7.7$ V. Which of the following *could be* the source of the problem? **(a)** R_1 open; **(b)** R_2 open; **(c)** R_C open; **(d)** R_E open.

5–36. You are troubleshooting a circuit of the form shown in Figure P5–15, which should be operating in the active region, and you measure the following voltages with respect to ground: $V_{CC} = 15$ V, $V_C = 15$ V, $V_E (= V_{R_E}) = 0$. All but one of the following *could be* the source of the problem. Identify the one condition that could *not* cause the problem. **(a)** R_1 open; **(b)** R_2 shorted; **(c)** R_C open; **(d)** R_E open.

5–37. If the transistor in Figure P5–1 is initially biased in the active region, indicate whether V_{CE} would *increase, decrease,* or *remain nearly the same* for each of the following troubles: **(a)** The 91-kΩ resistor is open; **(b)** the 91-kΩ resistor is shorted; **(c)** the 2-kΩ resistor is open; **(d)** the 2-kΩ resistor is shorted; **(e)** the transistor develops a short between collector and emitter; **(f)** temperature changes cause β_{dc} to increase.

5–38. For the current source of Figure P5–11, indicate whether I_C would *increase, decrease,* or *remain nearly the same* for each of the following changes: **(a)** The 6-V source decreases to 4 V; **(b)** the 2650-Ω resistance changes in resistance to 3.5 kΩ; **(c)** the 15-V power supply increases to 17 V; **(d)** the load resistance R_C increases by 20%, but the transistor remains in the active region; **(e)** the value of β_{dc} increases by 20%.

6

FIELD-EFFECT TRANSISTORS

OBJECTIVES

After completing this chapter, the reader should be able to:

- Draw a simplified layout representing the construction of a field-effect transistor (FET).
- Identify the major classifications of FETs (JFET versus IGFET, depletion mode versus enhancement mode, etc.).
- Identify the two types of FETs according to the polarities of the semiconductor materials.
- Draw the schematic symbols of the different types of FETs, and identify the three terminals.
- State the three operating regions of an FET, and identify conditions for each.
- Draw equivalent-circuit models for each of the three FET operating regions.
- Define the following terms for a depletion-mode FET: gate–source cutoff voltage, pinchoff voltage, zero-bias drain current, and transition voltage.
- Define the gate–source threshold voltage for an enhancement-mode IGFET.
- Sketch the form of the drain characteristics of a typical FET, and label all variables involved.
- Sketch the form of a transfer characteristic, and label the variables involved.
- State the relationships between drain current and gate–source voltage for both depletion-mode FETs and enhancement-mode FETs in the beyond-pinchoff region.
- State the relationships for conductance and resistance of an FET in the ohmic region.
- Discuss the similarities and differences between the operating characteristics of BJTs and FETs.
- Inspect the specifications sheets of typical FETs to determine pertinent data.
- Analyze some of the circuits in the chapter with PSPICE.

6–1 CLASSIFICATIONS

The second major class of transistor is the **field-effect transistor (FET).** Like the BJT, the FET is composed of semiconductor materials such as silicon. Many modern integrated circuits utilize FET technology, and discrete FETs are available as individual components.

The FET is often referred to as a **unipolar** transistor, because conduction occurs as a result of a single type of charge carrier (either holes *or* electrons) in a given unit. This is in contrast to the **bipolar** junction transistor, where conduction occurs as a result of both types of charge carriers in a given unit.

Although there are certain similarities between the BJT and the FET that will become evident as we study FET operation, there are significant differences. In its basic form, the BJT is considered as a **current-controlled device,** while the FET is a **voltage-controlled device.** The base control terminal of a BJT draws a moderate current from the signal source connected to it, so the effective input resistance is relatively low. In contrast, the corresponding control terminal of an FET (the *gate,* as we will see shortly) requires virtually no current, so the effective input resistance is very high. For the insulated gate type of FET, the input resistance may be as high as hundreds of megohms.

The classification scheme for FETs is quite complex and requires some study for mastery of the pertinent terminology. We will provide a brief overview of this structure, although the significance of the various terms will not be clear until the different types of units are studied in the chapter.

A "family tree" of the FET classification structure is shown in Figure 6–1. One major class of FET is the **junction field-effect transistor (JFET).** All JFETs

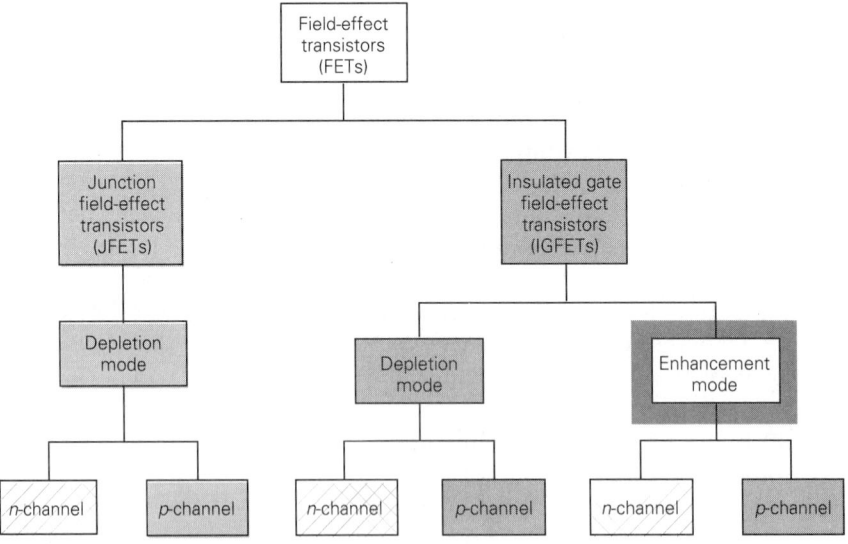

FIGURE 6–1
Family tree showing different classifications of FETs.

operate in what is called the **depletion mode,** in which conduction in the semiconductor is controlled by a **depletion layer.**

A second major class of FET is the **insulated gate field-effect transistor (IGFET).** The IGFET is further classified as either a **depletion-mode** device or an **enhancement-mode** device. A depletion-mode IGFET functions in the same manner as a JFET. An enhancement-mode IGFET, however, requires a process of channel enhancement in order to allow current flow.

Depletion-mode devices (either JFETs or IGFETs) are referred to as *normally on* devices, since they are capable of some conductivity in their basic forms with no control voltage. In contrast, enhancement-mode devices are referred to as *normally off* devices, since they will not conduct in their basic forms without the assistance of an enhancement voltage.

All the different FETs are available as either *n*-channel types or *p*-channel types. This last classification is similar to *npn* and *pnp* BJTs and relates to the polarities of the semiconductor material used in the device construction. To simplify the categorization somewhat, the developments in the next several sections assume only *n*-channel devices. Once the operation and general properties of these *n*-channel units are thoroughly understood, it is a straightforward process to infer the properties of the corresponding *p*-channel units simply by reversing all current directions and voltage polarities.

6–2 JUNCTION FETs

The coverage of FETs will begin with a discussion of the JFET. As indicated in the preceding section, we focus on the *n*-channel JFET.

N-Channel JFET

A simplified construction of an *n*-channel JFET is indicated by the sketch of Figure 6–2(a). A lightly doped bar of *n*-type semiconductor material (usually silicon)

FIGURE 6–2
Basic construction of an *n*-channel JFET and the schematic symbol.

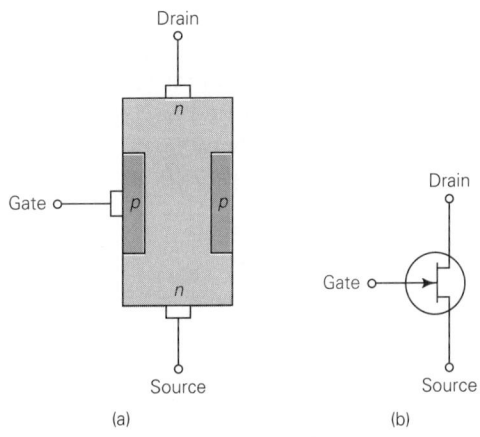

(a)

(b)

forms the **channel,** and contacts are placed at both ends. These ends are designated as the **source** and the **drain,** respectively. Two segments of heavily doped *p*-type semiconductor are embedded in the sides. The two *p*-type segments are internally connected and constitute the **gate.** Current can flow along the channel, and it is heavily influenced by the voltage at the gate through the creation of a **depletion layer.**

The interface between the *p*-type gate and the *n*-type channel constitutes a *pn* junction diode. In the normal operation of the *n*-channel JFET, however, the gate is always biased *negatively* (or possibly at zero) with respect to the source, so this diode remains reverse biased. Consequently, the only current flow in the gate will be the reverse current of the gate–channel diode.

The schematic symbol of the *n*-channel JFET is shown in Figure 6–2(b). The direction of the arrow can be remembered from the concept of the *pn* junction between the *p*-type gate and the *n*-type channel.

6–3 FET OPERATING REGIONS

There are three regions of operation for a JFET. These are referred to as the (1) **cutoff region,** (2) **ohmic region,** and (3) **beyond-pinchoff region.** There are both similarities and differences between these three regions and the three regions for a BJT. A comparison will be made after we have studied in detail the various FET properties.

Test Circuit

The circuit properties of the JFET will be introduced by showing the results of some possible experiments performed on a hypothetical *n*-channel JFET. The characteristics of the hypothetical JFET have been idealized somewhat for illustration, but they are representative of real-life devices.

Consider the test circuit of Figure 6–3. The circuit is connected in the **common-source configuration,** for which most data and operating parameters are determined. Adjustable dc voltage supplies are connected to both gate and drain terminals with respect to the source terminal. These bias supplies will be referred to as the **gate bias supply** and the **drain bias supply,** respectively. Note that the *gate* voltage is biased *negatively* with respect to the source, while the *drain* voltage is based *positively* with respect to the source.

Ohmic Region

The first experiment to be discussed involves varying the drain–source voltage V_{DS} over a very small positive range from zero up to, perhaps, 0.1 V or so. Refer to Figure 6–4 in the discussion that follows.

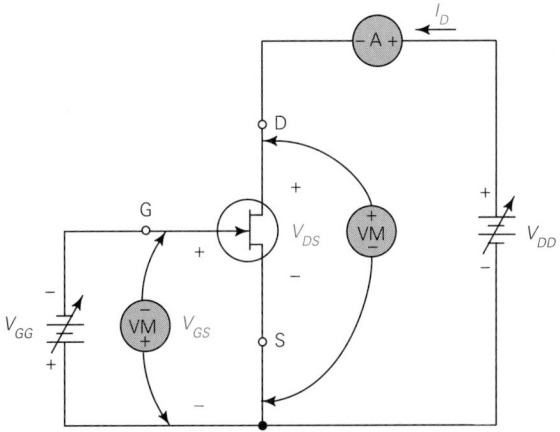

FIGURE 6–3
Test circuit for hypothetical *n*-channel JFET.

First, assume that the gate–source voltage is set to $V_{GS} = 0$. This could be achieved by setting the gate bias supply to zero or simply removing that bias supply and connecting the gate directly to the source. Assume now that the drain bias supply is adjusted in very small positive steps from zero and that I_D is measured as a function of V_{DS}. The current I_D increases approximately linearly with V_{DS}, as shown by the top curve in Figure 6–4. The resulting straight line, which starts at the origin, has the same form as for a passive resistor. The slope has the units of amperes/volt or siemens, so the slope represents the conductance.

FIGURE 6–4
Drain characteristics of hypothetical *n*-channel JFET in linear portion of ohmic region.

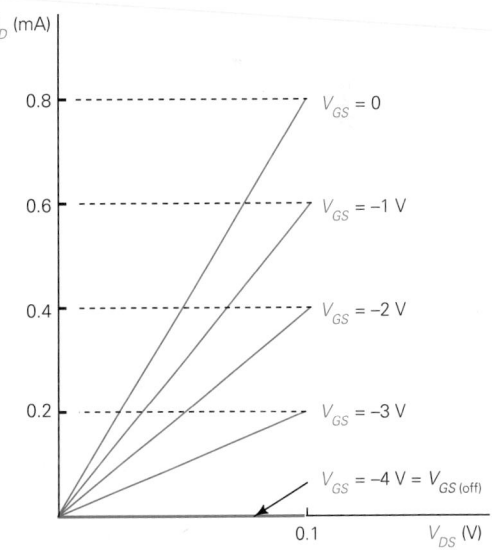

Thus, *the JFET is acting like a resistor in this region;* hence, the term *ohmic region* is quite appropriate.

Next, assume that we set $V_{GS} = -1$ V. (Note that the positive reference for V_{GS} is at the gate terminal, but the negative value takes care of the reverse voltage.) The curve of I_D versus V_{DS} is still a straight line, but it assumes a smaller slope, as shown by the line labeled $V_{GS} = -1$ V in Figure 6–4.

In Example 6–1 we will calculate the conductance and resistance for each line of our hypothetical JFET. For the moment, however, we will simply observe the trends.

The experiment is continued for $V_{GS} = -2$ V and $V_{GS} = -3$ V, and the slopes of the two lines continue to decrease as shown in Figure 6–4. In the ohmic region, the JFET assumes the role of a **voltage-controlled resistor.** For a fixed value of V_{GS}, the resistance is nearly constant for *small positive values* of V_{DS}. However, by varying V_{GS}, the resistance can be made to change.

The ohmic region actually extends farther to the right (more positive) than the range of V_{DS}, as we will see shortly. It also extends somewhat into the negative range of V_{DS}. However, the curves become nonlinear for increasing positive values or for negative values of V_{DS}, and the effective resistance is no longer a constant.

In the basic form shown, the assumption of a nearly constant resistance is generally valid only for small positive values of V_{DS}. By special feedback circuits, it is possible to extend the constant resistance both to negative values and more positive values of V_{DS}.

It might seem logical to introduce a mathematical model and an equivalent-circuit model for the ohmic region at this point. However, some of the terminology required in these models is best defined by first discussing the cutoff region and the beyond-pinchoff region. Consequently, models for the ohmic region will be delayed until Section 6–5. For the moment, remember that the JFET acts like a resistance in the ohmic region.

Cutoff Region

At some negative value of V_{GS}, the effective conduction width of the channel will be reduced completely to zero by depletion, and no further drain current can flow. The value of gate–source voltage at which drain current stops flowing is called the **gate–source cutoff voltage,** and it is denoted by $V_{GS(off)}$. For the characteristics of Figure 6–4, the value of the gate–source cutoff voltage is observed to be $V_{GS(off)} = -4$ V. This condition also corresponds to the beginning of the **cutoff region.** Further increases in the negative voltage applied to the gate have no effect, since the cutoff region has been reached, and no current flows in the channel. (This statement assumes that the magnitude of the voltage is not large enough to cause reverse junction breakdown.)

An ideal equivalent circuit for a JFET in the cutoff region is shown in Figure 6–5. This equivalent circuit has the same form as the model for a BJT in the cutoff region.

FIGURE 6–5
Ideal model of FET in *cutoff* region.

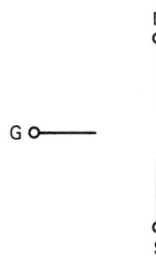

**EXAMPLE
6–1**

Determine the conductance and resistance values associated with each line for the hypothetical JFET characteristics of Figure 6–4.

Solutions
Let g_{DS} represent the conductance in siemens from drain to source, and let $r_{DS} = 1/g_{DS}$ represent the resistance in ohms. The slope g_{DS} in each case is determined by dividing the change in current along the line by the corresponding change in drain–source voltage. For the $V_{GS} = 0$ line, we have

$$g_{DS} = \frac{0.8 \times 10^{-3}\,\text{A}}{0.1\,\text{V}} = 8 \times 10^{-3}\,\text{S} = 8\,\text{mS} \qquad \textbf{(6–1)}$$

The resistance r_{DS} is

$$r_{DS} = \frac{1}{8 \times 10^{-3}} = 125\,\Omega \qquad \textbf{(6–2)}$$

For the $V_{GS} = -1$ V line, we have

$$g_{DS} = \frac{0.6 \times 10^{-3}\,\text{A}}{0.1\,\text{V}} = 6 \times 10^{-3}\,\text{S} = 6\,\text{mS} \qquad \textbf{(6–3)}$$

and

$$r_{DS} = \frac{1}{6 \times 10^{-3}} = 166.7\,\Omega \qquad \textbf{(6–4)}$$

The reader is invited to repeat this process for the other lines, and a short table summarizing these values follows:

V_{GS} (V)	g_{DS} (mS)	r_{DS} (Ω)
0	8	125
−1	6	166.7
−2	4	250
−3	2	500
−4	0	∞

Observe that as V_{GS} becomes more negative, the conductance decreases and the resistance increases. For $V_{GS} = V_{GS(off)}$ (−4 V in this case), the conductance becomes zero, and the resistance becomes infinite. This represents an open circuit between drain and source, since the cutoff region has been reached.

Beyond-Pinchoff Region

The next experiment to be discussed uses the same circuit as before (Figure 6–3) and follows the same procedure as the preceding one except for one major difference. The drain–source voltage will now be increased to a much larger value for each curve than for the ohmic region experiment. The complete set of curves so obtained are called the **output characteristics,** or the **drain characteristics.** The latter term is used most often in this book.

The form of the drain characteristics of the hypothetical transistor are shown in Figure 6–6. The scale is quite different here from that for Figure 6–4. The ohmic region of Figure 6–4 corresponds to a tiny portion of the curves near the origin in Figure 6–6. Observe that the overall drain characteristics do not increase linearly as they did for the small value of V_{DS}. Instead, their slopes decrease and eventually approach zero.

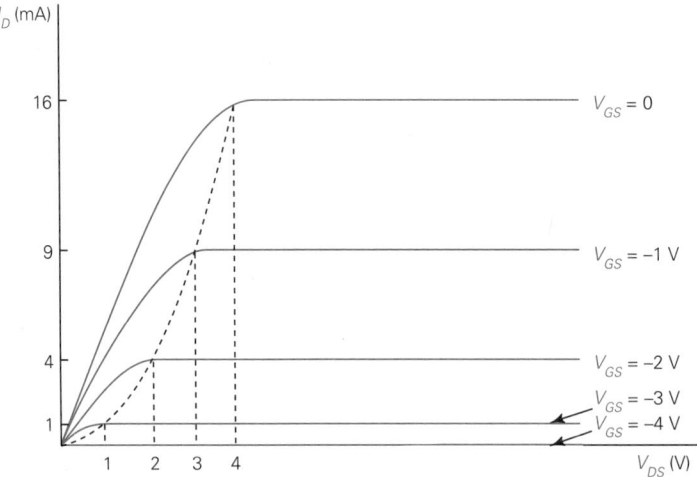

FIGURE 6–6
Drain characteristics of hypothetical *n*-channel JFET for full range of V_{DS}.

After the curves flatten out, there is very little additional change in I_D for increasing V_{DS}. This region is referred to as the **beyond-pinchoff region.** It is also called the **saturation region,** for reasons we will discuss. However, the latter designation could be confusing when the FET is compared with the BJT. Recall that saturation for a BJT corresponds to a collector-emitter voltage near zero, so the meaning of saturation in that case is quite different from what it is for an FET. We will use the term *beyond-pinchoff region* in this text.

A brief explanation of the phenomenon follows. As V_{DS} increases, the voltage between gate and drain (i.e., V_{GD}) becomes increasingly negative. This negative voltage results in a depletion region along the upper portion of the channel. When a depletion region was formed along the bottom of the channel by a negative value of V_{GS}, current eventually stopped flowing. In the present case, however, current has already been established, since V_{GS} is assumed to be more positive than $V_{GS(off)}$. The result is that the current flow becomes saturated and very little further increase can occur. The channel is said to be "pinched off," and a nearly constant value of current results for a fixed value of V_{GS}.

From Figure 6–6 it is noted that the curve corresponding to $V_{GS} = 0$ flattens out at about $V_{DS} = 4$ V. Recall, however, that $V_{GS(off)} = -4$ V. The equality between these magnitudes is more than a coincidence. The beyond-pinchoff region for an ideal JFET is reached when $V_{DS} = -V_{GS(off)}$ for $V_{GS} = 0$.

It is convenient for this purpose to define a quantity called the *pinchoff voltage*, and it is denoted by V_p. The pinchoff voltage is related to the cutoff voltage by

$$V_p = -V_{GS(off)} \tag{6–5}$$

Thus, for $V_{GS} = -4$ V, $V_p = 4$ V.

The transition from the ohmic region to the beyond-pinchoff region occurs for $V_{DS} = V_p$ along the curve corresponding to $V_{GS} = 0$. For other values of V_{GS}, however, the transition point is shifted back to the left as noted.

Let $V_{DS(tran)}$ represent the value of drain–source voltage for an ideal JFET at which this transition occurs. It is related to the pinchoff voltage V_p and the gate–source voltage by

$$\boxed{V_{DS(tran)} = V_p + V_{GS}} \tag{6–6}$$

When $V_{GS} = 0$, (6–6) reduces to $V_{DS(tran)} = V_p$, as already observed. Note that $V_p > 0$ and $V_{GS} < 0$ in (6–6) for an *n*-channel JFET. Note also that V_p is a constant for a given transistor, while the other two quantities vary with the operating curve.

EXAMPLE 6–2

Calculate the drain–source transition voltages for each characteristic curve of the hypothetical JFET whose output characteristics were given in Figure 6–6.

Solution

The gate–source cutoff voltage was determined earlier to be $V_{GS(off)} = -4$ V, and the pinchoff voltage was then noted to be $V_p = 4$ V. The values of the transition voltage $V_{DS(tran)}$ are determined from (6–6) as

$$V_{DS(tran)} = V_p + V_{GS} = 4 + V_{GS} \tag{6–7}$$

For $V_{GS} = 0$ we have

$$V_{DS(tran)} = 4 + 0 = 4 \text{ V} \qquad (6-8)$$

as expected. For $V_{GS} = -1$ V we have

$$V_{DS(tran)} = -1 + 4 = 3 \text{ V} \qquad (6-9)$$

The other results are tabulated as follows:

V_{GS} (V)	$V_{DS(tran)}$ (V)
0	4
−1	3
−2	2
−3	1
−4	0

The last case (cutoff) could be considered meaningless, but it is included to demonstrate that the transition point approaches $V_{DS(tran)} = 0$ as V_{GS} approaches $V_{GS(off)}$. The transition voltage is farthest to the right for $V_{GS} = 0$. These trends are readily observed in Figure 6–6.

The dashed curve superimposed on the characteristic curves of Figure 6–6 represents the locus of transition voltages. The region to the left of this curve is the ohmic region, and the region to the right is the beyond-pinchoff region.

6–4 CHARACTERISTICS IN BEYOND-PINCHOFF REGION

The forms of a few idealized n-channel JFET drain characteristic curves were shown in Figure 6–6. The reader may have observed the apparent nonlinear spacing between the curves; that is, the spacing between successive curves increases as V_{GS} is varied from cutoff to zero. This pattern is characteristic of FETs and is to be expected.

In this section a mathematical model for predicting the drain current as a function of the gate–source voltage applicable to the *beyond-pinchoff region* will be presented. The reader must understand that this equation is approximate and must be used with some caution in working with actual JFET characteristics. However, its form is sufficiently close to the actual behavior of typical JFET characteristics that it is widely used in analysis and design.

First, we define a parameter I_{DSS} known as the **zero-bias drain current.** Specifically, it is the value of I_D in the beyond-pinchoff region corresponding to $V_{GS} = 0$. For the characteristics of Figure 6–6, $I_{DSS} = 16$ mA.

The drain current in the beyond-pinchoff region follows a square-law functional relationship and can be closely approximated for many JFETs as

$$I_D = I_{DSS} \left[1 - \frac{V_{GS}}{V_{GS(\text{off})}} \right]^2 \tag{6–10}$$

For an n-channel JFET, $V_{GS(\text{off})}$ is negative, and V_{GS} is either negative or zero. Because of potential sign difficulties, some persons prefer to express (6–10) in the alternative form

$$I_D = I_{DSS} \left[1 - \frac{|V_{GS}|}{|V_{GS(\text{off})}|} \right]^2 \tag{6–11}$$

in which the magnitudes of V_{GS} and $V_{GS(\text{off})}$ are used. I_D and I_{DSS} are both positive for an n-channel JFET, so magnitude bars are not required for them.

Observe that V_{DS} does not appear in (6–10) and (6–11). The reason is that this model assumes no variation of I_D with V_{DS} in the beyond-pinchoff region. In practice, there is a slight increase in I_D as V_{DS} is increased, but this will be ignored in our idealized analysis.

Transfer Characteristic

We have displayed and discussed the drain characteristics of a hypothetical n-channel JFET. An alternative presentation, which is often more useful in designing circuits, is the **transfer characteristic.** The *transfer characteristic* is a plot of I_D versus V_{GS} in the beyond-cutoff region. The transfer characteristic of the hypothetical n-channel JFET considered earlier is shown in Figure 6–7. By comparing Figure 6–7 with Figure 6–6, the reader may readily verify that the values of V_{GS} and I_D at the circled points of Figure 6–7 correspond to the flat portions of the drain characteristics of Figure 6–6. Since a large number of individual drain characteristic curves could have been obtained in Figure 6–6 by assuming more values of V_{GS}, the continuous curve of Figure 6–7 represents the locus of all such values of V_{GS} and I_D. The square-law form of the drain current is rather evident from the curvature in Figure 6–7.

Reverse Form of Equation

The form of either (6–10) or (6–11) expresses I_D as a dependent function of the independent variable V_{GS}. However, suppose it is necessary to determine the value of V_{GS} that would produce a desired I_D. For this purpose it is necessary to solve for V_{GS} as a function of I_D.

To accomplish this task, the right-hand side of (6–10) is first expanded. The variable V_{GS} will appear both as a linear term and a second-degree term. The

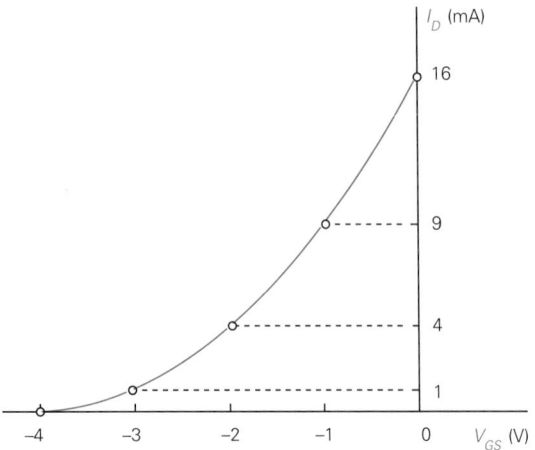

FIGURE 6–7
Transfer characteristic of hypothetical *n*-channel JFET considered in Figure 6–6.

resulting equation is a quadratic equation in the variable V_{GS}. The steps leading to the solution of V_{GS} will be left as an exercise for the reader (Problem 6–39). All quadratic equations have two solutions, but one solution in this case is simply a mathematical value that does not correspond to the physical limitations of the problem. The proper solution for I_D as a function of V_{GS} is the following equation:

$$V_{GS} = V_{GS(\text{off})}\left(1 - \sqrt{\frac{I_D}{I_{DSS}}}\right) \qquad \textbf{(6–12)}$$

Graphical Analysis

A normalized form of the beyond-pinchoff equation may readily be formulated, which permits a graphical approach to the analysis. First, both sides of (6–10) are divided by I_{DSS}, which results in

$$\frac{I_D}{I_{DSS}} = \left[1 - \frac{V_{GS}}{V_{GS(\text{off})}}\right]^2 \qquad \textbf{(6–13)}$$

For convenience, momentarily define the following normalized variables:

$$X = \frac{V_{GS}}{V_{GS(\text{off})}} \qquad \textbf{(6–14)}$$

$$Y = \frac{I_D}{I_{DSS}} \qquad \textbf{(6–15)}$$

The result of (6–13) may then be expressed in the simpler "pure" mathematical form

$$Y = (1 - X)^2 \qquad\qquad\qquad\text{(6–16)}$$

The result of (6–16) is shown in graphical form in Figure 6–8. The abscissa is labeled in terms of both $V_{GS}/V_{GS(\text{off})}$ and in terms of $|V_{GS}|/|V_{GS(\text{off})}|$, since both ratios are the same.

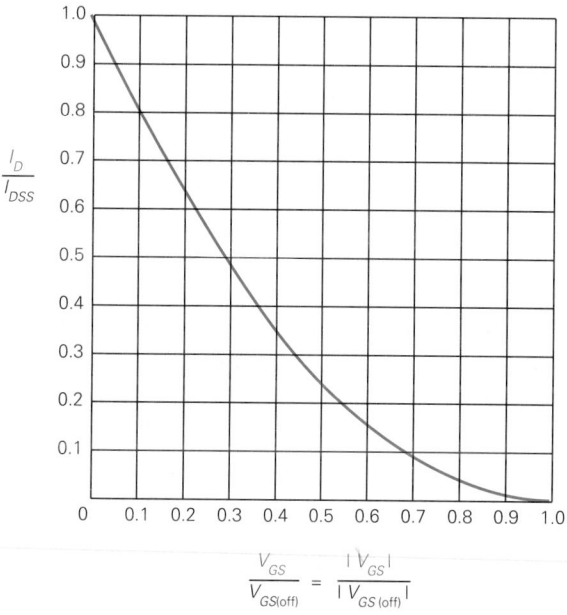

$$\frac{V_{GS}}{V_{GS(\text{off})}} = \frac{|V_{GS}|}{|V_{GS(\text{off})}|}$$

FIGURE 6–8
Normalized form of transfer characteristic.

This curve applies to both n-channel and p-channel devices as long as the quantities are appropriately interpreted. As we will see, it also applies to depletion-mode IGFETs (but not to enhancement-mode IGFETs). Because of the normalized nature of the variables, it is applicable for any values of $V_{GS(\text{off})}$ and I_{DSS}. Do not be confused by the fact that the abscissa is reversed in direction from the earlier form of the transfer characteristic curve. The right-hand side corresponds to cutoff, and the left-hand side corresponds to $V_{GS} = 0$.

EXAMPLE 6–3

Using the mathematical model for the drain current in the beyond-pinchoff region, verify that the circled values of I_D versus V_{GS} in Figure 6–7 (or the final values of I_D versus V_{DS} in Figure 6–6) represent the idealized values.

Solution

First, we note that $I_{DSS} = 16$ mA and $V_{GS(off)} = -4$ V, as determined previously. We will illustrate both forms of (6–10) and (6–11). The current I_D can be expressed in either of the following forms:

$$I_D = 16 \times 10^{-3} \left(1 - \frac{V_{GS}}{-4}\right)^2 \qquad (6\text{–}17)$$

or

$$I_D = 16 \times 10^{-3} \left(1 - \frac{|V_{GS}|}{4}\right)^2 \qquad (6\text{–}18)$$

where $|V_{GS(off)}| = 4$ V was used in (6–18).

From either (6–17) or (6–18), it is immediately noted that when $V_{GS} = 0$, $I_D = 16 \times 10^{-3}$ A $= 16$ mA, as expected.

Consider $V_{GS} = -1$ V and equation (6–17). The expression for I_D is

$$I_D = 16 \times 10^{-3} \left(1 - \frac{-1}{-4}\right)^2 = 16 \times 10^{-3} \left(1 - \frac{1}{4}\right)^2$$

$$= 9 \times 10^{-3} \text{ A} = 9 \text{ mA} \qquad (6\text{–}19)$$

Let's verify that (6–18) produces the same result.

$$I_D = 16 \times 10^{-3} \left(1 - \frac{1}{4}\right)^2 = 9 \times 10^{-3} \text{ A} = 9 \text{ mA} \qquad (6\text{–}20)$$

where $|V_{GS}| = 1$ V was used in (6–20).

Comparing the use of (6–17) and (6–18) in (6–19) and (6–20), respectively, we can see the possible sign difficulty in (6–17) and (6–19). Despite that, however, the form of (6–17) is usually more desirable when several separate algebraic manipulations are required. The point to remember is that the quantity in parentheses will always represent a numerical *difference* in the final analysis. In the remainder of the example, we use the form of (6–17). For $V_{GS} = -2$ V, we have

$$I_D = 16 \times 10^{-3} \left(1 - \frac{-2}{-4}\right)^2 = 16 \times 10^{-3} \left(1 - \frac{1}{2}\right)^2$$

$$= 4 \times 10^{-3} \text{ A} = 4 \text{ mA} \qquad (6\text{–}21)$$

For $V_{GS} = -3$ V,

$$I_D = 16 \times 10^{-3} \left(1 - \frac{-3}{-4}\right)^2 = 16 \times 10^{-3} \left(1 - \frac{3}{4}\right)^2$$

$$= 1 \times 10^{-3} \text{ A} = 1 \text{ mA} \qquad (6\text{–}22)$$

Finally, for $V_{GS} = -4$ V,

$$I_D = 16 \times 10^{-3} \left(1 - \frac{-4}{-4}\right)^2 = 16 \times 10^{-3}(1 - 1)^2 = 0 \qquad (6\text{–}23)$$

as we should expect. The preceding values are readily verified in either Figure 6–7 or 6–6.

EXAMPLE 6–4

Tests on a certain n-channel JFET in the beyond-pinchoff region are performed. When gate and source are connected together, $I_D = 10$ mA. When an adjustable dc source is connected between gate and source, the drain current ceases to flow when $V_{GS} = -2$ V. **(a)** Determine an equation for predicting I_D as a function of V_{GS} in the beyond-pinchoff region. **(b)** Determine the value of I_D when $V_{GS} = -1$ V.

Solution

(a) From the statement of the test, we surmise that $I_{DSS} = 10$ mA and $V_{GS(off)} = -2$ V. The equation for predicting I_D in the beyond-pinchoff region is

$$I_D = I_{DSS}\left[1 - \frac{V_{GS}}{V_{GS(off)}}\right]^2 = 0.01\left(1 - \frac{V_{GS}}{-2}\right)^2 = 0.01\left(1 + \frac{V_{GS}}{2}\right)^2 \quad \text{(6–24)}$$

(b) For $V_{GS} = -1$ V, we have

$$I_D = 0.01\left(1 - \frac{-1}{-2}\right)^2 = 0.0025 \text{ A} = 2.5 \text{ mA} \quad \text{(6–25)}$$

EXAMPLE 6–5

For the n-channel JFET of Example 6–4, determine the value of V_{GS} that produces $I_D = 5$ mA three ways: **(a)** by direct solution of the quadratic equation; **(b)** by use of the reverse equation as given by (6–12); **(c)** by use of the graph of Figure 6–8.

Solution

(a) The equation for I_D was stated in (6–24). We set $I_D = 5$ mA $= 5 \times 10^{-3}$ A, and this results in

$$5 \times 10^{-3} = 0.01\left(1 - \frac{V_{GS}}{-2}\right)^2 \quad \text{(6–26)}$$

This equation is simplified by dividing both sides by 0.01, and the right-hand side is expanded. This results in

$$0.5 = 1 + V_{GS} + \frac{V_{GS}^2}{4} \quad \text{(6–27)}$$

This equation is rearranged in standard quadratic form as

$$V_{GS}^2 + 4V_{GS} + 2 = 0 \quad \text{(6–28)}$$

Application of the quadratic formula yields

$$V_{GS} = \frac{-4 \pm \sqrt{16 - 8}}{2} = \frac{-4 \pm 2.828}{2} \quad \text{(6–29)}$$

When the two values are computed, one turns out to be $V_{GS} = -3.414$ V. This is an *impossible* physical solution, since this voltage is more negative than the pinchoff voltage and represents the cutoff region. This superfluous value is a mathematical solution but not a physically possible solution. The other solution is the correct one and is

$$V_{GS} = -0.586 \text{ V} \qquad \qquad \textbf{(6–30)}$$

When the preceding mathematical process is followed, the superfluous solution can always be determined by observing the physical limits of the problem. The correct solution is always the one involving *opposite* signs in the numerator of the quadratic solution, as in (6–29).

(b)   Substitution of the pertinent values in (6–12) results in

$$V_{GS} = -2 \left(1 - \sqrt{\frac{5 \times 10^{-3}}{10 \times 10^{-3}}} \right) = -2(1 - \sqrt{0.5}) \qquad \qquad \textbf{(6–31)}$$
$$= -2(0.293) = -0.586 \text{ V}$$

(c)   To use Figure 6–8, we first normalize the value of I_D by forming I_D/I_{DSS}. This value is $5 \times 10^{-3}/10 \times 10^{-3} = 0.5$. For an ordinate of 0.5, the corresponding abscissa is about $V_{GS}/V_{GS(\text{off})} = 0.29$. We then calculate V_{GS} as

$$V_{GS} = 0.29 \times (-2) = -0.58 \text{ V} \qquad \qquad \textbf{(6–32)}$$

Comparing the results of **(a)**, **(b)**, and **(c)**, we note that **(a)** and **(b)** produce exactly the same values, while **(c)** produces approximately the same result to the expected accuracy of the graph.

6–5   OHMIC-REGION CHARACTERISTICS

Early in the discussion of the JFET, we observed an idealized drain voltage–current characteristic in the ohmic region, and the reader may wish to refer to Figure 6–4 (p. 275). It was noted that for very small values of V_{DS}, the curves of I_D versus V_{DS} were straight lines whose slopes were a function of V_{GS}. In this region the FET behaves like a voltage-controlled resistor.

We now give a mathematical model for the ideal characteristics of these straight lines. Let g_{DS} represent the drain-to-source conductance in siemens corresponding to any one of these curves, and let g_{DSO} represent the conductance of the line representing $V_{GS} = 0$. The value of conductance is simply the slope of a particular line. It turns out that g_{DS} can be closely approximated for most JFETs as

$$g_{DS} = g_{DSO} \left[1 - \frac{V_{GS}}{V_{GS(\text{off})}} \right] \qquad \qquad \textbf{(6–33)}$$

The idealized value of g_{DSO} is

$$g_{DSO} = -\frac{2I_{DSS}}{V_{GS(off)}} \qquad (6\text{--}34)$$

At first glance, g_{DSO} may appear to have a negative value, but this is an illusion. The quantity I_{DSS} is positive, and $V_{GS(off)}$ is negative, so the latter negative value in consort with the negative sign in front results in a positive value of g_{DSO}. To delineate this point, an alternative form of g_{DSO} for an *n*-channel JFET is

$$g_{DSO} = \frac{2I_{DSS}}{|V_{GS(off)}|} \qquad (6\text{--}35)$$

Let $r_{DS} = 1/g_{DS}$ represent the resistance associated with a given line. This value may be expressed as

$$r_{DS} = \frac{r_{DSO}}{1 - V_{GS}/V_{GS(off)}} \qquad (6\text{--}36)$$

where

$$r_{DSO} = \frac{1}{g_{DSO}} = -\frac{V_{GS(off)}}{2I_{DSS}} = \frac{|V_{GS(off)}|}{2I_{DSS}} \qquad (6\text{--}37)$$

The equivalent circuit of the *n*-channel JFET in the *linear* portion of the ohmic region is shown in Figure 6–9. For the resistance to be assumed as a constant value, the value of V_{DS} must be limited to a small range, typically from 0 to about 0.1 V.

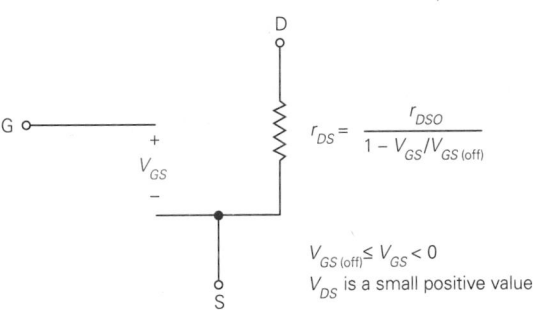

$$r_{DS} = \frac{r_{DSO}}{1 - V_{GS}/V_{GS(off)}}$$

$$V_{GS(off)} \leq V_{GS} < 0$$
V_{DS} is a small positive value

FIGURE 6–9
Ohmic-region model for *n*-channel JFET for small positive values of V_{DS}.

From the preceding relationships, it can be inferred that the maximum conductance and minimum resistance correspond to $V_{GS} = 0$. As the magnitude of V_{GS} increases (i.e., V_{GS} becomes more negative), the conductance decreases and the resistance increases.

One particular limiting case of the ohmic region corresponds to the use of an FET as an **analog switch.** In this type of circuit, the FET is alternately switched from the lowest resistance condition to the cutoff region. Some FETs are manufactured specifically for this purpose. More will be said about this application in Chapter 7.

EXAMPLE 6–6

Using the mathematical model for the conductance in the lower portion of the ohmic region, verify that the values of conductance computed in Example 6–1 (corresponding to the lines of Figure 6–4) represent the idealized values.

Solution

We note again that $I_{DSS} = 16$ mA and $V_{GS(off)} = -4$ V for the idealized JFET considered earlier. The value of g_{DSO} from (6–34) is

$$g_{DSO} = \frac{-2(16 \times 10^{-3})}{-4} = 8 \times 10^{-3}\,\text{S} = 8\,\text{mS} \qquad (6\text{–}38)$$

The conductance g_{DS} from (6–33) is

$$g_{DS} = 8 \times 10^{-3}\left(1 - \frac{V_{GS}}{-4}\right) \qquad (6\text{–}39)$$

For $V_{GS} = 0$, it is readily noted that

$$g_{DS} = 8 \times 10^{-3}\,\text{S} = 8\,\text{mS} \qquad (6\text{–}40)$$

For $V_{GS} = -1$ V, the conductance is

$$g_{DS} = 8 \times 10^{-3}\left(1 - \frac{-1}{-4}\right) = 8 \times 10^{-3}\left(1 - \frac{1}{4}\right)$$
$$= 6 \times 10^{-3}\,\text{S} = 6\,\text{mS} \qquad (6\text{–}41)$$

The reader is invited to repeat this process at $V_{GS} = -2$ V, -3 V, and -4 V, and the conductance values are 4 mS, 2 mS, and 0, respectively. These values are readily verified from the results of Example 6–1. Alternatively, we could have used (6–36) and (6–37) to compute the resistance instead of the conductance.

Before leaving this example, we need to bring the reader back to the "real world" again. In this example and several others earlier in the chapter, we have verified various relationships for the JFET with the hypothetical JFET characteristics provided. These characteristics were "rigged" to produce these characteristics exactly in the ideal forms. Seldom (if ever) do any actual characteristics work out so nicely. Most actual FET characteristics can be approximated by the models provided, but they should be interpreted as reasonable approximations rather than exact results. As we shall see, the actual parameters of most FETs of a given type vary widely from one device to another.

6–6 BEYOND-PINCHOFF DC MODEL

We will now consider a circuit model that can be used to show the dc operation of an *n*-channel JFET in the beyond-pinchoff region in a complete circuit. The basic form of this model is shown in Figure 6–10.

FIGURE 6–10
Beyond-pinchoff dc model for *n*-channel JFET.

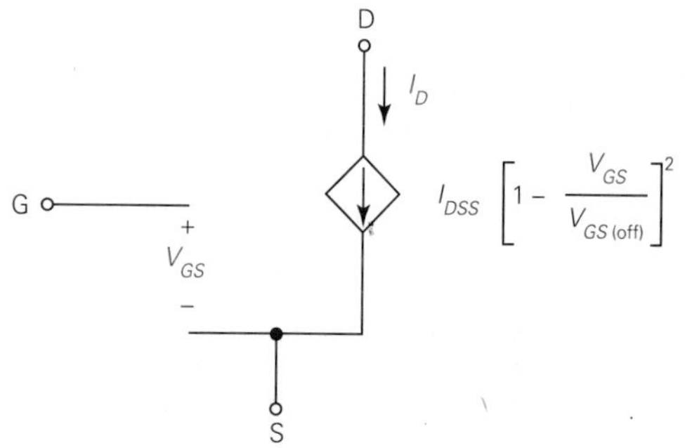

The drain side of this model consists of an ideal dependent current source whose value represents the drain current. This model assumes that the drain current remains constant as V_{DS} changes with the applicable region.

We immediately see that the drain circuit of the *n*-channel JFET has a striking similarity to the collector circuit active-region dc model of an *npn* BJT; that is, both devices act nearly like ideal current sources, and the current sources have the same direction. However, there are differences in the operation of the devices, as we will see.

The dc current source for the *BJT* is a function of the *base current,* while the dc current source for the *FET* is a function of the *gate–source voltage.* For the dc model, *we thus consider the BJT to be a current-controlled device, while we consider the FET to be a voltage-controlled device.* (Later we will see that the BJT can also be considered to be voltage controlled in amplifier applications.)

The value of the current source for the BJT was simply $\beta_{dc}I_B$, while the value of the current source in the FET is a somewhat more complex function of V_{GS}. Some significant differences between the BJT and the FET are evident on the input sides. The base side of a BJT looks into a forward-biased *pn* junction, in which a nearly constant voltage drop appears. Current must flow into the base circuit of a BJT, and the loading effect on the external circuit must be considered in applications. In contrast, *the input of an FET is a very large resistance, which for many applications can be assumed to be infinite.* This assumption has been made in Figure 6–10 and is noted by the open circuit between gate and source terminals. This means that in this model no current would flow into the gate terminal.

Actually, there is a small current (typically less than 1 nA) that flows in the gate of a JFET, and this could be represented by a very large resistance between gate and source. However, for most applications, this resistance can be ignored. For other classes of FETs to be considered later, the resistance can be as high as hundreds of megohms.

6–7 INSULATED GATE FIELD-EFFECT TRANSISTORS

The second broad category of FETs is the **insulated gate field-effect transistor (IGFET).** The most common type of IGFET is the **metal-oxide semiconductor field-effect transistor (MOSFET).** In practice, the terms *IGFET* and *MOSFET* are both used casually to refer to all transistors of this broad category. We use the term *IGFET* in most places in this book, since not all IGFETs are MOSFETs.

IGFETs can be classified in two ways according to their conducting status: (1) **depletion mode,** or "*normally on,*" and (2) **enhancement mode,** or "normally off." We will now discuss each category.

Depletion-Mode IGFET

The construction of an *n*-channel depletion mode IGFET is suggested by the sketch of Figure 6–11. As in the case of an *n*-channel JFET, there is a channel with *source* and *drain* contacts at the respective ends. A *p*-region called the **substrate** physically reduces the conducting path to a narrow channel. A thin layer of **silicon dioxide** (SiO_2) insulates the metallic gate from the channel in the common case of a MOSFET. Thus, the *pn* junction diode that exists in the JFET does not appear in the IGFET. The result of this insulation layer is that the gate current

FIGURE 6–11
Basic construction of an *n*-channel depletion-mode IGFET.

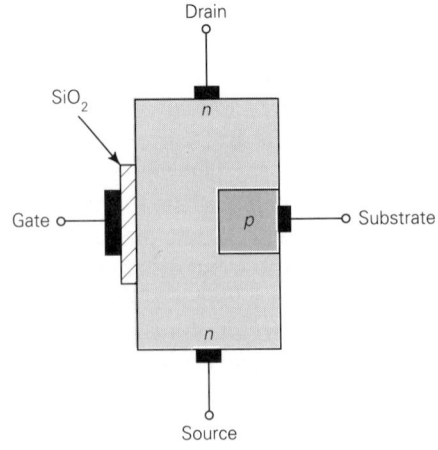

of the IGFET is even smaller than that of the JFET, which has already been indicated to be quite small.

As in the case of the JFET, the width of the conduction channel is controlled by the gate–source voltage. If the gate–source voltage in a depletion-mode IGFET is sufficiently negative, the channel is completely blocked and no current flows. This corresponds to the channel cutoff voltage.

As a general rule, the drain characteristics of an *n*-channel depletion-mode IGFET are very similar to those of the *n*-channel JFET except for one difference. In the *n*-channel JFET, V_{GS} is usually not permitted to assume positive values, since this would forward-bias the gate–channel junction. Because of the insulation layer, V_{GS} in an *n*-channel depletion-mode IGFET may assume positive values without affecting the high input resistance. Strictly speaking, this corresponds to enhancement-mode operation of the depletion-mode device. We will soon discuss devices specifically designed for enhancement-mode operation.

Although some IGFETs have the substrate terminal available as a fourth terminal, many devices are manufactured with this terminal connected internally to the source. The schematic diagram of an *n*-channel depletion-mode IGFET with substrate connected to source internally is shown in Figure 6–12.

FIGURE 6–12

Schematic symbol of an *n*-channel depletion-mode IGFET (substrate is connected to source internally).

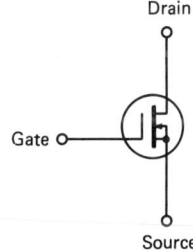

Drain

Gate

Source

Some typical drain characteristics of an *n*-channel depletion-mode IGFET are shown in Figure 6–13. Curves are shown for both negative and positive values of V_{GS} to illustrate that operation is possible for the IGFET with both polarities.

Enhancement-Mode IGFET

The construction of an *n*-channel enhancement-mode IGFET is suggested by the diagram of Figure 6–14. The *p*-type substrate completely separates the two *n*-type sections of the channel. The channel is thus blocked, and for that reason the enhancement-mode IGFET can be called a *normally off* IGFET.

When $V_{DS} > 0$, there is no conduction in an *n*-channel enhancement-mode IGFET for $V_{GS} \leq 0$. For conduction to occur, V_{GS} must assume some minimum *positive value*. A sufficient positive value of V_{GS} will attract free electrons from the *n*-type channel into the *p*-region in the vicinity of the gate. This results in an effective layer of *n*-type material that bridges across the substrate and permits

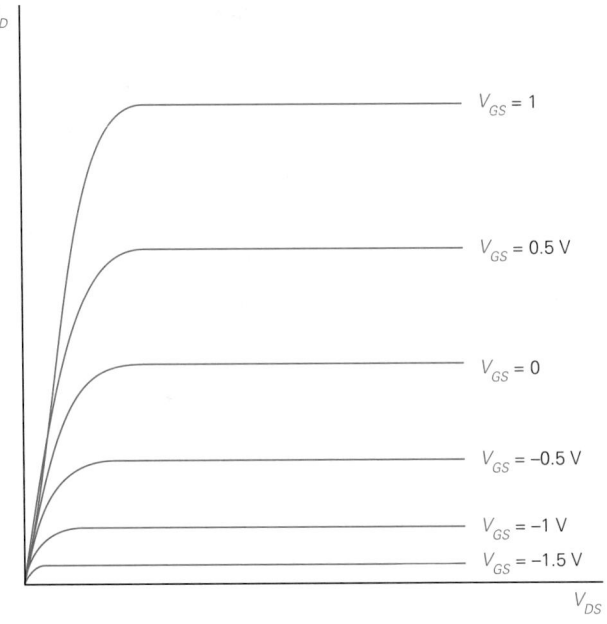

FIGURE 6–13
Typical drain characteristics of an *n*-channel depletion-mode IGFET showing operation for both negative and positive values of V_{GS}.

FIGURE 6–14
Basic construction of an *n*-channel enhancement-mode IGFET.

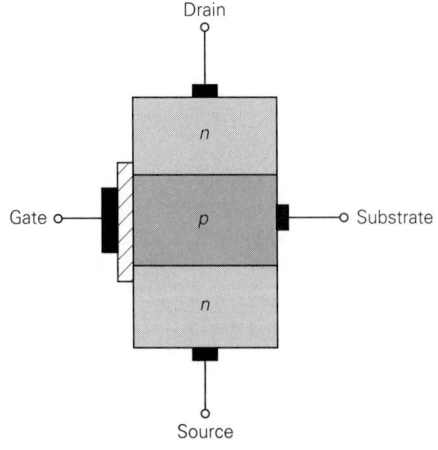

conduction. Thus conduction has been *enhanced* by the presence of a positive voltage.

Threshold Voltage

The minimum positive value of V_{GS} that results in drain current flow for an n-channel enhancement-mode IGFET is called the **threshold voltage** and is denoted as $V_{GS(th)}$. Thus $I_D = 0$ for $V_{GS} < V_{GS(th)}$. The threshold voltage for an n-channel enhancement-mode IGFET is analogous to the gate–source cutoff voltage for the n-channel depletion-mode IGFET and the JFET. However, note that for n-channel devices, $V_{GS(th)} > 0$ and $V_{GS(off)} < 0$.

The schematic symbol of an n-channel enhancement-mode IGFET is shown in Figure 6–15. The substrate is assumed to be connected to the source internally. The symbols for depletion-mode and enhancement-mode devices look somewhat similar, and the reader is encouraged to compare Figures 6–12 and 6–15 to observe the difference.

FIGURE 6–15

Schematic symbol of an n-channel en-hancement-mode IGFET (substrate is con-nected to source internally).

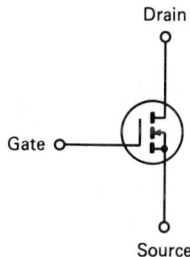

Drain

Gate

Source

6–8 ENHANCEMENT-MODE IGFET DRAIN CHARACTERISTICS

We discuss next the results of a possible experiment performed on a hypothetical n-channel enhancement-mode IGFET. As in the case of the JFET considered earlier in the chapter, the characteristics have been idealized somewhat.

Consider the test circuit shown in Figure 6–16 containing an n-channel enhancement-mode IGFET. If this circuit is compared with the JFET circuit of Figure 6–3, the main difference is that the gate voltage bias supply for the n-channel enhancement-mode device has its positive terminal connected to the gate rather than to ground.

Refer to the drain characteristics of Figure 6–17 in the discussion that follows. If $V_{GS} = 0$ (or a small positive voltage), $I_D = 0$ as V_{DS} is varied over a reasonable positive range because of the normally-off property. The IGFET is biased in the cutoff region until V_{DS} is increased beyond the threshold voltage $V_{GS(th)}$.

For this particular device, it is assumed that $V_{GS(th)} = 2$ V. Thus, for $V_{DS} > 2$ V, drain current is measured as V_{DS} is changed in positive increments.

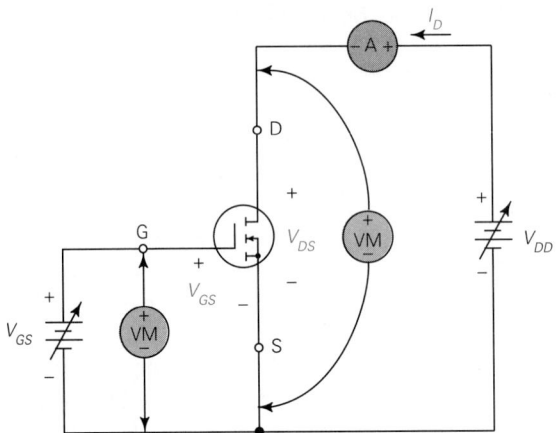

FIGURE 6–16
Test circuit for hypothetical *n*-channel enhancement-mode IGFET.

Beyond-Pinchoff Region

Observe in Figure 6–17 that the transition point from the ohmic region to the beyond-pinchoff region shifts farther to the right on the drain characteristics as V_{GS} increases. For depletion-mode IGFETs and JFETs, a convenient quantity called the pinchoff voltage was used in characterizing the transition point. However, there is no particular need to use such a definition for enhancement-mode IGFETs as the threshold voltage may be used directly in the formulation.

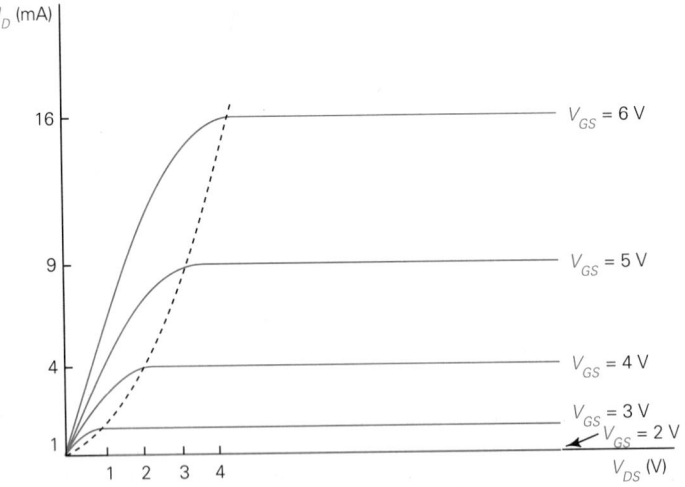

FIGURE 6–17
Drain characteristics of hypothetical *n*-channel enhancement-mode IGFET.

Let $V_{DS(\text{tran})}$ represent the transition voltage from the ohmic region to the beyond-pinchoff region. For an ideal n-channel enhancement-mode IGFET, this voltage is

$$V_{DS(\text{tran})} = V_{GS} - V_{GS(\text{th})} \qquad \textbf{(6–42)}$$

All the variables in (6–42) are positive quantities for an n-channel unit.

The curves of I_D versus V_{DS} for various values of V_{GS} are shown in Figure 6–17. The curves are similar in shape to those of the JFET and depletion-mode IGFET, but they correspond to a different range of V_{GS}. Like the JFET and the depletion-mode IGFET, the enhancement-mode IGFET possesses an ohmic region in which the device behaves like a voltage-controlled resistance. At small values of V_{DS}, the resistance is approximately constant. More will be said about the ohmic region later.

6–9 ENHANCEMENT-MODE IGFET MODELS

The drain current I_D as a function of the gate–source voltage V_{GS} for an n-channel enhancement-mode IGFET can be described very closely in the beyond-pinchoff region by the equation

$$I_D = K \left[\frac{V_{GS}}{V_{GS(\text{th})}} - 1 \right]^2 \qquad \textbf{(6–43)}$$

where $V_{GS(\text{th})}$ is the threshold voltage and K is a constant. The constant K is determined by matching the actual characteristics of a given unit to the mathematical model. In effect, it is a curve-fitting constant chosen to match the physical conditions. (Should we say a "fudge factor"?)

Transfer Characteristic

As for the JFET and depletion-mode IGFET, a **transfer characteristic** for the IGFET may also be displayed. The transfer characteristic of the hypothetical n-channel IGFET is shown in Figure 6–18. By comparing Figure 6–18 with Figure 6–17, the reader may readily verify that the values of V_{GS} and I_D at the circled points of Figure 6–18 correspond to the flat portions of the drain characteristics of Figure 6–17. Once again, the square-law form of the drain current is evident.

Beyond-Pinchoff dc Model

A dc circuit model of the n-channel enhancement-mode IGFET in the beyond-pinchoff region is shown in Figure 6–19. The form is the same as that of the n-channel JFET given in Figure 6–10 except for the expression for I_D. In both

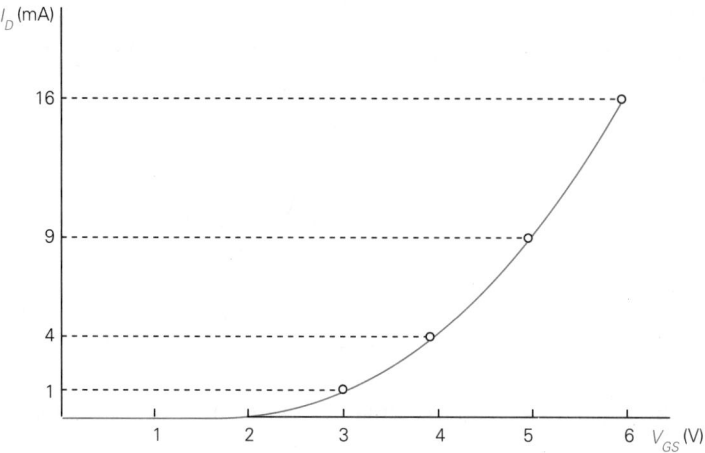

FIGURE 6–18
Transfer characteristic of hypothetical *n*-channel enhancement-mode IGFET.

models, the gate has been considered as an open circuit. In actual practice, this assumption is even more valid for the IGFET than for the JFET because of the insulated gate structure.

Reverse Equation

Determination of the value of V_{GS} that produces a given value of I_D in (6–43) is similar to the process considered in Section 6–4 for depletion-mode devices. It will be left as an exercise for the reader (Problem 6–40) to show that

$$V_{GS} = V_{GS(th)} \left(\sqrt{\frac{I_D}{K}} + 1 \right)$$

(6–44)

FIGURE 6–19
Beyond pinchoff model for *n*-channel enhance-ment-mode IGFET.

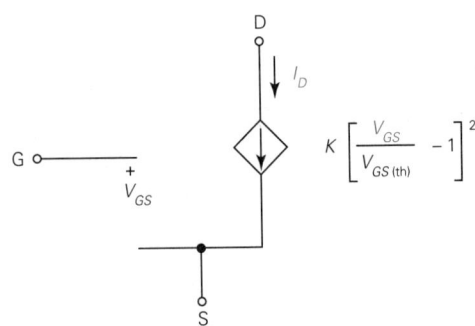

Ohmic-Region Characteristics

We now consider the ohmic-region mathematical model of the enhancement-mode IGFET. For $V_{GS} > V_{GS(th)}$ and very small values of V_{DS} (typically ≤ 0.1 V), the IGFET behaves approximately like a constant value of resistance. The conductance g_{DS} is of the form

$$g_{DS} = g_{DSO}\left[\frac{V_{GS}}{V_{GS(th)}} - 1\right] \tag{6-45}$$

The idealized value of g_{DSO} in this case is

$$g_{DSO} = \frac{2K}{V_{GS(th)}} \tag{6-46}$$

where K is the same constant appearing in (6–43).
    An expression for the resistance r_{DS} is

$$r_{DS} = \frac{1}{g_{DS}} = \frac{r_{DSO}}{\dfrac{V_{GS}}{V_{GS(th)}} - 1} \tag{6-47}$$

where

$$r_{DSO} = \frac{1}{g_{DSO}} = \frac{V_{GS(th)}}{2K} \tag{6-48}$$

A circuit model for the enhancement-mode IGFET in the ohmic region is shown in Figure 6–20.

FIGURE 6–20
Ohmic-region model for *n*-channel enhancement-mode IGFET for small positive values of V_{DS}.

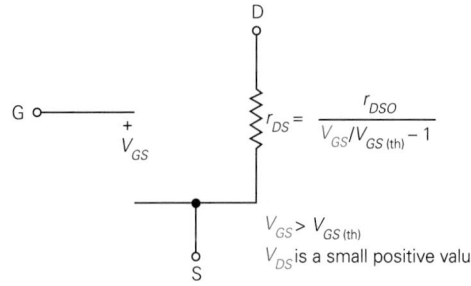

EXAMPLE 6–7

Calculate the drain–source transition voltage for each characteristic curve of the hypothetical enhancement-mode IGFET of Figure 6–17.

Solution

The threshold voltage was observed to be $V_{GS(th)} = 2$ V. An expression for transition voltage $V_{GS(tran)}$ is determined from (6–42) as

$$V_{DS(tran)} = V_{GS} - 2 \qquad (6\text{–}49)$$

For each value of V_{GS} on Figure 6–17, the transition voltage is easily calculated, and a short table of values follows.

V_{GS} (V)	$V_{DS(tran)}$ (V)
2	0
3	1
4	2
5	3
6	4

These trends are readily observed in Figure 6–17. The dashed curve superimposed on the characteristic curves represents the locus of transition voltages. The ohmic region is located to the left of this curve, and the beyond-pinchoff region is located to the right.

EXAMPLE 6–8

Using the mathematical model for the drain current in the beyond-pinchoff region, verify that the circled values of I_D versus V_{GS} in Figure 6–18 (or the final values of I_D versus V_{DS} in Figure 6–17) represent the idealized values.

Solution

Since $V_{GS(th)} = 2$ V, the form of the equation for I_D from (6–43) is

$$I_D = K\left(\frac{V_{GS}}{2} - 1\right)^2 \qquad (6\text{–}50)$$

This equation clearly predicts $I_D = 0$ when $V_{GS} = 2$ V, as it should. However, K must be determined before any subsequent values of I_D can be predicted.

In an actual application of the equation, one could select a known condition for I_D and V_{GS} to determine K. In our application, we will use the fact that $I_D = 1$ mA when $V_{GS} = 3$ V. When these values are substituted in (6–50), there results

$$1 \times 10^{-3} = K\left(\frac{3}{2} - 1\right)^2 = K(0.25) \qquad (6\text{–}51)$$

Solving for K, we obtain

$$K = 4 \times 10^{-3} \, \text{A} \qquad \textbf{(6–52)}$$

Note that K has the units of amperes.

The equation for I_D can be expressed as

$$I_D = 4 \times 10^{-3} \left(\frac{V_{GS}}{2} - 1 \right)^2 \qquad \textbf{(6–53)}$$

For $V_{GS} = 4$ V, we can determine I_D as

$$I_D = 4 \times 10^{-3} \left(\frac{4}{2} - 1 \right)^2 = 4 \times 10^{-3} \, \text{A} = 4 \, \text{mA} \qquad \textbf{(6–54)}$$

which is readily verified.

The reader is invited to show that the values for I_D of 9 mA and 16 mA are readily determined when $V_{GS} = 5$ V and 6 V, respectively. As was the case of the JFET considered earlier, these characteristics were idealized to fit the exact mathematical form.

In an actual IGFET, the value of K might depend somewhat on the point initially used to establish the curve-fitting process. In such a case, the best fit would probably be in the region in which operation of the IGFET is expected.

6–10 GENERAL OVERVIEW OF FETs

To simplify the development thus far in the chapter, we have considered only n-channel FETs. Thus, we have considered the n-channel JFET, the n-channel depletion-mode IGFET, and the n-channel enhancement-mode IGFET. All of these devices are also available as p-channel components. The construction of a given p-channel device is similar to that of the corresponding n-channel device except that all n-type semiconductor sections are replaced by p-type sections, and vice versa. All current directions and voltage polarities are reversed.

The schematic symbols of all the n-channel FETs and the corresponding p-channel devices are summarized in Figure 6–21. The left-hand column represents the various n-channel FETs, which have all been introduced earlier in the chapter. The right-hand column represents the corresponding symbols for the complementary p-channel FETs. In the symbols, if a given arrow is directed *into* the device (irrespective of whether it is on the left or right), the device is an n-channel unit. If the arrow is directed *out* of the device, it is a p-channel unit.

The positive direction of I_D is *into the drain* (and, of course, *out of the source*) for n-channel devices as we have seen throughout the chapter. By defining I_D as flowing *into the source and out of the drain* for p-channel devices, the value of I_D will be positive for normal operation. By performing this reversal in convention, the values of I_{DSS} and K will likewise be positive.

FIGURE 6–21
Summary of schematic symbols for the various FETs.

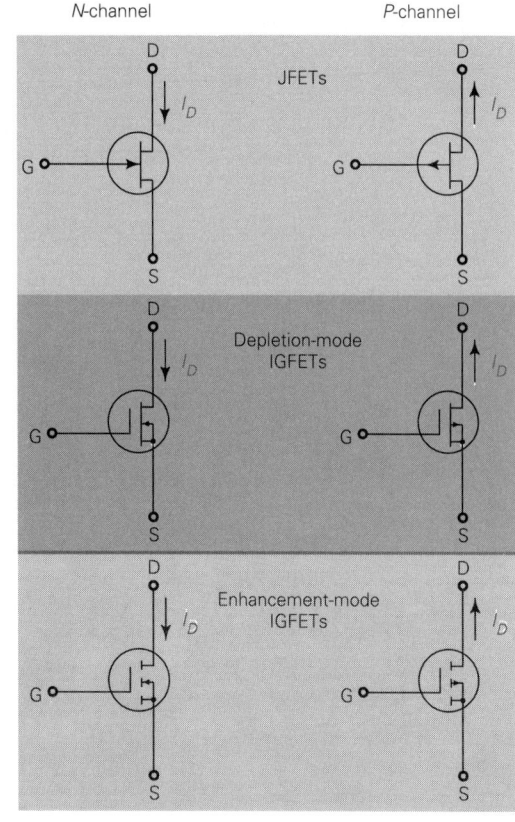

The reader will recall that when BJTs were studied, it was convenient to switch from V_{CE} and V_{BE} for *npn* transistors to V_{EC} and V_{EB} for *pnp* transistors, so that the resulting voltages were normally positive. Since we have just discussed reversing I_D for *p*-channel FETs, one might be led to conclude that we should switch references on the FET voltages in the same manner. Although this could be done, it is best to leave the pertinent FET voltages as V_{GS} and V_{DS} and let the signs assume whatever values are required, because reversing the conventions on the FET voltages creates some difficulties in working with the cutoff, pinchoff, and threshold voltage parameters and with identification of the beyond-pinchoff regions. Thus, the subscripts of the voltages V_{GS} and V_{DS} will be maintained, but the signs will be changed as required.

A summary of the direction and sign patterns is provided in Table 6–1. For this purpose, it is convenient to classify FETs as either the (1) **depletion-mode group,** or as the (2) **enhancement-mode group.** The depletion-mode group includes JFETs and depletion-mode IGFETs. As far as sign patterns and functional relationships are concerned, these two types of FETs behave very much alike. The enhancement-mode group consists of the enhancement-mode IGFETs.

TABLE 6-1
Summary of directions and polarities for different types of FETs in beyond-pinchoff (or possibly ohmic) regions

		Depletion-mode group (JFETs and depletion-mode IGFETs[a])					
		N-channel				P-channel	
I_D	$V_{GS(\text{off})}$	V_{GS}[a]	V_{DS}	I_D	$V_{GS(\text{off})}$	V_{GS}[a]	V_{DS}
↓	−	$V_{GS(\text{off})} < V_{GS} \leq 0$	+	↑	+	$0 \leq V_{GS} < V_{GS(\text{off})}$	−

$$I_D = I_{DSS}\left[1 - \frac{V_{GS}}{V_{GS(\text{off})}}\right]^2 = I_{DSS}\left[1 - \frac{|V_{GS}|}{|V_{GS(\text{off})}|}\right]^2$$

		Enhancement-mode group					
		N-channel				P-channel	
I_D	$V_{GS(\text{th})}$	V_{GS}	V_{DS}	I_D	$V_{GS(\text{th})}$	V_{GS}	V_{DS}
↓	+	$V_{GS} > V_{GS(\text{th})}$	+	↑	−	$V_{GS} < V_{GS(\text{th})}$	−

$$I_D = K\left[\frac{V_{GS}}{V_{GS(\text{th})}} - 1\right]^2 = K\left[\frac{|V_{GS}|}{|V_{GS(\text{th})}|} - 1\right]^2$$

[a] Depletion-mode IGFETs may also function with V_{GS} on other side of zero.

The left-hand column represents the *n*-channel devices, and the right-hand column represents the *p*-channel devices. As noted, I_D is defined in a direction such that it is positive. The directions of the arrows remind us of these positive directions (assuming that the drain is at the top).

Refer first to the left-hand-side (*n*-channel) block for the depletion-mode group. We note that $V_{GS(\text{off})}$ is negative (implying that V_p is positive) and V_{GS} normally assumes values in the range $V_{GS(\text{off})} < V_{GS} < 0$. W note also that V_{DS} is positive.

Corresponding patterns for the *p*-channel block on the right-hand side are opposite. Thus $V_{GS(\text{off})}$ is positive (implying that V_p is negative), the voltage V_{GS} normally assumes a value between 0 and the positive value $V_{GS(\text{off})}$, and V_{DS} normally assumes a negative value.

The equation for I_D in the beyond-pinchoff region is listed below the depletion-mode blocks in two ways. Both equations as given apply to both *n*-channel and *p*-channel units. The current I_D and the parameter I_{DSS} are both defined as positive, as noted. The quantities V_{GS} and $V_{GS(\text{off})}$ are both negative for an *n*-channel device and are both positive for a *p*-channel device. Thus, the ratio $V_{GS}/V_{GS(\text{off})}$ is positive for both cases. In the right-hand equation form, magnitudes are used, and the ratio is obviously positive.

Refer now to the left-hand-side block (*n*-channel) for the enhancement-mode group. We note that $V_{GS(th)}$ is positive and that V_{GS} normally assumes values in the range $V_{GS} > V_{GS(th)}$. It is also noted that V_{DS} is positive.

Corresponding inequalities for the *p*-channel block on the right-hand side are opposite. Thus $V_{GS(th)}$ is negative, the voltage V_{GS} normally assumes a lower value (more negative) than $V_{GS(th)}$, and V_{DS} assumes a negative value.

The equation for I_D in the beyond-pinchoff region is listed below the enhancement-mode blocks in two ways. Both equations as given apply to both *n*-channel and *p*-channel units. Since V_{GS} and $V_{GS(th)}$ both assume the same sign, the ratio $V_{GS}/V_{GS(th)}$ is always positive.

6–11 READING FET DATA SHEETS

A brief overview of some typical FET dc specifications is presented in this section. The particular FET group to be studied consists of the 2N4220 through the 2N4222 along with newer versions of these types (as indicated by the addition of A to the number). Refer to Appendix D for the data sheets on this group. These transistors are *n*-channel JFETs.

The absolute maximum ratings are listed near the top of the first data page. These data are similar in most ways to the corresponding BJT maximum ratings, and similar discussions of Section 4–6 are applicable here. The maximum values of the drain–source and drain–gate voltages are given as 30 V. The maximum gate–source voltage is indicated as −30 V. (Alternatively, this could have been stated as +30 V for the maximum source–gate voltage.) The maximum drain current is specified as 15 mA.

For a dc drain–source voltage V_{DS} and a dc drain current I_D, the power P_D dissipated in the FET is

$$P_D = V_{DS} I_D \qquad \qquad \textbf{(6–55)}$$

The maximum power dissipation at 25°C is listed as 300 mW, and the derating factor is 2 mW/°C.

Various electrical characteristics at 25°C are listed in the table. The **gate–source breakdown voltage** is denoted as $V_{(BR)GSS}$, and the minimum value is listed as −30 V. Earlier it was indicated that the maximum gate–source voltage rating was −30 V, so the reader may be confused by the use of the terms *minimum* and *maximum*.

This can be explained as follows: The *minimum* possible reverse breakdown voltage for the gate–source junction is 30 V. Therefore, if the maximum *gate–source* voltage is kept under 30 V, the junction should never encounter reverse breakdown. Thus, the absolute maximum ratings are our guide to safe design.

The normal mode of operation for the JFET is with the gate–channel reverse biased. The very small current that flows for this condition is called **gate reverse**

current, and it is denoted as I_{GSS}. A typical value at 25°C is -0.1 nA. The negative sign indicates that the current is flowing out of the gate. The small value confirms our earlier practical assumption that the gate is nearly an open circuit. At a temperature of 100°C, however, the maximum value increases to -100 nA.

The next item is the **gate–source cutoff voltage.** Minimum and typical values are not listed, but the maximum values range from -4 V for the 2N4220 to -8 V for the 2N4222. The corresponding range for the pinchoff voltage is 4 to 8 V.

In practice, the exact gate–source cutoff is somewhat difficult to measure accurately. For that reason, an additional block of data indicating the gate–source voltage for a reference small drain current is given.

Refer now to the block entitled **zero-gate–voltage drain current,** which is the same as the **zero-bias drain current.** The range of I_{DSS} depends on the transistor type and varies from 0.5 to 3 mA for a 2N4220 to a range of 5 to 15 mA for a 2N4222.

The block entitled **static drain–source ON resistance** corresponds to $V_{GS} = 0$ and was denoted as r_{DSO} in this chapter. This typical value ranges from 500 to 300 Ω, depending on the transistor type. The remaining data on the first page refer to small-signal or ac parameters and are considered later in the text.

Refer now to the second and third pages of the data. Several sets of drain characteristics, corresponding to different cutoff voltages, have been provided. In each case, the corresponding transfer characteristic is shown. These curves should be interpreted as *representative* of the possible types of characteristics that might be obtained for transistors in the group.

6–12 PSPICE EXAMPLES

PSPICE EXAMPLE 6–1

Develop a PSPICE program to determine the drain characteristics of the generic *n*-channel JFET model.

Solution

The PSPICE circuit used to determine the characteristics is shown in Figure 6–22(a), and the code is shown in Figure 6–22(b). Both V_{GS} and V_{DS} are to be varied so no values are included on their defining lines. As always, the analysis defaults to dc when there is no identifier provided for the source type.

The two lines used in providing the transistor identification are the following

```
J1 2 1 0 JTEST
.MODEL JTEST NJF
```

The letter J is the symbol of a JFET, and additional letters or numbers without a space may be added as appropriate. The order of the three integers that follow is very important, and they denote the *drain, gate,* and *source,* respectively. The name JTEST is an arbitrary name that must link with the model line.

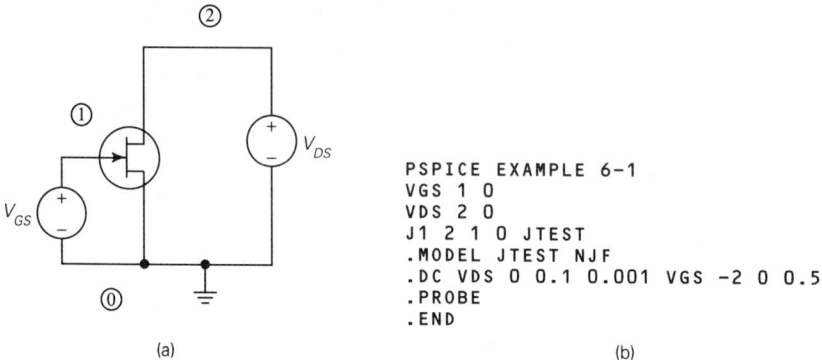

```
PSPICE EXAMPLE 6-1
VGS 1 0
VDS 2 0
J1 2 1 0 JTEST
.MODEL JTEST NJF
.DC VDS 0 0.1 0.001 VGS -2 0 0.5
.PROBE
.END
```

(a)   (b)

FIGURE 6–22
Circuit and code of PSPICE Example 6–1.

Following the .MODEL statement, the name JTEST is given, meaning that this name is attached to the model type to be provided. Finally, the generic n-channel JFET model is denoted by its required name NJF. Incidentally, the name of the corresponding p-channel JFET is PJF.

We will first focus on the characteristics in the ohmic region. This can be achieved by sweeping the drain–source voltage over a very narrow range. The

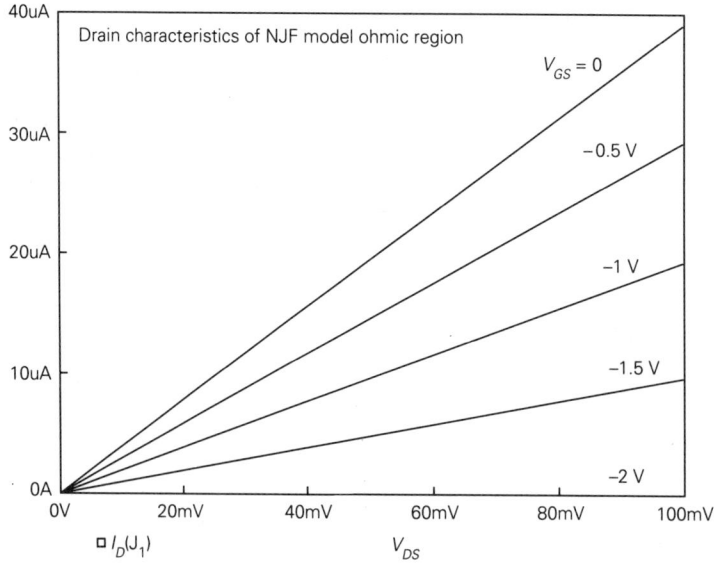

FIGURE 6–23
Drain characteristics of n-channel generic JFET model in ohmic region as determined in PSPICE Example 6–1.

gate–source cutoff voltage is known to be about −2 V, so V_{GS} is swept from −2 to 0 V in steps of 0.5 V.

The drain characteristics are shown in Figure 6–23. Note the general linear nature of the curves in this regon.

FIGURE 6–24

Code of PSPICE Example 6–1 modified for larger drain–source voltage sweeps.

```
PSPICE EXAMPLE 6-1
VGS 1 0
VDS 2 0
J1 2 1 0 JTEST
.MODEL JTEST NJF
.DC VDS 0 20 0.1 VGS -2 0 0.5
.PROBE
.END
```

Next, the program is modified to provide a much larger sweep of the drain–source voltage, and the code is shown in Figure 6–24. The corresponding drain characteristics are shown in Figure 6–25. Observe the flat behavior of the curves in the beyond-pinchoff region.

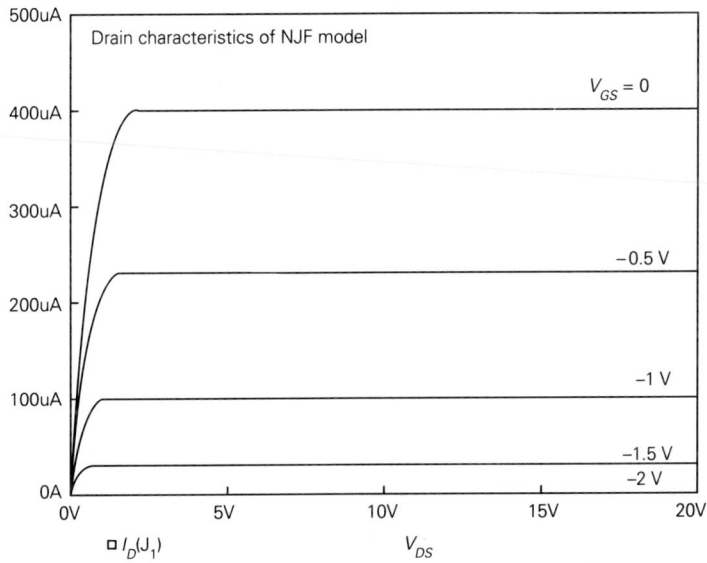

FIGURE 6–25

Drain characteristics of *n*-channel generic JFET over larger drain–source voltage range in PSPICE Example 6–1.

PROBLEMS

Drill Problems

6–1. The drain characteristics in the linear portion of the ohmic region for a certain idealized n-channel JFET are shown in Figure P6–1. Determine the values of conductance and resistance for each of the following values of V_{GS}: **(a)** 0; **(b)** -1 V; **(c)** -2 V.

FIGURE P6–1

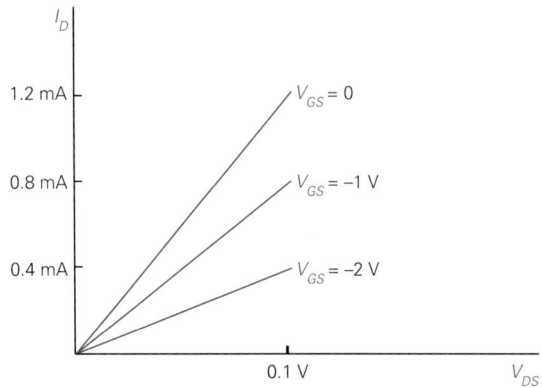

6–2. The drain characteristics in the linear portion of the ohmic region for a certain idealized n-channel JFET are shown in Figure P6–2. Determine the values of conductance and resistance for each of the following values of V_{GS}: **(a)** 0; **(b)** -0.5 V; **(c)** -1 V; **(d)** -1.5 V.

FIGURE P6–2

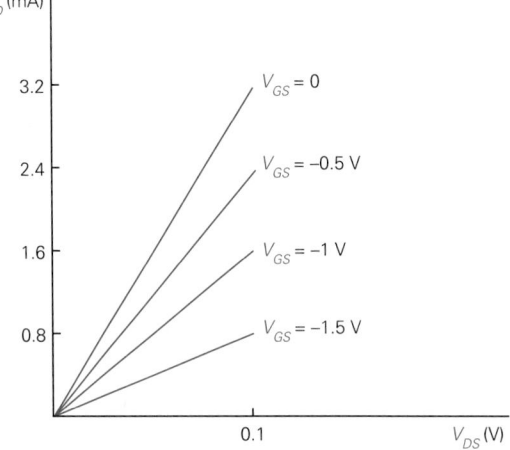

6–3. The overall drain characteristics of the idealized JFET of Problem 6–1 are shown in Figure P6–3. **(a)** Determine the gate–source cutoff voltage and the pinchoff volt-

FIGURE P6–3

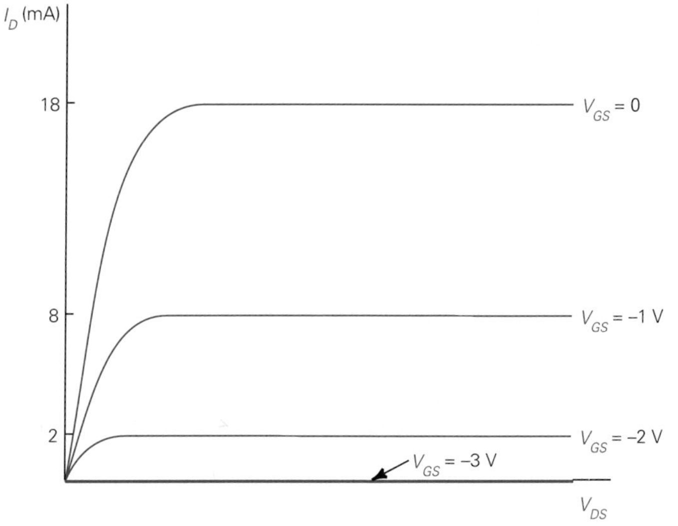

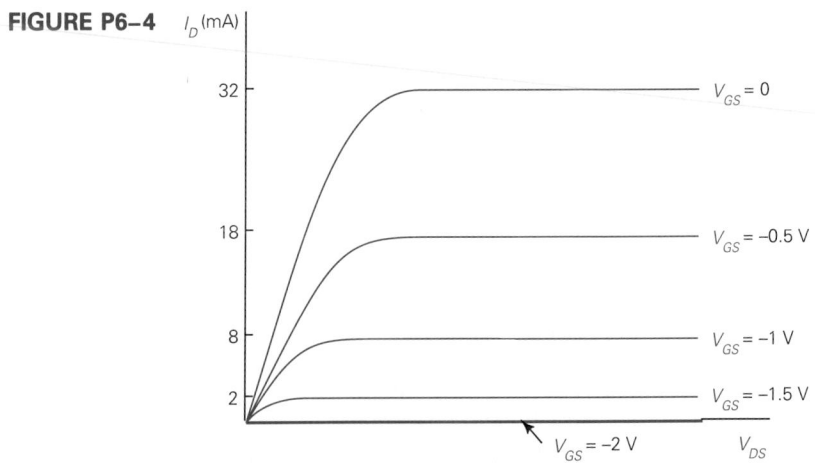

age. **(b)** Determine the transition voltage (from ohmic to beyond cutoff regions) for the following values of V_{GS}: 0, −1 V, −2 V.

6–4. The overall drain characteristics of the idealized JFET of Problem 6–2 are shown in Figure P6–4. **(a)** Determine the gate–source cutoff voltage and the pinchoff voltage. **(b)** Determine the transition voltage (from ohmic to beyond-pinchoff regions) for the following values of V_{GS}: 0, −0.5 V, −1 V, −1.5 V.

FIGURE P6–4

6–5. Refer to the idealized n-channel JFET characteristics of Figure P6–3. **(a)** Determine the zero-bias drain current. **(b)** Write an equation for I_D in the beyond-pinchoff region. **(c)** Verify that the result of **(b)** predicts the correct values of current for the curves of Figure P6–3.

6–6. Refer to the idealized n-channel JFET characteristics of Figure P6–4. **(a)** Determine the zero-bias drain current. **(b)** Write an equation for I_D in the beyond-pinchoff region. **(c)** Verify that the result of **(b)** predicts the correct values of current for the curves of Figure P6–4.

6–7. Using the results of Problem 6–5, determine for the given transistor the values of V_{GS} that produce the following drain current values (in beyond-pinchoff region): **(a)** 5 mA; **(b)** 10 mA; **(c)** 15 mA.

6–8. Using the results of Problem 6–6, determine for the given transistor the values of V_{GS} that produce the following drain current values (in beyond-pinchoff region): **(a)** 5 mA; **(b)** 12 mA; **(c)** 25 mA.

6–9. For the idealized JFET drain characteristics of Figure P6–3, construct the transfer characteristic.

6–10. For the idealized JFET drain characteristics of Figure P6–4, construct the transfer characteristic.

6–11. **(a)** Write an equation for the conductance in the linear portion of the ohmic region for the JFET of Problem 6–1. You may need to use some of the results of Problems 6–3 and 6–5. **(b)** Use the results of **(a)** to compute the conductance for each of the lines of Figure P6–1, and compare with the results of Problem 6–1.

6–12. **(a)** Write an equation for the conductance in the linear portion of the ohmic region for the JFET of Problem 6–2. You may need to use some of the results of Problems 6–4 and 6–6. **(b)** Use the results of **(a)** to compute the conductance for each of the lines of Figure P6–2, and compare with the results of Problem 6–2.

6–13. A certain n-channel JFET has a gate–source cutoff voltage of -5 V and a zero-bias drain current of 12 mA. Assume that the JFET is biased with a gate–source voltage of -2 V. Determine **(a)** the ideal drain–source resistance in the linear portion of the ohmic region, and **(b)** the ideal drain current in the beyond-pinchoff region.

6–14. A certain p-channel JFET has a gate–source cutoff voltage of 4 V and a zero-bias drain current of 1 mA. Assume that the JFET is biased with a gate–source voltage of 1 V. Determine **(a)** the ideal drain–source resistance in the linear portion of the ohmic region, and **(b)** the ideal drain current in the beyond-pinchoff region.

6–15. Draw a beyond-pinchoff dc model for the JFET of Problem 6–13.

6–16. Draw a beyond-pinchoff dc model for the JFET of Problem 6–14.

6–17. Assume that the drain characteristics of Figure P6–3 apply to some hypothetical n-channel depletion-mode IGFET. Calculate the idealized values of drain current in the beyond-pinchoff region for the following positive values of gate–source voltage: **(a)** 0.5 V; **(b)** 1 V.

6–18. Assume that the drain characteristics of Figure P6–4 apply to some hypothetical n-channel depletion-mode IGFET. Calculate the idealized values of drain current in the beyond-pinchoff region for the following positive values of gate–source voltage: **(a)** 0.5 V; **(b)** 1 V.

6–19. The drain characteristics of an idealized n-channel enhancement-mode IGFET are shown in Figure P6–19. **(a)** Determine the gate–source threshold voltage. **(b)** Determine the transition voltage (from ohmic to beyond-pinchoff region) for the following values of V_{GS}: 3 V, 4 V, 5 V, 6 V.

6–20. The drain characteristics of an idealized n-channel enhancement-mode IGFET are shown in Figure P6–20. **(a)** Determine the gate–source threshold voltage. **(b)** Determine the transition voltage (from ohmic to beyond-pinchoff region) for the following values of V_{GS}: 2 V, 2.5 V, 3.5 V, 4 V.

FIGURE P6–19

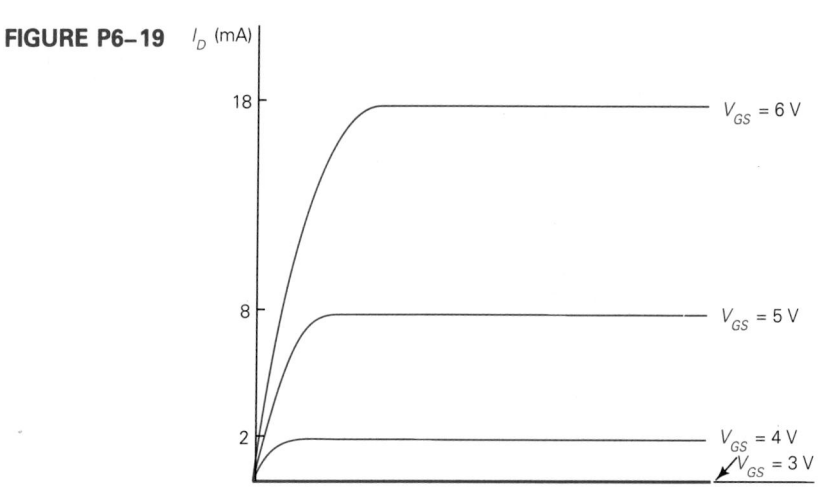

FIGURE P6–20

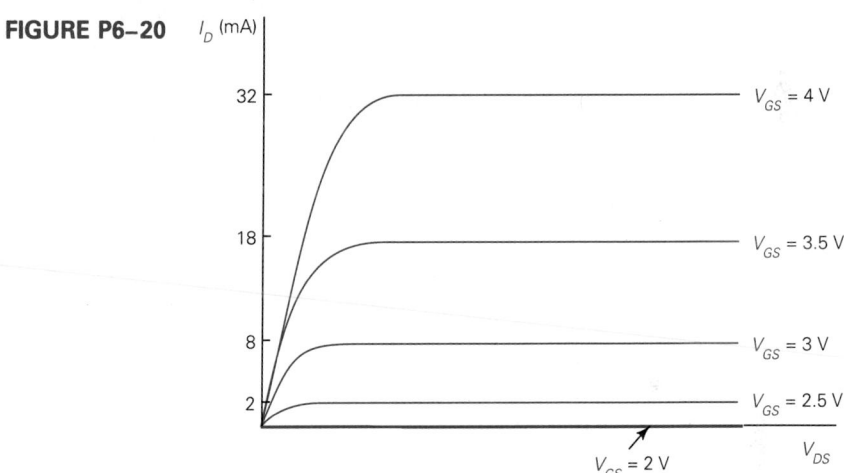

6–21. Refer to the idealized n-channel enhancement-mode characteristics of Figure P6–19.
(a) Determine the value of K. (b) Write an equation for I_D in the beyond-pinchoff region. (c) Verify that the result of (b) predicts the correct values of current for the curves of Figure P6–19.

6–22. Refer to the idealized n-channel enhancement-mode characteristics of Figure P6–20.
(a) Determine the value of K. (b) Write an equation for I_D in the beyond-pinchoff region. (c) Verify that the result of (b) predicts the correct values of current for the curves of Figure P6–20.

6–23. Using the results of Problem 6–21, determine for the given transistor the values of V_{GS} that produce the following values of drain current (in the beyond-pinchoff region):
(a) 5 mA; (b) 12 mA; (c) 20 mA.

6–24. Using the results of Problem 6–22, determine for the given transistor the values of V_{GS} that produce the following values of drain current (in the beyond-pinchoff region): **(a)** 5 mA; **(b)** 12 mA; **(c)** 25 mA.

6–25. For the idealized IGFET drain characteristics of Figure P6–19, construct the transfer characteristic.

6–26. For the idealized IGFET drain characteristics of Figure P6–20, construct the transfer characteristic.

6–27. A certain n-channel enhancement-mode IGFET starts to conduct when the gate–source voltage is 3 V. When the gate–source voltage is 4 V, the drain current is 5 mA. Determine the drain current when the gate–source voltage is 5 V.

6–28. A certain p-channel enhancement-mode IGFET starts to conduct when the gate–source voltage is −2 V. When the gate–source voltage is −3 V, the drain current is 6 mA. Determine the drain current when the gate–source voltage is −4 V.

6–29. Draw a beyond-pinchoff dc model for the IGFET of Problem 6–27.

6–30. Draw a beyond-pinchoff dc model for the IGFET of Problem 6–28.

6–31. The drain current in a certain n-channel JFET can be expressed as

$$I_D = 6 \times 10^{-3} \left(1 + \frac{V_{GS}}{2.3}\right)^2$$

Determine **(a)** the gate–source cutoff voltage, and **(b)** the zero-bias drain current.

6–32. The drain current in a certain p-channel JFET can be expressed as

$$I_D = 0.018 \left(1 - \frac{V_{GS}}{4.5}\right)^2$$

Determine **(a)** the gate–source cutoff voltage, and **(b)** the zero-bias drain current.

6–33. The drain current in a certain n-channel enhancement-mode IGFET can be expressed as

$$I_D = 0.012 \left(\frac{V_{GS}}{2.8} - 1\right)^2$$

Determine the threshold voltage.

6–34. The drain current in a certain p-channel enhancement-mode IGFET can be expressed as

$$I_D = 0.002(V_{GS} + 4)^2$$

Determine the threshold voltage.

6–35. A certain n-channel JFET has $V_{GS(off)} = -3$ V. Identify the region of operation (*cutoff, ohmic,* or *beyond pinchoff*) for each of the following conditions: **(a)** $V_{GS} = -2$ V, $V_{DS} = 12$ V; **(b)** $V_{GS} = -1$ V, $V_{DS} = 1$ V; **(c)** $V_{GS} = -4$ V, $V_{DS} = 12$ V; **(d)** $I_D = 5$ mA, $V_{DS} = 8$ V.

6–36. A certain p-channel JFET has $V_{GS(off)} = 4$ V. Identify the region of operation (*cutoff, ohmic,* or *beyond pinchoff*) for each of the following conditions: **(a)** $V_{GS} = 5$ V, $V_{DS} = -15$ V; **(b)** $V_{GS} = 2$ V, $V_{DS} = -12$ V; **(c)** $V_{GS} = 1$ V, $V_{DS} = -2$ V; **(d)** $I_D = 8$ mA, $V_{DS} = -10$ V.

6–37. A certain *n*-channel enhancement-mode IGFET has $V_{GS(th)} = 3$ V. Identify the region of operation (*cutoff, ohmic,* or *beyond pinchoff*) for each of the following conditions: **(a)** $V_{GS} = 2$ V, $V_{DS} = 10$ V; **(b)** $V_{GS} = 5$ V; $V_{DS} = 3$ V; **(c)** $V_{GS} = 5$ V, $V_{DS} = 1$ V.

6–38. A certain *p*-channel enhancement-mode IGFET has $V_{GS(th)} = -4$ V. Identify the region of operation (*cutoff, ohmic,* or *beyond pinchoff*) for each of the following conditions: **(a)** $V_{GS} = -7$ V, $V_{DS} = -2$ V; **(b)** $V_{GS} = -5$ V, $V_{DS} = -12$ V; **(c)** $V_{GS} = 0$ V, $V_{DS} = -10$ V.

Derivation Problems

6–39. Starting with (6–10) for I_D expressed in terms of V_{GS} for a depletion-mode FET, derive (6–12) for V_{GS} expressed in terms of I_D.

6–40. Starting with (6–43) for I_D expressed in terms of V_{GS} for an enhancement-mode FET, derive (6–44) for V_{GS} expressed in terms of I_D.

Troubleshooting Problems

6–41. While troubleshooting a circuit containing an *n*-channel JFET that *should* be operating in the beyond-pinchoff region, the following ground-referred voltages are measured: $V_S = 15$ V, $V_D = 36$ V, $V_G = 1$ V. What can you conclude about the *likely* region of operation for the transistor?

6–42. While troubleshooting a circuit containing a *p*-channel JFET that *should* be operating in the ohmic region, the following ground-referred voltages are measured: $V_D = 5$ V, $V_S = 30$ V, $V_G = 30$ V. What can you conclude about the *likely* region of operation for the transistor?

7

FET-RESISTOR CIRCUIT APPLICATIONS

OBJECTIVES

After completing this chapter, the reader should be able to:

- Construct a dc load line for the drain circuit of an FET.
- Demonstrate how changes in different transistor circuit parameters and variables affect the load line.
- Analyze or design a circuit in which an FET is used as a switch.
- Show how an FET biased in the ohmic region can be used as a voltage-controlled attenuator.
- Discuss the different approaches for biasing an FET and their relative stability.
- Analyze or design a self-bias circuit for an FET.
- Construct a bias line on the transfer characteristic.
- Discuss the operation of a self-plus-fixed bias circuit.
- Analyze or design a source-follower circuit.
- Analyze some of the circuits in the chapter with PSPICE.

7–1 OVERVIEW

We introduced the field-effect transistor (FET) in Chapter 6, and we developed the terminal characteristics in detail. We are now ready to begin the consideration of various circuits that use the FET.

In this chapter all circuits considered will use only transistors and resistors (along with dc bias sources). Primary focus will be on dc analysis and operating characteristics. However, many of these circuits can be used with time-varying inputs and outputs provided that it is the instantaneous total voltages and currents that are of interest.

313

Among the applications to be considered are FET switches, voltage-controlled resistors, current sources, and source-follower circuits. Such circuits find applications in interfacing between integrated circuits, in buffers and drivers, for switching and multiplexing, and in electronically adjustable-gain control circuits.

We consider the establishment of stable dc bias circuits for the different types of FETs. Both the analysis circuits and the design of existing circuits are discussed.

7–2 FET DRAIN–SOURCE LOAD LINE

We introduced the concept of the load line for a bipolar junction transistor (BJT) in Chapter 5. It is appropriate to begin the discussion of FET dc circuit analysis by showing a similar load line for a relatively simple circuit configuration. The transition from the linear ohmic region to the beyond-pinchoff region is more gradual for the FET than the transition from the saturation region to the active region for the BJT. The load line will help us to visualize the process better.

Consider the circuit of Figure 7–1(a) that contains an *n*-channel JFET, a resistance R_D in series with the drain, and two dc bias supplies. As it stands, the circuit is not necessarily intended to represent any useful application, but concepts inherent in this circuit will arise in practical application circuits later. Observe the directions of the polarities of the two bias sources for the *n*-channel JFET.

No current is assumed to flow in the gate. The equation for the gate–source portion of the circuit is simply

$$V_{GS} = -V_{GG} \tag{7–1}$$

Application of KVL for the drain–source loop in a counterclockwise direction yields

$$-V_{DD} + R_D I_D + V_{DS} = 0 \tag{7–2}$$

Solving for I_D, we obtain

$$I_D = \left(-\frac{1}{R_D}\right) V_{DS} + \frac{V_{DD}}{R_D} \tag{7–3}$$

This equation is of the form $y = mx + b$, and it is analogous to the collector–emitter loop equation of a BJT as given by (5–132).

Typical JFET drain characteristics are shown in Figure 7–1(b), and the linear equation of (7–3) may readily be superimposed on these curves. The vertical intercept is V_{DD}/R_D, the horizontal intercept is V_{DD}, and the slope of the load line is $-1/R_D$.

Most of the properties discussed in Chapter 5 for the BJT load line apply here. The Q-point is determined from the intersection of the curve corresponding to $V_{GS} = -V_{GG}$ with the load line. If V_{GS} is varied, movement of the Q-point takes

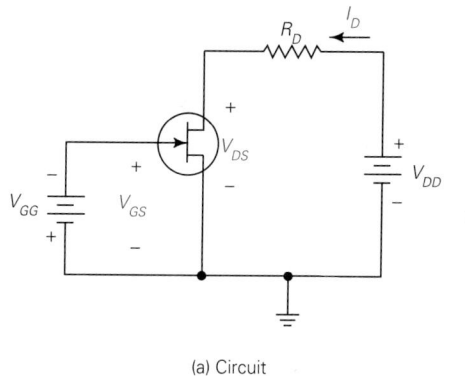

(a) Circuit

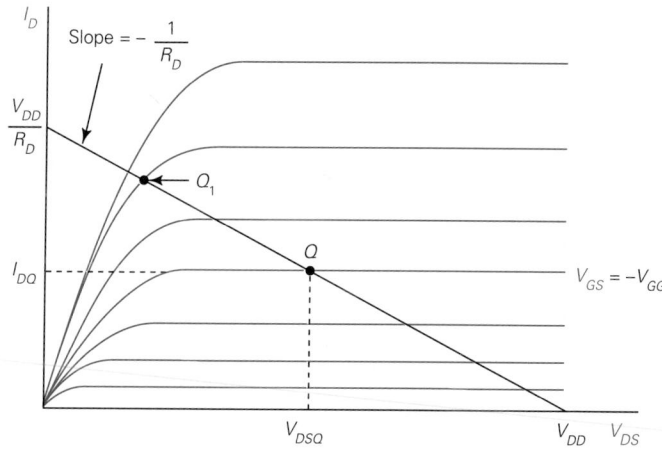

(b) Drain-source load line

FIGURE 7–1
JFET circuit and drain–source load line.

place along the load line. Once again, we see the **voltage control property** of the FET, which is in contrast to the **current control property** of the BJT. The subscript Q is added to any pertinent variables at the Q-point, as noted.

Suppose that V_{GS} were increased (made more positive) so that the Q-point moves to Q_1 in Figure 7–1(b). We are now out of the pinchoff region and into the ohmic region. This may not always be apparent from the circuit model if a computation of V_{DS} is made from the mathematical relationship. A check of the transition voltage $V_{DS(\text{tran})}$ would, of course, reveal that the beyond-pinchoff model would not be valid. However, the load-line interpretation helps us to visualize this process better.

If beyond-pinchoff operation is desired, make sure that the range of V_{GS} in the particular circuit is such that all applicable curves intersect with the load line in the flat region of the output characteristics. The calculated value of V_{DS} should always be greater than the pertinent transition voltage. If not, the beyond-pinchoff model is not valid, and some other approach must be used.

In applications involving the FET as a voltage-controlled resistor or as a switch, the load line should intersect the curves in the ohmic region. Furthermore, the range of V_{GS} (or other variables) must be such as to maintain operation in the ohmic region for that application.

A similar load line could be constructed for any of the FET types if the drain characteristics are known. The polarity of the gate–source voltage will depend, of course, on whether the device belongs to the depletion-mode group or the enhancement-mode group. For example, if the device is an n-channel enhancement-mode IGFET, the values of V_{GS} corresponding to the different output characteristics are positive and larger than $V_{GS(\text{th})}$.

EXAMPLE 7–1

The circuit of Figure 7–2(a) contains an n-channel JFET. The JFET drain characteristics are assumed to be idealized and equal to those considered in Figure 6–6. Verify analytically that operation is in the beyond-pinchoff region, and determine I_D and V_{DS}.

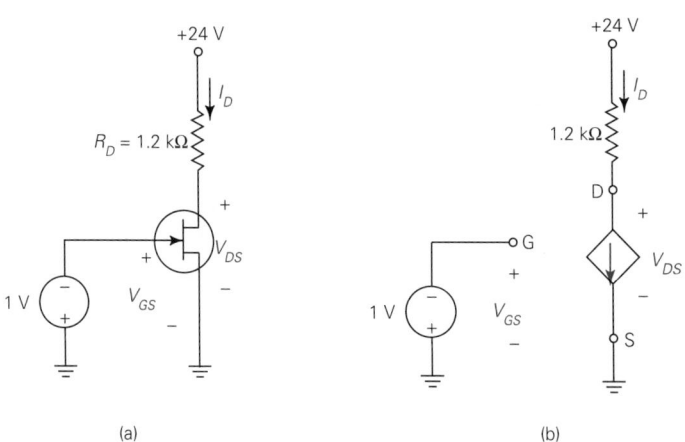

FIGURE 7–2
Circuit of Example 7–1 and equivalent-circuit model.

Solution
A brief review of Figure 6–6 indicates that $I_{DSS} = 16$ mA and $V_{GS(\text{off})} = -4$ V. Since $V_{GS} = -1$ V in this circuit, it is obvious that the FET is not in the cutoff region. Strictly speaking, we do not yet know whether the FET is operating in the beyond-pinchoff region or in the ohmic region. We will proceed on the assumption that the circuit is operating in the beyond-pinchoff region and, subsequently,

verify the validity of that assumption. An equivalent-circuit model is shown in Figure 7–2(b).

In the beyond-pinchoff region, the equation for I_D is

$$I_D = 16 \times 10^{-3} \left(1 - \frac{V_{GS}}{-4}\right)^2 \qquad (7\text{–}4)$$

For $V_{GS} = -1$ V, we have

$$I_D = 16 \times 10^{-3} \left(1 - \frac{-1}{-4}\right)^2 = 9 \times 10^{-3}\,\text{A} = 9\,\text{mA} \qquad (7\text{–}5)$$

For this particular value of V_{GS}, we could have determined I_D from Figure 6–6. The value of V_{DS} is

$$V_{DS} = 24 - (9 \times 10^{-3}) \times (1.2 \times 10^3) = 24 - 10.8 = 13.2\,\text{V} \qquad (7\text{–}6)$$

The transition voltage for $V_{GS} = -1$ V is $V_{DS(\text{tran})} = 3$ V. Since the calculated value of V_{DS} is much greater than the transition voltage, assumption of beyond-pinchoff region operation is valid.

EXAMPLE 7–2

For the circuit of Figure 7–2(a), assume that the drain resistance is changed to $R_D = 3$ kΩ. Attempt to repeat the analysis of Example 7–1.

Solution
The alert reader is probably aware that there is something different here, because of the term *attempt*. Proceeding with the questionable assumption of beyond-pinchoff operation, the calculated drain current would be the same as before (i.e., 9 mA). If a calculation of V_{DS} is attempted, the value obtained would be $V_{DS} = 24 - 9 \times 10^{-3} \times 3 \times 10^3 = 24 - 27 = -3$ V. This result is impossible! Not only is this value smaller (by 6 V) than the transition voltage of 3 V, but also it is negative.

The difficulty lies in the fact that operation has shifted into the ohmic region and that the beyond-pinchoff model is no longer valid. If the calculated value of V_{DS} is less than the transition voltage (i.e., if $V_{DS} < V_{DS(\text{tran})}$), the circuit is operating in the ohmic region.

The situation here is analogous to that of the BJT when saturation occurs. For FET circuits not designed specifically for switching purposes, however, the ohmic region may represent a substantial range of V_{DS}, and a simple assumption such as $V_{DS} \simeq 0$ may represent a gross error.

If operation is shifted far enough into the ohmic region to justify the constant resistance model given in Chapter 6, a complete solution may be possible. We consider this concept in Section 7–4. However, the drain characteristics are somewhat nonlinear in the transition region between the linear part of the ohmic region and the beyond-pinchoff region, and a simple analysis is not possible. Although nonlinear models for this region have been developed, their lack of utility does not justify further consideration here.

Based on the material developed thus far, about all we can say is that V_{DS} is somewhere between 0 and 3 V and I_D is less than 9 mA. We will show in Example 7–3 how a better solution and better visualization can be achieved through the use of the load line.

EXAMPLE 7–3

Construct load lines for the circuit conditions of **(a)** Example 7–1 and **(b)** Example 7–2.

Solution

The idealized output characteristics were originally given in Figure 6–6, and they are shown again in Figure 7–3. The horizontal scale of the latter figure has been extended somewhat because of the required range of V_{DS}.

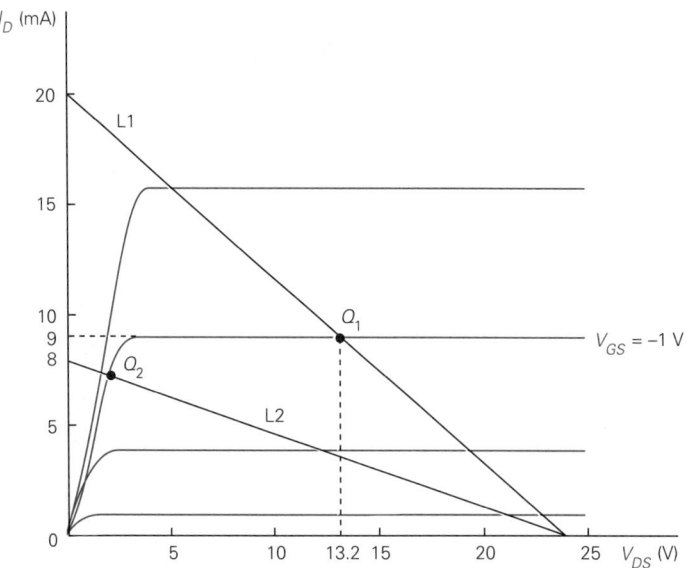

FIGURE 7–3
Drain characteristics and load lines of Example 7–2.

(a) In Example 7–1, $V_{DD} = 24$ V and $R_D = 1.2$ kΩ. The horizontal intercept is 24 V, and the vertical intercept is 24 V/1.2×10^3 Ω $= 20 \times 10^{-3}$ A $= 20$ mA. The resulting load line is L1 in Figure 7–3. The Q-point is determined from the intersection of L1 with the curve corresponding to $V_{GS} = -1$ V, and it is indicated as Q_1. The values of I_D and V_{DS} corresponding to the intersection are 9 mA and 13.2 V, respectively. These values are in agreement with the results of Example 7–1.

(b)   In Example 7–2, $V_{DD} = 24$ V and $R_D = 3$ kΩ. The horizontal intercept is again 24 V, but the vertical intercept is 24 V/3 × 10^3 Ω = 8 × 10^{-3} A = 8 mA. The resulting load line is L2 in Figure 7–3. The Q-point in this case is indicated as Q_2, and it is in the ohmic region. This portion of the ohmic region is somewhat nonlinear, and we can only make graphical estimates of I_D and V_{DS}. Reasonable estimates are $I_D \simeq 7.2$ mA and $V_{DS} \simeq 2.2$ V.

7–3   FET SWITCHES

FETs may be used as electronic switches in a manner quite similar to that of BJTs. In fact, well-designed FET switches have certain advantages over BJT switches. Although it was largely ignored in Chapter 5, a saturated BJT always exhibits a small offset $V_{CE(\text{sat})}$. In contrast, an FET acts as a purely passive resistor near the origin in the lower portion of the ohmic region.

A rather extensive line of special-purpose FETs is manufactured and marketed for the specific application of electronic switching. These units are referred to as **analog switches.** Most of them are integrated-circuit (IC) modules containing one or more FETs along with necessary drive and level conversion circuitry. The input control levels are often chosen to be compatible with standard digital logic circuits.

Consideration of these special-purpose ICs is not within the intended objective of this book. Instead, we will investigate some of the concepts of switching using basic FETs that can be used for limited noncritical switching functions. Most important for our purposes, they should help the reader to understand the general concept basic to more complex switching circuit configurations.

For any FET device used as a switch, the control signal must be capable of switching between two extremes: (1) cutoff region, and (2) ohmic region. In the cutoff region, the FET will act like an open circuit, and the resulting switch will be open. In the ohmic region, the FET will act like a resistor. The design must be such that the resulting value of V_{DS} in the ohmic region is very small compared with the source voltage, so that the FET approaches a short circuit. Said differently, the FET equivalent resistance in the ohmic region should be small compared with the net circuit series resistance.

Depending on whether the FET is a depletion-mode device or an enhancement-mode device and whether it is an n-channel or a p-channel device, a variety of possible control voltage levels and switched levels are possible. The cases that will be considered are intended as representative of the many possibilities.

N-Channel JFET Switch

Consider the n-channel JFET shown in Figure 7–4(a), in which a load is connected between the drain and a positive voltage V_{DD}. When the control input signal v_i is more negative than $V_{GS(\text{off})}$ (i.e., $v_i < V_{GS(\text{off})}$), the FET is at cutoff, and the equivalent circuit is shown in Figure 7–4(b).

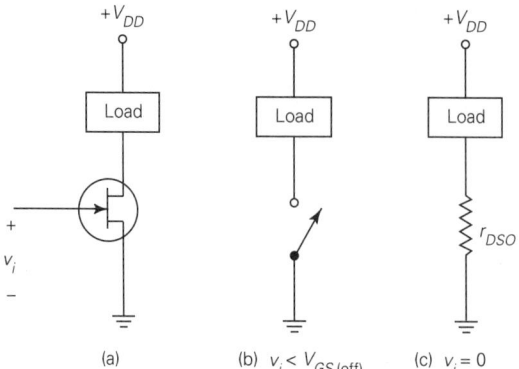

FIGURE 7–4
Electronic switch using an *n*-channel JFET to close switch with zero voltage.

Assume next that $v_i = 0$. Operation now shifts to the $V_{GS} = 0$ curve. If the circuit has been designed properly, the intersection of the load line with the $V_{GS} = 0$ curve will be in the ohmic region, and the FET will act like a small resistance r_{DSO}. The equivalent circuit is shown in Figure 7–4(c). Most of the supply voltage V_{DD} will appear across the load, although a small portion will appear across the FET.

It was assumed in the preceding development that the load line intersected the $V_{GS} = 0$ curve in the ohmic region, and that assumption is fundamental to the operation of the FET as a switch. With a poor design, it is possible that the intersection will occur in the beyond-pinchoff region, and the desired switching operation will not be achieved. Once again, we see the utility of the load line in visualizing the process.

N-Channel Enhancement-Mode IGFET Switch

The same switching arrangement considered before, but with an *n*-channel en-hancement-mode IGFET, is shown in Figure 7–5(a). The control input voltage v_i in this case consists of two nonnegative levels. If the control voltage is less than the threshold voltage (i.e., $v_i < V_{GS(th)}$), the FET is an open circuit as shown in Figure 7–5(b). The value $v_i = 0$ would be a good choice in this case.

If $v_i > V_{GS(th)}$, and if the load line is such that operation shifts to the ohmic region, the equivalent circuit is of the form shown in Figure 7–5(c). In practice, this control level should be sufficiently larger than $V_{GS(th)}$ to shift the value of V_{DS} to a very small voltage level.

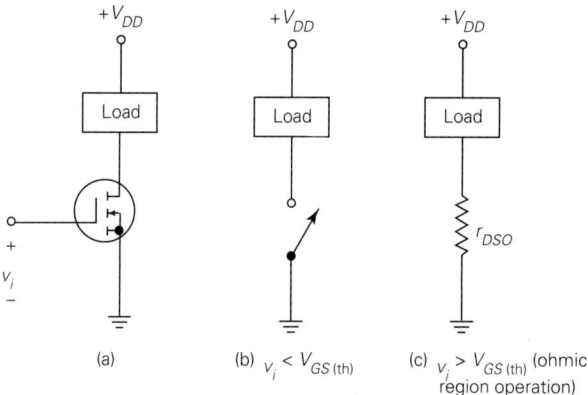

FIGURE 7–5
Electronic switch using an *n*-channel enhancement-mode IGFET to close switch with positive voltage.

EXAMPLE 7–4

It is desired to switch on and off the 6-kΩ load in the circuit of Figure 7–6. The FET characteristics are assumed to be those considered in Figure 6–6 and again in Figure 7–3. **(a)** Using a load line, show that the circuit is capable of operating as a switch. **(b)** Determine the required control levels for v_i. **(c)** Estimate the voltage v_L across the load when the switch is on.

FIGURE 7–6
Circuit of Example 7–4.

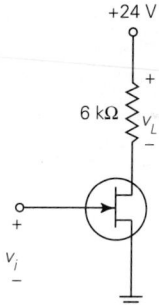

Solution
(a) The drain characteristics are shown again in Figure 7–7. To construct the load line, we first determine the horizontal and vertical intercepts. The horizontal intercept is 24 V, and the vertical intercept is $24 \text{ V}/6 \times 10^3 \ \Omega = 4 \times 10^{-3} \text{ A} = 4$ mA. The load line is then superimposed on the drain characteristics as shown in Figure 7–7. The intersection of the load line with the curves corresponding to the larger (less negative) values of V_{GS} is in the ohmic region, so operation as a switch is feasible.

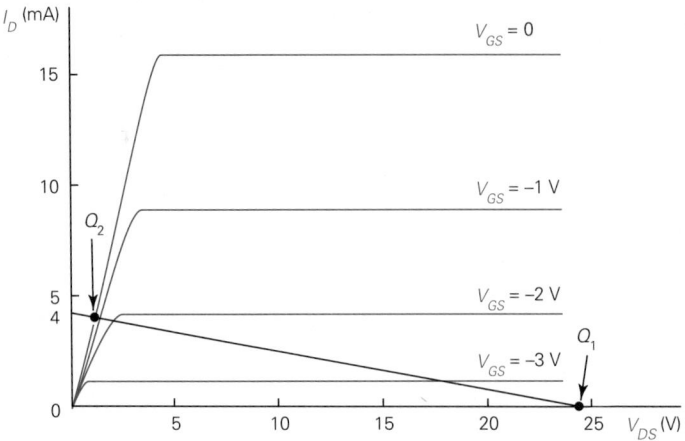

FIGURE 7–7
Drain characteristics of Example 7–4.

(b) For $v_i = V_{GS(off)} = -4$ V, the FET is at cutoff, as indicted by Q_1 in Figure 7–7. In practical FETs, cutoff is not as abrupt as suggested by these idealized curves. To ensure cutoff, a voltage more negative than $V_{GS(off)}$ would probably be used. (Of course, it cannot be as negative as the gate–source reverse breakdown voltage.)

For $v_i = 0$, operation shifts over to Q_2, and the FET is well into the ohmic region. One might be tempted to increase the gate drive above this level, but this should *not* be done for the JFET. The reason is that the gate–source diode would then be forward biased, and this would drastically alter its operation. Thus $v_i = 0$ is the usual maximum positive limit for the *n*-channel JFET.

(c) A reasonable estimate for V_{DS} at point Q_2 is $V_{DS} \approx 1$ V. The load voltage is then $V_L \approx 24 - 1 = 23$ V. Thus, we are not able to achieve perfect switching with this idealized "general-purpose" FET. Special-purpose analog switches could do much better.

7–4 VOLTAGE-CONTROLLED RESISTORS

We have observed that an FET acts as a voltage-controlled resistance in the ohmic region. This property may be used in a number of possible applications, such as voltage-controlled gain circuits, stabilization circuits, and tuning circuits. In this section, we discuss the concept in more detail and a representative example is shown.

In the same sense as for FET switches, we will choose to discuss the application of general-purpose FETs to this end. This should permit the reader to understand the concept of how the special-purpose devices work for this application.

Any circuit utilizing an FET for creating a voltage-controlled resistance must be designed so that operation is in the ohmic portion of the drain characteristics. Although the nonlinear portion may suffice for certain noncritical situations, the most exacting applications require operation in the linear portion of the ohmic region. In general, this means that the value of V_{DS} for the FET must be maintained at a relatively low level. If the signal level is very low, this condition is usually easy to achieve. If the signal level is moderate or large, a large drain resistance would be required to drop the voltage across the FET to a small level.

Refer to the expanded drain characteristics of Figure 7–8. These curves represent those of a typical *n*-channel JFET in the lower portion of the ohmic region, in which each curve has a nearly constant slope over a small range. The value 0.1 V is typical and may vary with different units. The load line of an applicable circuit should intersect the characteristic curves in the linear region as indicated.

FIGURE 7–8

Expanded scale of drain characteristics in lower part of ohmic region and load line for voltage-controlled resistance.

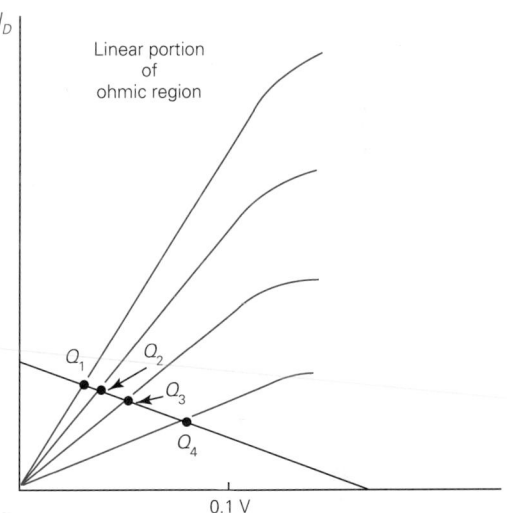

For the representative load line shown, the horizontal intercept is relatively small (<0.1 V), indicating a low-voltage application. By varying V_{GS}, operation may be shifted between the different Q-points shown. At each point, the resistance is different. Higher input voltages are possible as long as the load line intersects in the pertinent portion of the output characteristics.

EXAMPLE 7–5

The circuit of Figure 7–9 is to be used as a voltage-controlled **attenuator.** (**Attenuation** is the process of reducing a signal level, and an **attenuator** is a circuit that provides attenuation.) The FET characteristics in the ohmic region are assumed to be idealized and equal to those considered in Example 6–6. Let $A = v_o/v_i$

FIGURE 7–9

Voltage-controlled attenuator of Example 7–5.

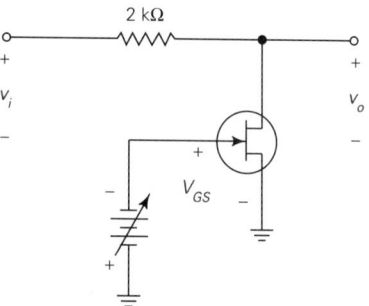

represent the signal transfer ratio. This transfer ratio is to be governed by the control voltage V_{GS}. The signal v_i is a small positive voltage with a peak value not exceeding 0.1 V. Compute the values of A corresponding to the following values of V_{GS}: **(a)** 0; **(b)** -1 V; **(c)** -2 V; **(d)** -3 V; **(e)** -4 V.

Solution

The hypothetical JFET was assumed to have linear characteristics for a range of V_{DS} up to about 0.1 V, and since the signal level will not exceed a peak level of 0.1 V, operation in the linear portion of the ohmic region should be ensured.

In Example 6–6, the conductance g_{DS} of this JFET was determined to be

$$g_{DS} = 8 \times 10^{-3} \left(1 - \frac{V_{GS}}{-4} \right) \tag{7–7}$$

The corresponding resistance $r_{DS} = 1/g_{DS}$ is

$$r_{DS} = \frac{1}{g_{DS}} = \frac{125}{1 - V_{GS}/(-4)} \tag{7–8}$$

The transfer, or output–input, ratio of the circuit in Figure 7–9 is determined by a simple application of the voltage-divider rule with r_{DS} representing the resistance of the JFET. We have

$$A = \frac{v_o}{v_i} = \frac{r_{DS}}{r_{DS} + 2000} \tag{7–9}$$

For each value of V_{GS}, the value of r_{DS} is computed from (7–8). This value is then substituted in (7–9) to determine A. A table of results follows.

	V_{GS} (V)	r_{DS} (Ω)	A
(a)	0	125	0.0588
(b)	-1	166.67	0.0769
(c)	-2	250	0.1111
(d)	-3	500	0.2
(e)	-4	∞	1

The voltage transfer ratio varies from 0.0588 to 1 as V_{GS} is varied from 0 to -4 V. This illustrates one way in which the level of a signal can be controlled electronically by an external control signal using an FET as a voltage-controlled resistor.

By using a voltage-controlled attenuator in conjunction with a fixed amplifier gain, the net effect is that of a voltage-controlled gain circuit. Such circuits are very important for achieving automatic gain adjustment in systems using feedback techniques.

7–5 DEPLETION-MODE FET BIAS CIRCUITS

In this section, we investigate some of the most common circuit configurations for establishing dc bias for depletion-mode FETs. Bias circuits are required for providing appropriate operating conditions with various applications, such as amplifiers.

For bias analysis we lump together JFETs and depletion-mode IGFETs, since the basic approaches in the circuit designs are virtually identical. (Strictly speaking, depletion-mode IGFETs can also be operated in the enhancement-mode, but that situation can be analyzed from the material given in Section 7–6.)

For ease in the development, *n*-channel units will be considered. The circuit configurations given can all be used for *p*-channel units if the polarities of all power supplies are reversed.

Fixed Bias

The most obvious way to establish an operating bias for a JFET is to use a separate fixed power supply. Consider the circuit of Figure 7–10 containing an *n*-channel JFET in a common-source configuration. The drain is connected through a drain resistance R_D to a positive dc voltage $+V_{DD}$. Since the required value of V_{GS} for

FIGURE 7–10
Fixed bias for an *n*-channel JFET using a separate power supply.

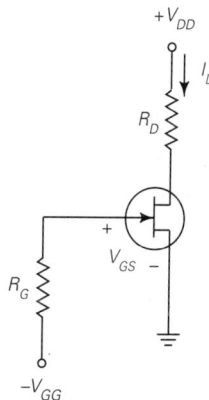

an n-channel JFET is negative, the gate is connected through a resistance R_G to a negative dc voltage $-V_{GG}$.

Until we study amplifier circuits, the reader may wonder why the negative dc voltage is not connected directly to the gate. The reason is that when an ac signal is applied to the input, there must be a resistance across which the input signal can be developed. Without R_G, the gate voltage can never vary and no signal amplification can occur. A similar property holds for the drain circuit with respect to R_D. We postpone full consideration to Chapter 11, but realize that these resistors will eventually be required.

The actual gate current in an FET is extremely small and will be neglected for this analysis. Therefore, the dc voltage drop across R_G is essentially zero, and the value of V_{GS} is

$$V_{GS} = -V_{GG} \tag{7-10}$$

Thus the gate–source voltage is simply equal to the value of the negative bias supply. For beyond-pinchoff operation, which is normal for most amplifier applications, the drain current is simply

$$I_D = I_{DSS} \left[1 - \frac{V_{GS}}{V_{GS(\text{off})}} \right]^2 \tag{7-11}$$

While the process seems straightforward as long as I_{DSS} and $V_{GS(\text{off})}$ are known, this circuit has only limited application in production situations. The values of I_{DSS} and $V_{GS(\text{off})}$ of a given type vary widely from one transistor to another, and it is usually impractical to adjust V_{GS} to compensate.

To illustrate the potential difficulty, let us return to the concept of the transfer characteristic curve introduced in Section 6–4. Consider the two transconductance curves T1 and T2 shown in Figure 7–11. Curve T1 has parameters $I_{DSS}^{(1)}$ and $V_{GS(\text{off})}^{(1)}$, and curve T2 has parameters $I_{DSS}^{(2)}$ and $V_{GS(\text{off})}^{(2)}$. Assume that a certain drain current $I_D^{(1)}$ has been established with a fixed bias $V_{GS} = -V_{GG}$ for an FET with the characteristic T1. Now assume that for the same bias voltage, the first FET is replaced with one that has the transfer characteristic T2. The drain current changes to a new value $I_D^{(2)}$, which is more than double the first value in the illustration. The basic parameters for a given FET type can vary by $4:1$ or more, so the potential stability of the operating point is very poor.

In summary, fixed bias circuits are practical only for limited applications, where it is feasible to adjust the bias for individual transistors or, in some applications, where the actual bias level is not critical.

Self-Bias

We now investigate a circuit that can be used to provide a degree of stabilization of the operating point while eliminating the necessity of providing a separate power supply for the gate bias. A portion of the drain power supply voltage is used to

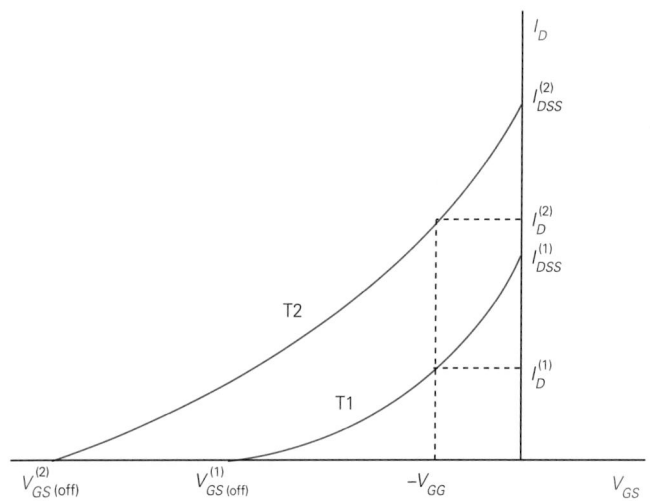

FIGURE 7–11
Illustration of shift in operating point with fixed bias.

establish bias, and because of the manner in which the bias is generated, it is referred to as **self-bias.**

Consider the circuit shown in Figure 7–12. As in the case of the fixed bias circuit, this circuit contains a drain resistance R_D and a gate resistance R_G. In addition, there is a source resistance R_S whose value is critical in establishing bias, as we will see shortly.

In the same fashion as for the fixed bias circuit, the resistance R_G serves as a load across which an input signal can be developed in an amplifier configuration.

FIGURE 7–12
Self-bias circuit for an *n*-channel JFET.

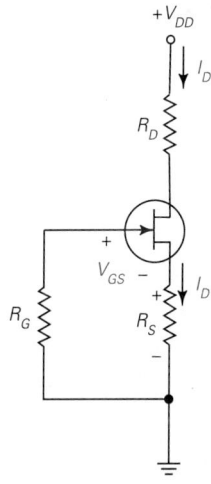

However, it also serves to provide a dc path between gate and ground, which is necessary in the self-bias circuit.

The value of R_G is usually quite large (typically, 100 kΩ or greater), and its exact value is usually not critical. However, R_S is much smaller (typically, <1 kΩ), and its value has a critical effect on the Q-point.

We will start at the gate terminal and form a clockwise loop through the source resistor to ground and back to the gate. As this loop is formed, we will recognize that the voltage across R_S is $R_S I_D$. However, since we are assuming no gate current in the ideal JFET, the voltage drop across R_G is zero, as noted earlier. The loop equation thus reads

$$V_{GS} + R_S I_D + 0 = 0 \qquad (7\text{--}12)$$

where the last 0 on the left represents the assumed zero drop across R_G.

There is, of course, a very small gate current, and, in order to neglect the drop across R_G, we are assuming that the product of R_G times this gate current is small compared with the other voltages. This condition is usually achieved by keeping R_G below a few megohms.

Solving for V_{GS} in (7–12), we have

$$\boxed{V_{GS} = -R_S I_D} \qquad (7\text{--}13)$$

Thus the gate-to-source voltage is simply the negative of the voltage drop across the source resistance. Since $R_S I_D$ is a positive value for an n-channel JFET, V_{GS} is a negative value, as required. We see why the circuit is called a self-bias circuit, since the FET generates its own bias through the voltage drop across R_S.

A different way of visualizing this process is as follows: The gate voltage with respect to ground is zero since the gate is connected through R_G to ground and there is negligible voltage drop across R_G. However, the FET source terminal voltage is now positive with respect to ground because of the drop across R_S. Lifting the source potential above ground while maintaining the gate potential at ground results in the same bias effect as maintaining the source terminal at ground and placing a negative bias voltage at the gate.

Assuming beyond-pinchoff–region operation, the drain current is given by

$$I_D = I_{DSS}\left[1 - \frac{V_{GS}}{V_{GS(\text{off})}}\right]^2 \qquad (7\text{--}14)$$

Let's inspect (7–13) and (7–14) to see what we have determined thus far. From (7–14), I_D is a function of V_{GS}. However, from (7–13), V_{GS} is, in turn, a function of I_D. It seems as if we are going around in circles! The key to the problem is that there are two equations, and as long as we have no more than two unknowns, we can solve the set of equations simultaneously.

Because of the uncertainty of the parameters and the fact that the expression for I_D itself is approximate, the resulting solution should be considered approximate at best. Nevertheless, there is value in the solution in predicting the approximate operating point when the circuit is given.

Assume for the moment that I_{DSS}, $V_{GS(off)}$, and R_S are known and that the two unknowns are V_{GS} and I_D. Simultaneous solution of (7–13) and (7–14) results in a quadratic equation, for which there are two possible solutions for each unknown. However, one each of the values represents a physically impossible situation and is rejected.

Self-Bias Solution

The solution that follows has been developed in a general manner that can be used for any depletion-mode device (i.e., JFETs and depletion-mode IGFETs). Let $|V_{GS(off)}|$ represent the magnitude of the cutoff voltage and let I_{DSS} represent the zero-bias–drain current. (We are defining I_{DSS} as a positive value for both n-channel and p-channel devices, so we do not need magnitude bars for it.) As a convenient parameter that simplifies the equation that follows, we define B as

$$B = \frac{|V_{GS(off)}|}{R_S I_{DSS}} \tag{7–15}$$

The drain current I_D for the self-bias circuit is given by

$$I_D = I_{DSS}\left(B + \frac{B^2}{2} - \sqrt{B^3 + \frac{B^4}{4}}\right) \tag{7–16}$$

for either an n-channel or a p-channel device. The magnitude of gate–source voltage is then

$$|V_{GS}| = R_S I_D \tag{7–17}$$

The value of V_{GS} is actually negative for an n-channel device and positive for a p-channel device.

EXAMPLE 7–6

The self-bias circuit shown in Figure 7–13 contains the hypothetical idealized n-channel JFET whose drain characteristics were given in Figures 6–6, 7–3, and 7–7. Determine analytically the values of I_D, V_{GS}, V_{R_S}, V_{R_D}, and V_{DS}.

Solution
The values of I_{DSS} and $V_{GS(off)}$ for this JFET were determined earlier to be $I_{DSS} = 16$ mA and $V_{GS(off)} = -4$ V. The analytical procedure involving equations (7–16) and (7–17) will be used to determine I_D. First, B from (7–15) is calculated as

$$B = \frac{4}{150 \times 16 \times 10^{-3}} = 1.6667 \tag{7–18}$$

FIGURE 7–13
Circuit of Example 7–6.

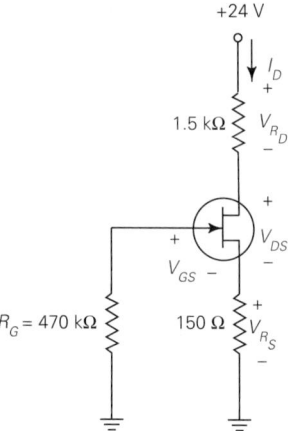

This quantity is substituted in (7–16), and I_D is determined as

$$I_D = 16 \times 10^{-3} \left[1.6667 + \frac{(1.6667)^2}{2} - \sqrt{(1.6667)^3 + \frac{(1.6667)^4}{4}} \right] \quad \textbf{(7–19)}$$

$$= 7.913 \times 10^{-3} \text{ A} = 7.913 \text{ mA}$$

The value of V_{GS} for an *n*-channel JFET is negative and is

$$V_{GS} = -R_S I_D = -150 \times 7.913 \times 10^{-3} = -1.187 \text{ V} \quad \textbf{(7–20)}$$

The voltage V_{R_S} across the source resistance has the same magnitude as V_{GS}, but the positive terminal is at the top as shown in Figure 7–13. Thus,

$$V_{R_S} = 1.187 \text{ V} \quad \textbf{(7–21)}$$

As indicated, it is assumed that the gate current is negligible and that there is no voltage drop across the 470-kΩ gate resistor. The voltage V_{R_D} across the drain resistance is

$$V_{R_D} = 1.5 \times 10^3 \times 7.913 \times 10^{-3} = 11.87 \text{ V} \quad \textbf{(7–22)}$$

The voltage V_{DS} across the FET is the supply voltage minus the drops across R_D and R_S.

$$V_{DS} = 24 - 11.87 - 1.187 = 10.94 \text{ V} \quad \textbf{(7–23)}$$

7–6 BIAS LINE

A great deal of insight on self-bias can be gained from a graphical approach. Assume for the moment that I_{DSS} and $V_{GS(\text{off})}$ are known. Consider the transfer characteristic displaying I_D as a function of V_{GS} in the beyond-pinchoff region, as

FIGURE 7–14
Bias line superimposed on transfer characteristic.

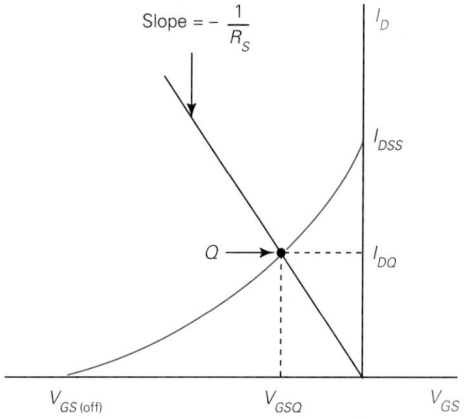

shown in Figure 7–14. From (7–13), we can solve for I_D in terms of V_{GS}, and we have

$$I_D = -\frac{1}{R_S} V_{GS} \qquad (7\text{–}24)$$

This equation is a straight-line equation of the form $y = mx + b$, in which the vertical intercept is $b = 0$. The straight line thus intersects the origin, and the slope is $m = -1/R_S$.

The straight line given by (7–24) can be superimposed on the transfer characteristic as shown in Figure 7–14. This line is called the **bias line,** and it has some similarities to the load line.

The intersection of the bias line with the transfer characteristic establishes the operating, or Q-point. The resulting values of V_{GS} and I_D at the Q-point are denoted as V_{GSQ} and I_{DQ} on Figure 7–14. When self-bias is employed, it is more convenient to work with a bias line on the transfer characteristic than with the load-line and drain characteristics.

The process that occurs in the circuit is that I_D settles at a value such that the value of $V_{GS} = -R_S I_D$ is the required voltage for causing I_D. This circuit represents a form of negative feedback, a major topic considered later in the book.

The self-bias circuit provides a moderate degree of stability to changes in transistor characteristics. To illustrate this concept, refer to Figure 7–15. Assume initially that a given FET has the transfer characteristic T_1 with parameters $V_{GS(\text{off})}^{(1)}$ and $I_{DSS}^{(1)}$. For a particular value of R_S, operation is at the point Q_1, corresponding to $V_{GS}^{(1)}$ and $I_D^{(1)}$, as shown.

Assume that the FET is replaced by one that has a different transfer characteristic T_2 with parameters $V_{GS(\text{off})}^{(2)}$ and $I_{DSS}^{(2)}$. If the circuit had employed fixed bias with a value $V_{GS}^{(1)}$, operation would shift all the way up to Q_3. However, with self-bias and the same source resistance, operation shifts to Q_2 instead. The values

FIGURE 7–15

Change in operating point with self-bias circuit
for different transistors.

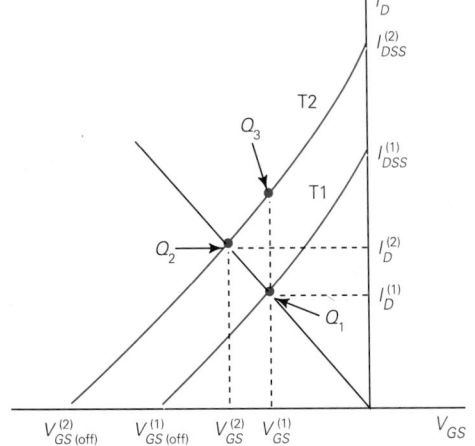

of V_{GS} and I_D at Q_2 are $V_{GS}^{(2)}$ and $I_D^{(2)}$, respectively. This represents a more moderate shift in the value of I_D than with fixed bias. There is, therefore, a degree of adjustment in the self-bias circuit that maintains a moderate degree of control on the operating-point current.

   Stabilization in the bias circuit usually requires that the drain current be kept within a reasonable range of variation. To achieve more stabilization with the basic self-bias circuit, it is necessary that the magnitude of the slope of the bias line be smaller. In this fashion the change in I_D for two different Q-points would be less. However, there is a practical limit to this process, as demonstrated next.

   Consider the two characteristics T_1 and T_2 in Figure 7–16. To minimize the variation in I_D, the bias line has been shown with a slope that has a small magnitude. This corresponds to a relatively large value of R_S. Although I_D has a higher degree of stability in this case, there are two problems with this approach: (1) The operating point for I_D is forced to assume a relatively small value. Later we will

FIGURE 7–16

Illustration of how more stabilization with self-bias forces operation to small values of I_D.

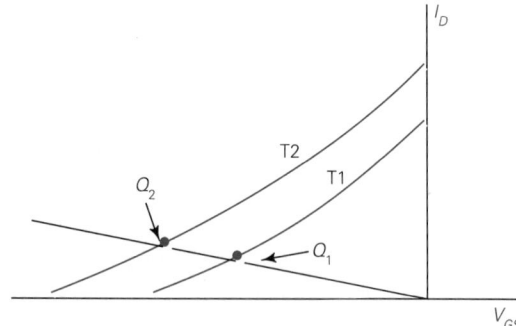

see that the resulting gain in amplifier applications will be relatively small as a consequence. (2) A larger dc voltage across R_S will result with this approach, and less of the supply voltage is available for the transistor and the load.

In conclusion, the self-bias circuit is very useful for less critical circuits, where a degree of stability is sufficient. By combining the self-bias circuit with fixed bias, a greater amount of stability can be achieved without the undesirable results discussed. This is covered in Section 7–8.

EXAMPLE 7–7

For the self-bias circuit of Example 7–6, determine the operating point using the bias line.

Solution

The transfer characteristic of the hypothetical JFET was given in Figure 6–7, and it is shown again in Figure 7–17. The bias line is a straight line through the origin with a slope $-1/R_S = -1/150 = -0.00667$ A/V. However, the easiest way to construct the bias line is to choose a convenient value of I_D, compute the corresponding voltage $V_{GS} = -150I_D$, and draw a straight line from the origin through this point. A convenient choice is $I_D = 10$ mA, in which case $V_{GS} = -150 \times 10^{-2} = -1.5$ V. Other points could have been chosen as well.

The intersection of this bias line with the transfer characteristic determines the operating point, and this is shown in Figure 7–17. It is obvious that the values of I_D and V_{GS} cannot be determined graphically to the degree of accuracy labeled. However, to show compatibility with the results of Example 7–6, the values obtained analytically have been labeled on the axes.

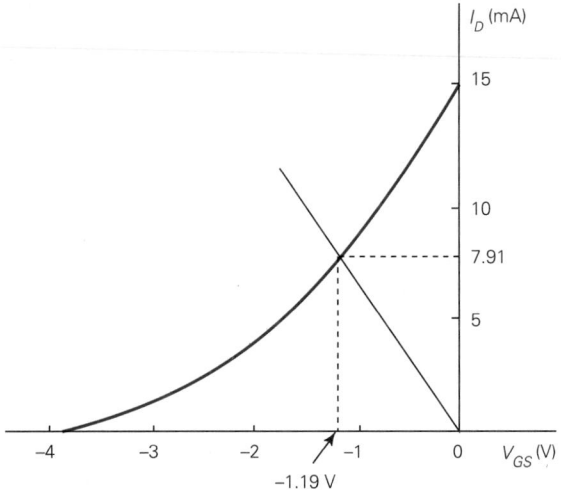

FIGURE 7–17
Bias-line analysis of Example 7–7.

7–7 DESIGN OF SELF-BIAS CIRCUIT

Consider the self-bias circuit of Figure 7–18 for an *n*-channel JFET. The design of this circuit involves the selection of a suitable operating point and the determination of the required source resistance R_S. Assume for the moment that the transconductance characteristics are known *exactly* (i.e., the values of $V_{GS(\text{off})}$ and I_{DSS} are known). For a desired operating-point drain current I_{DQ}, the value of V_{GSQ} that would produce this current is determined from either the transconductance graph or by solving the beyond-pinchoff equation.

FIGURE 7–18

Self-bias circuit for JFET with variable for design identified.

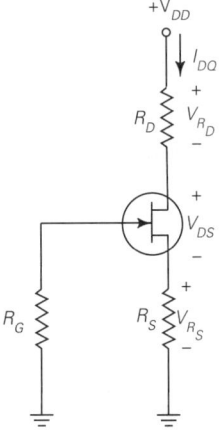

Because of the nonlinear relationship between V_{GS} and I_D, values of voltage and current along the curve do not follow an intuitively simple form. Some normalized convenient values that aid in dealing with this relationship are summarized in Table 7–1. For example, $I_D = 0.5 I_{DSS}$ when $V_{GS} = 0.2929 V_{GS(\text{off})}$, but $I_D =$

TABLE 7–1

Convenient gate–source voltage and drain current bias combinations

$V_{GS}/V_{GS(\text{off})}$	I_D/I_{DSS}
0	1
0.1340	0.75
0.25	0.5625
0.2929	0.5
0.5	0.25
0.75	0.0625
1	0

$0.25 I_{DSS}$ when $V_{GS} = 0.5 V_{GS(\text{off})}$. The normalized curve of Figure 6–8 is also useful for this purpose.

Once V_{GSQ} is determined, R_S is then calculated as

$$R_S = \frac{|V_{GSQ}|}{I_{DQ}} \tag{7-25}$$

Referring to Figure 7–18, the power supply voltage V_{DD} is divided into three parts: (1) V_{R_D}, (2) V_{DS}, and (3) V_{R_S}. The value of V_{R_S} is, of course, the magnitude of V_{GSQ}; that is,

$$V_{R_S} = |V_{GSQ}| \tag{7-26}$$

The remaining voltage is divided between V_{R_D} and V_{DSQ}, and R_D can be selected to achieve the required proportion. The required value of R_D is

$$R_D = \frac{V_{R_D}}{I_D} = \frac{V_{DD} - V_{DSQ} - V_{R_S}}{I_D} \tag{7-27}$$

The preceding procedure is fine *if* we know the exact parameters for a given FET. Suppose, however, that the design is to be performed for a particular type of FET in which the range, but not the exact values, of the parameter is known. Because the self-bias circuit has only a moderate degree of stability, one must accept the fact that the operating point will vary somewhat with different transistors. The simplicity of the self-bias circuit may justify its use as long as there is sufficient leeway in the specifications to accommodate the expected variations.

Assume that for a given FET, the minimum and maximum values of I_{DSS} are $I_{DSS}^{(1)}$ and $I_{DSS}^{(2)}$, respectively. Similarly, assume that the values of $V_{GS(\text{off})}$ corresponding to minimum and maximum magnitudes are $V_{GS(\text{off})}^{(1)}$ and $V_{GS(\text{off})}^{(2)}$, respectively. As a general rule, transistors with smaller values of I_{DSS} tend to correspond to smaller magnitudes of $V_{GS(\text{off})}$, and vice versa. As a reasonable assumption, two extreme transconductance characteristic curves can be constructed, as shown in Figure 7–19. The curve T_1 is an assumed lower bound, and T_2 is an assumed upper bound. All possible curves for the given FET are expected to fall in the bounded region, with some expected "tracking" of the characteristics.

Since R_S is a constant value, the bias line is unique and will intersect any applicable curve in the region. The range of possible drain currents will lie between $I_{DQ}^{(1)}$ and $I_{DQ}^{(2)}$, and the range of possible bias voltages will lie between $V_{GSQ}^{(1)}$ and $V_{GSQ}^{(2)}$.

If the self-bias circuit is to be acceptable in a production sense, the extreme range of possible values must produce acceptable dc operating-point values. This usually means investigating the effects of the various circuit voltages and currents as well as amplifier gain parameters (to be studied later). For a given power supply

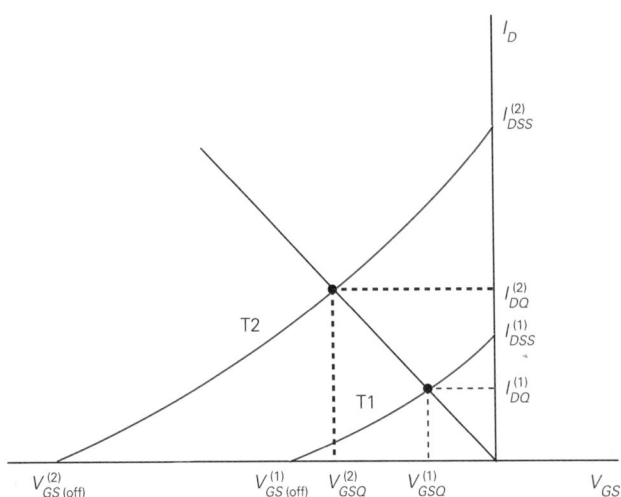

FIGURE 7–19
Range of operating points with self-bias circuit.

voltage and a particular drain resistance R_D, the voltage drops V_{R_D} and V_{R_S} will be minimum, and V_{DSQ} will be maximum at $I_{DQ}^{(1)}$. Conversely, V_{R_D} and V_{R_S} will be maximum, and V_{DSQ} will be minimum at $I_{DQ}^{(2)}$. Some detailed study along with a possible trial-and-error approach on the selection of R_S may be required to achieve an acceptable design.

For a p-channel JFET, the basic approach is the same, but the polarities of all voltages will be reversed. The preceding equations can therefore be modified to cover that situation if voltage magnitudes are used. For example, all voltages in (7–27) can be expressed as magnitudes, and the equation still applies. However, the two negative signs must be retained, since the magnitude of voltage across the FET is the magnitude of the power supply voltage minus the other two voltage magnitudes in the loop.

EXAMPLE 7–8

A certain n-channel JFET has the idealized characteristics given in Figures 6–6, 7–3, and 7–7. Design a self-bias circuit of the form of Figure 7–18 to achieve the following operating-point conditions: $I_{DQ} = 8$ mA and $V_{R_D} = V_{DSQ}$. The available power supply voltage is $V_{DD} = 15$ V.

Solution
Since $I_{DSS} = 16$ mA, $I_{DQ}/I_{DSS} = 8$ mA/16 mA $= 0.5$. The value of gate–source voltage required for this common operating-point condition is readily determined from Table 7–1. We note from the table that $V_{GS}/V_{GS(\text{off})} = 0.2929$ for $I_D/I_{DSS} = 0.5$. Since $V_{GS(\text{off})} = -4$ V, we have

$$V_{GSQ} = 0.2929 \times (-4) = -1.172 \text{ V} \qquad (7-28)$$

The required value of R_S is

$$R_S = \frac{1.172 \text{ V}}{8 \times 10^{-3} \text{ A}} = 146.5 \, \Omega \qquad (7\text{–}29)$$

The voltage V_{R_S} is

$$V_{R_S} = |V_{GSQ}| = 1.172 \text{ V} \qquad (7\text{–}30)$$

The dc voltage to be distributed between R_D and the drain–source drop is $15 - 1.172 = 13.828$ V. The specification calls for this voltage to be divided equally. Thus,

$$V_{R_D} = V_{DSQ} = \frac{13.828}{2} = 6.914 \text{ V} \qquad (7\text{–}31)$$

The drain resistance R_D is

$$R_D = \frac{V_{R_D}}{I_D} = \frac{6.914}{8 \times 10^{-3}} = 864.2 \, \Omega \qquad (7\text{–}32)$$

The value of the gate resistance R_G is not critical, and a range of possible values is usually acceptable. The value selected would probably depend on specifications concerning the minimum input resistance, which usually relates to the circuit as an amplifier. Since this question will be addressed later, we will indicate simply a typical standard value of 470 kΩ for the present purpose.

The circuit diagram with the calculated values of R_S and R_D is shown in Figure 7–20. Understand that these calculated values are *not* standard resistance values. In a situation such as this, one approach is to use the nearest standard values and hope that the shift in operating point is minimal. Another approach is to try to select an operating point such that all component values are standard. Example 7–9 will deal with this particular circuit after the component values are rounded to standard values.

FIGURE 7–20
Circuit design of Example 7–8 with idealized values.

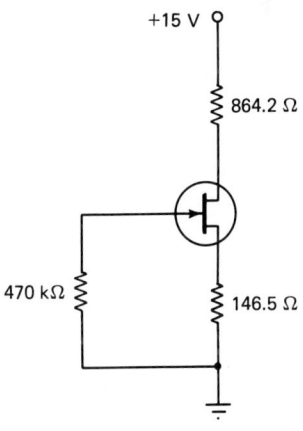

EXAMPLE 7–9

Assume that R_D and R_S in the design of Example 7–8 are rounded to the nearest standard values, as shown in Figure 7–21. Assuming the ideal JFET characteristics, determine the actual values of I_{DQ}, V_{GSQ}, V_{R_S}, and V_{DSQ}.

FIGURE 7–21
Circuit of Example 7–9 representing modification of design of Example 7–8 to standard values.

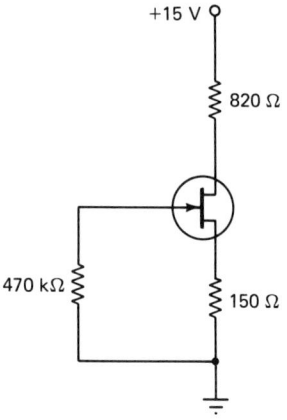

Solution
The first step is to determine the values of I_{DQ} and V_{GSQ} for the given transistor characteristics and the value of R_S. It turns out that the same transistor characteristics and the value of R_S were employed in Example 7–6, so we can use some of the results of that analysis. Reviewing the results of that problem, we have

$$I_{DQ} = 7.913 \text{ mA} \tag{7–33}$$

$$V_{GSQ} = -1.187 \text{ V} \tag{7–34}$$

$$V_{R_S} = 1.187 \text{ V} \tag{7–35}$$

The remaining variables are different than those of Example 7–6, since the power supply voltage and drain resistance are different. We have

$$V_{R_D} = 7.913 \times 10^{-3} \times 820 = 6.489 \text{ V} \tag{7–36}$$

$$V_{DS} = 15 - 6.489 - 1.187 = 7.324 \text{ V} \tag{7–37}$$

The various calculated values can be compared with the target design goals of Example 7–8. The most significant changes are in V_{R_D} and V_{DS}. The design goal was to achieve equal voltages. The value of V_{R_D} is less than its design goal, and V_{DS} is greater than its design goal.

EXAMPLE 7–10

Consider again the bias circuit design of Example 7–8. In this example we assume that the *exact* values of R_S and R_D calculated in Example 7–8 are employed. Obviously, this would produce the exact design goal operating point for the given exact transistor parameters. However, assume that the transistor production characteristics are subject to variation according to the following range:

$V_{GS(off)}$ variation: -6 to -2 V

I_{DSS} variation: 12 to 20 mA

The ideal values considered earlier of -4 V and 16 mA now represent the nominal midpoint values of the assumed range. Compute the range of operating-point conditions.

Solution
As discussed in the text, two extreme range transfer characteristic curves will be assumed. One corresponds to $V_{GS(off)} = -2$ V and $I_{DSS} = 12$ mA. The other corresponds to $V_{GS(off)} = -6$ V and $I_{DSS} = 20$ mA. All computations involving the first point will be denoted with a superscript (1), and similar computation at the second point can be denoted with a superscript (2).

The procedure of (7–15) and (7–16) can be employed to determine the operating-point current and gate–source voltage. The remaining calculations then proceed using basic circuit laws.

At the first point, the calculations are listed as follows:

$$B^{(1)} = \frac{2}{146.5 \times 12 \times 10^{-3}} = 1.1377 \qquad (7\text{–}38)$$

$$I_{DQ}^{(1)} = 12 \times 10^{-3} \left[1.1377 + \frac{(1.1377)^2}{2} - \sqrt{(1.1377)^3 + \frac{(1.1377)^4}{4}} \right]$$

$$= 4.915 \text{ mA} \qquad (7\text{–}39)$$

$$V_{GSQ}^{(1)} = -4.915 \times 10^{-3} \times 146.5 = -0.720 \text{ V} \qquad (7\text{–}40)$$

$$V_{R_S}^{(1)} = 0.720 \text{ V} \qquad (7\text{–}41)$$

$$V_{R_D}^{(1)} = 864.2 \times 4.915 \times 10^{-3} = 4.247 \text{ V} \qquad (7\text{–}42)$$

$$V_{DS}^{(1)} = 15 - 0.720 - 4.247 = 10.033 \text{ V} \qquad (7\text{–}43)$$

The reader is invited to perform similar computations for the other extreme condition. Such results, along with those computed previously, are summarized in the table that follows.

	Lower bound values[1]	Nominal values	Upper bound values[2]
$V_{GS(off)}$ (V)	-2	-4	-6
I_{DSS} (mA)	12	16	20
I_{DQ} (mA)	4.915	8	10.825
V_{GSQ} (V)	-0.720	-1.172	-1.586
V_{R_S} (V)	0.720	1.172	1.586
V_{R_D} (V)	4.247	6.914	9.356
V_{DSQ} (V)	10.033	6.914	4.058

While there is significant variation in the operating point, it is more stable than for fixed bias. For circuits in which this variation is acceptable, the self-bias circuit is a simple approach to the bias design.

7–8 FIXED-PLUS-SELF-BIAS CIRCUIT

An additional degree of stabilization in the drain current may be achieved by combining fixed bias with self-bias. If fixed bias alone were employed, a separate negative dc source would be required for an n-channel depletion-mode device, as noted. However, when the combination of fixed and self-bias is used, the fixed voltage required with an n-channel device is positive, and this voltage can be obtained from the drain supply through a voltage divider. This approach is referred to in many books as voltage-divider bias.

The basic form of the circuit is shown in Figure 7–22. Assuming negligible gate current, the dc voltage V_G at the gate with respect to ground is

$$V_G = \frac{R_2}{R_1 + R_2} V_{DD} \qquad (7\text{–}44)$$

Since this voltage is positive, it is opposite in direction to the required gate–source voltage. However, the drain current flowing through R_S produces a self-bias voltage $R_S I_D$, which turns out to be larger in magnitude than V_G. If the circuit is designed properly, the net value of V_{GS} will turn out to be negative, as we will see.

Starting at ground and moving clockwise around the gate–source loop, the KVL equation is

$$-V_G + V_{GS} + I_D R_S = 0 \qquad (7\text{–}45)$$

FIGURE 7–22
Self-plus-fixed-bias circuit for n-channel JFET.

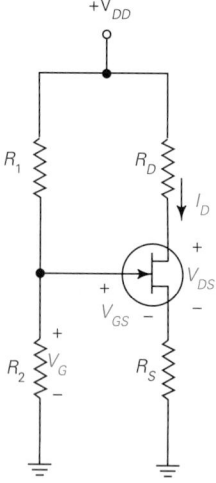

Solving for I_D, we obtain

$$I_D = -\frac{1}{R_S} V_{GS} + \frac{V_G}{R_S}$$ (7–46)

Consider that V_{GS} is the independent variable and I_D is the dependent variable. The result is a straight-line equation, and it represents the bias line for the fixed-plus-self-bias circuit.

To see how the higher degree of stabilization is achieved, refer to Figure 7–23. The two transfer characteristic curves T1 and T2 represent the extreme range situations, as considered earlier. The slope of the bias line is $-1/R_S$, and the horizontal intercept is determined as V_G. Because of the relatively large value of V_G and the fact that it is located to the right of the origin, the slope of the bias line may be reduced dramatically (by increasing R_S) in comparison to that of the fixed-bias circuit. The two extreme intersection points, corresponding to respective drain current values of $I_{DSS}^{(1)}$ and $I_{DSS}^{(2)}$, are much closer together than for the case of self-bias alone.

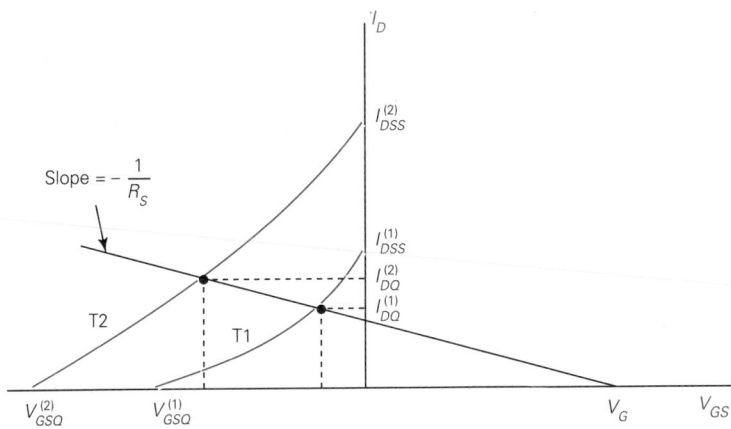

FIGURE 7–23
Range of operating points with self-plus-fixed-bias circuit.

7–9   ENHANCEMENT-MODE FET BIAS CIRCUITS

In the preceding sections, we discussed some of the common methods for establishing an operating point for JFETs and depletion-mode IGFETs. In this section, similar circuits for enhancement-mode IGFETs are discussed. While the basic approach is largely the same, the polarity of the required bias voltage changes certain conditions for the possible circuit configurations.

Fixed Bias

As in the case of depletion-mode devices, fixed bias may be employed with an enhancement-mode device. Consider the circuit of Figure 7–24(a), containing an n-channel IGFET in a common-source configuration. The drain is connected through a drain resistance R_D to a positive dc voltage $+V_{DD}$. In contrast to an n-channel depletion-mode device, which requires a negative value of V_{GS} for bias, the n-channel enhancement-mode device requires a positive bias voltage. In this circuit, the gate is connected through a resistance R_G to a positive bias voltage V_{GG}.

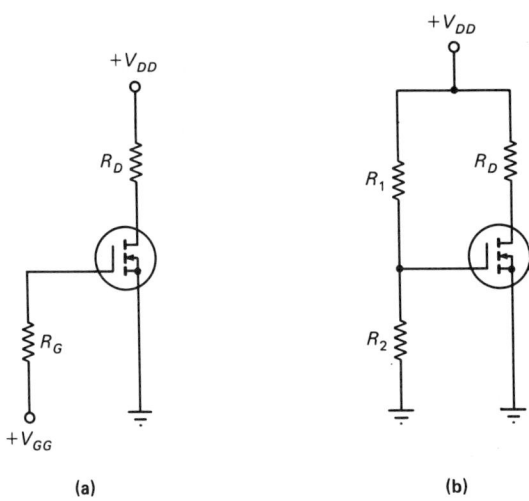

FIGURE 7–24
Fixed-bias circuits for n-channel enhancement-mode IGFET.

Since the drain and gate bias voltages have the same polarity, the drain power supply may be used to establish bias for both terminals. A circuit using this approach to achieve fixed bias for the n-channel IGFET is shown in Figure 7–24(b). Since the required value of V_{GS} is much smaller than V_{DD}, the voltage divider, consisting of R_1 and R_2, reduces the voltage to the appropriate level. Assuming no gate current, the Q-point gate–source voltage V_{GSQ} is

$$V_{GSQ} = \frac{R_2}{R_1 + R_2} V_{DD} \qquad (7\text{–}47)$$

The fixed-bias circuit for enhancement-mode IGFETs is thus simpler than for depletion-mode devices, since only one dc bias supply is required. However, it suffers from the same uncertainty in operating point because of the wide variation in transistor parameters.

Self-Plus-Fixed-Bias Circuit

The self-bias circuit that uses a source resistance does not work for enhancement-mode devices because the voltage drop across the source resistor has the wrong polarity to bias the gate. However, by combining the fixed-bias voltage-divider circuit with self-bias, a stable biasing circuit can be achieved.

The form of the biasing circuit is shown in Figure 7–25. Operation is similar to the corresponding circuit for depletion-mode devices, with one difference. For depletion-mode devices, the voltage drop across R_S is larger than the drop across R_2, so that the net value of V_{GS} is negative. For enhancement-mode devices, the voltage drop across R_2 is larger than the drop across R_S, and the net value of V_{GS} is positive.

FIGURE 7–25

Self-plus-fixed-bias circuit for *n*-channel enhancement-mode IGFET.

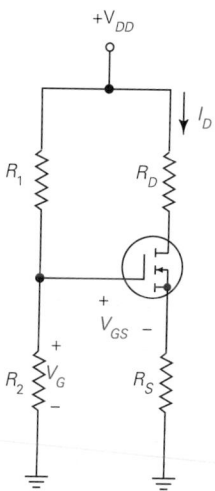

The dc voltage V_G at the gate with respect to ground is

$$V_G = \frac{R_2}{R_1 + R_2} V_{DD}$$ (7–48)

The current I_D produces a self-bias voltage $R_S I_D$ across R_S. A clockwise loop around the gate–source circuit results in

$$-V_G + V_{GS} + I_D R_S = 0$$

Solution for I_D yields

$$I_D = -\frac{1}{R_S} V_{GS} + \frac{V_G}{R_S}$$ (7–49)

This result is the same as (7–46) for depletion-mode devices. However, the applicable region of operation will be different for enhancement-mode devices, as we will see shortly.

Refer to Figure 7–26 for the discussion that follows. The two transfer characteristic curves T1 and T2 represent extreme range situations for the IGFET under consideration. Curve T1 has a threshold voltage $V_{GS(th)}^{(1)}$, and curve T2 has a threshold voltage $V_{GS(th)}^{(2)}$. The bias line intersects the two curves at $I_{DQ}^{(1)}$ and $I_{DQ}^{(2)}$, respectively. The variation in these operating-point currents is much less than the variation in characteristic curve parameters.

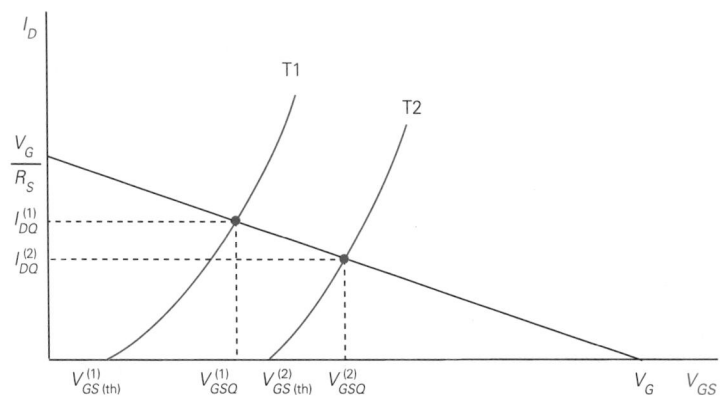

FIGURE 7–26
Range of operating points with self-plus-fixed-bias circuit.

7–10 FET CONSTANT CURRENT SOURCES

The FET closely approximates an ideal current source in the beyond-pinchoff region. It is therefore capable of being used for applications in which a constant current source is required.

We saw in Chapter 5 that when a BJT was used as a constant current source, the value of the current could be established to within a reasonably close value by appropriate circuit design. This is possible, because the base–emitter voltage drop does not vary considerably from one unit to another. For the FET, however, the wide variation in gate–source voltage for a given drain current makes it difficult to specify what the actual current will be for a general-purpose FET. The general-purpose FET is best suited for an application in which a current source is needed but for which the exact value either is not critical or can be adjusted as required.

Constant Current Diodes

Some FET circuits are specifically manufactured for the purpose of establishing constant current sources. The schematic diagram of such a device is shown in Figure 7–27. A bias resistor appears in the source lead, and the gate terminal is

FIGURE 7–27
FET constant current source diode.

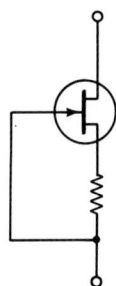

connected directly to one of the input leads. There is no provision for an external gate terminal, so the gate resistance is not required. Instead, the bias voltage developed across the source resistor is applied directly to the gate. Since there are only two external terminals, the device is a diode as far as external connections are concerned.

In a particular application, it is required that the gate–source voltage exceed a certain minimum positive voltage to ensure beyond-pinchoff operation. However, over a wide range of gate–source voltage, a constant current is established.

In a sense, the *constant current diode* provides the opposite type of regulation achieved by the *constant voltage zener diode* introduced in Chapter 2. The value of current for a given diode type is subject to some variation between different units.

7–11 DC SOURCE FOLLOWER

A basic BJT circuit considered in Chapter 5 was the dc **emitter follower.** A similar type of circuit exists for the FET, and it is called a **source follower.**

The basic form of a source follower intended for dc input and output is shown in Figure 7–28. For a given input v_i, a KVL equation around the gate–source loop reads

$$-v_i + v_{GS} + i_D R_S = 0 \qquad (7\text{--}50)$$

or

$$i_D = -\frac{1}{R_S} v_{GS} + \frac{v_i}{R_S} \qquad (7\text{--}51)$$

The output voltage is

$$v_o = R_S i_D \qquad (7\text{--}52)$$

An alternative interpretation is simply

$$\boxed{v_o = v_i - v_{GS} = v_i + |v_{GS}|} \qquad (7\text{--}53)$$

FIGURE 7–28
Basic form of an *n*-channel JFET dc
source follower.

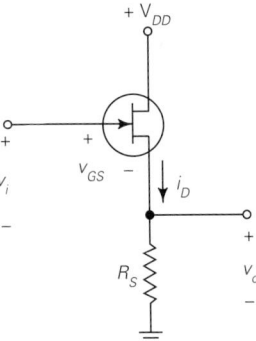

The form of (7–51) is a type of bias equation relating i_D and v_{GS}, which has been investigated earlier. However, it differs from earlier forms in that v_i is now a variable. For a given value of v_i, however, a bias line can be constructed. We consider next two possible values of v_i and the resulting bias lines.

Refer to Figure 7–29 in the discussion that follows. First consider a relatively small possible value of v_i, denoted as $v_i^{(1)}$. The bias line for this voltage is denoted as B1 on Figure 7–29, and the corresponding gate–source voltage is $v_{GS}^{(1)}$. From (7–53), we can deduce that the dc value of v_o will be larger than v_i by $|v_{GS}^{(1)}|$.

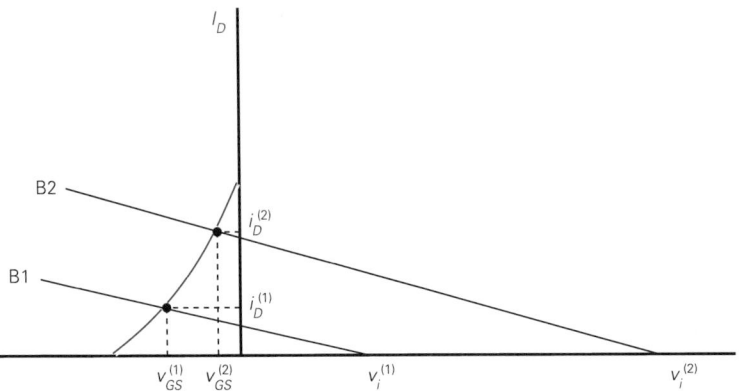

FIGURE 7–29
Bias lines for two possible input voltages with source follower.

Next, consider a larger input voltage $v_i^{(2)}$. The bias line is B2 on Figure 7–29, and the gate–source voltage is $v_{GS}^{(2)}$. The dc value of v_o is still larger than the input, but the difference is less at the larger voltage.

The output dc voltage thus tracks the dc input voltage over a certain range, but the total output voltage is larger than the input by the magnitude of gate–source voltage required to establish the operating point.

To avoid possible confusion later, it will be mentioned here that the change in output voltage is less than the change in input voltage and that the small-signal or ac voltage gain, a concept considered later, is less than 1. However, for the source-follower circuit with an *n*-channel JFET, the total dc output voltage is greater than the input dc voltage, as we have shown.

It is not as easy to predict the dc output voltage with the source follower as it was for the emitter follower. The reason is that the BJT base–emitter voltage drop is more predictable than the FET gate–source bias voltage. However, the FET offers a higher input impedance than the BJT.

The two "follower" circuits that we have considered thus far (i.e., the emitter follower and the source follower) are used primarily for isolation between a source and a load. With the BJT, the dc output voltage tracks the input voltage with a difference of about 0.7 V. With the JFET, the dc output voltage tracks the input voltage with a difference of v_{GS} for a given operating point. Thus, more accurate tracking is achieved when the magnitude of v_{GS} is very small.

7–12 PSPICE EXAMPLES

PSPICE EXAMPLE 7–1

Use PSPICE to perform the load-line analysis of Example 7–3.

Solution

To verify the results of the text example, it is necessary to modify the JFET model of PSPICE. It turns out that the ideal characteristics of the JFET assumed in the text can be obtained by altering two parameters in the NJF model file. The two parameters are VTO and BETA. The quantity VTO is what PSPICE uses to represent the gate–source cutoff voltage for a depletion-mode device and the threshold voltage for an enhancement-mode device.

The quantity BETA is defined by PSPICE as

$$\text{BETA} = \frac{I_{DSS}}{\text{VTO}^2} \tag{7-54}$$

Substituting VTO = -4 V and I_{DSS} = 16 mA for the text model, we have

$$\text{BETA} = \frac{16 \times 10^{-3}}{(4)^2} = 1 \times 10^{-3} \, \text{A/V}^2 \tag{7-55}$$

The circuit adapted to PSPICE is shown in Figure 7–30(a), and the code is shown in Figure 7–30(b). The two variables to be swept are V_{GS} and V_{DS}, so no values are inserted on these lines. The line beginning with J1 defines the connections for the JFET, and the model name provided is JTEXT. The .MODEL line matches the model name to the PSPICE model NJF. Without further information, the model would default to the form given at the end of Chapter 6. However, the model is modified by new values of VTO and BETA, as previously determined. For the .DC sweep, V_{DS} is swept from 0 to 25 V in steps of 0.1 V, and V_{GS} is swept from -4 V to 0 in steps of 1 V.

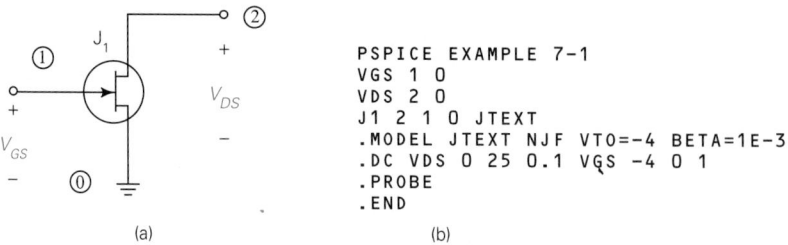

```
PSPICE EXAMPLE 7-1
VGS 1 0
VDS 2 0
J1 2 1 0 JTEXT
.MODEL JTEXT NJF VTO=-4 BETA=1E-3
.DC VDS 0 25 0.1 VGS -4 0 1
.PROBE
.END
```

(a)   (b)

FIGURE 7–30
Circuit and code for determining drain characteristics in PSPICE Example 7–1.

The drain characteristics obtained from .PROBE are shown in Figure 7–31. These can be compared with those of Figure 7–3.

The load lines can be added by the use of the "add trace" option in .PROBE. Along with the variables appearing on the circuit diagram, additional mathematical functions can be added to the plots with this option. To add a load line, we first review the equation describing it with I_D as the dependent variable, which was given in (7–3).

$$I_D = \left(-\frac{1}{R_D}\right) V_{DS} + \frac{V_{DD}}{R_D}$$
(7–56)

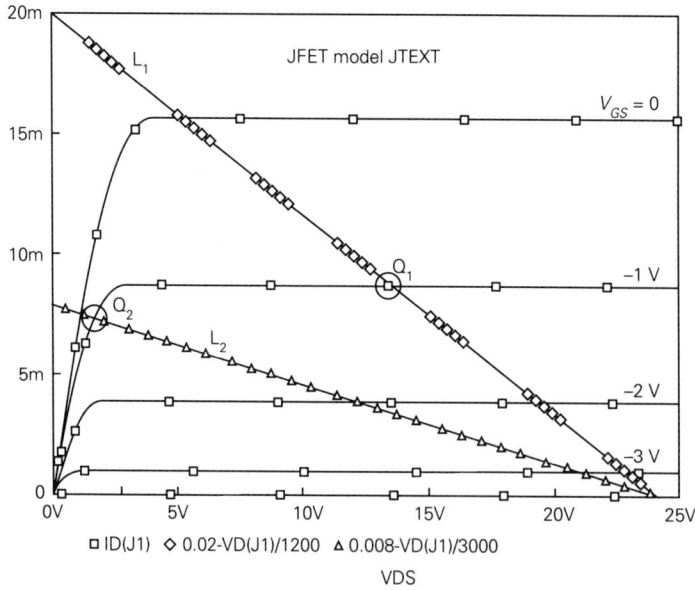

FIGURE 7–31
Drain characteristics of PSPICE Example 7–1 with two load lines.

In terms of the PSPICE variables, V_{DS} is the horizontal variable VDS, and I_D is the vertical variable ID (J1). First, the values $R_D = 1.2$ kΩ and $V_{DD} = 24$ V are substituted in (7–56), and the result after interchanging the terms is

$$I_D = 0.02 - \frac{V_{DS}}{1200} \qquad (7\text{–}57)$$

To form this equation, "add trace" is entered followed by 0.02 − VD (J1)/1200. This equation is labeled as L_1.

Next $R_D = 3$ kΩ and $V_{DD} = 24$ V are substituted, and the result is

$$I_D = 0.008 - \frac{V_{DS}}{3000} \qquad (7\text{–}58)$$

This equation is entered as 0.008 − VD (J1)/3000, and it is labeled as L_2. Both load lines are seen to be in agreement with those of Figure 7–3.

PSPICE EXAMPLE 7–2

Develop a PSPICE program to validate the operation of the voltage-controlled attenuator of Example 7–5, using the transfer function command.

Solution

The circuit adapted to PSPICE is shown in Figure 7–32(a), and the code is shown in Figure 7–32(b). Since the transfer function command works only for dc, a constant value of 0.1 V is assumed. (However, the actual circuit is intended for time-varying signals.) The special JFET model developed in the preceding example is used again.

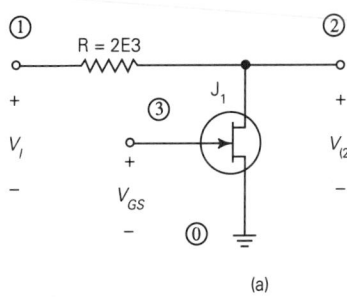

```
PSPICE EXAMPLE 7-2
.OPTIONS NOECHO NOPAGE NOMOD NOBIAS
VI 1 0 0.1
R 1 2 2E3
VGS 3 0
J1 2 3 0 JTEXT
.MODEL JTEXT NJF VTO=-4 BETA=1E-3
.STEP VGS 0 -4 -1
.TF V(2) VI
.END
```

(a)   (b)

FIGURE 7–32
Circuit and code of PSPICE Example 7–2.

The line following the .MODEL designation reads as follows:

```
.STEP VGS 0 -4 -1
```

This is the .STEP command that tells PSPICE to vary V_{GS} from 0 to −4 V in steps of −1 V.

The output data are shown in Figure 7–33. For each of the step values, the ratio $V(2)/V_I$, which is equivalent in the text notation to $A = v_0/v_i$, is computed. These values can be compared with those obtained in Example 7–5, and they are all in close agreement. For example, the first block of data is based on $V_{GS} = 0$, and the computer-generated value is 0.05891, which compares closely with the value of 0.0588 obtained in Example 7–5.

```
PSPICE EXAMPLE 7-2

****      CIRCUIT DESCRIPTION

************************************************************************

.OPTIONS NOECHO NOPAGE NOMOD NOBIAS

****      SMALL-SIGNAL CHARACTERISTICS

     V(2)/VI = 5.891E-02

     INPUT RESISTANCE AT VI = 2.125E+03

     OUTPUT RESISTANCE AT V(2) = 1.178E+02

****      SMALL-SIGNAL CHARACTERISTICS

     V(2)/VI = 7.711E-02

     INPUT RESISTANCE AT VI = 2.167E+03

     OUTPUT RESISTANCE AT V(2) = 1.542E+02

****      SMALL-SIGNAL CHARACTERISTICS

     V(2)/VI = 1.117E-01

     INPUT RESISTANCE AT VI = 2.251E+03

     OUTPUT RESISTANCE AT V(2) = 2.233E+02

****      SMALL-SIGNAL CHARACTERISTICS

     V(2)/VI = 2.033E-01

     INPUT RESISTANCE AT VI = 2.510E+03

     OUTPUT RESISTANCE AT V(2) = 4.066E+02

****      SMALL-SIGNAL CHARACTERISTICS

     V(2)/VI = 1.000E+00

     INPUT RESISTANCE AT VI = 1.000E+12

     OUTPUT RESISTANCE AT V(2) = 2.000E+03

          JOB CONCLUDED

          TOTAL JOB TIME          1.37
```

FIGURE 7–33
Computer printout of PSPICE Example 7–2.

PSPICE EXAMPLE 7-3

Develop a PSPICE model to verify the results of Example 7–6.

Solution

The circuit adapted to PSPICE is shown in Figure 7–34(a), and the code is shown in Figure 7–34(b). Once again, the special JFET model is employed. The .PRINT statement lists all the variables desired in the printout. Note that the gate–source voltage is V(1, 4) and the drain–source voltage is V(3, 4).

The computer printout is shown in Figure 7–35. The results are nearly in perfect agreement with those in Example 7–6.

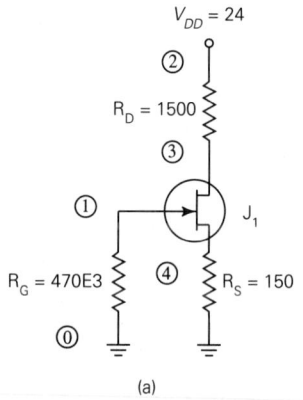

$V_{DD} = 24$

② $R_D = 1500$ ③ J_1

① $R_G = 470E3$ ④ $R_S = 150$

⓪

(a)

```
PSPICE EXAMPLE 7-3
.OPTIONS NOECHO NOPAGE NOMOD NOBIAS
RG 1 0 470E3
VDD 2 0 24
RD 2 3 1500
J1 3 1 4 JTEXT
RS 4 0 150
.MODEL JTEXT NJF VTO=-4 BETA=1E-3
.DC VDD 24 24 1
.PRINT DC ID(J1) V(1,4) V(RS) V(RD) V(3,4)
.END
```

(b)

FIGURE 7–34
Circuit and code of PSPICE Example 7–3.

```
PSPICE EXAMPLE 7-3

****       CIRCUIT DESCRIPTION

*******************************************************************

.OPTIONS NOECHO NOPAGE NOMOD NOBIAS

****       DC TRANSFER CURVES               TEMPERATURE = 27.000 DEG C

VDD        ID(J1)      V(1,4)     V(RS)      V(RD)       V(3,4)
2.400E+01  7.915E-03  -1.187E+00 1.187E+00  1.187E+01   1.094E+01



           JOB CONCLUDED

           TOTAL JOB TIME      1.10
```

FIGURE 7–35
Computer printout of PSPICE Example 7–3.

PROBLEMS

Drill Problems

7–1. The circuit of Figure P7–1(a) contains an *n*-channel JFET. The JFET characteristics are assumed to be idealized and are shown in Figure P7–1(b). Verify analytically that operation is in the beyond-pinchoff region, and determine I_D and V_{DS}.

FIGURE P7–1

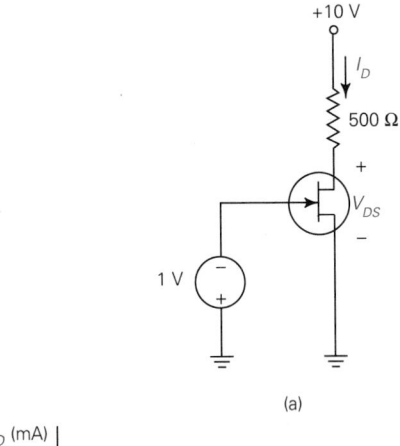

(a)

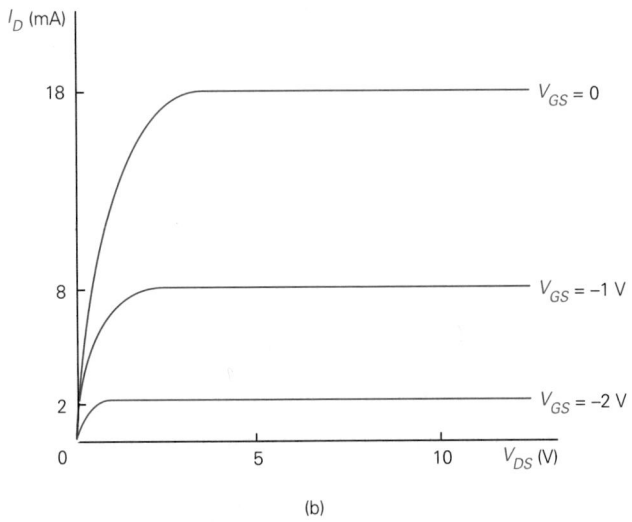

(b)

7–2. The circuit of Figure P7–2(a) contains an *n*-channel JFET. The JFET characteristics are assumed to be idealized and are shown in Figure P7–2(b). Verify analytically that operation is in the beyond-pinchoff region, and determine I_D and V_{DS}.

7–3. For the circuit of Figure P7–1(a), assume that the drain resistance is changed to $R_D = 2$ kΩ. Attempt to repeat the analysis of Problem 7–1.

7–4. For the circuit of Figure P7–2(a), assume that the drain resistance is changed to $R_D = 1.5$ kΩ. Attempt to repeat the analysis of Problem 7–2.

FIGURE P7–2

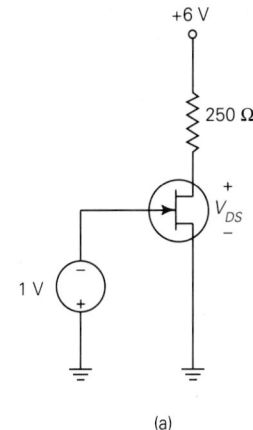

(a)

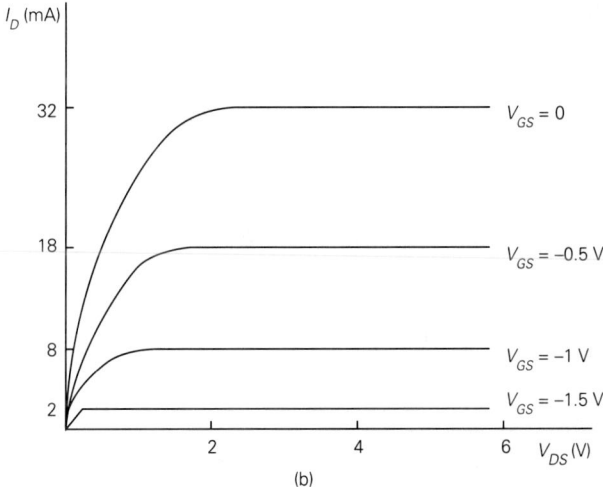

(b)

7–5. Construct a load line for the conditions of Problem 7–1.

7–6. Construct a load line for the conditions of Problem 7–2.

7–7. Construct a load line for the conditions of Problem 7–3.

7–8. Construct a load line for the conditions of Problem 7–4.

7–9. The self-bias circuit shown in Figure P7–9 contains the idealized n-channel JFET whose drain characteristics were given in Figure P7–1(b). Determine analytically the values of I_D, V_{GS}, V_{R_S}, V_{R_D}, and V_{DS}.

7–10. The self-bias circuit shown in Figure P7–10 contains the idealized n-channel JFET whose drain characteristics were given in Figure P7–2(b). Determine analytically the values of I_D, V_{GS}, V_{R_S}, V_{R_D}, and V_{DS}.

7–11. For the self-bias circuit of Problem 7–9, determine the operating point using the bias line.

FIGURE P7–9

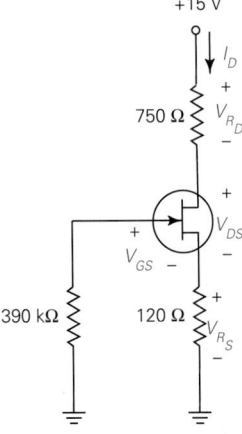

FIGURE P7–10

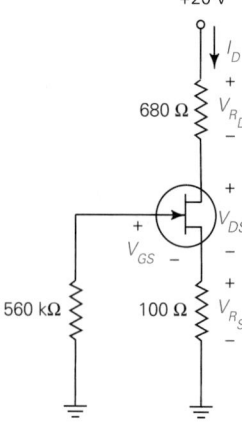

7–12. For the self-bias circuit of Problem 7–10, determine the operating point using the bias line.

Design Problems

7–13. It is desired to switch on and off the 10-kΩ load in the circuit of Figure P7–13. The FET characteristics are assumed to be those considered in Figure P7–1(b). **(a)** Using a load line, show that the circuit is capable of operating as a switch. **(b)** Determine the required control levels for v_i. **(c)** Estimate the voltage across the load when the switch is on.

7–14. Design a voltage-controlled attenuator, using the JFET whose characteristics were shown in Figure P7–2(b). The input voltage is a small positive voltage with a peak value of about 0.1 V. The desired transfer ratio $A = v_o/v_i$ is specified as 0.1 at $V_{GS} = -1$ V. After you have completed the design, determine A for the following values of V_{GS}: **(a)** 0; **(b)** −0.5 V; **(c)** −1.5 V; **(d)** −2 V.

FIGURE P7–13

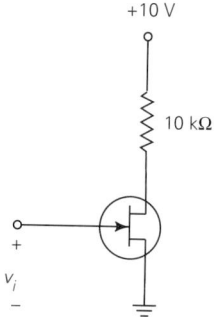

+10 V

10 kΩ

+

v_i

−

7–15. A certain *n*-channel JFET has the idealized characteristics given in Figure P7–1(b). Design a self-bias circuit of the form of Figure 7–18 to achieve the following operating point conditions: $I_{DQ} = 9$ mA, $V_{R_D} = V_{DSQ}$. The power supply voltage is 24 V. List the exact values of resistances as computed.

7–16. A certain *n*-channel JFET has the idealized characteristics given in Figure P7–2(b). Design a self-bias circuit of the form of Figure 7–18 to achieve the following operating point conditions: $I_{DQ} = 16$ mA, $V_{R_D} = V_{DSQ}$. The power supply voltage is 30 V. List the exact values of resistances as computed.

7–17. Assume in the design of Problem 7–15 that I_{DSS} is subject to a variation from 12 to 24 mA and that $V_{GS(off)}$ is subject to a variation from −2 to −4 V. Construct a table showing the possible range of circuit variables.

7–18. Assume in the design of Problem 7–16 that I_{DSS} is subject to a variation from 24 to 40 mA and that $V_{GS(off)}$ is subject to a variation from −1.5 to −3 V. Construct a table showing the possible range of circuit variables.

Troubleshooting Problems

7–19. If the transistor in Figure 7–18 is initially biased in the beyond-pinchoff region, indicate whether the ground-referred drain voltage V_D would *increase, decrease,* or *remain nearly the same* for each of the following troubles. **(a)** R_G shorted; **(b)** R_D shorted; **(c)** R_D open; **(d)** R_S shorted; **(e)** R_S open.

7–20. If the transistor in Figure 7–22 is initially biased in the beyond-pinchoff region, indicate whether V_D would *increase, decrease,* or *remain nearly the same* for each of the following troubles. **(a)** R_1 shorted; **(b)** R_1 open; **(c)** R_2 shorted; **(d)** R_2 open; **(e)** R_D shorted; **(f)** R_D open; **(g)** R_S shorted; **(h)** R_S open.

7–21. You are troubleshooting an FET circuit of the form shown in Figure 7–22. A comparison of dc voltages to *ground* specified by the equipment manual and those measured is as follows:

	Specified	Measured
V_{DD} (V)	24	23.9
V_D (V)	11	3.5
V_S (V)	1	2.5

Which one of the following *could* be the cause of the difficulty? **(a)** R_1 open; **(b)** R_2 open; **(c)** R_2 shorted; **(d)** R_S open; **(e)** R_D open.

7-22. You are troubleshooting an FET circuit of the form shown in Figure 7–22. A comparison of dc voltages to *ground* specified by the equipment manual and those measured is as follows:

	Specified	Measured
V_{DD} (V)	15	15.1
V_D (V)	8	14
V_S (V)	1	0.3

Which one of the following *could* be the cause of the difficulty? **(a)** R_D open; **(b)** R_1 open; **(c)** R_1 shorted; **(d)** R_2 open; **(e)** R_S open.

8

FREQUENCY RESPONSE

OBJECTIVES

After completing this chapter, the reader should be able to:

- Define the form of a sinusoidal waveform and its various properties, including peak value, peak-to-peak value, rms value, frequency, period, and phase angle.
- Calculate the capacitive reactance of a capacitor.
- Calculate the inductive reactance of an inductor.
- Construct and analyze the steady-state dc circuit model for a circuit containing capacitance and/or inductance.
- Construct and analyze the flat-band ac circuit model for a circuit containing capacitance and/or inductance.
- Define the concepts of amplitude response, phase response, and frequency response.
- Analyze the frequency response of simple RC low-pass and high-pass circuits.
- Analyze some of the circuits in the chapter with PSPICE.

8–1 OVERVIEW OF SINUSOIDAL AC CIRCUIT ANALYSIS

We introduced in Chapter 1 the three basic passive circuit parameters. These are **resistance, capacitance,** and **inductance.** As we have seen, the circuit element that exhibits resistance is called a **resistor.** In a similar fashion, the circuit element that exhibits capacitance is called a **capacitor,** and the circuit element that exhibits inductance is called an **inductor.** Depending on the application, an inductor is sometimes denoted as a **coil** or as a **choke.** A review of the three basic circuit parameters and their symbols is shown in Figure 8–1.

357

FIGURE 8–1

Three circuit parameters and their symbols.

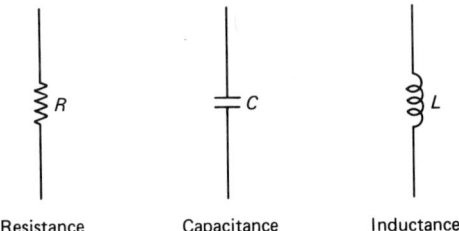

| | Resistance | Capacitance | Inductance |

Thus far in the book, the only passive element considered has been resistance. In all cases, the assumption has been that either the circuit has been operating at dc or, in the case of instantaneous voltages and currents, the effects of capacitance and inductance were negligible. However, we have now reached a point where the effects of capacitance and/or inductance must be considered in electronic circuits.

The manner in which capacitance and inductance is treated in basic electronic circuit analysis and design is through the concept of **sinusoidal steady-state ac circuit analysis.** This is an elegant body of circuit theory in which the source exciting the circuit is assumed to be sinusoidal in form, and the effects of all circuit elements to the input sinusoidal are determined. Although the ultimate application of the circuit may have nothing to do with a sinusoidal input, it is possible to infer many important properties of the circuit from the response to a sinusoid. The study of ac circuit theory utilizes certain mathematical forms of voltages and currents called **phasors** and the use of a so-called *j*-**operator.**

To accommodate readers who have not yet studied ac circuit theory, the use of the *j*-operator is deferred to Chapter 18. However, we will develop the concepts of ac analysis in this chapter in a manner that should allow a practical and intuitive approach to understanding the frequency-dependent properties of basic ac circuits. Readers who have already studied ac circuit theory using the *j*-operator may wish to fill in certain steps in this chapter.

Sinusoidal Function

The sinusoidal waveform is the most common function used in the analysis of electronic circuits. Consider the instantaneous voltage v shown in Figure 8–2. The equation of this waveform may be expressed as

$$v = V_p \sin \omega t \qquad \text{(8–1)}$$

The various quantities appearing in this equation are defined as follows:

V_p = *peak value* or *amplitude,* volts

ω = *angular frequency,* radians/second (rad/s)

   $= 2\pi f$

f = *cyclic frequency,* hertz (Hz)

t = independent variable *time,* seconds (s)

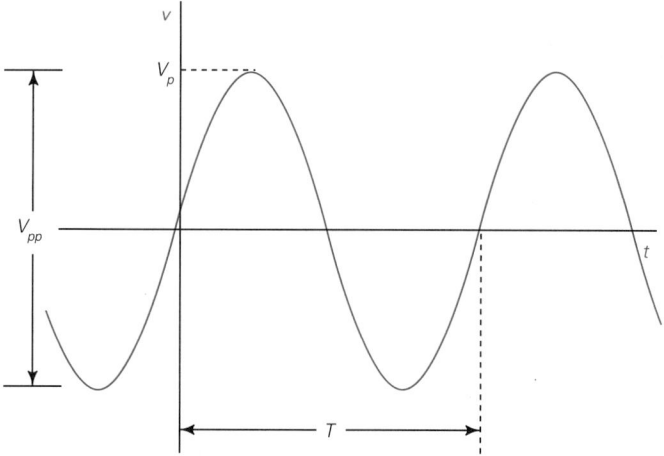

FIGURE 8–2
Sinusoidal voltage function.

Note that there are two definitions of frequency. The more common designation of frequency is the quantity f, which is the actual number of cycles of the waveform that occurs per second. The unit is hertz and 1 Hz = 1 cycle per second. However, it is the angular frequency ω that must appear as a multiplier of time in the equation to provide the correct dimensions for the angle of the sinusoidal function. In subsequent discussions, the term *frequency* without a modifier will be interpreted to be either the cyclic frequency f, or the reference could apply to either frequency form.

The *period T* is the duration of a full cycle in seconds. It is related to the cyclic frequency by

$$T = \frac{1}{f} \qquad \text{(8–2)}$$

In performing measurements on a sinusoid with an oscilloscope, it is often easier to measure the **peak-to-peak value** than the peak value. Let V_{pp} represent the peak-to-peak voltage. This quantity is related to the peak voltage simply as

$$V_{pp} = 2V_p \qquad \text{(8–3)}$$

Most of the various terms defined here are illustrated in Figure 8–2.

The function described by (8–1) and shown in Figure 8–2 is defined in terms of the basic sine function, and it passes through the origin with a positive slope. Many sinusoidal functions are shifted either to the left or to the right with respect to the assumed origin. For example, consider the voltage v shown in Figure 8–3.

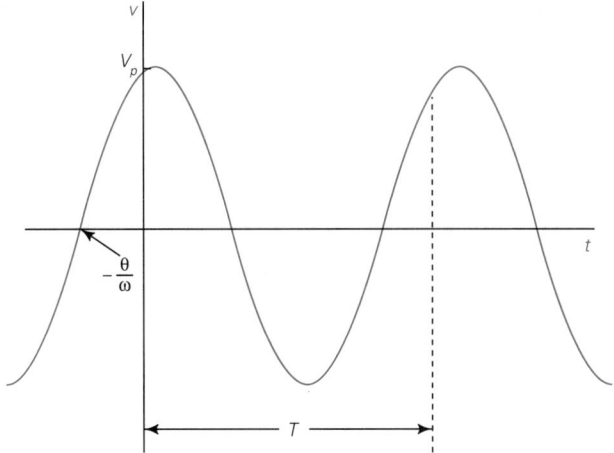

FIGURE 8–3
Sinusoidal voltage function with arbitrary phase angle.

This function can be expressed as

$$v = V_p \sin(\omega t + \theta) \tag{8–4}$$

The quantity θ in (8–4) represents an arbitrary phase shift, which for the example selected represents a shift to the left. Note that the time shift is θ/ω, since the units of the horizontal scale must be units of time. All other quantities in (8–4) have the same meaning as in (8–2).

The functions chosen thus far were voltages. However, currents appearing in steady-state ac circuits will have the same form. Thus, a typical instantaneous current will have the form

$$i = I_p \sin(\omega t + \phi) \tag{8–5}$$

where I_p is the peak value of the current and ϕ is some arbitrary phase angle.

The **sine function** has been chosen here as the basis for representation. However, the **cosine function** has the same basic shape, but it has a different time origin. Some books use the cosine function, but the practice in this book is to employ the sine function. Both the sine and cosine function, as well as shifts resulting from arbitrary phase angles, are collectively referred to as **sinusoidal functions.**

EXAMPLE 8–1   A certain sinusoidal voltage v has a peak value of 20 V and a frequency of 1000 Hz. The time origin is chosen in the form of Figure 8–2 (i.e., the function crosses $t = 0$ with a positive slope). **(a)** Determine the period. **(b)** Write an equation for the voltage.

Solution

(a) The period T is

$$T = \frac{1}{f} = \frac{1}{1000} = 0.001 \text{ s} = 1 \text{ ms} \tag{8–6}$$

(b) The radian frequency is $\omega = 2\pi f = 2\pi \times 1000 = 6283.2$ rad/s (to five significant figures). The function can be expressed as

$$v = 20 \sin 2\pi \cdot 1000t \tag{8–7a}$$
$$= 20 \sin 6283.2t \tag{8–7b}$$

Note the two alternative forms used in expressing the angles. The first form, with the 2π shown as a multiplier, is often used, since it allows the cyclic frequency to be immediately evident.

EXAMPLE 8–2

A certain current is described by the equation

$$i = 0.05 \sin 1600t \tag{8–8}$$

Determine **(a)** the peak value; **(b)** the radian frequency; **(c)** the cyclic frequency; and **(d)** the period.

Solution

(a) The peak value is determined by inspection as

$$I_p = 0.05 \text{ A} = 50 \text{ mA} \tag{8–9}$$

(b) The radian frequency is also determined by inspection as

$$\omega = 1600 \text{ rad/s} \tag{8–10}$$

(c) Since $\omega = 2\pi f$, the cyclic frequency is

$$f = \frac{\omega}{2\pi} = \frac{1600}{2\pi} = 254.65 \text{ Hz} \tag{8–11}$$

(d) The period is

$$T = \frac{1}{f} = \frac{1}{254.65} = 3.9270 \text{ ms} \tag{8–12}$$

Root Mean Square Value of Sinusoid

One very important quantity associated with a sinusoidal waveform is the **root-mean-square (rms)** value. For the moment, this quantity will be indicated by the subscript rms. The rms value of a voltage v will be denoted as V_{rms}, and the rms value of a current i will be denoted as I_{rms}.

The rms value of a sinusoidal voltage with peak value V_p is given by

$$V_{rms} = \frac{V_p}{\sqrt{2}} = 0.7071 \, V_p \qquad \text{(8–13)}$$

Similarly, the rms value of a sinusoidal current with peak value I_p is

$$I_{rms} = \frac{I_p}{\sqrt{2}} = 0.7071 \, I_p \qquad \text{(8–14)}$$

The significance of the rms value is that it is the quantity used when determining power produced by the ac waveform. If a sinusoidal voltage v exists across a resistance R, the power P is

$$P = \frac{V_{rms}^2}{R} \qquad \text{(8–15)}$$

Similarly, if a sinusoidal current i is flowing in a resistance R, the power P is

$$P = I_{rms}^2 R \qquad \text{(8–16)}$$

Thus, the rms voltage and current are treated like dc voltage and current for computing power in a resistance.

The expressions for power in (8–15) and (8–16) represent *average power,* which is the quantity most often desired in ac circuits. The *peak power* produced by the ac waveform is much greater. We defer detailed discussion of peak power to the point in the book where it is needed. For present developments, power computations with an ac waveform are considered to produce average power.

Many ac voltmeters measure the rms value of a sinusoidal waveform. It is important to recognize this fact when correlating oscilloscopic measurements with voltmeter measurements. For example, suppose that the peak-to-peak value of a certain sinusoidal voltage has been measured as 4 V. This means that the peak value is 2 V. An rms-measuring ac voltmeter applied to the waveform would ideally read 1.414 V. However, some ac voltmeters respond to different properties of the waveform, so it is important to understand the type of ac voltmeter used.

The rms value is also denoted as the **effective value** of the waveform, and some books use that definition. We use the rms designation in this book.

We have introduced the rms value to complete the treatment of the basic sine function here. For the next few chapters, however, very little use of the rms value will be required. However, it will be very significant in the treatment of power amplifiers and power supplies in Chapters 19 and 20.

A final note is that the formulas given by (8–13) and (8–14) apply only to a

sinusoidal function. In general, the relationship of the rms value to the peak value is different for different types of waveforms.

EXAMPLE 8–3

(a) Determine the rms value of the sinusoidal voltage of Example 8–1. (b) Determine the power dissipated in a 820-Ω resistor.

Solution

(a)   The peak value is 20 V, and the rms value is

$$V_{rms} = \frac{20}{\sqrt{2}} = 14.142 \text{ V} \qquad (8\text{–}17)$$

(b)   The power dissipated in a 820-Ω resistor is

$$P = \frac{V_{rms}^2}{R} = \frac{(14.142)^2}{820} = 0.2439 \text{ W} \qquad (8\text{–}18)$$

EXAMPLE 8–4

(a) Determine the rms value of the sinusoidal current of Example 8–2. (b) Determine the power dissipated in a 150-Ω resistor.

Solution

(a)   The peak value is 0.05 A, and the rms value in

$$I_{rms} = \frac{0.05}{\sqrt{2}} = 0.03536 \text{ A} \qquad (8\text{–}19)$$

(b)   The power dissipated in a 150-Ω resistor is

$$P = I_{rms}^2 R = (0.03536)^2 \times 150 = 0.1875 \text{ W} \qquad (8\text{–}20)$$

Spectral Analysis

The reason that the response of electronic circuits to sinusoidal waveforms is so important is the following fact: *Arbitrary waveforms can be considered to be composed of a sum of sinusoidal waveforms at different frequencies.* The process of determining the frequency components that compose a given waveform is called **spectral analysis.** It is also called **Fourier analysis** as a tribute to the great French mathematician Fourier, who developed the techniques that were later applied to spectral analysis.

To relate this concept to an area familiar to many readers, consider the common audio amplifier used in stereo music systems. The human ear is capable of responding to sinusoidal frequencies from about 20 Hz or so to nearly 20 kHz. Music waveforms are known to be composed of frequencies encompassing this range. Therefore audio amplifiers are tested with sinusoidal waveforms, and a good-quality amplifier should be capable of reproducing sinusoids equally well over that range. If the amplifier is defective and is not able to reproduce frequencies above 10 kHz, for example, there will be a loss of the higher-frequency notes, and the output signal will be degraded.

Steady-State Assumption

Consider a linear circuit with one or more sinusoidal sources of the same frequency. When the circuit is first excited by a sinusoid, there are certain *transient* effects occuring, which will eventually damp out, assuming that the circuit contains some resistance. The various voltages and currents in the circuit will then settle into what is called a **sinusoidal steady-state condition.**

A basic property of the sinusoidal steady-state for a *linear* constant parameter circuit is that *all voltages and currents are sinusoids that have the same frequency as the source*. Thus, all steady-state waveforms have the same basic sinusoidal shape. However, their peak values and phase angles will differ.

Since the various waveforms in the sinusoidal steady state are all sinusoids of the same frequency, only two quantities are required to specify each sinusoid: (1) peak value (or magnitude), and (2) phase angle. This is the basis of steady-state ac circuit analysis.

8–2 REACTANCE

A pure resistance presents the same opposition to current flow at all sinusoidal frequencies. Furthermore, the voltage across the resistance is in phase with the current through the resistance at all frequencies.

The effects of capacitance and inductance on ac current flow are somewhat more involved. Both capacitance and inductance provide opposition to ac current flow, but the manner and extent of the opposition are dependent on the frequency.

The opposition to ac current flow exhibited by either a capacitance or an inductance is called **reactance.** A reactance can be further classified as either a **capacitive reactance** or an **inductive reactance,** depending on which parameter is responsible. The unit of reactance is the same as the unit of resistance (i.e., the ohm). The general symbol for reactance is X.

If they oppose current flow, why don't we call these effects "capacitive resistance" and "inductive resistive"? The reason is that they describe a totally different kind of phenomenon. In pure resistance, the voltage and current are in phase, and all the power is dissipated in the resistor. For purely reactive elements, however, voltage and current are 90° out of phase, and there is no real power dissipation. Instead, the energy is alternately stored in the reactance and released back to the circuit. Therefore, the term *reactance* describes a completely different type of physical behavior from resistance.

Capacitive Reactance

The capacitive reactance of a capacitance C is denoted by X_C and it is determined as

$$X_C = \frac{1}{\omega C} \qquad \text{(8–21a)}$$

$$= \frac{1}{2\pi f C} \qquad \text{(8–21b)}$$

where alternate forms using f and ω are given. Strictly speaking, a more ''proper'' definition would include a negative sign for capacitive reactance, but in accordance with widespread usage at this level, it will be defined for convenience as a positive quantity.

The capacitive reactance is inversely proportional to both frequency and the capacitance. Thus, a larger capacitance exhibits a smaller reactance, and vice versa. The behavior of X_C as a function of frequency is illustrated in Figure 8–4. At very low frequencies, the capacitive reactance is very large (i.e., it presents a very large opposition to current flow). Conversely, at high frequencies, the capacitive reactance is very small.

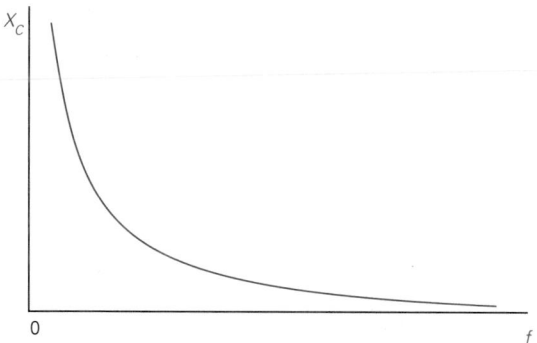

FIGURE 8–4
Capacitive reactance variation with frequency.

Inductive Reactance

The inductive reactance of an inductance L is denoted by X_L, and it is determined as

$$X_L = \omega L \qquad \text{(8–22a)}$$

$$= 2\pi f L \qquad \text{(8–22b)}$$

where alternative forms using f and ω are given. *The inductive reactance is directly proportional to both frequency and the inductance.* Thus, a larger inductance exhibits a larger reactance, and vice versa.

The behavior of X_L as a function of frequency is illustrated in Figure 8–5. At very low frequencies, the inductive reactance is very small (i.e., it presents a very small opposition to current flow). Conversely, at high frequencies, the inductive reactance is very large.

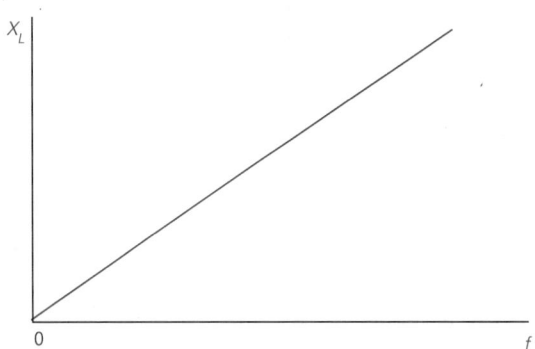

FIGURE 8–5
Inductive reactance variation with frequency.

EXAMPLE 8–5   Determine the reactance of a $0.01\text{-}\mu\text{F}$ capacitance at each of the following frequencies: **(a)** 10 Hz; **(b)** 1 kHz; **(c)** 100 kHz; **(d)** 10 MHz.

Solution
The expression for X_C is

$$X_C = \frac{1}{2\pi fC} = \frac{1}{2\pi \times 0.01 \times 10^{-6}f} = \frac{1.5915 \times 10^7}{f} \qquad \textbf{(8–23)}$$

where $C = 0.01 \ \mu\text{F}$ has been substituted and only f remains as a variable. This step was appropriate since several values of f are to be considered.

The remaining part of the problem amounts to substituting the four values of f and determining the four values of X_C. The different computations follow.

(a)   For $f = 10$ Hz,

$$X_C = \frac{1.5915 \times 10^7}{10} = 1.5915 \ \text{M}\Omega \qquad \textbf{(8–24)}$$

(b)   For $f = 1$ kHz,

$$X_C = \frac{1.5915 \times 10^7}{10^3} = 15.915 \ \text{k}\Omega \qquad \textbf{(8–25)}$$

(c) For $f = 100$ kHz,

$$X_C = \frac{1.5915 \times 10^7}{10^5} = 159.15 \; \Omega \tag{8-26}$$

(d) For $f = 10$ MHz,

$$X_C = \frac{1.5915 \times 10^7}{10^7} = 1.5915 \; \Omega \tag{8-27}$$

Observe the wide range of reactance as the frequency is varied. Observe how the reactance changes from a very large value at the lowest frequency to a very small value at the highest frequency.

EXAMPLE 8–6

Determine the reactance of a 50-mH inductance at each of the following frequencies: **(a)** 10 Hz; **(b)** 1 kHz; **(c)** 100 kHz; **(d)** 10 MHz.

Solution
The expression for X_L is

$$X_L = 2\pi f L = 2\pi \times 50 \times 10^{-3} f = 0.31416 f \tag{8-28}$$

where $L = 50$ mH has been substituted and only f remains as a variable. The different computations follow.

(a) For $f = 10$ Hz,

$$X_L = 0.31416 \times 10 = 3.1416 \; \Omega \tag{8-29}$$

(b) For $f = 1$ kHz,

$$X_L = 0.31416 \times 10^3 = 314.16 \; \Omega \tag{8-30}$$

(c) For $f = 100$ kHz,

$$X_L = 0.31416 \times 10^5 = 31.416 \; \text{k}\Omega \tag{8-31}$$

(d) For $f = 10$ MHz,

$$X_L = 0.31416 \times 10^7 = 3.1416 \; \text{M}\Omega \tag{8-32}$$

As in the case of the capacitive reactance of Example 8–5, we see a very wide range of inductive reactance as the frequency is varied. However, the inductive reactance changes from a very small value at the lowest frequency to a very large value at the highest frequency.

EXAMPLE 8–7

In a certain application requiring a capacitor, the exact value of the reactance is not critical, but it is desired that the reactance be no greater than 100 Ω at a frequency of 50 Hz. Determine the minimum value of capacitance required.

Solution
This problem has been stated in a practical fashion typical of many electronic circuit requirements. Before explaining the "no greater" and "minimum" qualifiers, we will first determine the exact value of C required to produce a reactance of exactly 100 Ω at 50 Hz.

In general, we have

$$X_C = \frac{1}{2\pi f C} \qquad (8\text{--}33)$$

Solving for C, we have

$$C = \frac{1}{2\pi f X_C} = \frac{1}{2\pi \times 50 \times 100} = 31.83 \ \mu\text{F} \qquad (8\text{--}34)$$

For the particular application, the exact value of X_C is not critical, so we need not be concerned about obtaining the exact value given by (8–34). This value of capacitance is fairly large, and the tolerances of standard capacitance values in that range are usually large. A larger standard value of capacitance would be acceptable for the application, since its reactance would be even smaller than the 100-Ω maximum value permitted. In conclusion, the value 31.83 μF represents the *minimum* value for the particular application, but a larger stock value would be acceptable.

8–3 STEADY-STATE DC MODELS FOR REACTANCE

Whenever dc analysis is employed in an electronic circuit containing reactive elements, the limiting forms of the reactance functions at dc lead to some simplified circuit models. For certain purposes, *dc may be considered as a special case of ac with zero frequency* (i.e., $f = 0$).

From the expression for X_C as given by (8–21) or from Figure 8–4, it is observed that X_C "approaches ∞" as f approaches 0 (i.e., X_C increases without limit as f approaches 0). An *infinite reactance* corresponds to an *open circuit*.

From the expression for X_L as given by (8–22) or from Figure 8–5, it is observed that $X_L = 0$ for $f = 0$. A *zero reactance* corresponds to a *short circuit*.

The dc steady-state models of an ideal capacitor and an ideal inductor are shown in Figure 8–6. To summarize, for *steady-state dc analysis, an ideal capacitor is represented by an open circuit, and an ideal inductor is represented by a short circuit.*

Under steady-state dc conditions, the capacitor will charge to a certain voltage and store that voltage until circuit conditions change. To compute the voltage stored on the capacitor, solve the dc circuit for the voltage across the open circuit representing the capacitor. This will be the voltage stored by the capacitor.

FIGURE 8–6
Steady-state dc equivalent-circuit models of ideal capacitor and inductor.

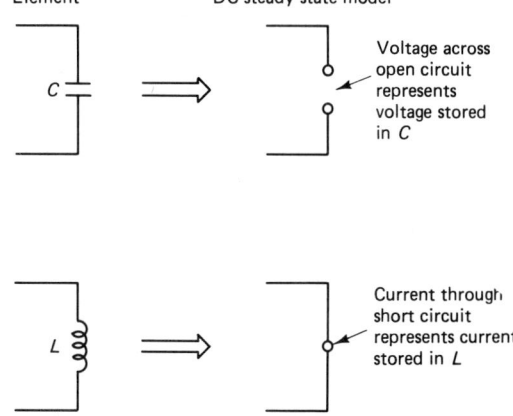

Element   DC steady-state model

Voltage across open circuit represents voltage stored in C

Current through short circuit represents current stored in L

In a similar fashion, the inductor will store magnetic flux, which is a result of current flowing through it. To compute the steady-state dc current flowing in the inductor, solve for the current through the short circuit representing the inductor. This will be the steady-state dc current flow for the inductor.

It should be stressed that these models apply after all transients damp out and steady-state conditions remain. When a reactive circuit is first excited by a dc voltage, current flow in a capacitor will occur until the capacitor is fully charged, and voltage can exist across an inductor until a full level of flux is obtained. The general solutions of such transients are covered in advanced circuits and network analysis texts. However, we consider certain important transient forms in Chapter 16 as they relate to nonlinear electronic circuits.

In addition to dc, these limiting models may be applied, with very little loss of accuracy, to analyze ac conditions at very low frequencies. The key to applying the limiting models is that all capacitive reactances must be large compared with all resistances, and all inductive reactances must be small compared with all resistances.

EXAMPLE 8–8

Determine the steady-state dc voltage across the capacitor in the simple *RC* circuit of Figure 8–7(a).

Solution
The steady-state dc equivalent circuit is shown in Figure 8–7(b). The capacitor is replaced by an open circuit. Since no current is flowing in the loop, the voltage across the capacitor is simply

$$V_C = 12 \text{ V} \qquad (8\text{–}35)$$

If this simple "quick result" bothers the reader, one can form a loop and write

$$-12 + 0 + V_C = 0 \qquad (8\text{–}36)$$

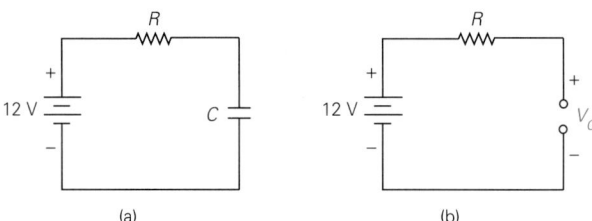

FIGURE 8–7
Circuit of Example 8–8.

in which the 0 term on the left represents the fact that the resistive voltage is zero, since the current is zero. The same result is readily obtained.

We need to remind ourselves that this model is applicable only after steady-state dc conditions are reached. When an uncharged capacitor is connected through a resistance to a dc source, current will flow in the circuit until the capacitor charges to a voltage level sufficient to terminate current flow in the capacitor (12 V in this case).

EXAMPLE 8–9 Determine the steady-state dc voltages across the three capacitors in the circuit of Figure 8–8(a).

Solution

The steady-state dc equivalent circuit is shown in Figure 8–8(b), in which all capacitors are replaced by open circuits. The right-hand loop is uncoupled from the left-hand loop by the open circuit representing C_3. This permits a simple application of the voltage-divider rule to the left-hand loop. We have

$$V_{C1} = \frac{2\,k\Omega}{3\,k\Omega + 2\,k\Omega + 1\,k\Omega} \times 60 = 20\ \text{V} \qquad \text{(8–37)}$$

and

$$V_{C2} = \frac{1\,k\Omega}{3\,k\Omega + 2\,k\Omega + 1\,k\Omega} \times 60 = 10\ \text{V} \qquad \text{(8–38)}$$

To determine V_{C3}, we note that the voltage across the 4.7-kΩ resistor is zero, since the current in the right-hand loop is zero. This forces V_{C3} to be the same as V_{C2}, and we have

$$V_{C3} = V_{C2} = 10\ \text{V} \qquad \text{(8–39)}$$

A more formal way to deduce the first part of (8–39) is the loop equation

$$-V_{C2} + V_{C3} + 0 = 0 \qquad \text{(8–40)}$$

where the 0 term on the left represents the voltage drop of zero for the 4.7-kΩ resistor.

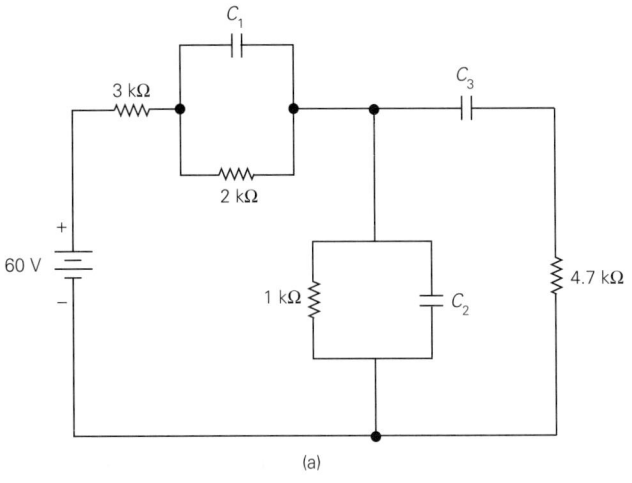

(a)

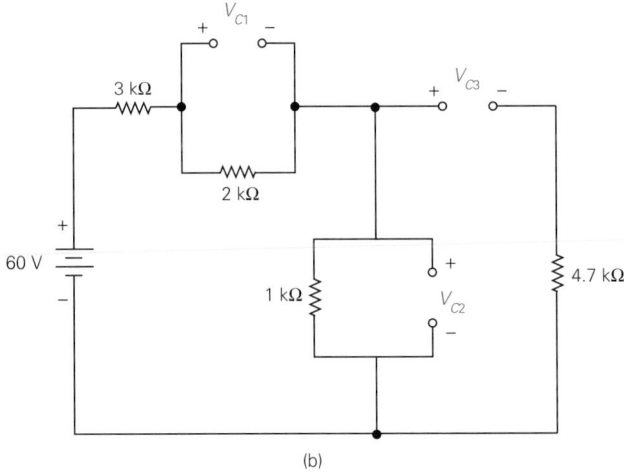

(b)

FIGURE 8–8
Circuit of Example 8–9.

EXAMPLE 8–10

The circuit of Figure 8–9(a) represents a complete single-stage RC-coupled BJT amplifier circuit, which is studied in detail in Chapter 10. Determine the steady-state dc voltages that will appear across the three capacitors. The function v_g represents an ac signal source, and by superposition, it can be modeled as a short-circuit for dc calculations.

Solution

A steady-state dc equivalent circuit, in which the three capacitors are represented by open circuits, is shown in Figure 8–9(b). We could, of course, represent the

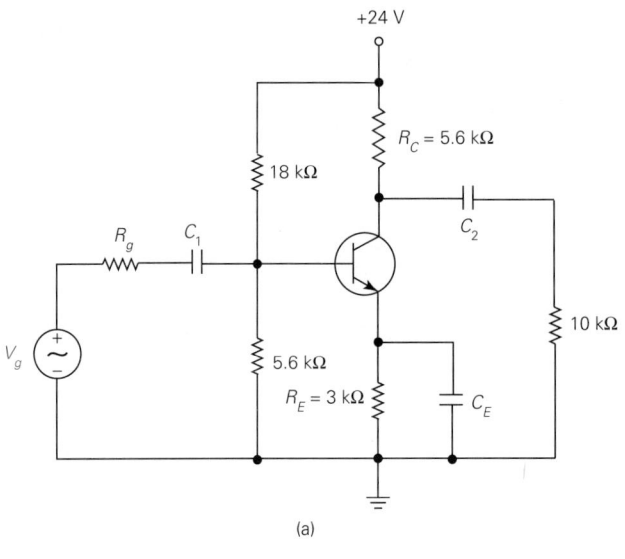

(a)

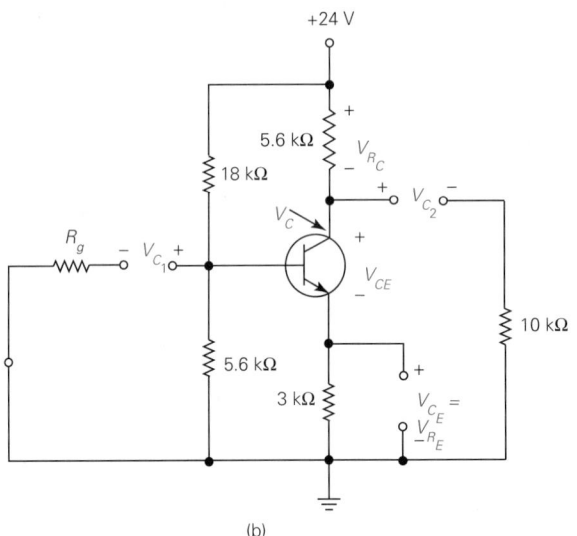

(b)

FIGURE 8–9
Circuit of Example 8–10.

transistor by its dc current source model, but we have had plenty of practice already on that type of model. Consequently, the transistor variables will be determined directly by the simplified formulas.

First, V_{TH} as a step for computing I_C is determined.

$$V_{TH} = \frac{5.6 \text{ k}\Omega}{5.6 \text{ k}\Omega + 18 \text{ k}\Omega} \times 24 \text{ V} = 5.695 \text{ V} \tag{8-41}$$

I_C is then determined as

$$I_C = \frac{5.695 - 0.7}{3 \times 10^3} = 1.66 \text{ mA} \tag{8-42}$$

where some rounding has been performed.

The three transistor voltages are

$$V_{R_E} = 5.695 - 0.7 = 5.00 \text{ V} \tag{8-43}$$
$$V_{R_C} = 5.6 \times 10^3 \times 1.66 \times 10^{-3} = 9.30 \text{ V} \tag{8-44}$$
$$V_{CE} = 24 - 9.30 - 5.00 = 9.70 \text{ V} \tag{8-45}$$

Up to this point, the analysis has been the same as performed back in Chapter 5. However, observe that the complete circuit is more involved than just the bias circuits considered in Chapter 5. Yet, the dc analysis on the transistor part of the circuit is the same.

The capacitors in the circuit serve to isolate the transistor dc circuit from the source on the left and the 10-kΩ load on the right. As we will see, the ac signal is effectively coupled through the various capacitors. However, the capacitors "block" the dc from the source and load. If the capacitors were not used in this circuit, the bias point would be affected drastically and adversely by the source and load.

Returning to the analysis, the dc voltage V_{C_1} is the same as the voltage across the 5.6-kΩ resistor and is

$$V_{C_1} = V_{TH} = 5.695 \text{ V} \tag{8-46}$$

The voltage V_{C_E} is the same as the voltage across the 3-kΩ resistor:

$$V_{C_E} = V_{R_E} = 5.00 \text{ V} \tag{8-47}$$

(Don't confuse V_{C_E}, the voltage across C_E, with V_{CE}, the collector–emitter voltage.) Finally, the voltage V_{C_2} is the same as the collector–ground voltage; that is,

$$V_{C_2} = V_C = V_{CE} + V_{R_E} = 9.70 + 5.00 = 14.70 \text{ V} \tag{8-48}$$

The analysis we have performed is helpful both in troubleshooting and in design. In troubleshooting, it leads to dc values to look for in checking the operation of a circuit. In design, it leads to the determination of the dc voltage ratings of capacitors in a circuit. In this context, the peak signal voltage should be added to the dc voltage to determine the maximum instantaneous voltage across a given capacitor. The voltage rating of the capacitor should then be selected to be somewhat larger than this value to provide additional leeway.

8–4 FLAT-BAND AC MODELS FOR REACTANCE

From the expression for X_C as given by (8–21) or from Figure 8–4, it is observed that X_C becomes very small as f increases without limit. Similarly, from the expression for X_L as given by (8–22) or from Figure 8–5, it is observed that X_L becomes very large as f increases without limit. Under certain conditions, X_C may be approximated as a short circuit and X_L may be approximated as an open circuit.

Many electrical and electronic circuits that have reactive elements are characterized by the fact that the following two conditions are met over a certain frequency range: (1) all capacitive reactances are very small compared with resistances and inductive reactances present, and (2) all inductive reactances are very large compared with resistances and capacitive reactances present. When these assumptions are valid, capacitive reactances may be approximated as short circuits, and inductive reactances may be approximated as open circuits. These are just the opposite conditions assumed in the steady-state dc models.

When these approximations are valid, the given circuit reduces to a resistive circuit. Although voltages and currents are varying with time, simple algebraic methods may be used to determine voltages and currents in the circuit on an instantaneous basis.

An important point is that *over the given frequency range where the approximations are valid, the peak value of any given voltage or current will be independent of the frequency of any sources.* We will use the term **flat-band** to describe the range of frequencies over which the limiting models are valid.

A summary of the flat-band ac models for capacitance and inductance is shown in Figure 8–10. In all subsequent developments in the book, a reference to *flat-band ac model* will refer to conditions in which these models are valid.

The reader is not expected to understand at this level when such models are valid. As we progress through the text, the judgment for assessing the validity of the models will develop. Ultimately, ac circuit theory will be used as an analytical way of scrutinizing the models for questionable situations. For the moment, simply

FIGURE 8–10
Steady-state flat-band ac models for ideal capacitor and inductor.

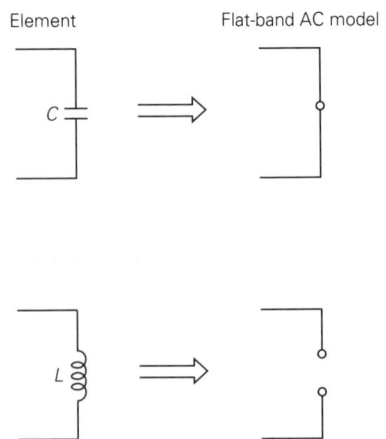

Element   Flat-band AC model

learn the models and how to work with them for analysis, troubleshooting, and design.

In this section and the preceding several sections, both capacitive and inductive reactances have been considered in order to provide a complete basic treatment. However, the particular order of development in the text leads first to amplifier circuits in which only capacitive effects are significant and the absence of inductance will be conspicuous in the next several chapters. Inductance will appear again in the applications of Chapter 17.

EXAMPLE 8–11

Consider the simple RC circuit of Figure 8–11(a). Assume that the frequency range of the time-varying signal v_g is in the flat-band region of the circuit. Determine the voltage v_o in terms of v_g.

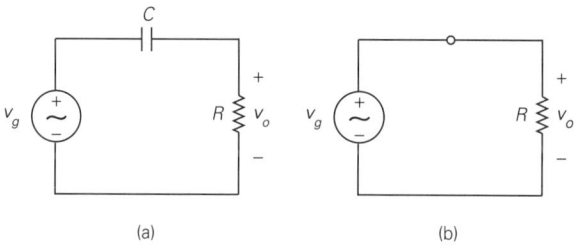

(a)   (b)

FIGURE 8–11
Circuit of Example 8–11.

Solution

The flat-band equivalent circuit is shown in Figure 8–11(b). The capacitor is replaced by a short circuit. Consequently, the voltage v_o in this circuit is simply the generator voltage v_g; that is,

$$v_o = v_g \qquad\qquad (8\text{–}49)$$

The circuit thus reduces to a trivial parallel connection of the source and load as far as the flat-band response is concerned. Understand that the frequency or frequencies of the source must be sufficiently high that the capacitive reactance is negligible.

EXAMPLE 8–12

In the circuit of Figure 8–12(a), assume that the frequency range of the time-varying signal v_g is in the flat-band region of the circuit. Determine the voltage v_o in terms of v_g.

Solution

The flat-band equivalent circuit is shown in Figure 8–12(b). The two capacitors are replaced by short circuits. The short around the 10-kΩ resistor effectively shunts all current around it, so that resistor has no effect in the flat-band region.

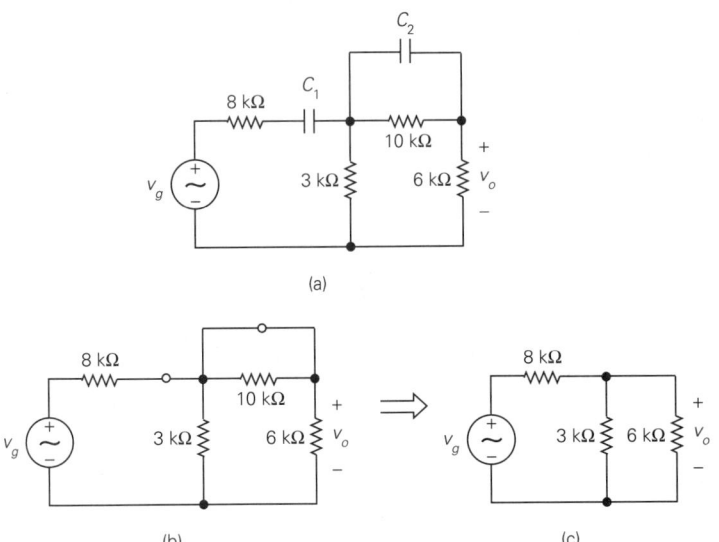

FIGURE 8–12
Circuit of Example 8–12.

Another way to deduce the same result is to recognize that the parallel combination of a short circuit and any element is a short circuit. Thus, the 3-kΩ resistor and the 6-kΩ resistor are effectively in parallel.

The circuit is redrawn in Figure 8–12(c) in a simplified form. By considering the parallel combination on the right, the voltage-divider rule may readily be employed. We have

$$v_o = \frac{3\text{ k}\Omega\|6\text{ k}\Omega}{3\text{ k}\Omega\|6\text{ k}\Omega + 8\text{ k}\Omega} \times v_g = \frac{2\text{ k}\Omega}{2\text{ k}\Omega + 8\text{ k}\Omega} \times v_g = \frac{1}{5}v_g \qquad \textbf{(8–50)}$$

8–5 DC-PLUS-FLAT-BAND AC RESPONSE

In many electronic circuits, one or more dc voltages are used to establish bias and other pertinent reference levels, but the ultimate goal is to process a time-varying signal. For this purpose it is necessary to consider separately a dc model and a time-varying or ac model. Although we are not yet ready to study time-varying models completely because of the frequency dependency of the reactances, in many cases it is the flat-band region models that are of primary interest. By combining the concepts of the preceding several sections, we can study both the dc and flat-band responses of certain circuits.

The focus in this section will therefore be on obtaining two models for a given circuit: (1) dc model, and (2) flat-band ac model. As we will see, the circuit form may change drastically as we convert from one form to the other.

The principle that allows separate models to be constructed for analysis is the principle of **superposition,** which was considered in Chapter 1. Strictly speaking, the principle of superposition applies only to linear circuits, and all electronic circuits have some degree of nonlinearity. However, as long as the behavior of the electronic devices can be represented by linear models over the range of operation, superposition may be applied with care. In this section, not all examples considered will contain electronic devices, so that the basic process may be developed. This will set the stage for working with linearized small-signal models for electronic devices in later chapters.

When the dc equivalent circuit is constructed, all ideal time-varying sources are replaced by short circuits (on paper). If the source contains an internal resistance, the resistance is retained in the model in the event that it could have some effect on various dc voltages and currents in the circuit. In many cases the circuit is designed so that internal resistances of time-varying sources do not affect dc operating points, but as a procedure at this point, we should consider any *possible* effects by keeping them in the models.

When the ac model is constructed, any dc sources are normally eliminated from the analysis. The possible drastic changes in configuation often confuse beginners, so the process will be illuminated.

AC Model for DC Source

Since a dc bias supply primarily establishes operating points, it is normally not considered as part of a signal. Therefore, it is simply modeled as a short circuit for ac signal analysis. This process is illustrated in Figure 8–13 in two ways. In Figure 8–13(a), the battery symbol is used, and it is replaced as a short circuit. In Figure 8–13(b), a dc connection to a ground-referred voltage is shown. In this case, the point is grounded in the model as shown.

FIGURE 8–13
Steady-state ac equivalent circuits of ideal battery and power supply connection.

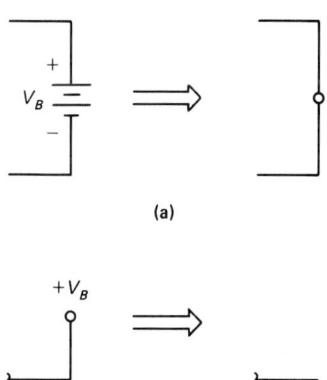

(a)

(b)

Two points need to be emphasized. First, this procedure with dc sources is simply a "paper model" and is not to be interpreted as a laboratory process. We obviously should not short any dc sources in the laboratory, since the consequences may be serious! In addition, the circuit will not function properly without the dc bias.

The second point is that we are considering situations at the moment where the signal is time varying and where dc sources represent bias or reference levels rather than signals. Later in the text, some direct-coupled amplifiers will be considered in which the signal itself could be a dc voltage.

The voltages and currents computed from the dc model represent the various dc levels in the circuit, and they would be the quantities measured with ideal dc instruments. Similarly, the voltages and currents computed from the ac model represent the time-varying quantities in the circuit. Depending on the type of time-varying signal, appropriate ac instruments could be used to measure these quantities. The sum of a given dc variable and the corresponding ac variable represents the instantaneous form of the pertinent voltage or current.

EXAMPLE 8–13

The circuit of Figure 8–14(a) contains an ac signal v_g and a 15-V power supply. Assume that all frequency components of v_g are in the flat-band region. A number of possible variables of interest are identified. (a) Construct a dc model, and compute the dc values of the variables. (b) Construct a flat-band ac model, and compute the ac values of the variables. (c) Write expressions for the total instantaneous variables.

Solution
Along with the analysis of the circuits, this problem will provide an opportunity to review the notational conventions given earlier in the text. Note that all variables except the signal source in the basic circuit of Figure 8–14(a) are labeled for total instantaneous form (i.e., lowercase symbols and uppercase subscripts).

(a) The dc model is shown in Figure 8–14(b), in which the two capacitors are replaced by open circuits and the source, v_g, is replaced by a short circuit. All variables are now labeled for dc analysis (i.e., uppercase symbols and uppercase subscripts).

The capacitors uncouple the dc portions of the circuit on the left and right, so by a simple inspection we have

$$I_A = 0 \tag{8–51}$$

$$I_D = 0 \tag{8–52}$$

$$V_B = 0 \tag{8–53}$$

The 15-V source, the 12-kΩ resistor, and the 6-kΩ resistor constitute a simple closed loop. The current through this loop is

$$I_B = I_C = \frac{15\text{ V}}{12\text{ k}\Omega + 6\text{ k}\Omega} = 0.8333\text{ mA} \tag{8–54}$$

FIGURE 8–14
Circuit of Example 8–13.

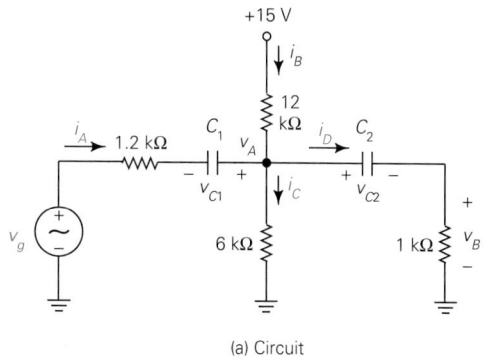

(a) Circuit

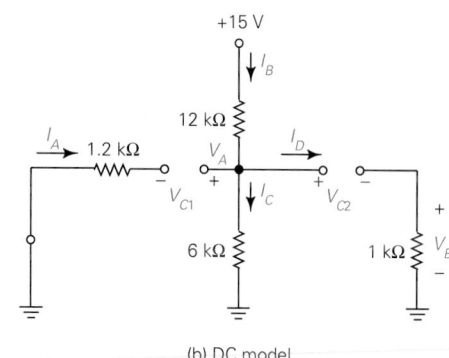

(b) DC model

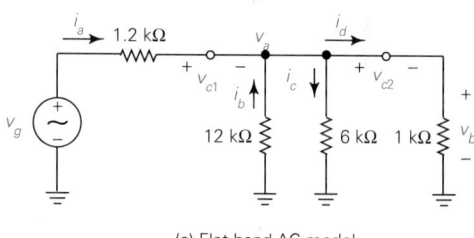

(c) Flat-band AC model

We could, of course, have "walked around the loop," but at this point in the book we assume that the reader has sufficient skill to analyze a simple loop such as this by a direct inspection process.

The voltage V_A is

$$V_A = 6 \times 10^3 \times 0.8333 \times 10^{-3} = 5 \text{ V} \qquad (8\text{–}55)$$

Alternatively, the voltage-divider rule could have been used.

Although the left- and right-hand portions of the circuits are uncoupled at dc, voltages will appear across the open circuits represented by the capacitors.

These voltages represent the dc values to which the capacitors will charge. It is observed that the two capacitor voltages are both equal to V_A, and we have

$$V_{C1} = V_{C2} = V_A = 5 \text{ V} \tag{8-56}$$

Note that the polarities were assumed in the directions resulting in positive values. With some practice, this can be done by intuition for many circuits.

(b) The flat-band ac model is shown in Figure 8–14(c). The two capacitors and the 15-V dc supply are replaced by short circuits. In the case of the dc supply, the circuit configuration is easier to visualize by turning the 12-kΩ resistor over and bringing it down to ground on the bottom. This process makes it obvious that three resistors are effectively in parallel as far as ac is concerned.

The variables are now labeled for ac analysis (i.e., lowercase symbols and lowercase subscripts). Note that the positive reference for i_b is opposite to that of the other currents on the right, because our original reference for i_B was from the power supply to the junction in the middle, and when this branch was inverted, it was necessary to keep the original reference direction.

Since the two capacitors are represented by short circuits, these two ac voltages may be written by a simple inspection as

$$v_{c1} = 0 \tag{8-57}$$

$$v_{c2} = 0 \tag{8-58}$$

The other variables may be determined by any standard resistive circuit analysis procedure. The approach used will be one of the "intuitive" methods of Chapter 1. The equivalent resistance R_{eq} seen by the source v_g is

$$R_{eq} = 1.2 \text{ k}\Omega + 12 \text{ k}\Omega\|6 \text{ k}\Omega\|1 \text{ k}\Omega$$
$$= 1200 \ \Omega + 800 \ \Omega = 2000 \ \Omega \tag{8-59}$$

The current i_a is then

$$i_a = \frac{v_g}{2000} = 5 \times 10^{-4} v_g \tag{8-60}$$

The voltages v_a and v_b are equal and may be determined as

$$v_a = v_b = 800 \times 5 \times 10^{-4} v_g = 0.4 v_g \tag{8-61}$$

Alternatively, the voltage-divider rule could have been employed. The three currents on the right are now determined as

$$i_b = \frac{-v_a}{12{,}000} = \frac{-0.4 v_g}{12{,}000} = -3.333 \times 10^{-5} v_g \tag{8-62}$$

$$i_c = \frac{v_a}{6000} = \frac{0.4 v_g}{6000} = 6.667 \times 10^{-5} v_g \tag{8-63}$$

$$i_d = \frac{v_a}{1000} = \frac{0.4 v_g}{1000} = 4 \times 10^{-4} v_g \tag{8-64}$$

Note that all ac voltages and currents are directly proportional to v_g and will have the same form. Thus, if v_g is a sinusoid of a particular frequency, all ac voltages and currents will be sinusoids of the same frequency.

(c) The total instantaneous variables are determined by combining the dc quantities plus the ac quantities. The various steps follow:

$$i_A = I_A + i_a \tag{8-65a}$$
$$= 0 + 5 \times 10^{-4} v_g = 5 \times 10^{-4} v_g \tag{8-65b}$$

$$i_B = I_B + i_b \tag{8-66a}$$
$$= 0.8333 \times 10^{-3} - 3.333 \times 10^{-5} v_g \tag{8-66b}$$

$$i_C = I_C + i_c \tag{8-67a}$$
$$= 0.8333 \times 10^{-3} + 6.667 \times 10^{-5} v_g \tag{8-67b}$$

$$i_D = I_D + i_d \tag{8-68a}$$
$$= 0 + 4 \times 10^{-4} v_g = 4 \times 10^{-4} v_g \tag{8-68b}$$

$$v_{C1} = V_{C1} + v_{c1} \tag{8-69a}$$
$$= 5 + 0 = 5 \text{ V} \tag{8-69b}$$

$$v_{C2} = V_{C2} + v_{c2} \tag{8-70a}$$
$$= 5 + 0 = 5 \text{ V} \tag{8-70b}$$

$$v_A = V_A + v_a \tag{8-71a}$$
$$= 5 + 0.4 v_g \tag{8-71b}$$

$$v_B = V_B + v_b \tag{8-72a}$$
$$= 0 + 0.4 v_g = 0.4 v_g \tag{8-72b}$$

Reviewing some of these results, the following observations are pertinent: (1) The capacitors uncouple the source and the load from the dc levels in the middle. Thus there is no dc current flow through the source v_g, and the load voltage v_B has only an ac component. (2) The voltages across the capacitors have no time-varying components (i.e., they have only dc levels). This is exactly what coupling capacitors should do; that is, they should be effective shorts for ac and thus should have no ac voltages across them.

8–6 FREQUENCY RESPONSE

As we prepare to study the properties of electronic amplifier circuits, it is necessary to understand the concept of **frequency response.** The frequency response of a system is a measure of the extent to which sinusoids of different frequencies are reproduced by the system.

Consider the block diagram of Figure 8–15 with the input v_i and output v_o. The block diagram could represent any type of linear electronic circuit. Assume that the input is a sinusoid with peak value V_{pi} of the form

$$v_i = V_{pi} \sin \omega t \tag{8-73}$$

FIGURE 8–15

Block diagram illustrating frequency response concept.

From previous discussions, if the circuit is linear, the steady-state output voltage is also a sinusoid and can be expressed as

$$v_o = V_{po} \sin(\omega t + \theta) \qquad (8\text{–}74)$$

where V_{po} is the peak value of the output voltage and θ is the phase shift introduced through the circuit. Thus, the output is a sinusoid of the same frequency as the input, but the peak value and angle are different from the input.

The process being discussed could be monitored on an oscilloscope, or an ac voltmeter could be used. On an oscilloscope, either the peak or the peak-to-peak value of each signal could be measured directly. With a voltmeter, the quantity measured would depend on the type of meter (i.e., peak reading meter, rms meter, etc.).

Irrespective of the type of instrument used, the values of V_{pi} and V_{po} may be determined. However, instead of working with the actual quantities in volts, it is the *ratio of output voltage to input voltage* that is most significant. Thus, if $V_{po}/V_{pi} = 0.5$, the output voltage has half the peak value of the input voltage at the given frequency. Let us denote the ratio simply as $|v_o|/|v_i|$ with the understanding that as long as the same property of each waveform is measured (peak, peak-to-peak, etc.), the ratio of the output quantity to the input quantity is the same.

The waveforms in Figure 8–16 illustrate how typical measurements of this type could be performed on a dual-trace oscilloscope. In Figure 8–16(a), the voltage v_i has a peak-to-peak value of 6 V, and v_o has a peak-to-peak value of 4 V at a certain frequency. The ratio $|v_o|/|v_i| = \frac{4}{6} = 0.667$ at that frequency.

Assume now that the frequency is changed to a new value, as shown in Figure 8–16(b). (The sweep rate has been changed to show two cycles here as in Figure 8–16[a].) The input voltage is maintained at a peak-to-peak value of 6 V. However, the output has now changed to a peak-to-peak value of 2 V. Thus, $|v_o|/|v_i| = \frac{2}{6} = 0.333$ at the new frequency. This process could be continued at as many frequencies as desired.

It is convenient to define the ratio of output voltage to input voltage at various frequencies by the function $A(f)$ in which

$$A(f) = \frac{|v_o|}{|v_i|} \qquad (8\text{–}75)$$

The quantity f represents the frequency, and the form $A(f)$ emphasizes that the function may vary with frequency. To simplify the notation, $A(f)$ will be expressed in many places simply as A.

FIGURE 8–16
Frequency response measurements on a dual-trace or dual-sweep oscilloscope.

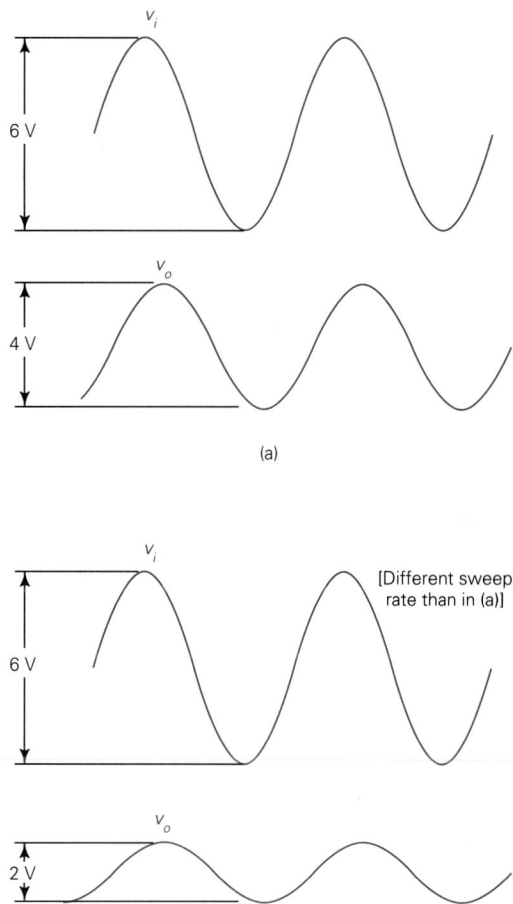

A plot of $A(f)$ as a function of frequency is called the **amplitude response function.** It is a measure of the relative weighting of the system to sinusoidal signals. Since complex waveforms can be considered to be composed of sinusoids, this information is very useful in characterizing an amplifier.

We will see in Chapter 9 that the same symbol A, along with the same definition of (8–75), will be used as the definition of amplifier gain. There is no conflict in these concepts, since amplifier gain is, in general, a frequency-dependent function. In this chapter, however, only passive circuits are being considered.

From the oscilloscopic waveforms of Figure 8–16, observe that there is also a phase shift, which varies with frequency. A function $\theta(f)$ providing a measure of the phase shift at various frequencies can be constructed. A plot of $\theta(f)$ as a function of frequency is called the **phase response function.** To simplify the notation, $\theta(f)$ will be expressed in many places simply as θ.

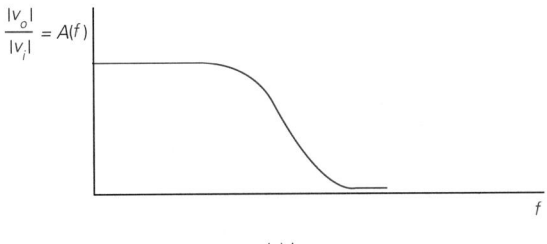

(a) Low-pass

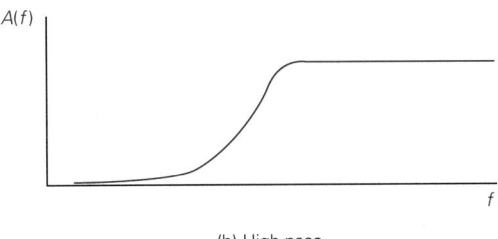

(b) High-pass

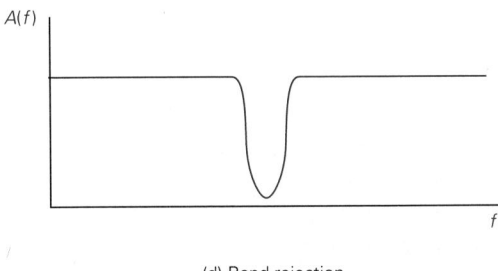

(c) Band-pass

(d) Band-rejection

FIGURE 8–17
Four important types of frequency response functions.

In practice, the amplitude response function is more widely used than the phase response function, although the latter function is important in some applications. Strictly speaking, the term *frequency response* refers to the combination of both the *amplitude response* and the *phase response*. In casual usage, however, the term *frequency response* is often used to mean simply the *amplitude response* whenever phase considerations are not important.

Some typical amplitude response functions of importance are shown in Figure 8–17. The function of Figure 8–17(a) is a **low-pass** function. It is characterized by the fact that dc and all frequencies lower than a certain reference frequency are passed with equal weighting. If the circuit involved is an amplifier, the level of the amplitude response in the low-frequency region will normally be greater than 1 as a result of the gain. If the circuit is a filter or a coupling network, the level in the low-frequency region may be 1 or less than 1. However, the important point is that all frequencies in that range are weighted equally. The low-frequency region of transmission is called the **passband.**

The function of Figure 8–17(b) is called a **high-pass** function. It is characterized by the fact that dc and all frequencies below a certain frequency are attenuated or reduced in amplitude, while higher frequencies are passed by the system. Thus, the passband for a high-pass filter is above the reference frequency.

The function of Figure 8–17(c) is called a **bandpass** function. Frequencies within a certain passband range are passed, while frequencies lower than and above this range are attenuated.

The function of Figure 8–17(d) is called a **band-rejection** function. Frequencies within a certain range are rejected, while frequencies lower than and above this range are passed.

8–7 SIMPLE *RC* CIRCUIT FORMS

We investigate the amplitude response forms of certain simple *RC* circuits in this section on a somewhat descriptive basis. The purpose is to emphasize certain kinds of low-frequency and high-frequency behavior that occur in electronic amplifiers. While the actual models of complex electronic circuits are quite involved, they can often be reduced to combinations of the forms to be discussed here. Thus, an understanding of the operation of these forms can lead to an understanding of similar effects in much more complex electronic circuits.

Certain useful formulas will be given with each of the circuits to be considered. These formulas may be readily derived using steady-state ac circuit analysis, and readers who have that background are strongly encouraged to verify the results. Readers who do not have a background in ac circuits should still be able to understand the significance of the results and to apply the formulas to practical problems.

Certain ''10% rules'' will be given. These guidelines are very useful in analyzing and designing circuits, and they provide useful intuitive approaches to circuit understanding. It should be stressed, however, that the rules of thumb provided should not be interpreted as absolute facts or requirements. Generally, they provide useful information, but when requirements are very exact, one must resort to an actual mathematical analysis for precise guidelines.

RC Low-Pass Model

Consider the circuit shown in Figure 8–18(a). Assume that v_i is a sinusoidal voltage source whose frequency can be varied. The steady-state output voltage v_o is measured across the capacitor. This arrangement thus allows a frequency response measurement to be performed.

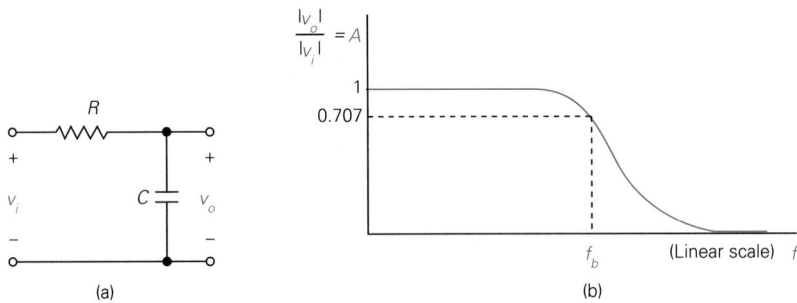

FIGURE 8–18
Simple *RC* low-pass circuit model and its amplitude response.

The dc steady-state model of the capacitor is an open circuit. Similarly, at very low frequencies, the capacitive reactance is very large, and the capacitor is acting nearly like an open circuit. Under these conditions, the output voltage is essentially the same as the input voltage, and $A \simeq 1$.

As the frequency continues to increase, the capacitive reactance decreases to a point where sufficient current is drawn from the source to cause a significant voltage drop across R. The output voltage then starts to drop.

At much higher frequencies, the capacitive reactance becomes very low. Almost all the input voltage is then dropped across R, and very little appears across C. This corresponds to the frequency range where, for some applications, the flat-band model (short circuit) for the capacitor might be used. However, for this application, we recognize that there will continue to be a small voltage across the capacitor. The drop in response as the frequency continues to increase is called **high-frequency rolloff.**

The form of $A(f)$ for this circuit is shown in Figure 8–18(b). This response is a *low-pass* type of function. The frequency scale here is assumed to be *linear*. Later we investigate the shape of such a curve when the scale is *logarithmic,* but let's not confuse the situation now.

Using ac circuit theory, it can be shown that the amplitude response $A(f)$ is given by

$$A(f) = \frac{X_C}{\sqrt{R^2 + X_C^2}} \qquad \text{(8–76)}$$

A convenient frequency to use in characterizing this response is that frequency at which $R = X_C$. For reasons that will not be clear until Section 9–10, we will define such a frequency as the **break frequency,** denoted as f_b.

The frequency f_b is determined by setting $X_C = R$. Letting $\omega = 2\pi f$ and $f = f_b$ yields

$$\frac{1}{2\pi f_b C} = R \qquad \text{(8–77)}$$

This leads to

$$\boxed{f_b = \frac{1}{2\pi RC}} \qquad \text{(8–78)}$$

By some algebraic manipulations, (8–76) can be rewritten as

$$A(f) = \frac{1}{\sqrt{1 + (f/f_b)^2}} \qquad \text{(8–79)}$$

It is readily shown that when $f = f_b$, the amplitude response is

$$\boxed{A(f_b) = \frac{1}{\sqrt{2}} \simeq 0.707} \qquad \text{(8–80)}$$

This frequency is identified on the curve of Figure 8–18(b).

At frequencies well below f_b, $A(f) \simeq 1$ and there is very little reduction in magnitude of a signal in passing through the circuit. Conversely, at frequencies well above f_b, there is considerable attenuation, and the output signal will be reduced significantly. In the vicinity of f_b, there is some attenuation of a signal, so f_b provides some measure of the transition from nearly flat response to the attenuation region. Some references define f_b as the **cutoff frequency,** but it is not a sharp cutoff characteristic for this simple circuit.

Note how the break frequency is inversely dependent on the RC product. Thus, an increase of either R or C will result in a reduction of the break frequency and a smaller passband region.

There are two separate points of view about this circuit that should be emphasized. First, the circuit has value as a simple low-pass filter, and it is often used for this purpose. As a low-pass filter, this circuit is about the most "elementary type" that can be implemented. Much sharper characteristics are available with more complex circuit structures, some of which are discussed later in the text. However, this simple circuit is widely used in minimal filtering applications where there is a wide band available between the desired signal and the undesired components to be attenuated.

The second point of view is that this circuit form arises frequently and naturally in electronic devices without intent on the part of the design. All electronic devices exhibit capacitance across the input, much of which arises from complex semiconductor properties. This capacitance, in conjunction with circuit resistance, results in an inherent frequency-limiting effect.

When flat transmission is desired through a circuit fitting the low-pass model, it is necessary that the signal frequency range of interest be well below f_b. If the capacitance is fixed (such as the input capacitance of an active device), this is achieved by ensuring that R be sufficiently small.

The actual amplitude response at any given frequency can be evaluated using (8–79). At a frequency $f = 0.1 f_b$, the value of $A(f)$ as determined from (8–79) is $A(0.1f_b) = 0.995$. This represents a drop in amplitude response of 0.5%. This particular frequency is sometimes used as a measure of the upper-frequency limit for negligible effect on the amplitude response. At the frequency $0.1 f_b$, the resistance is one-tenth the value of the capacitive reactance.

In many active circuits, the input circuit to the electronic device will contain a shunt capacitance in parallel with a resistance, and the driving circuit will contain a series resistance. The model for such a case is shown in Figure 8–19(a). The

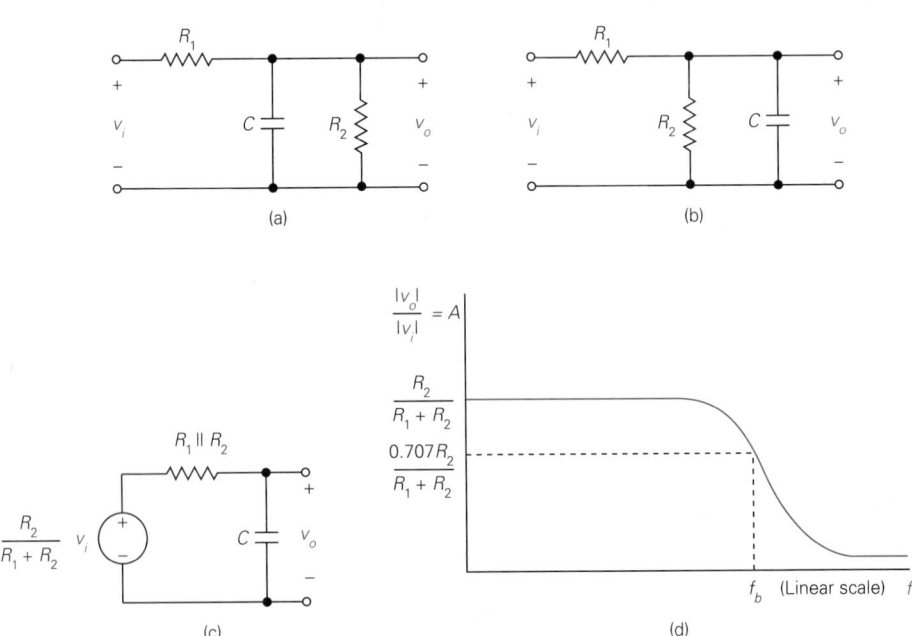

(a) (b)

(c) (d)

FIGURE 8–19

RC low-pass circuit containing both series and shunt resistors and simplification.

resistance R_2 has been deliberately shown on the right to assist the reader in seeing all possible points of view. We first redraw the circuit so that the capacitance C is on the right, as shown in Figure 8–19(b). A Thévenin equivalent circuit looking back from the capacitor is then determined as shown in Figure 8–19(c). The circuit is now in the form of the simpler one given in Figure 8–18(a).

The amplitude response for this circuit is shown in Figure 8–19(d). The low-frequency amplitude response is $R_2/(R_1 + R_2)$ because of the resistive voltage-divider reduction. The break frequency f_b is

$$f_b = \frac{1}{2\pi R_{eq} C} \qquad (8\text{–}81)$$

where R_{eq} is the parallel combination of R_1 and R_2; that is,

$$R_{eq} = R_1 \| R_2 = \frac{R_1 R_2}{R_1 + R_2} \qquad (8\text{–}82)$$

RC Coupling Circuit

Consider the *RC* circuit shown in Figure 8–20(a). At dc, the capacitor is an open circuit, and there is no coupling between input and output. Similarly, at very low frequencies, the capacitive reactance is very large, and it blocks most of the signal. As the frequency increases, the capacitive reactance decreases, thus allowing more of the signal to reach the output. At sufficiently high frequencies, the flat-band model of the capacitor (short-circuit) is valid, and the input signal is effectively coupled to the output.

The form of $A(f)$ for this circuit is shown in Figure 8–20(b). This response is a *high-pass* type of function. The drop in response at low frequencies is called **low-frequency rolloff.**

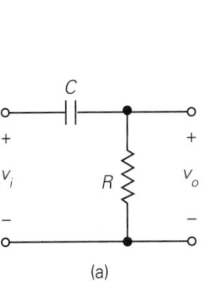
(a)

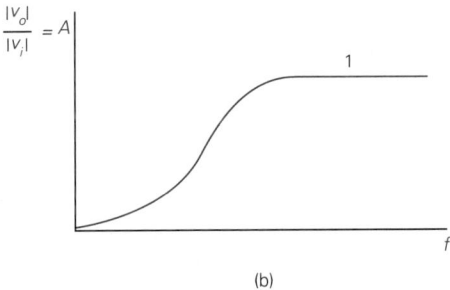
(b)

FIGURE 8–20
RC coupling circuit and form of the amplitude response.

Using ac circuit theory, it can be shown that the amplitude response $A(f)$ is given by

$$A(f) = \frac{R}{\sqrt{R^2 + X_C^2}} \qquad (8\text{--}83)$$

As in the case of the low-pass function, a *break frequency* f_b will be defined as the frequency at which $R = X_C$. The steps involved are exactly the same as for the low-pass case, so (8–77) and (8–78) are valid for the high-pass case. The result is exactly the same; that is,

$$\boxed{f_b = \frac{1}{2\pi RC}} \qquad (8\text{--}84)$$

By some algebraic manipulations, (8–83) can be expressed as

$$A(f) = \frac{1}{\sqrt{1 + (f_b/f)^2}} \qquad (8\text{--}85)$$

It is readily shown that when $f = f_b$, the amplitude response is

$$\boxed{A(f_b) = \frac{1}{\sqrt{2}} \simeq 0.707} \qquad (8\text{--}86)$$

Comparing low-pass and high-pass circuits, one can see that the expression for break frequency applies to both cases. For the *low-pass* case, the break frequency is at the *upper* end of the passband, while for the *high-pass* case, it is at the *lower* end. In both cases, the response has a level $1/\sqrt{2}$ at the break frequency. However, there is an important difference in the amplitude-response forms of (8–79) and (8–85). The difference is that f/f_b for the low-pass case is replaced by f_b/f for the high-pass case.

This circuit form is used extensively to pass a time-varying (or ac) signal while blocking a dc level. For distortion-free transmission, it is necessary that all frequencies comprising the signal fall in the flat-band region of the signal. The break frequency is inversely proportional to the RC product, so the lower range frequency response can be improved by increasing the RC product.

The actual amplitude response at any given frequency can be evaluated using (8–85). At a frequency $f = 10f_b$, the value of $A(f)$ as determined from (8–85) is $A(10f_b) = 0.995$. This particular frequency is sometimes used as a measure of the lower-frequency limit for negligible effect on the amplitude. In many design problems, R will be fixed and C must be selected for flat coupling. If the preceding criterion based on $10f_b$ is used, a capacitance C is calculated to establish f_b at one-tenth of the lowest applicable frequency.

In many practical coupling circuits, there will be both a source resistance and a load resistance. The applicable circuit model is shown in Figure 8–21(a).

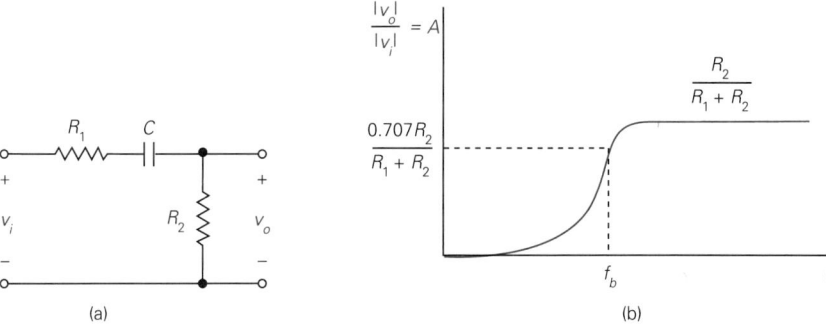

FIGURE 8–21
RC coupling network with attenuation in flat-band region.

The form of the amplitude response is shown in Figure 8–21(b). Note that the flat-band amplitude response is now $R_2/(R_1 + R_2)$, resulting from the resistive voltage-divider action.

The break frequency f_b is now

$$f_b = \frac{1}{2\pi R_{eq}C} \qquad (8\text{–}87)$$

where R_{eq} is the equivalent series resistance; that is,

$$R_{eq} = R_1 + R_2 \qquad (8\text{–}88)$$

RC Bypass Circuit

Consider the circuit shown in Figure 8–22 containing a resistance R in parallel with a capacitance C. This concept is best developed with a Norton model looking back from the capacitor. Although this form appears in a number of applications, one particular application arising in many electronic circuits is that of a **bypass circuit.** The concept is as follows: The resistor serves some dc function, such as establishing a transistor bias point. However, it is desired to pass the ac signal around the resistor, so that the resistance does not affect the signal. If the capacitor can be considered as a short circuit in the pertinent frequency range, the objective will be accomplished.

FIGURE 8–22
Basic form of an *RC* bypass circuit.

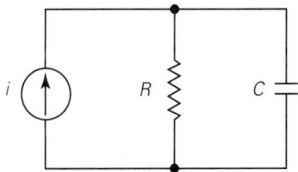

A useful rule of thumb applicable to this circuit is the following: *For most practical purposes, the capacitor can be considered to fully bypass the signal current at all ac frequencies in which the capacitive reactance is not greater than 10% of the net parallel resistance.* Expressed mathematically,

$$X_C \leq 0.1R \qquad \text{for bypass} \tag{8-89}$$

Note that it is the *net parallel resistance* that is used in this case. This rule is often misinterpreted in practical circuits whenever the *RC* combination shown is in parallel with some additional circuit whose equivalent resistance is not readily apparent.

Consider the circuit shown in Figure 8–23(a) in which a voltage source in series with R_1 appears on the left. Suppose that it is desired to bypass R_2 with the capacitor C. One might be tempted to base the rule on X_C and R_2 alone. However, it is necessary to determine the net equivalent parallel resistance. This is achieved by looking back to the left of R_2 and using a source transformation. The modified circuit is shown in Figure 8–23(b). The net parallel resistance R_{eq} is then

$$R_{eq} = R_1 \| R_2 = \frac{R_1 R_2}{R_1 + R_2} \tag{8-90}$$

Application of this rule yields

$$X_C \leq 0.1 R_{eq} \tag{8-91}$$

at the lowest frequency of interest.

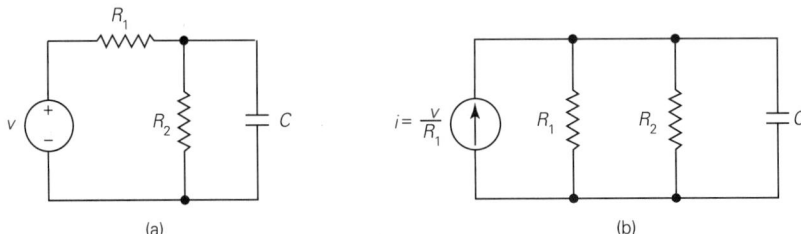

(a)   (b)

FIGURE 8–23
More complex bypass situation and equivalent circuit used to determine net parallel resistance.

EXAMPLE 8–14

A simple low-pass filter of the form of Figure 8–18(a) is to be designed to eliminate some high-frequency noise. Design the circuit to achieve a break frequency of 20 kHz.

Solution
The break frequency is given by

$$f_b = \frac{1}{2\pi RC} \tag{8-92}$$

The required RC product is then determined as

$$RC = \frac{1}{2\pi f_b} = \frac{1}{2\pi \times 20 \times 10^3} = 7.958 \times 10^{-6} \qquad (8\text{--}93)$$

This result indicates that any R and C whose product is equal to the value determined will achieve the objective. This situation illustrates a common occurrence in design. Since no other constraints are given, we are free to choose either R or C to be some common stock value and then determine the other value from the required product. It is usually better to select C as a stock value and determine R, since resistance is usually easier to adjust.

As a particular solution, we will choose $C = 0.001$ μF, in which case R is determined as

$$R = \frac{7.958 \times 10^{-6}}{C} = \frac{7.958 \times 10^{-6}}{0.001 \times 10^{-6}} = 7.958 \text{ k}\Omega \qquad (8\text{--}94)$$

However, many other combinations could be determined just as easily.

EXAMPLE 8–15

A certain signal transmission system can be modeled by the circuit of Figure 8–24. The 100-pF capacitance represents the net circuit shunt capacitance, which cannot be eliminated. The voltage v_g and resistance R_g represent the Thévenin model of a source used to drive the load. **(a)** Determine the break frequency f_b if $R_g = 10$ kΩ. **(b)** If the source is replaced with one having a much lower internal resistance, repeat the analysis based on $R_g = 50$ Ω.

FIGURE 8–24
Circuit of Example 8–15.

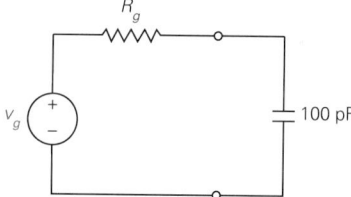

Solution
(a) The break frequency based on $R_g = 10$ kΩ is

$$f_b = \frac{1}{2\pi R_g C} = \frac{1}{2\pi \times 10^4 \times 100 \times 10^{-12}} = 159.15 \text{ kHz} \qquad (8\text{--}95)$$

(b) With $R_g = 50$ Ω, the break frequency now changes to

$$f_b = \frac{1}{2\pi \times 50 \times 100 \times 10^{-12}} = 31.83 \text{ MHz} \qquad (8\text{--}96)$$

Note the significant increase in the break frequency when the source resistance is reduced!

The actual usable bandwidth in each case would depend on how much deviation in the amplitude response would be acceptable. Recall that at the break frequency, the response is down to a level of 0.707. If, for example, the 0.1 f_b rule based on 0.5% degradation in amplitude response were used, the respective bandwidths in **(a)** and **(b)** would be 15.915 kHz and 3.183 MHz.

EXAMPLE 8–16

Determine the break frequency of the RC-coupling circuit of Figure 8–25.

FIGURE 8–25
Circuit of Example 8–16.

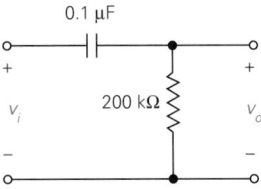

Solution
The break frequency is

$$f_b = \frac{1}{2\pi RC} = \frac{1}{2\pi \times 200 \times 10^3 \times 0.1 \times 10^{-6}} = 7.958 \text{ Hz} \qquad \textbf{(8–97)}$$

EXAMPLE 8–17

It is desired to couple two particular amplifiers together using a capacitor to block dc levels as shown in Figure 8–26(a). The ac Thévenin equivalent circuit at the output of the first stage is a voltage source in series with 2 kΩ, and the input resistance of the second stage is a 10-kΩ resistance. Determine the value of C required to achieve a break frequency of 10 Hz.

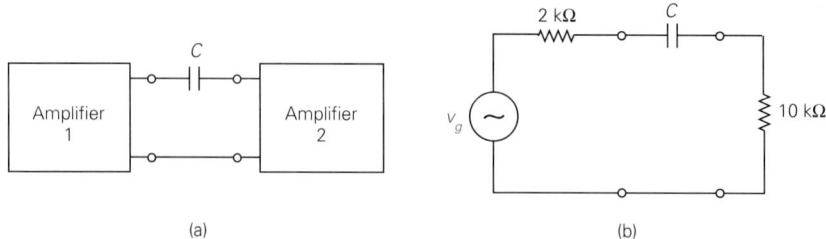

(a)   (b)

FIGURE 8–26
Circuit of Example 8–17.

Solution
An equivalent circuit at the point of coupling is shown in Figure 8–26(b). The net series resistance is

$$R_{eq} = 2 \text{ k}\Omega + 10 \text{ k}\Omega = 12 \text{ k}\Omega \qquad \textbf{(8–98)}$$

The break frequency is

$$f_b = \frac{1}{2\pi R_{eq}C} \tag{8–99}$$

Solving for C, we have

$$C = \frac{1}{2\pi R_{eq}f_b} = \frac{1}{2\pi \times 12 \times 10^3 \times 10} = 1.326 \ \mu F \tag{8–100}$$

8–8 BANDWIDTH AND RISE TIME

Many signals that must be transmitted or processed in modern electronic systems are digital or pulse-type signals. Such signals often have only two levels, representing a logical 1 and a logical 0, and perfect reproduction is not necessary as long as the two levels are distinguishable. In other pulse applications, it may be necessary to reproduce the pulse to the highest accuracy possible.

It turns out that the effect of a limited bandwidth is to impose a rise time on a pulse waveform. The basic concept will be developed with the RC low-pass circuit shown in Figure 8–27(a). Assume that the input is a dc voltage of magnitude

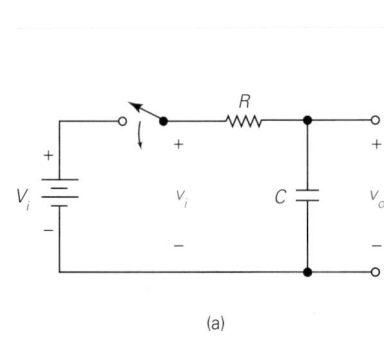

(a)

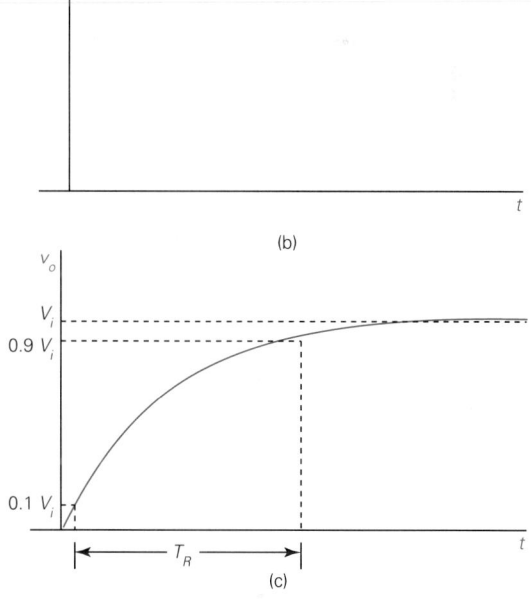

(b)

(c)

FIGURE 8–27
Low-pass RC circuit and its effect on a dc voltage switched on at $t = 0$.

V_i, which is switched in to the RC circuit at $t = 0$. The voltage v_i to the right of the switch then changes instantaneously from 0 to V_i at $t = 0$, as shown in Figure 8–27(b).

A fundamental property of capacitance is that a voltage across it cannot change instantaneously. Instead, it takes some time for the capacitor voltage to charge or discharge. For the circuit given, the voltage buildup across the capacitance has the form shown in Figure 8–27(c), and this is the circuit output voltage.

The capacitor voltage can be described by an **exponential function.** As a function of time t, the voltage can be expressed mathematically as

$$v_o = V_i(1 - e^{-t/RC}) \tag{8–101}$$

In this equation, $e = 2.718$ is the base for the natural system of logarithms. More will be said about exponential functions in the context of nonlinear circuits in Chapter 16, but the form of (8–101) will suffice for the moment.

Rise Time

The quantity **rise time** is a measure of the duration of the charging time in a circuit such as this. However, it is often difficult to determine precisely where the output voltage starts and where it reaches a final value. For that reason, the standard rise-time definition is based on the time between the point where the output voltage is at 10% of its final value and the point where the output voltage is at 90% of its final value. This is illustrated in Figure 8–27(c).

For the simple RC circuit, it can be shown from the basic properties of the exponential function that the rise time T_R is given by

$$\boxed{T_R = 2.2RC} \tag{8–102}$$

The rise time is thus directly proportional to both R and C.

A relationship between rise time and bandwidth will now be obtained. The break frequency f_b is

$$\boxed{f_b = \frac{1}{2\pi RC}} \tag{8–103}$$

If the product RC, expressed in terms of f_b, is determined from (8–103) and substituted in (8–102), the following equation is obtained:

$$\boxed{T_R = \frac{0.35}{f_b}} \tag{8–104}$$

Although this equation is based on the simple RC low-pass circuit, the equation can be used to estimate the approximate rise time of more complex circuits. Let us denote the low-pass bandwidth of any circuit of interest as B, in which case (8–104) as an estimate becomes

$$T_R = \frac{0.35}{B}$$ (8–105)

The results of this equation are very important and need to be emphasized. In general, *the rise time and bandwidth are inversely related*. Thus, to reproduce a perfect pulse without distortion, an infinite bandwidth is required.

Pulse Reproduction

The effects of finite bandwidth on a pulse train are illustrated in Figure 8–28. The waveform of Figure 8–28(a) represents some input pulse train v_i. (It could be digital data, for example.) The width of a given pulse is assumed to be T_p. When the bandwidth is relatively large, the output pulse train is quite close in form to the input, as shown in Figure 8–28(b). Some rounding at the edges occurs, but this may not even be apparent if the bandwidth is sufficiently large.

As the bandwidth is reduced, the output pulse train eventually assumes the form of Figure 8–28(c). The two-level character of the output is quite evident, but considerable rounding has occurred. Depending on the application, this type of output may or may not be acceptable.

If the bandwidth is reduced further, the output pulses are broadened and smeared to the point where it is nearly impossible to distinguish between different levels. This situation is illustrated in Figure 8–28(d).

From a quantitative point of view, the bandwidth required to reproduce a pulse train *accurately* should satisfy the inequality $B \gg 1/T_p$. However, for many forms of pulse data, where the levels must be distinguishable but where accurate pulse reproduction is not necessary, the following "coarse reproduction" estimate is useful:

$$B \simeq \frac{0.5}{T_p}$$ (8–106)

If the bandwidth is reduced much below the value $0.5/T_p$, the amplitude of the output pulses is degraded severely, and it will become nearly impossible to distinguish the levels involved.

EXAMPLE 8–18

A certain RC low-pass circuit of the form shown in Figure 8–27(a) has $R = 10$ kΩ and $C = 0.01$ μF. Calculate the rise time.

Solution
The rise time is

$$T_R = 2.2RC = 2.2 \times 10^4 \times 0.01 \times 10^{-6} = 220 \ \mu s$$ (8–107)

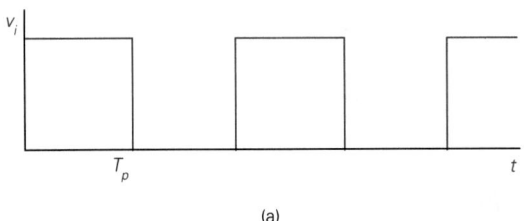

(a)

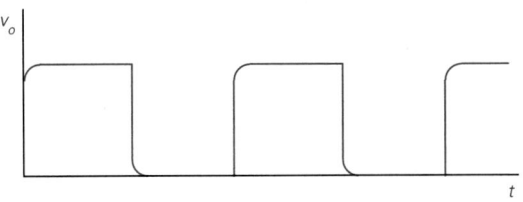

(b) Large bandwidth ($B \gg 1/T_p$)

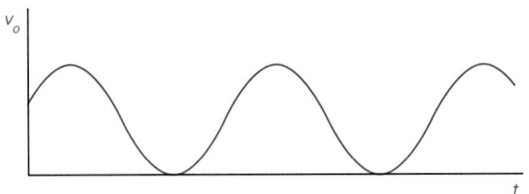

(c) Medium bandwidth ($B \approx 0.5/T_p$)

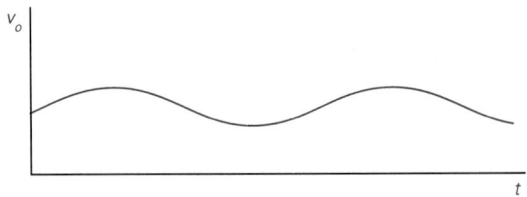

(d) Small bandwidth ($B \ll 1/T_p$)

FIGURE 8–28
Effects of finite bandwidth on pulse train.

EXAMPLE 8–19
A certain *RC* low-pass circuit of the form shown in Figure 8–27(a) has $f_b = 2$ kHz. Calculate the rise time.

Solution
The rise time is

$$T_R = \frac{0.35}{f_b} = \frac{0.35}{2 \times 10^3} = 175 \ \mu s \qquad (8\text{--}108)$$

EXAMPLE 8–20
A certain digital data signal has bits with width 2 μs. Rounding the data according to the "coarse bandwidth" estimate is acceptable for the particular application involved. Estimate the approximate bandwidth required for signal transmission.

Solution
The bandwidth estimate is given by

$$B \simeq \frac{0.5}{T_p} = \frac{0.5}{2 \times 10^{-6}} = 250 \ \text{kHz} \qquad (8\text{--}109)$$

Understand that this is an estimate rather than an exact equation.

8–9 PSPICE EXAMPLES

PSPICE EXAMPLE 8–1
For the circuit of Example 8–9 (Figure 8–8), determine the values of the steady dc voltages and currents using PSPICE.

Solution
Although the circuit contains capacitors, we are interested only in the steady-state dc values, and these are determined in PSPICE by a simple dc analysis. In fact, we could even leave out the capacitors if desired and replace them by open circuits, but our intent is to show what PSPICE does in this case.

The circuit adapted to PSPICE is shown in Figure 8–29(a), and the code is shown in. Figure 8–29(b). The computer printout is shown in Figure 8–30. The values of the three dc capacitor voltages are read as V(C1) = 20 V, V(C2) = V(C3) = 10 V. These results are in perfect agreement with the results of Example 8–9.

In summary, the steady-state dc values of a circuit containing capacitance and/or inductance may be determined from a .DC analysis with PSPICE. Capacitors are treated like open circuits, and inductors are treated like short circuits.

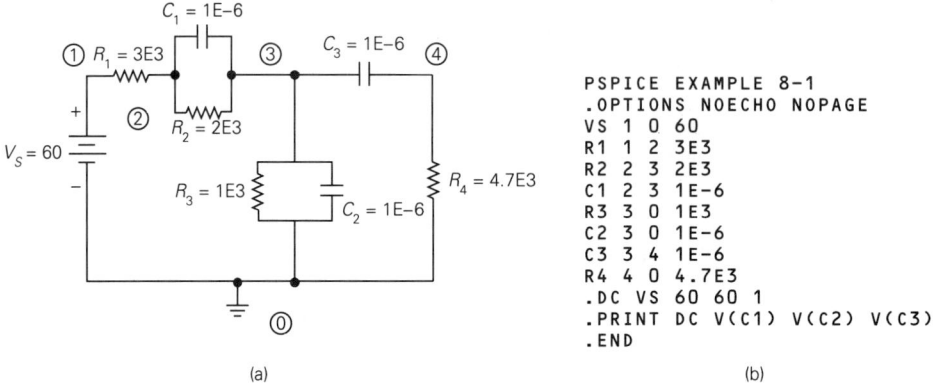

```
PSPICE EXAMPLE 8-1
.OPTIONS NOECHO NOPAGE
VS 1 0 60
R1 1 2 3E3
R2 2 3 2E3
C1 2 3 1E-6
R3 3 0 1E3
C2 3 0 1E-6
C3 3 4 1E-6
R4 4 0 4.7E3
.DC VS 60 60 1
.PRINT DC V(C1) V(C2) V(C3)
.END
```

(a)   (b)

FIGURE 8–29
Circuit and code of PSPICE Example 8–1.

```
PSPICE EXAMPLE 8-1

****        CIRCUIT DESCRIPTION

***********************************************************************

.OPTIONS NOECHO NOPAGE

****        DC TRANSFER CURVES                    TEMPERATURE = 27.000 DEG C

VS          V(C1)        V(C2)        V(C3)
6.000E+01   2.000E+01    1.000E+01    1.000E+01


            JOB CONCLUDED

            TOTAL JOB TIME        1.15
```

FIGURE 8–30
Computer printout of PSPICE Example 8–1.

PSPICE EXAMPLE 8–2

Solve the circuit of Example 8–12 (Figure 8–12) in the flat-band region using PSPICE.

Solution

We saw in the last example that a dc analysis was sufficient to determine the steady-state dc voltages and currents in the circuit. In contrast, it is necessary to employ an *ac analysis* to determine the flat-band behavior of a circuit. Most important, the frequency chosen should be sufficiently high that all capacitive reactances are small compared with circuit resistances, and all inductive reactances should be large compared with all circuit resistances. Since the values of

C_1 and C_2 in Example 8–12 are not specified, a number of combinations of capacitance and frequency values could be selected to achieve this objective.

As an arbitrary, yet simple, choice, the frequency will be selected as 10 kHz and the capacitances will be selected as 10 μF each. The reader is invited to show that the capacitive reactance of a 10-μF capacitor at 10 kHz is 1.59 Ω, which should be negligible in comparison to the circuit resistances.

In Example 8–12, the voltage v_o was determined in terms of the generator voltage v_g. With PSPICE an actual value must be assigned to v_g. The simplest choice is 1 V, and the resulting value for v_o should be the multiplier for v_g.

The circuit adapted to PSPICE is shown in Figure 8–31(a), and the code is shown in Figure 8–31(b). This is the first point at which an ac analysis with PSPICE is to be performed, so some explanation is needed.

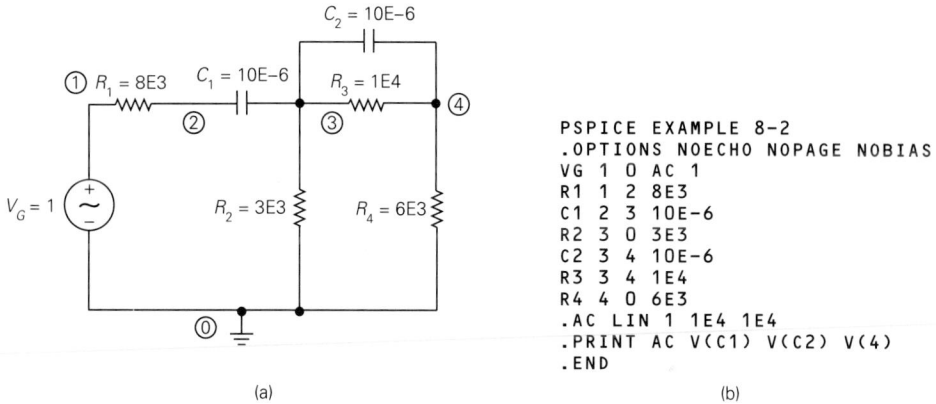

```
PSPICE EXAMPLE 8-2
.OPTIONS NOECHO NOPAGE NOBIAS
VG  1  0  AC  1
R1  1  2  8E3
C1  2  3  10E-6
R2  3  0  3E3
C2  3  4  10E-6
R3  3  4  1E4
R4  4  0  6E3
.AC LIN 1 1E4 1E4
.PRINT AC V(C1) V(C2) V(4)
.END
```

(a)   (b)

FIGURE 8–31
Circuit and code of PSPICE Example 8–2.

In general, ac analysis is based on the assumption of sinusoidal voltages and currents in the circuit. However, unless one is employing transient analysis, the sinusoids themselves are not observed in the solution. Instead, the values obtained are the ac **phasors,** which are represented in terms of a magnitude and an angle. Accompanying ac circuit courses should provide more of the analytical details, but we can start using the concepts at this point.

The line describing the generator reads as follows:

```
VG 1 0 AC 1
```

The first three entries have the same meaning as for a dc analysis. The fourth entry AC is required for PSPICE to perform an ac analysis. Without this entry, the analysis would default to dc. The last entry is the value of the ac voltage, which could be interpreted as either a peak value or an rms value depending on the desired objective. In this case, it is interpreted as a peak value. In the most general case, a sixth entry representing the phase angle of the ac voltage could be added. However, the program defaults to 0° when this entry is omitted, and that is the case here.

The next six lines describe the remainder of the circuit configuration, and the format here should be self-evident.

The first line of the control block reads

```
.AC LIN 1 1E4 1E4
```

The command .AC tells PSPICE to perform an ac analysis. The remaining part of the line provides information on the frequency or frequencies at which to perform the analysis. The command is designed for a frequency sweep in which a range of frequencies can be covered. However, our interest is one frequency and LIN (for "linear") is the best choice in this case. (Later, other options will be considered.)

The first number 1 means that only one frequency is to be used. The second number 1E4 represents the first frequency, and the third number 1E4 represents the last frequency, which, of course, is the same in this case, since only one frequency is being considered. PSPICE requires all three entries to satisfy the format.

The next line reads

```
.PRINT AC V(C1) V(C2) V(4)
```

This statement is almost exactly the same as for dc except, of course, that AC is substituted for DC. Our primary variable of interest is V(4), but for the sake of curiosity we will see what the voltages across the two capacitors are. Only the magnitudes of the ac voltages are requested at this time. Phase angles will be considered later.

The computer printout is shown in Figure 8–32. The value of V(4) is exactly

```
PSPICE EXAMPLE 8-2

****      CIRCUIT DESCRIPTION

*********************************************************************

.OPTIONS NOECHO NOPAGE NOBIAS

****      AC ANALYSIS                        TEMPERATURE = 27.000 DEG C

FREQ        V(C1)       V(C2)       V(4)
1.000E+04   1.592E-04   5.305E-05   2.000E-01


          JOB CONCLUDED

          TOTAL JOB TIME      1.32
```

FIGURE 8–32
Computer printout of PSPICE Example 8–2.

as expected, that is, 0.2 V (for an input voltage of 1 V). The two capacitor voltage magnitudes are not zero but are 0.1592 mV and 0.05305 mV, respectively. These voltages are negligible in comparison to the desired flat-band output voltage and may be considered to be zero for most practical purposes. If necessary, they could be made smaller by increasing either the frequency or the values of the capacitors.

PSPICE EXAMPLE 8–3

Use both ac and dc analysis modes of PSPICE to analyze the circuit of Example 8–13 (Figure 8–14).

Solution

The circuit adapted to PSPICE is shown in Figure 8–33(a), and the code is shown in Figure 8–33(b). A major theme of this problem is that both a dc analysis and an ac analysis are to be performed in the same program. The dc output will provide information on the dc steady-state circuit values, and the ac output will provide information on the flat-band values. As in PSPICE Example 8–2, the frequency and capacitance values must be chosen to provide negligible reactance values in comparison with circuit parameters. In this case, capacitance value of 100 μF along with a frequency of 1 kHz will be arbitrarily selected. The reader is invited to show that the reactance of a 100-μF capacitor at 1 kHz is about 1.59 Ω.

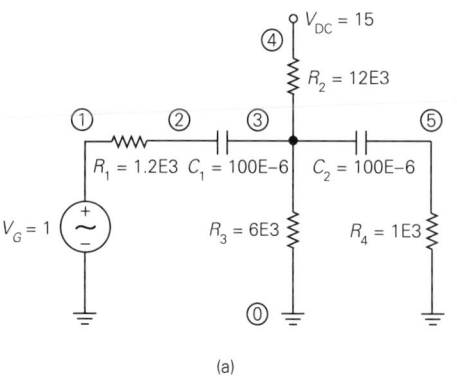

(a)

```
PSPICE EXAMPLE 8-3
.OPTIONS NOECHO NOPAGE NOBIAS
VG 1 0 AC 1
R1 1 2 1.2E3
C1 3 2 100E-6
VDC 4 0 DC 15
R2 4 3 12E3
R3 3 0 6E3
C2 3 5 100E-6
R4 5 0 1E3
.DC VDC 15 15 1
.PRINT DC I(R1) I(R2) I(R3) I(R4)
.PRINT DC V(C1) V(R3) V(C2) V(R4)
.AC LIN 1 1E3 1E3
.PRINT AC I(R1) I(R2) I(R3) I(R4)
.PRINT AC V(C1) V(R3) V(C2) V(R4)
.END
```

(b)

FIGURE 8–33
Circuit and code of PSPICE Example 8–3.

The circuit block starting with V_G and continuing through R_4 represents a complete description of the circuit in which both an ac source of 1 V and a dc source of 15 V are listed. This means that more than one type of source may appear in the same circuit code. We could have omitted DC from the 15-V line if desired, but it has been included for clarity. Incidentally, if both a dc and an ac analysis are to be performed with the same source, it is possible to include both descriptions on the same line, but in this case, there are two separate sources.

The control block contains both dc and ac commands. The .DC line informs PSPICE to perform a dc analysis for $V_{DC} = 15$ V. The .PRINT statements that follow list the various quantities that are to be printed.

The .AC line tells PSPICE to perform an ac analysis at a frequency of 1 kHz. The next two lines provide the print statements for certain magnitudes (but not phase angles).

The computer data are shown in Figure 8–34. The dc data can be compared with the values obtained in Example 8–13. The notation is somewhat different, but all values are in perfect agreement.

```
PSPICE EXAMPLE 8-3

****      CIRCUIT DESCRIPTION

***********************************************************************

.OPTIONS NOECHO NOPAGE NOBIAS

****      DC TRANSFER CURVES                TEMPERATURE = 27.000 DEG C


VDC        I(R1)        I(R2)        I(R3)        I(R4)
1.500E+01  0.000E+00    8.333E-04    8.333E-04    0.000E+00


****      DC TRANSFER CURVES                TEMPERATURE = 27.000 DEG C


VDC        V(C1)        V(R3)        V(C2)        V(R4)
1.500E+01  5.000E+00    5.000E+00    5.000E+00    0.000E+00


****      AC ANALYSIS                       TEMPERATURE = 27.000 DEG C

FREQ       I(R1)        I(R2)        I(R3)        I(R4)
1.000E+03  5.000E-04    3.333E-05    6.667E-05    4.000E-04


****      AC ANALYSIS                       TEMPERATURE = 27.000 DEG C

FREQ       V(C1)        V(R3)        V(C2)        V(R4)
1.000E+03  7.958E-04    4.000E-01    6.366E-04    4.000E-01



        JOB CONCLUDED
```

FIGURE 8–34
Computer printout of PSPICE Example 8–3.

The ac data can be compared with the flat-band ac data of Example 8–13 with the values from PSPICE interpreted as the multipliers of v_g. The values V(R3) and V(R4) represent v_A and v_B in the circuit, respectively, and they are in perfect agreement with Example 8–13. Note the small values for the two capacitor voltage magnitudes.

PROBLEMS

Drill Problems

8–1. A certain sinusoidal voltage v has a peak value of 5 V and a frequency of 2 kHz. The time origin is chosen in the form of Figure 8–2. **(a)** Determine the period. **(b)** Write an equation for the voltage.

8–2. A certain sinusoidal current i has a peak value of 120 mA and a frequency of 500 Hz. The time origin is chosen in the form of Figure 8–2. **(a)** Determine the period. **(b)** Write an equation for the current.

8–3. A certain voltage is described by the equation

$$v = 25 \sin 2\pi \times 5000t$$

Determine **(a)** the peak value; **(b)** the radian frequency; **(c)** the cyclic frequency; and **(d)** the period.

8–4. A certain current is described by the equation

$$i = 2 \cos 1000t$$

Determine **(a)** the peak value; **(b)** the radian frequency; **(c)** the cyclic frequency; and **(d)** the period.

8–5. **(a)** Determine the rms value of the voltage of Problem 8–1. **(b)** Determine the power dissipated in a 180-Ω resistor.

8–6. **(a)** Determine the rms value of the current of Problem 8–2. **(b)** Determine the power dissipated in a 200-Ω resistor.

8–7. **(a)** Determine the rms value of the voltage of Problem 8–3. **(b)** Determine the power dissipated in a 2.2-kΩ resistor.

8–8. **(a)** Determine the rms value of the current of Problem 8–4. **(b)** Determine the power dissipated in a 10-Ω resistor.

8–9. Determine the reactance of a 200-pF capacitor at each of the following frequencies: **(a)** 100 Hz; **(b)** 50 kHz; **(c)** 4 MHz.

8–10. Determine the reactance of a 10-μF capacitor at each of the following frequencies: **(a)** 50 Hz; **(b)** 20 kHz; **(c)** 1 MHz.

8–11. Determine the reactance of a 40-μH inductance at each of the following frequencies: **(a)** 200 Hz; **(b)** 100 kHz; **(c)** 5 MHz.

8–12. Determine the reactance of a 2-H inductance at each of the following frequencies: **(a)** 25 Hz; **(b)** 10 kHz; **(c)** 200 kHz.

8–13. Determine the steady-state dc voltage V_C across the capacitor in the circuit of Figure P8–13.

FIGURE P8–13

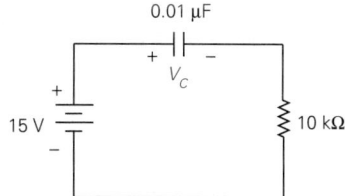

8–14. Determine the steady-state dc voltage V_C across the capacitor in the circuit of Figure P8–14.

FIGURE P8–14

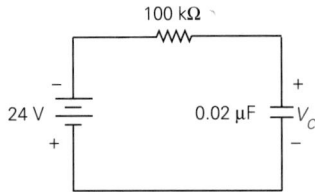

8–15. Determine the steady-state dc voltages V_{C1} and V_{C2} in the circuit of Figure P8–15.

FIGURE P8–15

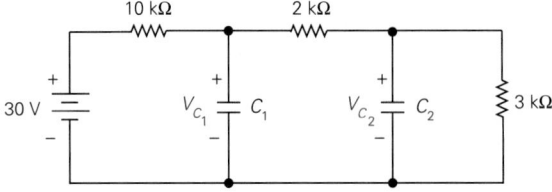

8–16. Determine the steady-state dc voltage V_C in the circuit of Figure P8–16.

FIGURE P8–16

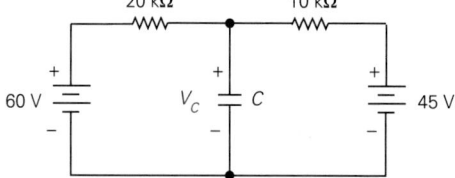

8–17. Determine the steady-state current I and voltage V_C for the circuit of Figure P8–17.

FIGURE P8–17

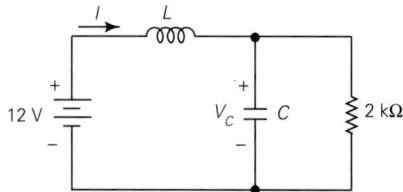

8–18. Determine the steady-state current I and voltages V_{C1} and V_{C2} for the circuit of Figure P8–18.

FIGURE P8–18

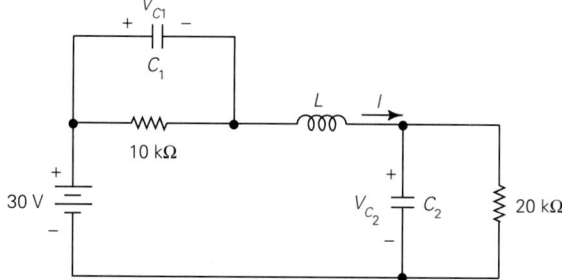

8–19. For the single-stage RC-coupled BJT amplifier of Figure P8-19, determine the steady-state dc voltages across the three capacitors. The voltage v_g is an ac signal source.

FIGURE P8–19

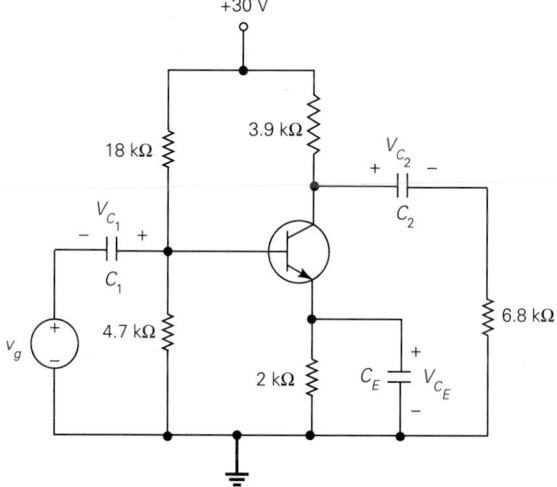

8–20. The circuit of Figure P8–20 represents an RC-coupled JFET amplifier. Determine the steady-state dc voltages across the three capacitors. The voltage v_g is an ac signal source.

FIGURE P8–20

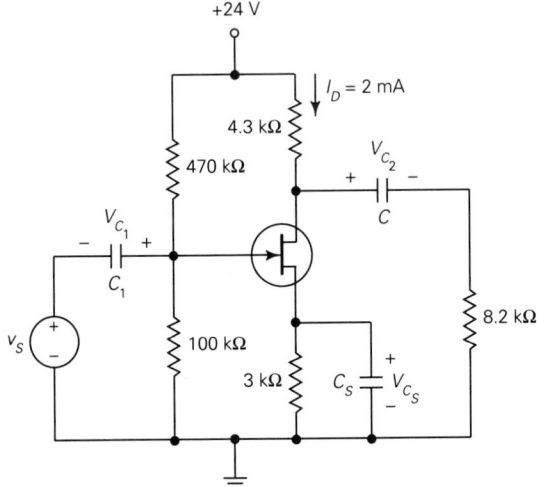

8–21. For the *RC* circuit of Figure P8–21, assume that the frequency range of the signal v_g is in the flat-band region. Determine the voltage v_o in terms of v_g.

FIGURE P8–21

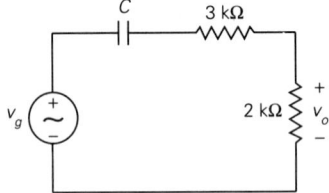

8–22. For the *RC* circuit of Figure P8–22, assume that the frequency range of the signal v_g is in the flat-band region. Determine the voltage v_o in terms of v_g.

FIGURE P8–22

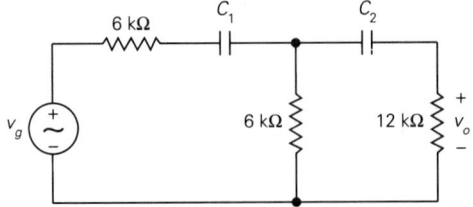

8–23. The circuit of Figure P8–23 contains a time-varying signal v_g and a 24-V dc source. Assume that all frequency components of v_g are in the flat-band region. A number of possible variables of interest are identified. **(a)** Construct a dc model, and compute the dc values of the variables. **(b)** Construct a flat-band model, and compute the ac values of the variables. **(c)** Write expressions for the total instantaneous variables.

FIGURE P8–23

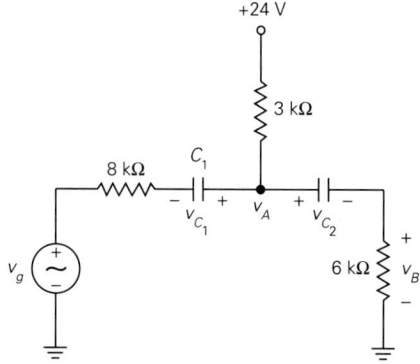

8–24. Repeat the analysis of Problem 8–23 for the circuit of Figure P8–24.

FIGURE P8–24

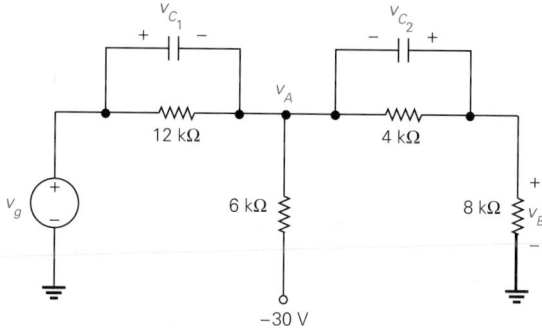

8–25. An *RC* low-pass circuit of the form of Figure 8–18(a) has $R = 10$ kΩ and $C = 0.001$ μF. Determine the break frequency.

8–26. An *RC* low-pass circuit of the form of Figure 8–18(a) has $R = 600$ Ω and $C = 200$ pF. Determine the break frequency.

8–27. An *RC* low-pass circuit of the form of Figure 8–19(a) has $R_1 = 12$ kΩ, $R_2 = 6$ kΩ, and $C = 0.01$ μF. Determine the break frequency.

8–28. An *RC* low-pass circuit of the form of Figure 8–19(a) has $R_1 = 200$ kΩ, $R_2 = 50$ kΩ, and $C = 2$ μF. Determine the break frequency.

8–29. An *RC* coupling network of the form of Figure 8–20(a) has $R = 20$ kΩ and $C = 0.01$ μF. Determine the break frequency.

8–30. An *RC* coupling network of the form of Figure 8–20(a) has $R = 470$ kΩ and $C = 50$ pF. Determine the break frequency.

8–31. An *RC* coupling network of the form of Figure 8–21(a) has $R_1 = 12$ kΩ, $R_2 = 24$ kΩ, and $C = 0.01$ μF. Determine the break frequency.

8–32. An *RC* coupling network of the form of Figure 8–21(a) has $R_1 = 150$ Ω, $R_2 = 220$ Ω, and $C = 100$ pF. Determine the break frequency.

8–33. A certain signal source has a Thévenin resistance of 600 Ω, and it is driving a capacitive load of 800 pF. Calculate the rise time.

8–34. A certain signal source has a Thévenin resistance of 2 kΩ, and it is driving a capacitive load of 50 pF. Calculate the rise time.

8–35. Calculate the rise time for the circuit of Problem 8–25.

8–36. Calculate the rise time for the circuit of Problem 8–26.

8–37. Calculate the rise time for the circuit of Problem 8–27.

8–38. Calculate the rise time for the circuit of Problem 8–28.

8–39. The measured rise time of a certain low-pass circuit is 20 μs. Estimate the bandwidth B.

8–40. The measured rise time of a certain low-pass circuit is 50 ns. Estimate the bandwidth B.

Derivation Problems

8–41. Starting with (8–76), fill in the steps required to express it in the form of (8–79).

8–42. Starting with (8–83), fill in the steps required to express it in the form of (8–85).

8–43. From the expression for v_o in (8–101), determine the time t_1 at which $v_o = 0.1\ V_i$ and the time t_2 at which $v_o = 0.9\ V_i$. The value of rise time is then $T_R = t_2 - t_1$. Show that the result is (8–102).

8–44. Fill in the steps leading from (8–102) to (8–104).

Design Problems

8–45. In a certain application requiring a capacitor, it is desired that the reactance be no greater than 250 Ω at a frequency of 20 Hz. Determine the minimum value of capacitance required.

8–46. In a certain application requiring a capacitor, it is desired that the reactance be no less than 500 Ω at a frequency of 2 kHz. Determine the maximum value of capacitance required.

8–47. In a certain application requiring an inductor, it is desired that the reactance be no less than 5 kΩ at a frequency of 200 kHz. Determine the minimum value of inductance required.

8–48. In a certain application requiring an inductor, it is desired that the reactance be no greater than 200 Ω at a frequency of 50 Hz. Determine the maximum value of inductance required.

8–49. A simple low-pass filter of the form of Figure 8–18(a) is to be designed to have a break frequency of 5 kHz. Determine the value of R required if the capacitance is selected as $C = 0.01\ \mu$F.

8–50. For the design of Problem 8–49, determine the value of C required if the resistance is selected as $R = 10$ kΩ.

8–51. A coupling network of the form of Figure 8–20(a) is to be designed to have a break frequency of 50 Hz. Determine the value of R required if the capacitance is selected as $C = 12\ \mu$F.

8–52. For the design of Problem 8–51, determine the value of C required if the resistance is selected as $R = 200$ kΩ.

Troubleshooting Problems

8–53. You are troubleshooting the circuit of Figure P8–15, and you measure the following voltages with a dc voltmeter: source voltage = 30 V, $V_{C1} = 5$ V, $V_{C2} = 0$. Which *two* of the following conditions could represent possible difficulties? **(a)** 10-kΩ resistor shorted; **(b)** C_1 shorted; **(c)** 2-kΩ resistor shorted; **(d)** 2-kΩ resistor open; **(e)** C_2 shorted; **(f)** 3-kΩ resistor shorted.

8–54. You are troubleshooting the circuit of Figure P8–16 and you measure the following voltages with a dc voltmeter: source voltages = 60 V and 45 V (as shown), V_C = 60 V. Which *two* of the following conditions could represent possible difficulties? **(a)** 20-kΩ resistor open; **(b)** 10-kΩ resistor open; **(c)** 20-kΩ resistor shorted; **(d)** 10-kΩ resistor shorted; **(e)** C shorted.

8–55. You are troubleshooting a simple low-pass filter circuit of the form shown in Figure 8–18(a). The circuit is not suppressing high-frequency noise to the extent expected. Which *two* of the following conditions could represent possible difficulties? **(a)** R too small; **(b)** R too large; **(c)** C too small; **(d)** C too large.

8–56. You are troubleshooting a circuit of the form shown in Figure 8–20(a), and the frequency of the source is in the specified range of flat-band frequencies. With an ac rms voltmeter, you measure v_g = 80 mV and v_o = 20 mV. Which *two* of the following conditions could represent possible difficulties? **(a)** R too small; **(b)** R too large; **(c)** C too small; **(d)** C too large.

9

INTRODUCTION TO AMPLIFIERS

OBJECTIVES

After completing this chapter, the reader should be able to:

- Define the voltage gain of an amplifier and discuss its physical significance.
- Explain various classifications of amplifiers such as noninverting or inverting, dc coupled or ac coupled, voltage or power, and so on.
- Construct the flat-band model of an amplifier.
- Define and explain the significance of the input and output resistances of an amplifier.
- Determine the net gain of an amplifier when the generator has internal resistance, and/or an external load resistance is connected to the output.
- Determine the net gain of a cascade of two or more amplifiers when there is interaction between the stages.
- Define and calculate decibel gain and loss quantities from the actual gain and loss values.
- Calculate actual gain and loss values from decibel gain and loss quantities.
- Calculate the decibel gain for a cascade of amplifier stages.
- Discuss the general concept of a Bode plot.
- Construct the Bode plot amplitude and phase approximations for a simple low-pass circuit.
- Construct the Bode plot amplitude and phase approximations for a simple high-pass circuit.
- Analyze some of the circuits in the chapter with PSPICE.

9–1 OVERVIEW

One of the most important applications of electronic devices is in the implementation of **amplifier** circuits. An **amplifier** is a circuit used to increase the level of a signal to meet some desired objective. For example, when a phonograph needle (stylus) moves along a record, tiny signals are generated at the output of the cartridge. By use of electronic amplifiers, these signals are increased to levels that may be connected to speakers from which audible sound is generated. A second example is a radio receiver, in which tiny electromagnetic signals generated in the antenna are amplified and processed to produce useful output.

A major objective of the next several chapters is to develop the various basic individual circuits from which composite amplifiers are formed. In many books, the reader is immediately exposed to the details of various amplifier circuits without thorough grounding in the general properties that are sought in such circuits. Consequently, beginning students are often bewildered with amplifier details and do not understand the significance of the major amplifier properties.

The approach in this book is to provide this chapter as a basis for general amplifier properties without regard to specific configurations or devices. Consequently, the circuit models developed in this chapter are quite general and may be applied to virtually all amplifiers to be considered later. Therefore, study the basic properties presented here carefully and make sure that you understand the significance of each. You will see these properties used extensively in later chapters as particular amplifiers using BJTs, FETs, and operational amplifiers are discussed.

9–2 AMPLIFIERS

An ideal amplifier is a circuit that increases the level of a signal while preserving the exact form. Consider the block diagram shown in Figure 9–1. The voltage v_i represents an assumed input signal, and the voltage v_o represents an assumed output signal. The output voltage is related to the input voltage in an ideal amplifier by the equation

$$v_o = Av_i$$

(9–1)

The quantity A represents the **voltage gain,** and it can be expressed as

$$A = \frac{v_o}{v_i}$$

(9–2)

for the ideal case.

FIGURE 9–1
Block diagram of amplifier with assumed input and output voltages.

The key to the definition of the ideal amplifier is that the *exact shape of the signal is preserved*. If the signal shape is altered, it is not possible to express input and output relationships exactly in the simple algebraic forms of (9–1) and (9–2).

Any change in the signal shape resulting from the amplifier is called **distortion.** An ideal amplifier is said to be **distortion-free.** All electronic devices introduce some amount of distortion, but a good design results in a distortion level so low that it produces negligible degradation for the intended application. It is appropriate, therefore, to use the simple algebraic forms of (9–1) and (9–2) in amplifiers that have very low distortion.

Although the output signal of an amplifier is normally larger than the input signal, it should be stressed that the amplifier is incapable of generating power. Rather, one or more dc power sources provide the power required for the unit to function properly. The small input signal acts as a control for the device and causes the power from the dc source to be converted to an output signal that has the form of the input signal. In performing signal analysis of an amplifier, it is customary to omit the dc power supplies in many cases, but their presence is understood for proper operation.

In the explanations that follow, we will discuss several methods for classifying amplifiers. Note the terms and meanings as they will arise throughout the remainder of the book.

Noninverting and Inverting

Amplifiers can be classified as either noninverting or inverting. If $A > 0$ (positive), the device is said to be a **noninverting amplifier,** and if $A < 0$ (negative), the device is said to be an **inverting amplifier.**

The transfer characteristic of an ideal *noninverting* amplifier is shown in Figure 9–2. It is a straight line through the origin with a slope A, which is assumed to be positive in this case.

The transfer characteristic of an ideal *inverting* amplifier is shown in Figure 9–3. It is also a straight line through the origin, but the slope A is assumed to be negative in this case.

FIGURE 9–2

Transfer characteristic of ideal *noninverting* amplifier.

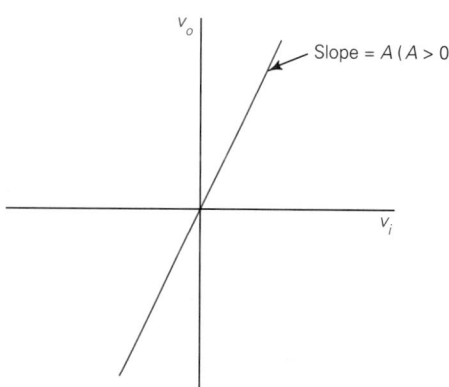

Slope = A ($A > 0$)

v_o

v_i

FIGURE 9–3
Transfer characteristic of ideal *inverting* amplifier.

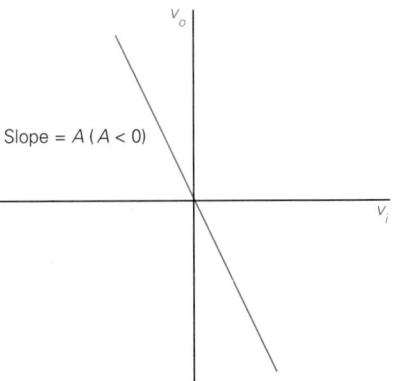

To illustrate the process of distortion-free transmission in a noninverting amplifier, consider the signal shown in Figure 9–4(a). The output of a noninverting amplifier with gain $A = 2$ is shown in Figure 9–4(b). All points on the curve in Figure 9–4(b) have exactly twice the levels of the points in (a). This gain is relatively low for an amplifier, but it was chosen for ease of illustration. Because the amplifier is noninverting, the input and output signals are said to be **in phase.**

FIGURE 9–4
Illustration of input and output waveforms for ideal *noninverting* amplifier.

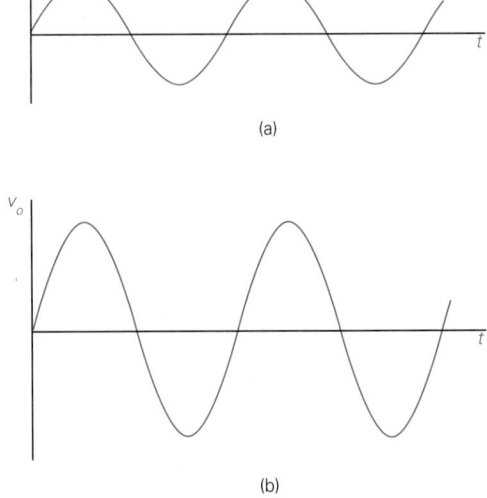

A similar illustration for an inverting amplifier is shown in Figure 9–5. The gain is now $A = -2$, and all points on the output waveform of Figure 9–5(b) have twice the magnitude but are inverted with respect to the input waveform of (a). Because of the inversion, the input and output waveforms are said to be **180° out of phase.** This simple inversion is usually not considered as a form of distortion,

FIGURE 9–5
Input and output waveforms for ideal *inverting* amplifier.

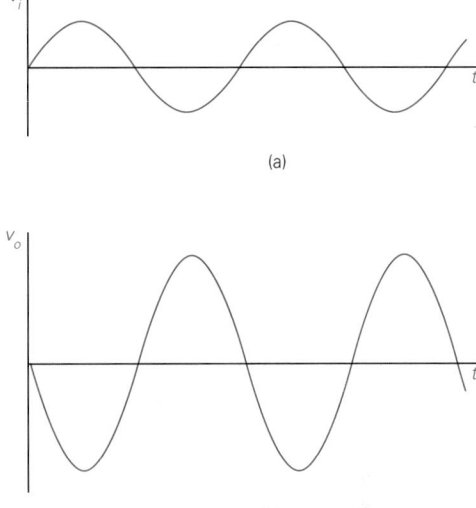

(a)

(b)

since the basic waveform shape is preserved. Many amplifier circuits exhibit the inversion property. In some applications, it is desirable, while in other situations inversion is unacceptable. For certain applications, it is immaterial whether the amplifier is inverting or noninverting.

AC Coupled and DC Coupled

Amplifiers may be classified as either **ac coupled** or **dc coupled.** An ac-coupled amplifier has one or more coupling circuits between stages to block dc levels while passing the time-varying signal on to subsequent stages. The most common coupling circuit is the *RC* coupling circuit considered in Chapter 8, but transformers are also used in some amplifiers.

Strictly speaking, the "dc" in the designation *dc amplifier* represents *directly coupled,* but the term *dc coupled* is widely used because of one of its basic properties; that is, it will amplify signals in frequency all the way down to dc because of the absence of coupling capacitors. The most common example of a dc amplifier is the operational amplifier, which will be studied in detail in Chapter 13 and subsequent chapters.

The major differences in the frequency responses between ac-coupled and dc-coupled amplifiers are shown in Figure 9–6. The plots given represent the amplitude response functions of the gains considered. The gain of the ac-coupled amplifier drops drastically at low frequencies, and there is no transmission at dc. However, the gain of the dc amplifier is flat all the way to dc. The ac amplifier exhibits a flat midfrequency range where the gain is essentially constant. Both types of amplifiers eventually show a decreasing gain characteristic as the frequency is increased. This high-frequency rolloff is caused primarily by shunt capacitive effects in the active elements providing the amplification.

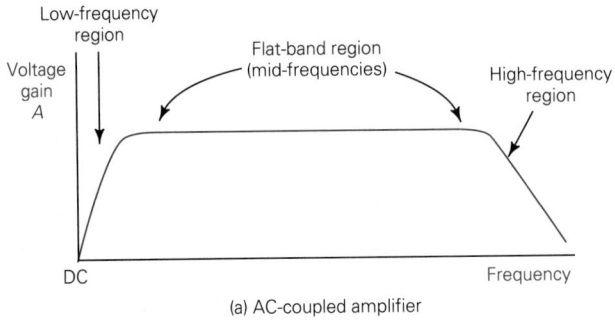

(a) AC-coupled amplifier

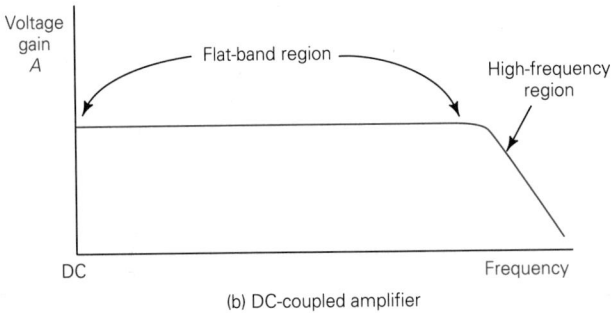

(b) DC-coupled amplifier

FIGURE 9–6
Frequency ranges for ac- and dc-coupled amplifiers.

Note again that the flat-band response of a dc-coupled amplifier extends all the way to dc, while the flat-band response of an ac-coupled amplifier starts at some frequency above dc and usually extends across a rather broad range of frequencies. When ac amplifiers are being discussed, the term *midband* is used to describe this broad range of frequencies. Although this midband varies considerably with the device and the circuit design, a typical range is from less than 100 Hz to more than 100 kHz.

The *high-frequency* region as shown in Figure 9–6 also varies considerably with the electronic device, the circuit design, and the type of configuration. It may range from 100 kHz or less to the range of tens of megahertz. Thus there are no absolute ranges of frequencies that can be associated with these definitions.

Untuned and Tuned Amplifiers

Amplifiers may be classified as untuned or tuned. An **untuned amplifier** is one in which no resonant circuits are employed. Most amplifiers designed for the frequency range below a few hundred kilohertz are untuned. Both frequency response curves of Figure 9–6 display untuned-type characteristics.

Tuned amplifiers are used primarily to create bandpass response curves that provide a uniform gain over a narrow band of frequencies. By employing resonant circuits consisting of inductance and capacitance combinations, the circuit capacitance actually becomes a part of the resonant function. In this fashion, operation at much higher frequencies is feasible. Tuned amplifiers are designed for operation well into the range of hundreds of megahertz. Most amplifiers operating above a few megahertz use tuned structures.

Common examples of tuned amplifiers appear in many stages of radio and television sets. The form of the amplitude response of a typical tuned amplifier is shown in Figure 9–7.

All amplifier circuits to be studied in Chapters 10 through 15 are untuned in the forms to be given, although some could be adapted to tuned forms. Examples of tuned amplifiers are considered in Chapter 18.

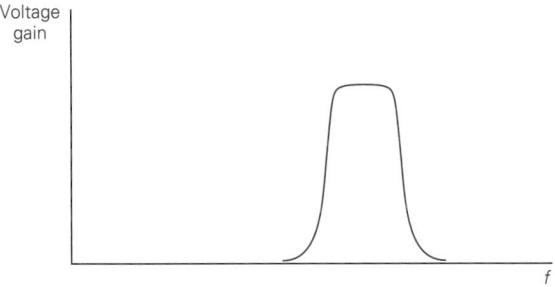

FIGURE 9–7
Typical form of the gain of a tuned amplifier.

Voltage and Power Amplifiers

Amplifiers are often classified as either **voltage amplifiers** or **power amplifiers.** The terms are somewhat ambiguous but have been established by long usage. The ambiguity arises because amplifiers classified as voltage amplifiers may also amplify power, and vice versa. It is the major function and intended application of the amplifier that determines the classification.

Voltage amplifiers are those amplifiers intended simply to raise the voltage level of a signal and for which there is no particular requirement for a specified amount of output power. The power level of the signal may be amplified in the process, but it is of secondary interest in the particular application. In many cases, the output of a voltage amplifier will be applied as the input to the next stage.

Power amplifiers are amplifiers designed to provide a specified output power to some given load. In some cases, the load may be an additional amplifier stage. However, most power amplifier circuits are used to provide power to an external load. A common example of a power amplifier is the output stage of an audio amplifier, which provides the substantial power required to drive the speakers.

This electrical power is in turn converted to acoustical power. The output stage of a commercial broadcast transmitter is a power amplifier, which provides power to the antenna.

There is no universal standard on the power level that characterizes a power amplifier. However, some references classify amplifiers providing an output power of 0.5 W or more as power amplifiers. Power amplifiers providing levels of hundreds of thousands of watts have been constructed.

Most of the amplifiers to be considered in the next several chapters could be classified primarily as voltage amplifiers. Power amplifiers are treated in Chapter 19.

The classifications that have been provided are only a few of the possible types employed, but they will suffice for this point in the book. Other classifications will be considered as the need arises and as the conceptual basis is established.

EXAMPLE 9–1

A certain inverting amplifier has a voltage gain $A = -20$. Determine the output signal voltage v_o when the input signal voltage has a value **(a)** $v_i = 0$; **(b)** $v_i = 0.5$ V; and **(c)** $v_i = -0.2$ V.

Solution
The input–output relationship is

$$v_o = -20v_i \qquad (9\text{–}3)$$

The three cases are analyzed as follows:

(a) For $v_i = 0$,

$$v_o = -20 \times 0 = 0 \qquad (9\text{–}4)$$

(b) For $v_i = 0.5$ V,

$$v_o = -20 \times 0.5 = -10 \text{ V} \qquad (9\text{–}5)$$

(c) For $v_i = -0.2$ V,

$$v_o = -20 \times (-0.2) = 4 \text{ V} \qquad (9\text{–}6)$$

Note that the output signal voltage is zero when the input signal voltage is zero. Note also that a positive input voltage results in an amplified negative output voltage, and vice versa.

EXAMPLE 9–2

The gain of a certain amplifier is to be determined from an oscilloscope measurement using a sine-wave input. The input and output waveforms are shown in Figure 9–8. The vertical sensitivities of the two channels are indicated. The time base reference is established by channel 1. Determine the gain.

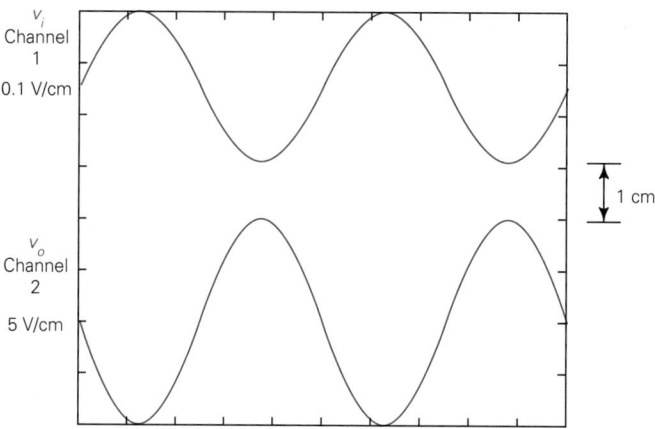

FIGURE 9–8
Waveforms of Example 9–2.

Solution
Gain determination on an oscilloscope is usually easier and more accurate if peak-to-peak values are used. Since the phase of channel 2 is established from the reference of channel 1, it is noted that the output is 180° out of phase with the input. Thus, the amplifier is an inverting unit.

Let V_{ipp} and V_{opp} represent the peak-to-peak values of the input and output voltages, respectively. We have

$$V_{ipp} = 0.1 \text{ V/cm} \times 3 \text{ cm} = 0.3 \text{ V} \tag{9-7}$$

and

$$V_{opp} = 5 \text{ V/cm} \times 4 \text{ cm} = 20 \text{ V} \tag{9-8}$$

The voltage gain is

$$A = -\frac{20}{0.3} = -66.7 \tag{9-9}$$

The peak-to-peak value of a quantity is always defined as a positive value, so the negative value in (9–9) was necessary to force the gain to be negative, as required.

9–3 AMPLIFIER FLAT-BAND SIGNAL MODEL

In this section we consider a signal model applicable to a wide variety of amplifier circuits. The model is useful in predicting the behavior of a single amplifier stage, and when combined with similar models for several stages, the analysis of a multistage amplifier may be easily performed.

The model is basically a linear model; that is, the amplifier is assumed to be distortion-free. It is applicable only to signals, and not any dc bias sources are shown. The restriction to the flat-band region means that the only passive elements to be considered are resistances; that is, all capacitive and inductive reactances are either short or open circuits in the applicable frequency ranges.

Consider the block diagram shown in Figure 9–9(a). The quantities v_i and i_i represent the input signal voltage and current, respectively. The quantity v_o represents the output signal voltage, and i_o represents the output current into any external load (not shown here).

FIGURE 9–9

Block diagram of linear amplifier and a flat-band signal model.

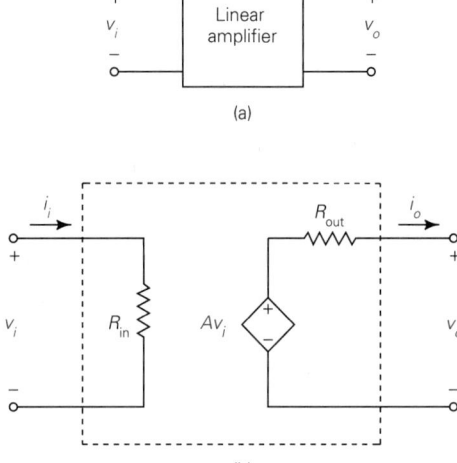

Input Resistance

The signal model for this amplifier is shown in Figure 9–9(b). The resistance R_{in} appearing across the input is called the **input resistance,** and it is related to v_i and i_i by

$$R_{\text{in}} = \frac{v_i}{i_i} \qquad\qquad (9\text{–}10)$$

The input resistance is the resistance "seen" by any signal source connected to the input. Its value is important in determining any loading on the signal connected to the input.

Output Resistance

The resistance R_{out} represents the **output resistance** of the amplifier. Basically, R_{out} is the Thévenin equivalent resistance as viewed from the output terminals. It is important in determining the loading of any external load connected to the output. It is not possible to write a simple equation such as (9–10) at the output, since there is a source in series with R_{out}.

Voltage Gain

The dependent voltage source Av_i in Figure 9–9(b) represents a signal voltage whose value is a constant times the input voltage. The quantity A represents the **open-circuit, or unloaded voltage, gain.** The term *open circuit* is necessary here, since the effect of any external load will modify the gain, as we will see. It is readily verified that if $i_o = 0$, $v_o = Av_i$, and the voltage gain under this open-circuit condition is A.

EXAMPLE 9–3

A certain amplifier is characterized by the following parameters:

$$R_{in} = 100 \text{ k}\Omega$$
$$R_{out} = 75 \text{ }\Omega$$
$$A = 60$$

Draw the flat-band signal model.

Solution
The model will have the same form as the circuit of Figure 9–9(b). For the particular parameters given, the circuit is shown in Figure 9–10.

FIGURE 9–10
Amplifier model of Example 9–3.

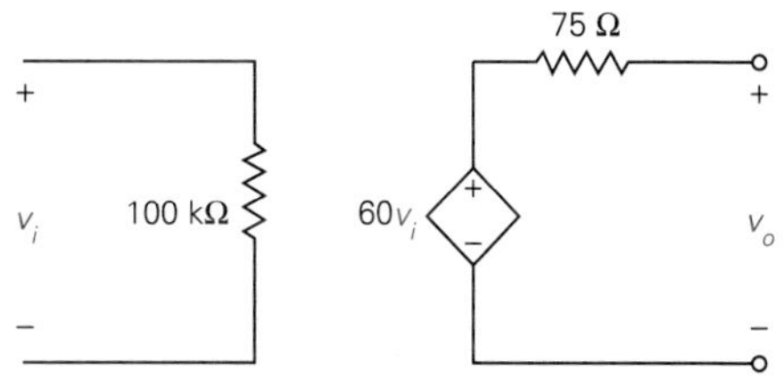

EXAMPLE 9–4

The circuit of Figure 9–11 represents the flat-band signal model of a certain amplifier. Determine the input resistance, the output resistance, and the open-circuit voltage gain.

Solution
We immediately note that the input and output resistances are

$$R_{in} = 20 \text{ k}\Omega \tag{9–11}$$

FIGURE 9-11
Amplifier model of Example 9-4.

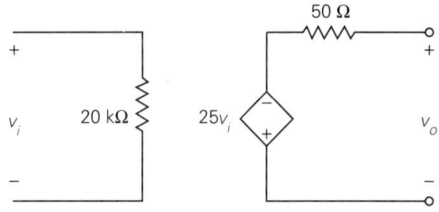

and

$$R_{\text{out}} = 50 \ \Omega \qquad\qquad\qquad (9\text{--}12)$$

However, when the given circuit is compared with the general model of Figure 9-9(b), we see that the polarity of the voltage source in Figure 9-11 is opposite to that of the general form. Since the positive references of both v_i and v_o are at the top, this means that the amplified voltage is inverted with respect to the reference; that is, this amplifier is an *inverting* form. Thus, the open-circuit voltage gain is

$$A = -25 \qquad\qquad\qquad (9\text{--}13)$$

An alternative way to show the dependent source is to keep the positive terminal of the source at the top and to label it as $-25 \ v_i$. However, the form given in Figure 9-11 is usually more convenient for inverting amplifiers.

9-4 AMPLIFIER, SOURCE, AND LOAD CONNECTIONS

We have seen that a linear amplifier can be represented in the flat-band frequency region by a model displaying input resistance, output resistance, and voltage gain. However, rarely does the amplifier stand alone in an application. A signal source must be connected to the input terminals, and in most cases some external load is connected to the output terminals. In general, the signal source contains an internal resistance that interacts with the input resistance of the amplifier to reduce the gain. Similarly, any external load connected to the output terminals will further act to reduce the gain.

In this section we consider a number of possible situations involving source and/or load connections. The best approach to emphasize to the reader is *not* to try to memorize the resulting formulas but, rather, to work them out as the need arises. It turns out that each case is a relatively simple exercise employing basic circuit laws. A pattern will emerge after a few cases are considered, and each subsequent situation can be dealt with from these principles.

The symbol A will be used to denote the voltage gain from the amplifier input terminals to the output terminals with no external load connected as in the preceding section. Under loaded conditions, subscripts will be added to represent various loaded gain parameters. The pattern should be clear as it is developed.

The term *signal source* is a popular term for describing the excitation to an amplifier, and we will use this term when appropriate. This would suggest the subscript *s* with this voltage indicated by v_s and its internal resistance as R_s. Indeed, many books use this terminology. There is, however, an ambiguity with this notation when FET amplifiers are considered, because one of the FET terminals is called the source, as we already know. To avoid this ambiguity, the source will be viewed as a generator, and the subscript *g* will be employed. In discussions, however, the terms *signal source* and *generator* will both be used, as appropriate.

Amplifier-Source Connection with No Loading

The most ideal situation to be considered is shown in Figure 9–12(a). The amplifier is excited by a signal source (or generator) v_g with zero internal resistance (i.e., the source is assumed to be an ideal voltage source). The output v_o is to be considered with no external load connected. In a practical sense, this means that whatever is used to monitor or respond to v_o presents no measurable loading.

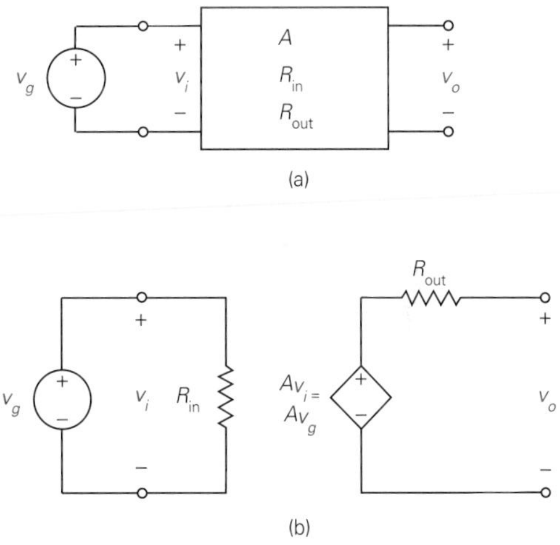

FIGURE 9–12
Amplifier connection with no loading at either port and circuit model.

(a)

(b)

The quantity A_{go} will be defined as the voltage gain between the generator v_g and the output voltage v_o; that is,

$$A_{go} = \frac{v_o}{v_g} \qquad (9\text{–}14)$$

The equivalent circuit model is shown in Figure 9–12(b). Since the input source is ideal, it is immediately obvious that the amplifier input voltage is the same as the generator voltage (i.e., $v_i = v_g$). The dependent source is thus $Av_i = Av_g$. The open-circuit output voltage is

$$v_o = Av_i = Av_g \qquad (9\text{--}15)$$

The gain between generator and output is

$$A_{go} = \frac{v_o}{v_g} = A \qquad (9\text{--}16)$$

This result should not come as any surprise, but it is worth emphasizing in the context of a complete amplifier-source connection. It says that the *composite amplifier circuit gain from signal generator to output is the same as the unloaded gain if the circuit is driven from an ideal voltage source and if there is no external load connected to the output.*

Amplifier with Input Loading

Consider the circuit of Figure 9–13(a), in which a generator v_g with internal resistance R_g excites an amplifier. However, the output has no external load connected. The equivalent-circuit model is shown in Figure 9–13(b).

In this case, the amplifier input voltage v_i differs from the generator voltage v_g because of the drop across R_g. By the voltage-divider rule, v_i is related to v_g by

$$v_i = \frac{R_{\text{in}}}{R_{\text{in}} + R_g} v_g \qquad (9\text{--}17)$$

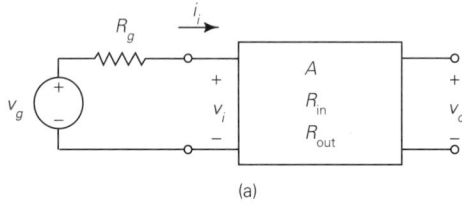

(a)

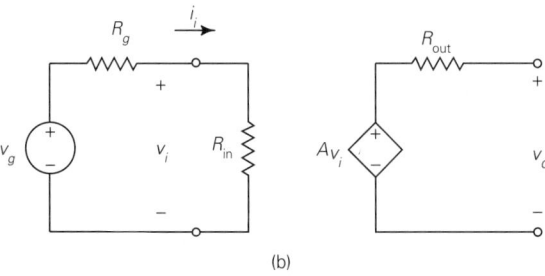

(b)

FIGURE 9–13
Amplifier connection with input loading and circuit model.

Since there is no loading at the output, the output voltage v_o is

$$v_o = Av_i \qquad (9\text{–}18)$$

To determine the gain from generator to output, we need a relationship between v_o and v_g. This is achieved by substituting v_i in terms of v_g from (9–17) in (9–18). The result is

$$v_o = \frac{AR_{\text{in}}}{R_{\text{in}} + R_g} v_g \qquad (9\text{–}19)$$

The gain A_{go} is

$$A_{go} = \frac{v_o}{v_g} = \frac{R_{\text{in}}}{R_{\text{in}} + R_g} A \qquad (9\text{–}20)$$

where a slight rearrangement was made, for reasons that will be clear later.

Amplifier with Output Loading

Consider the circuit of Figure 9–14(a), in which a signal source with zero internal resistance excites an amplifier. However, the output is connected to an external load R_L. The equivalent circuit model is shown in Figure 9–14(b).

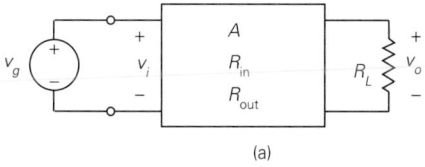

(a)

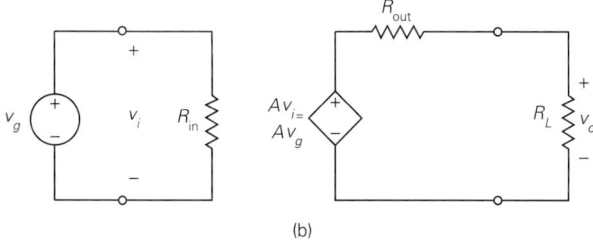

(b)

FIGURE 9–14
Amplifier connection with output loading and circuit model.

In this case the amplifier input voltage is the same as the generator voltage (i.e., $v_i = v_g$). The dependent source is $Av_i = Av_g$. However, the output voltage v_o is not equal to this dependent source, since a portion of this voltage is dropped across R_{out}.

By the voltage-divider rule, the output voltage v_o across the load is given by

$$v_o = \frac{R_L}{R_L + R_{out}} A v_i = \frac{R_L A v_g}{R_L + R_{out}}$$ (9–21)

The gain between generator and output is

$$A_{go} = \frac{v_o}{v_g} = A \frac{R_L}{R_L + R_{out}}$$ (9–22)

where a slight rearrangement was made, for reasons that will be clear later.

Amplifier with Loading at Input and Output

Consider the circuit of Figure 9–15(a), in which a generator v_g with internal resistance R_g excites an amplifier. Simultaneously, the output is connected to an external load R_L. The equivalent circuit is shown in Figure 9–15(b).

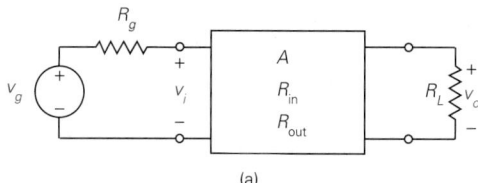

(a)

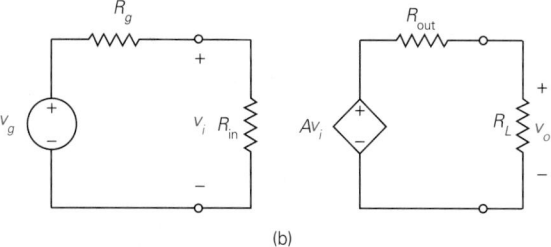

(b)

FIGURE 9–15
Amplifier connection with both input and output loading and circuit model.

By the voltage-divider rule, the amplifier input voltage is related to the generator voltage by

$$v_i = \frac{R_{in}}{R_{in} + R_g} v_g$$ (9–23)

Again, using the voltage-divider rule, the load voltage is related to the dependent voltage source by

$$v_o = \frac{R_L}{R_L + R_{out}} A v_i \qquad (9\text{--}24)$$

To obtain a relationship between v_o and v_g, the expression of (9–23) is substituted in (9–24), and we have

$$v_o = \frac{R_L}{R_L + R_{out}} A \frac{R_{in}}{R_{in} + R_g} v_g \qquad (9\text{--}25)$$

The gain between source and output is then determined as

$$A_{go} = \frac{v_o}{v_g} = \frac{R_{in}}{R_{in} + R_g} A \frac{R_L}{R_L + R_{out}} \qquad (9\text{--}26)$$

where a rearrangement was made, for reasons that will be clear shortly.

General Form

All the equations relating to the gain between generator voltage and output voltage of this section can be considered as variations of the following general equation:

$$A_{go} = \text{(input loading factor)} \times A \times \text{(output loading factor)} \qquad (9\text{--}27)$$

The input loading factor is simply the voltage-divider ratio representing the fraction of the open-circuit source or generator voltage appearing across the amplifier input; that is,

$$\text{input loading factor} = \frac{R_{in}}{R_{in} + R_g} \qquad (9\text{--}28)$$

When the amplifier is excited by an ideal voltage source, $R_g = 0$ and the input loading factor has a value of unity.

The output loading factor is simply the voltage-divider ratio representing the fraction of the amplifier open-circuit output voltage appearing across the load; that is,

$$\text{output loading factor} = \frac{R_L}{R_L + R_{out}} \qquad (9\text{--}29)$$

When the amplifier output is unloaded, R_L can be considered as approaching infinity ($R_L \to \infty$), and the output loading factor is unity. Thus, by appropriately setting the values of these factors, all expressions in this section can be considered as special cases of (9–27).

While the gain between the open-circuit generator voltage and the output is usually the primary parameter of interest, the gain between the amplifier input terminals and the output is also of interest. Let A_{io} represent this quantity; it is defined as

$$A_{io} = \frac{v_o}{v_i}$$

(9–30)

For this definition, the input loading factor does not have any effect, since the reference is to the voltage across the amplifier input terminals rather than the source voltage. Thus, for any of the configurations, A_{io} is given by

$$A_{io} = A \times \text{(output loading factor)}$$

(9–31)

Comparing (9–27) and (9–31), A_{go} and A_{io} are related by

$$A_{go} = \text{(input loading factor)} \times A_{io}$$

(9–32)

From (9–32) it can readily be deduced that $A_{go} \leq A_{io}$. The two quantities are equal when the amplifier is driven by an ideal voltage source; but when there is a nonzero source resistance, $A_{go} < A_{io}$.

The significance of A_{io} is that it is usually easier to measure it than A_{go} in the laboratory under operating conditions. When the source is connected, the voltage v_g is inside the source and cannot be directly measured. However, v_i and v_o can usually be directly measured under operating conditions.

Ideal Voltage Amplifier

Reviewing the general expression from (9–26), but considering the realistic limitations of practical circuits, let us determine those conditions that will maximize the gain A_{go}. First, the input loading factor should be made as close to unity as possible. This mandates that $R_{in} \gg R_g$. Second, the output loading factor should also be made as close to unity as possible. This mandates that $R_{out} \ll R_L$. Thus, *maximum voltage gain is achieved with an amplifier whose input resistance is*

FIGURE 9–16
Ideal voltage amplifier model.

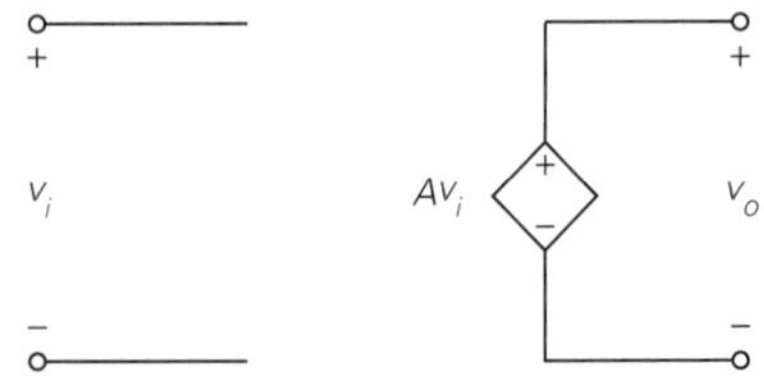

very large compared with the source resistance and whose output resistance is very small compared with the load resistance.

These conditions lead to the concept of the ideal voltage amplifier whose model is shown in Figure 9–16. This ideal amplifier model is characterized by $R_{in} = \infty$, which means that the input is essentially an open circuit. Conversely, $R_{out} = 0$, which means that the output dependent voltage source is an ideal voltage source with zero internal resistance. We will see that many practical circuit designs work toward these goals. Although they can never be truly achieved, many practical circuits come sufficiently close that the ideal model may be assumed.

EXAMPLE 9–5

The amplifier of Figure 9–17(a) has the following midband parameters: $A = 90$, $R_{in} = 6$ kΩ, and $R_{out} = 1.2$ kΩ. It is driven from a signal source with an internal resistance $R_g = 2$ kΩ, and a load $R_L = 2.4$ kΩ is connected across the output. Determine **(a)** $A_{go} = v_o/v_g$, and **(b)** $A_{io} = v_o/v_i$.

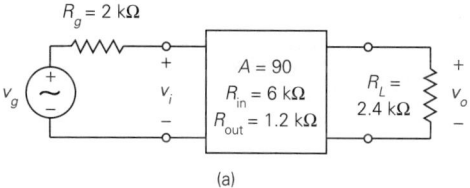

(a)

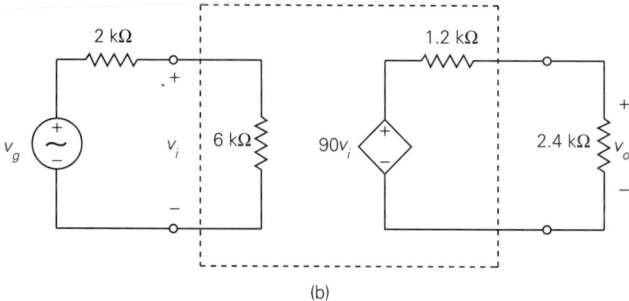

(b)

FIGURE 9–17
Circuit of Example 9–5.

Solution
The equivalent circuit of the composite amplifier with source and load connected is shown in Figure 9–17(b).

(a) The presence of both a generator resistance and a load resistance results in loading factors at both the input and the output. The gain between source and output is determined as follows:

$$A_{go} = \frac{v_o}{v_g} = \frac{6\,k\Omega}{6\,k\Omega + 2\,k\Omega} \times 90 \times \frac{2.4\,k\Omega}{2.4\,k\Omega + 1.2\,k\Omega} \tag{9-33}$$

$$= \frac{3}{4} \times 90 \times \frac{2}{3} = 45$$

The net effect of both factors is to reduce the voltage gain from the open-circuit value of 90 to a loaded gain of 45.

(b) To determine the gain between the amplifier input and the output, we consider only the output loading factor.

$$A_{io} = 90 \times \frac{2.4\,k\Omega}{2.4\,k\Omega + 1.2\,k\Omega} = 90 \times \frac{2}{3} = 60 \tag{9-34}$$

Note that $A_{io} > A_{go}$, as expected.

EXAMPLE 9-6
For the composite circuit of Example 9–5, assume that the signal voltage is $v_g = 0.2$ V. Determine **(a)** v_i, and **(b)** v_o.

Solution
(a) The voltage v_i is simply the fraction of v_g appearing across the 6-kΩ input resistance. Thus,

$$v_i = \frac{6\,k\Omega}{6\,k\Omega + 2\,k\Omega} \times 0.2 = 0.75 \times 0.2 = 0.15 \text{ V} \tag{9-35}$$

(b) From (9–33),

$$v_o = 45v_g = 45 \times 0.2 = 9 \text{ V} \tag{9-36}$$

As a check, since v_i is known, (9–34) can be used to determine v_o from v_i. We have

$$v_o = A_{io}v_i = 60v_i = 60 \times 0.15 = 9 \text{ V} \tag{9-37}$$

9-5 CASCADE CONNECTIONS OF AMPLIFIERS

In this section we consider the process of cascading two or more individual amplifier circuits. The effects of loading at the source and/or the output are then combined with the cascade assumption. The result is a procedure for working with various combinations of cascaded stages as well as source and load connections.

To provide as simple an approach as possible in the development that follows, only two amplifier stages will be considered. However, the pattern that emerges should allow the reader to extend the result to an arbitrary member of cascaded stages without any difficulty.

Cascade Arrangement without Input or Output Loading

Consider first the cascade arrangement of two amplifiers shown in Figure 9–18(a). The parameters of the two amplifiers are labeled on the respective blocks. In this case, the input is an ideal source of value v_g, and the output has no assumed external load.

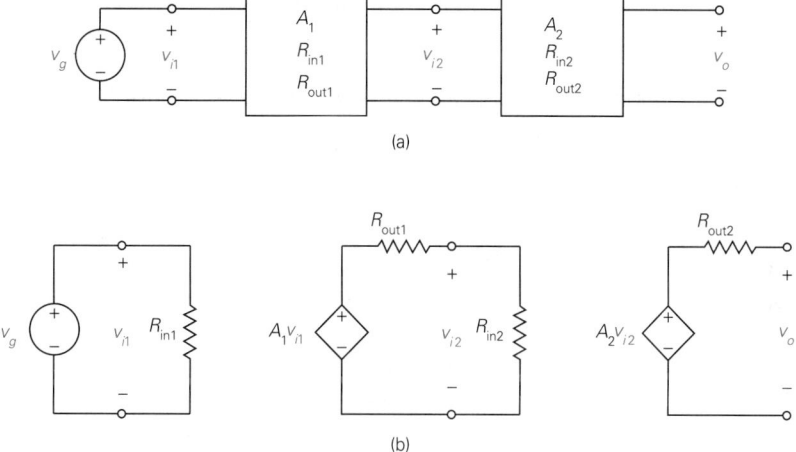

(a)

(b)

FIGURE 9–18
Cascade arrangement of two amplifiers without input or output loading and equivalent circuit.

An equivalent model for the arrangement is shown in Figure 9–18(b). Since there is no loading at the input, the voltage across the input of amplifier 1 is simply $v_{i1} = v_g$.

The voltage v_{i2} across the input to amplifier 2 is

$$v_{i2} = \frac{R_{in2}}{R_{in2} + R_{out1}} A_1 v_{i1} = \frac{R_{in2} A_1 v_g}{R_{in2} + R_{out1}} \tag{9–38}$$

Since there is no loading at the output, the output voltage v_o is simply the value of the dependent source on the right; that is,

$$v_o = A_2 v_{i2} \tag{9–39}$$

Substitution of (9–38) in (9–39) yields

$$v_o = \frac{A_2 R_{in2} A_1 v_g}{R_{in2} + R_{out1}} \tag{9–40}$$

Solving for the gain $A_{go} = v_o/v_g$ and rearranging in a particular order, we have

$$A_{go} = \frac{v_o}{v_g} = A_1 \frac{R_{in2}}{R_{in2} + R_{out1}} A_2 \tag{9–41}$$

This equation may be expressed as

$$A_{go} = A_1 \times (\text{interaction factor}) \times A_2 \qquad \text{(9–42)}$$

where

$$\text{interaction factor} = \frac{R_{in2}}{R_{in2} + R_{out1}} \qquad \text{(9–43)}$$

The interaction factor between the stages is simply the voltage-divider rule involving the input resistance to the second stage and the output resistance of the first stage.

Conditions for No Interaction

Assume in (9–43) that $R_{in2} \gg R_{out1}$. In this case the denominator may be approximated by R_{in2}, and the interaction factor is approximately $R_{in2}/R_{in2} = 1$. The gain A_{go} is then

$$A_{go} = A_1 A_2 \qquad \text{(9–44)}$$

This result is important and needs to be emphasized. *If the input resistance of a given amplifier is very large compared to the output resistance of the amplifier driving it, and if there is no loading at input and output, the composite voltage gain is simply the product of the individual gains.* As a practical rule of thumb, if the input resistance is at least 100 times as large as the output resistance of the stage driving it, the error in using (9–44) is no greater than 1%.

Once again we see the desirability of high input resistance and low output resistance for a voltage amplifier. Without these desirable conditions, the voltage gain of a cascade amplifier will be less than it is capable of achieving.

Interaction and Loading Combined

Consider now the cascade arrangement in Figure 9–19. In this case, the generator contains an internal resistance R_g, and the output of the second amplifier is connected to a load R_L.

The reader who has followed all the developments thus far should be able to write down the gain expression in this case. The composite gain is

$$A_{go} = \frac{v_o}{v_g} = (\text{input loading factor}) \times A_1 \times (\text{interaction factor})$$
$$\times A_2 \times (\text{output loading factor}) \qquad \text{(9–45)}$$

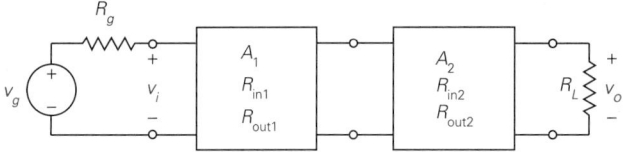

FIGURE 9–19
Cascade arrangement of two amplifiers with input and output loading.

The various factors are defined as

$$\text{input loading factor} = \frac{R_{in1}}{R_{in1} + R_g} \tag{9–46}$$

$$\text{interaction factor} = \frac{R_{in2}}{R_{in2} + R_{out1}} \tag{9–47}$$

$$\text{output loading factor} = \frac{R_L}{R_L + R_{out2}} \tag{9–48}$$

The reader should be able to generalize those results to more than two stages. In general, there will be a single input loading factor and a single output loading factor. However, each interface between stages will require an interaction factor. Thus, for three stages, there will be two interaction factors.

EXAMPLE 9–7

Consider the cascade arrangement shown in Figure 9–20. Determine the net gain $A_{go} = v_o/v_g$.

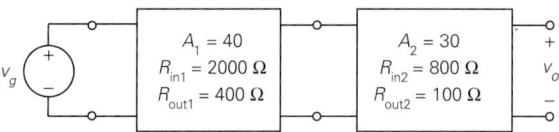

FIGURE 9–20
System of Example 9–7.

Solution
There is no loading at either input or output, so only an interaction factor need be considered. Using the procedure developed, we can write by inspection

$$A_{go} = \frac{v_o}{v_g} = 40 \times \frac{800}{800 + 400} \times 30 = 40 \times \frac{2}{3} \times 30 = 800 \tag{9–49}$$

If there were no interaction between stages, the net gain would be $A_1 A_2 = 40 \times 30 = 1200$.

EXAMPLE 9-8

Consider the cascade arrangement shown in Figure 9–21. The amplifiers are the same as in Example 9–7, but the source now contains an internal resistance and there is an external load connected. Determine the net gain $A_{go} = v_o/v_g$.

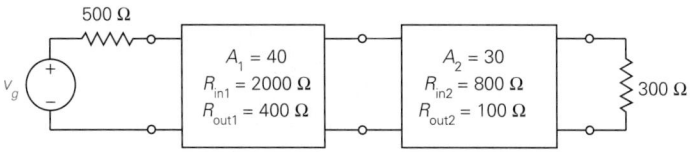

FIGURE 9–21
System of Example 9–8.

Solution

In this case we write

$$A_{go} = \frac{v_o}{v_g} = \frac{2000}{2000 + 500} \times 40 \times \frac{800}{800 + 400} \times 30 \times \frac{300}{300 + 100}$$

$$= \frac{4}{5} \times 40 \times \frac{2}{3} \times 30 \times \frac{3}{4} \qquad (9\text{–}50)$$

$$= 480$$

9-6 DECIBEL POWER GAIN AND LOSS

We develop the concepts of decibel power gain and loss in this section. **Decibel gain** is a method of characterizing the gain of a circuit by means of a logarithmic ratio. This approach was developed very early in the history of the electronics industry and has widespread usage in many application areas. A beginner may tend to question the value of the concept, but it is impossible to understand and interpret many of the specifications and properties of electronic devices without utilizing decibel quantities. After some proficiency is established in using decibel quantities, their utility becomes apparent.

Decibel Power Gain

The original, and perhaps the "purest," form of a decibel measure is that of the **decibel power gain.** To develop this concept, refer to the amplifier block diagram in Figure 9–22. Let P_{in} represent the power delivered to the input of the amplifier,

FIGURE 9–22

Block diagram for development of decibel power gain.

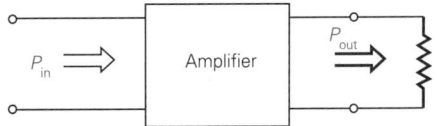

and let P_{out} represent the power delivered to the load. The *power gain* of the amplifier will be denoted as G, and it is defined as

$$G = \frac{P_{out}}{P_{in}} \qquad (9\text{--}51)$$

The *decibel power gain* will be denoted as G_{dB}, and it is defined as

$$G_{dB} = 10 \log_{10} G \qquad (9\text{--}52)$$

where $\log_{10}$ means the logarithm to the base 10. Even though G has no dimensions, it is customary to assign units to G_{dB}, and the units are **decibels (dB).** Thus, when G is specified, no units are given, but when G_{dB} is specified, the value given is stated in decibels, and this is a good way to recognize a decibel measure.

In effect, the decibel gain is a logarithmic conversion of the actual gain. Many applications of electronic signals have responses that are logarithmic in nature. It turns out that this logarithmic conversion provides a better way of relating these responses to signal levels.

The choice of the constant 10 in the definition of (9–52) was made to arrive at a convenient size of the individual logarithmic units. In fact, a logarithmic definition has been used without this factor, and the resulting unit was known as the **bel.** However, the bel was judged to be rather awkward because of its size, and the decibel was found to be more convenient. Consequently, virtually all logarithmic measures today are based on the decibel (and its variations), and that will be the basis in this text.

The quantity G in (9–51) will be assumed to be a positive real number. From some basic properties of the logarithmic function, the following conditions may be deduced:

1. If $G > 1$, G_{dB} is positive.
2. If $G < 1$, G_{dB} is negative.
3. If $G = 1$, $G_{dB} = 0$.

Thus the gain in decibels turns out to be positive when there is a true power gain ($G > 1$). However, when the output power is less than the input power ($G < 1$), the gain in decibels turns out to be negative. A negative value in decibels implies, therefore, that the output power is less than the input power. When input and output power levels are equal, there is no decibel gain (i.e., $G_{dB} = 0$).

Although the definition has been established here with an assumed amplifier, it is widely used for all types of transmission networks. In many passive circuits, the output power will be less than the input power (i.e., the signal is *attenuated* by the circuit). The decibel gain will thus turn out to be negative. In working with circuits having *attenuation*, it is often more convenient to work with loss factors. Let L represent the **power loss factor** associated with a circuit. It is defined as

$$L = \frac{P_{in}}{P_{out}} \tag{9-53}$$

Comparing (9-53) with (9-51), we see that

$$L = \frac{1}{G} \tag{9-54}$$

The **decibel power loss** will be denoted as L_{dB}, and it is defined as

$$L_{dB} = 10 \log_{10} L \tag{9-55}$$

As in the case of G_{dB}, L_{dB} is measured in decibels.
A basic property of logarithms is

$$\log_{10} \frac{1}{x} = -\log_{10} x \tag{9-56}$$

By substituting (9-54) in (9-55) and using (9-56), it is observed that

$$L_{dB} = -G_{dB} \tag{9-57}$$

Thus, while L and G have a reciprocal relationship as given by (9-54), L_{dB} and G_{dB} are simply related by negation.

Do not let these relationships confuse you, because the pattern is quite simple. If there is a true power gain ($G > 1$), the decibel power gain has a positive value, and there is very little need even to use a loss factor. However, in situations where there is a true loss ($G < 1$), one can use a decibel power gain with negative values of decibels, or the decibel loss function can be used, in which case the decibel values are positive.

EXAMPLE 9-9

A certain amplifier has a power gain $G = 1000$. Determine the decibel power gain G_{dB}.

Solution
The decibel power gain is

$$G_{dB} = 10 \log_{10} 1000 = 10 \times 3 = 30 \text{ dB} \tag{9-58}$$

EXAMPLE **9-10**	In a certain lossy transmission system, the input power is $P_{in} = 2$ W, and the output power is $P_{out} = 20$ mW. Determine **(a)** the power gain; **(b)** the decibel gain; **(c)** the power loss factor; and **(d)** the decibel loss.

Solution

(a) The power gain G is

$$G = \frac{P_{out}}{P_{in}} = \frac{0.02 \text{ W}}{2 \text{ W}} = 0.01 \tag{9-59}$$

(b) The decibel gain G_{dB} is

$$G_{dB} = 10 \log_{10} G = 10 \log_{10} 0.01 = 10 \times (-2) = -20 \text{ dB} \tag{9-60}$$

This is obviously not a true "gain," but it is perfectly proper to use the gain definitions. The conditions that $G < 1$ and $G_{dB} < 0$ imply an actual loss.

(c) The loss factor L is

$$L = \frac{P_{in}}{P_{out}} = \frac{2 \text{ W}}{0.02 \text{ W}} = 100 \tag{9-61}$$

(d) The decibel loss factor L_{dB} is

$$L_{dB} = 10 \log_{10} L = 10 \log_{10} 100 = 10 \times 2 = 20 \text{ dB} \tag{9-62}$$

It is more convenient to refer to a *loss of 20 dB than a gain of −20 dB*, but both references are correct.

Converting Decibel Power Gain to Power Gain

Suppose that the decibel gain (or loss) is given and we desire to determine the corresponding value of G (or L). We start with the definition of the decibel quantity, but rearrange it so that the unknown is on the left. For gain, we have

$$10 \log_{10} G = G_{dB} \tag{9-63}$$

First, both sides of (9-63) are divided by 10. This gives

$$\log_{10} G = \frac{G_{dB}}{10} \tag{9-64}$$

To extract G from the logarithmic function, both sides of (9-64) are raised as powers of 10. This step yields

$$10^{\log_{10} G} = 10^{G_{dB}/10} \tag{9-65}$$

By definition, however, the left side of (9-65) is simply G. Thus, we have

$$\boxed{G = 10^{G_{dB}/10}} \tag{9-66}$$

A similar process can be used to determine L from L_{dB}, and the result is

$$\boxed{L = 10^{L_{dB}/10}} \tag{9-67}$$

The reader is encouraged to perform the various steps required rather than memorizing the formulas.

EXAMPLE 9-11

A certain amplifier has a decibel power gain $G_{dB} = 26$ dB. Determine the value of G.

Solution
Since

$$10 \log_{10} G = 26 \tag{9-68}$$
$$\log_{10} G = 2.6 \tag{9-69}$$

and

$$G = 10^{2.6} = 398.1 \tag{9-70}$$

Most modern scientific calculators have functions of the form y^x and 10^x, each of which could be used to perform the computation of (9-70).

EXAMPLE 9-12

A certain transmission system has a decibel power loss $L_{dB} = 30$ dB. **(a)** Determine L directly from L_{dB}. **(b)** Convert to a power gain basis, and determine G.

Solution
(a) Since

$$10 \log_{10} L = 30, \tag{9-71}$$
$$\log_{10} L = 3 \tag{9-72}$$

and

$$L = 10^3 = 1000 \tag{9-73}$$

(b) The corresponding power gain is

$$G_{dB} = -L_{dB} = -30 \text{ dB} \tag{9-74}$$

We now have

$$10 \log_{10} G = -30 \tag{9-75}$$
$$\log_{10} G = -3 \tag{9-76}$$

and

$$G = 10^{-3} = 0.001 \tag{9-77}$$

We readily note that $G = 1/L$.

Although the gain approach and the loss approach are equally valid, it is more convenient to work with positive logarithms and ratios greater than 1. Thus, the gain approach is more convenient when there is a true gain, and the loss approach is more convenient when there is a true loss.

9–7 DECIBEL VOLTAGE AND CURRENT GAIN

As indicated in the preceding section, the original concept for decibel gain was based on power ratios. However, in modern usage, the definition has been extended to include both voltage and current gain. Here we will focus first on voltage gain, and the extension to current gain will then be evident.

To bring together the concepts of power and voltage gain, consider the block diagram of Figure 9–23. The input and output instantaneous voltages are denoted as v_i and v_o, respectively.

FIGURE 9–23
Block diagram for development of decibel voltage gain.

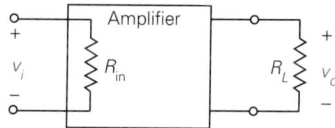

The input instantaneous power p_i is the power that is delivered to R_{in}, and it is

$$p_i = \frac{v_i^2}{R_{in}} \tag{9-78}$$

Similarly, the output instantaneous power p_o is the power delivered to R_L, and it is

$$p_o = \frac{v_o^2}{R_L} \tag{9-79}$$

The power gain is

$$G = \frac{p_o}{p_i} = \frac{v_o^2/R_L}{v_i^2/R_{in}} = \frac{R_{in}}{R_L} \left(\frac{v_o}{v_i}\right)^2 \tag{9-80}$$

Substituting $A = v_o/v_i$, we have

$$G = \frac{R_{in}}{R_L} A^2 \tag{9-81}$$

Assume momentarily that $R_{in} = R_L$; that is, the load and input resistances are equal. The power gain then reduces to

$$G = A^2 \quad \text{for } R_{in} = R_L \qquad (9\text{--}82)$$

Thus, *the power gain is simply the square of the voltage gain when the load and input resistances are equal.* This situation will be referred to as a **matched condition** in the development that follows.

Next, the decibel power gain is determined. This quantity is

$$G_{dB} = 10 \log_{10} G = 10 \log_{10} A^2 \qquad (9\text{--}83)$$

A fundamental property of logarithms is

$$\log_{10} x^2 = 2 \log_{10} x \qquad (9\text{--}84)$$

for any variable x. Using (9–84) in (9–83) results in

$$G_{dB} = 20 \log_{10} A \qquad (9\text{--}85)$$

Comparing this result with (9–52), we see that the decibel power gain can be determined directly from the voltage gain under matched conditions. However, the constant multiplier is now 20 instead of 10. Equation (9–85) thus provides an alternative way to compute the decibel power gain when the voltage gain is given and when matched conditions exist.

Suppose, however, that the system is not matched (i.e, $R_L \neq R_{in}$). Although decibel power gain could still be determined in some cases by calculating input and load power values or by using (9–81), the definition may not even make any sense in many cases. For example, we indicated earlier that an ideal voltage amplifier was characterized by infinite input resistance (i.e., an open circuit). In that case, the input power is zero, and the calculation of power gain would lead to an infinite value. Similarly, if the output voltage is measured with no external load, there is no output power.

To resolve this dilemma, a quantity we will define as the **decibel voltage gain** will be used. This quantity will be denoted as A_{dB} and it will be defined as

$$A_{dB} = 20 \log_{10} A \qquad (9\text{--}86)$$

where A is the voltage gain under consideration. The quantity A can be replaced by any of the voltage-gain definitions encountered previously (i.e., A_{go} and A_{io}). The definition is applied as given by (9–86) irrespective of whether input and load resistances are equal.

What is the significance of A_{dB} as compared to G_{dB}? The quantity G_{dB} represents a true power gain expressed as a decibel ratio. The quantity A_{dB}, in general, simply represents a voltage gain expressed in a "powerlike" logarithmic ratio.

Note that in the special matched case, the two quantities are equal; that is,

$$A_{dB} = G_{dB} \quad \text{for } R_L = R_{in}$$ (9–87)

In general, however, A_{dB} is simply a logarithmic expression of a voltage ratio rather than a true power ratio.

Converting Decibel Voltage Gain to Voltage Gain

The process of determining A from a given A_{dB} is nearly identical to that of the corresponding power quantities except for one constant. We start with

$$20 \log_{10} A = A_{dB}$$ (9–88)

First, both sides of (9–88) are divided by 20. This yields

$$\log_{10} A = \frac{A_{dB}}{20}$$ (9–89)

Both sides are raised as powers of 10, and we have

$$A = 10^{A_{dB}/20}$$ (9–90)

EXAMPLE 9–13

A certain amplifier has $A = 200$. Determine the decibel voltage gain.

Solution
The decibel voltage gain is

$$A_{dB} = 20 \log_{10} 200 = 20 \times 2.301 = 46.02 \text{ dB}$$ (9–91)

EXAMPLE 9–14

A certain amplifier has $A_{dB} = 40$ dB. Determine A.

Solution
Start with

$$20 \log_{10} A = 40$$ (9–92)

Division of both sides by 20 yields

$$\log_{10} A = 2$$ (9–93)

and

$$A = 10^2 = 100$$ (9–94)

9–8 DECIBEL GAIN FOR CASCADED SYSTEMS

One of the primary advantages of decibel forms is the ease of characterizing a system containing a cascade of individual stages. We develop the concept here on the basis of three stages, but the result can be readily extended to any arbitrary number.

Consider the system of Figure 9–24. For the initial phase of the development, we will assume that the three amplifiers are ideal (i.e., the input resistances are infinite, and the output resistances are zero). Thus, there are no loading or interaction factors to consider.

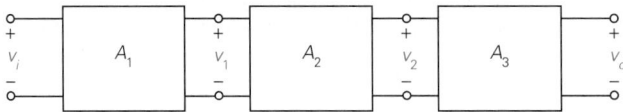

FIGURE 9–24
Cascade connection of three amplifiers.

A series of equations relating the outputs of the various stages to the inputs may be written. We have

$$v_1 = A_1 v_i \tag{9–95}$$

$$v_2 = A_2 v_1 \tag{9–96}$$

$$v_o = A_3 v_2 \tag{9–97}$$

To determine a relationship between v_o and v_i, (9–95) is substituted in (9–96), and this result is substituted in (9–97). The net result is

$$v_o = A_1 A_2 A_3 v_i \tag{9–98}$$

Finally, the net gain A is

$$A = \frac{v_o}{v_i} = A_1 A_2 A_3 \tag{9–99}$$

This result comes as no surprise, since a similar form was developed in Section 9–5 for two sections.

Next, the decibel voltage gain A_{dB} is determined. This quantity is

$$A_{dB} = 20 \log_{10} A = 20 \log_{10} A_1 A_2 A_3 \tag{9–100}$$

A fundamental property of logarithms is

$$\log_{10} xyz = \log_{10} x + \log_{10} y + \log_{10} z \tag{9–101}$$

for any variables x, y, and z. Application of (9–101) to (9–100) yields

$$A_{dB} = 20 \log_{10} A_1 + 20 \log_{10} A_2 + 20 \log_{10} A_3 \tag{9–102}$$

We now define

$$A_{1dB} = 20 \log_{10} A_1 \qquad (9\text{–}103)$$
$$A_{2dB} = 20 \log_{10} A_2 \qquad (9\text{–}104)$$
$$A_{3dB} = 20 \log_{10} A_3 \qquad (9\text{–}105)$$

The final form is then

$$\boxed{A_{dB} = A_{1dB} + A_{2dB} + A_{3dB}} \qquad (9\text{–}106)$$

Stated in words and generalizing, *the net decibel voltage gain of a cascade system of ideal amplifiers is the sum of the individual decibel gains.*

Suppose that the amplifiers are not ideal and that generator and output loading conditions are present. Decibel forms can still be applied provided that decibel measures for the loading and/or interaction effects are used. Since the basis of the analysis is decibel gain, the decibel measures of these factors must be treated as *negative* dB gains. Thus, their effects will be to reduce the overall gain.

EXAMPLE 9–15

Consider the cascade of three amplifiers in Figure 9–25. All amplifiers are assumed to be ideal. **(a)** Determine the net voltage gain $A = v_o/v_i$, and determine A_{dB} from A. **(b)** From the individual decibel gains, determine A_{dB}.

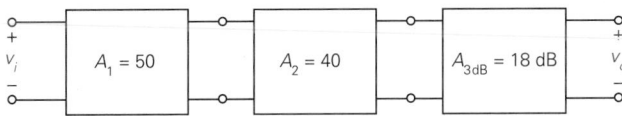

FIGURE 9–25
Cascade system of Example 9–15.

Solution
(a) Observe that the voltage gains of the first two stages are given, and the decibel gain of the third stage is given. The latter quantity must first be converted to a voltage gain. We have

$$20 \log_{10} A_{3dB} = 18 \qquad (9\text{–}107)$$

which leads to

$$A_{3db} = 10^{0.9} = 7.943 \qquad (9\text{–}108)$$

The net voltage gain A is

$$A = A_1 A_2 A_3 = 50 \times 40 \times 7.943 = 15{,}886 \qquad (9\text{–}109)$$

The net decibel voltage gain A_{dB} is

$$A_{dB} = 20 \log_{10} A = 20 \log_{10} 15{,}886 = 84.02 \text{ dB} \qquad (9\text{–}110)$$

(b) To work with all stages in decibels, we must first convert A_1 and A_2 to dB form. We have

$$A_{1dB} = 20 \log_{10} 50 = 33.98 \text{ dB} \qquad \textbf{(9–111)}$$

$$A_{2dB} = 20 \log_{10} 40 = 32.04 \text{ dB} \qquad \textbf{(9–112)}$$

The net decibel gain can then be determined as

$$A_{dB} = A_{1dB} + A_{2dB} + A_{3dB} = 33.98 + 32.04 + 18 = 84.02 \text{ dB} \quad \textbf{(9–113)}$$

which agrees with (9–110).

9–9 DECIBEL SIGNAL LEVELS

The basic application of decibel values is to express power or voltage gains as logarithmic ratios, as we have seen in the last several sections. In this context a decibel value defines a *ratio* of an output quantity to an input quantity.

In many segments of the electronics industry, it has become widespread practice to extend decibel measures to represent actual signal levels. However, it is necessary that a reference level be used in the definition, since a decibel measure is always a ratio. Such levels are denoted by adding an additional letter to the unit abbreviation dB. Several of the most common decibel-level definitions are summarized in Table 9–1.

TABLE 9–1
Some common dB-level definitions

Reference power level	Defining formula	Unit abbreviation
1 W	$10 \log_{10} \dfrac{P(\text{W})}{1 \text{ W}}$	dBW
1 mW	$10 \log_{10} \dfrac{P(\text{mW})}{1 \text{ mW}}$	dBm
1 fW	$10 \log_{10} \dfrac{P(\text{fW})}{1 \text{ fW}}$	dBf

For the three cases shown, the reference levels are 1 W, 1 mW, and 1 fW (10^{-15}W), respectively. The use of the 1-W reference is probably the most natural and is used in many higher-power communications-related areas. However, it is awkward for small-signal applications. The use of 1 mW is very common in many electronic small-signal applications and in telephone transmission systems. The use of 1 fW as a reference is relatively new and is being used in specifying antenna signal levels in communications receivers.

For each of the forms, the formula for computing the decibel signal level is the same as developed earlier in the book, with some additional stipulations. The power level must be expressed in the units given in parentheses for each case, and the corresponding reference is one unit for the particular case. The corresponding units are dBW, dBm, and dBf for the three cases, respectively.

Whenever there is confusion about whether a decibel value represents a signal level or a gain value, note the units. If dB is used without an additional letter, it is referring to a gain (or a loss). However, if it is accompanied by an additional letter, it represents a signal. Unfortunately for beginners, sometimes experienced personnel in a given area will use the basic term dB when they really mean one of the signal level units.

We will now investigate the advantage obtained from using decibel signal levels. For the development that follows, we use decibels referred to 1 mW as the basis because of its widespread usage.

Assume that the input power P_i and output power P_o are both expressed in milliwatts. For a power gain G, the system relationship is

$$P_o(\text{mW}) = GP_i(\text{mW}) \qquad (9\text{--}114)$$

Taking the logarithms of both sides of (9–114), multiplying by 10, and employing the logarithmic property of (9–101), we have

$$10 \log_{10}P_o(\text{mW}) = 10 \log_{10}GP_i(\text{mW}) = 10 \log_{10}G + 10 \log_{10}P_i(\text{mW}) \quad (9\text{--}115)$$

Let $P_i(\text{dBm})$ and $P_o(\text{dBm})$ represent the respective decibel levels of input and output expressed in dBm. Comparing (9–115) with the formulas of Table 9–1 and using the definition of G_{dB} as developed earlier in the chapter, we have

$$\boxed{P_o(\text{dBm}) = P_i(\text{dBm}) + G_{\text{dB}}} \qquad (9\text{--}116)$$

Stated in words, the result of (9–116) says that *the output level in dBm is the sum of the input level in dBm plus the decibel gain of the system.* This formula may be adapted to any of the decibel signal definitions of Table 9–1 simply by replacing dBm for the two power levels by the desired dB-level form. However, the input and output levels must be expressed in the same units. Note that the value of G_{dB} is not affected in any way by the input and output signal definitions, since it is based on a dimensionless ratio. Thus G_{dB} is expressed in decibels irrespective of the other units in the equation. This result often bothers students who have been taught that all quantities to be added must be expressed in the same units. In the present situation, both additive terms are really dimensionless, but the units added provide clarification of the process involved.

EXAMPLE 9–16   Assume that an amplifier with a power gain of 30 dB is excited by a signal source that supplies an input power level of 250 μW to the amplifier. Express the input in dBm, and determine the corresponding output level in dBm.

Solution

Let P_i(dBm) and P_o(dBm) represent the respective input and output power levels in dBm. First, P_i is expressed in mW as P_i(mW) = 0.25 mW. The value of P_i(dBm) is

$$P_i(\text{dBm}) = 10\log_{10}\frac{P_i(\text{mW})}{1\text{ mW}} = 10\log_{10}\frac{0.25}{1} = -6.02\text{ dBm} \qquad \text{(9–117)}$$

Note that this particular level is smaller than 1 mW, so the corresponding decibel level is negative. The dBm value of the output is

$$P_o(\text{dBm}) = P_i(\text{dBm}) + G_{\text{dB}} = -6.02 + 30 = 23.98\text{ dBm} \qquad \text{(9–118)}$$

9–10 BODE PLOTS

In general, amplifier gain will vary with frequency. The concept of frequency response was introduced in Chapter 8, and several representative frequency-dependent circuits containing reactances were provided. Amplifier circuits contain complex combinations of these reactance components, so the frequency response is an important consideration.

A widely employed method for displaying the amplitude response as a function of frequency is through the use of a **Bode plot,** in which A_{dB} versus frequency is shown. The method employs a **semilog** coordinate system. An example of semilog graph paper is shown in Figure 9–26. One scale is linear, and the other has a logarithmic form. Frequency is represented with the logarithmic scale, and the decibel amplitude response is represented with the linear scale. The particular sample shown has what is called by paper manufacturers "two cycles" on the logarithmic scale. Strictly speaking, it should be called "two decades," since each segment provides a 10 : 1 frequency range.

The use of Bode plots for involved frequency response functions is a major topic in more advanced circuit analysis books.* The treatment here will provide sufficient information to deal with certain amplifier functions to be encountered later in the text.

RC Low-Pass Model

Consider the simple *RC* low-pass circuit shown in Figure 9–27. This circuit was considered in Chapter 8, and some analysis was performed at that time. Recall that a so-called **break frequency f_b** was defined as

$$f_b = \frac{1}{2\pi RC} \qquad \text{(9–119)}$$

---

* See, for example, W. D. Stanley, *Network Analysis with Applications* (Englewood Cliffs, N.J.: Reston, 1985), Chap. 9.

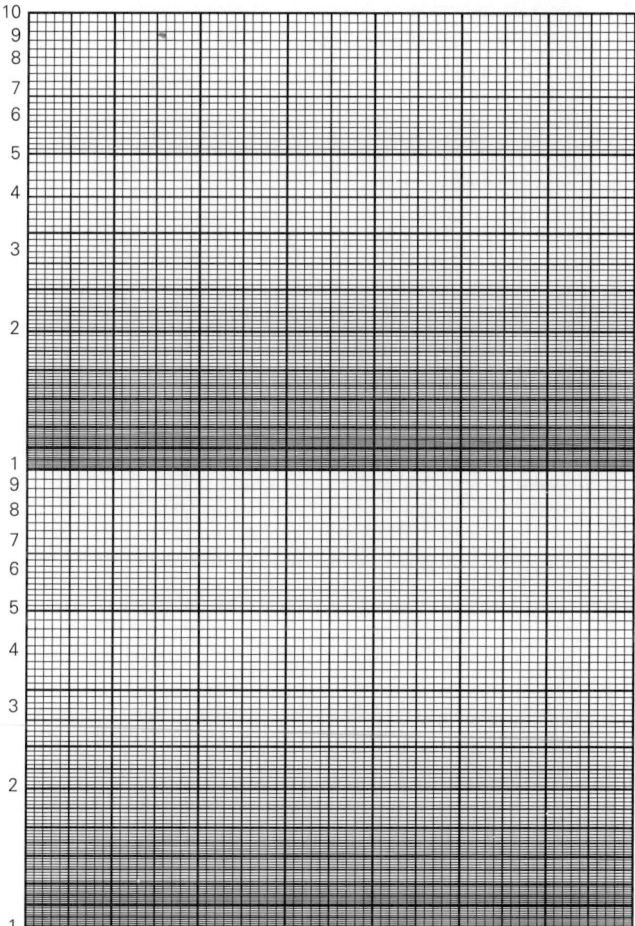

FIGURE 9–26
"Two-cycle" semilog graph scales.

The form of the Bode plot for the amplitude response A_{dB} is shown in Figure 9–28. The frequency scale is labeled as f/f_b, so the result can thus be extended to any frequency range. The exact curve is labeled as E. It starts at a level of 0 dB at very low frequencies and gradually starts to drop off. Eventually, it rolls

FIGURE 9–27
Simple RC low-pass circuit.

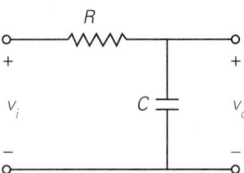

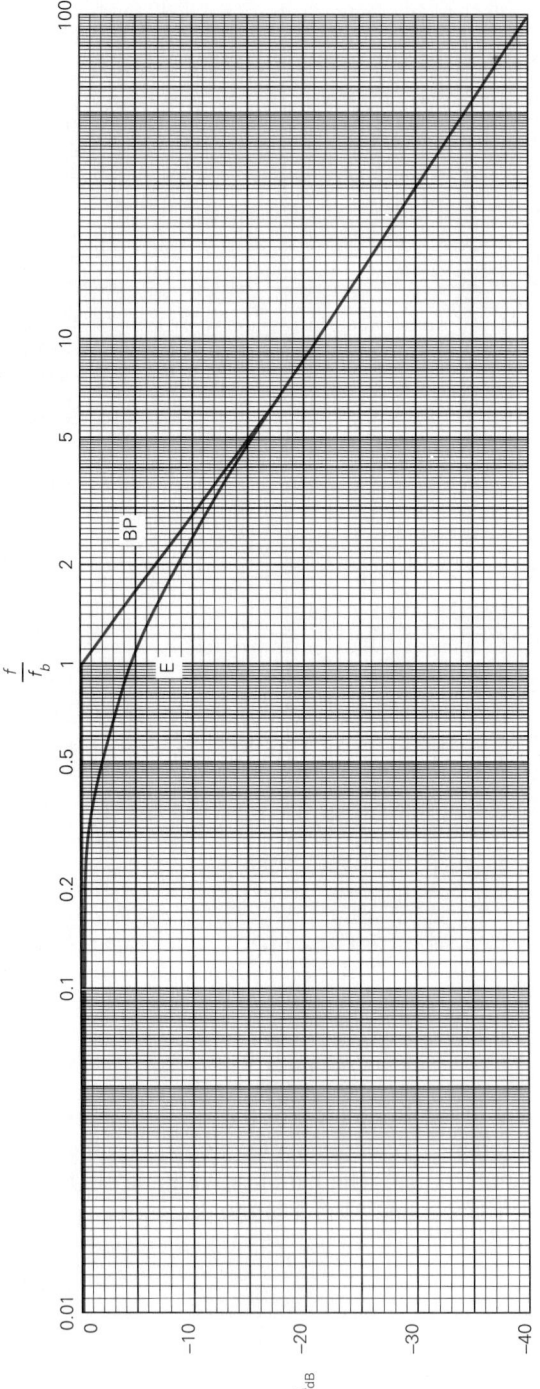

FIGURE 9–28
Bode plot of amplitude response for *RC* low-pass circuit.

off at a rate approximating a straight line on a logarithmic frequency scale. The actual response is down 3 dB at $f/f_b = 1$ (i.e., at $f = f_b$).

The curve labeled BP is the **Bode breakpoint amplitude** approximation. It is formed as a horizontal line with a level 0 dB until $f = f_b$, at which point it breaks downward as a straight line. Note the appropriateness of the term *break frequency*.

The slope of the straight-line approximation beyond the break frequency is -6 dB/octave or -20 dB/decade. An octave is a doubling of the frequency, and a decade is the multiplication of the frequency by 10. Thus, for each range in which the frequency is doubled, the response drops 6 dB, and in each range in which the frequency is increased by a factor of 10, the response drops 20 dB.

The reader may now compare the Bode plot amplitude functions of Figure 9–28 with the linear sketch of A for the same circuit shown in Figure 8–18(b) (p. 386). From this point on, the Bode plot forms will be used. In some cases the exact curve will be displayed, but in many cases the breakpoint approximation will suffice.

The phase shift θ for this circuit is shown in Figure 9–29 using a semilog scale again. The exact phase shift is the curve labeled E. There is also a Bode breakpoint approximation for the phase, and it is shown as the curve BP in Figure 9–29. This approximation assumes $0°$ up to $f/f_b = 0.1$, at which point the curve breaks to a slope of $-45°$ per decade. It passes through $-45°$ at $f/f_b = 1$ and reaches a level of $-90°$ at $f/f_b = 10$. The actual curve differs from the approximation by $5.73°$ at the points $f/f_b = 0.1$ and $f/f_b = 10$.

The curves shown were developed on the basis of a simple RC low-pass circuit. However, many amplifiers are dominated by a frequency response of this form. The major difference is that the low-frequency flat level will normally have a positive decibel level as opposed to the 0-dB level of the RC circuit. The curves of Figure 9–28 may be used in such a case simply by interpreting the dB level as the level relative to the maximum dB gain.

In its general form, any low-pass function displaying a -6-dB/octave high-frequency rolloff rate is referred to as a **one-pole low-pass** model. The term *pole* relates to mathematical definitions used in advanced network theory.

EXAMPLE 9–17

Consider the simple RC low-pass circuit of Figure 9–30(a). Sketch the Bode breakpoint approximations to the amplitude and phase response functions.

Solution

The break frequency is computed as

$$f_b = \frac{1}{2\pi RC} = \frac{1}{2\pi \times 10^4 \times 0.01 \times 10^{-6}} = 1592 \text{ Hz} \qquad \textbf{(9–120)}$$

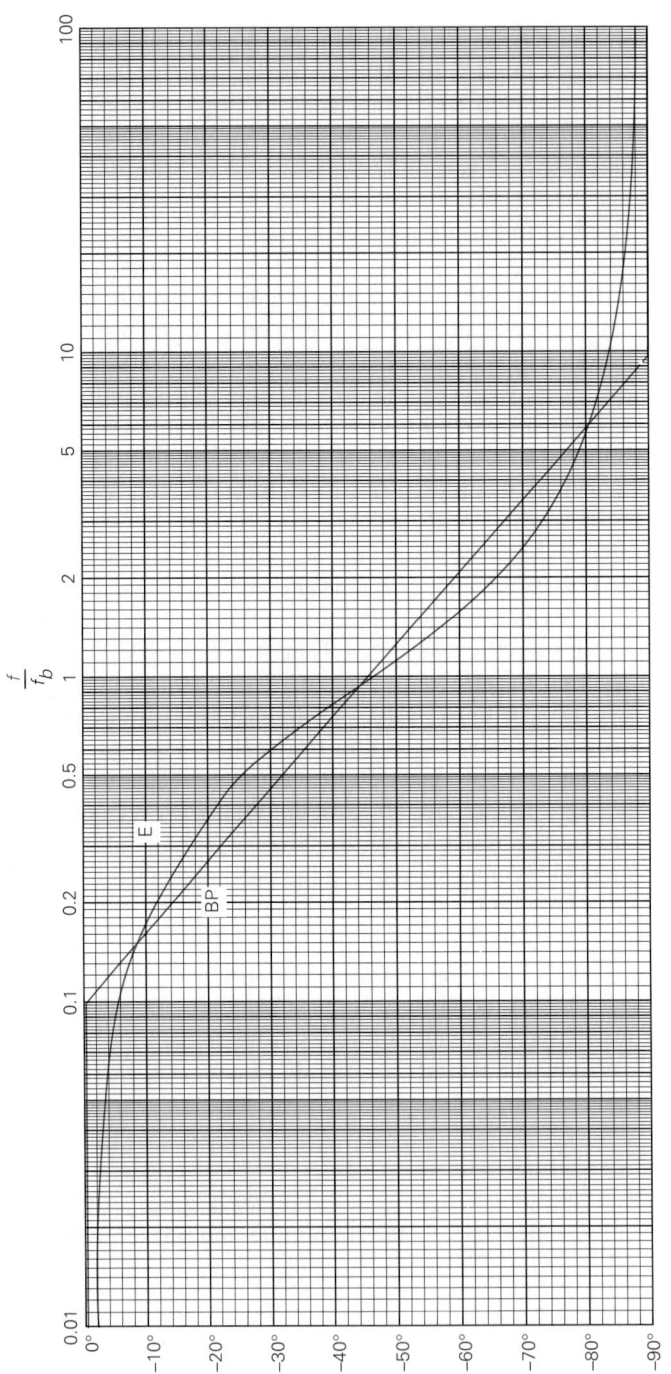

FIGURE 9-29
Bode plot of phase response for *RC* low-pass circuit.

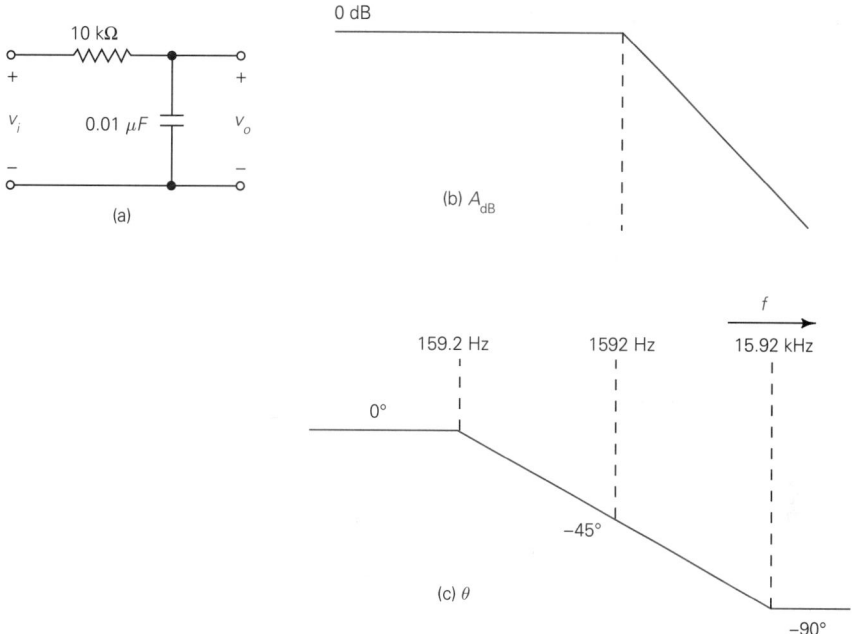

FIGURE 9–30
Circuit and Bode plot approximations of Example 9–17.

The breakpoint amplitude approximation is shown in Figure 9–30(b). The actual response differs from the approximation by 3 dB at 1592 Hz. The breakpoint phase approximation is shown in Figure 9–30(c).

EXAMPLE 9–18

A certain amplifier has a gain function of the form of the one-pole low-pass model. The low-frequency gain is 60 dB, and the break frequency is 10 kHz. Using the amplitude and phase curves of Figures 9-28 and 9-29, determine the amplitude response in dB and the phase response at each of the following frequencies: **(a)** 1 kHz; **(b)** 2 kHz; **(c)** 5 kHz; **(d)** 10 kHz; **(e)** 20 kHz; **(f)** 50 kHz; **(g)** 100 kHz.

Solution

At each frequency, the ratio $f/f_b = f/10^4$ is first determined so that the proper abscissa on the two curves may be used. The relative dB response is read from Figure 9–28, and the phase response is read from Figure 9–29. Note that the amplitude response as read from Figure 9–28 is subtracted from 60 dB to determine the actual response. It should be understood that the values as read from the curves will not have the accuracy that would be obtained from using the mathematical functions given in Chapter 8. The values listed below, although rounded, are stated with slightly more accuracy than can normally be read from the curves.

	f (kHz)	f/f_b	dB (relative)	A_{dB} (dB)	θ
(a)	1	0.1	≈ 0	≈ 60	$-5.7°$
(b)	2	0.2	-0.2	59.8	$-11.3°$
(c)	5	0.5	-1	59	$-26.6°$
(d)	10	1	-3	57	$-45°$
(e)	20	2	-7	53	$-63.4°$
(f)	50	5	-14	46	$-78.7°$
(g)	100	10	-20	40	$-84.3°$

RC High-Pass Model

Consider the simple *RC* high-pass circuit shown in Figure 9–31. As in the case of the low-pass circuit, this high-pass circuit was also discussed in Chapter 8. Let us now consider the form of the Bode plot for this circuit. Recall that the break frequency is

$$f_b = \frac{1}{2\pi RC} \tag{9–121}$$

FIGURE 9–31
Simple *RC* high-pass circuit.

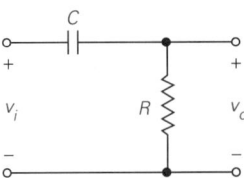

The form of the Bode plot for the amplitude response A_{dB} is shown in Figure 9–32. The frequency scale is labeled as f/f_b. The exact curve is labeled as E. One can never get to the origin on a logarithmic scale, but this response at zero frequency would be a negative infinite dB level. The slope is positive at low frequencies, but it gradually tapers off and flattens out. The actual response is -3 dB at $f/f_b = 1$ or $f = f_b$.

The curve labeled BP is the Brode breakpoint approximation to the high-pass function. It is formed as a straight line in the frequency range below f_b with a slope of $+6$ dB/octave or $+20$ dB/decade. At the frequency $f = f_b$, it reaches a 0-dB level and then breaks to a horizontal line.

The phase shift for this circuit is shown in Figure 9–33 using a semilog scale again. The exact phase shift is the curve labeled E. The Bode breakpoint approximation for the phase is labeled as BP. The approximation assumes $+90°$ up to $f/f_b = 0.1$, at which point the curve breaks to a slope of $-45°$ per decade. It passes through $+45°$ at $f/f_b = 1$. The curve reaches $0°$ at $f/f_b = 10$, at which point it breaks to a zero slope again.

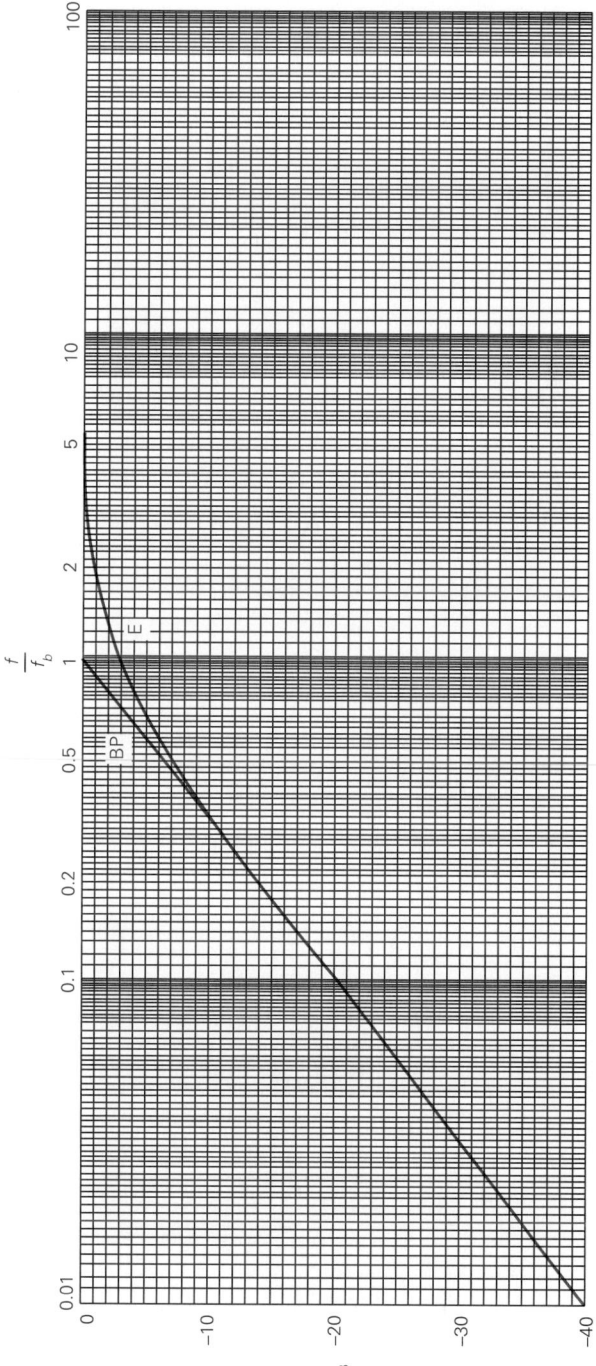

FIGURE 9–32
Bode plot of amplitude response for *RC* high-pass circuit.

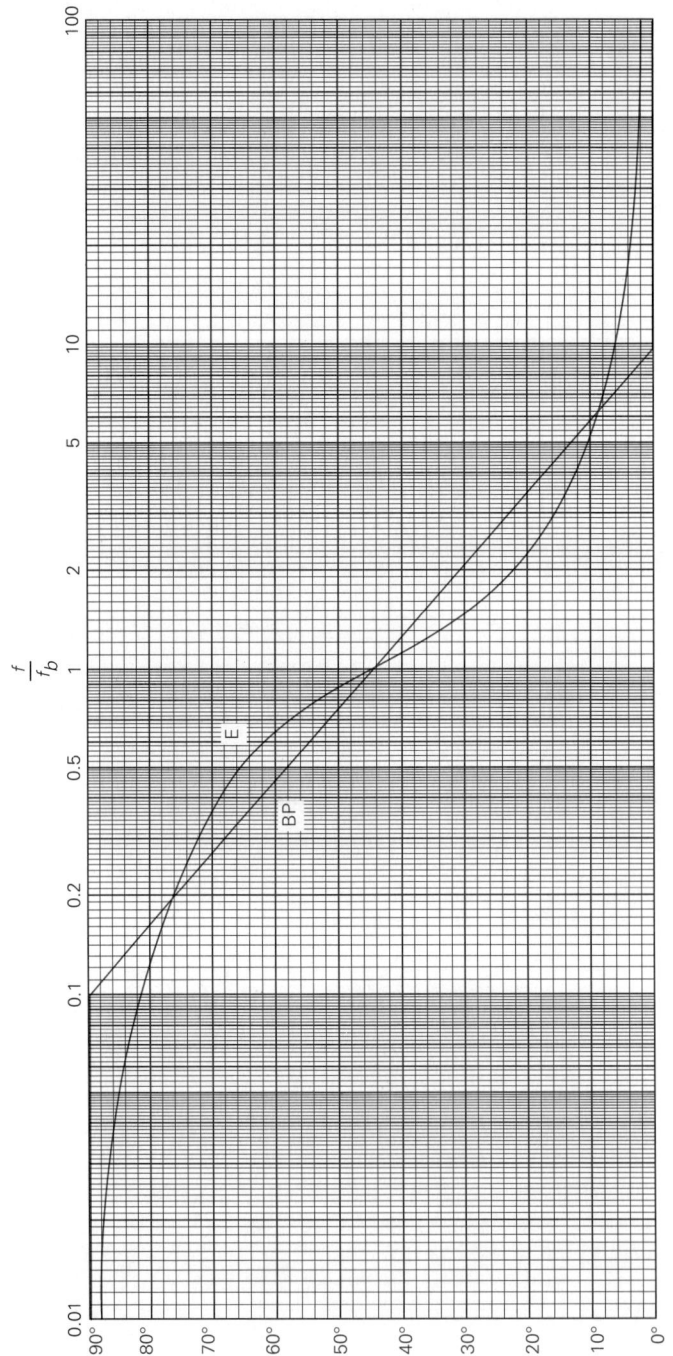

FIGURE 9-33
Bode plot of phase response for *RC* high-pass circuit.

EXAMPLE 9–19

Consider the simple *RC* high-pass circuit of Figure 9–34(a). Sketch the Bode breakpoint approximations to the amplitude and phase response functions.

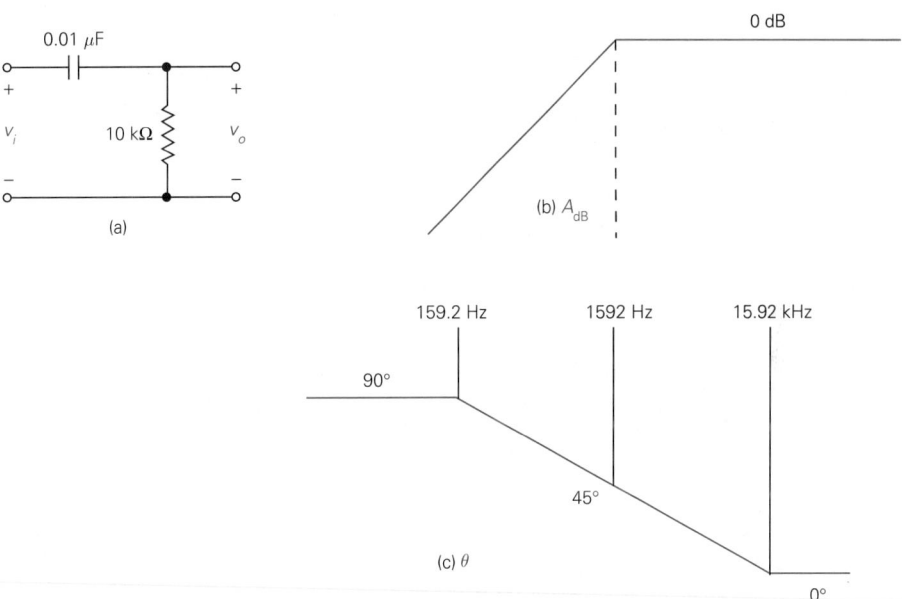

FIGURE 9–34
Circuit and Bode plot approximations of Example 9–19.

Solution

The break frequency is calculated as

$$f_b = \frac{1}{2\pi RC} = \frac{1}{2\pi \times 10^4 \times 0.01 \times 10^{-6}} = 1592 \text{ Hz} \qquad \text{(9–122)}$$

The breakpoint amplitude approximation is shown in Figure 9–34(b). The actual response differs from the approximation by 3 dB at 1592 Hz. The breakpoint phase approximation is shown in Figure 9–34(c).

The reader may find it instructive to compare the circuit and Bode plots of this example with those of Example 9–17. The element values and break frequencies of the two circuits are identical. However, the circuit of Example 9–17 is arranged in a low-pass form, while the circuit of this example is arranged in a high-pass form.

9-11 PSPICE EXAMPLES

PSPICE EXAMPLE 9-1

Develop a PSPICE model to analyze the circuit of Example 9–5 (Figure 9–17) for v_i and v_o.

Solution

The circuit adapted to PSPICE is shown in Figure 9–35(a), and the code is shown in Figure 9–35(b). The reader should be able to follow the code, since there are no new types of elements or commands here. The values of v_i and v_o are represented by V(2) and V(4), respectively.

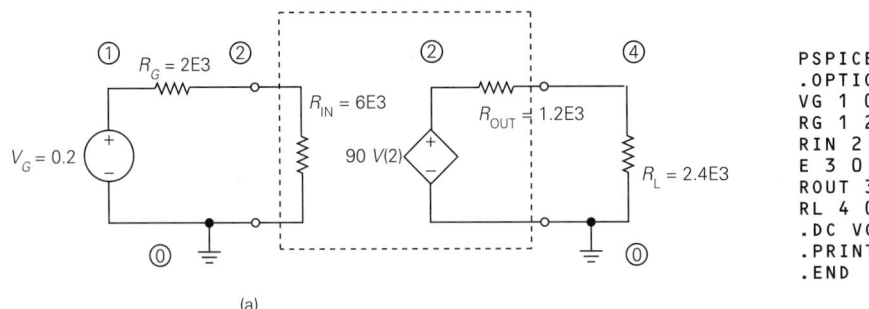

```
PSPICE EXAMPLE 9-1
.OPTIONS NOECHO NOPAGE
VG 1 0 0.2
RG 1 2 2E3
RIN 2 0 6E3
E 3 0 2 0 90
ROUT 3 4 1.2E3
RL 4 0 2.4E3
.DC VG 0.2 0.2 1
.PRINT DC V(2) V(4)
.END
```

(a)   (b)

FIGURE 9-35
Circuit and code of PSPICE Example 9–1.

The computer output is shown in Figure 9–36. The values are in exact agreement with those of Example 9–5.

```
PSPICE EXAMPLE 9-1

****        CIRCUIT DESCRIPTION

********************************************************************

.OPTIONS NOECHO NOPAGE

****        DC TRANSFER CURVES              TEMPERATURE = 27.000 DEG C


VG          V(2)        V(4)
2.000E-01   1.500E-01   9.000E+00


        JOB CONCLUDED

        TOTAL JOB TIME          1.04
```

FIGURE 9-36
Computer printout of PSPICE Example 9–1.

PSPICE EXAMPLE 9–2

Analyze the simple RC low-pass circuit of Example 9–17 (Figure 9–30) using PSPICE. **(a)** First, obtain some tabular data concerning the amplitude and phase response functions in the neighborhood of the 3-dB frequency. **(b)** Next, use .PROBE to obtain the response over the range from 10 Hz to 100 kHz.

Solution

(a) The circuit diagram adapted to PSPICE is shown in Figure 9–37(a), and the first code is shown in 9–37(b). Calculations made in the example indicated that the 3-dB break frequency was about 1592 Hz.

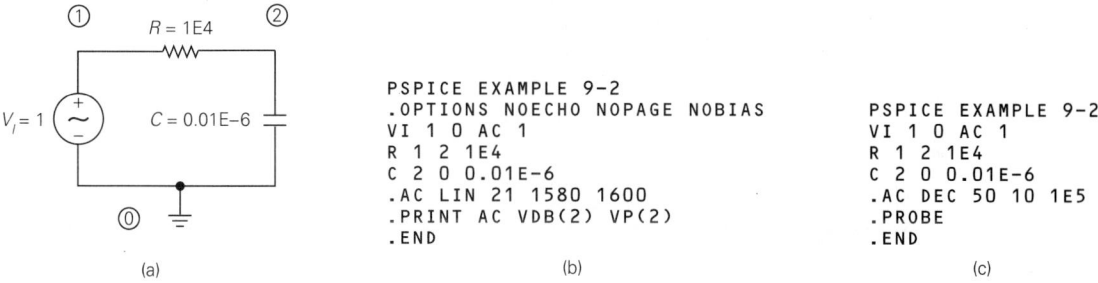

```
PSPICE EXAMPLE 9-2
.OPTIONS NOECHO NOPAGE NOBIAS
VI 1 0 AC 1
R 1 2 1E4
C 2 0 0.01E-6
.AC LIN 21 1580 1600
.PRINT AC VDB(2) VP(2)
.END
```

```
PSPICE EXAMPLE 9-2
VI 1 0 AC 1
R 1 2 1E4
C 2 0 0.01E-6
.AC DEC 50 10 1E5
.PROBE
.END
```

(a) (b) (c)

FIGURE 9–37
Circuit and two codes for PSPICE Example 9–2.

The sweep command reads as follows:

```
.AC LIN 21 1580 1600
```

The command .AC indicates an ac sweep, and LIN implies a *linear* sweep; that is, the frequencies are to be equally spaced. The number of points N is then determined from the beginning frequency f_1 and the ending frequency f_2 by

$$N = \frac{f_2 - f_1}{\Delta f} + 1 \qquad (9\text{–}123)$$

Note the addition of 1 to the ratio. This is necessary, since both a beginning value and an ending value are required. If this is confusing, think about stepping from 10 to 11 in steps of 0.5. The values are 10, 10.5, and 11, which means that there are 3 values even though $(11 - 10)/0.5 = 2$.

For the case at hand, a choice of sweeping from 1580 to 1600 Hz in steps of 1 Hz was made, and this results in 21 points from (9–123). Note in the code that the number of points 21 precedes the beginning and ending values for the sweep.

The print statement reads

```
.PRINT AC VDB(2) VP(2)
```

The addition of DB to V tells PSPICE to calculate the decibel level of V(2), which is defined as

$$VDB = 20 \log_{10}V \qquad\qquad (9\text{--}124)$$

Since the input voltage was 1 V, this means that VDB(2) is the gain or output–input ratio in decibels.

The addition of P to V tells PSPICE to calculate the phase angle of the output voltage. Since the input phase is by default 0°, the angle is the phase shift of the circuit.

The computer printout is shown in Figure 9–38. Note that the amplitude response is −3.012 dB and the angle is −45.01° at a frequency of 1592 Hz. This is within the normal range of roundoff error and validates the response of the circuit near the break frequency.

```
PSPICE EXAMPLE 9-2

****        CIRCUIT DESCRIPTION

*********************************************************************

.OPTIONS NOECHO NOPAGE NOBIAS

****        AC ANALYSIS                         TEMPERATURE = 27.000 DEG C


FREQ        VDB(2)      VP(2)
1.580E+03   -2.979E+00  -4.479E+01
1.581E+03   -2.982E+00  -4.481E+01
1.582E+03   -2.984E+00  -4.483E+01
1.583E+03   -2.987E+00  -4.485E+01
1.584E+03   -2.990E+00  -4.486E+01
1.585E+03   -2.992E+00  -4.488E+01
1.586E+03   -2.995E+00  -4.490E+01
1.587E+03   -2.998E+00  -4.492E+01
1.588E+03   -3.001E+00  -4.494E+01
1.589E+03   -3.003E+00  -4.495E+01
1.590E+03   -3.006E+00  -4.497E+01
1.591E+03   -3.009E+00  -4.499E+01
1.592E+03   -3.012E+00  -4.501E+01
1.593E+03   -3.014E+00  -4.503E+01
1.594E+03   -3.017E+00  -4.504E+01
1.595E+03   -3.020E+00  -4.506E+01
1.596E+03   -3.022E+00  -4.508E+01
1.597E+03   -3.025E+00  -4.510E+01
1.598E+03   -3.028E+00  -4.512E+01
1.599E+03   -3.031E+00  -4.513E+01
1.600E+03   -3.033E+00  -4.515E+01


        JOB CONCLUDED

    TOTAL JOB TIME              1.32
```

FIGURE 9–38
Computer printout of part (a) of PSPICE Example 9–2.

(b)   Based on part **(a),** we have seen that it is possible to obtain exact tabular data, but obviously this can become a very unwieldy response if a broad frequency range is desired. A much better way for a broad frequency range is to use .PROBE and to employ a logarithmic sweep. There are two types of logarithmic sweeps: (1) DEC, and (2) OCT. In the first form the sweep follows a 10-to-1 range in displaying the frequency, while in the second form, the sweep follows a 2-to-1 range.

The code for the broad frequency sweep is shown in Figure 9–37(c). A decade sweep was chosen over the range from 10 Hz to 100 kHz. The number 50 represents the number of points taken over *each* decade. These points become equally spaced on the logarithmic scale, so the exact values are a little complicated to predict. However, this is usually unnecessary as long as a large number of points is chosen.

The .PROBE decibel amplitude plot along with some additional labeling is shown in Figure 9–39. By the use of the "cursor" control, a point very close to the break frequency has been delineated. The corresponding phase response is shown in Figure 9–40. Because of the extrapolation properties of .PROBE, the closest points to the break frequency differ slightly for the two plots.

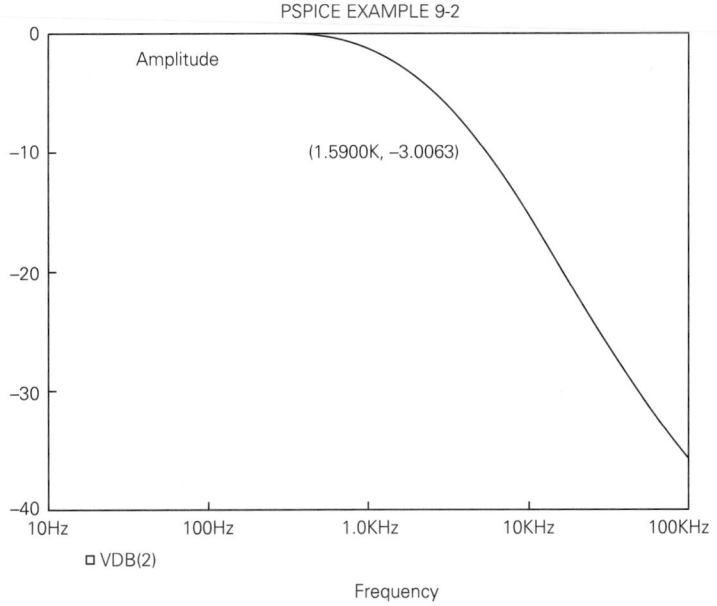

FIGURE 9–39
Decibel amplitude response of PSPICE Example 9–2.

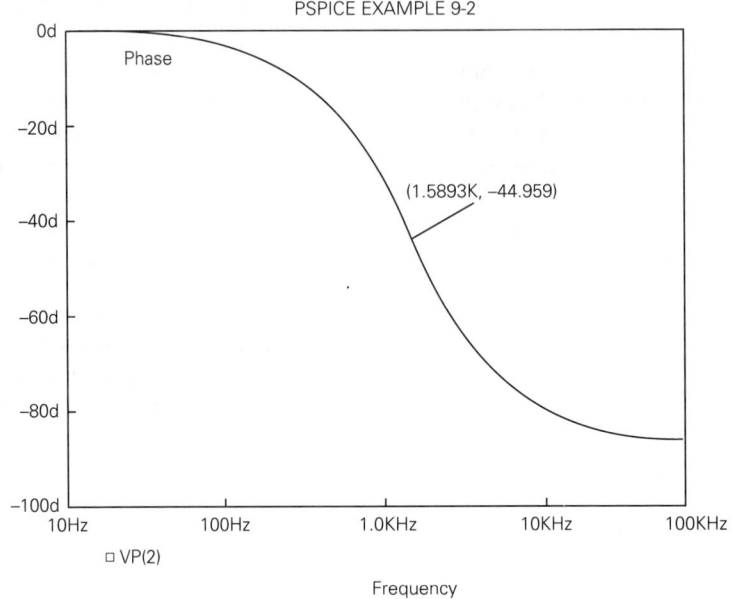

FIGURE 9–40
Phase response of PSPICE Example 9–2.

PSPICE EXAMPLE 9–3

Obtain amplitude and phase response curves of the simple RC high-pass circuit of Example 9–19 (Figure 9–34).

Solution
The circuit adapted to PSPICE is shown in Figure 9–41(a), and the code is shown in Figure 9–41(b). For this circuit, only the .PROBE plots are to be constructed. In this case, 100 points per decade are used.

The amplitude response with additional labels is shown in Figure 9–42. The corresponding phase response is shown in Figure 9–43.

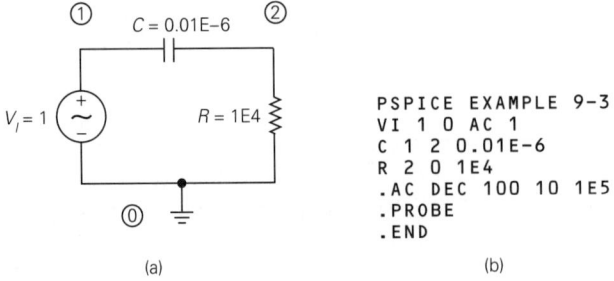

```
PSPICE EXAMPLE 9-3
VI 1 0 AC 1
C 1 2 0.01E-6
R 2 0 1E4
.AC DEC 100 10 1E5
.PROBE
.END
```

(a)   (b)

FIGURE 9–41
Circuit and code of PSPICE Example 9–3.

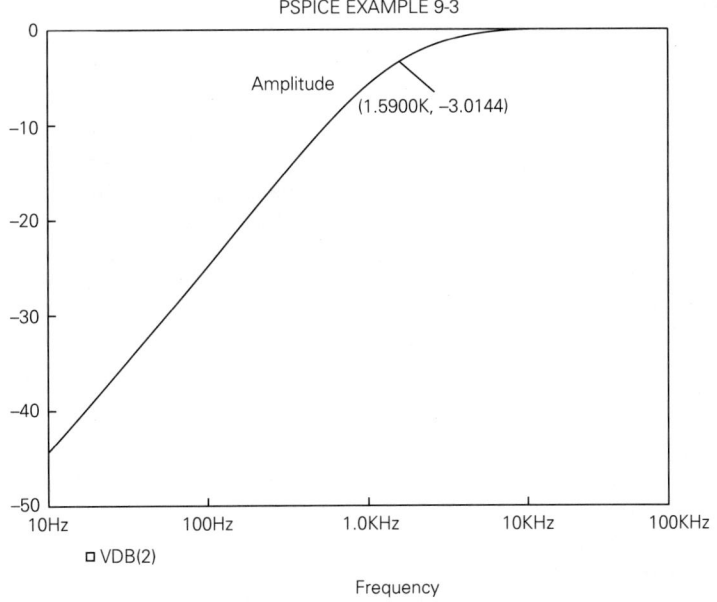

FIGURE 9–42
Decibel amplitude response of PSPICE Example 9–3.

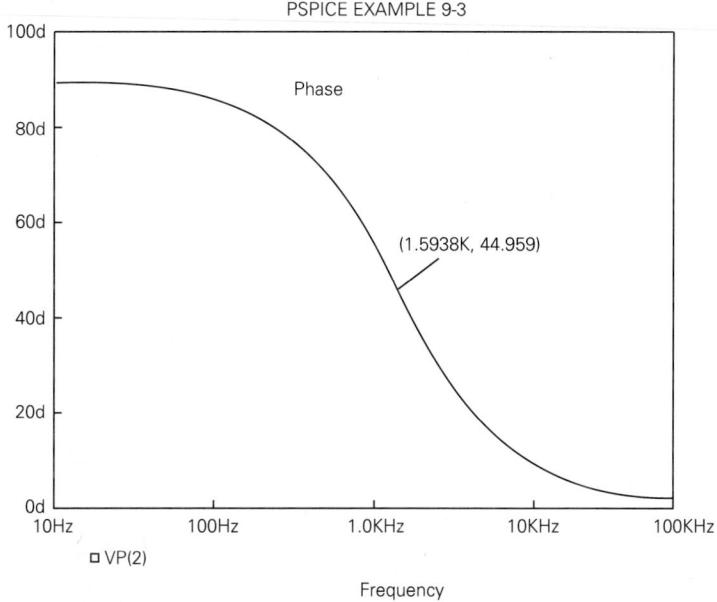

FIGURE 9–43
Phase response of PSPICE Example 9–3.

PROBLEMS

Drill Problems

9–1. A certain inverting amplifier has a voltage gain $A = -50$. Determine the output signal voltage v_o when the input signal voltage is **(a)** $v_i = 0$; **(b)** $v_i = 20$ mV; and **(c)** $v_i = -0.1$ V.

9–2. A certain noninverting amplifier has a voltage gain $A = 25$. Determine the output signal voltage v_o when the input signal voltage is **(a)** $v_i = 0$; **(b)** $v_i = 80$ mV; and **(c)** $v_i = -0.4$ V.

9–3. The gain of a certain amplifier is to be determined from an oscilloscope measurement. The peak-to-peak input voltage is 0.2 V, and the peak-to-peak output voltage is 8 V. The output is 180° out of phase with the input. Determine the voltage gain A.

9–4. The gain of a certain amplifier is to be determined from an oscilloscope measurement. The peak-to-peak input voltage is 0.5 V, and the peak-to-peak output voltage is 12 V. The output is in phase with the input. Determine the voltage gain A.

9–5. An oscilloscope is connected across the input terminals of a certain inverting amplifier with a sinusoidal signal. The peak-to-peak voltage is measured as 0.2 V. An ac voltmeter across the output reads 4 V rms. Determine the gain A.

9–6. An ac voltmeter is connected across the input terminals of a certain noninverting amplifier with a sinusoidal signal. The voltmeter reads 50 mV rms. The peak-to-peak value of the output voltage is determined by an oscilloscope to be 5 V. Determine the gain A.

9–7. A certain amplifier is characterized by the following parameters:

$$R_{in} = 200 \text{ k}\Omega$$
$$R_{out} = 50 \text{ }\Omega$$
$$A = 80$$

Draw the flat-band signal model.

9–8. A certain amplifier is characterized by the following parameters:

$$R_{in} = 500 \text{ k}\Omega$$
$$R_{out} = 200 \text{ }\Omega$$
$$A = -60$$

Draw the flat-band signal model.

9–9. The circuit of Figure P9–9 represents the flat-band signal model of a certain amplifier. Determine the input resistance, the output resistance, and the open-circuit voltage gain.

FIGURE P9–9

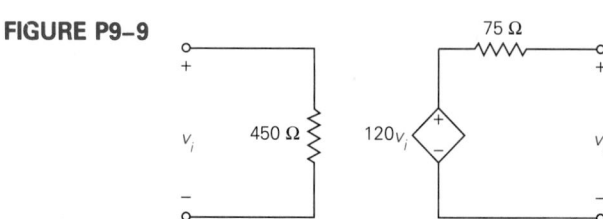

9–10. The circuit of Figure P9–10 represents the flat-band signal model of a certain amplifier. Determine the input resistance, the output resistance, and the open-circuit voltage gain.

FIGURE P9–10

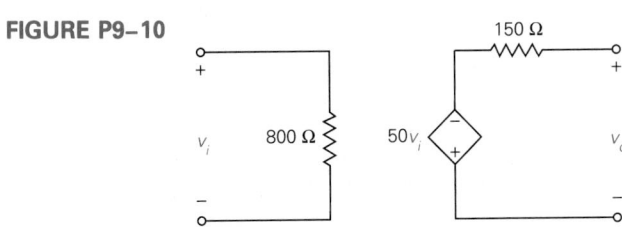

9–11. The circuit of Figure P9–11 represents the flat-band signal model of a certain amplifier with the output arranged in Norton equivalent circuit form. Determine the input resistance, the output resistance, and the open-circuit voltage gain.

FIGURE P9–11

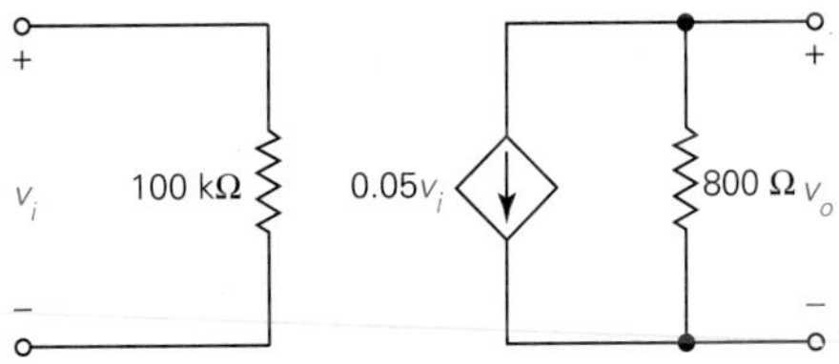

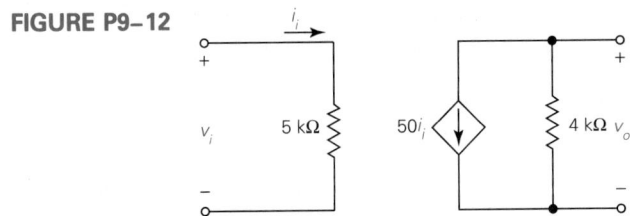

9–12. The circuit of Figure P9–12 represents the flat-band signal model of a certain amplifier with output arranged in a Norton equivalent circuit form and with input current expressed as the controlling variable. Determine the input resistance, the output resistance, and the open-circuit voltage gain.

FIGURE P9–12

9–13. An ideal voltage source signal is applied to the input of the amplifier of Problem 9–7, and there is negligible loading across the output. Determine the gain A_{go}.

9–14. An ideal voltage source signal is applied to the input of the amplifier of Problem 9–8, and there is negligible loading across the output. Determine the gain A_{go}.

9–15. A signal source with an internal resistance of 50 kΩ is applied to the input of the amplifier of Problem 9–7, and there is negligible loading across the output. Determine the gain A_{go}.

9–16. A signal source with an internal resistance of 100 kΩ is applied to the input of the amplifier of Problem 9–8, and there is negligible loading across the output. Determine the gain A_{go}.

9–17. An ideal voltage source signal is applied to the input of the amplifier of Problem 9–7, and a resistance of 150 Ω is connected across the output. Determine the gain A_{go}.

9–18. An ideal voltage source signal is applied to the input of the amplifier of Problem 9–8, and a resistance of 400 Ω is connected across the output. Determine the gain A_{go}.

9–19. A signal source with an internal resistance of 50 kΩ is applied to the input of the amplifier of Problem 9–7, and a resistance of 150 Ω is connected across the output. Determine the gain A_{go}.

9–20. A signal source with an internal resistance of 100 kΩ is applied to the input of the amplifier of Problem 9–8, and a resistance of 400 Ω is connected across the output. Determine the gain A_{go}.

9–21. Determine the net gain A_{go} of the cascade amplifier configuration of Figure P9–21.

FIGURE P9–21

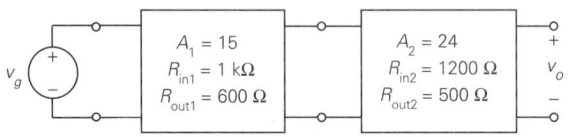

9–22. Determine the net gain A_{go} of the cascade amplifier configuration of Figure P9–22.

FIGURE P9–22

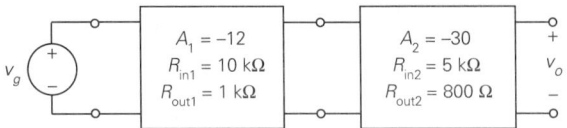

9–23. For the system of Problem 9–21, assume that the generator has an internal resistance of 200 Ω and that a 2-kΩ load is connected across the output. Determine the gain A_{go}.

9–24. For the system of Problem 9–22, assume that the generator has an internal resistance of 2 kΩ and that a 1-kΩ load is connected across the output. Determine the gain A_{go}.

9–25. A certain amplifier has a power gain $G = 8000$. Determine the decibel power gain G_{dB}.

9–26. A certain amplifier has a power gain $G = 500$. Determine the decibel power gain G_{dB}.

9–27. A certain amplifier has a decibel power gain G_{dB} = 48 dB. Determine G.

9–28. A certain amplifier has a decibel power gain G_{dB} = 33 dB. Determine G.

9–29. In a certain amplifier, the input power is 5 mW, and the output power is 2 W. Determine G and G_{dB}.

9–30. In a certain amplifier, the input power is 2 mW, and the output power is 1.2 W. Determine G and G_{dB}.

9–31. In a certain transmission system, the input power is 50 mW, and the output power is 2 mW. Determine G, G_{dB}, L, and L_{dB}.

9–32. In a certain transmission system, the input power is 200 mW, and the output power is 40 mW. Determine G, G_{dB}, L, and L_{dB}.

9–33. A certain amplifier has A = 120. Determine the decibel voltage gain A_{dB}.

9–34. A certain amplifier has A = 800. Determine the decibel voltage gain A_{dB}.

9–35. A certain amplifier has A_{dB} = 30 dB. Determine A.

9–36. A certain amplifier has A_{dB} = 68 dB. Determine A.

9–37. Consider the system of Problem 9–21 (Figure P9–21). Determine the decibel voltage gains of the two stages and the interaction factor, and use these results to determine the value of A_{dB} for the complete system.

9–38. Consider the system of Problem 9–22 (Figure P9–22). Determine the decibel voltage gains of the two stages and the interaction factor, and use these results to determine the value of A_{dB} for the complete system.

9–39. Express the following power levels in dBm: **(a)** 1 W; **(b)** 5 W; **(c)** 28 mW; **(d)** 60 μW; **(e)** 20 fW.

9–40. Determine the power in mW corresponding to the following dBm values: **(a)** 20 dBm; **(b)** 50 dBm; **(c)** 1 dBm; **(d)** −16 dBm; **(e)** −50 dBm.

9–41. Assume that an amplifier with a power gain of 50 dB is excited by a signal source that supplies an input power level of 60 μW to the amplifier. Express the input in dBm, and determine the corresponding output level in dBm.

9–42. The output power of a certain line amplifier is 2 W. It is connected to a transmission system that has a loss of 40 dB. Express the line input in dBm, and determine the corresponding line output level in dBm.

9–43. Consider the *RC* low-pass circuit of Figure P9–43. Sketch the Bode breakpoint approximations to the amplitude and phase response functions.

FIGURE P9–43

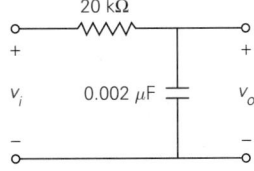

20 kΩ

v_i 0.002 μF v_o

9–44. Consider the *RC* low-pass circuit of Figure P9–44. Sketch the Bode breakpoint approximations to the amplitude and phase response functions.

9–45. Consider the simple *RC* high-pass circuit of Figure P9–45. Sketch the Bode breakpoint approximations to the amplitude and phase response functions.

9–46. Consider the simple *RC* high-pass circuit of Figure P9–46. Sketch the Bode breakpoint approximations to the amplitude and phase response functions.

FIGURE P9–44

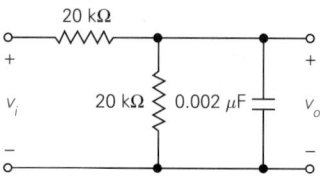

FIGURE P9–45

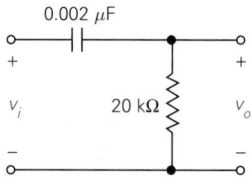

FIGURE P9–46

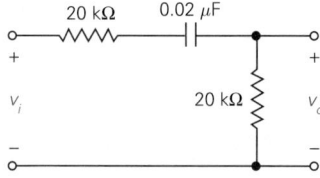

Derivation Problems

9–47. Consider an amplifier with open-circuit voltage gain A. Assume that it is connected to a generator whose resistance is matched to the amplifier input resistance (i.e., $R_g = R_{in}$). Show that

$$A_{go} = \tfrac{1}{2}A$$

9–48. Consider an amplifier with open-circuit voltage gain A. Assume that it is connected to a generator whose resistance is matched to the amplifier input resistance and that the output is connected to a matched load (i.e., $R_g = R_{in}$ and $R_{out} = R_L$). Show that

$$A_{go} = \tfrac{1}{4}A$$

9–49. Show that the power level in dBm is related to the power level in dBW by

$$P(\text{dBm}) = P(\text{dBW}) + 30$$

9–50. Show that the power level in dBf is related to the power level in dBW by

$$P(\text{dBf}) = P(\text{dBW}) + 150$$

Design Problems

9–51. An amplifier is to be selected to connect between a source with internal resistance of 50 Ω and a 2-kΩ load. The following criteria are specified:

Gain reduction due to input loading factor ≤ 0.1 dB

Gain reduction due to output loading factor ≤ 0.1 dB

Loaded voltage gain between source and load ≥ 40 dB under worst-case conditions of A and B

Specify the minimum value of A, the minimum value of R_{in}, and the maximum value of R_{out} for the amplifier to be selected.

9–52. Determine new design specifications for the amplifier of Problem 9–51 if the three criteria are changed as follows:

Input loading factor ≥ 0.9

Output loading factor ≥ 0.9

Loaded voltage gain between source and load ≥ 100 under worst-case conditions of A and B

9–53. Design an RC circuit whose Bode breakpoint amplitude approximation is as follows:

0 dB level from dc to 2 kHz
−6 dB/octave slope above 2 kHz

Select $C = 0.01 \ \mu F$.

9–54. Design an RC circuit whose Bode breakpoint amplitude approximation is as follows:

6 dB/octave slope below 500 Hz

0 dB above 500 Hz

Select $C = 0.1 \ \mu F$.

9–55. Design an RC circuit whose Bode breakpoint amplitude approximation is as follows:

−6 dB level from dc to 5 kHz

−6 dB/octave slope above 5 kHz

Select $C = 0.002 \ \mu F$.

9–56. Design an RC circuit whose Bode breakpoint amplitude approximation is as follows:

6 dB/octave slope from dc to 1 kHz

−6 dB level above 1 kHz

Troubleshooting Problems

9–57. You are testing an RC low-pass circuit of the form shown in Figure 9–27, and you find that the break frequency is lower than specified. Which *two* of the following could be possible troubles? **(a)** R too small; **(b)** R too large; **(c)** C too small; **(d)** C too large.

9–58. You are testing an RC high-pass circuit of the form shown in Figure 9–31, and you find that the break frequency is higher than specified. Which *two* of the following could be possible troubles? **(a)** R too small; **(b)** R too large; **(c)** C too small; **(d)** C too large.

10

BJT SMALL-SIGNAL AMPLIFIER CIRCUITS

OBJECTIVES

After completing this chapter, the reader should be able to:

- Explain the process by which amplification occurs in a BJT, and illustrate the process on a load line.
- Discuss the concept of a small-signal or ac model of a transistor.
- Define and be able to determine for a BJT the input dynamic resistance, the dynamic current gain, and the transconductance.
- Construct for a BJT the small-signal flat-band circuit models.
- Draw the schematic diagram for and explain the operation of a complete common-emitter amplifier using a stable bias design and with emitter resistance bypassed.
- Analyze the circuit of the preceding objective for gain, input resistance, and output resistance.
- Draw the schematic diagram for and explain the operation of a common-emitter amplifier with emitter feedback.
- Analyze the circuit of the preceding objective for gain, input resistance, and output resistance.
- Draw the schematic diagram for and explain the operation of a common-base amplifier.
- Analyze the circuit of the preceding objective for gain, input resistance, and output resistance.
- Draw the schematic diagram for and explain the operation of a common-collector amplifier.
- Analyze the circuit of the preceding objective for gain, input resistance, and output resistance.
- Analyze some of the circuits in the chapter with PSPICE.

10–1 OVERVIEW

We have considered general properties that apply to all amplifiers in the preceding two chapters in terms of gain, input and output resistances, and frequency response. We are now in a position to begin the implementation of specific types of amplifiers using various electronic devices.

In this chapter we return to the bipolar junction transistor (BJT) to study how amplification can be achieved with this device. We considered the dc properties of the BJT in Chapters 4 and 5, but all applications there were based either on the dc model or on total instantaneous signals. In this chapter we focus on the properties of small-signal amplifier circuits implemented with BJTs.

We will present a detailed treatment of BJT small-signal or ac parameters. These parameters are widely used to predict the performance of small-signal amplifiers. We develop small-signal models and study typical specifications.

The three basic amplifier configurations using a BJT are denoted respectively as the **common-emitter amplifier,** the **common-base amplifier,** and the **common-collector amplifier.** The latter circuit is also known as the **emitter-follower circuit,** and a dc version of this circuit was considered in Chapter 5. We study all of these basic amplifier circuits in detail in this chapter. In addition, an improved form of the common-emitter amplifier utilizing emitter feedback will be presented.

The basic amplifier configurations to be discussed represent some of the major circuit forms used in the formulation of linear integrated-circuit modules. To understand and fully use linear integrated circuits, it is important to comprehend the operation of the basic amplifier configurations. In addition, many high-frequency amplifiers still use these basic forms for implementation.

Most of the sections in this chapter contain a number of derivations. Among the formulas to be derived are voltage gain, input resistance, and output resistance for various amplifier configurations. Some of the derivations are straightforward, and others are a bit tricky. Depending on the objectives sought from the text, it may or may not be necessary to be able to reproduce the details of all the derivations. If you are taking a formal course using this text, the instructor will assist in identifying the expectations on derivations.

Even if you are not expected to be able to reproduce the details of the derivations, it is recommended that you at least study the general approaches taken. However, you can learn to use the results of the derivations intelligently if you understand the assumptions and limitations.

All the small-signal results derived in this chapter, as well as similar results derived in Chapter 11, are summarized and ''tied together'' in detail in Chapter 12.

10–2 SIMPLE SMALL-SIGNAL COMMON-EMITTER AMPLIFIER

The process of amplification with a BJT will be illustrated with the simple common-emitter ac-coupled circuit of Figure 10–1(a). This circuit has some serious practical limitations, as will be noted, but its simplicity serves to illustrate the amplification process more clearly than the more elaborate circuits to be developed later.

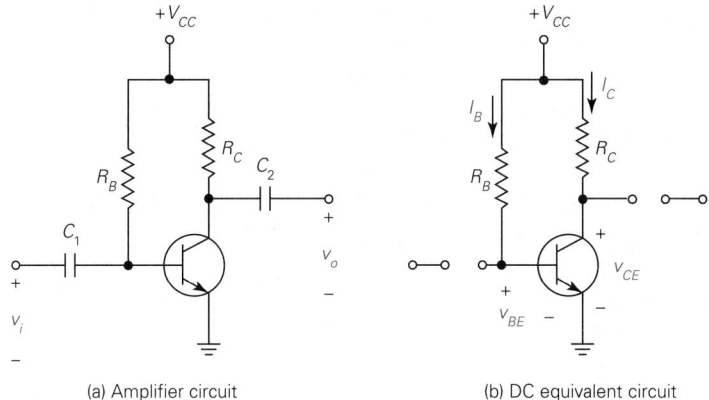

(a) Amplifier circuit   (b) DC equivalent circuit

FIGURE 10–1
Simple BJT common-emitter amplifier and dc equivalent circuit.

The voltage v_i represents an input signal voltage, and v_o represents the output signal voltage. The input voltage v_i could represent the voltage across an ideal voltage source, or it could represent the terminal voltage when there is an internal generator drop. From the work of Chapter 9, a loading factor could be used to account for loading effects between the signal generator and the amplifier, so we will consider only the amplifier input terminals now.

The output voltage v_o could represent a voltage across an open circuit (infinite resistance load), or there may be a load added for some application. As in the case of the input, a loading factor could be used when there is an external load connected. At this point only, the open-circuit ac voltage across the output will be considered.

The dc equivalent circuit is shown in Figure 10–1(b) based on the work of Chapter 8. The two capacitors are represented as open circuits, and this uncouples both the input signal and an output load from the dc path for the transistor. If v_i had been identified as an ac voltage source, the point to the left of the input capacitor would have been grounded. However, since it may or may not be a source, the point will be left uncommitted in this analysis, since it does not affect the dc circuit of the transistor.

The dc circuit was presented early in Chapter 5 to study basic transistor models. The base current is given by

$$I_B = \frac{V_{CC} - V_{BE}}{R_B} = \frac{V_{CC} - 0.7}{R_B} \qquad (10\text{–}1)$$

where $V_{BE} = 0.7$ V has been assumed. The collector current I_C is

$$I_C = \beta_{dc} I_B \qquad (10\text{–}2)$$

The collector-emitter voltage V_{CE} is

$$V_{CE} = V_{CC} - R_C I_C \qquad (10\text{–}3)$$

Before considering amplification, a property concerning V_{BE} needs to be emphasized. From a dc point of view, it is usually reasonable (as we have done many times) to assume that this voltage is a constant (i.e., $V_{BE} = 0.7$ V for many purposes). However, for amplification to occur, it is necessary that V_{BE} change when an input signal is applied. The amount by which it changes is usually quite small compared with the assumed constant value. In a typical application, V_{BE} might change from 0.69 V to 0.71 V or so, for example. When amplification is analyzed mathematically, these changes will have to be considered. However, the dc circuit may continue to assume a constant value in the process. Do not be disturbed by this apparent contradiction, since it is strictly a matter of focusing on the important effect for the purpose desired.

Load-Line Analysis

The load line introduced in Chapter 5 serves as a very useful concept for illustrating the basic amplification process qualitatively. Refer to Figure 10–2, in which some idealized collector characteristics for an *npn* transistor are shown. The particular values of V_{CE}, I_C, and I_B at the dc operating point (Q-point) are designated as V_{CEQ}, I_{CQ}, and I_{BQ}.

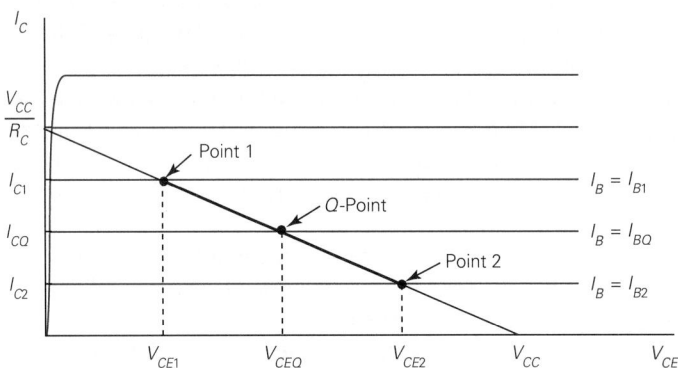

FIGURE 10–2
Load line for simple BJT common-emitter transistor amplifier.

In accordance with the load-line analysis of Chapter 5, an expression for I_C from (10–3) is determined as

$$I_C = \frac{V_{CC}}{R_C} - \frac{1}{R_C} V_{CE} \qquad \textbf{(10–4)}$$

The vertical intercept is V_{CC}/R_C, the horizontal intercept is V_{CC}, and the slope is $-1/R_C$.

The load line is shown as the straight line on the collector characteristics in Figure 10–2. The intersection of the load line with the curve $I_B = I_{BQ}$ defines the

Q-point, and the values of V_{CE} and I_C at that point are V_{CEQ} and I_{CQ}, respectively, as shown.

Amplification from Load Line

Next, we will study the process for amplification. The basic idea is to couple a time-varying signal to the base circuit in a manner that will cause the base–emitter voltage and base current to vary with the input signal. One might naively be tempted simply to connect the signal directly to the base. However, if the signal has a dc path to ground, the dc operating point would shift immediately (even with zero signal level), and this would adversely affect transistor operation. The basic rule for ac-coupled circuits is that *the input and output connection paths of the signal should not affect the dc operating point*. The input and output coupling capacitors achieve the required isolation.

Assume that the frequency range of the signal is sufficiently high that the flat-band models of the two capacitors (short circuits) are valid. However, assume that the frequency is sufficiently low that all the transistor high-frequency effects are negligible. We are thus assuming *midfrequency* operation of the circuit, which might be from about 100 Hz to 100 kHz or so in a typical case.

In the discussion that follows, refer both to the load line of Figure 10–2 and to the waveforms shown in Figures 10–3 and 10–4. It is customary in analyzing amplifiers to assume a sinusoidal input signal in many cases. The assumed sinusoidal input signal v_i is shown in Figure 10–3(a). The various other waveforms to be studied have been idealized in this development and are also shown as sinusoids. In actual practice *significant distortion* could appear in this simple amplifier, but it will not be considered at this point in the book.

The input signal v_i is coupled through the capacitor C_1 and adds algebraically to the Q-point base–emitter voltage. The resulting idealized instantaneous voltage v_{BE} is shown in Figure 10–3(b). This voltage varies up to some maximum value V_{BE1} on the positive half-cycle and down to a minimum value V_{BE2} on the negative half-cycle.

As v_{BE} varies about V_{BEQ}, i_B varies correspondingly about I_{BQ}. The resulting instantaneous current i_B is shown in Figure 10–3(c). This current varies up to some maximum value I_{B1} on the positive half-cycle and down to a minimum value I_{B2} on the negative half-cycle. Note that i_B is *in phase* with v_{BE}; that is, the positive half-cycle of v_{BE} corresponds to the positive half-cycle of i_B, and vice versa.

Next, refer to the load line in Figure 10–2. On the positive input half-cycle, operation shifts up to point 1, which corresponds to the intersection of the load line with the curve $I_B = I_{B1}$. On the negative input half-cycle, operation shifts down to point 2, which corresponds to the intersection of the load line with the curve $I_B = I_{B2}$. Note that compared to I_{CQ}, point 1 corresponds to a larger collector current I_{C1}, and point 2 corresponds to a smaller collector current I_{C2}. This certainly seems reasonable, since collector current increases with base current.

In contrast to collector current, note that point 1 corresponds to a smaller collector voltage V_{CE1}, and point 2 corresponds to a larger collector voltage V_{CE2}.

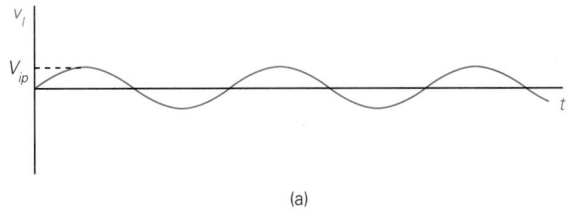

(a)

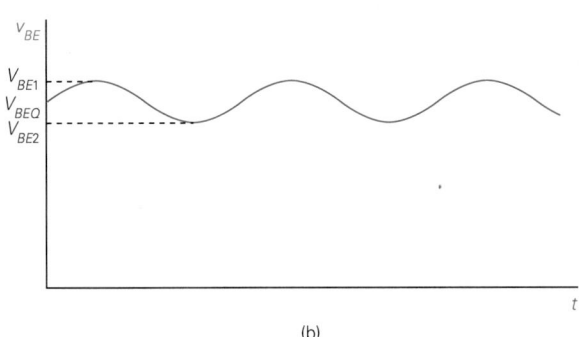

(b)

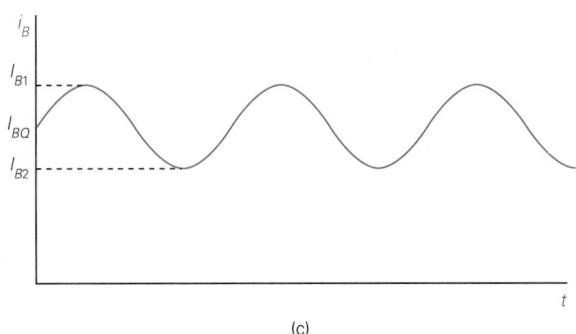

(c)

FIGURE 10–3
Idealized base waveforms of simple amplifier circuit. (Scale is different from that of Figure 10–4.)

The reason for this is that a larger drop across R_C occurs for a larger collector current, and the remaining voltage across the transistor is decreased. The opposite situation is true for a smaller collector current.

The various waveforms in the collector circuit are shown in Figure 10–4. Understand that the vertical scale is different here from that in Figure 10–3. It is necessary to use a different scale, since typical collector current and signal voltage levels could be many times larger than the corresponding base levels. The origin in Figure 10–4 is assumed to be aligned exactly with that of Figure 10–3; that is,

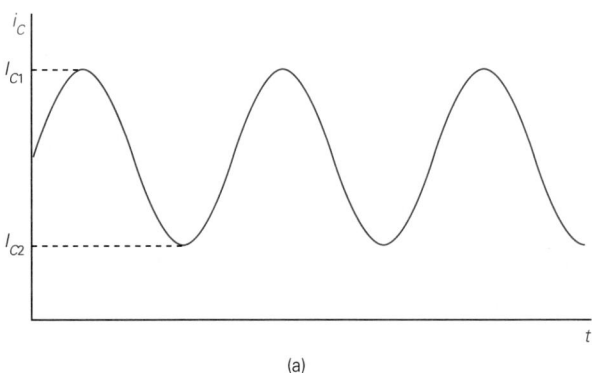

(a)

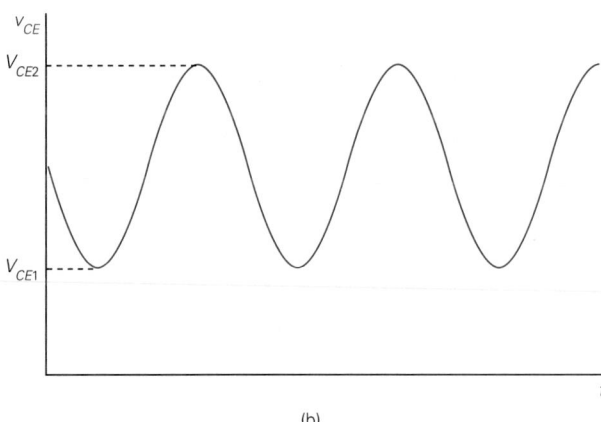

(b)

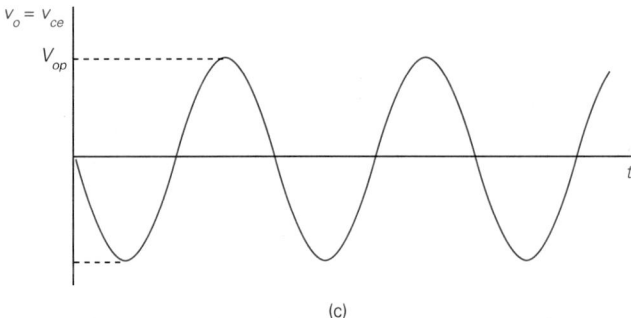

(c)

FIGURE 10–4
Idealized collector waveforms of simple amplifier circuit. (Scale is different from that of Figure 10–3.)

the first half-cycle of each waveform corresponds to the first half-cycle of the input voltage of Figure 10–3.

The instantaneous collector current waveform i_C is shown in Figure 10–4(a), and the limits I_{C1} and I_{C2} correspond to those shown on the load line. The instantaneous collector–emitter voltage v_{CE} is shown in Figure 10–4(b), and the limits V_{CE1} and V_{CE2} also correspond to those shown on the load line. Observe here how the variation in collector voltage is opposite to that of the collector current.

The output voltage waveform is shown in Figure 10–4(c). Comparing the output signal v_o with the input signal v_i of Figure 10–3(a), and assuming the much larger scale, two things may be concluded: (1) the input signal voltage has been amplified, and (2) the output signal is inverted with respect to the input signal (i.e., *the common-emitter amplifier is an inverting amplifier*).

From the load line and the figures, it appears that the Q-point in Figure 10–2 is at the "halfway" point; that is, the dc voltages are divided equally between the transistor and the load resistance R_C. This allows the output to swing symmetrically a maximum amount in both directions. If an input signal were sufficiently large to move operation over to the intersection with the I_C axis, saturation would occur. Conversely, if an input signal were sufficiently large to move operation over to the intersection with the V_{CE} axis, cutoff would occur.

EXAMPLE 10–1

Consider a simple BJT amplifier of the form shown in Figure 10–1 with waveforms as shown in Figures 10–3 and 10–4. Assume that for a specific set of operating conditions, the peak values of input and output voltages are measured as $V_{ip} = 12$ mV, and $V_{op} = 2.16$ V. Determine the voltage gain A between input and output terminals.

Solution

In general, the gain A may be expressed as

$$A = \frac{v_o}{v_i} \tag{10-5}$$

In measuring this gain for an ac amplifier, only the signal components at input and output are of interest. For sinusoidal waveforms, the peak values, the peak-to-peak values, or the rms values could be used as long as *both* input and output are expressed in the same forms. Oscilloscopic measurements are often best performed with peak-to-peak values.

In this example we are given the peak values of input and output voltages. We also know that the voltage gain of a common-emitter amplifier is inverting, so a negative sign will be added. Thus

$$A = -\frac{V_{op}}{V_{ip}} = -\frac{2.16 \text{ V}}{12 \times 10^{-3} \text{ V}} = -180 \tag{10-6}$$

10–3 SMALL-SIGNAL BJT MODELS

The load-line approach in the preceding section has provided some useful insight into the process of amplification for a BJT. If the collector characteristics were given for a particular transistor, it would be possible to determine various amplifier properties, such as gain, directly from the load line. Unfortunately, the collector characteristics vary so widely between individual transistors that such an approach is quite limited in practice.

The most meaningful approach for predicting amplifier performance with transistors is through the use of so-called small-signal models. These equivalent-circuit models are used to predict the levels of various *signal* voltages and currents within the transistor. These signals are the *changes* that occur in the dc levels when an input signal is applied. From the concept of superposition, the dc levels are not shown in the small-signal models. However, their presence is always understood in order to make the active devices function properly.

The adjective *small* in the term *small signal* refers to the assumption that the equivalent circuits are generally valid only when the shift of a given variable is small compared with the dc operating point. Most electronic devices are basically nonlinear, but the linear models are reasonably correct in a narrow region about a given operating point. However, when exacting requirements must be met in advanced applications, more complete nonlinear models may be required.

The term *ac model* will be used here to mean the same thing as *small-signal model*. The adjective *dynamic* is often used to describe an ac or small-signal quantity, while the term *static* could refer to a corresponding dc quantity.

Consider the *npn* BJT shown in Figure 10–5. No external connections are shown here, since the interest at this point is to focus on the transistor alone. In accordance with notation established in Chapter 5, the total instantaneous base current, base–emitter voltage, collector current, and collector–emitter voltage are identified. To review that notation in reference to this situation, the total base current i_B can be expressed as

$$i_B = I_B + i_b \qquad (10\text{–}7)$$

FIGURE 10–5
Bipolar junction transistor and variables used in establishing circuit models.

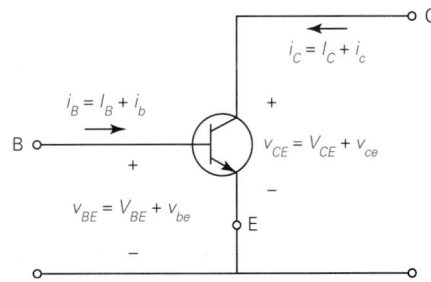

where I_B is the dc component and i_b is the small-signal or ac component. The total base–emitter voltage v_{BE} is

$$v_{BE} = V_{BE} + v_{be}$$ (10–8)

where V_{BE} is the dc component and v_{be} is the ac component. On the collector side, i_C is the total current and it is

$$i_C = I_C + i_c$$ (10–9)

where I_C is the dc component and i_c is the ac component. Finally, v_{CE} is the total collector–emitter voltage, and it is

$$v_{CE} = V_{CE} + v_{ce}$$ (10–10)

where V_{CE} is the dc component and v_{ce} is the ac component.

Dynamic Input Resistance

Refer to Figure 10–6(a), in which a typical BJT input characteristic is shown. As we have done on other occasions in graphical analysis, the subscript Q has been added to distinguish the dc operating point values from the abscissa and ordinate variables. Thus, the transistor is assumed to be biased at operating dc voltage and current V_{BEQ} and I_{BQ}, respectively.

An enlarged portion of the input characteristic is shown in Figure 10–6(b). Assume that the base–emitter voltage is changed by a positive amount ΔV_{BE} as shown. The base current changes by a positive amount ΔI_B. Although the curve

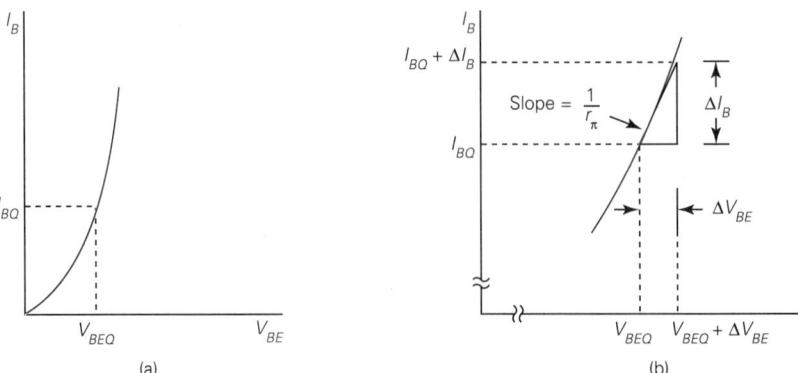

(a) (b)

FIGURE 10–6
Input BJT characteristic at operating point and changes.

is basically nonlinear, if the changes are small, the slope of the straight line between the two points will be nearly the same as the slope of the characteristic at the operating point. This slope can be expressed as

$$\text{slope} = \frac{\Delta I_B}{\Delta V_{BE}} \tag{10-11}$$

The slope has the dimensions of conductance (amperes/volt = siemens). It is more convenient to work with units of resistance. A resistance quantity will be defined as

$$r_\pi = \frac{1}{\text{slope}} = \frac{\Delta V_{BE}}{\Delta I_B} \tag{10-12}$$

The quantity r_π is defined as the **dynamic input resistance** of the BJT. It is clear that r_π will vary considerably as the operating point is changed, but at a given operating point, it can be approximated as a constant value.

For a *pnp* transistor, ΔV_{BE} in (10–11) is replaced by ΔV_{EB} to maintain a positive sense of the change. The important point to remember is that r_π is a positive quantity for both types of BJTs.

Dynamic Current Gain

The effect of a change in voltage and current at the base of a BJT has been represented in terms of a dynamic resistance. Let us now consider the corresponding effect at the collector.

Refer to Figure 10–7. Assume that the curve $I_B = I_{BQ}$ corresponds to the dc operating-point base current, and the corresponding collector current is $I_C = I_{CQ}$.

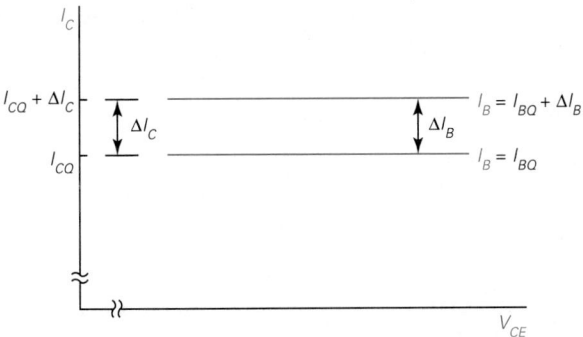

FIGURE 10-7
Output BJT characteristics at operating point and changes.

Assume that the base current increases by a positive value ΔI_B, which shifts operation to the curve corresponding to $I_B = I_{BQ} + \Delta I_B$. The corresponding change in the collector current is a positive shift ΔI_C, so the new collector current is $I_C + \Delta I_C$.

A dynamic flat-band current gain will now be defined. We use the term *flat band* to refer to the fact that none of the frequency-limiting aspects of the transistor are considered at this point. The flat-band dynamic current gain is denoted as β_o, and is defined as

$$\beta_o = \frac{\Delta I_C}{\Delta I_B}$$

(10–13)

Strictly speaking, this definition applies at a constant value of V_{CE}, but since the ideal curves are assumed to show no variation with V_{CE}, this point need not concern us at the moment.

How does β_o compare with β_{dc}? If the transistor characteristics were ideal, and if β_{dc} were constant for all active-region conditions, the two quantities would be equal. However, since β_{dc} is defined in terms of *total* collector and base currents, while β_o is defined for *small changes* about an operating point, the two quantities usually differ somewhat for realistic transistors. Manufacturers usually provide separate data for the two quantities. Usually, however, the values of β_{dc} and β_o for a given transistor tend to be reasonably close to each other.

Small-Signal Circuit Models

We have defined the input dynamic resistance and current gain in terms of changes. In a practical amplifier, these changes occur as a result of *signals*. Thus, the Δ-quantities used to define the parameters in this section become the ac or small-signal variables in an amplifier configuration.

Consider the BJT shown in Figure 10–8(a). Only signal or ac quantities are shown in this particular case, and all dc levels are understood. The first form of a flat-band ac model is shown in Figure 10–8(b). This model will be referred to as the *β-model,* for reasons that will be clear shortly.

The resistance r_π is connected between base and emitter, and it relates the ac base–emitter voltage to the ac base current by Ohm's law; that is,

$$v_{be} = r_\pi i_b$$

(10–14)

Typical values of r_π for small-signal transistors range from a few hundred ohms to several thousand ohms.

The process of ac current gain is accounted for by placing a dependent current source with the value $\beta_o i_b$ between collector and base. Note that $i_c = \beta_o i_b$ in the model.

If this ac equivalent circuit is compared with the corresponding dc equivalent circuit considered extensively earlier in the text, the main difference in form is

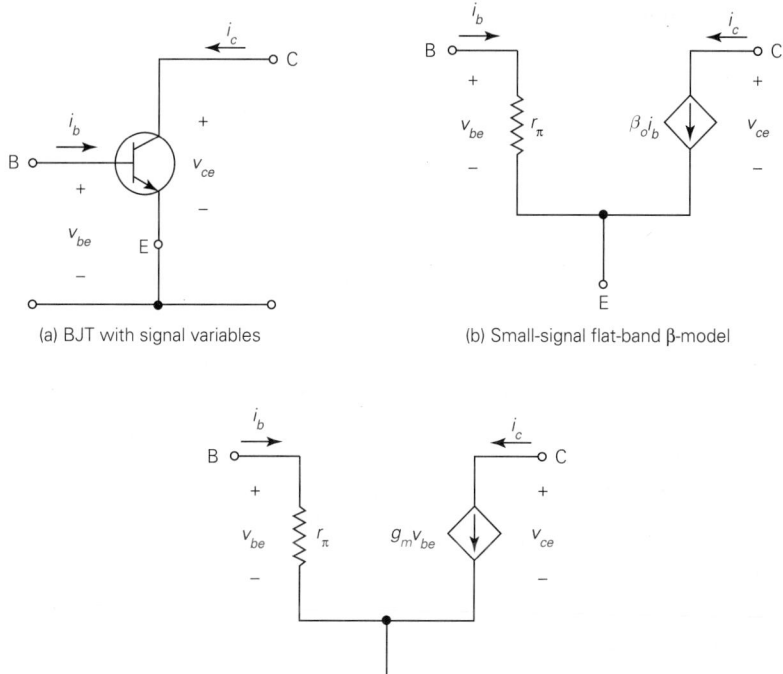

(a) BJT with signal variables

(b) Small-signal flat-band β-model

(c) Small-signal flat-band g_m-model

FIGURE 10–8
BJT and two flat-band small-signal models.

on the base side. The dc circuit employs a fixed voltage drop (usually 0.7 V), which is a reasonable assumption for dc analysis. However, for ac analysis, the dc drop is unimportant, but the dynamic resistance is most significant.

Although the ac collector–emitter signal voltage v_{ce} is labeled on the model, it is not possible to determine this quantity until external circuit parameters are connected. The collector acts nearly like a current source, and the voltage depends on how this current reacts to external circuit parameters.

In most applications of the BJT, it is more desirable to express the dependent current source in terms of the base–emitter voltage than the base current. The base current is readily expressed in terms of the base–emitter voltage as

$$i_b = \frac{v_{be}}{r_\pi} \tag{10–15}$$

The dependent current source is

$$\beta_o i_b = \frac{\beta_o v_{be}}{r_\pi} \tag{10–16}$$

The dependent source is now a **voltage-controlled current source.** A transconductance parameter g_m can then be defined as

$$g_m = \frac{\beta_o}{r_\pi} \qquad\qquad (10\text{--}17)$$

or

$$\boxed{\beta_o = g_m r_\pi} \qquad\qquad (10\text{--}18)$$

The model modified for v_{be} as the controlling variable is shown in Figure 10-8(c). This circuit will be referred to as the **g_m-model** in subsequent work.

Traditionally, the β-model has been used more than the g_m-model in basic texts, probably because of earlier conventions used in transistor modeling and because it correlates more closely with the dc model. However, the g_m-model has recently begun to emerge as a better model for a number of reasons that will be clearer later. Consequently, the latter model will serve as the basis for much of our work through the book.

It turns out that the g_m-model is a simplified form of a more elaborate circuit, called the **hybrid-pi model,** which is considered in Chapter 18. The hybrid-pi model can be readily employed to predict the high-frequency performance of transistors when frequency dependency is significant. Indeed, the subscript π in r_π refers to the hybrid-pi model notation. We will also see later that the g_m-model results in a closer comparison between BJT and FET small-signal models.

The parameters β_o and r_π at a given operating point vary considerably from one transistor to another for a given type. A typical range may be $4:1$ or greater. However, the parameter g_m tends to remain reasonably close for a given operating point. It can be shown from advanced semiconductor physics that for an assumed junction temperature of 25°C, the theoretical ideal value of g_m is $g_m = 38.9 I_C$. Because of the nonideal nature of the junction and the temperature variation, this constant has been rounded to several values in the literature. We choose the fairly common and more easily remembered value of 40. Consequently, the following reasonable estimate for g_m will be used here:

$$\boxed{g_m = 40 I_C} \qquad\qquad (10\text{--}19)$$

If I_C is given in amperes, g_m will be expressed in siemens. However, if I_C is expressed in milliamperes, g_m will be expressed in millisiemens.

From (10–17) and (10–18), we see that β_o, r_π, and g_m are all related. If r_π and β_o vary by as much as $4:1$ or more, how does g_m tend to be much closer to a constant value at a given operating point? The answer is determined by inspecting (10–17). The ratio of β_o to r_π tends to remain nearly the same at a given operating point. This means that a transistor having a higher value of β_o will normally have a higher value of r_π, and vice versa. However, the ratio tends to remain nearly the same.

One more point about the direction of the dependent current source needs to be made. In Chapter 4 we saw that the dependent current source in the dc model of a *pnp* transistor was opposite to that of an *npn* transistor because of the opposite direction of current flow. Our development of the amplifier circuit in this chapter has assumed an *npn* transistor, and the current source small-signal equivalent circuit has been observed to have the same direction as in the dc circuit. One might then be led to believe that the current source in the ac model for a *pnp* transistor should have the opposite direction of the current source in the *npn* case. This is *not* the case, however.

The current source in the ac model represents a *change* in total collector current rather than total current. In an *npn* transistor, a positive input signal causes an increase in positive base current and this increases positive collector current. In a *pnp* transistor, however, a positive input signal has the effect of decreasing the base current (whose positive direction is out of the base), and this decreases the collector current, whose positive direction is out of the collector. A decrease in positive current out of the collector is equivalent to an increase in signal current from collector to emitter. The final result is that the positive collector signal current is directed downward.

To summarize, *either of the small-signal equivalent circuits of Figure 10–8 applies to both npn and pnp transistors as given.*

EXAMPLE 10–2

Some fine dc measurements performed in the base circuit of a particular *npn* transistor with fixed collector–emitter voltage yield the following data:

V_{BE} (V)	I_B (μA)
0.65	100
0.66	150

Determine the approximate value of r_π at the operating point.

Solution

The first point to note is that V_{BE} is no longer assumed to be our "standard" value of 0.7 V. When small-signal models are being determined, it is necessary to measure carefully the actual voltage and changes about that value.

The concept of (10–12) is used for this purpose. The change ΔV_{BE} is given by

$$\Delta V_{BE} = 0.66 \text{ V} - 0.65 \text{ V} = 0.01 \text{ V} \qquad \textbf{(10–20)}$$

The corresponding change ΔI_B is

$$\Delta I_B = 150 \, \mu\text{A} - 100 \, \mu \text{ A} = 50 \, \mu\text{A} \qquad \textbf{(10–21)}$$

The value of r_π is

$$r_\pi = \frac{0.01\,\text{V}}{50 \times 10^{-6}\,\text{A}} = 200\,\Omega \qquad \text{(10–22)}$$

This example was formulated more as a learning process than as a practical procedure. The measurement of a small change in a dc voltage about a much larger dc voltage is difficult to achieve with any degree of accuracy. In practice, the parameter r_π is easier to determine using ac small-signal techniques.

EXAMPLE 10–3

Some dc measurements performed on a particular BJT with fixed collector–emitter voltage yield the following data:

I_B (μA)	I_C (mA)
100	9
120	11.2

Determine the approximate value of β_o at the operating point.

Solution

The concept of (10–13) is used for this purpose. The change ΔI_B is

$$\Delta I_B = 120\,\mu\text{A} - 100\,\mu\text{A} = 20\,\mu\text{A} \qquad \text{(10–23)}$$

The corresponding change ΔI_C is

$$\Delta I_C = 11.2\,\text{mA} - 9\,\text{mA} = 2.2\,\text{mA} \qquad \text{(10–24)}$$

The value of β_o is

$$\beta_o = \frac{2.2 \times 10^{-3}\,\text{A}}{20 \times 10^{-6}\,\text{A}} = 110 \qquad \text{(10–25)}$$

Incidentally, the value of β_{dc} could be computed at either of the two points. The reader is invited to show that neither value of β_{dc} is equal to β_o since the phenomena involved are different.

EXAMPLE 10–4

A certain transistor is operating at $I_C = 2$ mA, and it is known that $\beta_o = 120$. Determine **(a)** g_m, and **(b)** r_π.

Solution

(a) The value of g_m is estimated from (10–19) as

$$g_m = 40I_C = 40 \times 2 \times 10^{-3} = 80 \times 10^{-3}\,\text{S} = 80\,\text{mS} \qquad \text{(10–26)}$$

(b) From (10–18), we have

$$r_\pi = \frac{\beta_o}{g_m} = \frac{120}{80 \times 10^{-3}} = 1500 \ \Omega \tag{10–27}$$

EXAMPLE 10–5

For the transistor of Example 10–4 at the given operating point, construct the β-model and the g_m-model for small signals.

Solution
The models have the forms given in Figure 10–8, and they are shown in Figure 10–9 for the particular parameters of Example 10–4.

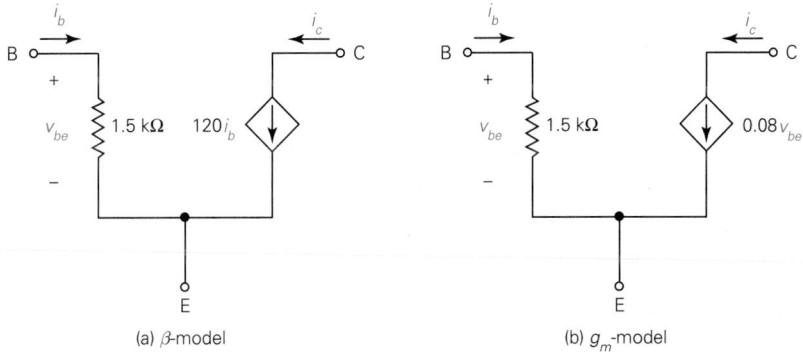

FIGURE 10–9
Small-signal models of Example 10–5.

EXAMPLE 10–6

A certain transistor is characterized by the fact that β_o remains nearly constant with the value $\beta_o \approx 100$ as I_C varies from 1 to 8 mA. For this assumption, determine g_m and r_π at the following values of I_C: **(a)** 1 mA; **(b)** 2 mA; **(c)** 4 mA; **(d)** 8 mA.

Solution
At any value of I_C, we have

$$g_m = 40I_C \tag{10–28}$$

The corresponding value of r_π is then determined as

$$r_\pi = \frac{\beta_o}{g_m} = \frac{100}{g_m} \tag{10–29}$$

For the given values of I_C, the results can be readily calculated and tabulated as follows:

	I_C (mA)	g_m (mS)	β_o	r_π
(a)	1	40	100	2.5 kΩ
(b)	2	80	100	1.25 kΩ
(c)	4	160	100	625 Ω
(d)	8	320	100	312.5 Ω

In practice, β_o does not remain constant as I_C varies, but it often varies little for a particular transistor over a reasonably wide range of collector current. Therefore, this example illustrates a very important point: *The value of r_π decreases very rapidly as the collector current increases. If β_o were exactly constant (as assumed in this example), the value of r_π would vary inversely with collector current.*

10–4 COMMON-EMITTER CIRCUIT WITH STABLE BIAS

The simple common-emitter circuit previously considered has provided insight into the amplification process. That basic circuit, however, is subject to a wide variation in the operating point as a result of transistor parameter variations and has only limited usefulness in practice.

We now consider a more practical common-emitter circuit employing stabilized bias. This circuit is analyzed with the small-signal model developed in the preceding section.

The basic form of the stabilized common-emitter ac-coupled amplifier using an *npn* transistor is shown in Figure 10–10(a). Effects of possible source resistance and external load resistance will not be considered at this time.

The dc circuit is shown in Figure 10–10(b). This circuit was developed in some detail in Chapter 5 and is the standard stabilized bias circuit for a BJT. Recall that if the bias circuit is designed properly, the collector current can be closely approximated as

$$I_C = \frac{V_{TH} - V_{BE}}{R_E} = \frac{V_{TH} - 0.7}{R_E} \tag{10–30}$$

where

$$V_{TH} = \frac{R_2 V_{CC}}{R_1 + R_2} \tag{10–31}$$

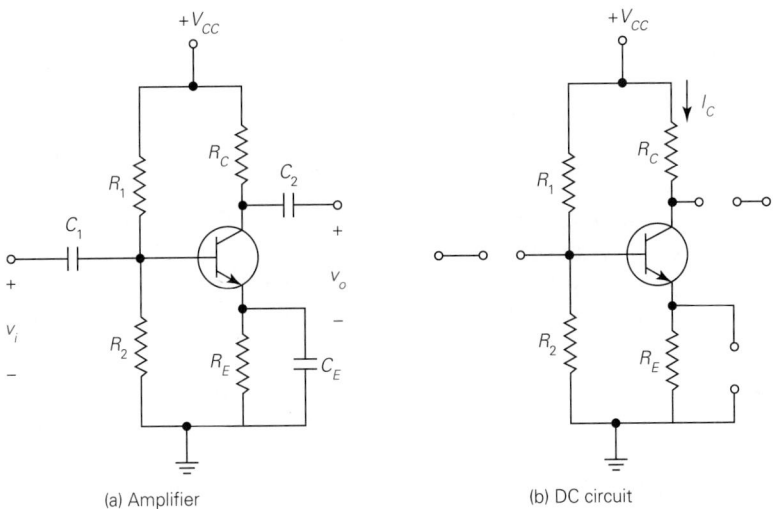

(a) Amplifier (b) DC circuit

FIGURE 10–10
Common-emitter amplifier with stabilized bias and dc equivalent circuit.

The corresponding *midfrequency* ac equivalent circuit for small signals is shown in Figure 10–11. This circuit combines some of the concepts of Chapter 8 with the ac signal model of Section 10–3. The actual transistor is replaced by the small-signal model between the three transistor terminals as noted. The value of g_m is determined from the dc operating-point current previously calculated and is approximated as

$$g_m = 40I_C \qquad (10\text{--}32)$$

The functions of C_1 and C_2 have already been discussed for the simpler amplifier, and they serve to couple the signal in and out while blocking dc from

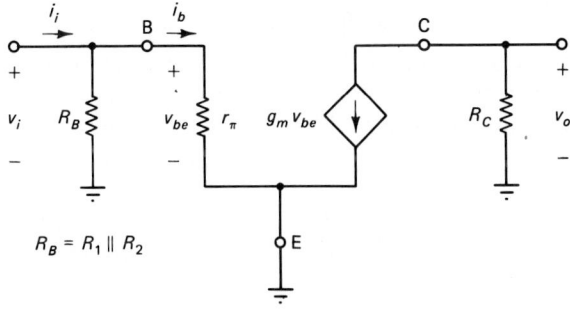

FIGURE 10–11
Small-signal midfrequency model of common-emitter amplifier.

flowing into the signal source or any external load. The function of C_E, however, is to bypass the signal around the emitter resistance R_E. Without C_E, the effect of R_E would be to reduce the gain considerably, as will be discussed in a later section. In the flat-band region of operation, C_E acts as a short, and all signal current flows directly from emitter to ground. Thus, R_E does not even appear in the midfrequency equivalent circuit.

In accordance with the principles discussed in Chapter 8, the dc bias supply is replaced by a short circuit to ground as far as signals are concerned. In effect, this brings both R_1 and the collector resistance R_C to ground. Since R_1 and R_2 appear as a simple parallel combination of two resistors, for convenience a resistance R_B is defined as

$$R_B = R_1 \| R_2 = \frac{R_1 R_2}{R_1 + R_2} \tag{10-33}$$

Note that two separate signal voltage labels v_i and v_{be} are shown on the diagram, even though they are equal in this case. The voltage v_i is the input signal voltage to the amplifier, and v_{be} is the signal base–emitter voltage that controls the collector current. In many configurations, these two voltages are not the same, but for the common-emitter configuration in the flat-band region with emitter resistance completely bypassed, $v_{be} = v_i$.

The voltage gain A will now be determined. The output ac voltage v_o must be determined in terms of the input ac voltage v_i. Since all of the available ac current flows into R_C, with the resulting positive flow into the lower terminal, we have

$$v_o = -g_m v_{be} R_C = -g_m v_i R_C \tag{10-34}$$

The voltage gain A is

$$A = \frac{v_o}{v_i} = -g_m R_C \tag{10-35}$$

Some other quantities of immediate interest for the amplifier are the input and output resistances. However, it is convenient to define two separate values for each of these quantities. The first value for each represents the quantities associated with the transistor itself without regard to external resistances. The second value for each represents the corresponding first value coupled with the effects of external resistances.

Let R'_{in} and R'_{out} represent the ac input and output resistances of the transistor *without the effects of external resistances*. From the equivalent circuit of Figure 10–11, we see that

$$R'_{in} = \frac{v_{be}}{i_b} = r_\pi \tag{10-36}$$

and

$$R'_{out} = \infty \qquad (10\text{–}37)$$

The result of (10–37) is based on the assumption of an ideal current source for the $g_m v_{be}$ generator. In practice, there is a large dynamic resistance in parallel with this current source, but it will be approximated as an infinite resistance.

Let R_{in} and R_{out} represent the input and output resistances of the *composite circuit*. We have

$$R_{in} = \frac{v_i}{i_i} = R_B \| r_\pi = R_1 \| R_2 \| r_\pi \qquad (10\text{–}38)$$

and

$$R_{out} = R_C \qquad (10\text{–}39)$$

The result of (10–38) indicates that the net input resistance of the complete amplifier circuit is simply the parallel combination of the three resistances R_1, R_2, and r_π. Although the transistor output is assumed to be an ideal current source, the result of (10–39) indicates that the output resistance of the composite circuit is the collector resistance R_C. Because of the typical moderate size of R_C, the assumption of (10–37) is usually valid, since the effect of the transistor dynamic resistance on R_{out} will often be insignificant.

The quantities R_{in} and R_{out} are generally the most meaningful of the input and output resistance quantities, since they predict the net effects of the complete circuit. However, when special attention is focused only on the transistor itelf, R'_{in} and R'_{out} are useful.

For this amplifier and other configurations to be discussed in this chapter and in Chapter 11, most of the effects of source and load interactions will be momentarily disregarded until the basic properties of the amplifiers themselves are established. The effects of source and load interactions, as well as combinations of the basic amplifier configurations, are considered in Chapter 12.

Recall the unstabilized common-emitter amplifier of Section 10–2, which was used to qualitatively analyze amplifier operation with a load line. If a small-signal ac circuit were constructed for that circuit, it would have exactly the same form as the stabilized circuit except that R_B in the simpler circuit represents one resistance instead of a parallel combination. This equivalence is due to the fact that the emitter stabilizing resistance R_E is bypassed for signals, and thus the emitter appears at signal ground.

**EXAMPLE
10–7**
The schematic diagram of a certain RC-coupled common-emitter amplifier is shown in Figure 10–12(a). **(a)** Determine the dc operating point current I_C and the three

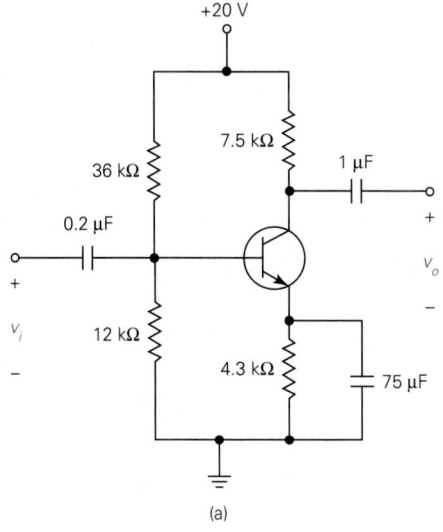

(a)

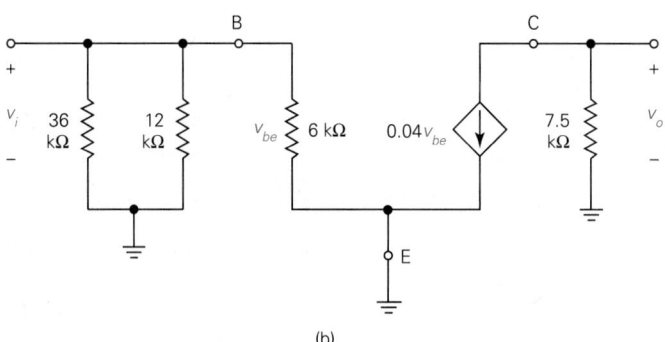

(b)

FIGURE 10–12
Common-emitter amplifier circuit and midfrequency model of Example 10–7.

voltages V_{R_E}, V_{R_C}, and V_{CE}. **(b)** Construct the midfrequency small-signal equiva-lent circuit. It is known that $\beta_o = 240$. **(c)** Determine the gain A and the input and output resistances R'_{in}, R'_{out}, R_{in}, and R_{out}.

Solution

(a) The first part is simply a rehash of the work of Chapter 5 but is a necessary part of a complete amplifier analysis. The steps involved in determining I_C follow.

$$V_{TH} = \frac{12 \text{ k}\Omega}{12 \text{ k}\Omega + 36 \text{ k}\Omega} \times 20 \text{ V} = 5 \text{ V} \qquad (10\text{–}40)$$

$$I_C = \frac{5 - 0.7}{4.3 \times 10^3} = 1 \text{ mA} \qquad (10\text{--}41)$$

Various voltages in the circuit are determined as follows:

$$V_{R_E} = 4.3 \times 10^3 \times 1 \times 10^{-3} = 4.3 \text{ V} \qquad (10\text{--}42)$$
$$V_{R_C} = 7.5 \times 10^3 \times 1 \times 10^{-3} = 7.5 \text{ V} \qquad (10\text{--}43)$$
$$V_{CE} = 20 - 4.3 - 7.5 = 8.2 \text{ V} \qquad (10\text{--}44)$$

(b) To construct the ac equivalent circuit we first compute g_m as

$$g_m = 40I_C = 40 \times 1 \times 10^{-3} = 0.04 \text{ S} \qquad (10\text{--}45)$$

Since β_o is given, r_π can be calculated as

$$r_\pi = \frac{\beta_o}{g_m} = \frac{240}{0.04} = 6 \text{ k}\Omega \qquad (10\text{--}46)$$

The midfrequency equivalent circuit is shown in Figure 10–12(b). For clarity, R_1 and R_2 are shown separately.

(c) The gain A is

$$A = \frac{v_o}{v_i} = -g_m R_C = -0.04 \times 7.5 \times 10^3 = -300 \qquad (10\text{--}47)$$

The input and output resistances of the transistor above are

$$R'_{in} = r_\pi = 6 \text{ k}\Omega \qquad (10\text{--}48)$$
$$R'_{out} = \infty \qquad (10\text{--}49)$$

The input and output resistances of the composite circuit are

$$R_{in} = R_1 \| R_2 \| r_\pi = 36 \text{ k}\Omega \| 12 \text{ k}\Omega \| 6 \text{ k}\Omega = 3.6 \text{ k}\Omega \qquad (10\text{--}50)$$
$$R_{out} = R_C = 7.5 \text{ k}\Omega \qquad (10\text{--}51)$$

It is not always necessary to construct the small-signal model to determine these various parameters, since the formulas can often be applied directly. However, the small-signal model provides some meaningful insight into the process involved, and it permits the desired properties to be determined directly from the circuit parameters.

Because of the uncertainties associated with transistor parameters and the approximations involved, these results should be interpreted as "good estimates" rather than exact values. In particular, the magnitude of the gain of an actual transistor circuit based on these conditions would tend to be somewhat lower than the calculated value of (10–47) because of the transistor dynamic output resistance in parallel with R_C and because g_m was rounded upward in earlier work.

10–5 COMMON-EMITTER AMPLIFIER WITH EMITTER FEEDBACK

The common-emitter amplifier of the preceding section is capable of assuming a very stable dc operating point, but its ac or signal characteristics have some serious limitations. The most serious limitation of the circuit is the degree of distortion present in the output when the input signal is nearly an ideal voltage source and when the desired output signal has an appreciable level.

   The reason for the serious distortion will be explained with Figure 10–13. Certain of the curves here will be exaggerated to demonstrate the situation fully.

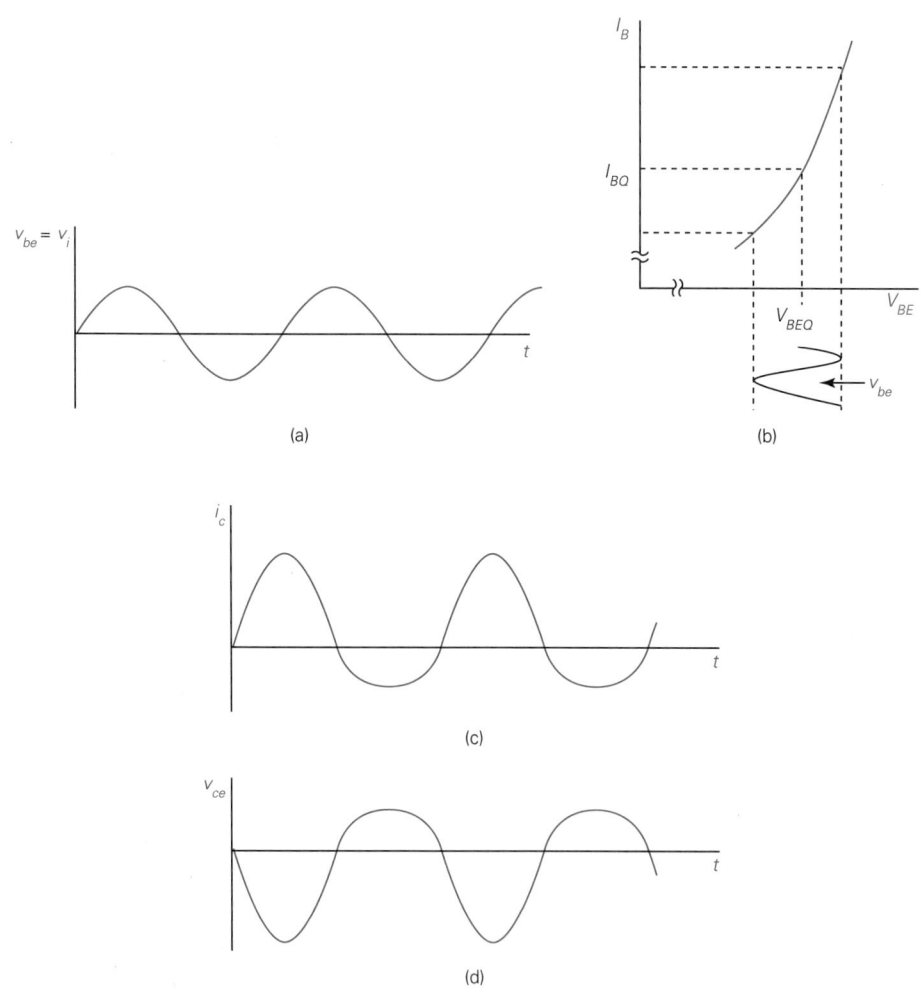

(a)

(b)

(c)

(d)

FIGURE 10–13
Illustration of distortion in common-emitter amplifier with voltage-driven input.

In Figure 10–13(a), an ideal sinusoidal signal is assumed across the amplifier input (i.e., the waveform is $v_{be} = v_i$). The nonlinear form of the input characteristic is shown in Figure 10–13(b), and the bias point corresponds to V_{BEQ} and I_{BQ}.

The sinusoidal voltage causes an equal swing on either side of V_{BEQ}. However, observe that the change in I_B is much greater on the positive half-cycle of v_{be} than on the negative half-cycle. If β_o is assumed to be constant over the operating range, this means that the positive peak collector signal current will be greater in magnitude than the corresponding negative peak as shown in Figure 10–13(c). The resulting collector voltage variation will be of the form shown in Figure 10–13(d). The signal is no longer a pure sinusoid, and it exhibits appreciable distortion.

An additional limitation of this circuit is the relatively wide variation in amplification that results from differences in transistor operating characteristics. Thus, while the circuit has been stabilized for dc, it is not particularly stable for ac.

What is needed is a way of stabilizing the ac characteristics in a manner similar to that of the dc circuit. A major factor in stabilizing the circuit for dc is the presence of the emitter resistance. However, in the amplifier configuration of Section 10–4, the emitter resistance was bypassed with a capacitor, so that its effect on ac was eliminated.

Suppose that the emitter bypass capacitor were completely eliminated, so that R_E functioned for both dc and ac. In that extreme case, the circuit would be very stable for dc and ac. Unfortunately, however, the gain would be reduced so low that the amplifier would be virtually useless. In effect, the stability imposed on the ac characteristics would be so extreme that almost no signal variation could occur.

A realistic and practically feasible compromise is to implement R_E with two separate resistances R_{E1} and R_{E2} and bypass one of the two resistances. This process is referred to as *emitter feedback,* and the resulting circuit is shown in Figure 10–14. For dc purposes, the circuit is still the stabilized bias circuit, and the required emitter resistance R_E for bias is simply

$$R_E = R_{E1} + R_{E2} \qquad\qquad (10\text{–}52)$$

The resistance R_{E2} is bypassed with the capacitor C_E, but R_{E1} is left unbypassed. In a typical situation, R_{E1} is much smaller than R_{E2}, so that only a small fraction of the total emitter resistance is left unbypassed. As we will see shortly, however, the relatively small value of R_{E1} will have a pronounced effect on the circuit operating characteristics.

Formulation of the small-signal midfrequency model follows the same procedure previously used with R_{E2} assumed to be bypassed completely by C_E. However, a major difference is that R_{E1} appears between the emitter terminal and ground as shown in Figure 10–15. The control voltage v_{be} is no longer equal to the input voltage v_i. Indeed, the input and output circuits are coupled not only

FIGURE 10–14
Common-emitter amplifier with emitter feedback.

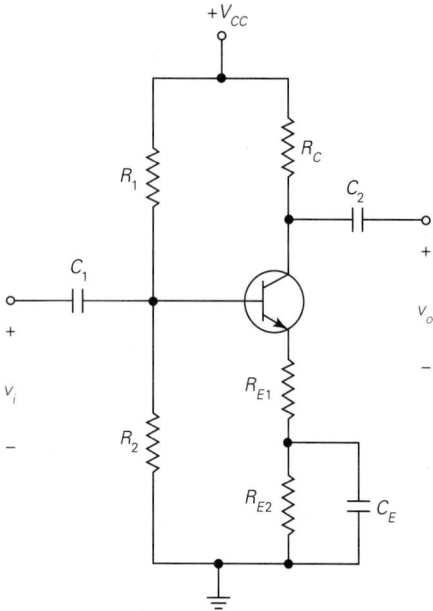

through g_m but also through R_{E1}. A KVL equation from ground to the base, through r_π, and back to ground through R_{E1}, reads

$$-v_i + v_{be} + R_{E1}i_e = 0 \qquad\qquad \textbf{(10–53)}$$

or

$$v_{be} = v_i - R_{E1}i_e \qquad\qquad \textbf{(10–54)}$$

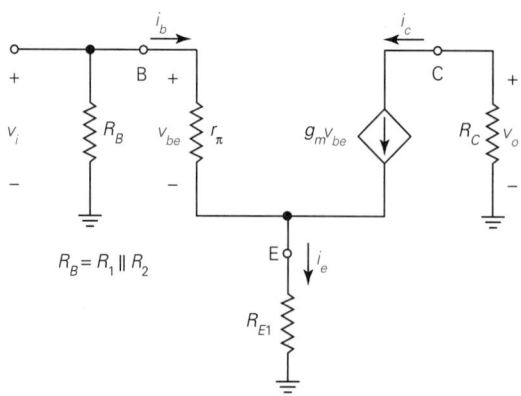

FIGURE 10–15
Small-signal midfrequency model of common-emitter amplifier with emitter feedback.

We will now briefly explain the process with reference to the model and (10–54). As v_i increases, v_{be} starts to increase, and the collector signal current $g_m v_{be}$ also starts to increase. This causes the emitter signal current i_e to increase, but this change tends to decrease v_{be} as noted by (10–54). The result is that the net increase in v_{be} is considerably less than when the signal voltage is applied directly between base and emitter.

Gain with Feedback

An expression for the gain A of the amplifier with emitter feedback will now be derived. Refer to the small-signal equivalent circuit of Figure 10–15. The collector signal current i_c is

$$i_c = g_m v_{be} \tag{10–55}$$

Since $v_{be} \neq v_i$, we must next substitute an expression for v_{be} in terms of v_i. Such an expression has already been determined and was given in (10–54). In practice, the signal emitter current i_e is very nearly equal to the signal collector current i_c (i.e., $i_e \simeq i_c$). Thus, (10–54) can be approximated as

$$v_{be} \simeq v_i - R_{E1} i_c \tag{10–56}$$

Substitution of (10–56) in (10–55) results in

$$i_c = g_m(v_i - R_{E1} i_c) = g_m v_i - g_m R_{E1} i_c \tag{10–57}$$

Note in (10–57) that i_c appears on both the left and the right sides of the equation. Regrouping and solving for i_c, we obtain

$$i_c = \frac{g_m v_i}{1 + g_m R_{E1}} \tag{10–58}$$

The output voltage v_o is then

$$v_o = -R_C i_C = \frac{-g_m R_C v_i}{1 + g_m R_{E1}} \tag{10–59}$$

Finally, the gain A is

$$A = \frac{v_o}{v_i} = \frac{-g_m R_C}{1 + g_m R_{E1}} \tag{10–60}$$

We will make comments on this result after deriving other parameters.

Input and Output Resistances

The input resistance will be derived in a two-step process. First, we will determine the resistance R'_{in}, in which case the effect of R_B will be ignored. The net resistance R_{in} will then be determined by combining R'_{in} with R_B.

For the circuit as given, the input current i_b can be expressed as

$$i_b = \frac{v_i - R_{E1}i_e}{r_\pi} \tag{10-61}$$

We next assume that $i_e \approx i_c$ and substitute the expression for i_c from (10–58) in (10–61). After a slight rearrangement, we have

$$i_b = \frac{v_i}{r_\pi}\left(1 - \frac{R_{E1}g_m}{1 + g_m R_{E1}}\right) = \frac{v_i}{r_\pi}\frac{1}{(1 + g_m R_{E1})} \tag{10-62}$$

The input resistance R'_{in} is then determined as

$$\boxed{R'_{in} = \frac{v_i}{i_b} = r_\pi(1 + g_m R_{E1})} \tag{10-63}$$

The net input resistance R_{in} is simply the parallel combination of R'_{in} and R_B; that is,

$$\boxed{R_{in} = R_B \| R'_{in} = R_B \| r_\pi(1 + g_m R_{E1})} \tag{10-64}$$

It can be shown that for the idealized model assumed, the output resistances R'_{out} and R_{out} are the same as for the amplifier circuit with emitter completely bypassed; that is,

$$\boxed{R'_{out} = \infty} \tag{10-65}$$

and

$$\boxed{R_{out} = R_C} \tag{10-66}$$

Comparison

We will now compare the common-emitter amplifier with the emitter resistance completely bypassed and with emitter feedback. A summary of the key formulas for the two circuits is provided in Table 10–1.

From the first line of the table, we observe that the magnitude of the gain is reduced by the factor $1/(1 + g_m R_{E1})$. This fact alone might be interpreted as a detrimental effect. However, as will be seen shortly, this is a small price to pay for the benefits that will result. A significant gain can still be achieved by a careful choice of R_{E1}. Note that the expression for gain with feedback reduces to the expression for gain without feedback when $R_{E1} = 0$.

From the second line of the table, note that the input resistance R'_{in} increases by a factor $(1 + g_m R_{E1})$. This is a very desirable effect and results in less loading

TABLE 10–1
Comparison of common-emitter properties with emitter resistance bypassed and with emitter feedback

	Emitter resistance bypassed	Emitter feedback
A	$-g_m R_C$	$\dfrac{-g_m R_C}{1 + g_m R_{E1}}$
R'_{in}	r_π	$r_\pi(1 + g_m R_{E1})$
R_{in}	$R_B \| r_\pi$	$R_B \| (1 + g_m R_{E1}) r_\pi$
R'_{out}	∞	∞
R_{out}	R_C	R_C

at the input. The net resistance R_{in} for both cases is the parallel combination of R_B and R'_{in}, so the increase in the net input resistance depends on the relationship between R_B and R'_{in}.

The output resistance based on the idealized model assumed is not affected by the feedback. It should be noted that if there is any shunt resistance associated with the dependent current source in the output, there will be a change in the output resistance resulting from feedback. However, that effect is not significant in most applications of this circuit and will be ignored.

Some of the major improvements resulting from feedback are not apparent from these equations. The most significant improvement is a drastic reduction in distortion. The increased input resistance at the base results in a more linear relationship between the input signal voltage and the input base current. Assuming a constant value of β_o, the resulting collector signal current is more nearly a replica of the input signal voltage.

A different way of explaining the same mechanism can be deduced by comparing the gain equations without and with feedback. We have
Without feedback:

$$A = -g_m R_C \tag{10–67}$$

With feedback:

$$A = \frac{-g_m R_C}{1 + g_m R_{E1}} \tag{10–68}$$

The transconductance g_m changes linearly with current (i.e., $g_m \simeq 40 I_C$). Without feedback the gain from (10–67) then varies around the operating point as the current changes, thus distorting the signal.

To observe the effect with feedback, let us momentarily assume that $g_m R_{E1} \gg 1$. With that assumption, (10–68) can be approximated as

$$A \approx \frac{-g_m R_C}{g_m R_{E1}} = \frac{-R_C}{R_{E1}} \tag{10–69}$$

The effect of g_m has been canceled out! Said differently, the denominator of (10–68) changes nearly in the same fashion as the numerator, so the net gain is stabilized. The approximation that $g_m R_{E1} \gg 1$ is not always valid, but even when it is not, the variation in the denominator of (10–68) definitely reduces the distortion.

Incidentally, the result of (10–69) provides a quick check of the approximate gain of the circuit without knowing the exact operating point. Although it may or may not be a very close approximation, depending on the product $g_m R_{E1}$, it does at least provide a ballpark estimate for rough checks.

To summarize, the common-emitter circuit with emitter feedback has less distortion and higher input resistance than the corresponding circuit with all the emitter resistance bypassed. It is also more stable from an ac point of view, and the gain is generally more predictable. The "price" that is paid for these improvements is a reduction in gain. As a general trade-off, the greater the sacrifice in gain, the greater are the improvements obtained.

EXAMPLE 10–8

The common-emitter amplifier of Example 10–7, which was shown in Figure 10–12(a), is modified to provide emitter feedback. The resulting circuit is shown in Figure 10–16. **(a)** Verify that the dc operating point is essentially the same as in Example 10–7. **(b)** Determine the midfrequency gain A and the input and output resistances R'_{in}, R'_{out}, R_{in}, and R_{out}.

Solution
(a) The only change in the dc circuit is that the original emitter resistance of 4.3 kΩ has now been split into two resistances of 3.9 kΩ and 390 Ω. Actually,

FIGURE 10–16
Amplifier circuit of Example 10–8.

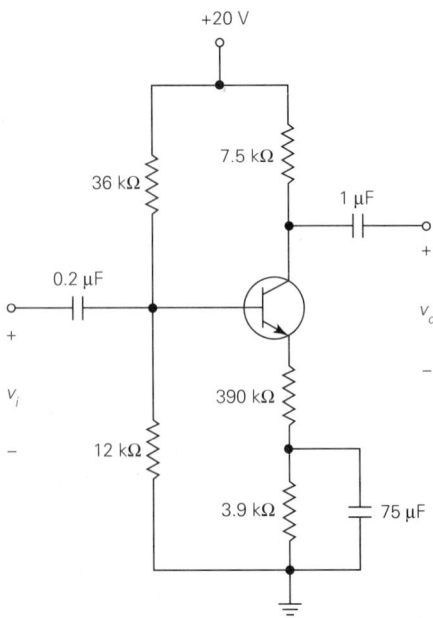

$3900 + 390 = 4290\ \Omega$, which appears to be $10\ \Omega$ short. However, both implementations have been based on assumed stock 5% resistance values, and when the uncertainty of the resistances and the approximations involved in the operating-point formulas are considered, the dc circuits are essentially equivalent. Thus, we assume that $I_C = 1$ mA and $g_m = 40$ mS as in Example 10–7.

(b) In the steps that follow, the results of Table 10–1 will be freely used. The unbypassed emitter resistance is $R_{E1} = 390\ \Omega$. In addition, we assume that $r_\pi = 6\ \text{k}\Omega$ as in Example 10–7. The gain A is

$$A = \frac{-g_m R_C}{1 + g_m R_{E1}} = \frac{-0.04 \times 7.5 \times 10^3}{1 + 0.04 \times 390} = \frac{-300}{1 + 15.6} = -18.1 \qquad \textbf{(10–70)}$$

The input resistance R'_{in} is

$$R'_{in} = r_\pi (1 + 0.04 \times 390) = 6 \times 10^3 (16.6) \approx 99.6\ \text{k}\Omega \qquad \textbf{(10–71)}$$

The net circuit input resistance is

$$R_{in} = R_B \| R'_{in} = 9\ \text{k}\Omega \| 99.6\ \text{k}\Omega = 8.25\ \text{k}\Omega \qquad \textbf{(10–72)}$$

The output resistances R'_{out} and R_{out} are the same as before; that is,

$$R'_{out} = \infty \qquad \textbf{(10–73)}$$

$$R_{out} = 7.5\ \text{k}\Omega \qquad \textbf{(10–74)}$$

A comparison of the amplifier properties without and with feedback is given below.

	Without feedback	With feedback
A	-300	-18.1
R'_{in} (kΩ)	6	99.6
R_{in} (kΩ)	4.5	8.25
R'_{out} (kΩ)	∞	∞
R_{out}(kΩ)	7.5	7.5

Several comments concerning these results are in order. The transistor input resistance R'_{in} is raised by the same factor as the gain is reduced ($1 + g_m R_{E1} = 16.6$). When this resistance is combined with R_B, the net input resistance is raised only from 4.5 to 8.25 kΩ, which seems like less of an improvement. However, this value of 8.25 kΩ is much more predictable and stable, since the effect of r_π is far less significant on the total input resistance when feedback is present. As discussed in the text, the distortion with the feedback circuit will be reduced dramatically, and this effect is not apparent from the data.

As one final check, let us see how closely (10–69) comes to predicting the actual gain. The ratio $-R_C/R_{E1}$ is

$$\frac{-R_C}{R_{E1}} = \frac{-7.5 \times 10^3}{390} = -19.23 \qquad \textbf{(10–75)}$$

This value differs from the actual gain by about 6%, so the utility of this result as a quick check is evident. Remember that this estimate always predicts a gain somewhat larger in magnitude than the actual gain.

10–6 COMMON-BASE AMPLIFIER

All BJT small-signal amplifier circuits considered up to this point have used the common-emitter configuration. However, it is also possible to implement amplifiers with either the base or the collector as the common terminal between the input and the output. Such configurations have properties that can be useful in certain applications. The **common-base** configuration is considered in this section, and the **common-collector** configuration is considered in the next section.

An ac-coupled common-base amplifier circuit is shown in Figure 10–17(a). This layout represents the usual manner in which the circuit is drawn. However, it will be useful at this point to show the circuit momentarily in a form that most easily relates to the common-emitter circuit, and this is provided in Figure 10–17(b). The dc circuit of the common-base configuration is equivalent to the common-emitter configuration. Thus, the operating point is established with R_1, R_2, R_E, and R_C exactly as for a common-emitter circuit.

The common-base circuit is different from the common-emitter circuit in two ways: (1) the signal is coupled to the emitter terminal by the capacitor C_1, and (2) the base terminal is established at signal ground by means of the capacitor C_B. However, the output signal is taken at the collector terminal in the same manner as in the common-emitter circuit.

Study the two forms in Figure 10–17 so that you can recognize the common-base circuit in either form. From this point on, we will use the form of Figure 10–17(a).

The midfrequency small-signal equivalent circuit of the common-base amplifier is shown in Figure 10–18. Note how the base terminal of the BJT is connected to ground and the control voltage v_{be} has its positive reference at ground. The input signal appears at the emitter terminal, and the dependent current source appears along the top.

The voltage gain A will now be derived. From Figure 10–18, the output voltage v_o is

$$v_o = -g_m v_{be} R_C \qquad \textbf{(10–76)}$$

However, the control voltage v_{be} is simply

$$v_{be} = -v_i \qquad \textbf{(10–77)}$$

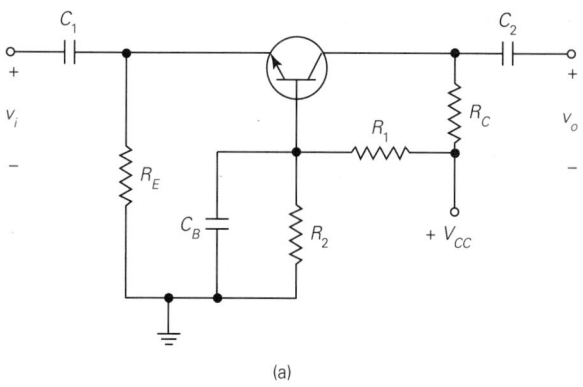

(a)

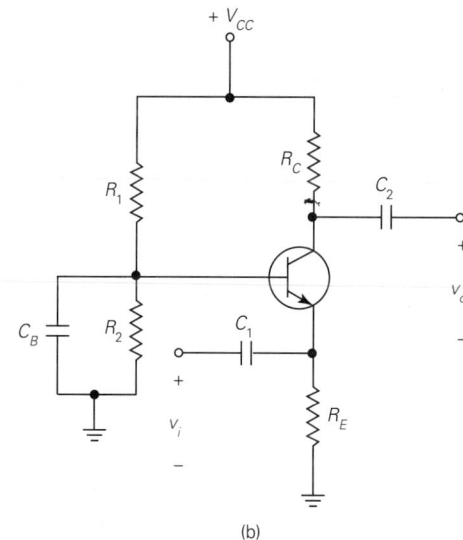

(b)

FIGURE 10–17
Common-base circuit drawn two ways.

FIGURE 10–18
Small-signal midfrequency model of common-base circuit.

Substitution of (10–77) in (10–76) yields

$$v_o = -g_m(-v_i)R_C = g_m R_C v_i \qquad \text{(10–78)}$$

The gain A is

$$\boxed{A = \frac{v_o}{v_i} = g_m R_C} \qquad \text{(10–79)}$$

This value is exactly the same as the magnitude of A for the common-emitter circuit. However, the gain of the common base is *noninverting*.

Next, the input resistance of the common-base amplifier will be derived. To that end, i_e is first calculated in terms of v_i. Application of KCL at the node to the right of the emitter terminal yields

$$-i_e + \frac{v_i}{r_\pi} - g_m v_{be} = 0 \qquad \text{(10–80)}$$

Substitution of $v_{be} = -v_i$ in (10–80) and rearrangement yield

$$v_i \left(\frac{1}{r_\pi} + g_m \right) = i_e \qquad \text{(10–81)}$$

The input resistance R'_{in} can be expressed as

$$R'_{in} = \frac{v_i}{i_e} = \frac{1}{(1/r_\pi) + g_m} \qquad \text{(10–82)}$$

The result of (10–82) is in the form of the reciprocal of the sum of two conductances, which is the formula for obtaining the parallel combination of two resistances. One resistance is r_π, and the other resistance is an equivalent resistance $1/g_m$. This quantity is defined in many texts as the dynamic emitter resistance r'_e, which is given by

$$\boxed{r'_e = \frac{1}{g_m}} \qquad \text{(10–83)}$$

However, we will simply refer to it as $1/g_m$.

The result of the preceding discussion is that R'_{in} can be expressed as the parallel combination of two resistances r_π and $1/g_m$; that is,

$$\boxed{R'_{in} = r_\pi \| \frac{1}{g_m}} \qquad \text{(10–84)}$$

The net circuit input resistance can be expressed as

$$R_{in} = R_E \| R'_{in} = R_E \| r_\pi \| \frac{1}{g_m} \qquad (10\text{--}85)$$

In many cases, the resistances R_E and r_π are much greater than $1/g_m$. In this common situation, the net input resistance is very close to

$$R_{in} \approx \frac{1}{g_m} \qquad (10\text{--}86)$$

This value is usually quite small compared to the input resistance of the common-emitter circuit.

The output resistances R'_{out} and R_{out} are

$$R'_{out} = \infty \qquad (10\text{--}87)$$

and

$$R_{out} = R_C \qquad (10\text{--}88)$$

The gain magnitude and the output resistance of the common-base circuit are the same as for the common-emitter circuit, but the input resistance is much lower. Because of this last property, the net voltage gain from source to output may be considerably lower than for the common-emitter circuit when the source resistance is appreciable because of the input loading factor. Thus, the low input resistance may be a serious limitation in many situations. For this reason, the common-base circuit finds only limited usage in low- and midfrequency applications.

The common-base circuit does offer some attributes at high frequencies that are not apparent from the low-frequency small-signal model. It turns out that there is an internal capacitance between the base and the collector of a transistor that tends to limit the upper frequency range in the common-emitter configuration. When the base is grounded for signals, this capacitance is isolated from the input, and the effect is reduced. By proper circuit design, it is possible to achieve stable amplifier gains at high frequencies in the common-base configuration.

EXAMPLE 10–9 Consider the *RC*-coupled common-base circuit shown in Figure 10–19. **(a)** Verify that the dc bias circuit is exactly the same as that of the common-emitter circuit of Example 10–7, which was shown in Figure 10–12(a). **(b)** Determine the mid-frequency gain A and the input and output resistances R'_{in}, R'_{out}, R_{in}, and R_{out}. Assume that the transistor has the same characteristics as in Example 10–7.

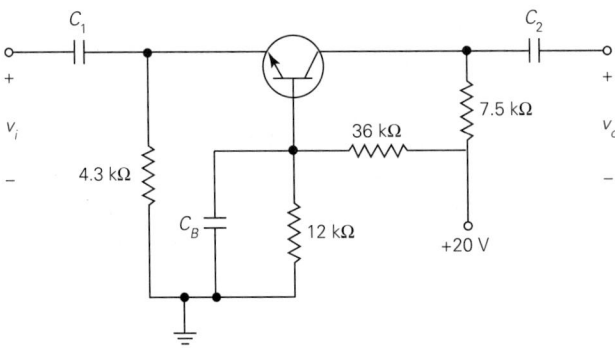

FIGURE 10–19
Common-base circuit of Example 10–9.

Solution

(a) By either rearranging the dc circuit form or comparing the two circuit diagrams of Figure 10–17, it is readily verified that the dc circuit is the same as in Figure 10–12(a). The dc collector current and the various dc voltages then are the same as in Example 10–7. The reader may wish to review the calculations from (10–40) through (10–46).

(b) Assuming that $g_m = 40$ mS, the gain A is

$$A = \frac{v_o}{v_i} = g_m R_C = 0.04 \times 7.5 \times 10^3 = 300 \qquad \text{(10–89)}$$

As in the preceding two examples, we will assume that $r_\pi = 6$ kΩ. The dynamic emitter resistance is $r'_e = 1/g_m = 1/0.04 = 25$ Ω. The transistor input resistance R'_{in} is

$$R'_{in} = r_\pi \| \frac{1}{g_m} = 6000 \| 25 = 24.9 \ \Omega \qquad \text{(10–90)}$$

The circuit input resistance R_{in} is

$$R_{in} = R_E \| R'_{in} = 4300 \| 24.9 \approx 24.8 \ \Omega \qquad \text{(10–91)}$$

Note how close this result is to $1/g_m = 25$ Ω, as given by the approximation of (10–86).

The output resistances are

$$R'_{out} = \infty \qquad \text{(10–92)}$$

and

$$R_{out} = 7.5 \ \text{k}\Omega \qquad \text{(10–93)}$$

10–7 COMMON-COLLECTOR AMPLIFIER

The third basic BJT configuration is the **common-collector amplifier,** which is also referred to as the **emitter-follower circuit.** The reader will recognize that we have already considered a dc-coupled version of this circuit in Chapter 5 as an isolation circuit. The circuit will now be considered in ac-coupled form and from the small-signal point of view.

The basic form of an ac-coupled common-collector circuit is shown in Figure 10–20. The dc biasing circuit is the same as for the common-emitter and common-base circuits except for one difference. There is no collector resistance in the common-collector circuit, and this establishes the collector at ground potential as far as ac signals are concerned (but obviously not for dc). The input signal is applied at the base terminal through a coupling capacitor, and the output signal is taken through a capacitor at the emitter terminal.

FIGURE 10–20
Common-collector (emitter-follower) ac-coupled amplifier circuit.

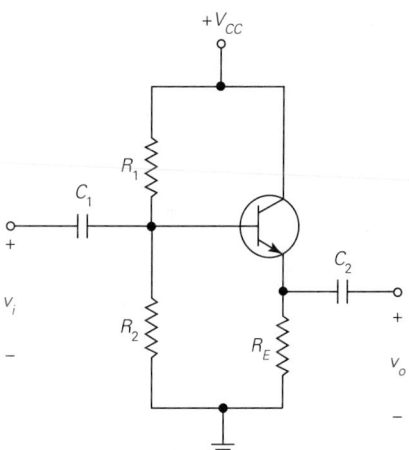

Determination of the operating point proceeds as follows: The voltage V_{TH} is

$$V_{\text{TH}} = \frac{R_2 V_{CC}}{R_1 + R_2}$$

(10–94)

For a stable bias design, the collector current is approximated as

$$I_C = \frac{V_{\text{TH}} - V_{BE}}{R_E} = \frac{V_{\text{TH}} - 0.7}{R_E}$$

(10–95)

The dc voltage V_{R_E} across the emitter resistance is approximately

$$V_{R_E} = R_E I_C \qquad (10\text{--}96)$$

The dc collector–emitter voltage V_{CE} is

$$V_{CE} = V_{CC} - R_E I_C \qquad (10\text{--}97)$$

The major difference between the dc circuits of the common-collector and common-emitter circuits is that the supply voltage V_{CC} is only divided into two parts for the common-collector circuit. Furthermore, since the output voltage is taken across the emitter, a logical division for many applications is about half of the available supply voltage across each of the two parts.

The midfrequency small-signal equivalent circuit of the common-collector circuit is first drawn as in Figure 10–21(a). Note how the collector terminal is grounded. It is best to redraw the circuit as in Figure 10–21(b) for the analysis that follows.

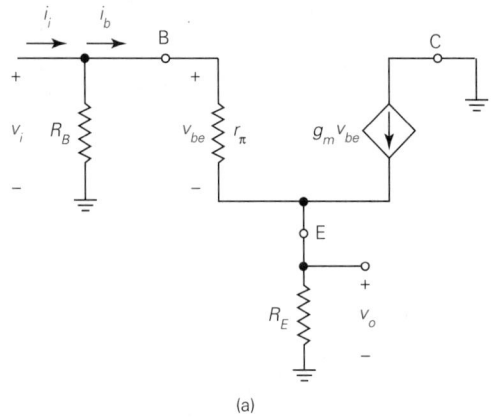

(a)

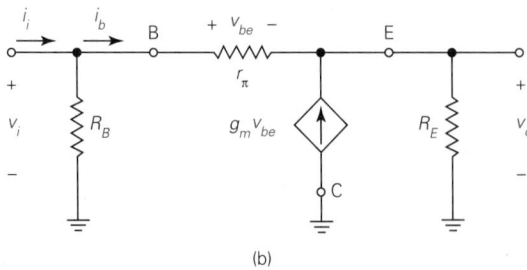

(b)

FIGURE 10–21
Small-signal midfrequency model of common-collector circuit as used to derive gain and input resistance.

The gain A will now be derived. A KVL loop around the complete circuit yields

$$-v_i + v_{be} + v_o = 0 \qquad \text{(10–98)}$$

or

$$v_{be} = v_i - v_o \qquad \text{(10–99)}$$

A KCL equation at the right-hand upper node reads

$$\frac{v_o}{R_E} + \frac{v_o - v_i}{r_\pi} - g_m v_{be} = 0 \qquad \text{(10–100)}$$

When the expression of v_{be} from (10–99) is substituted in (10–100), the resulting equation now contains only the variables v_i and v_o. After several steps of manipulation, the following result is obtained:

$$A = \frac{v_o}{v_i} = \frac{R_E/r_\pi + g_m R_E}{1 + R_E/r_\pi + g_m R_E} = \frac{R_E(1/r_\pi + g_m)}{1 + R_E(1/r_\pi + g_m)} \qquad \text{(10–101)}$$

For virtually any transistor, it turns out that $g_m \gg 1/r_\pi$. A very reasonable approximation for (10–101) is then

$$A = \frac{v_o}{v_i} = \frac{g_m R_E}{1 + g_m R_E} \qquad \text{(10–102)}$$

This expression will be retained as the basic gain equation for the common-collector configuration. However, it should be noted that in many applications, $g_m R_E \gg 1$, in which case A can be further approximated as

$$A \simeq 1 \qquad \text{(10–103)}$$

When this approximation is not used, the calculated gain turns out to be slightly less than unity.

Inasmuch as the voltage gain is slightly less than unity, one may logically question whether the common-collector circuit is really an "amplifier" or not. However, the virtue of the common-collector circuit is its isolation property. This will be evident from the results of input and output resistances.

The input resistance R'_{in} will now be derived. Referring to Figure 10–21(b), the base current i_b is

$$i_b = \frac{v_i - v_o}{r_\pi} \qquad \text{(10–104)}$$

From the gain equation of (10–102), v_o may be expressed in terms of v_i, and when this expression is substituted in (10–104), some simplification and rearrangement yield

$$i_b = \frac{1}{r_\pi} \frac{1}{(1 + g_m R_E)} v_i \qquad \text{(10–105)}$$

The transistor input resistance is

$$R'_{in} = \frac{v_i}{i_b} = r_\pi(1 + g_m R_E) \qquad \text{(10–106)}$$

The net circuit input resistance is

$$R_{in} = R_B \| R'_{in} = R_B \| r_\pi(1 + g_m R_E) \qquad \text{(10–107)}$$

Normally, the output resistance is determined with the input signal source replaced by a short circuit. Since we are assuming an ideal voltage source at the input at this time, the short circuit in the model would be in parallel with R_B and would, in effect, eliminate any effects of R_B. The result could be misleading, since many common-collector circuits are used in conjunction with high-resistance sources. To circumvent this difficulty and to permit the later inclusion of a source resistance, *the output resistance of the emitter follower will be determined with the source replaced by an open circuit.*

The circuit used to derive the output resistance is shown in Figure 10–22. As indicated, the input source is replaced by an open circuit, and this leaves R_B across the input. Although the circuit will not function without R_E, the small-signal model permits us to determine the resistance to the left of R_E, and this result can then be combined in parallel with R_E.

FIGURE 10–22
Midfrequency model used to derive R'_{out} for common-collector circuit.

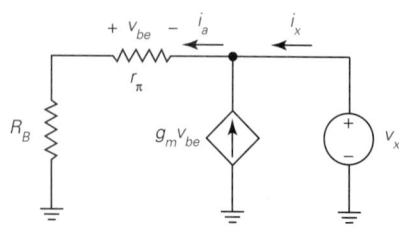

A fictitious source v_x is applied across the output, and the current i_x is to be determined. First, the current i_a is determined as

$$i_a = \frac{v_x}{r_\pi + R_B} \qquad \text{(10–108)}$$

The control voltage v_{be} is

$$v_{be} = -r_\pi i_a \qquad \text{(10–109)}$$

The current i_x is

$$i_x = i_a - g_m v_{be} \qquad \text{(10–110)}$$

Substitution of (10–109) in (10–110) yields

$$i_x = i_a + g_m r_\pi i_a = i_a(1 + g_m r_\pi) \qquad \text{(10–111)}$$

Next, (10–111) can be used to express i_a in terms of i_x, and this can be substituted in (10–108) to yield

$$\frac{i_x}{1 + g_m r_\pi} = \frac{v_x}{r_\pi + R_B} \qquad \text{(10–112)}$$

The output resistance R'_{out} is then determined as

$$R'_{\text{out}} = \frac{v_x}{i_x} = \frac{r_\pi + R_B}{1 + g_m r_\pi} \qquad \text{(10–113)}$$

The quantity $g_m r_\pi = \beta_o$ normally satisfies the inequality $\beta_o \gg 1$. With a slight approximation R'_{out} can be expressed as

$$R'_{\text{out}} = \frac{1}{g_m} + \frac{R_B}{\beta_o} \qquad \text{(10–114)}$$

If this signal source had been replaced by a short circuit, R'_{out} in (10–114) would simply be $1/g_m$, and this is the approximate value when the circuit is driven by an ideal voltage source.

The composite output resistance R_{out} is given by

$$R_{\text{out}} = R_E \| R'_{\text{out}} \qquad \text{(10–115)}$$

EXAMPLE 10–10   Consider the common-collector circuit shown in Figure 10–23. **(a)** Determine the operating-point dc collector current and circuit voltages. **(b)** Determine the gain A and input and output resistances. It is known that $\beta_o = 240$.

Solution
(a)   The Thévenin voltage V_{TH} is

$$V_{\text{TH}} = \frac{47 \text{ k}\Omega}{47 \text{ k}\Omega + 39 \text{ k}\Omega} \times 15 \text{ V} = 8.20 \text{ V} \qquad \text{(10–116)}$$

FIGURE 10–23

Common-collector circuit of Example 10–10.

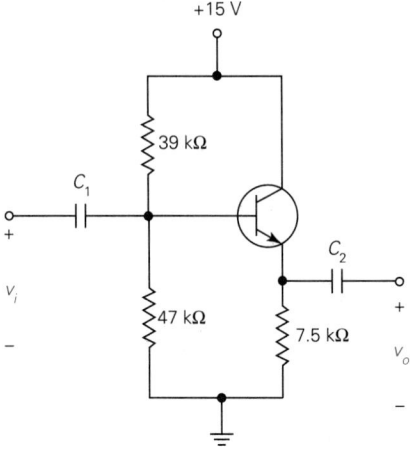

The collector current is

$$I_C = \frac{V_{TH} - V_{BE}}{R_E} = \frac{8.20 - 0.7}{7.5 \times 10^3} = 1 \text{ mA} \qquad \textbf{(10–117)}$$

The emitter dc voltage is

$$V_E = V_{R_E} = V_{TH} - 0.7 = 7.5 \text{ V} \qquad \textbf{(10–118)}$$

The collector–emitter voltage is

$$V_{CE} = 15 - 7.5 = 7.5 \text{ V} \qquad \textbf{(10–119)}$$

To the degree of accuracy of the approximations involved, the dc supply voltage is evenly distributed between the transistor and the emitter resistance.

(b) Since $I_C = 1$ mA, $g_m = 40 \times 10^{-3}$ S.
The value of the gain A is

$$A = \frac{g_m R_E}{1 + g_m R_E} = \frac{40 \times 10^{-3} \times 7.5 \times 10^3}{1 + 40 \times 10^{-3} \times 7.5 \times 10^3} = 0.997 \qquad \textbf{(10–120)}$$

This result is very close to unity.
The value of r_π is

$$r_\pi = \frac{\beta_o}{g_m} = \frac{240}{0.04} = 6 \text{ k}\Omega \qquad \textbf{(10–121)}$$

The input resistance R'_{in} is

$$R'_{in} = r_\pi(1 + g_m R_E) = 6 \times 10^3(301) = 1.81 \times 10^6 \; \Omega \qquad \textbf{(10–122)}$$

This large input resistance is one of the desirable properties of the common-collector circuit.

The output resistance R'_{out} with the *input open* is

$$R'_{\text{out}} = \frac{1}{g_m} + \frac{R_B}{\beta_o} = \frac{1}{0.04} + \frac{39 \times 10^3 \| 47 \times 10^3}{240} = 25 + \frac{21{,}314}{240} \qquad \textbf{(10–123)}$$

$$= 25 + 88.81 = 113.8 \ \Omega$$

This output resistance is quite small compared to the typical values obtained with the other configurations. Note again that the output resistance has been determined with the input terminals *open* rather than *shorted* as was the case for the other configurations. As we will see in Chapter 12, it will be much easier to incorporate the effect of a source resistance into the expression for output resistance from our assumed open-circuit condition.

The circuit input resistance is

$$R_{\text{in}} = R_B \| R'_{\text{in}} = R_1 \| R_2 \| R'_{\text{in}} = 39 \ \text{k}\Omega \| 47 \ \text{k}\Omega \| 1.81 \ \text{M}\Omega = 21 \ \text{k}\Omega \quad \textbf{(10–124)}$$

The circuit output resistance is

$$R_{\text{out}} = R_E \| R'_{\text{out}} = 7.5 \ \text{k}\Omega \| 113.8 \ \Omega = 112.1 \ \Omega \qquad \textbf{(10–125)}$$

10–8 BJT AC SPECIFICATIONS

Now that we have analyzed a number of basic amplifier configurations, it is appropriate to inspect some of the specifications pertinent to the analysis and design of such amplifiers. Refer again to Appendix D for the 2N3904 specifications sheets. We now observe certain of the ac or small-signal data values.

Some of the notation used in specifying small-signal data use the so-called **hybrid parameters.** The hybrid parameters were used extensively in the early developments of transistor theory and modeling. As the utilization of complete integrated circuits has increased, other methods have become more useful to the average technician and technologist. Consequently, hybrid parameters are included only as a supplement (Appendix C). However, since the data provided on specifications sheets still use the hybrid terminology, we must learn to recognize the symbols and convert the appropriate data to the symbols used in the text.

Refer to the section entitled "Dynamic Characteristics" on the second page of the 2N3904 specifications. Some of the data contained here will be considered in Chapter 18, when the complete hybrid-pi equivalent circuit is developed.

At this time, refer to the portion of the section entitled "Hybrid Parameters." The data provided here are based on $I_E = 1$ mA, $V_{CE} = 10$ V, and $f = 1$ kHz. This would correspond to midband operation in most cases.

The value of β_o is listed in the table as h_{fe}. Recall that β_{dc} was listed as h_{FE}, so there is some consistency between the corresponding symbols.

The range of h_{fe} is shown to be from 100 to 400 for the 2N3904. Thus, out of a large stock of these transistors, the value of h_{fe} (β_o) is expected to be within this wide range. This compares with a range of 100 to 300 for h_{FE} (β_{dc}) as noted on the first page of data. Thus, the range of the two quantities is similar, but the maximum value of h_{fe} can be greater.

The value of r_π is listed in the table as h_{ie}. The range of h_{ie} is shown to be from 1 to 10 kΩ for a 2N3904.

The quantities h_{re} and h_{oe} are neglected in our small-signal model. Their significance is discussed in Appendix C dealing with hybrid parameters. Considering the range of approximations involved with transistors, their effects are relatively small in most cases and are considered only in the most exacting analysis.

10–9 PSPICE EXAMPLES

PSPICE EXAMPLE 10–1

Devise a software test to determine the values of r_π, g_m, and β_o for the *npn* model of PSPICE at $IC = 1$ mA and $V_{CE} = 10$ V.

Solution

If we had no previous knowledge of the model, it would be necessary to step through various values of I_B to determine the exact value of I_B that would cause I_C to be 1 mA at $V_{CE} = 10$ V. However, that process was performed in PSPICE Example 4–1 to obtain the collector characteristics. From Figure 4–21, it appears that $I_B = 10$ μA is the approximate base current required, and we will use that value as a starting point.

The circuit used for this purpose is shown in Figure 10–24(a) and the code is shown in Figure 10–24(b). The scheme that we will use is to place a small ac

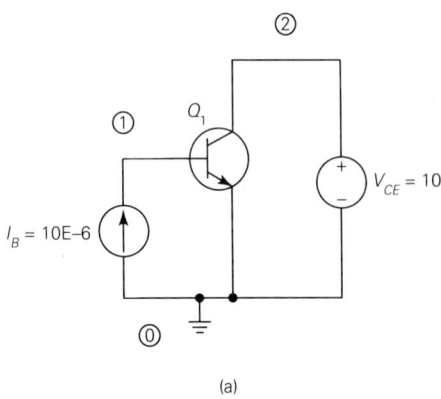

```
PSPICE EXAMPLE 10-1
.OPTIONS NOECHO NOPAGE NOMOD NOBIAS
IB 0 1 DC 10E-6 AC 1E-6
VCE 2 0 DC 10
Q1 2 1 0 QTEST
.MODEL QTEST NPN
.DC VCE 10 10 1
.PRINT DC IB(Q1) V(1) IC(Q1)
.AC LIN 1 1E3 1E3
.PRINT AC IB(Q1) V(1) IC(Q1)
.END
```

(a)   (b)

FIGURE 10–24
Circuit and code of PSPICE Example 10–1.

variation of the base current on top of the dc level. This is achieved through the following statement:

```
IB 0 1 DC 10E-6 AC 1E-6
```

Thus, the dc level is 10 μA (10E$-$6), and an ac level of 1 μA (1E$-$6) is added to this dc level. The dc collector voltage is set at 10 V.

To check the dc levels, the following print statement is used:

```
.PRINT DC IB(Q1) V(1) IC(Q1)
```

The quantity IB(Q1) is the dc base current of Q_1, and IC(Q1) is the corresponding collector current. The quantity V(1) is the dc base voltage with respect to ground.

The ac analysis is performed at a frequency of 1 kHz, which should be sufficiently low to avoid any high-frequency effects. The following ac print statement is used:

```
.PRINT AC IB(Q1) V(1) IC(Q1)
```

These are the same variables identified for the dc print, but in this case, the *ac* variables will be printed.

The computer printout is shown in Figure 10–25. The dc value of the base–emitter voltage is V(1) = 0.7742 V. The dc collector current is $IC(Q_1)$ = 1E$-$3 A = 1 mA, which is exactly the value desired.

Turning to the ac data, the ac value of the base current is 1 μA as established in the code, and the corresponding value of the ac collector current is $IC(Q_1)$ =

```
PSPICE EXAMPLE 10-1

****       CIRCUIT DESCRIPTION

******************************************************************

.OPTIONS NOECHO NOPAGE NOMOD NOBIAS

****       DC TRANSFER CURVES                TEMPERATURE = 27.000 DEG C

VCE        IB(Q1)     V(1)       IC(Q1)
1.000E+01  1.000E-05  7.742E-01  1.000E-03


****       AC ANALYSIS                       TEMPERATURE = 27.000 DEG C

FREQ       IB(Q1)     V(1)       IC(Q1)
1.000E+03  1.000E-06  2.586E-03  1.000E-04


           JOB CONCLUDED

           TOTAL JOB TIME          1.42
```

FIGURE 10–25
Computer printout of PSPICE Example 10–1.

1E$-$4 A. Thus, the value of β_o is

$$\beta_o = \frac{i_c}{i_b} = \frac{1 \times 10^{-4}}{1 \times 10^{-6}} = 100 \qquad \textbf{(10–126)}$$

Thus, for this somewhat idealized model, the ac β value is the same as the dc β values.

The value of r_π is determined from the ac values of base voltage and base current as follows:

$$r_\pi = \frac{v_{be}}{i_b} = \frac{\text{V(1)}}{\text{IB}(Q_1)} = \frac{2.586 \times 10^{-3}}{1 \times 10^{-6}} = 2586 \ \Omega \qquad \textbf{(10–127)}$$

The value of g_m is determined from the ac values of the collector current and base–emitter voltage

$$g_m = \frac{i_c}{v_{be}} = \frac{\text{IC}(Q_1)}{\text{V(1)}} = \frac{1 \times 10^{-4}}{2.586 \times 10^{-3}} = 38.7 \ \text{mS} \qquad \textbf{(10–128)}$$

This value compares very favorably with the value of $40I_C = 40 \times 10^{-3} = 40$ mS. (Remember that the value of $40I_C$ is simply a rounded approximation.) As an additional check, it can be readily verified that $g_m r_\pi = 38.7 \times 10^{-3} \times 2586 = 100 = \beta_o$.

PSPICE EXAMPLE 10–2

Develop a small-signal midfrequency model for the amplifier circuit of Example 10–7 (Figure 10–12[b]), and use the transfer function command to determine the voltage gain and overall input and output resistances.

Solution

The circuit adapted to PSPICE is shown in Figure 10–26(a), and the code is shown in Figure 10–26(b). Strictly speaking, the transfer function operation of PSPICE applies only to dc, but since the midfrequency model contains only resistances and linear components, it may be used with the model as given.

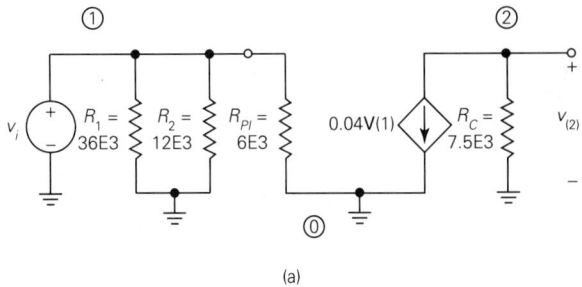

```
PSPICE EXAMPLE 10-2
.OPTIONS NOECHO NOPAGE NOBIAS
VI 1 0 1
R1 1 0 36E3
R2 1 0 12E3
RPI 1 0 6E3
GCE 2 0 1 0 0.04
RC 2 0 7.5E3
.TF V(2) VI
.END
```

(a)   (b)

FIGURE 10–26
Circuit and code of PSPICE Example 10–2.

The input signal is chosen as 1 V for convenience. Understand, however, that this value is meaningless for the actual circuit, since it would drive the transistor into saturation. However, this fictitious model will respond linearly to any software voltage level provided!

The model employs a voltage-controlled current source, for which the first letter must be G. This particular source is designated as GCE, and the code is as follows:

```
GCE 2 0 1 0 0.04
```

The first two integers 2 0 define the direction of current flow, which is from node 2 to node 0. The next two integers 1 0 define the nodes across which the controlling voltage is identified with the first node (1) representing the positive reference of that voltage. The last value 0.04 is the transconductance in siemens.

The transfer function command .TF identifies V(2) as the output voltage and VI as the input voltage. These values correspond, respectively, to v_o and v_i on the circuit diagram.

The computer printout is shown in Figure 10–27. The voltage gain is V(2)/VI = −300, which agrees exactly with the value in the text. The input resistance is 3600 Ω, and the output resistance is 7500 Ω. These values agree exactly with the text values for R_{in} and R_{out}, respectively. The values for R'_{in} and R'_{out} are not available directly, since a separate simulation without the presence of the circuit resistances would be required for this purpose.

```
PSPICE EXAMPLE 10-2

****        CIRCUIT DESCRIPTION

*********************************************************************

.OPTIONS NOECHO NOPAGE NOBIAS

****        SMALL-SIGNAL CHARACTERISTICS

V(2)/VI = -3.000E+02
INPUT RESISTANCE AT VI = 3.600E+03
OUTPUT RESISTANCE AT V(2) = 7.500E+03

    JOB CONCLUDED

    TOTAL JOB TIME          .38
```

FIGURE 10–27
Computer printout of PSPICE Example 10–2.

It is questionable whether we gain much from a computer simulation of this simple circuit configuration, but the next example will tell a different story.

PSPICE EXAMPLE 10–3	Develop a small-signal midfrequency model for the amplifier circuit of Example 10–8 (Figure 10–16), and use the transfer function command to determine the voltage gain and overall input and output resistances.

Solution

The general form of the midfrequency model with a portion of the emitter resistance bypassed was shown in Figure 10–15. This model adapted to PSPICE along with the element values of Example 10–8 are shown in Figure 10–28(a), and the code is shown in Figure 10–28(b).

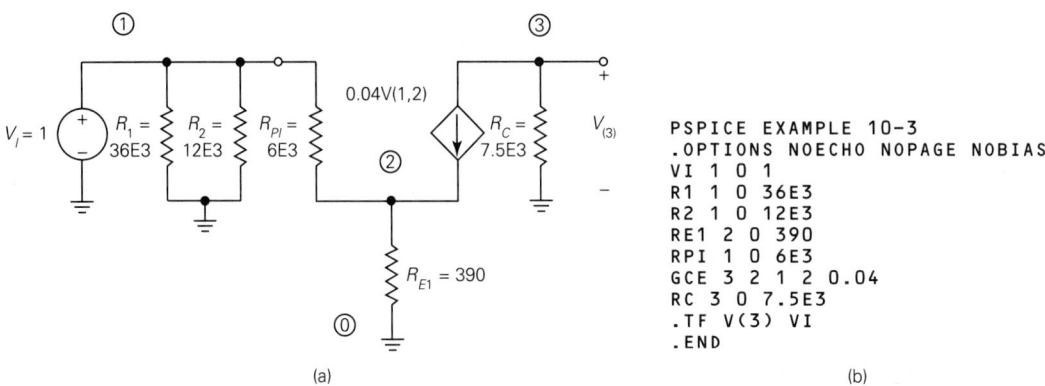

```
PSPICE EXAMPLE 10-3
.OPTIONS NOECHO NOPAGE NOBIAS
VI 1 0 1
R1 1 0 36E3
R2 1 0 12E3
RE1 2 0 390
RPI 1 0 6E3
GCE 3 2 1 2 0.04
RC 3 0 7.5E3
.TF V(3) VI
.END
```

(a)   (b)

FIGURE 10–28
Circuit and code of PSPICE Example 10–3.

The coding follows a pattern very similar to that of PSPICE Example 10–2 except for one significant difference. The controlling voltage in the present example is the voltage between nodes 1 and 2, and this results in a major difference in the overall behavior. A portion of the output current source flows through the emitter resistance R_{E1}, and the resulting voltage is a portion of the controlling voltage. Thus, the circuit contains *feedback,* a major topic to be studied later.

The computer output data are shown in Figure 10–29. The voltage gain is

```
PSPICE EXAMPLE 10-3

****        CIRCUIT DESCRIPTION

*****************************************************************

.OPTIONS NOECHO NOPAGE NOBIAS

****        SMALL-SIGNAL CHARACTERISTICS

V(3)/VI = -1.800E+01
INPUT RESISTANCE AT VI = 8.257E+03
OUTPUT RESISTANCE AT V(3) = 7.500E+03

    JOB CONCLUDED

    TOTAL JOB TIME           .39
```

FIGURE 10–29
Computer printout of PSPICE Example 10–3.

$V(3)/VI = -18$, which compares closely with the calculated value of -18.1 in the text. The input and output resistances are 8257 Ω and 7500 Ω, respectively. The first value compares very closely with the calculated value of 8250 Ω, and the second value is exactly the same as the text value.

PSPICE EXAMPLE 10–4

For the circuit of PSPICE Example 10–8 (Figure 10–16), prepare a program using the 2N2222A library model to simulate the complete circuit and determine various dc and ac voltages and currents directly from PSPICE.

Solution
The circuit adapted to PSPICE is shown in Figure 10–30(a), and the corresponding code is shown in Figure 10–30(b). While the program is probably the longest we have encountered thus far in the text, it should be straightforward to follow. Bear in mind that for a circuit this complex, it is unlikely that any two persons would use the exact same strategy and order of node numbering.

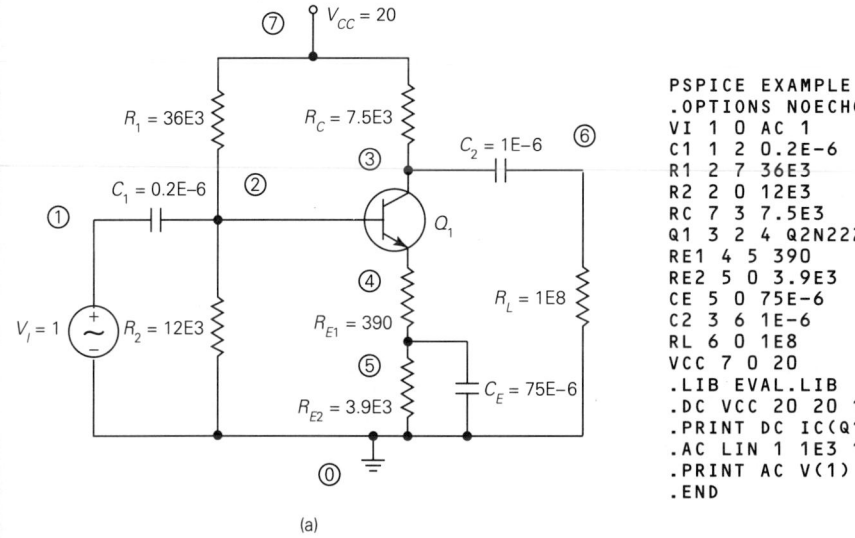

```
PSPICE EXAMPLE 10-4
.OPTIONS NOECHO NOPAGE NOBIAS NOMOD
VI  1 0 AC 1
C1  1 2 0.2E-6
R1  2 7 36E3
R2  2 0 12E3
RC  7 3 7.5E3
Q1  3 2 4 Q2N2222A
RE1 4 5 390
RE2 5 0 3.9E3
CE  5 0 75E-6
C2  3 6 1E-6
RL  6 0 1E8
VCC 7 0 20
.LIB EVAL.LIB
.DC VCC 20 20 1
.PRINT DC IC(Q1) V(4,0) V(RC) V(3,4)
.AC LIN 1 1E3 1E3
.PRINT AC V(1) V(2) V(5) V(3) V(6)
.END
```

(a)   (b)

FIGURE 10–30
Circuit and code of PSPICE Example 10–4.

To minimize the size of the output, four options have been employed, although any of these could be omitted if more data were required.

One point that should be emphasized is that even though dc voltages and currents will be determined based on the 20-V dc source, PSPICE will obtain a linear small-signal equivalent circuit for the ac analysis. Thus, the choice of 1 V for the ac input is still valid, but in the actual circuit this would represent an excessive drive. The resulting ac output voltage must then be interpreted in the light of the small equivalent circuit model.

The transistor in the text had an assumed value of $\beta_o = 240$, and no value of β_{dc} was specified. While it would be possible to modify the 2N2222A model to conform more exactly with the text model, as a matter of interest, we will use the model without alteration.

The computer output data are shown in Figure 10–31. A comparison of the dc values of various variables is provided in the following table:

	IC(mA)	V_{R_E}(V)	V_{R_C}(V)	V_{C_E}(V)
Text values	1	4.3	7.5	8.2
PSPICE values	0.9964	4.301	7.473	8.226

```
PSPICE EXAMPLE 10-4

****      CIRCUIT DESCRIPTION

**********************************************************************

.OPTIONS NOECHO NOPAGE NOBIAS NOMOD

****      DC TRANSFER CURVES                TEMPERATURE = 27.000 DEG C

VCC        IC(Q1)      V(4,0)      V(RC)       V(3,4)
2.000E+01  9.964E-04   4.301E+00   7.473E+00   8.226E+00


****      AC ANALYSIS                       TEMPERATURE = 27.000 DEG C

FREQ       V(1)        V(2)        V(5)        V(3)        V(6)
1.000E+03  1.000E+00   9.947E-01   5.044E-03   1.772E+01   1.772E+01


          JOB CONCLUDED

          TOTAL JOB TIME      1.70
```

FIGURE 10–31
Computer printout of PSPICE Example 10–4.

All of the PSPICE dc values agree with the text values to a tolerance well below 1%. This illustrates how immune a well-designed bias circuit is to parameter value changes even when the transistors are somewhat different.

The ac values for the different variables are shown at the bottom of the printout. Starting with V(1), which represents the input signal voltage, the various ac voltages can be traced from left to right in much the same manner as with an ac voltmeter in an actual circuit. The ac voltage at the base is V(2) = 0.9947 V.

Although this value differs from the input voltage by about 0.5%, remember that our test is at 1 kHz. If this amplifier must have a good low-frequency response (e.g., <100 Hz), this result would suggest that the value of C_1 should be increased. The ac voltage at the top of the bypass capacitor is $V(5) = 5.044$ mV. Theoretically, this value should approach zero as the value of C_E increases without limit. The value of the ac collector voltage is $V(3) = 17.72$ V, which is the same as the ac output voltage in view of the extremely large output resistance coupled with C_2. Once again, note that this voltage is a result of the linearized model, since the real circuit could not support an ac voltage this large.

The magnitude of the gain A is given by

$$|A| = \frac{V(6)}{V(1)} = \frac{17.72}{1} = 17.72 \qquad (10\text{--}129)$$

This value differs by only about 2% from the text magnitude value of 18.1. (The phase angle of the output voltage was not requested in the printout, but it would be close to $\pm 180°$.)

PSPICE EXAMPLE 10–5

For the circuit of PSPICE Example 10–4, modify the program to provide a transient response over two cycles of a sine wave input at 1 kHz. Reduce the input signal level so that the output is relatively distortion-free.

Solution

The circuit is essentially the same as shown in Figure 10–30(a) except for the signal level, and the new code is shown in Figure 10–32. A sinusoidal input of 0.1 V peak has been arbitrarily chosen. This level is representative of an actual level used in a real circuit of this nature. Note that the transient response is taken over an interval from $t = 0$ to $t = 2$ ms, which represents two cycles.

Using the available plot options of PROBE, traces of the input and output voltages are shown in Figure 10–33. The output voltage $V(6)$ is seen to be 180° out of phase with the input, which is expected for a common-emitter amplifier.

FIGURE 10–32
Code for PSPICE Example 10–5.

```
PSPICE EXAMPLE 10-5
.OPTIONS NOECHO NOPAGE NOBIAS NOMOD
VI 1 0 SIN 0 0.1 1E3
C1 1 2 0.2E-6
R1 2 7 36E3
R2 2 0 12E3
RC 7 3 7.5E3
Q1 3 2 4 Q2N2222A
RE1 4 5 390
RE2 5 0 3.9E3
CE 5 0 75E-6
C2 3 6 1E-6
RL 6 0 1E8
VCC 7 0 20
.LIB EVAL.LIB
.TRAN 2E-3 2E-3
.PROBE
.END
```

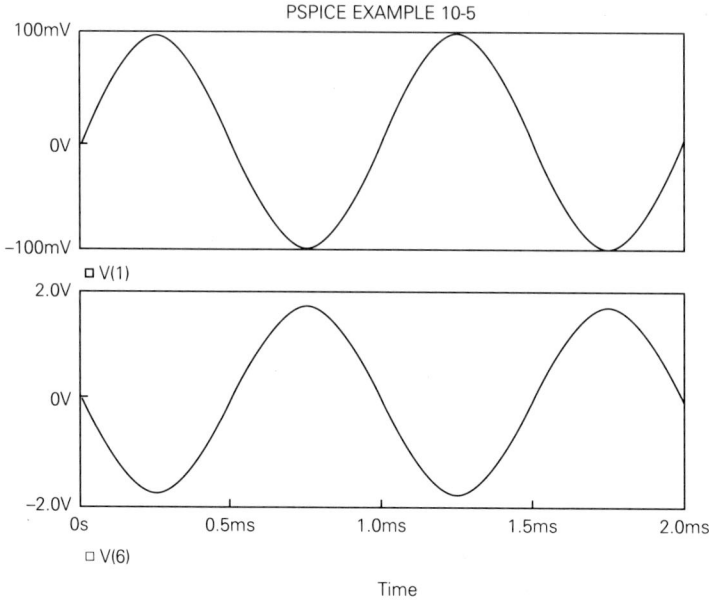

FIGURE 10–33
Input and output voltages of PSPICE Example 10–5.

The peak value of the output is about 1.8 V, which is also expected for an input peak value of 0.1 V, since the gain magnitude is about 18.

PSPICE EXAMPLE 10–6

To illustrate the phenomenon of distortion, repeat the transient analysis of PSPICE Example 10–5 with the input level increased to 1 V.

Solution
The code is not repeated, since the only change required is the ac source input, which is changed as follows:

```
VI 1 0 SIN 0 1 1E3
```

Plots of the collector voltage V(3) and the output voltage V(6) are shown in Figure 10–34. The collector voltage cannot rise above the power supply voltage without the presence of an inductor or a transformer, so it clips at an upper level of 20 V as shown. Conversely, the collector voltage cannot drop below the emitter dc voltage of about 4.3 V, so it clips at a lower level of that value. The first value corresponds to cutoff for the transistor, and the second value corresponds to saturation for the transistor.

The output voltage V(6) displays the same shape as V(3), but the coupling capacitor C_2 removes the dc level. Obviously, this circuit cannot handle a peak ac voltage of 1 V without severe distortion.

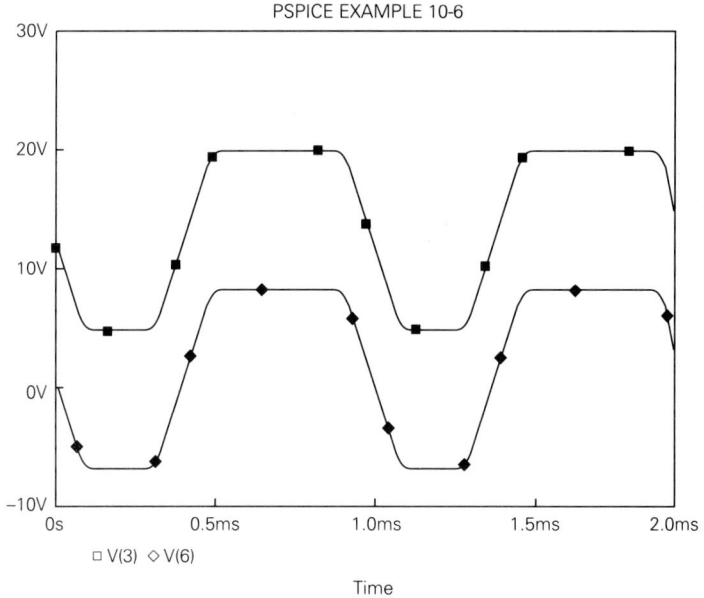

PSPICE EXAMPLE 10-6

□ V(3) ◇ V(6)

Time

FIGURE 10–34
Collector and output voltages of PSPICE Example 10–6.

PROBLEMS

Drill Problems

10–1. A simple BJT amplifier of the form shown in Figure 10–1 has waveforms as shown in Figures 10–3 and 10–4. The peak-to-peak values of input and output voltages are measured as V_{ipp} = 40 mV and V_{opp} = 2 V. Determine the voltage gain A.

10–2. A simple BJT amplifier of the form shown in Figure 10–1 has waveforms as shown in Figures 10–3 and 10–4. The rms values of input and output voltages are measured as $V_{i(\text{rms})}$ = 10 mV and $V_{o(\text{rms})}$ = 1.5 V. Determine the voltage gain A.

10–3. Input voltage–current data for a certain *npn* transistor are as follows:

V_{BE} (V)	I_B (μA)
0.68	120
0.70	200

Determine the approximate value of r_π.

10–4. Input voltage–current data for a certain *pnp* transistor are as follows:

V_{EB} (V)	I_B (μA)
0.64	90
0.65	110

Determine the approximate value of r_π.

10–5. Some dc measurements on a BJT with a fixed collector–emitter voltage yield the following data:

I_B (μA)	I_C (mA)
70	6.1
100	7.9

Determine the approximate value of β_o.

10–6. Some dc measurements on a BJT with fixed collector–emitter voltage yield the following data:

I_B (μA)	I_C (mA)
40	4.6
50	8.2

Determine the approximate value of β_o.

10–7. A certain *npn* BJT is biased at $I_C = 8$ mA, and it is known that $\beta_o = 150$. Determine **(a)** g_m, and **(b)** r_π.

10–8. A certain *pnp* BJT is biased at $I_C = 5$ mA, and it is known that $\beta_o = 200$. Determine **(a)** g_m, and **(b)** r_π.

10–9. For the transistor of Problem 10–7 at the given operating point, construct the β-model and the g_m-model for small signals.

10–10. For the transistor of Problem 10–8 at the given operating point, construct the β-model and the g_m-model for small signals.

10–11. The schematic diagram of a certain RC-coupled common-emitter amplifier is shown in Figure P10–11. **(a)** Determine the dc operating point current I_C and the three voltages V_{R_E}, V_{R_C}, and V_{CE}. **(b)** Construct the midfrequency small-signal equivalent circuit. It is known that $\beta_o = 200$. **(c)** Determine the gain A and the input and output resistances R'_{in}, R'_{out}, R_{in}, and R_{out}.

FIGURE P10–11

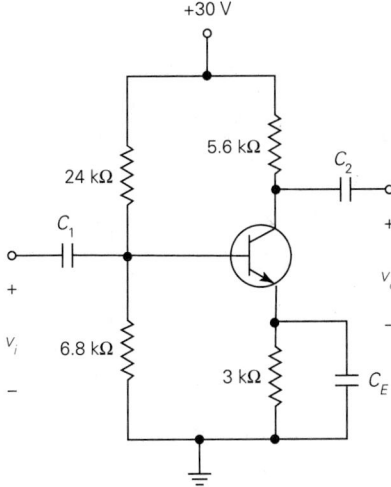

10–12. The schematic diagram of a certain *RC*-coupled common-emitter amplifier employing a *pnp* transistor and a positive bias supply is shown in Figure P10–12. **(a)** Determine the dc operating-point current I_C and the three voltages V_{R_E}, V_{R_C}, and V_{EC}. (Assume reference directions for first two voltages that will yield positive values.) **(b)** Construct the midfrequency small-signal equivalent circuit. It is known that $r_\pi = 4 \text{ k}\Omega$. **(c)** Determine the gain A and the input and output resistances R'_{in}, R'_{out}, R_{in}, and R_{out}.

FIGURE P10–12

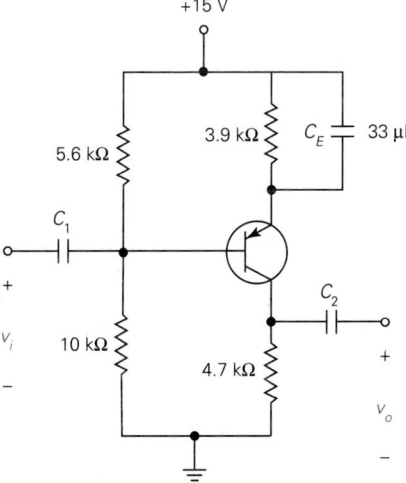

10–13. The common-emitter amplifier of Problem 10–11 is modified to provide emitter feedback, and the resulting circuit is shown in Figure P10–13. **(a)** Verify that the dc operating point is the same as in Problem 10–11. **(b)** Determine the midfrequency gain A and the input and output resistances R'_{in}, R'_{out}, R_{in}, and R_{out}.

FIGURE P10–13

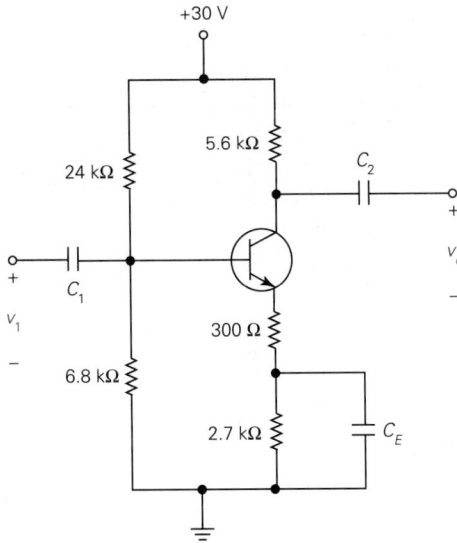

10–14. The common-emitter amplifier of Problem 10–12 is modified to provide emitter feedback, and the resulting circuit is shown in Figure P10–14. **(a)** Verify that the dc operating point is the same as in Problem 10–12. **(b)** Determine the midfrequency gain A and the input and output resistances R'_{in}, R'_{out}, R_{in}, and R_{out}.

FIGURE P10–14

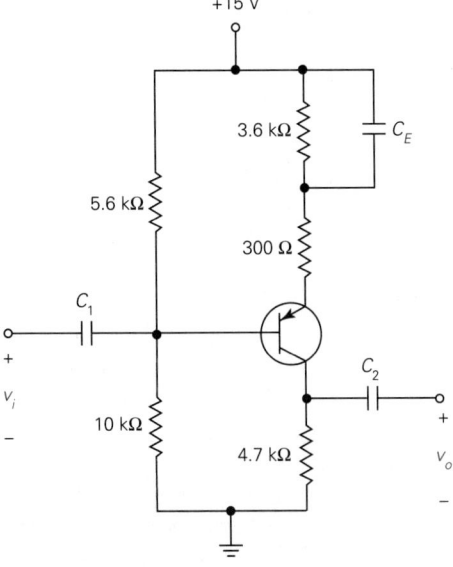

10–15. Consider the RC-coupled common-base circuit shown in Figure P10–15. **(a)** Verify that the dc bias circuit is exactly the same as that of the common-emitter circuit of Problem 10–11. **(b)** Determine the midfrequency gain A and the input and output resistances R'_{in}, R'_{out}, R_{in}, and R_{out}.

FIGURE P10–15

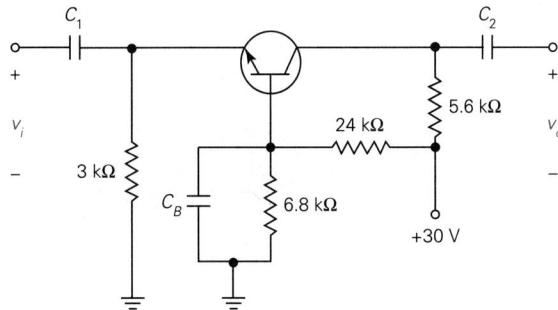

10–16. Consider the *RC*-coupled common-base circuit shown in Figure P10–16. **(a)** Verify that the dc bias circuit is the same as that of the common-emitter circuit of Problem 10–12. **(b)** Determine the midfrequency gain *A* and the input and output resistances R'_{in}, R'_{out}, R_{in}, and R_{out}.

FIGURE P10–16

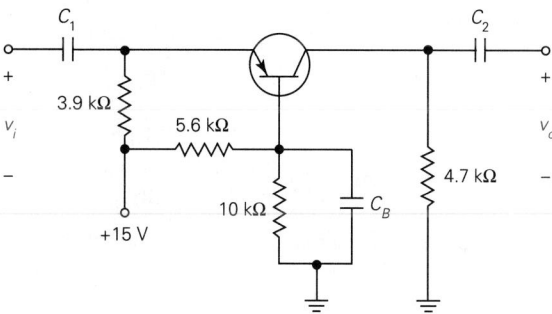

10–17. Consider the common-collector circuit shown in Figure P10–17. **(a)** Determine the dc operating-point collector current I_C and the voltages V_{R_E} and V_{CE}. **(b)** Determine the gain *A* and the input and output resistances R'_{in}, R'_{out}, R_{in}, and R_{out}.

FIGURE P10–17

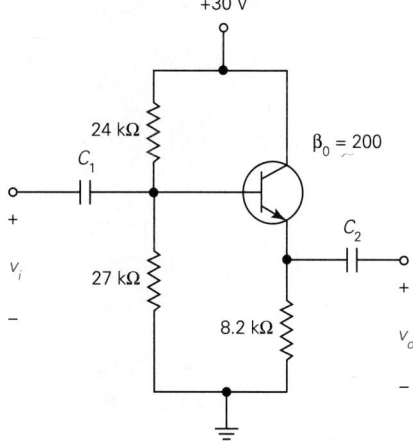

10–18. Consider the common-collector circuit shown in Figure P10–18. **(a)** Determine the dc operating-point collector current I_C and the voltages V_{R_E} and V_{EC}. **(b)** Determine the gain A and the input and output resistances R'_{in}, R'_{out}, R_{in}, and R_{out}.

FIGURE P10–18

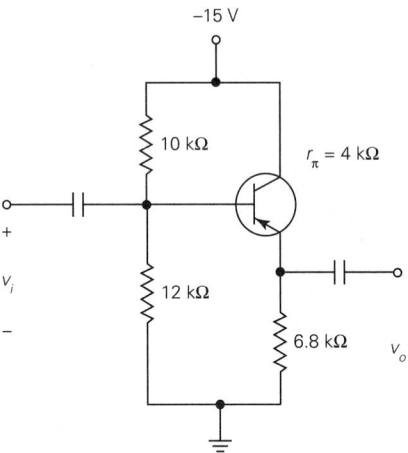

Derivation Problems

10–19. Consider the common-emitter circuit of Figure 10–10(a). Assume that the dc operating-point voltages are established as follows: $V_{CE} = V_{R_C} = V_{R_E}$. Show that the gain A is given by

$$A = -13.33 V_{CC}$$

10–20. Consider the common-emitter circuit of Figure 10–10(a). Assume that the dc operating-point voltages are established as follows: $V_{CEQ} = V_{R_C} = 2V_{R_E}$. Show that the gain A is given by

$$A = -16 V_{CC}$$

Design Problems

10–21. One possible approach to designing a common-emitter amplifier with feedback to achieve a specified gain is as follows: First, design a stable dc biasing circuit based on a desired I_C using the techniques of Chapter 5. One of the parameters determined from this process is R_E. Next, use (10–60) to determine the value of R_{E1} required to achieve the gain A. R_E can then be divided into two parts: R_{E1} and $R_{E2} = R_E - R_{E1}$. (In some cases, R_{E1} may be so small that the approximation $R_{E2} \approx R_E$ may be valid.) Assume that the following design constraints are given:

$$V_{CC} = 30 \text{ V}$$
$$V_{CE} = V_{R_C} = V_{R_E} = 10 \text{ V}$$
$$I_C = 1 \text{ mA}$$
$$A = -12$$

Assume $\beta_{dc}(\text{min}) = 50$ and $K = 20$. Perform the design by specifying all resistance values.

10–22. Repeat the design of Problem 10–21 if the dc voltage distribution and current are changed to the following:

$$V_{CE} = V_{R_C} = 12 \text{ V}$$
$$V_{R_E} = 6\text{V}$$
$$I_C = 2 \text{ mA}$$

Troubleshooting Problems

10–23. You are troubleshooting a common-emitter amplifier of the type shown in Figure P10–11 with an applied signal source. All dc levels seem satisfactory. The ac level at the base seems satisfactory, but the ac level at the collector is much too small. Which one of the following would likely be the trouble? **(a)** Collector resistance too small; **(b)** emitter resistance too large; **(c)** shorted emitter bypass capacitor; **(d)** open emitter bypass capacitor; **(e)** defective transistor.

10–24. You are troubleshooting a common-emitter amplifier of the type shown in Figure P10–11 with an applied signal source. The source has a dc path to ground through its output. The signal appears at the base, but there is no output signal. The dc voltages measured at the base and collector terminals are 0 and 30 V, respectively. Which one of the following would likely be the trouble? **(a)** Shorted collector resistor; **(b)** open 6.8-kΩ resistor; **(c)** shorted input coupling capacitor; **(d)** open emitter bypass capacitor; **(e)** shorted emitter resistor.

11

FET SMALL-SIGNAL AMPLIFIER CIRCUITS

OBJECTIVES

After completing this chapter, the reader should be able to:

- Explain the process by which amplification occurs in an FET and compare it with the process for a BJT.
- Define and be able to determine transconductance for both depletion-mode and enhancement-mode FETs.
- Construct for an FET the small-signal flat-band circuit model.
- Draw the schematic diagram for and explain the operation of a complete common-source amplifier with source resistance bypassed.
- Analyze the circuit of the preceding objective for gain, input resistance, and output resistance.
- Draw the schematic diagram for and explain the operation of a common-source amplifier with source feedback.
- Analyze the circuit of the preceding objective for gain, input resistance, and output resistance.
- Draw the schematic diagram for and explain the operation of a common-gate amplifier.
- Analyze the circuit of the preceding objective for gain, input resistance, and output resistance.
- Draw the schematic diagram for and explain the operation of a common-drain amplifier.
- Analyze the circuit of the preceding objective for gain, input resistance, and output resistance.
- Analyze some of the circuits in the chapter with PSPICE.

11-1 OVERVIEW

This chapter has essentially the same objectives for FET amplifiers as Chapter 10 did for BJT amplifiers. A detailed treatment of FET small-signal or ac parameters is considered, and a small-signal model is developed.

The three basic amplifier configurations using an FET are denoted as the **common-source amplifier,** the **common-gate amplifier,** and the **common-drain amplifier,** respectively. The latter circuit is also known as the **source-follower circuit,** and a dc version of this circuit is considered in Chapter 7. We will consider all of these basic amplifier circuits in some detail, in addition to an improved version of the common-source amplifier utilizing source feedback.

Each of the basic FET configurations has very close similarities to one of the basic BJT configurations. Because of these, the results of the various detailed derivations developed in Chapter 10 for BJT circuits can be adapted to the corresponding FET circuits. Except for a few of the more basic and simpler cases, which will be developed fully in this chapter, a common approach of this chapter is to adapt the results of Chapter 10 to FET circuits.

11-2 SMALL-SIGNAL FET MODELS

In developing the basic form of a BJT amplifier in Chapter 10, we used a load-line approach to convey a graphical description of the process. A similar approach could be used for an FET amplifier, but the basic concept is virtually the same. The major difference in such an approach for an FET would be that the variations in input signal would be directly related to changes in gate–source voltage rather than to changes in base current, which was the case for the BJT. We sidestep that approach for the FET and go directly to the small-signal models.

Consider the n-channel JFET shown in Figure 11–1. No external connections are shown here, since the intent at this point is to focus on the transistor alone. In accordance with earlier notation, the instantaneous gate–source voltage, drain current, and drain–source voltage are identified.

FIGURE 11–1

Junction FET and total instantaneous variables used in establishing circuit models.

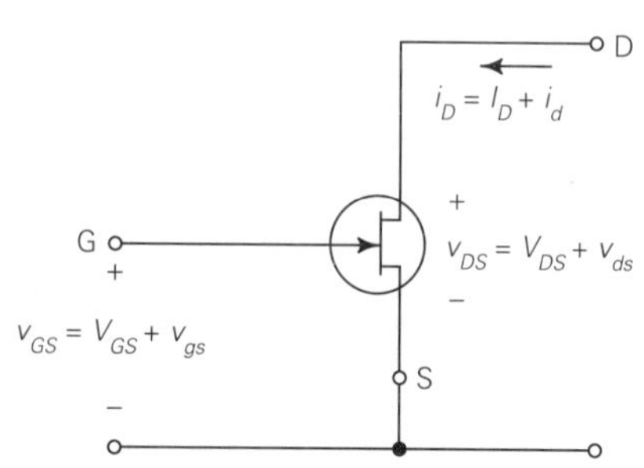

$i_D = I_D + i_d$

$v_{DS} = V_{DS} + v_{ds}$

$v_{GS} = V_{GS} + v_{gs}$

The gate–source voltage can be expressed as

$$v_{GS} = V_{GS} + v_{gs}$$ (11–1)

where V_{GS} is the dc component and v_{gs} is the ac component. The drain current is expressed as

$$i_D = I_D + i_d$$ (11–2)

where I_D is the dc component and i_d is the ac component. Finally, v_{DS} is the total drain–source voltage, and it is

$$v_{DS} = V_{DS} + v_{ds}$$ (11–3)

where V_{DS} is the dc component and v_{ds} is the ac component.

One noticeable difference between the FET and the BJT is that no input current is assumed for the FET. While the base current in a BJT is both sizable and significant to the operation of the device, the gate current is, for most applications, insigificant in an FET, and it serves no major function in the operation. At this point it will be assumed to be zero.

Transconductance

Refer to Figure 11–2, in which a portion of two curves of the drain characteristics of an FET are shown. Assume that the curve $V_{GS} = V_{GSQ}$ corresponds to the dc operating-point gate–source voltage, and the corresponding drain current is $I_D = I_{DQ}$. Assume that the gate–source voltage increases by a positive value ΔV_{GS},

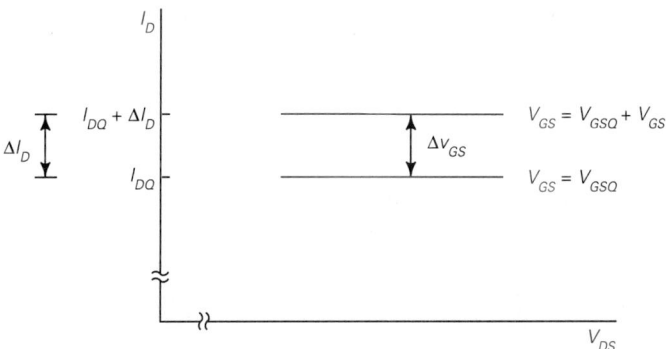

FIGURE 11–2
FET drain characteristics at operating point and changes.

which shifts operation to the curve corresponding to $V_{GS} = V_{GSQ} + \Delta V_{GS}$. The corresponding change in the drain current is a positive shift ΔI_D, so the new drain current is $I_{DQ} + \Delta I_D$.

A dynamic flat-band transconductance g_m is defined as

$$g_m = \frac{\Delta I_D}{\Delta V_{GS}}$$

(11–4)

A constant value for V_{DS} is assumed in this definition, but since the ideal curves are assumed to show no variation with V_{DS} in the beyond-pinchoff region, this point need not concern us now.

Recalling the procedure used for a BJT, we first determined β_o and then changed the reference to g_m. In the case of the FET, however, there is no β_o parameter since the input current is assumed to be zero. Thus, the only meaningful approach is that of the g_m parameter. This leads to another argument for the use of the g_m-model for a BJT since, as we will see shortly, it results in a very close similarity between BJT and FET circuit models.

Small-Signal Circuit Model

We will now develop a small-signal model for the FET. Consider the n-channel JFET shown in Figure 11–3(a). Only signal or ac quantities are shown in this particular case, and all dc levels are understood.

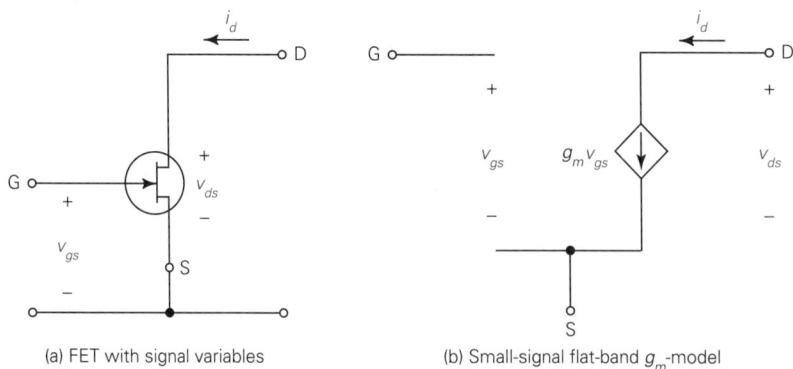

(a) FET with signal variables

(b) Small-signal flat-band g_m-model

FIGURE 11–3
FET and the flat-band small-signal g_m-model.

The flat-band small-signal or ac model is shown in Figure 11–3(b). This form is a g_m-model employing a voltage-controlled current source.

Let us now compare the g_m-model of the BJT from Figure 10–8(c) with the corresponding model of the FET from Figure 11–3(b). Basically, the forms have the same structure. In each case, a voltage across the input terminals controls a current in the output. This similarity results in a great deal of analysis flexibility in studying complex circuit configurations.

One noticeable difference is that the BJT contains a resistance r_π across the input while the ideal FET has an open circuit across the input. The dynamic input resistance of a BJT at the base is sufficiently low that the input signal current plays a significant role in the overall analysis of the circuit. However, an FET has a very high input resistance that, for many applications, may be considered as an open circuit or infinite resistance.

Depletion-Mode FET Transconductance

We considered the drain current I_D as a function of V_{GS} for a **depletion-mode FET** in Chapter 6, and it was shown to be

$$I_D = I_{DSS}\left[1 - \frac{V_{GS}}{V_{GS(\text{off})}}\right]^2 \qquad (11\text{–}5)$$

Recall that the classification *depletion-mode FET* includes both *JFETs* and *depletion-mode IGFETS*. An expression for g_m may be derived from (11–5) using differential calculus. The result will be stated here, and it is

$$g_m = g_{mo}\left[1 - \frac{V_{GS}}{V_{GS(\text{off})}}\right] \qquad (11\text{–}6)$$

where g_{mo} is the zero-bias value of g_m (i.e., the value of g_m at $V_{GS} = 0$). This quantity is

$$g_{mo} = \frac{-2I_{DSS}}{V_{GS(\text{off})}} \qquad (11\text{–}7)$$

The negative sign in (11–7) may be misleading since it arises in the mathematical formulation of the model. For an *n*-channel unit, $I_{DSS} > 0$ and $V_{GS(\text{off})} < 0$. The latter negative sign cancels the negative sign in front, resulting in a positive value of g_{mo}. For a *p*-channel unit, $I_{DSS} > 0$ and $V_{GS} > 0$, which would imply a negative value of g_m. However, recall that the positive direction of I_D is *into* the drain for an *n*-channel unit and *out* of the drain for a *p*-channel. It is more convenient to establish the same current directions in the small-signal models for both *n*-channel and *p*-channel units. To accomplish this, the direction of the small-signal model for the *p*-channel unit will also be considered into the drain, in which case the sign will be changed to a positive value.

If the preceding discussion seems confusing, it will be summarized very simply as follows: The direction of the small-signal current source for *both* *n*-channel and *p*-channel units may be assumed from drain to source as shown by

Figure 11–3(b). The value of g_m for n-channel units with negative gate–source voltage and p-channel units with positive gate–source voltage may be considered to be positive according to the following equation:

$$g_m = g_{mo}\left[1 - \frac{|V_{GS}|}{|V_{GS(off)}|}\right] \qquad (11\text{–}8)$$

where

$$g_{mo} = \frac{2|I_{DSS}|}{|V_{GS(off)}|} \qquad (11\text{–}9)$$

The variation of g_m with V_{GS} for a depletion-mode FET has the form shown in Figure 11–4. Observe that g_m varies from g_{mo} to zero as the gate–source bias voltage varies from zero to cutoff.

FIGURE 11–4
Variation of g_m with $|V_{GS}|$ for a depletion-mode FET.

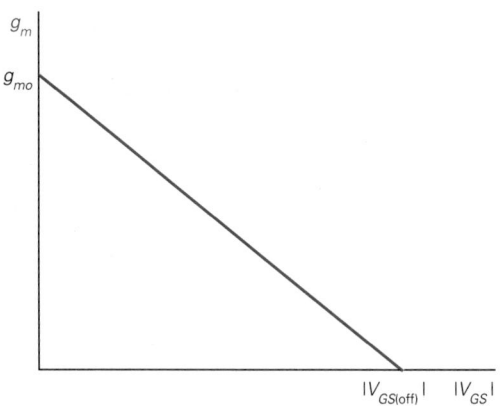

Enhancement-Mode FET Transconductance

We considered the drain current I_D as a function of V_{GS} for an n-channel enhancement-mode IGFET in Chapter 6, and it was shown to be

$$I_D = K\left[\frac{V_{GS}}{V_{GS(th)}} - 1\right]^2 \qquad (11\text{–}10)$$

An expression for g_m may be derived from (11–10) using differential calculus. The result will be stated here, and it is

$$g_m = \frac{2K}{V_{GS(th)}}\left[\frac{V_{GS}}{V_{GS(th)}} - 1\right] \qquad (11\text{–}11)$$

For an n-channel unit, all parameters and variables in (11–11) are positive in the beyond-pinchoff region, so g_m is necessarily positive. For a p-channel unit, the value of g_m as given literally by (11–11) in the beyond-pinchoff region would be negative. However, by the same type of argument previously used for depletion-mode devices, the direction of the small-signal current will be made to conform to that of n-channel units, and in the process, the value of g_m will be made to be positive.

In summary, the equivalent-circuit model for enhancement-mode IGFETs is the same form as that for depletion-mode FETs, and the value of g_m for both n-channel and p-channel enhancement-mode devices can be expressed as

$$g_m = \frac{2K}{|V_{GS(th)}|}\left[\frac{|V_{GS}|}{|V_{GS(th)}|} - 1\right] \qquad \textbf{(11–12)}$$

The quantity K has been defined all along as being positive, so no magnitude bars are required for it. The variation of g_m with V_{GS} for an enhancement-mode IGFET has the form shown in Figure 11–5.

FIGURE 11–5
Variation of g_m with $|V_{GS}|$ for an enhancement-mode IGFET.

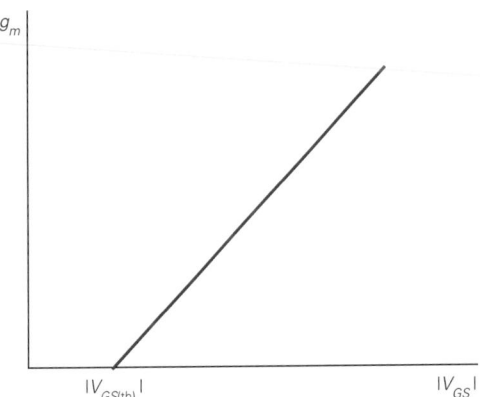

EXAMPLE 11–1

Some dc measurements performed on an n-channel JFET with fixed drain–source voltage yield the following data:

V_{GS} (V)	I_D (mA)
-2	8
-1.9	8.6

Determine the approximate value of g_m in the region of operation.

Solution

The concept of (11–4) is used for this purpose. The change ΔV_{GS} is

$$\Delta V_{GS} = -1.9 \text{ V} - (-2 \text{ V}) = 0.1 \text{ V} \qquad \textbf{(11–13)}$$

The corresponding change ΔI_D is

$$\Delta I_D = 8.6 \text{ mA} - 8 \text{ mA} = 0.6 \text{ mA} \qquad \textbf{(11–14)}$$

The value of g_m is

$$g_m = \frac{0.6 \times 10^{-3} \text{ A}}{0.1 \text{ V}} = 6 \times 10^{-3} \text{ S} = 6 \text{ mS} \qquad \textbf{(11–15)}$$

EXAMPLE 11–2

A certain idealized *n*-channel JFET with $I_{DSS} = 16$ mA and $V_{GS(\text{off})} = -4$ V was used as the basis for several examples in Chapters 6 and 7. **(a)** Determine g_{mo}. **(b)** Determine an expression for g_m as a function of V_{GS}. **(c)** Compute g_m at the following values of V_{GS}: 0; -1 V; -2 V; -3 V; -4 V.

Solution

(a)   The value of g_{mo} is determined from (11–9) as

$$g_{mo} = \frac{2|I_{DSS}|}{|V_{GS(\text{off})}|} = \frac{2 \times 16 \times 10^{-3} \text{ A}}{4 \text{ V}} = 8 \times 10^{-3} \text{ S} = 8 \text{ mS} \qquad \textbf{(11–16)}$$

(b)   The value of g_m as a function of V_{GS} is determined from (11–8) as

$$g_m = 8 \times 10^{-3} \left(1 - \frac{|V_{GS}|}{4} \right) \qquad \textbf{(11–17)}$$

(c)   The values of V_{GS} are substituted in (11–17), and the results are summarized as follows:

V_{GS} (V)	g_m (mS)
0	8
-1	6
-2	4
-3	2
-4	0

As expected, $g_m = g_{mo} = 8$ mS for $V_{GS} = 0$. The value of g_m then decreases linearly with increasing negative voltage until it reaches a value of zero at cutoff.

EXAMPLE 11–3

For the *n*-channel JFET of Example 11–2, construct the small-signal g_m-model at an operating point corresponding to $V_{GS} = -2$ V.

FIGURE 11–6
Small-signal model of Example 11–3.

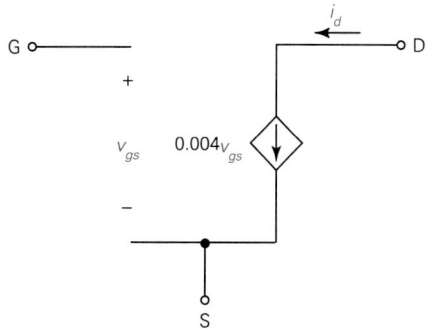

Solution
From the results of Example 11–2, the value of g_m at $V_{GS} = -2$ V is $g_m = 4$ mS. The small-signal model has the form of Figure 11–3(b), and it is shown in Figure 11–6.

11–3 COMMON-SOURCE AMPLIFIER CIRCUIT

The form of a practical common-source amplifier circuit using an *n*-channel JFET is shown in Figure 11–7(a). This circuit uses a combination of fixed plus self-bias. However, the analysis to be employed here can be adapted to the simpler case

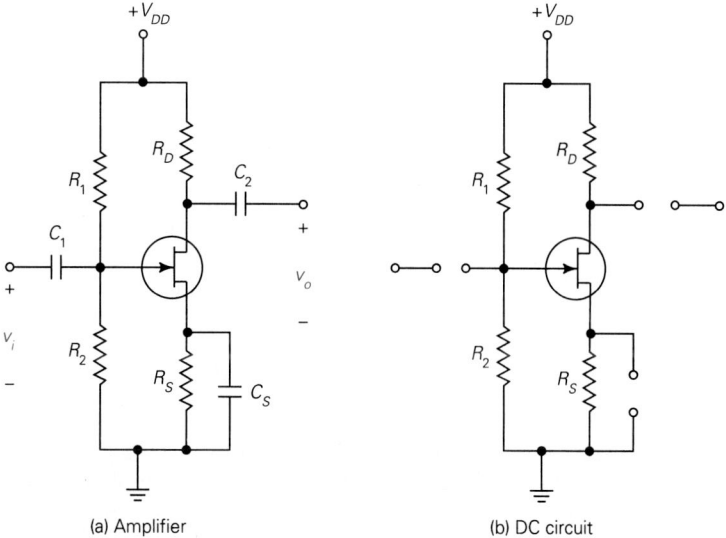

(a) Amplifier

(b) DC circuit

FIGURE 11–7
Common-source amplifier with stabilized bias and dc equivalent circuit.

of self-bias just by eliminating R_1 in the circuit model. Effects of possible source resistance and external load resistance will be considered later.

The dc circuit is shown in Figure 11–7(b). As a result of the fixed- and self-bias combination, a dc operating point for the transistor is established.

The midfrequency ac equivalent circuit for small signals is shown in Figure 11–8. The flat-band small-signal model replaces the transistor between the three terminals. The capacitors C_1 and C_2 are used for input and output coupling, and they act as short circuits in the midfrequency range. The function of C_S is to bypass signals around R_S. Without C_S, the gain would be reduced considerably. In the midfrequency region C_S appears as a short circuit.

FIGURE 11–8

Small-signal midfrequency model of common-source amplifier.

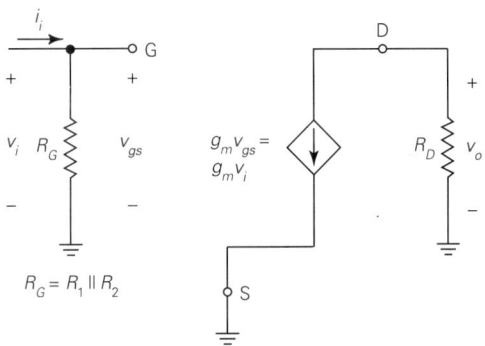

The resistances R_1 and R_D are both brought to ground for signal analysis purposes. Since R_1 and R_2 appear as a parallel combination, a resistance R_G is defined as

$$R_G = R_1 \| R_2 = \frac{R_1 R_2}{R_1 + R_2} \qquad \text{(11–18)}$$

The voltage gain A will now be determined. The output voltage v_o is expressed in terms of the input voltage as

$$v_o = -g_m v_{gs} R_D = -g_m v_i R_D \qquad \text{(11–19)}$$

where $v_{gs} = v_i$ was noted. The voltage gain A is

$$\boxed{A = \frac{v_o}{v_i} = -g_m R_D} \qquad \text{(11–20)}$$

As in the case of the BJT, input and output resistances of the transistor and of the transistor plus external circuitry will be determined. Let R'_{in} and R'_{out} represent the input and output resistances of the transistor itself. We have

$$\boxed{R'_{in} = \infty} \qquad \text{(11–21)}$$

$$\boxed{R'_{\text{out}} = \infty} \qquad\qquad \text{(11–22)}$$

The infinite input resistance of the transistor is a result of the assumed open circuit across the input, and the infinite output resistance is a result of the assumed ideal current source model.

Let R_{in} and R_{out} represent the input and output resistances of the composite circuit. We have

$$\boxed{R_{\text{in}} = \frac{v_i}{i_i} = R_G = R_1 \| R_2} \qquad\qquad \text{(11–23)}$$

and

$$\boxed{R_{\text{out}} = R_D} \qquad\qquad \text{(11–24)}$$

A comparison of the common-source amplifier of this section with the common-emitter amplifier of Section 10–4 will lead to a number of close similarities. With the exception of component subscripts, the gain expressions and output resistances have exactly the same forms. The input resistances differ only with respect to the presence of r_π for the BJT.

EXAMPLE 11–4   A certain ac-coupled common-source amplifier is shown in Figure 11–9(a). At the particular operating point established, $g_m = 20\ \text{mS}$. (a) Construct the midfrequency small-signal equivalent circuit. (b) Determine the voltage gain A and the input and output resistances R'_{in}, R'_{out}, R_{in}, and R_{out}.

Solution
(a)   Since we are given the value of g_m, it is not necessary in this case to determine the dc operating point, nor is it necessary to know R_S. The small-signal midfrequency model is shown in Figure 11–9(b). For clarity, R_1 and R_2 are shown separately.

(b)   The gain A is determined as

$$A = \frac{v_o}{v_i} = -g_m R_D = -0.02 \times 3 \times 10^3 = -60 \qquad\qquad \text{(11–25)}$$

The ideal input and output resistances of the transistor are

$$R'_{\text{in}} = \infty \qquad\qquad \text{(11–26)}$$
$$R'_{\text{out}} = \infty \qquad\qquad \text{(11–27)}$$

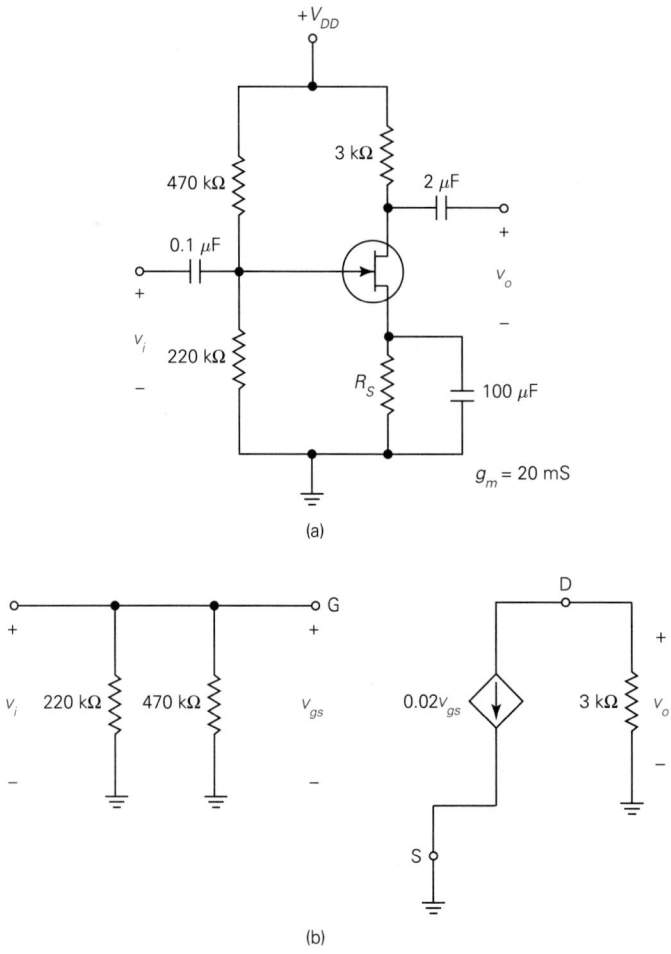

FIGURE 11–9
Common-source circuit and midfrequency model of Example 11–4.

The input and output resistances of the composite circuit are

$$R_{in} = 220\ k\Omega \| 470\ k\Omega = 150\ k\Omega \qquad (11\text{–}28)$$
$$R_{out} = 3\ k\Omega \qquad (11\text{–}29)$$

11–4 COMMON-SOURCE AMPLIFIER WITH SOURCE FEEDBACK

The common-source amplifier with source resistance completely bypassed is similar in many respects to the common-emitter amplifier with emitter resistance bypassed. Like the common-emitter amplifier, the common-source amplifier is subject to serious distortion except at the very smallest signal levels. Recall that the relationship between gate–source voltage and drain current is a second-degree function and that the transconductance varies with operating point. Thus, as the signal level varies, the gain varies, which results in a distorted output.

It turns out that the distortion may be reduced by a process very similar to that of the common-emitter amplifier. In the case of the FET, a portion of the source resistance is left unbypassed. This process is referred to as **source feedback,** and the circuit is shown in Figure 11–10.

FIGURE 11–10
Common-source amplifier with source feedback.

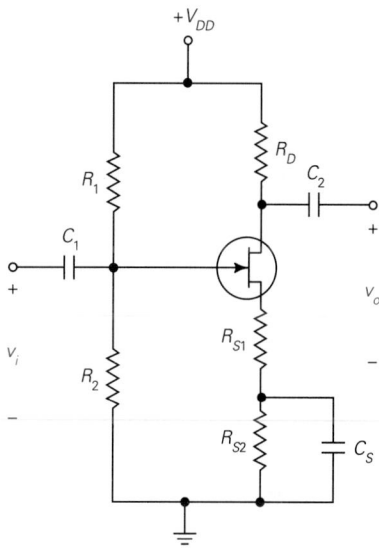

For dc bias circuit analysis, the net source resistance R_S is

$$R_S = R_{S1} + R_{S2} \tag{11–30}$$

For ac, the resistance R_{S2} is bypassed with the capacitor C_S, but R_{S1} is much smaller than R_{S2}, so that only a small fraction of the total source resistance is left unbypassed.

The small-signal midfrequency model of this amplifier is shown in Figure 11–11. The control voltage v_{gs} is no longer equal to the input voltage v_i. A KVL equation from ground to the gate, down to the source, and back to ground reads

$$-v_i + v_{gs} + R_{S1}i_d = 0 \tag{11–31}$$

or

$$v_{gs} = v_i - R_{S1}i_d \tag{11–32}$$

The process that occurs here is virtually the same as we encountered in the common-emitter with feedback. As v_i increases, v_{gs} starts to increase, and the

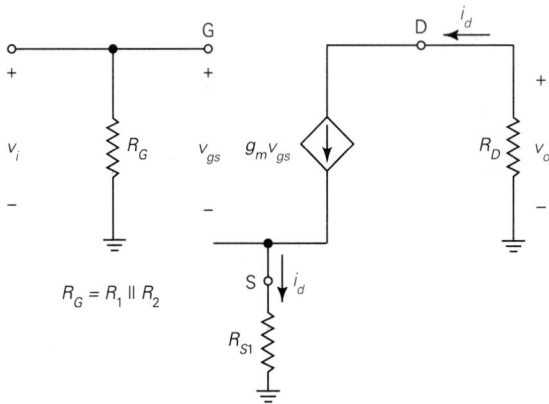

FIGURE 11–11
Small-signal midfrequency model of common-source amplifier with source feedback.

drain signal current $g_m v_{gs}$ also starts to increase. This causes the voltage across R_{S1} to increase, and this change tends to decrease v_{gs}, as noted by (11–32). The result is that the net increase in v_{gs} is considerably less than when the signal voltage is applied directly between gate and source.

Because of the very close similarity between the g_m-models of the BJT and the FET, many of the BJT results of Chapter 10 can readily be adapted to the corresponding FET circuits. Along with notational changes, it will be necessary to replace r_π in any circuit by an open circuit. This is achieved in various equations by allowing r_π to increase without limit (i.e., $r_\pi = \infty$).

An expression for the voltage gain of a BJT amplifier with emitter feedback was given by (10–60). By replacing R_C with R_D and R_{E1} with R_{S1}, the corresponding gain for the FET with source feedback can be determined. This result is

$$A_{io} = \frac{v_o}{v_i} = \frac{-g_m R_D}{1 + g_m R_{S1}} \qquad (11\text{–}33)$$

The input and output resistances of the transistor alone are assumed to be

$$R'_{in} = \infty \qquad (11\text{–}34)$$

and

$$R'_{out} = \infty \qquad (11\text{–}35)$$

The corresponding input and output resistances of the composite circuit are

$$R_{\text{in}} = \frac{v_i}{i_i} = R_G = R_1 \| R_2 \qquad \textbf{(11–36)}$$

and

$$R_{\text{out}} = R_D \qquad \textbf{(11–37)}$$

Comparison

A comparison between the common-source amplifier with the source resistance completely bypassed and the corresponding amplifier with source feedback can now be made. A summary of the key formulas for the two circuits is provided in Table 11–1. It may also be helpful for the reader to refer to Table 10–1 (p. 499), in which a similar comparison was made for the common-emitter amplifier.

TABLE 11–1
Comparison of common-source properties with emitter resistance bypassed and with emitter feedback

	Source resistance bypassed	source feedback
A_{io}	$-g_m R_D$	$\dfrac{-g_m R_D}{1 + g_m R_{S1}}$
R'_{in}	∞	∞
R_{in}	R_G	R_G
R'_{out}	∞	∞
R_{out}	R_D	R_D

The magnitude of the gain with source feedback is reduced by the factor $1/(1 + g_m R_{S1})$. Note that this expression reduces to the expression for gain without feedback when $R_{S1} = 0$.

The input resistance for the ideal assumptions employed is infinite in both cases. Strictly speaking, if the very large resistance between gate and source were shown, it would be multiplied by the factor $(1 + g_m R_{S1})$ in a manner analogous to the BJT. However, with the assumption of an open circuit from the outset, this factor does not appear. In practice, the external resistance R_G dominates the net circuit input resistance.

In some situations, $g_m R_{S1} \gg 1$. In this case, the gain A may be approximated as

$$A \simeq -\frac{g_m R_D}{g_m R_{E1}} = -\frac{R_D}{R_{E1}} \tag{11–38}$$

The gain for this assumption then becomes virtually independent of transistor parameters.

EXAMPLE 11–5

A certain common-source amplifier with source feedback is shown in Figure 11–12. At the particular operating point established, $g_m = 20$ mS. Determine the midfrequency voltage gain A and the input and output resistances R'_{in}, R'_{out}, R_{in}, and R_{out}.

FIGURE 11–12
Amplifier circuit of Example 11–5.

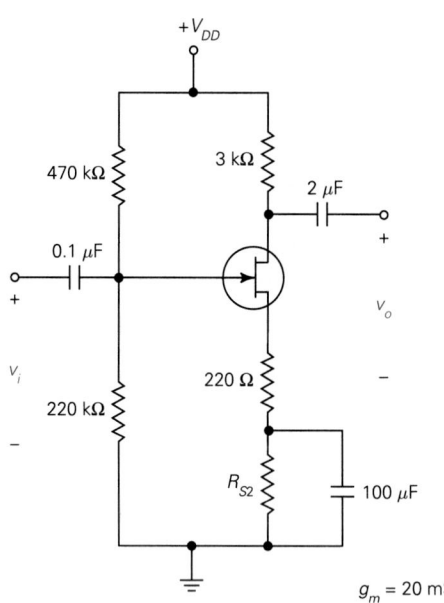

Solution
We will use the various results established in this section in the analysis. The gain A is

$$A = \frac{v_o}{v_i} = \frac{-g_m R_D}{1 + g_m R_{S1}} = \frac{-0.02 \times 3 \times 10^3}{1 + 0.02 \times 220} = \frac{-60}{5.4} = -11.11 \tag{11–39}$$

The transistor input and output resistances are

$$R'_{in} = \infty \tag{11–40}$$

and

$$R'_{out} = \infty \tag{11-41}$$

The net circuit input and output resistances are

$$R_{in} = 220 \text{ k}\Omega \| 470 \text{ k}\Omega = 150 \text{ k}\Omega \tag{11-42}$$

and

$$R_{out} = 3 \text{ k}\Omega \tag{11-43}$$

A check on the closeness of (11–38) will now be made. This ratio yields $-3 \times 10^3/220 = -13.64$. This result differs from the actual gain by about 23%.

11-5 COMMON-GATE AMPLIFIER

We will consider the **common-gate** amplifier in this section. The common-gate amplifier for an FET corresponds to the common-base amplifier for a BJT and has many similar properties.

An ac-coupled common-gate amplifier is shown in Figure 11–13(a). This layout represents the most common form. However, it is useful at this point to show the circuit in a form easily related to the common-source circuit, and this is shown in Figure 11–13(b). The dc circuit of the common-gate amplifier is equivalent to that of the common-source circuit. Thus, the operating point is established with R_1, R_2, R_S, and R_D as for a common-source circuit.

The signal is coupled to the source terminal by the capacitor C_1, and the gate terminal is established at signal ground by means of the capacitor C_G. However, the output signal is taken at the drain terminal in the same manner as in the common-source circuit.

A small-signal midfrequency model of the common-gate circuit is shown in Figure 11–14. Note how the gate terminal of the FET is connected to ground and how the control voltage v_{gs} has its positive reference at ground. The input signal appears at the source terminal, and the dependent current source appears along the top.

By comparing Figure 11–14 for the common-gate circuit with Figure 10–18 for the common-base circuit, the similarities in the circuit models are evident. The various circuit properties for the common-gate circuit may be determined from those of the common-base circuit by appropriate notational changes.

The voltage gain is determined from (10–79) by replacing R_C with R_D. We have

$$A = \frac{v_o}{v_i} = g_m R_D \tag{11-44}$$

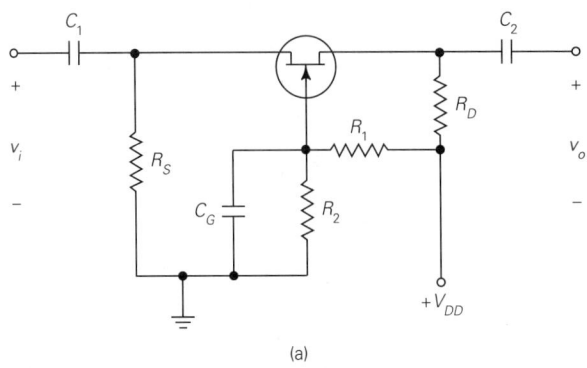

(a)

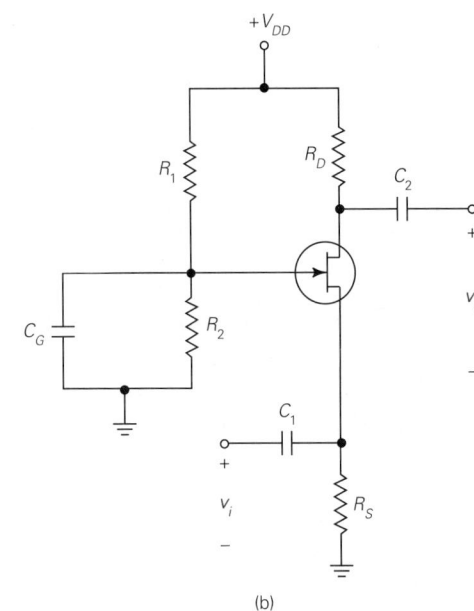

(b)

FIGURE 11–13
Common-gate circuit drawn two ways.

FIGURE 11–14
Small-signal midfrequency model of common-gate circuit.

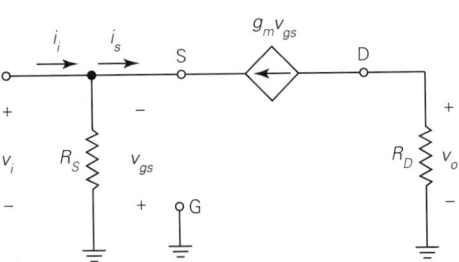

The transistor input resistance R'_{in} is determined from (10–82) by assuming $r_\pi = \infty$. This means that $1/r_\pi = 0$. The result is

$$R'_{in} = \frac{v_i}{i_s} = \frac{1}{g_m}$$

(11–45)

The corresponding output resistance R'_{out} is

$$R'_{out} = \infty$$

(11–46)

The composite circuit input and output resistances are now determined as

$$R_{in} = R_S \| \frac{1}{g_m}$$

(11–47)

and

$$R_{out} = R_D$$

(11–48)

EXAMPLE 11–6

A certain common-gate amplifier is shown in Figure 11–15. At the particular operating point established, $g_m = 0.008$ S. Determine the midfrequency values of A, R'_{in}, R'_{out}, R_{in}, and R_{out}.

Solution
We will apply the results established in this section directly in the analysis. The gain A is

$$A = \frac{v_o}{v_i} = g_m R_D = 0.008 \times 2 \times 10^3 = 16$$

(11–49)

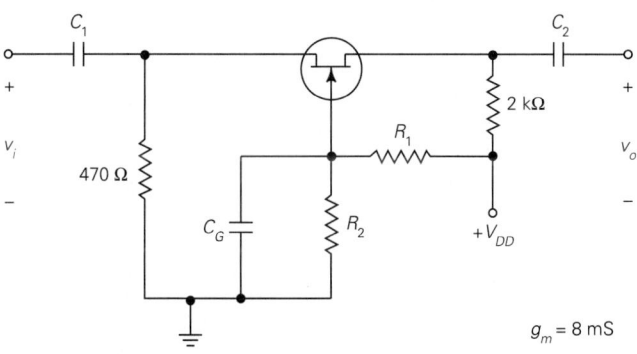

FIGURE 11–15
Common-gate circuit of Example 11–6.

The transistor input and output resistances are

$$R'_{in} = \frac{1}{g_m} = \frac{1}{0.008} = 125 \ \Omega \qquad \text{(11–50)}$$

and

$$R'_{out} = \infty \qquad \text{(11–51)}$$

The net circuit input and output resistances are

$$R_{in} = R_E \left\| \frac{1}{g_m} = 470 \right\| 125 = 98.7 \ \Omega \qquad \text{(11–52)}$$

and

$$R_{out} = R_D = 2 \ k\Omega \qquad \text{(11–53)}$$

11–6 COMMON-DRAIN AMPLIFIER

The third basic FET configuration is the **common-drain amplifier,** which is also referred to as the **source-follower circuit.** A dc-coupled version of this circuit was considered in Chapter 7. The circuit will now be considered in ac-coupled form and with the small-signal approach.

The basic form of an ac-coupled common-drain amplifier circuit using an *n*-channel JFET is shown in Figure 11–16. There is no drain resistance in the common-drain circuit, so the drain terminal is established at ground potential for ac signals.

FIGURE 11–16
Common-drain (source-follower) ac-coupled amplifier circuit employing an *n*-channel JFET.

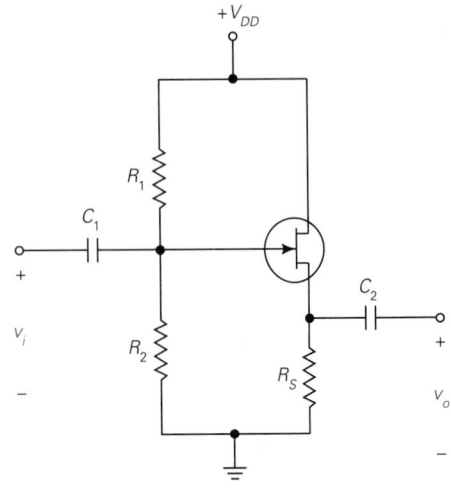

The midfrequency small-signal equivalent circuit is first drawn as in Figure 11–17(a). Note how the drain terminal is grounded. It is best to redraw the circuit as in Figure 11–17(b) for the discussion that follows.

A comparison of Figure 11–17(b) with Figure 10–21(b) reveals a very close similarity between the common-drain FET circuit and the common-collector circuit. Aside from terminology, the only real difference is that r_π for the BJT is replaced by an open circuit for the FET. As done earlier for several other cases, the results of the BJT analysis will be adapted to the FET case.

FIGURE 11–17
Small-signal midfrequency model of common-drain circuit.

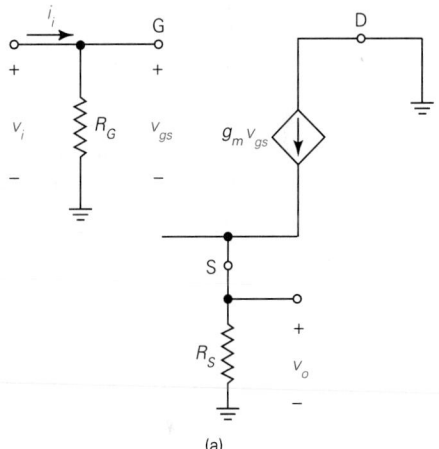

(a)

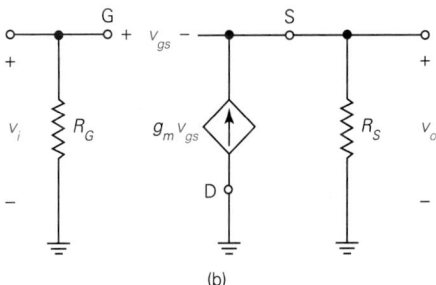

(b)

An expression for the voltage gain of the common-collector circuit was given in (10–102). By replacing R_E with R_S, the corresponding gain for the common-drain circuit is

$$A = \frac{v_o}{v_i} = \frac{g_m R_S}{1 + g_m R_S} \qquad (11\text{–}54)$$

We will retain this expression as the basic gain equation for the common-drain circuit. However, in many applications, $g_m R_S \gg 1$, in which case A may be further approximated as

$$\boxed{A \simeq 1} \tag{11-55}$$

The transistor input resistance is assumed to be

$$\boxed{R'_{in} = \infty} \tag{11-56}$$

The net circuit input resistance is

$$\boxed{R_{in} = R_G = R_1 \| R_2} \tag{11-57}$$

The output resistance of the common-collector circuit was given by (10–113), based on replacing the source by an open circuit. To adapt this expression to the common-drain circuit, R_B is replaced by R_G and r_π is allowed to increase without limit. When the latter process is performed, $r_\pi \gg R_G$ and $g_m r_\pi \gg 1$. The numerator can then be approximated as r_π, and the denominator can be approximated as $g_m r_\pi$. The ratio can then be closely approximated as $r_\pi / g_m r_\pi = 1/g_m$. The resulting value of R'_{out} is

$$\boxed{R'_{out} = \frac{1}{g_m}} \tag{11-58}$$

The assumed infinite gate–source resistance thus uncouples the effect of R_G on the output resistance in the ideal case.

The net circuit output resistance is given by

$$\boxed{R_{out} = R_S \| R'_{out} = R_S \| \frac{1}{g_m}} \tag{11-59}$$

EXAMPLE 11–7

A certain common-drain amplifier is shown in Figure 11–18. At the particular operating point established, $g_m = 12$ mS. Determine the midfrequency values of A, R'_{in}, R'_{out}, R_{in}, and R_{out}.

Solution

We will use the results established in this section in the analysis. The gain A is

$$A = \frac{v_o}{v_i} = \frac{g_m R_S}{1 + g_m R_S} = \frac{0.012 \times 2 \times 10^3}{1 + 0.012 \times 2 \times 10^3} = \frac{24}{25} = 0.96 \tag{11-60}$$

FIGURE 11–18
Common-drain circuit of Example 11–7.

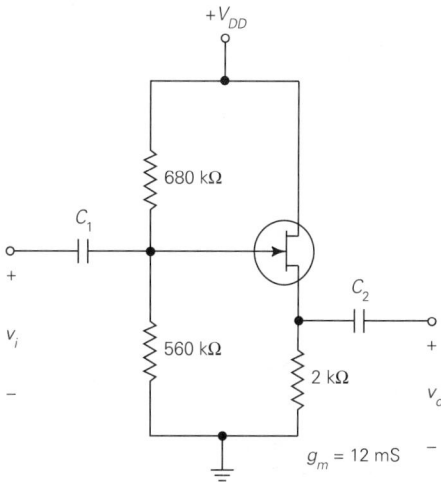

The transistor input and output resistances are

$$R'_{in} = \infty \tag{11-61}$$

and

$$R'_{out} = \frac{1}{g_m} = \frac{1}{0.012} = 83.3 \ \Omega \tag{11-62}$$

The net circuit input and output resistances are

$$R_{in} = 680 \ \text{k}\Omega \| 560 \ \text{k}\Omega = 307 \ \text{k}\Omega \tag{11-63}$$

$$R_{out} = \frac{1}{g_m} \| R_S = 83.3 \| 2000 = 80 \ \Omega \tag{11-64}$$

11–7 FET AC SPECIFICATIONS

We will now study some FET small-signal specifications. Refer to the first page of the 2N4220 through 2N4222 data sheets in Appendix D. The first block in the "small-signal characteristic" section is entitled "forward transfer admittance common source." The symbol used there is $|y_{fs}|$, and this is equivalent to g_m. The symbol y is based on a set of parameters called the "short-circuit admittance" parameters that have been employed in some references for FET specifications. The subscript fs refers to forward (gain) and common source. The value of $|y_{fs}|$ depends on the particular transistor type, but for each one, there is more than a 2 : 1 range in the possible value. The older unit of admittance "μmho" is used here.

We have assumed that the effective resistance in parallel with the output current source is infinite. When it is necessary to use an actual resistance in the model, the "output admittance common source" data may be useful. This admittance value is denoted as $|y_{os}|$, and only the maximum value is given for each transistor. The reciprocal of this quantity is the minimum value of the shunt resistance.

The three additional blocks provide information on three FET capacitances. Their significance is discussed in Chapter 18.

11–8 PSPICE EXAMPLE

PSPICE EXAMPLE 11–1 Develop a small-signal midfrequency model for the common-drain amplifier circuit of Example 11–7 (Figure 11–18), and use the transfer function command to determine the voltage gain and overall input and output resistances.

Solution
The circuit adapted to PSPICE is shown in Figure 11–19(a), and the code is shown in Figure 11–19(b). This model is based on the form of Figure 11–17(b), although the two resistors R_1 and R_2 have not been combined in the simulation. The code is similar to several earlier examples and requires little explanation. Note that the controlling voltage for the current source is the voltage between nodes 1 and 2, with the positive reference at 1.

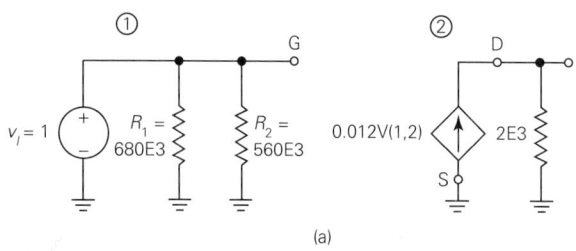

```
PSPICE EXAMPLE 11-1
.OPTIONS NOPAGE NOECHO NOBIAS
VI 1 0 1
R1 1 0 680E3
R2 1 0 560E3
G 0 2 1 2 0.012
RD 2 0 2E3
.TF V(2) VI
.END
```

(a)   (b)

FIGURE 11–19
Circuit and code of PSPICE Example 11–1.

The computer printout is shown in Figure 11–20. The voltage gain is V(2)/VI = 0.96, which agrees exactly with the text values. The input and output resistances are 307.1 kΩ and 80 Ω, respectively. The first value is almost exactly the same as the computed value of 307 kΩ, and the second value is exactly the same as the computed value of 80 Ω.

```
PSPICE EXAMPLE 11-1

****       CIRCUIT DESCRIPTION

*********************************************************************

.OPTIONS NOPAGE NOECHO NOBIAS

****       SMALL SIGNAL CHARACTERISTICS

V(2)/VI = 9.600E-01
INPUT RESISTANCE AT VI = 3.071E+05
OUTPUT RESISTANCE AT V(2) = 8.000E+01

    JOB CONCLUDED

    TOTAL JOB TIME          1.10
```

FIGURE 11–20
Computer printout of PSPICE Example 11–1.

PROBLEMS

Drill Problems

11–1. Some dc measurements performed on an *n*-channel JFET with fixed drain–source voltage yield the following data:

V_{GS} (V)	I_D (mA)
−2.5	4
−2.3	5.8

Determine the approximate value of g_m in the region of operation.

11–2. Some dc measurements performed on a *p*-channel JFET with fixed drain–source voltage yield the following data:

V_{GS} (V)	I_D (mA)
3	2
2.4	4.4

Determine the approximate value of g_m in the region of operation.

11-3. Some dc measurements performed on an n-channel enhancement-mode MOSFET yield the following data:

V_{GS} (V)	I_D (mA)
2.7	8
2.8	9.2

Determine the approximate value of g_m in the region of operation.

11-4. Some dc measurements on a p-channel enhancement-mode MOSFET yield the following data:

V_{GS} (V)	I_D (mA)
-3.6	11.5
-3.8	14.3

Determine the approximate value of g_m in the region of operation.

11-5. A certain idealized n-channel JFET has $I_{DSS} = 18$ mA and $V_{GS(off)} = -3$ V. **(a)** Determine g_{mo}. **(b)** Compute g_m at the following values of V_{GS}: 0; -1 V; -2 V; -3 V.

11-6. A certain idealized p-channel JFET has $I_{DSS} = 32$ mA and $V_{GS(off)} = 2$ V. **(a)** Determine g_{mo}. **(b)** Compute g_m at the following values of V_{GS}: 0; 1 V; 2 V.

11-7. For the n-channel JFET of Problem 11-5, construct the small-signal g_m-model at an operating point of $V_{GS} = -1$ V.

11-8. For the p-channel JFET of Problem 11-6, construct the small-signal g_m-model at an operating point of $V_{GS} = 1$ V.

11-9. Refer to the characteristics of the idealized n-channel enhancement-model IGFET of Problem 6-19 (Figure P6-19). **(a)** Determine an expression for g_m. **(b)** Compute g_m at the following values of V_{GS}: 3 V; 4 V; 5 V; 6 V.

11-10. Refer to the characteristics of the idealized n-channel enhancement-mode IGFET of Problem 6-20 (Figure P6-20). **(a)** Determine an expression for g_m. **(b)** Compute g_m at the following values of V_{GS}: 2 V; 2.5 V; 3 V; 3.5 V; 4 V.

11-11. A certain ac-coupled common-source amplifier is shown in Figure P11-11. At the particular operating point established, $g_m = 12$ mS. **(a)** Construct the midfrequency small-signal equivalent circuit. **(b)** Determine the voltage gain A and the input and output resistances R'_{in}, R'_{out}, R_{in}, and R_{out}.

11-12. A certain ac-coupled common-source amplifier is shown in Figure P11-12. At the particular operating point established, $g_m = 18$ mS. **(a)** Construct the midfrequency small-signal equivalent circuit. **(b)** Determine the voltage gain A and the input and output resistances R'_{in}, R'_{out}, R_{in}, and R_{out}.

11-13. A common-source amplifier with source feedback is shown in Figure P11-13. At the operating point, $g_m = 12$ mS. Determine the midfrequency voltage gain A and the input and output resistances R'_{in}, R'_{out}, R_{in}, and R_{out}.

FIGURE P11–11

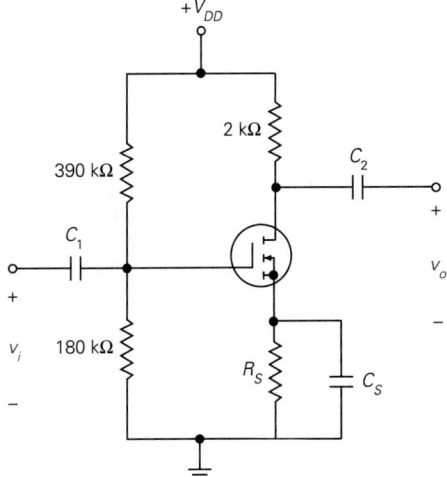

FIGURE P11–12

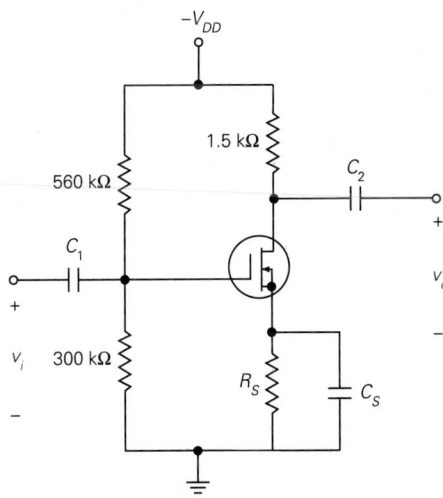

FIGURE P11–13

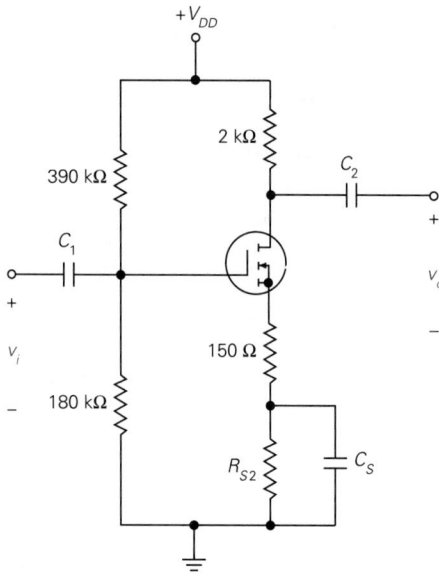

11–14. A common-source amplifier with source feedback is shown in Figure P11–14. At the operating point, $g_m = 18$ mS. Determine the midfrequency voltage gain A and the input and output resistances R'_{in}, R'_{out}, R_{in}, and R_{out}.

FIGURE P11–14

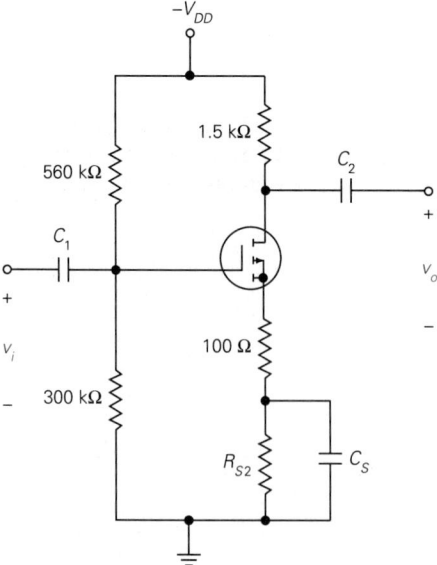

11–15. A certain common-gate amplifier is shown in Figure P11–15. At the operating point, $g_m = 15$ mS. Determine the midfrequency values of A, R'_{in}, R'_{out}, R_{in}, and R_{out}.

FIGURE P11–15

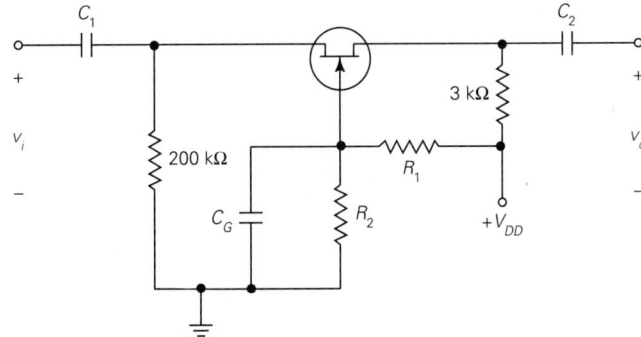

11–16. A certain common-gate amplifier is shown in Figure P11–16. At the operating point, $g_m = 10$ mS. Determine the midfrequency values of A, R'_{in}, R'_{out}, R_{in}, and R_{out}.

FIGURE P11–16

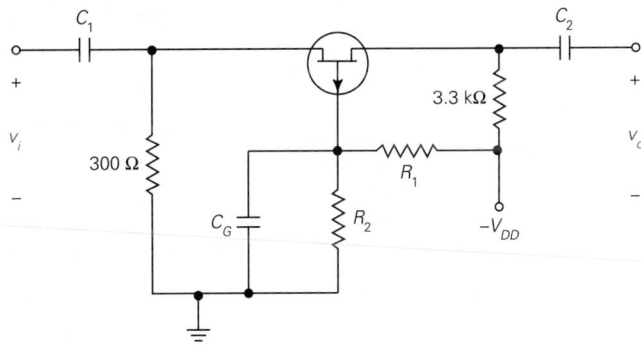

11–17. A certain common-drain amplifier is shown in Figure P11–17. At the operating point, $g_m = 12$ mS. Determine the midfrequency values of A, R'_{in}, R'_{out}, R_{in}, and R_{out}.

FIGURE P11–17

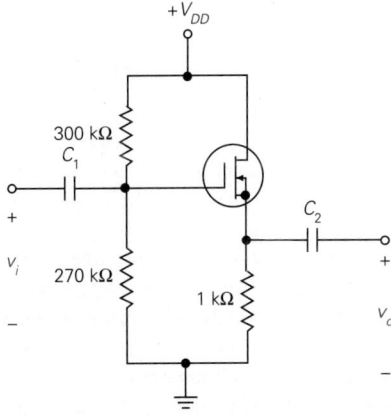

11–18. A certain common-drain amplifier is shown in Figure P11–18. At the operating point, $g_m = 18$ mS. Determine the midfrequency values of A, R'_{in}, R'_{out}, R_{in}, and R_{out}.

FIGURE P11–18

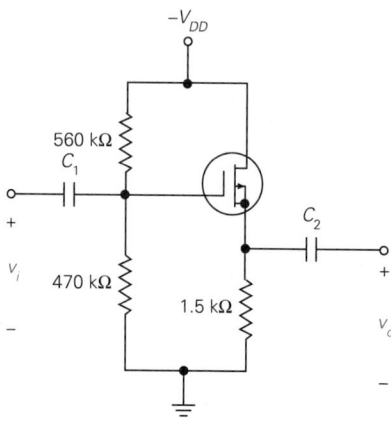

Derivation Problems

11–19. (Requires calculus.) Starting with (11–5), determine the derivative of I_D with respect to V_{GS}, and show that (11–6) is obtained using the definition of (11–7).

11–20. (Requires calculus.) Starting with (11–10), determine the derivative of I_D with respect to V_{GS}, and show that (11–11) is obtained.

Design Problems

11–21. The FET in a certain common-source amplifier circuit has $g_m = 20$ mS. Determine the value of drain resistance required to achieve a gain $A = -40$.

11–22. The FET in a certain common-gate amplifier circuit has $g_m = 30$ mS. Determine the value of drain resistance required to achieve a gain $A = 15$.

11–23. The FET in a certain common-source amplifier with source feedback has $g_m = 25$ mS. For proper drain–source bias voltage, the value of drain resistance is selected as 2 kΩ. Determine the value of unbypassed source resistance required to establish a gain $A = -12$.

11–24. The output resistance of a certain common-drain amplifier is required to be no greater than 200 Ω. Determine the approximate value of g_m required in the FET.

Troubleshooting Problems

11–25. You are troubleshooting a common-source amplifier of the type shown in Figure P11–11 using an oscilloscope and with an applied signal. With the input signal within the specified voltage level, the output signal is very distorted on the negative peak. Which one of the following *could* be the cause of the difficulty? **(a)** 180-kΩ resistor shorted; **(b)** 180-kΩ resistor open; **(c)** 2-kΩ resistor shorted; **(d)** 390-kΩ resistor open; **(e)** C_S open.

11–26. Suppose in Problem 11–25 that the output signal appears satisfactory on the negative peak but that it clips on the positive peak. Which one of the following *could* be the cause of the difficulty? **(a)** 180-kΩ resistor open; **(b)** 390-kΩ resistor shorted; **(c)** 2-kΩ resistor shorted; **(d)** R_S too small; **(e)** R_S too large.

12

MULTISTAGE AMPLIFIERS

OBJECTIVES

After completing this chapter, the reader should be able to:

- Compare the general properties of the common-emitter, common-base, and common-collector BJT amplifier circuits.
- Compare the general properties of the common-source, common-gate, and common-drain FET amplifier circuits.
- Indicate which of the FET configurations has similar characteristics to a given BJT configuration.
- Use tabulated results to determine voltage gain, input resistance, and output resistance for any of the basic configurations.
- Construct a standard amplifier model using the results of the preceding objective.
- Analyze any of the basic configurations with loading at the input and/or the output.
- Develop models for and analyze multistage amplifier configurations.
- Analyze some of the circuits in the chapter with PSPICE.

12–1 COMPARISON OF AMPLIFIER CIRCUITS

In the preceding two chapters we analyzed various BJT and FET single-stage amplifier circuits, and we derived various properties of these amplifiers. We are now in a position to review and compare the various properties and relationships.

As an initial qualitative comparison, Table 12–1 is provided for a general overview. This table is intended to provide some general guidelines as opposed to exact quantitative bounds. Consequently, the various ranges shown are intended as *typical* results rather than *exact* boundaries. Exceptions to these bounds are

TABLE 12–1
Relative comparison of different amplifier circuits

BJT Circuits

	R_{in}	R_{out}	Voltage Gain Magnitude	Current Gain Magnitude	Power Gain	Inverting (I) or Noninverting (N)
Common emitter	Medium	Medium	>1	>1	>1	I
Common base	Low	Medium	>1	<1	>1	N
Common collector	Medium to high	Low	<1	>1	>1	N

FET Circuits

	R_{in}	R_{out}	Voltage Gain Magnitude	Current Gain Magnitude	Power Gain	Inverting (I) or Noninverting (N)
Common source	Very high	Medium	>1	>1	>1	I
Common gate	Low to medium	Medium	>1	<1	>1	N
Common drain	Very high	Low to medium	<1	>1	>1	N

Typical ranges are:
Low $\quad\quad$ <100 Ω
Medium $\quad$ 100 Ω to 10 kΩ
High $\quad\quad$ 10 to 100 kΩ
Very high $\;$ >100 kΩ

fairly common and can probably be found in this book. Furthermore, the transistors are assumed to be small-signal types rather than large power units.

Two properties that have not been considered yet in the book, but that are included in the comparison, are **current gain** and **power gain.** The major gain parameter considered thus far has been **voltage gain,** and this will continue to be the dominant parameter. However, by relating output signal current to input signal current or output signal power to input signal power, current gain and power gain measures can be defined. We develop the quantitative relationships for current and power gains in the next section.

First, let us note the properties of the common-emitter amplifier. Based on the arbitrary ranges indicated, the common emitter has typical "medium" levels of input and output resistances. The voltage gain is inverting and its magnitude is normally greater than unity. Likewise, the current gain and power gain are normally greater than unity.

The common-base amplifier has a "low" input resistance and a "medium" output resistance. The voltage gain is noninverting and normally greater than unity. The power gain is also greater than unity but the current gain is less than unity. The latter result is due to the fact that collector current is less than emitter current.

The common-collector amplifier typically has a "medium" to "high" input resistance and a "low" output resistance. Its voltage gain is noninverting but is less than unity. (We often approximate the voltage gain as unity.) Its current gain and power gain are both greater than unity.

Reviewing the preceding results, only the common-emitter amplifier has all three gains (voltage, current, and power) greater than unity. However, all three configurations have power gains exceeding unity.

Turn now to the FET entries. In general, there is a strong association in gains and other results between the following corresponding amplifiers:

BJT		FET
Common emitter	$\longleftrightarrow$	Common source
Common base	$\longleftrightarrow$	Common gate
Common collector	$\longleftrightarrow$	Common drain

This pattern is immediately evident in the inverting and noninverting comparisons as well as the voltage gain, current gain, and power gain. Thus a common-source amplifier has properties very similar to a common-emitter amplifier, and so on. The major difference is in the relative level of certain resistances. For example, a common-emitter input resistance is classified as "medium," whereas a common-source input resistance is classified as "very high." This is a result of the fact that an FET has a much higher input resistance than a BJT.

12–2 AMPLIFIER SUMMARY

In this section the major results concerning both BJT and FET single-stage amplifier circuits in the midfrequency range are summarized and tables of pertinent formulas provided.

For BJT amplifiers, we have considered the three basic amplifier structures: *common emitter, common base,* and *common collector* (or *emitter follower*). For the common-emitter circuit, we have considered both the case with the emitter resistance bypassed and that with emitter feedback.

For FET amplifiers, we have also considered the three basic amplifier configurations: *common source, common gate,* and *common drain* (or *source follower*). For the common-source circuit, we have considered both the case with the source resistance bypassed and that with source feedback.

A summary of the basic schematic diagrams for the BJT amplifiers is shown in Figure 12–1, and a similar summary for the FET amplifiers is shown in Figure 12–2. None of these configurations yet show the effects of generator resistance or of external load resistance. These effects will be considered shortly.

FIGURE 12–1
Summary of BJT amplifier configurations.

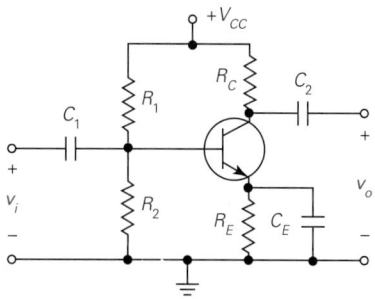

(a) Common emitter

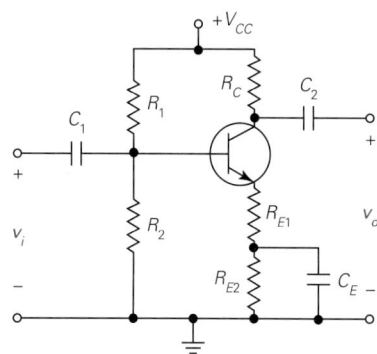

(b) Common emitter with
emitter feedback

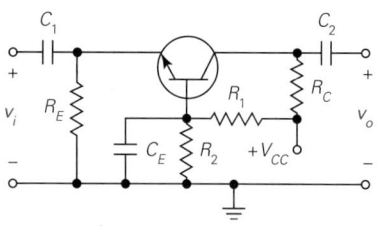

(c) Common base

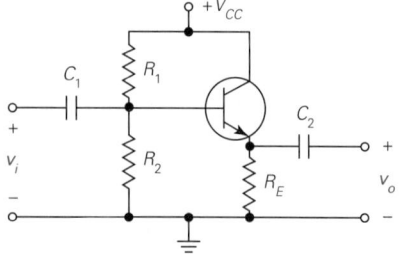

(d) Common collector

FIGURE 12–2
Summary of FET amplifier configurations.

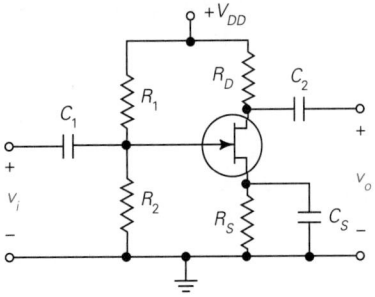

(a) Common source

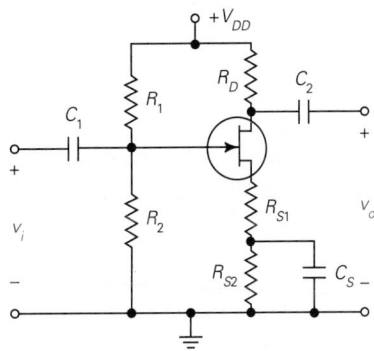

(b) Common source with
source feedback

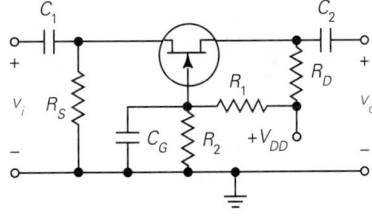

(c) Common gate

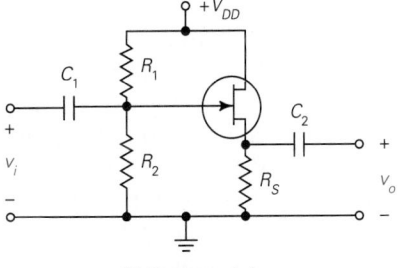

(d) Common drain

The standard basic amplifier form applicable to all cases is shown in Figure 12–3. It contains the following three parameters: input resistance R_{in}, output resistance R_{out}, and voltage gain $A = v_o/v_i$.

FIGURE 12–3

Standard model that can represent any of the basic amplifier configurations.

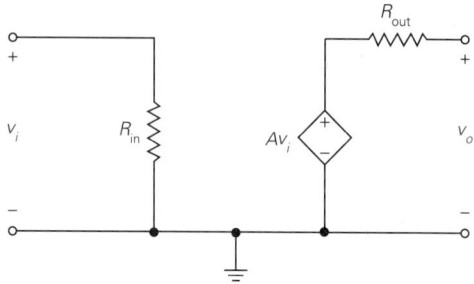

A general tabulation of the midfrequency small-signal gain and resistance relationships for all of the ac-coupled amplifier configurations is given in Table 12–2. For the complete amplifier circuits, R'_{in} and R'_{out} are no longer of prime interest, so the resistance parameters tabulated are R_{in} and R_{out}. For a given amplifier configuration, formulas for A, R_{in}, and R_{out} are provided. When these three parameters are computed, the basic form of Figure 12–3 may then represent the equivalent signal model for any of the basic circuits. The "generic" form of Figure 12–3 is particularly convenient when a cascade of several stages is analyzed.

The formulas given in Table 12–2 are the same as derived earlier except for two modifications. In the common-collector circuit, the input resistance depends on the actual load resistance, and the output resistance depends on the generator resistance. The formulas have been modified to allow these effects.

In the expression of R_{in} for the common-collector circuit, note the term R'_E. Assume that the output is connected to a load resistance R_L through a coupling capacitor. For midfrequency analysis, the value of R'_E is defined as

$$R'_E = R_E \| R_L = \frac{R_E R_L}{R_E + R_L} \tag{12–1}$$

Note that R_E appears without modification in the expression for A. The effect of R_L on A will be accounted for by a loading factor in the normal fashion.*

In the expression for R_{out} for the common-collector circuit, note that a parallel combination of R_B and the generator resistance R_g appears (i.e., $R_B \| R_g$). When the formula was derived in Chapter 10, the input terminals were left open since no generator resistance was considered. With an external generator resistance, the modification shown accounts for the effect of generator resistance on output resistance.

---

* Because of these additional coupling effects between generator and load resistances, a small error is introduced in common-collector loaded gain calculations. However, the results should be sufficiently close for most practical purposes.

TABLE 12–2
Summary of midfrequency small-signal amplifier relationships

BJT Circuits

	Common Emitter		Common Base	Common Collector
A	$-g_m R_C$	$\dfrac{-g_m R_C}{1 + g_m R_{E1}}$	$g_m R_C$	$\dfrac{g_m R_E}{1 + g_m R_E}$
R_{in}	$R_B \| r_\pi$	$R_B \| r_\pi (1 + g_m R_{E1})$	$R_E \| r_\pi \| \dfrac{1}{g_m}$	$R_B \| r_\pi (1 + g_m R'_E)$
R_{out}	R_C	R_C	R_C	$R_E \left\| \left(\dfrac{1}{g_m} + \dfrac{R_B \| R_g}{\beta_o} \right) \right.$
	R_E bypassed	Emitter feedback		$R'_E = R_E \| R_L$
	$R_B = R_1 \| R_2$			R_g = generator resistance

FET Circuits

	Common Source		Common Gate	Common Drain
A	$-g_m R_D$	$\dfrac{-g_m R_D}{1 + g_m R_{S1}}$	$g_m R_D$	$\dfrac{g_m R_S}{1 + g_m R_S}$
R_{in}	R_G	R_G	$R_S \left\| \dfrac{1}{g_m} \right.$	R_G
R_{out}	R_D	R_D	R_D	$R_S \left\| \dfrac{1}{g_m} \right.$
	R_S bypassed	Source feedback		
	$R_G = R_1 \| R_2$			

12–3 SINGLE STAGES WITH LOADING

In the actual application of most single-stage amplifier circuits, there will usually be loading effects at the input and the output that must be considered. A model depicting the interconnection is shown in Figure 12–4. The voltage v_g represents a signal generator (or source), and R_g represents its internal resistance. The resistance R_L represents some additional load resistance connected across the output. The basic amplifier form is shown again in the middle. The parameters for any one of the basic BJT or FET circuits apply to this standard form. Several formulas for additional gain measures are given below the circuit diagram of Figure 12–4. We will now discuss these formulas.

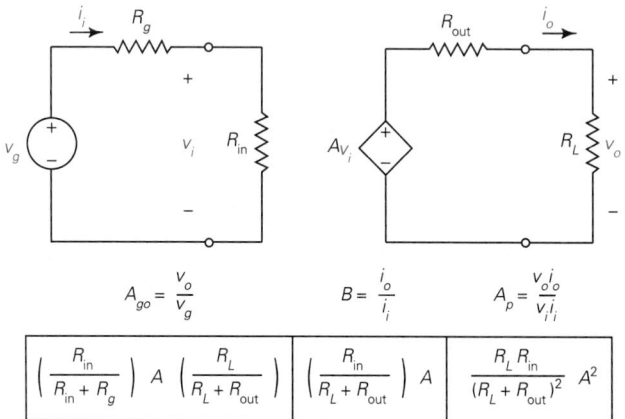

FIGURE 12–4
Model for connection of any single-stage amplifier with generator and load and additional gain relationships.

Loaded Voltage Gain

The loaded voltage gain will be defined as the ratio of the output voltage across an external load resistance R_L to the open-circuit generator voltage v_g. This parameter will be denoted as A_{go}. As we refer to Figure 12–4, the work of Chapter 9 enables us to write down this expression by inspection. We have

$$A_{go} = \frac{R_{in}}{R_{in} + R_g} A \frac{R_L}{R_L + R_{out}}$$ (12–2)

The values of A, R_{in}, and R_{out} for the various configurations are tabulated in Table 12–2.

Current Gain

The current gain for a given amplifier has not been considered up to this point in the text. However, it is a straightforward procedure to determine this gain in terms of the voltage gain and other pertinent amplifier parameters.

In this text the amplifier current gain is denoted as B, and it is defined as the ratio of the signal current i_o flowing out of the amplifier into a load to the input current i_i flowing into the amplifier. Referring to Figure 12–4, we have

$$B = \frac{i_o}{i_i}$$ (12–3)

The current i_o may be expressed in terms of v_o and R_L.

$$i_o = \frac{v_o}{R_L}$$ (12–4)

Similarly, i_i may be expressed in terms of v_i and R_{in}:

$$i_i = \frac{v_i}{R_{in}}$$ (12–5)

Substitution of (12–4) and (12–5) in (12–3) and simplification result in

$$B = \frac{v_o/R_L}{v_i/R_{in}} = \frac{v_o}{v_i}\frac{R_{in}}{R_L}$$ (12–6)

The ratio v_o/v_i under loaded output conditions can be expressed as

$$\frac{v_o}{v_i} = \frac{R_L}{R_L + R_{out}}A$$ (12–7)

Substitution of (12–7) in (12–6) results in

$$B = \frac{R_{in}}{R_L + R_{out}}A$$ (12–8)

We see that the current gain is proportional to the voltage gain, but it includes an additional resistance ratio.

Power Gain

There are a number of different definitions of power gain used in various electronic applications. For our purposes, we consider only the most intuitive form. It will be denoted as A_p, and it is defined as

$$A_p = \frac{P_L}{P_{in}}$$ (12–9)

where P_L is defined as the output signal power accepted by the load P_L, and P_{in} is the signal power accepted by the amplifier input.

The power P_L is

$$P_L = v_o i_o$$ (12–10)

where v_o is defined under loaded conditions. The input power P_{in} is

$$P_{in} = v_i i_i$$ (12–11)

The power gain is then

$$A_p = \frac{P_L}{P_{\text{in}}} = \frac{v_o i_o}{v_i i_i} = \frac{v_o}{v_i} \frac{i_o}{i_i}$$ (12–12)

Applying the relationship of (12–7) and the expression for $i_o/i_i = B$ from (12–8), we obtain

$$A_p = \frac{R_L R_{\text{in}}}{(R_L + R_{\text{out}})^2} A^2$$ (12–13)

The power gain is thus proportional to the square of the voltage gain.

EXAMPLE 12–1 A certain common-emitter circuit with generator and external load connected is shown in Figure 12–5(a). An anlysis is to be performed in the midfrequency range.

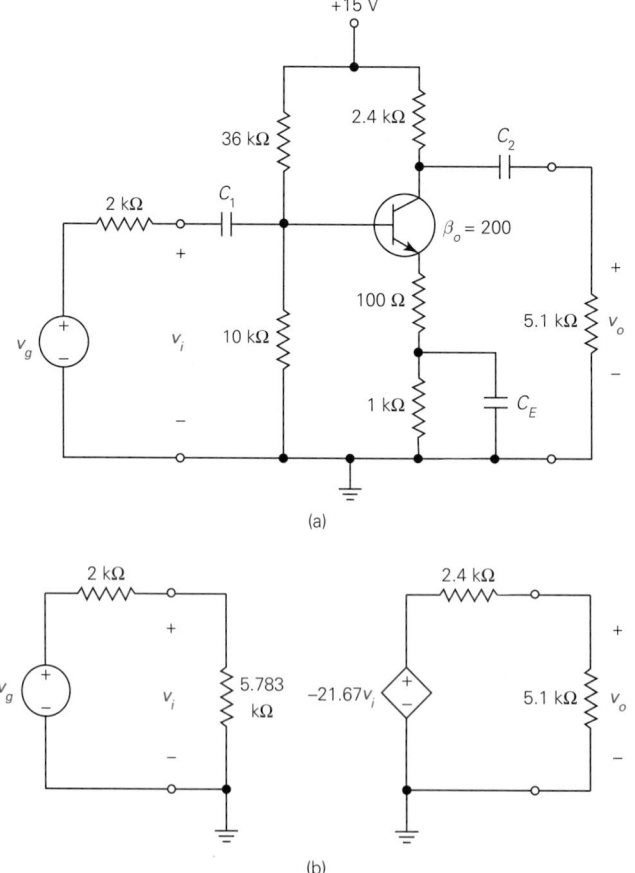

(a)

(b)

FIGURE 12–5
Circuit of Example 12–1 and midfrequency model.

(a) Determine the following properties of the amplifier: A, R_{in}, and R_{out}. **(b)** Draw a general standard equivalent circuit from the amplifier showing source and load connections. **(c)** From the model of **(b)**, determine A_{go}, B, and A_p.

Solution
(a) This part of the solution is a review of earlier work. First, the dc collector current must be determined. The voltage V_{TH} is

$$V_{TH} = \frac{10 \text{ k}\Omega}{10 \text{ k}\Omega + 36 \text{ k}\Omega} \times 15 = 3.261 \text{ V} \tag{12-14}$$

The net dc emitter resistance is $1 \text{ k}\Omega + 100 \text{ }\Omega = 1100 \text{ }\Omega$, and the dc current I_C is

$$I_C = \frac{3.261 - 0.7}{1100} = 2.33 \text{ mA} \tag{12-15}$$

The value of g_m is

$$g_m = 40I_C = 40 \times 2.33 \times 10^{-3} = 0.093 \text{ S} \tag{12-16}$$

The value of r_π is

$$r_\pi = \frac{\beta_o}{g_m} = \frac{200}{0.093} = 2.15 \text{ k}\Omega \tag{12-17}$$

Referring to Table 12–2, the three parameters of the basic amplifier circuit *without loading* are determined as follows:

$$A = \frac{-g_m R_C}{1 + g_m R_{E1}} = \frac{-0.093 \times 2.4 \times 10^3}{1 + 0.093 \times 100} = -21.67 \tag{12-18}$$

$$\begin{aligned} R_{in} &= R_B \| r_\pi (1 + g_m R_{E1}) \\ &= 36{,}000 \| 10{,}000 \| 2150(1 + 9.3) \\ &= 36{,}000 \| 10{,}000 \| 22{,}150 \text{ k}\Omega = 5.783 \text{ k}\Omega \end{aligned} \tag{12-19}$$

$$R_{out} = R_C = 2.4 \text{ k}\Omega \tag{12-20}$$

(b) The standard equivalent circuit depicting the amplifier properties is shown in Figure 12–5(b). If desired, the polarity of the dependent source could be reversed, and its value would then be $+21.67v_i$.

(c) The properties of the complete amplifier circuit are now determined as follows:

$$\begin{aligned} A_{go} &= \frac{R_{in}}{R_{in} + R_g} A \frac{R_L}{R_L + R_{out}} \\ &= \frac{5.783 \text{ k}\Omega}{5.783 \text{ k}\Omega + 2 \text{ k}\Omega} \times (-21.67) \times \frac{5.1 \text{ k}\Omega}{5.1 \text{ k}\Omega + 2.4 \text{ k}\Omega} \\ &= 0.743 \times (-21.67) \times 0.68 = -10.95 \end{aligned} \tag{12-21}$$

$$B = \frac{R_{in}}{R_L + R_{out}} A = \frac{5783}{5100 + 2400} \times (-21.67) = -16.71 \qquad \textbf{(12–22)}$$

$$A_P = \frac{R_L R_{in}}{(R_L + R_{out})^2} A^2 = \frac{5100 \times 5783}{(5100 + 2400)^2} \times (-21.67)^2 = 246.2 \qquad \textbf{(12–23)}$$

Note that the magnitude of the loaded voltage gain A_{go} is considerably less than that of the unloaded gain A. Note also that the magnitude of the current gain and the power gain of this amplifier are both greater than unity.

EXAMPLE 12–2 A certain common-collector circuit with source and external load connected is shown in Figure 12–6(a). An analysis is to be performed in the midfrequency range. **(a)** Determine the following parameters of the amplifier: A, R_{in}, and R_{out}. **(b)** Draw a general standard equivalent circuit for the amplifier showing source and load connections. **(c)** From the model of **(b)**, determine A_{go}, B, and A_p.

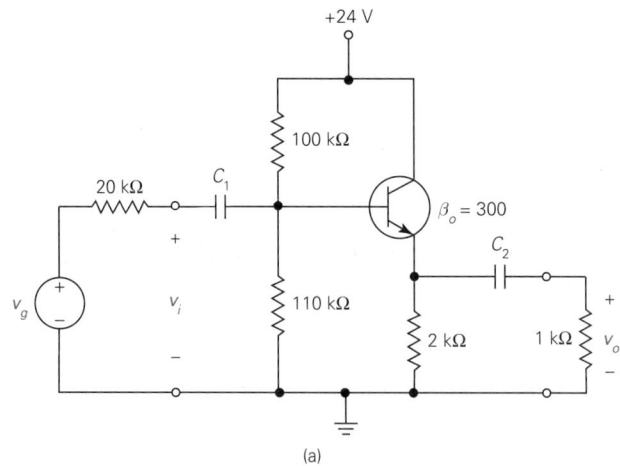

(a)

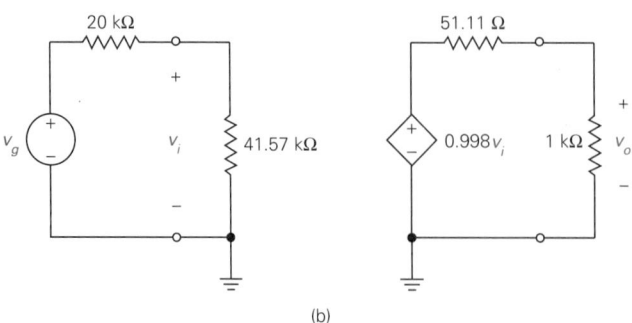

(b)

FIGURE 12–6
Circuit of Example 12–2 and midfrequency model.

Solution

(a) The dc collector current is first determined by the standard procedure. The voltage V_{TH} is

$$V_{TH} = \frac{110\text{ k}\Omega}{110\text{ k}\Omega + 100\text{ k}\Omega} \times 24 = 12.57\text{ V} \tag{12-24}$$

The collector current is

$$I_C = \frac{12.57 - 0.7}{2\text{ k}\Omega} = 5.94\text{ mA} \tag{12-25}$$

The value of g_m is

$$g_m = 40I_C = 40 \times 5.94 \times 10^{-3} = 0.238\text{ S} \tag{12-26}$$

The value of r_π is

$$r_\pi = \frac{\beta_o}{g_m} = \frac{300}{0.238} = 1261\ \Omega \tag{12-27}$$

Referring to Table 12–2, the amplifier parameters are now determined. The gain A is

$$A = \frac{g_m R_E}{1 + g_m R_E} = \frac{0.238 \times 2 \times 10^3}{1 + 0.238 \times 2 \times 10^3} = 0.998 \tag{12-28}$$

To determine R_{in}, it is first necessary to determine R'_E, which is

$$R'_E = R_E \| R_L = 2\text{ k}\Omega \| 1\text{ k}\Omega = 666.7\ \Omega \tag{12-29}$$

The input resistance is then

$$\begin{aligned}
R_{\text{in}} &= R_B \| r_\pi (1 + g_m R'_E) \\
&= 10^5 \| 110{,}000 \| 1261(1 + 0.238 \times 666.7) \\
&= 10^5 \| 110{,}000 \| 201{,}300 = 41{,}570\ \Omega
\end{aligned} \tag{12-30}$$

The output resistance is determined as

$$\begin{aligned}
R_{\text{out}} &= R_E \| \left(\frac{1}{g_m} + \frac{R_B \| R_g}{\beta_o} \right) \\
&= 2000 \| \left(\frac{1}{0.238} + \frac{10^5 \| 110{,}000 \| 20{,}000}{300} \right) \\
&= 2000 \| (4.202 + 48.246) \\
&= 51.11\ \Omega
\end{aligned} \tag{12-31}$$

(b) The standard equivalent circuit displaying the amplifier properties is shown in Figure 12–6(b).

(c)   The properties of the complete amplifier circuit will now be determined. The voltage gain A_{go} is

$$A_{go} = \frac{R_{in}}{R_{in} + R_g} A \frac{R_L}{R_L + R_{out}}$$

$$= \frac{41,570}{41,570 + 20,000} \times 0.998 \times \frac{1000}{1000 + 51.11} \qquad \textbf{(12–32)}$$

$$= 0.675 \times 0.998 \times 0.951 = 0.641$$

The current gain is

$$B = \frac{R_{in}}{R_L + R_{out}} A = \frac{41,570}{1000 + 51.11} \times 0.998 = 39.4 \qquad \textbf{(12–33)}$$

$$A_p = \frac{R_L R_{in}}{(R_L + R_{out})^2} A^2 = \frac{1000 \times 41,570}{(1000 + 51.11)^2} \times (0.998)^2 = 37.4 \qquad \textbf{(12–34)}$$

Note that while A is very close to unity, A_{go} is considerably less than unity. The reduction in net voltage gain is largely a result of the interaction at the input. Although the voltage gain is less than unity, both the current gain and the power gain are greater than unity.

EXAMPLE 12–3

A certain common-source circuit with source and external load connected is shown in Figure 12–7(a). An analysis is to be performed in the midfrequency range. **(a)** Determine the following properties of the amplifier: A, R_{in}, and R_{out}. **(b)** Draw a general standard equivalent circuit for the amplifier showing source and load connections. **(c)** From the model of **(b)**, determine A_{go}, B, and A_p.

Solution
(a)   Referring to Table 12–2, the amplifier properties without loading are determined as follows:

$$A = -g_m R_D = -12 \times 10^{-3} \times 2 \times 10^3 = -24 \qquad \textbf{(12–35)}$$

$$R_{in} = R_G = 200 \text{ k}\Omega \| 680 \text{ k}\Omega = 154.5 \text{ k}\Omega \qquad \textbf{(12–36)}$$

$$R_{out} = 2 \text{ k}\Omega \qquad \textbf{(12–37)}$$

(b)   The standard equivalent circuit showing the amplifier properties is given in Figure 12–7(b).

(c)   The properties of the complete amplifier will now be determined.

$$A_{go} = \frac{R_{in}}{R_{in} + R_g} A \frac{R_L}{R_L + R_{out}} = \frac{154.5 \text{ k}\Omega}{154.5 \text{ k}\Omega + 50 \text{ k}\Omega} \times (-24) \times \frac{5 \text{ k}\Omega}{5 \text{ k}\Omega + 2 \text{ k}\Omega} \qquad \textbf{(12–38)}$$

$$= 0.756 \times (-24) \times 0.714 = -13.0$$

$$B = \frac{R_{in}}{R_L + R_{out}} A = \frac{154.5 \text{ k}\Omega}{5 \text{ k}\Omega + 2 \text{ k}\Omega} \times (-24) = -530 \qquad \textbf{(12–39)}$$

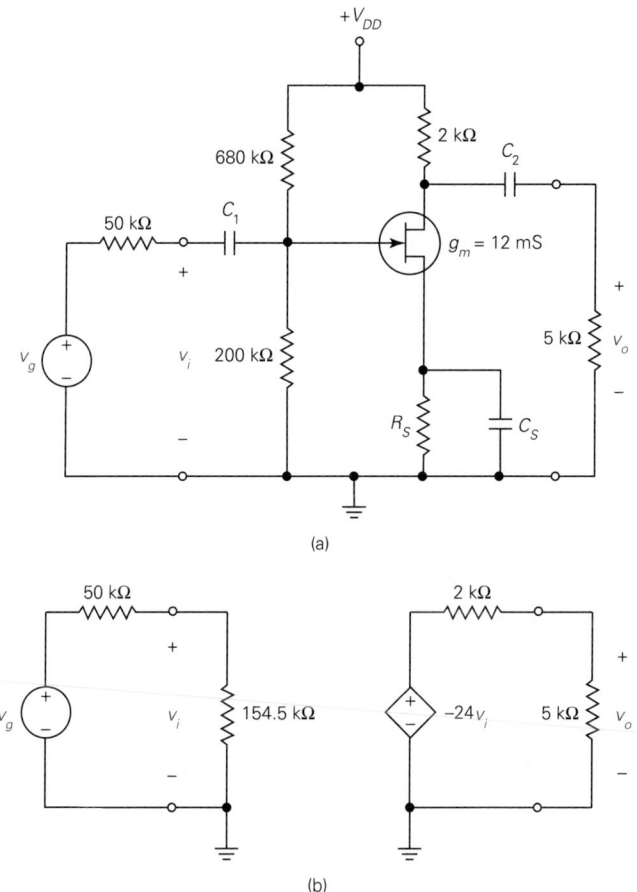

(a)

(b)

FIGURE 12–7
Circuit of Example 12–3 and midfrequency model.

$$A_p = \frac{R_L R_{in}}{(R_L + R_{out})^2} A^2 = \frac{5000 \times 154,500}{(5000 + 2000)^2} \times (-24)^2 = 9081 \qquad \textbf{(12–40)}$$

The very large values of current gain and power gain may be misleading. They are a result of the fact that since the input current is so small, the ratio of output current to input current is a very large value.

EXAMPLE 12–4   A certain common-gate circuit with source and external load connected is shown in Figure 12–8(a). An analysis is to be performed in the midfrequency range. **(a)** Determine the following properties of the amplifier: A, R_{in}, and R_{out}. **(b)** Draw a general standard equivalent circuit for the amplifier showing source and load connections. **(c)** From the model of **(b)**, determine A_{go}, B, and A_p.

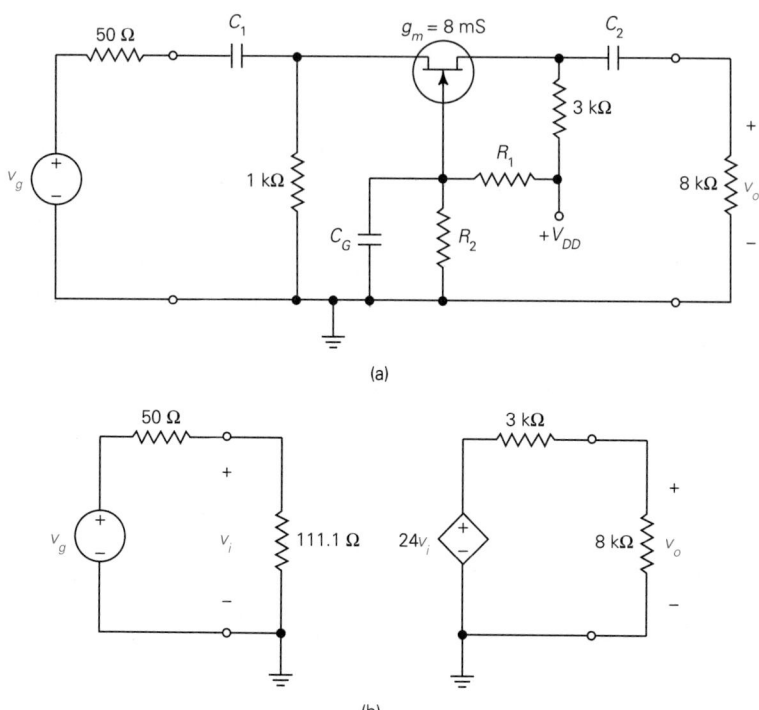

(a)

(b)

FIGURE 12–8
Circuit of Example 12–4 and midfrequency model.

Solution

(a) Referring to Table 12–2, the amplifier properties are determined as follows:

$$A = g_m R_D = 8 \times 10^{-3} \times 3 \times 10^3 = 24 \qquad (12\text{–}41)$$

$$R_{in} = R_S \Big\| \frac{1}{g_m} = 1\,k\Omega \Big\| \frac{1}{0.008} = 1\,k\Omega \| 125\,\Omega = 111.1\,\Omega \qquad (12\text{–}42)$$

$$R_{out} = R_D = 3\,k\Omega \qquad (12\text{–}43)$$

(b) The standard equivalent circuit showing the amplifier properties is given in Figure 12–8(b).

(c) The properties of the complete amplifier will now be determined.

$$
\begin{aligned}
A_{go} &= \frac{R_{in}}{R_{in} + R_g} A \frac{R_L}{R_L + R_{out}} \\
&= \frac{111.1}{111.1 + 50} \times 24 \times \frac{8\,k\Omega}{8\,k\Omega + 3\,k\Omega} \qquad (12\text{–}44) \\
&= 0.690 \times 24 \times 0.727 = 12.04
\end{aligned}
$$

$$B = \frac{R_{in}}{R_L + R_{out}} A = \frac{111.1}{8000 + 3000} \times 24 = 0.242 \qquad \textbf{(12–45)}$$

$$A_p = \frac{R_L R_{in}}{(R_L + R_{out})^2} A^2 = \frac{8000 \times 111.1}{(8000 + 3000)^2} \times (24)^2 = 4.23 \qquad \textbf{(12–46)}$$

A few comments are in order. First, even though the generator has a relatively low output resistance (50 Ω), there is still a significant loading effect at the input due to the low input resistance of the amplifier (111.1 Ω). If the generator output resistance had been much larger, all of the potential voltage gain could have been lost because of attenuation at the input. Thus the common-gate amplifier and its BJT counterpart, the common-base amplifier, require low-resistance driving sources to achieve any real gain. In high-frequency circuits, this can be accomplished with transformer coupling at the input.

A second comment is to note that the current gain is less than unity, a characteristic of both the common-gate and common-base circuits. However, the power gain is greater than unity as expected.

12–4 CASCADE AMPLIFIERS

In the preceding section, source and load connections were made to various single-stage amplifiers. In this section the process is extended to include a cascade of two or more amplifier stages.

Depending on the intended application, almost any conceivable cascade combination of amplifier stages is possible. In fact, BJT and FET active devices may be used together when the properties of each are to be exploited.

All the basic material required for this purpose has been covered in preceding chapters, and the individual stage properties have been summarized earlier in this chapter. The process of analysis involves first determining the properties of each individual stage. By determining interaction factors between stages, along with source and output loading factors, the complete properties of the composite amplifier can be determined.

The process is developed through several representative examples. Not all possible cases can be covered, but the general approaches for analyzing other circuits should evolve from these examples.

All feedback considered in this section is limited to feedback within a stage, such as the common emitter with emitter feedback, for example. We are not yet ready to deal with feedback around several stages.

EXAMPLE 12–5 The cascade amplifier of Figure 12–9(a) consists of two *npn* BJT common-emitter stages, each having emitter feedback. Determine for the composite circuit the net midfrequency gain between source voltage v_g and loaded output voltage v_o (i.e., $A_{go} = v_o/v_g$).

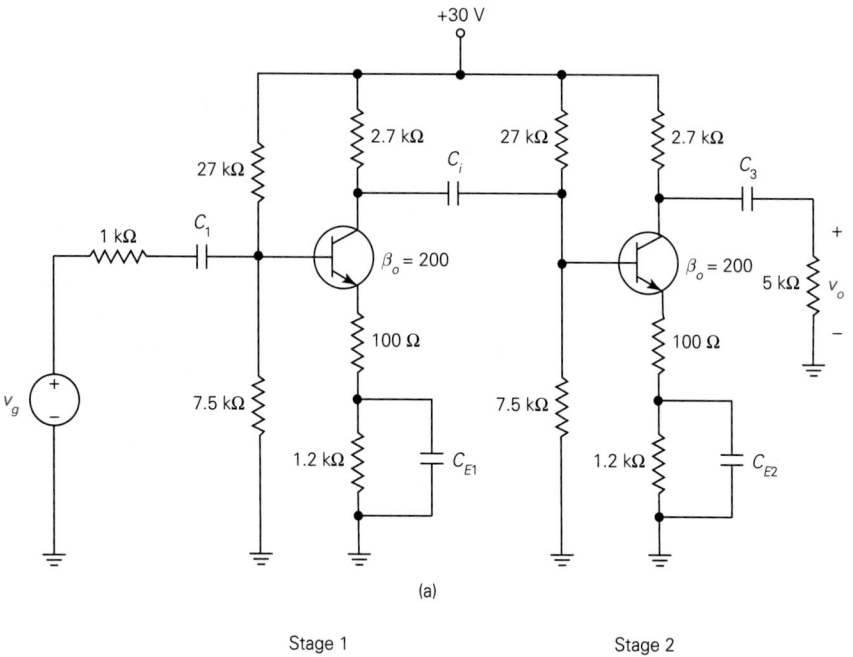

(a)

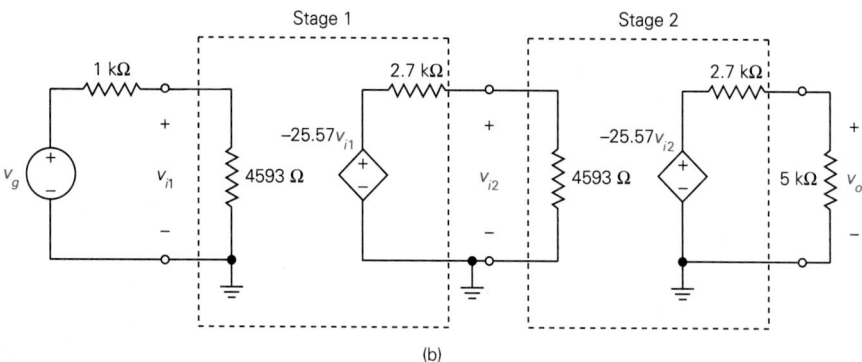

(b)

FIGURE 12–9
Circuit of Example 12–5 and midfrequency model.

Solution
It can be readily observed that the two common-emitter stages are identical. Thus, it will be necessary to analyze only one of the stages in detail. First, the dc operating point must be determined. The voltage V_{TH} is determined as

$$V_{TH} = \frac{7.5 \text{ k}\Omega}{7.5 \text{ k}\Omega + 27 \text{ k}\Omega} \times 30 = 6.52 \text{ V} \qquad (12\text{–}47)$$

The net dc emitter resistance is $1200 + 100 = 1300 \ \Omega$. The dc collector current is

$$I_C = \frac{6.52 - 0.7}{1300} = 4.48 \text{ mA} \qquad (12\text{–}48)$$

The value of g_m is

$$g_m = 40I_C = 40 \times 4.48 \times 10^{-3} = 0.179 \text{ S} \qquad \text{(12–49)}$$

The value of r_π is

$$r_\pi = \frac{\beta_o}{g_m} = \frac{200}{0.179} = 1117 \,\Omega \qquad \text{(12–50)}$$

The gain of each stage is

$$A = -\frac{g_m R_C}{1 + g_m R_{E1}} = -\frac{0.179 \times 2.7 \times 10^3}{1 + 0.179 \times 100} = -25.57 \qquad \text{(12–51)}$$

The input resistance of each stage is

$$R_{in} = R_B \| r_\pi (1 + g_m R_{E1}) = 27{,}000 \| 7500 \| 1117(18.9)$$
$$= 4593 \,\Omega \qquad \text{(12–52)}$$

The output resistance of each stage is

$$R_{out} = 2.7 \text{ k}\Omega \qquad \text{(12–53)}$$

It should be noted that the input resistance of the composite amplifier as seen by the source is the input resistance of the first stage (i.e., 4593 Ω). Furthermore, the output resistance looking back from the load is the output resistance of the second stage (i.e., 2.7 kΩ).

A block diagram depicting the interaction between stages, source, and load is shown in Figure 12–9(b). The total gain A_{go} may be expressed as follows:

$$A_{go} = \frac{4593}{4593 + 1000} \times (-25.57) \times \frac{4593}{4593 + 2700} \times (-25.57)$$

$$\times \frac{5000}{5000 + 2700} \qquad \text{(12–54)}$$

$$= 0.8212 \times (-25.57) \times 0.6298 \times (-25.57) \times 0.6494 = 219.6$$

Note that the net gain is noninverting since there is an even number of inversions.

EXAMPLE 12–6

The three-stage amplifier of Figure 12–10 consists of a common-source input stage, a common emitter with emitter feedback stage, and a common-collector output stage. Determine the net loaded gain $A_{go} = v_o/v_g$ in the midfrequency range.

Solution

Each of the three stages will be analyzed individually. The calculations are grouped separately as follows:

Common source:

$$A = -g_m R_D = -15 \times 10^{-3} \times 2.2 \times 10^3 = -33 \qquad \text{(12–55)}$$
$$R_{in} = R_G = 220 \text{ k}\Omega \| 680 \text{ k}\Omega = 166 \text{ k}\Omega \qquad \text{(12–56)}$$
$$R_{out} = 2.2 \text{ k}\Omega \qquad \text{(12–57)}$$

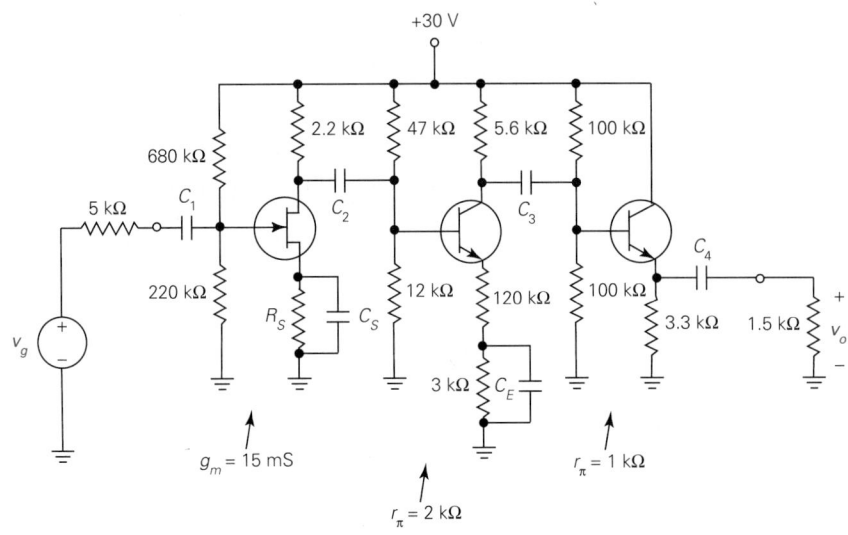

FIGURE 12–10
Circuit of Example 12–6.

Common emitter:

$$V_{\text{TH}} = \frac{12 \text{ k}\Omega}{12 \text{ k}\Omega + 47 \text{ k}\Omega} \times 30 = 6.102 \text{ V} \tag{12-58}$$

$$I_C = \frac{6.102 - 0.7}{3000 + 120} = 1.73 \text{ mA} \tag{12-59}$$

$$g_m = 40I_C = 40 \times 1.73 \times 10^{-3} = 0.0692 \text{ S} \tag{12-60}$$

$$A = \frac{-g_m R_C}{1 + g_m R_{E1}} = \frac{-0.0692 \times 5.6 \times 10^3}{1 + 0.0692 \times 120} = -41.65 \tag{12-61}$$

$$R_{\text{in}} = R_B \| r_\pi (1 + g_m R_{E1})$$
$$= 12{,}000 \| 47{,}000 \| 2000(1 + 0.0692 \times 120) = 6315 \ \Omega \tag{12-62}$$

$$R_{\text{out}} = R_C = 5.6 \text{ k}\Omega \tag{12-63}$$

Common collector:

$$V_{\text{TH}} = \frac{100 \text{ k}\Omega}{100 \text{ k}\Omega + 100 \text{ k}\Omega} \times 30 = 15 \text{ V} \tag{12-64}$$

$$I_C = \frac{15 - 0.7}{3.3 \times 10^3} = 4.33 \text{ mA} \tag{12-65}$$

$$g_m = 40I_C = 40 \times 4.33 \times 10^{-3} = 0.173 \text{ S} \tag{12-66}$$

$$A = \frac{g_m R_E}{1 + g_m R_E} = \frac{0.173 \times 3.3 \times 10^3}{1 + 0.173 \times 3{:}3 \times 10^3} = 0.998 \tag{12-67}$$

$$R'_E = R_E \| R_L = 3.3 \text{ k}\Omega \| 1.5 \text{ k}\Omega = 1.031 \text{ k}\Omega \tag{12-68}$$

$$\begin{aligned} R_{\text{in}} &= R_B \| r_\pi (1 + g_m R'_E) \\ &= 10^5 \| 10^5 \| 1000(1 + 0.173 \times 1031) = 39.1 \text{ k}\Omega \end{aligned} \tag{12-69}$$

Before computing R_{out}, it is necessary to determine β_o. We have

$$\beta_o = g_m r_\pi = 0.173 \times 10^3 = 173 \tag{12-70}$$

In computing R_{out}, the "generator" resistance R_g is the output resistance of the common emitter stage (i.e., 5.6 kΩ).

$$\begin{aligned} R_{\text{out}} &= R_E \left\| \left(\frac{1}{g_m} + \frac{R_B \| R_g}{\beta_o} \right) \right. \\ &= 3300 \left\| \left(\frac{1}{0.173} + \frac{50{,}000 \| 5600}{173} \right) \right. \\ &= 3300 \| (5.78 + 29.11) = 34.52 \ \Omega \end{aligned} \tag{12-71}$$

Although an equivalent circuit showing the parameters of the three stages could be readily constructed, that process will be sidestepped in this problem. Instead, an expression for the net gain will be written as follows:

$$\begin{aligned} A_{go} &= \frac{v_o}{v_g} \\ &= \frac{166 \text{ k}\Omega}{166 \text{ k}\Omega + 5 \text{ k}\Omega} \times (-33) \times \frac{6315}{6315 + 2200} \times (-41.65) \\ &\quad \times \frac{39.1 \text{ k}\Omega}{39.1 \text{ k}\Omega + 5.6 \text{ k}\Omega} \times 0.998 \times \frac{1500}{1500 \ \Omega + 34.52} \\ &= 0.9708 \times (-33) \times 0.7416 \times (-41.65) \times 0.8747 \times 0.998 \times 0.9775 \\ &= 844 \end{aligned} \tag{12-72}$$

12-5 CASCODE AMPLIFIER

A circuit that offers some significant advantages at high frequencies is the **cascode amplifier.** The basic form of the circuit utilizing JFETs is shown in Figure 12–11(a). The circuit may be considered as a common-source amplifier stage driving a common-gate amplifier stage. The input resistance of the common-gate stage acts as the load for the common-source stage.

A small-signal midfrequency equivalent circuit, after some slight approximations have been made, is shown in Figure 12–11(b). Neglecting gate current, the input resistance of the common-source stage is R_3. Since there is no source resistor for the common-gate stage, its input resistance is simply $1/g_{m2}$, as shown. The dependent current source $g_{m1}v_i$ for the first stage is effectively connected directly to this input resistance as shown.

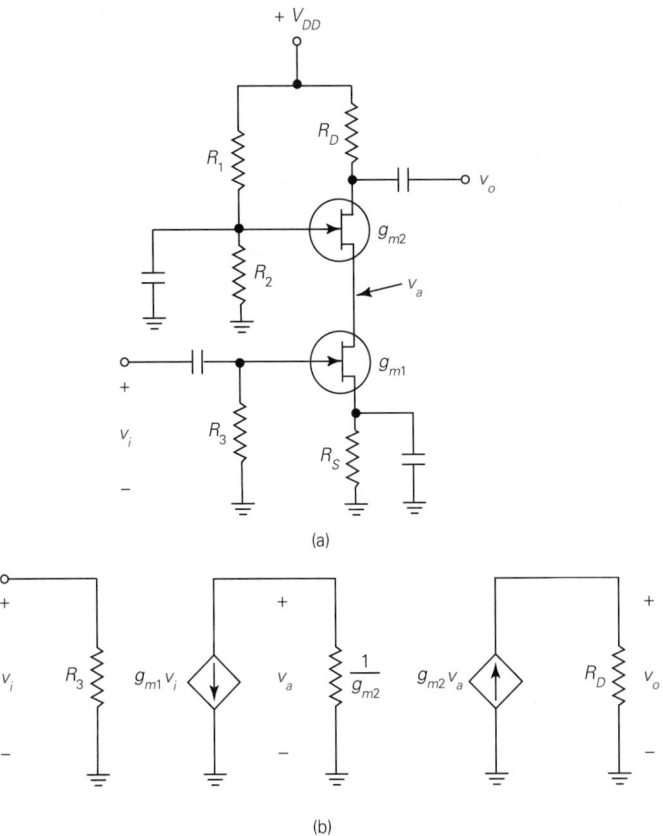

FIGURE 12–11
Cascode amplifier using JFETs and small-signal model.

The gate–source controlling voltage v_{gs} for the common-gate stage is $v_{gs} = -v_a$, where v_a is the signal at the junction point between stages with respect to ground. Thus, the dependent source in the second stage directed toward ground would be $g_{m2}v_{gs} = -g_{m2}v_a$. By reversing the current-source direction, the value of the current source is $g_{m2}v_a$ as shown.

An analysis of the circuit will now be performed. The voltage v_a is

$$v_a = -g_{m1}v_i \times \frac{1}{g_{m2}} = \frac{-g_{m1}}{g_{m2}}v_i \qquad (12\text{–}73)$$

The voltage v_o is given by

$$v_o = g_{m2}v_a R_D \qquad (12\text{–}74)$$

Substituting (12–73) in (12–74) results in

$$v_o = g_{m2}\left(\frac{-g_{m1}}{g_{m2}}v_i\right)R_D = -g_{m1}R_D v_i \qquad (12\text{–}75)$$

The gain A is

$$A = \frac{v_o}{v_i} = -g_{m1}R_D \qquad \textbf{(12–76)}$$

The expression for gain has the same form as that of a single-stage common-source circuit, but the FET transconductance is that of the common-source stage, and the drain resistance is that of the common-gate stage.

The prime virtue of this circuit is not evident from the midfrequency analysis. A concept known as the **Miller effect input capacitance** is developed in Chapter 18. This phenomenon is most pronounced in common-source (and common-emitter) stages and results in serious gain degradation at high frequencies. It turns out that the low input resistance of the common-gate circuit loads the common-source circuit so heavily that the Miller effect capacitance at the input of the cascode circuit is reduced dramatically. The cascode circuit thus offers the high input resistance and gain associated with the common-source circuit, but with improved high-frequency performance. Cascode circuits are often used in the radio-frequency stages of television sets.

12–6 PSPICE EXAMPLE

PSPICE EXAMPLE 12–1

Develop a complete dc and ac PSPICE simulation for the two-stage amplifier of Example 12–5 (Figure 12–9), and use it to determine the voltage gain of the complete amplifier at a frequency of 1 kHz.

Solution

The circuit adapted to PSPICE is shown in Figure 12–12(a), and the code is shown in Figure 12–12(b). Since the capacitor values were not specified in the original problem, representative values of 2 μF each have been assumed for C_1, C_2, and C_3, while values of 200 μF have been assumed for C_{E1} and C_{E2}. In general, much larger capacitors are required for emitter bypass capacitors than for coupling capacitors because of the much smaller equivalent resistance levels looking back into emitters.

The transistor model has been denoted as QMODEL, and it is the *npn* model with BF = 200 (representing β), as indicated in the example.

The computer printout is shown in Figure 12–13. We have chosen in this case to allow a printout of the bias data. These data provide values of all the dc node voltages, the source currents, and the total power dissipation. Unfortunately, the dc transistor currents are not printed, but they could be easily obtained with a dc analysis. However, we can easily check any dc currents by simple calculations from the available data. For example, the collector current of Q_1 can be calculated as

$$I_{C1} = \frac{V_{CC} - \text{V}(4)}{R_{C1}} = \frac{30 - 18.459}{2.7 \times 10^3} = 4.27 \text{ mA} \qquad \textbf{(12–77)}$$

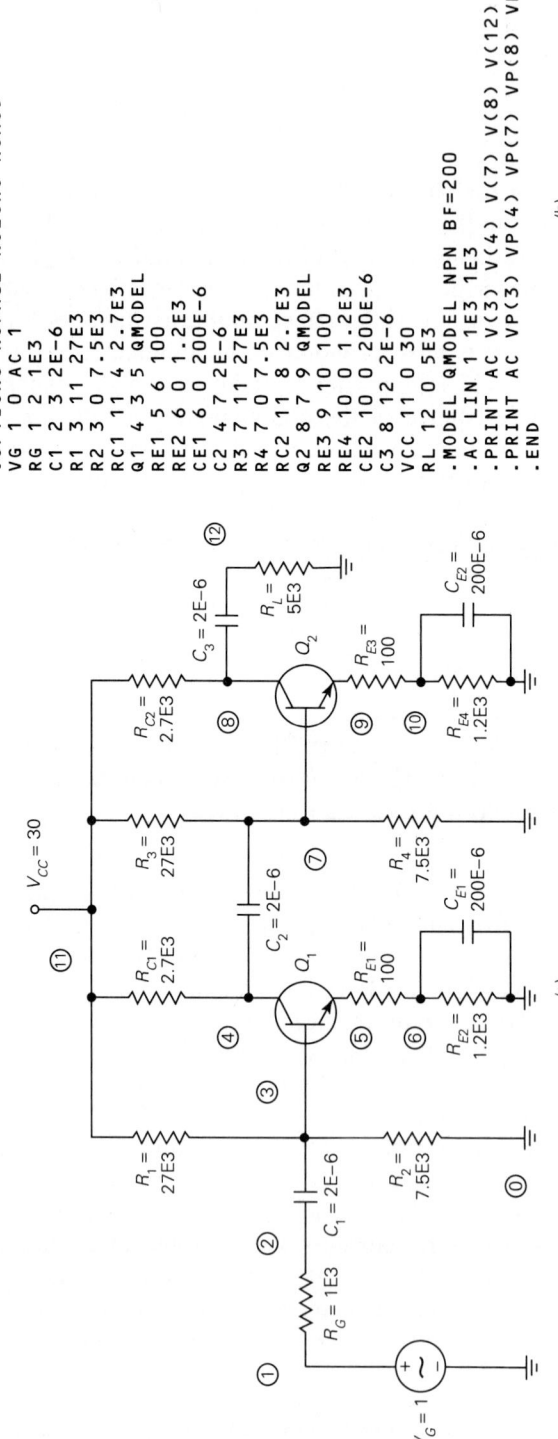

```
PSPICE EXAMPLE 12-1
.OPTIONS NOPAGE NOECHO NOMOD
VG 1 0 AC 1
RG 1 2 1E3
C1 2 3 2E-6
R1 3 11 27E3
R2 3 0 7.5E3
RC1 11 4 2.7E3
Q1 4 3 5 QMODEL
RE1 5 6 100
RE2 6 0 1.2E3
CE1 6 0 200E-6
C2 4 7 2E-6
R3 7 11 27E3
R4 7 0 7.5E3
RC2 11 8 2.7E3
Q2 8 7 9 QMODEL
RE3 9 10 100
RE4 10 0 1.2E3
CE2 10 0 200E-6
C3 8 12 2E-6
VCC 11 0 30
RL 12 0 5E3
.MODEL QMODEL NPN BF=200
.AC LIN 1 1E3 1E3
.PRINT AC V(3) V(4) V(7) V(8) V(12)
.PRINT AC VP(3) VP(4) VP(7) VP(8) VP(12)
.END
```

(b)

(a)

FIGURE 12–12

Circuit and code of PSPICE Example 12–1.

```
PSPICE EXAMPLE 12-1

****        CIRCUIT DESCRIPTION

***************************************************************

.OPTIONS NOPAGE NOECHO NOMOD

****        SMALL SIGNAL BIAS SOLUTION      TEMPERATURE = 27.000 DEG C

   NODE    VOLTAGE    NODE    VOLTAGE    NODE    VOLTAGE    NODE    VOLTAGE
   ( 1)    0.0000     ( 2)    0.0000     ( 3)    6.3963     ( 4)    18.4590
   ( 5)    5.5845     ( 6)    5.1549     ( 7)    6.3963     ( 8)    18.4590
   ( 9)    5.5845     (10)    5.1549     (11)    30.0000    (12)    0.0000

   VOLTAGE SOURCE CURRENTS
   NAME            CURRENT

   VG              0.000E+00
   VCC            -1.030E-02

   TOTAL POWER DISSIPATION     3.09E-01 WATTS


****        AC ANALYSIS                     TEMPERATURE = 27.000 DEG C

   FREQ       V(3)        V(4)        V(7)        V(8)        V(12)
   1.000E+03  8.214E-01   1.312E+01   1.312E+01   2.158E+02   2.158E+02

****        AC ANALYSIS                     TEMPERATURE = 27.000 DEG C

   FREQ       VP(3)       VP(4)       VP(7)       VP(8)       VP(12)
   1.000E+03  7.972E-01  -1.792E+02  -1.782E+02  1.928E+00   2.839E+00
```

FIGURE 12–13
Computer printout of PSPICE Example 12–1.

This differs from the predicted value of 4.48 mA by about 7.8%, which is the normal range of transistor uncertainty for a circuit such as this.

For the ac analysis, both the magnitude and phase angles of various voltages have been computed. These various quantities may be traced from input to output, and the reader is invited to carefully note the values involved. Note that the various phase angles are either very close to 0° or very close to −180°. Slight variations from 0° or ±180° are due to small capacitive reactances produced by the capacitors at 1 kHz.

The overall gain is determined by dividing the ac output voltage by the ac input voltage

$$A = \frac{V(12)}{V(1)} = \frac{215.8}{1} = 215.8 \qquad \textbf{(12–78)}$$

This value differs from the computed text value of 219.6 by about 1.7%. As has been pointed out on several other occasions, the measured output of 215.8 V is pure fiction for the circuit, but it is a convenient choice for gain determination with the linearized small-signal model.

PROBLEMS

Drill Problems

Note: In Problems 12–1 through 12–8, all analysis is to be performed in the midfrequency range. For each circuit, determine, A_{go}, B, and A_p.

12–1. See Figure P12–1.

FIGURE P12–1

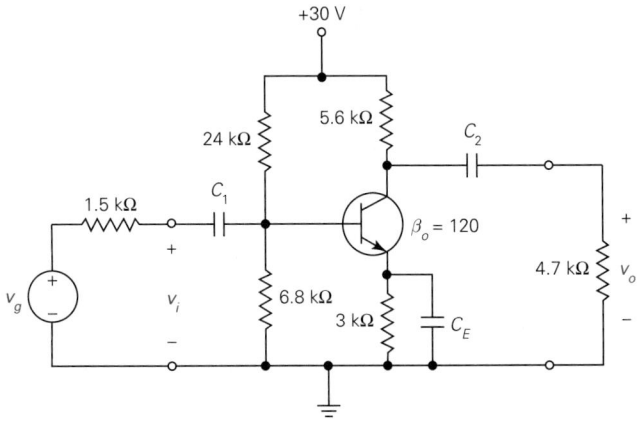

12–2. See Figure P12–2.

FIGURE P12–2

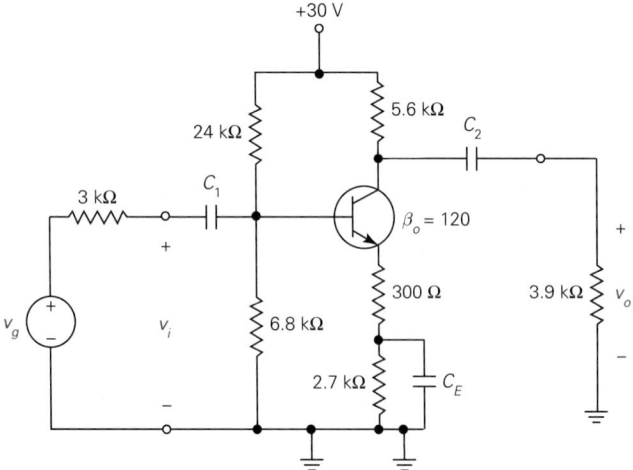

12–3. See Figure P12–3.

FIGURE P12–3

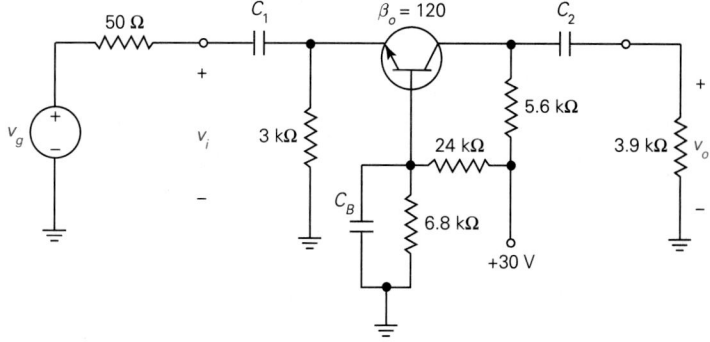

12–4. See Figure P12–4.

FIGURE P12–4

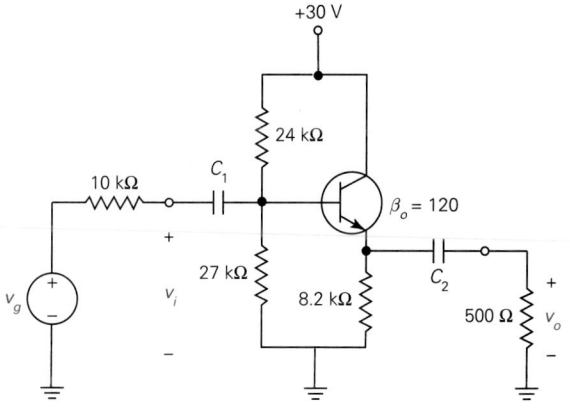

12–5. See Figure P12–5.

FIGURE P12–5

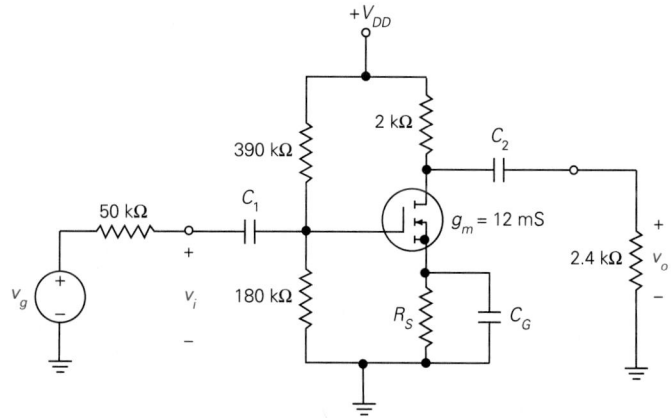

12–6. See Figure P12–6.

FIGURE P12–6

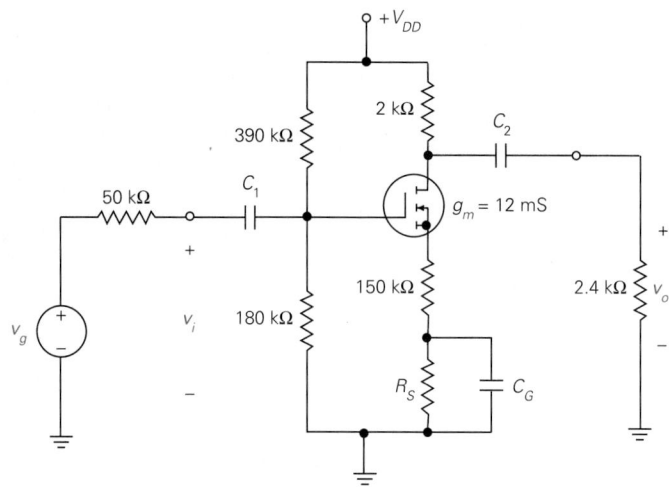

12–7. See Figure P12–7.

FIGURE P12–7

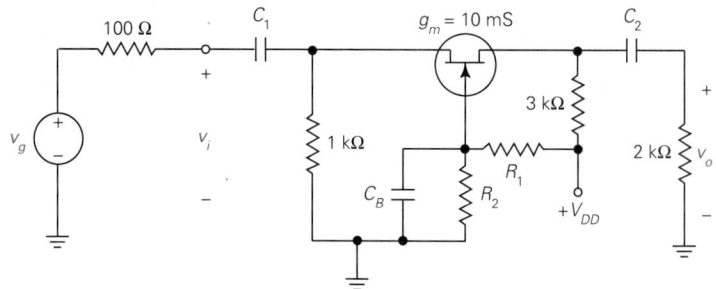

12–8. See Figure P12–8.

FIGURE P12–8

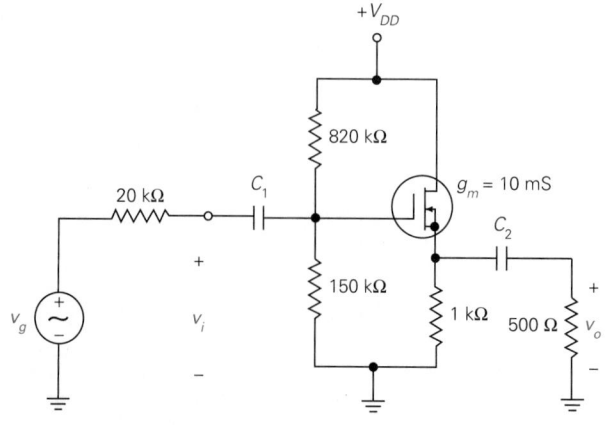

12–9. Determine the net loaded gain $A_{go} = v_o/v_g$ in the midfrequency range for the two-stage amplifier of Figure P12–9.

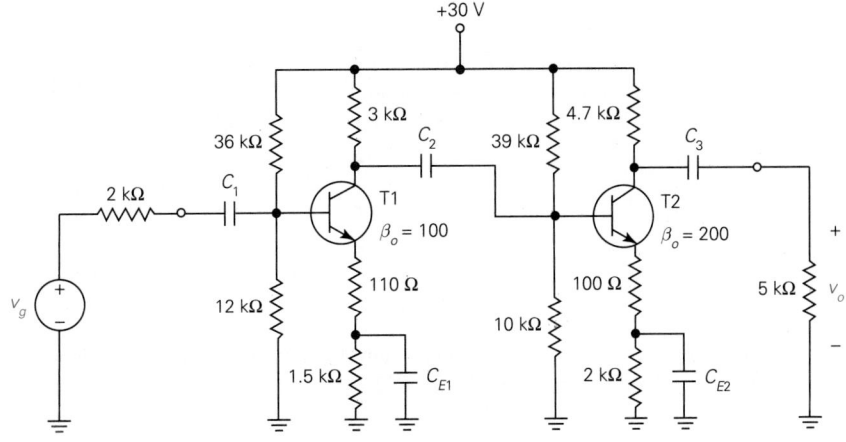

12–10. Determine the net loaded gain $A_{go} = v_o/v_g$ in the midfrequency range for the two-stage amplifier of Figure P12–10.

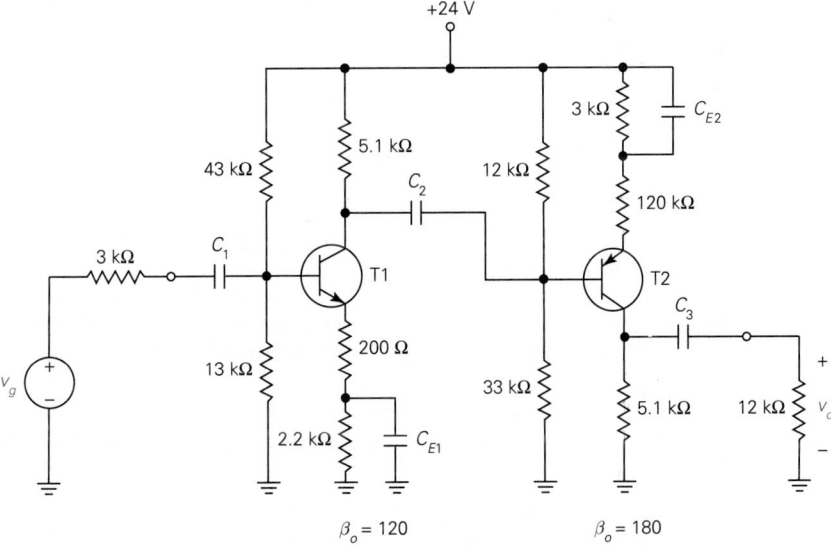

12–11. Determine the net loaded gain $A_{go} = v_o/v_g$ in the midfrequency range for the two-stage amplifier of Figure P12–11.

FIGURE P12–11

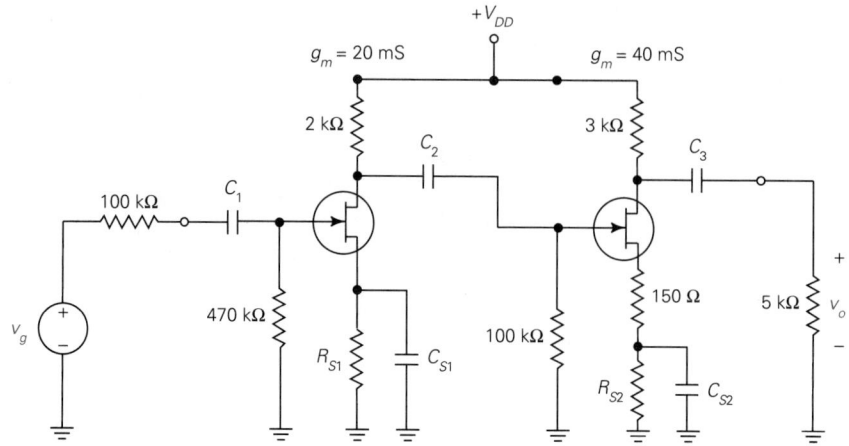

12–12. Determine the net loaded gain $A_{go} = v_o/v_g$ in the midfrequency range for the two-stage amplifier of Figure P12–12.

FIGURE P12–12

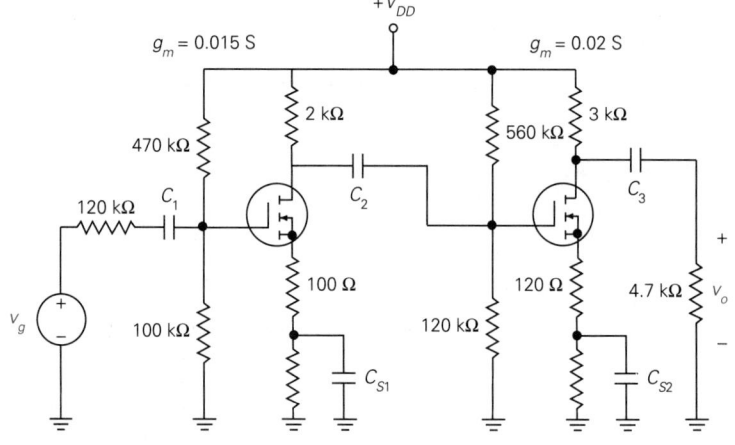

12–13. A cascode amplifier of the form shown in Figure 12–11(a) has $g_{m1} = 12$ mS, $g_{m2} = 10$ mS, and $R_D = 2$ kΩ. **(a)** Determine A. **(b)** If $v_i = 20$ mV, determine v_a and v_o.

12–14. A cascode amplifier of the form shown in Figure 12–11(a) has $g_{m1} = 20$ mS, $g_{m2} = 25$ mS, and $R_D = 1.5$ kΩ. **(a)** Determine A. **(b)** If $v_i = 50$ mV, determine v_a and v_o.

Troubleshooting Problems

12–15. You are troubleshooting the circuit of Figure P12–9 with an oscilloscope set for dc coupling. The dc levels at all points are close to specified values. A signal appears at the collector of T1, but there is no signal at the collector of T2. Which one of the following is the probable trouble? **(a)** C_{E2} open; **(b)** C_3 shorted; **(c)** C_2 shorted; **(d)** C_2 open; **(e)** defective transistor.

12–16. You are troubleshooting the circuit of Figure P12–9 with an oscilloscope set for dc coupling. The dc levels at all points are close to specified values. A signal of normal level appears at the collector of T1, but the signal level at the collector of T2 is much smaller than expected. Which one of the following is the probable trouble? **(a)** C_{E2} open; **(b)** C_3 shorted; **(c)** C_3 open; **(d)** C_2 open; **(e)** defective transistor.

13

DIFFERENTIAL AND OPERATIONAL AMPLIFIERS

OBJECTIVES

After completing this chapter, the reader should be able to:

- Explain the general operation of a discrete-component differential amplifier.
- Define differential input voltage.
- Define common-mode input voltage.
- Define differential gain.
- Define common-mode gain.
- Define common-mode rejection ratio.
- Explain the operation of and analyze a current mirror.
- Explain the operation of an active load.
- Explain the general operation of an operational amplifier.
- Show the manner in which dual power supplies are used to provide dc bias for an op-amp.
- Draw the schematic symbol for an op-amp, and explain the functions of the three signal terminals.
- Explain saturation in an op-amp, and estimate the approximate levels.
- Define input offset voltage for an op-amp.
- Define bias and offset currents in an op-amp.
- Discuss how the output dc offset for an op-amp can be minimized.
- Analyze some of the circuits in the chapter with PSPICE.

13–1 OVERVIEW

The differential amplifier is one of the most important circuits encountered in modern electronic systems. It represents a basic "building block" for many important linear and nonlinear integrated-circuit modules. It is an important key in the transition between discrete amplifier circuits and integrated circuits.

A major portion of this chapter is devoted to analyzing a two-transistor discrete-component differential amplifier. This detailed analysis is intended to assist the reader in comprehending the basic differential amplifier properties. Once this operation is understood, the transition to the integrated-circuit version may be achieved with minimum difficulty.

The last portion of the chapter is devoted to the consideration of the integrated-circuit operational amplifier. An operational amplifier is a high-gain dc-coupled differential amplifier. Emphasis at that point will be on the overall circuit modeling and operation rather than the detailed circuit elements. Typical operational amplifier specifications are investigated in detail.

13–2 DIFFERENTIAL AMPLIFIER

The simplest form of a discrete version of the basic differential amplifier circuit employing two *npn* bipolar junction transistors and two power supplies is shown in Figure 13–1. The differential amplifier normally has no coupling capacitors (i.e., it is **directly coupled**). It is therefore capable of processing frequencies all the way down to dc.

The circuit may have two input voltages (or **dual** inputs), and these voltages as referred to ground are designated as v_{i1} and v_{i2}, respectively. (When a single-input voltage is desired, one input may be connected to ground.) This version

FIGURE 13–1
Basic differential amplifier circuit.

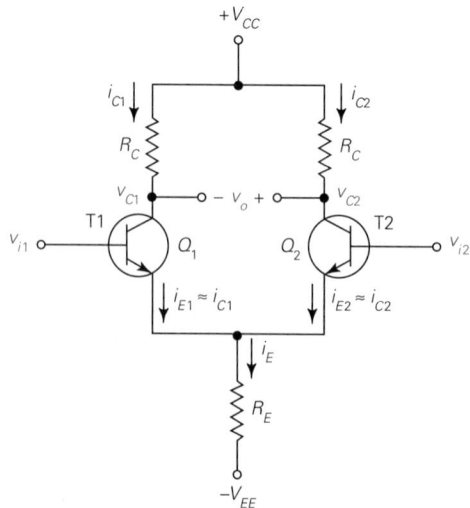

shown has a **double-ended output;** that is, both collector voltages v_{C1} and v_{C2} as referred to ground are possible outputs. Either one of the collector resistances may be omitted, and the output would be at the single collector with the resistance R_C. This latter form is denoted as a **single-ended output.** Because it is more general, the double-ended output form will be retained until the integrated-circuit form is introduced later.

Another possible output voltage of interest is the difference between the collector voltages, denoted as v_o, and it is defined as

$$v_o = v_{C2} - v_{C1} \qquad \textbf{(13–1)}$$

Some difficulty arises in utilizing v_o in this simplified circuit form since it is "floating" (i.e., it is *not* referred to ground). This problem is circumvented in the later integrated-circuit versions, so we will proceed to consider v_o without undue concern at this point.

The two emitters are connected together, and this point is connected through a common resistance R_E to a negative supply voltage $-V_{EE}$. The two transistor collector currents are represented as i_{C1} and i_{C2}, respectively. Although the bases are shown with no connection in this illustration, it is necessary that there be dc connections to ground for bias currents to flow in order for the circuit to function properly. Bias current flows from the positive terminal of the emitter power supply to ground and up to the bases through the external dc base connections.

It is assumed that the two transistors are matched; that is, corresponding dc and dynamic characteristics are essentially the same. Without such a match, the circuit may not function as intended. It is also assumed that β_{dc} for each transistor is very large. In the discussion that follows, it will be assumed in many steps that the two emitter currents i_{E1} and i_{E2} are nearly the same as the two collector currents (i.e., $i_{E1} \approx i_{C1}$ and $i_{E2} \approx i_{C2}$).

The operation of the differential amplifier is somewhat tricky and requires some detailed study for full comprehension. Before developing general equations, we move through a series of circuit diagrams with representative voltages and currents chosen to illustrate some of the key features of circuit operation. These results represent some typical (rather than general) results, but they should provide the reader with some insight about circuit operation.

The first step in this process is to consider only one of the transistors in the differential amplifier with the other completely removed. As we will see, the function is completely different with one transistor removed, and this will provide a way of comparing operation when both transistors are present.

Consider the circuit of Figure 13–2(a) containing only T1. The two power supply voltages are +15 V and −15 V. The base of T1 is momentarily grounded, so $v_{i1} = 0$. The emitter resistance is 14.3 kΩ, and the collector resistance is 8 kΩ.

A KVL loop will be formed by starting at the grounded base, moving over to the emitter, down through the emitter resistance to the negative power supply terminal, and back to ground again. This equation reads

$$v_{BE1} + 14.3 \times 10^3 i_{E1} - 15 = 0 \qquad \textbf{(13–2)}$$

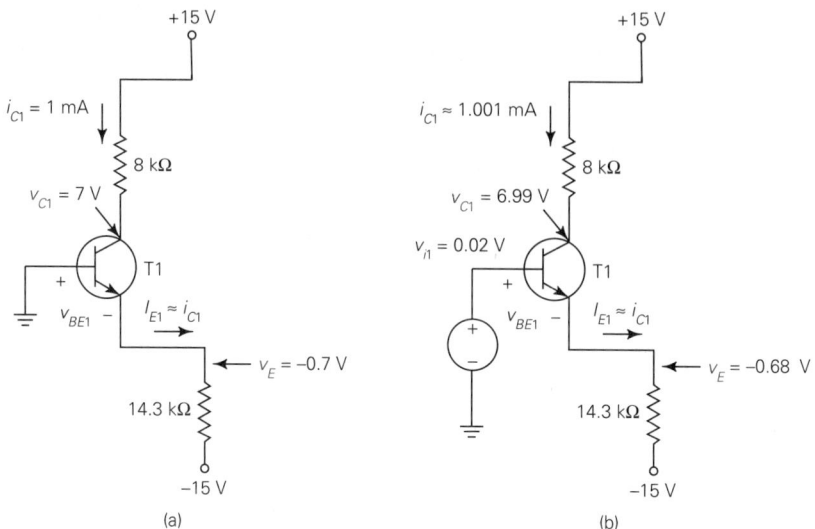

FIGURE 13–2

Illustration of representative values for one transistor in differential amplifier with other transistor removed.

We then solve for i_{E1} and assume that $v_{BE1} = 0.7$ V. The result is

$$i_{E1} \approx i_{C1} = \frac{15 - v_{BE1}}{14.3 \times 10^3} = \frac{15 - 0.7}{14.3 \times 10^3} = 1 \text{ mA} \tag{13-3}$$

The voltage drop across R_C is

$$v_{R_C} = 1 \times 10^{-3} \times 8 \times 10^3 = 8 \text{ V} \tag{13-4}$$

The collector voltage v_{C1} referred to ground is

$$v_{C1} = 15 - 8 = 7 \text{ V} \tag{13-5}$$

The emitter voltage is designated as v_E, and since $v_{BE1} = v_{B1} - v_E$, the value of v_E is

$$v_E = v_{B1} - v_{BE1} = 0 - v_{BE1} = -0.7 \text{ V} \tag{13-6}$$

where $v_{B1} = 0$ has been used. Finally, the collector–emitter voltage v_{CE1} is

$$v_{CE1} = v_{C1} - v_{E1} = 7 - (-0.7) = 7.7 \text{ V} \tag{13-7}$$

The various voltages and currents are labeled on Figure 13–2(a).

    The single-transistor circuit as shown can be viewed in two ways. First, it has the form of a common-emitter circuit with feedback (i.e., with unbypassed emitter resistance). However, the value of this emitter resistance is very large relative to the values normally left unbypassed. Consequently, the gain of the circuit as a common-emitter circuit is very low as we shall see shortly. In fact, the gain magnitude is less than unity.

A second interpretation of the single-transistor circuit is that it can be viewed as an emitter follower between the base and emitter. Although the emitter-follower circuit does not normally have a collector resistance, the presence of the collector resistance does not change the function of the circuit as an emitter follower at low frequencies except to limit the range of active-region operation (i.e., saturation occurs at a lower input voltage due to the drop across R_C).

With the preceding points of view in mind, assume that the base is connected to a very small dc voltage as shown in Figure 13–2(b). The value $v_{i1} = 20$ mV has been selected for illustration. Because of the emitter-follower action of the circuit, the emitter voltage "follows" the input, and it increases by about 20 mV. The new value of v_E is thus $v_E = -0.7 + 0.02 = -0.68$ V.

The new value of emitter current is computed most easily by first determining the voltage across the emitter resistance and employing Ohm's law. We have

$$i_{E1} \approx i_{C1} = \frac{-0.68 - (-15)}{14.3 \times 10^3} \approx 1.001 \text{ mA} \qquad \textbf{(13–8)}$$

The change in emitter current is very small and could hardly be sensed with normal instrumentation. The slight increase in collector current causes a slight increase in the drop across the 8-kΩ resistor, so the collector voltage also drops slightly. The new approximate collector voltage is $v_{C1} \approx 6.99$ V.

From the point of view of a common-emitter circuit, the slight changes in current and collector voltage are a result of the large unbypassed emitter resistance, which has reduced the gain to a very small value. However, from the emitter-follower point of view, the emitter voltage has changed approximately the same amount as the input, and the base–emitter voltage has remained essentially constant.

Now, let us add the second transistor T2 as shown in Figure 13–3 and return to the initial point of discussion. In this case, both bases are grounded again (i.e., $v_{i1} = v_{i2} = 0$). To restore the original operating point, we do have to make one change in a circuit parameter. For symmetrical conditions, $i_{C1} = i_{C2} = 1$ mA. The current through the common resistor is $i_{C1} + i_{C2} = 2$ mA. To preserve the same voltage drop, this resistor is changed from 14.3 to 7.15 kΩ. Thus, 2 mA flowing through 7.15 kΩ produces the same drop as 1 mA through 14.3 kΩ. All other points in the circuit have the same operating conditions as the single-transistor circuit of Figure 13–2.

To generalize these dc results for this basic configuration, the dc emitter current I_E corresponding to an emitter resistance R_E with both base terminals grounded and a power supply voltage $-V_{EE}$ is

$$I_E = \frac{V_{EE} - 0.7}{R_E} \qquad \textbf{(13–9)}$$

FIGURE 13–3
Representative differential amplifier voltages and
currents when $v_{i1} = v_{i2} = 0$.

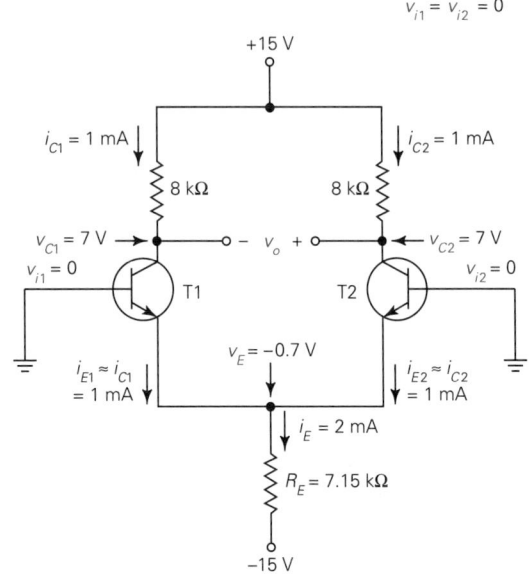

where $V_{BE} = 0.7$ V has been assumed. As we will see shortly, this current remains essentially constant for different values of v_{i1} and v_{i2}.

For $v_{i1} = 0$ and $v_{i2} = 0$ the emitter current divides equally between the transistors (i.e., $i_{E1} = i_{E2} \approx i_{C1} = i_{C2} = I_E/2$). However, for various conditions of v_{i1} and v_{i2}, i_{E1} and i_{E2} will be shown to change.

The voltage v_o between the two collectors is

$$v_o = v_{C2} - v_{C1} = 7 - 7 = 0 \qquad (13\text{--}10)$$

The reader should be reminded that the assumed base–emitter voltage of 0.7 V is approximate and is only a convenient assumption. However, once we assume this value as a reference, it will be necessary to follow variations up and down from that value, as we will shortly see.

Let us now assume that the small dc voltage of 20 mV is applied to the base of T1 (i.e., $v_{i1} = 20$ mV), as shown in Figure 13–4. However, we retain the ground connection for the base of T2 (i.e., $v_{i2} = 0$). In general, the two base–emitter voltages v_{BE1} and v_{BE2} can be expressed as

$$v_{BE1} = v_{i1} - v_E \qquad (13\text{--}11)$$

and

$$v_{BE2} = v_{i2} - v_E \qquad (13\text{--}12)$$

For the values of v_{i1} and v_{i2} of Figure 13–4, we have

$$v_{BE1} = 0.02 - v_E \qquad (13\text{--}13)$$
$$v_{BE2} = 0 - v_E = -v_E \qquad (13\text{--}14)$$

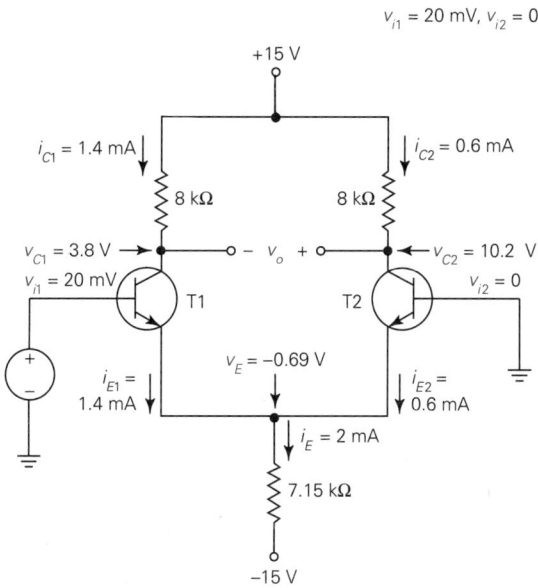

FIGURE 13–4
Representative differential amplifier voltages and currents when $v_{i1} = 20$ mV and $v_{i2} = 0$.

The effect of a positive voltage for v_{i1} is to cause v_E to become more positive. When there was only T1 in the circuit, v_E essentially "followed" v_{BE1}. However, when T2 is present, the effect of an increase in v_{BE1} is to cause a decrease in v_{BE2}. This can be inferred from (13–14) by noting that v_E is a negative quantity, and that when it becomes more positive, its magnitude decreases. Thus an increase in v_{BE1} is simultaneously accompanied by a decrease in v_{BE2}. An increase in v_{BE1} corresponds to an increase in i_{E1}, and a decrease in v_{BE2} corresponds to a decrease in i_{E2}.

The net current i_E through the emitter resistance is

$$i_E = i_{E1} + i_{E2}$$

(13–15)

The combination of the large emitter resistance and the negative power supply acts nearly like a constant current source as far as the emitters are concerned. Thus, i_E assumes nearly a constant value. This last property was also true when only T1 was present, but the single current in that case was the only variable. In this case, however, i_{E1} and i_{E2} both vary, but the sum remains nearly constant. Therefore, *an increase in current in one transistor is accompanied by a corresponding decrease in the other transistor, but the sum remains essentially constant.*

The combined effect of the two transistors working in opposition coupled with the constant current constraint is to prevent the emitter from simply following the base as it did for the single-transistor case. Instead, the emitter voltage changes enough to cause an increase in i_{E1} and a decrease in i_{E2} such that the sum of the currents remains the same. For $v_{i1} = 20$ mV, v_E changes by 10 mV. This has the effect of increasing v_{BE1} by 10 mV and decreasing v_{BE2} by 10 mV. Thus v_{BE1} increases to 0.71 V and v_{BE2} decreases to 0.69 V for the original assumed values.

The formulas for predicting various circuit variables will be developed later. For the conditions in Figure 13–4, it turns out that $i_{E1} \approx i_{C1} = 1.4$ mA and $i_{E2} \approx i_{C2} = 0.6$ mA. The two collector voltages may be readily verified to be $v_{C1} = 3.8$ V and $v_{C2} = 10.2$ V. The voltage v_o is

$$v_o = 10.2 - 3.8 = 6.4 \text{ V} \tag{13–16}$$

The changes in this case are rather pronounced and represent a totally different situation than when the single transistor was used. Although the emitter resistance is unbypassed, it does not suppress changes in signal voltages and currents as it did for the single-transistor case. In effect, the presence of the second transistor "pulling" in opposition to the first transistor permits a large gain to be achieved even with the presence of a large emitter resistance. The reduction in gain that would be caused by the large emitter resistance for a single transistor is effectively canceled out when two transistors are used.

Next, let us consider the effect when equal voltages are applied to both input terminals. Refer to Figure 13–5, in which $v_{i1} = v_{i2} = 20$ mV is assumed. In this case the emitter voltage v_E may easily follow the inputs since the two inputs are

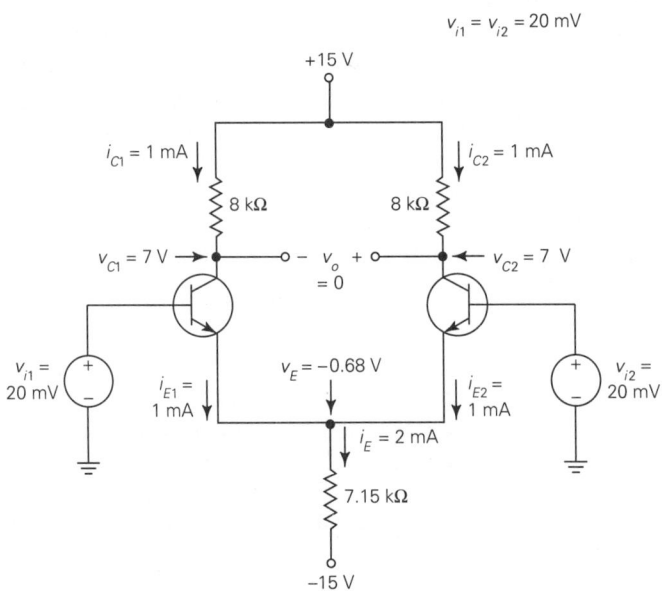

FIGURE 13–5
Representative differential amplifier voltages and currents when $v_{i1} = v_{i2} = 20$ mV.

equal. In a sense, the circuit here is acting almost like that of the single-transistor case of Figure 13–2. Indeed, there is even a very slight increase in collector current as was shown in Figure 13–2, but it will be ignored here. Because neither of the collector currents changes appreciably, the collector voltages are assumed to be the same as when the two inputs were grounded. The voltage v_o is

$$v_o = 7 - 7 = 0 \qquad\qquad (13\text{--}17)$$

The final situation to be considered for this circuit is shown in Figure 13–6. In this case, two different voltages are applied to the separate inputs. The specific values chosen are $v_{i1} = 60$ mV and $v_{i2} = 40$ mV. In this case, the "pulling" action between the transistors occurs as it did in the circuit of Figure 13–4. The effect is to cause the emitter voltage to shift upward by an amount equal to the midpoint between the two input voltages. Thus, v_E increases by 50 mV from its original value of -0.7 V to a value of -0.65 V. The reader can readily verify that $v_{BE1} = 0.71$ V and $v_{BE2} = 0.69$ V. These are the same as for the circuit of Figure 13–4.

The collector and emitter currents and the collector voltages are the same as for the circuit of Figure 13–4. The voltage v_o is obviously also the same (i.e., $v_o = 6.4$ V).

The conditions of Figures 13–4 and 13–6 produce identical results for the collector voltages and the voltage v_o. Yet in one case, $v_{i1} = 20$ mV and $v_{i2} = 0$, and in the other case, $v_{i1} = 60$ mV and $v_{i2} = 40$ mV. On the other hand, when

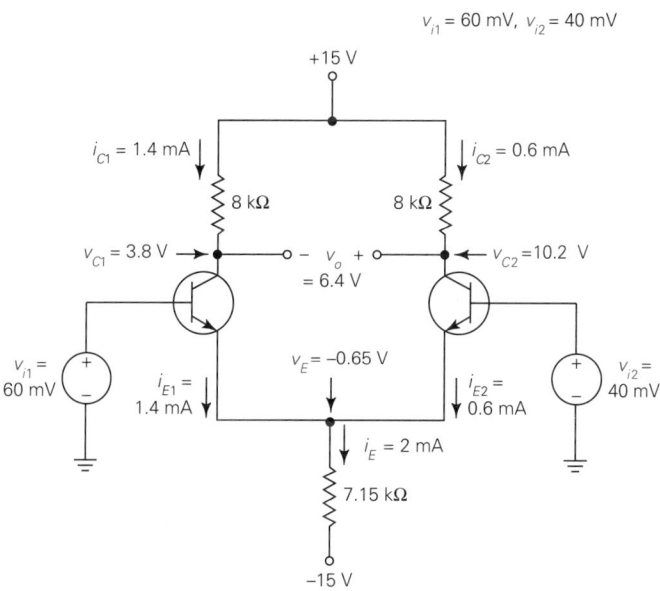

FIGURE 13–6
Representative differential amplifier voltages and currents when $v_{i1} = 60$ mV and $v_{i2} = 40$ mV.

we set $v_{i1} = v_{i2} = 20$ mV, the results were the same as when $v_{i1} = v_{i2} = 0$. *The implication is that the circuit is responding to the difference between the input voltages.* Thus, when $v_{i1} = 60$ mV and $v_{i2} = 40$ mV, the difference is 20 mV, which is the same as when $v_{i1} = 20$ mV and $v_{i2} = 0$. When the two input voltages are equal, the difference is zero, and there is virtually no change in the circuit variables.

In analyzing differential amplifiers, the term **differential input voltage** is a convenient definition. Representing this quantity as v_d, it is defined as

$$v_d = v_{i1} - v_{i2} \qquad\qquad \textbf{(13–18)}$$

It should be clear from the symmetry of the circuit that if any input conditions are reversed between the bases, the resulting collector and emitter conditions are similarly reversed. For example, if $v_{i1} = 0$ and $v_{i2} = 20$ mV, it is readily deduced that $i_{C1} = 0.6$ mA, $i_{C2} = 1.4$ mA, $v_{C1} = 10.2$ V, and $v_{C2} = 3.8$ V. The voltage v_o would then be $v_o = 3.8 - 10.2 = -6.4$ V.

When the two inputs are equal, there is very little change in circuit conditions, and the output v_o is essentially zero. When both inputs are equal, the amplifier is said to be excited with **common-mode input signals.** Thus, a differential amplifier responds primarily to differential input signals and offers very little gain to common-mode signals.

Why is this property important? It turns out that many spurious noise components and undesired pickup signals tend to be of the common-mode type. However, an actual signal may be applied as a differential input. Consequently, the differential amplifier tends to provide a high degree of suppression for common-mode pickup, and this is one of its many virtures.

Thus far, the reader has been guided through a series of voltage–current conditions for a differential amplifier for the purpose of establishing a "feel" for the process. Because no general equations for the circuit have been established, we are not yet in a position to work out example problems.

The next two sections are used to establish the general gain and transfer characteristics for the basic differential amplifier circuit. Examples considered then return to the particular amplifier configuration considered here.

13–3 CASCADE ANALYSIS OF DIFFERENTIAL AMPLIFIER

One particular interpretation of the differential amplifier using small-signal analysis is made in this section. This particular interpretation is limited in predicting the complete analysis of the circuit, but it does provide an interesting insight, and it ties in nicely with the work of Chapter 12.

Consider the two-transistor differential amplifier configuration of Figure 13–7(a) with a single input v_{i1} and with the second base grounded (i.e., $v_{i2} = 0$). The only output of interest for this purpose is v_{c2}, the signal portion of v_{C2}. Assume that the transistors are *identical* and have small-signal parameters g_m and r_π.

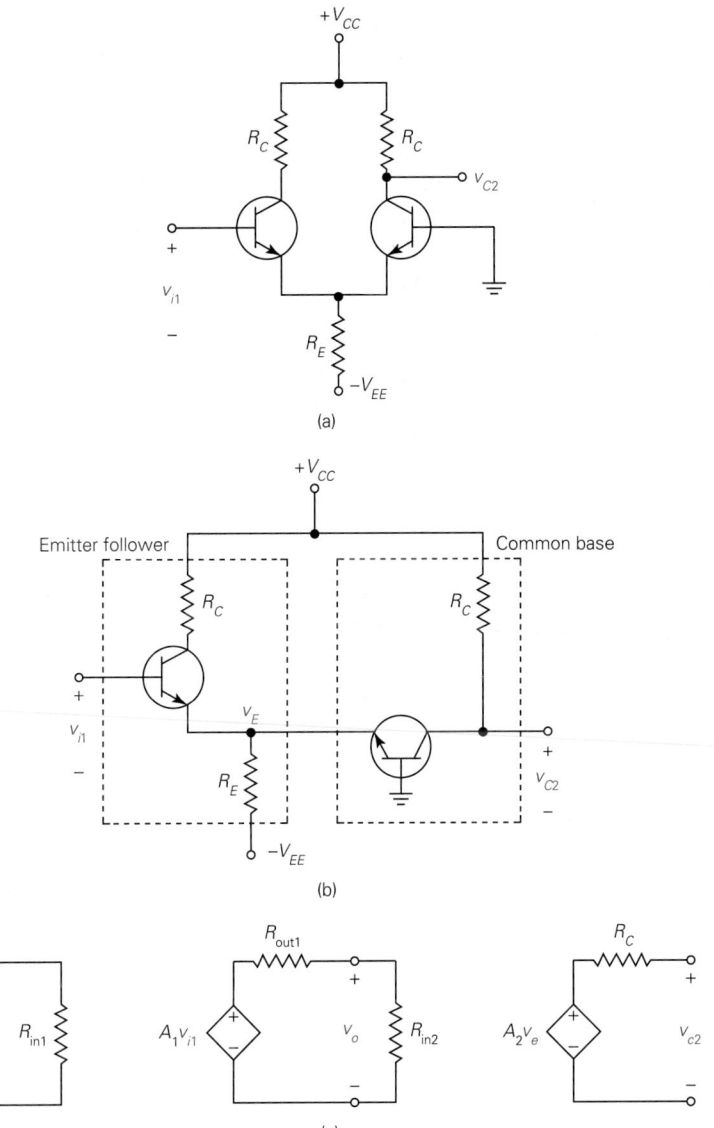

FIGURE 13–7
Interpretation of differential amplifier as emitter follower in cascade with common-base circuit.

One way to visualize this circuit is to consider it as an emitter follower in cascade with a common-base circuit, as shown in Figure 13–7(b). The emitter resistance could be considered as a part of either circuit, but it has been arbitrarily grouped with the emitter-follower circuit.

From the work of Chapter 12, the small-signal analysis of the complete circuit can be represented by a cascade arrangement in which the individual small-signal parameters of each circuit are identified. The composite form of such an arrangement is shown in Figure 13–7(c). The parameters here represent *signals,* so the instantaneous voltages v_E and v_{C2} have been replaced by small-signal voltages v_e and v_{c2}, respectively.

Because of the fact that the input resistance of the emitter follower is affected by the load of the common-base circuit, the parameters of the latter circuit are determined first. From the work of either Chapter 10 or the summary of Chapter 12, the input resistance of the common-base circuit, identified here as R_{in2}, is

$$R_{in2} = r_\pi \left\| \frac{1}{g_m} \simeq \frac{1}{g_m} \right. \tag{13–19}$$

where the assumption $r_\pi \gg 1/g_m$ has been made.

The gain A_2 of the common-base circuit is

$$A_2 = g_m R_C \tag{13–20}$$

The output resistance of the common-base circuit is simply R_C and is labeled as such on Figure 13–7(c).

The input resistance of the emitter-follower circuit is

$$R_{in1} = r_\pi (1 + g_m R'_E) \tag{13–21}$$

where

$$R'_E = R_E \| R_{in2} \tag{13–22}$$

Normally, $R_{in2} \ll R_E$, so (13–22) can be approximated as

$$R'_E \simeq R_{in2} = \frac{1}{g_m} \tag{13–23}$$

where (13–19) was employed. Substitution of (13–23) in (13–21) yields

$$R_{in1} = r_\pi \left(1 + g_m \frac{1}{g_m} \right) = 2r_\pi \tag{13–24}$$

The unloaded gain A_1 of the emitter-follower circuit is approximately

$$A_1 \simeq 1 \tag{13–25}$$

The output resistance R_{out1} of the common-collector circuit based on $v_{i1} = 0$ is

$$R_{\text{out1}} = R_E \left\| \frac{1}{g_m} \right. \simeq \frac{1}{g_m} \tag{13–26}$$

where the assumption $R_E \gg 1/g_m$ has been made.

From Figure 13–7 we see that there is neither a generator nor an output loading factor to consider. There is, however, an interaction factor between the stages. The net gain between v_{i1} and v_{c2} will be denoted as A_{12}, and it is

$$A_{12} = \frac{v_{c2}}{v_{i1}} = A_1 \frac{R_{\text{in2}}}{R_{\text{in2}} + R_{\text{out1}}} A_2 \tag{13–27}$$

Substitution of the various parameters in (13–27) results in

$$A_{12} = 1 \times \frac{1/g_m}{1/g_m + 1/g_m} g_m R_C = \frac{g_m R_C}{2} \tag{13–28}$$

This result is quite interesting in that it is one-half the gain of a common-base circuit (or one-half the magnitude of a common-emitter circuit). The factor of one-half can be thought of as arising from the interaction of the output resistance of the emitter-follower circuit and the input resistance of the common-base circuit. For equal transistor characteristics, these resistances are equal, and the signal voltage is halved at this point.

13–4 SMALL-SIGNAL ANALYSIS OF DIFFERENTIAL AMPLIFIER

In this section we perform a rather detailed small-signal analysis of the two-transistor differential amplifier. The derivations involved are about the most unwieldy of any in the text, so do not feel intimidated if you do not understand all the steps at first. Like many electronic circuit derivations, there are some "tricks" and some approximations involved in the various steps. Try to focus on the basic concepts involved and the significance of the results obtained. If you are using this text in a course, the instructor may choose to hold you responsible only for applying the results rather than deriving them.

Differential and Common-Mode Signals

Consider a differential amplifier of the form given in Figure 13–1 with input signals v_{i1} and v_{i2}. The **differential input voltage** v_d was defined earlier, but for completeness, the definition is repeated here as

$$\boxed{v_d = v_{i1} - v_{i2}} \tag{13–29}$$

We now define the **common-mode input voltage** v_{cm} as

$$v_{cm} = \frac{v_{i1} + v_{i2}}{2} \qquad\qquad \textbf{(13–30)}$$

The differential input voltage is the *difference* between the two input signals, and the common-mode input voltage is the *mean* or *average* of the two input signals. For example, if $v_{i1} = 8$ mV and $v_{i2} = 6$ mV, $v_d = 2$ mV and $v_{cm} = 7$ mV.

Equations (13–29) and (13–30) provide expressions for v_d and v_{cm} in terms of v_{i1} and v_{i2}. However, the latter two quantities can be treated as dependent variables, and by solving the two equations simultaneously, v_{i1} and v_{i2} can be expressed in terms of v_d and v_{cm}. The reader is invited to verify that the results of this process yield

$$v_{i1} = v_{cm} + \frac{v_d}{2} \qquad\qquad \textbf{(13–31)}$$

$$v_{i2} = v_{cm} - \frac{v_d}{2} \qquad\qquad \textbf{(13–32)}$$

For the numerical example considered earlier, substitution of $v_{cm} = 7$ mV and $v_d = 2$ mV yields $v_{i1} = 8$ mV and $v_{i2} = 6$ mV, as expected.

These forms may seem rather strange at this point. However, as we will see shortly, the effects of the amplifier on the quantities v_d and v_{cm} will be quite different, and some of the major properties of the differential amplifier can be most easily deduced from these forms.

Small-Signal Analysis

The small-signal equivalent circuit of the differential amplifier is shown in Figure 13–8. Note that the two input signals are represented in terms of differential and common-mode components.

The analysis will focus first on the determination of the emitter signal v_e. Once v_e is known, all other voltages and the currents may then be readily determined.

A KCL equation will now be written at the mode representing v_e. The currents can be expressed in terms of the various voltages. Assuming positive currents leaving the node, we have

$$\frac{v_e}{R_E} + \frac{v_e - v_{i1}}{r_\pi} + \frac{v_e - v_{i2}}{r_\pi} - g_m v_{be1} - g_m v_{be2} = 0 \qquad\qquad \textbf{(13–33)}$$

The final form of the equation should contain no variables other than v_e, v_d, and v_{cm}. The voltages v_{be1} and v_{be2} may be first expressed in terms of v_{i1} and v_{i2} by applying KVL to the two loops at the bottom of the circuit. After simplification, we have

$$v_{be1} = v_{i1} - v_e \qquad\qquad \textbf{(13–34)}$$

$$v_{be2} = v_{i2} - v_e \qquad\qquad \textbf{(13–35)}$$

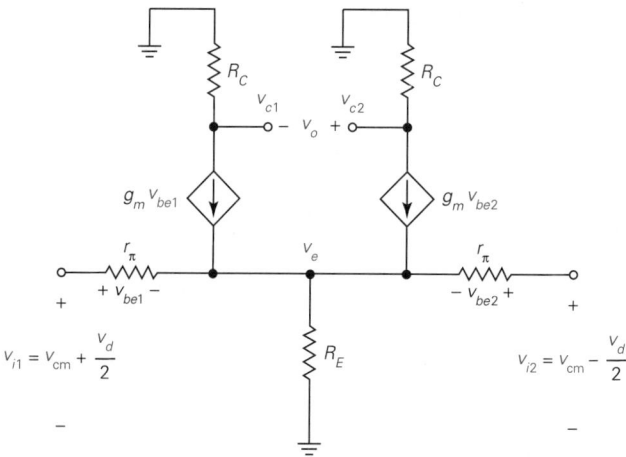

FIGURE 13–8
Small-signal equivalent circuit of differential amplifier.

These relationships for v_{be1} and v_{be2} may first be substituted in (13–33). Next, the expressions for v_{i1} and v_{i2} in terms of v_d and v_{cm} may be substituted. The expression of (13–33) then becomes

$$\frac{v_e}{R_E} + \frac{v_e}{r_\pi} - \frac{v_{cm}}{r_\pi} - \frac{v_d}{2r_\pi} + \frac{v_e}{r_\pi} - \frac{v_{cm}}{r_\pi} + \frac{v_d}{2r_\pi} - g_m v_{cm} - \frac{g_m v_d}{2} + g_m v_e$$

$$- g_m v_{cm} + \frac{g_m v_d}{2} + g_m v_e = 0 \quad \text{(13–36)}$$

Note that the terms involving v_d cancel. The terms involving v_e are grouped on the left, and those involving v_{cm} are grouped on the right. This results in

$$v_e \left(\frac{1}{R_E} + \frac{2}{r_\pi} + 2g_m \right) = \left(\frac{2}{r_\pi} + 2g_m \right) v_{cm} \quad \text{(13–37)}$$

Solving for v_e and simplifying the expression, we obtain

$$v_e = \frac{2(1 + g_m r_\pi) v_{cm}}{2(1 + g_m r_\pi) + r_\pi / R_E} \quad \text{(13–38)}$$

An approximation will be made to simplify this expression. However, a significant point to make at this time is that the emitter signal voltage is affected only by the common-mode input voltage; that is, the effect of the differential voltage cancels as far as the emitter signal voltage is concerned.

Next we determine v_{c1} and v_{c2}. Since the signal currents flow from ground through the two collector resistances, we can write

$$v_{c2} = -g_m R_C v_{be2} \quad \text{(13–39)}$$

$$v_{c1} = -g_m R_C v_{be1} \quad \text{(13–40)}$$

It is necessary to express v_{be1} and v_{be2} in terms of v_d and v_{cm}. By employing the relationships of (13–34) and (13–35) and substituting v_e from (13–38), the following results can be determined:

$$v_{be1} = v_{i1} - v_e \tag{13–41a}$$

$$= v_{cm} + \frac{v_d}{2} - \frac{2(1 + g_m r_\pi)v_{cm}}{2(1 + g_m r_\pi) + r_\pi/R_E} \tag{13–41b}$$

$$v_{be2} = v_{i2} - v_e \tag{13–42a}$$

$$= v_{cm} - \frac{v_d}{2} - \frac{2(1 + g_m r_\pi)v_{cm}}{2(1 + g_m r_\pi) + r_\pi/R_E} \tag{13–42b}$$

When the preceding expressions are simplified over a common denominator, they become

$$v_{be1} = \frac{(1 + g_m r_\pi + r_\pi/2R_E)v_d + (r_\pi/R_E)v_{cm}}{2(1 + g_m r_\pi + r_\pi/R_E)} \tag{13–43}$$

$$v_{be2} = \frac{-(1 + g_m r_\pi + r_\pi/2R_E)v_d + (r_\pi/R_E)v_{cm}}{2(1 + g_m r_\pi + r_\pi/R_E)} \tag{13–44}$$

The preceding two expressions may be substituted in (13–39) and (13–40) to permit the determination of v_{c1} and v_{c2} in terms of v_d and v_{cm}. The results are

$$v_{c2} = \frac{g_m R_C(1 + g_m r_\pi + r_\pi/2R_E)v_d - g_m R_C(r_\pi/R_E)v_{cm}}{2(1 + g_m r_\pi + r_\pi/R_E)} \tag{13–45}$$

$$v_{c1} = \frac{-g_m R_C(1 + g_m r_\pi + r_\pi/2R_E)v_d - g_m R_C(r_\pi/R_E)v_{cm}}{2(1 + g_m r_\pi + r_\pi/R_E)} \tag{13–46}$$

At this point, we have carried these bulky expressions far enough, and some reasonable approximations are in order. In practice, $g_m r_\pi = \beta_o \gg 1$ and $= \beta_o \gg r_\pi/2R_E$. With these expressions, (13–45), (13–46), and the expression for v_e can be simplified to

$$v_e \approx v_{cm} \tag{13–47}$$

$$v_{c2} \approx \frac{g_m R_C}{2} v_d - \frac{R_C}{2R_E} v_{cm} \tag{13–48}$$

$$v_{c1} \approx -\frac{g_m R_C}{2} v_d - \frac{R_C}{2R_E} v_{cm} \tag{13–49}$$

The two collector voltages are each a linear combination of the differential input voltage and of the common-mode input voltage. The differential output voltage v_o measured between the collectors is

$$v_o = v_{c2} - v_{c1} \tag{13–50}$$

When (13–48) and (13–49) are substituted in (13–50), the common-mode components cancel, and we have

$$v_o = g_m R_C v_d \tag{13–51}$$

Let

$$v_o = A'_d v_d \tag{13-52}$$

where A'_d is the gain factor relating the differential output voltage to the differential input voltage. This quantity is

$$A'_d = \frac{v_o}{v_d} = g_m R_C \tag{13-53}$$

This gain has the same value as for a single-stage common-base circuit and the same magnitude as for a common-emitter circuit.

Although the differential output voltage is theoretically free of any common-mode components, the differential amplifier is most often used in a single-ended form with a single output. For v_{c2} as the output, we can write

$$v_{c2} = v_{cd} + v_{ccm} \tag{13-54}$$

where v_{cd} is the differential component of the voltage v_{c2} and v_{ccm} is the common-mode component of the voltage v_{c2}. Comparing (13-54) with (13-48), we can write

$$v_{cd} = A_d v_d \tag{13-55}$$

and

$$v_{ccm} = A_{cm} v_{cm} \tag{13-56}$$

The quantity A_d is the differential gain relating the differential portion of the single-ended output to the differential input, and it is

$$A_d = \frac{v_{cd}}{v_d} = \frac{g_m R_C}{2} \tag{13-57}$$

The quantity A_{cm} is the common-mode gain relating the common-mode portion of the single-ended output to the common-mode input, and it is

$$A_{cm} = \frac{v_{ccm}}{v_{cm}} = -\frac{R_C}{2R_E} \tag{13-58}$$

The differential gain for a single-ended output is one-half the value for a full differential output. This result is seen to be in agreement with the cascade interpretation of Section 13-3.

Common-Mode Rejection Ratio

A quantity called the **common-mode rejection ratio (CMRR)** will now be defined. In general, it is the ratio of the differential gain to the common-mode gain; that is,

$$\text{CMRR} = \frac{|A_d|}{|A_{cm}|} \qquad \textbf{(13–59)}$$

where the magnitude signs indicate that inverting gains are converted to positive values in the definition. Substitution of (13–57) and (13–58) for the circuit just analyzed yields

$$\text{CMRR} = g_m R_E \qquad \textbf{(13–60)}$$

The criterion for a differential amplifier is that the value of CMRR should be as large as possible. Note that the CMRR varies linearly with the value of R_E employed.

EXAMPLE 13–1

Consider the differential amplifier of Section 13–2. Apply the small-signal relationships of this section to the conditions of Figure 13–4 to determine **(a)** the differential and common-mode input signals; **(b)** the differential gain (both for differential and single-ended output v_{c2}), common-mode gain, and CMRR; **(c)** the differential and common-mode output signals; and **(d)** the emitter signal voltage.

Solution

(a) The two input signals in Figure 13–4 are $v_{i1} = 20$ mV and $v_{i2} = 0$. The differential and common-mode input signals are

$$v_d = v_{i1} - v_{i2} = 20 - 0 = 20 \, \text{mV} \qquad \textbf{(13–61)}$$

and

$$v_{cm} = \frac{v_{i1} + v_{i2}}{2} = \frac{20 + 0}{2} = 10 \, \text{mV} \qquad \textbf{(13–62)}$$

(b) Before determining the various gains, we need to determine g_m. For $I_C = 1$ mA, the value of g_m is

$$g_m = 40 I_C = 40 \times 1 \times 10^{-3} = 40 \, \text{mS} \qquad \textbf{(13–63)}$$

The differential gain A'_d relating the full differential output voltage to the differential input voltage is

$$A'_d = g_m R_C = 0.04 \times 8 \times 10^3 = 320 \qquad \textbf{(13–64)}$$

The differential gain A_d relating the differential component of v_{c2} to the differential input voltage is

$$A_d = \frac{g_m R_C}{2} = \frac{320}{2} = 160 \qquad (13\text{--}65)$$

The common-mode gain A_{cm} is

$$A_{cm} = \frac{-R_C}{2R_E} = \frac{-8 \times 10^3}{2 \times 7.15 \times 10^3} = -0.559 \qquad (13\text{--}66)$$

The CMRR is

$$\text{CMRR} = \frac{|A_d|}{|A_{cm}|} = \frac{160}{0.559} = 286 \qquad (13\text{--}67)$$

(c) For the differential and common-mode components of **(a)**, the various output quantities are determined as follows:

$$v_o = A'_d v_d = 320 \times 0.02 = 6.4 \text{ V} \qquad (13\text{--}68)$$

$$v_{cd} = A_d v_d = 160 \times 0.02 = 3.2 \text{ V} \qquad (13\text{--}69)$$

$$v_{ccm} = A_{cm} v_{cm} = -0.559 \times 0.01 \approx -5.6 \text{ mV} \qquad (13\text{--}70)$$

$$v_{c2} = v_{cd} + v_{ccm} = 3.2 \text{ V} - 5.6 \text{ mV} \approx 3.2 \text{ V} \qquad (13\text{--}71)$$

(d) The emitter signal voltage v_e is

$$v_e = v_{cm} = 10 \text{ mV} \qquad (13\text{--}72)$$

Several comments concerning these results are in order. The value of v_o from Figure 13–4 is seen to be in agreement with (13–68). For the particular levels of v_{i1} and v_{i2} in this example, the common-mode component of v_{c2} is, for most practical purposes, negligible, and $v_{c2} \approx 3.2$ V as noted from (13–71). (Example 13–2 will illustrate a different situation, however.) The value of $v_{c2} = 3.2$ V is the *signal* voltage at the collector of T2. Consequently, this is recognized on Figure 13–4 as a change from the quiescent value of 7 V to a value of 10.2 V when the signal is applied. Similarly, the emitter signal voltage of 10 mV is recognized as a change in the emitter voltage from −0.7 to −0.69 V in Figure 13–4.

With this discrete-component circuit, the single-ended output thus contains a dc level, which would limit its use to ac coupling for most applications. However, we will later consider the use of dc-level translation circuitry to change the dc level so that dc coupling can be employed.

EXAMPLE 13–2

Consider again the differential amplifier of Section 13–2. However, assume now that the two input voltages are $v_{i1} = 2.01$ V and $v_{i2} = 1.99$ V. Determine **(a)** the differential and common-mode input signals; **(b)** the differential and common-mode output signals; and **(c)** the emitter signal voltage.

Solution

(a) The differential and common-mode input signals are

$$v_d = v_{i1} - v_{i2} = 2.01 - 1.99 = 0.02 \text{ V} = 20 \text{ mV} \qquad \textbf{(13–73)}$$

and

$$v_{cm} = \frac{2.01 + 1.99}{2} = 2 \text{ V} \qquad \textbf{(13–74)}$$

The differential input voltage is the same as in Example 13–1, but the common-mode input voltage is considerably larger.

(b) The gain values computed in Example 13–1 will be used freely here. The various output quantities are

$$v_o = A'_d v_d = 320 \times 0.02 = 6.4 \text{ V} \qquad \textbf{(13–75)}$$
$$v_{cd} = A_d v_d = 160 \times 0.02 = 3.2 \text{ V}$$
$$v_{ccm} = A_{cm} v_{cm} = -0.559 \times 2 = -1.12 \text{ V} \qquad \textbf{(13–76)}$$
$$v_{c2} = v_{cd} + v_{ccm} = 3.2 - 1.12 = 2.08 \text{ V} \qquad \textbf{(13–77)}$$

(c) The emitter signal voltage is

$$v_e = v_{cm} = 2 \text{ V} \qquad \textbf{(13–78)}$$

Comparing these results with those of Example 13–1, the common-mode component is seen to have a significant effect on v_{c2} in the present case.

13–5 SOME IMPORTANT INTEGRATED-CIRCUIT FUNCTIONS

Now that we have analyzed the operation of a discrete component differential amplifier, the next step is to discuss the evolution of that basic circuit to its integrated-circuit form. However, it is desirable to first observe a few basic functions that are achieved with integrated-circuit components. It turns out that some of the bias circuit and resistance loads for transistors are implemented with additional transistors. From a fabrication point of view, transistors are easier to realize in many cases than resistors and diodes. The resulting circuits appear to be overwhelmed with transistors, but many of these transistors are serving as active loads and bias sources.

Current Mirror

The first circuit that will be considered is the **current mirror.** This circuit has been used both in discrete component circuits and in integrated circuits.

Refer to Figure 13–9(a). The objective of the circuit is to establish a specified current I_C in the collector of the transistor. No particular connection is shown for

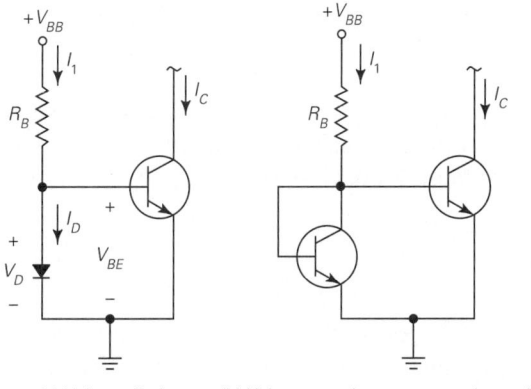

(a) Using a diode   (b) Using a transistor connected as a diode

FIGURE 13–9
Current mirror.

the collector terminal as it may be connected in a number of ways depending on the application. However, it is assumed that it is connected through some appropriate load to a proper bias circuit.

The base of the transistor is connected to the diode circuit on the left, which is returned through a resistance R_B to a positive voltage V_{BB}. This voltage may or may not be the voltage used to bias the collector of the transistor.

The current I_1 through the resistance is readily expressed as

$$I_1 = \frac{V_{BB} - V_D}{R_B} = \frac{V_{BB} - 0.7}{R_B} \qquad (13\text{--}79)$$

where $V_D = 0.7$ V has been assumed. Now assume that the transistor base–emitter diode has the same characteristics as that of the diode on the left; that is, the emitter current (and in turn the collector current) corresponding to a given base–emitter voltage is the same as the diode current for a given diode voltage. The two voltages are clearly the same since they are connected in parallel (i.e., $V_D = V_{BE}$). For identical characteristics, therefore, the collector current is forced to be essentially the same as the diode current, that is,

$$I_C \approx I_1 = \frac{V_{BB} - 0.7}{R_B} \qquad (13\text{--}80)$$

where the base current of the transistor has been neglected. Thus R_B is selected to produce a specified current on the left, and the *mirror image* of that current flows in the collector on the right.

In many integrated-circuit chips, it is easier to create a diode from a transistor. This is achieved by trying the base and collector terminals together, and this configuration is shown in Figure 13–9(b). The transistor on the left is thus functioning as a diode. On an integrated circuit, it is easier to approach identical characteristics than with discrete components since the two transistors are formed from the same semiconductor material. Thus the expression for I_C in Figure 13–9(b) is the same as in Figure 13–9(a).

Normally, $V_{BB} \gg V_{BE}$, so variations of V_{BE} with temperature have very little effect on I_B and I_C. The current established is primarily a function of V_{BB} and R_B, and variations of V_{BE} with temperature tend to be compensated for by the mirroring process.

Active Load

The current mirror circuit may be used to establish a resistive load to function in the same manner as a resistance. The resulting circuit is referred to as an **active load.**

One form of an active load that can be used as the collector resistance in a differential amplifier is shown in Figure 13–10(a). Note that the BJT is a *pnp* type, but as shown, it will serve as the collector load for an *npn* transistor. The corresponding form employing a transistor acting as a diode is shown in Figure 13–10(b). The latter form is used extensively in integrated circuits. Both forms are also referred to as **current sources.**

FIGURE 13–10
Active collector load obtained with a current mirror.

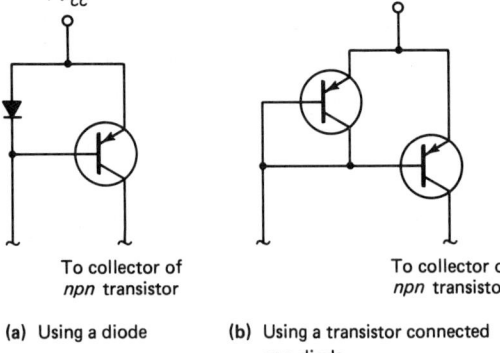

(a) Using a diode

(b) Using a transistor connected as a diode

The reason that these circuit forms behave like resistors is the fact that the dynamic collector resistance is very large; that is, the slope of the collector voltage versus current characteristic in the operating region is very large. The active load is particularly useful where a large dynamic resistance is required but where only a moderate dc voltage drop across the component is feasible.

Integrated-Circuit Differential Amplifier Stage

A representative integrated-circuit differential amplifier stage is shown in Figure 13–11. The circuit shown here does not represent any specific amplifier, but it is representative of a number of different types.

FIGURE 13–11
Typical integrated-circuit differential amplifier
stage.

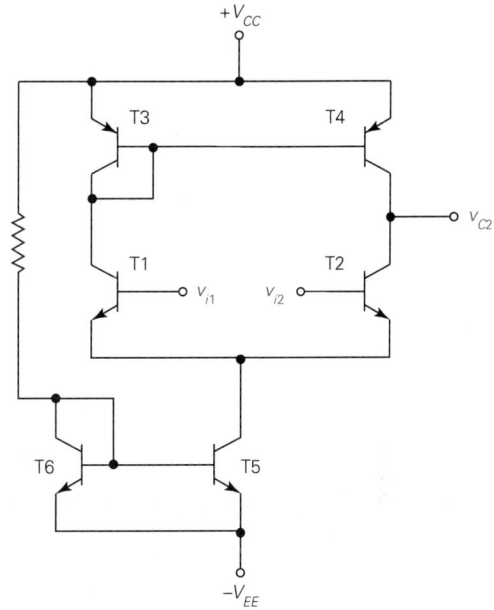

The *npn* transistors T1 and T2 represent the two input stages of the differential
amplifier. The *pnp* transistors T3 and T4 form an active load pair. Transistor T3
is functioning as the diode for T4, so the drop across T3 is constant. However,
T4 acts as a dynamic load for T2. Since only T2 has an active load, the output
v_{C2} is a single-ended type.

The transistor pair T5 and T6 act as a current source for the differential
pair (i.e., they replace the emitter resistance R_E). Because of the large dynamic
resistance compared with an actual practical resistor, the CMRR can be made to
be very large.

**EXAMPLE
13–3**

Determine the value of I_C established by the current mirror of Figure 13–12.

Solution
The transistor on the left is functioning as an active diode, and its collector current
establishes the collector current for the transistor on the right. Neglecting base
currents and assuming that $V_{BE} = 0.7$ V, we have

$$I_C = \frac{15 - 0.7}{5.1 \times 10^3} = 2.80 \text{ mA} \qquad \textbf{(13–81)}$$

Strictly speaking, this result is valid only if the transistor on the right is not
saturated. The voltage drop across the 2-kΩ resistor is readily determined as
$2.80 \times 10^{-3} \times 2 \times 10^3 = 5.60$ V. The collector–emitter voltage is $15 - 5.60 =$
9.40 V. Active-region operation is thus assured.

FIGURE 13–12
Circuit of Example 13–3.

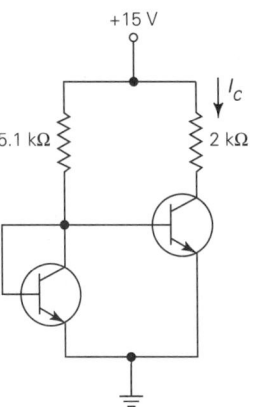

13–6 OPERATIONAL AMPLIFIER

We now turn our attention to the modern integrated-circuit **operational amplifier,** hereafter referred to frequently by the abbreviated term **op-amp.** From an external point of view, an op-amp may be considered as a high-gain dc-coupled differential amplifier. Depending on the model, an op-amp will generally contain the following circuits: (1) one or more cascade stages of basic differential amplifier circuits, (2) dc-level shifting or translation circuits to establish a reference-zero output voltage without coupling capacitors, (3) a compensation circuit to prevent instability and potential oscillations, and (4) a low-resistance output stage (often an emitter follower or source follower).

At this point in the development, it is not meaningful to continue working with the detailed circuit diagrams as we have done with the simpler two-transistor differential amplifier. Instead, the basic integrated-circuit operational amplifier will be accepted as a ''building block'' for many circuit functions, and a simplified approach to the modeling and layout will be assumed.

Before we develop this simplified approach, the actual schematic diagram of a typical operational amplifier will be observed, and this is shown in Figure 13–13. Do not be intimidated by this circuit diagram, since we will not analyze it in detail. Actually, most of the circuitry shown here either directly or indirectly establishes the operation of various differential amplifier stages, whose functions are the same as those of the much simpler discrete-component type considered earlier. If you inspect the various transistors closely, you will see a number of transistors functioning as active loads and as current sources. This entire operational amplifier is available in a small chip, at a low price, and a user does not need to understand all the details of the actual circuit to use it properly. However, the basic understanding of the differential amplifier pair considered earlier is important as an aid in the overall operation and utilization of the integrated-circuit chip.

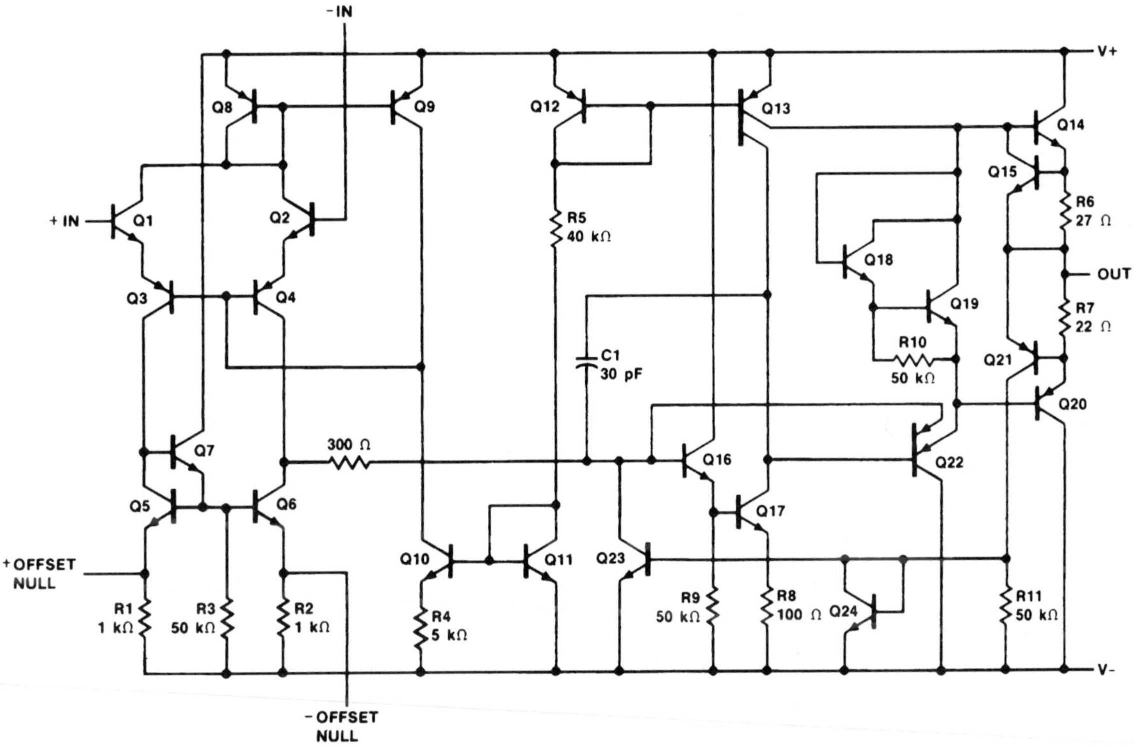

FIGURE 13–13
Schematic diagram of typical integrated-circuit operational amplifier. (Courtesy of Fairchild Semiconductor Corporation.)

Operational Amplifier Bias Connection

Most operational amplifiers require two dc bias supplies for full dynamic range operation. Normally, this is achieved with two separate dc power supplies. The most common arrangement is shown in Figure 13–14. One bias supply is established as a positive voltage of value V_{BB} with respect to ground, and one is

FIGURE 13–14
Dual power supply arrangement for biasing an op-amp.

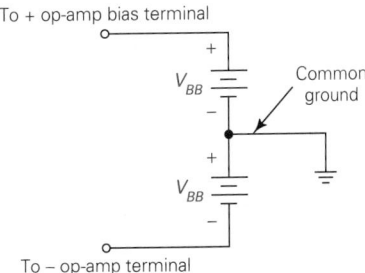

established as a negative voltage $-V_{BB}$ with respect to ground. The most common values are $+15$ V and -15 V. The operational amplifier provides terminals for the two voltages. However, the ground terminal itself is normally not connected to the op-amp. Rather, all ground connections are referred to the common point between the two power supplies, as shown in Figure 13–14. Referring to Figure 13–13, the two bias points on this particular op-amp are identified as V+ and V−.

Operational Amplifier Symbols and Conventions

The most widely employed symbol for the operational amplifier, both with bias supplies connected and without bias supplies connected, is shown in Figure 13–15. In Figure 13–15(a), the two bias connections are indicated as $+V_{BB}$ above and $-V_{BB}$ below the op-amp. Unless there is a particular reason for focusing on the power supply connections, it is customary in many op-amp circuit diagrams to omit the power supply connections. Thus, they will be omitted from most subsequent diagrams, but their presence must be understood for operation.

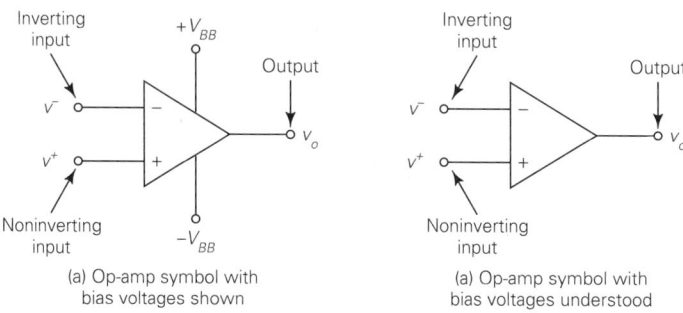

(a) Op-amp symbol with bias voltages shown

(a) Op-amp symbol with bias voltages understood

FIGURE 13–15
Operational amplifier symbol with power supply connections shown and with power sup-ply connections understood.

The op-amp symbol with the power supply connections omitted and under-stood is shown in Figure 13–15(b). Most integrated-circuit op-amps use a single-ended output, and it will be denoted here as v_o. Unlike the v_o variable used in the discrete-component circuit, however, this voltage is referred to ground. The earlier notation for inputs (v_{i1} and v_{i2}) will be dropped at this point for reasons that will be clear later. Instead, the two symbols v^+ and v^- will be used. (Do not confuse these lowercase instantaneous symbols with the op-amp power supply connections V+ and V− shown earlier on the representative op-amp schematic.)

The input labeled as (+) and denoted as v^+ is called the **noninverting input.** The input labeled as (−) and denoted as v^- is called the **inverting input.** The differential input voltage v_d is defined as

$$v_d = v^+ - v^-$$

(13–82)

The operational amplifier is characterized by a differential gain A_d, which is typically 100,000 or more at dc and very low frequencies. The ideal output voltage is related to the two input voltages as

$$v_o = A_d(v^+ - v^-) = A_d v_d \qquad \textbf{(13--83)}$$

Thus, the op-amp output responds to the difference between the two inputs. This is basically the same process as encountered for the simpler two-transistor differential amplifier. A major difference, however, is that A_d for the op-amp is considerably larger than for the simpler differential amplifier circuit.

It should be stressed that the + and − symbols at the op-amp input terminals have nothing to do with polarities of voltages at the inputs. Either input voltage may be positive or negative. Rather, it is the sense of the amplification that is indicated. A voltage of either polarity applied to the noninverting input will cause the output to shift in the direction of the input. However, a voltage of either polarity applied to the inverting input will cause the output to shift in the opposite direction of the input.

Transfer Characteristics

A typical transfer characteristic curve of an op-amp is shown in Figure 13–16. The output voltage v_o versus the differential input voltage v_d is displayed. The

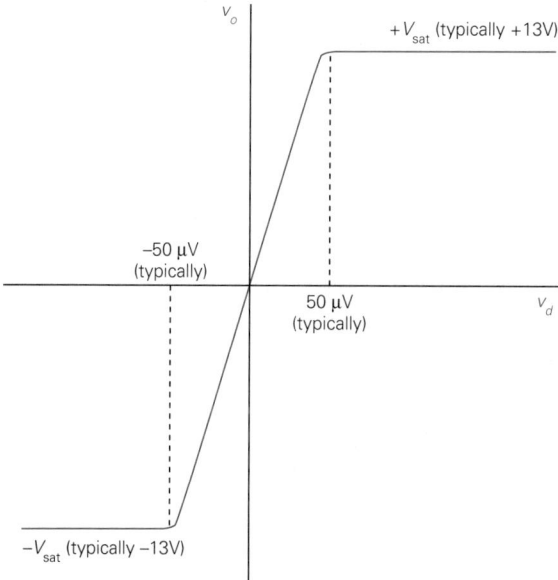

FIGURE 13–16
Typical transfer characteristic of an op-amp (different horizontal and vertical scales).

slope of the characteristic curve in the steep region is the differential gain A_d. Because of the very large gain, the horizontal scale here is quite different from the vertical scale. If the scales were the same, the steep line would appear nearly vertical. Note that the ideal output voltage is zero when the differential input voltage is zero. This is true even though there are no coupling capacitors in the op-amp. This dc cancellation is achieved by level-shifting circuitry and the dual power supplies.

Saturation

The curve of Figure 13–16 was based on assumed power supply voltages of ±15 V. It is observed that as the output voltage starts to approach either +15 V or −15 V, the characteristic curve starts to level off. Theoretically, the output could go as high as +15 V or as low as −15 V. In practice, however, certain transistors on the chip start to saturate before these levels are reached, and with most op-amps, the absolute power supply voltages are never quite reached.

The quantities $+V_{\text{sat}}$ and $-V_{\text{sat}}$ will be defined as the positive and negative saturation voltages, respectively. Typical values of V_{sat} range from 13 to 14 V, but they vary from one op-amp to another. Thus if $\pm V_{\text{sat}} = \pm 13$ V, *active-region* operation is from −13 to +13 V. As a good rule of thumb, when information is lacking, allow about 2 V margin or "backoff" from the two power supply limits. Thus for ±15-V power supplies, the output should be able to vary from about −13 to +13 V.

EXAMPLE 13–4 On the typical input–output characteristic curve of Figure 13–16, assume that the saturation level of 13 V is reached for a differential input voltage of 50 μV. Compute the differential gain A_d.

Solution
The differential gain is determined from the slope of the curve as

$$A_d = \frac{13 \text{ V}}{50 \times 10^{-6} \text{ V}} = 260{,}000 \qquad \textbf{(13–84)}$$

EXAMPLE 13–5 A certain op-amp is characterized by a dc differential gain of $A_d = 120{,}000$. What is the value of differential input voltage that would just barely cause the output to go into saturation if $V_{\text{sat}} = 13.5$ V?

Solution
The output v_o is related to the differential input by

$$v_o = A_d v_d \qquad \textbf{(13–85)}$$

Solving for v_d and substituting $v_o = 13.5$ V and $A_d = 120{,}000$, we have

$$v_d = \frac{v_o}{A_d} = \frac{13.5 \text{ V}}{120{,}000} = 112.5 \ \mu\text{V} \qquad \textbf{(13–86)}$$

13-7 OPERATIONAL AMPLIFIER SPECIFICATIONS

We will now inspect some typical operational amplifier specifications to see how pertinent data concerning its operating characteristics are determined. The op-amp selected for this purpose is the widely used 741. This particular op-amp was the first to employ internal stabilization and was introduced by Fairchild Semiconductor in the late 1960s. Variations of the 741 are now produced by a number of different manufacturers.

Refer to the μA741 data sheets in Appendix D for the discussion that follows. (The prefix "μA" is one used by Fairchild. Different manufacturers utilize prefixes with specific device numbers.)

The first page of the specifications provides data on the various packages and pin diagrams. The two power supply terminals are denoted as V+ and V−, respectively. Again, do not confuse these symbols with the v^+ and v^- symbols that we are using in the text for the instantaneous input signal variables.

Some terminals that we have not considered thus far are the "+offset null" and "−offset null" terminals. These are used when it is necessary to provide fine adjustments on the dc output level with no signal applied. They need not be connected for noncritical applications. The use of these terminals will be discussed later in this section.

The "absolute maximum ratings" are also given on the first page of the specifications. The maximum values of the supply voltages are listed as ±22 V for three variations of the 741 and ±18 V for the 741C. The maximum power dissipation depends on the type of package as noted.

The maximum values of the differential input voltage are ±30 V. However, the maximum range of either input voltage is ±15 V when power supplies of ±15 V are employed. From the footnote it is seen that these voltages cannot extend beyond the range of the power supply voltages when the op-amp uses lower supply voltages.

Following the schematic on the second page, the next several pages provide various electrical characteristics of different versions of the 741 both at fixed temperature ($T_A = 25°C$) and over a temperature range. Not all of these data are appropriate to be considered at this time, but some entries will be considered. To simplify the discussion that follows, only the 741 data will be discussed. As can be verified, data for the 741A, 741C, and 741E differ somewhat.

Input Offset Voltage

Ideally, the op-amp should have a dc output voltage of zero when the differential input voltage is zero. In practice, a small offset (either positive or negative) will appear at the output. This offset as it appears at the output turns out to be a function of the external dc gain established for the composite circuit, and it is more meaningful to refer the voltage to the input in specifications. Hence the name **input offset voltage** is used to denote it. By referring the voltage to the input, it is easier to predict the actual offset at the output from the given specification and the composite circuit gain.

The circuit model that can be used to predict the effect of dc input offset voltage is shown in Figure 13–17. The input offset voltage is shown as a dc voltage of value V_{IO} referred to the noninverting input. This effect will be ignored except for certain applications where it is critical. Scanning through the electrical characteristics, the input offset voltage for the 741 at 25°C is observed to have a typical value of 1 mV and a maximum value of 5 mV.

FIGURE 13–17
Circuit model used to show effect of dc input offset voltage.

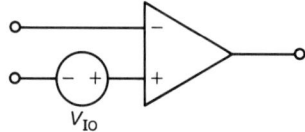

Input Bias and Offset Currents

The dc bias current used to establish the operating points of the op-amp input transistors must flow through any external circuit connected. For that reason it is necessary that *there be a dc path to ground at both op-amp input terminals for proper circuit operation.*

Let I_B^+ represent the bias current for the noninverting input, and let I_B^- represent the input bias current for the inverting input. Let I_{IB} represent the average input bias current, which is defined as

$$I_{IB} = \frac{I_B^+ + I_B^-}{2} \tag{13–87}$$

The value of this average bias current is denoted on specifications simply as "input bias current." For the 741 at 25°C, it has a typical value of 80 nA and a maximum value of 500 nA.

Another dc quantity of importance is the "input offset current," which will be denoted as I_{IO}, and it is defined as the magnitude of the difference between the two bias currents, that is,

$$I_{IO} = |I_B^+ - I_B^-| \tag{13–88}$$

If I_B^+ and I_B^- were exactly equal, the input offset current would be zero. In practice, however, they differ somewhat, so the difference is not zero.

For the 741 at 25°C, the typical value of I_{IO} is 20 nA, and the maximum value is 200 nA.

Offset Null Circuit

As noted, there is usually a small dc output voltage at the op-amp even when the differential input voltage is zero. In some applications, this is of no consequence.

However, when the op-amp is used to process small dc signals, this offset may be troublesome, and it is desirable to null it out.

On the eighth page of the specifications is a circuit diagram showing the method recommended by the manufacturer for nulling out this voltage. A 10-kΩ potentiometer is connected between pins 1 and 5 and the negative power supply as shown. The potentiometer is adjusted for a null at the output.

The actual nulling procedure is performed with the op-amp connected to the external circuit resistances expected for the particular application since the actual offset is dependent on the external connections.

Bias-Compensating Resistor

Bias currents flowing in any resistances connected to the two op-amp inputs will produce small dc voltages, whose combined effects appear at the output as a dc offset. The dc offset resulting from bias currents is in addition to that produced by the input offset voltage and adds algebraically to the latter.

The offset produced by the bias currents can be minimized, whenever possible, by forcing the effective dc resistances to ground from the two op-amp inputs to be equal. Since the two bias currents are usually in the same range of values, this condition will result in two nearly equal dc voltages at the two inputs, and since their effects on the output voltage are in opposite directions, there will be a near-cancellation process. Theoretically, if the two bias currents were equal, and if the resistances to ground were also equal, perfect cancellation would result.

Any resistance connected from an input to the output of the op-amp is considered as being connected to ground for the purpose of determining a bias resistance. This is true because the op-amp output presents a very low effective resistance to ground. External inputs are considered as being connected to ground through any source resistance present.

To illustrate the process of bias compensation, consider the inverting amplifier circuit of Figure 13–18(a). Full operation of this circuit is considered in detail

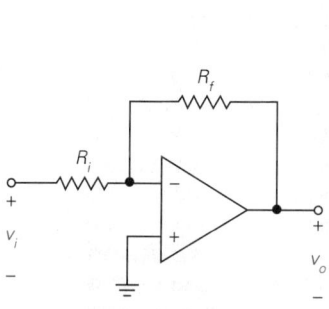

(a) Inverting amplifier without bias compensation resistor

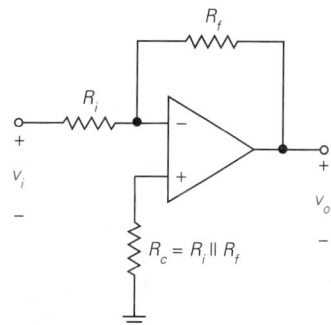

(b) Addition of bias compensation resistor R_c

FIGURE 13–18

Manner in which bias compensation resistor is added to one amplifier configuration.

in Chapters 14 and 15. For the moment, we are concerned only with the process of bias compensation.

The basic form of Figure 13–18(a) will work fine for many applications where small dc levels are not encountered. However, in applications where offset effects are to be minimized, a bias compensation resistance R_c may be added as shown in Figure 13–18(b). Assuming an ideal source at the input and considering it momentarily replaced by a short, the net resistance to ground from the inverting input is the parallel combination of R_i and R_f (i.e., $R_i \| R_f$). A bias-compensation resistance R_c having this value is connected between the noninverting input and ground as shown. This resistor does not affect the primary operation of the circuit, but it does minimize the bias current offset effect.

Input Resistance

The input resistance is referred to by the more general term *input impedance* and is denoted as Z_i. The minimum value for the 741 is indicated as 0.3 MΩ, and the typical value is 2 MΩ.

Differential Gain

The differential gain of the 741 is found on the line indicated as "large-signal voltage gain." The minimum value for the 741 can be interpreted as 50,000, and a typical value is 200,000. These are the values of gain at dc.

More information on gain is shown as curves on subsequent pages. On the sixth page of the specifications, a curve entitled "open-loop voltage gain as a function of supply voltage" is shown. As in the preceding listing, this is dc gain, and it is given in dB. The supply voltage is interpreted as both positive and negative (i.e., dual power supply voltages). For ±15 V, the gain is about 106 dB (or 200,000), which agrees with the typical value listed earlier.

On the ninth page of the specifications, a curve showing "open-loop frequency response" is given. The gain below 10 Hz is about 200,000, and this is in agreement with the typical value stated earlier. Observe, however, that this open-loop gain starts decreasing below 10 Hz and reaches a level of unity just above 1 MHz. The effect of this rolloff will be considered later.

Output Voltage Swing

On the seventh page of the specifications is a curve entitled "output voltage swing vs. supply voltage." The abscissa is interpreted as ± supply voltages. Note for bias supplies of ±15 V, the output voltage swing is given as a peak-to-peak value of 26 V. This corresponds to $+V_{sat} = 13$ V and $-V_{sat} = -13$ V, which is in agreement with our earlier rule of thumb.

Common-Mode Rejection Ratio

We defined CMRR earlier, but the definition is repeated here. It is

$$CMRR = \frac{|A_d|}{|A_{cm}|} \tag{13-89}$$

where A_d is the differential gain and A_{cm} is the common-mode gain. On most specifications, CMRR is given in dB, and this can be expressed as

$$CMRR(dB) = 20 \log_{10} CMRR \tag{13-90}$$

Values of CMRR are given on the third and fourth pages of the specifications. The minimum value is listed as 70 dB, and a typical value is listed as 90 dB. These correspond to absolute levels of 3162 and 31,623, respectively. Note how much larger these values are than the value computed in Example 13–1 for the single-stage differential amplifier.

It turns out that the CMRR varies with frequency. A curve displaying the variation with frequency is shown on the sixth page of the specifications.

13–8 PSPICE EXAMPLES

PSPICE EXAMPLE 13–1

Develop a PSPICE program to simulate the differential amplifier of Figure 13–3, and check the various dc voltages and currents when $v_{i1} = v_{i2} = 0$.

Solution

The circuit diagram adapted to PSPICE is shown in Figure 13–19(a), and the code is shown in Figure 13–19(b). Each transistor is the library model of the 2N2222A. The two inputs are nodes 1 and 2, respectively. Note that in the code for this example, values of zero have been listed.

The computer output data are shown in Figure 13–20. The two collector currents each have values of 0.9977 mA, which compare very closely with the text value of 1 mA. The current through R_E is 2.008 mA, which compares very closely with the text value of 2 mA. The output voltage is V(4, 3) = 0, as expected.

The two collector voltages each have values of 7.018 V, which compare closely with the text value of 7 V. Finally, the emitter voltage is −0.6435 V, which compares with −0.7 V.

PSPICE EXAMPLE 13–2

Use the transfer function command to determine the ratio of the differential output voltage to the input voltage v_{i1} for the differential amplifier circuit of PSPICE Example 13–1.

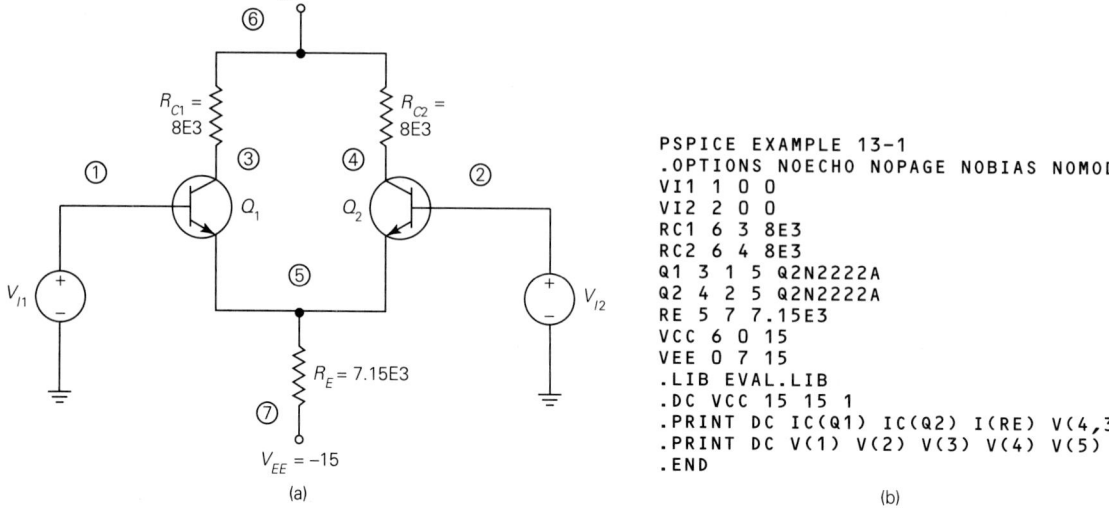

```
PSPICE EXAMPLE 13-1
.OPTIONS NOECHO NOPAGE NOBIAS NOMOD
VI1 1 0 0
VI2 2 0 0
RC1 6 3 8E3
RC2 6 4 8E3
Q1 3 1 5 Q2N2222A
Q2 4 2 5 Q2N2222A
RE 5 7 7.15E3
VCC 6 0 15
VEE 0 7 15
.LIB EVAL.LIB
.DC VCC 15 15 1
.PRINT DC IC(Q1) IC(Q2) I(RE) V(4,3)
.PRINT DC V(1) V(2) V(3) V(4) V(5)
.END
```

(a)   (b)

FIGURE 13-19
Circuit and code of PSPICE Example 13-1.

```
PSPICE EXAMPLE 13-1

****        CIRCUIT DESCRIPTION

*********************************************************************

.OPTIONS NOECHO NOPAGE NOBIAS NOMOD

****        DC TRANSFER CURVES                 TEMPERATURE = 27.000 DEG C


VCC          IC(Q1)        IC(Q2)      I(RE)        V(4,3)
1.500E+01    9.977E-04     9.977E-04   2.008E-03    0.000E+00


****        DC TRANSFER CURVES                 TEMPERATURE = 27.000 DEG C


VCC          V(1)          V(2)        V(3)         V(4)        V(5)
1.500E+01    0.000E+00     0.000E+00   7.018E+00    7.018E+00   -6.435E-01

            JOB CONCLUDED
            TOTAL JOB TIME                         .60
```

FIGURE 13-20
Computer printout of PSPICE Example 13-1.

Solution

The circuit diagram is the same as in PSPICE Example 13-1 (see Figure 13-19[a]). The code for determining the transfer function is shown in Figure 13-21. If only the transfer function is desired, and no output voltage values are requested, the input signal voltages may be left with zero values as in the preceding example.

FIGURE 13–21
Code for PSPICE Example 13–2.

```
PSPICE EXAMPLE 13-2
.OPTIONS NOECHO NOPAGE NOBIAS NOMOD
VI1 1 0 0
VI2 2 0 0
RC1 6 3 8E3
RC2 6 4 8E3
Q1 3 1 5 Q2N2222A
Q2 4 2 5 Q2N2222A
RE 5 7 7.15E3
VCC 6 0 15
VEE 0 7 15
.LIB EVAL.LIB
.TF V(4,3) VI1
.END
```

The transfer function line reads

```
.TF V(4, 3) VI1
```

The output variable of reference is V(4, 3), the voltage between the two collectors. Since only one input variable may be specified in a transfer function, VI1 has been selected. Understand, however, that the corresponding transfer function relating V(4, 3) to VI2 would simply be the negative of the one requested because of the circuit symmetry. Thus, we can obtain the desired information from this single transfer function.

The computer printout is shown in Figure 13–22. The transfer function is V(4, 3)/VI1 = 279.3, which can be interpreted as the differential gain. This value differs by about 12.7% from the value of 320 assumed in the text. However, the text assumptions were somewhat idealized and did not consider any dynamic collector resistance, which would lower the actual gain from the ideal value.

```
PSPICE EXAMPLE 13-2

****      CIRCUIT DESCRIPTION

********************************************************************

.OPTIONS NOECHO NOPAGE NOBIAS NOMOD

****      SMALL-SIGNAL CHARACTERISTICS

      V(4,3)/VI1 = 2.793E+02
      INPUT RESISTANCE AT VI1 = 9.236E+03
      OUTPUT RESISTANCE AT V(4,3) = 1.457E+04

      JOB CONCLUDED
      TOTAL JOB TIME            .83
```

FIGURE 13–22
Computer printout of PSPICE Example 13–2.

PSPICE EXAMPLE 13–3

Use PSPICE to determine the input–output characteristic of the library model for the 741 op-amp.

Solution

A block diagram depicting the external connections for the 741 library model is shown in Figure 13–23. There are five connections required, and they are denoted on the diagram as 1–5 in the following order:

1. Noninverting signal input
2. Inverting signal input
3. Positive power supply connection
4. Negative power supply connection
5. Output

FIGURE 13–23
PSPICE library model of 741 operational amplifier.

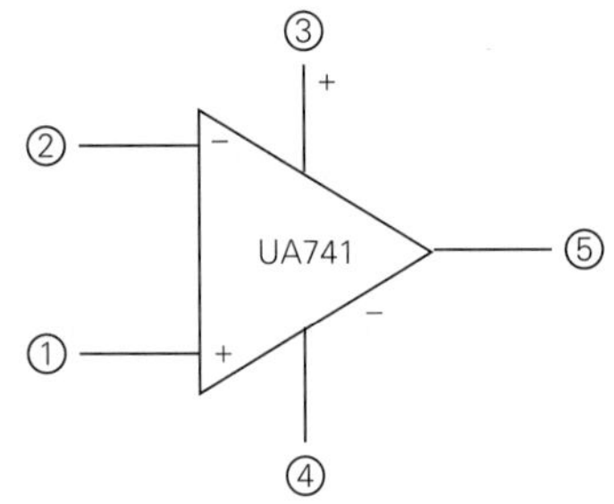

In an actual circuit connection, the node numbers may be quite different, but in the line describing the connection, it is very important that the *order* of the five terminals be the same as in this diagram. Note that unlike the simple VCVS model, the op-amp model requires power supplies to operate correctly.

The circuit diagram used for obtaining the input–output characteristic is shown in Figure 13–24(a), and the code is shown in Figure 13–24(b). A differential input voltage VDIFF has been connected across the two input terminals with the positive reference at the noninverting input. However, to provide dc paths to ground, a 1-Ω resistor has been connected to ground at each input. Power supply voltages of ±15 V have been connected to the respective power supply terminals.

The line identifying the op-amp and its connections reads as follows:

```
X 1 2 3 4 5 UA741
```

The quantity X identifies the op-amp, and additional letters or numbers could be added for further identification if there were more than one op-amp. The five numbers that follow identify the five connection nodes. In this particular example,

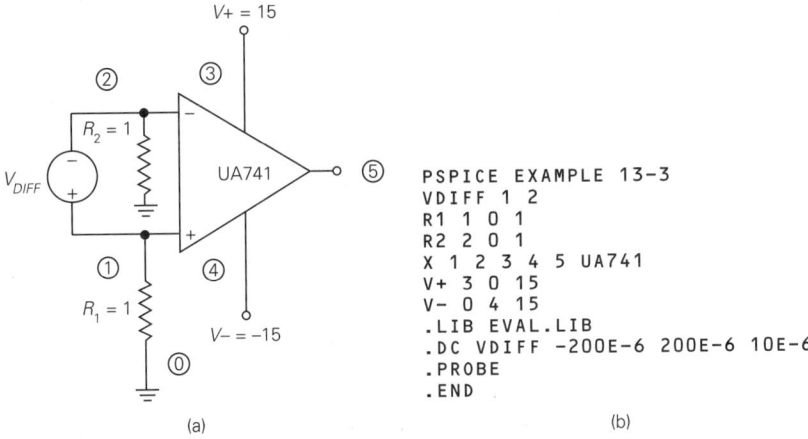

(a)

```
PSPICE EXAMPLE 13-3
VDIFF 1 2
R1 1 0 1
R2 2 0 1
X 1 2 3 4 5 UA741
V+ 3 0 15
V- 0 4 15
.LIB EVAL.LIB
.DC VDIFF -200E-6 200E-6 10E-6
.PROBE
.END
```

(b)

FIGURE 13–24
Circuit and code of PSPICE Example 13–3.

the numbers are exactly the same as in the illustration of Figure 13–23. However, remember for future examples that it is the *order* of the node numbers, rather than the actual node numbers, that defines the functions of the five terminals.

In the .DC sweep statement, the voltage VDIFF is swept from $-200~\mu V$ to $200~\mu V$ in steps of $1~\mu V$. The .PROBE operation is used to perform the plot.

The transfer characteristic is shown in Figure 13–25. Using the cursor of

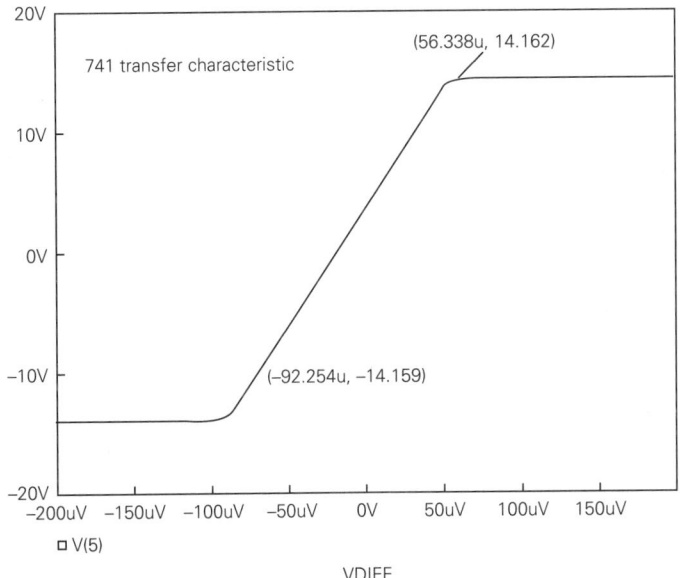

FIGURE 13–25
Transfer characteristic of PSPICE Example 13–3.

.PROBE, the approximate points at which saturation occurs were determined and are labeled. It is noted that saturation occurs near ±14.3 V. These values are somewhat greater than our conservative estimates of ±13 V. Note that there is an offset near the origin; that is, the output voltage is not zero when the input is zero. However, this effect is largely due to the open-loop conditions assumed here, and it would be reduced to nearly zero when the loop is closed with a well-designed feedback circuit.

An estimate of the open-loop gain can be made by determining the approximate slope of the transfer characteristic in the linear region. This value is approximated as

$$A = \frac{14.162 - (-14.159)}{56.338 \times 10^{-6} - (-92.254 \times 10^{-6})} = 190{,}600 \qquad \textbf{(13–91)}$$

This value is very close to the typical value of 200,000 provided in the data sheets.

PROBLEMS

Drill Problems

13–1. Consider the circuit of Figure P13–1, and let $V_{BB} = 12$ V, $R_E = 11.3$ kΩ, and $R_C = 10$ kΩ. Initially assume that both base terminals are grounded. **(a)** Determine the dc emitter current I_E. **(b)** At the dc operating point, determine the values of i_{C1}, i_{C2}, v_{C1}, v_{C2}, and v_o.

FIGURE P13–1

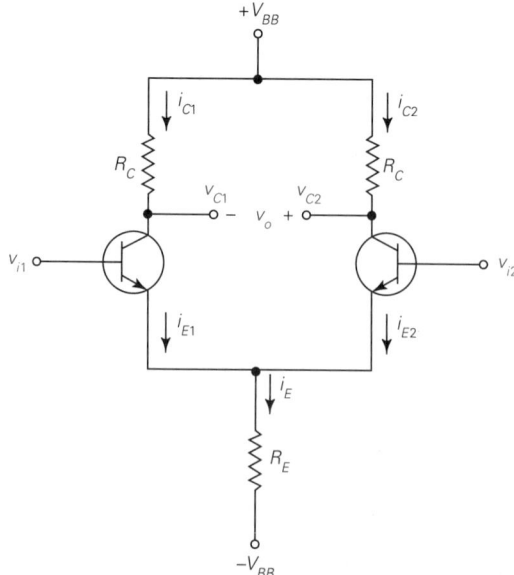

13–2. Consider the circuit of Figure P13–1, and let $V_{BB} = 8$ V, $R_E = 1.825$ kΩ, and $R_C = 2$ kΩ. Initially assume that both base terminals are grounded. **(a)** Determine the dc emitter current I_E. **(b)** At the dc operating point, determine the values of i_{C1}, i_{C2}, v_{C1}, v_{C2}, and v_o.

13–3. For the circuit of Problem 13–1, determine the values of **(a)** A'_d; **(b)** A_d; **(c)** A_{cm}; and **(d)** CMRR.

13–4. For the circuit of Problem 13–2, determine the values of **(a)** A'_d; **(b)** A_d; **(c)** A_{cm}; and **(d)** CMRR.

13–5. For the circuit of Problems 13–1 and 13–3, assume that the two inputs are $v_{i1} = 12$ mV and $v_{i2} = 0$. Determine **(a)** the differential and common-mode input signals; **(b)** the differential and common-mode output signals; and **(c)** the emitter signal voltage.

13–6. For the circuit of Problems 13–2 and 13–4, assume that the two inputs are $v_{i1} = 10$ mV and $v_{i2} = 0$. Determine **(a)** the differential and common-mode input signals; **(b)** the differential and common-mode output signals; and **(c)** the emitter signal voltage.

13–7. Repeat the analysis of Problem 13–5 if $v_{i1} = 2.012$ V and $v_{i2} = 2$ V.

13–8. Repeat the analysis of Problem 13–6 if $v_{i1} = 1.5$ V and $v_{i2} = 1.49$ V.

13–9. Repeat the analysis of Problem 13–5 if $v_{i1} = -1.8$ V and $v_{i2} = -1.79$ V.

13–10. Repeat the analysis of Problem 13–6 if $v_{i1} = -2$ V and $v_{i2} = -2.01$ V.

13–11. In the circuit of Figure P13–11, let $V_{BB} = 12$ V and $R_B = 5$ kΩ. Determine I_C.

FIGURE P13–11

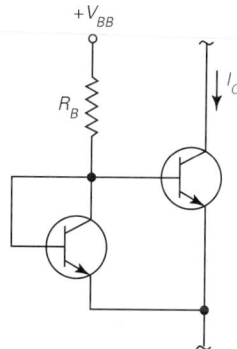

13–12. In the circuit of Figure P13–11, let $V_{BB} = 24$ V and $R_B = 4$ kΩ. Determine I_C.

13–13. In the circuit of Figure P13–11, let $V_{BB} = 20$ V. Determine the value of R_B required to establish a collector current $I_C = 5$ mA.

13–14. In the circuit of Figure P13–11, let $V_{BB} = 15$ V. Determine the value of R_B required to establish a collector current $I_C = 10$ mA.

13–15. The transfer characteristic of a certain op-amp, based on ±15-V power supplies, is shown in Figure P13–15. (Note that the horizontal and vertical scales are drastically different.) Determine **(a)** the saturation voltages, and **(b)** the differential gain.

13–16. The saturation voltages of a certain op-amp, based on ±15 V power supplies, are ±13.5 V. If saturation occurs for a differential input voltage of 50 μV, determine the differential gain.

FIGURE P13–15

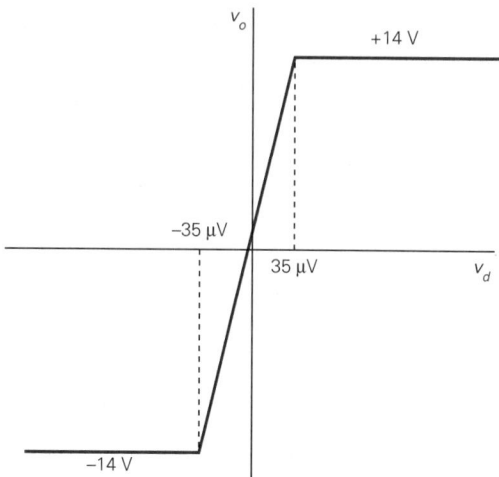

13–17. A certain op-amp has a dc differential gain of 240,000. Determine the value of differential input voltage that would just barely cause the output to go into saturation if $V_{sat} = 13.2$ V.

13–18. A certain op-amp has a dc differential gain of 106 dB. Determine the value of differential input voltage that would just barely cause the output to go into saturation if $V_{sat} = 14$ V.

13–19. A certain op-amp has input bias currents $I_B^+ = 110$ nA and $I_B^- = 100$ nA. Determine the average input bias current I_{IB}.

13–20. A certain op-amp has bias currents $I_B^+ = 40$ nA and $I_B^- = 36$ nA. Determine the average input bias current I_{IB}.

13–21. Determine the input offset current for the op-amp of Problem 13–19.

13–22. Determine the input offset current for the op-amp of Problem 13–20.

13–23. A certain op-amp has $A_d = 200,000$ and $A_{cm} = 50$. Determine the values of CMRR and CMRR(dB).

13–24. A certain op-amp has $A_d = 80,000$ and $A_{cm} = 12$. Determine the values of CMRR and CMRR(dB).

13–25. A certain op-amp has a specified differential gain of 120,000 and a common-mode rejection ratio of 100 dB. Determine the value of the common-mode gain A_{cm}.

13–26. A certain op-amp has a specified differential gain of 100 dB and a common-mode rejection ratio of 80 dB. Determine the value of the common-mode gain A_{cm}.

Derivation Problems

13–27. Starting with (13–29) and (13–30), derive (13–31) and (13–32).

13–28. Assume that the output v_o of a differential amplifier can be expressed in terms of the two inputs v_1 and v_2 as

$$v_o = A_1 v_1 + A_2 v_2$$

where A_1 and A_2 are gain factors relating to the two separate inputs. The output v_o can also be expressed in terms of differential input v_d and common-mode input v_{cm} as

$$v_o = A_d v_d + A_{cm} v_{cm}$$

where A_d and A_{cm} are the differential and common-mode gains and v_d and v_{cm} are defined as

$$v_d = v_1 - v_2$$

and

$$v_{cm} = \frac{v_1 + v_2}{2}$$

Show that

$$A_d = \frac{A_1 - A_2}{2}$$

and

$$A_{cm} = A_1 + A_2$$

13–29. Referring to Problem 13–28, show that

$$A_1 = A_d + \frac{A_{cm}}{2}$$

and

$$A_2 = -A_d + \frac{A_{cm}}{2}$$

13–20. Referring to Problem 13–28, assume that $A_1 = A$ and $A_2 = -A$. Determine A_d and A_{cm}, and comment on the significance of the result.

14

NEGATIVE FEEDBACK

OBJECTIVES

After completing this chapter, the reader should be able to:

- Discuss the general concept of negative feedback and its advantages.
- Draw a block diagram depicting a negative feedback system, and derive the equation relating closed-loop gain to open-loop gain and the feedback factor.
- Show how the closed-loop gain is very nearly dependent only on the feedback factor and virtually independent of the forward gain when the loop gain is very large.
- Show how the op-amp noninverting amplifier configuration fits the general feedback block diagram, and relate the various terms in the two cases.
- Derive the closed-loop gain expression for the noninverting amplifier using the feedback equation.
- Determine the gain, input resistance, and output resistance of the noninverting op-amp configuration.
- Determine the gain, input resistance, and output resistance of the inverting op-amp configuration.
- Analyze some of the circuits in the chapter with PSPICE.

14–1 NEGATIVE FEEDBACK

One of the most important developments in the history of electronics was the concept of **negative feedback.** The use of negative feedback started in the vacuum-tube era and continued on through the transistor era. It remains as important as ever in the application of modern integrated circuits.

To develop the concept of negative feedback we will use a block diagram form, from which the results may be adapted to specific devices and circuits.

FIGURE 14–1
Block diagram illustrating negative feedback concept.

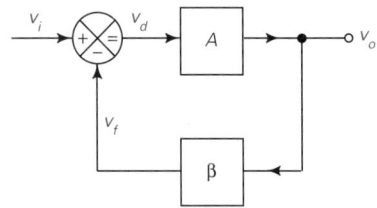

Consider the block diagram shown in Figure 14–1. The diagram contains three
important elements required in a negative feedback system. They are (1) the **open-
loop forward gain** A, (2) the **feedback factor** β, and (3) the **input difference circuit**
shown on the left.

The *open-loop forward gain* A represents the gain of the amplifier without
feedback. The *feedback factor* represents the fraction of the output that is fed
back to the input. The *input difference circuit* combines the feedback signal with
the input. Specifically, the feedback signal is *subtracted* from the input signal in
the model shown.

Assume that the input signal is v_i and the feedback signal is v_f. The output
of the difference circuit is a voltage v_d given by

$$v_d = v_i - v_f \tag{14–1}$$

This difference voltage is amplified by the gain A to produce the output voltage
v_o according to

$$v_o = A v_d \tag{14–2}$$

The feedback voltage v_f is proportional to the output voltage v_o and is

$$v_f = \beta v_o \tag{14–3}$$

When (14–3) is substituted in (14–1), there results

$$v_d = v_i - \beta v_o \tag{14–4}$$

Substitution of (14–4) in (14–2) results in

$$v_o = A(v_i - \beta v_o) = A v_i - A \beta v_o \tag{14–5}$$

This last result may seem strange since v_o is expressed in terms of both the
input v_i and itself. However, by regrouping all v_o terms on the left-hand side of
the equation, v_o can be expressed in terms of the input. The regrouping procedure
results in

$$v_o + A \beta v_o = A v_i \tag{14–6}$$

Let A_{CL} represent the *closed-loop gain* (i.e., the *gain with feedback*). It is
defined as

$$A_{CL} = \frac{v_o}{v_i} \tag{14–7}$$

Factoring out v_o on the left side of (14–6) and forming the ratio of (14–7), we obtain

$$A_{CL} = \frac{A}{1 + A\beta} \qquad \text{(14–8)}$$

The quantity $A\beta$ in the denominator has a special significance and is defined as

$$\text{loop gain} = A\beta \qquad \text{(14–9)}$$

The *loop gain* is the product of all gain factors around the loop.

The result of (14–8) is a most important equation and should be emphasized. Stated in words, it says that *the closed-loop gain with feedback is the open-loop forward gain divided by one plus the loop gain.*

Normally, the closed-loop gain is considerably less than the open-loop gain. What is then accomplished by the process? The answer is that several important improvements can be achieved through this trade-off. Among these are stability, reduction of distortion and noise, and predictability of gain.

EXAMPLE 14–1

A certain feedback amplifier can be modeled by the block diagram of Figure 14–1. The open-loop forward gain is $A = 100$, and the feedback is $\beta = 0.1$. Compute (a) the loop gain, and (b) the closed-loop gain A_{CL} with feedback.

Solution

(a) The loop gain is

$$A\beta = 100 \times 0.1 = 10 \qquad \text{(14–10)}$$

(b) The closed-loop gain is

$$A_{CL} = \frac{100}{1 + 10} = 9.09 \qquad \text{(14–11)}$$

Limiting Case

Consider the case when $A\beta \gg 1$, that is, when the loop gain is very large. In this case the unity term in the denominator of (14–8) may be neglected, and the closed-loop gain may be closely approximated as

$$A_{CL} \simeq \frac{A}{A\beta} = \frac{1}{\beta} \qquad \text{for } A\beta \gg 1 \qquad \text{(14–12)}$$

Simply stated, this result says that *when the loop gain is very large, the closed-loop gain is nearly equal to the reciprocal of the feedback factor.*

The implications of this result are most important and need to be emphasized. Usually, A is the gain of an active device that is difficult to control and hard to predict from one unit to another. On the other hand, β is usually the transfer ratio of a simple resistive network, which can be controlled to a high degree of accuracy. If the loop gain is made to be very large, the net gain will be virtually independent of the uncontrollable factor A and almost completely predictable from the controllable factor β.

What constitutes a large loop gain? There is no single answer for this question. However, as a general guideline, if $A\beta \geq 100$, the difference between the exact gain and the approximation of (14–12) is less than 1%.

EXAMPLE 14–2

In a certain feedback amplifier system, $A = 20,000$ and $\beta = 0.01$. Compute **(a)** the exact closed-loop gain using (14–8), and **(b)** the approximate closed-loop gain using (14–12).

Solution

(a) The loop gain is $A\beta = 20,000 \times 0.01 = 200$. The exact gain is

$$A_{CL} = \frac{20,000}{1 + 200} = 99.5 \tag{14–13}$$

(b) The approximate gain is

$$A_{CL} \simeq \frac{1}{\beta} = \frac{1}{0.01} = 100 \tag{14–14}$$

The results differ by about 0.5%, a result of the large loop gain.

EXAMPLE 14–3

To illustrate the stability of the closed-loop gain, consider the amplifier of Example 14–2. **(a)** Consider first that the amplifier gain drops to $A = 10,000$, and compute the closed-loop gain. **(b)** Next, assume that the amplifier gain increases to $A = 40,000$, and compute the closed-loop gain. **(c)** Compare the preceding results with those of Example 14–2.

Solution

(a) For $A = 10,000$ and $\beta = 0.01$, we have

$$A_{CL} = \frac{10,000}{1 + 10^4 \times 0.01} = 99.01 \tag{14–15}$$

(b) For $A = 40,000$ and $\beta = 0.01$, we have

$$A_{CL} = \frac{40,000}{1 + 4 \times 10^4 \times 0.01} = 99.75 \tag{14–16}$$

(c) The various results of Examples 14–2 and 14–3 are tabulated below. The last case, $A = \infty$, corresponds to the approximation of (14–12).

A	A_{CL}
10,000	99.01
20,000	99.5
40,000	99.75
∞	100

The stability of A_{CL} as A varies over a very wide range is quite obvious.

Variation in Gain

We have seen that the variation of closed-loop gain is much smaller than the variation in open-loop gain when negative feedback is employed. A useful formula for predicting the approximate fractional variation in closed-loop gain is given in this section.

Let ΔA represent a change in open-loop gain, and let ΔA_{CL} represent the corresponding change in closed-loop gain. An approximate relationship between ΔA and ΔA_{CL} may be derived using calculus, and it is stated as follows:

$$\frac{\Delta A_{CL}}{A_{CL}} = \frac{1}{1 + A\beta} \frac{\Delta A}{A} \qquad (14\text{--}17)$$

The quantity $\Delta A/A$ may be considered as the fractional change in open-loop gain, and $\Delta A_{CL}/A_{CL}$ may be considered as the corresponding fractional change in closed-loop gain. This interpretation of (14–17) may be stated as

$$\left(\begin{array}{c}\text{fractional change} \\ \text{in } A_{CL}\end{array}\right) = \frac{1}{1 + A\beta} \times \left(\begin{array}{c}\text{fractional change} \\ \text{in } A\end{array}\right) \qquad (14\text{--}18)$$

These fractional changes may also be expressed as percentage changes if desired. Whenever it is actually desired to compute ΔA_{CL}, the nominal value of A is used in the equation. This result is most useful for dealing with fractional and percentage changes.

Although (14–17) and (14–18) are approximations, they provide close estimates of the relationship between open-loop and closed-loop fractional gain changes for small and moderate changes in A. For drastic changes in A, such as considered in Example 14–3, the formula is less accurate in predicting the exact changes. However, the sense of the result is correct even for large gain changes. Basically, the sense of the formula is that the variation in closed-loop gain decreases as the loop gain increases.

EXAMPLE 14-4

A certain amplifier has a nominal open-loop gain of 20,000, and it is subject to a variation of ±10%. For $\beta = 0.01$, estimate the percentage variation in closed-loop gain.

Solution

The value of $\Delta A/A$ on a positive fractional basis is

$$\frac{\Delta A}{A} = 0.1 \qquad \text{(14-19)}$$

The corresponding value of $\Delta A_{CL}/A_{CL}$ is estimated as

$$\frac{\Delta A_{CL}}{A_{CL}} = \frac{1}{1 + 20,000(0.01)} \times 0.1 = \frac{1}{201} \times 0.1 \approx 0.0005 \qquad \text{(14-20)}$$

where some realistic rounding was employed. Expressed as a percentage, we can say that the expected variation in the closed-loop gain is about ±0.05%.

14-2 NONINVERTING OPERATIONAL AMPLIFIER CIRCUIT

The closed-loop **noninverting amplifier circuit** employing an op-amp is considered in this section. This amplifier circuit readily fits the feedback block diagram form considered in the preceding section and may be analyzed by the concepts developed there.

The basic circuit form as it is more commonly drawn is shown in Figure 14-2(a). The input signal, referred to ground, is applied to the noninverting input. A feedback circuit consisting of R_i and R_f is arranged as a voltage divider between output and input. The subscript i represents "input," and the subscript f represents "feedback."

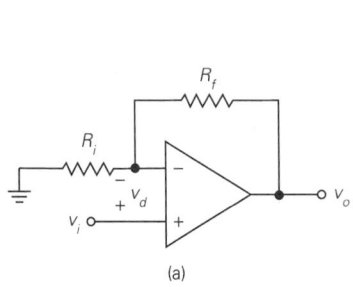
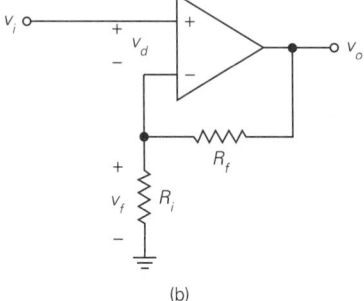

(a)   (b)

FIGURE 14-2
Noninverting amplifier drawn two ways.

An alternative way to draw the circuit to emphasize more clearly the feedback loop is shown in Figure 14–2(b). For this form, the inverting and noninverting inputs have been reversed in position.

The operational amplifier acts as both the difference circuit and the open-loop forward gain. We will momentarily change the notation of differential gain from A_d as defined in Chapter 13 to A, in order to conform to standard feedback notation. We have

$$v_o = Av_d = A(v_i - v_f) \qquad (14\text{–}21)$$

The feedback voltage is the fraction of the output voltage appearing across R_i. At this point, any current flowing into or out of the op-amp input terminals will be neglected. With this assumption, the voltage v_f is determined by the voltage-divider rule as

$$v_f = \frac{R_i}{R_i + R_f} v_o \qquad (14\text{–}22)$$

We immediately recognize that the feedback factor β is

$$\beta = \frac{R_i}{R_i + R_f} \qquad (14\text{–}23)$$

Application of (14–8) yields for the closed-loop gain A_{CL},

$$A_{\mathrm{CL}} = \frac{v_o}{v_i} = \frac{A}{1 + A\beta} = \frac{A}{1 + AR_i/(R_i + R_f)} \qquad (14\text{–}24)$$

Equation (14–24) can be manipulated algebraically to the form

$$A_{\mathrm{CL}} = \frac{(R_i + R_f)/R_i}{1 + (R_i + R_f)/AR_i} \qquad (14\text{–}25)$$

Assume next that A is very large. In this case the second term in the denominator of (14–25) will be small compared with unity, and A_{CL} may be approximated as

$$\boxed{A_{\mathrm{CL}} \simeq \frac{R_i + R_f}{R_i}} \qquad (14\text{–}26)$$

Alternatively, this is equivalent to assuming that the loop gain is very large, in which case $A_{\mathrm{CL}} \simeq 1/\beta$. The results are identical.

We see that the closed-loop gain of the noninverting amplifier circuit is essentially independent of the open-loop gain. However, the desired closed-loop gain is established by choosing appropriate values of R_i and R_f.

EXAMPLE 14–5

A certain noninverting op-amp amplifier circuit has $R_i = 1\ \text{k}\Omega$, $R_f = 99\ \text{k}\Omega$, and $A = 200{,}000$. Determine **(a)** β; **(b)** the loop gain; **(c)** the exact closed-loop gain; and **(d)** the approximate closed-loop gain based on the assumption that $A \to \infty$.

Solution

(a) The value of β is

$$\beta = \frac{R_i}{R_i + R_f} = \frac{1\ \text{k}\Omega}{1\ \text{k}\Omega + 99\ \text{k}\Omega} = 0.01 \qquad (14\text{–}27)$$

(b) The value of the loop gain is

$$\text{loop gain} = A\beta = 2 \times 10^5 \times 0.01 = 2000 \qquad (14\text{–}28)$$

(c) The exact gain A_{CL} can be determined from either (14–8) or (14–25). Using the first form, we have

$$A_{\text{CL}} = \frac{200{,}000}{1 + 2000} = 99.95 \qquad (14\text{–}29)$$

(d) The approximate gain based on the assumption that $A \to \infty$ is determined from (14–26) as

$$A_{\text{CL}} = \frac{R_i + R_f}{R_i} = \frac{1\ \text{k}\Omega + 99\ \text{k}\Omega}{1\ \text{k}\Omega} = 100 \qquad (14\text{–}30)$$

Alternatively, A_{CL} is simply

$$A_{\text{CL}} = \frac{1}{\beta} = \frac{1}{0.01} = 100 \qquad (14\text{–}31)$$

Note that the gain as determined in (14–30) and (14–31) differs from (14–29) by about 0.05%.

14–3 EFFECT OF FEEDBACK ON INPUT AND OUTPUT RESISTANCES

In the preceding section, the input resistance of the op-amp was assumed to be infinite, and the output resistance was assumed to be zero. In this section the effects of a finite input resistance and a nonzero output resistance will be considered. Because the two effects are different, and the values differ by several orders of magnitude, it is possible to focus on each effect individually while considering ideal conditions for the other effect. With this assumption, the mathematical expressions are greatly simplified with negligible loss in accuracy.

Effect on Input Resistance

The circuit model used to study the effect of feedback on finite input resistance in the noninverting amplifier is shown in Figure 14–3. The quantity r_d represents the **open-loop differential input resistance.** The differential gain is assumed again to be A. For this analysis, the output resistance is assumed to be zero, since it contributes very little effect to the input resistance.

FIGURE 14–3
Circuit model used to study effect of feedback input resistance.

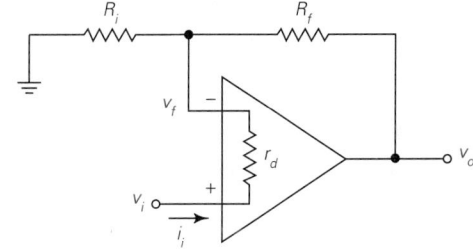

To determine the input resistance, a signal voltage v_i and an input current i_i are assumed. An expression must be determined for i_i in terms of v_i with all other variables eliminated.

From Figure 14–3 we can express i_i as

$$i_i = \frac{v_i - v_f}{r_d} \tag{14-32}$$

The feedback voltage v_f is

$$v_f = \beta v_o \tag{14-33}$$

However, from the work of Section 14–1, v_o can be expressed as

$$v_o = \frac{A}{1 + A\beta} v_i \tag{14-34}$$

Substitution of (14–34) in (14–33) yields

$$v_f = \frac{A\beta}{1 + A\beta} v_i \tag{14-35}$$

Next, v_f from (14–35) is substituted in (14–32).

$$i_i = \frac{1}{r_d}\left(v_i - \frac{A\beta}{1 + A\beta} v_i\right) = \frac{1}{r_d(1 + A\beta)} v_i \tag{14-36}$$

The input resistance R_{in} is then determined as

$$R_{in} = \frac{v_i}{i_i} = (1 + A\beta)r_d \qquad \textbf{(14–37)}$$

The actual circuit input resistance is thus (1 + Aβ) times the open-loop differential input resistance.

Effect on Output Resistance

The circuit model used to study the effect of feedback on nonzero output resistance in the noninverting amplifier is shown in Figure 14–4. The quantity r_o represents the **open-loop op-amp output resistance,** and, as before, A represents the assumed finite differential gain. However, in this analysis, the input resistance is assumed to be infinite since it contributes very little effect to the output resistance.

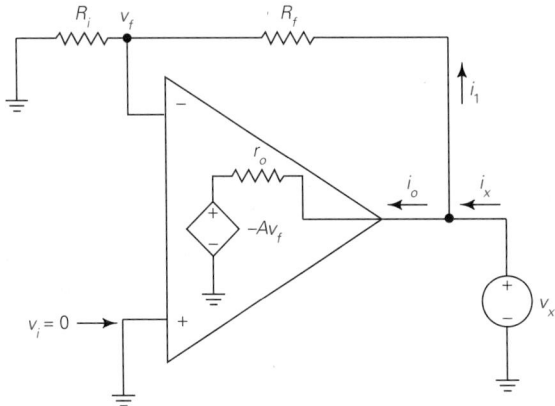

FIGURE 14–4
Circuit model used to study effect of feedback output resistance.

To determine the output resistance, the input signal v_i has been deenergized (replaced by a short), and a fictitious voltage generator v_x is applied to the output. The current i_x that would flow into the output is then determined.

The current i_x is composed of two components i_o and i_1 as shown. However, the current i_1 is so small compared with i_o in a typical case that we can assume that $i_x \approx i_o$. The value of i_o can then be determined as

$$i_o = \frac{v_x - (-Av_f)}{r_o} = \frac{v_x + Av_f}{r_o} \qquad \textbf{(14–38)}$$

The voltage v_f may be written as

$$v_f = \beta v_o = \beta v_x \tag{14-39}$$

Substitution of (14–39) in (14–38) yields

$$i_o \simeq i_x = \frac{v_x + A\beta v_x}{r_o} = \frac{1 + A\beta}{r_o} v_x \tag{14-40}$$

The output resistance R_{out} is then determined as

$$R_{\text{out}} = \frac{v_x}{i_x} = \frac{r_o}{1 + A\beta} \tag{14-41}$$

The actual circuit output resistance is thus the open-loop op-amp output resistance divided by $1 + A\beta$.

Voltage-Series Feedback

The noninverting amplifier circuit represents a type of feedback called **voltage-series feedback.** In this type of feedback, the variable sensed at the *output* is the *voltage* and the feedback is connected in *series* with the input. This is one of the most popular types of feedback arrangements, and its properties can be summarized as follows: *Voltage-series feedback causes the input resistance to increase by the factor* $1 + A\beta$, *and it causes the output resistance to decrease by the factor* $1/(1 + A\beta)$. These effects are desirable in an ideal voltage amplifier.

Voltage-series feedback is by no means the only type of feedback used. Several other forms will be encountered in this book. Depending on the form, the input and output resistances may be increased or decreased.

EXAMPLE 14–6

Consider the noninverting amplifier of Example 14–5, and assume that the open-loop op-amp input and output resistances are $r_d = 500$ kΩ and $r_o = 100$ Ω. Determine **(a)** the closed-loop input resistance, and **(b)** the closed-loop output resistance.

Solution
It was determined in Example 14–5 that $A\beta = 2000$.

(a) The closed-loop input resistance is

$$R_{\text{in}} = (1 + A\beta)r_d = (1 + 2000)500{,}000 \simeq 1 \times 10^9 \ \Omega \tag{14-42}$$

(b) The closed-loop output resistance is

$$R_{\text{out}} = \frac{r_o}{1 + A\beta} = \frac{100}{1 + 2000} \simeq 0.05 \ \Omega \tag{14-43}$$

14–4 EFFECT OF FEEDBACK ON BANDWIDTH

In this section the effect of negative feedback on the bandwidth of an amplifier is discussed. Certain results are stated here without proof. The actual development of these results is deferred to Chapter 18. At that point in the book, steady-state phasor analysis will be used to facilitate a proof.

The closed-loop gain A_{CL} as a function of the open loop gain A and feedback ratio β has been shown to be

$$A_{CL} = \frac{A}{1 + A\beta} \qquad\qquad (14\text{–}44)$$

Thus far, A has been assumed to be a constant value, which results in a constant value of A_{CL}, assuming, of course, that β is constant. In practice, the gain A will be frequency dependent. As we will see shortly, this results in a frequency-dependent closed-loop gain A_{CL}.

The high-frequency response of many amplifier circuits, particularly op-amps, is controlled over a wide frequency range by a dominating capacitance. (In op-amps, this dominating capacitive effect is established in the design process to achieve absolute stability.)

The Bode plot approximation of a typical open-loop gain response is shown by the upper curve in Figure 14–5. The value A_o represents the dc and very low frequency gain. The frequency f_o represents the 3-dB or "break" frequency in the open-loop gain response.

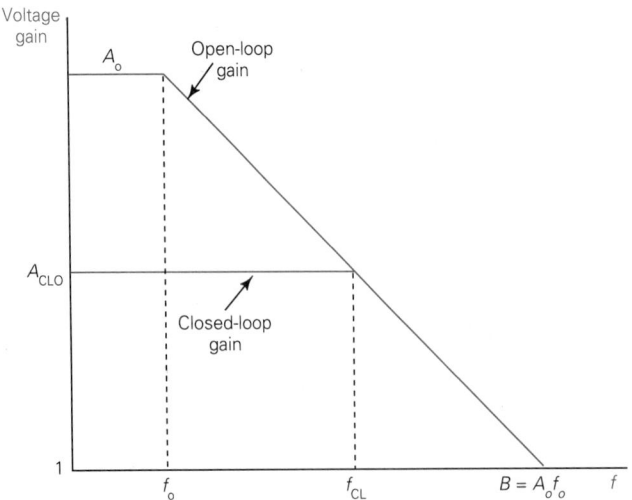

FIGURE 14–5
Open- and closed-loop amplitude response of amplifier with feedback.

Assume that this amplifier is used as the forward gain block in a negative feedback circuit. Assume that a closed-loop low-frequency gain A_{CLO} is established. Obviously, $A_{CLO} < A_o$ as shown on Figure 14–5. The quantity A_{CLO} is given by

$$A_{CLO} = \frac{A_o}{1 + A_o\beta} \tag{14–45}$$

It turns out that negative feedback causes the 3-dB closed-loop frequency to be greater than the corresponding open-loop frequency. Let f_{CL} represent the closed-loop 3-dB frequency. It will be shown in Chapter 18 with steady-state ac circuit analysis that f_{CL} and f_o are related by

$$\boxed{f_{CL} = (1 + A_o\beta)f_o} \tag{14–46}$$

Stated in words, *the closed-loop 3-dB frequency is* $(1 + A_o\beta)$ *times the open-loop 3-dB frequency.*

From Figure 14–5 it is noted that the Bode plot approximation of the closed-loop gain intersects the open-loop gain at the frequency f_{CL}. From this property, it can be inferred that *the bandwidth increases as the closed-loop gain decreases.*

Gain–Bandwidth Product

An interesting and meaningful property of the preceding development is the **gain–bandwidth product.** First, consider the product of the open-loop dc gain A_o and the open-loop 3-dB bandwidth f_o. This product is simply $A_o f_o$. Next, consider the product of the closed-loop dc gain A_{CLO} given by (14–45) and the closed-loop 3-dB bandwidth f_{CL} given by (14–46). This product is

$$A_{CLO}f_{CL} = \frac{A_o}{1 + A_o\beta}f_o(1 + A_o\beta) = A_o f_o \tag{14–47}$$

The product thus remains the same. In effect, gain is traded for more bandwidth. This property appears in many electronic circuits and represents one of the inherent constraints in electronic circuits.

The gain–bandwidth product is also called the **unity-gain frequency.** The reason is that it is the bandwidth resulting from a closed-loop gain of unity. This is observed in Figure 14–5 by noting that the open-loop gain intersects the unity gain ordinate at a frequency $A_o f_o$. For convenience, a quantity B is defined as

$$\boxed{B = A_o f_o = A_{CLO}f_{CL}} \tag{14–48}$$

Thus B is the gain–bandwidth product or the unity-gain frequency.

EXAMPLE 14–7

Consider the noninverting op-amp amplifier circuit of Example 14–5. The gain will be interpreted as the dc gain (i.e., $A_o = 200{,}000$). Assume that $f_o = 5$ Hz. Determine **(a)** the gain–bandwidth product, and **(b)** the closed-loop 3-dB frequency f_{CL}.

Solution

Since $\beta = 0.01$ in Example 14–5, $A_o\beta = 2000$.

(a) The gain–bandwidth product (or unity-gain frequency) is

$$B = A_o f_o = 200{,}000 \times 5 = 1 \times 10^6 \text{ Hz} = 1 \text{ MHz} \qquad \textbf{(14–49)}$$

(b) The closed-loop 3-dB frequency is

$$f_{CL} = f_o(1 + A\beta) = 5(1 + 2000) \simeq 10 \text{ kHz} \qquad \textbf{(14–50)}$$

14–5 INVERTING OPERATIONAL AMPLIFIER CIRCUIT

An op-amp circuit of equal importance with the noninverting amplifier is the **inverting amplifier circuit,** and its basic form is shown in Figure 14–6. (The power supply connections are understood but are not shown.) In this case the noninverting terminal is either grounded or connected through a bias compensation resistor to ground. For simplicity, it is shown here grounded. The signal is connected through a resistance R_i to the inverting input, and a feedback resistance R_f is connected between inverting input and output.

FIGURE 14–6
Basic form of inverting amplifier.

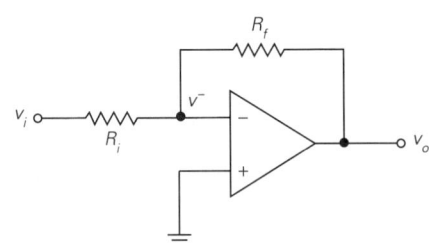

It turns out that β is the same in this circuit as in the noninverting amplifier. This can be seen by momentarily deenergizing the signal input point (v_i) to ground and noting that the feedback fraction β is simply

$$\beta = \frac{R_i}{R_i + R_f} \qquad \textbf{(14–51)}$$

However, the inverting amplifier differs from the noninverting amplifier in that the inverting amplifier employs **shunt feedback** at the input; that is, the feedback current is in parallel with the signal current. The result is that the circuit does not quite fit the standard block form given early in the chapter as well as the noninverting amplifier. Although the block form could be modified, we will employ a different approach here.

Assume that the voltage at the inverting input is v^- as shown in Figure 14–6. A KCL equation will be written at this node with currents expressed in terms of the voltages. It will be assumed that there is negligible current flowing into or out of the op-amp input terminals.

Summing the currents at the v^- mode, we have

$$\frac{v^- - v_i}{R_i} + \frac{v^- - v_o}{R_f} = 0 \qquad (14\text{–}52)$$

In addition, we have the op-amp equation according to which

$$v_o = A v_d = A(0 - v^-) = -A v^- \qquad (14\text{–}53)$$

Solving for v^- from (14–53) and substituting in (14–52), we obtain

$$v_o \left[\frac{1}{R_f} + \frac{1}{A} \left(\frac{1}{R_i} + \frac{1}{R_f} \right) \right] = \frac{-v_i}{R_i} \qquad (14\text{–}54)$$

After some manipulation, the closed-loop gain can be expressed as

$$A_{\text{CL}} = \frac{v_o}{v_i} = \frac{-R_f/R_i}{1 + 1/A\beta} \qquad (14\text{–}55)$$

where β was defined in (14–51).

Next, assume that $A\beta \gg 1$. In this case (14–55) reduces to

$$A_{\text{CL}} \simeq - \frac{R_f}{R_i} \qquad (14\text{–}56)$$

The closed-loop gain of the inverting amplifier under large loop gain conditions is a simple resistance ratio. The negative sign indicates the inherent sign inversion in the process. Once again, we see the beauty of the feedback process.

Input Resistance

Because of the shunt form of the feedback in the inverting amplifier, the input resistance *at the inverting input* approaches zero. This can be seen from (14–53) by first solving for v^-. We have

$$v^- = \frac{-v_o}{A} \qquad (14\text{–}57)$$

For any finite v_o, as A becomes very large, v^- approaches zero. This property is referred to as a *virtual ground;* that is, the voltage approaches zero, but it is not connected to the circuit ground. In fact, this point *must not* be connected to the actual circuit ground since it is necessary for a small voltage to exist across the input terminals for the circuit to function properly.

The external circuit input resistance R_{in} is the most meaningful property, and this resistance will be defined as the resistance seen at the point where v_i is connected. Since $v^- \approx 0$, the resistance seen by v_i is simply

$$R_{in} = R_i \qquad (14\text{--}58)$$

Output Resistance

As far as the output resistance is concerned, the inverting amplifier functions in the same manner as the noninverting circuit. This is readily observed by deenergizing the input signal and noting that the circuit structure is then the same as for the noninverting configuration. Thus the output resistance is

$$R_{out} = \frac{r_o}{1 + A\beta} \qquad (14\text{--}59)$$

EXAMPLE 14-8 A certain inverting op-amp amplifier circuit has $R_i = 1\ k\Omega$, $R_f = 24\ k\Omega$, and $A = 100{,}000$. Determine **(a)** β; **(b)** the loop gain; **(c)** the exact closed-loop gain; and **(d)** the approximate closed-loop gain based on the assumption that $A \to \infty$.

Solution
(a) The value of β is

$$\beta = \frac{R_i}{R_i + R_f} = \frac{1\ k\Omega}{1\ k\Omega + 24\ k\Omega} = 0.04 \qquad (14\text{--}60)$$

(b) The value of the loop gain is

$$\text{loop gain} = A\beta = 10^5 \times 0.04 = 4000 \qquad (14\text{--}61)$$

(c) The exact gain can be determined from (14–55) as

$$A_{CL} = \frac{-24\ k\Omega/1\ k\Omega}{1 + 1/4000} = -23.994 \qquad (14\text{--}62)$$

(d) The approximate gain based on the assumption that $A \to \infty$ is determined from (14–56) as

$$A_{CL} = \frac{-24\ k\Omega}{1\ k\Omega} = -24 \qquad (14\text{--}63)$$

The approximate gain differs from the exact gain by about 0.025%.

14–6 PSPICE EXAMPLES

PSPICE EXAMPLE 14–1

Develop a PSPICE model to simulate the noninverting op-amp amplifier of Examples 14–5 and 14–6, and use the transfer function command to determine the closed-loop gain, the input resistance, and the output resistance.

Solution

The circuit adapted to the PSPICE format is shown in Figure 14–7(a), and the code is shown in Figure 14–7(b). The internal structure of the op-amp model consists of three components: (1) the differential input resistance R_D, (2) the output resistance R_o, and (3) a VCVS with gain 2E5. The controlling voltage is the differential input voltage, which is denoted for PSPICE as V(1, 2).

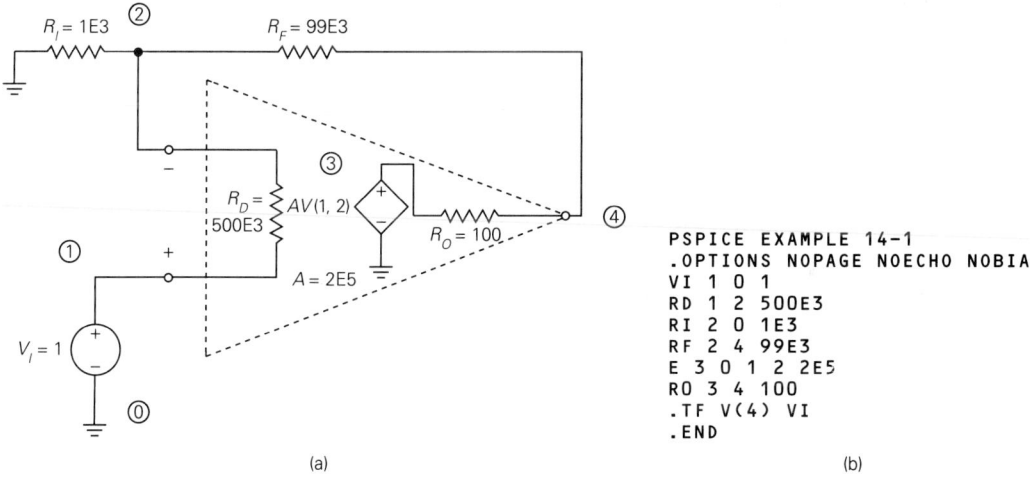

```
PSPICE EXAMPLE 14-1
.OPTIONS NOPAGE NOECHO NOBIAS
VI 1 0 1
RD 1 2 500E3
RI 2 0 1E3
RF 2 4 99E3
E 3 0 1 2 2E5
RO 3 4 100
.TF V(4) VI
.END
```

(a)   (b)

FIGURE 14–7
Circuit and code of PSPICE Example 14–1.

The computer printout is shown in Figure 14–8. The gain is V(4)/VI = 99.95, which agrees exactly with the result of Example 14–5(c). The input resistance is slightly less than 1×10^9 Ω, and the output resistance is slightly greater than 0.05 Ω. The values are very close to the theoretical values of Example 14–6.

PSPICE EXAMPLE 14–2

Develop a PSPICE model to simulate the inverting op-amp amplifier circuit of Example 14–8, and use the transfer function command to determine the closed-loop gain, the input resistance, and the output resistance.

```
PSPICE EXAMPLE 14-1

****        CIRCUIT DESCRIPTION

***************************************************************

.OPTIONS NOPAGE NOECHO NOBIAS

****        SMALL-SIGNAL CHARACTERISTICS

            V(4)/VI = 9.995E+01
            INPUT RESISTANCE AT VI = 9.995E+08
            OUTPUT RESISTANCE AT V(4) = 5.007E-02

            JOB CONCLUDED
            TOTAL JOB TIME                .83
```

FIGURE 14–8
Computer printout of PSPICE Example 14–1.

Solution

The circuit adapted to PSPICE is shown in Figure 14–9(a), and the code is shown in Figure 14–9(b). The internal structure of the op-amp model is similar in form to that of the preceding example except that the open-loop gain in the present case is 1E5. Since no values for differential input resistance and open-loop output resistance were specified in Example 14–8, the values from the preceding example have been used. Note in this case that the differential input voltage is V(0, 2).

The computer printout is shown in Figure 14–10. The closed-loop gain is V(4)/VI = −23.99, which is extremely close to the computed value of −23.994 in (14–62). The input resistance is 1000 Ω, which agrees exactly with the result of (14–58). The output resistance is indicated as 0.02504 Ω. A computation with (14–59) yields about 0.025 Ω, so the results are extremely close to each other.

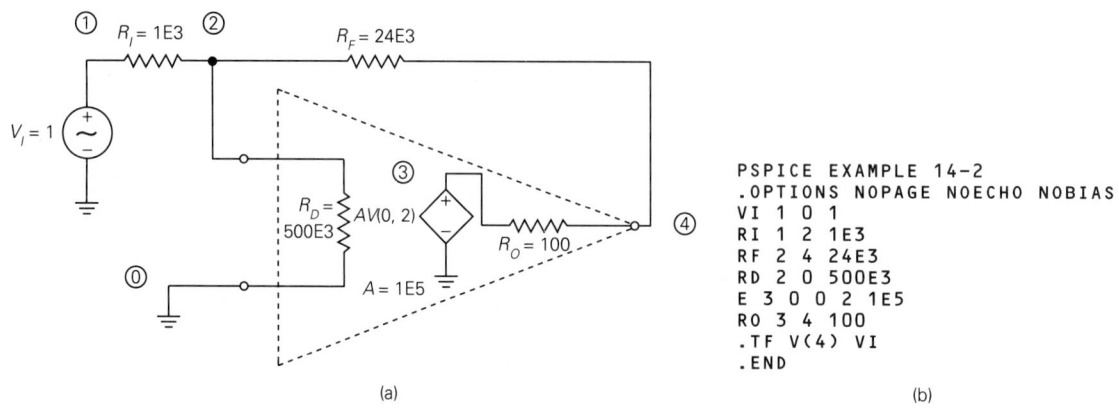

```
PSPICE EXAMPLE 14-2
.OPTIONS NOPAGE NOECHO NOBIAS
VI 1 0 1
RI 1 2 1E3
RF 2 4 24E3
RD 2 0 500E3
E 3 0 0 2 1E5
RO 3 4 100
.TF V(4) VI
.END
```

(a)   (b)

FIGURE 14–9
Circuit and code of PSPICE Example 14–2.

```
PSPICE EXAMPLE 14-2

****        CIRCUIT DESCRIPTION

************************************************************************

.OPTIONS NOPAGE NOECHO NOBIAS

****        SMALL-SIGNAL CHARACTERISTICS

            V(4)/VI = -2.399E+01
            INPUT RESISTANCE AT VI = 1.000E+03
            OUTPUT RESISTANCE AT V(4) = 2.504E-02

            JOB CONCLUDED
            TOTAL JOB TIME              .99
```

FIGURE 14-10
Computer printout of PSPICE Example 14-2.

PROBLEMS

Drill Problems

14-1. A certain feedback amplifier can be modeled by the block diagram of Figure 14-1. The open-loop forward gain is $A = 250$, and the feedback factor is $\beta = 0.08$. Compute **(a)** the loop gain, and **(b)** the closed-loop gain A_{CL} with feedback.

14-2. Repeat the computations of Problem 14-1 if $A = 500$ and $\beta = 0.05$.

14-3. In a certain feedback amplifier system, $A = 7500$ and $\beta = 0.02$. Compute **(a)** the exact closed-loop gain, and **(b)** the approximate closed-loop gain based on the assumption that $A \rightarrow \infty$.

14-4. Repeat the computations of Problem 14-3 if $A = 25,000$ and $\beta = 0.04$.

14-5. A certain feedback amplifier has a fixed feedback ratio $\beta = 0.05$. Determine the closed-loop gain A_{CL} for each of the following values of open-loop gain A: **(a)** 100; **(b)** 1000; **(c)** 10^4; **(d)** ∞.

14-6. A certain feedback amplifier has a fixed feedback ratio $\beta = 0.02$. Determine the closed-loop gain A_{CL} for each of the following values of open-loop gain A: **(a)** 100; **(b)** 1000; **(c)** 10^4; **(d)** ∞.

14-7. A certain amplifier has a nominal open-loop gain of 5000, and it is subject to a variation of $\pm 12\%$. For $\beta = 0.02$, estimate the percentage variation in closed-loop gain.

14-8. A certain amplifier has a nominal open-loop gain of 12,000, and it is subject to a variation of $\pm 20\%$. For $\beta = 0.02$, estimate the percentage variation in closed-loop gain.

14-9. A certain noninverting op-amp amplifier circuit has $R_i = 2$ kΩ, $R_f = 98$ kΩ, and $A = 120,000$. Determine **(a)** β; **(b)** the loop gain; **(c)** the exact closed-loop gain; and **(d)** the approximate closed-loop gain based on the assumption that $A \rightarrow \infty$.

14-10. Repeat the computation of Problem 14-9 if $R_i = 1$ kΩ, $R_f = 19$ kΩ, and $A = 50,000$.

14–11. Based on the assumption that $A \rightarrow \infty$ and $R_i = 1$ kΩ, compute the values of noninverting gain for each of the following values of R_f: **(a)** 9 kΩ; **(b)** 19 kΩ; **(c)** 49 kΩ; **(d)** 199 kΩ.

14–12. Based on the assumption that $A \rightarrow \infty$ and $R_f = 100$ kΩ, compute the values of noninverting gain for each of the following values of R_i: **(a)** 100 Ω; **(b)** 1 kΩ; **(c)** 10 kΩ; **(d)** 50 kΩ.

14–13. Consider the amplifier of Problem 14–9, and assume that the open-loop op-amp input and output resistances are $r_d = 400$ kΩ and $r_o = 60$ Ω. Determine **(a)** the closed-loop input resistance, and **(b)** the closed-loop output resistance.

14–14. Repeat the computations of Problem 14–13 for the amplifier of Problem 14–10 if $r_d = 1$ MΩ and $r_o = 50$ Ω.

14–15. Consider the noninverting op-amp circuit of Problem 14–9, and assume that the given gain is the dc gain (i.e., $A_o = 120{,}000$). Assume that $f_o = 20$ Hz. Determine **(a)** the gain–bandwidth product, and **(b)** the closed-loop 3-dB frequency f_{CL}.

14–16. Consider the noninverting op-amp circuit of Problem 14–10, and assume that the given gain is the dc gain (i.e., $A_o = 50{,}000$). Assume that $f_o = 10$ Hz. Determine **(a)** the gain–bandwidth product, and **(b)** the closed-loop 3-dB frequency f_{CL}.

14–17. A certain op-amp has a gain–bandwidth product $B = 2 \times 10^6$. Determine the 3-dB closed-loop bandwidth for each of the following noninverting closed-loop values of dc gain: **(a)** 10^3; **(b)** 100; **(c)** 10; **(d)** 1.

14–18. A certain op-amp connected as a noninverting amplifier has a closed-loop gain of 50 and a closed-loop 3-dB bandwidth of 30 kHz. **(a)** Determine the gain–bandwidth product. **(b)** If the open-loop 3-dB bandwidth is 7.5 Hz, determine the open-loop dc gain.

Derivation Problems

14–19. (Requires calculus.) Starting with (14–8), show that

$$\frac{dA_{CL}}{dA} = \frac{1}{(1 + A\beta)^2}$$

14–20. (Requires calculus.) Using the results of Problem 14–19 and employing differentials, derive (14–17).

Design Problems

14–21. It is desired to add feedback to a certain amplifier to improve the stability. The nominal value of the open-loop gain is 1000. Determine the value of the feedback factor β required to establish a closed-loop gain of 100.

14–22. It is desired to add feedback to a certain amplifier to improve the stability. The loop gain established for the amplifier is $A\beta = 20$. Determine the values of open-loop gain and feedback factor required to achieve a closed-loop gain of 12.

14–23. The open-loop gains of amplifiers in a certain system are subject to a $\pm16\%$ variation. It is desired to reduce this uncertainty with feedback. Determine the value of loop gain required to reduce the variation to about $\pm0.5\%$.

14–24. It is desired to achieve a net gain of 10,000 by connecting a number of amplifiers in cascade. Each amplifier has a gain of 200 ± 20. Feedback is to be employed in each stage so as to reduce the percentage variation to a level of about $\pm0.5\%$. Determine **(a)** the required closed-loop gain of each stage if all stages are to have equal gains, and **(b)** the number of stages required.

15

OPERATIONAL AMPLIFIER LINEAR CIRCUITS

OBJECTIVES

After completing this chapter, the reader should be able to:

- State the three assumptions and the associated implications used in the idealized analysis of op-amp circuits.
- Analyze an inverting amplifier circuit by employing the idealized assumptions.
- Design an inverting amplifier circuit to meet prescribed specifications.
- Analyze a noninverting amplifier circuit by employing the idealized assumptions.
- Design a noninverting amplifier circuit to meet prescribed specifications.
- Compare advantages and disadvantages of inverting and noninverting configurations.
- Analyze or design an op-amp summing circuit.
- Analyze or design an op-amp difference amplifier circuit.
- Analyze or design a current-controlled voltage source.
- Analyze or design a voltage-controlled current source.
- Analyze or design a current-controlled current source.
- Analyze some of the circuits in the chapter with PSPICE.

The material in this chapter and several other chapters is greatly expanded in W. D. Stanley, *Operational Amplifiers with Linear Integrated Circuits,* 3d Ed. (Columbus, OH: Charles E. Merrill, 1994).

15–1 SIMPLIFIED ANALYSIS OF OP-AMP CIRCUITS

The basis for a very simplified approach to the analysis of op-amp circuits will be established in this section. This simplified analysis is based on certain assumptions that may be made in a wide variety of applications. Designers select op-amps with specifications that justify these assumptions whenever possible. A wide variety of application circuits may be readily analyzed and designed around this approach.

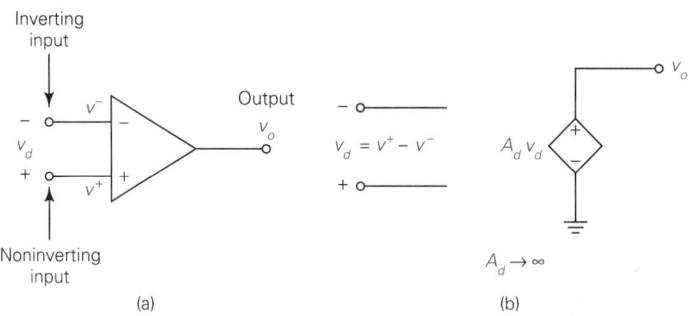

FIGURE 15–1
Op-amp symbol and ideal circuit model.

As a quick review of material previously considered, the symbol of an op-amp is shown in Figure 15–1(a). The two input terminals are known as the **noninverting input** (indicated by $+$) and the **inverting input** (indicated by $-$). The voltage at the noninverting input with respect to ground is indicated as v^+, and the voltage at the inverting input with respect to ground is indicated as v^-. The differential input voltage v_d is defined as the difference between these voltages in the following sense:

$$v_d = v^+ - v^- \qquad (15\text{–}1)$$

The output voltage with respect to ground is denoted as v_o. This voltage is related to the differential input voltage and the differential gain A_d by

$$v_o = A_d v_d = A_d(v^+ - v^-) \qquad (15\text{–}2)$$

It is assumed that the op-amp is powered by dual power supplies with a common ground established at the midpoint as discussed in Chapter 13. However, the power supply connections are not shown here.

A model depicting the ideal assumptions is shown in Figure 15–1(b). The ideal assumptions are as follows:

1. The input resistance of the op-amp is infinite. This is indicated on the model by the presence of an open circuit between the two input terminals.

2. The output resistance of the op-amp is zero. This is indicated on the model by the absence of a series resistance in the output circuit.
3. The open-loop differential gain A_d becomes infinite in the ideal case. This is indicated below the dependent source since it would be awkward to try to show an infinite value directly on the model schematic.

The implications of these three assumptions are the key to the simplified analysis and will now be discussed.

1. The assumed infinite input resistance means that the op-amp input current can be assumed to be zero.
2. The assumed zero output resistance means that the op-amp output voltage can be assumed to be independent of the load.
3. The implication of the assumed infinite gain is best seen by first solving for v_d in (15–2), in which case we have

$$v_d = \frac{v_o}{A_d} \qquad \text{(15–3)}$$

Assume that negative feedback forces v_o to be finite and in the linear operating region. If A_d is allowed to increase without limit, we have

$$\lim_{A_d \to \infty} v_d = \lim_{A_d \to \infty} \frac{v_o}{A} = 0 \qquad \text{(15–4)}$$

The differential voltage thus approaches zero and may be stated in either of the following ways:

$$\boxed{v_d = v^+ - v^- = 0} \qquad \text{(15–5)}$$

or

$$\boxed{v^+ = v^-} \qquad \text{(15–6)}$$

Stated in words, *the voltages at the two op-amp inputs are forced to be essentially equal under closed-loop stable conditions.*

A major qualification about the last implication is in order. Since the gain is never infinite, there must always be a small potential difference between these terminals to make the op-amp work. Thus, *the two input terminals must not be connected together.* However, in a variety of applications, it is acceptable to assume equal voltages at the two terminals since the difference in voltage will be very small compared with external circuit voltages.

EXAMPLE 15–1   Assume that in a certain op-amp closed-loop feedback circuit, the output is forced through feedback to be $v_o = 10$ V. Calculate the differential input voltage v_d for each of the following values of open-loop differential gain A_d: **(a)** 200,000; **(b)** 50,000; **(c)** 10,000.

Solution
This problem may seem backward in that we are given the output and gain and wish to determine the input. In a sense, however, this is the way in which many feedback circuits appear to work. Through a large loop gain, the output is forced to assume a value nearly independent of active-device characteristics. The input to the op-amp then assumes the value necessary to force the output to be the required value. Do not confuse the op-amp differential input voltage with the external input signal voltage, which is not given, but which is assumed to remain constant in this example.

The differential input voltage for each value of A_d is determined from (15–3) as

$$v_d = \frac{v_o}{A_d} = \frac{10}{A_d} \tag{15–7}$$

Each of the three values of A_d may be substituted in (15–7), and the results are summarized as follows:

	A_d	v_d
(a)	200,000	50 μV
(b)	50,000	200 μV
(c)	10,000	1 mV

Note that as the gain drops, the differential input voltage increases to maintain the assumed output voltage, and the assumption that $v_d = 0$ becomes less valid. Depending on the circuit, the assumption that $v_o = 10$ V will similarly become less valid, and the output will begin to depend somewhat on the open-loop gain.

15–2 INVERTING AMPLIFIER REVISITED

We begin the simplified analysis of op-amp circuits by considering again the inverting amplifier, of which a more exact analysis was given in Chapter 14. The analysis here employs the ideal assumptions of the preceding section.

The inverting amplifier with certain variables identified is shown in Figure 15–2. Since the noninverting input terminal is grounded, $v^+ = 0$. Based on the assumption that the differential input voltage is zero, we can therefore assume that $v^- = 0$. As discussed in Chapter 14, this situation is referred to as a **virtual ground**; that is, the inverting input is forced to assume ground potential even though it is not connected (and *should not be connected*) to ground.

FIGURE 15-2

Inverting op-amp amplifier circuit with certain variables labeled.

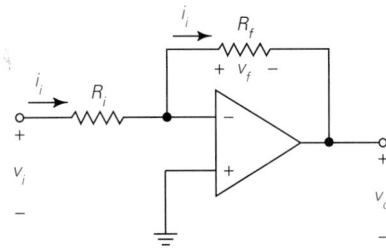

The existence of the virtual ground at the inverting input means that the input signal voltage v_i appears across R_i. The input current i_i is

$$i_i = \frac{v_i}{R_i} \qquad (15\text{-}8)$$

The current i_i cannot flow into the op-amp because of the assumed infinite input resistance, so it flows through R_f. A voltage v_f is produced across R_f given by

$$v_f = R_f i_i \qquad (15\text{-}9)$$

Substitution of (15-8) in (15-9) yields

$$v_f = R_f \frac{v_i}{R_i} = \frac{R_f}{R_i} v_i \qquad (15\text{-}10)$$

Since the assumed positive reference terminal of v_f is at ground potential, the output voltage v_o is simply

$$v_o = -v_f = -\frac{R_f}{R_i} v_i \qquad (15\text{-}11)$$

Let A_{CL} represent the closed-loop gain. It is determined as

$$A_{\mathrm{CL}} = \frac{v_o}{v_i} = -\frac{R_f}{R_i} \qquad (15\text{-}12)$$

This result was, of course, established in Chapter 14 by a more elaborate procedure after limiting conditions had been made. The procedure just given was much simpler and more intuitive.

The input resistance R_{in} of the noninverting amplifier is readily determined from (15-8) as

$$R_{\mathrm{in}} = \frac{v_i}{i_i} = R_i \qquad (15\text{-}13)$$

Do not confuse this input resistance of the composite circuit with the input resistance of the op-amp alone, which has been assumed to be infinite.

The output resistance R_{out} in the ideal case is simply

$$\boxed{R_{out} = 0}$$

(15–14)

Saturation

It must be understood that the gain equation of (15–12) is valid only for operation in the linear active region. For the common power supply voltages of $+15$ V and -15 V, the output saturation voltages are typically about $\pm V_{sat} = \pm 13$ V. Consequently, if the input should be large enough to cause the output to try to exceed these limits, the output will be clipped, and distortion will result.

Let V_{ip} represent the magnitude of the input at which saturation would start to occur. This value is

$$\boxed{V_{ip} = \frac{V_{sat}}{|A_{CL}|}}$$

(15–15)

based on symmetry in the two saturation voltages. The linear operating region for the input signal is $-V_{ip} < v_i < +V_{ip}$.

Simplified Design

It is possible to design useful amplifier circuits with the material just provided. A typical design would be based on a specification of the closed-loop gain A_{CL} and the minimum value of the input resistance R_{in}. The required gain constrains the ratio R_f/R_i as given by (15–12). The minimum input resistance constrains the minimum value of R_i.

In general, there may be a number of combinations of R_f and R_i that satisfy all the requirements. In deciding which values to use, the following points should be kept in mind: (1) Smaller values of resistance reduce the input resistance to the circuit and force the op-amp to supply more current to the feedback circuit. (2) Larger values of resistance tend to generate more thermal noise and cause possible offset effects due to bias currents.

As a general guideline, resistance values between a few thousand ohms and up to a few hundred thousand ohms tend to be best for most applications. However, this should be considered as a rough guideline rather than as an exact constraint.

EXAMPLE 15–2   Consider the inverting amplifier circuit of Figure 15–3. **(a)** Determine the closed-loop gain A_{CL}. **(b)** Determine the input resistance R_{in}. **(c)** For $+V_{sat} = \pm 13$ V, determine the input voltage level at which saturation occurs. **(d)** Determine the output voltage for each of the following input voltages: $v_i = 0$; 1 V; -1 V; 2 V; and 4 V.

FIGURE 15–3
Amplifier circuit of Example 15–2.

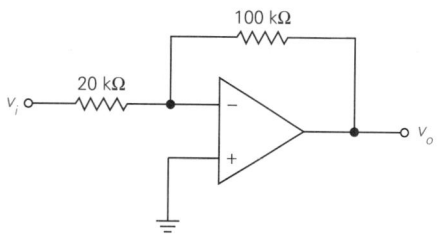

Solution
(a) The closed-loop gain is

$$A_{CL} = -\frac{R_f}{R_i} = -\frac{100,000}{20,000} = -5 \qquad (15\text{–}16)$$

(b) The input resistance is

$$R_{in} = R_i = 20\text{ k}\Omega \qquad (15\text{–}17)$$

(c) For $\pm V_{sat} = \pm 13$ V, the input voltage magnitude at which saturation starts to occur is

$$V_{ip} = \frac{13}{5} = 2.6\text{ V} \qquad (15\text{–}18)$$

The linear range in volts for the input is $-2.6 < v_i < 2.6$.

(d) For any value of input voltage in the linear region, the output voltage is

$$v_o = -5v_i \qquad (15\text{–}19)$$

The various output voltages for the first four cases are summarized as follows:

v_i (V)	v_o (V)
0	0
1	−5
−1	5
2	−10

Note that the output voltage is zero when the input is zero. For the other cases, the output magnitude is five times as large as the input, but is inverted in polarity. Since the op-amp is a dc-coupled device, the input and output could be dc levels, or they could represent certain properties (e.g., peak values) of ac signals.

Next, consider the last voltage (i.e., $v_i = 4$ V). If one simply substituted in (15–19), the output would appear to be −20 V. However, this is impossible since the output cannot extend outside the range from -13 to $+13$ V. The problem is that the peak input voltage as computed in (15–18) has been exceeded (i.e., saturation has occurred). For this condition, we can only say that $v_o = -13$ V. Note that negative saturation occurs for a positive input voltage, and vice versa.

**EXAMPLE
15–3**

Design an inverting op-amp amplifier circuit to meet the following specifications:

$$A_{CL} = -15$$
$$R_{in} \geq 10 \text{ k}\Omega$$

Use standard 5% resistance values, and keep all values below 200 kΩ to minimize noise and bias offset effects.

Solution
The gain specification requires that

$$\frac{R_f}{R_i} = 15 \qquad\qquad (15\text{–}20)$$

or

$$R_f = 15 \, R_i \qquad\qquad (15\text{–}21)$$

Since $R_{in} = R_i$, the second specification means that the smallest acceptable resistance would be 10 kΩ, which is a standard value. It turns out that $15 \times 10 \text{ k}\Omega = 150 \text{ k}\Omega$ is also a standard value. Thus one solution would be $R_i = 10 \text{ k}\Omega$ and $R_f = 150 \text{ k}\Omega$. However, to allow some extra leeway for tolerances on the input resistance, a better solution with standard values might be $R_i = 12 \text{ k}\Omega$ and $R_f = 180 \text{ k}\Omega$.

15–3 NONINVERTING AMPLIFIER REVISITED

Let us now analyze the noninverting amplifier configuration using the idealized assumptions of Section 15–1. The noninverting amplifier with certain variables identified is shown in Figure 15–4. The input voltage v_i is applied to the noninverting

FIGURE 15–4
Noninverting op-amp amplifier circuit with certain variables labeled.

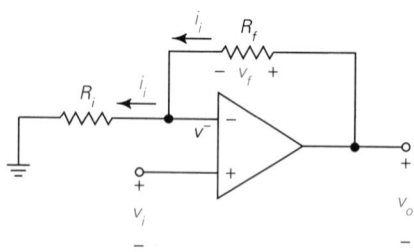

input, and this results in $v^+ = v_i$. Based on the assumption that the voltages at the two input terminals are forced to be the same, we can say that

$$v^- = v^+ = v_i \tag{15-22}$$

The voltage across the resistance R_i is thus v_i, and the current i_i is

$$i_i = \frac{v_i}{R_i} \tag{15-23}$$

The current i_i cannot flow out of the op-amp input, so it must be supplied by the op-amp output through R_f. A voltage v_f is produced across R_f given by

$$v_f = R_f i_i \tag{15-24}$$

Substitution of (15–23) in (15–24) yields

$$v_f = R_f \frac{v_i}{R_i} = \frac{R_f}{R_i} v_i \tag{15-25}$$

A KVL loop is now formed from ground to the inverting input, over to the output, and back to ground. This equation reads

$$-v_i - v_f + v_o = 0 \tag{15-26}$$

Solution for v_o and substitution of (15–25) for v_f result in

$$v_o = v_i + \frac{R_f}{R_i} v_i \tag{15-27}$$

The closed-loop voltage gain A_{CL} is then determined as

$$\boxed{A_{CL} = \frac{v_o}{v_i} = 1 + \frac{R_f}{R_i} = \frac{R_i + R_f}{R_i}} \tag{15-28}$$

where two equivalent forms have been given. As in the case of the inverting amplifier, the closed-loop gain expression agrees with the result obtained in Chapter 14.

Since the signal source looks directly into an assumed open circuit at the noninverting input, the circuit input resistance R_{in} in this case is assumed to be

$$\boxed{R_{in} = \infty} \tag{15-29}$$

The output resistance in this ideal case is simply

$$\boxed{R_{out} = 0} \tag{15-30}$$

Saturation

The discusstion concerning saturation as given in Section 15–2 for the inverting amplifier applies to the noninverting amplifier as well. The appropriate noninverting gain is used in (15–15) to predict the input at which saturation starts to occur. In the case of the noninverting amplifier, positive saturation would occur for a positive input voltage, and vice versa.

Voltage Follower

A special case of a noninverting amplifier is the voltage follower, whose basic form is shown in Figure 15–5. The result of (15–28) may be adapted to this case by setting $R_f = 0$ and $R_i = \infty$. The result is

$$\boxed{A_{CL} = 1} \tag{15–31}$$

FIGURE 15–5
Voltage-follower circuit.

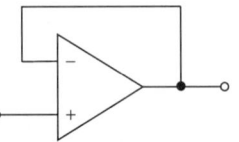

      The voltage follower thus has unity gain. However, since the ideal input resistance is infinite and the ideal output resistance is zero, it is very useful as a buffer. It is used to isolate a high-resistance source, in which minimal loading can be tolerated, from a load.

EXAMPLE 15–4

Consider the noninverting amplifier circuit of Figure 15–6. **(a)** Determine the closed-loop gain A_{CL}. **(b)** For $\pm V_{sat} = \pm 13$ V, determine the input voltage level at which saturation occurs. **(c)** Determine the output voltage for each of the following input voltages: $v_i = 0; 0.2$ V; -0.2 V; 0.5 V; 2 V.

FIGURE 15–6
Amplifier circuit of Example 15–4.

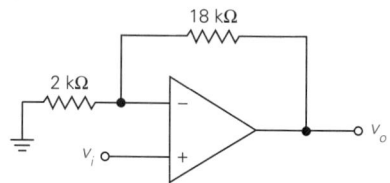

Solution

(a) The closed-loop voltage gain is

$$A_{CL} = 1 + \frac{R_f}{R_i} = 1 + \frac{18 \times 10^3}{2 \times 10^3} = 1 + 9 = 10 \tag{15–32}$$

(b) For $\pm V_{sat} = \pm 13$ V, the peak input magnitude is

$$V_{ip} = \frac{13}{10} = 1.3 \text{ V} \tag{15–33}$$

The linear range for the input is $-1.3 < v_i < +1.3$.

(c) For any value of input voltage in the linear region, the output voltage is

$$v_o = 10v_i \tag{15–34}$$

The various output voltages for the first four cases are summarized as follows:

v_i (V)	v_o (V)
0	0
0.2	2
−0.2	−2
0.5	5

Note that the output voltages have the same sign as the input voltages in this case. The last voltage, $v_i = 2$ V, exceeds the peak input permitted for linear operation, so saturation will occur. In this case, $v_o = 13$ V.

EXAMPLE 15–5

Design a noninverting op-amp amplifier circuit to achieve a gain $A_{CL} = 12$. Use standard 5% resistance values and keep all values below 100 kΩ in this particular design.

Solution

The gain specification requires that

$$1 + \frac{R_f}{R_i} = 12 \tag{15–35}$$

which simplifies to

$$\frac{R_f}{R_i} = 11 \tag{15–36}$$

or

$$R_f = 11R_i \tag{15–37}$$

A search through Appendix A results in the following acceptable combination:

$$R_i = 3 \text{ k}\Omega \quad \text{and} \quad R_f = 33 \text{ k}\Omega$$

15-4 COMPARISON OF AMPLIFIER CIRCUITS

We will now make a comparison of the inverting and noninverting amplifiers. The formulas for voltage gain, input resistance, and output resistance for the ideal cases are summarized in Table 15-1. For a given R_f/R_i ratio, the gain magnitude of the noninverting circuit is slightly higher than that of the inverting amplifier because of the additional unity term. However, this is neither an advantage nor a disadvantage since resistance ratios can always be adjusted easily.

TABLE 15-1
Summary of ideal relationships for inverting and noninverting amplifier circuits

	Inverting	Noninverting
A_{CL}	$-\dfrac{R_f}{R_i}$	$1 + \dfrac{R_f}{R_i}$
R_{in}	R_i	∞
R_{out}	0	0

The major advantage of the noninverting amplifier as compared to the inverting amplifier is the very high input resistance (assumed here as infinite) of the former. In designing inverting amplifiers, one must always select a value of R_i no smaller than the minimum acceptable input resistance, and this sometimes limits the use of an inverting amplifier.

On the other hand, the inverting amplifier has one advantage over the noninverting amplifier in the following sense: Inverting amplifiers can be cascaded to achieve noninverting gains if desired, provided that an even number of stages is used. However, one can never achieve inversion with a noninverting amplifier. Since many applications require inverting gains, the inverting amplifier circuit is essential.

Certain types of signal processing operations (e.g., summing) are more cumbersome to achieve with the noninverting amplifier. In fact, some operations (e.g., ideal integration) are theoretically impossible to achieve with the noninverting amplifier. The inverting amplifier is therefore more flexible in its possible range of applications.

The conclusion is that *both* the *noninverting* and *inverting* amplifiers are very important, and both have numerous applications in modern electronic circuits. Both the noninverting and inverting amplifier configurations may be considered in their basic forms as *voltage-controlled voltage sources*.

15-5 SUMMING AMPLIFIER CIRCUIT

The operational amplifier may be used to combine several signal voltages together while maintaining a common ground for each signal. The simplest designation for such a circuit is a **summing amplifier.** However, the circuit allows the relative weighting of each signal to be adjusted, so the circuit is also called a **linear combination circuit.** We use the first term in this book.

The basic form of a summing amplifier circuit is shown in Figure 15–7. The voltages $v_1, v_2, \ldots, v_n$ represent an arbitrary number of input voltages, each referred to circuit ground.

FIGURE 15–7
Summing amplifier circuit.

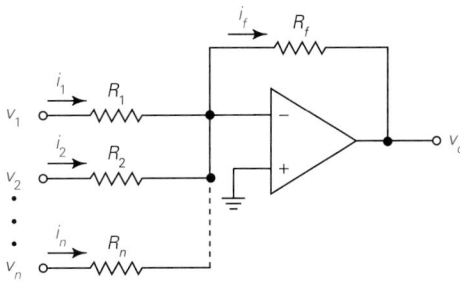

The inverting input terminal is forced to assume a virtual ground condition. Thus each input voltage appears across its particular input resistance. The various currents $i_1, i_2, \ldots, i_n$ may then be expressed as follows:

$$i_1 = \frac{v_1}{R_1} \tag{15–38}$$

$$i_2 = \frac{v_2}{R_2} \tag{15–39}$$

$$\vdots$$

$$i_n = \frac{v_n}{R_n} \tag{15–40}$$

By KCL, the current i_f may be expressed as

$$i_f = i_1 + i_2 + \cdots + i_n \tag{15–41a}$$

$$= \frac{v_1}{R_1} + \frac{v_2}{R_2} + \cdots + \frac{v_n}{R_n} \tag{15–41b}$$

where the various expressions for the currents were substituted in (15–41b).

The output voltage v_o may be expressed as

$$v_o = -R_f i_f \tag{15–42}$$

Substituting (15–41b) in (15–42), the output voltage is

$$v_o = -\frac{R_f}{R_1} v_1 - \frac{R_f}{R_2} v_2 - \cdots - \frac{R_f}{R_n} v_n \tag{15–43}$$

This last equation may be expressed as

$$v_o = A_1 v_1 + A_2 v_2 + \cdots + A_n v_n \qquad \text{(15–44)}$$

where

$$A_1 = -\frac{R_f}{R_1} \qquad \text{(15–45)}$$

$$A_2 = -\frac{R_f}{R_2} \qquad \text{(15–46)}$$

$$\cdot$$
$$\cdot$$
$$\cdot$$

$$A_n = -\frac{R_f}{R_n} \qquad \text{(15–47)}$$

The output voltage v_o is thus an algebraic sum of the n input voltages, with each input voltage weighted by a gain factor determined from a resistance ratio. The resistance R_f is common to all the gain factors, but each gain has the input resistance for the particular signal in the denominator, allowing individual adjustment for that gain.

As a by-product of this circuit, the gain factors of all the input signals are inverted. This may pose a minor inconvenience in some applications. However, it is readily circumvented by following this circuit with a simple inverting amplifier with a gain of -1. This causes the resulting sense of all the gain factors to be positive.

Because of the inversion property, a logical question is whether or not the noninverting amplifier may be used as a summing amplifier. The answer is that it can be used, but the circuit is not nearly as convenient to use. The beauty of the inverting amplifier is that the input currents sum at a virtual ground point, and each is independent of the others. With noninverting summing circuits, each input circuit affects all the other inputs, and the circuit design is somewhat awkward. We will restrict our consideration here to the inverting case since it is the most popular circuit. An additional inversion, when required, is a small price to pay for the convenience of this circuit.

In some cases it is desired that all combined signals be amplified by the same level. Assume that the desired output is

$$v_o = -A(v_1 + v_2 + \cdots + v_n) \qquad \text{(15–48)}$$

where $-A$ is the desired amplification. This is achieved by selecting $R_1 = R_2 = \cdots R_n = R$ and $R_f = AR$. The resulting circuit is shown in Figure 15–8.

FIGURE 15–8
Summing amplifier circuit with equal amplification for all signals.

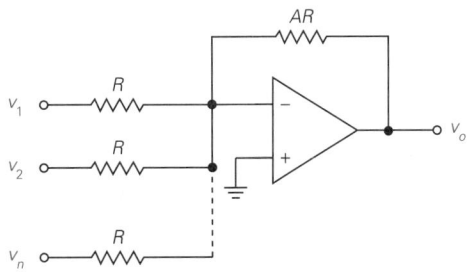

EXAMPLE 15–6 (a) For the circuit of Figure 15–9, determine an expression for the output voltage in terms of the three input voltages. (b) Determine the output voltage for the following input condition: $v_1 = 0.5$ V, $v_2 = 0.8$ V, $v_3 = -3$ V.

FIGURE 15–9
Circuit of Example 15–6.

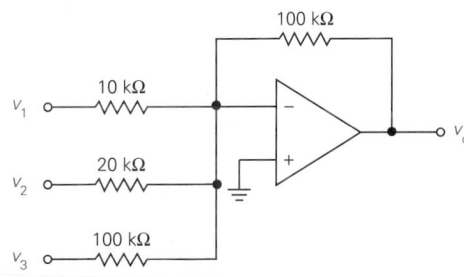

Solution
(a) The gain factors for the three input signals are determined as follows:

$$A_1 = -\frac{R_f}{R_1} = -\frac{100 \text{ k}\Omega}{10 \text{ k}\Omega} = -10 \qquad (15\text{–}49)$$

$$A_2 = -\frac{R_f}{R_2} = -\frac{100 \text{ k}\Omega}{20 \text{ k}\Omega} = -5 \qquad (15\text{–}50)$$

$$A_3 = -\frac{R_f}{R_3} = -\frac{100 \text{ k}\Omega}{100 \text{ k}\Omega} = -1 \qquad (15\text{–}51)$$

The output voltage is then expressed as

$$v_o = -10v_1 - 5v_2 - v_3 \qquad (15\text{–}52)$$

(b) For the three values of input voltages given in the problem statement, we have

$$v_o = -10(0.5) - 5(0.8) - 1(-3) = -5 - 4 + 3 = -6 \text{ V} \qquad (15\text{–}53)$$

Note that the net output effects of the first two voltages are in a negative sense, while the net effect of the third voltage on the output is in a positive sense. In the last case, a negative input multiplied by a negative gain factor produces a positive output sense.

EXAMPLE 15–7

It is desired to combine two signals v_1 and v_2 and produce an output v_o according to the following requirement:

$$v_o = -12v_1 - 6v_2$$

The input resistances at the two inputs must not be less than 10 kΩ. Design a circuit to achieve the objective. Use resistor values less than 200 kΩ.

Solution
The two gain specifications force the following constraints:

$$\frac{R_f}{R_1} = 12 \qquad (15\text{–}54)$$

and

$$\frac{R_f}{R_2} = 6 \qquad (15\text{–}55)$$

There are two gain constraints and three unknowns. This means that one resistance may be selected somewhat arbitrarily. However, the minimum input resistance of 10 kΩ establishes a lower bound at the input. Note that the *smallest* input resistance corresponds to the *highest* gain. One solution would be to select $R_1 = 10$ kΩ. The other two resistances would then be $R_f = 120$ kΩ and $R_2 = 20$ kΩ, all of which are standard values. In searching through Appendix A, however, a second possible solution is $R_1 = 15$ kΩ, $R_f = 180$ kΩ, and $R_2 = 30$ kΩ. Since the second solution provides a higher input resistance, it will be selected in this case. The circuit is shown in Figure 15–10.

FIGURE 15–10
Circuit design of Example 15–7.

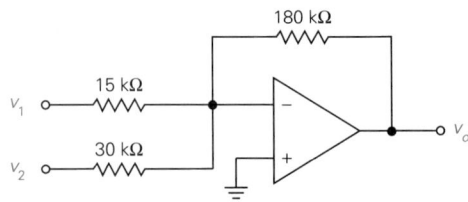

15–6 DIFFERENCE AMPLIFIER CIRCUIT

The operational amplifier may be used to form the difference between two signals and, if desired, amplify this difference at the same time. This circuit is called the **difference amplifier.**

FIGURE 15–11
Difference amplifier circuit.

FIGURE 15–11
Difference amplifier circuit.

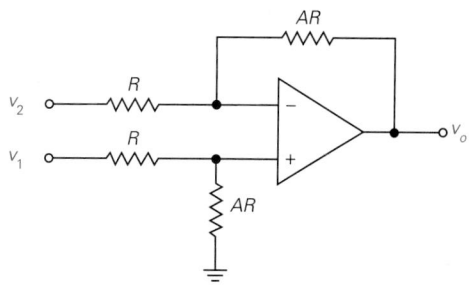

The form of a difference amplifier is shown in Figure 15–11. Note that in this form, two resistances have the value R, while the other two have the value AR. It is possible to create different weightings of the two signals by changing one of the resistances to a different value, but that situation will not be considered here. The circuit shown is the most common form.

The difference circuit with certain variables labeled to facilitate analysis is shown in Figure 15–12. The starting point for the analysis is to recognize that the voltage v^+ at the noninverting input is determined solely by v_1 since no other part of the circuit is connected there. By the voltage-divider rule, we have

$$v^+ = \frac{AR}{AR + R} v_1 = \frac{A}{A + 1} v_1 \qquad \textbf{(15–56)}$$

where the common factor R was canceled in the last form.

FIGURE 15–12
Difference amplifier labeled for analysis.

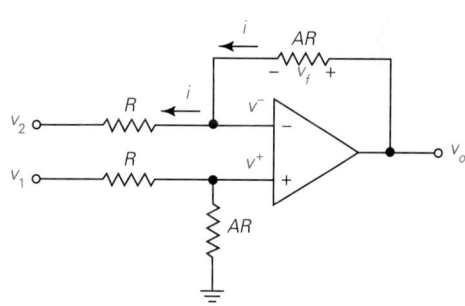

Assuming linear operation, the voltage v^- must be the same as v^+; that is,

$$v^- = v^+ = \frac{Av_1}{A + 1} \qquad \textbf{(15–57)}$$

A current i must flow between v^- and v_2. Assuming a positive reference for v^-, this current is

$$i = \frac{v^- - v_2}{R} \qquad \textbf{(15–58)}$$

Substitution of (15–57) in (15–58) results in

$$i = \frac{Av_1}{(A+1)R} - \frac{v_2}{R} \tag{15-59}$$

A voltage v_f is produced across the feedback resistance and is given by

$$v_f = ARi = \frac{A^2}{A+1}v_1 - Av_2 \tag{15-60}$$

Finally, the output voltage v_o is

$$v_o = v_f + v^- = \frac{A^2}{A+1}v_1 - Av_2 + \frac{Av_1}{A+1} \tag{15-61}$$

$$= \frac{A(A+1)}{A+1}v_1 - Av_2 \tag{15-62}$$

Simplification yields

$$\boxed{v_o = A(v_1 - v_2)} \tag{15-63}$$

The output voltage is thus A times the difference between the input voltages.

A special case for this circuit is when all four resistances are equal. In that case, the output voltage is simply the difference between the two input voltages.

EXAMPLE 15–8 (a) For the circuit of Figure 15–13, determine an expression for the output voltage in terms of the two input voltages. (b) Determine the output voltage when $v_1 = 2$ V and $v_2 = 1.9$ V

FIGURE 15–13
Circuit of Example 15–8.

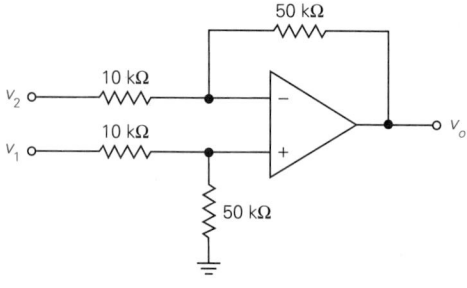

Solution
(a) The gain A is determined as

$$A = \frac{50\ \text{k}\Omega}{10\ \text{k}\Omega} = 5 \tag{15-64}$$

The output voltage can be expressed as

$$v_o = 5(v_1 - v_2) \tag{15-65}$$

(b) For $v_1 = 2$ V and $v_2 = 1.9$ V, we have

$$v_o = 5\,(2 - 1.9) = 5 \times 0.1 = 0.5 \text{ V} \tag{15-66}$$

15-7 CONTROLLED SOURCES

The noninverting and inverting amplifier circuits considered earlier in the chapter may be both considered as **voltage-controlled voltage sources (VCVS).** In each case, the controlling variable was the input voltage, and the controlled variable was the output voltage. In this section it will be shown how each of the three other types of dependent sources may be implemented with an op-amp.

Current-Controlled Voltage Source

The schematic diagram of the simplest type of **current-controlled voltage source (ICVS)** is shown in Figure 15–14. The circuit has essentially the form of an inverting amplifier. However, no input resistance is shown. Instead, the input variable is considered as the current i_i. Since the inverting input assumes a virtual ground level, the input current sees a very low input resistance. This is the ideal condition for a current-driven amplifier, which is in direct contrast to a voltage-driven amplifier.

The current i_i flows through the resistance R and produces a voltage across it of value Ri_i, with the positive terminal on the left. The output voltage is

$$\boxed{v_o = -Ri_i} \tag{15-67}$$

FIGURE 15–14
Current-controlled voltage source.

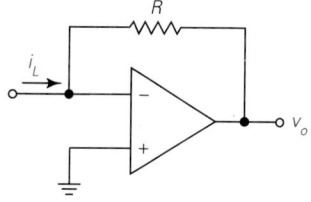

The output voltage is thus a constant times the input current, the desired situation for a current-controlled voltage source.

A constant that relates a dependent output voltage to an independent input current has the dimensions of ohms and is called the **transresistance R_m.** The transresistance for this circuit is simply

$$\boxed{R_m = R} \tag{15-68}$$

Voltage-Controlled Current Source

The schematic diagram of the simplest form of a **voltage-controlled current source** **(VCIS)** is shown in Figure 15–15. The circuit has the same form as an inverting amplifier. However, it differs from the inverting amplifier in that the output variable is the desired current i_L through the load resistance R_L.

FIGURE 15–15
VCIS for a floating load.

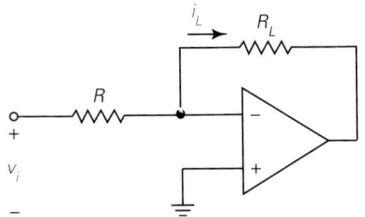

The current i_L is the same as the current established through the resistance R on the left and is

$$i_L = \frac{v_i}{R} \qquad \text{(15–69)}$$

The current is thus a constant times the input voltage and is independent of the resistance R_L. These are the requirements for a VCIS.

A constant that relates a dependent load current to an independent input voltage has the dimensions of siemens and is the **transconductance g_m**. The transconductance for the circuit is simply

$$g_m = \frac{1}{R} \qquad \text{(15–70)}$$

Since the op-amp output voltage is not of prime interest in this circuit, it is easy to overlook the fact that this output voltage must be maintained in the linear region of the op-amp. This limits the maximum value of the output resistance. The reader is invited to show (Problem 15–25) that linear operation requires that the following constraint be satisfied:

$$R_L i_L < V_{\text{sat}} \qquad \text{(15–71)}$$

Howland Current Source

The preceding current source requires that the load be floating with respect to ground. Whenever the terminal of the load must be connected to actual ground, the Howland current source can be used. This circuit is shown in Figure 15–16.

FIGURE 15–16
Howland VCIS for grounded load.

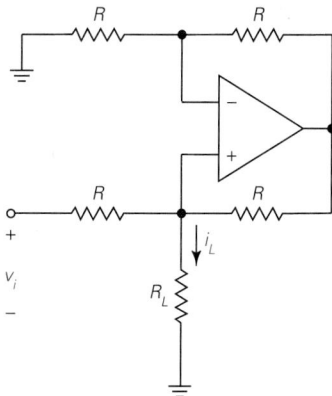

A guided exercise in the derivation of the input–output relationship is given as Problem 15–26. The basic relationship, which is certainly not obvious, is

$$i_L = \frac{v_i}{R} \qquad (15\text{–}72)$$

The transconductance is

$$g_m = \frac{1}{R} \qquad (15\text{–}73)$$

The reader is invited to show (Problem 15–27) that linear operation requires that the following constraint be satisfied:

$$R_L i_L < \frac{V_{\text{sat}}}{2} \qquad (15\text{–}74)$$

Current-Controlled Current Source

The schematic diagram of a **current-controlled current source (ICIS)** is shown in Figure 15–17. As in the case of the ICVS, the input resistance is very low, as required. A guided exercise in the derivation of the input–output relationship for this circuit is given as Problem 15–28. The basic relationship is

$$i_L = \left(1 + \frac{R_2}{R_1}\right) i_i \qquad (15\text{–}75)$$

FIGURE 15–17
ICIS.

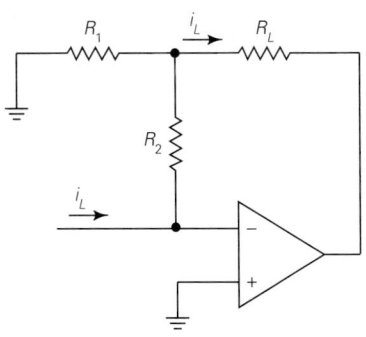

The current gain is denoted as β, and for this circuit it is

$$\beta = 1 + \frac{R_2}{R_1} \qquad (15\text{–}76)$$

For linear operation, it can be shown that the following requirement must be met:

$$\left[R_2 + R_L \left(1 + \frac{R_2}{R_1} \right) \right] |i_i| < V_{\text{sat}} \qquad (15\text{–}77)$$

EXAMPLE 15–9

It is desired to establish a dc current of 5 mA through a certain nonlinear floating resistive load. The circuit of Figure 15–18 is proposed. Assume that the nonlinear resistance is sufficiently small that saturation will not occur. Determine the value of R required.

FIGURE 15–18
Circuit of Example 15–9.

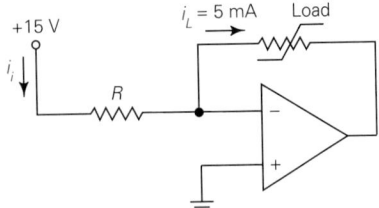

Solution
The circuit is a VCIS with a 15-V dc supply used as the controlling voltage. Since $i_L = i_i = v_i/R$, the value of R is

$$R = \frac{v_i}{i_i} = \frac{v_i}{i_L} = \frac{15 \text{ V}}{5 \times 10^{-3} \text{ A}} = 3 \text{ k}\Omega \qquad (15\text{–}78)$$

Some readers may question the need for using an op-amp since one could readily establish a fixed current through a simple series circuit consisting of a variable resistance and the nonlinear load. However, that type of circuit would require a readjustment after each change of the load resistance. The circuit shown will maintain a fixed load current of 5 mA for any value of load provided that linear operation is maintained.

EXAMPLE 15–10 Consider the Howland current source of Figure 15–19. **(a)** Determine the transconductance. **(b)** Determine the load current for $v_i = 2$ V and $v_i = 4$ V, assuming linear operation. **(c)** For $v_i = 4$ V, determine the range of R_L that will ensure linear operation. Assume that $\pm V_{\text{sat}} = \pm 13$ V.

FIGURE 15–19
Circuit of Example 15–10.

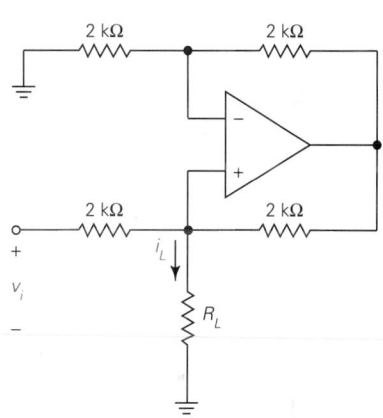

Solution
(a) The transconductance g_m is

$$g_m = \frac{1}{R} = \frac{1}{2 \times 10^3} = 0.5 \times 10^{-3}\,\text{S} \tag{15–79}$$

(b) For any input voltage v_i, the load current i_L in the linear region is

$$i_L = g_m v_i = 0.5 \times 10^{-3} v_i \tag{15–80}$$

For $v_i = 2$ V, i_L is

$$i_L = 0.5 \times 10^{-3} \times 2 = 1\,\text{mA} \tag{15–81}$$

For $v_i = 4$ V, i_L is

$$i_L = 0.5 \times 10^{-3} \times 4 = 2\,\text{mA} \tag{15–82}$$

(c) Linear operation is based on the constraint of (15–74). For $i_L = 2$ mA, we have

$$R_L \times 2 \times 10^{-3} < \frac{13}{2} \qquad (15\text{–}83)$$

This leads to the requirement that

$$R_L < 3250 \; \Omega \qquad (15\text{–}84)$$

Thus, the current through the load should remain fixed at 2 mA for all values of R_L less than 3250 Ω.

15–8 PSPICE EXAMPLES

PSPICE EXAMPLE 15–1

Using the 741 library model, write a PSPICE program to simulate the inverting amplifier of Example 15–2 (Figure 15–3), and use the transfer function command to investigate the properties of the circuit.

Solution

The circuit adapted to the PSPICE format is shown in Figure 15–20(a), and the code is shown in Figure 15–20(b). Various options have been employed to minimize the size of the output printout. The input is VI and the output is V(5). Carefully

FIGURE 15–20
Circuit and code of PSPICE Example 15–1.

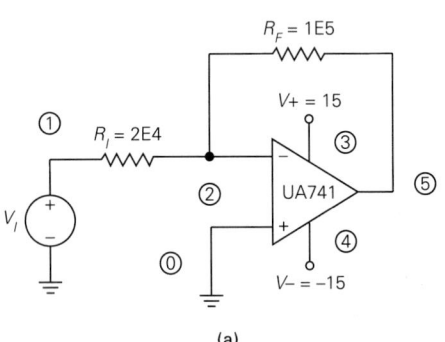

(a)

```
PSPICE EXAMPLE 15-1
.OPTIONS NOECHO NOPAGE NOBIAS NOMOD
VI 1 0
RI 1 2 2E4
RF 2 5 1E5
V+ 3 0 15
V- 0 4 15
X 0 2 3 4 5 UA741
.LIB EVAL.LIB
.TF V(5) VI
.END
```

(b)

note the order of the node connections to the 741 model. The two-power supplies are the standard ±15 V values.

The computer printout is shown in Figure 15–21. The voltage gain is V(5)/VI = −5, and the input resistance is 20 kΩ. These values are in perfect agreement with the results of Example 15–2. The output resistance is 4.645×10^{-3} Ω. This latter value was not calculated in Example 15–2, but the computer-generated value is very reasonable considering the fact that the output resistance of the circuit is expected to be very small.

```
PSPICE EXAMPLE 15-1

****      CIRCUIT DESCRIPTION

****************************************************************

.OPTIONS NOECHO NOPAGE NOBIAS NOMOD

****      SMALL-SIGNAL CHARACTERISTICS

     V(5)/VI = -5.000E+00
     INPUT RESISTANCE AT VI = 2.000E+04
     OUTPUT RESISTANCE AT V(5) = 4.645E-03

     JOB CONCLUDED
     TOTAL JOB TIME              1.75
```

FIGURE 15–21
Computer printout of PSPICE Example 15–1.

PSPICE EXAMPLE 15–2

Modify the program of PSPICE Example 15–1 to display a transient analysis when the input is a sinusoid with a peak value of 2 V at a frequency of 1 kHz.

Solution
The circuit configuration is exactly the same as in PSPICE Example 15–1, so reference can be made to Figure 15–20(a). The only difference is that VI will be changed from a dc voltage to a sinusoid.

The new code is shown in Figure 15–22. For a quick review of the source description, the second line now reads

```
VI 1 0 SIN 0 2 1E3
```

FIGURE 15–22
Code for PSPICE Example 15–2.

```
PSPICE EXAMPLE 15-2
VI 1 0 SIN 0 2 1E3
RI 1 2 2E4
RF 2 5 1E5
V+ 3 0 15
V- 0 4 15
X 0 2 3 4 5 UA741
.LIB EVAL.LIB
.TRAN 2E-3 2E-3
.PROBE
.END
```

This describes a sinusoidal source with no dc offset, a peak value of 2 V, and a frequency of 1 kHz.

The control line for the analysis type reads

```
.TRAN 2E-3 2E-3
```

This defines a transient analysis over an interval of 2 ms, which represents two cycles at 1 kHz. Finally, .PROBE is added to obtain a graphical printout.

Using various menu options of .PROBE, plots of the input and output voltages are shown in Figure 15–23. For a peak input voltage of 2 V, the peak output voltage is 10 V, as expected. However, because the gain is −5, the output voltage is inverted with respect to the input. These waveforms have the same forms as one would see on an oscilloscope with dual trace and with the time base reference established at the input.

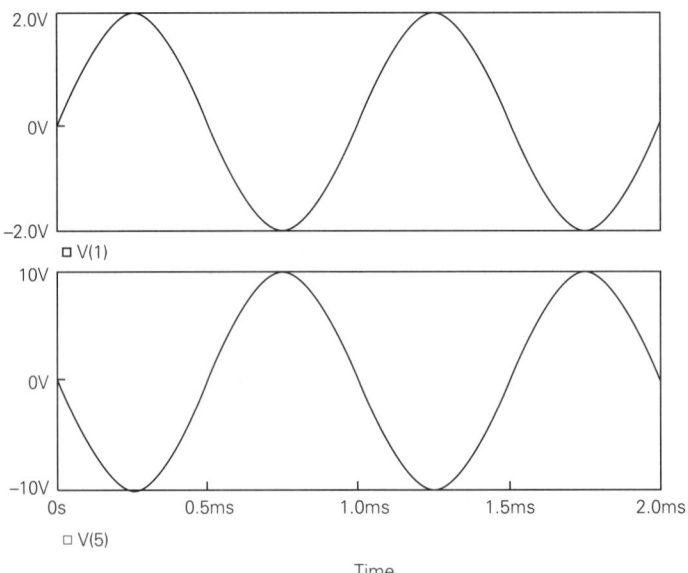

FIGURE 15–23
Input and output waveforms of PSPICE Example 15–2.

PSPICE EXAMPLE 15–3

Repeat the analysis of PSPICE Example 15–2 with the peak input voltage increased to 4 V.

Solution
The complete code will not be repeated since the only change is on the second line, which should now read

```
VI 1 0 SIN 0 4 1E3
```

For an input peak voltage of 4 V, the required peak output voltage would be 20 V if the circuit were linear in that region. This is impossible, however, since the power supply voltages are ±15 V and saturation will occur below these levels.

The input and output waveforms are shown in Figure 15–24. Observe that clipping of the output occurs at levels near ±15 V, representing the saturation voltages.

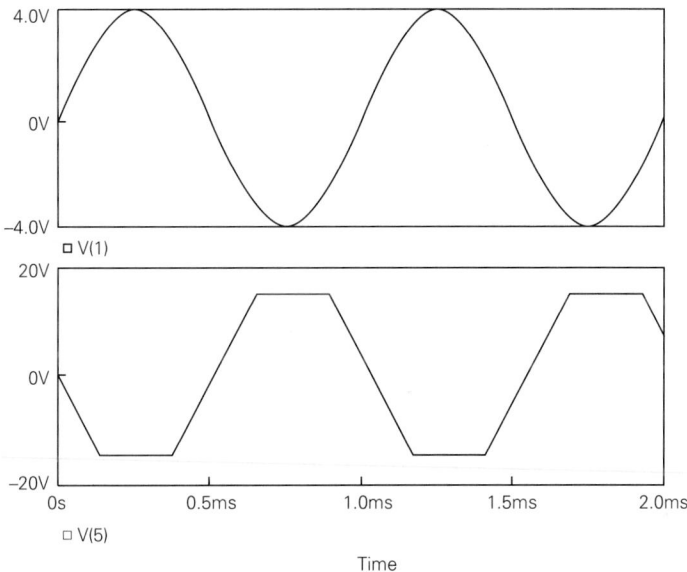

FIGURE 15–24
Input and output waveforms of PSPICE Example 15–3.

PSPICE EXAMPLE 15–4

Using the 741 library model, write a PSPICE program to simulate the noninverting amplifier of Example 15–4 (Figure 15–6), and use the transfer function command to investigate the properties of the circuit.

Solution
The circuit adapted to the PSPICE format is shown in Figure 15–25(a), and the code is shown in Figure 15–25(b). The format follows the same pattern as in PSPICE Example 15–1.

The output data are shown in Figure 15–26. The gain is $V(5)/VI = 9.999$, which differs insignificantly from the predicted value of 10. The input resistance is 2.068 GΩ, and the output resistance is 7.627 mΩ. A very large value of input resistance and a very small value of output resistance are expected for a noninverting amplifier.

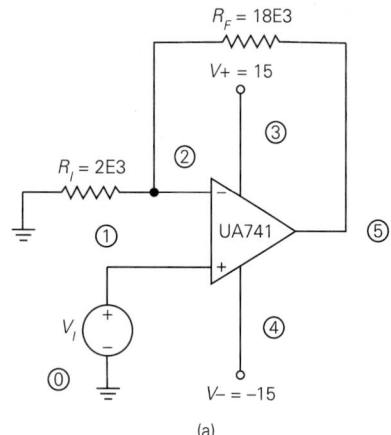

```
PSPICE EXAMPLE 15-4
.OPTIONS NOECHO NOPAGE NOBIAS NOMOD
VI 1 0
RI 0 2 2E3
RF 2 5 18E3
V+ 3 0 15
V- 0 4 15
X 1 2 3 4 5 UA741
.LIB EVAL.LIB
.TF V(5) VI
.END
```

(a)   (b)

FIGURE 15–25
Circuit and code of PSPICE Example 15–4.

```
PSPICE EXAMPLE 15-4

****          CIRCUIT DESCRIPTION

*********************************************************************

.OPTIONS NOECHO NOPAGE NOBIAS NOMOD

****          SMALL-SIGNAL CHARACTERISTICS

       V(5)/VI = 9.999E+00
       INPUT RESISTANCE AT VI = 2.068E+09
       OUTPUT RESISTANCE AT V(5) = 7.627E-03

       JOB CONCLUDED
       TOTAL JOB TIME                    1.75
```

FIGURE 15–26
Computer printout of PSPICE Example 15–4.

PROBLEMS

Drill Problems

15–1. Assume that in a certain op-amp closed-loop application, the output is forced through feedback to be $v_o = 12$ V. Calculate the differential input voltage v_d for each of the following values of open-loop differential gain A_d: **(a)** 10^5; **(b)** 20,000; **(c)** 5000.

15–2. Assume that in a certain op-amp closed-loop application, the output is forced through feedback to be $v_o = -10$ V. Calculate the differential input voltage v_d for each of the following values of open-loop differential gain A_d: **(a)** 120,000; **(b)** 40,000; **(c)** 8000.

15–3. The two input signal voltages for a certain op-amp circuit are $v^+ = 6.002$ V and $v^- = 6.000$ V. Determine the output voltage v_o if the differential gain is $A_d = 4000$.

15–4. The two input signal voltages for a certain op-amp circuit are $v^+ = 4.000$ V and $v^- = 4.003$ V. Determine the output voltage v_o if the differential gain is $A_d = 2500$.

15–5. Consider the amplifier circuit of Figure P15–5. **(a)** Determine the closed-loop gain A_{CL}. **(b)** Determine the input resistance R_{in}. **(c)** For $\pm V_{sat} = \pm 13$ V, determine the input voltage level at which saturation occurs. **(d)** Determine the output voltage for each of the following input voltages: $v_i = 0$; 0.2 V; −0.2 V; 0.5 V; 1 V.

FIGURE P15–5

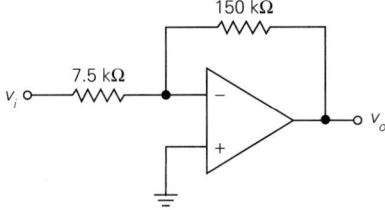

15–6. For the amplifier circuit of Figure P15–6, repeat the computations of Problem 15–5.

FIGURE P15–6

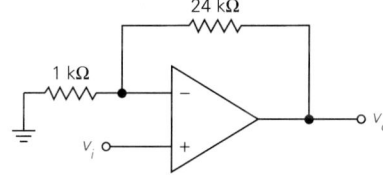

15–7. For the circuit of Figure P15–7, determine an expression for the output voltage in terms of the two input voltages.

FIGURE P15–7

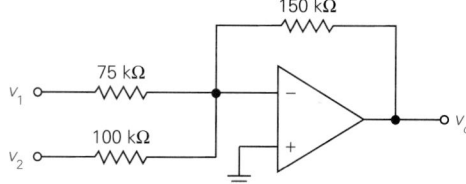

15–8. For the circuit of Figure P15–8, determine an expression for the output voltage in terms of the three input voltages.

FIGURE P15–8

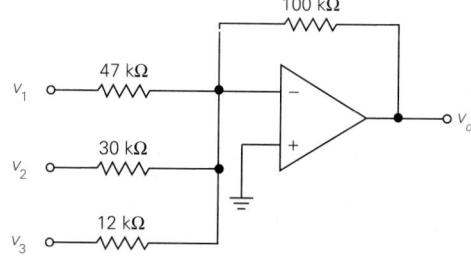

15–9. For the circuit of Problem 15–7 (Figure P15–7), determine the output voltage for each of the following combinations of input voltages: **(a)** $v_1 = 3$ V, $v_2 = 2$ V; **(b)** $v_1 = 3$ V, $v_2 = -4$ V; **(c)** $v_1 = 5$ V, $v_2 = -8$ V; **(d)** $v_1 = -5$ V, $v_2 = -8$ V. Assume that $\pm V_{\text{sat}} = \pm 13$ V.

15–10. For the circuit of Problem 15–8 (Figure P15–8), determine the output voltage for each of the following combinations of input voltages: **(a)** $v_1 = 4$ V, $v_2 = -4$ V, $v_3 = 2$ V; **(b)** $v_1 = 2$ V, $v_2 = 3$ V, $v_3 = -2$ V; **(c)** $v_1 = -1$ V, $v_2 = 2$ V, $v_3 = 1$ V. Assume that $\pm V_{\text{sat}} = \pm 13$ V.

15–11. For the circuit of Figure P15–11, determine an expression for the output voltage in terms of the two input voltages.

FIGURE P15–11

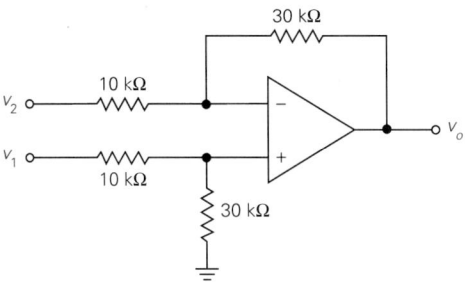

15–12. For the circuit of Figure P15–12, determine an expression for the output voltage in terms of two input voltages.

FIGURE P15–12

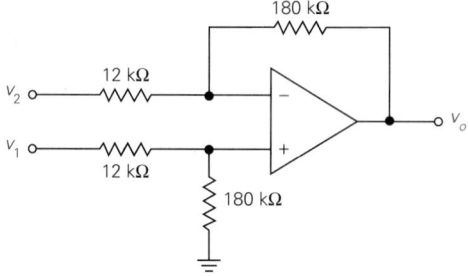

15–13. For the circuit of Problem 15–11 (Figure P15–11), determine the output voltage for each of the following combinations of input voltages: **(a)** $v_1 = 2$ V, $v_2 = 2$ V; **(b)** $v_1 = 2$ V, $v_2 = -2$ V; **(c)** $v_1 = -2$ V, $v_2 = 2$ V; **(d)** $v_1 = 5$ V, $v_2 = -5$ V. Assume that $\pm V_{\text{sat}} = \pm 13$ V.

15–14. For the circuit of Problem 15–12 (Figure P15–12), determine the output voltage for each of the following combinations of input voltages: **(a)** $v_1 = 2$ V, $v_2 = 2$ V; **(b)** $v_1 = 0.1$ V, $v_2 = -0.1$ V; **(c)** $v_1 = -0.1$ V, $v_2 = 0.1$ V; **(d)** $v_1 = 5$ V, $v_2 = -5$ V. Assume that $\pm V_{\text{sat}} = \pm 13$ V.

15–15. Consider the ICVS of Figure P15–15. **(a)** Determine the transresistance. **(b)** Determine the output voltage for each of the following values of input current i_i: 0; 2 mA; -2 mA; 10 mA. Assume that $\pm V_{\text{sat}} = \pm 13$ V.

15–16. Consider the ICVS of Figure P15–16. **(a)** Determine the transresistance. **(b)** Determine the output voltage for each of the following values of input current i_i: 0; 50 μA; 0.6 mA; -1 mA; -2 mA. Assume that $\pm V_{\text{sat}} = \pm 13$ V.

FIGURE P15–15

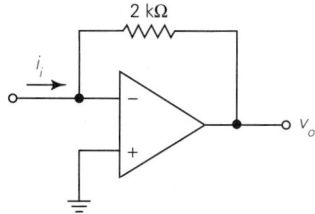

FIGURE P15–16

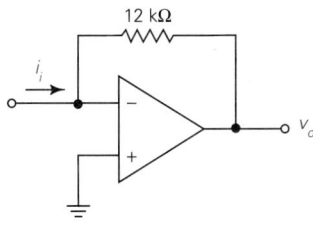

15–17. Consider the VCIS of Figure P15–17. Assume that $\pm V_{sat} = \pm 13$ V. **(a)** Determine the transconductance. **(b)** For $v_i = 2$ V, determine the maximum value of R_L for linear operation. **(c)** For $R_L = 10$ kΩ, determine the maximum value of v_i for linear operation.

FIGURE P15–17

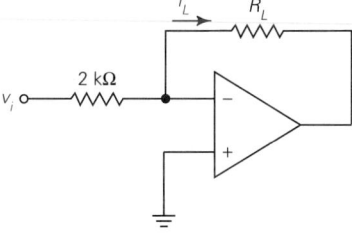

15–18. Consider the VCIS of Figure P15–18. Assume that $\pm V_{sat} = \pm 13$ V. **(a)** Determine the transconductance. **(b)** For $v_i = 5$ V, determine the maximum value of R_L for linear operation. **(c)** For $R_L = 40$ kΩ, determine the maximum value of v_i for linear operation.

FIGURE P15–18

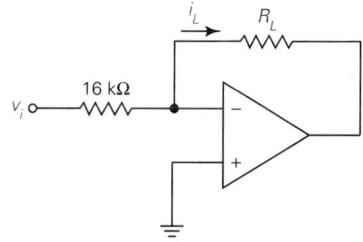

15–19. Consider a Howland current source of the form shown in Figure 15–16 with $R = 2$ kΩ. Assume that $\pm V_{sat} = \pm 13$ V. **(a)** Determine the transconductance. **(b)** For $v_i = 2$ V, determine the maximum value of R_L for linear operation. **(c)** For $R_L = 10$ kΩ, determine the maximum value of v_i for linear operation.

15–20. Consider a Howland current source of the form shown in Figure 15–16 with $R = 16$ kΩ. Assume that $\pm V_{sat} = \pm 13$ V. **(a)** Determine the transconductance. **(b)** For $v_i = 5$ V, determine the maximum value of R_L for linear operation. **(c)** For $R_L = 40$ kΩ, determine the maximum value of v_i for linear operation.

15–21. Consider the ICIS of Figure P15–21. **(a)** Determine the current gain β. **(b)** Determine the load current i_L if $i_i = 200$ μA, $R_L = 5$ kΩ, and $\pm V_{sat} = \pm 13$ V.

FIGURE P15–21

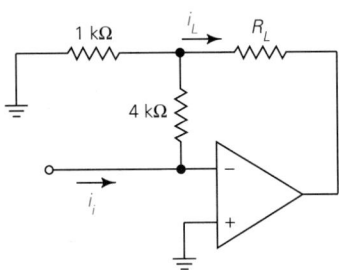

15–22. Consider the ICIS of Figure P15–22. **(a)** Determine the current gain β. **(b)** Determine the load current i_L if $i_i = 50$ μA. **(c)** Determine the load current i_L if $i_i = 100$ μA, $R_L = 1$ kΩ, and $\pm V_{sat} = \pm 13$ V.

FIGURE P15–22

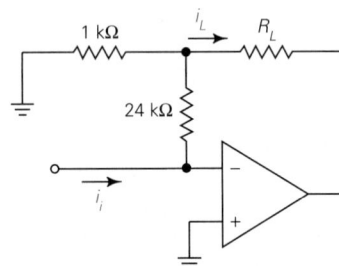

Derivation Problems

15–23. Show that the differential input voltage for an inverting amplifier configuration is given by

$$v_d = - \frac{R_f}{R_i} \frac{v_i}{A_d}$$

15–24. Show that the differential input voltage for a noninverting amplifier configuration is given by

$$v_d = \left(1 + \frac{R_f}{R_i}\right) \frac{v_i}{A_d}$$

15–25. For the VCIS of Figure 15–15, derive the constraint of (15–71).

15–26. The objective of this exercise is to derive the current relationship of the Howland current source as given by (15–72). On the circuit diagram of Figure 15–16, label the voltages v_o, v^+, and v^- at the appropriate points. Note that $v^- = [R/(R + R)]v_o = v_o/2$ and $v^+ = v^-$. Next, sum the currents leaving the noninverting terminal, and after appropriate substitution, derive (15–72).

15–27. For the Howland current source of Figure 15–16, derive the constraint of (15–74).

15–28. The objective of this exercise is to derive the current-gain relationship of an ICIS as given by (15–75). On the circuit diagram of Figure 15–17, label a current i_2 flowing up through R_2, a current i_1 flowing from ground through R_1, a voltage v_2 across R_2 (+ at bottom), and a voltage v_1 across R_1 (+ on left). The current i_2 flows through R_2 producing the voltage v_2. Since the inverting input is at virtual ground, $v_1 = v_2$. The voltage across R_1 results in the current i_1. Finally, i_L is determined from i_1 and i_2. Using this approach, derive (15–75).

15–29. For the circuit of Figure P15–29, show that the output voltage is given by

$$v_o = \frac{R_3 + R_4}{R_3} \left(\frac{R_2}{R_1 + R_2} v_1 + \frac{R_1}{R_1 + R_2} v_2 \right)$$

FIGURE P15–29

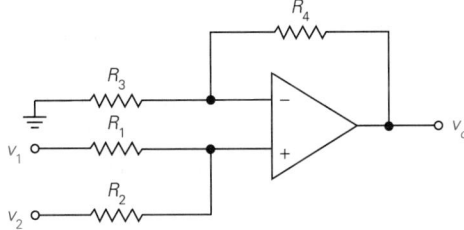

15–30. For the circuit of Figure P15–30, show that the output voltage is given by

$$v_o = \frac{R_2}{R_1 + R_2} \frac{R_3 + R_4}{R_3} v_1 - \frac{R_4}{R_3} v_2$$

FIGURE P15–30

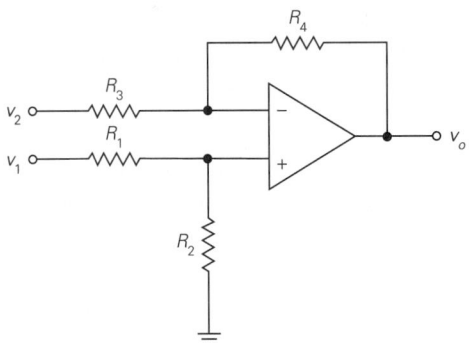

Design Problems

15–31. Design an inverting op-amp amplifier circuit to meet the following specifications:

$$A_{CL} = -12$$
$$R_{in} \geq 10\,k\Omega$$

Use standard 5% resistance values, and keep all values below 200 kΩ to minimize noise and bias offset effects.

15–32. Design an inverting op-amp amplifier circuit to meet the following specifications:

$$A_{CL} = -25$$
$$R_{in} \geq 20\,k\Omega$$

Use standard 5% resistance values, and keep all values below 1 MΩ to minimize noise and bias offset effects.

15–33. Design a noninverting op-amp amplifier circuit to achieve a gain $A_{CL} = 25$. Use standard 5% resistance values, and keep all values below 500 kΩ.

15–34. Design a noninverting op-amp amplifier circuit to achieve a gain $A_{CL} = 48$. Use standard 5% resistance values, and keep all values below 500 kΩ.

15–35. It is desired to combine two signals v_1 and v_2 to produce an output v_o according to the following requirement:

$$v_o = -5V_1 - 12v_2$$

The input resistance at the two inputs must not be less than 5 kΩ. Design a circuit to achieve the objective. Use standard 5% resistance values less than 200 kΩ.

15–36. It is desired to combine three signals v_1, v_2, and v_3 to produce an output v_o according to the following requirement:

$$v_o = -v_1 - 4v_2 - 12v_3$$

The input resistance at the two inputs must not be less than 10 kΩ. Design a circuit to achieve the objective. Use standard 5% resistance values less than 200 kΩ.

15–37. Design a ICVS to achieve a transresistance of 1.5 kΩ.

15–38. Design a VCIS for a floating load to achieve a transconductance of 1 mS.

15–39. Design a Howland current source for a grounded load to achieve a transconductance of 1 mS.

15–40. Design an ICIS to achieve a current gain of 28. Use standard 5% resistance values.

Troubleshooting Problems

15–41. You are troubleshooting an inverting amplifier circuit of the form shown in Figure 15–4. Specifications call for $v_o = -12$ V when $v_i = 1$ V. However, when v_i is a sine wave with a peak value of 1 V, the actual output is observed to be a square-wave oscillating between about -13 and $+13$ V. The frequency of the square wave is the same as that of the sine wave. Which *two* of the following *could* be the difficulty? **(a)** R_i too small; **(b)** R_i too large; **(c)** R_f too small; **(d)** R_f too large; **(e)** defective power supply; **(f)** defective op-amp.

15–42. You are troubleshooting a noninverting amplifier circuit of the form shown in Figure 15–4. Specifications call for $v_o = 12$ when $v_i = 1$ V. However, when a 1-V dc level is established at v_i, the following voltages are measured: $v_o = 5.5$ V, $v^- = 0.5$ V. The power supply voltages are correct at ± 15 V. Which *one* of the following *could* be the difficulty? **(a)** R_i too small; **(b)** R_i too large; **(c)** R_f too small; **(d)** R_f too large; **(e)** defective op-amp.

16

NONLINEAR ELECTRONIC CIRCUITS

OBJECTIVES

After completing this chapter, the reader should be able to:

- Discuss the major differences between linear and nonlinear electronic circuits.
- Analyze the operation of active rectifier circuits.
- Draw waveforms and write equations to describe capacitor charge and discharge.
- Analyze the operation of passive and active peak detector circuits.
- Analyze the operation of an envelope detector circuit.
- Analyze the operation of passive and active clamping circuits.
- Analyze the operation of comparator circuits implemented with a general-purpose op-amp.
- Analyze some of the circuits in the chapter with PSPICE.

16–1 OVERVIEW

For the past several chapters, most of our attention was directed toward the analysis, modeling, and design of various amplifier circuits. Ideal amplifiers are examples of **linear** circuits. Ideal linear amplifiers are characterized by the fact that the output waveform has the same form as the input waveform (i.e., no distortion is introduced).

In this chapter we observe some of the most common **nonlinear** electronic circuits. For this purpose a nonlinear circuit will be defined as one in which the output waveform does not have the same shape as the input waveform. Obviously, if an amplifier is nonlinear, it is an undesirable situation. However, there are numerous applications in which it is desired to reshape or change the form of a waveform or a signal in some predetermined manner. Thus, the nonlinear circuits to be considered here can be used to advantage in various waveshaping applications.

Along with passive nonlinear circuits, the op-amp will be shown to be quite useful in realizing precision nonlinear operations. When op-amps appear in applications, bear in mind that the feedback loop may be closed in some applications, while in others, the op-amp may be allowed to assume a saturation state. In the first case, the assumption of a zero differential input voltage will normally still be valid. In the latter case, however, the loop will often be broken, and the differential input voltage will no longer be forced to assume a nearly zero value. In designing such circuits, care must be taken that the maximum input differential voltage not be exceeded.

The nonlinear circuits of this chapter using op-amps may be classified as either **saturating** or **nonsaturating.** A **saturating circuit** is one in which the op-amp output becomes saturated for some portion of the normal range of the input signal. A **nonsaturating circuit** is one in which the op-amp output is prevented from becoming saturated for the normal range of the input signal.

It takes an appreciable amount of time for an op-amp to move into and out of saturation. Consequently, saturating circuits are considerably slower than nonsaturating circuits and may be used only at very low frequencies. However, their simplicity justifies their use in applications in which the speed of response is not critical.

16–2 ACTIVE RECTIFIER CIRCUITS

We considered several passive small-signal rectifier circuits in Chapter 3 to illustrate simple applications of diodes. These circuits were all characterized by certain errors between the input and the desired output as a result of forward-biased diode voltage drops. In this section we consider several rectifier circuits utilizing one or more op-amp circuits. The use of an op-amp with feedback can virtually eliminate the forward diode drop in the signal path. Rectifiers employing op-amps to eliminate the diode drop will be referred to as **active rectifier circuits.** Such circuits are also denoted elsewhere in the literature as *precision rectifier circuits*.

Noninverting Half-Wave Circuit

The simplest form of an active rectifier circuit is the noninverting half-wave rectifier circuit of Figure 16–1. The circuit has essentially the same form as a voltage follower except for the presence of the diode in the forward block. When v_i is positive, v_a will increase in a positive direction, and this will forward-bias the diode. The feedback loop will thus be closed, and this will force the noninverting output voltage to be the same as the input voltage v_i. However, the noninverting input is the same as the output, so $v_o = v_i$ for this condition. The voltage v_a will be clamped at about 0.7 V higher than v_i to maintain the feedback loop.

FIGURE 16–1
Noninverting active half-wave rectifier circuit.

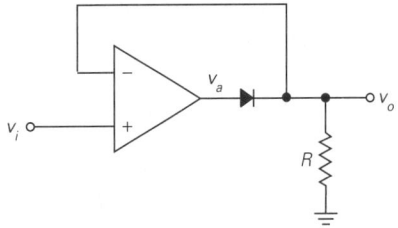

When v_i is negative, v_a will become negative, and this will open the loop. Consequently, v_a will assume the value of negative saturation (i.e., $v_a = -V_{sat}$). The voltage v_o is thus zero in this case.

A summary of the preceding conditions is as follows:

$$v_o = \begin{cases} v_i & \text{for } v_i > 0 \\ 0 & \text{for } v_i < 0 \end{cases} \qquad \begin{matrix} \textbf{(16–1a)} \\ \textbf{(16–1b)} \end{matrix}$$

Strictly speaking, v_i must be large enough to cause v_a to exceed the diode drop for (16–1a) to be valid, but since this value for v_i is of the order of microvolts, the condition of (16–1a) is quite accurate. The transfer characteristic for this rectifier is shown in Figure 16–2. The voltage v_a is given by

$$v_a = \begin{cases} v_i + 0.7 & \text{for } v_i > 0 \\ -V_{sat} & \text{for } v_i < 0 \end{cases} \qquad \begin{matrix} \textbf{(16–2a)} \\ \textbf{(16–2b)} \end{matrix}$$

FIGURE 16–2
Transfer characteristic curve of active noninverting rectifier circuit.

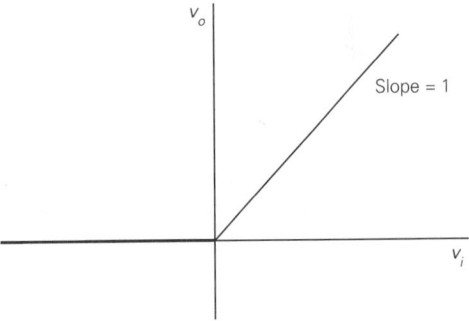

The primary disadvantage of this rectifier is that it is a *saturating circuit*, so its speed is somewhat limited.

Inverting Half-Wave Circuit

A nonsaturating rectifier can be achieved by employing two diodes in an inverting half-wave circuit. This circuit as arranged for active operation in the second quadrant is shown in Figure 16–3.

FIGURE 16–3
Inverting active half-wave rectifier circuit.

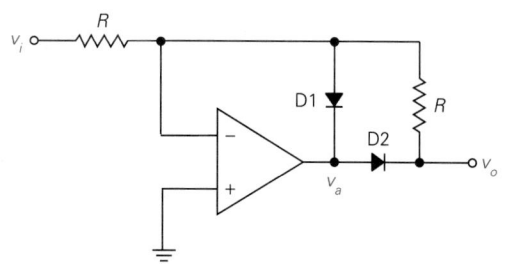

Refer to Figure 16–4(a) for the discussion that follows. When v_i is positive, v_a will become negative, and this will forward-bias D1. However, D2 will be reverse biased, which uncouples v_o from v_a. The feedback loop is thus closed around D1, and since the inverting input must assume a virtual ground for this condition, $v_o = 0$. The voltage v_a is clamped at 0.7 V below ground i.e., $v_a = -0.7$ V).

FIGURE 16–4
Two states in analyzing active rectifier circuit of Figure 16–3.

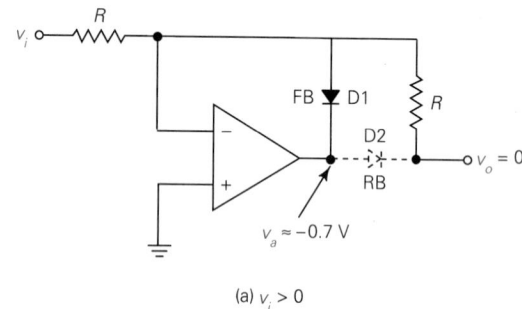

(a) $v_i > 0$

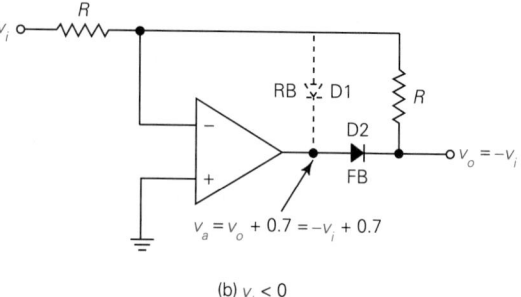

(b) $v_i < 0$

Refer now to Figure 16–4 (b). When v_i is negative, v_a will become positive, and this will reverse-bias D1. However, D2 will be forward biased for this condition. The feedback loop is thus closed around D2 with D1 open. Under this condition, the circuit is operating as an inverting amplifier with a diode added to

the forward gain block. The output v_o is thus forced to be $v_o = -(R/R)v_i = -v_i$. The voltage v_a assumes a voltage of 0.7 V above the level of v_o.
The preceding results for v_o are summarized as follows:

$$v_o = \begin{cases} 0 & \text{for } v_i > 0 \\ -v_i & \text{for } v_i < 0 \end{cases} \qquad \begin{matrix} \textbf{(16–3a)} \\ \textbf{(16–3b)} \end{matrix}$$

The transfer characteristic for this rectifier is shown in Figure 16–5.

FIGURE 16–5
Transfer characteristic curve of active inverting
rectifier circuit.

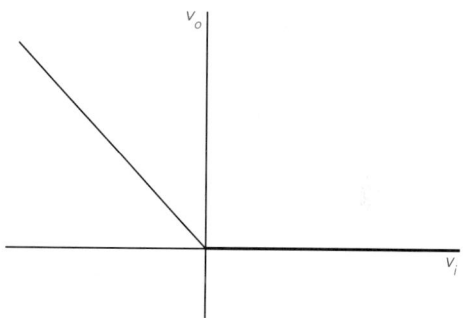

The two states for v_a are summarized as follows:

$$v_a = \begin{cases} -0.7\text{ V} & \text{for } v_i > 0 \\ -v_i + 0.7\text{ V} & \text{for } v_i < 0 \end{cases} \qquad \begin{matrix} \textbf{(16–4a)} \\ \textbf{(16–4b)} \end{matrix}$$

Although v_a switches abruptly between two levels as the input changes sign, it will remain out of saturation (unless v_i comes within 0.7 V of saturation). Consequently, this rectifier circuit is much faster than the noninverting circuit considered earlier.

EXAMPLE 16–1

For the rectifier circuit of Figure 16–1, determine v_o and v_a for $v_i =$ **(a)** 3 V, and **(b)** −3 V. Assume that $\pm V_{\text{sat}} = \pm 13$ V.

Solution

(a) When $v_i = 3$ V, the loop is closed through the diode. This corresponds to the first quadrant of Figure 16–2 and

$$v_o = 3\text{ V} \qquad \textbf{(16–5)}$$

The voltage v_a must be 0.7 V higher; that is,

$$v_a = 3.7\text{ V} \qquad \textbf{(16–6)}$$

(b) When $v_i = -3$ V, the loop is open and

$$v_o = 0 \qquad \textbf{(16–7)}$$

The op-amp output assumes negative saturation; that is,

$$v_a = -13 \text{ V} \qquad \qquad (16\text{-}8)$$

**EXAMPLE
16–2**

For the rectifier circuit of Figure 16–3, determine v_o and v_a for $v_i =$ **(a)** 3 V, and **(b)** −3 V.

Solution

(a) When $v_i = 3$ V, D1 is forward biased and D2 is reverse biased. The output is

$$v_o = 0 \qquad \qquad (16\text{-}9)$$

The voltage v_a is

$$v_a = -0.7 \text{ V} \qquad \qquad (16\text{-}10)$$

(b) When $v_i = -3$ V, D1 is reverse biased and D2 is forward biased. In this case

$$v_o = -v_i = -(-3) = 3 \text{ V} \qquad \qquad (16\text{-}11)$$

The voltage v_a is

$$v_a = -v_i + 0.7 = -(-3) + 0.7 = 3.7 \text{ V} \qquad \qquad (16\text{-}12)$$

16–3 CAPACITOR CHARGE AND DISCHARGE

Many nonlinear electronic circuits and timers use the charging and/or discharging of a capacitor to control the timing. It is, therefore, necessary to understand this process before undertaking a study of circuits using the effect.

Once a voltage is established across an ideal capacitor, the capacitor will hold or store the voltage until it is discharged. If the capacitor were ideal and had no losses, the voltage would remain indefinitely. Any practical capacitor will, of course, have some leakage, so it will eventually discharge. However, many capacitors are capable of holding a charge for a long period of time.

Capacitor Discharge

Although the order of presentation here may seem backward, it is easier mathematically to begin the discussion by considering the discharge of a capacitor that is already charged to some initial value. Consider the capacitor shown in Figure 16–6(a), which has been previously charged to a voltage V_o. The quantity v_C represents the instantaneous voltage, which will change with time once discharge is initiated. The capacitor is assumed to be lossless, so $v_C = V_o$ initially.

At some time, which we will conveniently designate as $t = 0$, assume that the resistor is connected across the capacitor. Current starts to flow through the resistor, and the initial stored energy starts to dissipate in the resistor.

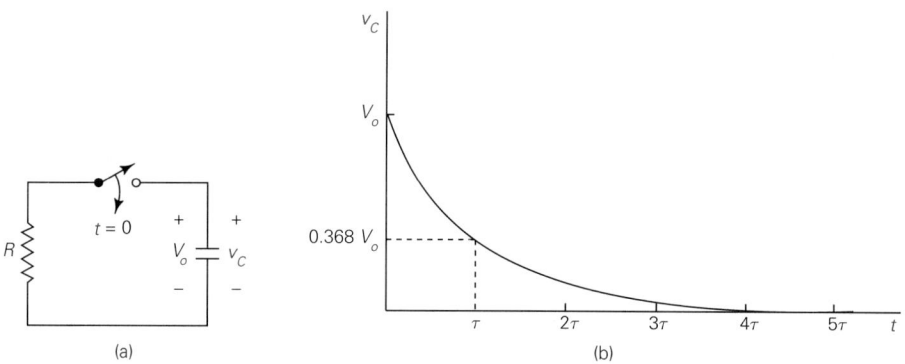

FIGURE 16–6
Initially charged capacitor and the discharge voltage waveform.

The exact mathematical form of the voltage v_C is given by

$$v_C = V_o e^{-t/\tau} = V_o e^{-t/RC} \qquad \textbf{(16–13)}$$

The quantity e represents the base of the natural system of logarithms. The value of e to four significant figures is 2.718. The quantity $-t/\tau = -t/RC$ is the exponent of e. The function e^x is available on all modern scientific calculators.

The form of the voltage discharge is shown in Figure 16–6(b). The initial value of the exponential function is unity, so the initial value of v_C is V_o, as required.

The quantity τ is defined as the **time constant** and is

$$\boxed{\tau = RC} \qquad \textbf{(16–14)}$$

After a time $t = \tau$, the voltage has discharged to about 36.8% of its initial value. In a time $t = 2\tau$, the voltage has discharged to about 13.5% of its initial value. Other values are readily determined with a calculator. In a time $t = 5\tau$, the voltage has discharged to a level less than 1% of its initial value.

Charging a Capacitor

Consider the circuit of Figure 16–7(a), and assume that the capacitor is initially uncharged. At $t = 0$, the switch is closed, and the capacitor is connected through a resistance to a dc voltage. Current begins to flow, and the voltage v_C across the capacitor starts to change. The exact mathematical form of the voltage v_C is given by

$$\boxed{v_C = V_B(1 - e^{-t/\tau})} \qquad \textbf{(16–15)}$$

where $\tau = RC$ was defined in (16–14).

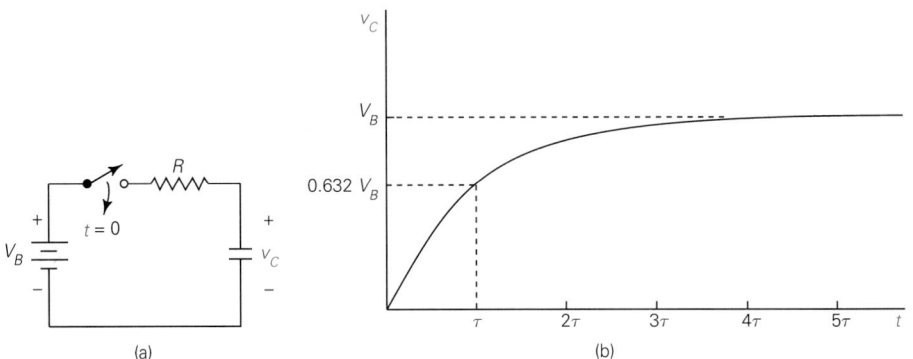

FIGURE 16–7
Initially charged capacitor and the charging voltage waveform.

The form of the voltage across the capacitor is shown in Figure 16–7(b). After a time $t = \tau$, the voltage will reach 63.2% of its final voltage. After a time $t = 2\tau$, the capacitor voltage will reach about 86.5% of its final value. In a time $t = 5\tau$, the voltage is over 99.3% of the final value. For most practical purposes, the capacitor can be thought of as reaching the final value in a time $t = 5\tau$.

Changing Levels

Thus far we have considered the complete discharge of a capacitor and the process of charging an initially uncharged capacitor. Next we consider the process of charging or discharging a capacitor from one level to a second level. Two possibilities must be considered: (1) discharging from a higher level to a lower level, and (2) charging from a lower level to a higher level.

The circuit to be considered is shown in Figure 16–8(a). The capacitor is assumed to be charged to a level V_o, and a dc voltage of value V_B is connected at $t = 0$.

$V_B < V_o$. If the applied voltage is smaller than the initial capacitor voltage, the capacitor will discharge toward the level of V_B. This situation is shown in Figure 16–8(b).

$V_B > V_o$. If the applied external voltage is larger than the initial capacitor voltage, the capacitor will charge toward the level of V_B. This situation is shown in Figure 16–8(c).

It turns out that the same equation can be used to describe both situations. The voltage v_C can be expressed as

$$v_C = V_B + (V_o - V_B)e^{-t/\tau}$$

(16–16)

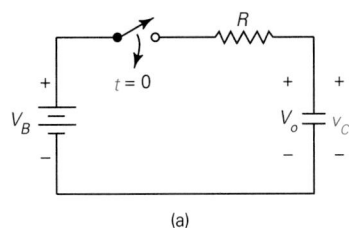

(a)

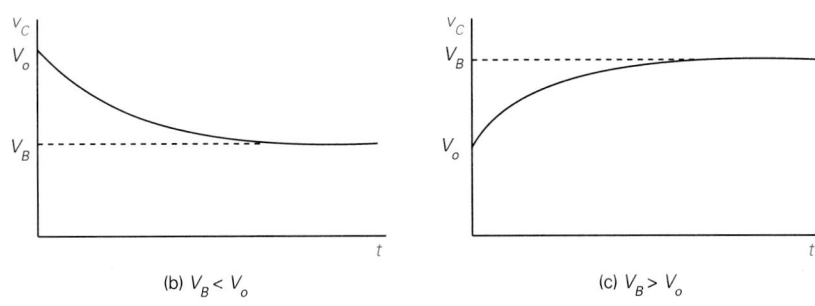

(b) $V_B < V_o$ (c) $V_B > V_o$

FIGURE 16–8
Initially charged capacitor, partial discharge to a lower level, and charging to a higher level.

Depending on which of the two cases is being considered, the factor of the exponential in (16–16) can be either positive or negative.

EXAMPLE 16–3   A portion of a certain electronic switching circuit is shown in Figure 16–9. Initially both switches are open. First, S1 is closed and the capacitor is charged from some external source not shown through the 100-Ω resistor. S1 is then opened and the voltage remains on the capacitor. At a later time, S2 is closed, and the capacitor discharges through the 100-kΩ resistor. **(a)** Determine the charging and discharging

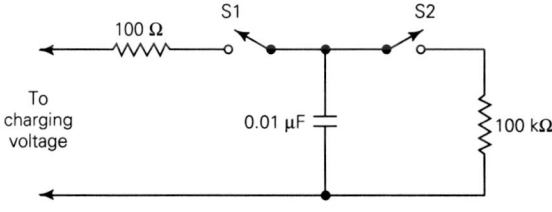

FIGURE 16–9
Circuit of Example 16–3.

time constants. **(b)** Based on the 5τ rule, determine the minimum times required for charging and discharging of the capacitor.

Solution
(a) Let τ_c and τ_d represent the charging and discharging time constants, respectively. For charging, the value of R is 100 Ω, and for discharging the value of R is 100 kΩ. The value of C is, of course, common for both. We have

$$\tau_c = 100 \times 0.01 \times 10^{-6} = 10^{-6}\,\text{s} = 1\,\mu\text{s} \tag{16–17}$$

and

$$\tau_d = 10^5 \times 0.01 \times 10^{-6} = 10^{-3}\,\text{s} = 1\,\text{ms} \tag{16–18}$$

(b) Let T_c and T_d represent the minimum times for charging and discharging, respectively, based on the 5τ rule. These values are

$$T_c = 5\tau_c = 5 \times 1\,\mu\text{s} = 5\,\mu\text{s} \tag{16–19}$$
$$T_d = 5\tau_d = 5 \times 1\,\text{ms} = 5\,\text{ms} \tag{16–20}$$

It will thus take about 1000 times as long to discharge the capacitor as to charge it because of the drastically different time constants. Understand that the 5τ rule is simply a convenient approximation.

16–4 PEAK DETECTOR CIRCUITS

An ideal **peak detector circuit** is one that extracts the peak value of a waveform and holds that value for some interval of time. A peak detector may be classified as either a **positive** peak detector or a **negative** peak detector, depending on the side of the waveform to which it responds. It may be further classified as a **passive** or an **active** peak detector, depending on whether or not an active device is used.

The action of an ideal positive peak detector is illustrated in Figure 16–10. Assume that the input waveform v_i of Figure 16–10(a) is first turned on at $t = 0$. The output v_o shown in Figure 16–10(b) follows the input up to the peak level V_{ip}. However, when the input drops below that level, the output remains at the value V_{ip} as shown. A negative peak detector would produce an output at the level $-V_{ip}$.

Passive Peak Detectors

The passive peak detector is the simplest circuit of this type, although it does introduce some error in the output level. The form of a passive positive peak detector is shown in Figure 16–11(a), and the corresponding negative peak detector is shown in Figure 16–11(b).

A brief analysis will now be made of the positive peak detector of Figure 16–11(a). On the positive half-cycle of the input, the capacitor charges through the diode resistance and any source resistance (not shown) to a voltage equal to

FIGURE 16–10
Waveforms of ideal peak detector circuit.

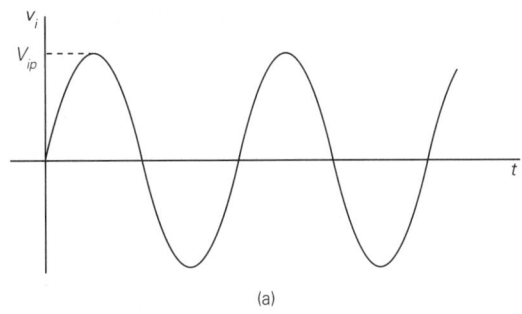

(a)

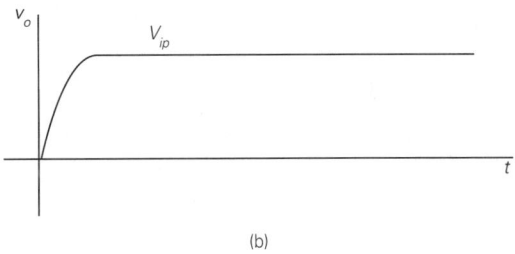

(b)

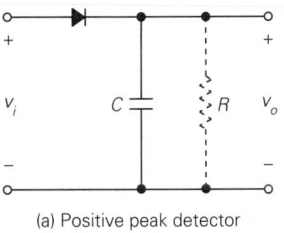

(a) Positive peak detector

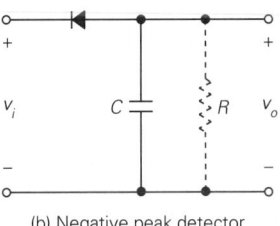

(b) Negative peak detector

FIGURE 16–11
Passive peak detector circuits.

the peak positive voltage minus the diode drop. It is assumed that the charging time constant is very small compared with the waveform period, which is usually the case. Assuming a drop of 0.7 V and a peak input voltage level of V_{ip}, the voltage v_o reaches a level of

$$v_o \simeq V_{ip} - 0.7 \qquad (16–21)$$

The output voltage is thus lower than the input peak by one forward-biased diode drop.

After the capacitor has been charged to its peak value and the input voltage starts to drop, the diode becomes reverse based. If there were no shunting resistance across the capacitor, the capacitor would remain at the voltage level of (16–21) indefinitely as long as the peak value of the input did not increase.

In practice, there is almost always some type of load across the capacitor. This is indicated by the resistance R in Figure 16–11. After all, if the output voltage is to serve any purpose, it must be connected to some additional circuit. In the normal type of application, the shunting resistance is much larger than the resistance of the charging part of the circuit. Consequently, the discharge time constant is very large compared to a period of the input voltage, and the capacitor will discharge only slightly during a cycle. At a point near the peak of the next cycle, the diode will turn on briefly and recharge the capacitor to the peak input level.

Even if there were no load at all on the capacitor, it might be desirable in some applications to provide a means to discharge the capacitor slowly. Consider the situation where it is desired to monitor the peak value of a waveform, but in which the peak might be expected to change over a period of time. If the peak value were to increase, the capacitor voltage will adjust to the new level quickly. However, if the peak value decreased, the output voltage could never readjust unless some means for discharge is provided.

The passive peak detector circuits are widely used in noncritical applications where some error in the exact peak level can be tolerated. When this circuit is used in instrumentation, it is possible to calibrate the output level to compensate for much of the error in the diode drop.

Active Peak Detector

In applications requiring high accuracy, an active peak detector employing an op-amp with negative feedback may be used. Op-amp active detectors may be classified as either **saturating** or **nonsaturating.** A saturating detector moves into and out of saturation once each cycle, while a nonsaturating detector always remains in the active region. Consequently, a nonsaturating peak detector is capable of operating at a much higher frequency. The treatment here will be limited to a nonsaturating circuit. The circuit arranged as a positive peak detector is shown in Figure 16–12. The analysis of this circuit is a little tricky and must be carefully studied.

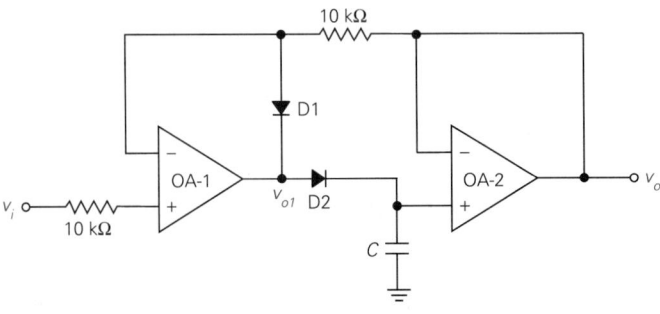

FIGURE 16–12
Active precision nonsaturating positive peak detector.

Assume first that the capacitor is uncharged and that the input v_i starts increasing from an initial zero level. The positive value of v_i causes the output v_{o1} of OA-1 to become positive, and this forward biases D2 and start to charge the capacitor C. OA-2 is connected as a voltage follower, and v_o thus follows the capacitor voltage. The voltage v_{o1} is held at 0.7 V above this voltage by D2, and this causes D1 to remain reverse biased.

Refer now to Figure 16–13 in which possible states of the circuit are shown. When v_i reaches its peak value (i.e., $v_i = V_{ip}$), the situation is shown in Figure 16–13(a). It is asumed here that the charging time constant is sufficiently small that the capacitor reaches its peak level at essentially the same time as the voltage reaches its peak value. The voltage v_{o1} is forced to be $v_{o1} = V_{ip} + 0.7$. The output is $v_o = V_{ip}$, and diode D1 is still reverse biased.

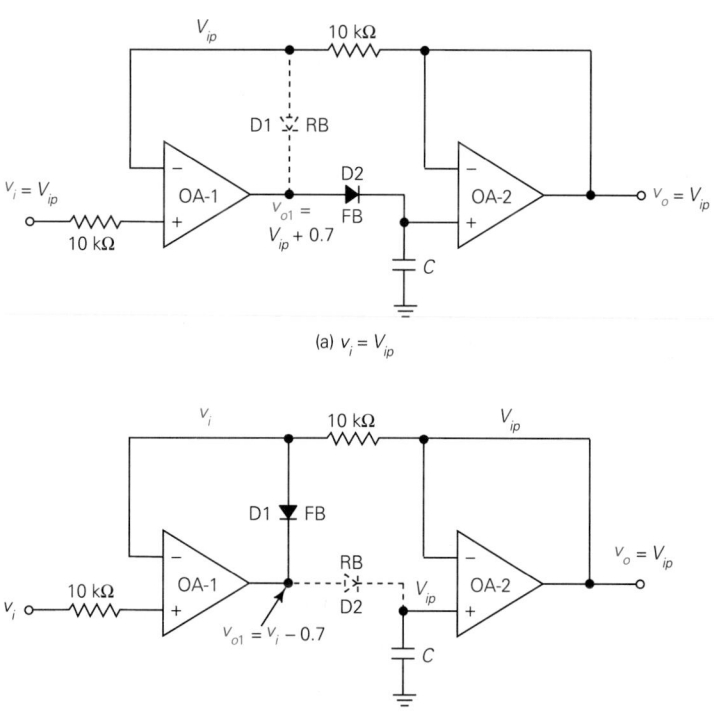

(a) $v_i = V_{ip}$

(b) $v_i < V_{ip}$ (after capacitor is charged to V_{ip})

FIGURE 16–13
Two states in analyzing active peak detector circuit of Figure 16–12.

Assume next that v_i decreases after the capacitor has been charged to V_{ip}. The voltage v_{o1} now starts to drop and diode D2 becomes reverse biased. However, this drop causes D1 to turn on. The two stages each have individual feedback loops now, and the inverting input of OA-1 is forced to assume a value v_i. The output v_{o1} will thus be clamped at 0.7 V lower than the input. This situation is shown in Figure 16–13(b).

As the circuit has been shown, the capacitor can only discharge through its internal leakage or through the noninverting input of OA-2. When it is desired to hold a peak voltage for a very long time, the circuit should employ an op-amp for OA-2 with an FET input and a capacitor with very low losses. In an application where it is desired to provide a nominal discharge path, a large value of resistance may be connected in parallel with the capacitor.

EXAMPLE 16–4

Consider an active peak detector of the form shown in Figure 16–12. Assume that a random signal varying from −6 to 6 V is applied at the input. Determine the values of v_o and v_{o1} for each of the following conditions: **(a)** $v_i = 6$ V on the first positive peak, and **(b)** $v_i = 2$ V after the first positive peak. Assume that the charging time constant is negligible compared with the time scale of observation.

Solution

(a) When $v_i = 6$ V, the capacitor charges quickly to 6 V, and we have

$$v_o = 6 \text{ V} \tag{16–22}$$

The voltage v_{o1} assumes the value

$$v_{o1} = 6 + 0.7 = 6.7 \text{ V} \tag{16–23}$$

(b) Once the capacitor is charged, it remains at the previous peak level, and

$$v_o = 6 \text{ V} \tag{16–24}$$

For $v_i = 2$ V, the loop around OA-1 forces the inverting input to have the same value (i.e., 2 V). The voltage v_{o1} is one diode drop below this value; that is,

$$v_{o1} = 2 - 0.7 = 1.3 \text{ V} \tag{16–25}$$

16–5 ENVELOPE DETECTOR

An important circuit widely employed in radio receivers is the **envelope detector.** A passive positive envelope detector circuit is shown in Figure 16–14(a). The circuit has the same form as that of a passive peak detector, but the discharge-time constant established for the circuit is what distinguishes the different type of operation.

To discuss the operation of the envelope detector, it is necessary to inspect the waveform of an **amplitude modulated (AM)** signal. A simplified AM signal is shown in Figure 16–14(b). In an AM signal, the amplitude of a high-frequency sinusoid (called a **carrier**) is varied (or *modulated*) with a lower-frequency signal. Both the positive and negative sides of the carrier then vary up and down in accordance with the lower-frequency signal. To enhance the illustration, only a few cycles of the higher-frequency signal are shown here. In practice, the frequency of the carrier may be 100 or more times as great as the lower-frequency modulating signal.

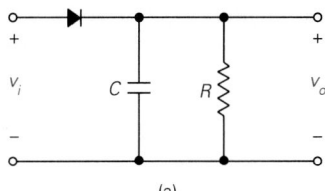

(a)

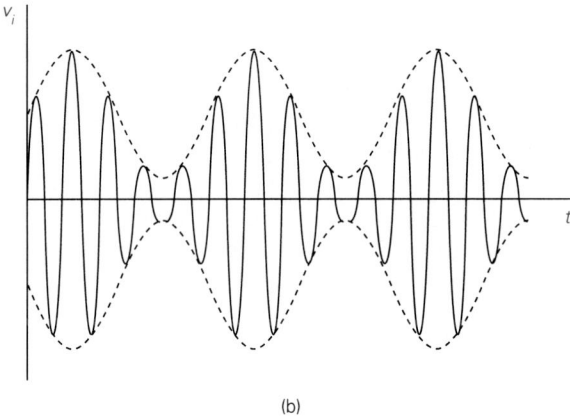

(b)

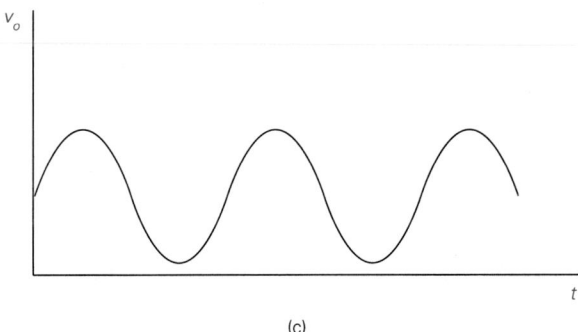

(c)

FIGURE 16–14
Passive envelope detector and waveforms.

Let T_c represent the period corresponding to the carrier, and let T_m represent the period of the modulating signal. The time constant $\tau = RC$ of the envelope detector is chosen to satisfy the following two inequalities:

$$\tau \gg T_c \qquad\qquad (16\text{–}26)$$
$$\tau \ll T_m \qquad\qquad (16\text{–}27)$$

It is possible to satisfy both inequalities since a typical value of T_m will be 100 or more times as great as T_c.

The inequality of (16–26) means that the voltage v_o cannot follow the high-frequency variations of the carrier. However, the inequality of (16–27) allows v_o to follow the envelope. The result is that the voltage v_o follows the positive envelope of the waveform reasonably well. There will be some error due to the diode drop and some slight discharge between "spurts" of the carrier. However, this basic circuit form has been used in millions of AM radios over the years.

The somewhat idealized form of the output is indicated as v_o in Figure 16–14(c). This signal contains a superfluous dc level, but it can be readily removed with an RC coupling circuit (not shown).

EXAMPLE 16–5

Consider a passive envelope detector of the form shown in Figure 16–14(a). The carrier frequency in a certain application is 500 kHz. Assume that the time constant τ is arbitrarily selected to satisfy the criterion $\tau = 20T_c$, where T_c is the carrier period. **(a)** If $C = 0.01$ μF, determine the value of R required. **(b)** For a modulating frequency of 2 kHz, show that $T_m \gg T_c$, where T_m is the modulating period.

Solution
(a) The carrier period is

$$T_c = \frac{1}{500 \times 10^3} = 2\ \mu\text{s} \tag{16–28}$$

Based on the criterion employed,

$$\tau = 20T_c = 20 \times 2\ \mu\text{s} = 40\ \mu\text{s} \tag{16–29}$$

Since $\tau = RC$ and $C = 0.01$ μF, we have

$$R = \frac{\tau}{C} = \frac{40 \times 10^{-6}}{0.01 \times 10^{-6}} = 4\ \text{k}\Omega \tag{16–30}$$

(b) The modulating signal period is

$$T_m = \frac{1}{2 \times 10^3} = 0.5\ \text{ms} = 500\ \mu\text{s} \tag{16–31}$$

Comparing the modulating signal period of 500 μs with the carrier period of 2 μs, the former is 250 times as large as the latter. Thus $T_m \gg T_c$.

16–6 CLAMPING CIRCUITS

A **clamping circuit** is one that shifts a signal either up or down about a specified reference level. A common application of a clamping circuit is that of a **dc restorer.** This application involves shifting the level of a signal that has been ac coupled in order to restore a dc level. The dc restorer circuit has been widely used in the video amplifiers of television receivers.

Passive Clamping Circuit

The simplest version of a clamping circuit is the **passive** version shown in Figure 16–15(a). On the first negative portion of the input signal, the diode conducts and charges the capacitor to the negative peak value minus one diode drop. The positive terminal of this voltage will be on the right. The diode then remains reverse biased as the input voltage changes in a positive direction. The voltage across the capacitor then adds to the input voltage and shifts the net output voltage upward.

FIGURE 16–15
Passive clamping circuit and typical waveforms.

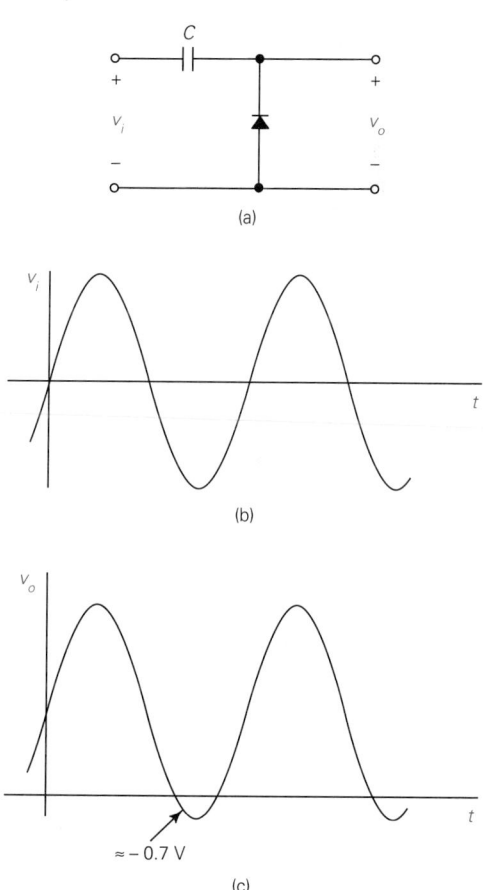

Waveforms illustrating this clamping action are shown in Figure 16–15(b) and (c). These waveforms apply *after* the initial clamping action has been established. The input waveform v_i of (b) has no dc level. However, the output waveform v_o of (c) is shifted upward. If the diode were perfect, the bottom level of the output would be exactly at a voltage level of zero. The diode drop causes the output to have a slight negative excursion as shown.

Active Clamping Circuit

The error resulting from the diode drop can be virtually eliminated with the active clamping circuit of Figure 16–16(a) using an op-amp. On the first negative half-cycle of the input waveform, the op-amp loop is closed through the diode, and the inverting input is forced to assume a virtual ground level. This forces the capacitor to charge to the negative peak of the input with the positive terminal on the right. When the input becomes more positive, the loop is broken, and the output of the op-amp goes into negative saturation. However, the capacitor voltage adds to the input voltage to produce an output clamped at zero volts.

Representative waveforms are shown in Figure 16–16(b) and (c). For the input v_i of (b), the output v_o will be of the form shown in (c).

FIGURE 16–16
Active clamping circuit and typical waveforms.

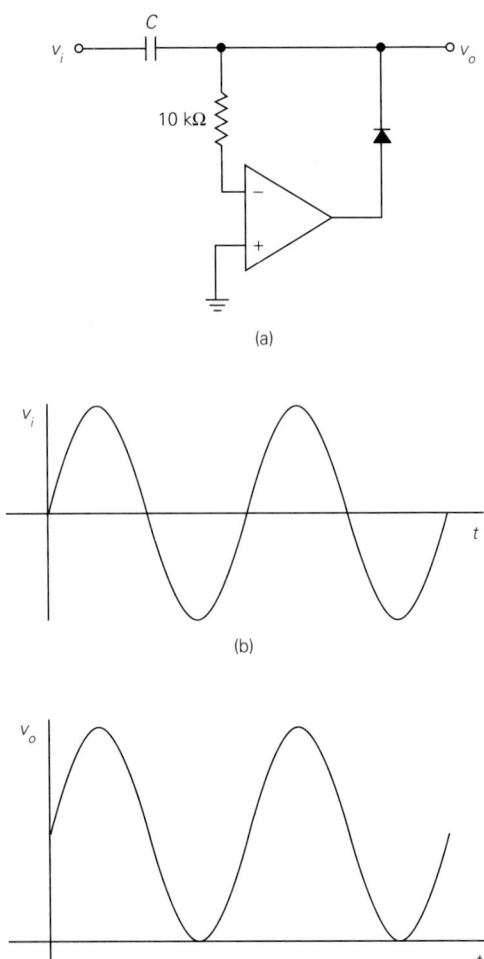

(a)

(b)

(c)

EXAMPLE 16–6

Consider an ac signal of the form of (b) in either Figure 16–15 or 16–16, and assume that it oscillates between −6 and 6 V. Determine the peak value V_{op} of the output for the following two circuits: **(a)** passive clamping circuit of Figure 16–15(a), and **(b)** active clamping circuit of Figure 16–16(a).

Solution

(a) For the circuit of Figure 16–15(a), the peak value of the output is

$$V_{op} = 2 \times 6 - 0.7 = 12 - 0.7 = 11.3 \text{ V} \tag{16–32}$$

(b) For the circuit of Figure 16–16(a), the peak value of the output is

$$V_{op} = 2 \times 6 = 12 \text{ V} \tag{16–33}$$

16–7 COMPARATOR CIRCUITS

A comparator circuit is one that provides one of two possible output levels in accordance with the range of an input signal. A variety of possible comparator types may be implemented with general-purpose op-amps, although their switching speeds are limited by the op-amp characteristics. Dedicated comparators are available in which the switching speed is optimized. The treatment here will be limited to the use of general-purpose op-amps as comparators since this approach will most clearly illustrate the principle of operation.

Noninverting Comparator

The simplest form of a noninverting op-amp comparator is shown in Figure 16–17(a). The op-amp is operated in an open-loop arrangement, and the inverting input is grounded. When the input voltage v_i exceeds V_{sat}/A_d (typically in the microvolt range), the output v_o moves into positive saturation and remains there for subsequent increases of the input.

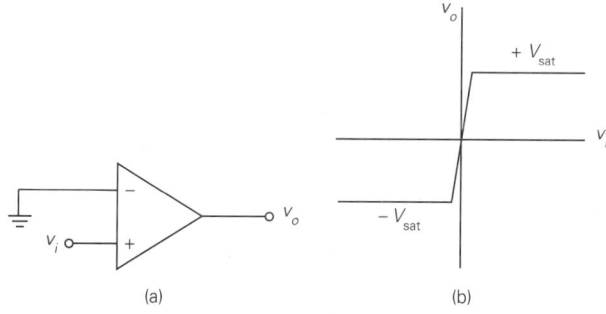

FIGURE 16–17
Op-amp noninverting comparator and its transfer characteristic.

The transfer characteristic curve is shown in Figure 16–17(b). Note that there is a small transition interval on either side of $v_i = 0$.

A comparator circuit may be used to convert a general time-varying signal into a sequence of positive and negative pulses. For example, consider the sinusoidal waveform of Figure 16–18(a). When $v_i > 0$ (by more than a few microvolts), the output is $v_o = V_{sat}$. The result is a square wave, as shown in Figure 16–18(b).

FIGURE 16–18

Sinusoidal input to noninverting comparator and form of the output (at low frequencies).

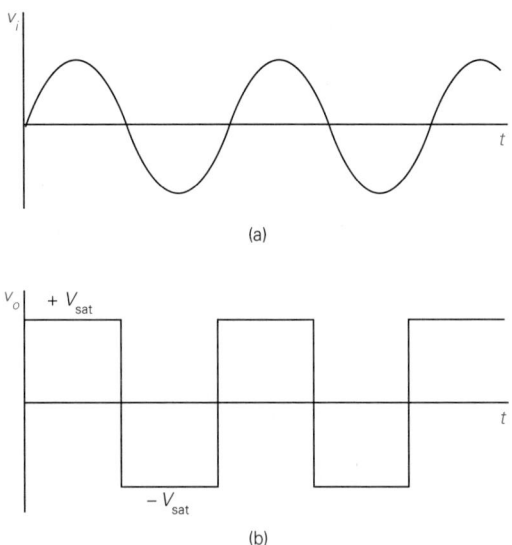

Once again, it should be stressed that the switching speed is limited in general-purpose op-amps, and a square wave generated by this process can only switch at a moderate rate. The square wave of Figure 16–18(b) is assumed to be at a sufficiently low frequency that the transition times are not visible.

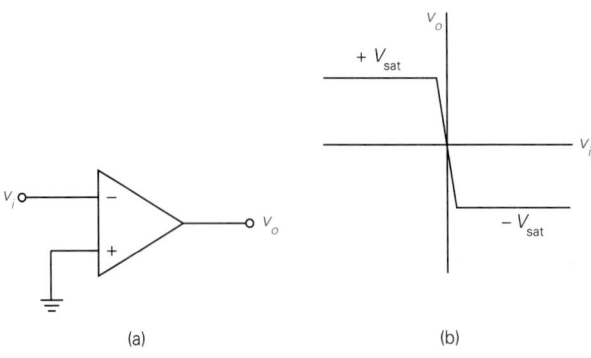

FIGURE 16–19

Op-amp inverting comparator and transfer characteristic.

Inverting Comparator

An inverting comparator can be implemented by grounding the noninverting input and applying the input signal to the inverting input as shown in Figure 16–19(a). In this case, a positive input causes the output to move into negative saturation and vice versa. The input–output characteristic is shown in Figure 16–19(b).

A sinusoidal input of the form of Figure 16–20(a) now results in an inverted square wave as shown in Figure 16–20(b). The discussion concerning switching speed for the noninverting comparator applies to the inverting comparator as well.

FIGURE 16–20
Sinusoidal input to inverting comparator and form of the output (at low frequencies).

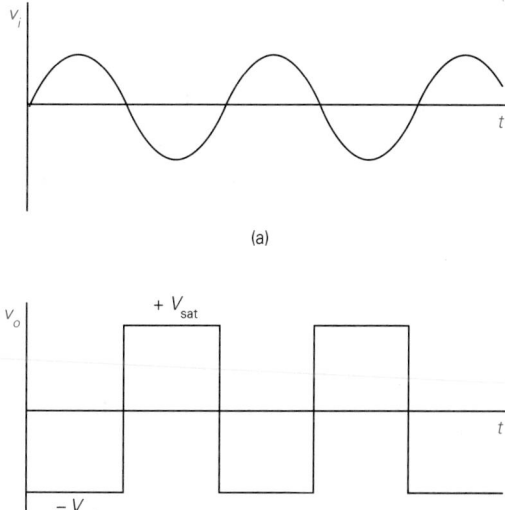

(a)

(b)

EXAMPLE 16–7

Assume for a given op-amp that $\pm V_{\text{sat}} = \pm 13$ V. If the differential gain is $A_d = 2 \times 10^5$, calculate the positive and negative input voltages at which saturation will be reached.

Solution
For $v_o = 13$ V, the value of v_i is

$$v_i = \frac{13}{A_d} = \frac{13}{2 \times 10^5} = 65 \ \mu V \qquad (16\text{–}34)$$

From symmetry, the corresponding negative voltage required to reach $v_o = -13$ V is

$$v_i = -65 \ \mu V \qquad (16\text{–}35)$$

16–8 PSPICE EXAMPLES

PSPICE EXAMPLE 16–1

Simulate a noninverting active half-wave rectifier circuit of the form shown in Figure 16–1, and obtain a transfer characteristic plot.

Solution

The circuit diagram using PSPICE notation is shown in Figure 16–21(a), and the code is shown in Figure 16–21(b). Library models of the 741 op-amp and of the 1N4148 diode have been employed. The input voltage is swept from −12 to 12 V in steps of 0.1 V.

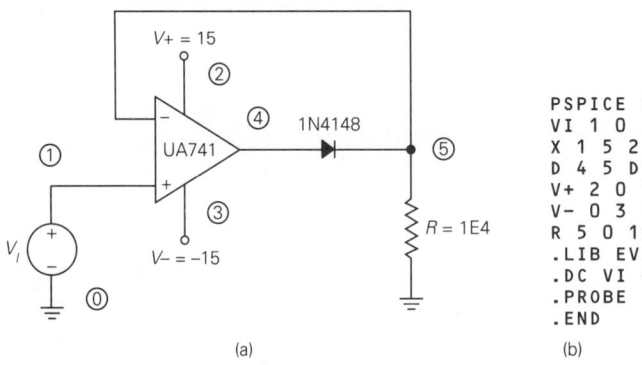

```
PSPICE EXAMPLE 16-1
VI 1 0
X 1 5 2 3 4 UA741
D 4 5 D1N4148
V+ 2 0 15
V- 0 3 15
R 5 0 1E4
.LIB EVAL.LIB
.DC VI -12 12 0.1
.PROBE
.END
```

(a)   (b)

FIGURE 16–21
Circuit and code of PSPICE Example 16–1.

The transfer characteristic is shown in Figure 16–22. This function is very linear over the given operating range.

PSPICE EXAMPLE 16–2

Consider the *RC* circuit of Figure 16–6 with $R = 1$ kΩ and $C = 1$ μF. Assume that the capacitor is initially charged to 100 V. Use transient analysis to obtain a plot of the capacitor discharge voltage.

Solution

The circuit adapted to the PSPICE format is shown in Figure 16–23(a), and the code is shown in Figure 16–23(b). The time constant is

$$\tau = RC = 1 \times 10^3 \times 1 \times 10^{-6} = 1 \text{ ms} \qquad (16\text{–}36)$$

An initial voltage is placed on the capacitor by the following line code:

```
C 1 0 1E-6 IC=100
```

The designation IC represents "initial condition." An initial voltage may be placed across a capacitor, or an initial current may be established in an inductor by this process.

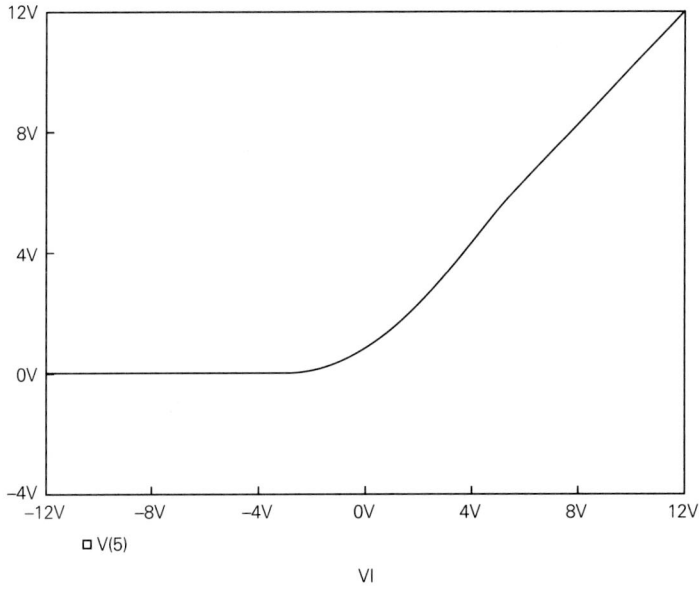

FIGURE 16–22
Transfer characteristic of PSPICE Example 16–1.

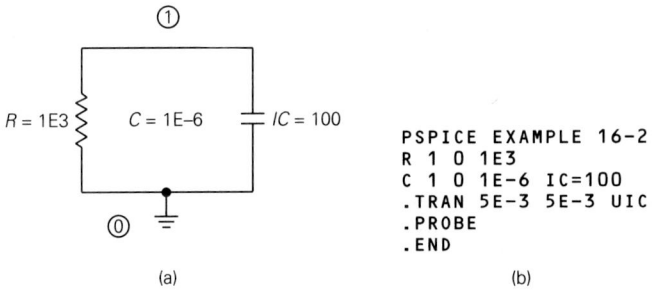

```
PSPICE EXAMPLE 16-2
R 1 0 1E3
C 1 0 1E-6 IC=100
.TRAN 5E-3 5E-3 UIC
.PROBE
.END
```

(a)   (b)

FIGURE 16–23
Circuit and code of PSPICE Example 16–2.

The transient statement reads

```
.TRAN 5E-3 5E-3 UIC
```

The analysis is to be performed from $t = 0$ to $t = 5$ ms, which represents five time constants. The control designation UIC means to "use the initial condition." Unless this code is placed on the transient line, the program will ignore the stated initial condition and assume 0.

The capacitor voltage obtained with .PROBE is shown in Figure 16–24. Note the marker at $t = 1.0004$ ms, which shows that the voltage is 36.785 V. This

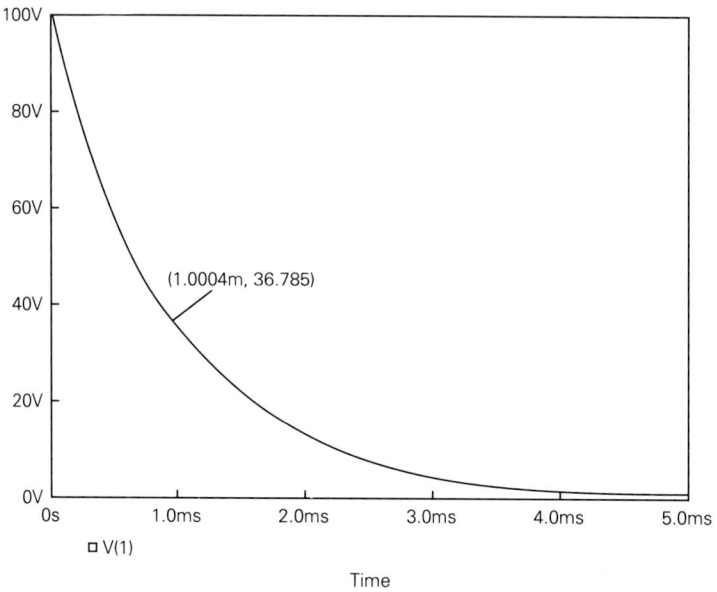

FIGURE 16–24
Discharge voltage waveform of PSPICE Example 16–2.

verifies that the voltage has decayed to about 36.8% of its initial value in one time constant. For most practical purposes, the capacitor may be assumed to be discharged by $t = 5$ ms.

PSPICE EXAMPLE 16–3

Consider again the resistance and capacitance values of PSPICE Example 16–2, but assume now that the capacitor is initially uncharged. For the circuit form of Figure 16–7 with $V_B = 100$ V, use transient analysis to plot the changing process for the capacitor.

Solution

The circuit adapted to PSPICE is shown in Figure 16–25(a), and the code is shown in Figure 16–25(b). It might appear that a dc voltage should be used as the input, but that approach could lead to some difficulty. The reason is that a dc analysis assumes a steady-state condition, which would mean that the capacitor would already be charged and would appear as an open circuit. Instead, the simulation will be based on a transient situation in which the voltage is applied to the circuit at $t = 0$.

The desired process can be accomplished with a piecewise linear (PWL) source. A PWL source is represented through a series of time and amplitude pairs

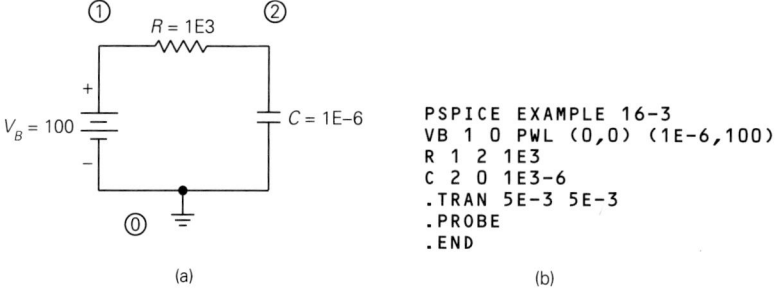

PSPICE EXAMPLE 16-3
```
VB 1 0 PWL (0,0) (1E-6,100)
R 1 2 1E3
C 2 0 1E3-6
.TRAN 5E-3 5E-3
.PROBE
.END
```

(a)   (b)

FIGURE 16–25
Circuit and code of PSPICE Example 16–3.

that have the following format:

$$\text{PWL } (t_1, v_1)(t_2, v_2) \cdots (t_n, v_n)$$

The parentheses and commas are optional but are recommended for clarity. Each pair of points describes a time t_K and a value v_K. The program assumes a straight line variation between any two adjacent points, and the voltage remains at the last value in the sequence.

To simulate a dc voltage switched on at $t = 0$, the first pair of values is (0, 0). The second pair of values chosen in this case is (1E-6, 100). This means that the voltage increases linearly to 100 V at $t = 1 \; \mu s$. This time must be small compared to the time constant of the circuit. The voltage will then remain at 100 V unless an additional pair of values follows in the statement. Thus, an adequate description of the function is

$$\text{PWL } (0, 0) (1E\text{-}6, 100)$$

This description follows the name of the voltage VB and the code connections.

The input and capacitor voltages are shown in Figure 16–26. Observe that the small rise time is not even visible on the time scale used. The capacitor voltage reaches a level of about 63.2 V in one time constant and can be considered to be fully charged in about five time constants.

Incidentally, it is possible to use a pure dc voltage for the input of a transient analysis such as this by clamping the capacitor voltage to a value of zero during the initial steady-state analysis. This can be achieved with an initial condition control statement (not to be confused with the initial condition placed within the function) that reads as follows:

```
.IC V(2) = 0
```

With this statement, a dc voltage of 100 V could have been employed, and it would be unnecessary to place an IC statement on the line defining the capacitance. We have chosen, however, to introduce the PWL function because of its versatility as a source.

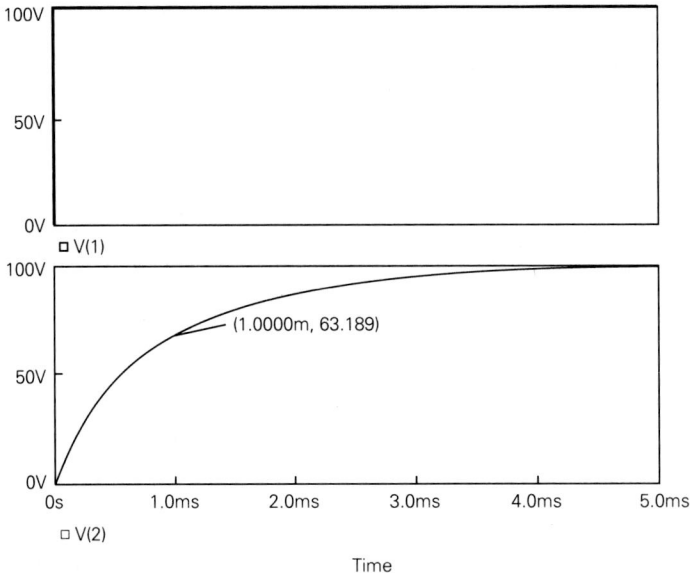

FIGURE 16–26
Waveforms of PSPICE Example 16–3.

PROBLEMS

Drill Problems

16–1. For the rectifier circuit of Figure 16–1, determine v_a and v_o for $v_i =$ **(a)** 8 V, and **(b)** −8 V. Assume that $\pm V_{\text{sat}} = \pm 13.5$ V.

16–2. For the rectifier circuit of Figure 16–1, determine v_a and v_o for $v_i =$ **(a)** 12 V, and **(b)** −12 V. Assume that $\pm V_{\text{sat}} = \pm 13$ V.

16–3. For the rectifier circuit of Figure 16–3, determine v_a and v_o for $v_i =$ **(a)** 8 V, and **(b)** −8 V. Assume that $\pm V_{\text{sat}} = \pm 13.5$ V.

16–4. For the rectifier circuit of Figure 16–3, determine v_a and v_o for $v_i =$ **(a)** 12 V, and **(b)** −12 V. Assume that $\pm V_{\text{sat}} = \pm 13$ V.

16–5. A certain discharge circuit is of the form shown in Figure 16–6(a) with $R = 220$ Ω and $C = 0.001$ μF. Using the 5τ rule, determine the approximate time required to discharge the capacitor.

16–6. Repeat the analysis of Problem 16–5 if $R = 220$ kΩ and $C = 0.1$ μF.

16–7. A certain charging circuit is of the form shown in Figure 16–7(a) with $R = 2$ MΩ and $C = 12$ μF. Using the 5τ rule, determine the approximate time required to charge the capacitor.

16–8. Repeat the analysis of Problem 16–7 if $R = 470$ Ω and $C = 220$ pF.

16–9. A certain discharge circuit is of the form shown in Figure 16–6(a) with $V_o = 100$ V and $\tau = 1$ ms. Use the exponential function on your calculator to determine the capacitor voltage at the following values of t: **(a)** 0; **(b)** 1 ms; **(c)** 2 ms; **(d)** 5 ms; **(e)** 10 ms.

16–10. A certain charging circuit is of the form shown in Figure 16–7(a) with $V_B = 100$ V and $\tau = 1$ ms. Use the exponential function on your calculator to determine the capacitor voltage at the following values of t: **(a)** 0; **(b)** 1 ms; **(c)** 2 ms; **(d)** 5 ms; **(e)** 10 ms.

16–11. Assume a passive positive peak detector of the form shown in Figure 16–11(a). Assume that a random signal varying from − 6 to 6 V is applied to the input. **(a)** Determine the value of v_o after the first positive peak of the input voltage has occurred. **(b)** Determine the peak reverse voltage V_{rp} that will appear across the diode.

16–12. Assume a passive negative peak detector of the form shown in Figure 16–11(b). Assume that a random signal varying from −6 to 6 V is applied to the input. **(a)** Determine the value of v_o after the first negative peak of the input voltage has occurred. **(b)** Determine the peak reverse voltage V_{rp} that will appear across the diode.

16–13. Assume an active positive peak detector of the form shown in Figure 16–12. Assume that a random signal varying from −4 to 4 V is applied to the input. Determine the values of v_o and v_{o1} for each of the following conditions: **(a)** $v_i = 4$ V on the first positive peak, and **(b)** $v_i = −1$ V after the first positive peak.

16–14. Assume an active negative peak detector based on the circuit of Figure 16–12, but with the two diodes reversed. Determine the values of v_o and v_{o1} for each of the following conditions: **(a)** $v_i = −4$ V on the first negative peak, and **(b)** $v_i = 4$ V after the first negative peak.

16–15. Consider a passive clamping circuit of the form shown in Figure 16–15(a), and assume an ac signal of the form shown in Figure 16–16(b) that oscillates between −8 and 8 V. **(a)** Determine the peak value V_{op} of the output voltage. **(b)** Determine the peak reverse voltage V_{rp} that will appear across the diode.

16–16. Consider a passive clamping circuit based on the form shown in Figure 16–15(a), but with the diode reversed. Assume an ac signal of the form shown in Figure 16–16(b) that oscillates between −8 and 8 V. Determine the upper and lower limits of the output voltage.

16–17. Consider an active clamping circuit of the form shown in Figure 16–16(a), and assume an ac signal of the form shown in Figure 16–16(b) that oscillates between −8 and 8 V. **(a)** Determine the peak value V_{op} of the output voltage. **(b)** Determine the peak reverse voltage V_{rp} that will appear across the diode. Assume that $\pm V_{sat} = \pm 13$ V.

16–18. Consider an active clamping circuit based on the form shown in Figure 16–16(a) but with the diode reversed. Assume an ac signal of the form shown in Figure 16–16(b) that oscillates between −8 and 8 V. Determine the upper and lower limits of the output voltage.

16–19. Consider an op-amp noninverting comparator circuit of the form shown in Figure 16–17(a), and assume that $\pm V_{sat} = \pm 13$ V. Determine v_o when $v_i =$ **(a)** 2 V, and **(b)** −2 V.

16–20. Consider an op-amp inverting comparator circuit of the form shown in Figure 16–19(a), and assume that $\pm V_{sat} = \pm 13$ V. Determine v_o when $v_i =$ **(a)** 2 V, and **(b)** −2 V.

Derivation Problems

16–21. Consider the circuit of Figure P16–21. **(a)** Determine equations for v_a and v_o in terms of v_i. **(b)** Sketch the transfer characteristic relating v_o to v_i.

FIGURE P16–21

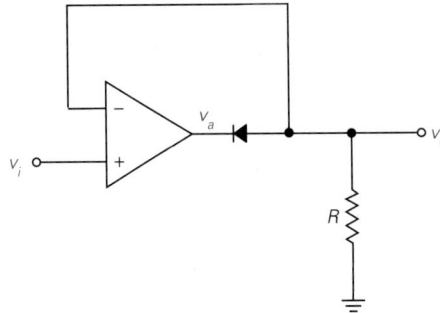

16–22. Consider the circuit of Figure P16–22. **(a)** Determine equations for v_a and v_o in terms of v_i. **(b)** Sketch the transfer characteristic relating v_o to v_i.

FIGURE P16–22

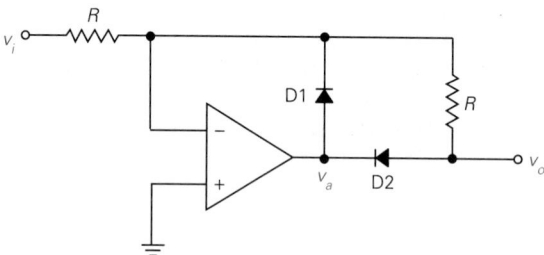

16–23. When $t \ll \tau$, the exponential function may be approximated as

$$e^{-t/\tau} \simeq 1 - \frac{t}{\tau}$$

Consider a capacitor discharge circuit as shown in Figure 16–6(a). Show that for $t \ll \tau$, the change in voltage Δv_C (from the initial value V_o) is

$$\Delta v_C \simeq - \frac{V_o t}{\tau}$$

16–24. Consider a capacitor charging circuit as shown in Figure 16–7(a). Show that for $t \ll \tau$, the voltage v_C can be approximated as

$$v_c \simeq \frac{V_B t}{\tau}$$

(See the approximation of Problem 16–23.)

Design Problems

16–25. A passive envelope detector circuit is to be designed to detect modulation on a 2-MHz carrier. Assume that the time constant is selected to satisfy the criterion $\tau = 40T_c$, where T_c is the carrier period. **(a)** If $C = 0.002 \; \mu F$, determine the value of R required. **(b)** For a modulating frequency of 5 kHz, show that $T_m \gg T_c$.

16–26. For the envelope detector of Example 16–5, assume that the time constant is to be selected as the geometric mean between the carrier period and the modulating period; that is, $\tau = \sqrt{T_c T_m}$. Selecting $C = 0.01 \; \mu F$ as in Example 16–5, determine the value of R required.

Troubleshooting Problems

16–27. You are troubleshooting an active rectifier circuit of the form shown in Figure 16–3, which is not working properly. When $v_i = -2$ V, $v_o = 2$ V, but when $v_i = 2$ V, $v_o = 2$ V. Which one of the following *could be* the difficulty? **(a)** D1 shorted; **(b)** D1 open; **(c)** D2 shorted; **(d)** D2 open.

16–28. If instead of the voltage indicated in Problem 16–27, suppose that when $v_i = 2$ V, $v_o = 0$ V, but when $v_i = -2$ V, $v_o = 0$ V. Which one of the following *could be* the difficulty? **(a)** D1 shorted; **(b)** D1 open; **(c)** D2 shorted; **(d)** D2 open.

17

OSCILLATOR CIRCUITS

OBJECTIVES

After completing this chapter, the reader should be able to:

- Discuss the general properties of oscillator circuits.
- Define the Barkhausen criterion.
- Recognize the form and analyze the following RC oscillators: Wien bridge oscillator and phase shift oscillator.
- Recognize the form and analyze the following LC oscillators: Colpitts oscillator, crystal oscillator, Hartley oscillator, and Clapp oscillator.
- Distinguish between the three types of multivibrators: (a) bistable, (b) monostable, and (c) astable.
- Recognize the form and analyze the op-amp square-wave generator.
- Analyze the operation of the 555 timer in both the astable and monostable modes.
- Discuss the operation of a unijunction transistor and its application as an oscillator.
- Analyze some of the circuits in the chapter with PSPICE.

17–1 OVERVIEW

An **oscillator** is a circuit that generates a periodic waveform of a predetermined form and frequency. All electronic circuits containing feedback are subject to generate oscillations if certain gain and phase shift conditions exist. In an amplifier circuit, these oscillations are undesirable. In an oscillator circuit, however, the design goal is to generate oscillations in a controlled manner.

Oscillators may be designed to produce a fixed single frequency, or they may be turnable over a range of frequencies. An oscillator may produce a continuous output, or it may be triggered off and on from an external signal. Output waveforms of oscillators include sinusoids, square waves, pulse waveforms, triangular waveforms, and many others.

Oscillations in a circuit are produced by **positive feedback.** This is in contrast to **negative feedback,** which is used to stabilize amplifiers and other circuits, as we have seen for the past several chapters. However, what is intended as negative feedback may turn out to be positive feedback because of unwanted phase shifts in the circuit. Sooner or later, everyone working with electronic circuits will encounter spurious oscillations occurring because of unintended positive feedback. Our focus throughout this chapter is on the *planned* use of positive feedback to establish and maintain oscillations of a specified type.

Barkhausen Criterion

The concept of the **Barkhausen criterion** as the basis for producing steady-state oscillations is developed in this section. The development is based on a simplified intuitive approach as opposed to a more rigorous mathematical basis.

Consider the block diagram of Figure 17–1 containing a forward gain block of value A and a feedback block of value β. This configuration differs from the negative feedback system of earlier chapters in two respects: (1) there is no subtraction or inversion of the input as was the case for negative feedback, and (2) there is no input provided for an external signal.

FIGURE 17–1
Block diagram of positive feedback loop.

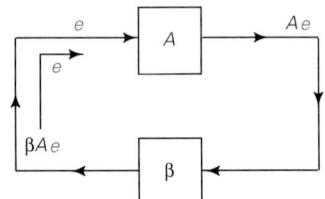

The diagram here, then, describes a **positive feedback loop,** for which self-sustained oscillations *may be* generated. Understand that we are observing only signal conditions in the loop. The energy required to sustain the oscillations is provided by the power supplies for the active components.

Assume that somehow a signal e appears at the input to the A block. This signal is amplified and appears at the output as Ae. This output is multiplied by β, and the output of the β-block is then $A\beta e$. To sustain the oscillation, this output must equal the value assumed at the beginning. Thus

$$A\beta e = e \qquad\qquad \textbf{(17–1)}$$

We now define the loop gain G_L as

$$\boxed{G_L = A\beta} \qquad\qquad \textbf{(17–2)}$$

Substituting this definition in (17–1) and canceling the common *e* factors, we determine the criterion for sustaining oscillations as

$$G_L = 1$$ **(17–3)**

This results in an expression for the Barkhausen criterion. Stated in words, it says that *to sustain oscillations in a loop, the net loop gain must be equal to unity.*

From the point of view of ac circuit theory, the Barkhausen criterion may be restated with the following two separate conditions:

Loop gain magnitude = 1 **(17–4a)**
Loop phase shift = $n \times 360°$ **(17–4b)**

where *n* is any integer. In practice, these two criteria are usually employed in analyzing the conditions for oscillation in practical circuits. Note that (17–4b) applies for any integer multiple of 360°, including 0°. Thus, if the loop gain magnitude is unity, oscillations will occur if the phase shift around the loop is 0°, 360°, −360°, 720°, and so on.

The preceding discussion justifies the process of sustaining an oscillation, but the skeptical reader may wonder how the process was started. From where did the assumed signal *e* arise?

Actually, it turns out that noise in the circuit is sufficient to start the process. In practical oscillators, the loop gain will be slightly greater than unity at very small-signal levels so that the oscillation, once started, will grow in amplitude. However, as the level grows, the loop gain drops to a level just sufficient to maintain oscillations.

It may be difficult to maintain the loop gain at exactly unity without some type of compensation. Some of the best oscillators employ both positive and negative feedback. The oscillations are established and maintained by the positive feedback. The negative feedback is used to control the amplitude of oscillations and to keep the loop gain maintained at exactly unity.

RC and LC Oscillators

Many of the standard oscillator circuits are classified as either *RC* or *LC* oscillators. The distinction lies in the types of components used to generate necessary phase shifts in the *β*-block of the loop. *RC* oscillators use resistance and capacitance to create the phase shift, while *LC* oscillators use inductance and capacitance.

As a general rule, *RC* oscillators are easier to design and have superior operating characteristics at relatively lower frequencies, while *LC* oscillators are superior at higher frequencies. There is no rigid boundary, and there are frequencies where either type would work as well as the other. As a rough guideline, *RC* oscillators work best below about 1 MHz or so.

EXAMPLE 17–1

A certain loop consists of an attenuator, an amplifier, and a phase shift circuit as shown in Figure 17–2. The output–input ratio of the attenuator is 0.04. The amplifier has a gain magnitude A and a phase shift θ at a certain frequency. At this reference frequency, the phase shift of the phase shift circuit is $-60°$, and its amplitude response is unity. What values of A and θ will produce oscillations in the circuit?

FIGURE 17–2

Loop of Example 17–1.

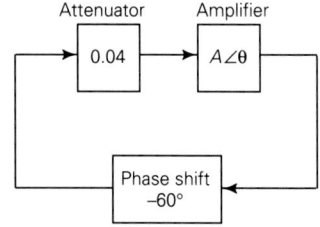

Solution

The magnitude-angle (or polar) form of the attenuator output–input ratio is denoted as $0.04\underline{/0°}$, and a similar form for the phase shift circuit is $1\underline{/-60°}$. The net loop gain is determined by multiplying the magnitudes and adding the angles; that is,

$$G_L = (0.04\underline{/0°})(A\underline{/\theta})(1\underline{/-60°})$$
$$= 0.04A\underline{/\theta - 60°} \tag{17–5}$$

Applying the Barkhausen criteria to the magnitude and angle, we have

$$0.04A = 1 \tag{17–6}$$

and

$$\theta - 60° = 0° \tag{17–7}$$

where $n = 0$ was chosen as the simplest case. These equations result in

$$A = 25 \tag{17–8}$$

and

$$\theta = 60° \tag{17–9}$$

Thus, oscillations will occur at the reference frequency if the amplifier gain is 25, and the amplifier phase shift is 60°.

The reader may be puzzled as to why we chose $n = 0$ in the phase constraint. It turns out that it does not really matter which *integer* value of n is selected since all the results are identical when the proper interpretation of the angle is made. For example, if 360° had been chosen in (17–7), the result would be 420°. However, an angle of 420° is equivalent to an angle of 60° with respect to the phasor rotation. Thus, the simplest form of the angle is usually chosen.

17–2 *RC* OSCILLATORS

Two of the most common types of *RC* oscillators are discussed in this section. As indicated earlier, *RC* oscillators are usually best for applications at relatively low frequencies.

The desired output waveforms from the circuits to be considered are sine waves. In practice, sine-wave oscillators will contain some distortion. The optimum design is one that minimizes the distortion and produces a nearly perfect sine wave.

Wien Bridge Oscillator

The most popular *RC* oscillator circuit employing the unity-gain feedback concept is the **Wien bridge oscillator,** whose basic form is shown in Figure 17–3. Any active device capable of producing a noninverting voltage gain $A = 3$ and negligible loading on the feedback circuit may be used. This gain requirement arises from the fact that the output voltage of the feedback circuit (on the left) is one-third of the input voltage (on the right) at the frequency of oscillation. At that frequency, the phase shift through the feedback circuit is zero. Thus the loop gain is $3 \times \frac{1}{3} = 1$ at the frequency of oscillation, and the Barkhausen criterion is met. Development of these results will be left as guided exercises (Problems 17–17 and 17–18).

The frequency of oscillation f_o is related to the circuit components by

$$f_o = \frac{1}{2\pi RC} \qquad\qquad \textbf{(17–10)}$$

FIGURE 17–3
Basic form of a Wien bridge oscillator.

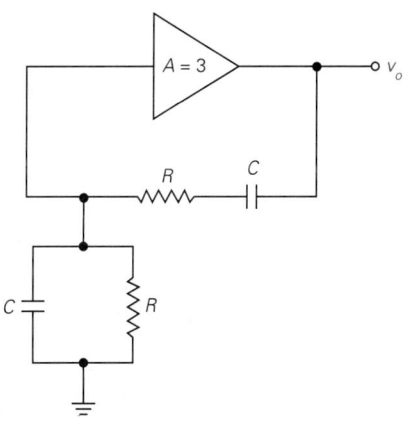

Note that two resistors have the same value R, and two capacitors have the same value C. If the R's and C's were different, the gain requirement and frequency of oscillation would be different than stated previously. Virtually all circuits built use equal values for the resistances and equal values for the capacitances.

One of the outstanding virtues of the Wien bridge oscillator is its ease of tuning. The two values of R can be realized by a ganged potentiometer, or the two values of C can be realized by a ganged capacitor. In either case, the particular parameter is changed by turning one control, and this provides a simple way to vary the frequency. Many general-purpose laboratory oscillators use the Wien bridge oscillator. If designed properly, the output waveform is nearly sinusoidal and contains very little distortion.

In practice, some means of stabilizing the gain to the exact required level is necessary. A number of ways utilizing such varied means as FET voltage-controlled resistors, diodes, thermistors, and nonlinear resistances have been employed. The process will be illustrated here with an incandescant lamp.

Consider the circuit of Figure 17–4. The required positive gain is achieved with a noninverting op-amp amplifier circuit. The gain of this amplifier is

$$A_{CL} = 1 + \frac{R_f}{R_i} \qquad \text{(17–11)}$$

To achieve $A_{CL} = 3$, it is necessary that $R_f = 2R_i$.

FIGURE 17–4
Wein bridge oscillator with stabilization using an incandescant lamp.

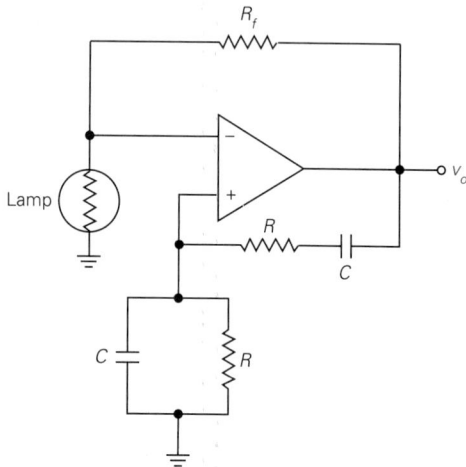

The gain can be stabilized by employing a nonlinear component as a portion of or all of either R_f or R_i. If the resistance *increases* with increasing output level, the nonlinear element should be part of or all of R_i. The gain will then be reduced if the output signal level tends to increase, since an increase in R_i corresponds to a lower gain. If the resistance *decreases* with increasing output level, the nonlinear element should be part of or all of R_f. In the circuit of Figure 17–4, it is assumed that the resistance of the incandescent lamp increases with the output level. This particular scheme is a very common one for Wien bridge oscillators.

Phase Shift Oscillator

Another popular *RC* oscillator is the **phase shift oscillator** shown in Figure 17–5. It turns out that the phase shift through this circuit at the desired frequency of oscillation is 180° so an inverting gain is required for this circuit. If there is negligible loading on the *RC* phase shift circuit; that is, if the input resistance to the amplifier can be assumed to be infinite, the magnitude of the output voltage of the phase shift circuit is $\frac{1}{29}$ of the input at the frequency of 180° phase shift. This establishes the requirement for the gain to be

$$-|A| = -29 \tag{17–12}$$

where the negative sign implies an inverting gain. The frequency f_o at which oscillations occur is

$$f_o = \frac{1}{2\pi\sqrt{6}\,RC} \tag{17–13}$$

Note that the three values of *R* are equal and the three values of *C* are equal.

FIGURE 17–5
Phase shift oscillator.

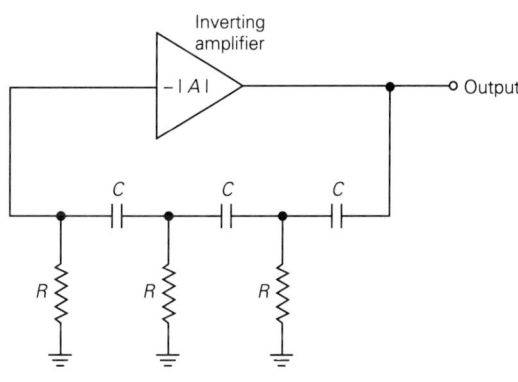

Inverting amplifier
$-|A|$
Output

A certain Wien bridge oscillator employs the circuit of Figure 17–4 with $R = 10$ kΩ and $C = 0.01$ μF (two values of each as shown). The feedback resistor is $R_f = 2$ kΩ, and a nonlinear lamp is used. Determine **(a)** the frequency of oscillation, and **(b)** the resistance of the lamp when stable oscillations are produced.

EXAMPLE 17–2

Solution

(a) The oscillation frequency is

$$f_o = \frac{1}{2\pi RC} = \frac{1}{2\pi \times 10^4 \times 0.01 \times 10^{-6}} = 1592 \text{ Hz} \qquad (17\text{--}14)$$

(b) For stable oscillations, $R_f = 2R_i$, or

$$R_i = \frac{R_f}{2} = \frac{2 \times 10^3}{2} = 1 \text{ k}\Omega \qquad (17\text{--}15)$$

This resistance will generally be greater than the "cold" resistance of the bulb as measured by an ohmmeter.

EXAMPLE 17–3   A certain Wien bridge oscillator is being designed to be tunable from 50 to 500 Hz with a dual potentiometer. Fixed values of 0.1 μF are employed for the two capacitors. Determine the range of resistances required to tune the specified range.

Solution
From (17–10), the resistance value is

$$R = \frac{1}{2\pi f_o C} = \frac{1}{2\pi \times 0.1 \times 10^{-6} f_o} \qquad (17\text{--}16)$$

Let R_{min} and R_{max} represent the minimum and maximum values of the resistance required. Since f_o appears in the denominator of (17–16), the *highest* frequency results in the *minimum* value of resistance, and vice versa. The resistance values are then calculated as follows:

$$R_{min} = \frac{1}{2\pi \times 0.1 \times 10^{-6} \times 500} = 3183 \ \Omega \qquad (17\text{--}17)$$

$$R_{max} = \frac{1}{2\pi \times 0.1 \times 10^{-6} \times 50} = 31.83 \text{ k}\Omega \qquad (17\text{--}18)$$

The two resistances must each be adjustable from 3183 Ω to 31.83 kΩ in order to tune the oscillator for the required frequency range. This can be achieved by two fixed resistances, each connected in series with a section of a dual potentiometer.

EXAMPLE 17–4   A certain phase shift oscillator is of the form shown in Figure 17–5. The resistance and capacitance values are $R = 10$ kΩ and $C = 0.01$ μF (three values of each). Determine the frequency of oscillation.

Solution
The frequency is

$$f_o = \frac{1}{2\pi\sqrt{6}RC} = \frac{1}{2\pi\sqrt{6} \times 10^4 \times 0.01 \times 10^{-6}} = 650 \text{ Hz} \qquad (17\text{--}19)$$

17–3 *LC* OSCILLATORS

In the frequency range from about 1 MHz or so up to several hundred megahertz, *LC* oscillators are more practical than *RC* oscillators. Over the years, quite a number of *LC* oscillators have been developed, and a complete treatment of this subject is best relegated to a textbook dealing with communications electronics. However, a few of the more popular *LC* oscillators are surveyed in this section.

Although it is possible to implement *LC* oscillators with op-amps in a limited frequency range, it turns out that the higher-frequency range where *LC* oscillators are best suited is the frequency range where op-amp usage is more limited. Consequently, most *LC* oscillators are implemented with BJTs and FETs.

Colpitts Oscillator

One of the most popular of the *LC* oscillators is the Colpitts oscillator, for which one implementation utilizing a BJT is shown in Figure 17–6. The resistors R_1, R_2, and R_E are used to establish the bias operating point, and C_E is used to bypass the emitter resistor as in an amplifier. The RF choke is an inductance whose reactance is large at the frequency of oscillation and would be an open circuit if it were not for losses. However, its losses create an effective resistance that, when coupled with any reflected resistance from the feedback circuit, acts as a dynamic load for the BJT. The reactance of C_3 is assumed to be small at the oscillation frequency.

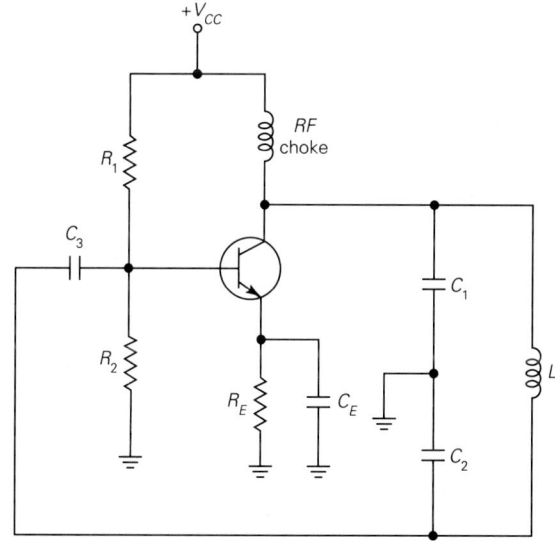

FIGURE 17–6
Colpitts oscillator employing a BJT.

The feedback circuit consists of L, C_1, and C_2. It turns out that the oscillation frequency f_o is given by

$$f_o = \frac{1}{2\pi\sqrt{LC_{\text{eq}}}}$$

(17-20)

where C_{eq} is the series equivalent capacitance of C_1 and C_2 as given by

$$C_{\text{eq}} = \frac{C_1 C_2}{C_1 + C_2}$$

(17-21)

EXAMPLE 17-5

A certain Colpitts oscillator of the form shown in Figure 17–6 has $C_1 = 100$ pF, $C_2 = 500$ pF, and $L = 60$ μH. Determine the frequency of oscillation.

Solution

The equivalent capacitance from (17–21) is

$$C_{\text{eq}} = \frac{C_1 C_2}{C_1 + C_2} = \frac{100 \times 10^{-12} \times 500 \times 10^{-12}}{600 \times 10^{-12}} = 83.33 \text{ pF}$$

(17-22)

The oscillation frequency from (17–20) is

$$f_o = \frac{1}{2\pi\sqrt{LC_{\text{eq}}}} = \frac{1}{2\pi\sqrt{60 \times 10^{-6} \times 83.3 \times 10^{-12}}} = 2.251 \text{ MHz}$$

(17-23)

Crystal Oscillator

A crystal oscillator is a circuit that employs a **crystal** to establish the oscillation frequency. The most common type of crystal is the **quartz crystal.** It has been included with LC oscillators because the equivalent circuit has the form of an LC circuit.

Quartz is a substance that exhibits a property called the **piezoelectric effect.** When a mechanical stress is applied across the crystal, a voltage is developed at the frequency of the mechanical vibration. Conversely, when a voltage is applied across the crystal, it vibrates at the frequency of the voltage.

The schematic symbol of a crystal is shown in Figure 17–7(a), and an equivalent-circuit model is shown in Figure 17–7(b). The crystal exhibits both a series resonant frequency and a parallel resonant frequency. The parallel resonant frequency is higher than the series resonant frequency.

An oscillator circuit employing a crystal is shown in Figure 17–8. This particular circuit is a modified Colpitts oscillator, and the parallel resonant mode of the crystal is utilized.

FIGURE 17–7
Schematic symbol for a quartz crystal
and an equivalent-circuit model.

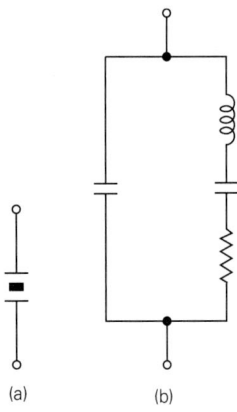

(a)   (b)

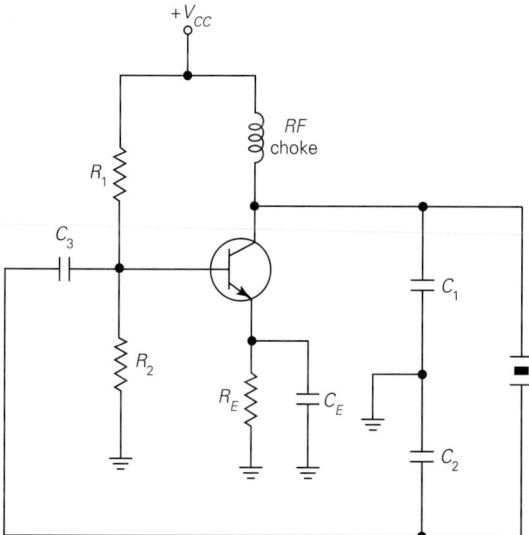

FIGURE 17–8
Crystal oscillator based on Colpitts circuit.

The primary advantage of a crystal oscillator is the excellent frequency stability provided. Commercial broadcast transmitters utilize crystal oscillators with the crystal kept in a temperature stabilized oven to provide additional stability in the operating frequency.

Hartley Oscillator

A **Hartley oscillator** is shown in Figure 17–9. The primary difference between this circuit and the Colpitts oscillator is that the phase shift circuit of the Hartley

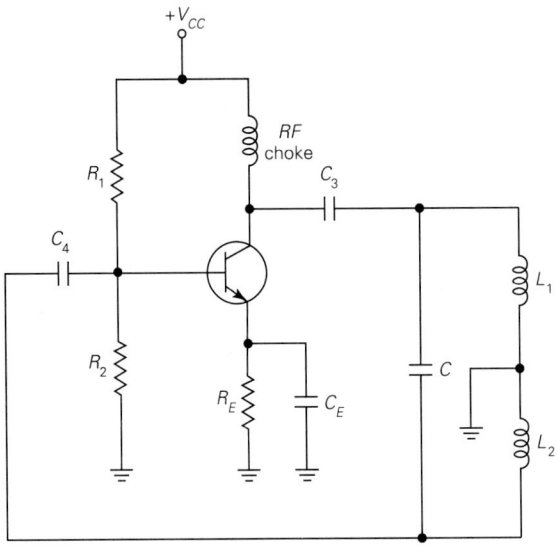

FIGURE 17–9
Hartley oscillator employing a BJT.

oscillator consists of two inductances, denoted by L_1 and L_2, and a single capacitor C.

The frequency of oscillation f_o of the Hartley oscillator is given by

$$f_o = \frac{1}{2\pi\sqrt{L_{eq}C}}$$ (17–24)

where L_{eq} is the equivalent inductance of the series combination of L_1 and L_2 as given by

$$L_{eq} = L_1 + L_2$$ (17–25)

EXAMPLE 17–6 A certain Hartley oscillator of the form shown in Figure 17–9 has $L_1 = 80\ \mu\text{H}$, $L_2 = 10\ \mu\text{H}$, and $C = 100\ \text{pF}$. Determine the frequency of oscillation.

Solution
The equivalent inductance from (17–25) is

$$L_{eq} = L_1 + L_2 = 80 + 10 = 90\ \mu\text{H}$$ (17–26)

The oscillation frequency from (17–24) is

$$f_o = \frac{1}{2\pi\sqrt{L_{eq}\,C}} = \frac{1}{2\pi\sqrt{90 \times 10^{-6} \times 100 \times 10^{-12}}} = 1.678\,\text{MHz} \quad \textbf{(17–27)}$$

Clapp Oscillator

An *LC* oscillator based on the Colpitts circuit, but with some improvement in performance, is the Clapp oscillator shown in Figure 17–10. A capacitor C_3 is placed in series with L, and it is chosen to be somewhat smaller than either C_1 or C_2. The frequency of oscillation is given by an expression of the form of (17–20), but C_{eq} in this case is determined from

$$\frac{1}{C_{eq}} = \frac{1}{C_1} + \frac{1}{C_2} + \frac{1}{C_3} \quad \textbf{(17–28)}$$

If $C_3 \ll C_1$ and $C_3 \ll C_2$, then to a first approximation

$$C_{eq} \approx C_3 \quad \textbf{(17–29)}$$

The capacitance C_3 is affected much less by transistor input and output capacitances than either C_1 or C_2. Consequently, the frequency of oscillation is

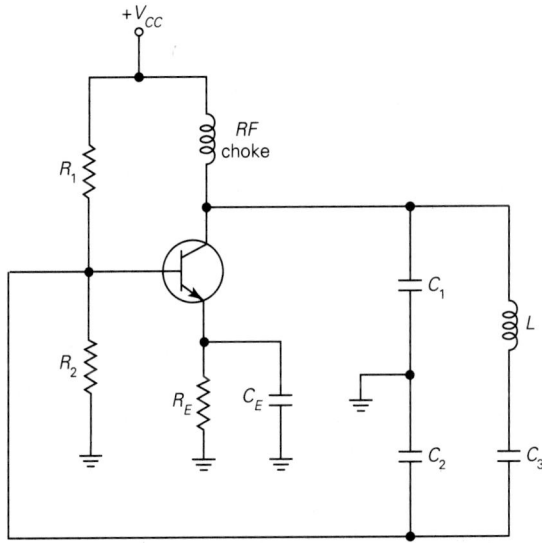

FIGURE 17–10
Clapp oscillator employing a BJT.

more stable with respect to uncontrollable circuit capacitances than with the Colpitts oscillator.

EXAMPLE 17-7

A certain Clapp oscillator of the form shown in Figure 17–10 has $C_1 = 220$ pF, $C_2 = 1000$ pF, $C_3 = 20$ pF, and $L = 50$ μH. Determine the frequency of oscillation.

Solution

The equivalent capacitance is determined from (17–28) with the following steps:

$$\frac{1}{C_{eq}} = \frac{1}{220 \times 10^{-12}} + \frac{1}{1000 \times 10^{-12}} + \frac{1}{20 \times 10^{-12}} \tag{17-30}$$

This leads to

$$C_{eq} = 18.00 \text{ pF} \tag{17-31}$$

The frequency of oscillation is then

$$f_o = \frac{1}{2\pi\sqrt{LC_{eq}}} = \frac{1}{2\pi\sqrt{50 \times 10^{-6} \times 18 \times 10^{-12}}} = 5.305 \text{ MHz} \tag{17-32}$$

17–4 SQUARE-WAVE OSCILLATORS

All the oscillator circuits considered up to this point have as their major objective a sinusoidal output waveform. However, there are many applications in which a nonsinusoidal waveform is desired. The most common nonsinusoidal signal is the **square wave**. A **bipolar square wave** is one that oscillates between positive and negative levels, and a **unipolar square wave** is one that oscillates between zero and a level of one polarity (either positive or negative).

Multivibrators

The term **multivibrator** is one that has long-standing usage in the electronics industry and dates back to the vacuum-tube era. It refers to certain circuits used for pulse and/or square-wave generation. There are three types of multivibrators: (1) **bistable** (or **flip-flop**), (2) **monostable** (or **one-shot**), and (3) **astable** (or **free-running**).

The **bistable multivibrator** has two stable states and will remain in either state until it receives a proper signal, in which case it will change states. Bistable multivibrators are used as memory devices in digital systems.

The **monostable multivibrator** has one stable state. Upon receiving a proper trigger, it will change states for a specified interval of time, but it will then return to the original stable state.

The **astable multivibrator** has no stable states. Consequently, it oscillates back and forth between states at a rate determined by the circuit parameters. It is, therefore, capable of generating a square wave in the process.

Op-Amp Square-Wave Generator

As a simple example of an astable multivibrator, consider the op-amp square-wave generator of Figure 17–11. As shown, the circuit produces an output square wave that oscillates between the positive and negative saturation levels. By the use of zener diodes, lower output magnitude levels can be established. However, the analysis here will be based on the two levels $+V_{sat}$ and $-V_{sat}$.

FIGURE 17–11
Square-wave generator using a single op-amp.

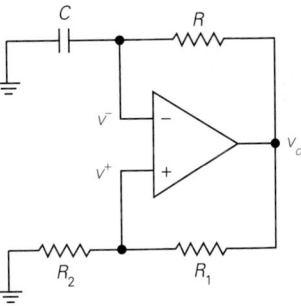

To analyze a multivibrator, it is usually necessary first to assume that steady-state conditions exist. The initial assumptions on levels may seem strange, but by moving through a complete cycle, these initial conditions may be verified.

In addition to the circuit diagram of Figure 17–11, refer to the waveforms of Figure 17–12 in the discussion that follows. Assume that a transition occurs at $t = 0$ with v_o assuming the positive state (i.e., $v_o = +V_{sat}$). By the voltage-divider rule, the voltage v^+ will be a fraction of v_o. We will define a feedback voltage V_f as

$$V_f = \frac{R_2}{R_1 + R_2} V_{sat} \tag{17–33}$$

Thus, $v^+ = V_f$ when $v_o = V_{sat}$ as shown in Figure 17–12(a) and (b), respectively. For reasons that will be clear later, assume that $v^- = -V_f$ at $t = 0$ as indicated in Figure 17–12(c). Note that v^- is the voltage across the capacitor at any given time.

The top portion of the circuit consists of a capacitance initially charged to a level $-V_f$, a resistance R, and a positive voltage $+V_{sat}$. The voltage across the capacitance will start climbing toward the level $+V_{sat}$ as indicated on Figure 17–12(c). However, when it reaches and slightly exceeds the level $+V_f$, the

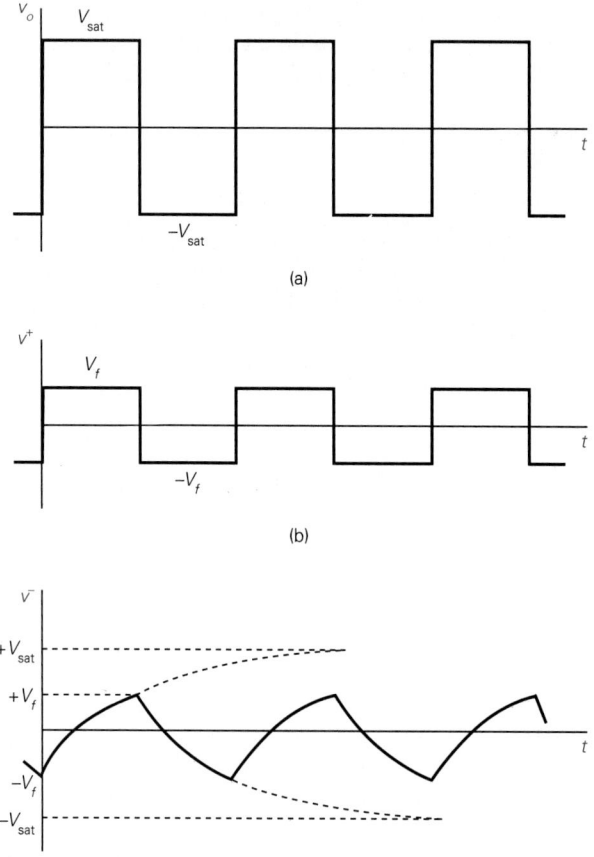

FIGURE 17–12
Waveforms of op-amp square-wave generator.

differential voltage $v_d = v^+ - v^-$ becomes negative. This will cause the output v_o to abruptly change states, in which case $v_o = -V_{sat}$. In turn, the voltage v^+ changes to $v^+ = -V_f$. The capacitor voltage now starts discharging toward a level $-V_{sat}$ as shown in Figure 17–12(c). When it reaches the level $-V_f$, the circuit changes state again, and a full cycle is complete.

The derivation of the expression for a period is left as a guided exercise at the end of the chapter (Problem 17–19). The period T can be determined from the equation as

$$T = 2RC \ln \left(1 + \frac{2R_2}{R_1} \right) \tag{17–34}$$

where "ln" refers to the natural logarithm of the quantity on the right. Note that the result is independent of the actual value of V_{sat}. The frequency of oscillation f_o is

$$f_o = \frac{1}{T} \tag{17–35}$$

EXAMPLE 17–8

A certain op-amp multivibrator of the form shown in Figure 17–11 has $R = 200$ kΩ, $C = 0.01\ \mu F$, $R_1 = 3$ kΩ, and $R_2 = 1$ kΩ. Determine the period and frequency of oscillation.

Solution
The period is determined from (17–34) as

$$T = 2 \times 2 \times 10^5 \times 0.01 \times 10^{-6} \times \ln\left(1 + \frac{2 \times 1 \times 10^3}{3 \times 10^3}\right)$$

$$= 4 \times 10^{-3} \times \ln 1.667 \tag{17–36}$$
$$= 4 \times 10^{-3} \times 0.5110 = 2.044 \times 10^{-3}\ s = 2.044\ ms$$

The frequency is

$$f_o = \frac{1}{2.044 \times 10^{-3}} = 489.2\ Hz \tag{17–37}$$

EXAMPLE 17–9

Design an op-amp square-wave generator of the form shown in Figure 17–11 to generate a 1-kHz square-wave signal.

Solution
The required period T is

$$T = \frac{1}{f_o} = \frac{1}{10^3} = 10^{-3}\ s \tag{17–38}$$

We thus have the requirement that

$$2RC \ln\left(1 + \frac{2R_2}{R_1}\right) = 10^{-3} \tag{17–39}$$

There are four parameter values to select and only one equation to satisfy. Consequently, three of the four values can be selected arbitrarily.

As a reasonable approach, first consider R_1 and R_2 since the ratio R_2/R_1 appears in the logarithmic function. Refer to Figure 17–12 and equation (17–33). If $R_2 \ll R_1$, then $V_f \ll V_{sat}$, and the timing waveform barely crosses zero in both directions. This causes the period to be more sensitive to the op-amp open-loop gain and offset effects. At the other extreme, if $R_2 \gg R_1$, V_f is nearly as large as V_{sat}, and this results in a poorly defined point of triggering on the upper part of the exponential curve. Thus these extreme cases should be avoided. Although there is no optimum point, a ratio of R_2/R_1 between roughly about $\frac{1}{9}$ and 1 should suffice. These two limits correspond to $V_f = 0.1V_{sat}$ and $0.5V_{sat}$.

As an arbitrary choice in this range, we will select $R_2/R_1 = \frac{1}{2}$, which can be achieved with $R_1 = 20$ kΩ and $R_2 = 10$ kΩ (or other values with the same ratio). This corresponds to $V_f = V_{sat}/3$. When $R_2/R_1 = \frac{1}{2}$ is substituted in (17–39), the equation reduces to

$$2RC \ln (1 + 2 \times \tfrac{1}{2}) = 10^{-3} \qquad\qquad \textbf{(17–40)}$$

This simplifies to

$$RC = \frac{10^{-3}}{2 \ln 2} = \frac{10^{-3}}{2(0.6931)} = 0.7213 \times 10^{-3} \text{ s} \qquad\qquad \textbf{(17–41)}$$

We have now reduced the number of unknowns to two. Any RC product satisfying (17–41) will theoretically work. However, in the capacitance range involved, it is easier to adjust R than C. Consequently, the best approach is to select C and calculate the required value of R.

Let's try $C = 0.01$ μF, which is a common standard value. With this choice, R can be calculated as

$$R = \frac{0.7213 \times 10^{-3}}{0.01 \times 10^{-6}} = 72.13 \text{ kΩ} \qquad\qquad \textbf{(17–42)}$$

While this value of R is not a standard one, it can easily be achieved with a potentiometer alone or with a fixed resistance plus a potentiometer. In fact, the final design of a circuit such as this is best adjusted in the lab.

The resulting design based on Figure 17–11 is summarized as follows:

$$R_1 = 20 \text{ kΩ}$$
$$R_2 = 10 \text{ kΩ}$$
$$R = 72.13 \text{ kΩ}$$
$$C = 0.01 \text{ μF}$$

17–5 555 TIMER

Integrated circuits that generate square waves and perform other multivibrator functions are readily available. Probably the most popular of such circuits is the widely used 555 timer, which was first introduced by Signetics, Inc.

A block diagram of the 555 timer is shown in Figure 17–13. It consists of two comparators, a control flip-flop (bistable multivibrator), a discharge transistor (T_1), a reset transistor (T_2), and an output stage. It can be operated with a dc voltage ($+V_{CC}$) between +5 and +18 V.

A simplified description of the logic is as follows: (1) If the voltage at the threshold terminal (pin 6) slightly exceeds $2V_{CC}/3$, the output (pin 3) assumes a

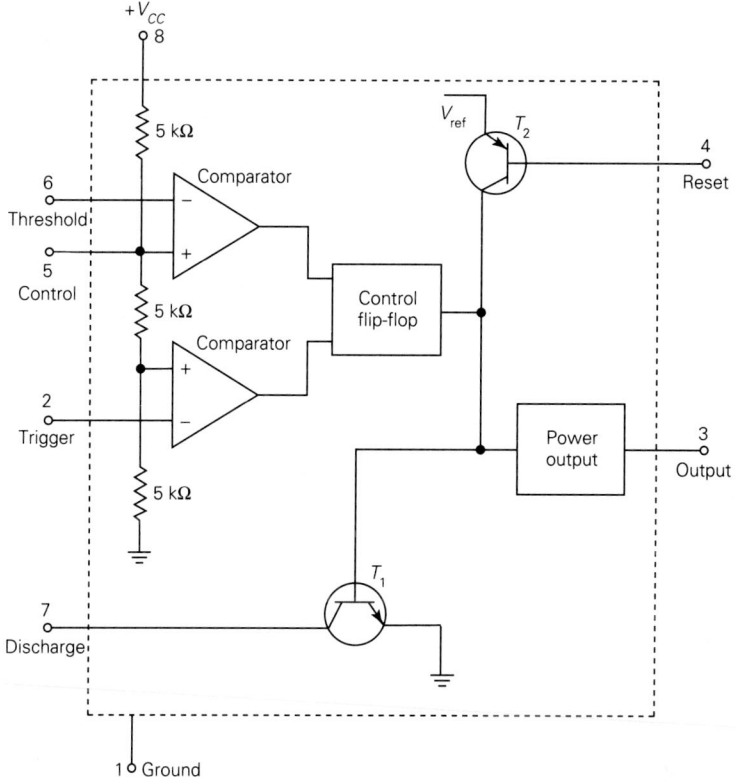

FIGURE 17–13
Block diagram of 555 timer IC.

low state (close to zero); and (2) if the trigger input (pin 2) drops slightly below $V_{CC}/3$, the output will assume a high state (close to V_{CC}). When the output is in a high state, the discharge transistor is at cutoff, and when the output is in a low state, the discharge transistor is saturated.

Astable Operation

The 555 connected for basic astable operation is shown in Figure 17–14. The only external components required for operation are R_A, R_B, C, and the 0.01-μF bypass capacitor.

Equivalent circuits for the timing process are shown in Figure 17–15, and pertinent waveforms are shown in Figure 17–16. Assume just prior to $t = 0$ that the capacitor voltage has dropped to a level slightly less than $V_{CC}/3$. The output v_o then assumes a high stage (i.e., $v_o = V_H$). When the output is in the high state, the discharge transistor (pin 7) is off. The equivalent circuit is shown in Figure

FIGURE 17–14
Astable multivibrator connection for 555.

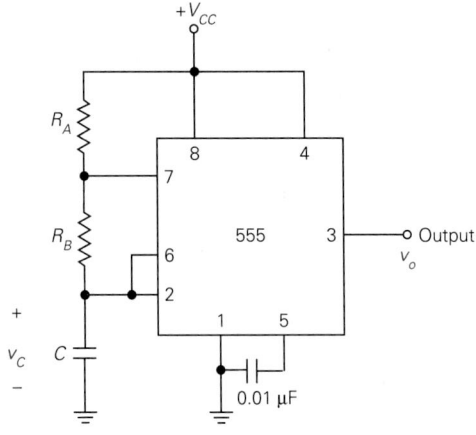

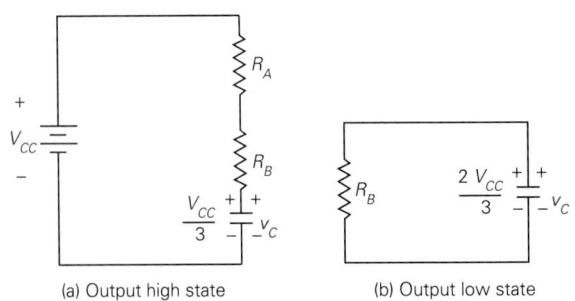

(a) Output high state   (b) Output low state

FIGURE 17–15
Timing equivalent circuits in 555 astable mode.

17–15(a). Based on this equivalent circuit, the capacitor voltage will rise exponentially toward V_{CC} (to which level it never reaches).

When v_C slightly exceeds $2V_{CC}/3$, the output v_o will assume the low state. When the output is in the low state, the discharge transistor becomes saturated, which effectively brings pin 7 to a ground level. The equivalent circuit then assumes the form shown in Figure 17–15(b). Based on this equivalent circuit, the capacitor voltage will discharge exponentially toward zero (which it never reaches).

When v_C drops slightly below $V_{CC}/3$, the output again assumes a high state. One complete cycle has now been completed.

Observe that the time constant during charging (output high level) is different from the time constant during the discharge process (output low level). Let τ_H represent the charging time constant, and let τ_L represent the discharge time constant. We have

$$\tau_H = (R_A + R_B)C \qquad \qquad (17\text{–}43)$$

and

$$\tau_L = R_B C \qquad \qquad (17\text{–}44)$$

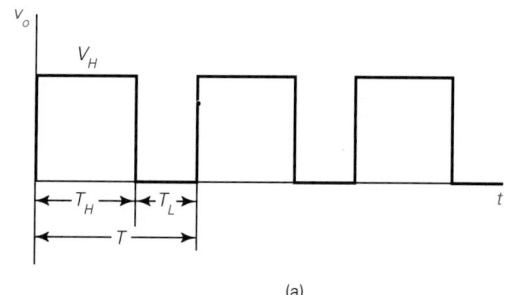

(a)

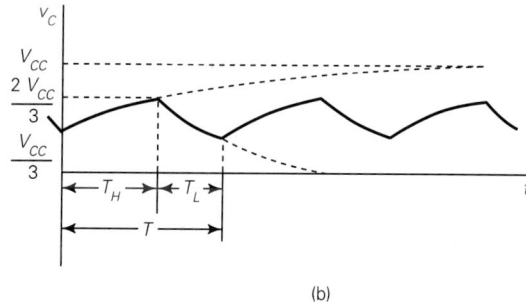

(b)

FIGURE 17–16
Astable 555 waveforms.

The total period T is composed of two intervals T_H and T_L as shown in Figure 17–16. It can be shown that

$$T_H = \tau_H \ln 2 = 0.693(R_A + R_B)C \qquad \text{(17–45)}$$

and

$$T_L = \tau_L \ln 2 = 0.693R_BC \qquad \text{(17–46)}$$

The total period is then

$$T = T_H + T_L = 0.693(R_A + 2R_B)C \qquad \text{(17–47)}$$

The frequency $f_o = 1/T$ is

$$f_o = \frac{1.443}{(R_A + 2R_B)C} \qquad \text{(17–48)}$$

Note from (17–45) and (17–46) that $T_H > T_L$; that is, the high interval is always longer than the low interval. Depending on the application, this property may or may not be significant. By choosing R_A to be small compared with R_B, the difference in the two intervals may be made to be very small. However, there is a limit to how small R_A can be made. During the low output interval, pin 7 assumes ground potential, and the power supply voltage is directly across R_A. The power dissipated in R_A during this interval is V_{CC}^2/R_A, and the power rating of the resistor will limit the minimum value of R_A.

A convenient parameter in characterizing the period is the **duty cycle *d*.** It will be defined as

$$d = \frac{T_H}{T} \times 100\% = \frac{T_H}{T_H + T_L} \times 100\% \qquad (17\text{-}49)$$

Stated in words, the duty cycle is the percentage of the period in which the output assumes the high state.

By substituting (17–45) and (17–46) in (17–49), the duty cycle can be expressed as

$$d = \frac{R_A + R_B}{R_A + 2R_B} \times 100\% \qquad (17\text{-}50)$$

Monostable Operation

The 555 connected for monostable operation is shown in Figure 17–17. The stable state is with the output voltage low. The trigger circuit consisting of a diode, a resistance R_t, and a capacitance C_t, is connected to the trigger input terminal (pin

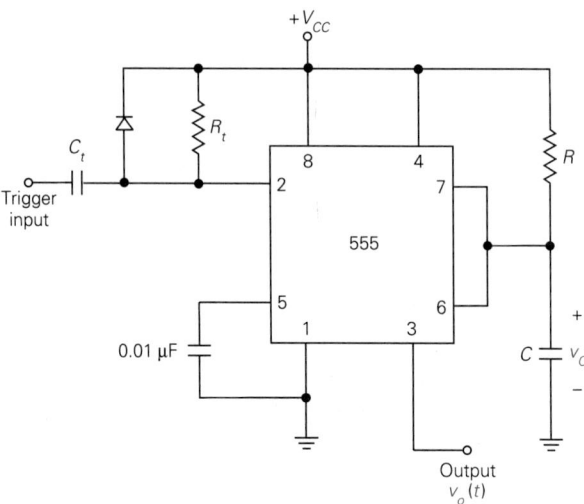

FIGURE 17–17
Monostable multivibrator connection for 555.

2). The timing circuit consists of a single resistance R and a capacitance C. The time constant τ is

$$\tau = RC \qquad (17\text{--}51)$$

An equivalent circuit for the timing process is shown in Figure 17–18, and pertinent waveforms are shown in Figure 17–19. Assume that the output v_o is initially in its low state. The discharge transistor is thus on, and this causes v_C to be at ground potential.

FIGURE 17–18
Timing equivalent circuit in 555 mono-stable mode.

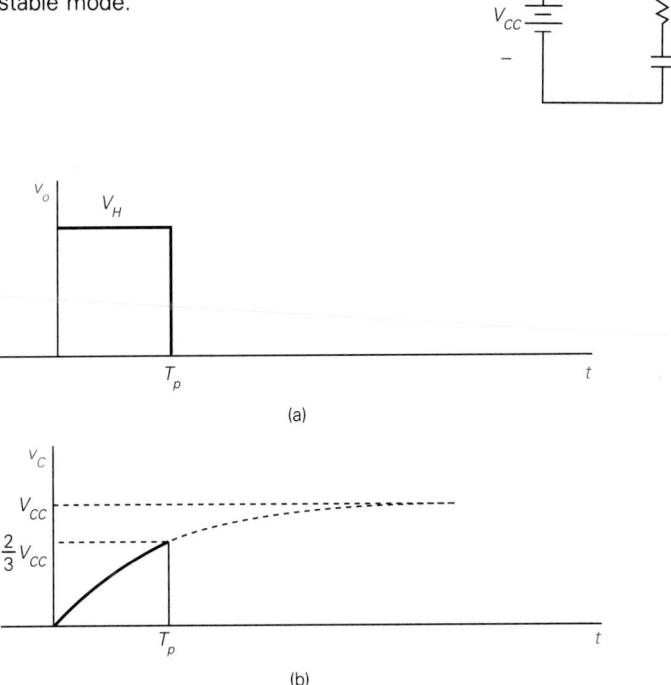

FIGURE 17–19
Monostable 555 waveforms.

At $t = 0$, a negative pulse of sufficient magnitude to drop the voltage at the trigger input below $V_{CC}/3$ is received. The output then assumes the high state (i.e., $v_C = V_H$), as shown in Figure 17–19. The discharge transistor now assumes an open-circuit condition, and the timing circuit of Figure 17–18 is applicable. The capacitor voltage charges exponentially toward a final value of V_{CC} (which it will never reach). Note that v_C is also the voltage applied to the upper comparator. When this voltage reaches a level slightly greater than $2V_{CC}/3$, the output drops to the low state, and the discharge transistor is turned on. The capacitor is quickly discharged through the transistor, and the circuit is returned to its stable state.

The pulse width T_p is the time required for the voltage across the capacitor to rise from zero to $2V_{CC}/3$. It can be shown that T_p is

$$\boxed{T_p = \tau \ln 3 = RC \ln 3 = 1.1RC} \qquad \text{(17–52)}$$

The exact value of the trigger circuit time constant $R_t C_t$ is not critical, but it should be small compared with the pulse width T_p.

EXAMPLE 17–10

A 555 timer connected for astable operation as shown in Figure 17–14 has $R_A = 1\ \text{k}\Omega$, $R_B = 4\ \text{k}\Omega$, and $C = 0.02\ \mu\text{F}$. Determine the high and low timing intervals, the total period, the frequency, and the duty cycle.

Solution

The high and low timing intervals are determined from (17–45) and (17–46) as

$$\begin{aligned} T_H &= 0.693(R_A + R_B)C \\ &= 0.693(10^3 + 4 \times 10^3) \times 0.02 \times 10^{-6} = 69.3\ \mu\text{s} \end{aligned} \qquad \text{(17–53)}$$

and

$$\begin{aligned} T_L &= 0.693R_B C \\ &= 0.693 \times 4 \times 10^3 \times 0.02 \times 10^{-6} = 55.4\ \mu\text{s} \end{aligned} \qquad \text{(17–54)}$$

The total period T is

$$T = 69.3 + 55.4 = 124.7\ \mu\text{s} \qquad \text{(17–55)}$$

The frequency f_o is

$$f_o = \frac{1}{124.7 \times 10^{-6}} = 8.019\ \text{kHz} \qquad \text{(17–56)}$$

The duty cycle d is

$$d = \frac{T_H}{T} \times 100\% = \frac{69.3\ \mu\text{s}}{124.7\ \mu\text{s}} \times 100\% = 55.6\% \qquad \text{(17–57)}$$

EXAMPLE 17–11

A 555 timer connected for monostable operation as shown in Figure 17–17 has $R = 10\ \text{k}\Omega$ and $C = 0.1\ \mu\text{F}$. Determine the pulse width.

Solution

From (17–52), the pulse width T_p is

$$T_p = 1.1RC = 1.1 \times 10^4 \times 0.1 \times 10^{-6} = 1.1\ \text{ms} \qquad \text{(17–58)}$$

EXAMPLE 17-12   Design a 555 astable circuit to produce a 500-Hz square wave with a duty cycle of 60%.

Solution

The pertinent design equations involve T_H and T_L as given by (17–45) and (17–46). Since T_H involves both R_A and R_B, while T_L involves only R_B, it is easier to work with the latter timing interval first. The period T is given by

$$T = \frac{1}{f_o} = \frac{1}{500} = 2 \text{ ms} \tag{17-59}$$

The values of T_H and T_L are given by

$$T_H = 0.6T = 0.6 \times 2 = 1.2 \text{ ms} \tag{17-60}$$
$$T_L = T - T_H = 2 - 1.2 = 0.8 \text{ ms} \tag{17-61}$$

From (17–46), we have the requirement that

$$0.693 R_B C = 0.8 \times 10^{-3} \tag{17-62}$$

or

$$R_B C = 1.154 \times 10^{-3} \tag{17-63}$$

The value of C will be arbitrarily selected as $C = 0.1 \ \mu$F. The value of R_B is then determined as

$$R_B = \frac{1.154 \times 10^{-3}}{0.1 \times 10^{-6}} = 11.54 \text{ k}\Omega \tag{17-64}$$

From (17–45), we have the requirement that

$$0.693(R_A + R_B)C = 1.2 \times 10^{-3} \tag{17-65}$$

or

$$(R_A + R_B)C = 1.732 \times 10^{-3} \tag{17-66}$$

Substituting $C = 0.1 \ \mu$F leads to

$$R_A + R_B = \frac{1.732 \times 10^{-3}}{0.1 \times 10^{-6}} = 17.32 \text{ k}\Omega \tag{17-67}$$

Finally, $R_B = 11.54$ kΩ is substituted in (17–67), and this leads to

$$R_A = 17.32 - 11.54 = 5.78 \text{ k}\Omega \tag{17-68}$$

To summarize, the circuit will be of the form shown in Figure 17–14 with the following parameter values:

$$R_A = 5.78 \text{ k}\Omega$$
$$R_B = 11.54 \text{ k}\Omega$$
$$C = 0.1 \ \mu\text{F}$$

17–6 UNIJUNCTION TRANSISTOR AND OSCILLATOR

Because of its importance in generating nonsinusoidal waveforms, the operation of a unijunction transistor has been delayed to the present chapter. Along with the principle of operation, application of this device to a representative oscillator circuit is considered.

Unijunction Transistor

A diagram suggesting the layout of a **unijunction transistor (UJT)** is shown in Figure 17–20(a). The UJT has two semiconductor regions and three external terminals. The three external terminals consist of one emitter terminal and two base terminals. The p region is heavily doped, but the n region is lightly doped. Because of the light doping in the emitter region, the resistance between the two bases is relatively high when the emitter is open.

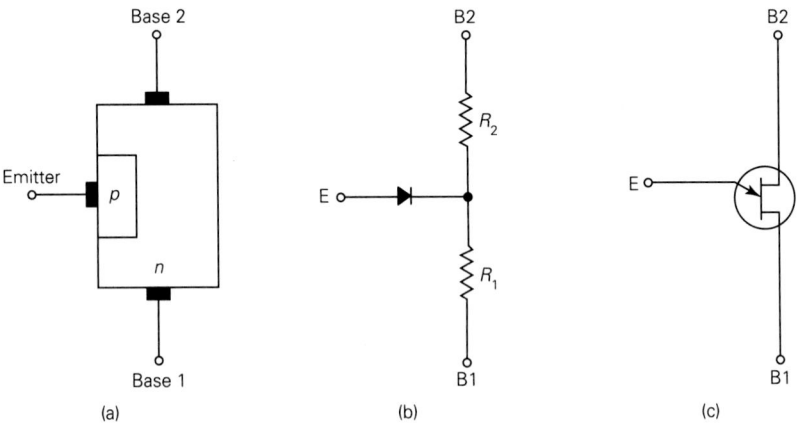

FIGURE 17–20
UJT and schematic symbol.

An equivalent circuit when the emitter is open is shown in Figure 17–20(b), and the schematic symbol is shown in Figure 17–20(c). The net resistance between the two bases when the emitter is open is called the **interbase resistance** and is denoted as R_{BB}. It is observed from the circuit diagram that

$$R_{BB} = R_1 + R_2 \tag{17–69}$$

Assume that a voltage V_{BB} is connected between the two bases as shown in Figure 17–21. The voltage V_1 across R_1 will be

$$V_1 = \frac{R_1}{R_1 + R_2} V_{BB} = \eta V_{BB} \tag{17–70}$$

FIGURE 17–21
Connection of a voltage between the two
bases of a UJT.

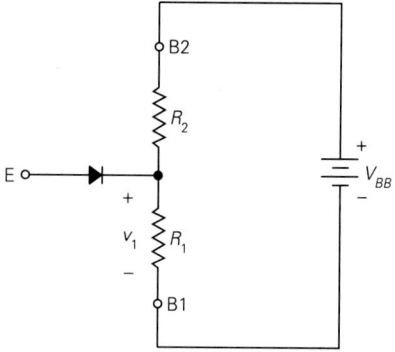

where η is the voltage-divider ratio preceding V_{BB}. The quantity η is called the
intrinsic standoff ratio. It is the fraction of the total voltage between the two bases
that appears across R_1. A typical value of η is from about 0.5 to 0.8.

To illustrate the action of a UJT, consider the circuit of Figure 17–22. Along
with a voltage V_{BB} connected between the two bases, an adjustable voltage V_{EE}
is connected between emitter and the lower base. When V_E is less than ηV_{BB}, the
emitter diode is reverse biased and the high resistance between the two bases is
unchanged. However, when v_E is increased slightly above the level ηV_{BB}, holes
from the heavily doped p region are injected into the lower half of the UJT and
create a strong conducting path. The net effect is to decrease the resistance R_1
to a very low value. The voltage v_E likewise drops to a low value.

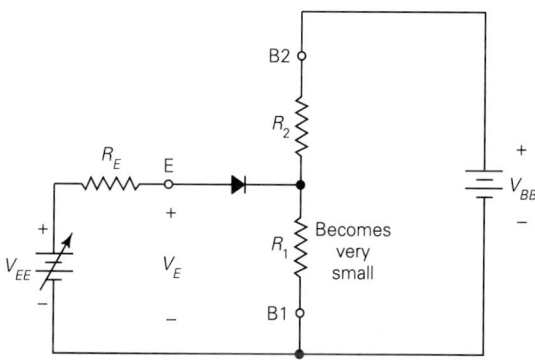

FIGURE 17–22
Circuit for illustrating UJT operations.

Once conduction is initiated, the UJT remains in a heavily conducting state
even if V_{EE} is reduced over a reasonable range. The UJT will drop out of the
strong conduction state if the emitter current is reduced below a specified **valley
current.**

UJT Relaxation Oscillator

The schematic diagram of a sawtooth relaxation oscillator circuit utilizing a UJT is shown in Figure 17–23. Assume initially that the voltage across the capacitor is zero. The UJT will then be in the open state, and the voltage across the capacitor will charge exponentially toward V_{BB}. However, when this voltage slightly exceeds ηV_{BB}, strong conduction occurs between emitter and B1. This discharges the capacitor voltage rapidly down to a level at which the emitter current is reduced below the valley current. At this point, the resistance between emitter and B1 changes back to the value R_1. This completes one cycle, and the process is repeated. The output voltage across the emitter then approximates a sawtooth waveform, as shown.

FIGURE 17–23

UJT sawtooth oscillator.

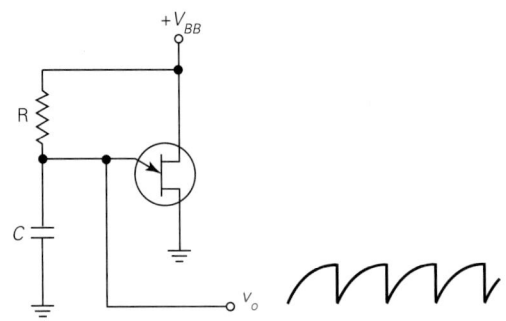

17–7 PSPICE EXAMPLES

PSPICE EXAMPLE 17–1

The circuit of Figure 17–24(a) represents a phase shift oscillator whose R and C values were as determined in Example 17–4. This particular form is convenient for use with an op-amp, and the gain is established by the ratio R_F/R_1. Develop a PSPICE simulation of the circuit.

Solution

The code using the 741 library model is shown in Figure 17–24(b). Oscillator circuits can be tricky to simulate since they do not have any external inputs. In a real circuit, the presence of thermal noise and other natural fluctuations such as the warm-up process are sufficient to start the oscillations. Such disturbances are not present in the simulation so it is necessary to "jump-start" the circuit.

Several points in the circuit could be used, but as an arbitrary choice, the output voltage V(7) was clamped at −1 V by the statement

```
.IC V(7)= -1
```

The output waveform for two cycles is shown in Figure 17–25. It can be observed that the waveform has not completely settled in this interval. The peak

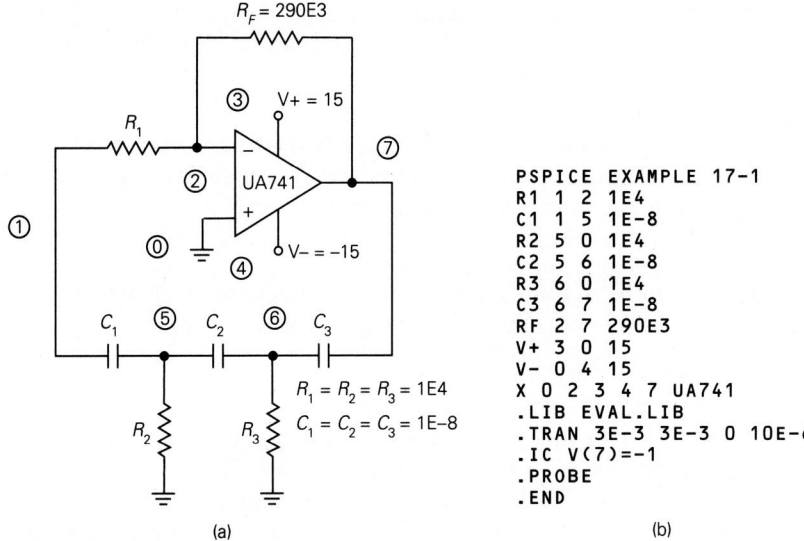

(a)

(b)

```
PSPICE EXAMPLE 17-1
R1 1 2 1E4
C1 1 5 1E-8
R2 5 0 1E4
C2 5 6 1E-8
R3 6 0 1E4
C3 6 7 1E-8
RF 2 7 290E3
V+ 3 0 15
V- 0 4 15
X 0 2 3 4 7 UA741
.LIB EVAL.LIB
.TRAN 3E-3 3E-3 0 10E-6
.IC V(7)=-1
.PROBE
.END
```

FIGURE 17–24
Circuit and code of PSPICE Example 17–1.

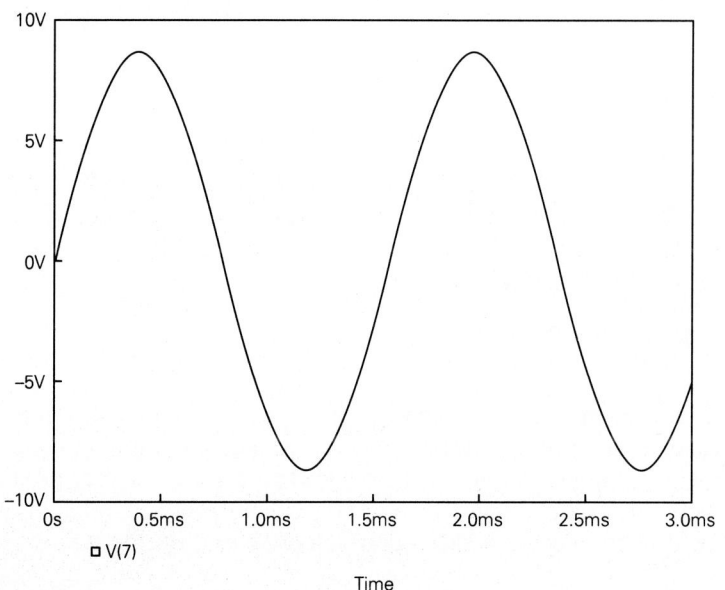

□ V(7)

Time

FIGURE 17–25
Output waveform of PSPICE Example 17–1.

value of the waveform obtained will depend on the value of the initial condition and on where it is placed. In an actual circuit, the amplitude will settle at that value that maintains a loop gain of unity.

The period of the waveform was measured as $T = 1.5916$ ms, which corresponds to a frequency of 628.3 Hz. This differs from the predicted frequency of 650 Hz by about 3.3%, which may result from the fact that the oscillation has not fully settled at the point of measurement.

PSPICE EXAMPLE 17–2

Develop a PSPICE program to simulate the op-amp square-wave generator design of Example 17–9. Use the LF411 op-amp library model.

Solution

The circuit diagram is shown in Figure 17–26(a), and the code is shown in Figure 17–26(b). When this program was prepared, the 741 op-amp was first used. The rise and fall times of the square wave were degraded, and the effective period increased because of the nonzero additional times required for the transition intervals. The 741 was then replaced by the 411, which has superior switching characteristics.

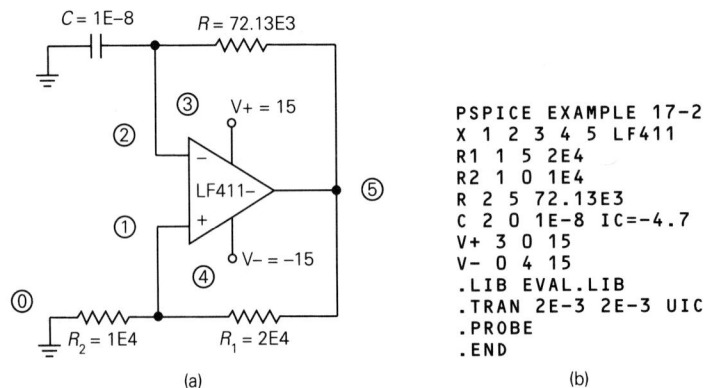

FIGURE 17–26
Circuit and code of PSPICE Example 17–2.

Once again, it is necessary to establish an initial condition in the circuit. In this case, a value of the initial capacitor voltage corresponding approximately to $-VF$ was used. This should make the corresponding waveforms start at the same time reference as in Figure 17–12. The value used was −4.7 V, which is about one-third of the negative saturation voltage based on the voltage divider R_1 and R_2.

The waveforms of the output voltage and the noninverting input voltage are shown in Figure 17–27. The period is 1 ms, corresponding to a frequency of 1 kHz, as expected.

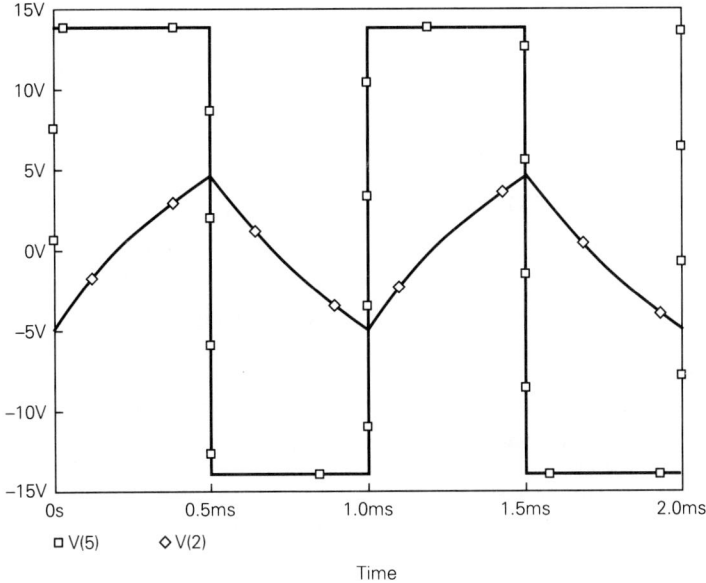

FIGURE 17–27
Waveforms of PSPICE Example 17–2.

PROBLEMS

Drill Problems

17–1. For the loop of Figure P17–1, determine the values of A and θ that will produce oscillations.

FIGURE P17–1

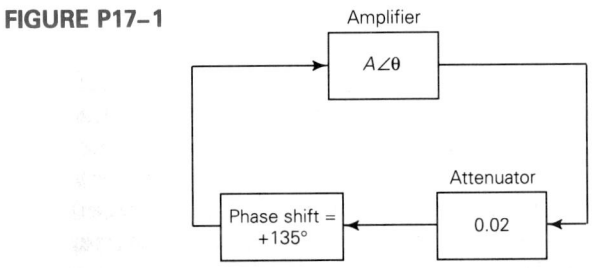

17–2. For the loop of Figure P17–2, determine the values of the attenuation ratio A (output–input ratio) and phase shift θ that will produce oscillations.

FIGURE P17–2

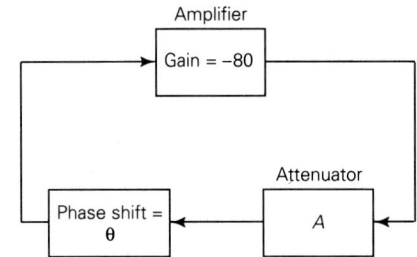

17–3. A certain Wien bridge oscillator employs the circuit of Figure 17–4 with $R = 20$ kΩ, $C = 0.005$ μF, and $R_f = 1$ kΩ. Determine **(a)** the frequency of oscillation, and **(b)** the resistance of the lamp when stable oscillations are produced.

17–4. Repeat the analysis of Problem 17–3 if $R = 100$ kΩ, $C = 0.2$ μF, and $R_f = 2.4$ kΩ.

17–5. A certain phase shift oscillator has $R = 200$ kΩ and $C = 0.2$ μF (three values of each). Determine the frequency of oscillation.

17–6. A certain phase shift oscillator has $R = 5.1$ kΩ and $C = 0.001$ μF (three values of each). Determine the frequency of oscillation.

17–7. A Colpitts oscillator of the form shown in Figure 17–6 has $C_1 = 50$ pF, $C_2 = 400$ pF, and $L = 20$ μH. Determine the frequency of oscillation.

17–8. Repeat the analysis of Problem 17–7 if $C_1 = 220$ pF, $C_2 = 0.001$ μF, and $L = 250$ μH.

17–9. A Hartley oscillator of the form shown in Figure 17–9 has $L_1 = 50$ μH, $L_2 = 10$ μH, and $C = 68$ pF. Determine the frequency of oscillation.

17–10. Repeat the analysis of Problem 17–9 if $L_1 = 2$ mH, $L_2 = 400$ μH, and $C = 0.001$ μF.

17–11. An op-amp multivibrator of the form shown in Figure 17–11 has $R = 10$ kΩ, $C = 0.01$ μF, $R_1 = 3.9$ kΩ, and $R_2 = 1.5$ kΩ. Determine the period and frequency of oscillation.

17–12. An op-amp multivibrator of the form shown in Figure 17–11 has $R = 120$ kΩ, $C = 0.2$ μF, $R_1 = 4.7$ kΩ, and $R_2 = 2$ kΩ. Determine the period and frequency of oscillation.

17–13. A 555 timer connected for astable operation as shown in Figure 17–14 has $R_A = 10$ kΩ, $R_B = 62$ kΩ, and $C = 0.1$ μF. Determine the high and low timing intervals, the total period, the frequency, and the duty cycle.

17–14. Repeat the analysis of Problem 17–13 if $R_A = 200$ Ω, $R_B = 820$ Ω, and $C = 0.001$ μF.

17–15. A 555 timer connected for monostable operation as shown in Figure 17–17 has $R = 200$ kΩ and $C = 2$ μF. Determine the pulse width.

17–16. Repeat the analysis of Problem 17–15 if $R = 150$ Ω and $C = 0.001$ μF.

Derivation Problems

17–17. (Requires steady-state ac phasor analysis.) Consider the Wien bridge oscillator of Figure 17–3. Let $\overline{V}_o$ represent the feedback circuit input voltage phasor (which is also the amplifier output voltage), and let $\overline{V}_1$ represent the feedback circuit output voltage phasor. Show that

$$\overline{V}_1 = \frac{j\omega RC\overline{V}_o}{1 - \omega^2 R^2 C^2 + j3\omega RC}$$

17–18. (Requires steady-state ac phasor analysis.) **(a)** Use the results of Problem 17–17, and obtain an expression for the loop gain G_L for the Wien bridge oscillator. **(b)** Apply the Barkhausen criterion to the results of **(a)** to derive the requirements for oscillation [i.e., $A = 3$ and $f_o = 1/(2\pi RC)$].

17–19. Consider the op-amp square-wave generator of Figure 17–11 along with the waveforms of Figure 17–12. **(a)** Referring to the process of capacitor charging between two states discussed in Section 16–3, write an equation for v^- based on the given initial value and the final value that *would* result if the capacitor were allowed to completely charge to V_{sat}. **(b)** From the expression obtained in **(a)**, set $v^- = V_f$, and determine the time t_1 corresponding to half of a cycle. **(c)** By the symmetry involved, determine the period T.

17–20. For the op-amp square-wave generator of Figure 17–11, show that if R_2 is selected to satisfy the condition that $R_2 = 0.86R_1$, the expression for the period reduces to

$$T = 2RC$$

Design Problems

17–21. A Wien bridge oscillator is to be designed to operate at a fixed frequency of 1 kHz. If C is selected as 0.01 μF, determine the value of R required.

17–22. Assume that in the design of Problem 17–21, the value of R is selected as 10 kΩ. Determine the value of C required.

17–23. A certain Wien bridge oscillator is being designed to be tunable from 200 to 1200 Hz. Fixed values of 0.2 μF are employed for the two capacitors. Determine the range of resistance required.

17–24. A certain Wien bridge oscillator is being designed to be tunable from 100 Hz to 1 kHz using a dual potentiometer, from which the two sections can be varied from zero to 10 kΩ. Fixed values of resistance R_f will be connected in series with the sections to provide the minimum resistance required. Determine the values of R_f and C.

17–25. The resistance R (in ohms) of a certain incandescent lamp is assumed to increase linearly with the voltage across it and can be approximated as

$$R = 100 + 300V_p$$

where V_p is the peak voltage across the lamp. The lamp is used to stabilize a Wien bridge oscillator of the form shown in Figure 17–4. **(a)** Determine the value of R_f required in a certain design in order to produce an oscillator peak output voltage of 9 V. **(b)** In a different design, the feedback resistor R_f is set to 1.6 kΩ. Determine the peak value of the oscillator output voltage.

17–26. The resistance R (in ohms) of a certain incandescant lamp is assumed to be a quadratic function of the voltage across it and can be approximated as

$$R = 200 + 50V_p^2$$

where V_p is the peak voltage across the lamp. The lamp is used to stabilize a Wien bridge oscillator of the form shown in Figure 17–4. **(a)** Determine the value of R_f required in a certain design in order to produce an oscillator peak output voltage of 6 V. **(b)** In a different design the feedback resistor R_f is set to 1.3 kΩ. Determine the peak value of the oscillator output voltage.

17–27. A phase shift oscillator is to be designed to operate at a fixed frequency of 1 kHz. If C is selected as 0.01 μF, determine the value of R required.

17–28. Assume that in the design of Problem 17–27, the value of R is selected as 10 kΩ. Determine the value of C required.

17–29. Design an op-amp square-wave generator of the form shown in Figure 17–11 to generate a 500-Hz square wave. In this particular design, select $R_1 = R_2$ and $C = 0.02 \ \mu F$.

17–30. Perform a new design of the circuit of Problem 17–29 by selecting the feedback voltage as $V_f = 0.1V_{sat}$. Choose $C = 0.02 \ \mu F$.

17–31. Design an op-amp square-wave generator of the form shown in Figure 17–11 to generate a 2-kHz square wave. In this particular design, select the feedback voltage as $V_f = 0.4V_{sat}$, and select $C = 0.01 \ \mu F$.

17–32. Perform a new design of the circuit of Problem 17–31 by selecting $R_1 = 4R_2$. Choose $C = 0.01 \ \mu F$.

17–33. Design a 555 astable circuit to produce a 1-kHz square wave with a duty cycle of 65%. Select $C = 0.02 \ \mu F$.

17–34. Perform a new design of the circuit of Problem 17–33 to achieve a duty cycle of 52%. Select $C = 0.02 \ \mu F$.

17–35. Design a 555 monostable circuit to produce a pulse width of 400 μs. Select $C = 0.01 \ \mu F$.

17–36. Design a 555 monostable circuit to produce a pulse width of 30 s. Select $C = 50 \ \mu F$.

Troubleshooting Problems

17–37. You are troubleshooting a Wien bridge oscillator of the type shown in Figure 17–4. The problem is that oscillation has stopped. Which *three* of the following *could be* the problem? **(a)** R too large; **(b)** R too small; **(c)** C too large; **(d)** C too small; **(e)** R_f too large; **(f)** R_f too small; **(g)** lamp shorted; **(h)** lamp open; **(i)** defective op-amp.

17–38. You are troubleshooting a Colpitts oscillator of the type shown in Figure 17–6. The problem is that the frequency is too low. Which *three* of the following *could be* the problem? **(a)** L too large; **(b)** L too small; **(c)** C_1 too large; **(d)** C_1 too small; **(e)** C_2 too small; **(f)** C_2 too large; **(g)** shorted RF choke; **(h)** R_1 too small; **(i)** R_2 too large.

18

HIGH-FREQUENCY EFFECTS IN AMPLIFIERS

OBJECTIVES

After completing this chapter, the reader should be able to:

- State the phasor forms for sinusoidal voltages and currents.
- Construct phasor impedance models for circuit components.
- Apply phasor analysis to determine steady-state voltages and currents in ac circuits.
- Explain the Miller effect, and determine the effective input capacitance of an inverting amplifier.
- Analyze a series resonant circuit to determine resonant frequency, bandwidth, and Q.
- Analyze a parallel resonant circuit to determine resonant frequency, bandwidth, and Q.
- Analyze a simplified tuned amplifier and discuss its major advantages with respect to a low-pass amplifier.
- Define and discuss the significance of the gain–bandwidth product of a transistor.
- Draw the schematic diagram for and define the parameters contained in the hybrid-pi model for a BJT.
- Draw the schematic diagram for and define the parameters contained in the hybrid-pi model for an FET.
- Sketch the waveform of a transistor current pulse, and identify and define the following parameters: delay time, rise time, turn-on time, storage time, fall time, and turn-off time.
- Define and discuss the significance of the gain–bandwidth product of an operational amplifier.
- Determine the closed-loop 3-dB bandwidth for either a noninverting or an inverting amplifier configuration.

- Discuss the significance of the slew rate parameter for an op-amp.
- Determine an op-amp output rise time produced by the slew rate effect.
- Determine the slew rate limiting frequency.
- Compare the relative effects of finite gain–bandwidth product and finite slew rate.
- Analyze some of the circuits in the chapter with PSPICE.

18–1 AC CIRCUIT ANALYSIS

A major portion of basic circuit theory is devoted to **steady-state ac circuit analysis.** This technique involves the representation of sinusoidal voltages and currents by **phasors.** Phasor voltages and currents associated with passive circuit elements are related to each other by **impedances** or **admittances.**

It is assumed by this point in the book that most readers will have studied basic ac circuit theory in some detail. The treatment here is intended to facilitate the use of that theory in analyzing certain types of electronic circuits. However, readers without a background in ac circuits should still be able to use the results given here in working with electronic circuits.

Phasor Form for Sinusoids

Consider the sinusoidal voltage v shown in Figure 18–1 with peak value V_p and a leading phase angle θ. This voltage may be expressed as

$$v = V_p \sin(\omega t + \theta) \tag{18–1}$$

where $\omega = 2\pi f = 2\pi/T$. Reviewing the terminology introduced in Chapter 8, ω is the angular frequency in radians/second, f is the repetition frequency in hertz, and T is the period in seconds.

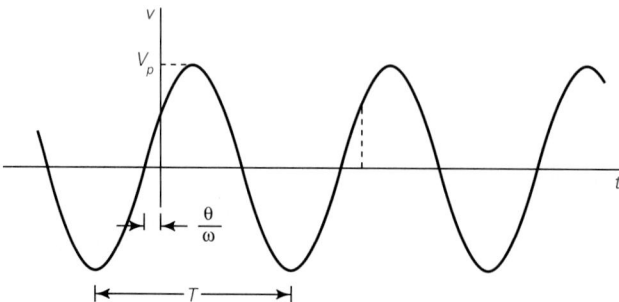

FIGURE 18–1
Instantaneous sinusoidal voltage with arbitrary phase angle.

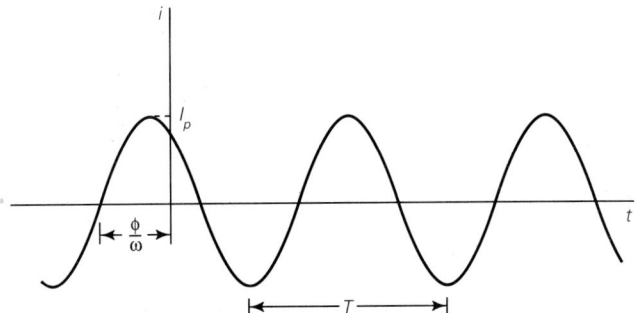

FIGURE 18–2
Instantaneous sinusoidal current with arbitrary phase angle.

Consider now the sinusoidal current i of Figure 18–2 with peak value I_p and a leading phase angle ϕ. This current may be expressed as

$$i = I_p \sin(\omega t + \phi) \tag{18-2}$$

A **phasor** is a vector quantity that represents a sinusoidal voltage or current in the so-called **complex plane.** This quantity has both length and angular orientation. The phasor representing the instantaneous voltage v is denoted as $\overline{V}$, and the phasor representing the instantaneous current i is denoted as $\overline{I}$.

The phasor form of $\overline{V}$ is expressed as

$$\boxed{\overline{V} = V_p \underline{/\theta}} \tag{18-3}$$

and is shown as one of the two quantities in Figure 18–3. The value V_p is the phasor length or **magnitude,** and θ is the angle measured in a counterclockwise sense from the positive x axis.

The phasor form of $\overline{I}$ is expressed as

$$\boxed{\overline{I} = I_p \underline{/\phi}} \tag{18-4}$$

FIGURE 18–3
Phasor forms of sinusoids of Figures 18–1 and 18–2.

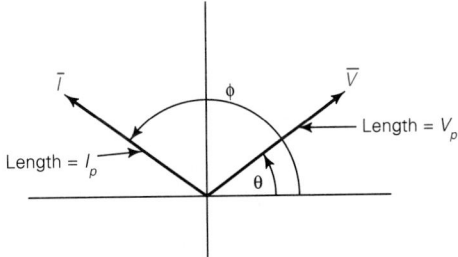

and is the other quantity of Figure 18–3. The value I_p is the length of this phasor, and ϕ is the angle measured from the positive x axis. In the terminology of complex numbers, the x axis is called the **real axis,** and the y axis is called the **imaginary axis.**

Impedance

One of the most important properties of steady-state ac phasor analysis is that the algebraic properties of resistive circuits may be extended to circuits containing capacitance and inductance by use of the impedance concept. For any **passive** linear circuit consisting of combinations of resistance, capacitance, and inductance, the impedance $\overline{Z}$ is defined as

$$\overline{Z} = \frac{\overline{V}}{\overline{I}} \qquad (18–5)$$

In general, impedance is a complex quantity having both magnitude and angle. However, it is best represented in **rectangular form.**

The rectangular form of an impedance may be expressed as

$$\overline{Z} = R + jX \qquad (18–6)$$

The quantity R is the real part of $\overline{Z}$, and it is the net **resistance.** The quantity X is the imaginary part of $\overline{Z}$, and it is the net **reactance.** The quantity $j = \sqrt{-1}$ is referred to as the j-operator. When displayed as a vector quantity, the real part represents the projection on the x, or **real,** axis, and the quantity preceded by j represents the projection on the y, or **imaginary,** axis.

An alternative way to characterize impedance is through the **admittance** Y. The admittance is

$$\overline{Y} = \frac{\overline{I}}{\overline{V}} = \frac{1}{\overline{Z}} \qquad (18–7)$$

Admittance can also be expressed in terms of real and imaginary parts in the form

$$\overline{Y} = G + jB \qquad (18–8)$$

where G is called the **conductance** and B is called the **susceptance.**

Each of the three circuit parameters may be represented in terms of a complex impedance whenever ac circuit analysis is employed. Refer to Figure 18–4 in the discussion that follows. The instantaneous notational forms for voltage and current, along with the circuit elements, are shown in the left-hand column. The instantaneous voltage–current relationship for a resistance obeys Ohm's law as

FIGURE 18–4

Resistance, capacitance, and inductance and their phasor impedances.

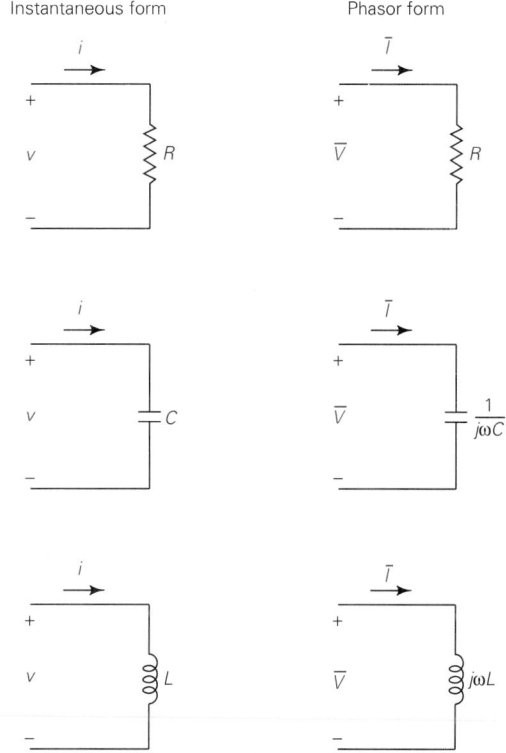

was seen in Chapter 1. However, the instantaneous voltage–current relationships for capacitance and inductance require differential and integral calculus, and these forms are not considered in this book. On the other hand, the steady-state phasor forms, which require only algebraic relationships with complex numbers, are given in the second column.

The first parameter shown is resistance R. The phasor relationship is

$$\overline{V} = R\overline{I} \tag{18–9}$$

The resistive impedance $\overline{Z}_R$ is simply

$$\overline{Z}_R = \frac{\overline{V}}{\overline{I}} = R \tag{18–10}$$

The next quantity is capacitance C. The phasor voltage–current relationship for a capacitance is

$$\overline{V} = \frac{1}{j\omega C}\overline{I} \tag{18–11}$$

The capacitive impedance $\bar{Z}_C$ is

$$\bar{Z}_C = \frac{1}{j\omega C} \qquad (18\text{--}12)$$

A basic property of the j-operator is that $1/j = -j$. Thus (18–12) can be expressed in the alternative form

$$\bar{Z}_C = \frac{-j}{\omega C} \qquad (18\text{--}13)$$

Thus, the *phasor impedance of a capacitor is a purely imaginary negative function.* Note that $\bar{Z}_C$ is simply $-j$ multiplied by the capacitive reactance as defined in Chapter 8.

The final quantity is inductance L. The phasor voltage–current relationship for an inductance is

$$\bar{V} = j\omega L \bar{I} \qquad (18\text{--}14)$$

The inductive impedance $\bar{Z}_L$ is

$$\bar{Z}_L = j\omega L \qquad (18\text{--}15)$$

Thus, *the phasor impedance of an inductor is a purely imaginary positive function.* Note that $\bar{Z}_L$ is simply j multiplied by the inductive reactance as defined in Chapter 8.

A positive j-operator corresponds to a positive (counterclockwise) rotation of 90°, and a negative j-operator corresponds to a negative (clockwise) rotation of 90°. This means that the sinusoidal voltage and current associated with an inductance or a capacitance have a 90° phase difference. The voltage and current for a resistance are, of course, in phase. Sinusoidal waveforms depicting these conditions, along with typical phasor forms, are shown in Figure 18–5. Thus, *the voltage across a pure capacitor lags the current by 90°, and the voltage across a pure inductor leads the current by 90°.*

Since the impedance of a pure capacitance or a pure inductance has only reactance, inductance and capacitance are referred to as **reactive elements.**

EXAMPLE 18–1 Express the complex form of the capacitive impedance of a 2-μF capacitor at a frequency of 1 kHz.

Solution
The capacitive impedance $\bar{Z}_C$ is

$$\bar{Z}_C = \frac{-j}{\omega C} = \frac{-j}{2\pi \times 10^3 \times 2 \times 10^{-6}} = -j79.58 \ \Omega \qquad (18\text{--}16)$$

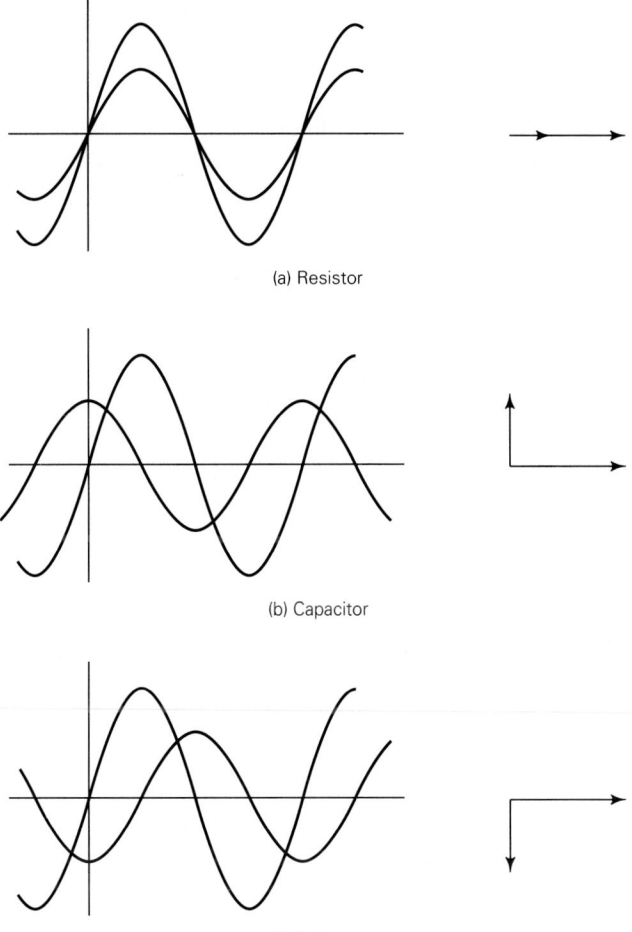

(a) Resistor

(b) Capacitor

(c) Inductor

FIGURE 18–5
Instantaneous and phasor forms for R, C, and L.

EXAMPLE	Express the complex form of the inductive impedance of a 3-mH inductor at a
18–2	frequency of 2 kHz.

Solution
The inductive impedance $\overline{Z}_L$ is

$$\overline{Z}_L = j\omega L = j2\pi \times 2 \times 10^3 \times 3 \times 10^{-3} = j37.70\ \Omega \qquad (18\text{–}17)$$

18–2 MILLER EFFECT

An important phenomenon known as the **Miller effect** is derived and discussed in this section. This effect is most significant in a high-gain inverting amplifier, and it results in a severe degradation of the high-frequency response.

Consider the block shown in Figure 18–6(a) representing an inverting amplifier with gain $-|A|$. (The gain has been indicated in this manner to emphasize the fact that it is an inverting gain.) The amplifier may represent any type of circuit in which gain and inversion occur; that is, it could be a BJT amplifier, an FET amplifier, or an op-amp. The input and output phasor voltages with respect to ground are denoted as $\overline{V}_i$ and $\overline{V}_o$, respectively. The output voltage is related to the input voltage by

$$\overline{V}_o = -|A|\overline{V}_i \tag{18–18}$$

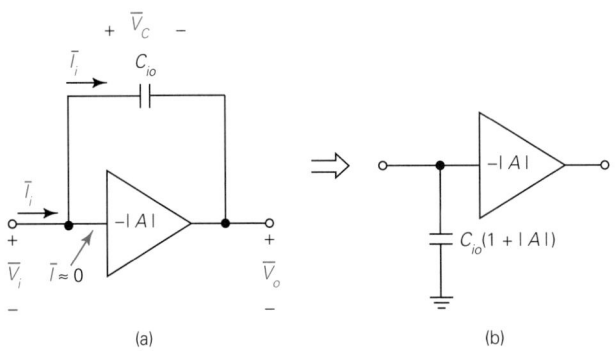

(a)　　　　　　　　(b)

FIGURE 18–6
Inverting amplifier used to develop Miller effect and equivalent capacitance *as referred to input.*

It will be assumed for this development that the input impedance of the amplifier is sufficiently high that the input current is negligible. With this assumption, we need only concentrate on the current drawn by the capacitance. The assumed input current $\overline{I}_i$ must then flow into the capacitance.

To determine the capacitive current, it is necessary to know the voltage across the capacitance. Let $\overline{V}_C$ represent this phasor voltage. A loop is formed by starting from ground, moving up to the left-hand side of the capacitance, moving through the capacitance to $\overline{V}_o$, and then back to ground. The KVL equation reads

$$-\overline{V}_i + \overline{V}_C + \overline{V}_o = 0 \tag{18–19}$$

or

$$\overline{V}_C = \overline{V}_i - \overline{V}_o \tag{18–20}$$

Substitution of (18–18) in (18–20) results in

$$\overline{V}_C = \overline{V}_i - (-|A|\overline{V}_i) = (1 + |A|)\overline{V}_i \tag{18–21}$$

The voltage across the capacitance is thus $1 + |A|$ times the input signal voltage. The current $\bar{I}_i$ is expressed as

$$\bar{I}_i = \frac{\bar{V}_C}{1/j\omega C_{io}} = j\omega C_{io}\bar{V}_C \qquad \text{(18–22)}$$

Substitution of (18–21) in (18–22) yields

$$\bar{I}_i = j\omega C_{io}(1 + |A|)\bar{V}_i \qquad \text{(18–23)}$$

Since the input current is purely capacitive in nature, $\bar{I}_i$ must be related to $\bar{V}_i$ by an equation of the form

$$\bar{I}_i = j\omega C_{eq}\bar{V}_i \qquad \text{(18–24)}$$

where C_{eq} is the equivalent capacitance seen at the input.

Comparing (18–23) with (18–24), we see that the equivalent input capacitance is

$$\boxed{C_{eq} = (1 + |A|)C_{io}} \qquad \text{(18–25)}$$

The capacitance between input and output is thus multiplied by $1 + |A|$. This is a direct result of the fact that the voltage across the capacitance is multiplied by $1 + |A|$.

The equivalent circuit *referred to the input of the amplifier* is shown in Figure 18–6(b). An equivalent capacitance of value $(1 + |A|)C_{io}$ to ground can be considered to exist across the input of the amplifier.

A typical amplifier will also have an actual shunt input capacitance C_{ig} to ground as shown in Figure 18–7(a). This capacitance simply adds to the Miller

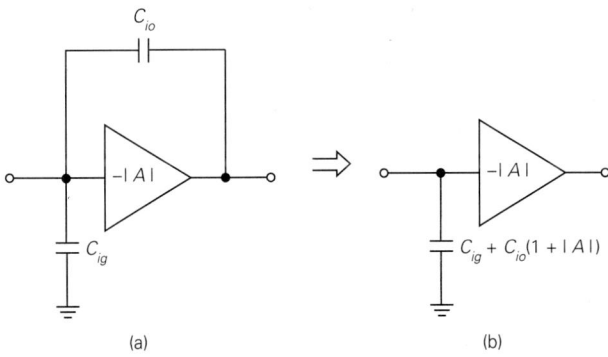

(a)   (b)

FIGURE 18–7
Combination of Miller effect capacitance and shunt input capacitance *as referred to input.*

effect capacitance so that the total equivalent capacitance C_{eq} is now

$$C_{eq} = C_{ig} + C_{io}(1 + |A|) \qquad (18–26)$$

An equivalent circuit *referred to the input* is shown in Figure 18–7(b).

Since the 3-dB upper frequency of a circuit with shunt capacitance is inversely proportional to the value of the capacitance, the enlarged capacitance seen at the input acts to reduce significantly the upper operating frequency range. This phenomenon is a serious limitation of an inverting amplifier with capacitive coupling between input and output.

EXAMPLE 18–3

A certain inverting amplifier has a gain $A = -99$. The amplifier has an input shunt capacitance of 8 pF and a capacitance between input and output terminals of 2 pF. Determine the net equivalent input capacitance.

Solution
From the data provided $C_{ig} = 8$ pF and $C_{io} = 2$ pF. The equivalent capacitance is

$$C_{eq} = C_{ig} + (1 + |A|)C_{io} = 8 + (1 + 99)2 = 8 + 200 = 208 \text{ pF} \qquad (18–27)$$

Note that the Miller effect capacitance alone is 200 pF, and this is 100 times as great as the actual input–output capacitance of 2 pF.

EXAMPLE 18–4

Assume that the amplifier of Example 18–3 is driven from a source whose internal resistance is 2 kΩ. Determine the 3-dB break frequency f_b.

Solution
Since the net capacitance represents a shunt effect, the circuit model is a basic low-pass form. Recall from earlier work that the 3-dB break frequency for such a circuit is

$$f_b = \frac{1}{2\pi RC} \qquad (18–28)$$

Substituting $R = 2$ kΩ and $C = C_{eq} = 208$ pF, we have

$$f_b = \frac{1}{2\pi \times 2 \times 10^3 \times 208 \times 10^{-12}} = 382.6 \text{ kHz} \qquad (18–29)$$

18–3 RESONANT CIRCUITS

As a necessary prelude to the introduction of tuned amplifier circuits, the basic properties of series and parallel resonant circuits are discussed in this section. The properties discussed are limited to those essential for a basic understanding of tuned amplifiers.

Series Resonant Circuit

In basic ac circuit texts, resonant circuits usually assume a single resistance. However, to make the work here applicable to some actual electronic circuits, the models assumed will contain several typical circuit resistances.

Consider the circuit of Figure 18–8(a) having a voltage source with internal resistance R_g and a load R_L. A series combination of a capacitance C and an inductance L is connected between the signal terminals and the load.

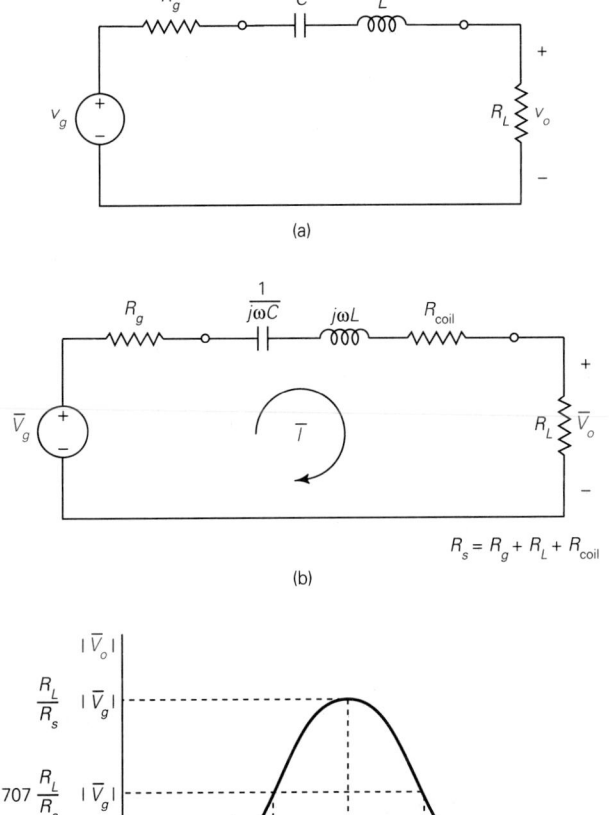

FIGURE 18–8
Series resonant circuit and general form of response.

In addition to the source and load resistances, the inductance (or "coil") also contains a certain resistance R_{coil}. This coil resistance normally cannot be determined from a simple dc ohmmeter check since the actual resistance of the coil typically varies with frequency. The coil resistance may be determined indirectly from the "quality factor" of the inductance with a so-called Q-meter, as will be discussed.

The phasor model of the circuit with the coil resistance indicated is shown in Figure 18–8(b). There are thus three series resistances in this circuit. For convenience in some later steps, we will define R_s as the net resistance; that is,

$$R_s = R_g + R_L + R_{coil} \qquad (18\text{--}30)$$

The phasor impedance Z of the series circuit is

$$\overline{Z} = R_g + \frac{1}{j\omega C} + j\omega L + R_{coil} + R_L = R_s + j\left(\omega L - \frac{1}{\omega C}\right) \qquad (18\text{--}31)$$

The phasor current $\overline{I}$ is

$$\overline{I} = \frac{\overline{V}_g}{\overline{Z}} = \frac{\overline{V}_g}{R_s + j(\omega L - 1/\omega C)} \qquad (18\text{--}32)$$

Assume that the desired output voltage is $\overline{V}_o$, which is given by

$$\overline{V}_o = R_L \overline{I} = \frac{R_L \overline{V}_g}{R_s + j(\omega L - 1/\omega C)} \qquad (18\text{--}33)$$

The magnitude of $\overline{V}_o$, indicated by $|\overline{V}_o|$, is given by

$$|\overline{V}_o| = \frac{R_L |\overline{V}_g|}{\sqrt{R_s^2 + (\omega L - 1/\omega C)^2}} \qquad (18\text{--}34)$$

where $|\overline{V}_g|$ is the magnitude of $\overline{V}_g$.

The variation of $|\overline{V}_o|$ as a function of the cyclic frequency $f = \omega/2\pi$ is shown in Figure 18–8(c). At low frequencies, $|\overline{V}_o|$ is small due to the large series capacitive reactance. At high frequencies, $|\overline{V}_o|$ is small due to the large series inductive reactance. However, there is a frequency at which the reactive effects cancel. This is the frequency at which the quantity in parentheses in (18–34) is zero. Denoting this radian frequency as ω_o and the corresponding cyclic frequency as $f_o = \omega_o/2\pi$, we have

$$\omega_o L - \frac{1}{\omega_o C} = 0 \qquad (18\text{--}35)$$

which leads to

$$\omega_o = \frac{1}{\sqrt{LC}} \qquad (18\text{--}36)$$

and

$$f_o = \frac{1}{2\pi\sqrt{LC}} \tag{18-37}$$

The frequency f_o is called the **series resonant frequency.** At the series resonant frequency, the impedance magnitude has its *minimum value,* the current has its *maximum value,* and the voltage across R_L has its *maximum value.* At f_o, the voltage across R_L from (18–34) is

$$|\overline{V}_o| = \frac{R_L}{R_s}|\overline{V}_g| \tag{18-38}$$

The response curve for $|\overline{V}_o|$ thus has a **bandpass form.**

The relative width of the resonant curve is an important design parameter since it determines the **selectivity** of the circuit. The selectivity is a relative measure of the circuit's ability to pass a desired bandpass signal while rejecting others outside the desired passband. If a circuit is too selective, it will reject a portion of the desired signal, while if it is not sufficiently selective, it will pass undesired signals.

Let f_1 and f_2 represent frequencies on the lower and upper sides of resonance at which the response drops by 3 dB. The relative magnitude level at these points is $1/\sqrt{2}(= 0.707)$ times the peak value. Let B represent the 3-dB bandwidth defined by

$$B = f_2 - f_1 \tag{18-39}$$

The selectivity of the circuit is often specified in terms of the Q (or "quality factor"). For this purpose, Q is defined as

$$Q = \frac{f_o}{B} \tag{18-40}$$

As Q increases, the ratio of f_o to B increases and the relative selectivity increases.

To distinguish the Q of the series circuit from that of the parallel circuit to be studied later, let Q_s represent the Q of the series circuit. It is shown in basic ac circuit texts that Q_s may be expressed in either of the following three ways:

$$Q_s = \frac{\omega_o L}{R_s} = \frac{1}{\omega_o R_s C} = \frac{1}{R_s}\sqrt{\frac{L}{C}} \tag{18-41}$$

where $\omega_o = 2\pi f_o$ and R_s was defined in (18–30).

Coil Q

In practice, the ac resistance of a coil is often determined indirectly from a measurement with a Q-meter. Let Q_{coil} represent a Q parameter for the coil itself, and it is defined by

$$Q_{coil} = \frac{\omega_o L}{R_{coil}} \qquad \text{(18–42)}$$

If (18–42) is compared with the first form of (18–41), it can be deduced that if R_{coil} were the only circuit resistance, Q_{coil} would be identical with Q_s. In practice, however, $R_s > R_{coil}$ and $Q_s < Q_{coil}$. Thus the circuit Q can never be greater than the coil Q but will almost always be smaller. Since R_{coil} is usually less predictable and adjustable than the other series resistances, the most desirable situation is to have the highest possible value of Q_{coil} and adjust the circuit Q with external resistances.

Parallel Resonant Circuit

Consider the circuit of Figure 18–9(a), which has a voltage source with internal resistance R_g and a load R_L. A parallel combination of a capacitance C and an inductance L is connected across the load.

To simplify the analysis of the circuit, a source transformation is first applied to v_g and R_g, and this series combination is converted to an equivalent parallel combination of a current source $i_g = v_g/R_g$ and the resistance R_g. Incidentally, many electronic circuit models actually start with the current source parallel form, as we will see later, but the development here started with the voltage source form to show the equivalence.

In addition to the source and load resistances, the inductance also contains a resistance R_{coil} as shown in Figure 18–9(b). The natural form of this resistance is a series parameter, but it is awkward to analyze the circuit with this form since the remainder of the circuit is a parallel structure.

The key to this dilemma is to replace the series RL combination by an equivalent parallel RL combination. This approach is most meaningful if $Q_{coil} \geq 10$. For this condition, and to within an error not exceeding 1% in the vicinity of resonance, the inductance is unchanged, but the equivalent *parallel* coil resistance R'_{coil} is approximately

$$R'_{coil} \simeq Q^2_{coil} R_{coil} \qquad \text{(18–43)}$$

The net parallel equivalent circuit is shown in phasor form in Figure 18–9(c). For convenience, we will define the net parallel resistance R_p as

$$R_p = R_g \| R_L \| R'_{coil} \qquad \text{(18–44)}$$

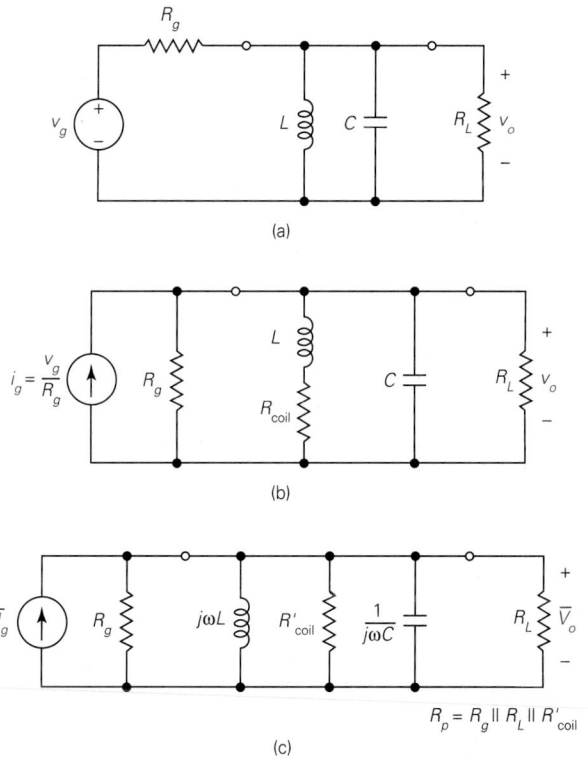

(a)

(b)

(c)

$$R_p = R_g \parallel R_L \parallel R'_{coil}$$

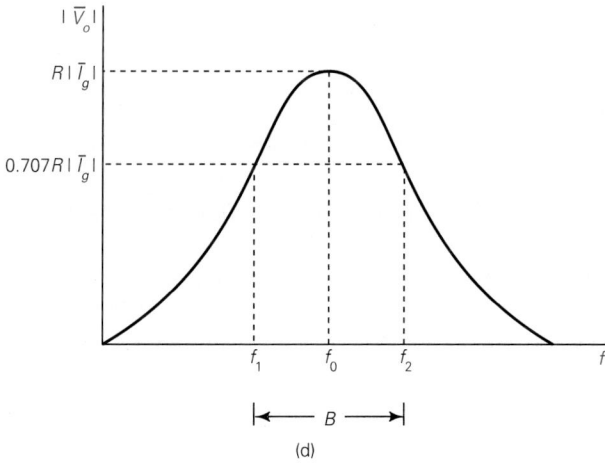

(d)

FIGURE 18–9
Parallel resonant circuit and general form of response.

The phasor admittance $\overline{Y}$ of the parallel circuit is

$$\overline{Y} = G_p + j\omega C + \frac{1}{j\omega L} = G_p + j\left(\omega C - \frac{1}{\omega L}\right) \qquad (18\text{--}45)$$

where

$$G_p = \frac{1}{R_p} \qquad (18\text{--}46)$$

The phasor voltage $\overline{V}_o$ is

$$\overline{V}_o = \frac{\overline{I}_g}{\overline{Y}} = \frac{\overline{I}_g}{G_p + j(\omega C - 1/\omega L)} \qquad (18\text{--}47)$$

The magnitude of $\overline{V}_o$, indicated by $|\overline{V}_o|$, is given by

$$|\overline{V}_o| = \frac{|\overline{I}_g|}{\sqrt{G_p^2 + (\omega C - 1/\omega L)^2}} \qquad (18\text{--}48)$$

where $|\overline{I}_g|$ is the magnitude of $\overline{I}_g$.

The variation of $|\overline{V}_o|$ as a function of f is shown in Figure 18–9(d). At low frequencies, $|\overline{V}_o|$ is small because of the shunting effect of the small inductive reactance. At high frequencies, $|\overline{V}_o|$ is small because of the shunting effect of the small capacitive reactance. However, there is a resonant frequency at which the reactive effects cancel. This frequency is determined by setting the quantity in parentheses in the denominator of (18–48) to zero. Denoting the resonant frequency as f_o, it is readily shown to be

$$f_o = \frac{1}{2\pi\sqrt{LC}} \qquad (18\text{--}49)$$

Thus the expression for the parallel resonant frequency is the same as for the series resonant frequency.

At the **parallel resonant frequency,** the impedance magnitude has its *maximum* value, and the voltage across the resonant circuit has its *maximum* value. At f_o, the voltage across the resonant circuit from (18–48) is

$$|\overline{V}_o| = \frac{|\overline{I}_g|}{G_p} = R_p|\overline{I}_g| \qquad (18\text{--}50)$$

As in the case of the series circuit, we may define B and Q as

$$B = f_2 - f_1 \qquad (18\text{--}51)$$

and

$$Q = \frac{f_o}{B} \qquad (18\text{--}52)$$

However, the expressions for determining the Q from the circuit elements are different for the parallel resonant circuit. Let Q_p represent the parallel circuit Q. It is shown in basic ac circuit texts that Q_p may be expressed in either of the following three ways:

$$Q_p = \frac{R_p}{\omega_o L} = \omega_o R_p C = R_p \sqrt{\frac{C}{L}} \qquad (18\text{--}53)$$

where $\omega_o = 2\pi f_o$ and R_p was defined in (18–44).

As in the case of the series RLC circuit, the maximum possible value of the net circuit Q is limited by Q_{coil}. If the driving source were an ideal current source ($R_g = \infty$), and if the load were an open circuit ($R_L = \infty$), $Q_p = Q_{\text{coil}}$. In general, however, $Q_p < Q_{\text{coil}}$.

Effect of Q on Response

For either the series or the parallel resonant circuits, the relative effect of the circuit Q on the response shape is illustrated in Figure 18–10. It is assumed here that Q_1 is the smallest of the three values and Q_3 is the largest. The smallest value

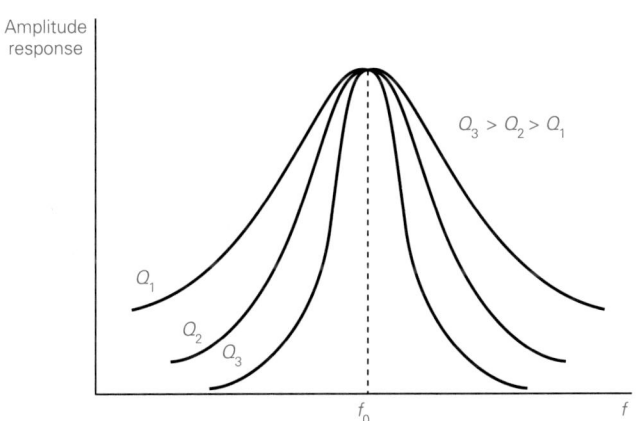

FIGURE 18–10
Effect of Q on resonance curves.

772 | HIGH-FREQUENCY EFFECTS IN AMPLIFIERS

Q_1 corresponds to a relatively broad bandpass response. At the other extreme, Q_3 corresponds to a very sharp, narrow bandpass response.

EXAMPLE 18–5

A certain electronic amplifier utilizes a series resonant circuit whose form may be represented by the model of Figure 18–8(a). The generator resistance is 250 Ω and the load resistance is 500 Ω. The reactive elements are $L = 5$ mH and $C = 100$ pF. **(a)** Determine the resonant frequency. **(b)** At the resonant frequency, the Q of the coil is measured as 30. Determine the net circuit Q at that frequency. **(c)** Determine the 3-dB bandwidth B.

Solution

(a) The resonant frequency f_o is

$$f_o = \frac{1}{2\pi\sqrt{LC}} = \frac{1}{2\pi\sqrt{5 \times 10^{-3} \times 100 \times 10^{-12}}} = 225.1 \text{ kHz} \qquad \textbf{(18–54)}$$

(b) To determine the net circuit Q, we must first determine R_{coil}. The coil Q is related to its resistance by

$$Q_{coil} = \frac{\omega_o L}{R_{coil}} \qquad \textbf{(18–55)}$$

Solving for R_{coil}, we have

$$R_{coil} = \frac{\omega_o L}{Q_{coil}} = \frac{2\pi \times 225.1 \times 10^3 \times 5 \times 10^{-3}}{30} = 235.7 \ \Omega \qquad \textbf{(18–56)}$$

The net series resistance R_s is

$$R_s = R_g + R_L + R_{coil} = 250 + 500 + 235.7 = 985.7 \ \Omega \qquad \textbf{(18–57)}$$

The net series Q is

$$Q_s = \frac{\omega_o L}{R_s} = \frac{2\pi \times 225.1 \times 10^3 \times 5 \times 10^{-3}}{985.7} = 7.174 \qquad \textbf{(18–58)}$$

The net circuit Q is much less than the Q of the inductance alone.

(c) The net Q is related to the bandwidth B by

$$Q_s = \frac{f_o}{B} \qquad \textbf{(18–59)}$$

The 3-dB bandwidth B is

$$B = \frac{f_o}{Q_s} = \frac{225.1 \times 10^3}{7.174} = 31.38 \text{ kHz} \qquad \textbf{(18–60)}$$

EXAMPLE 18–6

A certain electronic amplifier utilizes a parallel resonant circuit whose form may be represented by the model of Figure 18–9(a). The generator resistance is 5 kΩ, and the load resistance is 10 kΩ. The reactive elements are $L = 8$ μH and $C = 125$ pF. **(a)** Determine the resonant frequency. **(b)** At the resonant frequency, the Q of the coil is measured as 50. Determine the values of the series resistance and the equivalent parallel resistance. **(c)** Determine the net circuit Q at the resonant frequency. **(d)** Determine the 3-dB bandwidth B.

Solution

(a) The resonant frequency f_o is

$$f_o = \frac{1}{2\pi\sqrt{LC}} = \frac{1}{2\pi\sqrt{8\times10^{-6}\times125\times10^{-12}}} = 5.033 \text{ MHz} \qquad \textbf{(18–61)}$$

(b) The coil Q is related to its series resistance R_{coil} by

$$Q_{\text{coil}} = \frac{\omega_o L}{R_{\text{coil}}} \qquad \textbf{(18–62)}$$

Solving for R_{coil}, we have

$$R_{\text{coil}} = \frac{\omega_o L}{Q_{\text{coil}}} = \frac{2\pi\times5.033\times10^6\times8\times10^{-6}}{50} = 5.060 \ \Omega \qquad \textbf{(18–63)}$$

The equivalent parallel resistance R'_{coil} is

$$R'_{\text{coil}} = Q^2_{\text{coil}}R_{\text{coil}} = (50)^2 \times 5.060 = 12.65 \text{ k}\Omega \qquad \textbf{(18–64)}$$

(c) The net parallel resistance R_p is

$$R_p = R_g \| R_L \| R'_{\text{coil}} = 5 \text{ k}\Omega \| 10 \text{ k}\Omega \| 12.65 \text{ k}\Omega = 2.638 \text{ k}\Omega \qquad \textbf{(18–65)}$$

The net circuit Q is

$$Q_p = \frac{R_p}{\omega_o L} = \frac{2638}{2\pi\times5.033\times10^6\times8\times10^{-6}} = 10.43 \qquad \textbf{(18–66)}$$

(d) The 3-dB bandwidth is

$$B = \frac{f_o}{Q} = \frac{5.033\times10^6}{10.43} = 482.6 \text{ kHz} \qquad \textbf{(18–67)}$$

18–4 TUNED AMPLIFIER

Tuned amplifiers employ series or parallel resonant circuits to create a bandpass characteristic over the frequency range of interest. Shunt capacitive effects, which severely limit the upper frequency range of low-pass amplifiers, become an integral part of the resonant circuit and are thus used to advantage in tuned amplifiers.

With careful design and adjustment, tuned amplifiers employing discrete components may be used to several hundred megahertz.

A full treatment of tuned amplifiers is beyond the scope of this text and is more appropriate for a text on communications or high-frequency electronics. The treatment here will be limited to the basic concepts involved. This should allow the reader insight into the more complex circuits encountered in the field.

A simplified representative tuned amplifier employing a BJT and a parallel resonant circuit is shown in Figure 18–11(a). The input of this particular circuit as shown is similar to any standard BJT amplifier, although the circuit could be driven by another tuned amplifier. The resistance R_L represents any load resistance, and C_{in} represents the net shunt capacitance of the circuit as viewed by the collector. This could represent the Miller effect capacitance looking into any additional stage plus all shunt capacitance across the circuit (including collector-to-ground capacitance of the BJT).

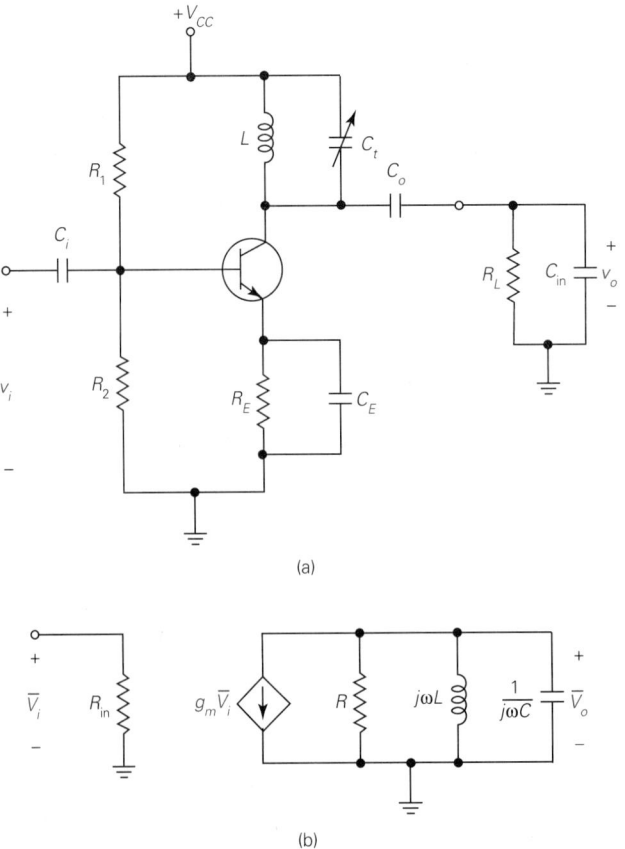

(a)

(b)

FIGURE 18–11
Typical tuned amplifier stage and a simplified equivalent circuit.

The inductance L and the capacitance C_t represent the discrete components constituting a parallel resonant circuit. Note that C_t is adjustable (i.e., it is a "tuning" capacitor). Seldom is it possible to predict with high accuracy the exact value of capacitance required for this capacitor because of the various stray circuit capacitances. Consequently, C_t is adjusted under dynamic circuit conditions to achieve parallel resonance at the exact center frequency desired. An alternative approach is to employ a fixed capacitance and an adjustable inductance. In receivers, the process of tuning a capacitance or inductance to an exact frequency is known as **alignment.**

The capacitances C_i, C_E, and C_o are all assumed to be very large compared with C_t and C_{in}. This means that these first three capacitors may be assumed to be short circuits in the frequency range where resonance occurs.

With the preceding assumptions, a simplified small-signal phasor model for the circuit is shown in Figure 18–11(b). Note that the variables and components are replaced by their phasor and impedance equivalents. The resistance R_{in} represents the parallel combination of R_1, R_2, and r_π; that is,

$$R_{\text{in}} = R_1 \| R_2 \| r_\pi \qquad \textbf{(18–68)}$$

The capacitance C represents the combination of the tuning capacitance C_t and the circuit shunt capacitance; that is,

$$C = C_t + C_{\text{in}} \qquad \textbf{(18–69)}$$

Finally, the resistance R represents the net parallel combination of the load resistance R_L and the parallel equivalent R'_{coil} of the coil resistance; that is,

$$R = R_L \| R'_{\text{coil}} \qquad \textbf{(18–70)}$$

Analysis of this circuit is very similar to that of the parallel resonant circuit considered in the preceding section and differs only with respect to the fact that a gain can be achieved with this circuit. The admittance $\overline{Y}$ of the parallel circuit is

$$\overline{Y} = G + j\omega C + \frac{1}{j\omega L} \qquad \textbf{(18–71)}$$

$$= G + j\left(\omega C - \frac{1}{\omega L}\right) \qquad \textbf{(18–72)}$$

where $G = 1/R$. The voltage $\overline{V}_o$ is

$$\overline{V}_o = \frac{-g_m \overline{V}_i}{\overline{Y}}$$

$$= \frac{-g_m \overline{V}_i}{G + j(\omega C - 1/\omega L)} \qquad \textbf{(18–73)}$$

The voltage gain has a frequency-dependent nature, so we will denote it as $\overline{A}$. It is

$$\overline{A} = \frac{\overline{V}_o}{\overline{V}_i} = \frac{-g_m}{G + j(\omega C - 1/\omega L)} \tag{18-74}$$

The resonant frequency f_o is readily shown to be

$$f_o = \frac{1}{2\pi\sqrt{LC}} \tag{18-75}$$

At resonance, the gain is

$$A = \frac{-g_m}{G} = -g_m R \tag{18-76}$$

A sketch of the gain magnitude as a function of frequency is shown in Figure 18–12. The form here is the same as that of the resonant circuits considered earlier. However, the magnitude of the gain at f_o is the same as that of a low-pass common-emitter amplifier. In effect, the use of the tuned circuit has resulted in the same gain at f_o as would occur in the midband for a low-pass type of amplifier.

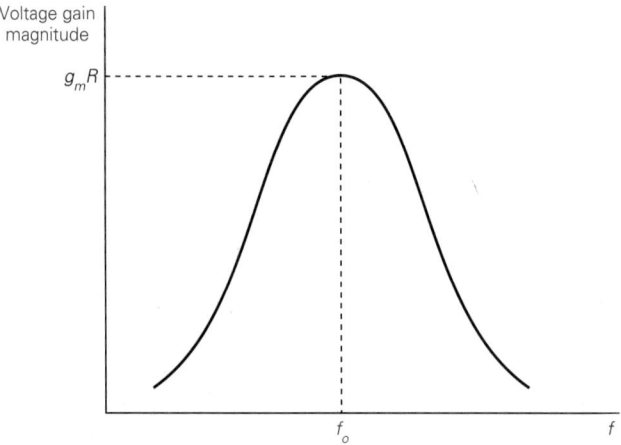

FIGURE 18–12
Voltage gain magnitude of tuned amplifier.

Expressions for the Q for this amplifier are the same as those for the parallel resonant circuit of the preceding section, but they will be repeated here. Letting Q_p represent this Q, we have

$$Q_p = \frac{R}{\omega_o L} = \omega_o RC = R\sqrt{\frac{C}{L}} \tag{18-77}$$

While the concept shown here is generally applicable more or less to all tuned amplifiers, there are many variations in the circuit forms. Although this circuit employed parallel resonance, some employ series resonance, and some multistage tuned amplifiers employ a combination of both. In many cases the desired Q is easier to achieve with one form than the other.

An additional process used in many tuned amplifiers is that of impedance transformation. It turns out that the input and output resistance levels of many active devices, sources, and loads limit the value of Q directly attainable with realistic values of inductance and capacitance. By employing an impedance transformation, however, these resistance levels may be transformed to a range permitting the attainment of a desired Q with practical values of L and C.

Impedance transformation may be achieved with a transformer, a tapped inductance, or with a capacitive-divider circuit. A representative example of a tuned amplifier employing a tapped inductance and an FET is shown in Figure 18–13.

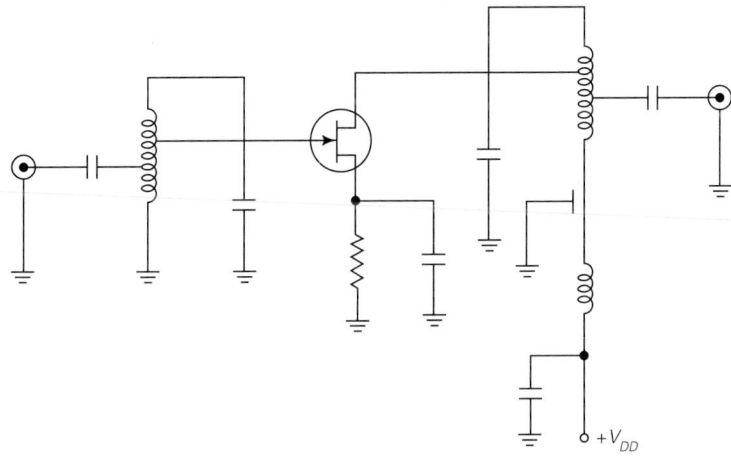

FIGURE 18–13
Typical high-frequency amplifier employing tapped inductors for impedance matching.

EXAMPLE 18–7 An amplifier of the form shown in Figure 18–11(a) has a center frequency of 2 MHz and a 3-dB bandwidth of 200 kHz. The inductance is 50 μH, and the input capacitance C_{in} is estimated as 40 pF. The load resistance is 10 kΩ, and at the operating point, $g_m = 0.02$ S. Determine **(a)** the value of C_t required for correct tuning; **(b)** the net shunt resistance R; and **(c)** the gain A at resonance.

Solution
(a) Resonance results from an inductance of 50 μH and a net capacitance C. The resonance formula is

$$f_o = \frac{1}{2\pi\sqrt{LC}} \tag{18–78}$$

Solving for C and substituting values for f_o and L, we obtain

$$C = \frac{1}{4\pi^2 f_o^2 L} = \frac{1}{4\pi^2 (2 \times 10^6)^2 \times 50 \times 10^{-6}} = 126.7 \text{ pF} \qquad \textbf{(18–79)}$$

Of this total capacitance, about 40 pF is supplied by input and transistor capacitance. The tuning capacitance C_t is

$$C_t = 126.7 - 40 = 86.7 \text{ pF} \qquad \textbf{(18–80)}$$

(b) The net circuit Q is

$$Q_p = \frac{f_o}{B} = \frac{2 \times 10^6}{200 \times 10^3} = 10 \qquad \textbf{(18–81)}$$

The net Q is related to R by

$$Q_p = R\omega_o C \qquad \textbf{(18–82)}$$

Solving for R, we have

$$R = \frac{Q_p}{\omega_o C} = \frac{10}{2\pi \times 2 \times 10^6 \times 126.7 \times 10^{-12}} = 6281 \text{ }\Omega \qquad \textbf{(18–83)}$$

Note that the net shunt resistance is less than the load resistance of 10 kΩ. This is a result of the fact that R is a parallel combination of the load resistance, the equivalent parallel resistance of the coil, and any other shunt resistance (e.g., the dynamic output resistance of the BJT).

(c) The net gain A_o at resonance is

$$A_o = -g_m R = -0.02 \times 6281 = -126 \qquad \textbf{(18–84)}$$

18–5 TRANSISTOR GAIN–BANDWIDTH PRODUCT

In earlier chapters the small-signal value of the short-circuit ac current gain β was assumed to be independent of frequency and was denoted as β_o. The actual behavior of β as a function of frequency follows a one-pole roll-off behavior quite closely.

Refer to the curve shown in Figure 18–14 in which logarithmic scales are assumed for both variables. At relatively low frequencies, $\beta = \beta_o$ as assumed earlier in the book. The value of β drops to $0.707\beta_o$ at a frequency f_β, which corresponds to the break frequency of a Bode representation. The value of β continues to drop above f_β as shown.

At a frequency $f = f_T$, the value of β is $\beta = 1$ as shown. The frequency f_T is called the **transition frequency.** It can be shown from the one-pole roll-off model that

$$f_T = \beta_o f_\beta \qquad \textbf{(18–85)}$$

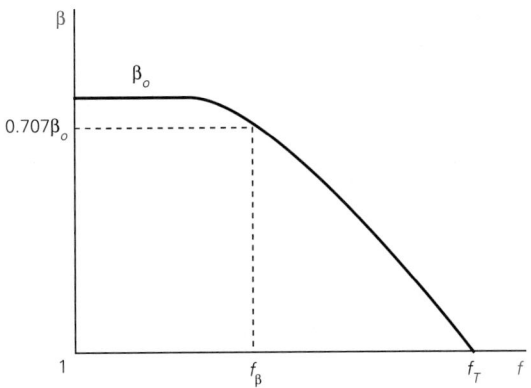

FIGURE 18–14
Variation of β with frequency for a BJT.

Since β_o is the low-frequency gain and f_β is a bandwidth parameter, the frequency f_T is also called the **gain–bandwidth product.**

   Typical values of f_T for general-purpose transistors may be determined from inspecting the 2N3903/2N3904 specifications in Appendix D. On the third data page of Appendix D, it is noted that the minimum value for the 2N3903 is f_T = 250 MHz, and the minimum value for the 2N3904 is f_T = 300 MHz. Transistors designed specifically for microwave applications have values of f_T exceeding 1 GHz.

EXAMPLE 18–8
The minimum value of f_T for a 2N3904 was noted to be 300 MHz. Determine the value of f_β for each of the following values of β_o: **(a)** 100, and **(b)** 400.

Solution
An expression for f_β is determined from (18–85) to be

$$f_\beta = \frac{f_T}{\beta_o} \qquad \text{(18–86)}$$

(a)   For β_o = 100, we have

$$f_\beta = \frac{300 \times 10^6}{100} = 3 \text{ MHz} \qquad \text{(18–87)}$$

(b)   For β_o = 400, we have

$$f_\beta = \frac{300 \times 10^6}{400} = 0.75 \text{ MHz} \qquad \text{(18–88)}$$

Note that for a given f_T, the bandwidth varies inversely with the low-frequency gain.

18–6 HYBRID-PI TRANSISTOR MODELS

When serious analysis of high-frequency effects in a transistor is necessary or when a high-frequency amplifier is designed, a model displaying the high-frequency effects is required. The most convenient and widely employed circuit is the so-called **hybrid-pi model.** The hybrid-pi models for both a BJT and an FET are discussed in this section.

BJT Hybrid-Pi Model

The form of the hybrid-pi model of the BJT is shown in Figure 18–15. First, note that if $r_x = 0$, $C_\pi = 0$, $C_\mu = 0$, and $r_d = \infty$, the circuit reduces to the simple small-signal g_m-model for the BJT, which was used extensively in early chapters. For the frequency range we have considered thus far, the effects of these parameters are usually not significant.

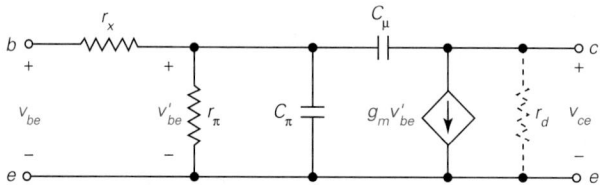

FIGURE 18–15
Hybrid-pi model of bipolar junction transistor.

A forward-biased *pn* junction exhibits a capacitive effect as a result of temporary charge storage in the semiconductor material. This capacitance for the base-emitter junction is referred to as **diffusion capacitance** and is denoted as C_π.

A reverse-biased *pn* junction exhibits a capacitive effect arising from the depletion layer between conducting regions. This capacitance for the base–collector junction is referred to as **depletion** or **junction** capacitance and is denoted as C_μ.

There is always some resistance between the active base region and the external contact. This resistance is referred to as the **base spreading resistance** and is denoted by r_x. Note that when the base spreading resistance is present, the control voltage v'_{be} for the dependent current source is not the same as v_{be} because of the drop across r_x.

The resistance r_d is shown as an optional quantity to use when it is necessary to consider the actual slope of the collector characteristics in the active region. In tuned amplifiers, this effective resistance will tend to lower the overall Q of the circuit.

FET Hybrid-Pi Model

As in the case of the BJT, there is a hybrid-pi model for an FET. The form of the hybrid-pi model for an FET is shown in Figure 18–16.

The hybrid-pi model of the FET contains three capacitances. The capacitance C_{gs} represents the **gate–source capacitance**, C_{gd} represents the **gate–drain capacitance**, and C_{ds} represents the **drain–source capacitance**. When the dynamic effects of the drain–source resistance are to be considered, r_d may be included. Note that this model reduces to the simpler FET model when the three capacitances are assumed to be zero.

FIGURE 18–16
Hybrid-pi model of an FET.

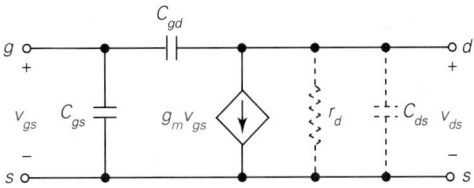

18–7 TRANSISTOR SWITCHING TIMES

In various transistor switching applications considered early in the text, the transistors were assumed to switch instantaneously between appropriate levels. In actual practice, a transistor requires some time to change between levels, and this limitation is important in high-speed pulse and switching applications. Various transistor capacitances act to slow down the switching process.

Consider the transistor circuit of Figure 18–17(a). Assume initially that $i_B = 0$ and $i_C = 0$. Assume that through some external switching arrangement (not shown), an ideal pulse of base current is applied to the transistor as shown in Figure 18–17(b). If the transistor were ideal, the collector current would be a pulse of the same shape as the base current, but would have a value $\beta_{dc} i_B$. The ideal collector current would start and stop at the same times as the base current.

The actual form of the collector current is as shown in Figure 18–17(c). On the leading edge of the collector current pulse are two time parameters. These are called the **delay time** and the **rise time** and are denoted by t_d and t_r, respectively. The *delay time* t_d is the time between the point where the input pulse is initiated and the point where the collector current reaches 10% of its final level. The *rise time* t_r is the time between the 10% and 90% points on the leading edge of the current pulse. The **turn-on time** t_{on} is the sum of the delay time and the rise time, that is,

$$t_{on} = t_d + t_r \qquad \textbf{(18–89)}$$

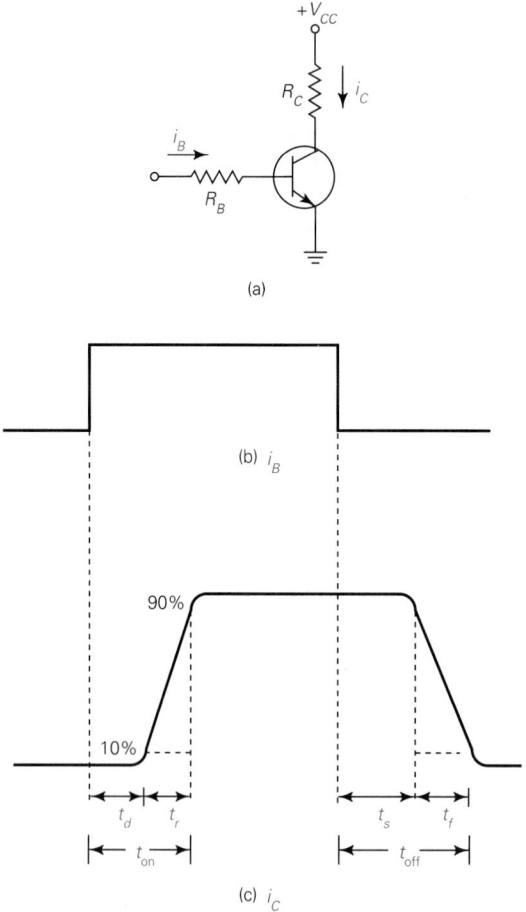

FIGURE 18–17
Transistor switching current waveforms.

On the trailing edge of the collector current pulse are two additional time parameters. These are called the **storage time** and the **fall time** and are denoted by t_s and t_f, respectively. The *storage time* t_s is the time between the point where the base current pulse drops to zero and the point where the collector current drops to 90% of its maximum value. The storage time is due to charge carriers being trapped in the depletion region. This time is usually longer than the other time parameters.

The *fall time* t_f is the time required for the collector to drop from 90% of its peak value down to 10% of its peak value. The *turn-off time* t_{off} is the sum of the storage time and the fall time, that is,

$$t_{\text{off}} = t_s + t_f \qquad \qquad \textbf{(18–90)}$$

Typical values of these various parameters for general-purpose transistors may be noted by referring to the third and fourth pages of the 2N3903/2N3904 data in Appendix D. The maximum values of t_d and t_r are both listed as 35 ns. The maximum value of t_s for the 3904 is listed as 200 ns. (The value for the 3903 is 175 ns.) The maximum value of t_f is listed as 50 ns.

EXAMPLE 18–9

Based on the data provided for the 2N3904 transistor, calculate **(a)** the turn-on time, and **(b)** the turn-off time.

Solution
The values of t_d, t_r, t_s, and t_f were summarized in the last paragraph preceding this example.

(a) The turn-on time t_{on} is

$$t_{\text{on}} = t_d + t_r = 35 + 35 = 70 \text{ ns} \qquad \textbf{(18–91)}$$

(b) The turn-off time t_{off} is

$$t_{\text{off}} = t_s + t_f = 200 + 50 = 250 \text{ ns} \qquad \textbf{(18–92)}$$

Since the individual time parameters were given as maximum values, the computed turn-on and turn-off times are also maximum values.

18–8 OPERATIONAL AMPLIFIER HIGH-FREQUENCY LIMITATIONS

Operational amplifiers are generally most useful in the frequency range below about 1 MHz or so. While a few special-purpose op-amps are available with usable frequency capabilities well above 10 MHz, these units are relatively expensive.

In general, there are two separate effects that contribute to the high-frequency degradation of op-amps. These are (1) the **finite bandwidth effect,** and (2) the **slew rate effect.** Each effect will be considered separately in the developments that follow.

Finite Bandwidth Effect

A discussion was given in Chapter 14 concerning the effect of feedback on the closed-loop bandwidth, and some results were stated at that time without proof. In this section, we expand these results and present more detailed considerations of the effects.

We have seen that an op-amp has a very large open-loop dc gain, typically of the order of 10^5 or greater. However, the amplitude response of the open-loop response starts to decrease at a relatively low frequency, often less than 10 Hz.

The variation in open-loop gain as a function of frequency for a typical op-amp is shown by the Bode plot in Figure 18–18. The level A_o represents the dc (and very-low-frequency) open-loop gain, and f_o represents the 3-dB open-loop, or break, frequency. If the op-amp were operated as an open-loop amplifier, the 3-dB bandwidth would be only f_o.

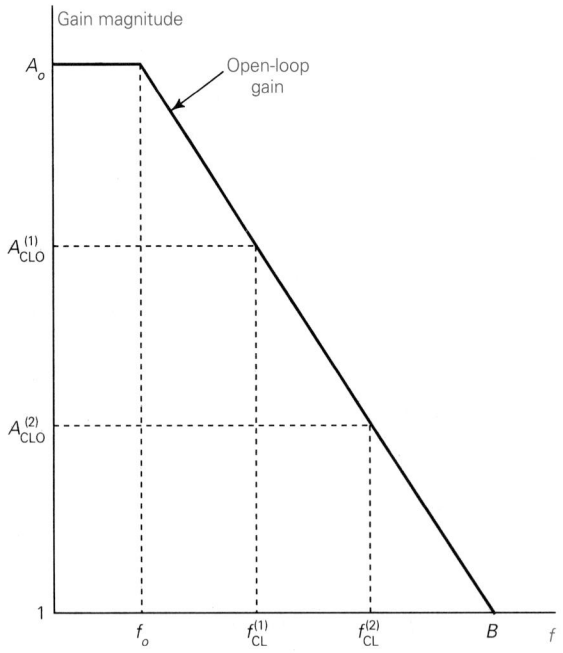

FIGURE 18–18
Manner in which closed-loop gain and break frequency depend on open-loop gain of an op-amp.

Although there are many frequency-limiting effects inside the op-amp, most modern op-amps are of the **internally compensated** variety. Such op-amps have included in their circuitry deliberate frequency shaping so as to attain a one-pole roll-off rate (−6 dB/octave) over the frequency range where the open-loop gain is greater than 1. It turns out that from more advanced feedback theory, this condition ensures that the op-amp will be stable for all closed-loop configurations. Unstabilized op-amps are available, in which the user adds a specifically tailored stabilizing circuit, but we do not consider these here.

The mathematical form of the open-loop differential gain $\overline{A}$ as a function of frequency for internally stabilized op-amps can be closely approximated by a one-pole roll-off function, which is of the form

$$\overline{A} = \frac{A_o}{1 + j(f/f_o)} \qquad (18\text{--}93)$$

where A_o is the dc and very-low-frequency gain and f_o is the break frequency. The closed-loop gain A_{CL} expressed in terms of the open-loop differential gain is

$$\overline{A}_{CL} = \frac{\overline{A}}{1 + \overline{A}\beta} \qquad (18\text{--}94)$$

When (18–93) is substituted in (18–94) and some rearrangement is made, the following result is obtained:

$$\overline{A}_{CL} = \frac{A_o/(1 + A_o\beta)}{1 + jf/f_o(1 + A_o\beta)} \qquad (18\text{--}95)$$

This result may be expressed as

$$\overline{A}_{CL} = \frac{A_{CLO}}{1 + j(f/f_{CL})} \qquad (18\text{--}96)$$

in which the new dc gain A_{CLO} is

$$A_{CLO} = \frac{A_o}{1 + A_o\beta} \qquad (18\text{--}97)$$

and the new break frequency f_{CL} is

$$f_{CL} = f_o(1 + A_o\beta) \qquad (18\text{--}98)$$

These results were, of course, stated in Chapter 14 without proof. Based on the one-pole roll-off model, the results can readily be derived.

The manner in which closed-loop gain and closed-loop bandwidth are related for an op-amp is illustrated in Figure 18–18. Consider first the closed-loop gain $A_{CLO}^{(1)}$. The point where this gain level intersects the open-loop gain magnitude curve is $f_{CL}^{(1)}$, and this value represents the 3-dB closed-loop bandwidth in this case. Consider next the smaller closed-loop gain $A_{CLO}^{(2)}$. This level intersects the open-loop gain magnitude curve at $f_{CL}^{(2)}$, so the 3-dB closed-loop bandwidth is larger at this smaller gain.

Eventually, unity-gain level is reached, and the 3-dB frequency corresponding to this level is denoted as B. The value of B is called the **unity-gain frequency,** and it is the frequency corresponding to the intersection of the open-loop magnitude response and a gain level of unity.

It can be shown from the one-pole frequency response model that B is related to the dc gain A_o and 3-dB open-loop frequency f_o by

$$B = A_o f_o \tag{18–99}$$

From this relationship, an alternative definition of B is the *gain–bandwidth product*. Thus, *unity-gain frequency and gain–bandwidth product will be considered to be equivalent specifications.*

Simplified Bandwidth Forms

The closed-loop bandwidth has been expressed in terms of the open-loop gain and the feedback factor. In practice, however, these quantities may not be readily apparent from a circuit diagram. Instead, some forms will now be given in which the closed-loop bandwidth is expressed in terms of the closed-loop gain and the unity-gain frequency.

The gain–bandwidth product is

$$A_{\text{CLO}} f_{\text{CL}} = A_o f_o = B \tag{18–100}$$

Solving for f_{CL} and expressing A_{CLO} simply as a *desired* real gain A_{CL}, we have

$$f_{\text{CL}} = \frac{B}{A_{\text{CL}}} \qquad (\textit{noninverting amplifier}) \tag{18–101}$$

The value of β for an inverting amplifier is the same as for a noninverting amplifier, but the gain is different. Specifically,

$$\frac{1}{\beta} = \frac{R_i + R_f}{R_i} = 1 + \frac{R_f}{R_i} = 1 + |A_{\text{CL}}| \tag{18–102}$$

where $|A_{\text{CL}}|$ is the magnitude of the *desired* real gain. The corresponding bandwidth relationship is

$$f_{\text{CL}} = \frac{B}{1 + |A_{\text{CL}}|} \qquad (\textit{inverting amplifier}) \tag{18–103}$$

For other types of op-amp circuits, the relationship between closed-loop bandwidth and open-loop unity-gain frequency involves a quantity called the **noise gain.*** The two relationships given here will suffice for the purposes of this book.

---

* W. D. Stanley, *Operational Amplifiers with Linear Integrated Circuits,* 3d Ed. (Columbus, Ohio: Charles E. Merrill, 1994).

In all cases, the general inverse relationship between desired gain and attainable closed-loop bandwidth tends to be the case. In fact, some books simply state a formula such as

$$\text{closed-loop bandwidth} = \frac{\text{unity-gain frequency}}{\text{closed-loop gain}}$$

This formula may always be used qualitatively for estimation purposes, but it is strictly correct only for the case of the noninverting amplifier.

EXAMPLE 18–10

A certain op-amp has a unity-gain frequency B = 2 MHz. Determine the value of closed-loop 3-dB bandwidth B_{CL} for each of the following values of *closed-loop noninverting gain:* **(a)** 1000; **(b)** 100; **(c)** 10; **(d)** 1.

Solution
From (18–101), the relationship used is

$$f_{CL} = \frac{B}{A_{CL}} = \frac{2 \times 10^6}{A_{CL}} \qquad \text{(18–104)}$$

The four values of A_{CL} are then substituted in (18–104), and the four values of f_{CL} are determined. The resulting values of f_{CL} are summarized in the second column of the table in Example 18–11. Comments will be made at the conclusion of Example 18–11.

EXAMPLE 18–11

For the op-amp of Example 18–10, determine the closed-loop 3-dB bandwidth for the following values of *closed-loop inverting gain:* **(a)** −1000; **(b)** −100; **(c)** −10; **(d)** −1.

Solution
From (18–103), the relationship used is

$$f_{CL} = \frac{B}{1 + |A_{CL}|} = \frac{2 \times 10^6}{1 + |A_{CL}|} \qquad \text{(18–105)}$$

The four values of A_{CL} are substituted in (18–105), and the four values of f_{CL} are determined. The resulting values of f_{CL} are summarized in the third column of the table that follows.

	$|A_{CL}|$	Noninverting f_{CL}	Inverting f_{CL}
(a)	1000	2 kHz	1998 Hz
(b)	100	20 kHz	19.8 kHz
(c)	10	200 kHz	181.9 kHz
(d)	1	2 MHz	1 MHz

Observe the manner in which the 3-dB frequency increases as the desired gain level decreases. For moderate and large gains, there is negligible difference between the closed-loop 3-dB frequencies for the noninverting and inverting configurations. However, at very low gain, there is a significant difference.

Slew Rate Effect

To provide the internal compensation we discussed, a capacitor is added to the op-amp at a critical point in the circuitry. This capacitor in conjunction with the current capability of the stage driving it will limit the rate at which the output of the op-amp can change. This limitation is given in specifications as a **slew rate** parameter.

The slew rate, denoted as S, is the maximum rate of change of the op-amp *output* voltage. The basic units are volts per second. However, because of the large value involved, it is customary to specify the slew rate in volts per microsecond (V/μs). A typical value of slew rate is 0.5 V/μs for a general-purpose op-amp.

The manner in which the slew rate effect exhibits itself in pulse-type situations is illustrated in Figure 18–19. It is assumed in this development that rise time effects due to finite bandwidth are negligible.

Assume an input pulse V_i shown in Figure 18–19(a) in which there is a rise time t_i. (Only the leading edge is shown.) The input height is V_i. The closed-loop

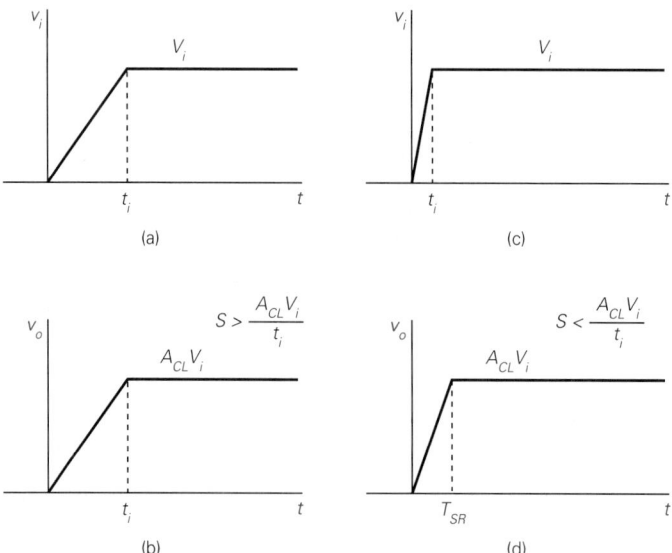

FIGURE 18–19
Effects of slew rate for two different input rise times.

gain of the amplifier is A_{CL}, so the desired output is a pulse whose height should be $A_{\text{CL}} V_i$. The required rate of change of the output voltage v_o is $A_{\text{CL}} V_i/t_i$. If the slew rate of the op-amp equals or exceeds this rate of change, the pulse will be reproduced without distortion. This situation is shown in Figure 18–19(b).

Next, assume an input voltage having a shorter rise time t_i as shown in Figure 18–19(c). The rate of change of the desired output voltage is much greater than for the first voltage and is assumed to exceed the slew rate of the op-amp. Since the op-amp output cannot change faster than S, the output will now be of the form shown in Figure 18–19(d). Consequently, the rise time will be degraded.

In Figure 18–19(d), the actual time required for the pulse to reach full output is a function of the slew rate and is denoted as T_{SR}. For a given *output voltage change V_o*, the value of T_{SR} is

$$T_{\text{SR}} = \frac{V_o}{S}$$

(18–106)

If V_o is expressed in volts and S is expressed in volts per microsecond, T_{SR} will be expressed in microseconds.

Slew Rate Limiting Frequency

The slew rate concept has been developed on the basis of a pulse-type waveform. However, for complex waveforms (e.g., music, speech, etc.), the effect of the slew rate is to limit the usable frequency range to those frequencies in which the output can follow the maximum rate of change required.

Let f_{SR} represent the frequency of a sinusoidal waveform at which the op-amp output can track the required signal. Using calculus, it can be shown that

$$f_{\text{SR}} = \frac{S}{2\pi V_o}$$

(18–107)

where V_o is the peak value of the output sinusoid. In using (18–107), S should be expressed in the basic units of volts per second instead of volts per microsecond. The relationship of (18–107) provides an estimate of the highest operating frequency for complex signals. The frequency will be denoted as the **slew rate limiting frequency.**

Finite Bandwidth and Slew Rate

We have discussed finite bandwith and slew rate as two separate isolated phenomena. In practice, of course, both effects occur simultaneously, and the combined frequency limitation of the op-amp will result from both effects.

Although both effects are always present, the following general guidelines should be noted: (1) *At very low output signal levels, the finite bandwidth effect is usually the dominating effect;* and (2) *at very high output signal levels, the slew rate effect is usually the dominating effect.* In general, however, calculations should be performed to ascertain the degree of each effect.

EXAMPLE 18–12

From 741 data in Appendix D, it can be determined that the slew rate is 0.5 V/μs. Determine the time required for an output pulse to reach its full level for each of the following op-amp output voltages: **(a)** 0.1 V; **(b)** 1 V; **(c)** 10 V.

Solution
The pertinent relationship is (18–106), and since S is given in V/μs, the time T_{SR} will be expressed directly in microseconds. We have

$$T_{SR}(\mu s) = \frac{V_o}{0.5} \tag{18–108}$$

The three values of V_o may be substituted in (18–108), and the results are summarized in the following table:

	V_o (V)	T_{SR} (μs)
(a)	0.1	0.2
(b)	1	2
(c)	10	20

The output rise time is thus directly proportional to the peak value of the output voltage. Note that the gain does not enter into these results since we are given the output voltage in each case. Thus, for slew rate purposes, it is immaterial whether an output voltage of 10 V arose from an input voltage of 10 V and a gain of 1 or an input voltage of 1 V and a gain of 10.

EXAMPLE 18–13

For the 741 op-amp with a slew rate of 0.5 V/μs, determine the highest operating frequency with analog-type signals for each of the following peak output voltages: **(a)** 0.1 V; **(b)** 1 V; **(c)** 10 V.

Solution
The pertinent relationship is (18–107). In this case we should first express the slew rate as $S = 0.5 \times 10^6$ V/s. When this value is substituted in (18–107), we have

$$f_{SR} = \frac{0.5 \times 10^6}{2\pi V_o} = \frac{79,577}{V_o} \tag{18–109}$$

The three values of V_o may be substituted in (18–109), and the results are summarized in the following table:

	V_o (V)	f_{SR} (kHz)
(a)	0.1	796
(b)	1	79.6
(c)	10	7.96

The highest operating frequency is thus inversely proportional to the peak value of the output voltage.

18–9 PSPICE EXAMPLE

PSPICE EXAMPLE 18–1

Prepare a PSPICE program to simulate the series resonant circuit of Example 18–5, and obtain a decibel amplitude response in the vicinity of resonance.

Solution

The circuit adapted to the PSPICE format is shown in Figure 18–20(a), and the code is shown in Figure 18–20(b). The input voltage is chosen as 1 V. An octave sweep is used in this case over the frequency range of 112.5 to 450 kHz. This represents a range from one octave below to one octave above the predicted center frequency of 225 kHz.

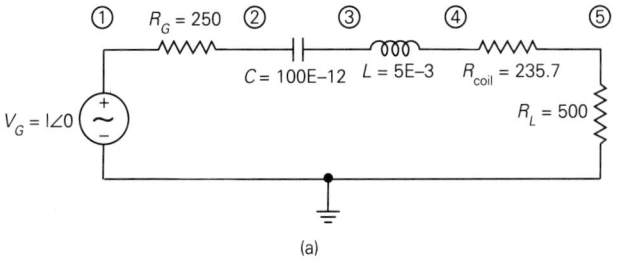

```
PSPICE EXAMPLE 18-1
VG 1 0 AC 1
RG 1 2 250
C 2 3 100E-12
L 3 4 5E-3
RCOIL 4 5 235.7
RL 5 0 500
.AC OCT 100 1.125E5 4.5E5
.PROBE
.END
```

(a)   (b)

FIGURE 18–20
Circuit and code of PSPICE Example 18–1.

The decibel amplitude response is shown in Figure 18–21. With the .PROBE marker, the center frequency was determined to be 225 kHz, as expected. The level of −5.8956 dB represents the ratio of output voltage to input voltage at resonance, which is determined by the voltage-divider ratio 500/(500 + 235.7 +

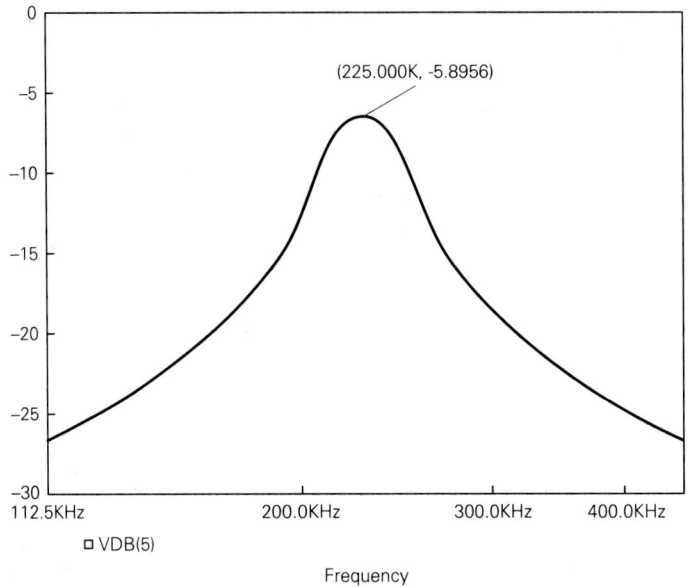

FIGURE 18–21
Decibel amplitude response for the output voltage of PSPICE Example 18–1.

$250) = 500/985.7 = 0.50725$. The corresponding decibel value is -5.896 dB, which agrees almost exactly with the computer result.

PROBLEMS

Drill Problems

18–1. Express the complex form of the capacitive impedance of a 0.01-μF capacitor at a frequency of 5 kHz.

18–2. Express the complex form of the capacitive impedance of a 200-pF capacitor at a frequency of 3 MHz.

18–3. Express the complex form of the inductive impedance of a 50-mH inductor at a frequency of 8 kHz.

18–4. Express the complex form of the inductive impedance of a 200-μH inductor at a frequency of 6 MHz.

18–5. A certain inverting amplifier has a gain $A = -50$. The amplifier has an input shunt capacitance of 20 pF and a capacitance between input and output terminals of 3 pF. Determine the net equivalent input capacitance.

18–6. A certain inverting amplifier has a gain $A = -80$. The amplifier has an input shunt capacitance of 40 pF and a capacitance between input and output terminals of 5 pF. Determine the net equivalent input capacitance.

18–7. Assume that the amplifier of Problem 18–5 is driven from a source whose internal resistance is 1 kΩ. Determine the 3-dB break frequency f_b.

18–8. Assume that the amplifier of Problem 18–6 is driven from a source whose internal resistance is 500 Ω. Determine the 3-dB break frequency f_b.

18–9. Determine the resonant frequency corresponding to $L = 80$ mH and $C = 0.001$ μF.

18–10. Determine the resonant frequency corresponding to $L = 50$ μH and $C = 30$ pF.

18–11. A certain resonant circuit has a resonant frequency of 5 MHz and a 3-dB bandwidth of 200 kHz. Determine the circuit Q.

18–12. A certain resonant circuit has a resonant frequency of 12 MHz and the circuit Q is 50. Determine the 3-dB bandwidth.

18–13. The Q of a certain 20-mH inductance is 50 at a frequency of 250 kHz. Determine the series resistance of the coil.

18–14. The Q of a certain 40-μH inductance is 80 at a frequency of 5 MHz. Determine the series resistance of the coil.

18–15. A certain coil with a Q of 12 at a certain frequency has a series resistance of 20 Ω. Determine the equivalent parallel resistance.

18–16. A certain coil with a Q of 20 at a certain frequency has an equivalent parallel resistance of 2 kΩ. Determine the series resistance.

18–17. A certain electronic amplifier uses a series resonant circuit whose form may be represented by the model of Figure 18–8(a). Both the generator and load resistances are 50 Ω. The reactive elements are $L = 40$ μH and $C = 50$ pF. **(a)** Determine the resonant frequency. **(b)** At the resonant frequency, the Q of the coil is measured as 50. Determine the net circuit Q at that frequency. **(c)** Determine the 3-dB bandwidth B.

18–18. A certain electronic amplifier uses a parallel resonant circuit whose form may be represented by the model of Figure 18–9(a). The generator resistance is 8 kΩ, and the load resistance is 16 kΩ. The reactive elements are $L = 2$ mH and $C = 0.003$ μF. **(a)** Determine the resonant frequency. **(b)** At the resonant frequency, the Q of the coil is measured as 30. Determine the values of the series resistance and the equivalent parallel resistance. **(c)** Determine the net circuit Q at the resonant frequency. **(d)** Determine the 3-dB bandwidth B.

18–19. An amplifier of the form shown in Figure 18–11(a) has a center frequency of 5 MHz and a 3-dB bandwidth of 250 KHz. The inductance is 10 μH, and resonance at the center frequency is achieved when the tuning capacitance is $C_t = 80$ pF. The load resistance is 12 kΩ, and at the operating point, $g_m = 0.03$ S. **(a)** Estimate the circuit shunt capacitance C_{in}. **(b)** Determine the net shunt resistance R. **(c)** Determine the gain A at resonance.

18–20. An amplifier of the form shown in Figure 18–11(a) has an inductance of 20 μH, and the circuit shunt capacitance C_{in} is estimated as 12 pF. **(a)** Determine the resonant frequency when the tuning capacitance is set at $C_t = 40$ pF. **(b)** For the condition of **(a)**, the 3-dB bandwidth is measured as 200 kHz. Determine the net shunt resistance R. **(c)** If $g_m = 5$ mS, determine the gain A at resonance.

18–21. The value of f_T for a certain transistor is 500 MHz. Determine the value of f_B for each of the following values of β_o: **(a)** 50, and **(b)** 200.

18–22. The value of f_B for a certain transistor is 4 MHz, and the value of β_o is 200. Determine the value of f_T.

18–23. A certain BJT has the following parameters:

> Base–emitter diffusion capacitance = 18 pF

> Base–collector depletion capacitance = 2 pF

Base spreading resistance = 50 Ω

Dynamic base–emitter resistance = 2 kΩ

Dynamic collector–emitter resistance = 100 kΩ

For I_C = 2 mA, draw a complete hybrid-pi equivalent circuit model.

18–24. A certain FET has the following parameters:

Gate–source capacitance = 8 pF

Gate–drain capacitance = 4 pF

Drain–source capacitance = 1 pF

Dynamic drain resistance = 200 kΩ

g_m = 8 mS

Draw a complete hybrid-pi equivalent circuit model.

18–25. A certain BJT has the following switching parameters:

Delay time = 40 ns

Rise time = 30 ns

Storage time = 120 ns

Fall time = 45 ns

Determine **(a)** the turn-on time, and **(b)** the turn-off time.

18–26. A certain BJT has the following parameters:

Turn-on time = 100 ns

Turn-off time = 240 ns

The rise time is 50% longer than the delay time, and the storage time is three times as long as the fall time. Determine the **(a)** delay time; **(b)** rise time; **(c)** storage time; and **(d)** fall time.

18–27. A certain op-amp has a unity-gain frequency B = 1.2 MHz. Determine the value of closed-loop 3-dB bandwidth for each of the following values of *closed-loop noninverting gain*: **(a)** 500; **(b)** 50; **(c)** 5; **(d)** 1.

18–28. A certain op-amp has a unity-gain frequency B = 4 MHz. Determine the value of closed-loop 3-dB bandwidth for each of the following values of *closed-loop noninverting gain*: **(a)** 1000; **(b)** 100; **(c)** 10; **(d)**1.

18–29. For the op-amp of Problem 18–27, determine the closed-loop 3-dB bandwidth for the following values of *closed-loop inverting gain*: **(a)** −500; **(b)** −50; **(c)** −5; **(d)** −1.

18–30. For the op-amp of Problem 18–28, determine the closed-loop 3-dB bandwidth for the following values of *closed-loop inverting gain*: **(a)** −1000; **(b)** −100; **(c)** −10; **(d)** −1.

18–31. A certain op-amp has a slew rate of 2 V/μs. Determine the time required for an output pulse to reach its full level for each of the following op-amp output voltages: **(a)** 0.2 V; **(b)** 4 V; **(c)** 12 V.

18–32. A certain op-amp has a slew rate of 40 V/μs. Determine the time required for an output pulse to reach its full level for each of the following op-amp output voltages: **(a)** 0.1 V; **(b)** 1 V; **(c)** 10 V.

18–33. For the op-amp of Problem 18–31, determine the highest operating frequency with analog-type signals for each of the following peak output voltages: **(a)** 0.2 V; **(b)** 4 V; **(c)** 12 V.

18–34. For the op-amp of Problem 18–32, determine the highest operating frequency with analog-type signals for each of the following peak output voltages: **(a)** 0.1 V; **(b)** 1 V; **(c)** 10 V.

Derivation Problems

18–35. Consider the series combination of an inductance L and resistance R_{coil} as shown in a portion of the circuit of Figure 18–9(b). **(a)** Write an expression for the impedance $\overline{Z}$ of the series combination. **(b)** Write an expression for the admittance $\overline{Y}$. Rationalize the denominator by multiplying both numerator and denominator by the complex conjugate of the denominator. **(c)** From the result of **(b)**, draw an equivalent parallel circuit showing the values of the equivalent parallel inductance L' and the parallel resistance R'. **(d)** Show that if $Q_{coil} \gg 1$, $L' \approx L$ and $R' \approx Q_{coil}^2 R_{coil}$.

18–36. Consider the equivalent parallel resonant circuit of Figure 18–9(c). Show that the net Q of the circuit may be determined from the expression

$$\frac{1}{Q} = \frac{1}{Q_g} + \frac{1}{Q_L} + \frac{1}{Q_{coil}}$$

where $Q_g = \omega_o R_g C$, $Q_L = \omega_o R_L C$, and Q_{coil} is the coil Q.

18–37. Starting with (18–94), substitute the expression of (18–93), and carry out the details leading to (18–98).

18–38. (Requires calculus.) Assume that the output v_o of an op-amp is expressed as

$$v_o = V_o \sin \omega t$$

(a) Applying differential calculus, determine the rate of change of the output, that is,

$$\frac{dv_o}{dt}$$

(b) Determine the maximum value of the result of **(a)**, that is, $dv_o/dt]_{max}$.
(c) To determine the highest frequency at which the op-amp can track the required output, set the result of **(b)** equal to the slew rate S. Let f_{SR} represent the frequency involved, and show that the result of (18–107) is obtained.

Design Problems

18–39. A certain tuned amplifier using series resonance is being designed. The basic circuit can be modeled by the circuit form of Figure 18–8(b). The inductance is selected as $L = 8$ mH. Determine the value of capacitance required for a center frequency of 150 kHz.

18–40. A certain tuned amplifier using parallel resonance is being designed. The basic circuit can be modeled by the circuit form of Figure 18–9(c). The inductance is selected as $L = 12$ μH. Determine the net value of capacitance required for a center frequency of 4 MHz.

18–41. An op-amp is being selected for a noninverting amplifier application requiring a closed-loop gain of 30. If the closed-loop 3-dB bandwidth is required to be at least 50 kHz, determine the minimum acceptable value of the gain–bandwidth product of the op-amp selected.

18–42. An op-amp is being selected for an inverting amplifier application requiring a closed-loop gain of −7. If the closed-loop 3-dB bandwidth is required to be at least 100 kHz, determine the minimum acceptable value of the gain–bandwidth product of the op-amp selected.

18–43. For the op-amp of Problem 18–41, assume that the slew rate limiting frequency is also specified as 50 kHz. If the maximum input voltage is 0.2 V, determine the minimum acceptable slew rate of the op-amp selected.

18–44. For the op-amp of Problem 18–42, assume that the slew rate limiting frequency is also specified as 100 kHz. If the maximum input voltage is 0.5 V, determine the minimum acceptable slew rate of the op-amp selected.

Troubleshooting Problems

18–45. You are troubleshooting a tuned amplifier of the type shown in Figure 18–11(a) and find that the inductor L is open. The original inductor had a value of 40 μH, but the only readily available inductor has a value of 20 μH. While waiting for an exact replacement, you decide to temporarily replace the faulty inductor with the available one. To compensate for the smaller inductance, the capacitance of the tuning capacitor will be changed, and the circuit will be realigned. Indicate which one of the following represents the expected required change in C_t. (*Hint:* Remember that a portion of the capacitance is C_{in}.) **(a)** C_t must be more than doubled; **(b)** C_t must be increased some, but not doubled; **(c)** C_t must be decreased to less than half of the original value; **(d)** C_t must be decreased some, but not to half of the original value.

18–46. In the situation of Problem 18–45, assume that the net shunt resistance has the same value at the center frequency as in the original situation. Indicate what change (if any) you would expect in the bandwidth B: **(a)** increase; **(b)** remain the same; **(c)** decrease.

19

POWER AMPLIFIERS

OBJECTIVES

After completing this chapter, the reader should be able to:

- Explain how a power amplifier differs from a small-signal amplifier.
- Express the mathematical form of and sketch the hyperbola of maximum power dissipation.
- Define class A, B, and C amplifiers and discuss their relative efficiencies.
- Discuss the balance between power supply input power, signal output power, and internal power dissipation.
- Analyze a transformer-coupled class A power amplifier for input power, output power, internal power dissipation, and efficiency.
- Analyze a class B push-pull amplifier for input power, output power, internal power dissipation, and efficiency.
- Discuss the operation of a class C amplifier.
- Draw an electrical analog of the heat flow process in a transistor.
- State and apply the equation relating temperature difference, power dissipation, and thermal resistance of a transistor.
- Apply the derating factor to determine maximum power dissipation at a particular temperature.
- Discuss the significance of various power transistor specifications.
- Analyze some of the circuits in the chapter with PSPICE.

19–1 OVERVIEW

A **power amplifier** is an amplifier circuit in which the primary objective is to deliver power to an external load. There is no precise boundary between the so-called "small-signal" amplifiers that we have studied thus far and power amplifiers since all amplifiers are capable of delivering some output power. However, a characterizing feature of most power amplifiers is that a specified power level must be delivered to some external load. For example, in an audio amplifier, the power amplifier stage delivers electrical power to the speakers, and the speakers in turn convert the electrical power to sound. Other examples of power amplifiers are the amplifiers used to provide drive for motors and the output stages of broadcast transmitters, which supply power to the antennas.

A second characterizing feature of most power amplifiers is the level of the required power. Earlier in the book, the level of 0.5 W was indicated as a common dividing line between small-signal amplifiers and power amplifiers. Actually, there are some power amplifiers that operate below this level (e.g., small portable radios with miniature speakers). However, the output power levels of some power amplifiers can be as high as hundreds of kilowatts.

Two factors that are of major importance in power amplifiers are (1) efficiency, and (2) heat transfer or thermal conductivity. Efficiency may not be important in a small-signal amplifier operating at 10 mW, but it is extremely important in a power amplifier operating at 10 kW. Similarly, if adequate heat transfer is not provided, power amplifier transistors may be quickly destroyed.

19–2 HYPERBOLA OF MAXIMUM POWER DISSIPATION

Since power dissipation is a major factor in a power transistor, an important concept concerning its operation may be depicted graphically. For this purpose, a BJT will be used, although an FET could be used as well.

Refer to Figure 19–1, in which the coordinate axes of I_C and V_{CE} are shown. The actual transistor output characteristic curves are not shown here since they would add confusion to the curve to be developed. However, their presence is understood.

For this purpose, assume a dc collector–emitter voltage V_{CE} and a dc collector current I_C. Neglecting any small base power, the total power P_D dissipated in the transistor is given by

$$\boxed{P_D = V_{CE}I_C} \qquad \text{(19–1)}$$

For a given transistor, heat transfer conditions, and temperature, there is a maximum value of P_D specified for operation. Indicating this value as $P_{D(\text{max})}$, we can express (19–1) as

$$V_{CE}I_C = P_{D(\text{max})} \qquad \text{(19–2)}$$

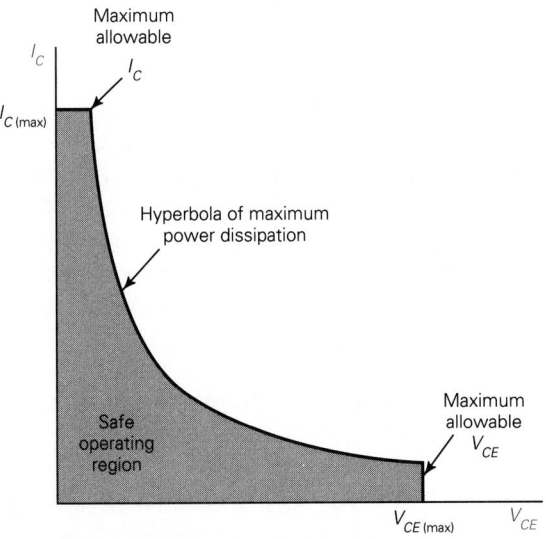

FIGURE 19–1

Hyperbola of maximum power dissipation.

If $P_{D(max)}$ is a constant, then V_{CE} and I_C may be considered as variables. This means that as V_{CE} increases, the maximum allowable value of I_C must decrease, and vice versa. The curve is of the form $xy = constant,$ and this is a **hyperbola.** The resulting curve is called the **hyperbola of maximum power dissipation,** and its form is shown in Figure 19–1.

The hyperbola is not the only limit that must be observed. There is a maximum value of V_{CE}, indicated here as $V_{CE(max)}$, and there is also a maximum value of I_C, indicated here as $I_{C(max)}$. Thus, the safe operating region for a transistor is that region bounded by the hyperbola of maximum power dissipation and the maximum respective values of voltage and current. For any specified operating point, it is absolutely essential that voltage, current, and power all be below their maximum respective values.

EXAMPLE 19–1

A certain power transistor has a maximum allowable power dissipation $P_{D(max)} =$ 30 W at 25°C. The maximum allowable collector–emitter voltage is $V_{CE(max)} = 50$ V, and the maximum allowable collector current is $I_{C(max)} = 2$ A. **(a)** Determine the maximum allowable collector current when $V_{CE} = 25$ V. **(b)** Determine the maximum allowable collector current when $V_{CE} = 6$ V. **(c)** Determine the maximum allowable collector–emitter voltage when $I_C = 1.5$ A. **(d)** Determine the maximum allowable collector–emitter voltage when $I_C = 0.25$ A. All calculations are to be based on 25°C.

Solution

The first point to note is that the product $V_{CE(max)}I_{C(max)} = 50 \text{ V} \times 2 \text{ A} = 100 \text{ W}$, and this exceeds $P_{D(max)}$ by a factor of $100/30 = 3.33$! This means that the transistor cannot operate at maximum collector–emitter voltage and maximum collector current simultaneously. Referring to Figure 19–1, note that when V_{CE} is near $V_{CE(max)}$, I_C is quite small. Similarly, when I_C is near $I_{C(max)}$, V_{CE} is quite small.

For a given value of V_{CE}, we will first calculate I_C based on the maximum power rating. If $I_C \leq I_{C(max)}$, the calculated current is the maximum allowable value for that particular value of V_{CE}. If $I_C > I_{C(max)}$, we must use $I_{C(max)}$ rather than the calculated current as the acceptable value. An analogous process is used with V_{CE} when I_C is given.

(a) For $V_{CE} = 25 \text{ V}$, the value of I_C corresponding to a dissipation of 30 W is

$$I_C = \frac{30 \text{ W}}{25 \text{ V}} = 1.2 \text{ A} \qquad\qquad \text{(19–3)}$$

Since this value is less than 2 A, the maximum current allowed for $V_{CE} = 25 \text{ V}$ is 1.2 A.

(b) For $V_{CE} = 6 \text{ V}$, the value of I_C corresponding to 30 W dissipation is

$$I_C = \frac{30 \text{ W}}{6 \text{ V}} = 5 \text{ A} \qquad\qquad \text{(19–4)}$$

However, this value *exceeds* $I_{C(max)}$. Consequently, the maximum current allowed for $V_{CE} = 6 \text{ V}$ is $I_{CE(max)}$ (i.e., 2 A).

(c) For $I_C = 1.5 \text{ A}$, the value of V_{CE} corresponding to 30 W dissipation is

$$V_{CE} = \frac{30 \text{ W}}{1.5 \text{ A}} = 20 \text{ V} \qquad\qquad \text{(19–5)}$$

Since this value is less than 50 V, the maximum voltage allowed for $I_C = 1.5 \text{ A}$ is 20 V.

(d) For $I_C = 0.25 \text{ A}$, the value of V_{CE} corresponding to a dissipation of 30 W is

$$V_{CE} = \frac{30 \text{ W}}{0.25 \text{ A}} = 120 \text{ V} \qquad\qquad \text{(19–6)}$$

However, this value *exceeds* $V_{CE(max)}$. Consequently, the maximum voltage allowed for $I_C = 0.25 \text{ A}$ is $V_{CE(max)}$ (i.e., 50 V).

Reviewing the four parts of the problem, the conditions of **(a)** and **(c)** represent *power-limiting situations*, since it was the power rating rather than the voltage or current rating that established the limit. The condition of **(b)** represented a *current-limiting situation* since it was the maximum current rating that established the limit. Finally, the condition of **(d)** represented a *voltage-limiting condition*.

19–3 CLASSIFICATION OF AMPLIFIERS

We will now consider a particular classification scheme for amplifiers that has been widely used since the early days of vacuum tube amplifiers. This classification scheme, which is very important in dealing with power amplifiers, relates to the *conduction angle* of active operation of the device.

It is customary for this purpose to assume an input sinusoidal signal. Although a sinusoidal signal rarely represents an actual useful signal in an application of an amplifier, it is widely used as a standard test signal. The response of the amplifier to a sinusoidal test signal is a meaningful way to provide specifications for the unit, and the properties resulting from other waveforms may be inferred from the sinusoidal response.

A **full cycle** of the input signal refers to the sinusoid over a time duration equal to one period. The angle involved is 360°. A **half-cycle** of the input signal refers to the sinusoid over a time duration equal to one half-cycle. The angle involved is 180°.

Class A Amplifier

A **class A amplifier** is one in which the active device conducts for a full-cycle (or 360°) of the input signal. All amplifiers considered in the earlier chapters on transistors and op-amps have been assumed to be of the class A type. Most small-signal amplifiers are operated as class A amplifiers.

Class B Amplifier

A class B amplifier is one in which each active device conducts for one-half cycle (or 180°C) of the input signal. Normally, class B amplifiers employ two active devices, with each device conducting on alternate half-cycles. The combination of the two devices then results in full-angle conduction. Strictly speaking, there is no conduction in a class B amplifier without a signal applied so that power dissipation occurs only when the amplifier is excited.

Class AB Amplifier

Some amplifiers deviate slightly from the class B condition and provide a slight forward bias with no signal applied. In this case, conduction for each active device occurs for somewhat more than a half-cycle. Such amplifiers are classified as class AB. Except for the slight forward bias, class AB amplifiers are very close to class B amplifiers in their various properties.

Class C Amplifier

A class C amplifier is one in which the active device conducts for less than one-half cycle. Conduction thus occurs in short pulses. A class C amplifier is basically a nonlinear circuit and imparts significant distortion to an input signal in its basic form. However, by using one or more tuned circuits, a useful output may be derived for specialized applications, such as in radio-frequency circuits.

Efficiency

The efficiency of amplifiers increases with the alphabetical listing of the amplifier type. Thus, a class A amplifier has the lowest efficiency, and a class C amplifier has the highest efficiency. Certain idealized values of efficiencies will be stated in subsequent sections.

19–4 POWER RELATIONSHIPS

It is essential to understand the manner in which power is distributed in a power amplifier. Actually, power is a time-varying function, and it is possible to describe all power relationships on an instantaneous basis. However, it is more customary to assume a standard signal waveform and develop the various power relationships on an average basis. The standard waveform used for this basis is the sinusoidal function. Thus, if an amplifier is rated at 20 W output, the specification usually means that the average power supplied to some specified load is 20 W based on an assumed sinusoidal signal.

The following quantities are defined:

P_S: average power delivered by power supply to amplifier (also called the **dc input power**)

P_O: output signal power delivered to load (also called the **ac output power**)

P_D: power dissipated by amplifier

The conservation of power for the circuit requires that

$$P_S = P_O + P_D \qquad \textbf{(19–7)}$$

Stated in words, the power delivered by the power supply must equal the sum of output signal power delivered to the load plus any power dissipated in the amplifier.

The equation as given applies only to the active device. Obviously, any bias circuitry involved will consume power. Similarly, small-signal amplifiers used to amplify the signal to a level sufficient to drive the power amplifier stage will also consume power. Our focus now is on the power output stage only.

Efficiency

The efficiency η of the power amplifier can be expressed as

$$\eta = \frac{P_O}{P_S} \times 100\% \qquad \textbf{(19–8)}$$

Greater efficiency is thus obtained as more of the dc input power is converted to ac output power and less of the power is dissipated in the amplifier. Depending on the amplifier type and the manner in which it is operated, efficiency may range from a few percent to a theoretical level approaching 100% for very ideal conditions. Most practical power amplifiers, however, operate well below 100% efficiency.

EXAMPLE 19–2

A certain power amplifier delivers 12 W to an external load. The power delivered by the power supply is 20 W. Determine **(a)** the internal power dissipated by the power amplifier, and **(b)** the efficiency.

Solution

(a) Since the dc input power is 20 W and the ac output power is 12 W, the internal power dissipation P_D is

$$P_D = P_S - P_O = 20 - 12 = 8 \text{ W} \qquad \text{(19–9)}$$

(b) The efficiency is

$$\eta = \frac{P_O}{P_S} \times 100\% = \frac{12}{20} \times 100\% = 60\% \qquad \text{(19–10)}$$

19–5 CLASS A POWER AMPLIFIERS

For reasons of operational comparison and historical perspective, a transformer-coupled class A power amplifier will be discussed even though it is limited in modern usage. The circuit form is shown in Figure 19–2. As shown, the circuit does not have a stabilized bias circuit and is thus impractical. However, this simplification enhances the load-line analysis and the analytical analysis to be given later in this section.

FIGURE 19–2

Simplified transformer-coupled class A power amplifier.

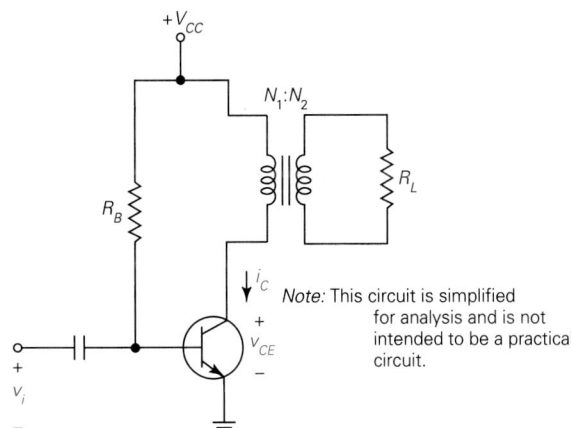

Note: This circuit is simplified for analysis and is not intended to be a practical circuit.

The transformer serves two purposes: (1) it allows the full dc supply voltage to be applied to the collector (in practice, there is some resistance in the transformer windings, but it will be neglected in this analysis), and (2) the turns ratio $N_1 : N_2$ of the transformer can be selected to provide an optimum impedance transformation between the transistor and the load.

Assuming negligible resistance in the transformer, the quiescent dc collector–emitter voltage is simply $V_{CEQ} = V_{CC}$. A *dc load line* may be constructed as shown in Figure 19–3. Since the assumed resistance of the transformer is zero, the slope of the dc load line is infinite and is a vertical line as shown.

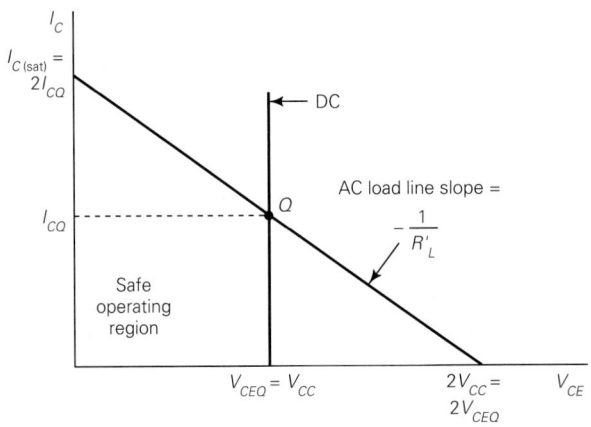

FIGURE 19–3
Load lines for class A power amplifier based on optimum conditions.

When an input signal is applied, the action of the transformer becomes significant. An effective dynamic resistance R'_L is coupled back to the primary and is given by

$$R'_L = \left(\frac{N_1}{N_2}\right)^2 R_L \qquad (19\text{–}11)$$

The primary voltage of the transformer will then vary up and down about V_{CEQ} because of transformer action. This process may be illustrated with an *ac load line* as shown in Figure 19–3. Thus a transformer-coupled amplifier may be characterized by the combination of a *dc load line and an ac load line*.

The optimum primary condition for class A operation is when V_{CE} reaches a level of $2V_{CC}$ on a positive half-cycle and 0 on a negative half-cycle. Similarly, I_C should reach $2I_{CQ} = I_{C(\text{sat})}$ on a positive half-cycle and zero on a negative half-cycle. The inverse magnitude of the slope of the ac load line is given by

$$\frac{2V_{CC}}{2I_{CQ}} = \frac{V_{CC}}{I_{CQ}} = \frac{V_{CEQ}}{I_{CQ}} = R'_L \qquad (19\text{–}12)$$

From the maximum power transfer theorem, maximum output power is obtained when

$$\left(\frac{N_1}{N_2}\right)^2 R_L = R'_L = \frac{V_{CEQ}}{I_{CQ}} \tag{19-13}$$

Power and Efficiency

Assume now that the instantaneous collector current i_c is varying sinusoidally about a Q-point according to

$$i_C = I_{CQ} + kI_{CQ} \sin \omega t \tag{19-14}$$

The quantity k is a **drive factor** and lies in the range $0 \le k \le 1$. When $k = 0$, there is no signal present, and when $k = 1$, the signal has the maximum allowable variation without distortion, assuming idealized characteristics.

Assume that the collector–emitter voltage is of the form

$$\begin{aligned} v_{CE} &= V_{CQ} - kV_{CQ} \sin \omega t \\ &= V_{CC} - kV_{CC} \sin \omega t \end{aligned} \tag{19-15}$$

Note that the signal component of V_{CE} is chosen to be 180° out of phase with the signal component of i_C as required.

The instantaneous power p_S delivered by the power supply is

$$\begin{aligned} p_S &= V_{CC} i_C = V_{CC}(I_{CQ} + kI_{CQ} \sin \omega t) \\ &= V_{CC} I_{CQ} + kV_{CC} I_{CQ} \sin \omega t \end{aligned} \tag{19-16}$$

The average or dc power P_S delivered by the power supply is simply the constant part of (19–16) since the sinusoidal portion has an average value of zero. Hence

$$P_S = V_{CC} I_{CQ} \tag{19-17}$$

This average power is independent of the drive, so the class A power amplifier consumes the same dc power with or without a signal.

The instantaneous output power p_O is also the power accepted by the transformer primary (assuming a loseless transformer), and it is given by

$$p_O = i_C(V_{CC} - v_{CE}) = kV_{CEQ} I_{CQ} \sin \omega t + k^2 V_{CEQ} I_{CQ} \sin^2 \omega t \tag{19-18}$$

It can be shown that the average output power P_O can be determined from (19–18) to be

$$P_O = \frac{k^2 V_{CEQ} I_{CQ}}{2} = \frac{k^2 V_{CC} I_{CQ}}{2} = \frac{k^2 V_{CC}^2}{2R'_L} \tag{19-19}$$

The maximum output power $P_{O(max)}$ occurs when $k = 1$ and is

$$P_{O(max)} = \frac{V_{CEQ} I_{CQ}}{2} = \frac{V_{CC} I_{CQ}}{2} = \frac{V_{CC}^2}{2R'_L} \tag{19-20}$$

The efficiency η is

$$\eta = \frac{P_O}{P_S} \times 100\% = \frac{k^2 V_{CC} I_{CQ}/2}{V_{CC} I_{CQ}} \times 100\% = \frac{k^2}{2} \times 100\% \qquad \textbf{(19–21)}$$

The maximum efficiency η_{max} occurs when $k = 1$ and is

$$\boxed{\eta_{\text{max}} = 50\%} \qquad \textbf{(19–22)}$$

To simplify the analysis, recall that the form of the amplifier chosen was highly simplified. When stable bias is added, the maximum peak output will be reduced, and this lowers the efficiency. Consequently, the 50% value should be considered as an ideal upper bound. Practical class A amplifiers operate with somewhat lower efficiencies.

Internal Dissipation

The average power P_D dissipated in the transistor is

$$P_D = P_S - P_O = V_{CC} I_{CQ} - \frac{k^2 V_{CC} I_{CQ}}{2} = V_{CC} I_{CQ}\left(1 - \frac{k^2}{2}\right) \qquad \textbf{(19–23)}$$

The maximum dissipation $P_{D(\text{max})}$ occurs for $k = 0$ and is

$$\boxed{P_{D(\text{max})} = V_{CC} I_{CQ} = 2 P_{O(\text{max})}} \qquad \textbf{(19–24)}$$

Thus the maximum internal dissipation in a class A transformer-coupled amplifier is twice the maximum output power. The major results of this section, along with results to be developed in Section 19–6, are summarized in Table 19–1.

We have only considered a transformer-coupled class A amplifier in this section. It is possible to use an RC-coupled amplifier as a class A power amplifier, but the efficiency is even less. The theoretical maximum efficiency of a class A RC-coupled amplifier is 25%, and practical amplifiers of this type have efficiencies well below this value.

EXAMPLE 19–3   A certain idealized class A transformer-coupled power amplifier has $V_{CC} = 24$ V, $I_{CQ} = 0.5$ A, and it is used to drive a speaker whose dynamic resistance is $R_L = 8 \ \Omega$. Based on the simplified assumptions of the model of Figure 19–2, determine **(a)** V_{CEQ}; **(b)** R'_L; **(c)** N_1/N_2; **(d)** P_S; **(e)** $P_{O(\text{max})}$; and **(f)** $P_{D(\text{max})}$.

Solution

(a)   Assuming a lossless transformer, the Q-point collector–emitter voltage is

$$V_{CEQ} = V_{CC} = 24 \text{ V} \qquad \textbf{(19–25)}$$

(b) The dynamic resistance R'_L reflected to the primary should satisfy

$$R'_L = \frac{V_{CEQ}}{I_{CQ}} = \frac{24 \text{ V}}{0.5 \text{ A}} = 48 \ \Omega \tag{19-26}$$

(c) The turns ratio should satisfy the equation

$$\left(\frac{N_1}{N_2}\right)^2 R_L = R'_L \tag{19-27}$$

or

$$\frac{N_1}{N_2} = \sqrt{\frac{R'_L}{R_L}} = \sqrt{\frac{48}{8}} = 2.449 \tag{19-28}$$

(d) The average power delivered by the power supply is

$$P_S = V_{CC}I_{CQ} = 24 \times 0.5 = 12 \text{ W} \tag{19-29}$$

(e) The maximum value of the output power is

$$P_{O(max)} = \frac{V_{CC}I_{CQ}}{2} = \frac{24 \times 0.5}{2} = 6 \text{ W} \tag{19-30}$$

(f) The maximum internal dissipation is

$$P_{D(max)} = 2P_{O(max)} = 2 \times 6 = 12 \text{ W} \tag{19-31}$$

TABLE 19–1
Summary of Certain Power Relationships[a]

	Class A (with transformer)	Class B
P_O	$\dfrac{k^2 V_{CC}I_{CQ}}{2}$	$\dfrac{k^2 V_{CC}I_{C(sat)}}{2}$
P_S	$V_{CC}I_{CQ}$	$\dfrac{2}{\pi} k V_{CC}I_{C(sat)}$
P_D	$2P_{O(max)}\left(1 - \dfrac{k^2}{2}\right)$	$P_{O(max)}\left(\dfrac{4k}{\pi} - k^2\right)$
η	$\dfrac{k^2}{2} \times 100\%$	$\dfrac{\pi}{4} k \times 100\%$
$P_{O(max)}$[b]	$\dfrac{V_{CC}I_{CQ}}{2}$	$\dfrac{V_{CC}I_{C(sat)}}{2}$
$P_{S(max)}$	$V_{CC}I_{CQ}$	$\dfrac{4}{\pi} P_{O(max)}$
$P_{D(max)}$	$2P_{O(max)}$	$\dfrac{4}{\pi^2} P_{O(max)}$
η_{max}	50%	78.5%

[a] k is a "drive" factor $0 \le k \le 1$.
[b] $P_{D(max)}$ occurs at a different value of k than do $P_{O(max)}$ and $P_{S(max)}$.

19–6 CLASS B AMPLIFIER

A class B amplifier is one in which conduction in each transistor occurs for 50% of a cycle. To obtain full-cycle coverage without tuned circuits, therefore, it is necessary that two transistors be employed.

Earlier class B amplifiers employed transformers to couple from the output stages to the load. Most modern circuits utilize direct coupling between a low-impedance output stage and the load. The treatment here will focus on a very popular circuit of this type, namely the **emitter-follower push-pull amplifier.** This circuit is widely employed, both in discrete form and as the output stage of many integrated circuits.

Emitter-Follower Push-Pull Amplifier

The basic form of the circuit is shown in Figure 19–4. Some refinements will be made on this circuit later to make it better, but the basic form shown here will suffice for the discussion that follows.

FIGURE 19–4
Basic form of a class B emitter-follower push-pull amplifier. (Circuit as given is subject to crossover distortion.)

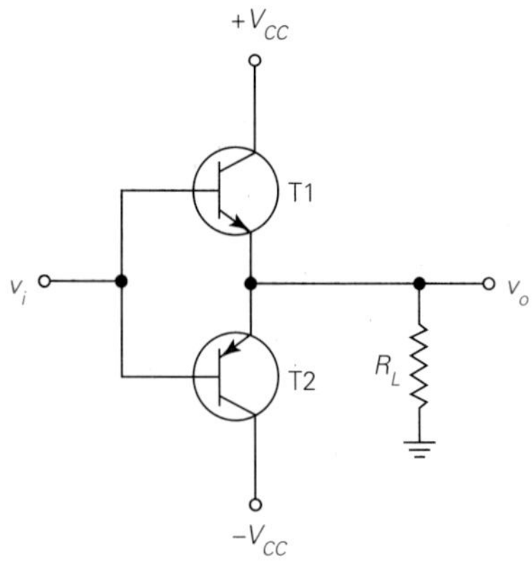

Note that one transistor is an *npn* unit, and the other is a *pnp* unit. To minimize distortion, their characteristics should be identical (except, of course, for opposite polarities). A pair of transistors with matched characteristics and opposite polarities is said to be a **complementary-symmetry matched pair.**

Assume initially that $v_i = 0$. Neither base–emitter diode has any forward bias, so there is no conduction in either transistor, and $v_o = 0$. Thus, in this basic circuit, there is no power dissipation with no signal applied.

Now assume that v_i is a positive voltage well above 0.7 V. The *npn* transistor T1 is driven into conduction, but the *pnp* transistor T2 is reverse biased. The input signal is thus coupled to the load but suffers a loss of about 0.7 V due to the base–emitter drop of T1.

Next, assume that v_i is a negative voltage well below -0.7 V. The base–emitter junction of T2 is now forward biased, but T1 is reverse biased. Conduction between source and load is now through T2. Once again, the output tracks the input, but with a loss of about 0.7 V in magnitude.

Power amplification is achieved by the fact that the required power at the input is very small, but the transistors are capable of delivering considerable power to a load R_L. Thus, although the voltage is not amplified, the power level can be increased dramatically.

The basic circuit as given suffers from a phenomenon called **crossover distortion.** This distortion arises from the **deadband** between about -0.7 and $+0.7$ V when neither transistor is conducting. The manner in which crossover distortion affects a sinusoidal input signal is illustrated in Figure 19–5.

FIGURE 19–5
Crossover distortion in basic push-pull amplifier of Figure 19–4.

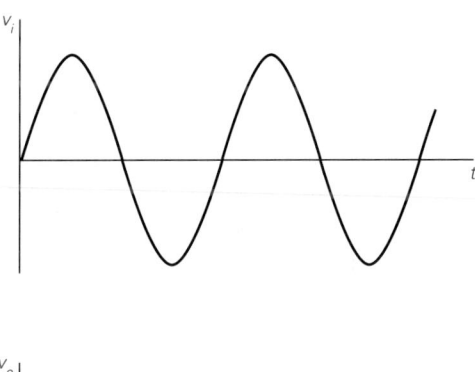

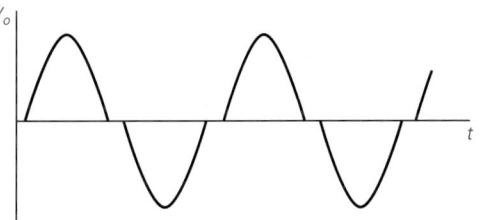

Crossover distortion is eliminated by placing a forward bias on each of the two base–emitter junctions. There are several ways to do this, one of which employs two diodes and dual power supplies and is shown in Figure 19–6. Assume first that $v_i = 0$. The base of T1 will then be 0.7 V above ground, and the base of T2 will be at 0.7 V below ground. The two transistors will barely be starting to conduct at $v_o = 0$.

Full conduction for T1 is now established at a very low positive signal level, and full conduction for T2 is now established at a very low negative signal level.

FIGURE 19–6
Class B emitter-follower push-pull amplifier with circuit used to eliminate crossover distortion.

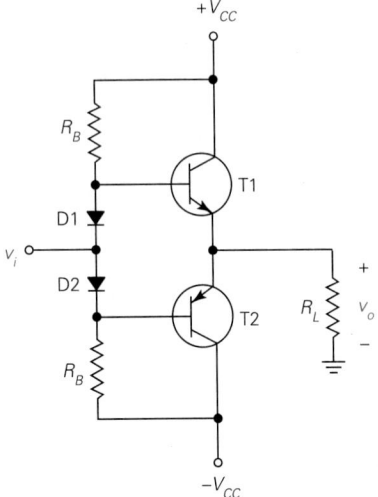

In effect, the base of T1 tracks v_i, but it is at a level of 0.7 V above v_i. Conversely, the base of T2 tracks v_i, but it is at a level of 0.7 V below v_i. Crossover distortion is virtually eliminated and $v_o = v_i$ without the 0.7-V loss.

Maximum effectiveness in this circuit is achieved when the characteristics of the diodes match those of the base–emitter transistor junctions. On integrated circuit chips, this is achieved by using transistors as diodes. This process is illustrated in Figure 19–7.

FIGURE 19–7
Integrated-circuit power amplifier stage
with transistors used for diodes.

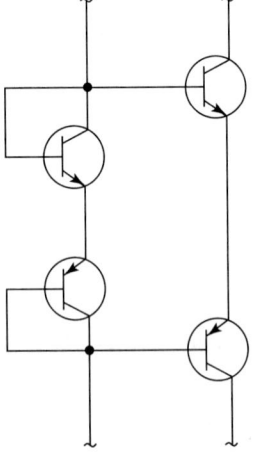

Incidentally, the emitter-follower push-pull amplifier may be biased with a single power supply if ac coupling is employed at the output. This form of the circuit is shown in Figure 19–8. Depending on the manner in which the circuit is driven, ac coupling may or may not be required at the input.

FIGURE 19–8
Class B amplifier operated with single
power supply.

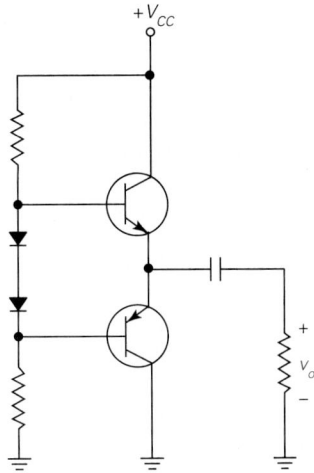

Power and Efficiency

The various power and efficiency relationships of the class B amplifier will now
be developed. A sinusoidal signal will be assumed, and the circuit form of Figure
19–6 employing dual power supplies will be used.

   If the saturation drops of the two transistors are neglected, the output voltage
v_o can swing all the way down to $-V_{CC}$ and all the way up to $+V_{CC}$. The output
voltage v_o can then be expressed as

$$v_o = kV_{CC} \sin \omega t \qquad (19\text{–}32)$$

where k is a "drive factor" in the range $0 \le k \le 1$. The instantaneous output
current i_o can be expressed in a similar form as

$$i_o = kI_{C(\text{sat})} \sin \omega t \qquad (19\text{–}33)$$

where $I_{C(\text{sat})}$ is the transistor saturation current occurring at the peak voltage V_{CC}.
The peak load voltage and peak load current are related by

$$\frac{V_{CC}}{I_{C(\text{sat})}} = R_L \qquad (19\text{–}34)$$

   The rms values of the output voltage and current are $kV_{CC}/\sqrt{2}$ and $kI_{C(\text{sat})}/$
$\sqrt{2}$, respectively. The output load power P_O can then be expressed as

$$P_O = \frac{kV_{CC}}{\sqrt{2}} \frac{kI_{C(\text{sat})}}{\sqrt{2}} = \frac{k^2 V_{CC} I_{C(\text{sat})}}{2} \qquad (19\text{–}35)$$

The maximum value of this power corresponds to $k = 1$ and is denoted as $P_{O(\text{max})}$.
We have

$$P_{O(\text{max})} = \frac{V_{CC} I_{C(\text{sat})}}{2} \qquad (19\text{–}36)$$

The instantaneous source power p_S from *either* one of the two sources over an interval of *one-half cycle* is

$$p_S = V_{CC}i_O = kV_{CC}I_{C(\text{sat})} \sin \omega t \qquad (19\text{–}37)$$

It can be shown using calculus that the average value P_S of the source power determined over a half-cycle interval, which is then the total average power due to the symmetry, is

$$P_S = \frac{2}{\pi} kV_{CC}I_{C(\text{sat})} \qquad (19\text{–}38)$$

A significant point to note is that $P_S = 0$ when $k = 0$. This means that the amplifier consumes no power in the absence of a signal. (Of course, power is consumed in the bias circuit and any class A driver circuits, etc.)

The maximum value of P_S is denoted as $P_{S(\text{max})}$, and it corresponds to $k = 1$ in (19–38).

$$P_{S(\text{max})} = \frac{2}{\pi} V_{CC}I_{C(\text{sat})} \qquad (19\text{–}39)$$

The efficiency η is

$$\eta = \frac{P_O}{P_S} \times 100\% = \frac{k^2 V_{CC}I_{C(\text{sat})}/2}{2kV_{CC}I_{C(\text{sat})}/\pi} \times 100\% = \frac{\pi}{4}k \times 100\% \qquad (19\text{–}40)$$

The maximum efficiency $\eta_{\max}$ occurs at $k = 1$ and is

$$\boxed{\eta_{\max} = \frac{\pi}{4} \times 100\% = 78.5\%} \qquad (19\text{–}41)$$

The class B amplifier is thus theoretically capable of achieving an efficiency of 78.5% at full output. This compares with the theoretical maximum of 50% for a class A amplifier. In addition, the ideal class B amplifier consumes no power with no input signal while the class A amplifier consumes power on a continual basis.

Practical class B amplifiers, like class A amplifiers, operate at efficiencies below the theoretical maximum level.

Internal Dissipation

The average internal power dissipation P_D may be expressed as

$$P_D = P_S - P_O = \frac{2}{\pi} kV_{CC}I_{C(\text{sat})} - \frac{k^2}{2} V_{CC}I_{C(\text{sat})} \qquad (19\text{–}42)$$

It can be shown using calculus that the maximum value of the power dissipation occurs when $k = 2/\pi = 0.637$. The maximum value of dissipation $P_{D(\text{max})}$ corresponding to this value of k is

$$\boxed{P_{D(max)} = \frac{2}{\pi^2} V_{CC} I_{C(sat)}}$$ (19–43)

Using (19–36), the maximum power dissipation may be expressed in terms of the maximum output power as

$$\boxed{P_{D(max)} = \frac{4}{\pi^2} P_{O(max)} \approx 0.405\, P_{O(max)}}$$ (19–44)

Thus the maximum power dissipation for the circuit is about 40% of the maximum output power. This is a considerable improvement over the class A circuit.

The major results for this section, along with the results of Section 19–5, are summarized in Table 19–1.

EXAMPLE 19–4

A certain class B push-pull emitter-follower power amplifier is biased by two power supplies with values ± 12 V, and it is used to drive a speaker whose dynamic resistance is $R_L = 8\ \Omega$. Based on the assumptions in this section, determine (a) $I_{C(sat)}$; (b) $P_{S(max)}$; (c) $P_{O(max)}$; (d) $P_{D(max)}$.

Solution

(a) From (19–34), the peak current $I_{C(sat)}$ is determined as

$$I_{C(sat)} = \frac{V_{CC}}{R_L} = \frac{12\ \text{V}}{8\ \Omega} = 1.5\ \text{A}$$ (19–45)

(b) The peak power delivered by the power supply is

$$P_{S(max)} = \frac{2}{\pi} V_{CC} I_{C(sat)} = \frac{2}{\pi} \times 12 \times 1.5 = 11.46\ \text{W}$$ (19–46)

(c) The maximum output power is

$$P_{O(max)} = \frac{V_{CC} I_{C(sat)}}{2} = \frac{12 \times 1.5}{2} = 9\ \text{W}$$ (19–47)

(d) The maximum internal power dissipated is

$$P_{D(max)} = \frac{4}{\pi^2} P_{O(max)} = \frac{4}{\pi^2} \times 9 = 3.648\ \text{W}$$ (19–48)

19–7 CLASS C AMPLIFIER

In a class C power amplifier, conduction occurs for less than 50% of a full cycle. Because of the low conduction cycle, the use of class C amplifiers is primarily limited to narrow-band signals in which a high-Q resonant circuit can be used to restore the output signal to a full conduction cycle. Primary application of class C amplifiers is in communications circuits such as radio transmitters and radio receivers. Signals in these systems tend to have all their frequency components in a narrow portion of the frequency spectrum, and the resonant effect can be used to advantage in high-efficiency power applications.

A simplified class C amplifier employing a BJT is shown in Figure 19–9. The output of the circuit is very similar to that of a tuned amplifier as described in Section 18–4. An LC resonant circuit is employed with the Q selected for the appropriate bandwidth. However, the feature that establishes the circuit as a class C amplifier is the presence of a negative voltage $-V_{BB}$ in the base circuit. The presence of this negative voltage means that the base of the BJT is at a negative voltage with respect to ground, and conduction in the transistor can only occur whenever the input voltage exceeds this negative voltage by about 0.7 V.

FIGURE 19–9
Basic form of a class C amplifier.

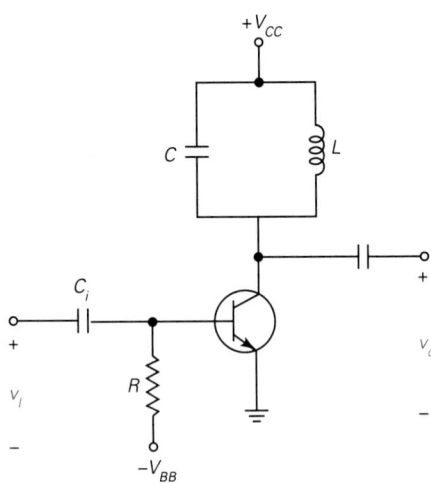

In the discussion that follows, refer to the waveforms of Figure 19–10 as a supplement to the schematic diagram. For discussion purposes, assume a sinusoidal input v_i as shown in Figure 19–10(a). The transistor begins conducting when the input signal exceeds the dc base bias by about 0.7 V. Conduction continues until the base voltage drops below the cut-in voltage. The result is a sequence of collector current pulses as shown in Figure 19–10(b).

Although collector current flow consists of short pulses, the high-Q resonant circuit produces a "flywheel effect." Energy stored in the capacitor on one half-cycle is related to the inductor on the next half-cycle, and the resulting voltage across the tuned circuit is sinusoidal in form. The result is that the collector voltage v_C is centered at V_{CC} but oscillates above and below this level as shown in Figure 19–10(c). After coupling through the output capacitor, the output waveform is a sinusoid oscillating about zero voltage.

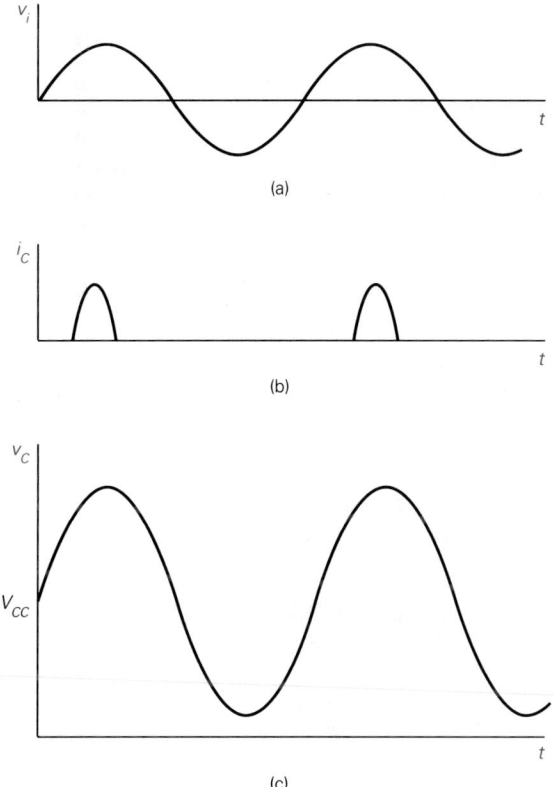

FIGURE 19–10
Waveforms in a class C amplifier.

Optimum efficiency conditions occur when the transistor is driven to full saturation on the peak of the input. Under this condition, the voltage v_C varies all the way from zero to $2V_{CC}$, and the output voltage v_O is a sinusoid with peak value V_{CC}. Efficiency in this limiting case approaches 100%.

19–8 THERMAL ANALYSIS

Because of the large power levels in power amplifiers, a considerable amount of heat may have to be dissipated. Some important insight into this process may be obtained from certain thermal models and the associated analysis.

Consider the model shown in Figure 19–11. Assume a physical situation in which two temperatures T_A and T_B are separated by some thermal path. Assume that $T_A > T_B$. Heat will then flow from T_A in the direction of T_B. The amount of heat flow is directly proportional to the difference in temperatures and inversely proportional to a quantity called the **thermal resistance.**

FIGURE 19-11
Basic form of thermal model for heat
flow analysis.

FIGURE 19-11
Basic form of thermal model for heat
flow analysis.

The situation may be analyzed by an analogy to Ohm's law. The quantities involved, units, and the corresponding electrical variables are summarized in Table 19-2. The form of the heat flow equation is

$$P = \frac{T_A - T_B}{\Theta_{AB}}$$ (19–49)

where Θ_{AB} is the thermal resistance between points A and B. This equation is analogous to $I = V/R$. Note that the "flow variable" for the thermal system is power, representing the amount of heat flow in watts. The units for thermal resistance are degrees Celsius per watt (°C/W).

TABLE 19-2
Thermal Quantities and the Associated Electrical Analogies

Quantity	Symbol	Units	Abbreviation of units	Electrical analog
Temperature	T	Degrees Celsius	°C	Voltage
Heat flow	P	Watts	W	Current
Thermal resistance	Θ	Degrees Celsius per watt	°C/W	Resistance

A thermal model describing heat flow in a transistor is shown in Figure 19–12. The following quantities are defined:

T_J = transistor junction temperature

T_C = transistor case temperature

T_A = ambient or air temperature outside transistor

Θ_{JC} = thermal resistance between junction and case

Θ_{CA} = thermal resistance between case and air

Θ_{JA} = net thermal resistance between junction and air = $\Theta_{JC} + \Theta_{CA}$

P_D = power dissipated by transistor

FIGURE 19-12
Thermal model for transistor.

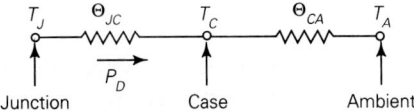

The power P_D dissipated by the transistor is converted to heat, and this heat must be transferred to the outside. By the Ohm's law analogy, this power must satisfy the equation

$$P_D = \frac{T_J - T_A}{\Theta_{JA}} \tag{19–50}$$

When (19–50) is solved for T_J, the following result is obtained:

$$T_J = T_A + \Theta_{JA}P_D \tag{19–51}$$

For a given transistor, there is a maximum value permitted for the junction temperature T_J. From (19–51), for a given thermal resistance, T_J increases as either the ambient temperature T_A increases or as the power dissipation P_D increases. The specified maximum value of power dissipation $P_{D(max)}$ is normally based on $T_A = 25°C$, and it represents the value that would result in the maximum allowable junction temperature.

If $T_A > 25°C$, the only way to maintain a balance on the right-hand side of (19–51) is to reduce P_D since Θ_{JA} is fixed. This is accomplished through a derating factor, for which the concept was briefly introduced back in Chapter 4.

To keep the notation from getting too bulky, the subscript "max" in the power equation that follows will be omitted, but the power level will be understood to be the maximum value. Let $P_D(T_A = 25°C)$ represent the maximum transistor power dissipation at an ambient temperature of 25°C. The value of maximum power dissipation at a temperature T_A will be denoted as $P_D(T_A)$, and it is

$$\boxed{P_D(T_A) = P_D(T_a = 25°C) - (T_A - 25)D_{JA}} \tag{19–52}$$

where D_{JA} is the junction-to-air derating factor measured in watts/°C. Inspection of (19–52) reveals that there is a simple relationship between the thermal resistance and the derating factor, which is

$$\Theta_{JA} = \frac{1}{D_{JA}} \tag{19–53}$$

Heat Sinks

Power transistors are almost always operated with **heat sinks.** A heat sink is a quantity of metal having a relatively large surface area and high heat conductivity. It facilitates heat transfer to the outside air and allows the transistor to operate at a cooler case temperature. This results in the ability of the transistor to operate at higher values of power dissipation, as we will see shortly.

Heat Flow with Heat Sink

We now consider the effect of a heat sink on the thermal circuit. Refer to Figure 19–13. The heat sink provides a separate parallel path for power to flow to the

FIGURE 19–13
Effect of heat sink on thermal model.

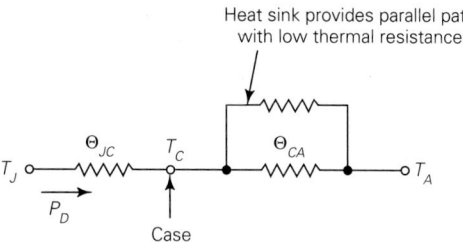

ambient environment. If the heat sink had infinite area, the shunt path would be a short circuit and the case and ambient temperatures would be the same. In practice, however, T_C will always be larger than T_A, but a good heat sink reduces the difference significantly.

A separate specification for power transistors is usually given based on $T_C = 25°C$. The junction temperature may be expressed as a function of the case temperature as

$$T_J = T_C + \Theta_{JC}P_D \qquad (19\text{--}54)$$

The maximum power dissipation at a temperature T_C may now be expressed in terms of the power dissipation when $T_C = 25°C$ and a junction-to-case derating factor D_{JC}. As in the previous case, the subscript "max" will be omitted from $P_D(T_C)$, but the value is understood to be the maximum value. We have

$$\boxed{P_D(T_C) = P_D(T_C = 25°C) - (T_C - 25)D_{JC}} \qquad (19\text{--}55)$$

The thermal resistance Θ_{JC} is related to D_{JC} by

$$\Theta_{JC} = \frac{1}{D_{JC}} \qquad (19\text{--}56)$$

EXAMPLE 19–5   The specifications of some actual power transistors will be studied in Section 19–9. The thermal specifications of these particular transistors may be summarized as follows:

Total power dissipation at $T_C = 25°C$: 30 W
Derate above 25°C: 0.24 W/°C
Total power dissipation at $T_A = 25°C$: 2 W
Derate above 25°C: 0.016 W/°C

(a) Determine Θ_{JC}, Θ_{JA}, and Θ_{CA}. **(b)** In a certain application with a heat sink, the case temperature is $T_C = 35°C$. Determine the maximum allowable power dissipation.

Solution
(a) The value of Θ_{JC} is

$$\Theta_{JC} = \frac{1}{D_{JC}} = \frac{1}{0.24} = 4.167°C/W \tag{19-57}$$

The value of Θ_{JA} is

$$\Theta_{JA} = \frac{1}{D_{JA}} = \frac{1}{0.016} = 62.5°C/W \tag{19-58}$$

From the circuit model analogy of Figure 19–12, the value of Θ_{CA} is

$$\Theta_{CA} = \Theta_{JA} - \Theta_{JC} = 62.5 - 4.167 = 58.3°C/W \tag{19-59}$$

Note that Θ_{CA} is considerably larger than Θ_{JC}. This is a common characteristic of power transistors.

(b) For $T_C = 35°C$, the maximum allowable power dissipation is

$$\begin{aligned} P_D(T_C) &= P_D(T_C = 25) - (35 - 25)D_{JC} \\ &= 30 - 10 \times 0.24 = 27.6 \text{ W} \end{aligned} \tag{19-60}$$

Although this transistor is considered as a 30-W transistor, it should be noted that the allowable power dissipation will be less than 30 W unless the case temperature can be maintained at 25°C.

19–9 POWER TRANSISTOR SPECIFICATIONS

Earlier in the book, some typical specifications of small-signal transistors were studied. Now that we have studied the operation of power amplifiers, it seems appropriate to consider some typical specifications of power transistors.

A wide variety of power transistors are available, with power levels ranging from less than 1 W to thousands of watts. The transistors selected for our review are the TIP29 and TIP30 units along with several variations. The specifications are given in Appendix D, so the reader should refer to the two-page set for the discussion that follows.

The TIP29 and TIP30 are a complementary pair, with the first being an *npn* unit and the second being a *pnp* unit. Several variations of each are indicated with the suffix A, B, or C following the transistor type.

First, study the transistor package arrangement as shown. In addition to the normal emitter, base, and collector terminals, there is also a larger tab shown at the top of the transistor and internally connected to the collector. The purpose of this connection is to provide an effective means for heat transfer out of the transistor.

Refer now to the "maximum ratings." The maximum collector–emitter voltage rating with base open is indicated as V_{CEO}, and the value varies from 40 to 100 V, depending on the suffix. The value of the collector–base voltage V_{CB} is the same in each case as V_{CEO}. Strictly speaking, these values are correct as given for the *npn* units, but they must be interpreted as magnitudes for the *pnp* units, since the actual reverse-biased values are negative in that sense. The maximum reverse voltage V_{EB} between emitter and base is 5 V for all units. Again, this is the magnitude for the *pnp* units.

The continuous maximum collector current rating is 1 A, and the peak current rating is 3 A. The maximum base current is 0.4 A.

It is customary to provide two sets of power dissipation data for power transistors. One set is based on $T_C = 25°C$, and the other is based on $T_A = 25°C$. In each case, the maximum power dissipation and the derating factor are provided. To achieve the maximum power dissipation rating, it would be necessary to maintain the case temperature at 25°C, and this is difficult to acheve in practice. The power rating is then an upper bound in a practical situation. The particular data given here were used in Example 19–5.

Some data concerning "thermal characteristics" are provided in a separate table. The notation there differs slightly from the notation of Section 19–8, but the meanings should be clear. In fact, these particular values were determined from the derating factors in Example 19–5.

Refer now to the "electrical characteristics" data. The collector–emitter maximum voltage values indicated earlier are now repeated as "sustaining voltages" based on a pulse test. The quantity I_{CEO} represents the collector-to-emitter current with the base open, and I_{CES} represents the collector-to-emitter current with the base shorted to the emitter. The maximum value of I_{CEO} is indicated as 0.3 mA, but the maximum value of I_{CES} is 200 μA = 0.2 mA. The quantity I_{EBO} is the emitter-to-base current with the collector open. The maximum value of this quantity is 1 mA. All of the preceding current values should be interpreted as magnitudes for the *pnp* units.

Some data concerning $h_{FE}(\beta_{dc})$ are provided in the table, but more extensive data on this parameter are provided from the curves on the second page. The typical range of h_{FE} at 25°C is around 100 to 150 or so for collector currents less than 1 A.

Returning to the table, note that the maximum value of $V_{CE(sat)}$ is 0.7 V. The maximum value of $V_{BE(on)}$ is 1.3 V, which is considerably larger than the "standard" value of 0.7 V assumed for small signal units.

19–10 PSPICE EXAMPLES

PSPICE EXAMPLE 19–1 Prepare a PSPICE simulation of the class B emitter-follower amplifier shown in Figure 19–14(a). Obtain a transient response over two cycles, and measure the distortion using the .FOUR control.

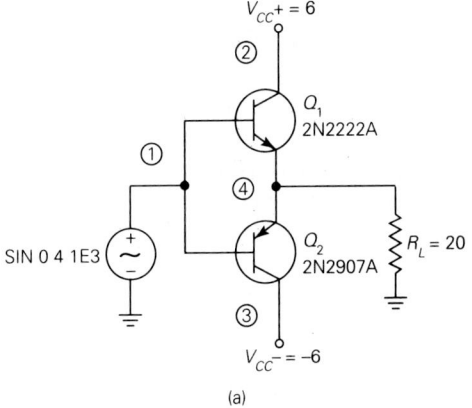

```
PSPICE EXAMPLE 19-1
.OPTIONS NOECHO NOPAGE NOMOD NOBIAS
VI 1 0 SIN 0 4 1E3
Q1 2 1 4 Q2N2222A
Q2 3 1 4 Q2N2907A
VCC+ 2 0 6
VCC- 0 3 6
RL 4 0 20
.LIB EVAL.LIB
.TRAN 2E-3 2E-3 0 10E-6
.PROBE
.FOUR 1E3 V(4)
.END
```

(a)   (b)

FIGURE 19–14
Circuit and code of PSPICE Example 19–1.

Solution

This amplifier circuit requires a matched pair consisting of an *npn* and a *pnp* transistor. The evaluation version of PSPICE contains a 2N2222A *npn* transistor and a 2N2907A *pnp* transistor, in which the characteristics are nearly identical but reversed in polarity. However, it should be stressed that these transistors *are not intended for power amplifier applications* since their power dissipation ratings are quite limited. Nevertheless, we can use these to demonstrate the concept even though the result will be a "very small power amplifier."

The code is shown in Figure 19–14(b). The input signal is a sinusoid with a peak value of 4 V and a frequency of 1 kHz. If the transistors were perfect, the output peak voltage would also be 4 V, which corresponds to an rms value of $4/\sqrt{2} = 2.828$ V. This would produce a power in the 20-Ω load resistor of $(2.828)^2/20 = 0.4$ W.

A new control statement that has not been encountered thus far in the text is the one that reads

```
.FOUR 1E3 V(4)
```

The term FOUR represents **Fourier analysis,** which is the process of determining the frequency components of a signal. These components are also referred to as the **spectrum.** If the amplifier were perfect, the only frequency component in the output would be the 1-kHz input signal. However, any nonlinear effects of the amplifier will introduce distortion components that have frequencies that are integer multiples of the input frequency. These frequencies are referred to as **harmonics,** and the resulting distortion is called **harmonic distortion.** The mathematical process of Fourier analysis is beyond the scope of this text, but the use of PSPICE for this purpose is very straightforward.

Returning to the control statement, the value 1E3 represents the input frequency of 1 kHz, which is called the **fundamental frequency** in Fourier terminology. In order to perform a spectral analysis, the transient response must be run for a time at least as long as the period of the fundamental, which is 1 ms in this case. In general, if there is a significant change during the transient interval, it should be run long enough to settle into steady-state conditions. Since there are no external capacitors in this circuit, it should settle quickly, and 2 ms has been used as the transient interval. The quantity V(4), which is the output, represents the variable to be analyzed.

When the program is first run, .PROBE permits the observation of various circuit variables, and the output voltage is shown in Figure 19–15. The effects of crossover distortion are immediately evident.

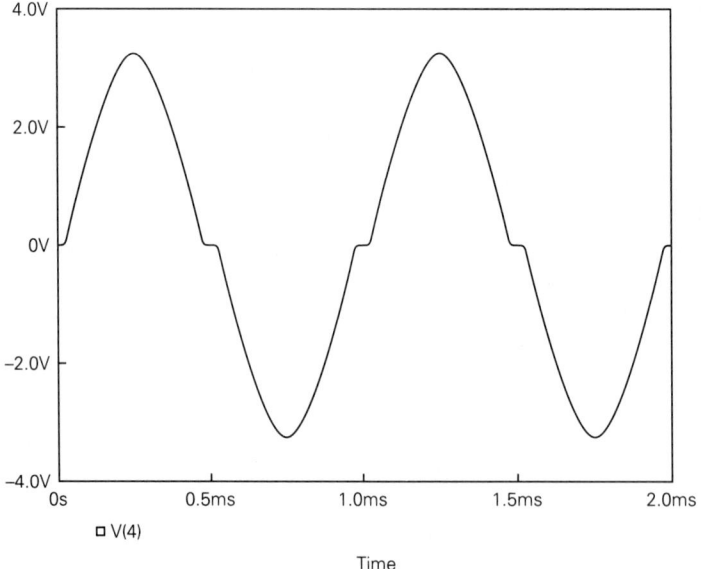

FIGURE 19–15
Output waveform of PSPICE Example 19–1.

To study the Fourier components, the output file can be printed, and the results are shown in Figure 19–16. The harmonic labeled as number 1 is the fundamental, and its peak value is 2.965 V. The values of the various harmonics through the ninth appear in the same column. Notice that the even-numbered harmonics are much smaller than the odd-numbered harmonics. In an ideal push-pull amplifier, the even-numbered harmonics would be zero.

```
PSPICE EXAMPLE 19-1

****      CIRCUIT DESCRIPTION

*********************************************************************

.OPTIONS NOECHO NOPAGE NOMOD NOBIAS

****      FOURIER ANALYSIS              TEMPERATURE = 27.000 DEG C

FOURIER COMPONENTS OF TRANSIENT RESPONSE V(4)

DC COMPONENT =   3.156523E-02
HARMONIC  FREQUENCY    FOURIER   NORMALIZED    PHASE     NORMALIZED
   NO       (HZ)      COMPONENT  COMPONENT    (DEG)     PHASE (DEG)

    1     1.000E+03   2.965E+00  1.000E+00   2.591E-03   0.000E+00
    2     2.000E+03   6.878E-03  2.320E-03  -8.983E+01  -8.983E+01
    3     3.000E+03   2.993E-01  1.010E-01   1.800E+02   1.800E+02
    4     4.000E+03   8.176E-03  2.758E-03  -8.976E+01  -8.976E+01
    5     5.000E+03   1.553E-01  5.237E-02   1.800E+02   1.800E+02
    6     6.000E+03   7.096E-03  2.393E-03  -8.965E+01  -8.966E+01
    7     7.000E+03   9.093E-02  3.067E-02  -1.800E+02  -1.800E+02
    8     8.000E+03   5.695E-03  1.921E-03  -8.959E+01  -8.959E+01
    9     9.000E+03   5.428E-02  1.831E-02  -1.799E+02  -1.799E+02
TOTAL HARMONIC DISTORTION =  1.193169E+01 PERCENT

     JOB CONCLUDED
     TOTAL JOB TIME                   8.79
```

FIGURE 19–16
Distortion analysis of amplifier in PSPICE Example 19–1.

It is assumed here that harmonics above the ninth contribute very little and, based on this assumption, the total harmonic distortion is calculated by PSPICE to be about 11.93%. This is a large level of distortion and is caused primarily by the crossover effect. We will see how to eliminate that problem in the next example.

PSPICE EXAMPLE 19–2

The circuit of PSPICE Example 19–1 was redesigned with diodes on the input as shown in Figure 19–17(a). Repeat the analysis of PSPICE Example 19–1.

Solution
The code for this circuit is shown in Figure 19–17(a). The input signal and simulation time are the same as in the preceding example.

The output voltage obtained from .PROBE is shown in Figure 19–18. There is a marked improvement in the waveform, and the crossover effects have been virtually eliminated.

The Fourier analysis results are shown in Figure 19–19. Among other improvements, the fundamental component is 3.727 V, which is not much below the input peak of 4 V. This corresponds to an rms value of $3.727/\sqrt{2} = 2.635$ V, which produces a power of $(2.635)^2/20 = 0.347$ W in the load.

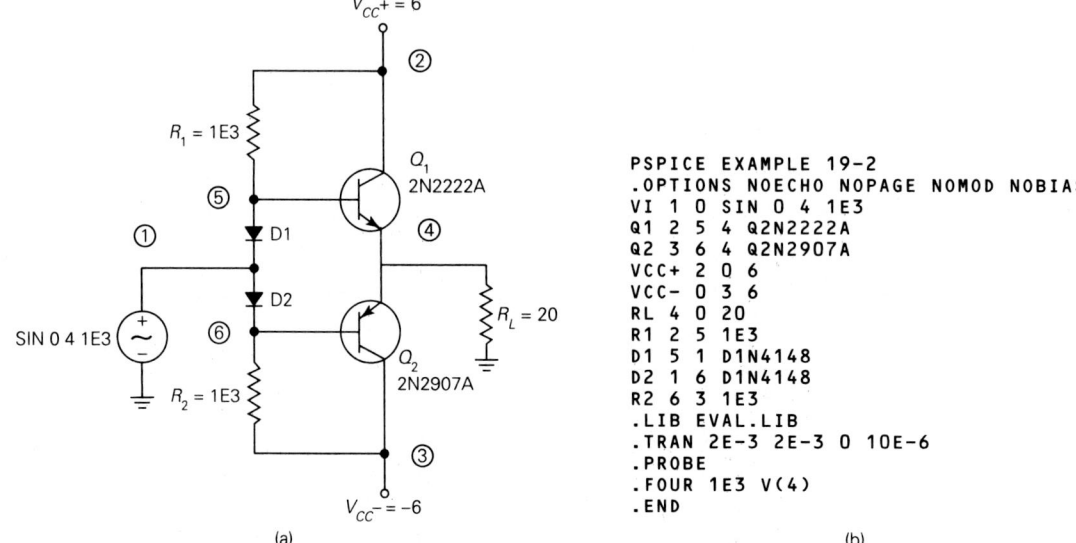

(a)

```
PSPICE EXAMPLE 19-2
.OPTIONS NOECHO NOPAGE NOMOD NOBIAS
VI 1 0 SIN 0 4 1E3
Q1 2 5 4 Q2N2222A
Q2 3 6 4 Q2N2907A
VCC+ 2 0 6
VCC- 0 3 6
RL 4 0 20
R1 2 5 1E3
D1 5 1 D1N4148
D2 1 6 D1N4148
R2 6 3 1E3
.LIB EVAL.LIB
.TRAN 2E-3 2E-3 0 10E-6
.PROBE
.FOUR 1E3 V(4)
.END
```

(b)

FIGURE 19–17
Circuit and code of PSPICE Example 19–2.

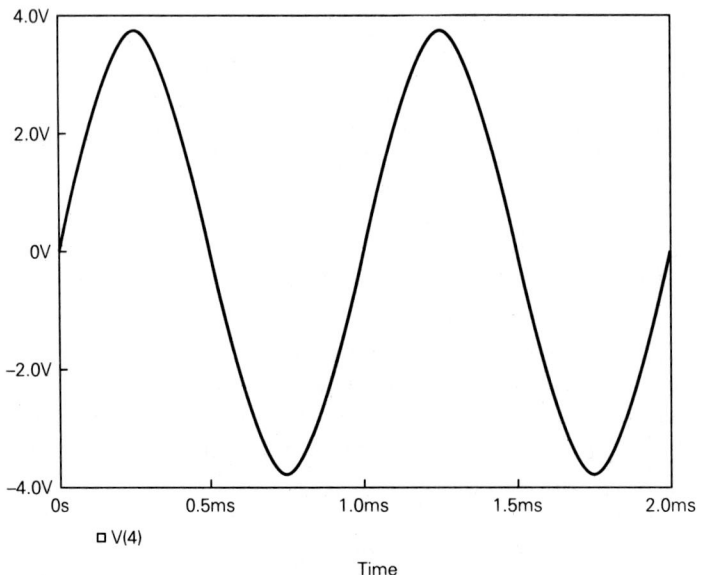

□ V(4)

Time

FIGURE 19–18
Output waveform of PSPICE Example 19–2.

```
PSPICE EXAMPLE 19-2

****        CIRCUIT DESCRIPTION

*********************************************************************

.OPTIONS NOECHO NOPAGE NOMOD NOBIAS

****        FOURIER ANALYSIS                    TEMPERATURE = 27.000 DEG C

FOURIER COMPONENTS OF TRANSIENT RESPONSE V(4)

DC COMPONENT =   2.795671E-02
HARMONIC  FREQUENCY   FOURIER    NORMALIZED    PHASE     NORMALIZED
   NO       (HZ)     COMPONENT   COMPONENT    (DEG)     PHASE (DEG)

    1     1.000E+03  3.727E+01   1.000E+00   -7.756E-04   0.000E+00
    2     2.000E+03  1.163E-02   3.121E-03    8.999E+01   8.999E+01
    3     3.000E+03  5.663E-03   1.520E-03   -1.796E+02  -1.796E+02
    4     4.000E+03  4.300E-03   1.154E-03   -8.996E+01  -8.996E+01
    5     5.000E+03  1.484E-02   3.982E-03   -1.798E+02  -1.798E+02
    6     6.000E+03  1.997E-03   5.358E-04    9.006E+01   9.006E+01
    7     7.000E+03  3.227E-03   8.659E-04   -1.792E+02  -1.792E+02
    8     8.000E+03  1.448E-03   3.884E-04   -8.988E+01  -8.987E+01
    9     9.000E+03  4.454E-03   1.195E-03   -1.793E+02  -1.793E+02

TOTAL HARMONIC DISTORTION =   5.643599E-01  PERCENT

        JOB CONCLUDED
        TOTAL JOB TIME                    10.05
```

FIGURE 19–19
Distortion analysis of amplifier in PSPICE Example 19–2.

The most dramatic achievement is observed from the total harmonic distortion, which has been reduced to about 0.564% (about one-half of 1%). This value could be reduced even further by careful adjustment of the resistances R_1 and R_2 to adjust the forward bias of the diodes or by selecting diodes that match more closely the base–emitter junctions of the transistors.

PROBLEMS

Drill Problems

19–1. A certain power transistor has a maximum allowable power dissipation of 50 W at 25°C. The maximum allowable collector–emitter voltage is 80 V, and the maximum allowable collector current is 4 A. Determine the maximum allowable collector current **(a)** when $V_{CE} = 10$ V, and **(b)** when $V_{CE} = 25$ V. Determine the maximum allowable collector–emitter voltage **(c)** when $I_C = 2$ A, and **(d)** when $I_C = 0.5$ A. All calculations are to be based on 25°C.

19–2. A certain power transistor has the following specifications at 25°C: $P_{D(\max)} = 10$ W, $V_{CE(\max)} = 40$ V, $I_{C(\max)} = 2$ A. Determine the maximum allowable collector current **(a)** when $V_{CE} = 4$ V, and **(b)** when $V_{CE} = 20$ V. Determine the maximum allowable collector–emitter voltage **(c)** when $I_C = 0.2$ A, and **(d)** when $I_C = 0.5$ A.

826 | POWER AMPLIFIERS

19–3. A certain idealized class A transformer-coupled power amplifier has $V_{CC} = 32$ V, $I_{CQ} = 0.5$ A, and it is used to drive a load $R_L = 4$ Ω. Based on the simplified assumptions in the text, determine **(a)** V_{CEQ}; **(b)** R'_L; **(c)** N_1/N_2; **(d)** P_S; **(e)** $P_{O(max)}$; and **(f)** $P_{D(max)}$.

19–4. Repeat the analysis of Problem 19–3 if $V_{CC} = 12$ V, $I_{CQ} = 2$ A, and $R_L = 96$ Ω.

19–5. A certain class B push-pull emitter-follower power amplifier is biased by two power supplies with values ± 16 V, and it is used to drive a load $R_L = 4$ Ω. Based on the assumptions of this section, determine **(a)** $I_{C(sat)}$; **(b)** $P_{S(max)}$; **(c)** $P_{O(max)}$; and **(d)** $P_{D(max)}$.

19–6. Repeat the analysis of Problem 19–5 if the two power supply voltages are ± 6 V and $R_L = 96$ Ω.

19–7. A certain power transistor has the following specifications:

Total power dissipation at $T_C = 25°C$: 50 W

Derate above 25°C: 0.4 W/°C

Total power dissipation at $T_A = 25°C$: 4 W

Derate above 25°C: 20 mW/°C

(a) Determine Θ_{JC}, Θ_{JA}, and Θ_{CA}. **(b)** If the transistor is used without a heat sink and the ambient temperature is 30°C, determine the maximum allowable power dissipation. **(c)** If the transistor is used with a heat sink and the case temperature is 40°C, determine the maximum allowable power dissipation.

19–8. A certain power transistor has the following specifications:

Total power dissipation at $T_C = 25°C$: 10 W

Derate above 25°C: 0.1 W/°C

Total power dissipation at $T_A = 25°C$: 1.5 W

Derate above 25°C: 10 mW/°C

(a) Determine Θ_{JC}, Θ_{JA}, and Θ_{CA}. **(b)** If the transistor is used without a heat sink and $T_A = 32°C$, determine the maximum allowable power dissipation. **(c)** If the transistor is used with a heat sink and $T_C = 38°C$, determine the maximum allowable power dissipation.

Derivation Problems

19–9. (Requires calculus.) Consider the expression of the source power p_S as given by (19–37). Determine the average value of this function over the interval from $t = 0$ to $t = T/2$ and show that the result is given by (19–38).

19–10. (Requires calculus.) Consider the expression for the internal power dissipation P_D as given by (19–42). **(a)** Show that the maximum value occurs for $k = 2/\pi$. **(b)** Show that the maximum value of P_D is the quantity of (19–43).

Design Problems

19–11. Some preliminary design on a class B push-pull emitter-follower power amplifier is to be performed. The desired maximum output power is 4 W into a resistive load of 8 Ω. Based on a sinusoidal waveform, determine **(a)** the minimum value of the power dissipation rating of the transistors to be selected; **(b)** the maximum value of collector current; and **(c)** the maximum value of power delivered by the power supplies.

19–12. Repeat Problem 19–11 if the desired maximum output power is 25 W into a resistive load of 10 Ω.

Troubleshooting Problems

19–13. You are troubleshooting a push-pull amplifier of the type shown in Figure 19–6. Some dc voltages are measured as follows:

> Positive power supply: +15 V
>
> Negative power supply: −15 V
>
> Base of T1: 0.7 V
>
> Base of T2: −0.7 V

With a sine wave applied to the input, the output is a half-wave rectified version of the input (i.e., the negative segments are missing). Which one of the following is the most likely trouble? **(a)** T1 defective; **(b)** T2 defective; **(c)** upper bias resistor open; **(d)** lower bias resistor open; **(e)** lower bias resistor shorted.

19–14. You are troubleshooting a class C amplifier of the type shown in Figure 19–9. All dc voltages appear to be normal, and an oscilloscopic display of v_i appears normal. However, the transistor collector voltage, while periodic, is not sinusoidal. Which *two* of the following could represent a possible trouble or troubles? **(a)** Defective transistor; **(b)** insufficient negative bias; **(c)** incorrect value of C; **(d)** incorrect value of L; **(e)** R shorted; **(f)** incorrect value of C_i.

20

POWER SUPPLIES

OBJECTIVES

After completing this chapter, the reader should be able to:

- Apply the voltage and current relationships for an ideal transformer.
- Draw the schematic diagram for and analyze the operation of a half-wave rectifier.
- Draw the schematic diagram for and analyze the operation of a full-wave rectifier with a center-tapped transformer.
- Draw the schematic diagram for and analyze the operation of a full-wave bridge rectifier.
- Discuss the need for and analyze the operation of a ripple filter in a power supply.
- Discuss the importance of the peak reverse voltage for a diode, and determine the value for each of the rectifier circuits studied.
- Explain the operation of an IC regulator circuit.
- Discuss the operation of a silicon-control rectifier (SCR), a diac, and a triac.
- Analyze some of the circuits in the chapter with PSPICE.

20–1 OVERVIEW

As we have seen throughout the text, dc voltages of various levels are required to properly bias various electronic devices such as transistors and op-amps. However, the commercial power distribution system is an ac system. The commercial power system in the United States has as its standard level close to 120 V rms at a frequency of 60 Hz. This voltage level must be changed and converted to dc before it can be used for electronic device operation.

A circuit that provides the required level change and converts the ac to dc is called a **power supply.** Virtually all electronic equipment operated from the ac line (e.g., radios, television sets, VCRs, etc.) has one or more power supply circuits.

A block diagram displaying the major components of a power supply is given in Figure 20–1. The circuit shows a grounded plug, a switch, and a fuse, all of which are essential for safety. To simplify subsequent diagrams, these components will be omitted, but they should be considered as requirements in a practical power supply.

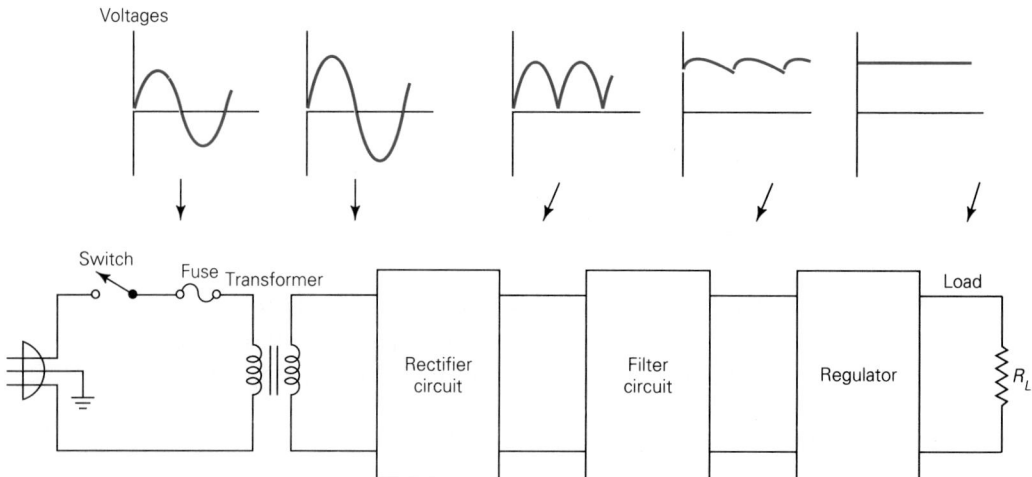

FIGURE 20–1
Block diagram of complete power supply circuit.

The transformer is used to change the level of the voltage. Depending on the final power supply requirements, the voltage level may be increased or decreased. Note the ac waveforms shown above the transformer on both sides.

The **rectifier circuit** is used to convert the alternating transformer output voltage to a voltage having only one polarity. However, this voltage will still vary considerably over the range of a cycle as shown. This widely varying voltage of one polarity is called **pulsating dc.**

The **filter circuit** provides a substantial smoothing effect on the pulsating dc. The result is a voltage having a relatively small variation as shown. This variation is called **ripple.**

Most modern power supplies employ a **regulator.** The regulator provides further smoothing on the voltage, and it establishes and maintains a predetermined output voltage level. This last process is quite important since there are many causes of possible output voltage variation in a typical power supply.

Much older equipment designed before the advent of the modern regulator circuit is still in service. In those units, the filter circuit was designed to provide all the smoothing required. Consequently, filter circuits in such older equipment were much more elaborate than is required with modern regulator circuits.

20–2 IDEAL TRANSFORMERS

At several earlier points in the text, transformers have appeared in certain amplifier circuits. A brief discussion of the property used was given in each situation, but a lengthy discussion of transformer properties was not provided. However, before considering the operation of power supplies, it is necessary to take a more detailed look at the properties of transformers.

A transformer is a circuit consisting of two or more coils wound on a common magnetic core. The coils are usually electrically isolated from each other, but the magnetic fields are coupled together. Because of the magnetic coupling, energy may be transferred between the coils through the magnetic field. In the process, the voltage may be increased ("stepped up") or decreased ("stepped down"). The coils of the transformers may be tightly coupled or loosely coupled depending on the application. Virtually all power supply transformers are tightly coupled. Tight coupling is achieved through a ferromagnetic core and close windings.

Tightly coupled transformers such as used in power supplies may be closely approximated by the **ideal transformer** model. The schematic symbol for an ideal transformer is shown in Figure 20–2. The coil on the left is called the **primary winding,** and the coil on the right is called the **secondary winding.** The primary winding is the one normally connected to the ac line, and the secondary winding is the one from which the working voltage for the power supply is obtained.

FIGURE 20–2

Schematic symbol for ideal transformer with primary and secondary labeled.

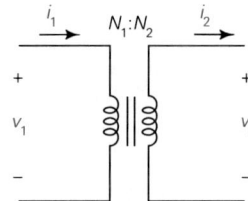

The voltage–current relationships for the ideal transformer to be given here apply to either instantaneous values, peak values, or rms values provided that *all* variables are expressed in the *same forms.* For the moment, we leave the exact forms unspecified. The primary voltage and current are then denoted simply as v_1 and i_1, and the secondary voltage and current are denoted as v_2 and i_2 in Figure 20–2. Obviously, the transformer must be connected to other parts of a circuit for all the variables to be present.

The ideal transformer is characterized by a **turns ratio,** denoted as $N_1 : N_2$ on Figure 20–2. Ideally, N_1 is the number of primary turns and N_2 is the number of secondary turns. In a practical transformer, it is usually impossible to determine the actual number of turns, so it is the *ratio* that is important. For example, assume that $N_1 = 500$ and $N_2 = 2000$. The turns ratio of such a transformer would be $N_1/N_2 = 1:4$. In actual practice, transformer specifications are often stated in terms of the intended working voltages. For example, a transformer might be specified as operating at 120 V primary/24 V secondary. In this case the turns ratio could be assumed to be $N_1/N_2 = 120/24 = 5/1$ and $N_1 : N_2 = 5:1$.

The voltage–current relationships of the ideal transformer are as follows:

$$\frac{v_2}{v_1} = \frac{N_2}{N_1} \qquad \text{(20–1)}$$

and

$$\frac{i_2}{i_1} = \frac{N_1}{N_2} \qquad \text{(20–2)}$$

Thus the voltage changes in direct proportion to the turns ratio, while the current changes in reverse proportion to the turns ratio. This means a "stepup" from primary to secondary requires that $N_2 > N_1$, while a "stepdown" requires that $N_2 < N_1$.

If (20–1) and (20–2) are multiplied together, we obtain

$$\frac{v_2 i_2}{v_1 i_1} = \frac{N_2 N_1}{N_1 N_2} = 1 \qquad \text{(20–3)}$$

Equation (20–3) simply tells us that the total power is not changed by the transformer. If the voltage is increased, the current is decreased, but the product remains the same. In an actual transformer, some power is always dissipated internally, so the output power is less than the input power.

EXAMPLE 20–1

An ideal transformer has a turns ratio of 4 : 1 from primary to secondary. The primary voltage is the ac line voltage whose rms value will be assumed to be 120 V. For a particular secondary load, the primary rms current is 2 A. Determine the secondary rms voltage and current.

Solution
Since the primary variables are specified as rms values, the corresponding secondary values will also be determined as rms values. By convention, uppercase symbols will be used for rms values.

The voltages are related by

$$\frac{V_2}{V_1} = \frac{N_2}{N_1} = \frac{1}{4} \qquad \text{(20–4)}$$

from which we obtain

$$V_2 = \frac{1}{4} V_1 = \frac{1}{4} \times 120 = 30 \text{ V rms} \qquad (20\text{--}5)$$

The currents are related by

$$\frac{I_2}{I_1} = \frac{N_1}{N_2} = 4 \qquad (20\text{--}6)$$

from which we determine that

$$I_2 = 4I_1 = 4 \times 2 = 8 \text{ A rms} \qquad (20\text{--}7)$$

Thus, when the voltage is decreased from 120 to 30 V, the current is increased from 2 to 8 A.

20–3 HALF-WAVE RECTIFIER (UNFILTERED)

In working with small-signal diodes early in the book, a forward drop of 0.7 V was always assumed. Power supply levels vary considerably, and for large currents, typical of many power supplies, diode drops may be somewhat greater. In addition, transformer losses will also be present, and these are difficult to predict exactly.

In view of more uncertain diode and transformer voltage losses, we take a simplified approach in the analysis of this rectifier circuit and others to follow. The simplified approach will be to neglect all voltage drops in the discussion of the basic operation of the circuit, and this should allow the reader to focus on the principle of operation without undue ''clutter'' resulting from losses and unwieldy notation. Bear in mind, however, that the actual output voltages will be somewhat smaller than the values predicted from the idealized assumptions, and it will be necessary to allow for this in a practical application.

The simplest rectifier power supply circuit is the unfiltered **half-wave rectifier** circuit, whose basic form is shown in Figure 20–3. Let $V_{1(\text{rms})}$ represent the rms value of the primary voltage v_1, and let $V_{2(\text{rms})}$ represent the rms value of the secondary voltage v_2. These voltages are related by

$$V_{2(\text{rms})} = \frac{N_2}{N_1} V_{1(\text{rms})} \qquad (20\text{--}8)$$

FIGURE 20–3
Schematic diagram of basic half-wave rectifier circuit.

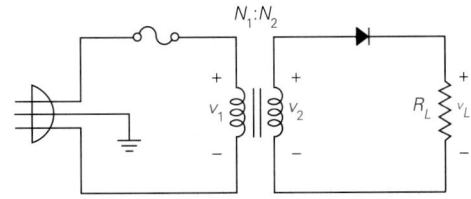

Let V_{2p} represent the peak value of the secondary voltage, which is related to the secondary rms voltage by

$$V_{2p} = \sqrt{2}\, V_{2(rms)} \qquad\qquad\qquad \textbf{(20–9)}$$

Refer to the waveforms of Figure 20–4 in the discussion that follows. The voltage v_2 is shown in Figure 20–4(a).

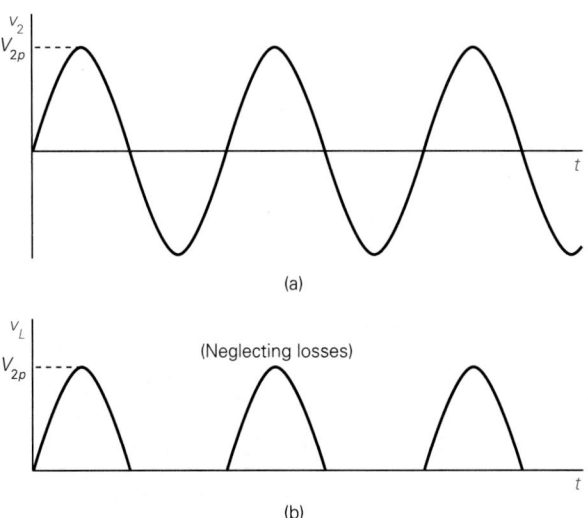

(a)

(b)

FIGURE 20–4
Waveforms of half-wave rectifier circuit.

When the secondary voltage v_2 is positive at the top of the transformer secondary winding with respect to the bottom, the diode will conduct. However, when the secondary voltage has the opposite polarity, the diode will be open and no conduction will occur. The load voltage v_L thus has the form of Figure 20–4(b). If there were no losses, the peak load voltage *would be* V_{2p}.

The load voltage as shown is pulsating dc. It is certainly not pure dc, but it does have a dc component. It has only limited value as shown, but there are some industrial and commercial processes for which pulsating dc is used. Let $V_{L(dc)}$ represent the dc value of this waveform. It can be shown with calculus that this voltage has the value

$$\boxed{V_{L(dc)} = \frac{V_{2p}}{\pi} = 0.318 V_{2p}} \qquad\qquad \textbf{(20–10)}$$

In practice, the value will always be smaller than (20–10) because of transformer and diode losses.

For a resistive load R_L, the dc load current $I_{L(dc)}$ is

$$I_{L(dc)} = \frac{V_{L(dc)}}{R_L}$$

(20–11)

**EXAMPLE
20–2**

A certain unfiltered half-wave rectifier circuit of the form shown in Figure 20–3 is powered by the 120-V rms ac power system, and the turns ratio of the transformer is $N_1 : N_2 = 6 : 1$. Determine **(a)** the rms secondary voltage; **(b)** the peak secondary voltage; **(c)** the dc load voltage; and **(d)** the dc load current if $R_L = 5\ \Omega$. Neglect any diode and transformer losses.

Solution

(a) The rms secondary voltage $V_{2(rms)}$ is

$$V_{2(rms)} = \frac{N_2}{N_1} V_{1(rms)} = \frac{1}{6} \times 120 = 20\text{ V}$$

(20–12)

(b) The peak secondary voltage V_{2p} is

$$V_{2p} = \sqrt{2}\, V_{2(rms)} = \sqrt{2} \times 20 = 28.28\text{ V}$$

(20–13)

(c) The dc load voltage $V_{L(dc)}$ is

$$V_{L(dc)} = \frac{V_{2p}}{\pi} = \frac{28.28}{\pi} = 9.00\text{ V}$$

(20–14)

(d) The dc load current $I_{L(dc)}$ is

$$I_{L(dc)} = \frac{V_{L(dc)}}{R_L} = \frac{9.00}{5} = 1.8\text{ A}$$

(20–15)

20–4 FULL-WAVE RECTIFIER (UNFILTERED)

The basic form of a **full-wave rectifier** circuit is shown in Figure 20–5. The secondary winding of the transformer is center tapped in this case. As will be seen shortly, only one-half of the secondary will conduct at a given time. For convenience, the

FIGURE 20–5

Schematic diagram of basic full-wave rectifier circuit.

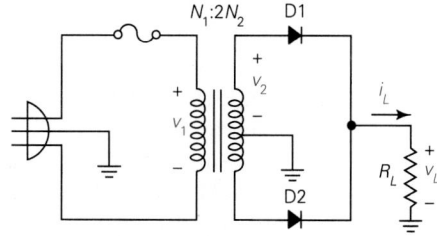

total number of secondary turns will be denoted as $2N_2$, with N_2 representing the number of turns on half of the secondary.

Refer to Figures 20–6 and 20–7 in the discussion that follows. Let v_2 represent the voltage across the upper half of the secondary. The rms value of this voltage is

$$V_{2(rms)} = \frac{N_2}{N_1} V_{1(rms)} \qquad (20–16)$$

The peak value is

$$V_{2p} = \sqrt{2}\, V_{2(rms)} \qquad (20–17)$$

FIGURE 20–6
Manner in which conduction occurs on alternate half-cycles in full-wave rectifier circuit.

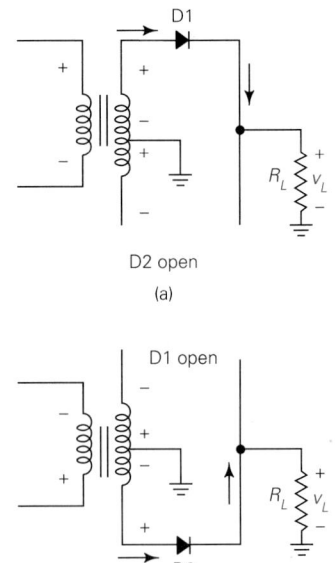

When v_2 is positive as defined, the upper diode D1 conducts and the lower diode D2 is open as shown in Figure 20–6(a). The first half-cycle of v_2 as shown in Figure 20–7(a) is then transferred to the load, and the load current returns through ground and the center tap of the transformer.

When v_2 as defined is negative, the voltage on the lower half of the transformer is positive at the bottom with respect to the center tap. Under this condition, D2 conducts and D1 is open. From Figure 20–6(b), note that the direction of this voltage is still positive at the top of the load with respect to ground as shown. Current again flows to ground and back to the center tap of the transformer.

The form of v_L is shown in Figure 20–7(b). Once again, rectification has taken place, but now it is **full-wave rectification.**

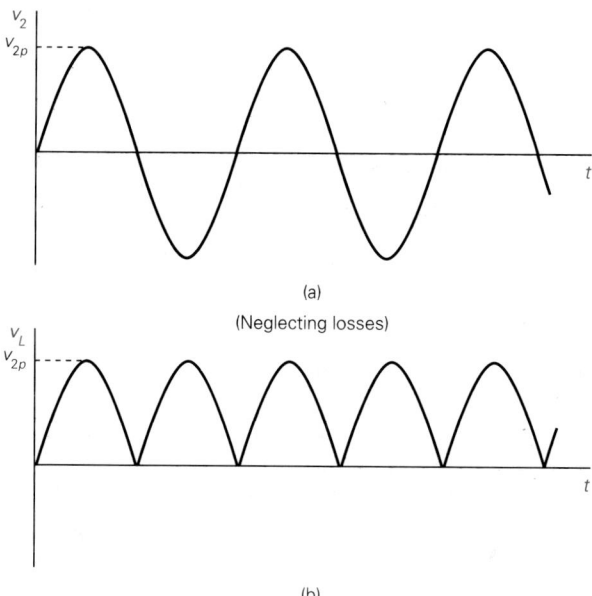

(a)

(Neglecting losses)

(b)

FIGURE 20–7
Waveforms in full-wave rectifier circuit.

As in the case of the half-wave rectifier, this voltage is pulsating dc. However, its dc value $V_{L(dc)}$ has twice the value as for the half-wave rectifier and is

$$V_{L(dc)} = \frac{2}{\pi} V_{2p} = 0.637 V_{2p} \qquad (20\text{–}18)$$

EXAMPLE 20–3   A certain unfiltered full-wave rectifier circuit of the form of Figure 20–5 is powered by the 120-V rms ac power system, and the turns ratio of primary to one-half of the secondary is $N_1 : N_2 = 6 : 1$. (Equivalently, $N_1 : 2N_2 = 3 : 1$.) Determine **(a)** the rms secondary voltage across one-half of the transformer; **(b)** the peak secondary voltage; **(c)** the dc load voltage; and **(d)** the dc load current if $R_L = 5\ \Omega$. Neglect any diode and transformer losses.

Solution
(a)   The rms voltage across one-half the secondary is

$$V_{2(rms)} = \frac{N_2}{N_1} V_{1(rms)} = \frac{1}{6} \times 120 = 20 \text{ V} \qquad (20\text{–}19)$$

(b)   The peak voltage across one-half of the secondary is

$$V_{2p} = \sqrt{2}\, V_{2(rms)} = \sqrt{2} \times 20 = 28.28 \text{ V} \qquad (20\text{–}20)$$

(c)
$$V_{L(\text{dc})} = \frac{2}{\pi} V_{2p} = \frac{2}{\pi} \times 28.28 = 18.00 \text{ V} \qquad \text{(20–21)}$$

(d) The dc load current is

$$I_{L(\text{dc})} = \frac{V_{L(\text{dc})}}{R_L} = \frac{18.00}{5} = 3.6 \text{ A} \qquad \text{(20–22)}$$

20–5 BRIDGE RECTIFIER (UNFILTERED)

Another type of full-wave rectifier circuit is the so-called **bridge rectifier,** whose basic form is shown in Figure 20–8. Although four diodes are required, the circuit has the advantage that no center tap is required on the transformer, and the entire secondary voltage is available for rectification. Another advantage is the lower peak reverse voltage requirement, and this will be discussed in Section 20–7. Because of the popularity of the circuit, diode bridge packages containing the four required diodes are available as stock items.

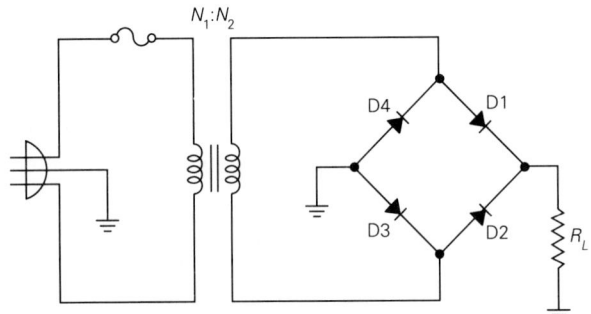

FIGURE 20–8
Schematic diagram of full-wave bridge rectifier circuit.

Operation of the circuit can be explained with the diagram of Figure 20–9. Assume first that the top of the transformer is positive with respect to the bottom. As shown in Figure 20–9(a), diodes D1 and D3 are forward biased, while D2 and D4 are reverse biased. Current flows out of the top of the transformer through D1, through the load to ground, up from ground through D3, and back to the lower terminal of the transformer.

Assume next that the bottom of the transformer secondary is positive with respect to the top as shown in Figure 20–9(b). Diodes D2 and D4 are now forward biased, while diodes D1 and D3 are reverse biased. Current flows out of the bottom of the transformer through D2, through the load to ground, up from ground through D4, and back to the upper terminal of the transformer.

FIGURE 20–9
Manner in which conduction occurs on alternate half-cycles in bridge rectifier circuit.

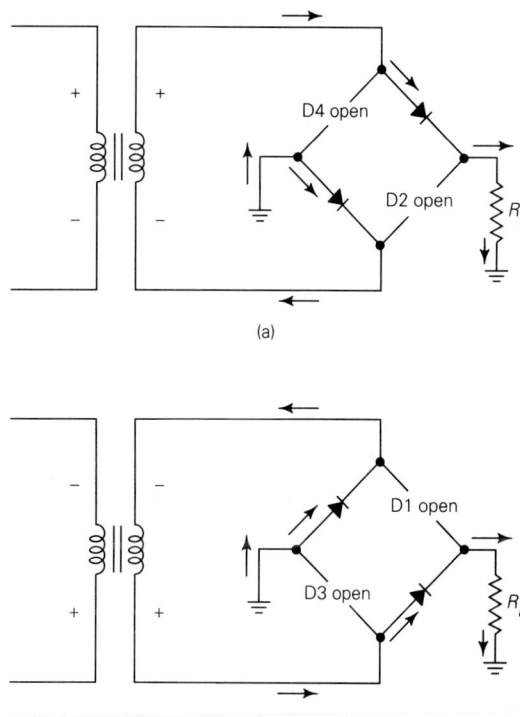

(a)

(b)

As can be seen, current flows through the load in the same direction on both half-cycles. The waveforms in this case are essentially the same as in Figure 20–7.

The various relationships for the bridge circuit will now be given. The secondary rms voltage is

$$V_{2(rms)} = \frac{N_2}{N_1} V_{1(rms)}$$
(20–23)

The peak secondary voltage is

$$V_{2p} = \sqrt{2}\, V_{2(rms)}$$
(20–24)

The dc load voltage is

$$V_{L(dc)} = \frac{2V_{2p}}{\pi} = 0.637V_{2p}$$
(20–25)

It should be noted that when losses are considered, there are *two* forward-biased diode drops on each half-cycle to be considered in the bridge rectifier.

EXAMPLE
20–4

A certain unfiltered bridge rectifier circuit of the form of Figure 20–8 is powered by the 120-V rms ac power system, and the turns ratio is $N_1 : N_2 = 6 : 1$. Determine **(a)** the rms secondary voltage; **(b)** the peak secondary voltage; **(c)** the dc load voltage; and **(d)** the dc load current if $R_L = 5\ \Omega$. Neglect any diode and transformer losses.

Solution

(a) The rms secondary voltage is

$$V_{2(\text{rms})} = \frac{1}{6} \times 120 = 20 \text{ V} \tag{20–26}$$

(b) The peak secondary voltage is

$$V_{2p} = \sqrt{2} \times 20 = 28.28 \text{ V} \tag{20–27}$$

(c) The dc load voltage is

$$V_{L(\text{dc})} = \frac{2}{\pi} \times 28.28 = 18.00 \text{ V} \tag{20–28}$$

(d) The dc load current is

$$I_{L(\text{dc})} = \frac{18.00}{5} = 3.6 \text{ A} \tag{20–29}$$

The reader has likely observed a great deal of redundancy between Examples 20–2, 20–3, and 20–4 because of the similarities. However, the values should help the reader to understand the differences between the different circuit forms.

20–6 FILTER CIRCUITS

The rectifier circuits considered thus far convert ac to pulsating dc. As noted, pulsating dc has only limited value in practice. The large output variations in pulsating dc can be converted to very small variations by means of a **filter circuit.** The resulting small variations are referred to as **ripple.** We will explain the action of filter circuits qualitatively in this section and give a detailed quantitative analysis in Section 20–8.

Capacitor Filter

Since modern power supplies use IC regulators, which provide additional smoothing, a very widely used power supply filter is a **single-capacitor filter,** shown in Figure 20–10. The resistance R_L represents any load. This could be an actual load or an equivalent load seen from a regulator input. The particular rectifier circuit shown in this case is the basic full-wave rectifier.

FIGURE 20–10
Full-wave rectifier with capacitor filter.

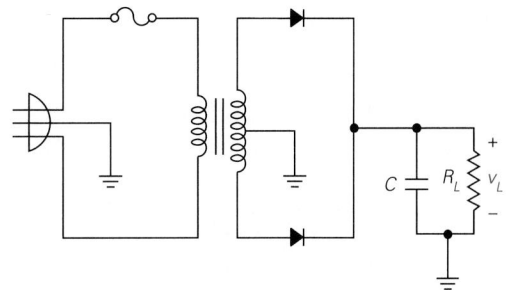

Refer to the waveforms of Figure 20–11. The dashed curve in Figure 20–11(a) shows the pulsating dc that *would* appear across the load *if* the capacitor were not present. The capacitor, however, charges to the peak value of this voltage very soon after the circuit is energized. If there were no load present, this voltage would remain at the peak value.

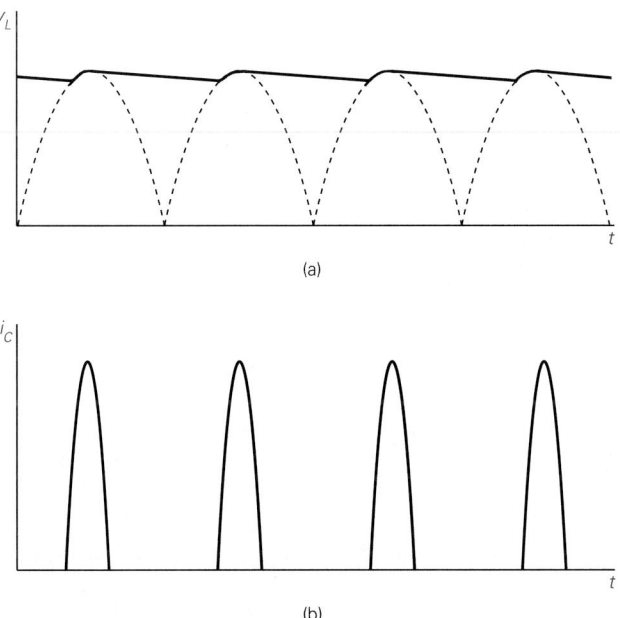

FIGURE 20–11
Waveforms of full-wave rectifier with capacitor filter.

The presence of a load causes the voltage on the capacitor to decay slowly over the interval of a half-cycle for a full-wave rectifier (or a full-cycle in the case of a half-wave rectifier). The decay has been exaggerated here for clarity.

At some point in each half-cycle, the rectified voltage exceeds the decaying voltage v_L. At that point, and depending on the appropriate half-cycle, one of the two diodes turns on again, and the capacitor is recharged quickly to the peak voltage. The form of the charging current is shown in Figure 20–11(b). The peak value of this charging current is rather large and is an important factor in selecting the diode ratings for a power supply.

Other Filters

As noted earlier, the use of modern IC regulators has reduced the requirements for filtering with reactive elements. However, in special designs and in earlier equipment, more elaborate filters are used.

One filter that found widespread usage in earlier equipment is the **inductive-input *LC* filter,** for which a single section is shown in Figure 20–12. Additional sections could be added for further filtering.

FIGURE 20–12
Single section of an inductive-input *LC* filter.

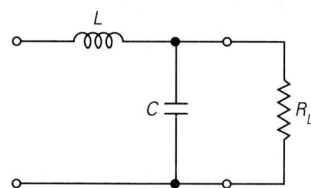

A second earlier popular filter was the **capacitive input *LC* filter,** for which a single section is shown in Figure 20–13. As in the previous case, additional sections may be added.

FIGURE 20–13
Single section of a capacitive-input *LC* filter.

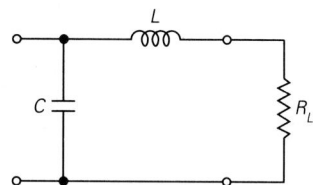

Another filter was the ***RC* filter,** for which a typical circuit form is shown in Figure 20–14. In a sense, the single capacitor filter is a special case of the *RC* filter in which there is one capacitance, and the load serves as the resistance.

FIGURE 20–14
Typical *RC* filter.

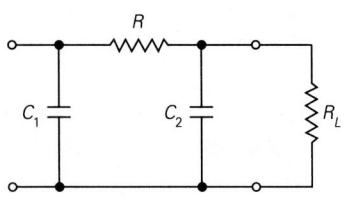

20–7 PEAK REVERSE VOLTAGE

A very important rating that must be considered in all power supply rectifier circuits is the **peak reverse voltage (PRV)** rating of the diodes. The peak reverse voltage of a diode is the peak voltage across the diode when it is reverse biased. As we will see shortly, this voltage may be larger than any of the obvious circuit voltages. Each of the major circuits previously considered will now be analyzed to determine the peak reverse voltage.

Half-Wave Rectifiers

First consider the unfiltered half-wave rectifier circuit of Figure 20–15(a). Operation is assumed to be at the point in the cycle where the diode is reverse biased ("off") and the secondary voltage is at a peak negative value $-V_{2p}$. Since no current is flowing in the load at this point, there is no voltage across the resistor, and the value of the peak reverse voltage V_{pr} is simply

$$\boxed{V_{pr} = V_{2p}} \qquad \text{(20–30)}$$

Thus, with no filter, *the peak reverse diode voltage of a half-wave rectifier is equal to the peak voltage across the transformer secondary.*

FIGURE 20–15
Analysis of diode peak reverse voltage in two half-wave rectifier circuits.

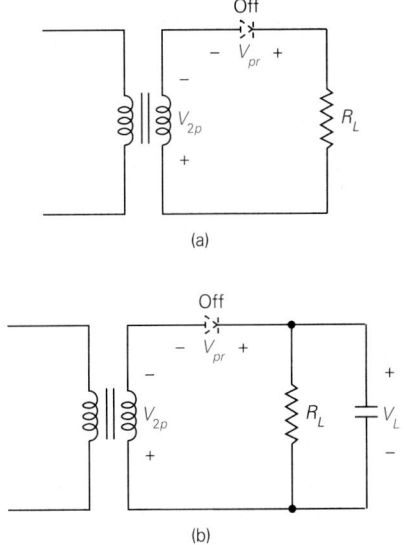

(a)

(b)

Next, consider the case where there is a capacitor filter as shown in Figure 20–15(b). The voltage across the capacitor is the load voltage V_L. A direct application of KVL to the loop yields

$$V_{pr} = V_{2p} + V_L \tag{20-31}$$

Although V_L varies slightly during a cycle because of the discharge, for very low ripple and for the purpose of this analysis, it will be assumed that $V_L = V_{2p}$. Thus the maximum peak reverse voltage is simply

$$V_{pr} = 2V_{2p} = 2V_L \tag{20-32}$$

The presence of the capacitor thus results in a peak reverse voltage across the diode equal to twice the dc output voltage.

Full-Wave Rectifier

Consider now the full-wave rectifier with no filter of Figure 20–16. Operation is assumed to be at the point in the cycle where D1 is reverse biased, D2 is forward biased, and the secondary voltage across each transformer half is V_{2p} with the polarities shown. Neglecting the small forward drop across D2, a KVL loop will be formed around the secondary. We have

$$V_{2p} + V_{2p} - V_{pr} = 0 \tag{20-33}$$

or

$$\boxed{V_{pr} = 2V_{2p}} \tag{20-34}$$

A similar analysis for the opposite half-cycle results in the same reverse voltage for the other diode.

Before commenting on this result, it can readily be demonstrated that the presence of a filter across the load does not affect the value in this case. *Thus, the peak reverse voltage of a full-wave rectifier is twice the peak voltage across each half or, equivalently, the voltage across the entire transformer secondary.*

FIGURE 20–16

Analysis of diode peak reverse voltage in full-wave rectifier circuit.

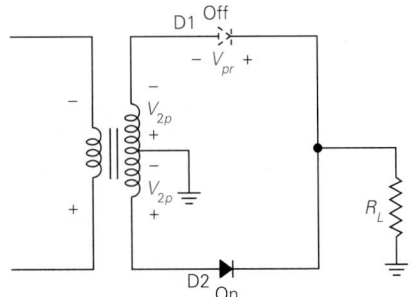

Bridge Rectifier

Finally, consider the bridge rectifier circuit of Figure 20–17. Operation is assumed to be at the point in the cycle where D1 and D3 are reverse biased and D2 and D4 are forward biased. The secondary voltage is at a peak V_{2p} with the positive terminal at the bottom.

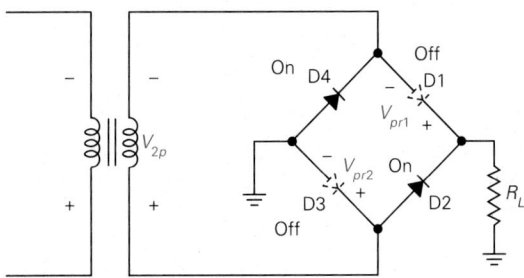

FIGURE 20–17
Analysis of diode peak reverse voltage in bridge rectifier circuit.

Two reverse voltages, V_{pr1} and V_{pr2}, must be considered in this case. A KVL loop through the transformer, down through D1 and D2, and back to the transformer will be formed to determine V_{pr1}. Neglecting the small forward drop of D2, this loop equation reads

$$V_{2p} - V_{pr1} = 0 \qquad (20\text{--}35)$$

or

$$V_{pr1} = V_{2p} \qquad (20\text{--}36)$$

From the symmetry of the circuit, it can be readily inferred that

$$\boxed{V_{pr2} = V_{pr1} = V_{2p}} \qquad (20\text{--}37)$$

A similar analysis for the opposite half-cycle results in the same reverse voltage for all four diodes.

As in the case of the basic full-wave rectifier, the presence of a filter on the load does not affect the value in this case. *Thus the peak reverse voltage of a bridge rectifier is simply the peak voltage across the secondary of the transformer.*

Let us now compare the bridge rectifier with the basic full-wave rectifier circuit. *For a given dc output voltage, the peak reverse voltage rating of the diodes in a full-wave rectifier must be twice as large as for a bridge rectifier.* This fact provides another distinct advantage of a bridge rectifier.

The various peak reverse voltage relationships developed in this section are summarized in Table 20–1.

TABLE 20-1
Summary of peak reverse voltages for different rectifier configurations

Rectifier Type	Diode Peak Reverse Voltage[a]
Half-wave (unfiltered)	V_{2p}
Half-wave (capacitor filter)	$2V_{2p}$
Full-wave rectifier with secondary center tap (with or without filtering)	$2V_{2p}$
Full-wave bridge rectifier (with or without filtering)	V_{2p}

[a] V_{2p} is the peak voltage across the complete transformer secondary for half-wave rectifiers and the full-wave bridge rectifier. It is the voltage across half of the transformer secondary for the full-wave rectifier with center tap.

EXAMPLE 20-5

Determine the peak reverse voltages for the diodes in each of the four following rectifier circuits: **(a)** unfiltered half-wave rectifier with $V_{2p} = 30$ V; **(b)** filtered half-wave rectifier with $V_L = 30$ V; **(c)** filtered full-wave rectifier (using a center-tapped transformer) with $V_L = 30$ V; **(d)** filtered bridge rectifier with $V_L = 30$ V.

Solution
The relationships of this section are readily applied, and the following short table can be constructed.

	Circuit	V_{pr} (V)
(a)	Unfiltered half-wave rectifier	30
(b)	Filtered half-wave rectifier	60
(c)	Filtered full-wave rectifier	60
(d)	Filtered bridge rectifier	30

20-8 ANALYSIS OF RIPPLE

As discussed, the output of a filter circuit does contain some ripple. We now investigate the nature of this ripple to determine how it relates to various circuit parameters. The analysis will be limited to the case of a single capacitor across a resistive load, since this is the most important case.

An enlarged view of the ripple is shown in Figure 20-18(a). The exact form of the discharge for a constant resistance load is an exponential function. In practice, however, the actual ripple level of realistic filters is so low that some simplifying approximations may be made.

The following approximations have been found to greatly simplify the analysis of the ripple without creating significant errors (assuming that the ripple level

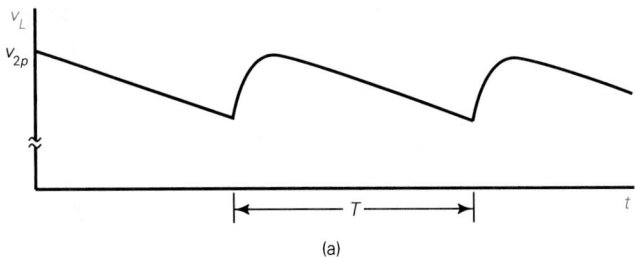

(a)

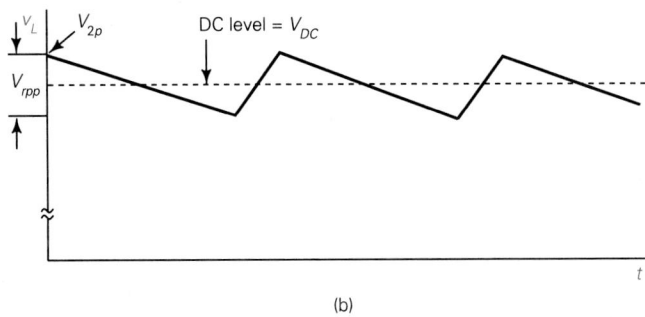

(b)

FIGURE 20–18
Waveforms used in analysis of ripple.

is relatively low): (1) The ripple waveform is approximated by a triangular wave-form as shown in Figure 20–18(b). (2) The load current is assumed to be constant for the purpose of calculating the change in voltage across the capacitor. This change in voltage may then be used to compute the ripple level and the actual dc output. (Note that the actual dc level is at the midpoint of the triangular waveform.) (3) The short charging time at the beginning of a half-cycle will be neglected in comparison with the discharge time.

A point to note from the waveforms is that the period T of the ripple for a full-wave rectifier is one-half the period of the ac line voltage. In contrast, the period of the ripple for a half-wave rectifier is the actual period of the ac line voltage. In order not to derive two sets of formulas, we will employ the symbol $f = 1/T$ as the frequency of the ripple. Thus, we have the following relationships for the commercial 60-Hz power system:

$$f = 60 \text{ Hz (half-wave)} \tag{20–38a}$$
$$= 120 \text{ Hz (full-wave)} \tag{20–38b}$$

First, a fundamental circuit principle must be established. Assume that a capacitor C is being charged from a constant current source I. Let Δv represent the change in voltage across the capacitor in a time Δt. It is established in basic circuit theory that

$$\Delta v = \frac{1}{C} I \Delta t \tag{20–39}$$

Let V_{rpp} represent the peak-to-peak ripple voltage, and let I_L represent the dc load current at the output of the filter. Referring to Figure 20–18(b) and applying (20–39), we have

$$V_{rpp} = \frac{I_L T}{C} = \frac{I_L}{fC} \qquad \text{(20–40)}$$

From (20–40), the following points are readily deduced: (1) The ripple level increases with an increase in the load current, and (2) the ripple level decreases with increasing frequency or with increasing capacitance.

From the frequency dependency, it should be clear that a full-wave rectifier is easier to filter than a half-wave rectifier since its frequency is twice as great. Generally, the full-wave rectifier is almost always preferred over the half-wave rectifier.

Since the ripple decreases with increased capacitance, one might be tempted to suggest that the capacitance could be increased to any size required for a given ripple level. In practice, however, there is an important trade-off that must be considered. As the capacitance increases, the initial surge of charging current when the power supply is first turned on increases. This surge is limited only by diode and transformer resistance and can be very large. In some circuits, a surge resistance is placed in the circuit to limit the value of this current.

The actual dc output voltage V_{dc} can be determined by locating the midpoint of the ripple. For a peak secondary voltage V_{2p}, the dc output can be expressed as

$$V_{dc} = V_{2p} - \frac{V_{rpp}}{2} = V_{2p} - \frac{I_L}{2fC} \qquad \text{(20–41)}$$

Ripple Factor

Frequently, the rms ripple voltage is used to express the ripple level. Let V_r represent this value. It can be shown that this rms value is related to the peak-to-peak ripple by

$$V_r = \frac{V_{rpp}}{2\sqrt{3}} = \frac{I_L}{2\sqrt{3}fC} \qquad \text{(20–42)}$$

The ripple factor r is defined as

$$r = \frac{\text{rms ripple voltage}}{\text{dc output voltage}} \qquad \text{(20–43a)}$$

$$= \frac{V_r}{V_{dc}} \qquad \text{(20–43b)}$$

EXAMPLE 20-6   A certain full-wave power supply with a single capacitor filter of the type shown in Figure 20–10 is powered by the 120-V rms, 60-Hz system. The peak value of the output voltage under loaded conditions is 30 V. (The peak transformer output voltage across half the secondary is somewhat greater to compensate for losses.) The capacitance is $C = 2000$ μF. For a load current of 2 A, determine **(a)** the peak-to-peak ripple voltage; **(b)** the rms ripple voltage; **(c)** the dc load voltage; and **(d)** the ripple factor.

Solution
(a)   The peak-to-peak ripple voltage is

$$V_{rpp} = \frac{I_L}{fC} = \frac{2}{120 \times 2000 \times 10^{-6}} = 8.33 \text{ V} \tag{20–44}$$

(b)   The rms ripple voltage is

$$V_r = \frac{V_{rpp}}{2\sqrt{3}} = \frac{8.33}{2\sqrt{3}} = 2.41 \text{ V} \tag{20–45}$$

(c)   The dc output voltage is

$$V_{dc} = V_{2p} - \frac{V_{rpp}}{2} = 30 - \frac{8.33}{2} = 25.83 \text{ V} \tag{20–46}$$

(d)   The ripple factor is

$$r = \frac{V_r}{V_{dc}} = \frac{2.41}{25.83} = 0.093 \tag{20–47}$$

20-9 REGULATED POWER SUPPLIES

The various power supply circuits considered thus far all suffer from the fact that the dc output voltage is not regulated. At the front end, the commercial ac supply voltage is subject to a wide variation, and this variation is in turn transferred to the output dc voltage. Coupled with this are transformer losses and diode drops. As load requirements change, voltage variations also occur. Consequently, these basic power supply circuits are used alone only in applications where the expected variation of the output dc voltage is not critical.

The modern approach is to use a **voltage regulator** circuit to establish a nearly precision dc output voltage. A voltage regulator is an electronic control circuit capable of establishing and maintaining a nearly constant dc output voltage as load requirements vary and as the input voltage varies.

For moderate degrees of regulation and where variations are not too great, zener diodes can be used directly for regulation. The approach for analyzing and designing zener circuits was considered in Chapter 3, and that material is applicable to the present discussion. However, the approach here will be to focus on the more comprehensive voltage regulator circuit. As a part of such a circuit, zener diodes are often used. However, the variation of the current in the zener will be far less than expected for the complete power supply, and zeners can be used to establish references within regulator circuits.

Basic Regulator Circuit

The operation of a basic regulator circuit implemented with an op-amp is illustrated in Figure 20–19. The op-amp is connected in the noninverting configuration with the feedback resistor connected to the dc output voltage. A zener reference voltage V_Z is also obtained from the output voltage using a dropping resistor R_Z. The reference voltage V_Z is less than the required load voltage V_L. For a zener current I_Z, the resistance R_Z must satisfy

$$R_Z = \frac{V_L - V_Z}{I_Z} \tag{20–48}$$

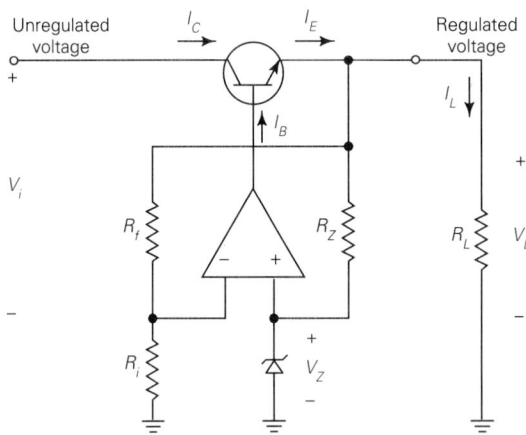

FIGURE 20–19
Voltage regulator implemented with an op-amp and a pass transistor.

Assume for the moment that the transistor operates in the active linear region. The nonlinear base–emitter junction is a part of the forward block of a feedback loop. The feedback network consisting of R_f and R_i forces the output voltage V_L to assume the value.

$$V_L = \left(1 + \frac{R_f}{R_i}\right) V_Z \tag{20–49}$$

The output of the op-amp assumes a voltage equal to V_L plus the transistor base–emitter drop. For the current levels involved, the base–emitter drop is typically greater than the 0.7-V drop assumed for small-signal transistors.

The op-amp need only supply the base current required to drive the transistor. However, the transistor and collector currents will be considerably greater. Because of the action of the transistor in allowing a much greater load current than its base drive current, it is referred to as a **pass transistor.**

In a typical application with moderate to large load currents, the current through the feedback circuit and the zener bias current will both be considerably smaller than the load current I_L. Thus, a reasonable approximation in many cases is

$$I_E \simeq I_L \tag{20-50}$$

The collector current I_C can then be expressed as

$$I_C = I_E - I_B \tag{20-51}$$

As a further approximation, I_B will be neglected in comparison with I_E in (20–51), which results in the simplification

$$I_C \simeq I_E \simeq I_L \tag{20-52}$$

Thus, in subsequent computations, it will be assumed that collector current, emitter current, and load current are all essentially the same. Although these assumptions are not always necessary, they do simplify the subsequent analysis.

Based on the preceding assumptions, the base current I_B can be expressed as

$$I_B \approx \frac{I_L}{\beta_{dc}} \tag{20-53}$$

where β_{dc} is the dc transistor current gain. The op-amp must be capable of providing this value of base drive. Where the required load current is considerably greater than the available op-amp drive current, a transistor compound connection may be used to achieve a very large value of β_{dc}.

Efficiency

Let V_i represent the value of the unregulated voltage at the input to the regulator. In order to ensure linear active region operation for the pass transistor, it is necessary that $V_i > V_L$. Because of the typical larger saturation voltage of a power transistor and the necessity to provide some leeway for line variations, ripple, and other factors, V_i is often chosen to be at least 2 or 3 V greater than V_L.

The input power P_i can be closely approximated as

$$P_i \simeq V_i I_L \tag{20-54}$$

The output load power P_L is

$$P_L = V_L I_L \tag{20-55}$$

The pass transistor must dissipate a power P_D given by the difference between the preceding values, that is,

$$P_D = P_i - P_L = V_i I_L - V_L I_L = (V_i - V_L) I_L \tag{20-56}$$

Let η represent the efficiency as a percentage. It is given by

$$\eta = \frac{P_L}{P_i} \times 100\% \qquad (20\text{–}57)$$

Substitution of (20–54) and (20–55) in (20–57) results in

$$\eta = \frac{V_L}{V_i} \times 100\% \qquad (20\text{–}58)$$

Current Limiting

To prevent the possibility of drawing excessive load current, a current-limiting transistor may be added to the regulator as shown in Figure 20–20. The combination of the pass transistor and the current-limiting transistor in Figure 20–20 replaces the pass transistor in Figure 20–19. All connections to the left, right, and below are the same as before.

FIGURE 20–20

Addition of current-limiting transistor (T2) to regulator circuit.

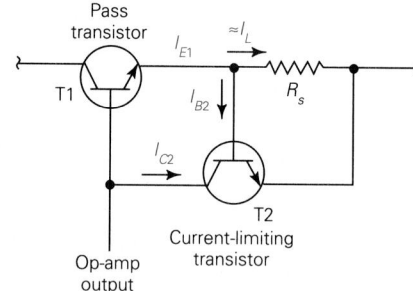

The concept behind the limiting actions is as follows: When the current I_L is small, the base–emitter voltage of T2, which is given by $R_s I_L$, is insufficient to forward-bias T2, and thus $I_{C2} = 0$. The limiter T2 has no effect for this condition. As I_L increases, the voltage $R_s I_L$ becomes large enough to forward-bias the base–emitter junction of T2, and base current flows. This results in collector current for T2, as indicated by I_{C2} on Figure 20–20. This current has to be supplied from the op-amp and is "robbed" from the base supply of T1. This effect tends to reduce the collector and emitter currents of T1, and the result is to effectively limit further increases in the load current. The resistance R_s is selected so that the product of R_s and the maximum load current is the expected cut-in voltage for T2.

Integrated-Circuit Regulators

This development has explained the basic operation of the regulator utilizing an op-amp. In actual practice, a wide variety of IC chips are available, in which the op-amp, reference zener, and pass transistor are all part of the given circuit. Some are designed for fixed dc output voltages such as $+5$, $+15$, and -15 V. Others are "programmable" by providing suitable resistors (i.e., R_i and R_f in Figure 20–19).

The wiring layout of a typical IC regulator used for a fixed dc output voltage is shown in Figure 20–21. This particular regulator is a three-terminal device having an input, an output, and a common ground. It is assumed that a reasonable amount of filtering has been provided for the input since this voltage must not drop below some minimum level for regulation to be sustained. Most regulators also recommend additional capacitors at input and output as shown.

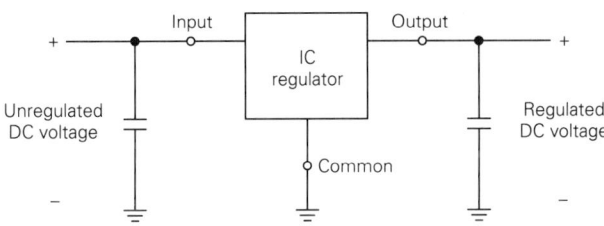

FIGURE 20–21
Connections for fixed positive voltage regulator.

Switching Regulator Power Supply

One of the newest concepts in regulated power supplies is that of the **switching regulator power supply.** The circuit is quite complex and will be discussed here from a block diagram point of view.

Refer to Figure 20–22, representing a typical switching power supply block diagram. The input ac line voltage is first rectified with a conventional rectifier circuit, and some filtering is employed. This voltage is applied to a high-frequency inverter that is controlled by a control circuit. The output of this inverter is a series of high-frequency pulses, for which a typical frequency (20 kHz) is indicated. The width of the pulses is controlled by the modulator in the control circuit. This higher-frequency voltage is rectified and filtered to produce the dc output voltage as shown.

The regulation action of the circuit is a function of the feedback signal applied to the control voltage. If the output dc voltage starts to drop, the sense of the comparator is such as to increase the width of the modulated pulses. Since these pulses control the high-frequency inverter pulses, the result is to increase the dc output to compensate. The opposite situation occurs if the dc output increases. The output signal from the modulator is called a **pulse-wide modulated (PWM)** signal.

The primary advantage of switching regulator power supplies is their relatively high efficiencies. Efficiencies may run as high as 80% in many cases, which is considerably higher than the typical efficiencies of standard regulator circuits.

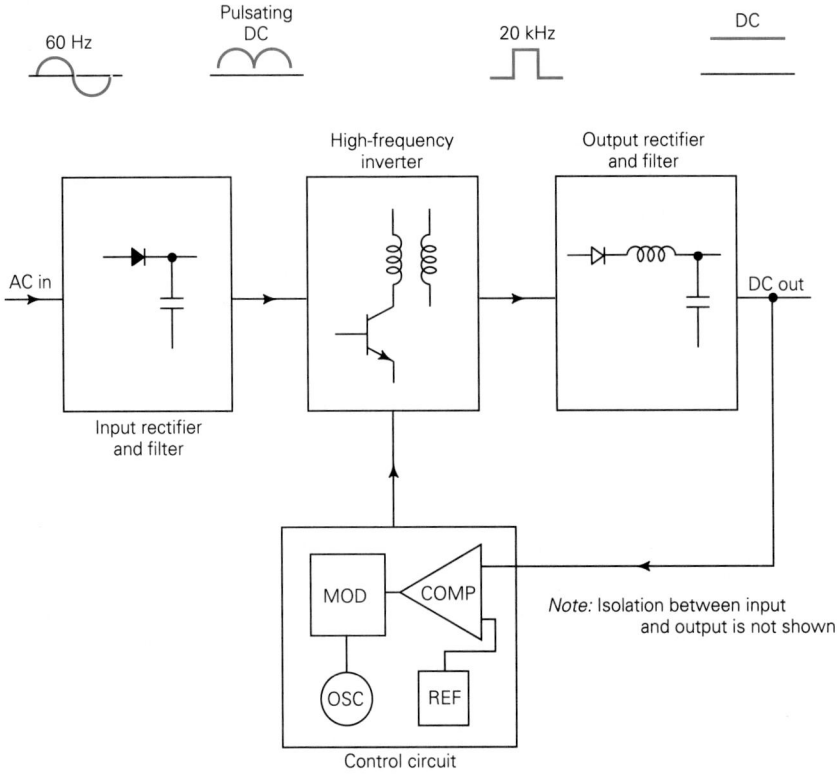

FIGURE 20–22
Block diagram of switching power supply.

EXAMPLE 20–7

Assume an op-amp regulator of the type shown in Figure 20–19 with $R_i = 2$ kΩ, $R_f = 3.357$ kΩ, and $V_Z = 5.6$ V. The unregulated voltage is approximately 20 V.

(a) Determine V_L. **(b)** If the maximum value of load current is 1 A and $\beta_{dc} = 100$ for T1, determine the maximum op-amp output current required for base drive I_B. **(c)** Determine the load power, the approximate input power, and the efficiencies under peak-load conditions. **(d)** Determine the maximum power dissipation $P_{D(\text{max})}$ for the pass transistor.

Solution

(a)   The value of the load voltage is

$$V_L = \left(1 + \frac{R_f}{R_i}\right) V_Z = \left(1 + \frac{3357}{2000}\right)5.6 = 15 \text{ V} \qquad \textbf{(20–59)}$$

(b)   For $I_L = 1$ A and $\beta_{dc} = 100$, the op-amp must supply a base drive of

$$I_B = \frac{I_L}{\beta_{dc}} = \frac{1 \text{ A}}{100} = 10 \text{ mA} \qquad \textbf{(20–60)}$$

(c) The load power is

$$P_L = V_L I_L = 15 \times 1 = 15 \text{ W} \qquad \textbf{(20–61)}$$

The approximate input power is

$$P_i = V_i I_L = 20 \times 1 = 20 \text{ W} \qquad \textbf{(20–62)}$$

The efficiency is

$$\eta = \frac{15 \text{ W}}{20 \text{ W}} \times 100\% = 75\% \qquad \textbf{(20–63)}$$

(d) The maximum power dissipation occurs under full-load conditions and is

$$P_{D(\text{max})} = 20 - 15 = 5 \text{ W} \qquad \textbf{(20–64)}$$

20–10 THYRISTORS

Thyristors are semiconductor devices that use internal feedback to produce latching action. Thyristors operate only in the switching mode. Generally, they are used to control large amounts of power, such as in lighting systems, heaters, motors, and so on. Because of their importance in power systems, a discussion of their properties has been included in this chapter.

Silicon-Control Rectifier

A **silicon-control rectifier (SCR)** is a thyristor employing four layers of semiconducting materials. The construction is suggested by the diagram of Figure 20–23(a). The three external connections are denoted as the **cathode,** the **anode,** and the **gate.**

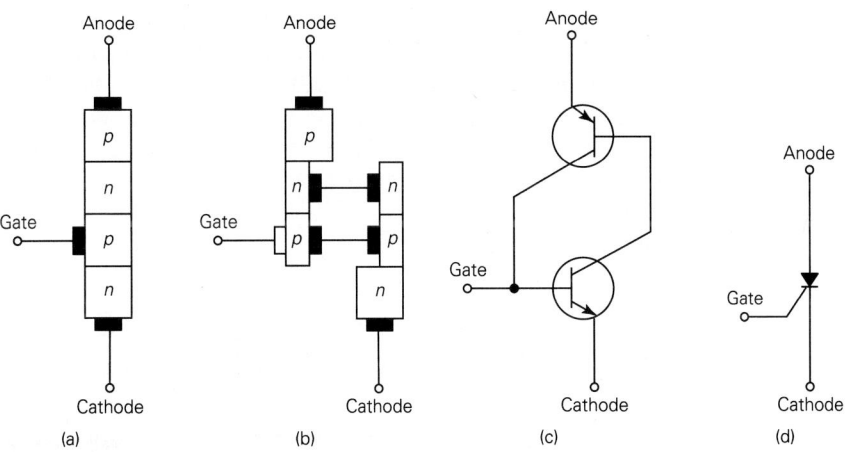

FIGURE 20–23
Construction of an SCR and the schematic symbol.

The SCR may be visualized in the manner shown in Figure 20–23(b). It is equivalent to the connection of a *pnp* transistor and an *npn* transistor as shown in Figure 20–23(c). The corresponding schematic symbol is shown in Figure 20–23(d).

The SCR will act as an open circuit unless either one of two possible conditions is satisfied: (1) A positive trigger pulse is received at the gate, or (2) the anode–cathode voltage exceeds a certain minimum voltage called the **breakover voltage.** Most SCRs are designed to function with a trigger input.

Assume that the voltage at the anode is positive with respect to the cathode, but there is no base current flowing in either transistor. This is the initial condition encountered with the SCR, and it acts as an open circuit under this condition as long as the anode–cathode voltage is less than the breakover voltage.

Assume now that a positive pulse is received at the gate terminal. The base–emitter junction of the *npn* transistor will be forward biased, and current will flow in its collector through the base–emitter junction of the *pnp* transistor. This will result in current flow in the *pnp* transistor collector, and this current flow adds to the initial base current of the *npn* transistor. The result is a regenerative action that rapidly changes the SCR to a nearly short-circuit condition. The SCR thus acts as a closed switch under this condition.

Even though the base pulse is removed, the SCR will remain in the closed condition until the current through the SCR drops to a small value. The transistors will come out of saturation when the current is reduced below the **holding current.** It then remains in the open-circuit condition until a new pulse is received.

Half-Wave Control

A typical control circuit using an SCR to control the voltage over a half-cycle is shown in Figure 20–24. The resistance R_1 is used to limit the current, and R_2 is used to establish the trigger level for the SCR. If the SCR is turned on near the beginning of a positive half-cycle, conduction in the load R_L will occur for nearly 180°. Conversely, if the SCR is turned on only near the peak of the half-cycle, conduction will occur for only about 90°. By adjustment of the variable arm for R_2, the conduction angle can thus be varied, and the power delivered to the load can be adjusted. During the negative half-cycle of the input, the SCR is turned off. The diode prevents this negative ac voltage from being applied to the SCR gate.

FIGURE 20–24
Half-wave SCR control circuit.

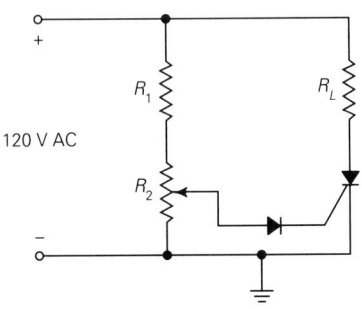

Diac

A **diac** is equivalent to the parallel connection of two four-layer semiconductor devices in opposite directions, and the schematic diagram is shown in Figure 20–25(a). It is a **bidirectional** thyristor in that current can flow in either direction.

FIGURE 20–25
Schematic symbol for diac and equivalent switching circuit.

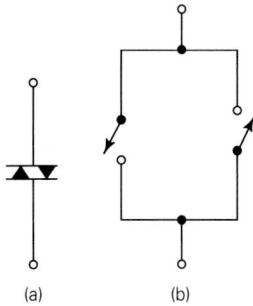

(a)   (b)

The diac acts like an open circuit until the voltage in either direction across it tries to exceed the breakover voltage. In that event one of two parallel diodes changes state and conducts strongly, approximating a short circuit in the ideal case. Once the diac starts to conduct, the only way it can be stopped is for the current through it to drop below the rated holding current of the device. Operation can be visualized as two switches in parallel as shown in Figure 20–25(b).

Triac

The **triac** is a bidirectional device with a control terminal. The schematic symbol is shown in Figure 20–26(a). It can be visualized as a parallel connection of two SCRs, as shown in Figure 20–26(b). The normal mode of operation for the triac is to turn it on with a trigger at the gate. When the voltage across the triac has one polarity, a positive trigger is required, but when the voltage has the opposite polarity, a negative trigger is required.

FIGURE 20–26
Schematic symbol for triac and its equivalence as two SCRs.

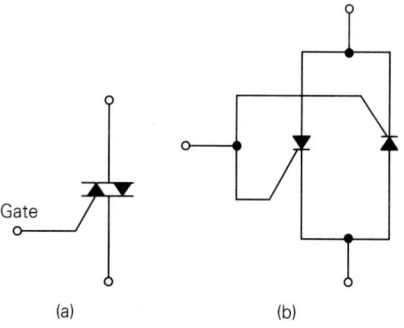

Gate

(a)   (b)

20–11 PSPICE EXAMPLES

PSPICE EXAMPLE 20–1

Develop a PSPICE program to simulate the unfiltered half-wave rectifier circuit of Example 20–2, and obtain input and output voltage waveforms.

Solution

The circuit diagram adapted to PSPICE is shown in Figure 20–27(a), and the code is shown in Figure 20–27(b). This is the first point in the text where a transformer has been simulated, so some discussion is required. In general, a transformer contains two or more inductances, each of which has a self-inductance measured in henries (H). For an ideal or nearly ideal transformer, such as we are assuming here, the inductance ratio is directly proportional to the turns ratio squared; that is,

$$\frac{L_1}{L_2} = \left(\frac{N_1}{N_2}\right)^2 \tag{20–65}$$

where L_1 and L_2 are the inductance values with N_1 and N_2 turns, respectively. To obtain a turns ratio of $6:1$, the required inductance ratio must be

$$\frac{L_1}{L_2} = \left(\frac{6}{1}\right)^2 = 36 \tag{20–66}$$

Thus, L_1 must be 36 times as great as L_2. Arbitrary values of $L_1 = 36$ H and $L_2 = 1$ H have been chosen.

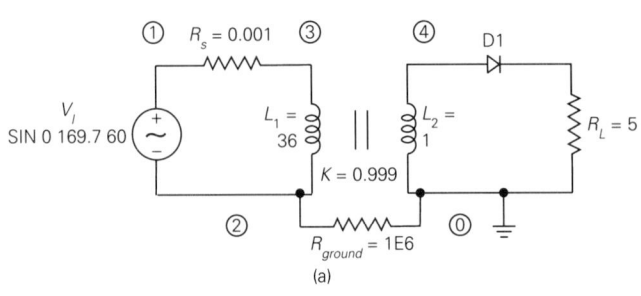

```
PSPICE EXAMPLE 20-1
VI 1 2 SIN 0 169.7 60
RS 1 3 0.001
L1 3 2 36
L2 4 0 1
K L1 L2 0.999
RGROUND 2 0 1E6
D1 4 5 DMODEL
RL 5 0 5
.LIB EVAL.LIB
.MODEL DMODEL D
.TRAN 33.3E-3 33.3E-3 0 10E-6
.PROBE
.END
```

(b)

FIGURE 20–27
Circuit and code for PSPICE Example 20–1.

To force the two inductors to act as a transformer, it is necessary to couple the magnetic flux from one inductor with the other. The degree of coupling is measured by a coefficient of coupling K, in which $0 \le K \le 1$. When $K = 0$, there is no coupling, and at the other extreme, $K = 1$ represents one requirement of an ideal transformer. A value such as $K = 0.999$ should suffice to approximate a nearly ideal transformer.

The portion of the program defining the transformer consists of three lines that read as follows:

```
L1 3 2 36
L2 4 0 1
K L1 L2 0.999
```

The first two lines define the node connections and inductance values and are self-explanatory. The last line begins with K, which indicates a coefficient of coupling, and this is followed by the names of the inductors involved. The last entry is the coefficient of coupling, which is 0.999.

All PSPICE circuit elements must have a dc path to ground. On the secondary, one side of the circuit is grounded. In its basic form as given, however, the primary would not have a dc path to ground. To achieve the requirement, a resistance of 1 MΩ is connected between the lower primary terminal and the grounded secondary terminal.

One other peculiarity is that PSPICE will not accept a loop that would be a short-circuit at dc. With the original circuit and the ideal inductance assumed, the primary loop would be a short at dc. The problem is circumvented by the addition of a small resistance, which in this case is 0.001 Ω.

Note that the peak value of the input voltage is set at 169.7 V, which corresponds to an rms voltage of $169.7/\sqrt{2} = 120$ V. The frequency is, of course, 60 Hz. The generic diode model is used, and a transient analysis is performed over two cycles.

The input and output waveforms are shown in Figure 20–28. When PSPICE was allowed to set its own time steps, the output appeared to be a bit "ragged," and the input peak of 169.7 V was not evident. The transient statement was adjusted for a maximum time step of 10E-6 = 10 μs, and the results improved dramatically. The peak output voltage is 27.063 V, which differs by about 4.3% from the value calculated in Example 20–2. This difference is due primarily to the fact that the diode drop was ignored in Example 20–2.

PSPICE EXAMPLE 20–2

Develop a PSPICE program to simulate the unfiltered full-wave rectifier circuit of Example 20–3, and obtain the output voltage waveform.

Solution

The circuit adapted to PSPICE is shown in Figure 20–29(a), and the code is shown in Figure 20–29(b). Many of the features of this circuit are similar to those of the preceding example, and the discussion given there applies here. However, there is an additional complication in this example produced by the presence of the center-tapped secondary.

To create a center-tapped winding, it is necessary to use two inductors in

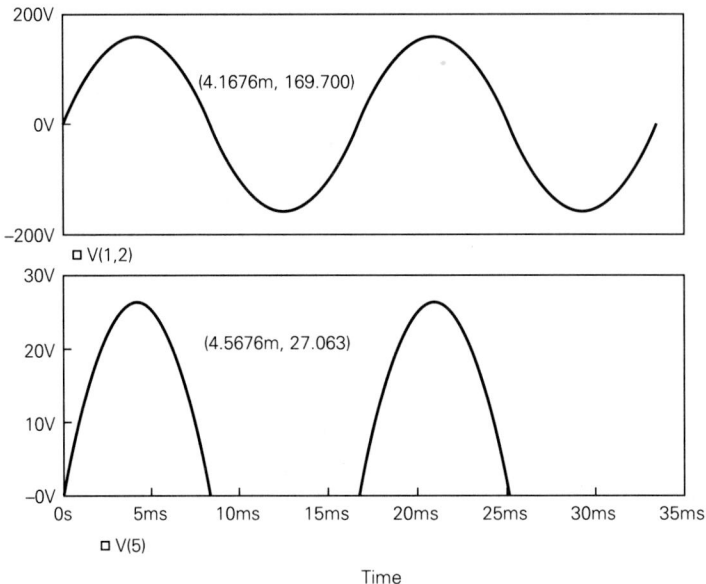

FIGURE 20–28
Input and output waveforms of PSPICE Example 20–1.

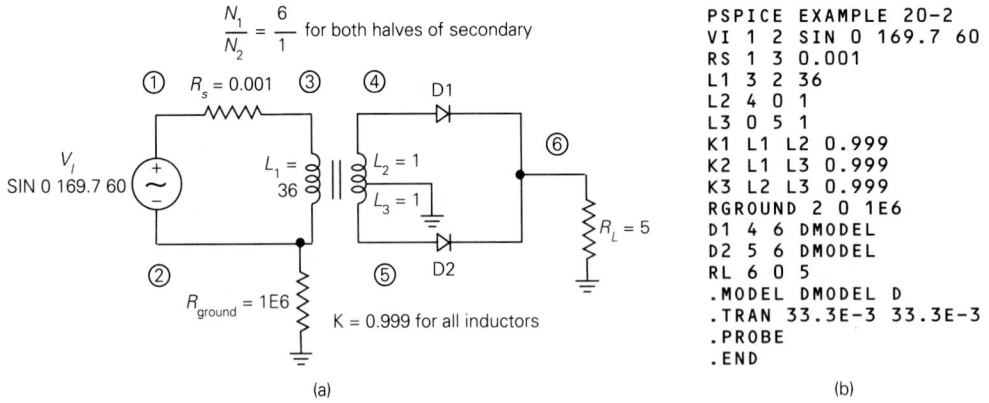

(a)

```
PSPICE EXAMPLE 20-2
VI 1 2 SIN 0 169.7 60
RS 1 3 0.001
L1 3 2 36
L2 4 0 1
L3 0 5 1
K1 L1 L2 0.999
K2 L1 L3 0.999
K3 L2 L3 0.999
RGROUND 2 0 1E6
D1 4 6 DMODEL
D2 5 6 DMODEL
RL 6 0 5
.MODEL DMODEL D
.TRAN 33.3E-3 33.3E-3
.PROBE
.END
```

(b)

FIGURE 20–29
Circuit and code of PSPICE Example 20–2.

series and to couple them with a *K* statement. Likewise, the primary must be separately coupled to each of the secondary inductors.

To accomplish the preceding, six lines of code are required. The first three define the primary inductance L_1 and the two secondary inductances L_2 and L_3. The ratio of primary inductance to *each* of the secondary inductances must be 36:1, which will establish turns ratios of 6:1 for each half. Three coefficients of coupling are provided on the next three lines. These represent the coupling between L_1 and L_2, between L_1 and L_3, and between L_2 and L_3. Care must be used in a situation such as this to ensure that the two secondary voltages aid each other, and this is achieved by defining the nodes in the same order. Thus, node 4 precedes node 0 in L_2 and node 0 precedes node 5 in L_3.

The output voltage waveform is shown in Figure 20–30. The peak output voltage is about 27.063 V, which differs from the computed value of 28.28 V by about 4.3%. As in the preceding example, the diode drop is the major source for the difference.

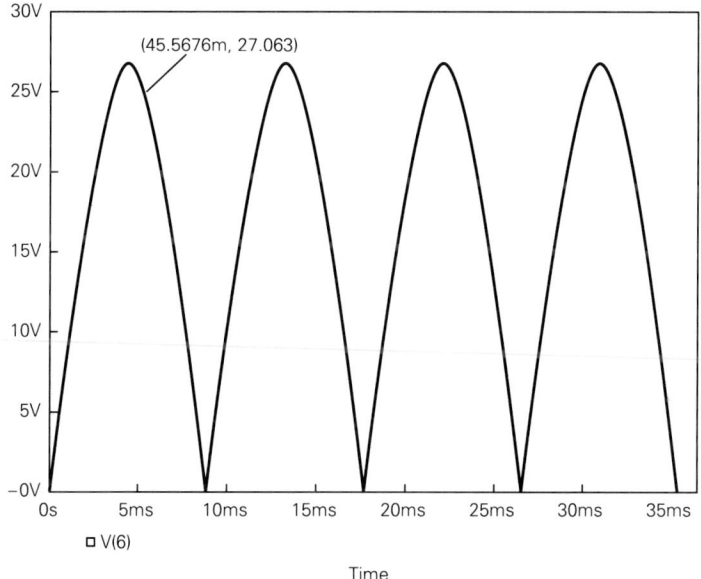

FIGURE 20–30
Output waveform of PSPICE Example 20–2.

PSPICE EXAMPLE 20–3

Assume in the full-wave rectifier circuit of PSPICE Example 20–2 that the 5-Ω load is replaced by a 50-Ω load and that a 2000-μF filter capacitor is placed across the load as shown in Figure 20–31(a). Modify the PSPICE program to accommodate these changes, and obtain a plot of the output voltage.

Solution
The new code is shown in Figure 20–31(b). The primary changes from the preceding example are the change in R_L and the addition of C. To minimize the transient interval and expedite plotting, some initial runs were taken to determine the approximate voltage at the beginning of a cycle. A value of about 25.2 V was determined, and this was established as an initial condition for C.

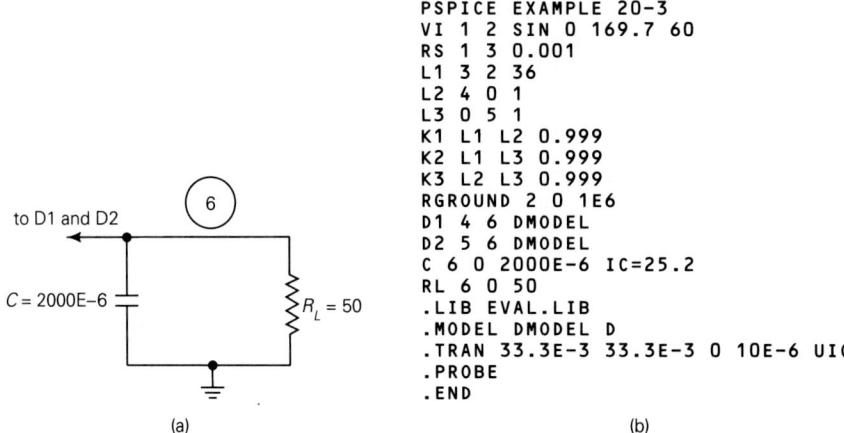

```
PSPICE EXAMPLE 20-3
VI 1 2 SIN 0 169.7 60
RS 1 3 0.001
L1 3 2 36
L2 4 0 1
L3 0 5 1
K1 L1 L2 0.999
K2 L1 L3 0.999
K3 L2 L3 0.999
RGROUND 2 0 1E6
D1 4 6 DMODEL
D2 5 6 DMODEL
C 6 0 2000E-6 IC=25.2
RL 6 0 50
.LIB EVAL.LIB
.MODEL DMODEL D
.TRAN 33.3E-3 33.3E-3 0 10E-6 UIC
.PROBE
.END
```

(a)   (b)

FIGURE 20–31
Capacitor filter and load for PSPICE Example 20–3.

The plot of the output voltage over two cycles of the input voltage is shown in Figure 20–32. Note that the full-wave rectification process, in effect, creates four cycles of the output during this interval.

The dc output voltage can be estimated by taking the midpoint between the maximum and minimum voltages. The value obtained is $(25.836 + 24.484)/2 =$

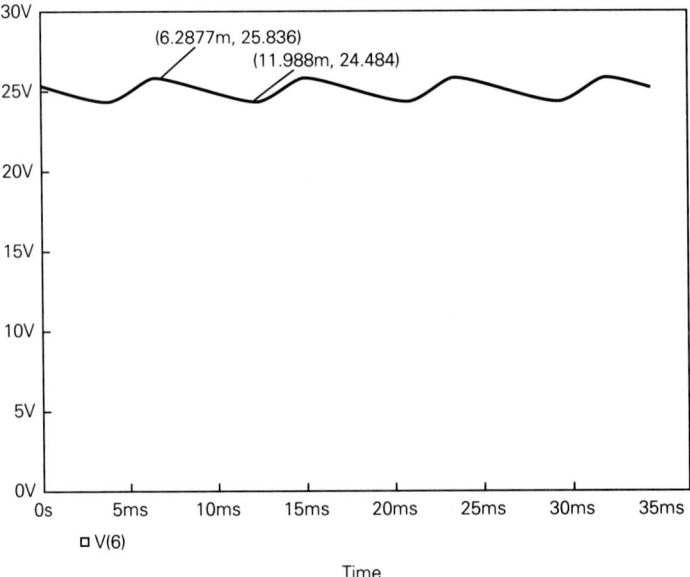

FIGURE 20–32
Output waveform of PSPICE Example 20–3.

25.16 V. The peak-to-peak ripple voltage is about $25.836 - 24.484 = 1.35$ V. However, equation (20–40) predicts a ripple of about 2.1 V. While this result is in the right ballpark, it illustrates the difficulty in predicting precisely certain values in a power rectifier circuit. The use of a voltage regulator can circumvent this problem.

The ripple level obtained is too large for most applications. Additional filtering and/or a regulator could be added to reduce significantly the ripple level and to establish a more precise output voltage.

PROBLEMS

Drill Problems

20–1. An ideal transformer has a turns ratio of 5:1 from primary to secondary. Assume a primary ac line voltage of 120 V rms. For a particular secondary load, the primary current is 3 A. Determine the secondary rms voltage and current.

20–2. An ideal transformer has a turns ratio of 1:4 from primary to secondary. Assume a primary ac line voltage of 120 V rms. For a particular secondary load, the primary current is 8 A. Determine the secondary rms voltage and current.

20–3. An ideal transformer has a turns ratio of 8:1 from primary to secondary. Assume a primary ac line voltage of 120 V rms. A certain load draws a secondary current of 4 A rms. Determine the rms values of the primary current and the secondary voltage.

20–4. The primary of an ideal transformer is connected to the 120-V rms ac line voltage. A secondary 20-Ω load draws an rms current of 2 A. Determine the rms secondary voltage, the rms primary current, and the turns ratio.

20–5. A certain unfiltered half-wave rectifier circuit of the form of Figure 20–3 is powered by the 120-V rms ac power system. The turns ratio is $N_1 : N_2 = 8 : 1$. Determine **(a)** the rms secondary voltage; **(b)** the peak secondary voltage; **(c)** the dc load voltage; and **(d)** the dc load current if $R_L = 6 \, \Omega$. Neglect any diode and transformer losses.

20–6. Repeat the analysis of Problem 20–5 if $N_1 : N_2 = 1 : 4$ and $R_L = 50 \, \Omega$.

20–7. A certain unfiltered full-wave rectifier circuit of the form of Figure 20–5 is powered by the 120-V rms ac power system. The turns ratio is $N_1 : N_2 = 8 : 1$. Determine **(a)** the rms secondary voltage; **(b)** the peak secondary voltage; **(c)** the dc load voltage; and **(d)** the dc load current if $R_L = 6 \, \Omega$. Neglect any diode and transformer losses.

20–8. Repeat the analysis of Problem 20–7 if $N_1 : N_2 = 1 : 4$ and $R_L = 50 \, \Omega$.

20–9. A certain unfiltered bridge rectifier circuit of the form of Figure 20–8 is powered by the 120-V rms ac power system. The turns ratio is $N_1 : N_2 = 8 : 1$. Determine **(a)** the rms secondary voltage; **(b)** the peak secondary voltage; **(c)** the dc load voltage; and **(d)** the dc load current if $R_L = 6 \, \Omega$. Neglect any diode and transformer losses.

20–10. Repeat the analysis of Problem 20–9 if $N_1 : N_2 = 1 : 4$ and $R_L = 50 \, \Omega$.

20–11. Determine the peak reverse voltages for the diodes in the rectifier circuits of the following problems: **(a)** 20–5; **(b)** 20–7; **(c)** 20–9.

20–12. Determine the peak reverse voltages for the diodes in the rectifier circuits of the following problems: **(a)** 20–6; **(b)** 20–8: **(c)** 20–10.

20–13. Determine the peak reverse voltages for the diodes in each of the four following rectifier circuits: **(a)** unfiltered half-wave rectifier with $V_{2P} = 24$ V; **(b)** filtered half-wave rectifier with $V_L = 24$ V; **(c)** unfiltered full-wave rectifier (using a center-tapped transformer) with $V_{2p} = 24$ V; **(d)** unfiltered bridge rectifier with $V_{2p} = 24$ V.

20–14. Determine the peak reverse voltages for the diodes in each of the four following rectifier circuits: **(a)** unfiltered half-wave rectifier with $V_{2P} = 400$ V; **(b)** filtered half-wave rectifier with $V_L = 400$ V; **(c)** filtered full-wave rectifier (using a center-tapped transformer) with $V_L = 400$ V; **(d)** filtered bridge rectifier with $V_L = 400$ V.

20–15. A certain half-wave rectifier circuit employs a single capacitor filter and $C = 600$ μF. It is powered by the 120-V rms, 60-Hz ac power system. The dc load current is 180 mA. Determine **(a)** the peak-to-peak ripple voltage, and **(b)** the rms ripple voltage.

20–16. Repeat the analysis of Problem 20–15 if the circuit is a full-wave rectifier. All other conditions are unchanged.

20–17. A certain full-wave power supply with a single capacitor filter of the type shown in Figure 20–10 is powered by the 120-V rms, 60-Hz ac power system. The peak value of the output voltage under loaded conditions is 36 V and $C = 2000$ μF. For a load current of 3 A, determine **(a)** the peak-to-peak ripple voltage; **(b)** the rms ripple voltage; **(c)** the dc load voltage; and **(d)** the ripple factor.

20–18. A certain full-wave power supply with a single capacitor filter uses a bridge circuit and is powered by the 120-V, 60-Hz ac power system. The peak value of the output voltage under loaded conditions is 600 V and $C = 100$ μF. For a load current of 200 mA, determine **(a)** the peak-to-peak ripple voltage; **(b)** the rms ripple voltage; **(c)** the dc load voltage; and **(d)** the ripple factor.

20–19. An op-amp regulator of the type shown in Figure 20–19 has $R_i = 10$ kΩ, $R_f = 14.24$ kΩ, and $V_Z = 3.3$ V. The unregulated voltage is approximately 12 V. **(a)** Determine V_L. **(b)** If the maximum value of load current is 0.5 A and $\beta_{dc} = 80$ for the transistor, determine the maximum op-amp output current required for base drive I_B. **(c)** Determine the load power, the approximate input power, and the efficiency under peak load conditions. **(d)** Determine the maximum power dissipation $P_{D(max)}$ for the pass transistor.

20–20. An op-amp regulator of the type shown in Figure 20–19 has $R_i = 10$ kΩ, $V_Z = 5.1$ V, and $V_L = 20$ V. The unregulated voltage is approximately 28 V. **(a)** Determine R_f. **(b)** If the maximum value of load current is 250 mA and $\beta_{dc} = 50$ for the transistor, determine the maximum op-amp output current required for base drive I_B. **(c)** Determine the load power, the approximate input power, and the efficiency under peak load conditions. **(d)** Determine the maximum power dissipation $P_{D(max)}$ for the pass transistor.

Derivation Problems

20–21. (Requires calculus.) Consider a half-wave rectified voltage as defined by

$$v(t) \begin{cases} = V_p \sin \omega t & 0 < t < \dfrac{T}{2} \\[2mm] = 0 & \dfrac{T}{2} < t < T \end{cases}$$

Prove that the dc value V_{dc} is given by

$$V_{dc} = \frac{V_p}{\pi}$$

20–22. (Requires calculus.) Consider a full-wave rectified voltage as defined by

$$v(t) = |V_p \sin \omega t| \qquad 0 < t < T$$

Prove that the dc value V_{dc} is given by

$$V_{dc} = \frac{2V_{dc}}{\pi}$$

(*Hint:* In the first half-cycle, $v(t)$ is the same as in Problem 20–21, and the integral for the complete cycle is double the value for the first half-cycle.)

Design Problems

20–23. In a certain industrial process, an unfiltered half-wave rectifier circuit of the form of Figure 20–3 is to be employed and is to be powered by the 120-V ac power system. In this initial design stage, idealized components will be assumed, and losses will be neglected. A dc current of 5 A in a 4-Ω load is required. Determine the turns ratio $N_1 : N_2$.

20–24. Repeat the design of Problem 20–23 if an unfiltered bridge rectifier of the form of Figure 20–8 is employed. (In this case, N_2 is the number of turns on the complete secondary.)

20–25. In a certain rectifier circuit employing a single-capacitor filter, the dc output voltage is to be 30 V at a load current of 800 mA. At that load current, the ripple factor is not to exceed 0.04. Determine the minimum value of capacitance required if a full-wave rectifier is used. The circuit is powered by the 120-V rms, 60-Hz ac power system.

20–26. Determine the minimum value of capacitance required in Problem 20–25 if a half-wave rectifier is used.

20–27. An unregulated full-wave rectifier circuit employing a center-tapped transformer and a single capacitor filter is to be designed for a particular application. In this initial design step, idealized components will be assumed, and losses will be neglected. The design specifications are

dc load voltage = 20 V at a load current of 0.5 A

Ripple factor = 0.05

Determine **(a)** the capacitance required, and **(b)** the rms voltage required across each half of transformer secondary (assuming the 120-V, 60-Hz ac power system).

20–28. An unregulated full-wave bridge rectifier circuit employing a single-capacitor filter is to be designed for a particular application. In this initial design step, idealized components will be assumed, and losses will be neglected. The design specifications are

dc load voltage = 300 V at a load current of 100 mA

Ripple factor = 0.02

Determine **(a)** the capacitance required, and **(b)** the rms voltage required across transformer secondary (assuming the 120-V, 60 Hz ac power system).

20–29. You are to perform a preliminary design of a voltage regulator circuit of the form shown in Figure 20–19. The desired dc load voltage is 12 V, and a 5.1-V zener diode is to be used as a reference voltage. The maximum load current is specified as 1 A. Zener bias current is to be established as 20 mA. **(a)** Determine the value of R_Z required. **(b)** Determine possible values of R_i and R_f. **(c)** Assume that the value of β_{dc} for the pass transistor is subject to a range from 50 to 200. Determine the maximum output drive current for the op-amp. **(d)** If the maximum unregulated voltage is 16 V, determine the maximum power dissipation of the pass transistor.

20–30. Based on the circuit form of Figure 20–19, draw a new schematic diagram of a regulator circuit for regulating a negative dc voltage.

Troubleshooting Problems

20–31. You are troubleshooting a circuit of the form shown in Figure 20–3. The measured ac value for v_2 is within the acceptable range as given by the manual, but there is no dc load voltage. The following voltages are measured on separate dc and ac scales of a voltmeter:

v_2 (V)	dc	ac (rms)
v_2 (V)	0	23.5
v_L (V)	0	23.5

Which one of the following is a possible trouble? **(a)** Diode open; **(b)** diode shorted; **(c)** shorted load; **(d)** open transformer.

20–32. You are troubleshooting a circuit of the form shown in Figure 20–5. The specified value of v_L is about 24 V dc, but the dc value read by a voltmeter is only about 11.8 V. The ac rms voltage read across half the secondary is measured as 26.2 V and across the full secondary as 52.4 V. Which one of the following is a possible trouble? **(a)** One diode open; **(b)** two diodes open; **(c)** one diode shorted; **(d)** two diodes shorted.

20–33. A power supply of the form of Figure 20–10 is used to power an amplifier circuit. There is an excessive level of hum accompanying the signal. Which one of the following could be the cause of the difficulty? **(a)** One or more shorted diodes; **(b)** one or more open diodes; **(c)** half of transformer secondary shorted; **(d)** open C; **(e)** shorted C.

20–34. You are troubleshooting a circuit of the form of Figure 20–10, for which there is no dc output voltage. After performing ohmmeter checks on the diodes, you discover that both are open and appear to have experienced excessive current flow. Which one of the following could be the possible trouble? **(a)** Load disconnected; **(b)** half of transformer secondary open; **(c)** half of transformer secondary shorted; **(d)** open C; **(e)** shorted C.

21

ACTIVE
FILTERS

OBJECTIVES

After completing this chapter, the reader should be able to:

- Define a filter, and discuss its general characteristics.
- Explain the differences between active and passive filters and the advantages and disadvantages of each type.
- Sketch the amplitude response of an ideal low-pass, a high-pass, a band-pass, and a band-rejection filter.
- Discuss the differences among a Butterworth, a Chebyshev, and a maximally flat time-delay filter.
- Apply to a filter the procedures for frequency scaling and for impedance scaling.
- Determine the number of poles required in a low-pass or high-pass Butterworth filter to meet specified filtering requirements.
- Design a low-pass and a high-pass Butterworth active filter using unity-gain sections.
- Apply curves to determine the response of a two-pole band-pass filter as a function of center frequency and bandwidth or Q.
- Design a two-pole band-pass active filter using a multiple-feedback circuit and a single op-amp.
- Discuss the concept of a state-variable filter, show its basic circuit form, and design one to achieve either a low-pass, a high-pass, a band-pass or a band-rejection circuit.
- Analyze some of the circuits in the chapter with PSPICE.

21-1 INTRODUCTION

Before we discuss the specific principles of active filters, it is instructive to consider the role that active filters play within the general framework of filter theory and design. Electrical filters have been an important part of the evolution of the electrical field from the beginning. Indeed, it would have been impossible to achieve many of the outstanding technological accomplishments without the use of electrical filters.

Because of the importance of filters, much research and development has been performed in the areas of filter theory, design, and implementation. Many books and thousands of articles have been written on the subject. Therefore, one chapter in a book must necessarily assume a very modest profile in dealing with a subject so vast. Treatment here will consider some key aspects of filter usage and workable design material. Although much of the underlying theoretical basis will necessarily be omitted, useful design procedures will be developed within the chapter. After completing this chapter, the reader should have acquired both an appreciation of how filters work in general as well as proficiency in actually designing certain common types of specific filters to meet required specifications.

Section 21-2 is a general discussion of the various classifications of filters on a somewhat broader level than the remainder of the chapter, to show the varied types of filters and their qualitative characteristics.

We focus primarily on low-pass and high-pass filters of the widely employed Butterworth types. Means for predicting their performance and for determining the complexity required to meet specific filtering requirements are covered.

We discuss two-pole band-pass response functions and their relationships to basic *RLC* resonance curves, then cover the design of such circuits using a single op-amp with multiple feedback.

We also discuss the state-variable filter and its evolution. Finally we present techniques for designing a state-variable filter to achieve a low-pass, a high-pass, a band-pass, or a band-rejection filter.

21-2 FILTER CLASSIFICATIONS

For our purposes, a **filter** is defined as any circuit that produces a prescribed frequency response characteristic, of which the most common objective is to pass certain frequencies while rejecting others. Some circuits classified as filters have objectives other than frequency response criteria, but our considerations here will be limited to the common amplitude and phase filter interpretations.

Filters may be classified in a number of different ways, and like most other classification schemes, filter classes often contain ambiguities and contradictions. The classifications to be given here are simplified as much as possible to convey some of the major categories within the space limitations.

Passive Filters versus Active Filters

The first classification to be considered is that of **passive filters** versus **active filters.** Passive filters consist of combinations of resistance, capacitance, and inductance. These RLC passive filters were the major types used for many years, and such filter technology still represents the dominant form above the audio frequency range. Passive RLC structures are capable of achieving relatively good filter characteristics in applications ranging from the audio frequency range to portions of the radio frequency range.

A problem occurs with passive RLC filters at the lower end of the audio frequency range. Inductance values increase as the required frequency decreases, creating several problems. First, inductors are somewhat imperfect devices because of internal losses, but these losses increase markedly in the very large inductance range required at lower frequencies. These heavy losses degrade significantly the quality factor (Q) for each coil, and the associated filter responses have large deviations from their desired forms. Second, the actual physical sizes of the large inductance values limit their usefulness. Third, the costs of such inductances are certainly not trivial considerations.

Active filters consist of combinations of resistance, capacitance, and one or more active devices (such as op-amps) employing feedback. Such filters are theoretically capable of achieving the same responses as passive RLC filters. Because inductances are not required, the difficulties associated with them at low frequencies are thus eliminated. Indeed, active RC filters operating at very low frequencies can be readily implemented. Active-filter frequency response characteristics can be made to approach the ideal forms very closely, and their costs and physical sizes are very reasonable.

Although active filters are capable of circumventing most of the low-frequency limitations of passive filters, they introduce a few disadvantages and problems of their own. Because they are active, power is required to make them operate properly. A closely related fact is that active components in general are less reliable than passive circuits. Finally, active filters employ feedback, and there is always a possibility of instability.

In spite of these limitations, the use of active filters has grown, and they now play a very prominent role in filter technology. Active filters are superior to passive filters in the majority of low- and very-low-frequency applications. There is a range of frequencies in which arguments could be made for either passive or active filters, depending on the various design constraints. However, as the region of application extends into the so-called radio frequency (RF) range, passive RLC filters take the lead. As a general rule with some probable exceptions, the practical limit of most RC active filters is perhaps 100 kHz or so, and most active filters are used well below that frequency.

From the title of this chapter, it is obvious that our interest is directed toward active RC filters because such devices fit within the objectives of this book. Passive RLC filters are considered in various books devoted to network synthesis and filter theory and design. However, many of the concepts that will be developed here can be related to passive filters. The next several classification schemes, for example, apply equally well to both passive and active filters.

Band Classifications

From the point of view of the amplitude response, most filters can be classified as **low-pass, high-pass, band-pass,** or **band-rejection.** It is helpful to define some ideal block characteristics to represent these different filter types, and such models are shown in Figure 21–1.

The amplitude response for the low-pass ideal block characteristic form is shown in Figure 21–1(a). The quality f_c represents the **cutoff frequency.** In this ideal case, the amplitude response for $f < f_c$ is unity, so frequencies in this range

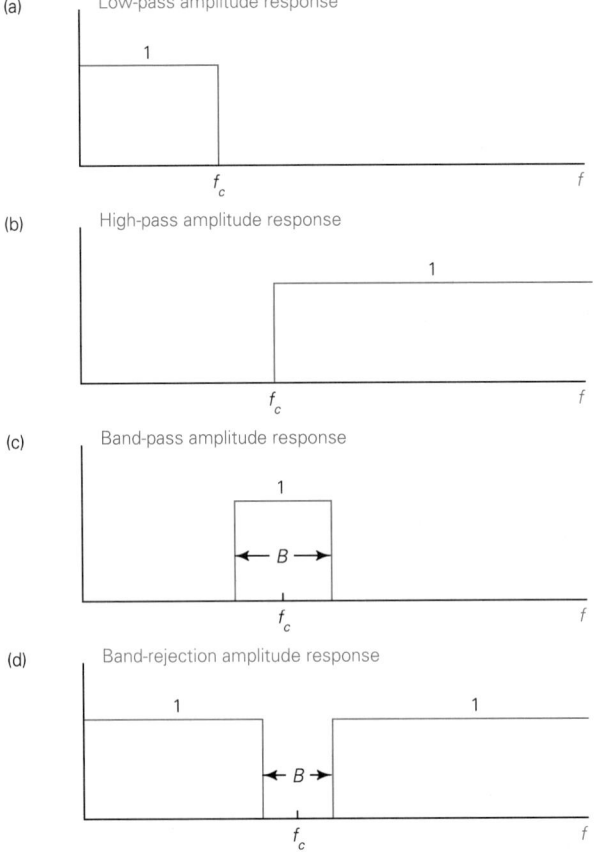

FIGURE 21–1
Ideal block form of the amplitude response for (a) low-pass, (b) high-pass, (c) band-pass, and (d) band-rejection filters.

are passed by the filter. However, for $f > f_c$, the amplitude response is zero, so frequencies in this range are completely eliminated by the filter.

The high-pass ideal block characteristic is shown in Figure 21–1(b). Observe that its nature is inverted with respect to the low-pass filter; that is, frequencies above the cutoff frequency f_c are passed by the filter, and lower frequencies are rejected.

The band-pass ideal block characteristic is shown in Figure 21–1(c). The band-pass amplitude response is characterized by a center frequency f_o and a bandwidth B. Frequencies falling within the band-pass region are passed by the filter, and components either below the lower band edge or above the upper band edge are rejected.

The band-rejection ideal block characteristic is shown in Figure 21–1(d). This filter passes all frequencies except those within a certain band-rejection region. Parameters f_o and B are used for this filter, but in this case they refer to the center and width, respectively, of the rejection band.

For each of the ideal filter forms, we can define a **pass band** and a **stop band.** The **pass band** is the range of frequencies that is transmitted through the filter, and the **stop band** is the range of frequencies that is rejected.

Actual filters do not possess the ideal block characteristics. Using a low-pass filter for reference, the form of the amplitude response for a realistic filter is illustrated in Figure 21–2. For such nonideal characteristics, it is useful to define three regions: (1) **pass band,** (2) **stop band,** and (3) **transition band.** The exact boundaries between these three regions are somewhat arbitrary. Typically, the amplitude response is specified to be within a certain range from unity in the pass band, and the response is specified to be below a certain level in the stop band. The connecting region between these levels then identifies the transition band.

As a general rule, the flatter the pass band and the lower the stop band, the more complex the filter. This basic trade-off has led to much research and development in filter theory through the years. The ideal block filter characteristic

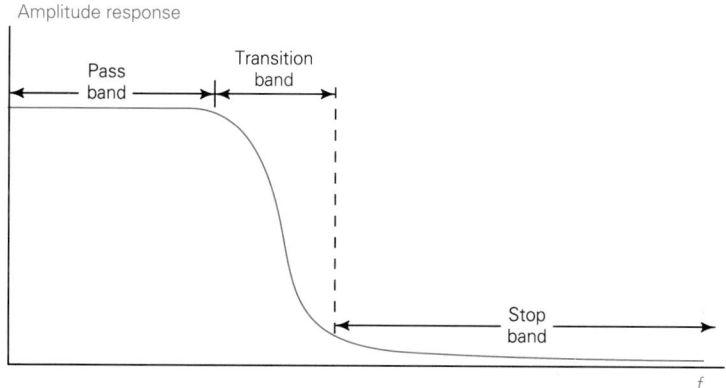

FIGURE 21–2
Representative amplitude response of realistic low-pass filter.

can never be reached, but as it is approached, the complexity of the associated filter designs increases markedly. Fortunately, the vast majority of filter applications can be met with realistic filter characteristics. The typical approach is for the systems designer to determine the minimum level of filtering required, and a filter meeting or exceeding the specifications is then determined.

Butterworth Response

The form of the **Butterworth amplitude characteristic** is illustrated in Figure 21–3. The Butterworth amplitude response is also referred to as a **maximally flat amplitude response** because of the mathematical structure of its development. The response always decreases as the frequency increases. The Butterworth function will be one of the primary forms developed in depth, so a detailed discussion will be delayed to Section 21–4.

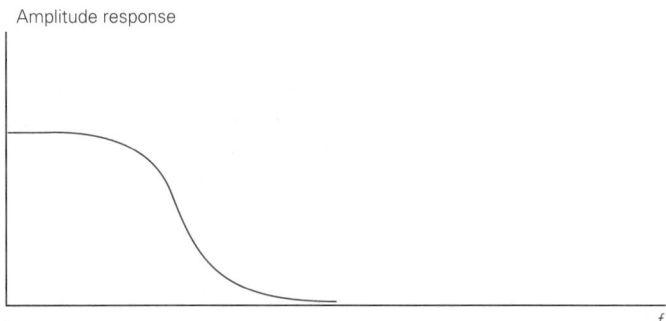

FIGURE 21–3
General form of amplitude response of Butterworth filter.

Chebyshev Response

The form of one particular **Chebyshev amplitude characteristic** is illustrated in Figure 21–4. The Chebyshev response is referred to as an **equiripple** response because the pass band is characterized by a series of ripples that have equal maximum levels and equal minimum levels. The number of ripples is a function of the number of reactive elements in the design. The response shown is one particular example within the Chebyshev class. Chebyshev filters have a sharper slope than Butterworth filters and are thus capable of achieving more attenuation in the stop band for a given number of reactive elements. However, their time delay and phase characteristics are less ideal than those of the Butterworth filter, and they tend to exhibit a ringing effect with transient signals.

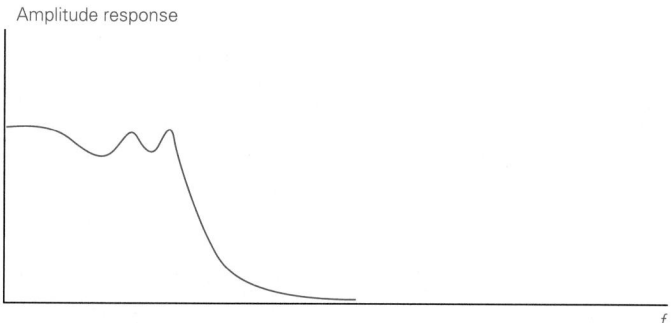

FIGURE 21–4
Form of amplitude response of a particular Chebyshev filter.

Maximally Flat Time-Delay Response

A different approach to the approximation problem is that of the **maximally flat time-delay (MFTD)** filter. With the MFTD filter, the phase response is optimized so that all frequency components have nearly constant time delay through the filter. The general form of the associated amplitude response for a low-pass case is illustrated in Figure 21–5. At first glance, this response resembles the Butterworth characteristic in that the response always decreases as the frequency increases. Compared with the Butterworth response, however, the MFTD amplitude response is not as constant in the pass band, and the attenuation is not as high in the stop band. MFTD filters are used in phase-sensitive applications, where constant time delay is very important but the attenuation requirements are moderate.

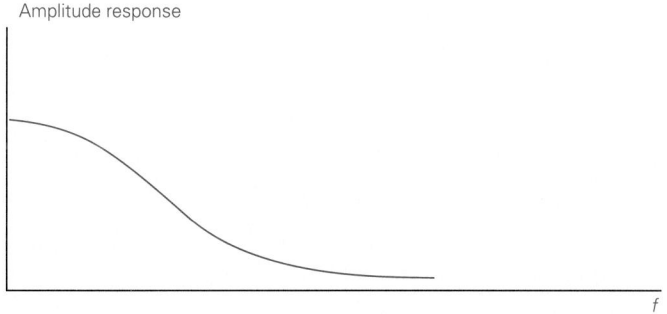

FIGURE 21–5
General form of amplitude response of maximally flat time-delay filter.

Comparison

There are various other approximations, but these three types are among the most widely employed. Within the group of these three filter characteristics, the Chebyshev amplitude response has the sharpest rate of attenuation increase above cutoff, but its phase and time-delay characteristics are the poorest. In contrast, the MFTD filter has the most ideal time-delay and phase characteristics, but its amplitude response is the poorest. The Butterworth filter is a reasonable compromise between these extremes, and as a result, it is a very popular choice.

The major emphasis on low-pass and high-pass filter design in this text will utilize Butterworth filter characteristics. Although such forms cannot solve all filtering problems, they are certainly capable of solving many. Their ease of design and implementation makes them most suitable for the level intended for this book.

Butterworth filters are capable of achieving any basic form of filtering, but the most common applications are in low-pass and high-pass forms. For band-pass and band-rejection functions, we consider a class of active *RC* circuits that have operating characteristics equivalent to those of passive *RLC* resonant circuits.

21-3 FREQUENCY AND IMPEDANCE SCALING

Before considering specific filter designs, we will discuss a widely employed method for dealing with general filter design procedures. Actual filter design requirements vary over a considerable range of frequency and impedance levels, so it would be impossible to tabulate sufficient data for all possible cases. When specific design formulas are used to determine element values, they are often cumbersome and unwieldy.

The most widely used method for tabulating filter design data is based on providing element values for a **normalized circuit.** The normalized circuit is developed on the basis of very simplified cutoff or center frequencies and convenient impedance levels. Typically, most of the element values in a normalized design will be in the neighborhood of 1 Ω, 1 F, and so on. Obviously, such element values are quite unrealistic, but they serve as the normalized prototype from which a workable design can be developed. The techniques for converting a normalized design to an actual design are straightforward and can be readily mastered. In fact, with some practice, the conversion can be done almost intuitively.

Conversion of a normalized design to a realistic design can be achieved through the process of two scaling operations: (1) **frequency scaling** and (2) **impedance scaling.** Frequency scaling is used to change the frequency of the normalized design to the required frequency of the actual design. Impedance scaling is used to change element values to more realistic or workable values. Each of the procedures will now be discussed in detail.

Frequency Scaling

Frequency scaling in an *RC* active filter may be most easily achieved by changing all filter capacitance values in the same proportion *or* by changing all filter resistance values in the same proportion (but not both). The basic rule to remember is that the frequency changes in *inverse* proportion to the change in either *C* or *R*. For example, if all capacitor values in an *RC* active low-pass filter are halved, the cutoff frequency is doubled. In turn, if all resistance values are halved, the cutoff frequency is again doubled. Thus, the process of changing *either* the capacitance values *or* the resistance values may be used to shift the frequency range of an *RC* filter.

When the capacitance values are changed, the impedance level in the modified frequency range is the same as the impedance level in the original circuit in the corresponding frequency range. However, when resistance values are changed, the impedance level is modified. If the resistances are increased, the impedance level in the new frequency range increases in direct proportion to the resistance changes. The change in impedance level may cause difficulty in some applications, but it may result in an advantage in others.

Note that *all filter* capacitances or *all filter* resistances must be changed in the same proportion. The modifier *filter* is used here because active *RC* circuits will typically contain some resistances used only for establishing gains and for bias compensation. Such resistances need not necessarily be scaled. The criterion is whether a resistance is truly part of the filter network itself or simply a resistance value used to establish an amplifier gain or to compensate for bias.

Impedance Scaling

Impedance scaling is the process of changing the relative impedance levels of all components in the filter by a specified amount. This process is performed for the purpose of acquiring realistic and workable element values for the circuit. In the most basic form, to be discussed now, impedance scaling does not alter the frequency scale. (When the resistance values are changed to perform frequency scaling, the impedance level also changes as a secondary effect, but the primary process of impedance scaling leaves the frequency response unchanged.)

Impedance scaling is performed by changing all filter resistance values in a common proportion *and* by changing all filter capacitance values in the opposite proportion. The impedance level change is directly proportional to the resistance change, or it is inversely proportional to the capacitance change. For example, if all filter resistances in an *RC* active filter are doubled and all filter capacitances are halved, the new filter will have exactly the same frequency response as the old filter, but the impedance level at a given frequency will be twice as great as before. Note that because resistance and capacitance values are changed in opposite directions, the frequency change that would have resulted from changing only one parameter is effectively canceled by the change in the other parameter.

Impedance level scaling is required to ensure that element values fall within practical limits for implementation. In fact, with some care, it is often possible to select an impedance scaling factor that will force one or more component values to assume readily available standard values, as will be demonstrated.

The earlier discussion about resistance values associated only with op-amp operation for frequency scaling holds equally true for impedance scaling. Thus, resistance values used only for establishing closed-loop op-amp gains and in bias compensating circuits can be adjusted independently of the filter resistances.

Having stated the impedance scaling rule in its general form, we will give one modification. Active *RC* filters consisting of several isolated stages have the advantage that the impedance level of one stage does not affect the impedance level of other stages. This permits the impedance level of one stage to be adjusted independently of other stages. Thus, it is possible to employ one impedance scaling factor for one stage and a different factor for a second stage. Such a procedure could permit a better set of component values in certain cases. Bear in mind that if different impedance scaling levels are selected, *all pertinent components in a given stage must be modified by the impedance scaling constant for that particular stage*. One can, of course, use the same impedance scaling factor for the entire filter, but the possibility of different constants offers a degree of additional flexibility.

Most normalized filter design data are tabulated on the basis of radian frequencies. The reason for such tabulation is that radian frequencies are related more easily to the mathematical techniques from which most of these filter designs are obtained. However, most real-life design specifications are given in terms of cyclic frequency in hertz, so one must be careful about the 2π factor. This author has observed experienced designers implementing unworkable designs because they momentarily overlooked this basic concept.

Procedure for Scaling

Assume that the normalized design has some specific reference radian frequency $\bar{\omega}_r$ of importance. This frequency is typically the cutoff frequency of a low-pass or a high-pass filter, or it might be the center frequency of a band-pass filter. Often, the normalized design will be based on a reference radian frequency of 1 rad/s. However, the notation and approach here will provide some generality to the concept.

Assume that there is a cyclic reference frequency f_r in the actual filter which is to correspond to $\bar{\omega}_r$ in the normalized design. Let $\omega_r = 2\pi f_r$ represent the radian frequency corresponding to f_r. A frequency scaling constant K_f is defined as

$$K_f = \frac{\omega_r}{\bar{\omega}_r} = \frac{2\pi f_r}{\bar{\omega}_r} \qquad (21\text{--}1)$$

Note that the normalized frequency is specified from the beginning in radian form, whereas the final design is initially specified in cyclic form because these forms are commonly encountered in practice. Irrespective of how they arise, however, the frequency scaling constant must have the same units in both numerator and denominator.

After the frequency scaling constant is determined, it is helpful to note in which direction the component values must change. For example, if the actual design is to be higher in frequency than the normalized design, either capacitance or resistance values will be reduced. By this intuitive approach, it is easier to identify what is happening with K_f rather than relying on a rigid formula.

We have seen that frequency scaling can be performed by changing either capacitance or resistance values. In practice, changing the capacitances is used more often for the process of converting to a higher frequency (the most common case) because the resulting capacitance values are smaller. Resistance changes are most often used to create tunable filter types, because it is usually much easier to vary resistances than capacitances. The procedures to be discussed emphasize changing the capacitance for frequency scaling, but the possibility of varying the resistance should also be recognized.

An impedance scaling constant K_r will be used in the scaling operation. This constant will be defined as

$$K_r = \frac{\text{impedance level of final circuit}}{\text{impedance level of normalized circuit}} \qquad \textbf{(21–2)}$$

When one is starting with normalized designs, K_r will almost always be larger than unity. However, one may have occasion to take a workable circuit and convert it to a lower impedance level, in which case K_r could be less than unity. The intuitive concept to remember is that the circuit having the highest impedance level will have the largest values of resistances and the smallest values of capacitances, and vice versa.

To convert a normalized active filter design to a realistic design, use the following recommended sequence of steps:

1. Perform the frequency scaling operation first by dividing all filter capacitances by K_f. (Alternately, all resistance values could be divided by K_f in the event that resistance changes are to be used.) Remember the intuitive point that a higher frequency corresponds to smaller element values and vice versa.
2. Select an impedance scaling constant K_r. There is no single way that is best for this procedure, as it will depend on the various constraints. In some cases, a value of K_r may be selected such that one or more final element values turn out to be readily available in stock components. From various design examples considered later, some general guidelines will evolve.
3. Multiply all resistance values by K_r and divide all capacitance values by K_r. Remember the intuitive point: A higher impedance level corresponds to larger resistance values and smaller capacitance values and vice versa. Note that the frequency scaling process is completely specified but that

the impedance scaling process leaves some choice in the matter. For this reason, frequency scaling should normally be performed first.

The following two steps are not always necessary, but they provide additional flexibility to ease the overall design process:

4. For filters with more than one section, different impedance scaling constants may be selected for the different sections if desired.
5. For any active filter, resistances that are used only to set gain or attenuation may be scaled completely independently of any other impedance scaling for either that or other sections. However, a common scale factor must necessarily be used for all resistances in a particular network.

21–4 BUTTERWORTH FILTER RESPONSES

During the brief introduction to a number of the most common filter approximations in Section 21–2, it was indicated that the Butterworth form was one of the most widely employed filter types. In this section, we discuss the Butterworth form in some detail and present quantitative data concerning the relative performance.

The emphasis in this section is on relating the actual filter performance to the complexity or order of the approximation. In this manner, it will be possible to determine if given specifications can be achieved with a Butterworth filter and to determine the particular filter complexity required. The actual design of such filters will be considered in later sections. It should be noted at the outset that the performance specifications apply equally well to both passive and active Butterworth filters, so the material presented in this section has broader scope than just active filters.

Number of Poles

A common parameter used in specifying filters is the **order** of the filter. In advanced mathematical terminology, the term *order* is synonymous with the **number of poles.** For our purposes, the order or number of poles is the number of nonredundant reactive elements in the circuit. Both inductors and capacitors are reactive elements, but because active filters contain no inductors, the order or number of poles will be the number of nonredundant capacitors. None of the circuit diagrams considered later will contain redundant capacitors, so the order or number of poles will be the number of capacitors in the circuit. (This does not include capacitors external to the filter network, such as power supply bypass capacitors and op-amp compensating capacitors.)

For a given filter type, the performance generally becomes closer to the ideal block characteristic as the number of poles increases. Thus, a higher-order filter will have a flatter pass-band response and a lower stop-band response (more attenuation) compared with one of lower order. However, the ideal block characteristic can never be fully attained.

Butterworth functions for low-pass and high-pass designs will be considered. Band-pass and band-rejection forms will be developed in later sections using a different approach. Next, we present the mathematical forms for the low-pass and high-pass Butterworth functions.

Low-Pass Butterworth

The Butterworth functions utilize a convenient reference frequency at which the amplitude response drops to $1/\sqrt{2}$ of its maximum pass-band level. This corresponds to the response being down 3.01 dB, and this value is usually rounded to 3 dB. This frequency will be referred to as the **cutoff frequency,** but it should be understood that it is not an abrupt cutoff. Let f_c represent the cutoff frequency, and let n represent the order (number of poles) of the approximation. The amplitude response $A(f)$ of the Butterworth low-pass function is given by

$$A(f) = \frac{1}{\sqrt{1 + (f/f_c)^{2n}}} \qquad \textbf{(21–3)}$$

The maximum value of $A(f)$ occurs at $f = 0$, and it has been established at unity for convenience; that is, $A(0) = 1$. With some passive filters, the maximum value is less than unity, and the filter is said to have a **flat loss.** Conversely, some active filters have a maximum value greater than unity. One can readily incorporate these differences into (21–3) by putting a constant in the numerator, but because it is the relative response that is of primary interest, the simpler form will be assumed here.

Let $A_{dB}(f)$ represent the decibel form of the response relative to the maximum level. Because the maximum level of (21–3) is unity, the decibel form can be expressed as

$$A_{dB}(f) = 20 \log_{10}\left[\frac{1}{\sqrt{1 + (f/f_c)^{2n}}}\right] \qquad \textbf{(21–4a)}$$

or

$$A_{dB}(f) = -10 \log_{10}[1 + (f/f_c)^{2n}] \qquad \textbf{(21–4b)}$$

where some basic properties of the logarithmic function were used in converting from (21–4a) to (21–4b).

High-Pass Butterworth

The cutoff frequency f_c for high-pass Butterworth filters has essentially the same meaning as for low-pass filters, except, of course, that it is at the low end of the passband. The response is down by 3.01 dB at this point. The amplitude response $A(f)$ of the Butterworth high-pass function is given by

$$A(f) = \frac{1}{\sqrt{1 + (f_c/f)^{2n}}} \qquad \textbf{(21–5)}$$

The decibel form $A_{dB}(f)$ of the high-pass response can be expressed as

$$A_{dB}(f) = 20 \log_{10} \frac{1}{\sqrt{1 + (f_c/f)^{2n}}} \tag{21-6a}$$

or

$$A_{dB}(f) = -10 \log_{10} \left[1 + \left(\frac{f_c}{f} \right)^{2n} \right] \tag{21-6b}$$

Comparing (21–6a) and (21–6b) with (21–4a) and (21–4b), we note that the high-pass form is derived from the low-pass form by simply replacing f/f_c with f_c/f. This transformation "inverts" the low-pass form and converts it to the high-pass form.

Amplitude Response Curves

Plots of the stop-band amplitude response for Butterworth filters with orders 2 through 7 are given in Figure 21–6. All the curves start at the -3 dB level, which corresponds to $f = f_c$. The response curves are shown as *relative* decibel output based on a reference dc gain of unity as given by (21–4a) and (21–4b). In the event that the filter has a flat loss or gain, the curves still apply, but a constant decibel level could be either added to or subtracted from the curves given. Alternately, the curves could simply be interpreted as the decibel response *relative* to the maximum response, which is usually the most meaningful approach.

Attenuation in a positive sense corresponds to gain in a negative sense. Thus, if a relative attenuation of 20 dB is specified, one could look for a level of -20 dB on the given curves.

The abscissa, labeled "Normalized frequency," is interpreted differently for low-pass and high-pass filters. For low-pass filters, the abscissa is interpreted as f/f_c. For example, *normalized frequency* = 1 corresponds to $f = f_c$, and *normalized frequency* = 10 corresponds to $f = 10f_c$. For high-pass filters, the abscissa is interpreted as f_c/f. In this case, *normalized frequency* = 10 corresponds to $f = 0.1f_c$. While the curves are shown only for a range of one decade above cutoff, this is the region where most specifications are directed, and the curves are most useful in evaluating relative filter performance.

EXAMPLE 21–1

A low-pass filter is desired for a given application. The specifications are as follows:

1. Relative attenuation (or loss) ≤ 3 dB for $f \leq 800$ Hz
2. Relative attenuation (or loss) ≥ 23 dB for $f \geq 2$ kHz

Specify the minimum number of poles for a Butterworth filter that will satisfy the requirements.

Solution
These specifications are given in a typical manner. The requirement of (1) specifies a "pass band" in which the response drops no more than 3 dB. The requirement

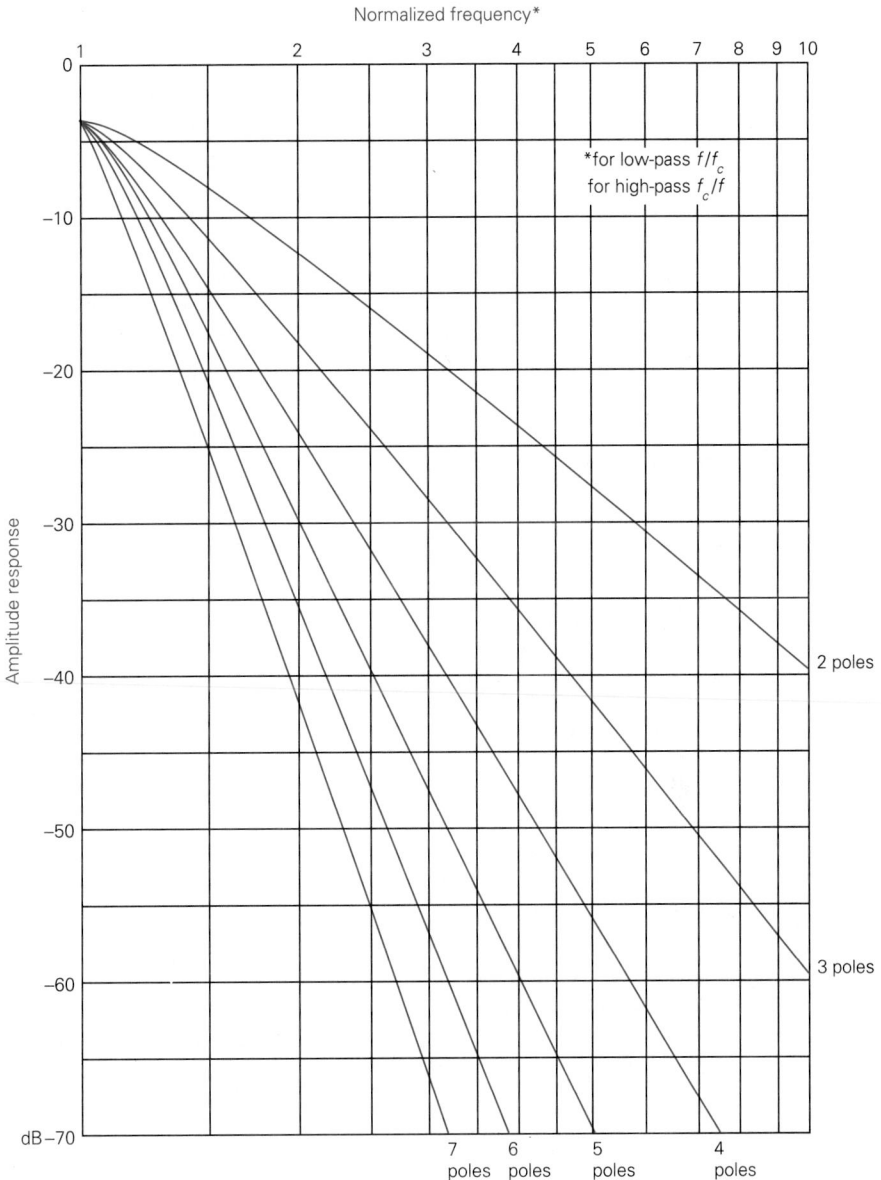

FIGURE 21-6
Stop-band amplitude response of Butterworth filters.

of (2) specifies a ''stop band'' in which the response is required to be at or below a certain level. The region between (1) and (2) can then be interpreted as a ''transition band.''

The specifications as given are not very demanding, because a relatively wide transition band is provided. The curves of Figure 21–6 may be readily used.

We interpret the cutoff frequency to be $f_c = 800$ Hz. The actual frequency $f = 2$ kHz corresponds to a normalized frequency $f/f_c = 2000/800 = 2.5$. At an abscissa of 2.5 on the normalized frequency scale, we drop down to determine the response curves that will achieve the required attenuation. The response of a two-pole function is down by only about 16 dB at this frequency, so it is inadequate. However, a three-pole response is down by about 24 dB at the normalized frequency of 2.5, so it more than meets the specifications. The minimum number of poles required in the Butterworth filter is *three*.

Before leaving this problem, we should note that the relative amplitude response curve for any three-pole Butterworth filter will meet specifications as given for any frequencies having the same ratios as those given. For example, if the frequency of specification (1) had been 5 kHz and the frequency of specification (2) had been 12.5 kHz, a three-pole Butterworth filter would again be the correct solution, because 12.5 kHz/5 kHz = 2.5 is the same normalized frequency as determined in the problem. The utility of the normalized frequency concept is that the various results apply to any frequency combinations having the same relative ratio.

EXAMPLE 21–2

For a three-pole Butterworth filter with a cutoff frequency of 800 Hz as determined in Example 21–1, determine the relative amplitude response at the following frequencies: dc, 400 Hz, 800 Hz, 2 kHz, 4 kHz, and 8 kHz.

Solution

Since the 3-dB frequency was established as 800 Hz, the normalized frequency is $f/f_c = f/800$. For all but dc and 400 Hz, the normalized frequency is calculated, and the response is read from Figure 21–6. For dc, the relative amplitude response is always 0 dB. For 400 Hz, the function of (21–4b) can be used to compute the response. In fact, this equation could be used to calculate the response at any frequency, and this approach is more accurate than the graphical approach. However, the graphical approach is usually sufficient for routine analysis and design. The results are summarized as follows:

f	$f/800$	Relative amplitude response (dB)
dc	0	0
400 Hz	0.5	−0.067
800 Hz	1	−3
2 kHz	2.5	−24
4 kHz	5	−42
8 kHz	10	−60

EXAMPLE 21–3

A high-pass filter is desired for a given application. The specifications are as follows:

1. Relative attenuation $\leq$ 3 dB for $f \geq$ 500 Hz
2. Relative attenuation $\geq$ 46 dB for $f \leq$ 125 Hz

Specify the minimum number of poles for a Butterworth filter that will satisfy the requirements.

Solution

The manner in which the specifications are given suggests that the 3-dB cutoff frequency can be established at 500 Hz. For a high-pass filter, the frequency of 125 Hz corresponds to an inverted normalized frequency of $f_c/f = 500/125 = 4$. At a normalized frequency of 4, the smallest number of poles satisfying the specification is *four*. The attenuation at this point is slightly greater than 48 dB, so the specification is met with some reserve.

21–5 LOW-PASS AND HIGH-PASS UNITY-GAIN DESIGNS

Design data for certain low-pass and high-pass Butterworth active filters are presented in this section. These active circuits utilize what are referred to as **finite gain realizations;** that is, the gain of the active device (op-amp) is established at a relatively low finite level by feedback. In particular, the gain in the circuits of this section will be set at unity by using voltage followers for the active stages.

All of the designs to be presented have a normalized 3-dB radian cutoff frequency of $\omega_c = 1$ rad/s. This value represents the defined upper edge of the pass band for low-pass designs and the lower edge of the pass band for high-pass designs.

The unity-gain designs are based on combinations of two-pole and three-pole sections. All even-numbered realizations employ only two-pole sections, and odd-numbered realizations employ one three-pole section and as many two-pole sections as required to realize the required numbers of poles.

Low-Pass Designs

The forms of the normalized two-pole and three-pole unity-gain sections for the low-pass designs are shown in Figure 21–7. The filter resistances have been established as 1 Ω in the normalized designs. However, as noted, the bias compensating resistances are set at their optimum values, but they do not affect the filter characteristics. It is common practice not to display units on schematics of normalized designs, but all component values are assumed to be in their basic units, that is, ohms and farads. Note that the op-amps are functioning as voltage followers.

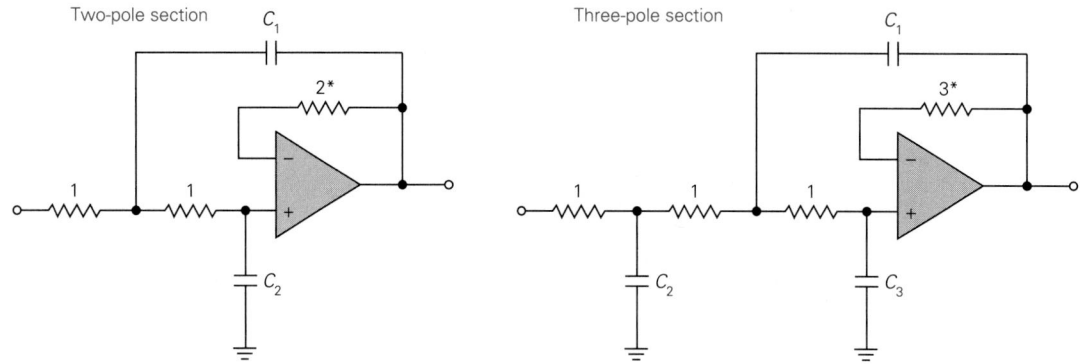

*These resistors are used for bias compensation only
and do not contribute to the filter response.

FIGURE 21–7
Normalized unity-gain low-pass Butterworth active filter designs. (See Table 21–1 for capacitance values.)

The values of the capacitances are given in Table 21–1 for different low-pass filter orders (number of poles). The units for these normalized capacitances are **farads.** Rows containing only C_1 and C_2 represent two-pole sections, and rows containing C_1, C_2, and C_3 represent *three-pole* sections. Do not make the mistake of using the form of a three-pole section when only C_1 and C_2 are given because the filter will not work properly.

TABLE 21–1
Capacitance values for low-pass Butterworth active filter designs

Poles	C_1	C_2	C_3
2	1.414	0.7071	
3	3.546	1.392	0.2024
4	1.082	0.9241	
	2.613	0.3825	
5	1.753	1.354	0.4214
	3.235	0.3089	
6	1.035	0.9660	
	1.414	0.7071	
	3.863	0.2588	
7	1.531	1.336	0.4885
	1.604	0.6235	
	4.493	0.2225	

The various sections are connected in cascade to implement the composite filter. Because each section contains a voltage follower at the output, theoretically the sections could be connected in any order. In practice, however, some sections are more peaked than others and more easily overloaded. Such sections are best located closer to the output. The data are arranged in the order for achieving the best implementation; that is, the section in the top row for a given number of poles should be connected at the input, and at the other extreme, the bottom row should be connected at the output section.

It should be strongly emphasized that *a higher-order Butterworth filter is not a cascade of lower-order Butterworth filters*. For example, consider a four-pole Butterworth filter consisting of two two-pole sections. Observe from Table 21–1 that the capacitors in the two sections are all different and that none are equal to the values for a two-pole Butterworth filter. (Because of the symmetry of the mathematics involved, it does turn out that one of the three sections for a six-pole filter is the same as a two-pole Butterworth filter, but this is a "mathematical coincidence.")

High-Pass Designs

The two-pole and three-pole sections for the high-pass filters are shown in Figure 21–8. For the high-pass filters, the capacitors are all normalized to a level of 1F. The resistance values for the circuits are tabulated in Table 21–2. The various procedures for determining the data and implementing the circuits discussed earlier for the low-pass filter designs apply here as well.

The extent to which the final active design matches the theoretical form depends on both the closeness of the element values to their ideal values and the performance of the op-amp in the frequency range involved. Polystyrene capaci-

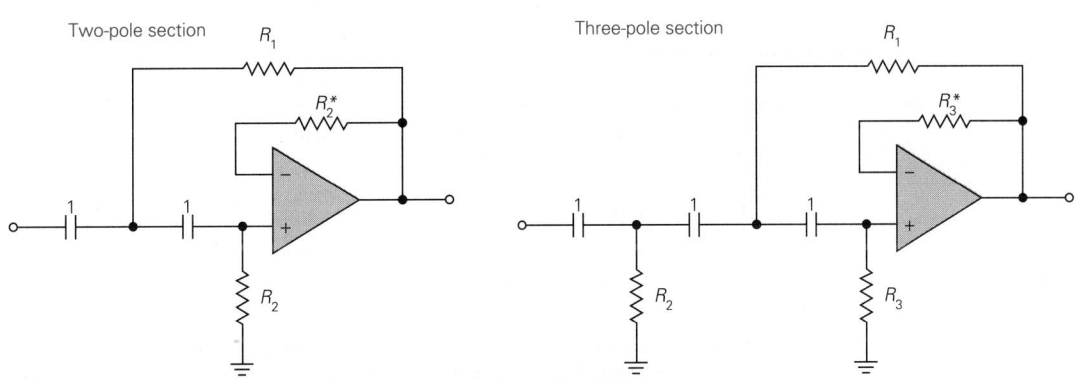

*These resistors are for bias compensation only
and do not contribute to the filter response.

FIGURE 21–8
Normalized unity-gain high-pass Butterworth active filter designs. (See Table 21–2 for resistance values.)

TABLE 21–2
Resistance values for high-pass Butterworth active filter designs

Poles	R_1	R_2	R_3
2	0.7072	1.414	
3	0.2820	0.7184	4.941
4	0.9242	1.082	
	0.3827	2.614	
5	0.5705	0.7386	2.373
	0.3091	3.237	
6	0.9662	1.035	
	0.7072	1.414	
	0.2589	3.864	
7	0.6532	0.7485	2.047
	0.6234	1.6038	
	0.2226	4.494	

tors are recommended when possible, but high-quality mica and Mylar® capacitors may also be used. Do not use electrolytic or other polarized capacitors.

The closed-loop bandwidth of the op-amp and the slew rate should be sufficient for the frequency range involved. In a sense, the high-pass filter is really a sort of band-pass filter because the finite bandwidth and slew rate of the op-amp will cause a roll-off at high frequencies. This factor should be considered in the overall design for any particular application.

The use of the design data for low-pass and high-pass filters is now illustrated with several examples.

EXAMPLE 21–4

Design a low-pass active filter to meet the specifications of Example 21–1.

Solution

In Example 21–1, the requirements called for a 3-dB cutoff frequency of 800 Hz, and it was determined that the specifications could be met with a three-pole Butterworth filter. The normalized data are obtained from Table 21–1, and one three-pole section of the form given in Figure 21–7 is used. The normalized circuit diagram with element values given is shown in Figure 21–9(a).

Conversion to the proper frequency range and a realistic impedance level is achieved in a two-step process. The frequency scaling constant K_f is determined such that 1 rad/s converts to 800 Hz. We have

$$K_f = \frac{2\pi \times 800 \text{ rad/s}}{1 \text{ rad/s}} = 5026.548 \qquad \textbf{(21–7)}$$

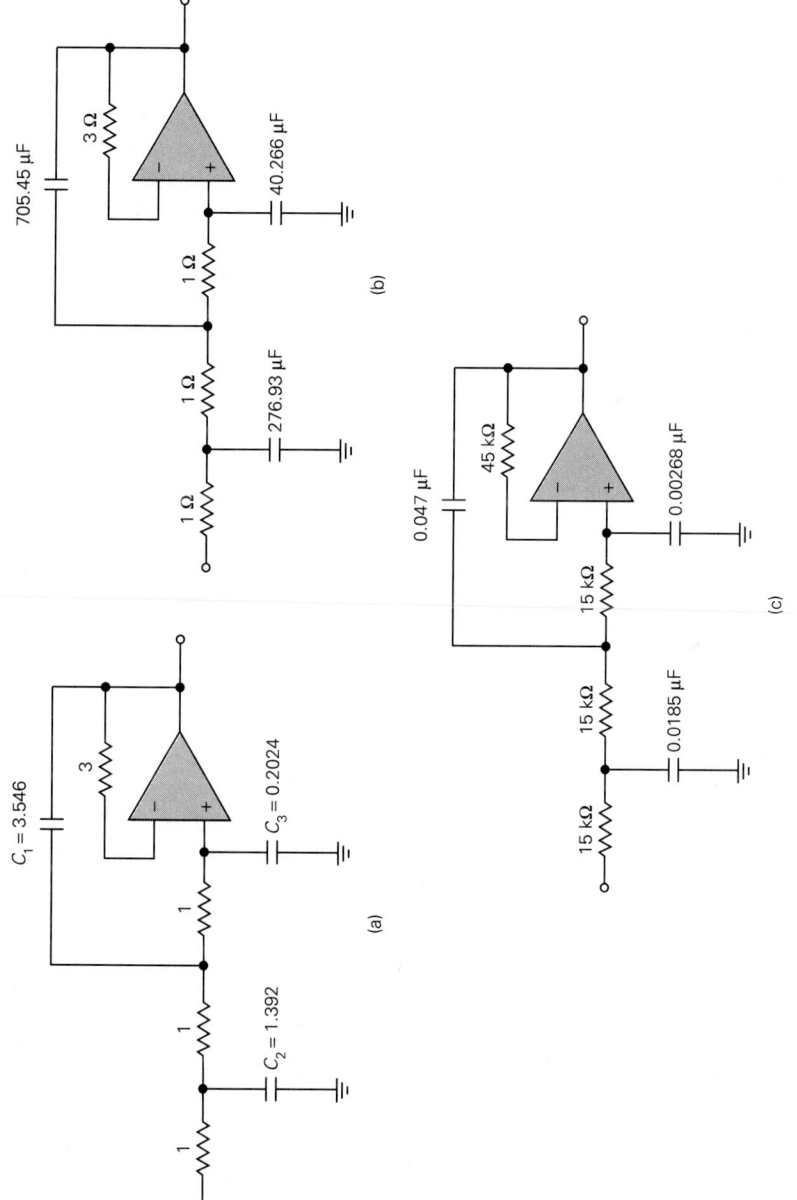

FIGURE 21-9
Circuit design of Example 21-4.

For the first step, the circuit is scaled to the proper frequency range by dividing all capacitance values by K_f, and the corresponding circuit is shown in Figure 21–9(b).

The next step in the process is to perform an impedance scaling that will yield realistic values of both resistances and capacitances. There is no single impedance factor K_r that is best for this purpose. For example, one might try to achieve available standard values whenever possible. However, some of the values required may not always correspond to standard values, and some "tweaking" or combining of several component values may be necessary. Some trial and error is often required, and a keen intuition is useful. The latter capability develops with experience.

After some trial and error, a constant $K_r = 15,000$ was selected. All resistance values are multiplied by K_r, and capacitance values are divided by K_r. The circuit obtained is shown in Figure 21–9(c). The advantage of this design is that the filter resistance values are standard and the three calculated capacitance values are all very close to standard values. For noncritical applications, one could start with the standard values of 0.047, 0.018, and 0.0027 μF for C_1, C_2, and C_3, respectively. Additional "tweaking" could be performed if necessary.

EXAMPLE 21–5

Design a five-pole, low-pass, active Butterworth filter with a 3-dB cutoff frequency of 2 kHz.

Solution

The normalized data are obtained from Table 21–1, and the forms of the two sections are given in Figure 21–7. Observe that one two-pole section and one three-pole section are required. The complete normalized schematic diagram is shown in Figure 21–10(a).

The frequency scaling constant K_f is determined such that 1 rad/s converts to 2 kHz. We have

$$K_f = \frac{2\pi \times 2 \times 10^3 \text{ rad/s}}{1 \text{ rad/s}} = 12,566.37 \qquad (21\text{–}8)$$

All capacitor values are first divided by K_f to convert the basic circuit to the proper frequency range, and the resulting circuit is shown in Figure 21–10(b).

As indicated earlier in the chapter, different impedance scaling constants may be used for different sections of an active filter provided that all element values in a given section are modified by the same constant. After some trial and error, an impedance scaling constant $K_r = 27,000$ was selected for the first section, and a constant $K_r = 22,000$ was selected for the second section. Resistance values are multiplied, capacitance values are divided by their respective constants, and the final circuit is shown in Figure 21–10(c).

Observe that the capacitance values in the first stage are quite close to the standard values of 0.0051, 0.0012, and 0.0039 μF and that the capacitance values in the second stage are close to the standard values of 0.012 and 0.0011 μF.

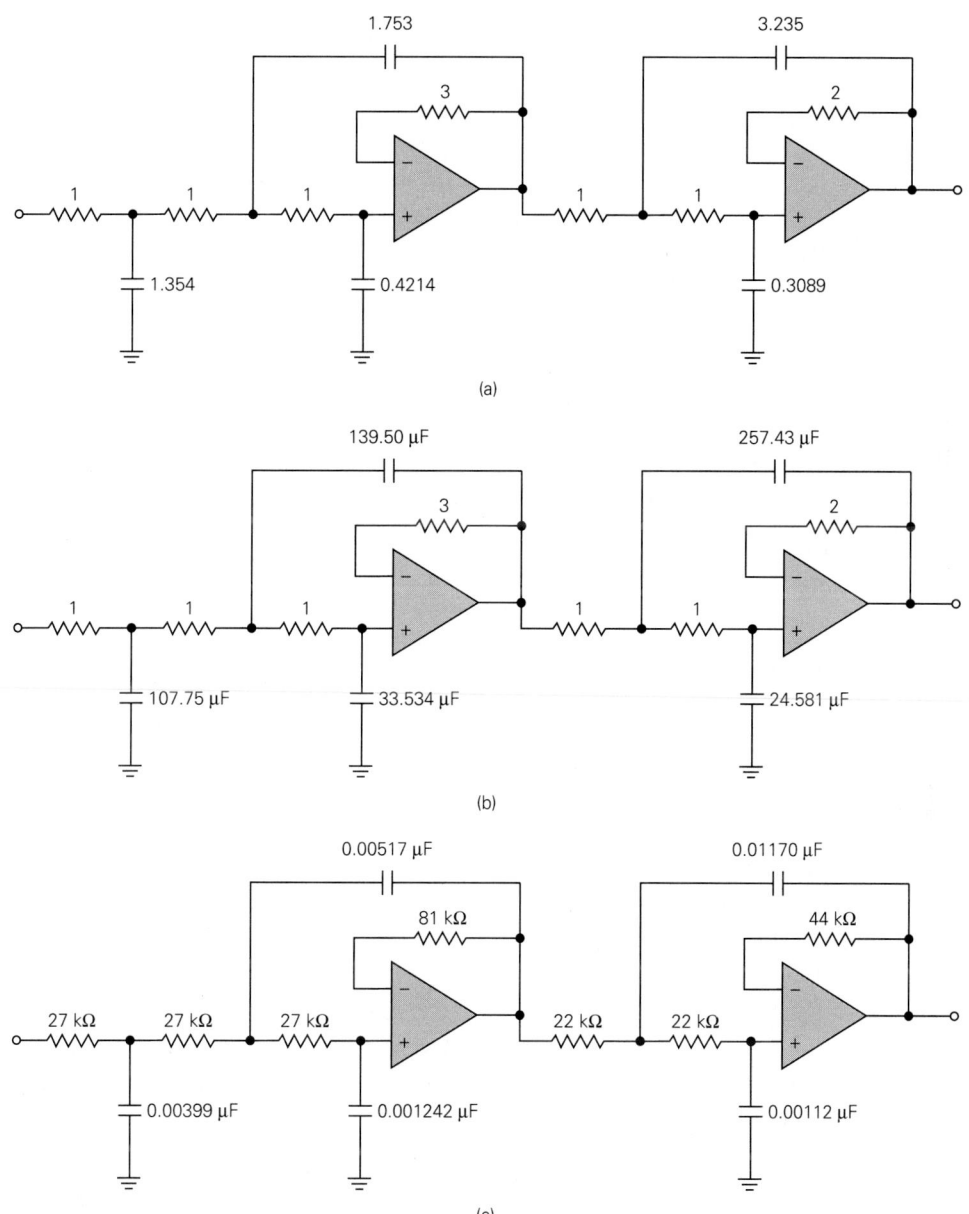

FIGURE 21–10
Circuit design of Example 21–5.

EXAMPLE 21-6

Design a two-pole, high-pass, active Butterworth filter with a 3-dB cutoff frequency of 1 kHz.

Solution

The normalized data are obtained from Table 21–2, and one two-pole section of the form given in Figure 21–8 is used. The normalized circuit diagram with element values given is shown in Figure 21–11(a).

The frequency scaling constant K_f is

$$K_f = 2\pi \times 10^3 = 6283.185 \qquad (21-9)$$

Conversion to the proper frequency range is obtained by dividing the two capacitances by K_f, and the result is shown in Figure 21–11(b).

For this case, one of the two resistance values is twice the other value. This suggests the possibility of choosing a scaling constant such that the two final resistance values are 10 kΩ and 20 kΩ, which are two of the 5% values that are in the right proportion. We thus require that $1.414 \times K_r = 20,000$, which yields $K_r = 14,144$. Dividing the two equal capacitance values by K_r results in required

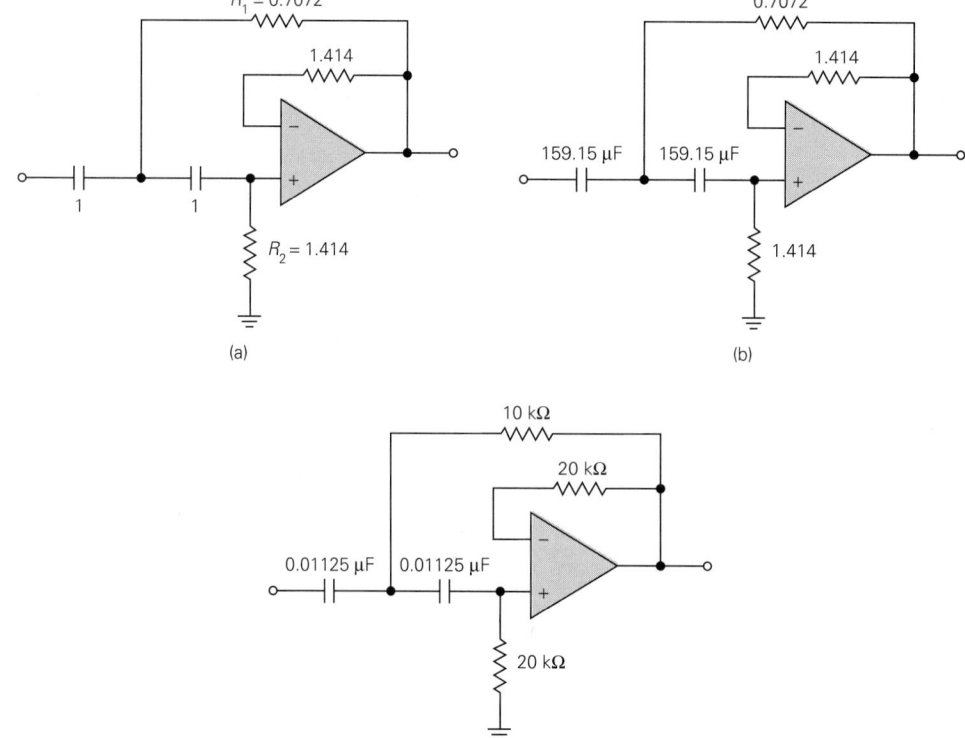

(a)

(b)

(c)

FIGURE 21–11
Circuit design of Example 21–6.

capacitance values of $0.01125 \, \mu\text{F}$, which is close to $0.011 \, \mu\text{F}$. The resulting circuit is shown in Figure 21–11(c).

21–6 TWO-POLE BAND-PASS RESPONSE

Before considering any specific active band-pass filter designs, it is desirable to investigate the value of the amplitude response forms that can be obtained. In general, band-pass filters are among the most complex of all filter types, and much effort has been directed toward developing the myriad of band-pass filters for various system requirements. Because the bulk of band-pass filter requirements are in frequency ranges in which op-amps are not well suited, band-pass active filters are more limited in their ability to compete with passive designs. Nevertheless, there are some low-frequency applications for band-pass filters in which active designs are significantly superior to passive forms.

The most common band-pass active filters are two-pole forms. Although limited compared to passive higher-order designs such as those used in high-frequency communications systems, they are capable of achieving good results in less demanding applications.

The two-pole forms we will consider are mathematically equivalent to certain configurations involving series or parallel *RLC* resonant circuits. This is a good example of how active *RC* circuits can perform operations equivalent to passive *RLC* circuits.

Amplitude Response

The amplitude response $A(f)$ of any two-pole band-pass filter can be expressed as

$$A(f) = \frac{A_o}{\sqrt{1 + Q^2 \left(\dfrac{f}{f_o} - \dfrac{f_o}{f} \right)^2}} \tag{21–10}$$

where A_o is the maximum gain within the band, and f_o is called the **geometric center frequency.** (In passive *RLC* circuits, f_o is called the **resonant frequency.**) The parameter Q is a measure of the selectivity or sharpness of the filter, and its quantitative meaning is discussed next. As was done with other forms earlier in the chapter, it is convenient to divide by A_o and work with the relative amplitude response. The relative decibel response $A_{\text{dB}}(f)$ corresponding to (21–10) can be expressed as

$$A_{\text{dB}}(f) = 20 \log_{10} \frac{A(f)}{A_o} = 20 \log_{10} \left[\frac{1}{\sqrt{1 + Q^2 \left(\dfrac{f}{f_o} - \dfrac{f_o}{f} \right)^2}} \right] \tag{21–11a}$$

or

$$A_{dB}(f) = -10 \log \left[1 + Q^2 \left(\frac{f}{f_o} - \frac{f_o}{f} \right)^2 \right] \qquad \text{(21–11b)}$$

The general form of $A(f)$ in (21–10) is illustrated in Figure 21–12 as it would appear on a *linear* scale. Observe that the rolloff on the high-frequency side occurs at a slower rate than on the low-frequency side. In other words, a point at which the response is down from the maximum by a specified amount is farther away from the center frequency on the high side than on the low side.

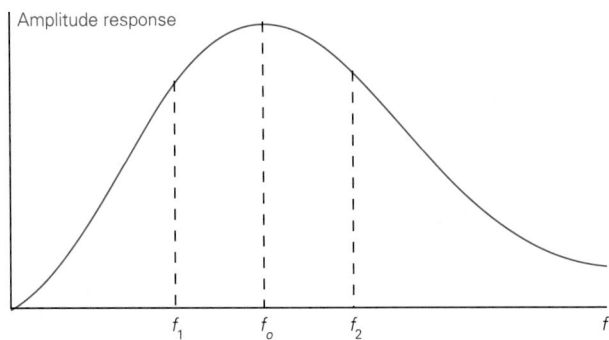

FIGURE 21–12
Illustration of band-pass filter response on linear scales.

Bandwidth and Q

Let f_1 and f_2 represent frequencies on the low and high sides, respectively, at which the response is $1/\sqrt{2}$ times the peak response (-3.01 dB down). The bandwidth B is defined as

$$B = f_2 - f_1 \qquad \text{(21–12)}$$

The frequencies f_1 and f_2 have **geometric** symmetry about the center frequency f_o. This property means that the following relationship is satisfied:

$$f_o = \sqrt{f_1 f_2} \qquad \text{(21–13)}$$

The parameter Q is related to the center frequency and bandwidth by

$$Q = \frac{f_o}{B} \qquad \text{(21–14)}$$

As Q is increased, the filter becomes more selective; that is, the 3-dB bandwidth is smaller for a given center frequency. At higher values of Q, the frequencies f_1 and f_2 are approximately the same distance on either side of f_o, and this narrow-

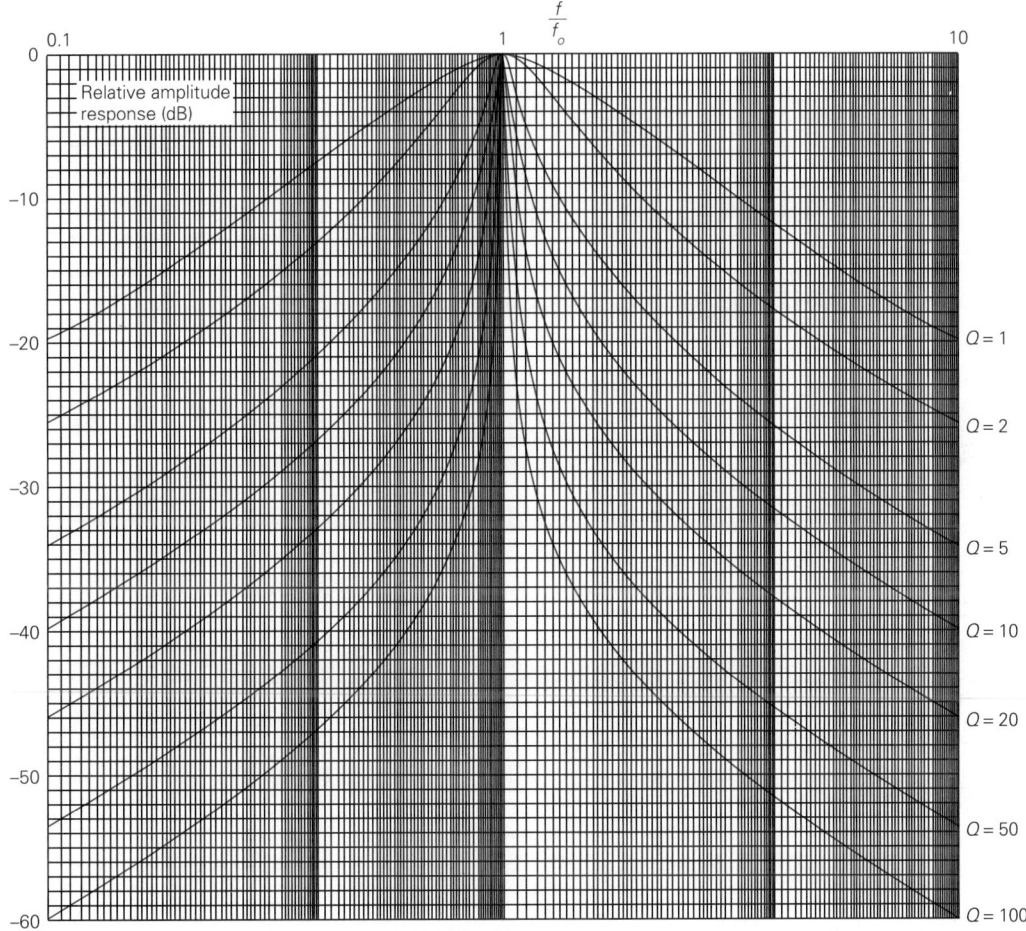

FIGURE 21–13
Two-pole band-pass amplitude response curves.

band response approximates **arithmetic** symmetry. However, the true correct condition is that of geometric symmetry as given by (21–13).

Band-Pass Response Curves

A family of curves for the amplitude response of the two-pole band-pass response is given in Figure 21–13. Observe that the horizontal scale is the normalized frequency f/f_o and that the scale is **logarithmic** in form. The response curves are symmetrical on this logarithmic scale. However, on a linear scale, the curves are somewhat skewed as observed qualitatively in Figure 21–12. Note that the relative amplitude response is shown over a frequency range from $0.1\,f_o$ to $10\,f_o$.

Observe the marked effect of the parameter Q on the amplitude response. At the lowest value of Q shown ($Q = 1$), the response drops off very slowly on either side of f_o. At the other extreme, for the highest value of Q shown ($Q = 100$), the response drops off very sharply on either side of f_o.

The geometric center frequency f_o is a very convenient parameter to use in analyzing and designing two-pole band-pass filters. However, because it is at the geometric center rather than the arithmetic center, confusion sometimes arises when it is desired to locate the two 3-dB band edges precisely. It can be shown that the arithmetic center between the two 3-dB frequencies is at a frequency $f_o\sqrt{1 + (1/2Q)^2}$, which is larger than f_o. The two 3-dB frequencies f_1 and f_2 are then given by

$$f_1 = f_o\sqrt{1 + \left(\frac{1}{2Q}\right)^2} - \frac{B}{2} \tag{21-15}$$

$$f_2 = f_o\sqrt{1 + \left(\frac{1}{2Q}\right)^2} + \frac{B}{2} \tag{21-16}$$

EXAMPLE 21-7

A certain two-pole band-pass filter response is desired with a geometric center frequency of 2 kHz and a 3-dB bandwidth of 400 Hz. **(a)** Calculate Q, f_1, and f_2. **(b)** Determine the relative decibel amplitude response at each of the following frequencies: 200 Hz, 400 Hz, 1 kHz, 1.5 kHz, 2 kHz, 3 kHz, 4 kHz, 10 kHz, and 20 kHz.

Solution

(a) We readily calculate Q as

$$Q = \frac{f_o}{B} = \frac{2000 \text{ Hz}}{400 \text{ Hz}} = 5 \tag{21-17}$$

From (21-15) and (21-16), the two band-edge frequencies f_1 and f_2 are calculated as

$$f_1 = 2000\sqrt{1 + \left(\frac{1}{2 \times 5}\right)^2} - \frac{400}{2} = 1810 \text{ Hz} \tag{21-18}$$

$$f_2 = 2000\sqrt{1 + \left(\frac{1}{2 \times 5}\right)^2} + \frac{400}{2} = 2210 \text{ Hz} \tag{21-19}$$

Observe that the arthmetic center between the two 3-dB frequencies is 2010 Hz, which is close to 2 kHz. The difference between the geometric and arithmetic center frequencies becomes quite small as Q increases.

(b) The relative decibel response can be computed with (21–11b), or the curves can be used if Q and the frequency range permit. Because $Q = 5$ is one of the curves provided and all frequencies are in the range from $0.1f_o$ to $10f_o$, this approach will be used. A convenient tabulation of the data follows:

f	f/f_o	Relative amplitude response (dB)
200 Hz	0.1	−33.9
400 Hz	0.2	−27.6
1 kHz	0.5	−17.6
1.5 kHz	0.75	−9.8
2 kHz	1	0
3 kHz	1.5	−12.6
4 kHz	2	−17.6
10 kHz	5	−27.6
20 kHz	10	−33.9

21–7 MULTIPLE-FEEDBACK BAND-PASS FILTER

In this section we present a two-pole band-pass filter utilizing one operational amplifier. Based on the theoretical development leading to this circuit, it is commonly referred to as a **multiple-feedback band-pass filter.** The multiple-feedback filter is capable in theory of achieving band-pass filter designs of the form discussed in Section 21–6 and is thus operationally equivalent to an RLC resonant circuit. As we will see, however, this circuit is best suited to designs utilizing relatively low values of Q.

The basic form of the normalized circuit is shown in Figure 21–14. The circuit is normalized to have a geometric center frequency $\omega_o = 1$ rad/s. Observe that the normalized resistance values are functions of the desired circuit Q. The Q parameter is a ratio of two frequency values, and both frequencies are changed by the same ratio during a scaling operation. Thus, *the desired Q is established directly in the normalized design, and it remains unchanged during the scaling process.*

FIGURE 21–14
Normalized form of band-pass filter using single operational amplifier.

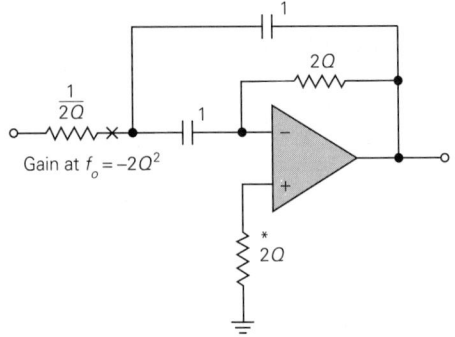

Gain at $f_o = -2Q^2$

*This resistor is used for bias compensation only.

From the data provided on the figure, it should be observed that the gain at the geometric center frequency is inverting and that its magnitude increases with the square of Q. This marked increase of gain as Q increases may pose serious problems at moderate to high values of Q because of possible overload of the op-amp as well as bandwidth and slew rate limitations.

Achieving a Gain of One

A partial solution to the problem is to place an attenuator network ahead of the filter to reduce the overall gain to a nominal level. The normalized form of a possible circuit is shown in Figure 21–15. The actual derivation of the normalized

FIGURE 21–15
Normalized attenuator network used in band-pass filter of Figure 21–14 to establish unity gain at f_o. (This circuit replaces the first resistor.)

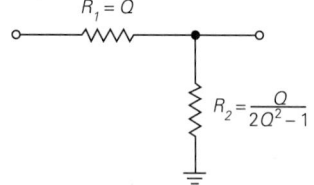

$R_1 = Q$

$R_2 = \dfrac{Q}{2Q^2 - 1}$

design will be left as an exercise for the reader, but two criteria have been employed in the circuit:

1. The open-circuit voltage is $1/2Q^2$ times the input voltage on the left. This factor exactly cancels the magnitude of the gain of the filter at f_o, so the *overall* magnitude of the gain at f_o is unity.
2. The net Thévenin resistance looking to the left in the circuit of Figure 21–15 must equal the value of the left-hand resistance (normalized value $1/2Q$) in Figure 21–14. This requirement results in the same resonant frequency and Q as for the basic design of Figure 21–14.

The attenuator network replaces the first resistance in the circuit of Figure 21–14. The complete normalized design is shown in Figure 21–16. Because the

FIGURE 21–16
Normalized band-pass filter using single op-amp with attenuator network used to establish unity gain at f_o.

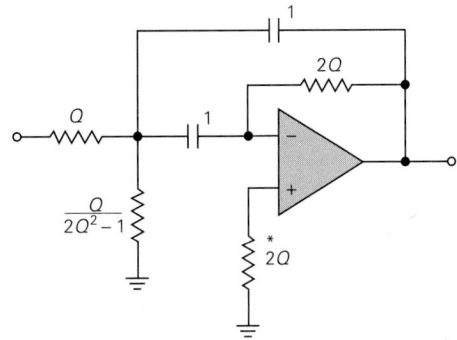

*This resistor is used for bias compensation only.

attenuator network becomes an integral part of the complete design, it must be scaled by the same factor as all other resistances in the scaling process.

Although the use of this circuit improves the design, the spread of element values and the associated sensitivity of this circuit still make it difficult to use for high values of Q. As an approximate rule of thumb, this circuit should be limited to Q values less than about 20 or so. Where higher values of Q are required, the state-variable filter of Section 21–8 is recommended.

EXAMPLE 21–8

Design a band-pass filter using a single op-amp to meet the specifications of Example 21–7. Include an attenuator network to achieve unity gain at f_o.

Solution
In Example 21–7 the requirements called for a geometric center frequency f_o = 2000 Hz and $Q = 5$. The basic form of the circuit is obtained from Figure 21–16, and as applied to the Q values specified here, the normalized circuit is given in Figure 21–17(a).

First, the frequency scaling operation will be performed. The response of the normalized design at 1 rad/s must correspond to the response of the actual design at 2 kHz. The constant K_f is thus determined as

$$K_f = \frac{2\pi \times 2 \times 10^3 \text{ rad/s}}{1 \text{ rad/s}} = 12,566.37 \qquad \textbf{(21–20)}$$

The capacitor values are then divided by K_f, and the corresponding circut is shown in Figure 21–17(b).

Although a number of possible solutions could be investigated for an appropriate impedance scaling constant, a brief trial-and-error approach led to the selection of the standard value of 0.01 μF for the capacitors. The impedance scaling constant K_r must be determined such that $79.577 \times 10^{-6}/K_r = 0.01 \times 10^{-6}$ F, which leads to $K_r = 7957.7$. The resistance values are multiplied by K_r, and the capacitance values are changed to 0.01 μF, as shown in Figure 21–17(c). As a starting point, the standard resistance values of 39 kΩ, 820 Ω, and 82 kΩ could

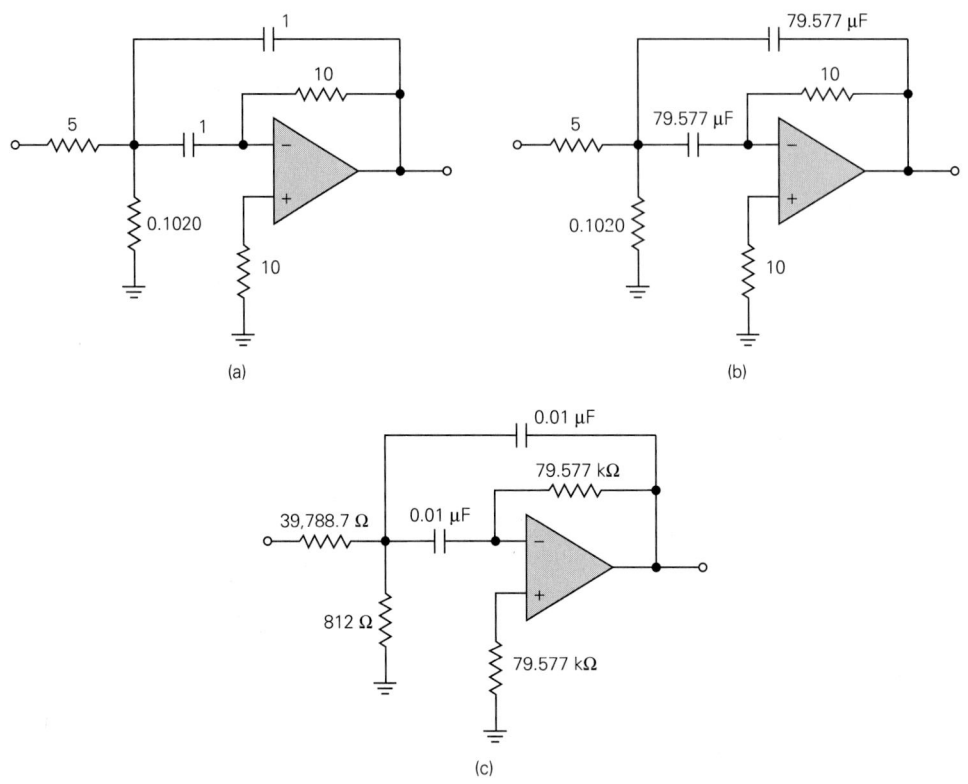

FIGURE 21–17
Circuit design of Example 21–8.

be used. The directions of these values will contribute to increased gain, so an additional fixed resistance could be added in series with the 39-kΩ resistance if necessary.

21–8 STATE-VARIABLE FILTER

The **state-variable filter** is one of the most versatile of all the active filter types. The same structure can be used for implementing a low-pass, a high-pass, or a band-pass filter; and with the addition of a summing circuit, it can even be used for a band-rejection filter.

The evolution of the state-variable filter is rather interesting and should be mentioned. The earliest applications of operational amplifiers were in analog computers, which have been used for many years to solve differential equations and to simulate physical systems. A technique evolved for "programming" a differential equation by a certain systematic layout and interconnection of various integrators and summers. As operational amplifiers became available in integrated-circuit form, the concept of realizing a filter by "programming" the corresponding differential equation was conceived. Thus, the state-variable filter represents essentially the same implementation strategy as that of simulating the applicable differential equation on an analog computer.

The term *state-variable* is related to a form of system analysis called *state-variable theory,* which provides a systematic means of formulating the differential equations of large systems. As part of such a formulation, a state diagram showing the interconnection of all the mathematical operations may be constructed. The state diagram is, in reality, a mathematical form of an analog computer simulation. The reader, of course, need not understand all this background to utilize state-variable designs properly.

Although it is theoretically possible to form state-variable sections of any order, most designs are based on two-pole sections. The two-pole section is less sensitive to parameter variations than those of higher order. Further, it is possible to create sections of any order by cascading sections of lower order. Incidentally, there is no need to form a one-pole section because any first-order function can be created with a simple passive RC circuit.

Two-Pole Circuit Form

There are several variations of the two-pole state-variable circuit, all of which are functionally equivalent. The form that we emphasize here is shown in Figure 21–18 in normalized form. Observe that the normalized values of all passive components except R_Q are unity.

Several element values have been labeled for purposes of clarity and explanation. The two resistances denoted as R_t and the two capacitances denoted as C_t will be collectively referred to as the **four tuning elements.** Note that the two tuning capacitances have identical values and that the two tuning resistances have identical values. When frequency scaling, impedance scaling, or tuning is performed, the two capacitors *or* the two resistors must be varied together. One common approach to implementing frequency-adjustable units is to switch in or out fixed capacitance values (two at a time) and to employ ganged variable resistors (two on a shaft) for continuous frequency adjustment over a range.

The value of R_Q establishes the required Q for the circuit. The meaning of Q for band-pass and band-rejection circuits has been discussed at length, but the interpretation for low-pass and high-pass forms will be given later.

The impedance level of the voltage divider network containing R_Q and the 1-Ω resistance on the left may be adjusted independently of all other resistances in the circuit provided that these two resistances remain in exactly the same proportion. After the initial form of a circuit is designed, one may desire to alter

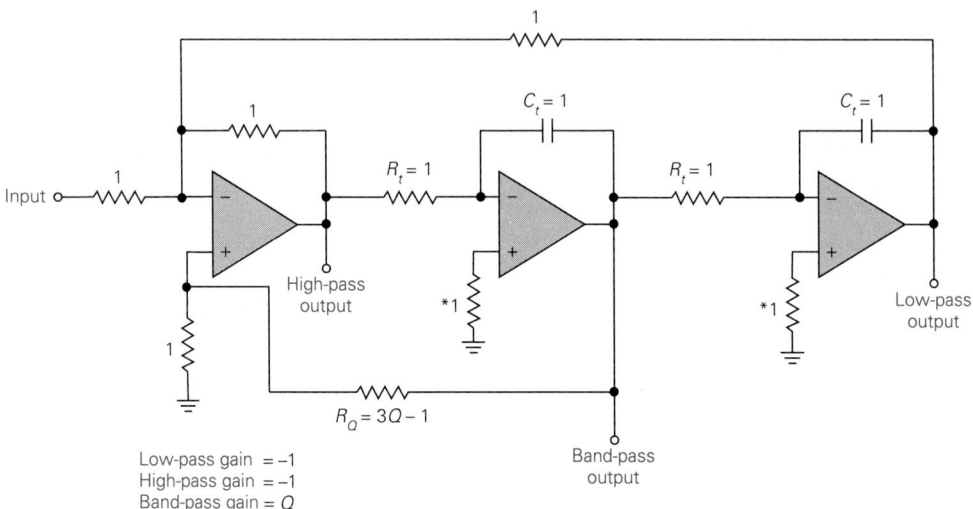

Low-pass gain = −1
High-pass gain = −1
Band-pass gain = Q

* These resistors are used for bias compensation only
and do not contribute to the filter response.

FIGURE 21–18
State-variable normalized filter.

the impedance level here so that the optimum bias compensating resistance is seen from both terminals.

The impedance level of the three 1-Ω resistances connected to the inverting input of the left-hand summing networks may be adjusted independently of other parts of the circuit. Again, it is necessary that these three components be scaled in exactly the same proportion.

The reader should understand that it may not be necessary to use different impedance scaling constants at different points, and the simplest approach "on paper" is to use the same constant. However, flexibility in the choice of element values as well as the realization of optimum bias resistance levels may result from exploiting the possibility of different levels for different parts of the circuit. Each of the different circuit forms will now be discussed.

Band-Pass Form

The input signal is applied on the extreme left, and the output is taken to the right of the first integrator as noted. The circuit can be assumed to be normalized to a geometric center frequency $\omega_o = 1$ rad/s. The desired bandpass Q is established by setting $R_Q = 3Q - 1$. Note that the gain at ω_o is Q, so care must be taken not to overload any amplifier at high values of Q.

Low-Pass and High-Pass Forms

For Butterworth two-pole low-pass and high-pass filters, the normalized circuit has a 3-dB frequency $\omega_c = 1$ rad/s. Because the only types of low-pass and high-pass characteristics being considered in this book are Butterworth forms, the 1 rad/s value identifies the upper 3-dB frequency for the low-pass case and the lower 3-dB frequency for the high-pass case. However, it should be emphasized that for other types of filter characteristics (for example, Chebyshev, maximally flat time-delay, and so on), the cutoff frequency of the normalized section may not be 1 rad/s. Such a discussion is not within the intended objective, but we caution anyone wanting to extend the material in this book to other filter forms.

Although the Q parameter has a direct physical meaning for band-pass filters and, as we will see, for band-rejection filters, its meaning for low-pass and high-pass filters is more subtle and not as easily associated with common physical characteristics. The value of Q for such cases relates to the relative damping of the transfer function. The Q value or values must be used in the two-pole sections as specified, but there is no simple, observable relationship as for band-pass and band-rejection functions.

For either two-pole low-pass or high-pass Butterworth filters, the value of Q is $Q = 1/\sqrt{2} = 0.7071$, so the corresponding value of the resistance is $R_Q = 3Q - 1 = 3(0.7071) - 1 = 1.1213$. Thus, the same value of R_Q establishes the two-pole section as either a low-pass or a high-pass Butterworth filter. If a low-pass form is desired, the output is taken on the extreme right, as shown in Figure 21–18, and if a high-pass form is desired, the output is taken to the right of the summing circuit on the left. Observe from the information associated with the figure that the maximum pass-band gain of both forms is −1.

Band-Rejection Form

The band-rejection form is obtained by summing the output of the low-pass section to the output of the high-pass section. Thus, an additional op-amp is required as shown in Figure 21–19. Because the gains of the low-pass and high-pass sections are both −1, the additional inversion for both inputs of the summing circuit results in a net gain of +1 for the band-rejection circuit.

The geometric center frequency f_o is now the geometric center of the rejection band, and f_1 and f_2 now represent the points on either side of the center at which the response is 3 dB below the levels of the high- and low-frequency flat bands. These concepts are illustrated in Figure 21–20. The Q parameter is still defined as

$$Q = \frac{f_o}{B} \qquad\qquad \textbf{(21–21)}$$

However, the frequency values f_o and B now refer to a rejection band rather than a pass band.

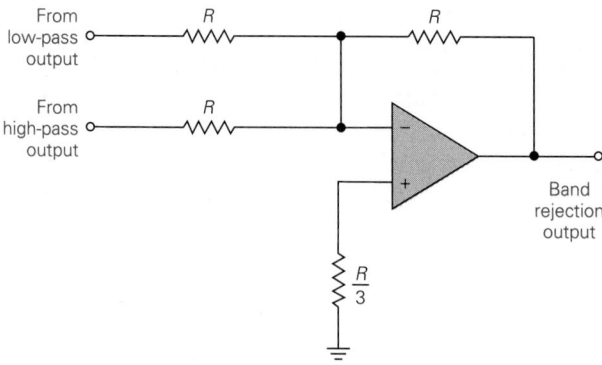

FIGURE 21–19
Summing circuit used to create band-rejection response in state-variable filter.

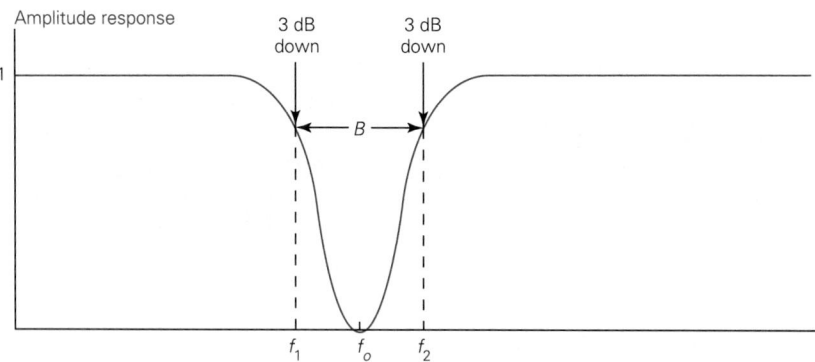

FIGURE 21–20
General form of amplitude response of band-rejection filter.

**EXAMPLE
21–9**

Design a state-variable filter that can serve as either a low-pass or a high-pass Butterworth two-pole filter if the desired 3-dB cutoff frequency is 1 kHz.

Solution
For a two-pole Butterworth filter, $Q = 0.7071$, and the value of R_Q in Figure 21–18 is $R_Q = 1.1213$. All other normalized values are the same as in Figure 21–18. The frequency of 1 rad/s in the normalized circuit must convert to 1 kHz in the final circuit, so the constant K_f is

$$K_f = \frac{2\pi \times 1000 \text{ rad/s}}{1 \text{ rad/s}} = 6283.185 \qquad \textbf{(21–22)}$$

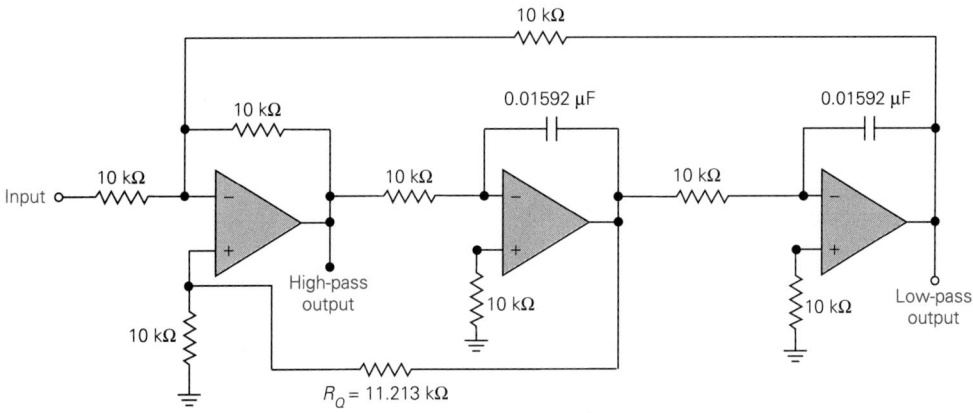

FIGURE 21–21
Circuit design of Example 21–9.

The two 1-F capacitors in Figure 21–18 are then divided by K_f, and the resulting values are each 159.2 μF.

Selection of the impedance scaling constant is somewhat arbitrary, but after some trial and error, a choice was made to set $K_r = 10^4$. All the 1-Ω resistances then convert to 10 kΩ, the value of R_Q becomes 11.213 kΩ, and the two capacitors assume values of 0.01592 μF. The final circuit diagram is shown in Figure 21–21. Standard capacitance values of 0.016 μF should suffice for the capacitors, and a standard value of 11 kΩ could be used for R_Q.

The circuit as designed could be used as either a low-pass filter or a high-pass filter in accordance with where the output is taken. As a pass-band filter, however, this design would be quite limited, as $Q = 0.7071$, and the selectivity would be far too low for most applications. This example illustrates that although the state-variable filter can be used for several types of filter functions, the Q value required for one type may be quite different from that required for other types.

EXAMPLE 21–10   Design a state-variable two-pole band-pass filter with a geometric center frequency of 500 Hz and a 3-dB bandwidth of 10 Hz.

Solution
The required value of Q is

$$Q = \frac{f_o}{B} = \frac{500}{10} = 50 \qquad \textbf{(21–23)}$$

The normalized design of Figure 21–18 is used as the starting point, with $R_Q = 3Q - 1 = 3 \times 50 - 1 = 149$. The normalized frequency of 1 rad/s must convert to 500 Hz, so the frequency scaling constant is

$$K_f = \frac{2\pi \times 500 \text{ rad/s}}{1 \text{ rad/s}} = 3141.593 \qquad \text{(21–24)}$$

The capacitance values are divided by K_f, and the new value of each is 318.31 μF.

After some trial and error, the resistance scaling constant was selected as $K_r = 2 \times 10^4$. All resistance values except R_Q then become 20 kΩ, and the two values of C become 0.01592 μF. The value of the Q adjustment resistance is $R_Q = 2.98$ MΩ. The design is shown in Figure 21–22. The 20-kΩ resistances are standard values, and a value of 0.016 μF should suffice for C. A 3-MΩ resistance could be used as a starting point for R_Q, but some delicate adjustments might be needed in view of the large value of Q. Depending on the acceptable tolerance, the next smallest standard value of resistance in series with a variable resistance could be used if necessary.

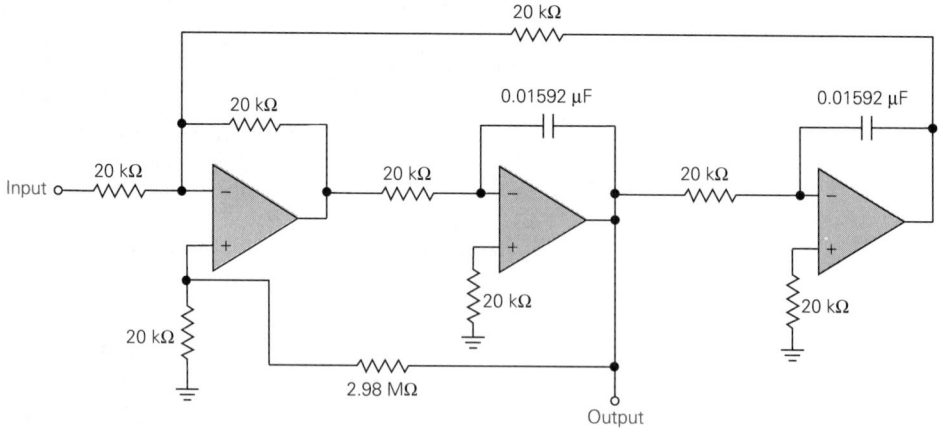

FIGURE 21–22
State-variable band-pass filter design of Example 21–10.

The gain of this circuit at 500 Hz is 50, so care must be taken not to overload any stages. In high Q circuits such as this, it may be necessary to attenuate the input signal if it is too high. If this is necessary, an isolation stage between the attenuator and the first stage may be required to avoid the undesirable interaction between the attenuator and the gain constants of the summing circuit.

EXAMPLE 21–11

An undesired interfering component at 500 Hz is to be eliminated in a system, and a band-rejection filter is desired. Design a state-variable two-pole filter to accomplish the purpose if the width between 3-dB points in the rejection band is to be set at 10 Hz.

Solution

The required value of Q is

$$Q = \frac{500}{10} = 50 \qquad (21\text{–}25)$$

Because the center frequency and the bandwidth are the same as in Example 21–10, the design of that example can be used as the basis for this band-rejection filter. All that is required is to sum the low-pass and high-pass outputs, and the complete circuit is shown in Figure 21–23. The same point about the high gain made in Example 21–10 applies here.

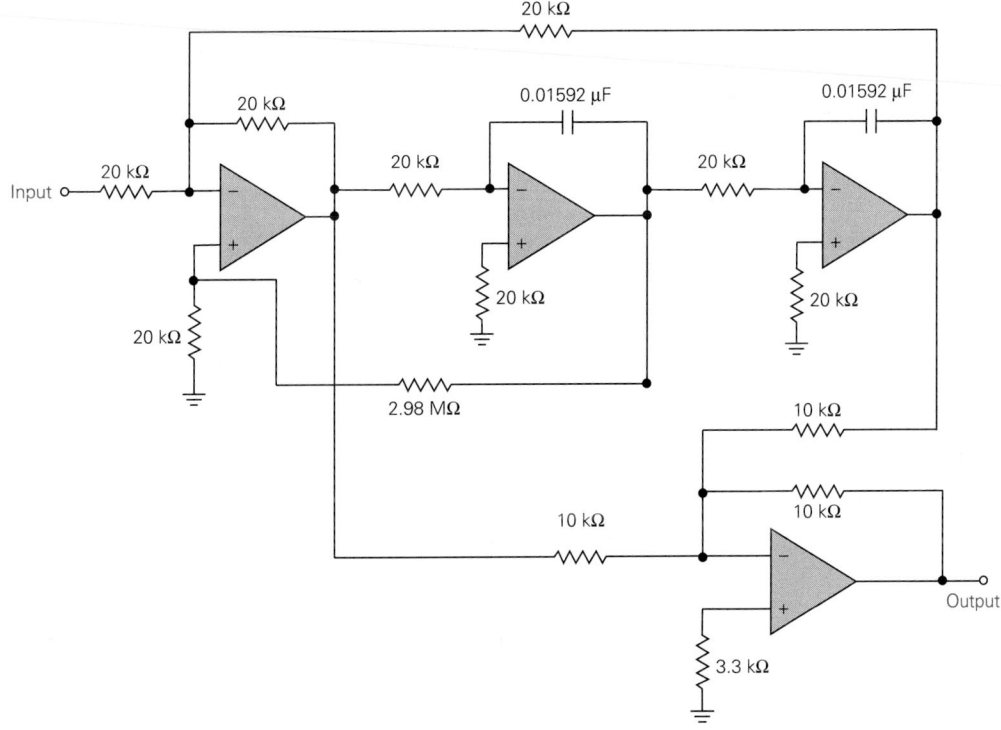

FIGURE 21–23

State-variable band-rejection filter design of Example 21–11.

In a notch filter such as this, the actual level of attenuation at the notch frequency (500 Hz) is a critical function of the parameter tolerances. Consequently, some trimming or adjustments of the element values may be required in order to obtain optimum rejection at 500 Hz.

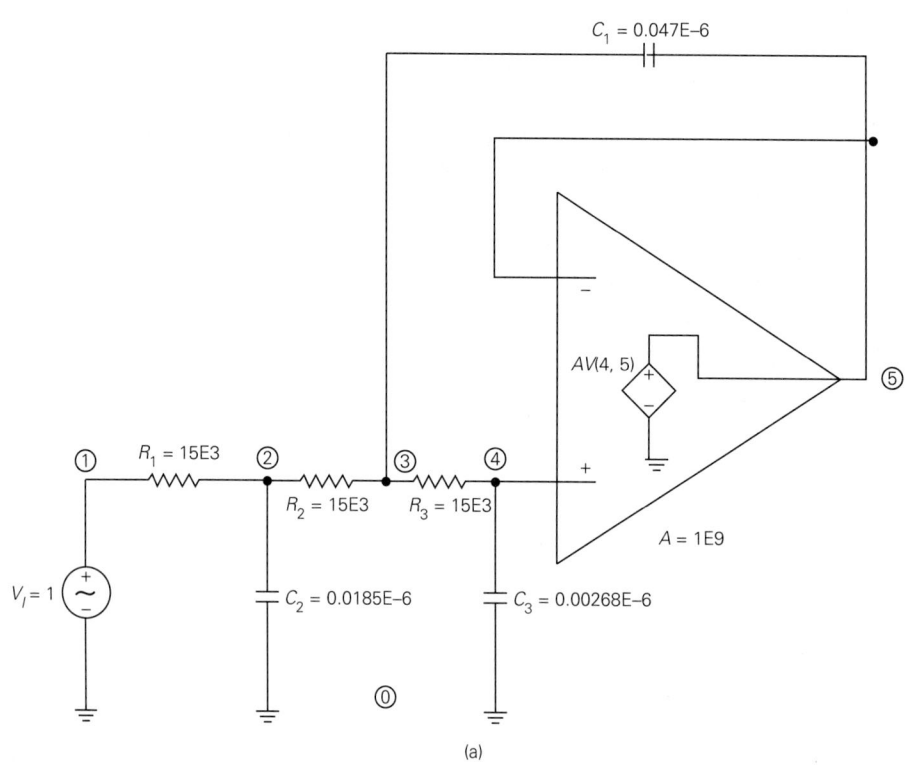

(a)

```
PSPICE EXAMPLE 21-1
VI 1 0 AC 1
R1 1 2 15E3
R2 2 3 15E3
R3 3 4 15E3
C1 3 5 0.047E-6
C2 2 0 0.0185E-6
C3 4 0 0.00268E-6
E 5 0 4 5 1E9
.AC DEC 50 100 5E3
.PROBE
.END
```

(b)

FIGURE 21–24
Circuit and code for PSPICE Example 21–1.

21–9 PSPICE EXAMPLES

The frequency response of an active filter may be readily displayed with PSPICE. The examples that follow illustrate this process with several of the designs of this chapter.

PSPICE EXAMPLE 21–1

Use PSPICE to model the low-pass active filter design of Example 21–4 (Figure 21–9(c)), and obtain a decibel amplitude response.

Solution

The circuit modified for PSPICE and the accompanying code are shown in Figure 21–24. The nearly ideal op-amp model is used in this simulation.

The decibel amplitude response is shown in Figure 21–25. The response follows the expected behavior.

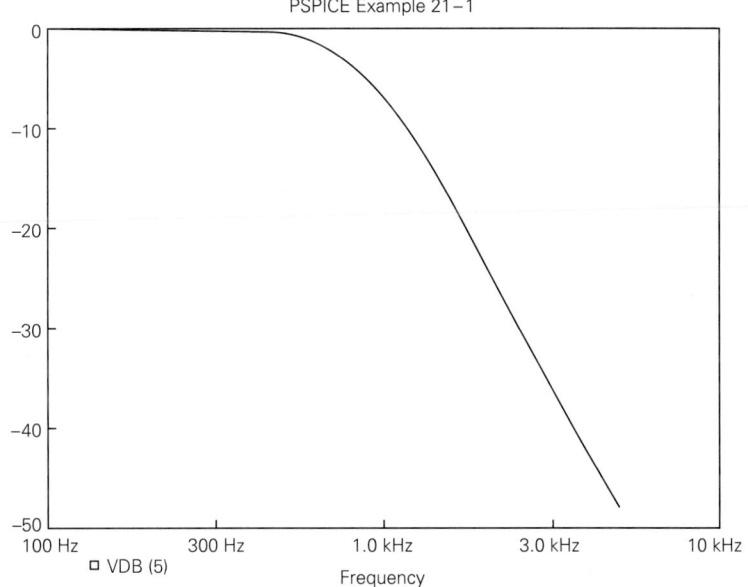

FIGURE 21–25
Decibel amplitude response of PSPICE Example 21–1.

PSPICE EXAMPLE 21–2

Use PSPICE to model the band-pass active filter design of Example 21–8 (Figure 21–17(c)), and obtain a decibel amplitude response.

Solution

The circuit modified for PSPICE and the code are shown in Figure 21–26. As in the preceding example, the nearly ideal op-amp model is used. The decibel amplitude response is shown in Figure 21–27.

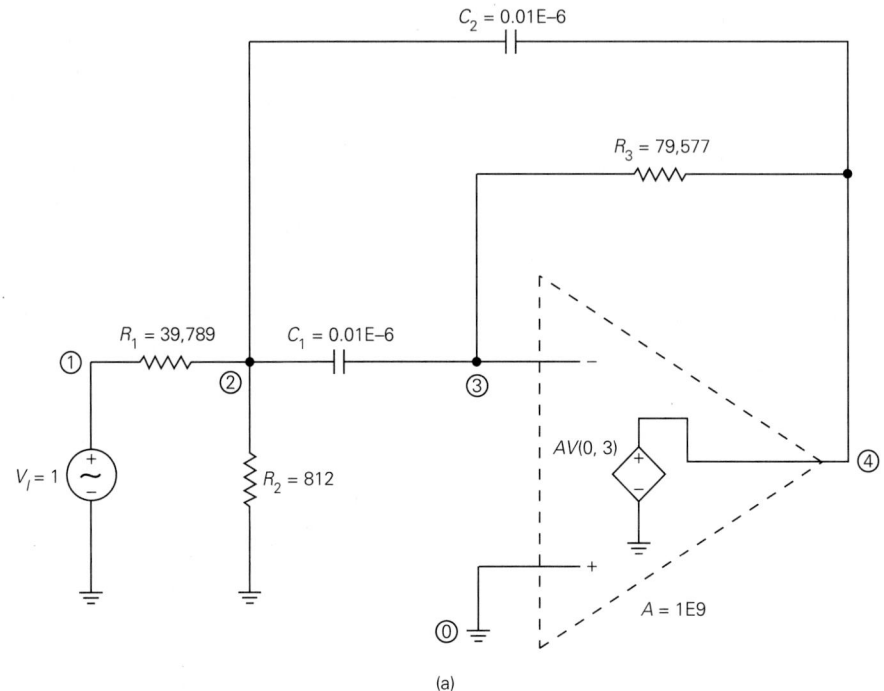

(a)

```
PSPICE EXAMPLE 21-2
VI 1 0 AC 1
R1 1 2 39789
R2 2 0 812
C1 2 3 0.01E-6
C2 2 4 0.01E-6
R3 3 4 79577
E 4 0 0 3 1E9
.AC DEC 50 200 2E4
.PROBE
.END
```

(b)

FIGURE 21–26
Circuit and code for PSPICE Example 21–2.

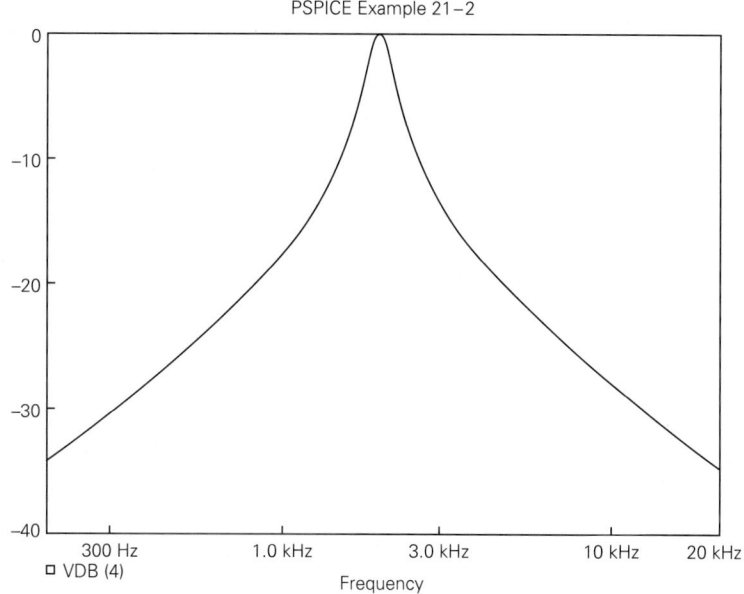

FIGURE 21–27
Decibel amplitude response of PSPICE Example 21–2.

PSPICE EXAMPLE 21–3

Use PSPICE to model the low-pass state-variable active filter design of Example 21–9 (Figure 21–21), and obtain a decibel amplitude response.

Solution

The circuit modified for PSPICE and the code are shown in Figure 21–28. Because three op-amps are required, the use of a subcircuit is warranted. The subcircuit employs a nearly ideal op-amp model. The decibel amplitude response is shown in Figure 21–29.

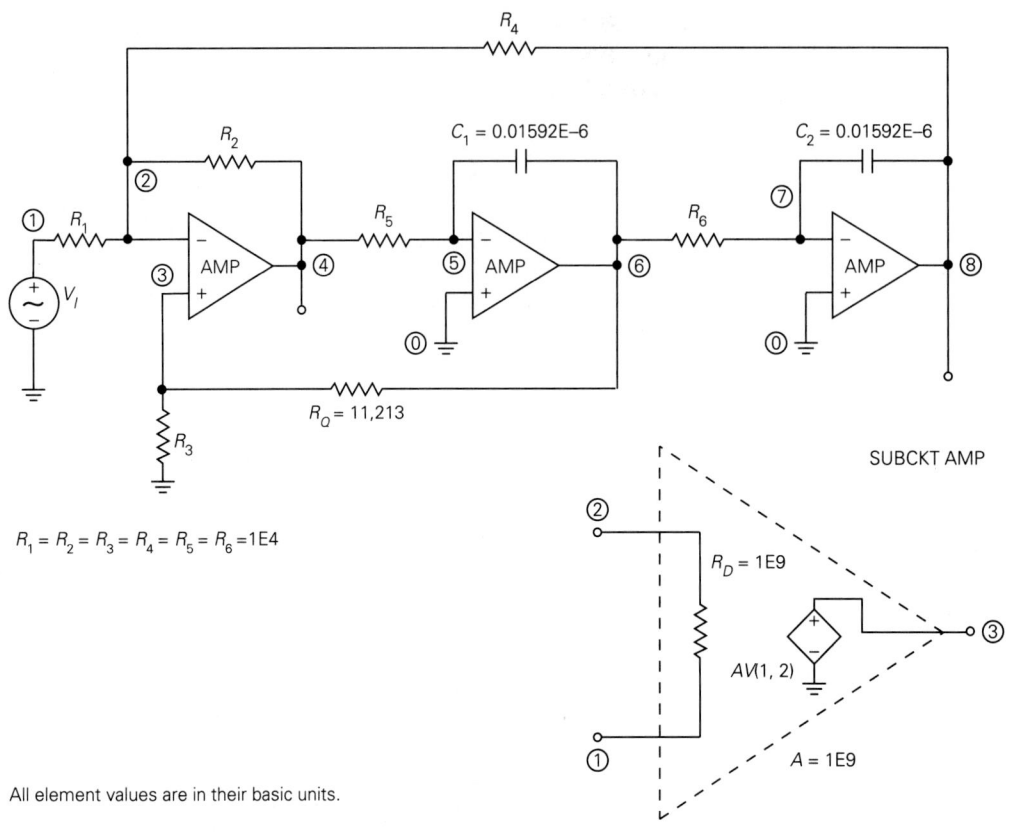

(a)

All element values are in their basic units.

$R_1 = R_2 = R_3 = R_4 = R_5 = R_6 = 1E4$

```
PSPICE EXAMPLE 21-3
VI 1 0 AC 1
R1 1 2 1E4
R2 2 4 1E4
R3 3 0 1E4
R4 2 8 1E4
R5 4 5 1E4
R6 6 7 1E4
RQ 3 6 11213
C1 5 6 0.01592E-6
C2 7 8 0.01592E-6
X1 3 2 4 AMP
X2 0 5 6 AMP
X3 0 7 8 AMP
.SUBCKT AMP 1 2 3
RD 1 2 1E9
E 3 0 1 2 1E9
.ENDS AMP
.AC DEC 50 100 1E4
.PROBE
.END
```

(b)

FIGURE 21–28

Circuit and code of PSPICE Example 21–3.

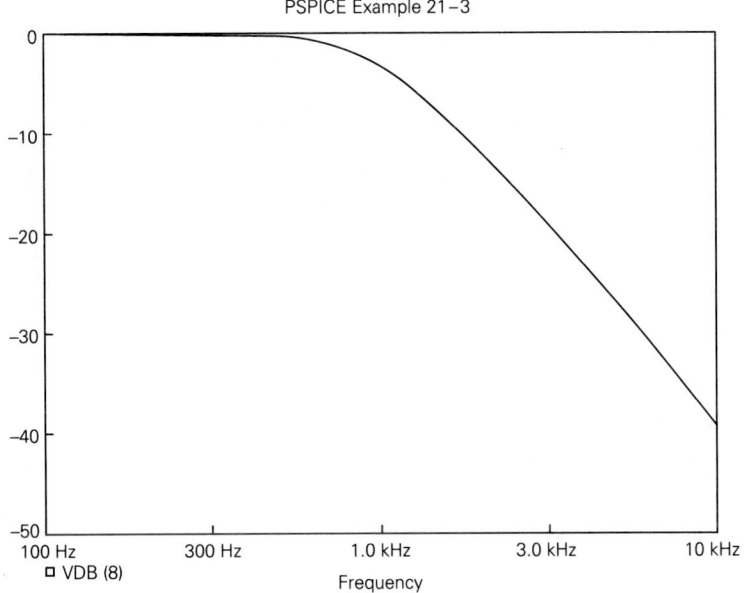

PSPICE Example 21–3

FIGURE 21–29
Decibel amplitude response of PSPICE Example 21–3.

PSPICE EXAMPLE 21–4

Use PSPICE to model the band-pass state-variable active filter design of Example 21–10 (Figure 21–22), and obtain a decibel amplitude response.

Solution

The circuit modified for PSPICE and the code are shown in Figure 21–30. The same subcircuit employed in the preceding example is used here.

In order to provide a convenient 0-dB reference level, the input voltage is chosen as 0.02 V because the gain is 50. This choice establishes a reference level of $50 \times 0.02 = 1$ V at the center of the passband. The decibel amplitude response is shown in Figure 21–31.

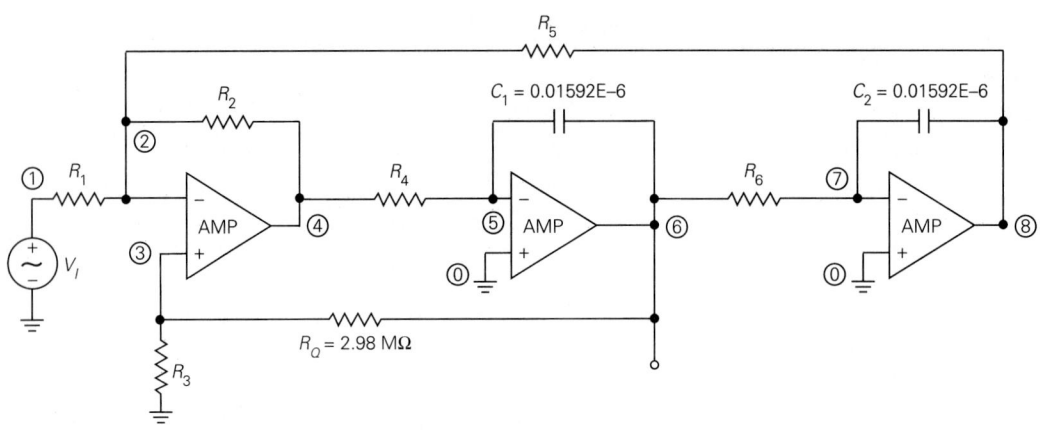

$R_1 = R_2 = R_3 = R_4 = R_5 = R_6 = 2E4$

All element values are in their basic units.

(a)

```
PSPICE EXAMPLE 21-4      X1 3 2 4 AMP
VI 1 0 AC 0.02          X2 0 5 6 AMP
R1 1 2 2E4             X3 0 7 8 AMP
R2 2 4 2E4            .SUBCKT AMP 1 2 3
R3 3 0 2E4            RD 1 2 1E9
R4 2 8 2E4            E 3 0 1 2 1E9
R5 4 5 2E4           .ENDS AMP
R6 6 7 2E4           .AC LIN 100 400 600
RQ 3 6 2.98E6        .PROBE
C1 5 6 0.01592E-6    .END
C2 7 8 0.01592E-6
```

(b)

FIGURE 21–30
Circuit and code of PSPICE Example 21–4.

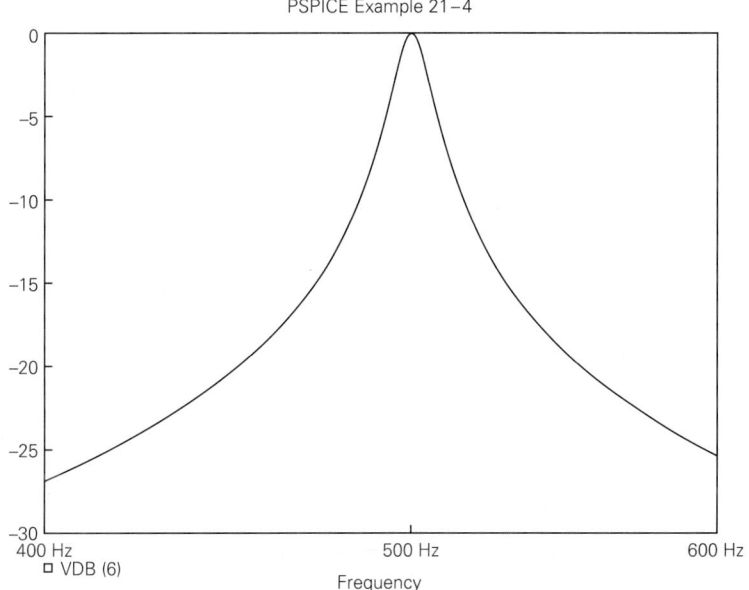

FIGURE 21–31
Decibel amplitude response of PSPICE Example 21–4.

PSPICE EXAMPLE 21–5

Use PSPICE to model the band-rejection state-variable active filter design of Example 21–11 (Figure 21–23), and obtain a decibel amplitude response.

Solution
The circuit modified for PSPICE and the code are shown in Figure 21–32. Once again, the nearly ideal op-amp model is represented through a subcircuit, and in this case, four are required. The decibel amplitude response is shown in Figure 21–33.

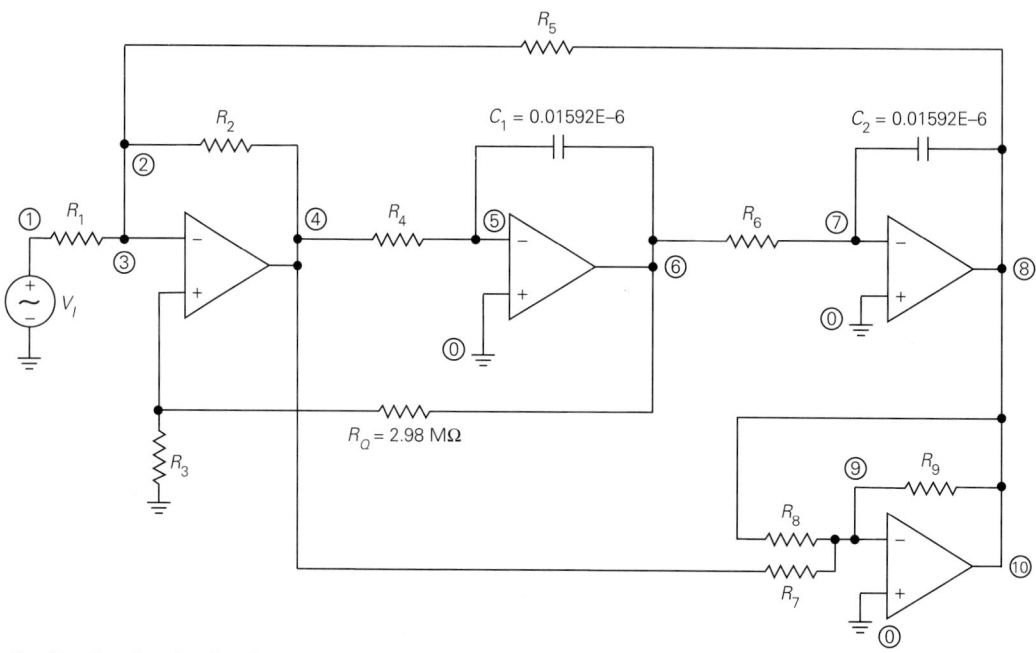

$R_1 = R_2 = R_3 = R_4 = R_5 = R_6 = R_7 = R_8 = R_9 = 2E4$

All element values are in their basic units.

(a)

```
PSPICE EXAMPLE 21-5
VI 1 0 AC 1
R1 1 2 2E4
R2 2 4 2E4
R3 3 0 2E4
R4 2 8 2E4
R5 4 5 2E4
R6 6 7 2E4
R7 4 9 2E4
R8 8 9 2E4
R9 9 10 2E4
RQ 3 6 2.98E6
C1 5 6 0.01592E-6
C2 7 8 0.01592E-6
X1 3 2 4 AMP
X2 0 5 6 AMP
X3 0 7 8 AMP
X4 0 9 10 AMP
.SUBCKT AMP 1 2 3
RD 1 2 1E9
E 3 0 1 2 1E9
.ENDS AMP
.AC LIN 201 450 550
.PROBE
.END
```

(b)

FIGURE 21–32
Circuit and code of PSPICE Example 21–5.

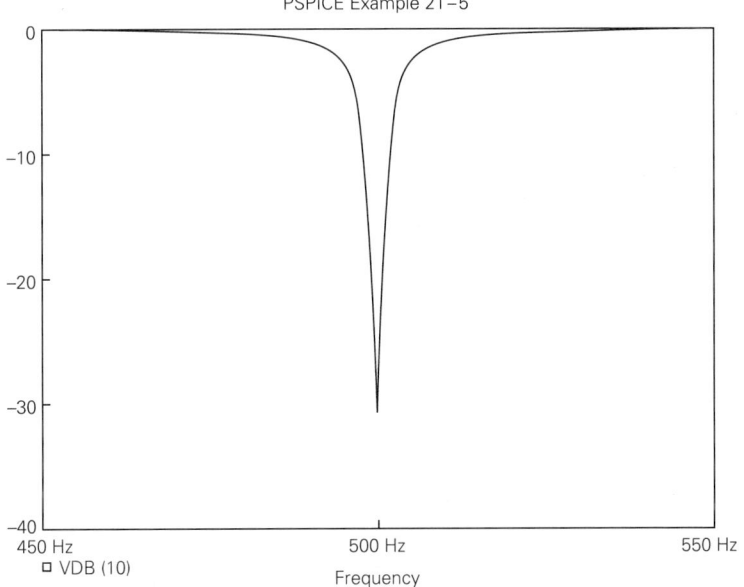

FIGURE 21-33
Decibel amplitude response of PSPICE Example 21-5.

PROBLEMS

Drill Problems

21-1. A 6-pole low-pass Butterworth filter has a 3-dB cutoff frequency of 2 kHz.
    a. Use Figure 21-6 to determine the decibel response at the following frequencies: 3 kHz, 4 kHz, 5 kHz, 6 kHz, 7 kHz.
    b. Use an appropriate equation in Section 21-4 to compute the decibel response at each of the preceding frequencies.

21-2. A 7-pole low-pass Butterworth filter has a 3-dB cutoff frequency of 8 kHz.
    a. Use Figure 21-6 to determine the decibel response at the following frequencies: 12 kHz, 16 kHz, 24 kHz, 32 kHz.
    b. Use an appropriate equation in Section 21-4 to compute the decibel response at each of the preceding frequencies.

21-3. A 4-pole high-pass Butterworth filter has a 3-dB cutoff frequency of 600 Hz.
    a. Use Figure 21-6 to determine the decibel response at the following frequencies: 300 Hz, 150 Hz, 100 Hz, 80 Hz.
    b. Use an appropriate equation in Section 21-4 to compute the decibel response at each of the preceding frequencies.

21-4. A 3-pole high-pass Butterworth filter has a 3-dB cutoff frequency of 1.5 kHz.
    a. Use Figure 21-6 to determine the decibel response at the following frequencies: 1 kHz, 500 Hz, 300 Hz, 150 Hz.
    b. Use an appropriate equation in Section 21-4 to compute the decibel response at each of the preceding frequencies.

21–5. A two-pole band-pass filter has a center frequency of 1.5 kHz and a 3-dB bandwidth of 300 Hz. Determine:
  a. The Q
  b. The exact 3-dB frequencies f_1 and f_2
  c. Use Figure 21–13 to determine the decibel response at 150 Hz and 15 kHz.
  d. Use an appropriate equation from Section 21–6 to compute the decibel response at the preceding two frequencies.
21–6. A two-pole band-pass filter has a center frequency of 800 Hz and a Q of 20. Determine:
  a. The 3-dB bandwidth
  b. The exact 3-dB frequencies f_1 and f_2
  c. Use Figure 21–13 to determine the decibel response at 80 Hz and 8 kHz.
  d. Use an appropriate equation from Section 21–6 to compute the decibel response at the preceding two frequencies.

Derivation Problems
21–7. Show that the amplitude response of low-pass Butterworth filters well above cutoff (that is, $f \gg f_c$) decreases by about $6n$ dB per octave. (An octave is a doubling of the frequency.)
21–8. Show that the amplitude response of low-pass Butterworth filters well above cutoff (that is, $f \gg f_c$) decreases by $20n$ dB per decade. (A decade corresponds to multiplying the frequency by a factor of 10.)

Design Problems
21–9. Design a two-pole low-pass Butterworth active filter using a unity-gain section to achieve a 3-dB frequency of 1 kHz. Select the two filter resistances as 10 kΩ each.
21–10. Design a two-pole low-pass Butterworth active filter using a unity-gain section to achieve a 3-dB frequency of 200 Hz. Select the two filter resistances as 30 kΩ each.
21–11. Repeat the unity-gain design of Problem 21–9 if the largest capacitor is selected as 0.01 μF.
21–12. Repeat the unity-gain design of Problem 21–10 if the largest capacitor is selected as 0.02 μF.
21–13. A low-pass filter is desired for a given application. The specifications are as follows:

  1. Relative attenuation ≤ 3 dB for $f \leq 1$ kHz
  2. Relative attenuation ≥ 35 dB for $f \geq 4$ kHz

Specify the minimum number of poles for a Butterworth filter that will satisfy the requirements.
21–14. A low-pass filter is desired for a given application. The specifications are as follows:

  1. Relative attenuation ≤ 3 dB for $f \leq 1$ kHz
  2. Relative attenuation ≥ 23 dB for $f \geq 2$ kHz

Specify the minimum number of poles for a Butterworth filter that will satisfy the requirements.
21–15. A high-pass filter is desired for a given application. The specifications are as follows:

  1. Relative attenuation ≤ 3 dB for $f \geq 1$ kHz
  2. Relative attenuation ≥ 40 dB for $f \leq 200$ Hz

Specify the minimum number of poles for a Butterworth filter that will satisfy the requirements.

21–16. A high-pass filter is desired for a given application. The specifications are as follows:

1. Relative attenuation $\leq$ 3 dB for $f \geq$ 1 kHz
2. Relative attenuation $\geq$ 30 dB for $f \leq$ 400 Hz

Specify the minimum number of poles for a Butterworth filter that will satisfy the requirements.

21–17. Design a unity-gain filter to meet the specifications of Problem 21–13. Select the filter resistances as 10 kΩ each.

21–18. Design a unity-gain filter to meet the specifications of Problem 21–14. Select the filter resistances as 10 kΩ each.

21–19. Perform a new unity-gain design to meet the specifications of Problem 21–13 if the largest capacitance is selected as 0.01 μF.

21–20. Perform a new unity-gain design to meet the specifications of Problem 21–14 if the largest capacitance in *each section* is selected as 0.01 μF.

21–21. Design a unity-gain filter to meet the specifications of Problem 21–15. Select the two capacitances as 0.01 μF each.

21–22. Design a unity-gain filter to meet the specifications of Problem 21–16. Select the two capacitances *for each section* as 0.01 μF each.

21–23. Design a band-pass filter using a single op-amp to achieve a center frequency of 1 kHz and a 3-dB bandwidth of 100 Hz. The capacitances are to be selected as 0.01 μF each.

21–24. Design a band-pass filter using a single op-amp to achieve a center frequency of 500 Hz and a 3-dB bandwidth of 25 Hz. The capacitances are to be selected as 0.02 μF each.

21–25. Design a state-variable Butterworth active filter that can be used as either a low-pass or a high-pass filter with a 3-dB frequency of 2 kHz. Select the filter-tuning resistances at 10 kΩ each. Show the output connections for low-pass or high-pass.

21–26. Repeat the design of Problem 21–25 if the capacitances are selected as 0.01 μF each.

21–27. Design a state-variable two-pole band-pass filter with a geometric center frequency of 2 kHz and a 3-dB bandwidth of 20 Hz. Select the filter-tuning resistances to be 10 kΩ each.

21–28. Using the result of Problem 21–27, design a band-rejection filter with the geometric center frequency of rejection at 2 kHz and a rejection bandwidth of 20 Hz.

22

ANALOG-TO-DIGITAL AND DIGITAL-TO-ANALOG CONVERSION

OBJECTIVES

After completing this chapter, the reader should be able to:

- Define *analog signal, quantization, quantized variable, discrete-time signal, sampled-data signal,* and *digital signal.*
- State Shannon's sampling theorem, and discuss its general significance.
- Define and explain the significance of *aliasing error.*
- Explain the concept and advantages of *multiplexing.*
- Explain the purposes of digital-to-analog (D/A) and analog-to-digital (A/D) converters.
- State the form of a digital word with *n* bits.
- Determine the normalized value of the analog representation of a digital word.
- Define the terms *full-scale voltage* and *one least-significant bit* for a D/A converter.
- Draw the schematic diagrams of several representative D/A converter circuits, and discuss their operation.
- Explain the general operation of and compare the relative speeds of *counter, successive approximation,* and *flash* A/D converters.
- Explain the general characteristics of and differences between *unipolar encoding* and *bipolar encoding.*
- Explain the difference between *rounding* and *truncation.*
- Sketch the forms of the quantization characteristics and of the normalized D/A conversion characteristics for both unipolar and bipolar encoding.
- Discuss the general specifications for an A/D converter, and determine various step error sizes from the specifications.

- Discuss the concept of an analog switch and its general operation.
- Discuss the operation of a voltage-to-frequency converter and its applications.
- Discuss the operation of a frequency-to-voltage converter and its applications.
- Analyze some of the circuits in the chapter with PSPICE.

22–1 INTRODUCTION

The major objectives of this chapter are to develop the theoretical and operational basis for data conversion requirements and to provide details on some of the actual circuits and systems used in data conversion applications. The major types of data conversion processes we will consider are analog-to-digital and digital-to-analog conversion, as reflected in the chapter title.

In Section 22–2, we establish the theoretical basis for representing a signal with a finite number of samples using Shannon's sampling theorem. We also give various definitions in that section.

We present some actual circuit details of representative digital-to-analog converters and then give a general explanation of the operation of and comparison between the different types of analog-to-digital converters. And we explore in detail the specifications required in selecting the various converters to meet prescribed specifications.

As compared to earlier chapters, the reader will find fewer detailed circuit schematic diagrams in this chapter. Instead, we show more of the units in block diagram form. Many of the circuits we have considered thus far are those that can be implemented readily with one or more op-amps and a few components. With the exception of the most elementary types of digital-to-analog converters, however, this procedure is neither desirable nor feasible for most of the circuits in this chapter. Instead, most users will select a complete module off the shelf based on the specified operational performance and acceptable error. It would serve little purpose to try to present complete, detailed schematic diagrams of circuits such as the integrated-circuit analog-to-digital converters. The block diagram approach is much better and explains the general operation.

22–2 SAMPLING AND DATA CONVERSION

As digital technology has advanced in sophistication and components have decreased in cost, many systems that were previously all analog in form have incorporated digital subsystems in their designs. Digital equipment such as microprocessors permit very complex control and computational functions to be achieved through software with associated flexibility and ease of adjustment.

When digital components are used in systems that are predominantly analog in form, it is often necessary to convert data from one form to the other. The processes of analog-to-digital and digital-to-analog conversion are required for this purpose. Before these specific operations are considered, however, it is essential to understand the basic forms of the data representations and the assumptions required for conversion. The definitions that follow are useful in establishing the various conversion processes.

Definitions

An **analog signal** is a signal that is defined over a continuous range of time and in which the amplitude may assume a continuous range of values. The term *analog* apparently originated in the field of analog computation, but it has assumed a very widespread connotation in actual usage.

Quantization describes the process of representing a variable by a finite set of distinct values. A **quantized variable** is one that may assume only distinct values.

A **discrete-time signal** is a function that is defined only at a particular set of values of time. Thus, the independent variable time is quantized. If, however, the amplitude of a discrete-time signal is permitted to assume a continuous range of values, the function is referred to as a **sampled-data signal.** A sampled-data signal could arise from sampling an analog signal at discrete values of time.

A **digital signal** is a function in which both time and amplitude are quantized. A digital signal may always be represented by a sequence of **words** in which each word has a finite number of **bits** (binary digits).

Analog-to-Digital Conversion System

A simplified diagram of a possible data conversion system is shown in Figure 22–1. Some of the preceding definitions will be clarified by a discussion of the operation of this system.

The input to the system is assumed to be an analog signal. It is first processed by a low-pass filter, the function of which will be discussed later. The signal is then sampled at a rate f_s as determined by the clock rate. In the context employed,

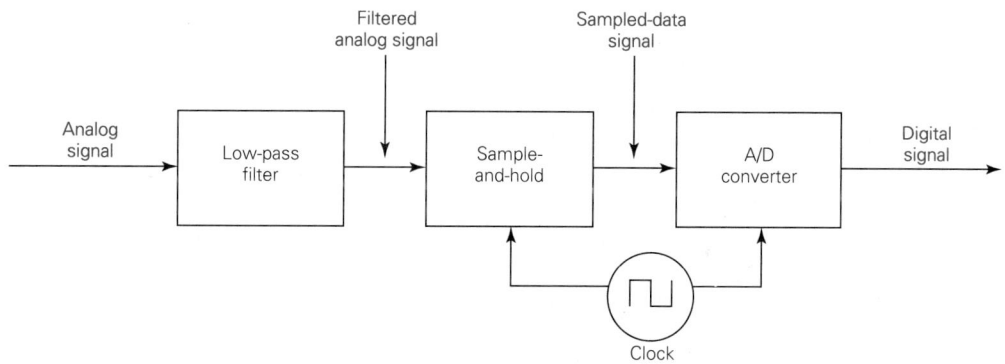

FIGURE 22–1
Block diagram of analog-to-digital conversion system.

1 hertz is considered as 1 sample/second. The sampling period $T = 1/f_s$ is the time interval between successive samples.

The sample-and-hold circuit senses the level of the analog signal once per sampling period and "freezes" that value until the corresponding point in the next sampling period. The signal at the output of the sample-and-hold circuit is a sampled-data analog signal because it can assume any analog level, but new samples are obtained only at discrete intervals of time.

The analog-to-digital (A/D) converter converts the constant analog signal to a digital word as represented by a finite number of bits. Assume that each binary word has n bits. The possible number of levels m is given by

$$m = 2^n \qquad \qquad (22\text{--}1)$$

Only a finite number of levels can be represented by the conversion process. Consequently, amplitude quantization will occur in the A/D converter, with each analog sample being assigned to one of the m possible digital levels by some strategy, of which several common forms will be discussed. The resulting digital values are thus quantized both in time and in amplitude.

The difference between the analog signal amplitude and the nearest standard digital level is called **quantization error** (also called **quantization noise**). In theory, the quantization error can be made arbitrarily small by selecting a sufficient number of bits for each word, but there are practical limits with actual circuits. However, analog systems also contain noise and spurious disturbances, so there are ultimate limits of signal quality that can be attained with both types of systems.

Shannon's Sampling Theorem

A fundamental question of great importance concerns the minimum sampling rate that must be used with the system. In other words, because only a finite number of samples of the analog signal can be taken in a finite time, will information be lost as a result of the sampling operation? The key to this question is **Shannon's sampling theorem,** which establishes the theoretical basis for all discrete sampling operations applied to analog signals.

Assume that the signal has a spectrum containing frequency components extending from near dc to some upper frequency f_u. Shannon's sampling theorem states that the signal can be reproduced by samples taken at intervals no greater than $1/2f_u$ apart. The associated sampling rate f_s must then satisfy

$$f_s \geq 2f_u \qquad \qquad (22\text{--}2)$$

For example, an audio signal extending from dc to 20 kHz could theoretically be reconstructed by taking uniformly spaced samples at a rate of 40,000 samples/second.

In practice, a rate somewhat greater than the theoretical minimum is always used. Operation at the ideal minimum sampling rate would require perfect filters for data recovery. The actual sampling rate is often chosen to be 3 or 4 (or more) times the highest frequency in order to ease the burden of signal recovery.

Aliasing Error

A signal sampled at a rate lower than $2f_s$ will suffer from **aliasing error,** a phenomenon in which frequencies appear to be different from their true values and the signal can never be correctly recovered. A very common example of aliasing is the apparently backward rotation of the wagon wheels in western movies resulting from the inability of the frame rate of the camera to sample the rotation of the wheels at a sufficiently high rate.

Because it is not always possible to define the exact frequency content of random data signals, a common practice is to pass the analog signal through a low-pass filter before sampling, as was indicated in Figure 22–1. The filter characteristics are chosen to reject (or at least to strongly attenuate) all components at frequencies equal to or greater than half the sampling rate. Without such a filter, spurious input components outside the frequency range of the signal could be shifted into the signal frequency range through the sampling and aliasing processes.

The digital words at the output of the conversion system are assumed to be transferred to some type of digital system for which processing is desired. Depending on the application, this system could be a general-purpose computer, a microprocessor, or some special-purpose digital hardware. Computational, communications, or control functions could be performed with the data.

Digital-to-Analog Conversion System

At some point in the overall system, it may be necessary to convert the train of digital data back to analog form. A simplified block diagram of a possible data conversion system for this purpose is shown in Figure 22–2. The sequence of digital words is first converted to analog form by a digital-to-analog converter. Depending on whether serial or parallel forms of the digital words are available, some buffering of the data may be required at the input. In some cases, a holding circuit is used to maintain the restored analog level for the duration of a sampling period. The converted signal has abrupt discontinuities and presents a "staircase"

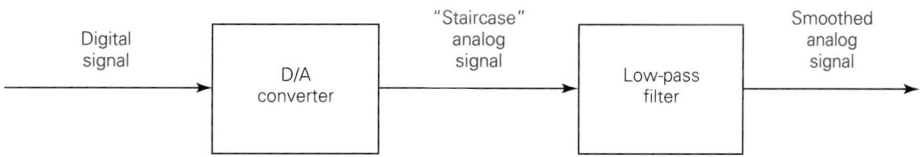

FIGURE 22–2
Block diagram of digital-to-analog conversion system.

type of pattern. The low-pass filter provides a smoothing of the signal by removing high-frequency content.

Multiplexing

The conversion systems of Figures 22–1 and 22–2 imply one input channel and one output channel in each case. However, one of the most important benefits of data conversion systems is that of **multiplexing.** (Specifically, the focus here will be on **time-division multiplexing,** but because it is the only form discussed, the modifier *time-division* will be omitted in subsequent discussions.)

Because the sampling theorem permits a signal to be defined in terms of a discrete set of samples, it is possible to accommodate a number of different data signals with the same processing system by the use of multiplexing. For example, one computer can simultaneously monitor and control a large number of different operations in an industrial process. Each signal is sampled in order, and the computer deals only with that particular operation during the time the sample is being processed. As long as each signal is sampled at a sufficiently high rate, one system effectively performs the functions of many separate systems.

Multiplexers are circuits designed to perform the sequential operation, and analog multiplexing units will be discussed in Section 22–8. However, it is possible to employ separate analog-to-digital conversion modules followed by a **digital multiplexer.** A **demultiplexer** is a unit used for separating a group of multiplexed signals into individual data channels.

EXAMPLE 22–1

A certain data processing system of the form indicated in Figure 22–1 is to be used to process an analog signal whose frequency content ranges from near dc to 2.5 kHz. Assume that the actual sampling rate is selected to be 60% greater than the theoretical minimum. **(a)** Determine the sampling rate and the sampling period. **(b)** If 8-bit words and real-time processing are employed, determine the maximum bit width if a given word is to be transferred serially during the conversion interval of the next word.

Solution

(a) The minimum sampling rate is twice the highest frequency. However, the actual sampling rate f_s should be 1.6 times the minimum rate (60% greater). Thus,

$$f_s = 1.6 \times 2 \times 2500 = 8000 \text{ Hz} \qquad (22\text{–}3)$$

The sampling period is

$$T = \frac{1}{8000} = 125 \ \mu s \qquad (22\text{–}4)$$

(b) The A/D converter must be capable of converting each analog value to a digital word in a time not exceeding 125 μs if real-time processing is employed.

In this time interval, 8 bits from the preceding converted word must be transferred serially from the A/D converter to the remainder of the system. If τ is the bit width, the maximum value is

$$\tau = \frac{125 \ \mu s}{8} = 15.625 \ \mu s \qquad (22\text{–}5)$$

With multiplexed systems, a number of words must be processed during a sampling period if real-time processing is employed. The corresponding conversion time, word lengths, and bit widths must be reduced in inverse proportion to the number of channels.

22–3 DIGITAL-TO-ANALOG CONVERSION

Digital-to-analog conversion is the process of converting a digital word to an analog voltage or current level, with the magnitude and sign of the analog level representing the binary value of the digital word in some sense. A circuit that performs this function is called a **digital-to-analog converter,** which will hereafter be referred to by the common designation **D/A converter.** (Another designation is **DAC.**)

Some of the most common circuit design strategies utilized in D/A converters are presented in this section. Although many D/A converter chips are readily available, the initial emphasis is on establishing the basic principles of operation.

Digital Word

For this discussion, assume a digital word $\hat{X}$ consisting of n bits. The quantity $\hat{X}$ will be represented as

$$\hat{X} = b_1 b_2 b_3 \cdots b_n \qquad (22\text{–}6)$$

where b_1 is the **most significant bit (MSB),** b_2 is the next most significant bit, and so on. The last bit b_n is the **least significant bit (LSB).** A given bit can assume only the two possible states 1 or 0. For example, consider the 4-bit number 1010. We have $b_1 = 1$, $b_2 = 0$, $b_3 = 1$, and $b_4 = 0$. The MSB is 1, and the LSB is 0.

The actual analog value corresponding to a given digital word can be established at any practical level desired. Furthermore, some encoding schemes involve analog levels of both polarities, whereas others are based on a single polarity. These various encoding schemes will be considered in due course, but for the moment, assume that the digital word is defined as a fractional value with the MSB representing the analog weight $0.5 = 2^{-1}$. The next MSB then represents a weight of $0.25 = 2^{-2}$, and so on. Finally, the LSB represents a weight $(0.5)^n = 2^{-n}$.

Normalized Analog Level

Let X represent the decimal (or analog) representation of the binary word $\hat{X}$. The actual value of X can be expressed as

$$X = b_1 2^{-1} + b_2 2^{-2} + b_3 2^{-3} + \cdots + b_n 2^{-n} \tag{22–7}$$

As the number of bits increases, the numerical value approaches unity. Theoretically, it never quite reaches unity, but it can be made to be arbitrarily close if you choose a sufficiently large number of bits. We will refer to the form of X as given in (22–7) as a **normalized level** because it is based on a maximum limiting value of unity.

When a digital word such as $\hat{X}$ in (22–6) is assumed to represent a number less than unity, as given by (22–7), a binary point should precede the digital value. However, as long as this normalized or fractional form is understood, the binary point will be omitted in subsequent work.

A D/A converter must provide an analog output voltage that is a weighted combination of the bits in the digital word. A binary 0 for a given bit should produce no contribution to the output, while a binary 1 should produce a contribution, with the weight a function of the relative significance of the bit.

Linear Combination D/A Circuit

The first D/A circuit that we consider is based on the op-amp summing circuit of Figure 22–3. A reference voltage $-V_r$ is simultaneously applied as the input to an array of switches. In practice, electronic switches are used, and they are controlled by the logic levels of the various bits. (Electronic analog switches are discussed in Section 22–8.) Thus, a logic level of 1 for a given bit connects a given resistance to the source $-V_r$, and a logic level of 0 connects the resistance to ground. In the diagram of Figure 22–3, it is convenient to represent the electronic switching operation by "mechanical" switches.

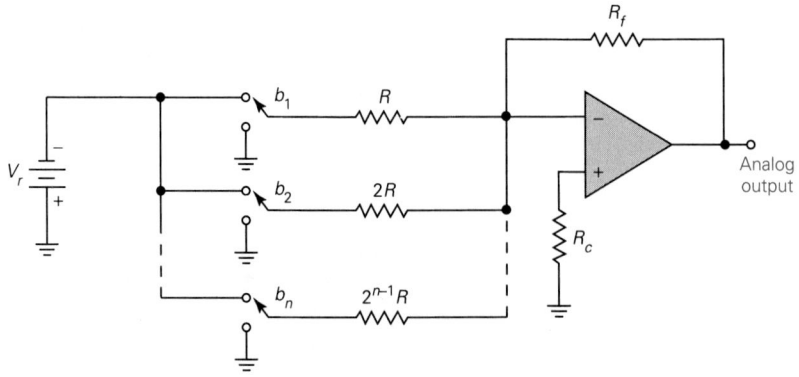

FIGURE 22–3
Digital-to-analog converter implemented with linear combination circuit.

Observe the binary pattern of the various input resistances. The gain for the b_1 signal path is $-R_f/R$, the gain for the b_2 signal path is $-R_f/2R$, and so on. Finally, the gain for the b_n signal path is $-R_f/2^{n-1}R$. Thus, the weighting of the various bits follows a binary pattern with the MSB receiving the greatest weight, and so on. In view of the inversion of the summer, coupled with the negative values of V_r, the output voltage v_o is either positive or zero and can be expressed as

$$v_o = \frac{R_f}{R} b_1 V_r + \frac{R_f}{2R} b_2 V_r + \cdots + \frac{R_f}{2^{n-1}R} b_n V_r \qquad \text{(22–8)}$$

where each b_k multiplier assumes a value of either 0 or 1.

After some factoring and rearrangement, the expression of (22–8) can be simplified to

$$v_o = \frac{2R_f V_r}{R}(b_1 2^{-1} + b_2 2^{-2} + b_3 2^{-3} + \cdots + b_n 2^{-n}) \qquad \text{(22–9)}$$

Observe that the quantity in parentheses is the analog normalized value of the binary word as given by (22–7), and it approaches a maximum value of unity. The factor $2R_f V_r/R$ is a scale-setting factor, and it allows the normalized value of the binary word to be converted to any absolute analog level within the operating limits of the op-amp. This quantity can be defined as the full-scale voltage V_{fs}:

$$V_{fs} = \frac{2R_f V_r}{R} \qquad \text{(22–10)}$$

Note that the term *full-scale* is actually a misnomer as the output voltage never quite reaches the value V_{fs}, but it is a convenient reference value.

The output voltage step between successive levels is often referred to as the value of 1 LSB. This value is determined by setting $b_n = 1$ and all other $b_k = 0$ in (22–9). We have

$$1 \text{ LSB} = \frac{2R_f V_r}{R} 2^{-n} = \frac{V_{fs}}{2^n} \qquad \text{(22–11)}$$

Although this D/A converter circuit is about the simplest type to understand conceptually, it presents some difficulties in practice. Observe that the largest value of input resistance is 2^{n-1} times the smallest value of input resistance. For 4 bits, this ratio is 8, which does not present any difficulties. However, for 8 bits, this ratio is 128, and for 16 bits, the ratio is 32,768! Finding acceptable resistance values with such large ratios is very difficult, if not impossible. Further, variations in resistance due to temperature changes and other factors are very difficult to control with large resistance spreads. In practice, therefore, this D/A converter is limited to noncritical applications where a small number of bits is used.

R-2R D/A Circuit

One of the most widely employed conversion circuits is the **R-2R D/A converter,** whose basic form is shown in Figure 22–4. The designation "R-2R" is rather obvious in that only two resistance values are employed in the input summing circuit, so the problem with an excessive resistance spread is circumvented with the circuit. Special resistance packages containing the R-2R ladder are commercially available.

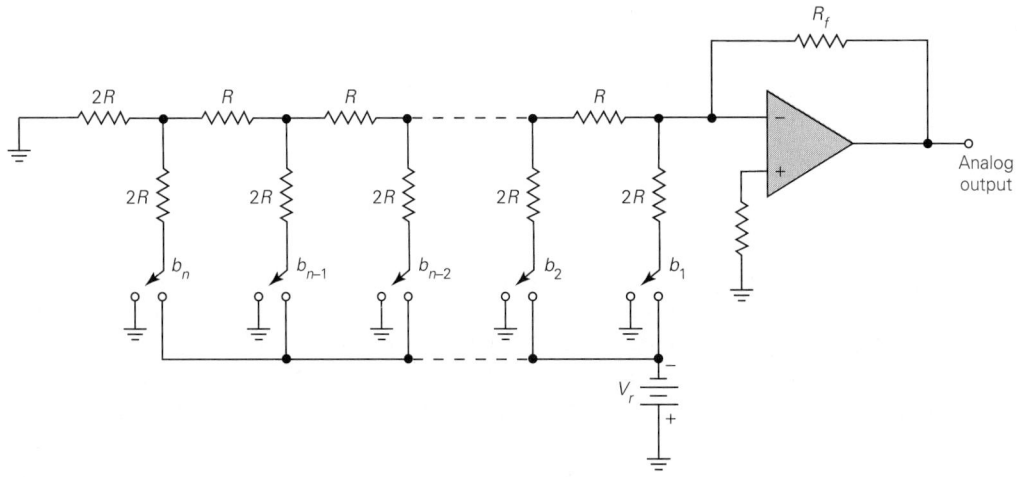

FIGURE 22–4
R-2R digital-to-analog converter.

Because of the interaction of the different branches in the R-2R ladder, it is not immediately evident that the various bit values combine in the proper ratios. Correct circuit verification is best achieved by successive application of Thévenin's theorem. First, the input ladder circuit is replaced by the equivalent circuit shown in Figure 22–5. Because the single voltage source of value $-V_r$ is connected to all the b_k branches, the circuit is equivalent to one having n separate voltage sources connected to the n branches. The source for the kth branch is represented as $-b_k V_r$, where b_k is the bit value (0 or 1) for that particular branch.

Starting with the points a and a', we determine a Thévenin equivalent circuit looking back to the left. This step is illustrated in Figure 22–6(a). The open-circuit voltage is $(1/2) \times b_n V_r/2$ in the direction shown, and the equivalent Thévenin resistance is $2R \| 2R = R$, as shown on the right.

Next, the branch corresponding to b_{n-1} is connected as shown in Figure 22–6(b), and the Thévenin equivalent circuit looking back from points b and b' is determined. The best way to determine the open-circuit voltage at this point is by superposition. Note that the b_{n-1} contribution will have twice the value of the b_n contribution, as expected. The resistance looking back again is $2R \| 2R = R$,

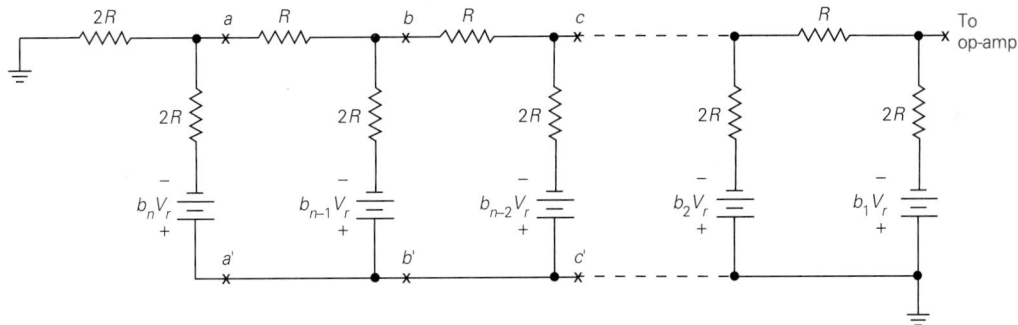

FIGURE 22–5
Equivalent circuit model of R-2R ladder used in analysis, with effect of reference voltage shown in each branch.

as shown on the right. This pattern continues as illustrated for points c and c', as shown in Figure 22–6(c).

After the Thévenin transformation is performed n times, the net equivalent circuit looking back from the op-amp inverting terminal is as shown in Figure 22–6(d). This equivalent circuit may then be considered to be applied to the inverting op-amp input, and the basic gain equation for an inverting amplifier may be used. Since the Thévenin voltage is negative with respect to ground and the gain is negative, the output voltage v_o is positive, and it may be expressed as

$$v_o = \frac{R_f}{R}\left(\frac{b_1 V_r}{2} + \frac{b_2 V_r}{4} + \cdots + \frac{b_n V_r}{2^n}\right) \qquad \textbf{(22–12)}$$

After some factoring and rearrangement, the expression of (22–12) can be simplified to

$$v_o = \frac{R_f V_r}{R}(b_1 2^{-1} + b_2 2^{-2} + \cdots + b_n 2^{-n}) \qquad \textbf{(22–13)}$$

Observe that the quantity in parentheses is the normalized value of the binary word as given by (22–7). The factor $R_f V_r / R$ is a scale-setting factor, and it allows the normalized value of the binary word to be converted to any absolute analog level within the operating limits of the op-amp. As in the case of the previous circuit, this quantity can be defined as the full-scale voltage V_{fs}:

$$V_{fs} = \frac{R_f V_r}{R} \qquad \textbf{(22–14)}$$

The output voltage step between successive levels, which is also denoted as the value of 1 LSB, is determined by setting $b_n = 1$ and all other $b_k = 0$ in (22–13). We have

$$1\ \text{LSB} = \frac{R_f V_r}{R} 2^{-n} = \frac{V_{fs}}{2^n} \qquad \textbf{(22–15)}$$

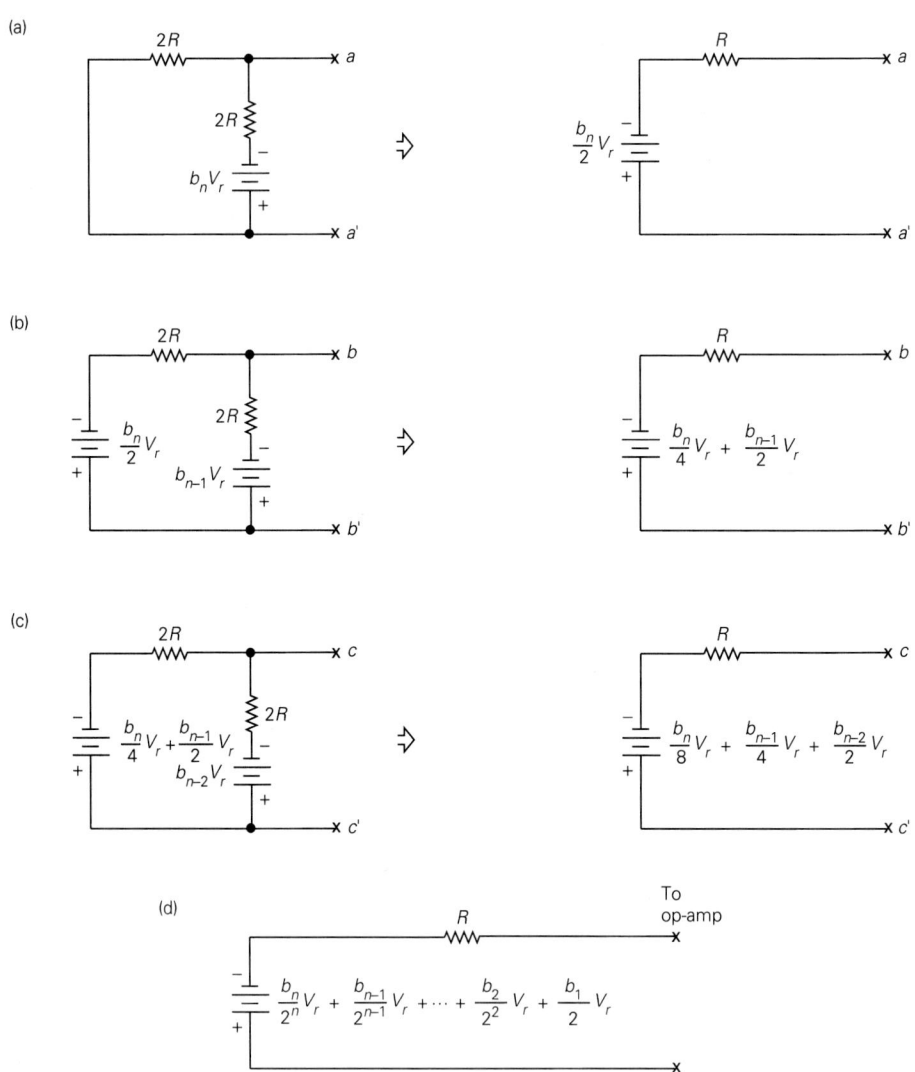

FIGURE 22–6

Successive Thévenin transformations applied to R-2R ladder network.

Comparing (22–15) and (22–14) with (22–11) and (22–10), we note that for given values of R_f, R, and V_r, the full-scale and incremental voltage values for the R-2R D/A converter are one-half the values of those for the first D/A converter considered.

Multiplying D/A Converter

Observe in (22–9) and (22–13) that the output voltage for any particular digital word is directly proportional to the value of the reference voltage V_r. Although many D/A converters employ a fixed stabilized reference voltage, some converters allow the user to apply an external reference voltage to achieve a multiplication effect. Such a unit is designated as a **multiplying digital-to-analog converter (MDAC).** A number of possible applications can be achieved with these converters, including gain control, modulation, and multiplication operations.

Current-Weighted D/A Converter

The R-2R D/A converter previously considered used a voltage source as the basis for the relative weighting. Another widely employed technique uses the relative weighting of constant current sources. This type of unit is referred to as a **current-weighted digital-to-analog converter (IDAC).** Although there are a number of circuits, the concept is illustrated in Figure 22–7. Note that a variation of the R-2R ladder network is used. The individual current sources are typically implemented with an array of bipolar junction transistors (BJTs) and a precision biasing circuit. For an arbitrary bit b_k, the corresponding transistor represents a constant current source when $b_k = 1$, and the transistor is open when $b_k = 0$. If the value of current is I, the corresponding current source may then be represented as $b_k I$.

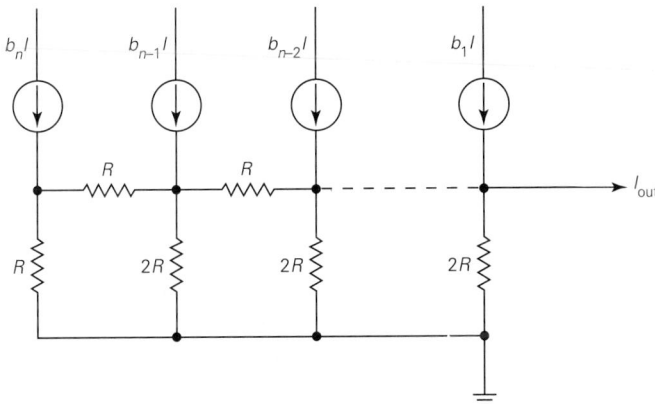

FIGURE 22–7
Basic form of current-weighted D/A converter.

Verification of the operation of this circuit can be achieved by a process very similar to the earlier development of the R-2R ladder with a voltage source reference, and the reader is invited to perform such an analysis. The results of successively reducing the circuit and representing the output in terms of a Norton equivalent circuit are shown in Figure 22–8. The value of the Norton current source I_N is

$$I_N = I[b_1 + b_2 2^{-1} + b_3 2^{-2} + \cdots + b_n 2^{-(n-1)}] \qquad \textbf{(22–16)}$$

A slight rearrangement of (22–16) yields

$$I_N = 2I[b_1 2^{-1} + b_2 2^{-2} + \cdots + b_n 2^{-n}] \tag{22–17}$$

The quantity in brackets is the analog normalized value of the binary word as given by (22–7), and it approaches a maximum value of unity.

The actual output current I_{out} will depend on the equivalent load resistance to which the circuit is attached. If an op-amp inverting current-to-voltage converter is used, the virtual ground permits the entire value of I_N to flow into the current-to-voltage conversion resistance. However, if the resistance is connected directly across the circuit, an appropriate circuit analysis using the model of Figure 22–8 connected to the load must be performed to determine the actual output level. Converters utilizing the current-switching concept will likely be obtained as a package by the user, so the manufacturer's data concerning the output connections should be closely followed.

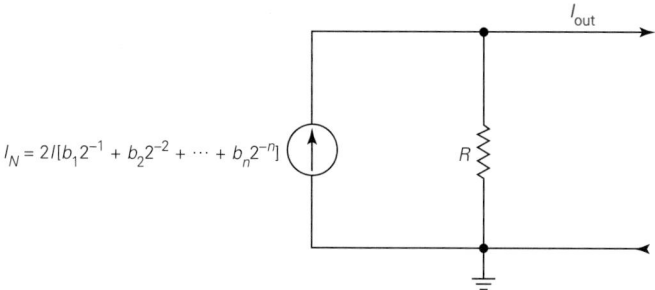

FIGURE 22–8
Norton equivalent circuit of current-weighted D/A converter in Figure 22–7.

An advantage of current-weighted D/A converters is their speed. By biasing the transistors in a linear region of operation, speed problems resulting from saturation can be avoided.

EXAMPLE 22–2 Consider the R-2R 4-bit converter shown in Figure 22–9 in which the resistances in the R-2R ladder and the dc voltage V_r are fixed, but in which R_f is variable. Determine the separate values of R_f that would be required to achieve *each* of the following conditions: **(a)** the value of 1 LSB at the output is 0.5 V; **(b)** a binary input of 1000 results in an analog output of 6 V; **(c)** full-scale output voltage is 10 V; **(d)** the *actual* maximum output voltage is 10 V.

Solution
(a) From (22–15), we set $R = 10$ kΩ, $V_r = 10$ V, and $n = 4$, and we determine R_f such that the value of 1 LSB = 0.5 V. We have

$$\frac{R_f \times 10 \times 2^{-4}}{10^4} = 0.5$$

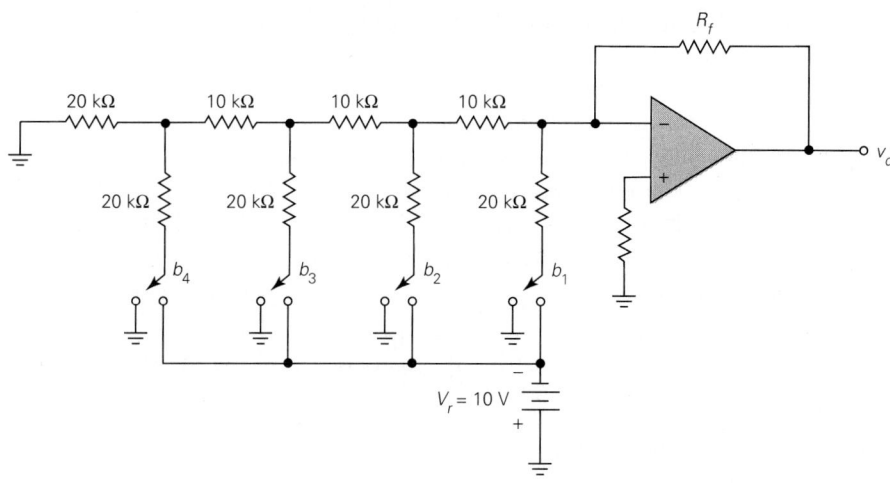

FIGURE 22–9
Four-bit D/A converter.

or

$$R_f = \frac{10^4}{10 \times 2^{-4}} \times 0.5 = 8000 \ \Omega \qquad \text{(22–18)}$$

(b) With a binary value of 1000, the output voltage is determined from (22–13) by setting $b_1 = 1$ and $b_2 = b_3 = b_4 = 0$. Substituting the values of R and V_r, and setting the result equal to 6 V, we have

$$\frac{R_f \times 10 \times 2^{-1}}{10^4} = 6$$

or

$$R_f = \frac{10^4}{10 \times 2^{-1}} \times 6 = 12 \ \text{k}\Omega \qquad \text{(22–19)}$$

(c) From (22–14), we substitute the given parameters and require that $V_{fs} = 10$ V; that is,

$$\frac{R_f \times 10}{10^4} = 10$$

or

$$R_f = \frac{10^4}{10} \times 10 = 10 \ \text{k}\Omega \qquad \text{(22–20)}$$

(d) Recall that the so-called full-scale voltage is never quite reached, so in **(c),** the maximum voltage is actually less than 10 V. In this part, however, the actual output voltage maximum is to be 10 V, and this value will be achieved when the input word has a value of 1111. Setting $b_1 = b_2 = b_3 = b_4 = 1$ in (22–13), we require that

$$\frac{R_f \times 10}{10^4}(2^{-1} + 2^{-2} + 2^{-3} + 2^{-4}) = 10$$

or

$$R_f = \frac{10^4}{10 \times 0.9375} \times 10 = 10.667 \text{ k}\Omega \qquad \textbf{(22–21)}$$

In this case, the "unreachable" full-scale voltage is 10.667 V.

22–4 ANALOG-TO-DIGITAL CONVERSION

Analog-to-digital conversion is the process of converting a sample of an analog signal to a digital word, with the value of the digital word representing the magnitude and sign of the analog level in some sense. A circuit that performs this function is called an **analog-to-digital converter,** which will hereafter be referred to by the common designation **A/D converter.** (Another designation is **ADC.**)

We considered some of the common D/A converter design concepts in the preceding section. The reason that D/A converters were considered first is that the D/A conversion process is notably simpler in both concept and implementation than the process of A/D conversion. Furthermore, most A/D converters employ a D/A converter as one component within the system, so one should understand the process of D/A conversion before attempting to analyze an A/D converter.

Although it is possible for a clever individual to implement an A/D converter using a D/A converter and various logic chips, there is very little incentive for this process with the technology available today. It should be quite difficult for an individual to implement in a short time anything more than the most rudimentary form of an A/D converter, so the use of available A/D converter packages is highly recommended.

In view of their complexity, A/D converters have traditionally been rather expensive. Some of the earlier units, which were implemented using discrete components, cost thousands of dollars. As integrated circuit technology has advanced, however, the price has decreased markedly. Certain A/D converter models are available for less than $10; however, the majority are in the range from tens of dollars to hundreds of dollars.

A/D Converter Types

Some of the most common design strategies used in A/D converters are discussed in this section, including the following types: (1) **counter** (or **ramp**); (2) **successive**

approximation; and (3) **flash.** Because of the complex interaction of the analog and digital portions of an A/D converter, the discussion will be aimed at a block diagram level rather than a detailed circuit operation level. Because the objective here is to convey understanding of how available circuits work rather than to provide information on building a circuit, this approach is felt to be the optimum one.

In discussions of the different circuits, reference will be made to an analog sample at the input. As suggested in Section 22–2, many data conversion systems employ a sample-and-hold circuit at the input to hold the analog value during the conversion process. It is actually possible in some cases to apply the analog signal directly to an A/D converter, but the sample-and-hold circuit is normally used in most data acquisition systems. The various types of A/D converters will next be considered individually.

Counter A/D Converter

A block diagram of a counter A/D converter is shown in Figure 22–10. Assume that the counter is initially set at zero before conversion is started. A sample of the analog signal appears at one input to the comparator, and the counting process is initiated. Each successive step of the counter causes the digital word at the output to advance one level in the binary sequence. Each of these successive digital words is converted back to analog form by the D/A converter, and the output is compared with the analog sample. At the first point at which the D/A converter exceeds the level of the analog sample, the comparator changes states, and by means of the logic gate shown, the counter is inhibited. The value of the digital word at the counter output represents the desired digital representation of

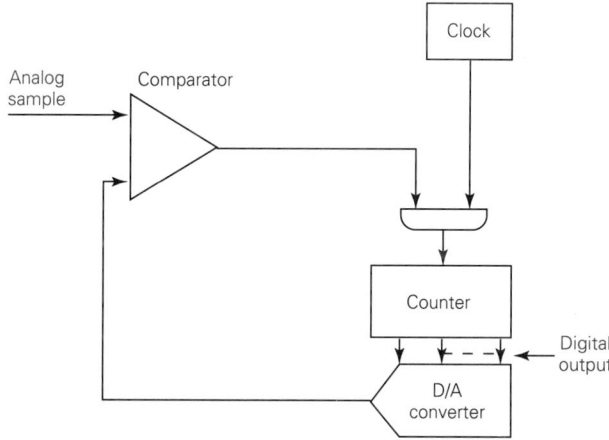

FIGURE 22–10
Block diagram of counter A/D converter.

the analog signal. Depending on the conversion strategy desired, an additional offset voltage in the feedback circuit may be required to establish the exact point for the comparator at which the counter is inhibited. Further discussion of this concept as it relates to A/D specifications will be given in Section 22–5.

Although the counter A/D converter is probably the easiest type to understand conceptually, it is relatively slow for a general-purpose converter. As the counter changes state, settling time must be allowed for the D/A converter, and a comparison must be performed at each step. Sufficient conversion time must be allowed during each conversion cycle for the counter to assume $2^n - 1$ states. As the number of bits increases, the number of states required increases exponentially. For example, if $n = 12$, conversion time must be allowed for $2^{12} - 1 = 4095$ counts. Obviously, only the highest level of the analog signal will necessitate all 4095 counts before the conversion is completed, but with the simplest circuit logic, a total time appropriate for the largest possible value must be allowed.

Certain variations of the counter A/D converter have been devised that make use of the property that a typical analog signal level will not normally change drastically from one sample to the next. With this logic, an up-down counter is used, and the counter follows the change of the analog signal. This type of counter A/D converter is referred to as a **tracking converter.** The tracking converter is faster than the basic counter converter, but the circuit logic is more complex.

Successive Approximation A/D Converter

The most widely employed general-purpose A/D converter is the successive approximation type, for which a block diagram is shown in Figure 22–11. Much of the operation of a successive approximation A/D converter depends on the control logic. Indeed, in some systems employing microprocessors, the control logic is generated by software.

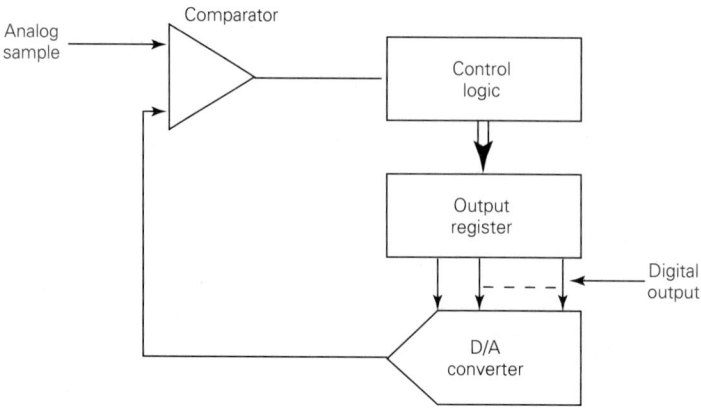

FIGURE 22–11
Block diagram of successive approximation A/D converter.

As in the case of the counter A/D converter, a sample of the analog signal appears at one input to the comparator. The digital word being generated is converted to analog form and compared with the analog sample, and the completion of conversion is triggered by a change in the state of the comparator. However, the logic for the successive approximation type is quite different from that of the counter type, as we will see.

The successive approximation logic is illustrated by the flow chart shown in Figure 22–12. Starting with all zeros, a first "trial" digital value is generated by changing the MSB to 1. The D/A converter converts this to an analog value, and this level is compared with the input analog sample. If the analog sample is less than the trial value, the assumed bit of 1 is rejected and replaced by 0. If, however, the input analog sample is greater than or equal to the trial value, the assumed

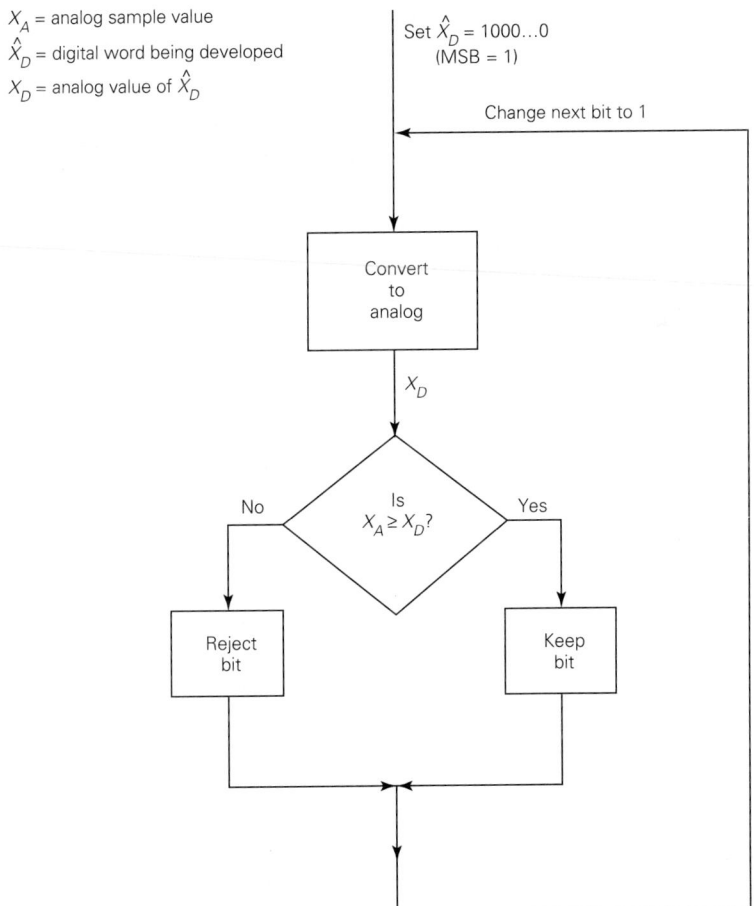

FIGURE 22–12

Flow chart of successive approximation A/D converter.

bit of 1 is retained. The MSB remains at the value established in this part as subsequent comparisons are made.

Focus is next directed to the second most significant bit of the trial value, and this value is changed to 1. The trial value is converted to analog form and compared with the input sample. If the analog sample is less than the converted value, the second assumed bit level of 1 is rejected and replaced by 0. However, if the analog sample is greater than or equal to the converted value, the assumed second bit of 1 is retained.

The preceding process is repeated at all bit locations. The digital word appearing in the output register at the end of the process is the converted digital value. As pointed out for the ramp A/D converter, it may be necessary to add a bias level somewhere in the loop to establish transition points appropriate for a given conversion code.

From the preceding discussion, it is seen that for n bit words, only n comparisons are required for the successive approximation A/D converter. Consequently, the successive approximation type can perform a conversion in a much shorter time than the counter type and is thus capable of operating at higher conversion rates.

Flash A/D Converter

The flash (or parallel) A/D converter has the shortest possible conversion time of any available types. As we will see, however, its use is somewhat limited by practical and economic reasons.

A block diagram of the flash converter is shown in Figure 22–13. For $m = 2^n$ possible values, $m - 1 = 2^n - 1$ comparators are required. The analog sample is simultaneously applied as one input to all comparators. The other input to each comparator is a dc voltage whose level is a function of the location on the resistive voltage divider network on the left. The choices of the resistance $R/2$ at the bottom and $3R/2$ at the top are related to transitions midway between quantization levels in accordance with unipolar encoding, and this will be discussed in Section 22–5.

For a given analog sample level, all comparators below a certain point in the ladder will have one particular state, and those above that point will have the opposite state. This pattern of states is applied to a decoder circuit, which produces the required digital word in accordance with the input pattern. A portion of the decoder circuit could be a read-only memory (ROM), in which the input controls the address, and the output represents the digital value stored in the given location.

The short conversion time is evidenced by the fact that all the comparisons are performed in parallel, so the conversion time is based on the time of one complete comparison. The limitation, however, is the large number of comparators required. For example, an 8-bit converter requires $2^8 - 1 = 255$ comparators. However, a 16-bit converter would require 65,535 comparators! Practical implementations of flash A/D converters appear to have been limited to 8 bits, and these command premium prices; however, conversion times less than 20 ns, corresponding to conversion rates exceeding 50 MHz, are apparently possible with flash converters.

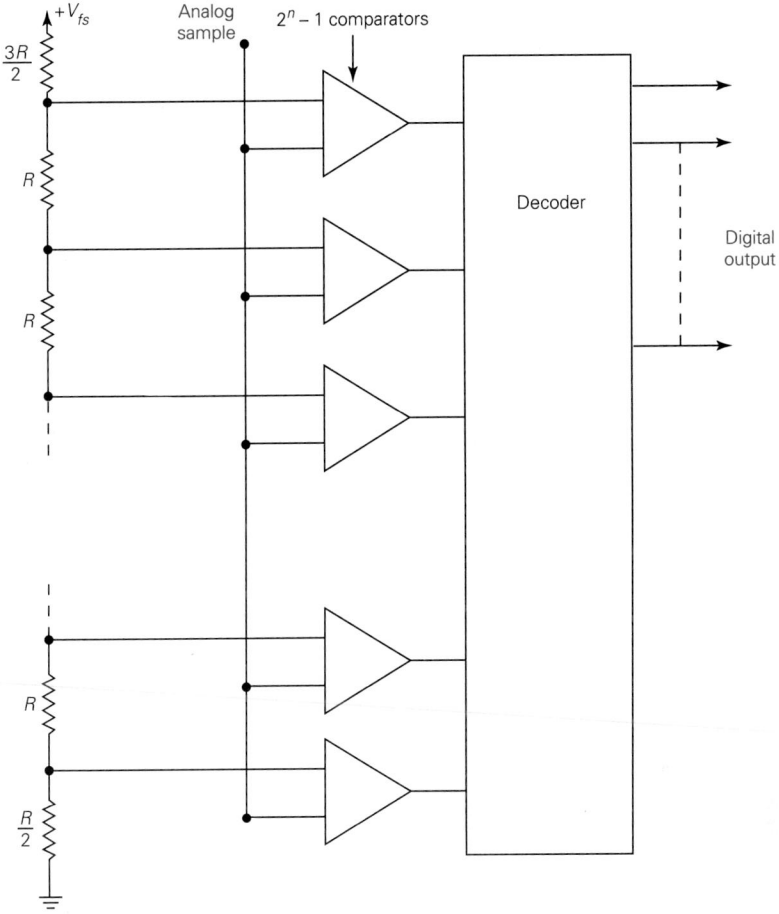

FIGURE 22–13
Block diagram of flash (or parallel) A/D converter.

Comparison

Comparing the three converters considered, we see that the conversion time for each is proportional to the number of comparisons required, and these results are summarized as follows:

Type	Comparisons
Counter (basic form)	$2^n - 1$
Successive approximation	n
Flash	1

The flash type is obviously the fastest, but it is the most expensive (by far), and the number of bits available is limited. The successive approximation type is readily available with a variety of different word lengths and conversion rates, and it is the most widely employed of the types considered here.

EXAMPLE 22–3

Assume that a certain form of integrated-circuit technology allows a comparator operation, amplifier settling, and the associated logic to be performed in about 1 μs. If all other times are assumed to be negligible, compute the approximate conversion times and data conversion rates for the three A/D converters considered in this section for the following word lengths: **(a)** 4 bits; **(b)** 8 bits; and **(c)** 16 bits.

Solution

Recognize that the assumption is based on one type of representative technology and should not be interpreted as any general result. The purpose is to provide a reasonable and meaningful comparison of the relative speeds of the different converter types. Let τ represent the time required for one comparison, and let T_c represent the total conversion time. The following formulas will be assumed:

$$\text{Counter: } T_c = (2^n - 1)\tau \tag{22–22}$$

$$\text{Successive approximation: } T_c = n\tau \tag{22–23}$$

$$\text{Flash: } T_c = \tau \tag{22–24}$$

The conversion rate f_c is determined as

$$f_c = \frac{1}{T_c} \tag{22–25}$$

The results for different values of n are readily calculated and are summarized as follows:

4 bits	T_c	f_c
Counter	15 μs	66.7 kHz
Successive approximation	4 μs	250 kHz
Flash	1 μs	1 MHz

8 bits	T_c	f_c
Counter	255 μs	3.922 kHz
Successive approximation	8 μs	125 kHz
Flash	1 μs	1 MHz

16 bits	T_c	f_c
Counter	65.535 ms	15.259 Hz
Successive approximation	16 μs	62.5 kHz
Flash	1 μs	1 MHz

Observe the extreme range in conversion times and frequencies as the number of bits increases.

22–5 CONVERTER SPECIFICATIONS

The emphasis in this section is directed toward the analysis and interpretation of A/D and D/A converter specifications. The available package units employ many of the various circuit design techniques that have been discussed in the preceding sections. However, the user is often not able to ascertain exactly how the various components within the unit perform their operations. Instead, he or she is concerned more with relating the given specifications to the requirements of the particular system.

Manufacturers differ somewhat in the manner in which they give specifications. The material given here is not intended to represent any particular manufacturer's approach, but it is intended to provide enough background to deal with a number of possible approaches.

Normalized Value

The discussion will utilize the concept of a **normalized value**, which was introduced in Section 22–3. The normalized value can be expressed as

$$\text{Normalized value} = \frac{\text{actual value}}{\text{full-scale value}} \qquad (22\text{--}26)$$

Conversely, the actual value can be expressed as

$$\text{Actual value} = \text{normalized value} \times \text{full-scale value} \qquad (22\text{--}27)$$

The relationships of (22–26) and (22–27) can be applied to either analog or digital values, depending on whether A/D conversion or D/A conversion is being performed. The "full-scale value" will be either a reference voltage or a current, depending on the converter.

Number of Levels

Let n represent the number of bits in each word. The number of levels m is then given by

$$m = 2^n \qquad (22\text{--}28)$$

In some of the discussion that follows, the value $n = 4$ will be used for illustration. This value gives $m = 2^4 = 16$ words, which is sufficiently large to illustrate the general trend but small enough to show the results graphically.

Table 22–1 provides information concerning various binary encoding schemes to be considered for 4 bits. The 16 possible digital words expressed in

TABLE 22–1
Natural binary numbers, decimal values, and unipolar and bipolar offset values for 4 bits

Natural Binary Number	Decimal Value[a]	Unipolar Normalized Decimal Value[b]	Bipolar Offset Normalized Decimal Value[b]
1111	15	$\frac{15}{16} = 0.9375$	$\frac{7}{8} = 0.875$
1110	14	$\frac{14}{16} = 0.875$	$\frac{6}{8} = 0.75$
1101	13	$\frac{13}{16} = 0.8125$	$\frac{5}{8} = 0.625$
1100	12	$\frac{12}{16} = 0.75$	$\frac{4}{8} = 0.5$
1011	11	$\frac{11}{16} = 0.6875$	$\frac{3}{8} = 0.375$
1010	10	$\frac{10}{16} = 0.625$	$\frac{2}{8} = 0.25$
1001	9	$\frac{9}{16} = 0.5625$	$\frac{1}{8} = 0.125$
1000	8	$\frac{8}{16} = 0.5$	0
0111	7	$\frac{7}{16} = 0.4375$	$-\frac{1}{8} = -0.125$
0110	6	$\frac{6}{16} = 0.375$	$-\frac{2}{8} = -0.25$
0101	5	$\frac{5}{16} = 0.3125$	$-\frac{3}{8} = -0.375$
0100	4	$\frac{4}{16} = 0.25$	$-\frac{4}{8} = -0.5$
0011	3	$\frac{3}{16} = 0.1875$	$-\frac{5}{8} = -0.625$
0010	2	$\frac{2}{16} = 0.126$	$-\frac{6}{8} = -0.75$
0001	1	$\frac{1}{16} = 0.0625$	$-\frac{7}{8} = -0.875$
0000	0	0	$-\frac{8}{8} = -1$

[a] Decimal value for integer representation of binary number.
[b] Decimal value with binary point understood on left-hand side of binary number.

natural binary and their corresponding decimal integer values are shown in the left-hand columns. Normalization of the values of the binary words to a range less than 1 is achieved by adding a binary point to the left of the values as given in the table. However, to simplify the notation, the binary point will usually be omitted, but it will be understood in all discussions in which the normalized form is assumed.

We consider two common encoding relationships between the analog and digital values. These are called (1) **unipolar encoding,** and (2) **bipolar offset encoding.** Each process will be discussed separately.

Unipolar Encoding

The unipolar representation is based on the assumption that the analog signal is either positive or zero, but never negative. Let X_u represent the normalized unipolar analog signal, and let $\hat{X}_u$ be the corresponding digital binary representation. In view of the normalized form, the analog level is always bounded by $0 \leq X_u < 1$. Note that the analog function is assumed never to equal unity, although it can approach that limit arbitrarily close.

For $n = 4$, the 16 possible values of X_u are shown in the third column of Table 22–1. The binary value 0000 corresponds to true decimal 0, but the binary value 1111 corresponds to the decimal value $15/16 = 0.9375$, which is less than unity, in accordance with the preceding discussion. The binary value 1000 corresponds to the decimal value 0.5.

Let ΔX_u represent the normalized step size, that is, the difference between successive levels. The step size is also the decimal value corresponding to 1 LSB. The normalized step size for unipolar encoding in general is

$$\Delta X_u = 2^{-n} \tag{22-29}$$

The largest unipolar normalized decimal value $X_u(\max)$ differs from unity by one step size and is

$$X_u(\max) = 1 - 2^{-n} \tag{22-30}$$

Observe that $X_u(\max)$ approaches 1 arbitrarily close as n increases without limit. The digital word whose MSB is 1 and whose other bits are 0s corresponds to the normalized decimal value 0.5. This point will be conveniently referred to as the **half-scale level.** In addition, the value $X_u = 1$ will be referred to as the **full-scale level** even though it is never quite reached.

Bipolar Offset Encoding

The bipolar offset representation is based on the assumption that the analog signal can assume both positive and negative values with nearly equal peak magnitudes. Let X_b represent the normalized bipolar analog signal, and let $\hat{X}_b$ represent the corresponding digital binary representation. For the bipolar case, the analog function is bounded by $-1 \le X_b < 1$. Note that the analog function can assume the value -1, but it can never quite reach the value $+1$.

For $n = 4$, the 16 possible values of X_b are shown in the fourth column of Table 22–1. The binary value 0000 corresponds to the decimal value -1, but the binary value 1111 corresponds to the decimal value $7/8 = 0.875$, which is less than unity. The binary value 1000 corresponds to a decimal value of 0.

Let ΔX_b represent the normalized step size, which is also the value of 1 LSB. Because the normalized range for bipolar offset encoding is assumed to have twice the range as for unipolar encoding, the step size is twice as big and is

$$\Delta X_b = 2^{-n+1} \tag{22-31}$$

The largest bipolar normalized decimal value $X_b(\max)$ differs from $+1$ by one step size and is

$$X_b(\max) = 1 - 2^{-n+1} \tag{22-32}$$

The digital word whose MSB is 1 and whose other bits are 0s corresponds to the decimal value 0.

From the fourth column of Table 22–1 for 4 bits, it is observed that 8 binary values correspond to negative decimal levels, 7 binary values correspond to positive decimal levels, and 1 binary value corresponds to decimal 0. In general, there are $m/2 = 2^{n-1}$ binary words representing negative decimal values, $2^{n-1} - 1$

binary words corresponding to positive decimal values, and one word corresponding to decimal zero.

One significant property of the bipolar offset representation is that it can be readily converted to the digital twos-complement form. To convert to twos-complement form, the MSB of each bipolar offset word is replaced by its logical complement. In fact, the bipolar form is sometimes denoted as the "modified twos-complement" representation. The close relationship between bipolar offset and twos-complement representations provides advantages in interfacing between analog and digital systems where arithmetic operations are to be performed on the digital data.

Before proceeding further, we present some discussion about the step size differences between unipolar and bipolar offset representations. We have defined the normalized ranges for unipolar as 0 to 1 and for bipolar as -1 to 1. Consequently, the step size for bipolar has twice the value as that for unipolar. This interpretation seems natural and in accordance with the manner in which such converters would be applied. However, the internal operation of A/D and D/A converters is generally based on a unipolar concept, and the bipolar operation is achieved by adding a fixed bias level. Thus, a given converter designed to operate with an analog signal level from 0 V to 10 V in unipolar would operate from -5 V to $+5$ V in bipolar if no relative weighting changes were made in the circuitry. To produce this shift, a fixed bias level is added in the circuit.

Because of the actual circuit operational form, some manufacturers prefer to specify error as a function of the peak-to-peak signal range of the converter without regard to whether it is to be used for unipolar or bipolar encoding. This approach is more natural from the standpoint of the circuit operation, but the normalized approach given earlier is more natural from the standpoint of the user who may or may not understand the circuit details.

Although confusing, these differences need not cause any difficulty, provided that we are careful. For unipolar encoding, there should be no ambiguity because the two approaches yield identical results. For bipolar offset encoding, however, we will define the full-scale signal range of the converter as one-half the peak-to-peak signal range, and the normalized value will be defined over the range from -1 to $+1$. The percent error computed using this approach may appear to be twice the value specified by some manufacturers, but it is a result of the difference in the definition and interpretation.

Unipolar Quantization Characteristic

The exact correspondence between the quantized decimal values and the range of input analog levels is represented by a **quantization characteristic curve.** Either **truncation** or **rounding** may be employed. With **truncation,** the encoded value of an analog sample is assigned to the **next lower** quantized level. With **rounding,** the encoded value of an analog sample is assigned to the **nearest** quantized level. For example, assume that two successive digital words correspond to analog levels of 3.1 and 3.2 V, respectively. With truncation, samples of 3.14 and 3.16 V would both be assigned the digital word corresponding to 3.1 V. However, with rounding, the value of 3.14 V would be assigned to the digital word corresponding to 3.2 V. The average error is less with rounding than with truncation, but there are some situations in which truncation is preferred. Rounding will be assumed in subsequent developments unless stated otherwise.

The quantization characteristic of an ideal 4-bit A/D converter employing **rounding** and **unipolar** encoding is shown in Figure 22–14. The actual quantizer relationship is the "staircase" function, and the straight line represents an ideal linear relationship of equality betwen the output and the input. For a normalized

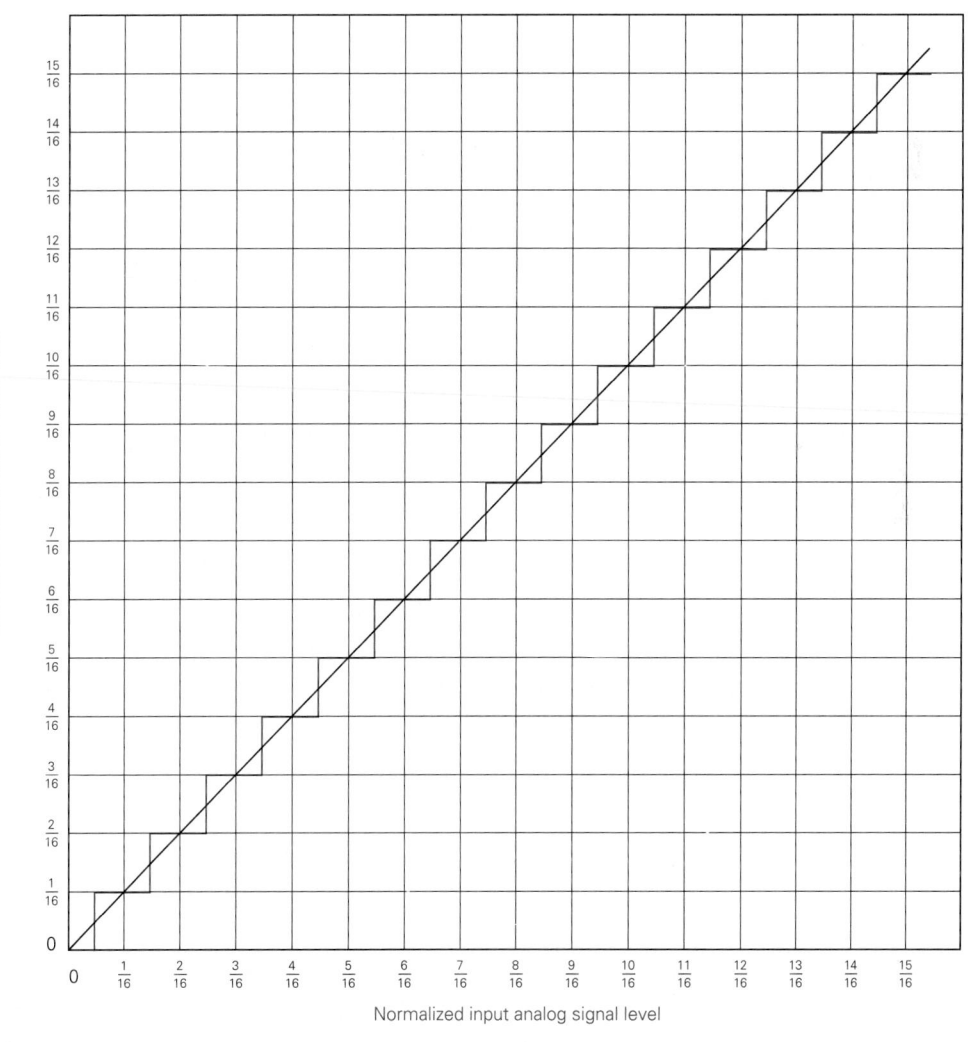

Normalized input analog signal level

FIGURE 22–14
Unipolar quantization characteristic for 4-bit A/D converter.

input analog signal x in the range $x < 1/32$, the digital word is 0000. However, for $1/32 < x < 3/32$, the digital word is 0001. The pattern continues to the maximum range $x > 29/32$, in which the output assumes the binary value 1111.

The vertical difference between the quantizer characteristic and the straight line represents the **quantization error.** For specific values where the curves intersect, there is no quantization error. If we assume that the peak normalized input signal is 31/32, the peak quantization error is 1/32, which is one-half the value of 1 LSB.

In general, for unipolar encoding and rounding, the normalized peak quantization error can be expressed as

$$\text{Normalized quantization error} = \pm 2^{-(n+1)} \quad \text{(22–33)}$$

The actual quantization error in appropriate units is

$$\text{Quantization error} = \pm 2^{-(n+1)} \times \text{full-scale value} \quad \text{(22–34)}$$

Typical full-scale values for common A/D converters are 2.5, 5, 10, and 20 V. The percentage quantization error is

$$\text{Percentage quantization error} = \pm 2^{-(n+1)} \times 100\% \quad \text{(22–35)}$$

The value of the percentage error is the same for either normalized or actual values.

Bipolar Quantization Characteristic

The quantization characteristic of an ideal 4-bit A/D converter employing **rounding** and **bipolar** encoding is shown in Figure 22–15. For a normalized input analog signal x in the range $x < -15/16$, the digital word is 0000. For $-15/16 < x < -13/16$, the digital word is 0001. In the mid-range, for $-1/16 < x < 1/16$, the digital word is 1000. Finally, for $x > 13/16$, the digital word is 1111.

The peak error for the bipolar case is now 1/16 on a normalized basis. As explained earlier, this value is a result of the fact that the peak-to-peak range is assumed to be twice as large as for unipolar encoding.

With bipolar offset encoding and rounding, the normalized peak quantization error can be expressed as

$$\text{Normalized quantization error} = \pm 2^{-n} \quad \text{(22–36)}$$

The actual quantization error is

$$\text{Quantization error} = \pm 2^{-n} \times \text{full-scale value} \quad \text{(22–37)}$$

Recall that the full-scale value for bipolar offset encoding is defined as one-half the peak-to-peak range in this development. The percentage quantization error is

$$\text{Percentage quantization error} = \pm 2^{-n} \times 100\% \quad \text{(22–38)}$$

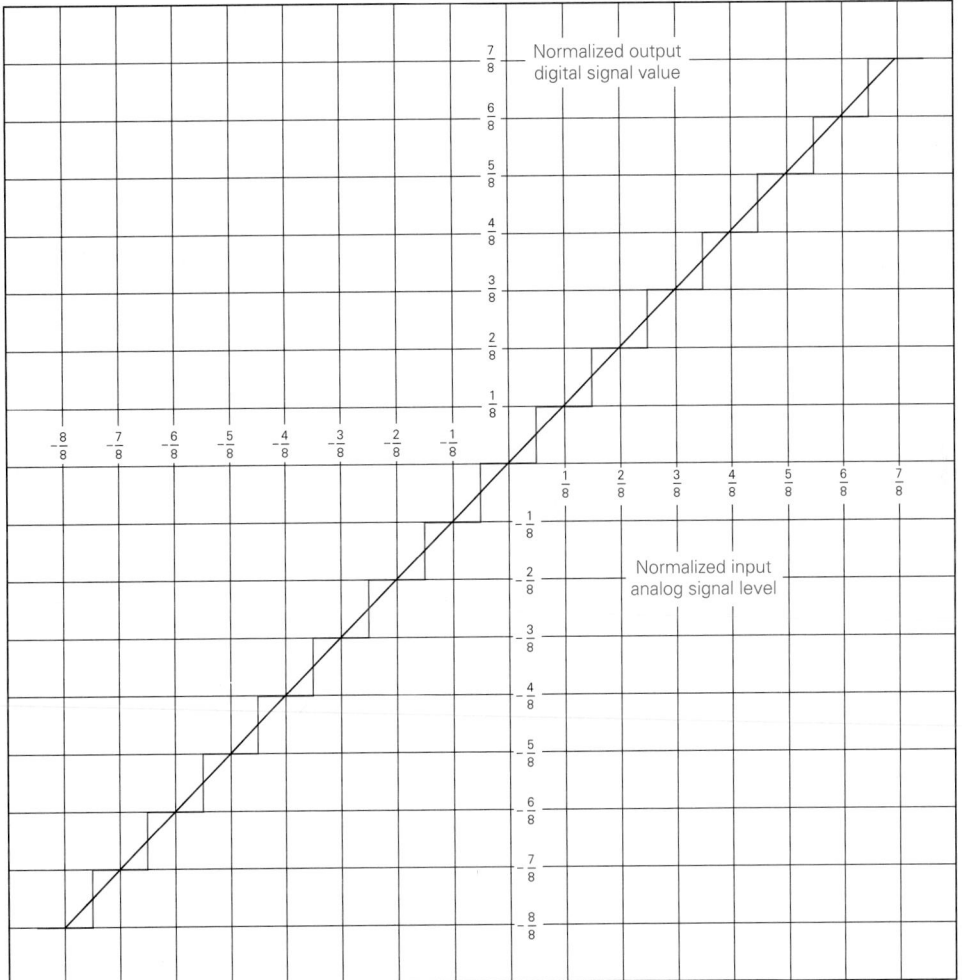

FIGURE 22–15
Bipolar offset quantization characteristic for 4-bit A/D converter.

This percentage is referred to as the peak value rather than the peak-to-peak value.

D/A Unipolar Conversion Characteristic

Having observed the quantization characteristics for A/D converters, we next investigate the corresponding characteristics for D/A converters. Using 4-bit words for illustration again, the normalized characteristics for unipolar encoding are shown in Figure 22–16. The 16 digital word values ranging from 0000 to 1111 are indicated along the horizontal scale, and the corresponding normalized

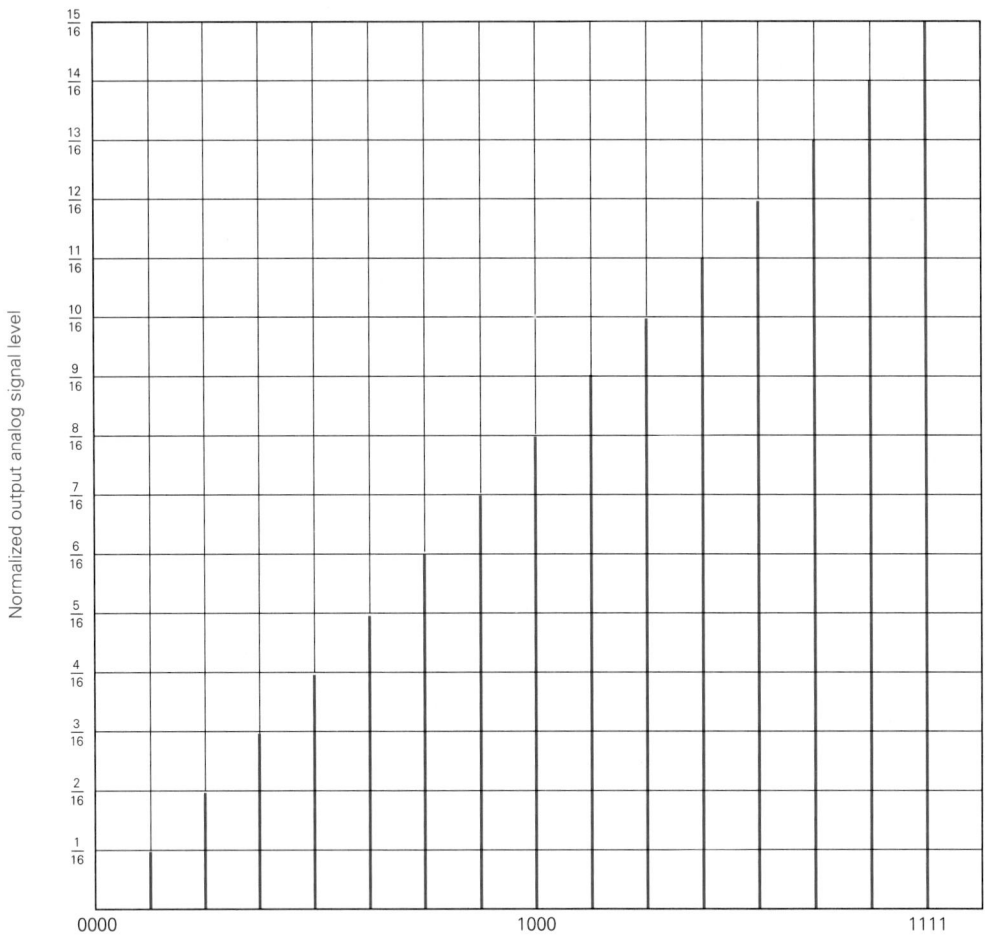

FIGURE 22–16
Normalized unipolar D/A conversion characteristic.

analog values are shown on the vertical scale. Unlike the A/D conversion, the D/A conversion process is unique; that is, there is a one-to-one correspondence between a given digital word and an analog value once the D/A circuit is established. However, there are gaps between levels, arising from the existing quantized states.

Observe that binary value 0000 corresponds to an analog level of zero, binary 0001 corresponds to a normalized analog level of 1/16, and so on. Finally, the binary value 1111 corresponds to a normalized analog level of 15/16.

In view of the abrupt changes between successive levels, filtering and smoothing of the restored analog signal are normally employed. In many systems, a holding circuit is employed to maintain a given analog level until the next conversion is performed. Further smoothing by a low-pass filter is then used.

D/A Bipolar Conversion Characteristic

The normalized D/A characteristics for bipolar offset encoding with 4 bits are shown in Figure 22–17. Observe that binary 0000 now corresponds to a normalized analog level of -1, binary 0001 corresponds to a normalized analog level of $-7/8$, and so on. At the mid-range, binary 1000 corresponds to an analog level of 0. Finally, binary 1111 corresponds to an analog example of 7/8.

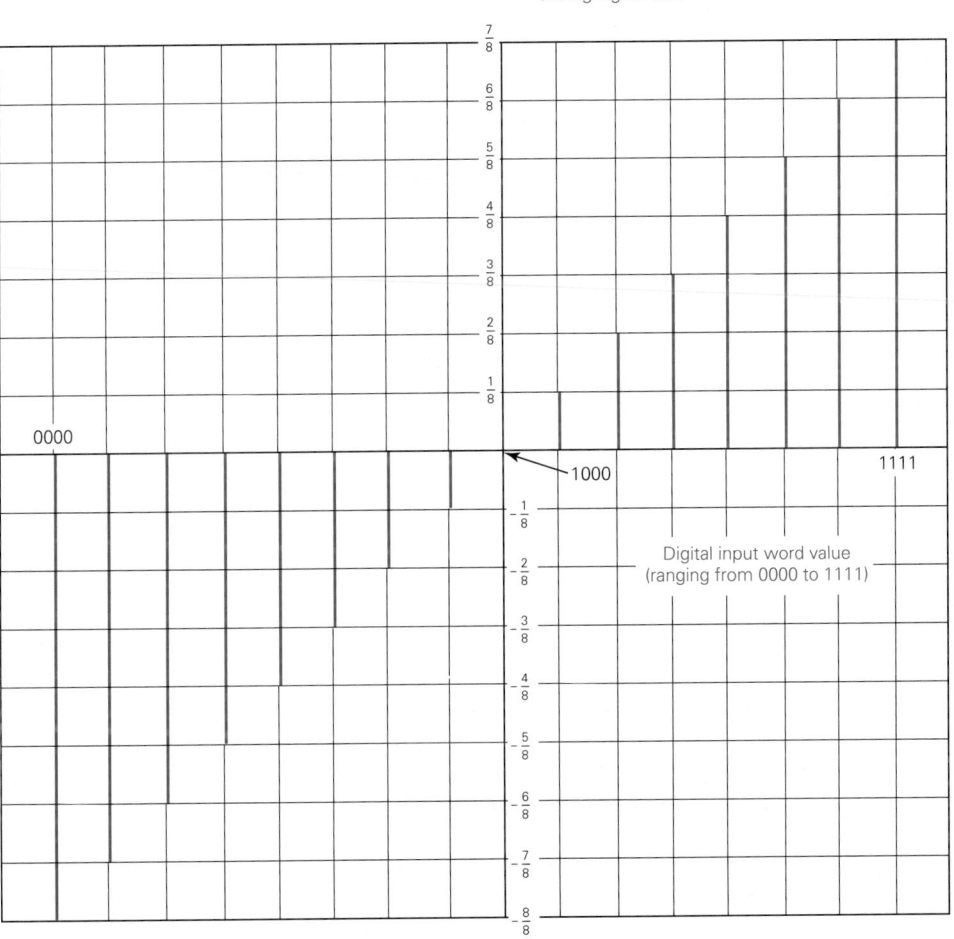

FIGURE 22–17
Normalized bipolar offset D/A conversion characteristic.

EXAMPLE 22–4

A certain 8-bit A/D converter connected for unipolar encoding (with rounding) has a full-scale voltage of 10 V. Determine the following quantities:

(a) Normalized step size

(b) Actual step size in volts

(c) Normalized maximum quantized level

(d) Actual maximum quantized level in volts

(e) Normalized peak quantization error

(f) Actual peak quantization error in volts

(g) Percentage quantization error

Solution

The desired quantities may be determined from the unipolar relationships of this section. The full-scale voltage is 10 V. The various results are developed in the steps that follow.

(a) The normalized unipolar step size is

$$\Delta X_u = 2^{-n} = 2^{-8} = 0.003906 \tag{22–39}$$

(b) Actual step size = $0.003906 \times 10 = 39.06$ mV $\tag{22–40}$

(c) The normalized maximum quantized level is

$$X_u(\max) = 1 - 2^{-n} = 1 - 2^{-8} = 0.9961 \tag{22–41}$$

(d) Actual maximum quantized voltage = $0.9961 \times 10 = 9.961$ V $\tag{22–42}$

(e) Normalized peak quantization error =

$$\pm 2^{-(n+1)} = \pm 2^{-9} = \pm 0.001953 \tag{22–43}$$

(f) Actual peak quantization error = $\pm 0.001953 \times 10 = \pm 19.53$ mV $\tag{22–44}$

(g) Percentage quantization error =

$$\pm 2^{-(n+1)} \times 100\% = \pm 2^{-9} \times 100\% = \pm 0.1953\% \tag{22–45}$$

EXAMPLE 22–5

The 8-bit A/D converter of Example 22–4 is next connected for bipolar offset encoding with the same peak-to-peak voltage range as in Example 22–4. (This range is achieved by adding a positive internal bias of 5 V to the signal before conversion.) The converter is now intended for signals in the range from −5 V to just under +5 V. Repeat all the calculations of Example 22–4 for this case.

Solution

For bipolar offset encoding, the normalized range is assumed to be from -1 to $+1$. The full-scale voltage is now assumed to be 5 V. The results are developed in the steps that follow.

(a) The normalized bipolar step size is

$$\Delta X_b = 2^{-n+1} = 2^{-7} = 0.007813 \qquad (22\text{–}46)$$

(b) Actual step size $= 0.007813 \times 5 = 39.06$ mV $\qquad (22\text{–}47)$

Comparing (22–46) with (22–39) and (22–47) with (22–40), we note that the normalized step size for bipolar is twice the value as for unipolar, but the actual step size is the same. The equality of the actual step sizes is a result of the same peak-to-peak voltage range, but the differences in the normalized steps result from the way the normalized range is defined.

(c) The normalized maximum quantized level is

$$X_b(\text{max}) = 1 - 2^{-n+1} = 1 - 2^{-7} = 0.9922 \qquad (22\text{–}48)$$

(d) Actual maximum quantized voltage $= 0.9922 \times 5 = 4.9610$ V $\qquad (22\text{–}49)$

(e) Normalized peak quantization error $= \pm 2^{-n} = \pm 2^{-8} = \pm 0.003906$ $\qquad (22\text{–}50)$

(f) Actual peak quantization error $= \pm 0.003906 \times 5 = \pm 19.53$ mV $\qquad (22\text{–}51)$

(g) Percentage quantization error $= \pm 2^{-n} \times 100\% = \pm 0.3906\%$ $\qquad (22\text{–}52)$

Comparing (22–50), (22–51), and (22–52) with (22–43), (22–44), and (22–45), respectively, we note that the actual peak quantization error for bipolar is the same value as for unipolar, but the normalized and percentage quantization errors for bipolar are twice the values for unipolar. The percentage comparison is based on the peak (rather than the peak-to-peak) value. Thus, while the **actual** error is the same, the **relative** error is twice as big.

22–6 ADC0801-0805 A/D CONVERTERS

This section introduces one group of popular economical integrated-circuit A/D converters. The group begins with the ADC0801 and continues through the ADC0805. We will refer to any one of the group or the entire group simply by the number 801. These units are manufactured by National Semiconductor Corporation.

The 801 units are all 8-bit successive approximation A/D converters. The basic operation of all units is similar, but the different identification numbers relate to the accuracy specifications. These A/D converters are compatible with the 8080 microprocessor and its subsequent modifications. Thus, they are ideal for applications in which sampled data are to be processed with a microprocessor.

The 801 converters are designed to operate with a 5-V peak-to-peak analog signal range. Consequently, if the analog signal has a larger range, it is necessary to attenuate the signal before conversion. Likewise, bipolar offset encoding requires the addition of a 2.5-V reference to the analog signal, which would have to be limited from −2.5 to 2.5 V.

The 801 units have a clock generator on the chip. The speed of the clock can be varied within some limits. Accuracy of the process is assured at clock frequencies up to 640 kHz. It takes a minimum of 66 clock periods to perform a conversion. This means that the minimum conversion time is about $66 \times 1/(640 \times 10^3) = 103 \ \mu s$. This conversion time corresponds to a sampling frequency of about 9697 Hz. On the other hand, the maximum number of clock periods is 73, which corresponds to a conversion time of about $73 \times (1/640 \times 10^3) = 114 \ \mu s$. The sampling frequency in this case is about 8767 Hz.

Free-Running Connection

The 801 converters may be synchronized with an external signal (e.g., a microprocessor), or they may be operated in a free-running mode. A typical free-running connection, based on information provided in data sheets, is shown in Figure 22–18. The choice of $V_{CC} = 5.12$ V results in convenient values for the various steps, as will be demonstrated in Example 22–6. (The absolute maximum value permitted for V_{CC} is 6.5 V.) The circuit as shown in connected for application with a unipolar analog signal ranging from zero to about 5.12 V.

To ensure that the conversion process starts, the WR and INTR controls are momentarily grounded. This is achieved with the switch on the left. Note that the circuit contains both an A (analog) ground and a D (digital) ground. However, these two separate grounds should be connected together.

The voltage appearing at the $V_{ref}/2$ terminal is half the supply voltage. Thus, for $V_{CC} = 5.12$ V, a voltage of 2.56 V appears at this terminal.

The 8 bits representing the converted digital words are denoted as DB0 through DB7. The most significant bit is DB7, and the least significant bit is DB0.

EXAMPLE 22–6

Consider the 801 A/D converter connection of Figure 22–18. **(a)** Let v_1, v_2, v_3, . . . , v_8 represent the eight actual analog (or decimal) unipolar voltage levels corresponding to the eight digital words 00000001, 00000010, 00000100, . . . , 10000000. Calculate these voltages. **(b)** Let v_{t1}, v_{t2}, v_{t3}, . . . , v_{t8} represent the eight corresponding actual unipolar lower transition voltages at which the previous digital words will be generated. Calculate these voltages.

Solution

(a) For each of the eight voltages being calculated in this part, only one bit has a level of 1, and the other seven have a level of 0. Starting with v_1, this voltage is simply the value corresponding to 1 LSB and is

$$v_1 = 5.12 \times 2^{-8} = 20 \ \text{mV} \qquad (22\text{–}53)$$

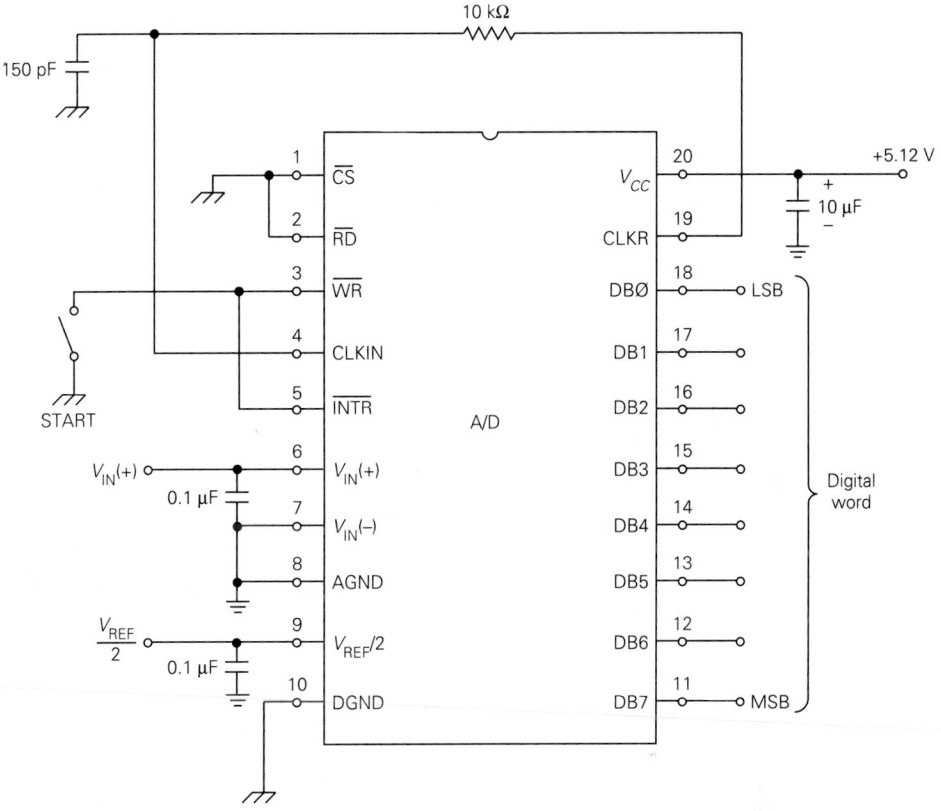

FIGURE 22–18
Typical free-running connection for 801 A/D converters. (Courtesy of National Semiconductor, Inc.)

The value of v_2 is

$$v_2 = 5.12 \times 2^{-7} = 2v_1 = 40 \text{ mV} \qquad \textbf{(22–54)}$$

The value of v_3 is

$$v_3 = 5.12 \times 2^{-6} = 2v_2 = 80 \text{ mV} \qquad \textbf{(22–55)}$$

The pattern continues in this fashion until all eight values are calculated. These values are summarized in the first row of the table following part **(b).**

(b) The transition on the low side for each of the voltages calculated in **(a)** occurs at a level of 1/2 LSB below the given voltage level. Thus, in general

$$v_{tk} = v_k - \frac{1}{2}\text{LSB} = v_k - 10 \text{ mV} \qquad \textbf{(22–56)}$$

The eight values are readily calculated and are given on the second row of the table that follows.

	$k = 1$	2	3	4	5	6	7	8
v_k	20 mV	40 mV	80 mV	160 mV	320 mV	640 mV	1.28 V	2.56 V
v_{tk}	10 mV	30 mV	70 mV	150 mV	310 mV	630 mV	1.27 V	2.55 V

22–7 DAC0806-0808 D/A CONVERTERS

An introduction to an economical group of integrated-circuit D/A converters is given in this section. This group consists of the DAC0806, the DAC0807, and the DAC0808 converters. We will refer to any one or the group simply by the number 806. These converters are manufactured by National Semiconductor Corporation, and complete data sheets appear in data manuals.

The 806 units are 8-bit converters and can interface directly with TTL, DTL, or CMOS logic levels. Dual power supply biasing is normally used, and the required ranges are from ±4.5 V to ±18 V.

Typical Connection

A typical circuit diagram of a D/A conversion system utilizing an 806 converter is shown in Figure 22–19. The particular scheme shown provides a unipolar positive output voltage. The output current is converted to a voltage through the current-controlled voltage source below the converter.

Assume that $R_{14} = R_{15} = R$. The full-scale voltage V_{fs} for this configuration is given by

$$V_{fs} = \frac{R_f}{R} V_{\text{REF}} \tag{22–57}$$

In practice, one of the two resistances can be adjusted to provide a more exact scaling relationship.

EXAMPLE 22–7

Consider the D/A connection of Figure 22–19, and assume that $R_{14} = R_{15} = R = R_f = 10$ kΩ and $V_{\text{REF}} = 5.12$ V. Determine the full-scale voltage.

Solution
The full-scale voltage is

$$V_{fs} = \frac{R_f}{R} V_{\text{REF}} = \frac{10 \text{ k}\Omega}{10 \text{ k}\Omega} \times 5.12 \text{ V} = 5.12 \text{ V} \tag{22–58}$$

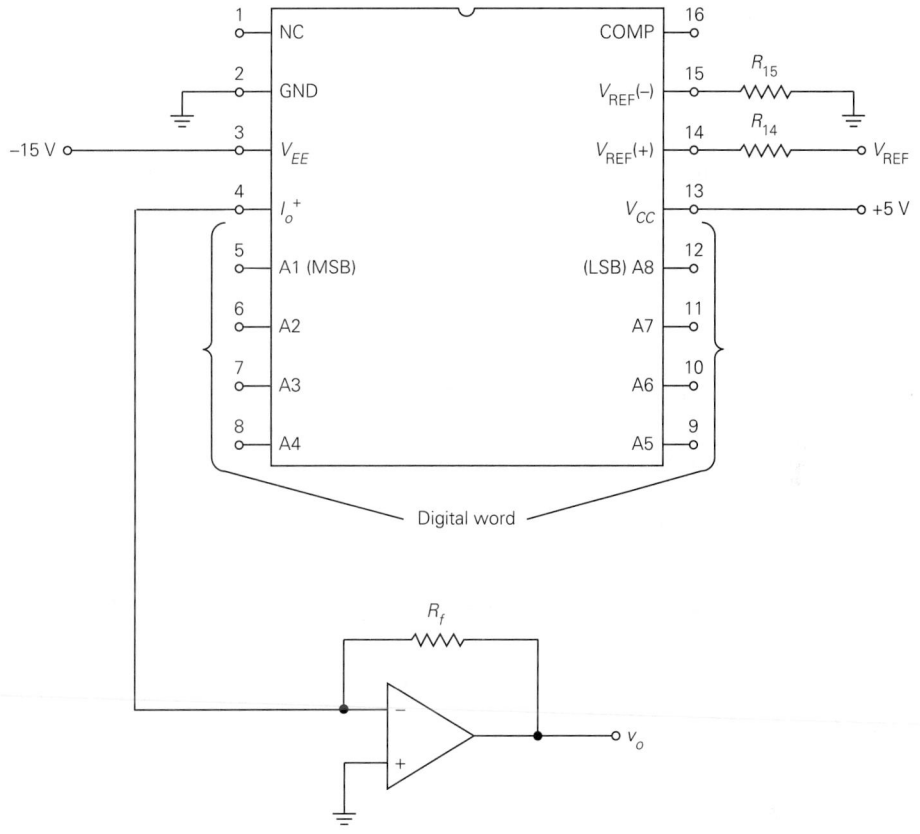

FIGURE 22–19
Typical connection for 806 D/A converter. (Courtesy of National Semiconductor, Inc.)

This particular circuit design matches the A/D design of Example 22–6 in the sense that the output analog levels correspond to the mid-range values of the inputs to the A/D converter.

22–8 ANALOG SWITCHES AND MULTIPLEXERS

Analog switches are components in which an "open" or a "closed" condition between two points can be achieved by means of a control signal without any moving parts. An ideal analog switch would perform electronically the same operation as a mechanical relay. Practical analog switches exhibit a small forward resistance in the "on" state and some leakage in the "off" state. Earlier in the text, references were made to electronic switches in several applications.

Most analog switches utilize field effect transistor (FET) technology. Some employ only junction field effect transistors (JFETs), and others consist of hybrid combinations of JFETs and metal oxide semiconductor field effect transistors (MOSFETs). Still others employ complementary (CMOS) forms. The detailed consideration of the different forms of analog switching is the subject of at least one complete book.

An **analog multiplexer** is a combination of several switches for the purpose of transmitting or processing samples of two or more analog signals in sequential order (serial form) on the same line. Although such a unit could be implemented with individual switches, multiplexers are available containing a number of switches with common characteristics.

For a given type of analog switch or multiplexer, there is a certain range of input signal in which the unit will behave properly as a switch. Consequently, a switch or multiplexer must be carefully selected to provide a proper range of operation for a given signal condition. Other parameters of concern are the switching times (both off and on), "on" resistance, "off" leakage current, and the control signal requirements.

P-Channel JFET Switch

The switch concept will be illustrated with one representative type, a P-channel JFET. The basic form of such a switch is illustrated in Figure 22–20(a). This simplified form is intended to function only with either positive signals or with a virtual ground form, as will be illustrated.

FIGURE 22–20
JFET P-channel analog switch with off and on models.

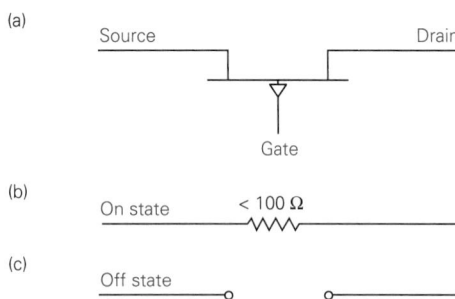

(a)

Source   Drain

Gate

(b) On state $< 100\ \Omega$

(c) Off state

The P-channel type is a useful form because it is compatible with standard transistor-transistor logic (TTL) levels. A TTL 1 applied to the gate turns off (opens) the switch for certain signal levels, and a TTL 0 applied to the gate turns on (closes) the switch. Simplified circuit forms for on and off conditions are illustrated in Figures 22–20(b) and (c). The "on" resistance is typically less than $100\ \Omega$.

4-Channel Multiplexer

A 4-channel multiplexer composed of *P*-channel switches connected in an appropriate multiplexing circuit is illustrated in Figure 22–21. This circuit employs the concept of summing at the virtual ground level, an acceptable condition to ensure proper operation of the JFET switches. Observe the extra JFET connected in the feedback circuit. This unit adds a resistance to the feedback resistance approximately equal to the value contributed by each FET in the input circuit, and the result is a more accurate gain realization. Channel selection is achieved by a digital circuit in which a TTL 0 is sequentially applied to each of the four gates while 1s are applied to the other three at a given time.

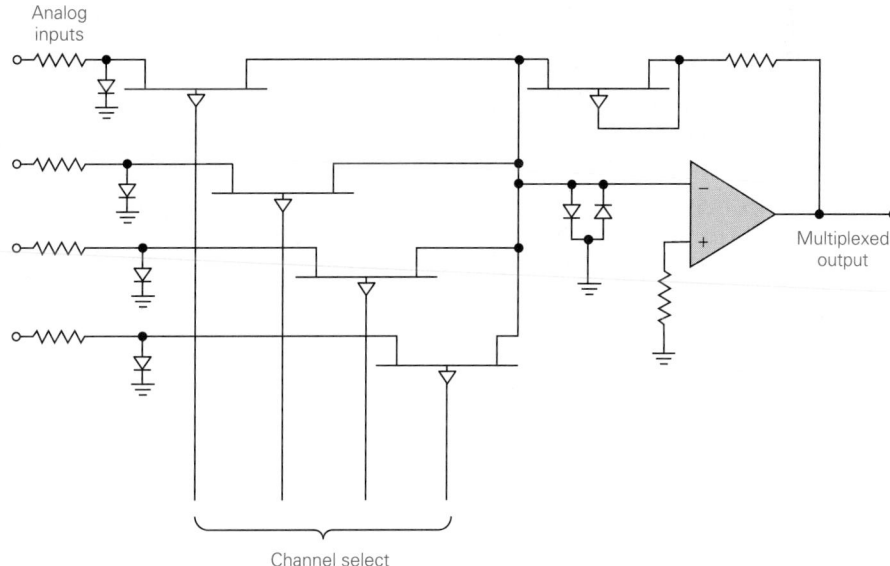

FIGURE 22–21
Circuit utilizing 4-channel JFET *P*-channel multiplexer.

22–9 VOLTAGE AND FREQUENCY CONVERSIONS

Two special types of data conversion circuits are discussed in this section. They are the (1) **voltage-to-frequency (V/F) converter** and (2) the **frequency-to-voltage (F/V) converter.** Some available chips can perform either operation, but the functions are discussed separately in this section. The emphasis here is on the external operating characteristics rather than on detailed circuit considerations.

Voltage-to-Frequency Converter

A V/F converter produces an output signal whose instantaneous frequency is a function of an external control voltage. The output signal may be a sine wave, a square wave, or a pulse train. For the latter case, the instantaneous frequency is often interpreted as the number of pulses generated per unit time.

The resulting frequency variation with signal is a form of frequency modulation (FM), and some of the technology developed in communications may be applied to V/F conversion. The term *voltage-controlled oscillator* (*VCO*) originated with FM, and this term is synonymous with *V/F converter*.

The frequency of many oscillator circuits can be varied by application of a control voltage at certain points. However, any circuit that is to be considered seriously for use as a V/F converter must display the proper characteristics; thus, it requires some careful design.

The ideal conversion characteristic of a V/F converter is shown in Figure 22–22. The input control voltage is shown on the horizontal scale, and the frequency deviation is shown on the vertical scale. The frequency deviation is the difference between the actual frequency and the frequency with no external control signal applied. The ideal curve is linear. For example, assume that a V/F converter has a frequency of 1 kHz with no signal applied. If an input signal of 1 V causes the frequency to shift to 1050 Hz, a signal of 2 V should cause the frequency to shift to 1100 Hz if the curve is linear. The first case corresponds to a deviation of 50 Hz, and the second case corresponds to a deviation of 100 Hz.

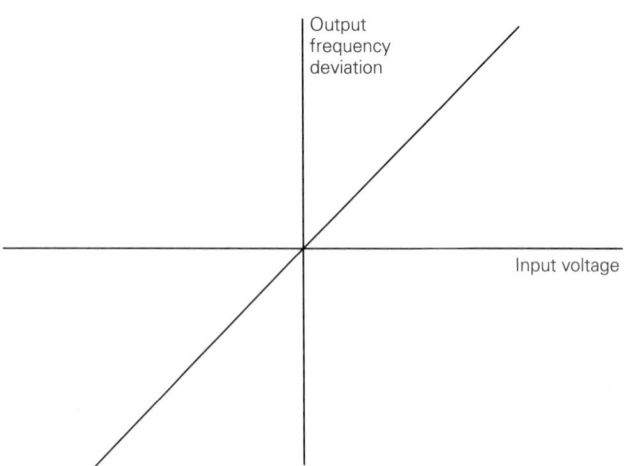

FIGURE 22–22
Ideal V/F converter characteristic.

Frequency-to-Voltage Converter

An F/V converter produces an output voltage whose amplitude is a function of the frequency of the input signal. Like the V/F converter, the unit may respond to a sine wave, a square wave, or a pulse train. Such a circuit is essentially an FM discriminator or detector.

The ideal conversion characteristic of an F/V converter is shown in Figure 22–23. The input frequency deviation (referred to some reference center frequency) is shown on the horizontal scale. The output voltage is shown on the vertical scale. The ideal curve is again linear. For example, assume that the output voltage is zero with an input frequency of 1 kHz. If an input frequency of 1050 Hz (corresponding to a 50-Hz deviation) causes the output voltage to be 1 V, an input frequency of 1100 Hz (a 100-Hz deviation) should cause the output voltage to be 2 V if the conversion is linear.

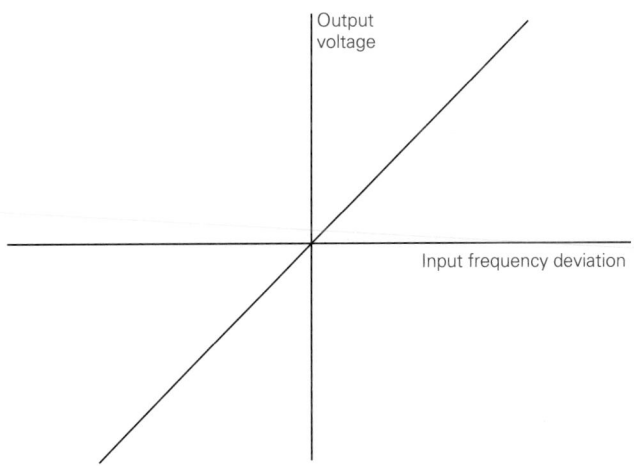

FIGURE 22–23
Ideal F/V converter characteristic.

Applications

In one sense, V/F conversion with a pulse train can be thought of as a form of A/D conversion. This interpretation is based on the fact that pulses assume only two states, and the number of pulses produced per unit of time is a function of the signal level. Voltage-to-frequency conversion is particularly useful for conversion systems in which the data are to be transmitted over some reasonable distance, as the output is in a modulated form. Thus, there is no need to transmit a dc level, and various ac coupling methods, such as transformers and ac amplifiers, may be employed. At the receiving end, an F/V converter matched to the V/F converter at the sending end is used to convert the data back to analog form.

22-10 PSPICE EXAMPLE

Most of the circuits discussed in this chapter are purchased as off-the-shelf subsystems and do not always lend themselves to the type of circuit modeling that we have employed in earlier chapters. However, the following example illustrates one case of a specific circuit configuration.

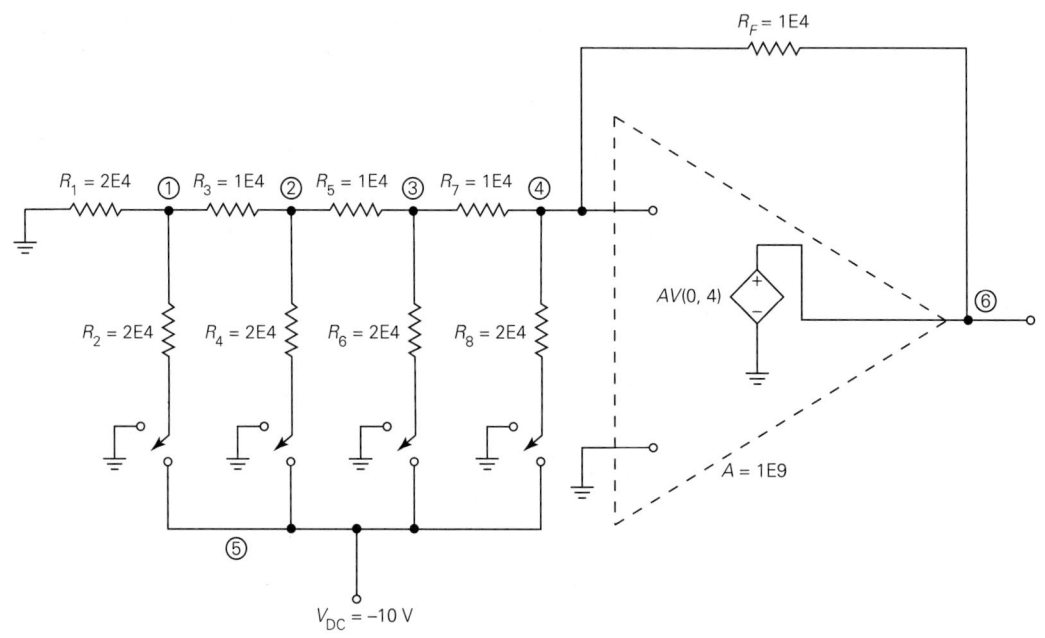

All element values are in their basic units.

(a)

```
PSPICE EXAMPLE 22-1
.OPTIONS NOECHO NOPAGE
R1 0 1 2E4
R2 1 5 2E4
R3 1 2 1E4
R4 2 5 2E4
R5 2 3 1E4
R6 3 5 2E4
R7 3 4 1E4
R8 4 5 2E4
RF 4 6 1E4
VDC 0 5 DC 10
E 6 0 0 4 1E9
.DC VDC 10 10 1
.PRINT DC V(6)
.END
```

(b)

FIGURE 22-24
Circuit and code of PSPICE Example 22-1.

PSPICE EXAMPLE 22-1

Develop a PSPICE model for the R-2R 4-bit A/D converter of Example 22–2 with $V_{fs} = 10$ V, and use it to determine the exact output voltage for the input digital word 1111.

Solution

The circuit adapted to the PSPICE format is shown in Figure 22–24(a), and the code is shown in Figure 22–24(b). The feedback resistance is selected as RF = 10 kΩ in accordance with the analysis of Example 22–2. Note that because all bits have the value 1, all switch connections are to the −10-V bus. A nearly ideal model of the op-amp is used in this example.

The results of the simulation for the output voltage $V(6)$ are shown in Figure 22–25. This voltage is 9.375 V, which is exactly as predicted.

```
PSPICE EXAMPLE 22-1

****        CIRCUIT DESCRIPTION

*****************************************************************************

.OPTIONS NOECHO NOPAGE

****        DC TRANSFER CURVES              TEMPERATURE =    27.000 DEG C


VDC          V(6)
1.000E+01    9.375E+00



     JOB CONCLUDED
     TOTAL JOB TIME                    1.27
```

FIGURE 22-25
Computer printout for PSPICE Example 22–1.

PROBLEMS

Drill Problems

22-1. The basic sampling rate for the compact disc (CD) recording industry has been set at 44.1 kHz.

   a. Determine the highest frequency of the analog signal that could theoretically be reconstructed from this sampling rate.

   b. Determine the maximum possible conversion time of the A/D converter if real-time processing is used.

22-2. The highest frequency contained in a certain analog signal is 4 kHz. It is to be sampled and converted to digital form.

   a. Determine the sampling rate if it is chosen to be 50% greater than the theoretical minimum.

b. Determine the maximum possible conversion time of the A/D converter if real-time processing is used.

22–3. A certain data processing system of the form shown in Figure 22–1 is to be used to process an analog signal whose frequency content ranges from near dc to 800 Hz. Assume that the actual sampling rate is selected to be 25% greater than the theoretical minimum rate.
   a. Determine the sampling rate and the sampling period.
   b. Determine the maximum conversion time of the A/D converter if real-time processing is employed.
   c. If 16-bit words are used, determine the maximum bit width if a given word is to be transferred serially during the conversion interval of the next word.

22–4. Assume in the system of Problem 22–3 that the sampling rate is increased to a rate 50% greater than the theoretical minimum rate and that 8-bit words are used. Repeat the analysis of Problem 22–3.

22–5. Assume that in the system of Problem 22–3, *four* data channels, each with frequencies from near dc to 800 Hz, are to be sampled and multiplexed. Repeat the calculations of parts **(b)** and **(c)** if a single A/D converter is used.

22–6. Assume that in the system of Problem 22–4, *eight* data channels, each with frequencies from near dc to 800 HZ, are to be sampled and multiplexed. Repeat the calculations of parts **(b)** and **(c)** if a single A/D converter is used.

22–7. The number of bits used for each sample in the CD recording industry is 16 bits. Determine the maximum number of levels that can be encoded.

22–8. In a certain A/D conversion system, it is desired to encode the signal into 2048 possible levels. Determine the number of bits required for each digital sample.

22–9. Consider the 4-bit D/A converter of Example 22–2 (Figure 22–9) with $V_r = 10$ V and $R_f = 10$ kΩ. Determine:
   a. Full-scale voltage   **b.** Value of 1 LSB

22–10. Consider the 4-bit D/A converter of Example 22–2 (Figure 22–9) with $V_r = 5$ V and $R_f = 12$ kΩ. Determine:
   a. Full-scale voltage   **b.** Value of 1 LSB

22–11. For the D/A converter of Problem 22–9, determine the output voltage for each of the following input digital words:
   a. 0000   **b.** 0001   **c.** 1000   **d.** 1010   **e.** 1111

22–12. For the D/A converter of Problem 22–10, determine the output voltage for each of the digital words of Problem 22–11.

22–13. Consider the 5-bit D/A converter of Figure P22–13 with $V_r = 10$ V and $R_f = 10$ kΩ. Determine:
   a. Number of possible output levels
   b. Full-scale voltage
   c. Value of 1 LSB

22–14. Consider the 5-bit D/A converter of Figure P22–13 with $V_r = 5$ V and $R_f = 12$ kΩ. Determine:
   a. Full-scale voltage   **b.** Value of 1 LSB

22–15. For the D/A converter of Problem 22–13, determine the output voltage for each of the following input digital words:
   a. 00001   **b.** 00010   **c.** 00100   **d.** 10000   **e.** 11111

22–16. For the D/A converter of Problem 22–14, determine the output voltage for each of the digital words of Problem 22–15.

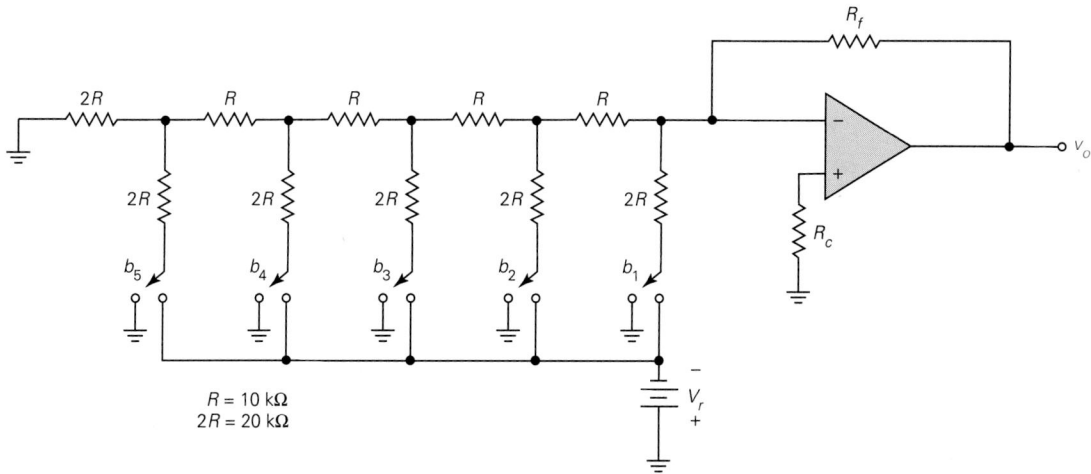

FIGURE P22–13

22–17. Assume that a certain form of integrated-circuit technology allows a comparator operation, amplifier settling, and the associated logic to be performed in about 0.2 μs. If all other times are assumed to be negligible, compute the approximate conversion times and data conversion rates for 6-bit words with the following A/D converter types:

    a. Counter   **b.** Successive approximation   **c.** Flash

22–18. Repeat the analysis of Problem 22–17 for 12-bit words.

22–19. A certain 16-bit A/D converter connected for unipolar encoding (with rounding) has a full-scale voltage of 10 V. Determine the following quantities:

    a. Normalized step size
    b. Actual step size in volts
    c. Normalized maximum quantized level
    d. Actual maximum quantized level in volts
    e. Normalized peak quantization error
    f. Actual peak quantization error in volts
    g. Percentage quantization error

22–20. The 16-bit A/D converter of Problem 22–19 is next connected for bipolar offset encoding with the same peak-to-peak voltage range as in Problem 22–19. The converter is now intended for signals in the range from −5 to just under +5 V. Repeat all the calculations of Problem 22–19 for this case.

Derivation Problems

22–21. The form of a current-weighted D/A converter for 4 bits is shown in Figure P22–21. Starting with the terminals a–a' and continuing to the right, show that the final Norton equivalent circuit looking back from the output is a current source I_N in parallel with R, in which the value of the current source is

$$I_N = 2I(b_1 2^{-1} + b_2 2^{-2} + b_3 2^{-3} + b_4 2^{-4})$$

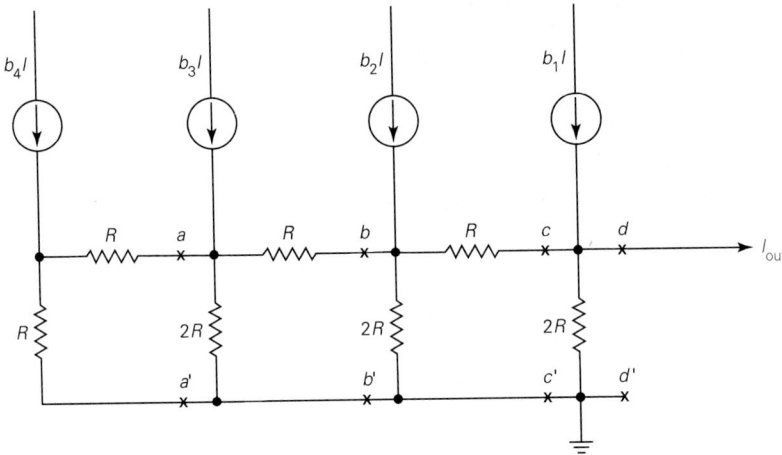

FIGURE P22–21

Hint: Because the current sources are assumed to be ideal, for analysis purposes, the circuit may be redrawn with each current source considered to be in parallel with the corresponding resistance to ground from the junction point.

22–22. An alternate form of a D/A converter utilizing an R-2R network is shown in Figure P22–22 for the case of 4 bits. This form has the advantage that various currents in the network are summed at a virtual ground point. Show that the output voltage is given by

$$v_o = \frac{R_f V_r}{R}(b_1 2^{-1} + b_2 2^{-2} + b_3 2^{-3} + b_4 2^{-4})$$

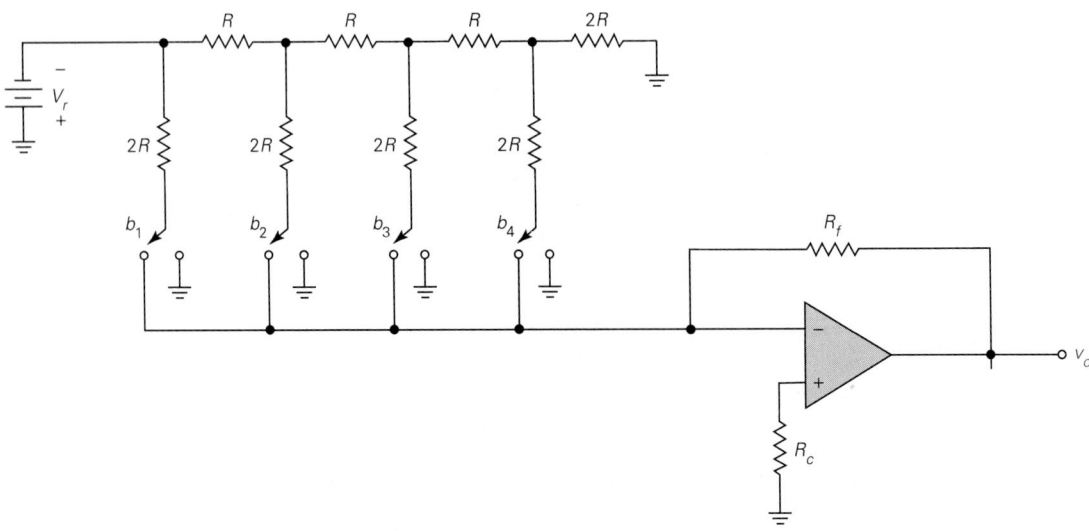

FIGURE P22–22

22–23. Consider the flash A/D converter of Figure 22–13, and assume 4 bits. Starting with the lowest comparator, let $v_1, v_2, \ldots, v_{15}$ represent the dc voltages at the 15 comparator inputs.

    a. Calculate the 15 voltages.

    b. Let $V_1, V_2, \ldots, V_{15}$ represent the corresponding normalized voltages. Determine these values.

    c. Show that the values of **(b)** correspond to the transition points for unipolar encoding with rounding.

    d. If a truncation quantization strategy were desired, what changes in the resistive network would be required?

22–24. Assume an A/D converter connected for bipolar encoding with rounding. Assume that the signal is symmetrical and that its peak-to-peak range is exactly equal to the range of the A/D converter. For the purpose of this problem, a quantity signal-to-noise ratio (SNR) will be defined as the ratio of the peak signal voltage to the peak quantization noise voltage.

    a. Show that the SNR is

$$\text{SNR} = 2^n$$

    b. Show that the corresponding value in decibels is nearly equal to

$$\text{SNR(dB)} = 6n \text{ dB}$$

One significance of this result is that for each additional bit, the signal-to-quantization ratio increases by about 6 dB.

Design Problems

22–25. Consider the 4-bit D/A converter of Example 22–2 (Figure 22–9) with $V_r = 10$ V and R_f variable. Determine the separate values of R_f that would be required to achieve each of the following conditions:

    a. The value of 1 LSB at the output is 0.3 V.

    b. A binary input of 1000 results in an analog output of 5 V.

    c. The full-scale output voltage is 12 V.

    d. The *actual* maximum output voltage is 12 V.

22–26. Consider the 4-bit D/A converter of Example 22–2 (Figure 22–9) with $R_f = 15$ kΩ and V_r variable. Determine the separate values of V_r required to satisfy each of the stated requirements in the four parts of Example 10–27.

22–27. Consider the 5-bit D/A converter of Figure P22–13 with $V_r = 10$ V and R_f variable. Determine the separate values of R_f that would be required to achieve *each* of the following conditions:

    a. The value of 1 LSB at the output is 200 mV.

    b. A binary input of 10000 results in an analog output of 4 V.

    c. The full-scale output voltage is 12 V.

    d. The *actual* maximum output voltage is 12 V.

22–28. Consider the 5-bit D/A converter of Figure P22–13 with $R_f = 15$ kΩ and V_r variable. Determine the separate values of V_r required to satisfy *each* of the stated requirements in Problem 22–27.

22–29. An A/D converter is to be selected in conjunction with the design of a data processing system. The following properties and specifications are given:

    1. Input analog signal: 0–10 V

    2. Rounding encoding strategy

3. Percent error due to quantization $\leq 0.1\%$ of full-scale voltage

For the A/D converter selected, determine:

a. Full-scale voltage

b. Minimum number of bits required

22–30. An A/D converter is to be selected in conjunction with the design of a data processing system. The following properties and specifications are given:

1. Input analog signal range: -10 to 10 V
2. Rounding encoding strategy
3. Maximum quantization error ≤ 5 mV

For the A/D converter selected, determine the minimum number of bits required.

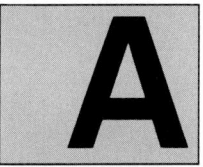

A

RESISTOR VALUES

A–1 TYPICAL 5% AND 10% VALUES

The most commonly used resistors in electronic circuits are the low-power, axial-lead, carbon-composition, and film resistors with tolerances of ±5% and ±10%. The values in the range 10 Ω to 10 MΩ are listed below. All values listed are available in ±5% tolerances, but only the *boldface* values are available in ±10% tolerances.

Ω	Ω	kΩ	kΩ	kΩ	MΩ
10	**100**	**1**	**10**	**100**	**1**
11	110	1.1	11	110	1.1
12	**120**	**1.2**	**12**	**120**	**1.2**
13	130	1.3	13	130	1.3
15	**150**	**1.5**	**15**	**150**	**1.5**
16	160	1.6	16	160	1.6
18	**180**	**1.8**	**18**	**180**	**1.8**
20	200	2.0	20	200	2.0
22	**220**	**2.2**	**22**	**220**	**2.2**
24	240	2.4	24	240	2.4
27	**270**	**2.7**	**27**	**270**	**2.7**
30	300	3.0	30	300	3.0
33	**330**	**3.3**	**33**	**330**	**3.3**
36	360	3.6	36	360	3.6
39	**390**	**3.9**	**39**	**390**	**3.0**
43	430	4.3	43	430	4.3
47	**470**	**4.7**	**47**	**470**	**4.7**

(continued)

Ω	Ω	kΩ	kΩ	kΩ	MΩ
51	510	5.1	51	510	5.1
56	**560**	**5.6**	**56**	**560**	**5.6**
62	620	6.2	62	620	6.2
68	**680**	**6.8**	**68**	**680**	**6.8**
75	750	7.5	75	750	7.5
82	**820**	**8.2**	**82**	**820**	**8.2**
91	910	9.1	91	910	9.1

A-2 TYPICAL 1% VALUES

Values in the range 100 to 1000 Ω are listed as follows. Values in other ranges are obtained by adding zeros or a decimal point to these values.

100	147	215	316	464	681
102	150	221	324	475	698
105	154	226	332	487	715
107	158	232	340	499	732
110	162	237	348	511	750
113	165	243	357	523	768
115	169	249	365	536	787
118	174	255	374	549	806
121	178	261	383	562	825
124	182	267	392	576	845
127	187	274	402	590	866
130	191	280	412	604	887
133	196	287	422	619	909
137	200	294	432	634	931
140	205	301	442	649	953
143	210	309	453	665	976

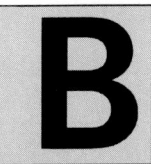

B

CAPACITOR VALUES

The patterns of capacitor values tend to track those of 5% and 10% resistor values as far as the digits are concerned. However, these patterns are somewhat more complex and depend on capacitor types and tolerances. A few digit combinations used for capacitors are different than for resistors. The following values should be interpreted as *typical* rather than *general*.*

pF	pF	pF	pF	μF	μF	μF	μF	μF	μF	μF
5	50	500	5000		0.05	0.5	5	50	500	5000
—	51	510	5100		—	—	—	—	—	—
—	56	560	5600		0.056	0.56	5.6	56	—	5600
—	—	—	6000		0.06	—	6	—	—	6000
—	62	620	6200		—	—	—	—	—	—
—	68	680	6800		0.068	0.68	6.8	—	—	—
—	75	750	7500		—	—	—	75	—	—
—	—	—	8000		—	—	8	80	—	—
—	82	820	8200		0.082	0.82	8.2	82	—	—
—	91	910	9100		—	—	—	—	—	—
10	100	1000		0.01	0.1	1	10	100	1000	10000
—	110	1100		—	—	—	—	—	—	
12	120	1200		0.012	0.12	1.2	—	—	—	
—	130	1300		—	—	—	—	—	—	
15	150	1500		0.015	0.15	1.5	15	150	1500	
—	160	1600		—	—	—	—	—	—	
18	180	1800		0.018	0.18	1.8	18	180	—	
20	200	2000		0.02	0.2	2	20	200	2000	
22	220	2200		—	0.22	2.2	22	—	—	

(continued)

pF	pF	pF	pF	μF	μF	μF	μF	μF	μF	μF
24	240	2400		—	—	—	—	240	—	
—	250	2500		—	0.25	—	25	250	2500	
27	270	2700		0.027	0.27	2.7	27	270	—	
30	300	3000		0.03	0.3	3	30	300	3000	
33	330	3300		0.033	0.33	3.3	33	330	3300	
36	360	3600		—	—	—	—	—	—	
39	390	3900		0.039	0.39	3.9	39	—	—	
—	—	4000		0.04	—	4	—	400	—	
43	430	4300		—	—	—	—	—	—	
47	470	4700		0.047	0.47	4.7	47	—	—	

* The element values in this table were extracted from David A. Bell, *Electronic Devices and Circuits*, 3rd Ed. (Englewood Cliffs, NJ: Reston/Prentice Hall, 1986).

HYBRID
PARAMETERS

Within the text, the modeling of small-signal transistor behavior has been achieved primarily with the hybrid-pi equivalent circuit because of its simplicity in the midfrequency range and because it can be easily adapted to high-frequency models. However, many books and data manuals use references to the classical hybrid parameter equivalent circuit. This model was heavily utilized in much of the earlier transistor work.

It turns out that three separate hybrid parameter equivalent circuits can actually be developed, corresponding to common emitter, common base, and common collector. However, all cases can be developed with the common-emitter configuration, and most data manuals provide numerical values based solely on that configuration. Consequently, the treatment here will be limited to the common-emitter situation.

Consider the common-emitter configuration shown in Figure C–1. The dc circuit used to establish an operating point is not shown but is obviously under-

FIGURE C–1
BJT common-emitter configuration with signal variables labeled.

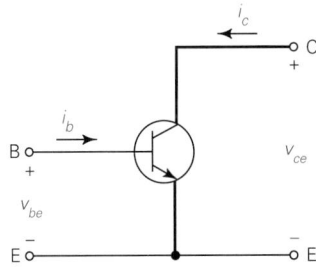

stood. Only the small-signal or ac variables are shown. The pertinent dynamic variables are as follows:

v_{be}: base–emitter signal voltage

i_b: base signal current

v_{ce}: collector–emitter signal voltage

i_c: collector signal current

The hybrid parameters as commonly applied to BJTs are based on assuming i_b and v_{ce} as the *independent* variables and v_{be} and i_c as the *dependent* variables. The resulting equations are

$$v_{be} = h_{ie}i_b + h_{re}v_{ce} \qquad \text{(C–1)}$$

$$i_c = h_{fe}i_b + h_{oe}v_{ce} \qquad \text{(C–2)}$$

The quantities h_{ie}, h_{re}, h_{fe}, and h_{oe} are the **hybrid parameters** for the common-emitter configuration. The second subscript e in each case refers to common-emitter. The subscripts $i, r, f,$ and o refer to *input, reverse, forward,* and *output*, respectively.

To understand the physical significance of the parameters, assume in (C–1) and (C–2) that the total collector–emitter voltage is held constant so that the signal voltage is $v_{ce} = 0$. The two equations then reduce to

$$v_{be} = h_{ie}i_b \qquad \text{for } v_{ce} = 0 \qquad \text{(C–3)}$$
$$i_c = h_{fe}i_b \qquad \text{for } v_{ce} = 0 \qquad \text{(C–4)}$$

The quantities h_{ie} and h_{fe} may then be expressed as

$$h_{ie} = \frac{v_{be}}{i_b}\bigg]_{v_{ce}=0} \qquad \text{(C–5)}$$

and

$$h_{fe} = \frac{i_c}{i_b}\bigg]_{v_{ce}=0} \qquad \text{(C–6)}$$

From these expressions, it is observed that h_{ie} is an input resistance and that h_{fe} is a forward current gain. Both are determined under conditions of an "ac short" (but not a total short, obviously) on the output.

Next, assume that the total input base current is held constant so that the signal current is $i_b = 0$. Equations (C–1) and (C–2) then reduce to

$$v_{be} = h_{re}v_{ce} \qquad \text{for } i_b = 0 \qquad\qquad \textbf{(C–7)}$$

$$i_c = h_{oe}v_{ce} \qquad \text{for } i_b = 0 \qquad\qquad \textbf{(C–8)}$$

The quantities h_{re} and h_{oe} may then be expressed as

$$h_{re} = \frac{v_{be}}{v_{ce}}\bigg]_{i_b=0} \qquad\qquad \textbf{(C–9)}$$

$$h_{oe} = \frac{i_c}{v_{ce}}\bigg]_{i_b=0} \qquad\qquad \textbf{(C–10)}$$

The parameter h_{re} is observed to be a reverse feedback ratio, and h_{oe} is an output conductance. Both are determined under conditions of an "ac open" (but not a total open) at the input.

C–1 EQUIVALENT CIRCUIT

Equation (C–1) expresses v_{be} as the sum of two voltages, while (C–2) expresses i_c as the sum of two currents. The first case can be interpreted as a series connection, while the second can be interpreted as a parallel connection.

A hybrid parameter equivalent circuit based on the preceding interpretations is shown in Figure C–2. From the preceding discussion and the circuit diagram, it can be inferred that h_{ie} has the dimensions of ohms, h_{oe} has the dimensions of siemens, and h_{re} and h_{fe} are dimensionless. Note that the resistance at the output is labeled as $1/h_{oe}$ (which would be in ohms), as is customary.

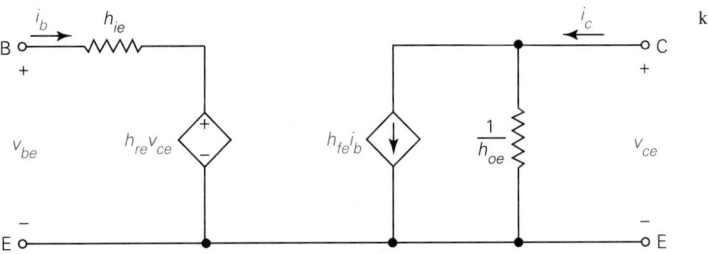

FIGURE C–2
Hybrid-parameter equivalent circuit based on common-emitter configuration.

C–2 RELATIONSHIP TO HYBRID-PI

By comparing the circuit of Figure C–2 with the low-frequency hybrid-pi model and other parameters of Chapter 10, some relationships between certain quantities may be established. First, h_{ie} is the input dynamic resistance, which was defined as r_π in Chapter 10. Second, h_{fe} is the same as the flat-band current gain, which was defined as β_o in Chapter 10.

In the midfrequency hybrid-pi model of Chapter 10, the following two assumptions were made: (1) Reverse feedback from collector back to base was ignored, and (2) the dependent output current source was assumed to be ideal. Because of these assumptions, there are no parameters in Chapter 10 that correspond directly with h_{re} and h_{oe}. The hybrid-pi model can be modified to take these effects into consideration, but considering the approximate nature of the parameters, such efforts are mandated in only the most exacting situations.

A summary of the comparison is as follows:

Hybrid parameters	Chapter 10 and later
h_{ie}	r_π
h_{fe}	β_o
h_{ie}	Effect is neglected
h_{oe}	Assumed to be zero

C–3 AMPLIFIER ANALYSIS

The use of the hybrid parameters for analyzing a typical amplifier circuit will now be illustrated. Consider the common-emitter circuit of Figure C–3(a). Assume operation in the midfrequency range where external capacitances are represented as shorts.

The net small-signal equivalent circuit based on replacing the transistor by its hybrid equivalent circuit is shown in Figure C–3(b). Note that the input voltage is $v_i = v_{be}$ and that the output voltage is $v_o = v_{ce}$.

A KVL loop equation around the input reads

$$-v_i + h_{ie}i_b + h_{re}v_{ce} = 0 \qquad \textbf{(C–11)}$$

Solving for i_b, there results

$$i_b = \frac{v_i - h_{re}v_{ce}}{h_{ie}} = \frac{v_i - h_{re}v_o}{h_{ie}} \qquad \textbf{(C–12)}$$

where the substitution $v_{ce} = v_o$ was made.

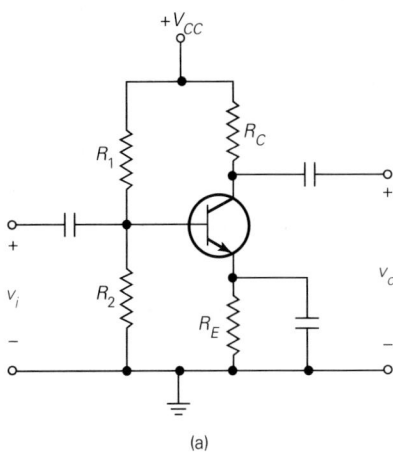

(a)

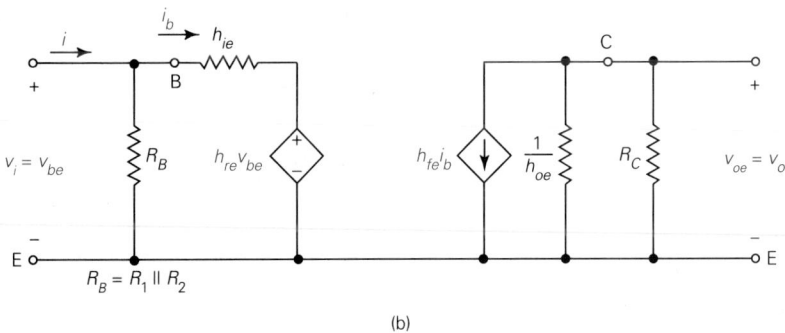

$R_B = R_1 \| R_2$

(b)

FIGURE C–3
Common-emitter amplifier and small-signal equivalent circuit using hybrid parameters.

The output voltage v_o can be determined from the output portion of the circuit as

$$v_o = -\left(R_C \| \frac{1}{h_{oe}}\right) h_{fe}i_b = -\frac{R_C h_{fe}i_b}{1 + R_C h_{oe}} \qquad \text{(C–13)}$$

The result of (C–12) is substituted in (C–13), and the equation is manipulated to obtain an expression for v_o. The result is

$$v_o = \frac{-h_{fe}R_C v_i}{h_{ie} + R_C(h_{ie}h_{oe} - h_{re}h_{fe})} \qquad \text{(C–14)}$$

The gain A is obtained as

$$A = \frac{v_o}{v_i} = \frac{-h_{fe}R_C}{h_{ie} + R_C(h_{ie}h_{oe} - h_{re}h_{fe})} \qquad \text{(C–15)}$$

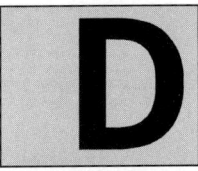

DATA SHEETS

Data sheets for the particular devices discussed in the text are reproduced in this appendix. Appreciation is expressed to the following organizations for permission to reproduce the specific data sheets indicated:

1. Pages 978 through 989: Copyright of Motorola, Inc. Used by permission.
2. Pages 990 through 995: Reprinted with permission of GE Semiconductor, Research Triangle Park, NC.
3. Pages 996 through 1007: Copyright Fairchild Semiconductor Corporation. Used by permission.

1N4001
thru
1N4007

GENERAL-PURPOSE RECTIFIERS

. . . subminiature size, axial lead mounted rectifiers for general-purpose low-power applications.

**LEAD MOUNTED
SILICON RECTIFIERS**

50-1000 VOLTS
DIFFUSED JUNCTION

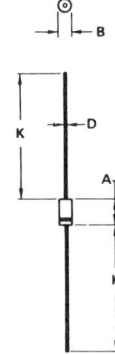

*MAXIMUM RATINGS

Rating	Symbol	1N4001	1N4002	1N4003	1N4004	1N4005	1N4006	1N4007	Unit
Peak Repetitive Reverse Voltage Working Peak Reverse Voltage DC Blocking Voltage	V_{RRM} V_{RWM} V_R	50	100	200	400	600	800	1000	Volts
Non-Repetitive Peak Reverse Voltage (halfwave, single phase, 60 Hz)	V_{RSM}	60	120	240	480	720	1000	1200	Volts
RMS Reverse Voltage	$V_{R(RMS)}$	35	70	140	280	420	560	700	Volts
Average Rectified Forward Current (single phase, resistive load, 60 Hz, see Figure 8, T_A = 75°C)	I_O				1.0				Amp
Non-Repetitive Peak Surge Current (surge applied at rated load conditions, see Figure 2)	I_{FSM}				30 (for 1 cycle)				Amp
Operating and Storage Junction Temperature Range	T_J, T_{stg}				−65 to +175				°C

*ELECTRICAL CHARACTERISTICS

Characteristic and Conditions	Symbol	Typ	Max	Unit
Maximum Instantaneous Forward Voltage Drop (i_F = 1.0 Amp, T_J = 25°C) Figure 1	v_F	0.93	1.1	Volts
Maximum Full-Cycle Average Forward Voltage Drop (I_O = 1.0 Amp, T_L = 75°C, 1 inch leads)	$V_{F(AV)}$	−	0.8	Volts
Maximum Reverse Current (rated dc voltage) T_J = 25°C T_J = 100°C	I_R	0.06 1.0	10 50	μA
Maximum Full-Cycle Average Reverse Current (I_O = 1.0 Amp, T_L = 75°C, 1 inch leads	$I_{R(AV)}$	−	30	μA

*Indicates JEDEC Registered Data.

MECHANICAL CHARACTERISTICS

CASE: Transfer Molded Plastic

MAXIMUM LEAD TEMPERATURE FOR SOLDERING PURPOSES: 350°C, 3/8" from case for 10 seconds at 5 lbs. tension

FINISH: All external surfaces are corrosion-resistant, leads are readily solderable

POLARITY: Cathode indicated by color band

WEIGHT: 0.40 Grams (approximately)

NOTES:
1. POLARITY DENOTED BY CATHODE BAND.

2. LEAD DIAMETER NOT CONTROLLED WITHIN "F" DIMENSION.

DIM	MILLIMETERS		INCHES	
	MIN	MAX	MIN	MAX
A	5.97	6.60	0.235	0.260
B	2.79	3.05	0.110	0.120
D	0.76	0.86	0.030	0.034
K	27.94	−	1.100	−

CASE 59-04
(Does not meet DO-41 outline)

1.5KE6.8,A thru 1.5KE200,A
See Page 4-74

1N746 thru 1N759
1N957A thru 1N986A
1N4370 thru 1N4372

GLASS ZENER DIODES
500 MILLIWATTS
2.4-110 VOLTS

Designers Data Sheet

500-MILLIWATT HERMETICALLY SEALED
GLASS SILICON ZENER DIODES

- Complete Voltage Range — 2.4 to 110 Volts
- DO-35 Package — Smaller than Conventional DO-7 Package
- Double Slug Type Construction
- Metallurgically Bonded Construction
- Nitride Passivated Die

Designer's Data for "Worst Case" Conditions

The Designer's Data sheets permit the design of most circuits entirely from the information presented. Limit curves — representing boundaries on device characteristics — are given to facilitate "worst case" design.

MAXIMUM RATINGS

Rating	Symbol	Value	Unit
DC Power Dissipation @ $T_L \leq 50^oC$, Lead Length = 3/8''	P_D		
*JEDEC Registration		400	mW
*Derate above $T_L = 50^oC$		3.2	mW/^oC
Motorola Device Ratings		500	mW
Derate above $T_L = 50^oC$		3.33	mW/^oC
Operating and Storage Junction Temperature Range	T_J, T_{stg}		^oC
*JEDEC Registration		−65 to +175	
Motorola Device Ratings		−65 to +200	

*Indicates JEDEC Registered Data.

MECHANICAL CHARACTERISTICS

MAXIMUM LEAD TEMPERATURE FOR SOLDERING PURPOSES: 230^oC, 1/16'' from case for 10 seconds

FINISH: All external surfaces are corrosion resistant with readily solderable leads.

POLARITY: Cathode indicated by color band. When operated in zener mode, cathode will be positive with respect to anode.

MOUNTING POSITION: Any

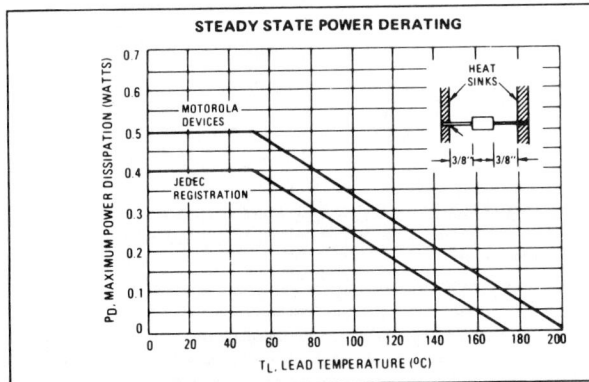

STEADY STATE POWER DERATING

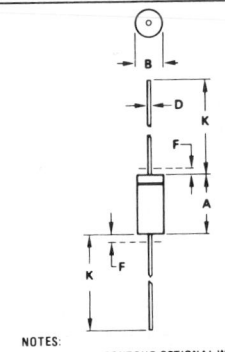

NOTES:
1. PACKAGE CONTOUR OPTIONAL WITHIN A AND B. HEAT SLUGS, IF ANY, SHALL BE INCLUDED WITHIN THIS CYLINDER, BUT NOT SUBJECT TO THE MINIMUM LIMIT OF B.
2. LEAD DIAMETER NOT CONTROLLED IN ZONE F TO ALLOW FOR FLASH, LEAD FINISH BUILDUP AND MINOR IRREGULARITIES OTHER THAN HEAT SLUGS.
3. POLARITY DENOTED BY CATHODE BAND.
4. DIMENSIONING AND TOLERANCING PER ANSI Y14.5, 1973.

DIM	MILLIMETERS		INCHES	
	MIN	MAX	MIN	MAX
A	3.05	5.08	0.120	0.200
B	1.52	2.29	0.060	0.090
D	0.46	0.56	0.018	0.022
F	—	1.27	—	0.050
K	25.40	38.10	1.000	1.500

All JEDEC dimensions and notes apply.

CASE 299-02
DO-204AH
(DO-35)

1N746 thru 1N759, 1N957A thru 1N986A, 1N4370 thru 1N4372

ELECTRICAL CHARACTERISTICS ($T_A = 25°C$, $V_F = 1.5$ V max at 200 mA for all types)

Type Number (Note 1)	Nominal Zener Voltage V_Z @ I_{ZT} (Note 2) Volts	Test Current I_{ZT} mA	Maximum Zener Impedance Z_{ZT} @ I_{ZT} (Note 3) Ohms	*Maximum DC Zener Current I_{ZM} (Note 4) mA		Maximum Reverse Leakage Current $T_A = 25°C$ I_R @ $V_R = 1$ V µA	$T_A = 150°C$ I_R @ $V_R = 1$ V µA
1N4370	2.4	20	30	150	190	100	200
1N4371	2.7	20	30	135	165	75	150
1N4372	3.0	20	29	120	150	50	100
1N746	3.3	20	28	110	135	10	30
1N747	3.6	20	24	100	125	10	30
1N748	3.9	20	23	95	115	10	30
1N749	4.3	20	22	85	105	2	30
1N750	4.7	20	19	75	95	2	30
1N751	5.1	20	17	70	85	1	20
1N752	5.6	20	11	65	80	1	20
1N753	6.2	20	7	60	70	0.1	20
1N754	6.8	20	5	55	65	0.1	20
1N755	7.5	20	6	50	60	0.1	20
1N756	8.2	20	8	45	55	0.1	20
1N757	9.1	20	10	40	50	0.1	20
1N758	10	20	17	35	45	0.1	20
1N759	12	20	30	30	35	0.1	20

Type Number (Note 1)	Nominal Zener Voltage V_Z (Note 2) Volts	Test Current I_{ZT} mA	Maximum Zener Impedance (Note 3) Z_{ZT} @ I_{ZT} Ohms	Z_{ZK} @ I_{ZK} Ohms	I_{ZK} mA	*Maximum DC Zener Current I_{ZM} (Note 4) mA		Maximum Reverse Current I_R Maximum µA	Test Voltage Vdc 5% V_R	10%
1N957A	6.8	18.5	4.5	700	1.0	47	61	150	5.2	4.9
1N958A	7.5	16.5	5.5	700	0.5	42	55	75	5.7	5.4
1N959A	8.2	15	6.5	700	0.5	38	50	50	6.2	5.9
1N960A	9.1	14	7.5	700	0.5	35	45	25	6.9	6.6
1N961A	10	12.5	8.5	700	0.25	32	41	10	7.6	7.2
1N962A	11	11.5	9.5	700	0.25	28	37	5	8.4	8.0
1N963A	12	10.5	11.5	700	0.25	26	34	5	9.1	8.6
1N964A	13	9.5	13	700	0.25	24	32	5	9.9	9.4
1N965A	15	8.5	16	700	0.25	21	27	5	11.4	10.8
1N966A	16	7.8	17	700	0.25	19	37	5	12.2	11.5
1N967A	18	7.0	21	750	0.25	17	23	5	13.7	13.0
1N968A	20	6.2	25	750	0.25	15	20	5	15.2	14.4
1N969A	22	5.6	29	750	0.25	14	18	5	16.7	15.8
1N970A	24	5.2	33	750	0.25	13	17	5	18.2	17.3
1N971A	27	4.6	41	750	0.25	11	15	5	20.6	19.4
1N972A	30	4.2	49	1000	0.25	10	13	5	22.8	21.6
1N973A	33	3.8	58	1000	0.25	9.2	12	5	25.1	23.8
1N974A	36	3.4	70	1000	0.25	8.5	11	5	27.4	25.9
1N975A	39	3.2	80	1000	0.25	7.8	10	5	29.7	28.1
1N976A	43	3.0	93	1500	0.25	7.0	9.6	5	32.7	31.0
1N977A	47	2.7	105	1500	0.25	6.4	8.8	5	35.8	33.8
1N978A	51	2.5	125	1500	0.25	5.9	8.1	5	38.8	36.7
1N979A	56	2.2	150	2000	0.25	5.4	7.4	5	42.6	40.3
1N980A	62	2.0	185	2000	0.25	4.9	6.7	5	47.1	44.6
1N981A	68	1.8	230	2000	0.25	4.5	6.1	5	51.7	49.0
1N982A	75	1.7	270	2000	0.25	1.0	5.5	5	56.0	54.0
1N983A	82	1.5	330	3000	0.25	3.7	5.0	5	62.2	59.0
1N984A	91	1.4	400	3000	0.25	3.3	4.5	5	69.2	65.5
1N985A	100	1.3	500	3000	0.25	3.0	4.5	5	76	72
1N986A	110	1.1	750	4000	0.25	2.7	4.1	5	83.6	79.2

NOTE 1. TOLERANCE AND VOLTAGE DESIGNATION

Tolerance Designation

The type numbers shown have tolerance designations as follows:

1N4370 series: ±10%, suffix A for ±5% units.
1N746 series: ±10%, suffix A for ±5% units.
1N957 series: suffix A for ±10% units,
suffix B for ±5% units.

Voltage Designation

To designate units with zener voltages other than those listed, the Motorola type number should be modified as shown below. Unless otherwise specified, the electrical characteristics other than the nominal voltage (V_Z) and test voltage for leakage current will conform to the characteristics of the next higher voltage type shown in the table.

EXAMPLE: 1N746 series, 1N4370 series variations

EXAMPLE: 1N957 series variations

Matched Sets for Closer Tolerances or Higher Voltages

Series matched sets make zener voltages in excess of 100 volts or tolerances of less than 5% possible as well as providing lower temperature coefficients, lower dynamic impedance and greater power handling ability.

For Matched Sets or other special circuit requirements, contact your Motorola Sales Representative.

1N746 thru 1N759, 1N957A thru 1N986A, 1N4370 thru 1N4372

NOTE 2. ZENER VOLTAGE (V_Z) MEASUREMENT

Nominal zener voltage is measured with the device junction in thermal equilibrium at the lead temperature of $30^\circ C \pm 1^\circ C$ and 3/8'' lead length.

NOTE 3. ZENER IMPEDANCE (Z_Z) DERIVATION

Z_{ZT} and Z_{ZK} are measured by dividing the ac voltage drop across the device by the ac current applied. The specified limits are for $I_Z(ac) = 0.1 \, I_Z(dc)$ with the ac frequency = 60 Hz.

NOTE 4. MAXIMUM ZENER CURRENT RATINGS (I_{ZM})

Maximum zener current ratings are based on the maximum voltage of a 10% 1N746 type unit or a 20% 1N957 type unit. For closer tolerance units (10% or 5%) or units where the actual zener voltage (V_Z) is known at the operating point, the maximum zener current may be increased and is limited by the derating curve.

APPLICATION NOTE

Since the actual voltage available from a given zener diode is temperature dependent, it is necessary to determine junction temperature under any set of operating conditions in order to calculate its value. The following procedure is recommended:

Lead Temperature, T_L, should be determined from:

$$T_L = \theta_{LA}P_D + T_A$$

θ_{LA} is the lead-to-ambient thermal resistance ($^\circ C/W$) and P_D is the power dissipation. The value for θ_{LA} will vary and depends on the device mounting method. θ_{LA} is generally 30-$40^\circ C/W$ for the various clips and tie points in common use and for printed circuit board wiring.

The temperature of the lead can also be measured using a thermocouple placed on the lead as close as possible to the tie point. The thermal mass connected to the tie point is normally large enough so that it will not significantly respond to heat surges generated in the diode as a result of pulsed operation once steady-state conditions are achieved. Using the measured value of T_L, the junction temperature may be determined by:

$$T_J = T_L + \Delta T_{JL}$$

ΔT_{JL} is the increase in junction temperature above the lead temperature and may be found from Figure 1 for dc power.

$$\Delta T_{JL} = \theta_{JL}P_D$$

For worst-case design, using expected limits of I_Z, limits of P_D and the extremes of $T_J(\Delta T_J)$ may be estimated. Changes in voltage, V_Z, can then be found from:

$$\Delta V = \theta_{VZ}\Delta T_J$$

θ_{VZ}, the zener voltage temperature coefficient, is found from Figures 3 and 4.

Under high power-pulse operation, the zener voltage will vary with time and may also be affected significantly by the zener resistance. For best regulation, keep current excursions as low as possible.

Surge limitations are given in Figure 6. They are lower than would be expected by considering only junction temperature, as current crowding effects cause temperatures to be extremely high in small spots, resulting in device degradation should the limits of Figure 6 be exceeded.

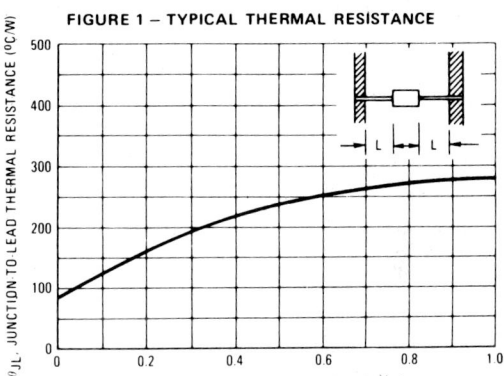

FIGURE 1 – TYPICAL THERMAL RESISTANCE

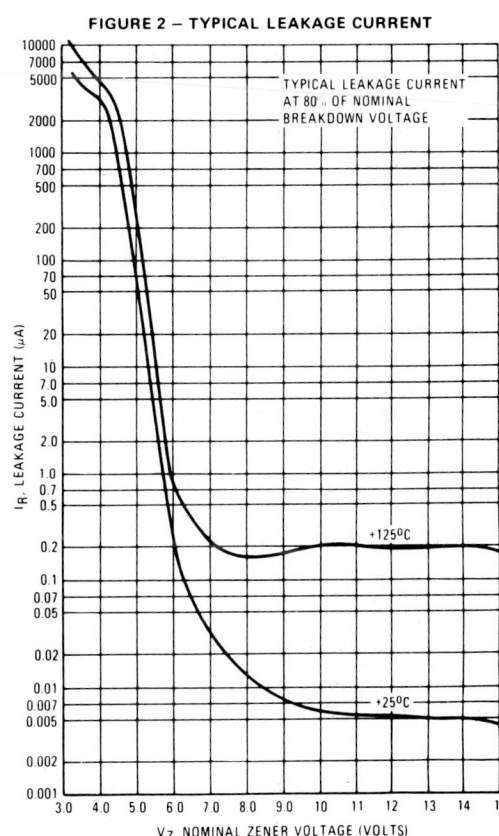

FIGURE 2 – TYPICAL LEAKAGE CURRENT

1N746 thru 1N759, 1N957A thru 1N986A, 1N4370 thru 1N4372

FIGURE 3 — TEMPERATURE COEFFICIENTS
(-55°C to +150°C temperature range; 90% of the units are in the ranges indicated.)

a — RANGE FOR UNITS TO 12 VOLTS

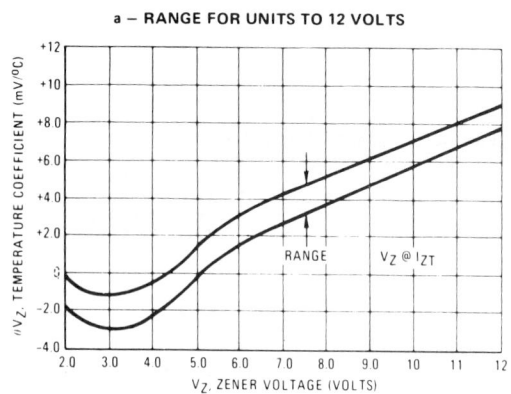

b — RANGE FOR UNITS 12 TO 100 VOLTS

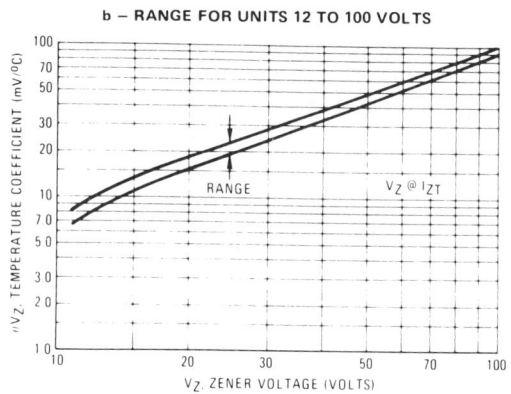

FIGURE 4 — EFFECT OF ZENER CURRENT

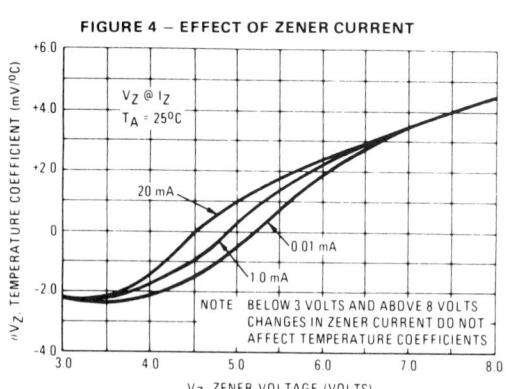

FIGURE 5 — TYPICAL CAPACITANCE

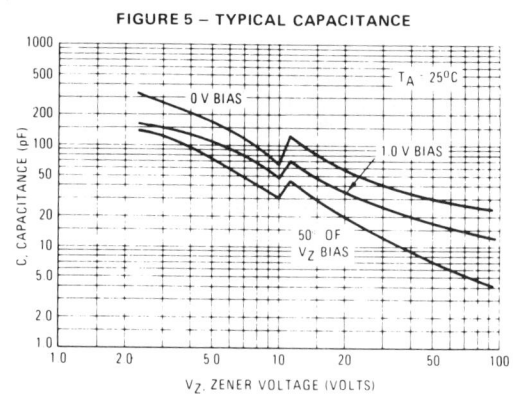

FIGURE 6 — MAXIMUM SURGE POWER

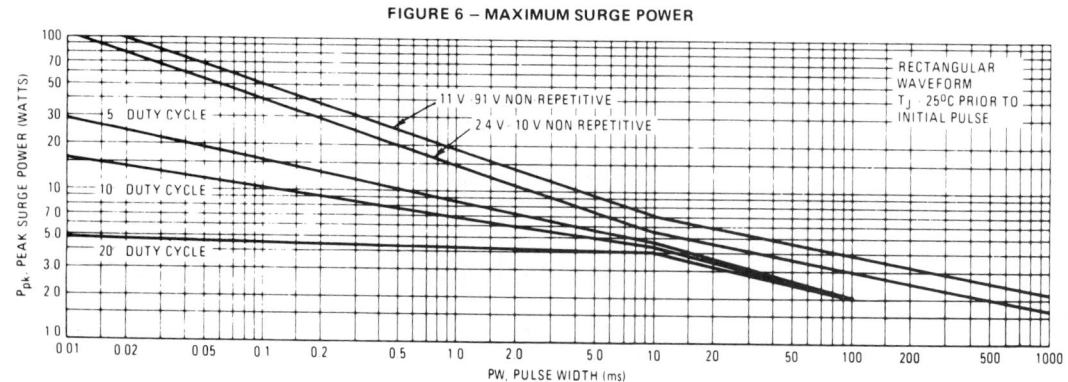

This graph represents 90 percentil data points.
For worst case design characteristics, multiply surge power by 2/3.

1N746 thru 1N759, 1N957A thru 1N986A, 1N4370 thru 1N4372

FIGURE 7 — EFFECT OF ZENER CURRENT ON ZENER IMPEDANCE

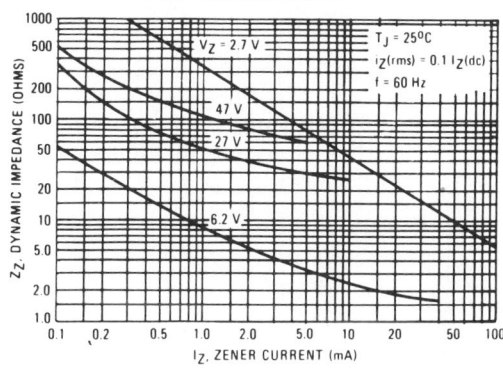

FIGURE 8 — EFFECT OF ZENER VOLTAGE ON ZENER IMPEDANCE

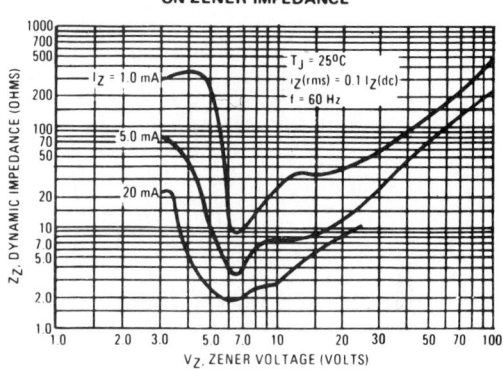

FIGURE 9 — TYPICAL NOISE DENSITY

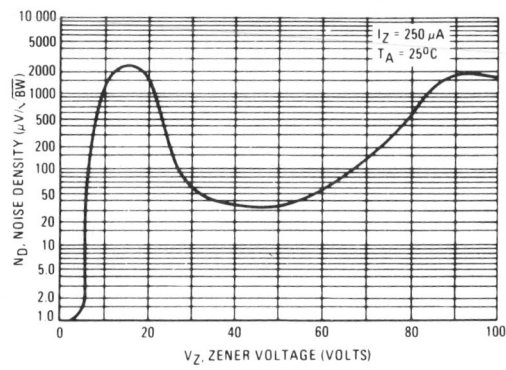

FIGURE 10 — NOISE DENSITY MEASUREMENT METHOD

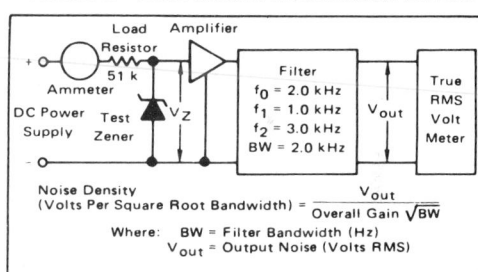

Noise Density
(Volts Per Square Root Bandwidth) $= \dfrac{V_{out}}{\text{Overall Gain }\sqrt{BW}}$

Where: BW = Filter Bandwidth (Hz)
         V_{out} = Output Noise (Volts RMS)

The input voltage and load resistance are high so that the zener diode is driven from a constant current source. The amplifier is low noise so that the amplifier noise is negligible compared to that of the test zener. The filter bandpass is known so that the noise density can be calculated from the formula shown.

FIGURE 11 — TYPICAL FORWARD CHARACTERISTICS

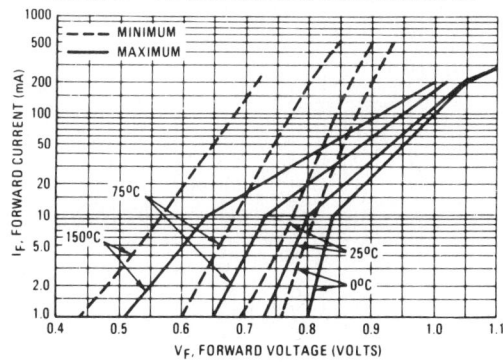

(Copyright of Motorola, Inc. Used by permission.)

983

1N746 thru 1N759, 1N957A thru 1N986A, 1N4370 thru 1N4372

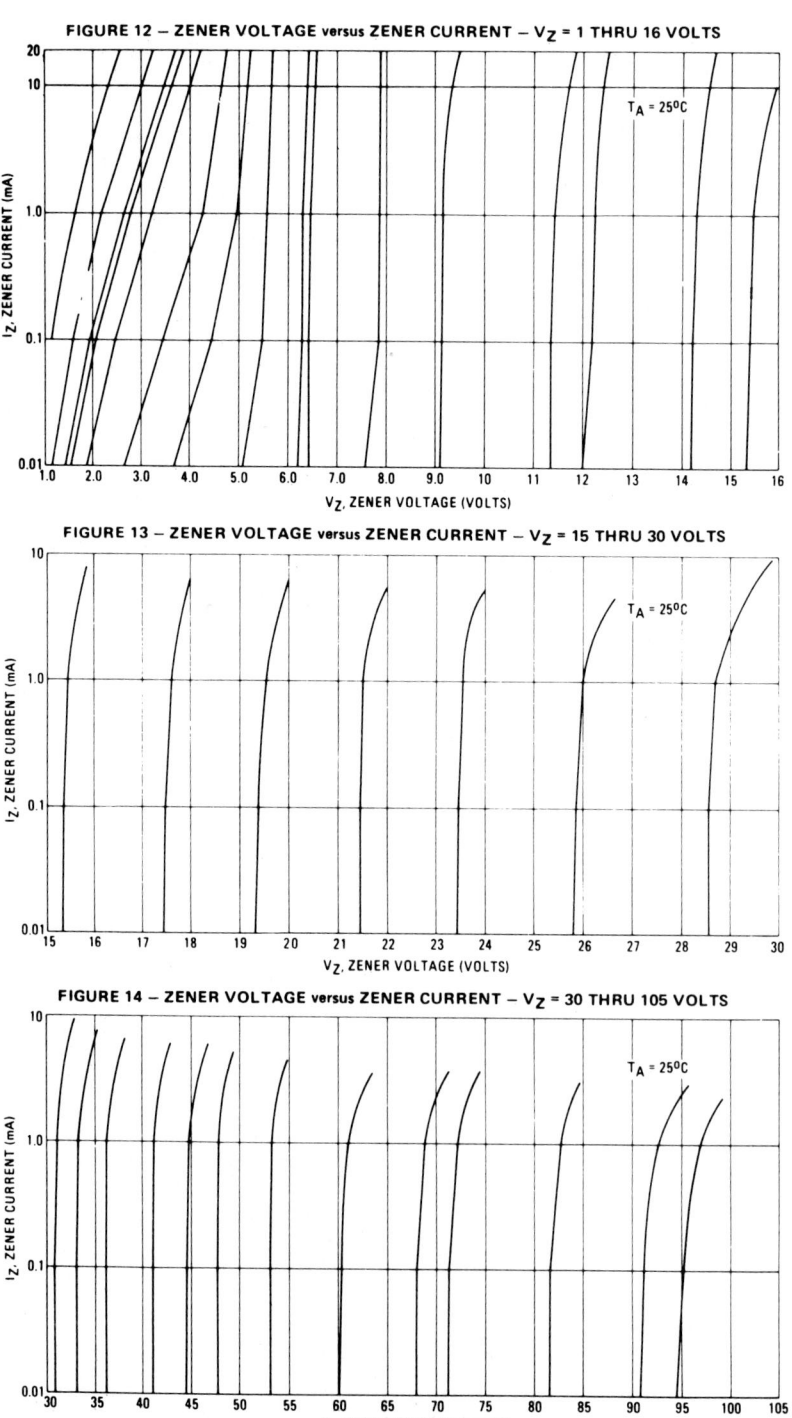

FIGURE 12 — ZENER VOLTAGE versus ZENER CURRENT — V_Z = 1 THRU 16 VOLTS

FIGURE 13 — ZENER VOLTAGE versus ZENER CURRENT — V_Z = 15 THRU 30 VOLTS

FIGURE 14 — ZENER VOLTAGE versus ZENER CURRENT — V_Z = 30 THRU 105 VOLTS

984

2N4220
thru
2N4222

2N4220,A
thru
2N4222,A

CASE 20-03, STYLE 3
TO-72 (TO-206AF)

JFET
LOW-FREQUENCY, LOW NOISE

N-CHANNEL — DEPLETION

MAXIMUM RATINGS

Rating	Symbol	Value	Unit
Drain-Source Voltage	V_{DS}	30	Vdc
Drain-Gate Voltage	V_{DG}	30	Vdc
Gate-Source Voltage	V_{GS}	-30	Vdc
Drain Current	I_D	15	mAdc
Total Device Dissipation @ $T_A = 25°C$ Derate above 25°C	P_D	300 2	mW mW/°C
Junction Temperature Range	T_J	175	°C
Storage Channel Temperature Range	T_{stg}	-65 to $+200$	°C

ELECTRICAL CHARACTERISTICS ($T_A = 25°C$ unless otherwise noted.)

Characteristic		Symbol	Min	Typ	Max	Unit
OFF CHARACTERISTICS						
Gate-Source Breakdown Voltage ($I_G = -10\ \mu Adc$, $V_{DS} = 0$)		$V_{(BR)GSS}$	-30	—	—	Vdc
Gate Reverse Current ($V_{GS} = -15$ Vdc, $V_{DS} = 0$) ($V_{GS} = -15$ Vdc, $V_{DS} = 0$, $T_A = 150°C$)		I_{GSS}	 — —	 — —	 -0.1 -100	nAdc
Gate Source Cutoff Voltage ($I_D = 0.1$ nAdc, $V_{DS} = 15$ Vdc)	2N4220,A 2N4221,A 2N4222,A	$V_{GS(off)}$	 — — —	 — — —	 -4 -6 -8	Vdc
Gate Source Voltage ($I_D = 50\ \mu Adc$, $V_{DS} = 15$ Vdc) ($I_D = 200\ \mu Adc$, $V_{DS} = 15$ Vdc) ($I_D = 500\ \mu Adc$, $V_{DS} = 15$ Vdc)	2N4220,A 2N4221,A 2N4222,A	V_{GS}	 -0.5 -1.0 -2.0	 — — —	 -2.5 -5.0 -6.0	Vdc
ON CHARACTERISTICS						
Zero-Gate-Voltage Drain Current* ($V_{DS} = 15$ Vdc, $V_{GS} = 0$)	2N4220,A 2N4221,A 2N4222,A	I_{DSS}	 0.5 2.0 5.0	 — — —	 3.0 6.0 15	mAdc
Static Drain-Source On Resistance ($V_{DS} = 0$, $V_{GS} = 0$)	2N4220,A 2N4221,A 2N4222,A	$r_{DS(on)}$	 — — —	 500 400 300	 — — —	Ohms
SMALL-SIGNAL CHARACTERISTICS						
Forward Transfer Admittance Common Source* ($V_{DS} = 15$ Vdc, $V_{GS} = 0$, $f = 1.0$ kHz)	2N4220,A 2N4221,A 2N4222,A	$\|y_{fs}\|$	 1000 2000 2500	 — — —	 4000 5000 6000	μmhos
Output Admittance Common Source ($V_{DS} = 15$ Vdc, $V_{GS} = 0$, $f = 1.0$ kHz)	2N4220,A 2N4221,A 2N4222,A	$\|y_{os}\|$	 — — —	 — — —	 10 20 40	μmhos
Input Capacitance ($V_{DS} = 15$ Vdc, $V_{GS} = 0$, $f = 1.0$ MHz)		C_{iss}	—	4.5	6.0	pF
Reverse Transfer Capacitance ($V_{DS} = 15$ Vdc, $V_{GS} = 0$, $f = 1.0$ MHz)		C_{rss}	—	1.2	2.0	pF
Common-Source Output Capacitance ($V_{DS} = 15$ Vdc, $V_{GS} = 0$, $f = 30$ MHz)		C_{osp}	—	1.5	—	pF

MOTOROLA SEMICONDUCTORS

SMALL-SIGNAL DEVICES

2N4220 thru 2N4222, 2N4220A thru 2N4222A

ELECTRICAL CHARACTERISTICS (continued) (T_A = 25°C unless otherwise noted.)

Characteristic		Symbol	Min	Typ	Max	Unit
FUNCTIONAL CHARACTERISTICS						
Noise Figure		NF				dB
(V_{DS} = 15 Vdc, V_{GS} = 0, R_S = 1.0 megohm,	2N4220A		—	—	2.5	
f = 100 Hz)	2N4221A		—	—	2.5	
	2N4222A		—	—	2.5	

*Pulse Test: Pulse Width = 630 ms, Duty Cycle = 10%.

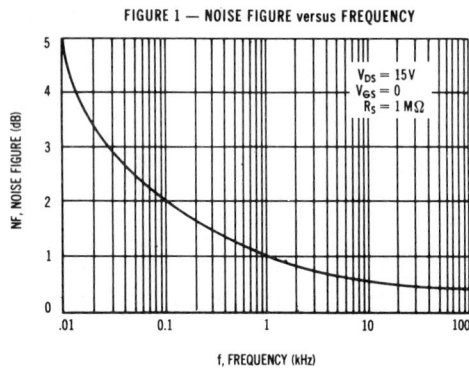

FIGURE 1 — NOISE FIGURE versus FREQUENCY

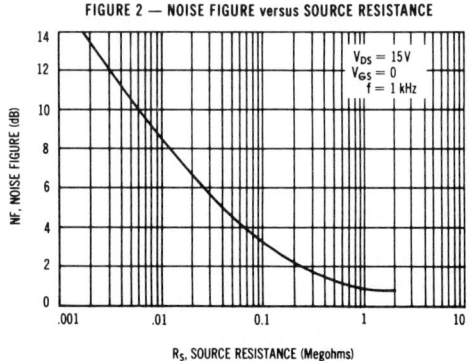

FIGURE 2 — NOISE FIGURE versus SOURCE RESISTANCE

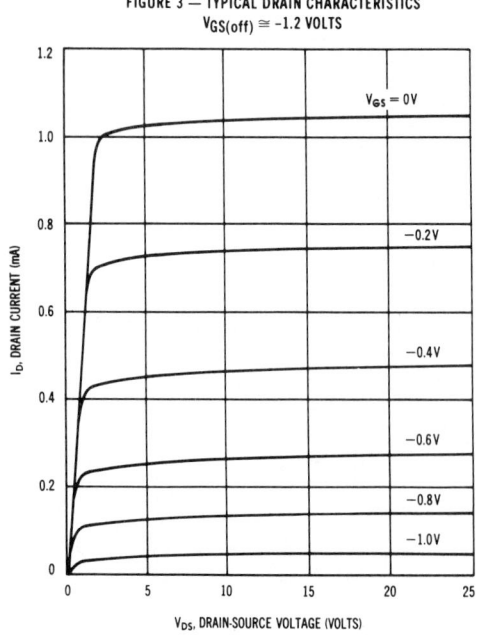

FIGURE 3 — TYPICAL DRAIN CHARACTERISTICS
$V_{GS(off)} \cong$ −1.2 VOLTS

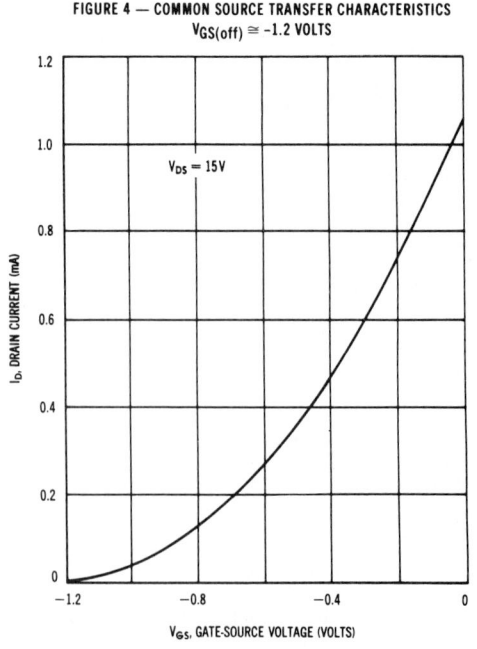

FIGURE 4 — COMMON SOURCE TRANSFER CHARACTERISTICS
$V_{GS(off)} \cong$ −1.2 VOLTS

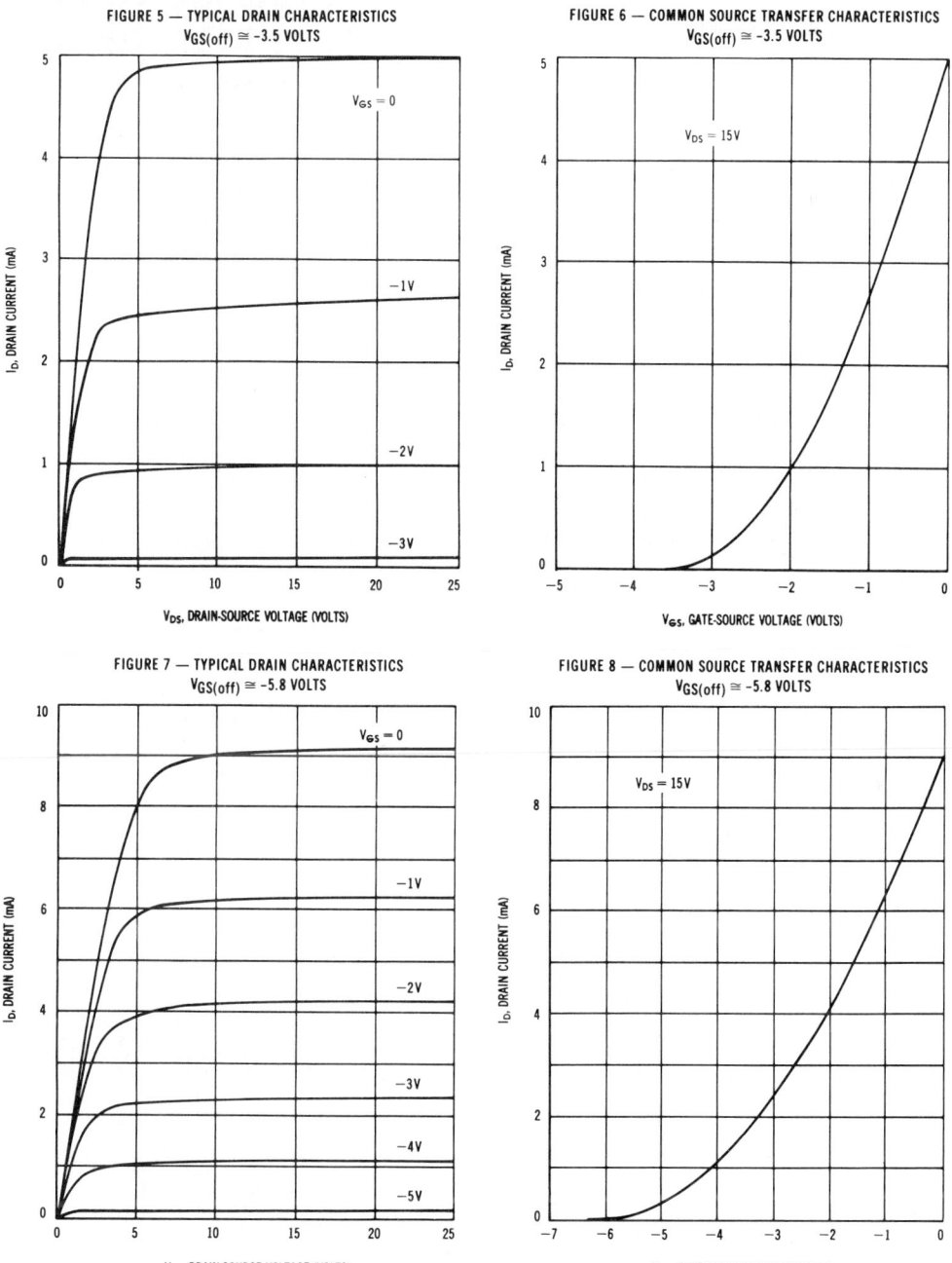

FIGURE 5 — TYPICAL DRAIN CHARACTERISTICS
$V_{GS(off)} \cong -3.5$ VOLTS

FIGURE 6 — COMMON SOURCE TRANSFER CHARACTERISTICS
$V_{GS(off)} \cong -3.5$ VOLTS

FIGURE 7 — TYPICAL DRAIN CHARACTERISTICS
$V_{GS(off)} \cong -5.8$ VOLTS

FIGURE 8 — COMMON SOURCE TRANSFER CHARACTERISTICS
$V_{GS(off)} \cong -5.8$ VOLTS

NOTES: 1. Graphical data is presented for dc conditions. Tabular data is given for pulsed conditions (Pulse Width = 630 ms, Duty Cycle = 10%). Under dc conditions, self heating in higher I_{DSS} units reduces I_{DSS} (See Figure 10).

2. Figures 8, 9, 10: Data taken in a standard printed circuit with a TO-18 type socket mounting and 1/4'' lead length.

COMPLEMENTARY SILICON PLASTIC POWER TRANSISTORS

... designed for use in general purpose amplifier and switching applications. Compact TO-220 AB package. TO-66 leadform also available.

1 AMPERE

POWER TRANSISTORS COMPLEMENTARY SILICON

40-60-80-100 VOLTS
30 WATTS

MAXIMUM RATINGS

Rating	Symbol	TIP29 / TIP30	TIP29A / TIP30A	TIP29B / TIP30B	TIP29C / TIP30C	Unit
Collector-Emitter Voltage	V_{CEO}	40	60	80	100	Vdc
Collector-Base Voltage	V_{CB}	40	60	80	100	Vdc
Emitter-Base Voltage	V_{EB}	5.0				Vdc
Collector Current — Continuous Peak	I_C	1.0				Adc
Peak		3.0				
Base Current	I_B	0.4				Adc
Total Power Dissipation @ T_C = 25°C	P_D	30				Watts
Derate above 25°C		0.24				W/°C
Total Power Dissipation @ T_A = 25°C	P_D	2.0				Watts
Derate above 25°C		0.016				W/°C
Unclamped Inductive Load Energy (See Note 3)	E	32				mJ
Operating and Storage Junction Temperature Range	T_J, T_{stg}	−65 to +150				°C

THERMAL CHARACTERISTICS

Characteristic	Symbol	Max	Unit
Thermal Resistance, Junction to Case	$R_{\theta JC}$	4.167	°C/W
Thermal Resistance, Junction to Ambient	$R_{\theta JA}$	62.5	°C/W

ELECTRICAL CHARACTERISTICS (T_C = 25°C unless otherwise noted)

Characteristic		Symbol	Min	Max	Unit
OFF CHARACTERISTICS					
Collector-Emitter Sustaining Voltage (1)	TIP29, TIP30	$V_{CEO(sus)}$	40	—	Vdc
(I_C = 30 mAdc, I_B = 0)	TIP29A, TIP30A		60	—	
	TIP29B, TIP30B		80	—	
	TIP29C, TIP30C		100	—	
Collector Cutoff Current		I_{CEO}			mAdc
(V_{CE} = 30 Vdc, I_B = 0)	TIP29, TIP29A, TIP30, TIP30A		—	0.3	
(V_{CE} = 60 Vdc, I_B = 0)	TIP29B, TIP29C, TIP30B, TIP30C		—	0.3	
Collector Cutoff Current		I_{CES}			µAdc
(V_{CE} = 40 Vdc, V_{EB} = 0)	TIP29, TIP30		—	200	
(V_{CE} = 60 Vdc, V_{EB} = 0)	TIP29A, TIP30A		—	200	
(V_{CE} = 80 Vdc, V_{EB} = 0)	TIP29B, TIP30B		—	200	
(V_{CE} = 100 Vdc, V_{EB} = 0)	TIP29C, TIP30C		—	200	
Emitter Cutoff Current		I_{EBO}			mAdc
(V_{BE} = 5.0 Vdc, I_C = 0)			—	1.0	
ON CHARACTERISTICS (1)					
DC Current Gain		h_{FE}			—
(I_C = 0.2 Adc, V_{CE} = 4.0 Vdc)			40	—	
(I_C = 1.0 Adc, V_{CE} = 4.0 Vdc)			15	75	
Collector-Emitter Saturation Voltage		$V_{CE(sat)}$	—	0.7	Vdc
(I_C = 1.0 Adc, I_B = 125 mAdc)					
Base-Emitter On Voltage		$V_{BE(on)}$	—	1.3	Vdc
(I_C = 1.0 Adc, V_{CE} = 4.0 Vdc)					
DYNAMIC CHARACTERISTICS					
Current Gain — Bandwidth Product (2)		f_T	3.0	—	MHz
(I_C = 200 mAdc, V_{CE} = 10 Vdc, f_{test} = 1 MHz)					
Small-Signal Current Gain		h_{fe}	20	—	—
(I_C = 0.2 Adc, V_{CE} = 10 Vdc, f = 1 kHz)					

(1) Pulse Test: Pulse Width ⩽ 300 µs, Duty Cycle ⩽ 2.0%.
(2) $f_T = h_{fe} \cdot f_{test}$
(3) This rating based on testing with L_C = 20 mH, R_{BE} = 100 Ω, V_{CC} = 10 V, I_C = 1.8 A, P.R.F. = 10 Hz.

STYLE 1
PIN 1 BASE
2 COLLECTOR
3 EMITTER
4 COLLECTOR

DIM	MILLIMETERS MIN	MILLIMETERS MAX	INCHES MIN	INCHES MAX
A	14.60	15.75	0.575	0.620
B	9.65	10.29	0.380	0.405
C	4.06	4.82	0.160	0.190
D	0.64	0.89	0.025	0.035
F	3.61	3.73	0.142	0.147
G	2.41	2.67	0.095	0.105
H	2.79	3.93	0.110	0.155
J	0.36	0.56	0.014	0.022
K	12.70	14.27	0.500	0.562
L	1.14	1.39	0.045	0.055
N	4.83	5.33	0.190	0.210
Q	2.54	3.04	0.100	0.120
R	2.04	2.79	0.080	0.110
S	1.14	1.39	0.045	0.055
T	5.97	6.48	0.235	0.255
U	0.00	1.27	0.000	0.050
V	1.14	—	0.045	—
Z	—	2.03	—	0.080

**CASE 221A-02
TO-220AB**

TIP29, TIP29A, TIP29B, TIP29C, NPN, TIP30, TIP30A, TIP30B, TIP30C, PNP

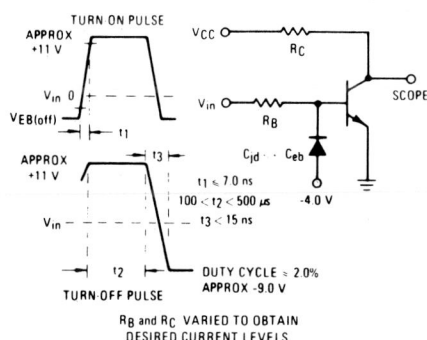

FIGURE 1 – DC CURRENT GAIN

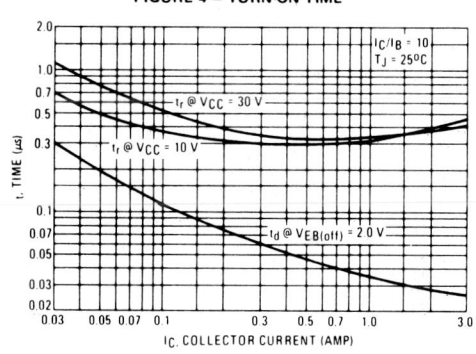

FIGURE 2 – TURN-OFF TIME

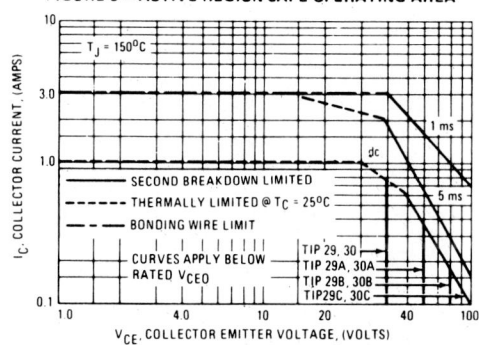

FIGURE 3 – SWITCHING TIME EQUIVALENT CIRCUIT

FIGURE 4 – TURN-ON TIME

FIGURE 5 – ACTIVE REGION SAFE OPERATING AREA

There are two limitations on the power handling ability of a transistor: average junction temperature and second breakdown. Safe operating area curves indicate I_C-V_{CE} operation; i.e., the transistor must not be subjected to greater dissipation than the curves indicate.

The data of Figure 5 is based on $T_{J(pk)} = 150°C$; T_C is variable depending on conditions. Second breakdown pulse limits are valid for duty cycles to 10% provided $T_{J(pk)} \leq 150°C$. At high case temperatures, thermal limitations will reduce the power that can be handled to values less than the limitations imposed by second breakdown.

(Copyright of Motorola, Inc. Used by permission.)

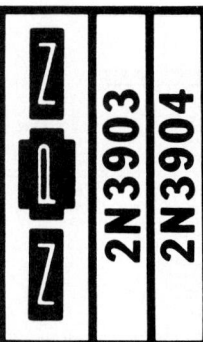

NPN

2N3903
2N3904

Silicon Transistors

The General Electric 2N3903 and 2N3904 are silicon NPN planar epitaxial transistors designed for general purpose switching and amplifier applications.

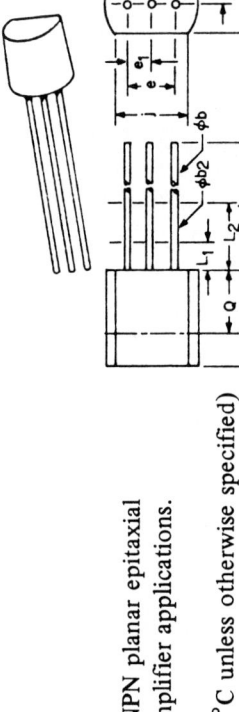

1. EMITTER
2. BASE
3. COLLECTOR

TO-92

SYMBOL	MILLIMETERS		INCHES		NOTES
	MIN.	MAX.	MIN.	MAX.	
A	4.320	5.330	.170	.210	
ϕb	.407	.550	.016	.022	1,3
$\phi b2$	.407	.482	.016	.019	3
ϕD	4.450	5.200	.175	.205	
E	3.180	4.190	.125	.165	
e	2.410	2.670	.095	.105	
e_1	1.150	1.395	.045	.055	
j	3.430	4.320	.135	.170	
L	12.700	–	.500	–	1,3
L_1	–	1.270	–	.050	3
L_2	6.350	–	.250	–	3
Q	2.920	–	.115	–	3
s	2.030	2.670	.080	.105	2

NOTES:
1. THREE LEADS
2. CONTOUR OF PACKAGE UNCONTROLLED OUTSIDE THIS SIDE.
3. (THREE LEADS) $\phi b2$ APPLIES BETWEEN L_1 AND L_2. ϕb APPLIES BETWEEN L_2 AND 12.70 MM (.500") FROM THE SEATING PLANE. DIAMETER IS UNCONTROLLED IN L_1 AND BEYOND 12.70 MM (.500") FROM SEATING PLANE.

absolute maximum ratings: ($T_A = 25°C$ unless otherwise specified)

VOLTAGES

Collector to Emitter	V_{CEO}	40	Volts
Collector to Base	V_{CBO}	60	Volts
Emitter to Base	V_{EBO}	6	Volts

CURRENT

Collector	I_C	200	mA

DISSIPATION

Total Power $T_A \leq 25°C$	P_T	350	m Watts
Derate Factor $T_A > 25°C$	P_T	2.8	mW/°C

TEMPERATURE

Operating	T_J	-55°C to +135°C	°C
Storage	T_{STG}	-55°C to +135°C	°C
Lead (1/16" ± 1/32" from case for 10 sec.)	T_L	+230°C	°C

*electrical characteristics: ($T_A = 25°C$ unless otherwise specified)

STATIC CHARACTERISTICS		SYMBOL	MIN.	MAX.	UNITS
Collector-Emitter Breakdown Voltage		$V_{(BR)CEO}$	40	—	Volts
($I_C = 1mA$, $I_B = 0$)					
Collector-Base Breakdown Voltage		$V_{(BR)CBO}$	60	—	Volts
($I_C = 10\mu A$, $I_E = 0$)					
Emitter-Base Breakdown Voltage		$V_{(BR)EBO}$	6	—	Volts
($I_E = 10\mu A$, $I_C = 0$)					
Collector Cutoff Current		I_{CEV}	—	50	nA
($V_{CE} = 30V$, V_{EB} (off) = 3V)					
Base Cutoff Current		I_{BEV}	—	50	nA
($V_{CE} = 30V$, V_{EB} (off) = 3V)					
Forward Current Transfer Ratio					
($V_{CE} = 1V$, $I_C = 100\mu A$)	2N3903	h_{FE}	20	—	
	2N3904	h_{FE}	40	—	
($V_{CE} = 1V$, $I_C = 1mA$)	2N3903	h_{FE}	35	—	
	2N3904	h_{FE}	70	—	
($V_{CE} = 1V$, $I_C = 10mA$)	2N3903	†h_{FE}	50	150	
	2N3904	†h_{FE}	100	300	
($V_{CE} = 1V$, $I_C = 50mA$)	2N3903	†h_{FE}	30	—	
	2N3904	†h_{FE}	60	—	
($V_{CE} = 1V$, $I_C = 100mA$)	2N3903	†h_{FE}	15	—	
	397 2N3904	†h_{FE}	30	—	

(Reprinted with permission of GE Semiconductor Research, Triangle Park, NC)

2N3903
2N3904

STATIC CHARACTERISTICS (Continued)

	Device	SYMBOL	MIN.	MAX.	UNITS
Collector-Emitter Saturation Voltage					
($I_C = 10mA$, $I_B = 1mA$)		$†V_{CE(sat)}$	—	.200	Volts
($I_C = 50mA$, $I_B = 5mA$)		$†V_{CE(sat)}$	—	.300	Volts
Base-Emitter Saturation Voltage					
($I_C = 10mA$, $I_B = 1mA$)		$†V_{BE(sat)}$	.65	.85	Volts
($I_C = 50mA$, $I_B = 5mA$)		$†V_{BE(sat)}$	—	.95	

DYNAMIC CHARACTERISTICS

	Device	SYMBOL	MIN.	MAX.	UNITS
Collector-Base Capacitance					
($V_{CB} = 5V$, $I_E = O, f = 1\,MHz$)		C_{cb}	—	4	pF
Emitter-Base Capacitance					
($V_{EB} = .5V$, $I_C = O, f = 1\,MHz$)		C_{eb}	—	8	pF
Current – Gain – Bandwidth Product					
($V_{CE} = 20V$, $I_E = 10mA, f = 100\,MHz$)	2N3903	f_T	250	—	MHz
	2N3904	f_T	300	—	MHz
Noise Figure					
($I_E = 100\mu A$, $V_{CE} = 5V$, $R_G = 1\,kHz$)	2N3903	NF	—	6	dB
BW = 15.7 kHz	2N3904	NF	—	5	dB
Turn-On Delay		t_d	—	35	ns
Collector Current Rise Time		t_r	—	35	ns
($I_C = 10mA$, $I_{B1} = 1mA$, V_{BE} (off) = .5V)					
($R_L = 275\Omega$)					
Storage Delay Time	2N3903	t_s	—	175	ns
	2N3904	t_s	—	200	ns

Collector Current Fall Time
($I_C = 10mA, I_{B1} = I_{B2} = 1mA$)
($R_L = 275\Omega, V_{CC} = 3V$)
Hybrid Parameters
($I_E = 1mA, V_{CE} = 10V, f = 1KHz$)

		Min	Max	Units
t_f		—	50	ns
h_{fe}	2N3903	50	200	
h_{fe}	2N3904	100	400	
h_{ie}	2N3903	.5	8	kΩ
h_{ie}	2N3904	1	10	kΩ
h_{re}	2N3903	.1	5	X10^{-4}
h_{re}	2N3904	.5	.8	X10^{-4}
h_{oe}		1.0	40	μmhos

†Pulse width ≤ 300 μsec., Duty Cycle ≤ 2%.
*JEDEC Registered Parameters.

SWITCHING TIME EQUIVALENT TEST CIRCUITS

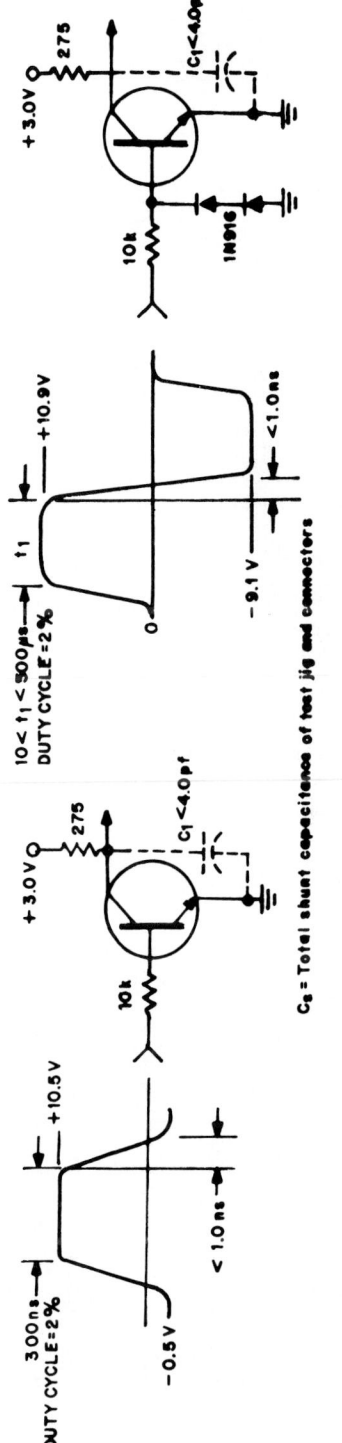

1. TURN-ON TIME TEST CIRCUIT t_d AND t_r

2. TURN-OFF TIME TEST CIRCUIT t_s AND t_f

C_s = Total shunt capacitance of test jig and connectors

(Reprinted with permission of GE Semiconductor Research, Triangle Park, NC)

993

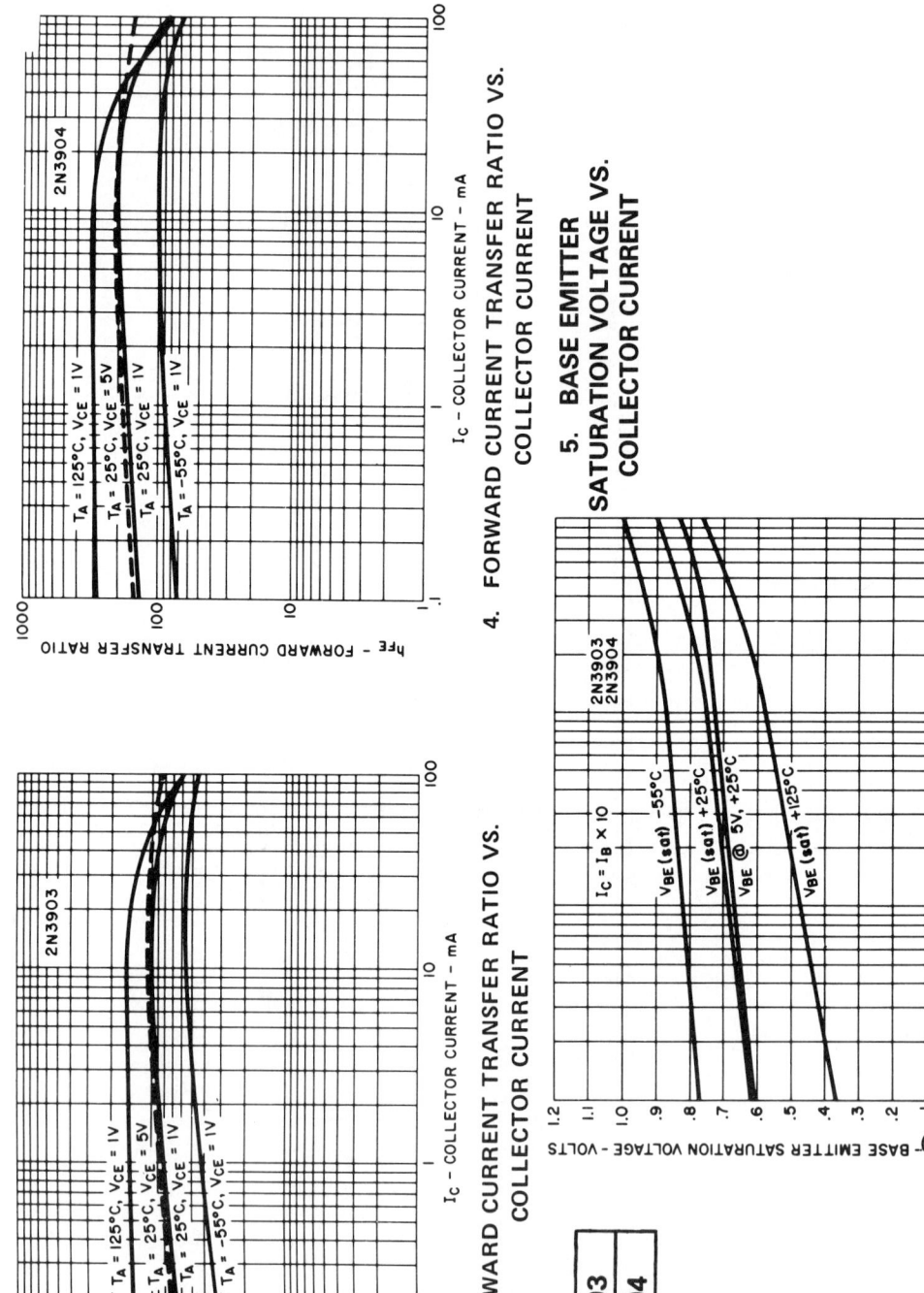

3. FORWARD CURRENT TRANSFER RATIO VS.
COLLECTOR CURRENT

4. FORWARD CURRENT TRANSFER RATIO VS.
COLLECTOR CURRENT

5. BASE EMITTER
SATURATION VOLTAGE VS.
COLLECTOR CURRENT

2N3903
2N3904

(Reprinted with permission of GE Semiconductor Research, Triangle Park, NC)

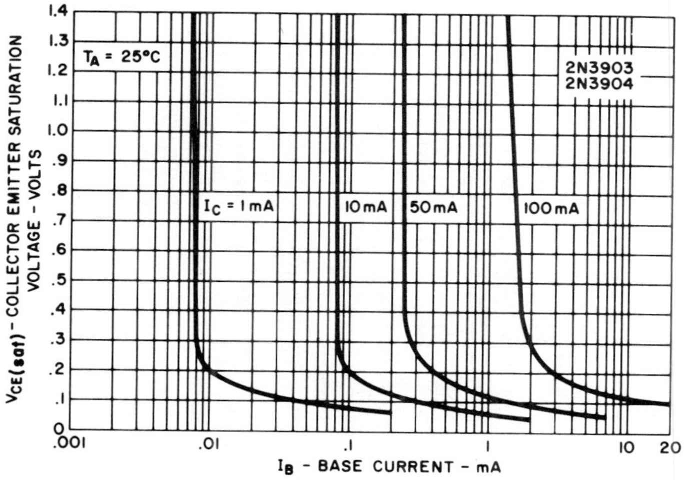

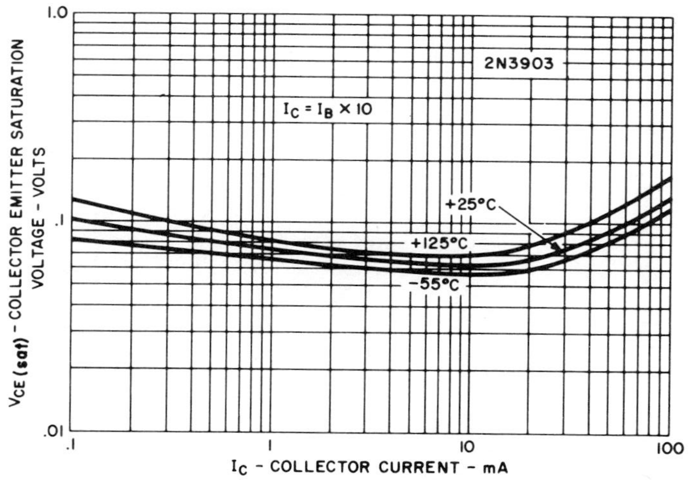

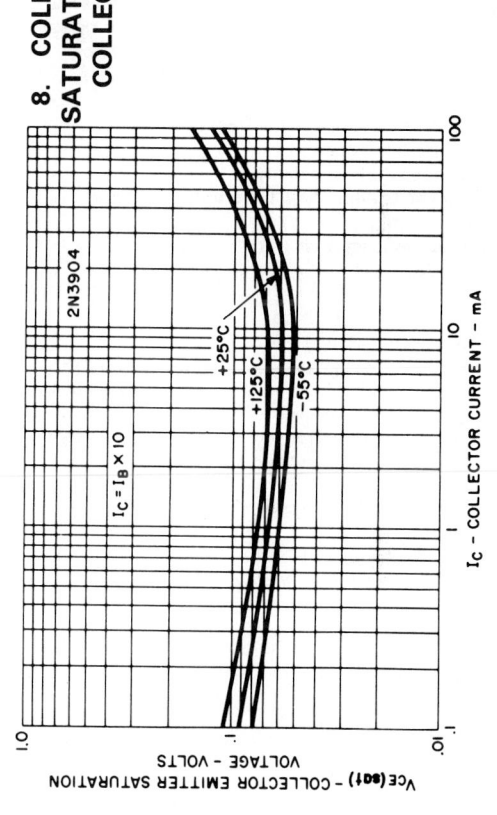

7. COLLECTOR EMITTER SATURATION
VOLTAGE VS. COLLECTOR CURRENT

2N3903

$I_C = I_B \times 10$

+25°C
+125°C
−55°C

$V_{CE(sat)}$ − COLLECTOR EMITTER SATURATION
VOLTAGE − VOLTS

I_C − COLLECTOR CURRENT − mA

8. COLLECTOR EMITTER
SATURATION VOLTAGE VS.
COLLECTOR CURRENT

2N3904

$I_C = I_B \times 10$

+25°C
+125°C
−55°C

$V_{CE(sat)}$ − COLLECTOR EMITTER SATURATION
VOLTAGE − VOLTS

I_C − COLLECTOR CURRENT − mA

6. COLLECTOR EMITTER SATURATION
VOLTAGE VS. BASE CURRENT

2N3903
2N3904

$T_A = 25°C$

$I_C = 1mA$

10mA

50mA

100mA

$V_{CE(sat)}$ − COLLECTOR EMITTER SATURATION
VOLTAGE − VOLTS

I_B − BASE CURRENT − mA

2N3903
2N3904

(Reprinted with permission of GE Semiconductor Research, Triangle Park, NC)

A Schlumberger Company

µA741
Operational Amplifier

Description

The µA741 is a high performance monolithic operational amplifier constructed using the Fairchild Planar Epitaxial process. It is intended for a wide range of analog applications. High common mode voltage range and absence of latch up tendencies make the µA741 ideal for use as a voltage follower. The high gain and wide range of operating voltage provide superior performance in integrator, summing amplifier, and general feedback applications.

- **No Frequency Compensation Required**
- **Short Circuit Protection**
- **Offset Voltage Null Capability**
- **Large Common Mode And Differential Voltage Ranges**
- **Low Power Consumption**
- **No Latch Up**

Absolute Maximum Ratings

Storage Temperature Range	
Metal Can and Ceramic DIP	−65°C to +175°C
Molded DIP and SO-8	−65°C to +150°C
Operating Temperature Range	
Extended (µA741AM, µA741M)	−55°C to +125°C
Commercial (µA741EC, µA741C)	0°C to +70°C
Lead Temperature	
Metal Can and Ceramic DIP (soldering, 60 s)	300°C
Molded DIP and SO-8 (soldering, 10 s)	265°C
Internal Power Dissipation[1, 2]	
8L-Metal Can	1.00 W
8L-Molded DIP	0.93 W
8L-Ceramic DIP	1.30 W
SO-8	0.81 W
Supply Voltage	
µA741A, µA741, µA741E	± 22 V
µA741C	± 18 V
Differential Input Voltage	± 30 V
Input Voltage[3]	± 15 V
Output Short Circuit Duration[4]	Indefinite

Notes

1. $T_{J\ Max}$ = 150°C for the Molded DIP and SO-8, and 175°C for the Metal Can and Ceramic DIP.
2. Ratings apply to ambient temperature at 25°C. Above this temperature, derate the 8L-Metal Can at 6.7 mW/°C, the 8L-Molded DIP at 7.5 mW/°C, the 8L-Ceramic DIP at 8.7 mW/°C, and the SO-8 at 6.5 mW/°C.
3. For supply voltages less than ± 15 V, the absolute maximum input voltage is equal to the supply voltage.
4. Short circuit may be to ground or either supply. Rating applies to 125°C case temperature or 75°C ambient temperature.

Connection Diagram
8-Lead Metal Package
(Top View)

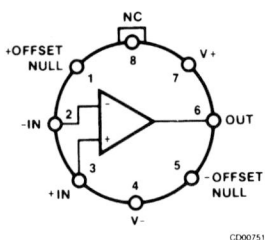

Lead 4 connected to case.

Order Information

Device Code	Package Code	Package Description
µA741HM	5W	Metal
µA741HC	5W	Metal
µA741AHM	5W	Metal
µA741EHC	5W	Metal

Connection Diagram
8-Lead DIP and SO-8 Package
(Top View)

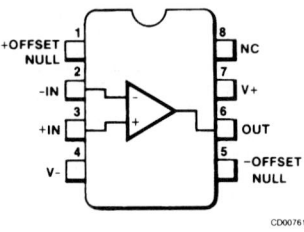

Order Information

Device Code	Package Code	Package Description
µA741RM	6T	Ceramic DIP
µA741RC	6T	Ceramic DIP
µA741SC	KC	Molded Surface Mount
µA741TC	9T	Molded DIP
µA741ARM	6T	Ceramic DIP
µA741ERC	6T	Ceramic DIP
µA741ETC	9T	Molded DIP

Equivalent Circuit

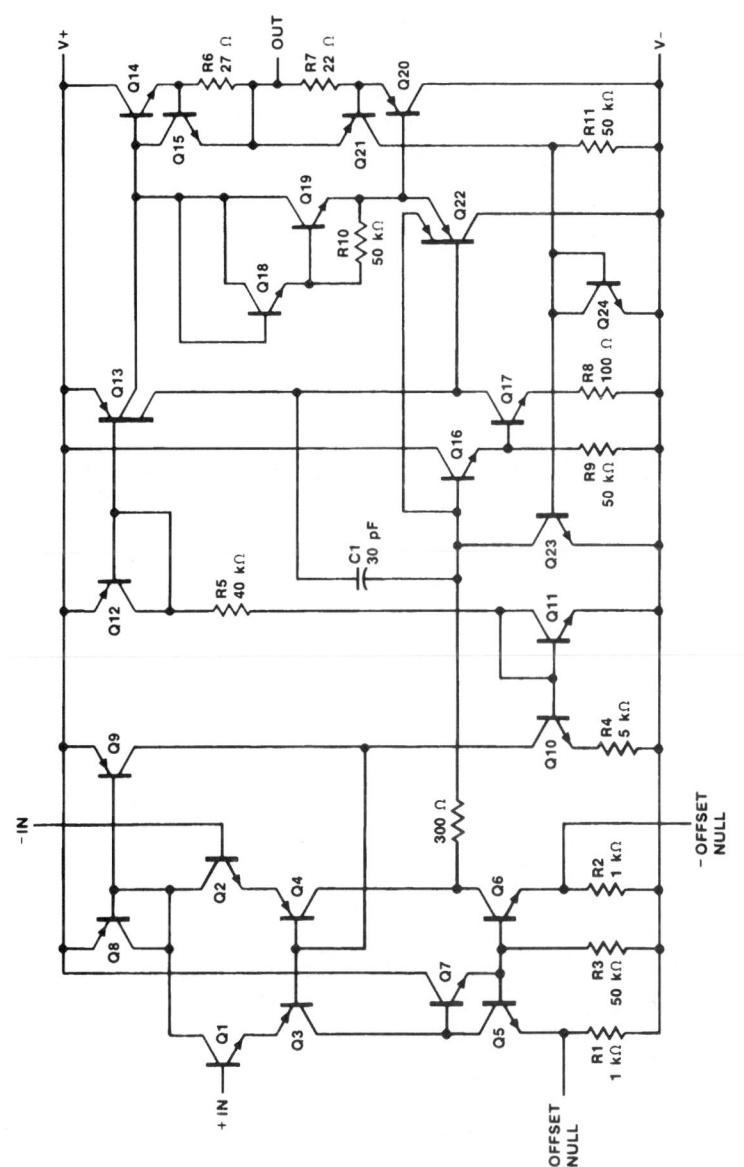

B00035IF

μA741 and μA741C
Electrical Characteristics $T_A = 25°C$, $V_{CC} = \pm 15$ V, unless otherwise specified.

Symbol	Characteristic	Condition	μA741			μA741C			Unit
			Min	Typ	Max	Min	Typ	Max	
V_{IO}	Input Offset Voltage	$R_S \leqslant 10$ kΩ		1.0	5.0		2.0	6.0	mV
$V_{IO\ adj}$	Input Offset Voltage Adjustment Range			±15			±15		mV
I_{IO}	Input Offset Current			20	200		20	200	nA
I_{IB}	Input Bias Current			80	500		80	500	nA
Z_I	Input Impedance		0.3	2.0		0.3	2.0		MΩ
I_{CC}	Supply Current			1.7	2.8		1.7	2.8	mA
P_c	Power Consumption			50	85		50	85	mW
CMR	Common Mode Rejection		70			70	90		dB
V_{IR}	Input Voltage Range		±12	±13		±12	±13		V
PSRR	Power Supply Rejection Ratio	$V_{CC} = \pm 5.0$ V to ±18 V		30	150		30	150	μV/V
I_{OS}	Output Short Circuit Current			25			25		mA
A_{VS}	Large Signal Voltage Gain	$R_L \geqslant 2.0$ kΩ, $V_O = \pm 10$ V	50	200		20	200		V/mV
V_{OP}	Output Voltage Swing	$R_L = 10$ kΩ	±12	±13		±12	±14		V
		$R_L = 2.0$ kΩ	±10	±13		±10	±13		
TR	Transient Response — Rise time	$V_I = 20$ mV, $R_L = 2.0$ kΩ, $C_L = 100$ pF, $A_V = 1.0$		0.3			0.3		μs
	Transient Response — Overshoot			5.0			5.0		%
BW	Bandwidth			1.0			1.0		MHz
SR	Slew Rate	$R_L \geqslant 2.0$ kΩ, $A_V = 1.0$		0.5			0.5		V/μs

(*Copyright Fairchild Semiconductor Corporation, 1987. Used by permission.*)

μA741 and μA741C (Cont.)
Electrical Characteristics Over the range of −55°C ≤ T_A ≤ +125°C for μA741, 0°C ≤ T_A ≤ +70°C for μA741C, unless otherwise specified.

Symbol	Characteristic	Condition	μA741			μA741C			Unit
			Min	Typ	Max	Min	Typ	Max	
V_{IO}	Input Offset Voltage	R_S ≤ 10 kΩ		1.0	6.0			7.5	mV
$V_{IO\ adj}$	Input Offset Voltage Adjustment Range			±15			±15		mV
I_{IO}	Input Offset Current							300	nA
		T_A = +125°C		7.0	200				
		T_A = −55°C		85	500				
I_{IB}	Input Bias Current							800	nA
		T_A = +125°C		0.03	0.5				μA
		T_A = −55°C		0.3	1.5				
I_{CC}	Supply Current	T_A = +125°C		1.5	2.5				mA
		T_A = −55°C		2.0	3.3				
P_c	Power Consumption	T_A = +125°C		45	75				mW
		T_A = −55°C		60	100				
CMR	Common Mode Rejection	R_S ≤ 10 kΩ	70	90					dB
V_{IR}	Input Voltage Range		±12	±13					V
PSRR	Power Supply Rejection Ratio			30	150				μV/V
A_{VS}	Large Signal Voltage Gain	R_L ≥ 2.0 kΩ, V_O = ±10 V	25			15			V/mV
V_{OP}	Output Voltage Swing	R_L = 10 kΩ	±12	±14					V
		R_L = 2.0 kΩ	±10	±13		±10	±13		

μA741A and μA741E
Electrical Characteristics $T_A = 25°C$, $V_{CC} = \pm 15$ V, unless otherwise specified.

Symbol	Characteristic		Condition	Min	Typ	Max	Unit
V_{IO}	Input Offset Voltage		$R_S \leqslant 50\ \Omega$		0.8	3.0	mV
I_{IO}	Input Offset Current				3.0	30	nA
I_{IB}	Input Bias Current				30	80	nA
Z_I	Input Impedance		$V_{CC} = \pm 20$ V	1.0	6.0		MΩ
P_c	Power Consumption		$V_{CC} = \pm 20$ V		80	150	mW
PSRR	Power Supply Rejection Ratio		$V_{CC} = +10$ V, -20 V to $V_{CC} = +20$ V, -10 V, $R_S = 50\ \Omega$		15	50	μV/V
I_{OS}	Output Short Circuit Current			10	25	40	mA
A_{VS}	Large Signal Voltage Gain		$V_{CC} = \pm 20$ V, $R_L \geqslant 2.0$ kΩ, $V_O = \pm 15$ V	50	200		V/mV
TR	Transient Response	Rise time	$A_V = 1.0$, $V_{CC} = \pm 20$ V, $V_I = 50$ mV, $R_L = 2.0$ kΩ, $C_L = 100$ pF		0.25	0.8	μs
		Overshoot			6.0	20	%
BW	Bandwidth			0.437	1.5		MHz
SR	Slew Rate		$V_I = \pm 10$ V, $A_V = 1.0$	0.3	0.7		V/μs

The following specifications apply over the range of $-55°C \leqslant T_A \leqslant +125°C$ for the μA741A, and $0°C \leqslant T_A \leqslant +70°C$ for the μA741E.

Symbol	Characteristic	Condition	Min	Typ	Max	Unit
V_{IO}	Input Offset Voltage				4.0	mV
$\Delta V_{IO}/\Delta T$	Input Offset Voltage Temperature Sensitivity				15	μV/°C
V_{IO} adj	Input Offset Voltage Adjustment Range	$V_{CC} = \pm 20$ V	10			mV
I_{IO}	Input Offset Current				70	nA

Symbol	Parameter	Conditions				Units
$\Delta I_{IO}/\Delta T$	Input Offset Current Temperature Sensitivity				0.5	nA/°C
I_{IB}	Input Bias Current				210	nA
Z_I	Input Impedance	$V_{CC} = \pm 20$ V		0.5		MΩ
P_c	Power Consumption	$V_{CC} = \pm 20$ V	μA741A	−55°C		165	mW
				+125°C		135	
			μA741E			150	
CMR	Common Mode Rejection	$V_{CC} = \pm 20$ V, $V_I = \pm 15$ V, $R_S = 50\ \Omega$		80	95	dB
I_{OS}	Output Short Circuit Current			10	40	mA
A_{VS}	Large Signal Voltage Gain	$V_{CC} = \pm 20$ V, $R_L \geqslant 2.0$ kΩ, $V_O = \pm 15$ V		32		V/mV
		$V_{CC} = \pm 5.0$ V, $R_L \geqslant 2.0$ kΩ, $V_O = \pm 2.0$ V		10		
V_{OP}	Output Voltage Swing	$V_{CC} = \pm 20$ V	$R_L = 10$ kΩ	±16		V
			$R_L = 2.0$ kΩ	±15		

(Copyright Fairchild Semiconductor Corporation, 1987. Used by permission.)

1001

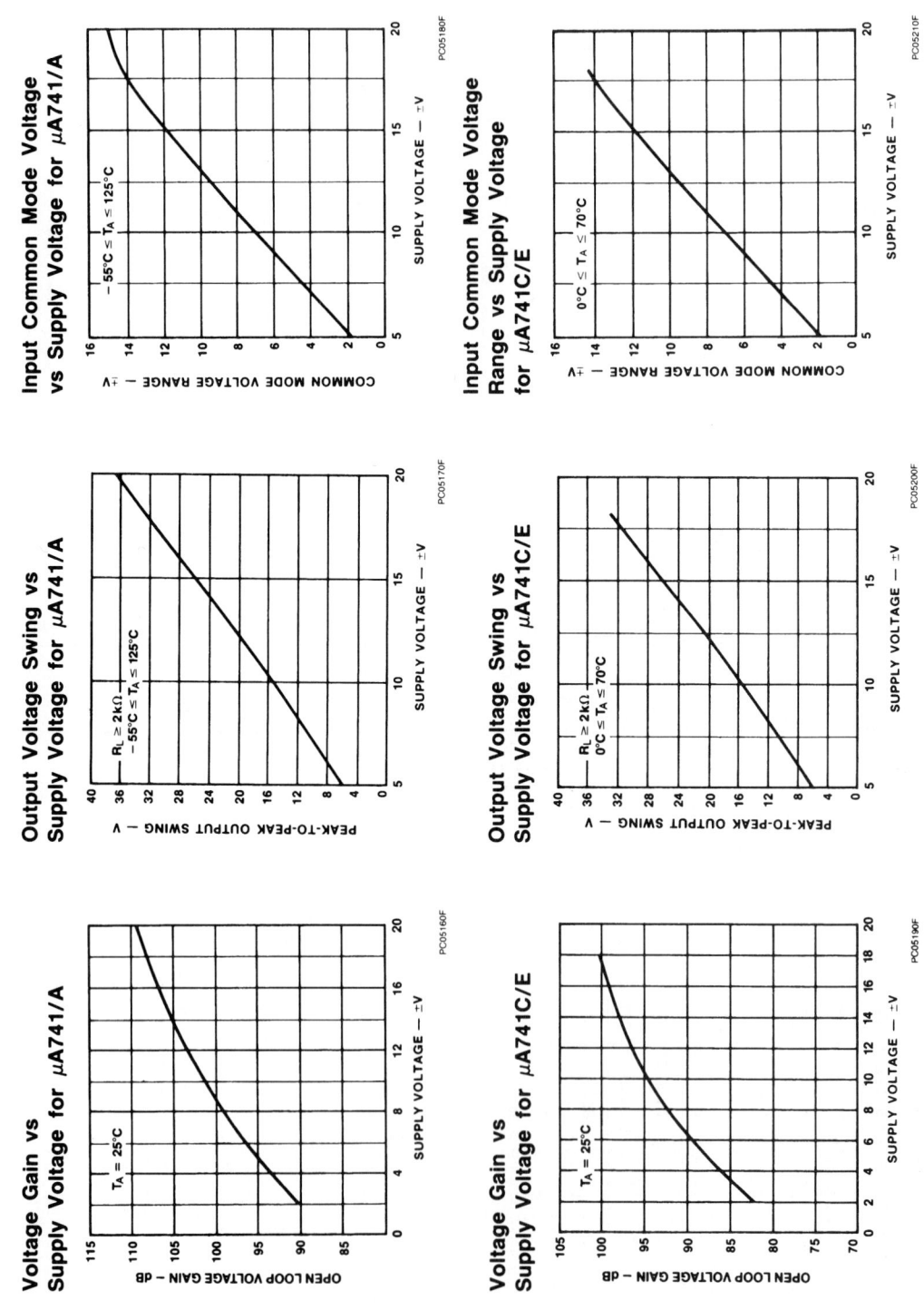

Voltage Gain vs
Supply Voltage for μA741/A

$T_A = 25°C$

OPEN LOOP VOLTAGE GAIN – dB

SUPPLY VOLTAGE — ±V

PC05160F

Output Voltage Swing vs
Supply Voltage for μA741/A

$R_L \geq 2k\Omega$
$-55°C \leq T_A \leq 125°C$

PEAK-TO-PEAK OUTPUT SWING — V

SUPPLY VOLTAGE — ±V

PC05170F

Input Common Mode Voltage
vs Supply Voltage for μA741/A

$-55°C \leq T_A \leq 125°C$

COMMON MODE VOLTAGE RANGE — ±V

SUPPLY VOLTAGE — ±V

PC05180F

Voltage Gain vs
Supply Voltage for μA741C/E

$T_A = 25°C$

OPEN LOOP VOLTAGE GAIN – dB

SUPPLY VOLTAGE — ±V

PC05190F

Output Voltage Swing vs
Supply Voltage for μA741C/E

$R_L \geq 2k\Omega$
$0°C \leq T_A \leq 70°C$

PEAK-TO-PEAK OUTPUT SWING — V

SUPPLY VOLTAGE — ±V

PC05200F

Input Common Mode Voltage
Range vs Supply Voltage
for μA741C/E

$0°C \leq T_A \leq 70°C$

COMMON MODE VOLTAGE RANGE — ±V

SUPPLY VOLTAGE — ±V

PC05210F

Common Mode Rejection Ratio vs Frequency for μA741C/E

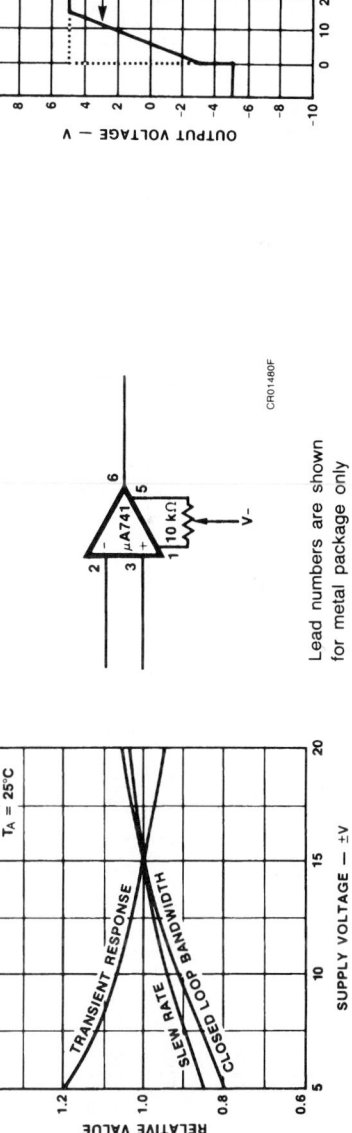

PC05241F

Voltage Follower Large Signal Pulse Response for μA741C/E

Transient Response Test Circuit for μA741C/E

CR01470F

Lead numbers are shown for metal package only

Voltage Offset Null Circuit for μA741C/E

CR01480F

Lead numbers are shown for metal package only

Transient Response for μA741C/E

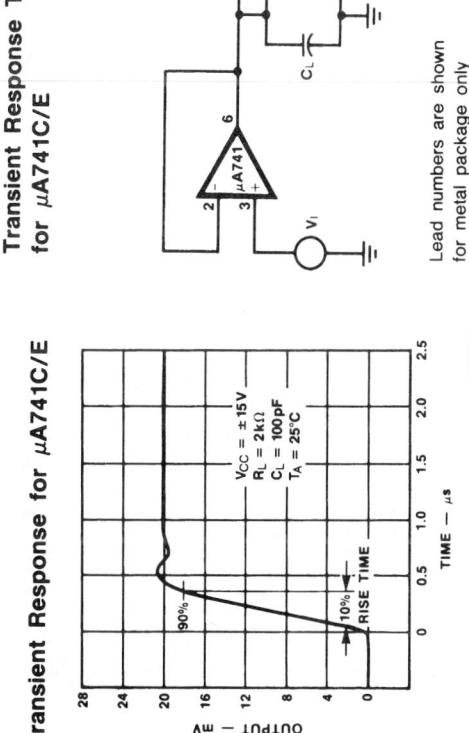

PC05220F

Frequency Characteristics vs Supply Voltage for μA741C/E

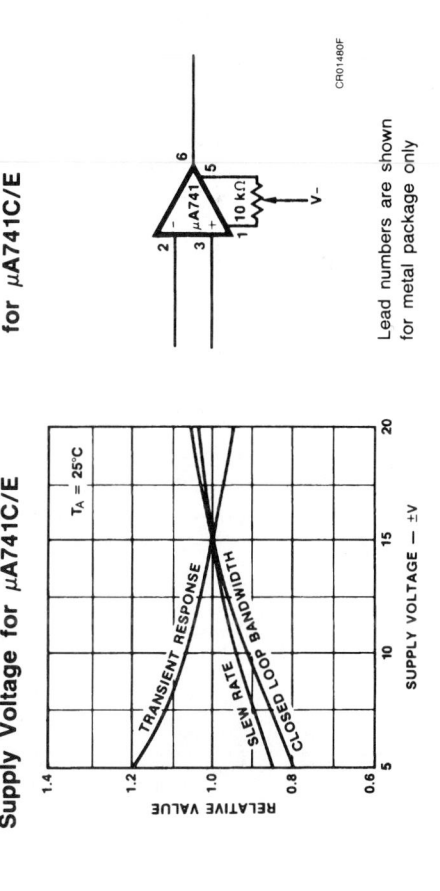

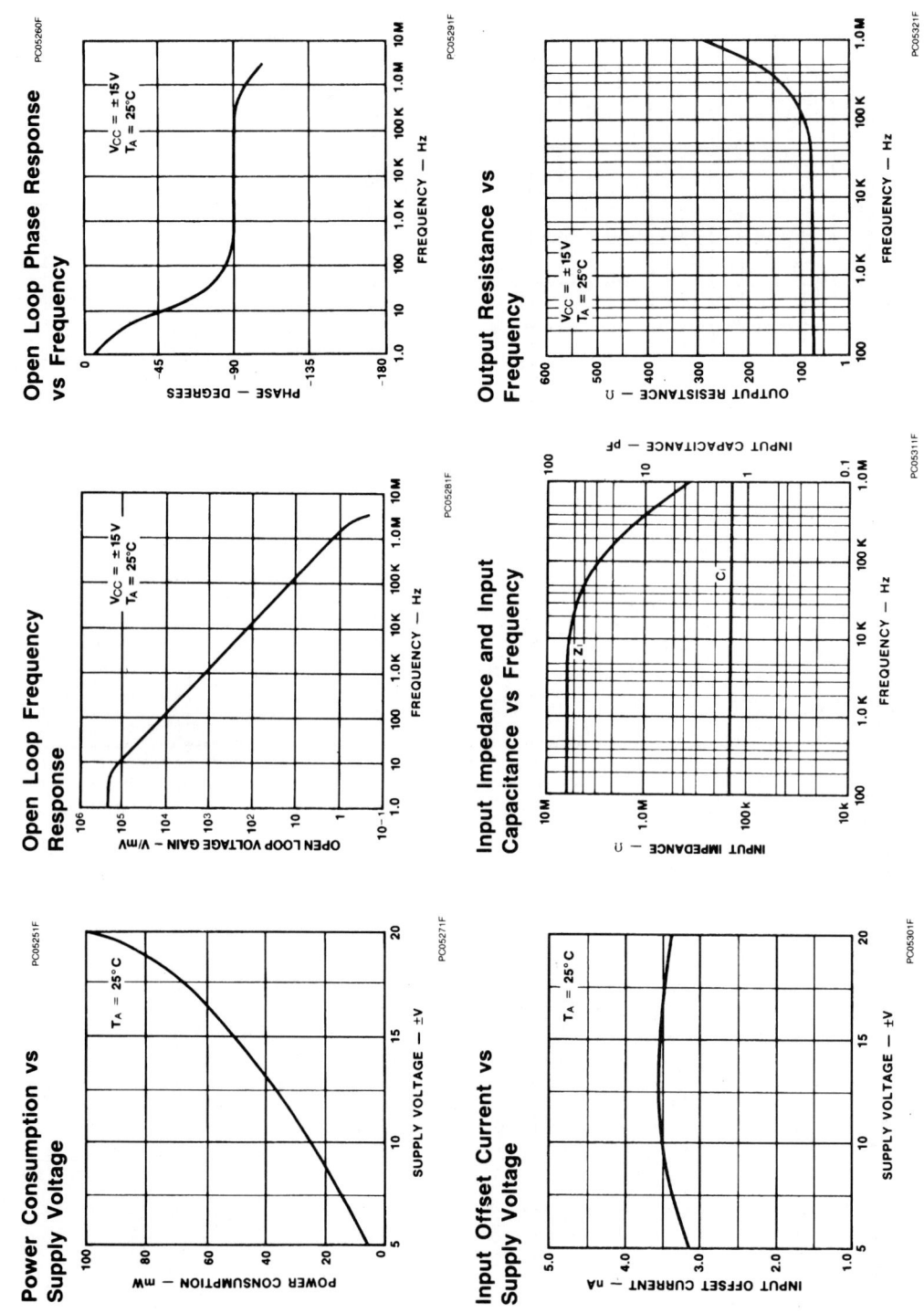

Open Loop Phase Response vs Frequency

PC05260F

PHASE — DEGREES

FREQUENCY — Hz

$V_{CC} = \pm15V$
$T_A = 25°C$

Output Resistance vs Frequency

PC05321F

OUTPUT RESISTANCE — Ω

FREQUENCY — Hz

$V_{CC} = \pm15V$
$T_A = 25°C$

Open Loop Frequency Response

PC05281F

OPEN LOOP VOLTAGE GAIN — V/mV

FREQUENCY — Hz

$V_{CC} = \pm15V$
$T_A = 25°C$

Input Impedance and Input Capacitance vs Frequency

PC05311F

INPUT CAPACITANCE — pF

INPUT IMPEDANCE — Ω

FREQUENCY — Hz

Z_i

C_i

Power Consumption vs Supply Voltage

PC05251F

POWER CONSUMPTION — mW

SUPPLY VOLTAGE — ±V

$T_A = 25°C$

Input Offset Current vs Supply Voltage

PC05301F

INPUT OFFSET CURRENT — nA

SUPPLY VOLTAGE — ±V

$T_A = 25°C$

PC05291F

PC05271F

1004

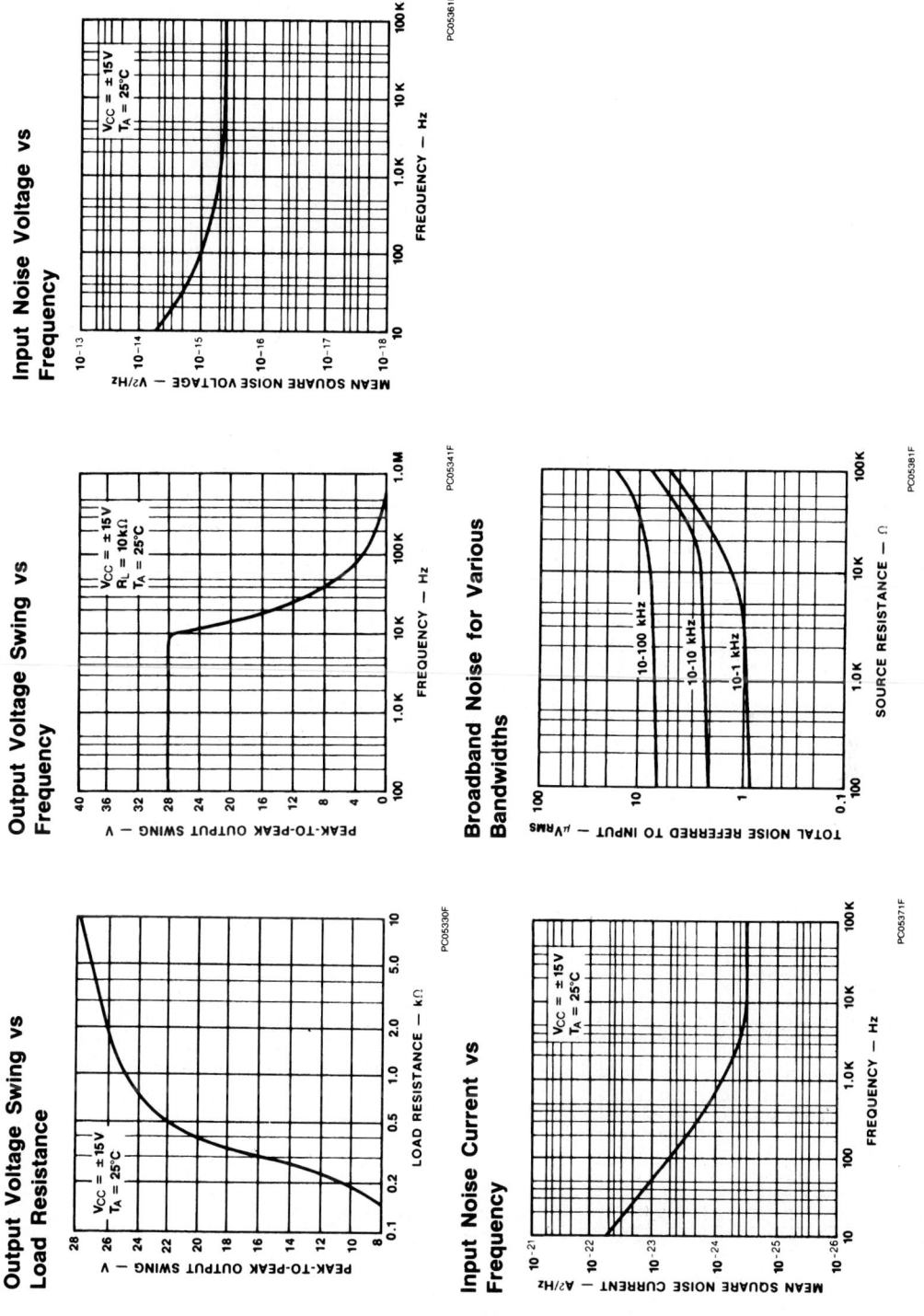

Output Voltage Swing vs Load Resistance

PEAK-TO-PEAK OUTPUT SWING — V

LOAD RESISTANCE — kΩ

$V_{CC} = \pm 15V$
$T_A = 25°C$

PC05330F

Output Voltage Swing vs Frequency

PEAK-TO-PEAK OUTPUT SWING — V

FREQUENCY — Hz

$V_{CC} = \pm 15V$
$R_L = 10kΩ$
$T_A = 25°C$

PC05341F

Input Noise Voltage vs Frequency

MEAN SQUARE NOISE VOLTAGE — V^2/Hz

FREQUENCY — Hz

$V_{CC} = \pm 15V$
$T_A = 25°C$

PC05361F

Input Noise Current vs Frequency

MEAN SQUARE NOISE CURRENT — A^2/Hz

FREQUENCY — Hz

$V_{CC} = \pm 15V$
$T_A = 25°C$

PC05371F

Broadband Noise for Various Bandwidths

TOTAL NOISE REFERRED TO INPUT — μV_{RMS}

SOURCE RESISTANCE — Ω

10-100 kHz
10-10 kHz
10-1 kHz

PC05381F

1005

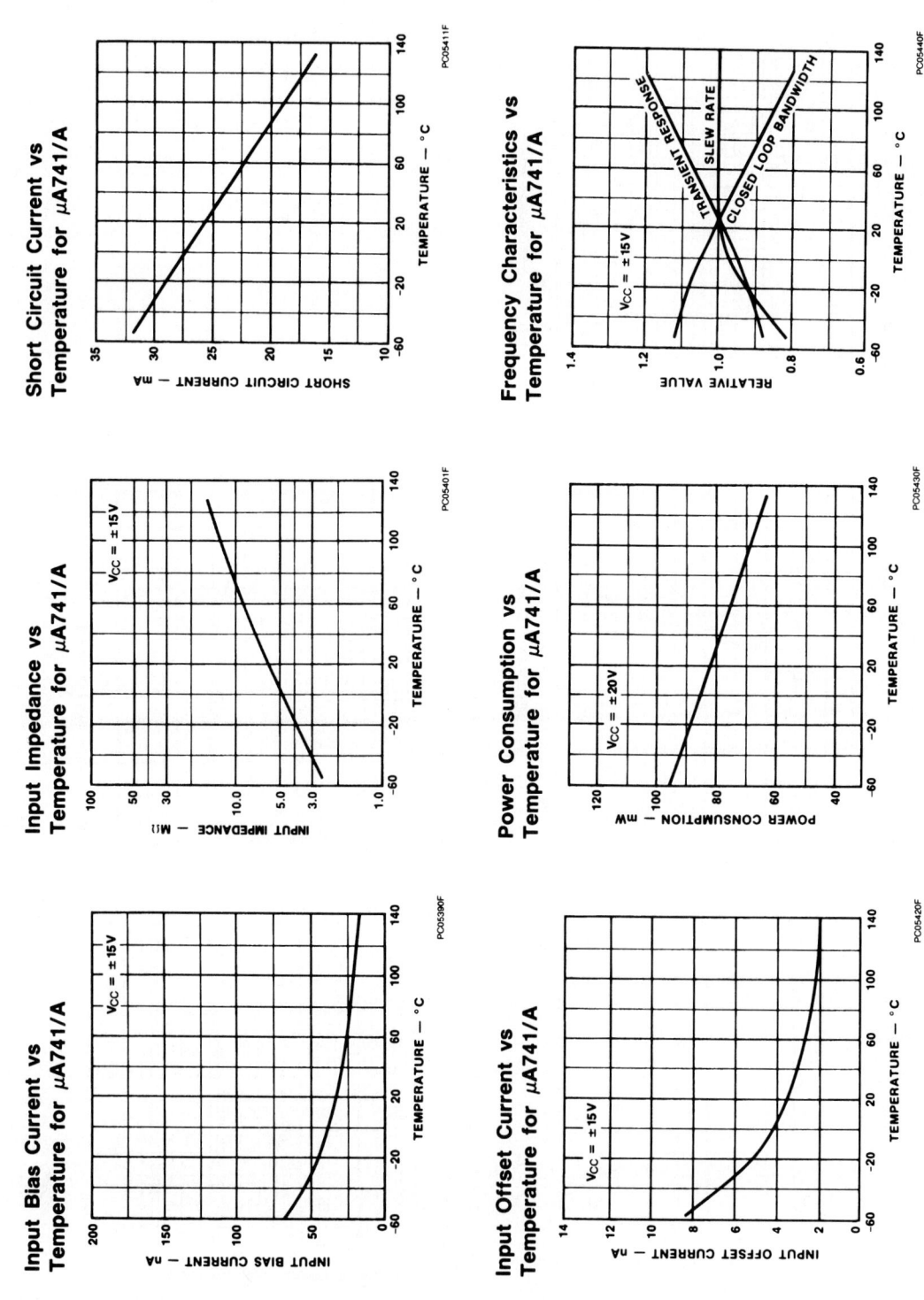

Short Circuit Current vs Temperature for μA741/A

SHORT CIRCUIT CURRENT — mA

TEMPERATURE — °C

PO05411F

Input Impedance vs Temperature for μA741/A

$V_{CC} = \pm 15V$

INPUT IMPEDANCE — MΩ

TEMPERATURE — °C

PO05401F

Input Bias Current vs Temperature for μA741/A

$V_{CC} = \pm 15V$

INPUT BIAS CURRENT — nA

TEMPERATURE — °C

PO05390F

Frequency Characteristics vs Temperature for μA741/A

$V_{CC} = \pm 15V$

TRANSIENT RESPONSE

SLEW RATE

CLOSED LOOP BANDWIDTH

RELATIVE VALUE

TEMPERATURE — °C

PO05440F

Power Consumption vs Temperature for μA741/A

$V_{CC} = \pm 20V$

POWER CONSUMPTION — mW

TEMPERATURE — °C

PO05430F

Input Offset Current vs Temperature for μA741/A

$V_{CC} = \pm 15V$

INPUT OFFSET CURRENT — nA

TEMPERATURE — °C

PO05420F

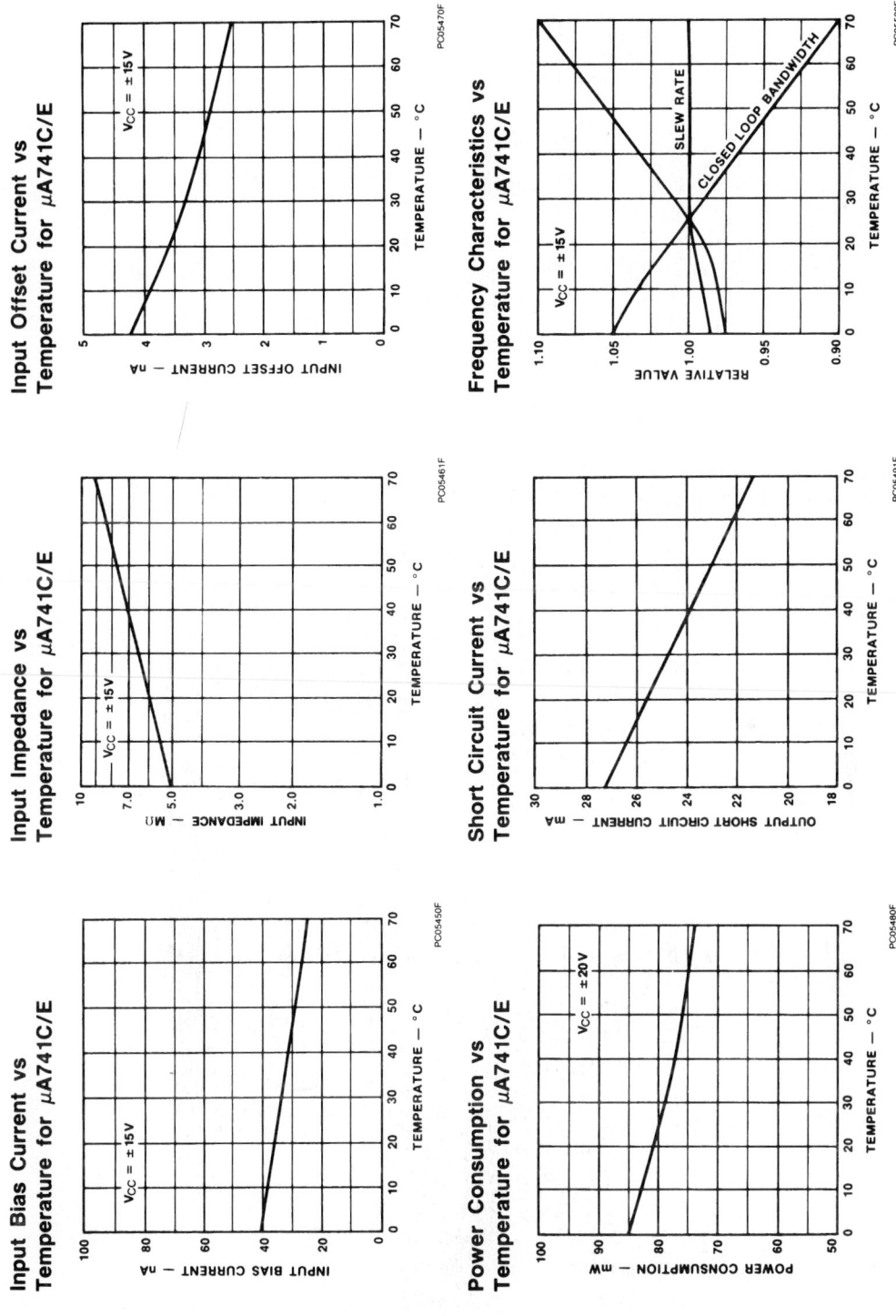

Input Offset Current vs Temperature for µA741C/E

INPUT OFFSET CURRENT — nA

$V_{CC} = \pm 15V$

TEMPERATURE — °C

PC05470F

Frequency Characteristics vs Temperature for µA741C/E

RELATIVE VALUE

$V_{CC} = \pm 15V$

SLEW RATE

CLOSED-LOOP BANDWIDTH

TEMPERATURE — °C

PC05500F

Input Impedance vs Temperature for µA741C/E

INPUT IMPEDANCE — MΩ

$V_{CC} = \pm 15V$

TEMPERATURE — °C

PC05461F

Short Circuit Current vs Temperature for µA741C/E

OUTPUT SHORT CIRCUIT CURRENT — mA

TEMPERATURE — °C

PC05491F

Input Bias Current vs Temperature for µA741C/E

INPUT BIAS CURRENT — nA

$V_{CC} = \pm 15V$

TEMPERATURE — °C

PC05450F

Power Consumption vs Temperature for µA741C/E

POWER CONSUMPTION — mW

$V_{CC} = \pm 20V$

TEMPERATURE — °C

PC05480F

ELEMENTS OF PSPICE

E–1 GENERAL DISCUSSION

The original Spice program was developed at the University of California at Berkeley in the 1970s. Spice stands for *Simulation Program with Integrated Circuit Emphasis*. This program, along with several variations and improvements, has become an industry and university standard over the years.

PSpice was introduced by MicroSim Inc.* in the 1980s as a version of Spice specifically tailored for use on microcomputers. It has received widespread use in colleges and universities as an aid in studying electrical and electronic circuit analysis and design. Its popularity is in part due to the liberal policy of MicroSim in that they supply a student version to faculty members and permit them to reproduce an unlimited number of copies for student distribution.

The student version of PSpice is limited to 10 transistors, which is much less than the professional version. However, most of the general features of the program are available in the student version. For teaching purposes, the 10-transistor limitation rarely causes a problem since few student problems will involve circuits with more complexity. All PSpice examples given in this text can be solved with the student version.

A PSpice file can be broken into four consecutive parts, as illustrated in Figure E–1. They are (1) the **title;** (2) the **circuit block;** (3) the **control block;** and (4) an **end** statement.

The *title* must be included, but it can be any name that you wish to assign to the circuit. Normally it will consist of one line only, but if necessary, it can be extended.

---

* MicroSim Corporation, 20 Fairbanks, Irvine, CA 92718.

FIGURE E–1
Form of a PSpice file.

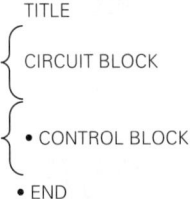

The *circuit block* is usually the longest block (except for very simple circuits), and it contains a complete description of the circuit in a special format that the program will translate to the appropriate mathematical formulation for solution. As a general rule, there will be *one line for each branch* (although more than one line could be used if necessary).

The *control block* provides a series of one or more lines that provides instruction to the program on (1) the type of analysis to be performed and (2) the nature and form of the output results. All lines in this block must begin with a period (.).

The *end* statement must appear as the last line in the file. Like the lines in the control block, it should begin with a period (.). Thus, the correct form is .END.

E–2 EXAMPLE OF PSPICE CIRCUIT CODE

The process of analyzing a circuit is illustrated with a simple dc circuit example. The circuit to be analyzed is shown in Figure E–2(a). The circuit is purely resistive, and it contains one dc voltage source and one dc current source. Although many circuit variables can be determined with PSpice, the desired variables in this case will be the node voltages and the resistive branch currents.

The circuit must be properly labeled for analysis, and this process is shown in Figure E–2(b). First, all nodes must be numbered. A ground node is selected and it is numbered as "0." When an actual working circuit is analyzed, the ground node should normally be selected to conform to the actual circuit ground.

In the present circuit, the bus along the bottom is chosen as the ground. In this text. PSpice nodes are enclosed by circles for clarity. Other nodes are labeled as "1," "2," and "3" as shown.

There is no particular order required on the node numbering. They could just as easily have been numbered as 3, 2, 1 from left to right, or as 1, 3, 2 for that matter. However, some logical pattern is recommended as an aid in checking the circuit. This author tends to follow a left-to-right and bottom-to-top pattern when possible.

Note in the circuit form of Figure E–2(b) and the code of Figure E–2(c) that various labels and values have been changed somewhat in form. First, the various subscripts on parameter names are written on the same levels as the first letter in the code. Subscripts may not be entered on the computer in the same form as

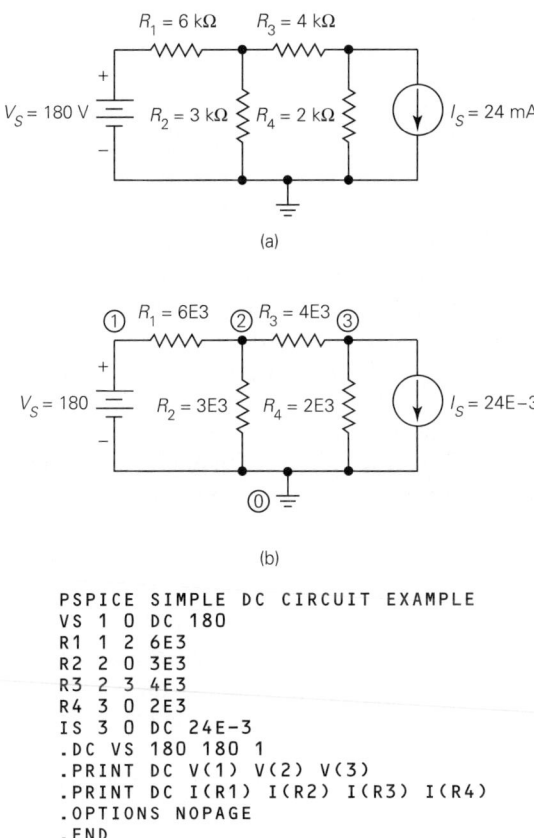

(a)

(b)

```
PSPICE SIMPLE DC CIRCUIT EXAMPLE
VS 1 0 DC 180
R1 1 2 6E3
R2 2 0 3E3
R3 2 3 4E3
R4 3 0 2E3
IS 3 0 DC 24E-3
.DC VS 180 180 1
.PRINT DC V(1) V(2) V(3)
.PRINT DC I(R1) I(R2) I(R3) I(R4)
.OPTIONS NOPAGE
.END
```

(c)

FIGURE E–2
Example used to illustrate PSpice code.

in a textbook, so all entries must be at the same level. For example, V_s must be entered on the computer as VS, and R_1 must be entered as $R1$, and so forth.

Although there are various prefixes available in PSpice for entering values, to simplify this abbreviated treatment, *we will always enter values in their basic units*. When numbers are very small or very large, they will be expressed in scientific or engineering notation form using floating-point values, and the basic units will be understood. For example, a resistance of value $2\,M\Omega$ will be expressed as $2 \times 10^6\,\Omega$, and the proper format we will employ for entering it on the computer will be 2E6. As a second example, a capacitor of value $0.5\,\mu F$ will be expressed as $0.5 \times 10^{-6}\,F$, and the proper format for computer entry will be 0.5E-6.

Now let us turn our attention to the actual circuit description (or *code*), which is shown in (c) of Figure E–2. The first line is the *title*, which has been given the name PSPICE SIMPLE DC CIRCUIT EXAMPLE.

The next six lines constitute the *circuit block,* and a complete description of the circuit is given. Note that there are six branches so that there is one line for each branch.

The first line in the circuit block reads

```
VS 1 0 DC 180
```

There are five separate items on the line. The first item identifies the type of circuit element in the branch. The symbol *V* as the first letter is the code for a voltage source and subsequent letters or numbers (up to 7) *without a space* are additional identifying information. (Since there is only one voltage source in this example, we could actually get by with the symbol *V*.) One or more spaces following VS identify the end of the first item, and this space pattern will continue after additional items.

The next two items in the voltage source description are the integers 1 0, and they identify the two nodes to which the element is connected. *For a voltage source, the order is important in that the first node listed is the positive node of the voltage source.* Thus, if the two integers had been entered in the order 0 1, it would tell the program that the voltage had its positive terminal at ground.

The next item in the line is the designation DC. This tells the computer that the source VS is a dc source. As you might guess, AC would be the designation for an ac phasor source; many other types of sources are also available in PSpice.

The last item is the quantity 180, and this is the value of the dc voltage source in volts. Incidentally if the node order had been expressed as 0 1, a value of -180 would, in effect, still create a positive value at node 1. (We have the algebraic negative of a negative!) However, that form is awkward and not recommended unless necessary for some reason.

The second line in the circuit block reads

```
R1 1 2 6E3
```

There are four separate items on this line. Once again, the first item identifies the type of circuit element in the branch. The symbol *R* as the first letter is the code for a resistance, and the 1 corresponds to the subscript.

The next two items in the description are integers 1 2, and they identify the two nodes to which the resistor is connected. The order of the two nodes represents the reference direction of assumed current flow, which in this case is from node 1 to node 2. As always in circuit analysis, if the current is actually flowing from node 2 to node 1, the current value would turn out to be negative.

As a general rule, if the desired output data are to be voltages only, the order of the nodes for *R, L,* and *C* is arbitrary.

The last item is the value 6E3, and this represents a resistance of $6 \times 10^3 \ \Omega$ or 6 kΩ.

The next three lines describe the remaining three resistive branches, and the code in each case follows the format of *R*1.

The last line in the circuit block reads

```
IS 3 0 DC 24E-3
```

As in the case of the voltage source, the first letter *I* of the first item identifies the element as a current source, and the letter *S* is the subscript. (Since there is only one current source, *I* would have been sufficient.)

The next two items in the current source description are the integers 3 0, and they identify the two nodes to which the element is connected. As in the case of the voltage source, the order is important for a current source.

In the case of a current source, the first node listed is the node at which the current enters the source, and the second node listed is the node at which the current leaves the source. Thus, the direction of flow in this case is from 3 to 0 for the source. If the order of the numbers were reversed, the computer would assume that the current source would be directed upward.

The next item on the line is the designation DC which, of course, identifies the source as a dc source. Finally, the value 24E-3 is the current source value in amperes.

We now direct our attention to the *control* block. The first line reads

```
.DC VS 180 180 1
```

This line is a bit clumsy looking, and it represents one of the eccentricities of PSpice when a single-point analysis is desired. Basically, dc analysis was intended for sweeping an input voltage or current over a range of values. When analysis at a single source value is desired, it is necessary to "fool" the program so that the "sweep" turns out to be a single value.

The first two items represent the type of analysis .DC and the name of the source that is to be "swept," which has been chosen as *VS*. (Since there are two sources in this circuit, *IS* could also have been selected for this purpose.) The three data values are given in order as 180 180 1. The first value represents the beginning value of the sweep, which is 180 V. The second value represents the ending value of the sweep, which is also 180 V in this case. The third value is the increment between successive sweeps, and for this single-point case, any nonzero value would satisfy the computer. The simplest choice is the value 1.

The next line reads

```
.PRINT DC V(1) V(2) V(3)
```

The command .PRINT tells the program that it should provide an output to a printer, and DC indicates that it is the dc data that are desired, which is the only type of data in this example. The next three items represent desired output variables, which for this line are the voltages at nodes 1, 2, and 3, respectively.

A single print statement can handle up to five different output variables in a block form. In this case, there are seven desired output variables (three node voltages and four branch currents). Although two of the currents could have been put on the same line with the voltages, for convenience in the output data, all currents have been entered in a second .PRINT statement. As long as no more than five output variables are to be grouped together, a single .PRINT statement will suffice.

On the second .PRINT line, note the manner in which the currents are listed.

$I(R1)$ represents the current through $R1$; $I(R2)$ represents the current through $R2$, and so forth.

The next line reads

```
.OPTIONS NOPAGE
```

This statement is optional and could have been eliminated without affecting the analysis. In general, when a program requires more than one page of printout, much of the header information concerning title, date, time, and so forth, is repeated page after page. Unless there is a reason for wanting to have this information on each separate page, the use of this option will generally reduce the volume of paper and provide a more compact package of data. Our practice will be to use it extensively throughout.

All the printer output for this example fits on one page, and it is shown in Figure E–3. The title of the program (PSPICE SIMPLE DC CIRCUIT EXAMPLE) appears near the top of the page, and the remainder of the code appears under CIRCUIT DESCRIPTION. In the event that you try to run any of the text programs listed in this manner, do not forget that you will need to supply a title line as the

```
PSPICE SIMPLE DC CIRCUIT EXAMPLE

****       CIRCUIT DESCRIPTION

***************************************************************************************
VS 1 0 DC 180
R1 1 2 6E3
R2 2 0 3E3
R3 2 3 4E3
R4 3 0 2E3
IS 3 0 DC 24E-3
.DC VS 180 180 1
.PRINT DC V(1) V(2) V(3)
.PRINT DC I(R1) I(R2) I(R3) I(R4)
.OPTIONS NOPAGE
.END

****       DC TRANSFER CURVES              TEMPERATURE =   27.000 DEG C


VS          V(1)        V(2)        V(3)
1.800E+02   1.800E+02   3.300E+01   -2.100E+01


****       DC TRANSFER CURVES              TEMPERATURE =   27.000 DEG C


VS          I(R1)       I(R2)       I(R3)       I(R4)
1.800E+02   2.450E-02   1.100E-02   1.350E-02   -1.050E-02


        JOB CONCLUDED
        TOTAL JOB TIME             4.23
```

FIGURE E–3
Output data of simple dc circuit example.

first line in the code. The remainder of the circuit description is exactly the same as that of the code in Figure E–2.

For each .PRINT statement, there is a collection of data listed under the heading of DC TRANSFER CURVES. As explained earlier, this heading refers to a possible sweep of the input voltage, and it does not quite describe the situation for a single-point analysis. The designation of a temperature refers to the default temperature (27°C) for semiconductor and temperature-dependent parameter variation. This designation should be ignored for the present analysis.

The first line of data provides a listing of the input voltage (because it was the source that was "swept") and all node voltages referred to ground. All values are expressed in floating-point form in their basic units. Note that $V(1) = VS = 180$ V, which was obvious from the circuit diagram. The other voltages are $V(2) = 33$ V and $V(3) = -21$ V. Note that the latter voltage is negative with respect to ground.

The second line of data is keyed to the second .PRINT statement, and the value of VS is repeated. The next four values are the various branch currents in amperes, and they are read as $I(R1) = 2.45 \times 10^{-2}$ A $= 24.5$ mA; $I(R2) = 1.1 \times 10^{-2}$ A $= 11$ mA; $I(R3) = 1.35 \times 10^{-2}$ A $= 13.5$ mA; $I(R4) = -1.050 \times 10^{-2}$ A $= -10.5$ mA

The last two lines tell us that the analysis is concluded and the time required in seconds. Thus, the analysis took 4.23 s.

E–3 ELEMENT CODES

A general summary of some of the most common circuit models and their PSpice codes is given in this section. For the most part, these codes are explained through examples in the various figures provided in this section.

Passive Elements

The forms of the circuit models and codes for (a) resistance (R), (b) capacitance (C), and (c) inductance (L) are provided through examples in Figure E–4. In each case, arbitrary numbers for the nodes, subscripts, and values have been assumed in a way that no single value appears more than once in the example. This author has seen examples in some texts where the number for the node might, for example, be the same as a parameter value, and this might lead to possible misinterpretation. Basically, the forms for all three passive elements are the same except for the first letter $(R, C,$ or $L)$.

Independent dc and ac Voltage Sources

The forms and codes for independent dc and phasor ac voltage sources are shown in Figure E–5. Basically they follow the same format except that one more entry, representing the angle of the voltage source, might appear as the last item on the

$R_3 = 100\ \Omega$

⑧ —/\/\/\— ⑨

R3 8 9 100

Resistor R_3 connected between nodes 8 and 9 with a value of 100 Ω.

(a)

$C_7 = 0.39\ \mu F$

③ —| |— ⑥

C7 3 6 0.39E–6

Capacitor C_7 connected between nodes 3 and 6 with a value of 0.39 μF.

(b)

$L_4 = 1.5\ mH$

⑦ —00000— ⑥

L4 7 6 1.5E–3

Inductor L_4 connected between nodes 7 and 6 with a value of 1.5 mH.

(c)

FIGURE E–4
PSpice codes for three passive elements.

FIGURE E–5
PSpice codes for independent dc and ac voltage
sources.

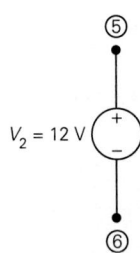

$V_2 = 12\ V$

V2 5 6 DC 12

Voltage source V_2 is connected between
nodes 5 and 6 with the positive terminal
at node 5. It is a dc source with value 12 V.

(a)

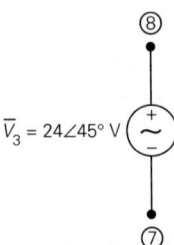

$\overline{V}_3 = 24\angle 45°\ V$

V3 8 7 AC 24 45

Voltage source $\overline{V}_3$ is connected between nodes 8 and 7
with the positive terminal at node 8. It is an ac source
with value 24 V and angle 45°.

(b)

ac source code. If this value is omitted, PSpice will assume a value of 0°. The first entry in each case must be "V." For the two node numbers, the first one represents the positive reference terminal of the source. Following the node connections, either "DC" or "AC" should be entered. If this quantity is omitted, PSpice defaults to "DC."

One point about the ac magnitude should be noted. If the desired calculations do not involve power, the magnitude may be interpreted as either the peak or the rms value, as desired, and the resulting output will be interpreted the same way. If power computations are to be performed, however, the magnitude should be the rms value. For example, if the peak value of a phasor voltage is 10 V and power computations are to be performed, the magnitude of this voltage should be entered as $10/\sqrt{2} = 7.071$ V.

Independent dc and ac Current Sources

The forms and code for independent dc and phasor ac current sources are shown in Figure E–6. As in the case of voltage sources, their formats are similar except for the possible additional entry for the angle of an ac source. The first entry in each case is the letter I. For the two-node numbers, the current source is assumed to be flowing from the first node to the second node. Once again, either "DC" or "AC" should be entered.

The point concerning peak or rms values for voltage sources applies equally to current sources. Thus, if power computations are desired, current source values should be entered as rms values.

FIGURE E–6
PSpice codes for independent dc and ac current sources.

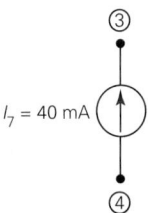

③

$I_7 = 40$ mA

④

I7 4 3 DC 40E–3
Current source I_7 is connected between nodes 4 and 3 with current flow from 4 to 3. It is a dc source with value 40 mA.

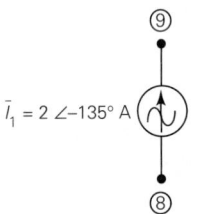

⑨

$\bar{I}_1 = 2 \angle{-135°}$ A

⑧

I1 8 9 AC 2 –135
Current source $\bar{I}_1$ is connected between nodes 8 and 9 with current flow from 8 to 9. It is an ac source with magnitude 2 A and angle –135°.

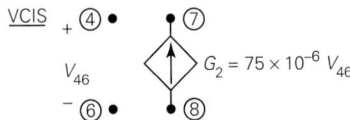

$$\text{VCVS} \quad +②• \qquad •⑥$$

$$V_{23} \qquad \boxed{+ \atop -} \ E_1 = 13.5\,V_{23}$$

$$-③• \qquad •⑦$$

E1 6 7 2 3 13.5

The dependent voltage source E_1 is connected between nodes 6 and 7 with the positive terminal at node 6. It is controlled by the voltage between nodes 2 and 3 (positive reference at 2) and the voltage gain is 13.5.

(a)

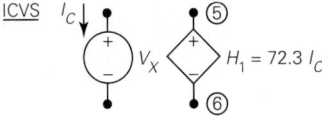

$$\text{VCIS} \quad +④• \qquad •⑦$$

$$V_{46} \qquad \Diamond\ G_2 = 75 \times 10^{-6}\,V_{46}$$

$$-⑥• \qquad •⑧$$

G2 8 7 4 6 75E–6

The dependent current source G_2 is connected between nodes 8 and 7 with current flow from 8 to 7. It is controlled by the voltage between nodes 4 and 6 (positive reference at 4) and the transconductance is 75 µS.

(b)

$$\text{ICVS} \quad I_C\downarrow$$

$$\bigcirc\ V_X \quad \Diamond\ H_1 = 72.3\,I_C$$

$$⑥$$

H1 5 6 VX 72.3

The dependent voltage source H_1 is connected between nodes 5 and 6 with the positive terminal at node 5. It is controlled by the current I_C flowing through the source V_X and the transresistance is 72.3 Ω.

(c)

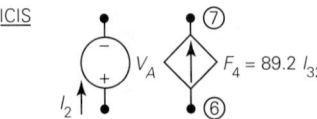

$$\text{ICIS} \qquad •⑦$$

$$\bigcirc\ V_A \quad \Diamond\ F_4 = 89.2\,I_{32}$$

$$I_2\uparrow \qquad •⑥$$

F4 6 7 VA 89.2

The dependent current source F_4 is connected between nodes 6 and 7 with current flow from 6 to 7. It is controlled by the current I_2 flowing through the source V_A, and the current gain is 89.2.

(d)

FIGURE E–7
PSpice models and codes for dependent sources.

Dependent Sources

PSpice permits the four forms for dependent sources considered in Section 1–4 to be modeled. These forms are summarized in Figure E–7. Since *V* and *I* are used for independent sources, other symbols are used for dependent sources. These symbols are listed as follows:

E   voltage-controlled voltage source (VCVS)
G   voltage-controlled current source (VCIS)
H   current-controlled voltage source (ICVS)
F   current-controlled current source (ICIS)

For voltage-controlled sources (VCVS and VCIS), there are four integer numbers following the source designation. The first two integers specify the nodes to which the source is connected, with the first representing the positive reference for a voltage source and the node at which the current enters the source for a current source. The third and fourth integers represent the nodes that define the controlling voltage, with the third representing the positive reference for the voltage.

For current-controlled sources (ICVS and ICIS), there are only two integer numbers following the source designation, and they specify the connection nodes in the same manner as for the voltage-controlled sources. However, following the node number is the specification of a voltage source through which the controlling current flows. This voltage source may be a dummy source having a value of zero.

The last entry is the controlling factor ("gain") for the particular type of source. The following quantities are represented by a third value:

VCVS   voltage gain A (no dimensions)
VCIS   transconductance g_m (siemens)
ICVS   transresistance r_m (ohms)
ICIS   current gain β (no dimensions)

The form of the source (dc, ac, or otherwise) is not specified for a dependent source since the form will have the same nature as the controlling variable.

Study the four examples in Figure E–7 carefully so that the forms may be recognized when they appear in the text.

ANSWERS TO SELECTED ODD-NUMBERED PROBLEMS

Chapter 1

1–1. **(a)** 15 W delivered, **(b)** 24 W absorbed

1–3. 4.667 mA, 58.8 mW

1–5. 36 V, 432 mW

1–7. 15 kΩ, 60 mW

1–9. 0.5 mS

1–11. 1 kΩ

1–13. 6 V, 20 mA, 11 mA, 1.636 kΩ

1–15. 15 V, 3 Ω

1–17. 50 kΩ

1–19. **(a)** 3 A, **(b)** 12 V, 9 V, 15 V, **(c)** total power = 150 W

1–21. **(a)** 2 A, **(b)** 4 V, 8 V, 10 V, **(c)** −16 V

1–23. **(a)** 24 V, **(b)** 1 A, 2 A, 3 A

1–25. **(a)** 20 V, **(b)** 4 A, 10 A, **(c)** 5 A

1–27. 12 V, 8 V

1–29. 24 mA, 12 mA

1–31. R: 30 kΩ 60 kΩ 18 kΩ 12 kΩ
     I: 2 mA   0.667 mA   1.333 mA   1.333 mA
     V: 60 V   40 V   24 V   16 V

1–33. 18 V

1–35. 12 V

1–37. 3 A in parallel with 8 Ω

1–39. 8 V in series with 4 Ω

1–41. Ideal 6-V voltage source

1–43. Ideal 40-μA current source

1–45. 9-V source in series with 4 kΩ

1–47. 14-V source in series with 7 Ω

1–49. 8-V source in series with 5 kΩ

1–59. 5.556 Ω

1–61. $R_1 = 450\ \Omega$, $R_2 = 55.56\ \Omega$

1–63. (c)

Chapter 2

2–1. (a) 8 mA, 0, 16 V,   (b) 7.65 mA, 0.7 V, 15.3 V

2–3. −16 V

2–5. 7.662 mA, 0.677 V, 15.323 V

2–7. 7.68 V

2–9. 7.7 V

2–11. 2.9 V

2–13. 20.77 mA

2–15. (a) A, B, G, E, D,   (b) 100 mA

2–17. (a) 150 Ω,   (b) 23.33 mA to 16.67 mA

2–19. Diode open (or possibly connected backward)

Chapter 3

3–1. (a) 19.3 V,   (b) 19.3 mA,   (c) 20 V

3–3. (a) 24.3 V,   (b) 24.3 mA,   (c) 30 V

3–5. (a) 20 V,   (b) 19.3 mA,   (c) 20 V

3–7. (a) 25 V,   (b) 24.3 mA,   (c) 25 V

3–9. (a) $v_o = 0$ for $v_s < 0.7$ V
$= 0.6667v_s - 0.4667$ for $v_s > 0.7$ V
(b) 19.53 V, 19.53 mA, 30 V

3–11. $v_o = -0.7$ V for $v_i < -0.933$ V
$= 0.75v_i$ for $v_i > -0.933$ V

3–13. (a) $v_o = v_i$ for $v_i < 6.7$ V
$= 6.7$ V for $v_i > 6.7$ V
(b) −12 V, −5 V, 5 V, 6.7 V,   (c) 23.3 mA

3–15. (a) 12.7 V, −12.7 V
(b) $v_o = v_i$ for -12.7 V $< v_i < 12.7$ V
$= -12.7$ V for $v_i < -12.7$ V
$= 12.7$ V for $v_i > 12.7$ V
(c) 17.3 mA,   (d) 24.7 V

3–17. (a) $v_o = v_i$ for -0.7 V $< v_i < 6.2$ V
$= -0.7$ V for $v_i < -0.7$ V
$= 6.2$ V for $v_i > 6.2$ V
(b) 6.2 V, 69 mA

3–19. (a) $v_o = -6.9$ V for $v_i < -6.9$ V
$= v_i$ for -6.9 V $< v_i < 6.9$ V
$= 6.9$ V for $v_i > 6.9$ V
(b) 6.9 V, 65.5 mA

3–21. (a) 100 mA, 175 mA, 75 mA,   (b) 125 mA, 175 mA, 50 mA

3–23. (a) 100 mA, 111.1 mA, 11.1 mA,   (b) 100 mA, 129.6 mA, 29.6 mA,
(c) 88.2 mA, 129.6 mA, 41.4 mA

3–25. 1465 Ω, 30 V

3–27. 857.5 Ω, 30.7 V

3–29. Use 5.1-V zener diodes; $R = 1105$ Ω

3–31. $R = 1.72 \text{ k}\Omega$

3–33. $141.4 \ \Omega$, 70 mA

3–35. Either the diode or the 1-kΩ resistor open

3–37. Either the 500-Ω resistor shorted or the 1500-Ω resistor open

Chapter 4

4–1. **(a)** Cutoff, **(b)** active, **(c)** saturation, **(d)** active, **(e)** cutoff, **(f)** saturation

4–3. **(a)** -14 V, **(b)** -14.3 V, **(c)** 0.7 V, **(d)** 14.3 V, **(e)** 13 V, **(f)** -0.7 V

4–5. 250, 10.04 mA

4–7. 3.6 mA, 3.63 mA

4–9. 200

4–13. 1.9 mA

4–15. 20 μA

4–17. 336 mW

4–19. Calculated value ≈ 0.47 V, which compares with 0.48 V on curve

4–21. **(a)** 0.98, **(b)** 0.99, **(c)** 0.995, **(d)** 0.9998

4–27. **(a)** Cutoff, **(b)** active, **(c)** saturation, **(d)** cutoff, **(e)** saturation, **(f)** active

4–29. Transistor is saturated.

Chapter 5

5–1. 6 mA, 75 μA, 7.525 V

5–3. **(a)** 0, 0, 0, 12 V, **(b)** 25.3 μA, 2.022 mA, 2.047 mA, 7.956 V, **(c)** 58.2 μA, 4.659 mA, 4.718 mA, 2.681 V, **(d)** 91.2 μA, 6 mA, 6.091 mA, 0

5–5. **(a)** 5.295 mA, 13.41 V, **(b)** 10.59 mA, 2.818 V, **(c)** 12 mA, 0

5–7. **(a)** 5 mA, 0, **(b)** 3.667 mA, 4 V, **(c)** 2.554 mA, 7.339 V

5–9. **(a)** 9.32 mA, 14.68 V, **(b)** 9.32 mA, 5.36 V, **(c)** 8 mA, 0

5–11. **(a)** 2 mA, **(b)** 4850 Ω

5–13. **(a)** $v_i - 0.7$, **(b)** 0.7 V $< v_i <$ 24.7 V, **(c)** 1.582 mA

5–15. **(a)** 1.90 mA, 1.90 mA, 8.18 V, 6.28 V, 9.54 V, **(b)** 38.1 μA

5–17. 1.19 mA, 1.19 mA, 5.10 V, 4.62 V, 5.28 V

5–19. **(a)** Load-line equation: $4000I_C + V_{CE} = 20$; $V_{CEQ} = 12$ V, $I_{CQ} = 2$ mA
(b) 5 mA, 25 μA

5–21. **(a)** 4.9 V, **(b)** 0.1 V

5–23. **(a)** 5 V, **(b)** 0, **(c)** NOR (combination of OR and NOT)

5–27. 4708 Ω

5–29. Load is connected between 15-V supply and collector. Emitter is biased through R_s to 15-V supply. Required value of emitter resistance is $R_E = 14.3$ kΩ. R_s is not critical as long as $R_s i_B \ll 0.7$. A reasonable value is $R_s = 2$ kΩ.

5–31. $R_C = 10$ kΩ, $R_E = 4$ kΩ, $R_1 = 48.25$ kΩ, $R_2 = 12.37$ kΩ

5–33. $R_C = 10$ kΩ, $R_E = 4$ kΩ, $R_1 = 12.06$ kΩ, $R_2 = 2975$ Ω

5–35. **(b)**

5–37. **(a)** Increase, **(b)** decrease, **(c)** decrease, **(d)** increase, **(e)** decrease, **(f)** decrease

Chapter 6

6–1. (a) 12 mS, 83.3 Ω,   (b) 8 mS, 125 Ω,   (c) 4 mS, 250 Ω

6–3. (a) −3 V, 3V,   (b) 3 V, 2 V, 1 V

6–5. (a) 18 mA,   (b) $18 \times 10^{-3} \left(1 - \dfrac{V_{GS}}{-3}\right)^2$

6–7. (a) − 1.419 V,   (b) −0.764 V,   (c) −0.261 V

6–11. (a) $12 \times 10^{-3} \left(1 - \dfrac{V_{GS}}{-3}\right)$

6–13. (a) 347 Ω,   (b) 4.32 mA

6–17. (a) 24.5 mA,   (b) 32 mA

6–19. (a) 3 V,   (b) 0, 1 V, 2 V, 3 V

6–21. (a) 18×10^{-3} A,   (b) $18 \times 10^{-3} \left(\dfrac{V_{GS}}{3} - 1\right)^2$

6–23. (a) 4.581 V,   (b) 5.449 V,   (c) 6.162 V

6–27. 20 mA

6–31. (a) −2.3 V,   (b) 6 mA

6–33. 2.8 V

6–35. (a) Beyond pinchoff,   (b) ohmic,   (c) cutoff,   (d) beyond pinchoff

6–37. (a) Cutoff,   (b) beyond pinchoff,   (c) ohmic

6–41. Cutoff

Chapter 7

7–1. 8 mA, 6 V

7–3. Operation moves to the ohmic region.

7–7. $I_D \approx 4.7$ mA, $V_{DS} \approx 0.4$ V

7–9. 8.164 mA, −0.9796 V, 0.9796 V, 6.123 V, 7.897 V

7–13. (b) $v_i = 0$ and −3 V,   (c) 0.3 V

7–15. $R_S = 97.6$ Ω, $R_D = 1285$ Ω

7–17.

	I_{DQ}	V_{R_S}	V_{R_D}	V_{DSQ}
Lower bound:	6 mA	0.586 V	7.71 V	15.70 V
Upper bound:	12 mA	1.171 V	15.42 V	7.409 V

7–19. (a) Nearly the same,   (b) increase,   (c) decrease,   (d) decrease,   (e) increase

7–21. (b)

Chapter 8

8–1. (a) 0.5 ms,   (b) $5 \sin 4000\pi t$

8–3. (a) 25 V,   (b) $10,000\pi$ rad/s,   (c) 5 kHz,   (d) 0.2 ms

8–5. (a) 3.536 V,   (b) 69.44 mW

8–7. (a) 17.678 V,   (b) 142.0 mW

8–9. (a) 7.958 MΩ,   (b) 15.92 kΩ,   (c) 198.9 Ω

8–11. (a) 50.27 mΩ,   (b) 25.13 Ω,   (c) 1257 Ω

8–13. 15 V

8–15. 10 V, 6 V

8–17. 6 mA, 12 V

8–19. $V_{C1} = 6.211$ V, $V_{C2} = 19.25$ V, $V_{C_E} = 5.51$ V

8–21. $0.4v_g$

8–23. **(a)** $V_{C1} = V_{C2} = V_A = 24$ V, $V_B = 0$, **(b)** $v_{c1} = 0$, $v_{c2} = 0$, $v_a = v_b = 0.2v_g$, **(c)** $v_{C1} = 24$ V, $v_{C2} = 24$ V, $v_A = 24 + 0.2v_g$, $v_B = 0.2v_g$

8–25. 15.92 kHz

8–27. 3979 Hz

8–29. 795.8 Hz

8–31. 442.1 Hz

8–33. 1.056 μs

8–35. 22 μs

8–37. 88 μs

8–39. 17.5 kHz

8–45. 31.83 μF

8–47. 3.979 mH

8–49. 3183 Ω

8–51. 265.3 Ω

8–53. (e), (f)

8–55. (a), (c)

Chapter 9

9–1. **(a)** 0, **(b)** -1 V, **(c)** 5 V

9–3. -40

9–5. -56.57

9–9. 450 Ω, 75 Ω, 120

9–11. 100 kΩ, 800 Ω, -40

9–13. 80

9–15. 64

9–17. 60

9–19. 48

9–21. 240

9–23. 160

9–25. 39.03 dB

9–27. 63.10×10^3

9–29. 400, 26.02 dB

9–31. 0.04, -13.98 dB, 25, 13.98 dB

9–33. 41.58 dB

9–35. 31.62

9–37. 47.60 dB

9–39. **(a)** 30 dBm, **(b)** 36.99 dBm, **(c)** 14.47 dBm, **(d)** -12.22 dBm, **(e)** -106.99 dBm

9–41. Output level = 37.78 dBm

9–43. Break frequency = 3979 Hz

9–45. Break frequency = 3979 Hz

9–51. 102.33, 4318 Ω, 23.16 Ω

9–53. Circuit form of Figure 9–27; R = 7.958 kΩ

9–55. Circuit form of Figure 8–19(a) with $R_1 = R_2$ = 31.83 kΩ

9–57. (b), (d)

Chapter 10

10–1. −50

10–3. 250 Ω

10–5. 60

10–7. (a) 0.32 S, (b) 468.75 Ω

10–11. (a) 1.974 mA, 5.923 V, 11.057 V, 13.020 V, (c) −442.3, 2532 Ω, ∞, 1713 Ω, 5600 Ω

10–13. (b) −17.91, 62.52 kΩ, ∞, 4885 Ω, 5600 Ω

10–15. (b) 442.3, 12.60 Ω, ∞, 12.55 Ω, 5600 Ω

10–17. (a) 1.852 mA, 15.182 V, 14.818 V, (b) 0.998, 1.643 MΩ, 77.03 Ω, 12.608 kΩ, 76.32 Ω

10–21. R_C = 10 kΩ, R_1 = 48.25 kΩ, R_2 = 28.16 kΩ, R_{E1} = 808 Ω, R_{E2} = 9192 Ω

10–23. (d)

Chapter 11

11–1. 9 mS

11–3. 12 mS

11–5. (a) 12 mS, (b) 12 mS, 8 mS, 4 mS, 0

11–9. (a) $12 \times 10^{-3} \left(\dfrac{V_{GS}}{3} - 1 \right)$, (b) 0, 4 mS, 8 mS, 12 mS

11–11. (b) −24, ∞, ∞, 123.2 kΩ, 2 kΩ

11–13. −8.571, ∞, ∞, 123.2 kΩ, 2 kΩ

11–15. 45, 66.67 Ω, ∞, 50 Ω, 3 kΩ

11–17. 0.923, ∞, 83.33 Ω, 142.1 kΩ, 76.92 Ω

11–21. 2 kΩ

11–23. 126.7 Ω

11–25. (b)

Chapter 12

12–1. −88.89, −50.70, 10,233

12–3. 36.32, 0.5822, 105.7

12–5. −9.311, −671.8, 8794

12–7. 5.714, 0.5455, 6.545

12–9. 270.8

12–11. 346.5

12–13. (a) -24, (b) -24 mV, -0.48 V
12–15. (d)

Chapter 13

13–1. (a) 1 mA, (b) 0.5 mA, 0.5 mA, 7 V, 7 V, 0
13–3. (a) 200, (b) 100, (c) -0.4425, (d) 226
13–5. (a) 12 mV, 6 mV, (b) 2.4 V, 1.2 V, -2.66 mV, (c) 6 mV
13–7. (a) 12 mV, 2.006 V, (b) 2.4 V, 1.2 V, -0.8877 V, (c) 2.006 V
13–9. (a) -10 mV, -1.795 V, (b) -2 V, -1 V, 0.7943 V, (c) -1.795 V
13–11. 2.26 mA
13–13. 3860 Ω
13–15. (a) $\pm$ 14 V, (b) 400,000
13–17. 55 μV
13–19. 105 nA
13–21. 10 nA
13–23. 4000, 72.04 dB
13–25. 1.2

Chapter 14

14–1. (a) 20, (b) 11.91
14–3. (a) 49.67, (b) 50
14–5. (a) 16.67, (b) 19.61, (c) 19.96, (d) 20
14–7. 0.119%
14–9. (a) 0.02, (b) 2400, (c) 49.98, (d) 50
14–11. (a) 10, (b) 20, (c) 50, (d) 200
14–13. (a) 960 MΩ, (b) 0.025 Ω
14–15. (a) 2.4 MHz, (b) 48.02 kHz
14–17. (a) 2 kHz, (b) 20 kHz, (c) 200 kHz, (d) 2 MHz
14–21. 0.009
14–23. 31

Chapter 15

15–1. (a) 120 μV, (b) 600 μV, (c) 2.4 mV
15–3. 8 V
15–5. (a) -20, (b) 7.5 kΩ, (c) 0.65 V, (d) 0, -4 V, 4 V, -10 V, -13 V
15–7. $-2v_1 - 1.5v_2$
15–9. (a) -9 V, (b) 0, (c) 2 V, (d) 13 V
15–11. $3(v_1 - v_2)$
15–13. (a) 0, (b) 12 V, (c) -12 V, (d) 13 V
15–15. (a) 2 kΩ, (b) 0, -4 V, 4 V, -13 V
15–17. (a) 0.5 mS, (b) 13 kΩ, (c) 2.6 V
15–19. (a) 0.5 mS, (b) 6.5 kΩ, (c) 1.3 V
15–21. (a) 5, (b) 1 mA

15–31. $R_i = 15$ kΩ, $R_f = 180$ kΩ
15–33. $R_i = 10$ kΩ, $R_f = 240$ kΩ
15–35. $R_1 = 24$ kΩ, $R_2 = 10$ kΩ, $R_f = 120$ kΩ
15–37. $R = 1.5$ kΩ
15–39. $R = 1$ kΩ
15–41. (a), (d)

Chapter 16

16–1. **(a)** 8.7 V, 8 V, **(b)** −13.5 V, 0
16–3. **(a)** −0.7 V, 0, **(b)** 8.7 V, 8 V
16–5. 1.1 μs
16–7. 2 min
16–9. **(a)** 100 V, **(b)** 36.79 V, **(c)** 13.53 V, **(d)** 0.674 V, **(e)** 4.54×10^{-3} V
16–11. **(a)** 5.3 V, **(b)** 11.3 V
16–13. **(a)** 4 V, 4.7 V, **(b)** 4 V, −1.7 V
16–15. **(a)** 15.3 V, **(b)** 15.3 V
16–17. **(a)** 16 V, **(b)** 29 V
16–19. **(a)** 13 V, **(b)** −13 V
16–21. $v_o = v_i$ for $v_i < 0$
           $= 0$ for $v_i > 0$
16–25. **(a)** 10 kΩ
16–27. (b)

Chapter 17

17–1. 50, −135°
17–3. **(a)** 1592 Hz, **(b)** 500 Ω
17–5. 1.625 Hz
17–7. 5.338 MHz
17–9. 2.492 MHz
17–11. 114.1 μs, 8764 Hz
17–13. 4.99 ms, 4.30 ms, 9.29 ms, 108 Hz, 53.7%
17–15. 0.440 s
17–21. 15.92 kΩ
17–23. 3979 Ω to 663.1 Ω
17–25. **(a)** 2 kΩ, **(b)** 7 V
17–27. 6497 Ω
17–29. 45.51 kΩ
17–31. $R_2 = 10$ kΩ, $R_1 = 15$ kΩ, $R = 29.51$ kΩ
17–33. $R_A = 21.65$ kΩ, $R_B = 25.25$ kΩ

17–35. 36.36 kΩ

17–37. (f), (h), (i)

Chapter 18

18–1. $-j3183$ Ω

18–3. $j2513$ Ω

18–5. 173 pF

18–7. 920.0 kHz

18–9. 17.79 kHz

18–11. 25

18–13. 628.3 Ω

18–15. 2880 Ω

18–17. **(a)** 3.559 MHz, **(b)** 7.587, **(c)** 469.1 kHz

18–19. **(a)** 21.32 pF, **(b)** 6283 Ω, **(c)** -188.5

18–21. **(a)** 10 MHz, **(b)** 2.5 MHz

18–25. **(a)** 70 ns, **(b)** 165 ns

18–27. **(a)** 2.4 kHz, **(b)** 24 kHz, **(c)** 240 kHz, **(d)** 1.2 MHz

18–29. **(a)** 2395 Hz, **(b)** 23.53 kHz, **(c)** 200 kHz, **(d)** 600 kHz

18–31. **(a)** 0.1 μs, **(b)** 2 μs, **(c)** 6 μs

18–33. **(a)** 1.592 MHz, **(b)** 79.58 kHz, **(c)** 26.53 kHz

18–39. 140.7 pF

18–41. 1.5 MHz

18–43. 1.885 V/μs

18–45. (a)

Chapter 19

19–1. **(a)** 4 A, **(b)** 2 A, **(c)** 25 V, **(d)** 80 V

19–3. **(a)** 32 V, **(b)** 64 Ω, **(c)** 4, **(d)** 16 W, **(e)** 8 W, **(f)** 16 W

19–5. **(a)** 4 A, **(b)** 40.74 W, **(c)** 32 W, **(d)** 12.97 W

19–7. **(a)** 2.5°C/W, 50°C/W, 47.5°C/W, **(b)** 3.9 W, **(c)** 44 W

19–11. **(a)** 0.8106 W for each transistor, **(b)** 1 A, **(c)** 5.093 W

19–13. (b)

Chapter 20

20–1. 24 V, 15 A

20–3. 0.5 A, 15 V

20–5. **(a)** 15 V, **(b)** 21.21 V, **(c)** 6.752 V, **(d)** 1.125 A

20–7. **(a)** 15 V, **(b)** 21.21 V, **(c)** 13.50 V, **(d)** 2.251 A

20–9. Same answers as Problem 20–7

20–11. **(a)** 21.21 V, **(b)** 42.43 V, **(c)** 21.21 V

20–13. **(a)** 24 V, **(b)** 48 V, **(c)** 48 V, **(d)** 24 V

20–15. **(a)** 5 V, **(b)** 1.443 V

20–17. **(a)** 12.5 V, **(b)** 3.608 V, **(c)** 29.75 V, **(d)** 0.1213

20–19. (a) 8 V,   (b) 6.25 mA,   (c) 4 W, 6 W, 66.7%,   (d) 2 W
20–23. 2.707
20–25. 1604 μF
20–27. (a) 1203 μF,   (b) 15.37 V
20–29. (a) 345 Ω,   (b) R_i = 10 kΩ, R_f = 13.53 kΩ,   (c) 20 mA,   (d) 4 W
20–31. (b)
20–33. (d)

Chapter 21
 21–1. (b) −21.2 dB, −36.1 dB, −47.8 dB, −57.3 dB, −65.3 dB
 21–3. (b) −24.1 dB, −48.2 dB, −62.3 dB, −70 dB
 21–5. (a) 5,   (b) 1357 Hz, 1657 Hz,   (d) −33.89 dB
 21–9. C_1 = 0.0225 μF, C_2 = 0.01125 μF
21–11. R = 22.5 kΩ, C_2 = 0.005 μF
21–13. 3 poles
21–15. 3 poles
21–17. C_1 = 0.0564 μF, C_2 = 0.0222 μF, C_3 = 0.00322 μF
21–19. R = 56.4 kΩ, C_2 = 0.00393 μF, C_3 = 571 pF
21–21. R_1 = 4488 Ω, R_2 = 11.43 kΩ, R_3 = 78.64 kΩ
21–23. Feedback R = 318.3 kΩ, Input R = 159.2 kΩ, Shunt R = 800 Ω
21–25. C_t = 0.00796 μF, R_Q = 11.213 kΩ with 10-kΩ shunt R
21–27. C_t = 0.00796 μF, R_Q = 2.99 MΩ with 10-kΩ shunt R

Chapter 22
 22–1. (a) 22.05 kHz,   (b) 22.68 μs
 22–3. (a) 2 kHz, 0.5 ms,   (b) 0.5 ms,   (c) 31.25 μs
 22–5. (b) 0.125 ms,   (c) 7.8125 μs
 22–7. 65,536
 22–9. (a) 10 V,   (b) 0.625 V
22–11. (a) 0,   (b) 0.625 V,   (c) 5 V,   (d) 6.25 V,   (e) 9.375 V
22–13. (a) 32,   (b) 10 V,   (c) 0.3125 V
22–15. (a) 0.3125 V,   (b) 0.625 V,   (c) 1.25 V,   (d) 5 V,   (e) 9.6875 V
22–17. (a) 12.6 μs, 79.37 kHz,   (b) 1.2 μs, 833.3 kHz,   (c) 0.2 μs, 5 MHz
22–19. (a) 15.26 × 10^{-6},   (b) 152.6 μV,   (c) 0.999985,   (d) 9.99985 V,   (e) 7.63 × 10^{-6},   (f) 76.3 μV,   (g) 7.63 × 10^{-4}%
22–23. (b) 0.03125, 0.09375, 0.15625, . . . , 0.90625,   (d) all resistances would have the value R
22–25. (a) 4.8 kΩ,   (b) 10 kΩ,   (c) 12 kΩ,   (d) 12.8 kΩ
22–27. (a) 6.4 kΩ,   (b) 8 kΩ,   (c) 12 kΩ,   (d) 12.387 kΩ
22–29. (a) 10 V,   (b) 9 bits

INDEX

1031